Springer Science+Business Media, LLC

Applied Mathematical Sciences

1. *John:* Partial Differential Equations, 4th ed.
2. *Sirovich:* Techniques of Asymptotic Analysis.
3. *Hale:* Theory of Functional Differential Equations, 2nd ed.
4. *Percus:* Combinatorial Methods.
5. *von Mises/Friedrichs:* Fluid Dynamics.
6. *Freiberger/Grenander:* A Short Course in Computational Probability and Statistics.
7. *Pipkin:* Lectures on Viscoelasticity Theory.
8. *Giacaglia:* Perturbation Methods in Non-linear Systems.
9. *Friedrichs:* Spectral Theory of Operators in Hilbert Space.
10. *Stroud:* Numerical Quadrature and Solution of Ordinary Differential Equations.
11. *Wolovich:* Linear Multivariable Systems.
12. *Berkovitz:* Optimal Control Theory.
13. *Bluman/Cole:* Similarity Methods for Differential Equations.
14. *Yoshizawa:* Stability Theory and the Existence of Periodic Solution and Almost Periodic Solutions.
15. *Braun:* Differential Equations and Their Applications, 3rd ed.
16. *Lefschetz:* Applications of Algebraic Topology.
17. *Collatz/Wetterling:* Optimization Problems.
18. *Grenander:* Pattern Synthesis: Lectures in Pattern Theory, Vol. I.
19. *Marsden/McCracken:* Hopf Bifurcation and Its Applications.
20. *Driver:* Ordinary and Delay Differential Equations.
21. *Courant/Friedrichs:* Supersonic Flow and Shock Waves.
22. *Rouche/Habets/Laloy:* Stability Theory by Liapunov's Direct Method.
23. *Lamperti:* Stochastic Processes: A Survey of the Mathematical Theory.
24. *Grenander:* Pattern Analysis: Lectures in Pattern Theory, Vol. II.
25. *Davies:* Integral Transforms and Their Applications, 2nd ed.
26. *Kushner/Clark:* Stochastic Approximation Methods for Constrained and Unconstrained Systems.
27. *de Boor:* A Practical Guide to Splines, Rev. ed.
28. *Keilson:* Markov Chain Models—Rarity and Exponentiality.
29. *de Veubeke:* A Course in Elasticity.
30. *Sniatycki:* Geometric Quantization and Quantum Mechanics.
31. *Reid:* Sturmian Theory for Ordinary Differential Equations.
32. *Meis/Markowitz:* Numerical Solution of Partial Differential Equations.
33. *Grenander:* Regular Structures: Lectures in Pattern Theory, Vol. III.
34. *Kevorkian/Cole:* Perturbation Methods in Applied Mathematics.
35. *Carr:* Applications of Centre Manifold Theory.
36. *Bengtsson/Ghil/Källén:* Dynamic Meteorology: Data Assimilation Methods.
37. *Saperstone:* Semidynamical Systems in Infinite Dimensional Spaces.
38. *Lichtenberg/Lieberman:* Regular and Chaotic Dynamics, 2nd ed.
39. *Piccini/Stampacchia/Vidossich:* Ordinary Differential Equations in $\mathbf{R}^n$.
40. *Naylor/Sell:* Linear Operator Theory in Engineering and Science.
41. *Sparrow:* The Lorenz Equations: Bifurcations, Chaos, and Strange Attractors.
42. *Guckenheimer/Holmes:* Nonlinear Oscillations, Dynamical Systems, and Bifurcations of Vector Fields.
43. *Ockendon/Taylor:* Inviscid Fluid Flows.
44. *Pazy:* Semigroups of Linear Operators and Applications to Partial Differential Equations.
45. *Glashoff/Gustafson:* Linear Operations and Approximation: An Introduction to the Theoretical Analysis and Numerical Treatment of Semi-Infinite Programs.
46. *Wilcox:* Scattering Theory for Diffraction Gratings.
47. *Hale et al:* An Introduction to Infinite Dimensional Dynamical Systems—Geometric Theory.
48. *Murray:* Asymptotic Analysis.
49. *Ladyzhenskaya:* The Boundary-Value Problems of Mathematical Physics.
50. *Wilcox:* Sound Propagation in Stratified Fluids.
51. *Golubitsky/Schaeffer:* Bifurcation and Groups in Bifurcation Theory, Vol. I.
52. *Chipot:* Variational Inequalities and Flow in Porous Media.
53. *Majda:* Compressible Fluid Flow and System of Conservation Laws in Several Space Variables.
54. *Wasow:* Linear Turning Point Theory.
55. *Yosida:* Operational Calculus: A Theory of Hyperfunctions.
56. *Chang/Howes:* Nonlinear Singular Perturbation Phenomena: Theory and Applications.
57. *Reinhardt:* Analysis of Approximation Methods for Differential and Integral Equations.
58. *Dwoyer/Hussaini/Voigt (eds):* Theoretical Approaches to Turbulence.
59. *Sanders/Verhulst:* Averaging Methods in Nonlinear Dynamical Systems.

(continued following index)

George R. Sell Yuncheng You

Dynamics of Evolutionary Equations

With 19 Illustrations

Springer

George R. Sell
School of Mathematics
University of Minnesota
Minneapolis, MN 55455, USA
sell@math.umn.edu

Yuncheng You
Department of Mathematics
University of South Florida
Tampa, FL 33620-5700, USA
you@math.usf.edu

Mathematics Subject Classification (2000): 35K55, 58Fxx, 58D25, 34G20

Library of Congress Cataloging-in-Publication Data
Sell, George R, 1937–
Dynamics of evolutionary equations / George R. Sell, Yuncheng You.
p. cm. — (Applied mathematical sciences ; 143)
Includes bibliographical references and indexes.
DOI 10.1007/978-1-4757-5037-9

1. Differentiable dynamical systems. 2. Evolution equations. I. You, Yuncheng. II. Title III. Applied mathematical sciences (Springer-Verlag New York, Inc.) ; v. 143
QA1.A647 Vol. 143
[QA614.8]
510s—dc21
[515′.35] 00-056314

Printed on acid-free paper.

Production managed by Terry Kornak; manufacturing supervised by Jerome Basma.
Photocomposed copy prepared by the authors.

9 8 7 6 5 4 3 2 1

www.springer.com/mycopy

With great fondness and gratitude, we dedicate
this volume to our parents.

George P. Sell and Alice O. (Roecker) Sell
and
You Qiwen and Wei Yinmei

PREFACE

The theory and applications of infinite dimensional dynamical systems have attracted the attention of scientists for quite some time. Dynamical issues arise in equations that attempt to model phenomena that change with time. The infinite dimensional aspects occur when forces that describe the motion depend on spatial variables, or on the history of the motion. In the case of spatially dependent problems, the model equations are generally partial differential equations, and problems that depend on the past give rise to differential-delay equations. Because the nonlinearities occurring in thse equations need not be small, one needs good dynamical theories to understand the longtime behavior of solutions.

Our basic objective in writing this book is to prepare an entreé for scholars who are beginning their journey into the world of dynamical systems, especially in infinite dimensional spaces. In order to accomplish this, we start with the key concepts of a *semiflow* and a *flow*. As is well known, the basic elements of dynamical systems, such as the theory of attractors and other invariant sets, have their origins here.

In the applications to partial differential equations, for example, the properties of a semiflow serve as a precise statement of the notion of a well-posed problem, which is a central feature in the study of reaction diffusion equations, nonlinear wave equations, and the Navier-Stokes equations.This concept serves as a road map for finding proper solutions in order to drive into the inner city of dynamics of partial differential equations.

Since a time-varying solution of a partial differential equation can be viewed as a trajectory, or curve, in some Banach space W, this suggests that one should rewrite the equation of motion of this solution as an equation in W. The resulting equation is called an evolutionary equation, be it linear or nonlinear. The main approach in this volume is built around the theory of evolutionary equations. (See Chapters 3 and 4.)

Chapter 4 is an especially important feature of this work. Many aspects of evolutionary equations are collected here, perhaps for the first time, in book form. One should read this chapter on two levels: as a basic introduction and as a reference source. A good approach is to read it more than once, where one goes deeper as the need arises.

The basic applications to the semiflow theory of the Navier-Stokes equations and other partial differential equations are in Chapters 5 and 6. Several aspects of the modern theories of dynamical systems to linear and nonlinear evolutionary equations, such as the perturbation theory of a saddle point, the reduction principle and center manifold, periodic orbits and invariant manifolds, and inertial manifolds appear in Chapters 7 and 8.

Chapter 1 is a brief essay on the Evolution of Evolutionary Equations, with emphasis on the theory of the longtime dynamics of the solutions of these equations. We have chosen to use some poetic license to keep this chapter short. As a result, we do not include an exhaustive list of references to the literature in this chapter. Additional references do appear in the Commentary sections elsewhere in this volume.

There are a number of general readings that are relevant. First there is the pioneering set of lecture notes by Henry (1981) on dynamical issues of nonlinear partial differential equations. Next there is the book by Temam (1988), which contains an encyclopedic treatment of many applications arising in mechanics and physics.The monograph by Hale (1988) contains valuable information on nonlinear dynamics in infinite dimensions, with applications to partial differential equations and differential equations with time delays. For a very good treatment of the dynamics of functional differential equations, see Hale and Verduyn Lunel (1993). The book by Pazy (1983) contains an excellent treatment of the linear theory of semigroups on Banach spaces. Background information on metric space theory, the geometry of Hilbert spaces, and the theory of linear operators can be found in Naylor and Sell (1982). An extensive bibliography on dynamical systems is available on the World Wide Web; see Sell (2000). The references in this bibliography are updated from time to time.

Acknowledgments

We are grateful to the many students who have attended our lectures while we were developing the notes for this volume. We also appreciate the helpful comments and corrections on the manuscript received from these young scholars. In addition we are most appreciative of the kind suggestions received from our colleagues, and especially from Anatoli Babin, John Ball, Ciprian Foias, Jack Hale, Michael Jolly, Klaus Kirchgässner, John Mallet-Paret, Sergei Pilyugin, Victor A. Pliss, Andreas Prohl, Hal Smith, and Edriss Titi, about various aspects of this book.

CONTENTS

1
THE EVOLUTION OF EVOLUTIONARY EQUATIONS

"May you live in exciting times!" This traditional Chinese saying aptly describes the environment surrounding the basic developments in mathematics, physics, and chemistry over the past four centuries. From the founding of the European Academies of Sciences during the era of Peter the Great and Napoleon, to the founding of the National Science Foundation in the United States of America during the presidency of Harry Truman, governments have realized the importance of scientific research.[1] To trace the implications of this academic research, as it affects the evolution of evolutionary equations, we begin in 1601 in Prague with the appointment of Johannes Kepler (1571 - 1630) to the position of Imperial Mathematician of the Holy Roman Empire, after the death of his predecessor, Tycho Brahe (1546 - 1601).

Newtonian Modeling: It was Kepler's early work on planetary motion which attracted the attention and respect of Brahe, who in turn invited Kepler to join his research staff. As Brahe's successor, Kepler had access to Brahe's very extensive records and observations of planetary motion. Kepler's goal was to derive a good mathematical model for planetary motion in our solar system. He succeeded!

In his 1609 paper *Astronomia Nova,* he derives two laws of planetary motion: (1) each planet travels in an elliptical orbit with the sun at a focus and (2) each planet sweeps out equal areas in equal times when traveling in its orbit. Then in 1619 he published *Harmonice Mundi,* in which he presents a third law: (3) the 3:2 rule relating the mean distance between

[1] Also, the emigration of some major scientists, such as Daniel Bernoulli (1700 - 1782) and Leonhard Euler (1707 - 1783) to St Petersburg, and Albert Einstein (1879 - 1955) and John von Neumann (1903 - 1957) to Princeton, accelerated the growth and international impact of science during this period.

the planet and the sun with the period of the motion. These three Keplerian Laws gave astronomers a new and unexpected paradigm for the study of the motion of asteroids and comets, as well as planetary motion.

Why was this considered to be a "good" model? Before answering, we must emphasize that any model must be measured by the standards of its time. Brahe's observations preceded the invention of the telescope, so one cannot fault the Keplerian model for lack of better experimental data.

The Keplerian Laws, which are fully valid for the 2-body problem, are only approximations to the planetary motion in the celestial mechanical N-body model of the solar system. It is a fact that Kepler, like Brahe before him, was very meticulous in his work, so much so that one wonders whether he would have found the three Keplerian Laws if the astronomical data of Brahe had been obtained with the more accurate telescopic observations. Only with the telescope did astronomers have the technology to see that the planetary motions do in fact deviate somewhat from a true elliptical orbit. The major importance of the Keplerian Laws is the seminal role they played later in the time of Isaac Newton (1643 - 1727) with the birth of classical mechanics.

The idea of formulating mathematical models of the solar system, in terms of differential equations representing the laws of motion, began to take hold in the scientific community around the time of Galileo Galilei (1564 - 1642). The mathematicians[2] of that day were trying to understand the basic relationships between force, momentum, displacement, and mass. Because of his extensive experimental work with pendula and inclined planes, Galileo was instrumental in the development of what is now called classical mechanics. It is at this point that Newton, a professor at Trinity College in Cambridge, enters the scene.

Newton, like Galileo, was searching for universal principles which could be used to explain the physical world around him. In this process, he arrived at a set of three laws, the Newtonian Laws of Motion. The first law, which is a reformulation of the Galilean concept of uniform motion, states that a body in motion remains in motion until a force acts on the body. The second law, $F = ma$, equates the force with the rate of change of momentum; and the third law states that for each action there is an equal and opposite reaction. These three laws are as insightful as they are simple. Even today, in the aftermath of more recent mechanical theories, such as quantum mechanics and relativistic mechanics, the Newtonian Laws are very widely used. For example, the momentum equation, which arises in the Navier-Stokes model of fluid flow, is a reformulation of the second law of Newton.

It was Newton's belief that the Keplerian Laws could be derived from the Newtonian Laws of Motion and the existence of a force caused by a gravitational field acting between the sun and the planets. By using the three laws

[2]During the time of Galileo, and for a long time thereafter, the science of physics was viewed as a part of mathematics.

of motion and the Keplerian 3:2 rule in the case of a single planet revolving about the sun, Newton found that the centripetal force acting on the planet was given by the inverse-square law, i.e., the force is inversely proportional to the square of the distance between the planet and the sun. However, the great achievement of Newton was to prove the truth of the opposite implication. That is, he succeeded in showing that, by using his Laws of Motion, together with the assumption of a gravitational force given by the inverse-square formula, one can derive the three Keplerian Laws as consequences. It was this theorem which gave birth to the Newtonian concept of the universal law of gravity. This work of Newton on mechanics appeared in 1687 in his masterpiece: *Philosophiae Naturalis Principia Mathematica,* or the Principia, for short.

It is very hard to overstate the importance of Newton's contributions to the advancement of science. The Principia, which includes the Newtonian model of mechanics (the three laws of motion, the law of universal gravity, and the inverse square law and the Keplerian model), as well as the beginnings of the differential and integral calculus, is probably the most important and the most significant treatise on mathematics ever written. While many other major successes in mathematical modeling were to follow, no other single achievement would have the same impact on the history of man's attempt to understand the world about us. It was this work of Newton, and the simultaneous discovery of the calculus by Newton and Gottfried Wilhelm Leibniz (1646 - 1716), that fully established the role of mathematics as the principal tool for modeling the laws of nature.

The Newtonian laws of motion for the N-body problem of celestial mechanics enable one to describe the dynamics of the problem in terms of the solutions of a system of ordinary differential equations. For the full problem, one has a three-dimensional (3D) position vector and a 3D velocity vector for each body. Thus for the three-dimensional problem, the equations of motion are described by a $6N$-dimensional system of ordinary differential equations. In the planar problem, where the N bodies are restricted to a plane, the equations of motion are described by a $4N$-dimensional system of ordinary differential equations.

However, there are some conservation laws for these problems which effectively reduce the dimension of the phase space for the equations of motion. In particular, the time derivatives of: the center of mass, the linear momentum, the angular momentum, and the energy are all zero. One has six conservation laws for the planar problem, and ten laws for the full three-dimensional problem. As a result, the reduced dimension for the planar problem is $4N - 6$; and for the full problem it is $6N - 10$, see Meyer and Hall (1992) and Siegel and Moser (1971). In particular, the 2-body problem is described by a system of ordinary differential equations in the plane $\mathbb{R}^2$.

In the celestial mechanical model of the solar system, where $N \geq 10$, the complexity of the equations of motion have defied all attempts at trying to find explicit formulae for the solutions, except in a few very special cases. Nevertheless, this model for the solar system is so good and the analysis of the solutions is so accurate that this has led to a very high degree of predictability of the position of the planets. As a matter of fact, on two occasions, this predictability has enabled astronomers to locate new planets. How did this happen?

The processes leading up to the discovery of Neptune (in 1846) and Pluto (in 1930) both began with the observations that the predicted positions of the then known planets were deviating from the actual positions in a way which could not be explained on the basis of the gravitational field of the sun and the known planets alone. This led in turn to the idea that one might postulate the existence of a new planet and then use its gravitational field to rederive the predicted positions. By adjusting the parameters (e.g., mass and position) of the new planet, one could try to reduce the deviations to zero. In other words, one seeks to use the deviations themselves to locate the unknown planet.

As it happens, the mass of the planet Pluto appears to be too small to explain fully the previously observed deviations between the predicted and actual orbits of Uranus and Neptune. Does that imply the existence of yet another planet, a Planet X? That is not known, and because of the long 248 Earth-year-period for Pluto's orbit, it may be too early to answer this question. However, the methodology for finding a tenth planet is now in place. Time will tell.

Birth of Dynamical Systems: The 3-body problem, in particular, presented a major challenge to the mathematical world. Of special interest was the satellite problem, for example, the Sun-Earth-Moon system, where the third body has a relatively small mass when compared to the two major bodies. As the efforts to find explicit formulae for the general solutions fell short, greater interest was placed on new qualitative methods for the analysis of the dynamics of the solutions. Furthermore, these new methods grew in importance as researchers turned to the issues of longtime dynamics, such as the stability of the solar system.

Certainly among the most important advances in this area are two works of Henri Poincaré (1854 - 1912): his 1890 paper *Sur le problème des trois corps et les équations de la dynamique* and the 1892 treatise *Les Méthodes Nouvelles de la Mécanique Céleste* I-II-III. One of the most interesting features of these works was the realization of the possibility of an instability in the N-body problem (where $N \geq 3$) owing to intersections of the stable and unstable manifolds of a periodic orbit. Poincaré's works are highly significant. Not only did he win the prestigious King Oscar Prize, see Goroff (1993), but more importantly, Poincaré, along with Alexander Mikhailovich Lyapunov (1857 - 1918) and George David Birkhoff (1884 - 1944), emerged as a co-founder of a new area: dynamical systems.

The issue of stability arises, in one way or another, in all mathematical models. It is omnipresent and multifaceted. Whether a given dynamical feature is stable or not, depends on the context, or point of view. The different meanings of the word stability come from the point of view. In the van der Pol equation, for example, there is an (unstable) source, within the (stable) global attractor, and the source is a stable dynamical feature of the global attractor. The major work on stability theory appears in the 1892 paper by Lyapunov, *Problème géneral de la stabilité du mouvement.* This important study, which coincided with the advances in celestial mechanics noted above, is a very significant development in the evolution of evolutionary equations, for several reasons.

First, this is the work in which Lyapunov presented his theory of a generalized energy functional, now called a Lyapunov function, which can be used to study the stability of certain systems of differential equations without first solving for the solutions. This theory is a precursor of the LaSalle Invariance Principle and Morse structures for dissipative evolutionary equations. Unlike the N-body problem, in which the total energy is constant along solutions, in dissipative problems, the energy can vary along solutions, but it is typically ultimately bounded. This is a common feature of those dynamical systems that have a global attractor.

Second, a theory of characteristic exponents, now called Lyapunov exponents, for time-varying linear differential equations is developed in this work. Based on contributions of Lyapunov and his followers, it is now appreciated that the theory of Lyapunov exponents offers a good framework for finding upper bounds for the dimension of an attractor of an evolutionary equation. While there are now several theories of dimension which are applicable to this study, the theory of Lyapunov dimension plays a unique role because of the strong analytical tools it brings to the problem.

It is noteworthy that both of these theories, which Lyapunov had developed for applications to finite systems of ordinary differential equations, have meaningful extensions to the infinite dimensional world of dynamical systems. The work of Lyapunov is as important today as it was when it first appeared in 1892.

The concept of a dynamical system, as we know it today, was developed by G D Birkhoff in the early part of the twentieth century. His theory of minimal sets, recurrence, nonwandering sets, central motions, transitivity, and the foundations of Hamiltonian systems forms the basis of many advances. Much of this material appears in his 1927 book *Dynamical Systems.* An especially important contribution is his well-known ergodic theorem, see Birkhoff (1931a,b). This theorem, which is a pillar for the theory of statistical mechanics, also serves as a bridge for the use of related functional-analytic techniques in the theory of dynamical systems.

The issues studied by Poincaré, Lyapunov, and Birkhoff all fit within the general theory of finite dimensional systems of ordinary differential equations. At a later time, other researchers would show that some of the techniques developed by the Founders do extend to selected infinite dimensional problems. During the early period of dynamical systems there were, of course, other advances. Two of these are especially noteworthy. First, there are the extensions of the Birkhoff theory to the question of the existence of invariant measures and ergodic measures, for compact, invariant sets in a dynamical system, see, for example, Krylov and Bogoliubov (1937). Second, there are two methods for the construction of invariant manifolds for nonlinear problems: (1) the Hadamard (1901) method and (2) the Lyapunov (1892) - Perron (1928, 1930b) method.

Infinite Dimensional Challenge: Not surprisingly, the theory of the longtime dynamical properties of solutions of infinite dimensional evolutionary equations generated by partial differential equations was slower in coming than the finite dimensional counterpart. Among other issues, the early researchers encountered additional difficulties, not seen on the theory of ordinary differential equations, in sorting out which partial differential equations problems had good solutions. A major step in resolving this was the concept of a well-posed problem proposed by Jacques Hadamard (1865 - 1963). It is very likely that in formulating this concept, Hadamard was influenced by the concurrent developments in the dynamics of ordinary differential equations, see, for example, Hadamard (1901). At a later time, the concept of a well-posed problem would play a central role in the definition of a semiflow generated by an evolutionary equation.

The development of the theories of the longtime dynamics for linear and nonlinear evolutionary equations generated by partial differential equations is one of the major triumphs of the area of functional analysis. This area of mathematics began with the works of Henri Lebesgue (1875 - 1941), who presented his new definition of the integral at the beginning of the twentieth century. Owing to the good limit theorems for the Lebesgue integral, this concept quickly replaced the Riemann integral in mathematical analysis. David Hilbert (1862 - 1943) used the theory of Lebesgue to analyze solutions of integral equations. In so doing, he built the basis for the abstract theory of Hilbert spaces, a term which was later coined by von Neumann (1930). In 1932, Stefan Banach (1892 - 1945) published a beautiful volume on *Théorie des Opérations Linéaries*. The concept of a Banach space is derived from this work.

Let us return to the concept of a well-posed problem in the context of a parabolic, or hyperbolic, partial differential equation. In each of these problems, unlike the case of an elliptic partial differential equation, one encounters an Initial Value Problem (IVP), or, as it is sometimes called, a Cauchy problem. This suggests that a time-varying solution of the IVP can be viewed as a trajectory, or curve, in some Banach space, which is the phase space for the problem. The equation of motion of this trajectory in the Banach space is given by a linear or nonlinear evolutionary equation. Loosely speaking, an evolutionary equation is an ordinary differential equation on a Banach space. This simple observation, by some researcher unknown to the authors, gave rise to the study of the dynamics of solutions of partial differential equations.

Nevertheless, the issue of the proper definition of a solution of an evolutionary equation is more complicated in the infinite dimensional setting. One might have a strong solution, which is an absolutely continuous function that satisfies the evolutionary equation almost everywhere in time; or one might have a mild solution, which is a solution given by an integral equation, the variation of constants formula.[3] In finite dimensions, these concepts are the same, but they differ in the infinite dimensional setting.

The raison d'être for Hilbert spaces and Banach spaces is the study of linear operators, both bounded and unbounded. Some of the early applications of this study were in the analysis of solutions of linear partial differential equations. In the case of linear evolutionary equations, an operator calculus is needed to study the solutions. Such a calculus was developed by means of a linear semigroup of bounded linear operators and its infinitesimal generator, see Hille and Phillips (1948, 1957). For some linear problems, such as the Stokes problem, the semigroup is analytic. This permits one to introduce a tower of Banach spaces, which in turn offers a good framework for the analysis of the nonlinear problems.

For those evolutionary equations generated by a nonlinear system of partial differential equations, there are basically two approaches for the study of solutions: (1) a methodology based on the theory of mild and strong solutions of the nonlinear problem, and (2) a methodology based on a theory of weak and strong solutions. The mild-strong approach builds on the variation of constants formula

$$u(t) = e^{-At}u_0 + \int_0^t e^{-A(t-s)}F(u(s))\,ds,$$

which defines the mild solution $u(t)$ in the Banach space H, where $u_0 \in H$, e^{-At} is a linear semigroup, and $-A$ is the infinitesimal generator. The nonlinear term $F = F(u)$ includes the nonlinearity in the underlying partial differential equation. In the case where F is a suitable Lipschitz continuous mapping of the phase space H into itself, then the proofs of the existence of mild solutions and the properties of these solutions follow the ordinary differential equation paradigm.

However, a serious complication occurs, as in the Navier-Stokes equations or the Cahn-Hilliard equation, when the nonlinear term in the underlying partial differential equation contains spatial derivatives of the unknown solution. In this case, the evolutionary equation is not well-defined on an L^2-space. One needs to set the problem in a space of functions with greater spatial regularity. However, since the image $v = F(u)$ will have less spatial regularity than u, there is another difficulty.

One would be at an impass here, except for confluence of two very important developments. First, there is the notion of a tower of Banach spaces

[3]See Sections 4.2, 4.6, and 4.7 for more details.

which arise in the case where the linear semigroup is analytic. Second, there are the imbedding theorems of Sergei L Sobolev (1908 - 1989) and the applications to the Sobolev spaces $W^{m,p}$, see Sobolev (1938, 1950). What follows from these two theories is a calculus for the study of the nonlinear terms appearing in many partial differential equations. For example, the inertial term $F = F(u)$ arising in the Navier-Stokes equations is a mapping $F : W^{2,2} \to W^{1,2}$ with two continuous Fréchet derivatives.

The alternate weak-strong approach for solutions, was discovered by Jean Leray (1906 - 1998) in his three papers on the Navier-Stokes equations, one in 1933 and two in 1934. The concept of a weak solution is based on the observation that any bounded set in H, where H is a Hilbert space or a reflexive Banach space, has compact closure in the weak topology on H.

A simplified description of Leray's approach is to begin with the construction of a sequence of approximate solutions for a given IVP for the Navier-Stokes equations. The next step is to use properties of the linear and nonlinear terms in the equations to show that the given sequence is in a bounded set and, therefore, that there is a subsequence that converges weakly. The limit of this subsequence is shown to be a weak solution. With a variation of this argument, one shows that if the initial datum has greater spatial regularity, then the weak solution is a strong solution, at least for time in some finite interval.

While the theory of Leray was formulated for unbounded domains in $\mathbb{R}^3$, Hopf (1951) showed that a similar theory was valid on suitable bounded domains in $\mathbb{R}^3$. Owing to the good properties of the Stokes operator on a bounded domain, the Hopf theory is especially important for the study of the longtime dynamics of the Navier-Stokes equations.

The study of differential equations with time delays, or more generally functional differential equations, is a rather recent development in the theory of evolutionary equations. The basic impetus for obtaining a good theory for these problems was the simple, yet insightful, observation by Hale (1963) and Krasovskii (1963) that the initial value problem is well-posed only when the initial condition is in a suitable function space (e.g., the space of continuous real-valued functions) defined over the delay interval. While it is then a straightforward issue to generate solutions, it should be noted that the resulting theory behind the dynamical features can have all the complexity seen in the case of partial differential equations, see Hale and Verduyn Lunel (1993).

Just the Beginning: By the 1930s, the basic theory of dynamical systems was well in place, and the basic studies, which at a later time would lead to a theory of flows and semiflows for the infinite dimensional evolutionary equations arising in partial differential equations, had begun. During the period 1930 - 1970 there were many major developments in the study of the longtime dynamics of systems of ordinary differential equations, including perturbation theory for invariant manifolds, bifurcation theory, exponential dichotomies and hyperbolic structures, the Pliss reduction principle (center manifold), the Kolmogorov-Arnold-Moser theory, skew products flows for nonautonomous problems, Morse-Smale dynamical systems, the structural stability program, the role of symmetries, and index theory.

By the 1970s, the dynamical theories for dissipative partial differential equations, such as reaction diffusion equations, the Navier-Stokes equations, and the Cahn-Hilliard equation, were coming to fruition. In this area and during the subsequent 30 years, one finds the development of existence theories and dimension theories for global attractors and inertial manifolds, the use of smooth and discrete-valued Lyapunov functions to find Morse-Smale structures and Poincaré-Bendixson theories, and the use of exponential trichotomies and hyperbolic structures for the perturbation theory of invariant manifolds, for example.

The year 1970 is an approximate date of the merger of finite dimensional and infinite dimensional dynamical systems. Since that time, this has become a united subject, the Dynamics of Evolutionary Equations. Other major developments in longtime dynamics which date from the time of this merger include the Melnikov method, singular perturbations, random dynamical systems, almost periodic and almost automorphic dynamics, and approximation dynamics. The subject of the Dynamics of Evolutionary Equations is only at its beginning. While it is not possible to predict the future, we sincerely hope that this volume will be helpful for scholars working in these areas and in some of the newer areas of dynamics, such as global climate modeling, numerical simulation of longtime dynamics, and control theory in time-varying media.

2
DYNAMICAL SYSTEMS: BASIC THEORY

The basic concept underlying the study of dynamics in infinite dimensional spaces is that of a semiflow, or as it is sometimes called, a semigroup. This semiflow is a time-dependent action on the ambient space, which we assume to be a complete metric space W, for example, a Banach space or a Fréchet space. One should think of the semiflow as a mechanism for describing the solutions of an underlying evolutionary equation. This evolutionary equation is oftentimes the abstract formulation of a given partial differential equation or, sometimes, an ordinary differential equation with time delays. In this chapter we will examine some basic properties of semiflows. Our principal objective is to describe the longtime dynamics in terms of the invariant sets, the limit sets, and the attractors of the semiflow. A comprehensive theory of global attractors is included here. Later in this volume, we will develop the connections between the semiflow and the underlying evolutionary equation.

Since one objective of this chapter is to lay the foundations for widespread applicability to the study of solutions of partial differential equations, we have formulated the basic concepts (such as the definition of a semiflow and of Lyapunov stability) to apply in problems, such as the Navier-Stokes equations, which may fail to be continuous at $t = 0$. We hope that this presentation, which has several novel features, will be a useful introduction to the longtime dynamics of infinite dimensional problems.

The basic theory of semiflows and attractors for infinite dimensional semiflows is presented in Section 2.3. The theory presented here does have overlap with other works, perhaps the closest to the point of view developed herein being Hale (1988). Other references include Babin and Vishik (1992), Ball (1997), Conley (1978), Ladyzhenskaya (1991), Sell (1971), Temam (1988), and Vishik (1992). Key features of our theory include the Stability Theorem 23.10, which gives a characterization of an attractor of a

κ-contracting semiflows in terms of uniform asymptotic stability, and the Robustness of Attractors Theorem 23.14. The latter result is the basic principle underlying the many theorems in the literature on the robustness (or the upper semicontinuous dependence on parameters) of attractors. We also present a new proof of the basic theorem concerning the existence of a global attractor (see Theorem 23.12).

The very definition of an attractor is a subtle issue, which we have dealt with carefully. Some authors have used various minimality, or maximality, or connectedness, properties in their definitions. As a result, this can lead to difficulties in comparing theories in different papers. Since all these concepts are, in fact, consequences of more basic features of an attractor, we feel that they should not be included as a part of the basic definition. Even though the definition of an attractor we use here is weaker than that used by others, we are able to develop a theory with the same richness found in other works. Comparisons between our theory and the theories of other researchers are presented in Section 2.7.

In Section 2.1 we present the basic dynamical concepts such as a semiflow, an invariant set, an alpha and omega limit set, and a hull. Section 2.2 is concerned with the concepts of compact and κ-contracting semiflows. These latter two concepts form the underpinnings of the theory of dissipative semiflows. As noted above, the basic theory of attractors is presented in Section 2.3, and a thumbnail sketch of the entire theory is contained in Section 2.3.7. We invite the reader to read this summary at an early stage. Sections 2.4 and 2.5 contain a number of related dynamical concepts which are useful in such problems as longtime dynamics in the presence of finite-time blowup, dynamics for equations with time-varying coefficients, and dynamics of singular perturbations. Section 2.7 includes a brief introduction to the literature. An extensive bibliography on dynamical systems appears in Sell (2001) on the World Wide Web.

Throughout this chapter we will let W denote a complete metric space. The distance between two points u and v in W will be denoted by $d(u,v) = d_W(u,v)$, where $d = d_W$ is a metric on W. Recall that if W is a Banach space, then the standard metric on W is given by $d(u,v) = \|u - v\|$, where $\|\cdot\|$ is the norm on W. In the case of a Fréchet space W, we will use an invariant metric d, where $d(u,v) = d(u-v,0)$, for all $u, v \in W$.

2.1. Semiflows and Nonlinear Semigroups.

Let M be a subset of a complete metric space W, and let $\mathbb{R} = (-\infty, \infty)$ and $\mathbb{R}^+ = [0,\infty)$. A mapping $\sigma = \sigma(u,t)$, where $\sigma : M \times [0,\infty) \to M$, is said to be a **semiflow** on M, provided the following hold:

(1) $\sigma(w,0) = w$, for all $w \in M$.

(2) The **semigroup property** holds, i.e.,

$$\sigma(\sigma(w,s),t) = \sigma(w,s+t), \qquad \text{for all } w \in M, \text{ and } s,t \in \mathbb{R}^+. \tag{21.1}$$

(3) The mapping $\sigma : M \times (0, \infty) \to M$ is continuous.

If in addition, the mapping $\sigma : M \times [0, \infty) \to M$ is continuous, we will say that the semiflow[4] σ is **continuous at** $t = 0$. A situation, which is weaker than continuity at $t = 0$ in the sense used here, and which occurs in some semiflows, arises when the semiflow has the property that for each $w \in M$, the mapping of $[0, \infty)$ into M given by $t \to \sigma(w, t)$ is continuous. (See Chapter 6 for the weak solutions of the 2D Navier-Stokes equations.) Another instructive example is presented at the end of this chapter.

As we will see in the next chapter, a prototype of a semiflow is a C_0-semigroup of linear operators on a Banach space W, and this semiflow is continuous at $t = 0$. In this case, the mapping $u \to \sigma(u, t)$ is a bounded linear operator on W. If the mapping $u \to \sigma(u, t)$ is not linear, then the semiflow is sometimes referred to as a **nonlinear semigroup**. On occasion we will write σ in the form $\sigma(u, t) = S(t)u$. In this notation, the semigroup property (21.1) becomes

$$S(s)S(t)u = S(s + t)u, \qquad \text{for all } s, t \geq 0. \tag{21.2}$$

It may happen that for each $t \geq 0$, the mapping $S(t)$ is a one-to-one mapping of M onto M with a continuous inverse $S(t)^{-1}$. In this case, we set $S(-t) \stackrel{\text{def}}{=} S(t)^{-1}$, for $t > 0$. As a result, (21.2) holds for all $u \in M$ and all $s, t \in \mathbb{R}$. In this case, the dynamical system will be referred to as a **flow**, or **(nonlinear) group**.

There do exist discrete versions of these concepts. For a fixed number $\tau > 0$, define

$$\tau Z^+ = \{n\tau : n = 0, 1, \cdots\} \quad \text{and} \quad \tau Z = \{n\tau : n = 0, \pm 1, \cdots\}.$$

A mapping $\sigma(\cdot, t) : M \to M$, for $t \in \tau Z^+$, is said to be a **discrete semiflow** on M, provided that σ is continuous, $\sigma(w, 0) = w$, for all $w \in M$, and Equation (21.1) holds for all $w \in M$ and $s, t \in \tau Z^+$. A **discrete flow** is defined similarly, but now with τZ replacing τZ^+. The prototype of a discrete semiflow arises when one begins with a semiflow $\sigma(w, t)$ on M, where $t \in [0, \infty)$, and then restricts time t to satisfy $t \in \tau Z^+$, for some $\tau > 0$. Another example is the Poincaré map generated by a system of ordinary differential equations with periodic coefficients. More generally, a mapping $\sigma(u, n\tau)$ is a discrete semiflow on M if and only if there is a continuous mapping $T : M \to M$ such that $\sigma(u, n\tau) = T^n(u)$, with $T^0(u) = u$, for all $u \subset M$ and $n \in Z^+$.

Let Λ be a metric space. We will say that S_λ, for $\lambda \in \Lambda$, is a **continuous family of semiflows** on M, provided that $S_\lambda(t)u = \sigma(\lambda, u, t)$, and the mapping

$$\sigma : \Lambda \times M \times [0, \infty) \to M$$

---

[4]Notice that *continuity at* $t = 0$ is a statement of joint continuity in (u, t) at each point $(u_0, 0)$.

satisfies the following conditions:

(1) the restriction mapping

$$\sigma : \Lambda \times M \times (0, \infty) \to M \tag{21.3}$$

is continuous;

(2) for each $\lambda \in \Lambda$, the mapping $S_\lambda(t)$ is a semiflow on M.

We will say the semiflow $S_0(t)$ is **imbedded into** a continuous family of semiflows S_λ, for $\lambda \in \Lambda$, provided that there is a $\lambda_0 \in \Lambda$ such that

$$S_{\lambda_0}(t)u = S_0(t)u, \qquad \text{for } u \in M \text{ and } t \in [0, \infty).$$

Lemma 21.1 (Continuity Lemma). *Let $S_\lambda(t)$ be a continuous family of semiflows on $M \subset W$, for $\lambda \in \Lambda$. Then the following hold:*

(1) *For any convergent sequences u_n and λ_n with limits $u_n \to u \in M$ and $\lambda_n \to \lambda \in \Lambda$, one has*

$$\sup_{\tau \le t \le T} d(\sigma(\lambda_n, u_n, t), \sigma(\lambda, u, t)) \to 0, \qquad \text{as } n \to \infty, \tag{21.4}$$

for any τ and T with $0 < \tau \le T < \infty$.

(2) *For any compact sets K in M and K_0 in Λ, the set*

$$N = \{S_\lambda(t)u : u \in K,\ \lambda \in K_0,\ \tau \le t \le T\}$$

is compact in M, for any τ and T with $0 < \tau \le T < \infty$.

(3) *Let K be a compact set in M with the property that, for some $\lambda_0 \in \Lambda$, one has $S_{\lambda_0}(t)K \subset K$, for all $t \ge 0$. Then for any τ and T with $0 < \tau \le T < \infty$ and any $\epsilon > 0$, there is a neighborhood $O = O(\lambda_0)$ of λ_0 in Λ and a $\delta > 0$, such that if $d(u, K) \le \delta$, then for all $\lambda \in O$, one has*

$$d(S_\lambda(t)u, K) < \epsilon, \qquad \text{for } \tau \le t \le T.$$

If in addition, each semiflow $S_\lambda(t)$, for $\lambda \in \Lambda$, is continuous at $t = 0$, then the three properties above remain valid with $\tau = 0$.

Proof. Item (1) follows immediately from the fact that any continuous function is uniformly continuous on a compact set. For Item (2) we let $v_n = S_{\lambda_n}(t_n)u_n$ be a sequence in N. By using the compactness of $[\tau, T]$, K, and K_0, we can extract subsequences, which we relabel as t_n, λ_n, and u_n, so that the limits $t = \lim t_n$, $\lambda = \lim \lambda_n$ and $u = \lim u_n$ exist. One then has

$$d(S_\lambda(t)u, S_{\lambda_n}(t_n)u_n) \le d(S_\lambda(t)u, S_\lambda(t_n)u) + d(S_\lambda(t_n)u, S_{\lambda_n}(t_n)u_n).$$

Now $d(S_\lambda(t)u, S_\lambda(t_n)u) \to 0$, as $n \to \infty$, because of the continuity in t. From (21.4), we obtain $d(S_\lambda(t_n)u, S_{\lambda_n}(t_n)u_n) \to 0$, as $n \to \infty$. Item (3) is an immediate consequence of the continuity property (21.3). If the semiflows $S_\lambda(t)$ are continuous at $t = 0$, then the argument above remains valid with $\tau = 0$. □

2.1.1 Invariant Sets. Let σ be a semiflow on $M \subset W$. For any $u \in M$ the **(positive) trajectory through** u is defined as the set

$$\gamma^+(u) \stackrel{\text{def}}{=} \{S(t)u : t \geq 0\},$$

and the mapping $t \to S(t)u$ of $\mathbb{R}^+$ into M is referred to as the **(positive) motion** through u. If σ is a flow on M, then the **(full) trajectory through** u is the set

$$\gamma(u) \stackrel{\text{def}}{=} \{S(t)u : t \in \mathbb{R}\}.$$

The **(full) motion** through u is the mapping $t \to S(t)u$ of $\mathbb{R}$ into M. For any set $K \subset M$ we define

$$S(t)K = \sigma(K, t) \stackrel{\text{def}}{=} \{S(t)u : u \in K\},$$

and the **trajectories through** K are given by

$$\gamma^+(K) \stackrel{\text{def}}{=} \{S(t)K : t > 0\} = \{S(t)u : u \in K, t \geq 0\}$$

and $\gamma(K) \stackrel{\text{def}}{=} \{S(t)K : t \in \mathbb{R}\}$. Note that "trajectories" are sets while "motions" are mappings.

A set $K \subset M$ is said to be **positively invariant** if $S(t)K \subset K$, for all $t \geq 0$, and K is said to be an **invariant** set if $S(t)K = K$, for all $t \geq 0$. For example, the ambient space M, as well as the trajectory $\gamma^+(B)$, where $B \subset M$, are positively invariant sets. A **stationary motion** (where $S(t)u = u$, for all $t \geq 0$) as well as the trajectory of a **periodic motion** (where $S(t+\tau)u = S(t)u$, for all $t \geq 0$ and some $\tau > 0$) generate examples of invariant sets. Notice that for any set B in M one has

$$\gamma^+(S(t)B) = S(t)\gamma^+(B), \qquad \text{for all } t \geq 0. \tag{21.5}$$

There is a very useful characterization of an invariant set, which is given in the next lemma. We will say that a continuous mapping $\phi^u : \mathbb{R} \to M$ is a **globally defined motion** through u with range in M if $\phi^u(0) = u$ and one has

$$S(t)\phi^u(\tau) = \phi^u(\tau + t), \qquad \text{for all } \tau \in \mathbb{R} \text{ and } t \in \mathbb{R}^+. \tag{21.6}$$

Notice that (21.6) implies that $S(t)u = \phi^u(t)$, for $t \geq 0$. The restriction of a globally defined motion ϕ^u to $\mathbb{R}^- = (-\infty, 0]$ is referred to as a **negative continuation** of the motion through u. There is a related concept which arises when one has a continuous mapping $\phi^u : [-T, \infty) \to M$ that satisfies $\phi^u(0) = u$ and (21.6) is valid for $\tau \in [-T, \infty)$ and $t \geq 0$, where $T > 0$.

In this case the restriction $\phi^u : [-T, 0] \to M$ is referred to as a **partial negative continuation** of the motion through u.

For any set $K \subset M$, we define $W^u(K)$, the **unstable set (associated with)** K, as the collection of all points $w \in W$ such that there is a negative continuation $\phi^w(t)$, where $\phi^w(0) = w$ and $\text{dist}(\phi^w(t), K) \to 0$, as $t \to -\infty$. Note that for every set K in M, the unstable set $W^u(K)$ is always an invariant set, although in some cases, $W^u(K)$ may be the empty set.

Not every point $u \in M$ needs to have a globally defined motion passing through it. Also note that we have not addressed the issue of the uniqueness of a negative continuation. An example of a nonunique negative continuation occurs in the semiflow T^0 constructed in Example 25.2. However, it is known that, under certain assumptions, which hold for many partial differential equations, a negative continuation is unique whenever it exists, see Bardos and Tartar (1973) or Temam (1988).

Lemma 21.2. *Let σ be a semiflow on $M \subset W$, and let K be a set in M. Then K is an invariant set if and only if every $u \in K$ has a globally defined motion ϕ^u passing through u with $\phi^u(t) \in K$ for all $t \in \mathbb{R}$.*

Proof. First assume that for every $u \in K$ there is a globally defined motion ϕ^u passing through u with $\phi^u(t) \in K$ for all $t \in \mathbb{R}$. Then by using (21.6), we see that for each $\tau \geq 0$ and any $u \in K$ one has $S(\tau)u = \phi^u(\tau) \in K$, so that $S(\tau)K \subset K$. Since $u = S(\tau)\phi^u(-\tau)$, one has $K \subset S(\tau)K$. Hence K is invariant.

Next assume that K is invariant. Since $S(1)K = K$, we see that for each $u \in K$ there is a $v^1 \in K$ with $S(t)v^1 \in K$, for $0 \leq t \leq 1$, and $S(1)v^1 = u$. By iterating this and using induction, one finds a sequence $v^k \in K$ with $S(t)v^k \in K$, for $0 \leq t \leq 1$, and $S(1)v^k = v^{k-1}$, for $k = 1, 2, \cdots$, where $v^0 = u$. We now define $\phi^u : (-\infty, 0] \to K$ by $\phi^u(t) = S(t+k)v^k$, for $-k < t \leq -k+1$ and $k = 1, 2, \cdots$. The final step, which is left as an exercise, is to show that (21.2) implies (21.6). □

The following result, while very simple, has profound consequences. We leave the proof as an exercise.

Lemma 21.3. *Let N be an invariant set for a semiflow σ on $M \subset W$, and assume that $K = \text{Cl}_M N$ is a compact set. Then K is an invariant set for σ.*

2.1.2 Limit Sets. The limit sets of a semiflow, namely, the omega limit set and the alpha limit set, are important concepts arising in the study of longtime dynamics. Let σ be a semiflow on $M \subset W$. For any set $B \subset M$ we define the the **(positive) hull** $H^+(B)$ and the **omega limit set** $\omega(B)$ as follows:

$$H^+(B) \stackrel{\text{def}}{=} \text{Cl}_M \gamma^+(B) \quad \text{and} \quad \omega(B) \stackrel{\text{def}}{=} \cap_{\tau \geq 0} H^+(S(\tau)B).$$

Since $\omega(B)$ is the intersection of closed sets, we see that $\omega(B)$ is always a closed set in M. Note that $\omega(B)$ may be empty, even when B is not empty.

If $\sigma(u,t) = S(t)u$ is a flow on $M \subset W$, then the study of the alpha limit sets, i.e., the behavior as $t \to -\infty$, is very much like the study of omega limit sets, i.e., the behavior as $t \to +\infty$. Indeed, if one reverses time by defining $\hat{S}(t)$ by $\hat{S}(t)u = S(-t)u$, for $t \in \mathbb{R}$, then the alpha limit sets of S are precisely the omega limit sets of $\hat{S}$. The definitions and the theories are the same.

On the other hand, if $S(t)u$ is a semiflow on $M \subset W$, then several complications arise because one is not able to use the time-reversal trick. In order to describe the dynamics on the alpha limit sets in this case, we will use the globally defined motions of σ, see (21.6). Before doing this though, we note the following:

(1) The globally defined motions of σ need not be defined for each $u \in M$. This will mean that the alpha limit set $\alpha(u)$ need not be defined for every $u \in M$.
(2) Since the time t-mapping, $u \to S(t)u$, need not be one-to-one, for $t > 0$, a given point $u \in M$ may have more than one globally defined motion passing through it.
(3) Another complication arises when M is a positively invariant set for a semiflow σ on N, where $M \subset N \subset W$. One may have a point $u \in M$ with a globally defined motion $\phi^u : \mathbb{R} \to N$ with $\phi^u(t) \not\subset M$, for some $t < 0$. Thus globally defined motions will depend on the ambient space.

With these points in mind, we let $\sigma(u,t) = S(t)u$ be a semiflow on $M \subset W$, and let B be a set in M. For each $t \leq 0$ we define $\hat{S}_M(t)B$ to be the set of all points $\phi^u(t)$, where $u \in B$ and $\phi^u : \mathbb{R} \to M$ is a globally defined motion through u with range in M. The **negative trajectory** through B and the **negative hull** are defined as

$$\gamma_M^-(B) \stackrel{\text{def}}{=} \bigcup_{t \leq 0} \hat{S}_M(t)B \quad \text{and} \quad H_M^-(B) \stackrel{\text{def}}{=} \text{Cl}_M \gamma_M^-(B).$$

The **alpha limit set** is defined by

$$\alpha_M(B) \stackrel{\text{def}}{=} \bigcap_{\tau \leq 0} H_M^-(\hat{S}_M(\tau)B).$$

The **trajectory** $\gamma_M(B)$ and the **hull** $H_M(B)$ are defined by

$$\gamma_M(B) \stackrel{\text{def}}{=} \gamma_M^-(B) \cup \gamma^+(B) \quad \text{and} \quad H_M(B) \stackrel{\text{def}}{=} H_M^-(B) \cup H^+(B).$$

Similarly, we define the negative hull $H^-(\phi^u) = H_M^-(\phi^u)$ and the alpha limit set $\alpha(\phi^u) = \alpha_M(\phi^u)$, for a single negative continuation ϕ^u with $\phi^u(t) \in M$, for all $t \leq 0$. In particular, with $\phi_\tau^u(t) = \phi^u(\tau + t)$, one has

$$H^-(\phi^u) = \text{Cl}_M\{\phi^u(t) : t \leq 0\} \qquad \text{and} \qquad \alpha(\phi^u) = \cap_{\tau \leq 0} H^-(\phi_\tau^u).$$

Let us return to the three issues raised above.[5] First note that if B contains a point u with no globally defined motion ϕ^u passing through it with range in M, then $\hat{S}_M(0)B \subsetneqq B$. Next note that if B is a set in M with the property that for no point $u \in B$ does there exist a globally defined motion ϕ^u through u with range in M, then the sets $\hat{S}(\tau)B$, $\gamma_M^-(\hat{S}_M(\tau)B)$, and $H_M^-(\hat{S}_M(\tau)B)$ are empty for each $\tau \leq 0$, and $\alpha_M(B)$ is empty as well. Also note that if σ is a semiflow on $N \subset W$, and if M is a positively invariant set in N, then for every set B in M one has

$$\gamma_M^-(B) \subset \gamma_N^-(B), \qquad H_M^-(B) \subset H_N^-(B), \qquad \text{and} \qquad \alpha_M(B) \subset \alpha_N(B).$$

The following result, which gives a characterization of the limit sets, is very useful.

Lemma 21.4 (Characterization Lemma). *Let σ be a semiflow on $M \subset W$, and let B be a subset of M. Then the omega limit set $\omega(B)$ is characterized as the set of all $v \in M$ for which one has*

$$v = \lim_{n\to\infty} S(t_n)u_n \tag{21.7}$$

for some sequences $u_n \in B$ and $t_n \in \mathbb{R}^+$, with $t_n \to \infty$, as $n \to \infty$. The alpha limit set $\alpha_M(B)$ is characterized as the set of all $v \in M$ for which there are sequences $v_n \in M$, $u_n \in B$, and $t_n \to \infty$ with $v = \lim_{n\to\infty} v_n$ and $v_n = \phi^{u_n}(-t_n)$ for some globally defined motion ϕ^{u_n} passing through u_n with range in M.

Proof. We will prove this for the omega limit set only. Assume that v satisfies (21.7) for appropriate sequences u_n and t_n, and let $v_n = S(t_n)u_n$. Since $t_n \to \infty$, one has

$$v_n \in \gamma^+(S(\tau)B) \subset H^+(S(\tau)B), \qquad \text{for all } t_n \geq \tau \text{ and all } \tau \geq 0.$$

Since $H^+(S(\tau)B)$ is a closed set and since $t_n \to \infty$, one has

$$v = \lim_{n\to\infty} v_n = \lim_{n\to\infty} S(t_n)u_n \in H^+(S(\tau)B), \qquad \text{for all } \tau \geq 0,$$

which implies that $v \in \omega(B)$. On the other hand, assume that $v \in \omega(B)$. Then one has $v \in H^+(S(n)B)$, for all $n \geq 1$. Hence there is a $v_n \in \gamma^+(S(n)B)$ such that $d(v, v_n) \leq \frac{1}{n}$, for all $n \geq 1$. That is to say, one has $v_n = S(n)S(\tau_n)u_n$ for some sequences $\tau_n \geq 0$ and $u_n \in B$. Consequently, (21.2) implies that

$$v = \lim_{n\to\infty} v_n = \lim_{n\to\infty} S(n)S(\tau_n)u_n = \lim_{n\to\infty} S(n+\tau_n)u_n.$$

Since $(n + \tau_n) \to \infty$, as $n \to \infty$, (21.7) holds. □

---

[5]This is the bad news. The good news is still to come.

Lemma 21.5. *Let σ be a semiflow on $M \subset W$ and let K be a nonempty, compact, invariant set in M. Then for every nonempty set B in K, the following are nonempty, compact, invariant sets in K: the alpha limit set $\alpha_K(B)$, the hull $H_K(B)$, and the omega limit set $\omega(B)$.*

Furthermore, if ϕ^u is a negative continuation, where $H_M^-(\phi^u)$ is a nonempty compact set, then the alpha limit set $\alpha_M(\phi^u)$ is a nonempty, compact, invariant set in M.

Proof. Since the compact set K is invariant, it follows that for every point $u \in K$ there is a globally defined motion ϕ^u passing through u with range in K. It follows that for each $\tau \leq 0$ the sets $H_K^-(\hat{S}_K(\tau)B)$ and $H^+(S(-\tau)B)$ are nonempty and compact. By using the monotonicity of $H^+(S(t)B)$, for $t \geq 0$, and the finite intersection property, we see that $\omega(B)$ is nonempty and compact in K. Similarly, each of the sets $\alpha_K(B)$ and $H_K(B)$ are nonempty, compact subsets of K.

In order to show that $\omega(B)$ is invariant, we first fix $\tau \geq 0$. Next let $v \in \omega(B)$ be fixed so that $v = \lim_{n\to\infty} S(t_n)u_n$, for sequences $t_n \to \infty$ and $u_n \in B$, see (21.7). We seek to find a $u \in \omega(B)$ that satisfies $v = S(\tau)u$. Since K is compact, it follows that there are subsequences of $(t_n - \tau)$ and u_n, which we relabel as $(t_n - \tau)$ and u_n, such that the limit $u = \lim_{n\to\infty} S(t_n - \tau)u_n$ exists. Since $(t_n - \tau) \to \infty$, it follows from the Characterization Lemma 21.4 that $u \in \omega(B)$. By using the continuity of the mapping $w \to S(\tau)w$ and the semigroup property (21.2) one finds that

$$\begin{aligned} S(\tau)u &= S(\tau) \lim_{n\to\infty} S(t_n - \tau)u_n = \lim_{n\to\infty} S(\tau)S(t_n - \tau)u_n \\ &= \lim_{n\to\infty} S(t_n)u_n = v. \end{aligned}$$

The remainder of the argument is left as an exercise. □

2.1.3 Semiflows on Function Spaces. Let W be a given Banach space and let $C(\mathbb{R}, W)$ denote the collection of all continuous functions $f : \mathbb{R} \to W$. For each $f \in C(\mathbb{R}, W)$ we define the **translate** f_τ by

$$f_\tau(t) = f(\tau + t), \qquad \text{for all } \tau, t \in \mathbb{R}.$$

We will assume that $C(\mathbb{R}, W)$ is given with the topology of uniform convergence on bounded sets, see Appendix B. In this case $C(\mathbb{R}, W)$ is a metric space and the mapping

$$(f, \tau) \to \sigma(f, \tau) \stackrel{\text{def}}{=} f_\tau$$

is a continuous mapping of $C(\mathbb{R}, W) \times \mathbb{R}$ into $C(\mathbb{R}, W)$. Since $\sigma(f, 0) = f$ and since (21.1) is satisfied, for all $s, t \in \mathbb{R}$, we see that σ is a flow on $C(\mathbb{R}, W)$. Similarly the translation mapping $\tau \to f_\tau$ of $\mathbb{R}^+$ into $C(\mathbb{R}^+, W)$ defines a semiflow on $C(\mathbb{R}^+, W)$.

Let us now turn to $C = C(V \times \mathbb{R}, W)$, where V and W are Banach spaces and C has the topology of uniform convergence on bounded sets, see Appendix B. For any $f \in C$ and $\tau \in \mathbb{R}$, the **translate** f_τ is defined by

$$f_\tau(v,t) = f(v, \tau + t), \qquad \text{for } v \in V \text{ and } \tau, t \in \mathbb{R}.$$

It is easily verified that the mapping $(f,\tau) \to \sigma(f,\tau) \stackrel{\text{def}}{=} f_\tau$ is a continuous mapping of $C \times \mathbb{R}$ into C. Since $\sigma(f,0) = 0$ and since (21.1) is satisfied, for all $s, t \in \mathbb{R}$, we see that σ is a flow on C. Furthermore, the space $C_{\text{Lip}} = C_{\text{Lip}}(V \times \mathbb{R}, W)$ is an invariant subspace of C. Similarly the mapping $(f,\tau) \to f_\tau$ is a semiflow on $C(V \times \mathbb{R}^+, W)$, and $C_{\text{Lip}}(V \times \mathbb{R}^+, W)$ is a positively invariant subspace of $C(V \times \mathbb{R}^+, W)$.

2.1.4 Ordinary Differential Equations on Banach Spaces. The fact that the theory described above was formulated for a semiflow σ defined on a subset M contained in W, takes into account a natural phenomenon found in the theory of differential equations on Banach spaces. To be more precise, let $f \in C_{\text{Lip}}(W \times \mathbb{R}, W)$ and consider the (ordinary) differential equation $w' = f(w,t)$ on the Banach space W. More specifically we consider the **Initial Value Problem**,

$$\partial_t w = f(w,t), \qquad w(t_0) = w_0, \tag{21.8}$$

where $w_0 \in W$. A **classical solution of** (21.8) is a pair (ϕ, I), where I is an open interval in $\mathbb{R}$ with $t_0 \in I$ and the function ϕ is a differentiable mapping of I into W that satisfies $\phi(t_0) = w_0$ and $\frac{d\phi}{dt} = f(\phi(t), t)$, for $t \in I$. Let $(\phi_i, I_i), i = 1, 2$, be two solutions of (21.8). We will say that ϕ_2 is a **proper extension of** ϕ_1 if $I_1 \subsetneqq I_2$ and $\phi_1(t) = \phi_2(t)$ for $t \in I_1$. A solution (ϕ, I) of (21.8) is said to be **maximally defined** if ϕ has no proper extension that is a solution of (21.8). The following theorem is the special case of a result proved in Section 4.6.

Theorem 21.6. *For every $f \in C_{\text{Lip}} = C_{\text{Lip}}(W \times \mathbb{R}, W)$, $t_0 \in \mathbb{R}$, and $w_0 \in W$, there is a unique maximally defined solution (ϕ, I) of (21.8), with $I = (\alpha, \omega)$ and $t_0 \in I$. Furthermore, the following statements are valid:*

(1) *Either $\omega = \infty$, or one has $\|\phi(t)\| \to \infty$, as $t \to \omega^-$.*
(2) *Either $\alpha = -\infty$, or one has $\|\phi(t)\| \to \infty$, as $t \to \alpha^+$.*

Moreover, the mapping $(f, w_0, t_0, t) \to \phi(t)$ is a continuous mapping of Σ into W, where

$$\Sigma \stackrel{\text{def}}{=} \{(f, w_0, t_0, t) \in C_{\text{Lip}} \times W \times \mathbb{R} \times \mathbb{R} : t \in I\},$$

and Σ is an open set in $C_{\text{Lip}} \times W \times \mathbb{R} \times \mathbb{R}$.

Let us now restrict to autonomous differential equations $w' = f(w)$, i.e., the vector field f is in $C_{\text{Lip}}(W, W)$ and does not depend on time t. For every

$w_0 \in W$, we let $\sigma(w_0, t)$ denote the unique, maximally defined solution of the Initial Value Problem

$$\partial_t w = f(w), \qquad w(0) = w_0,$$

given by the last theorem, and let $I(w_0) = (\alpha(w_0), \omega(w_0))$ be the associated interval of definition of $\sigma(w_0, t)$. Next define

$$M = M(f) \stackrel{\text{def}}{=} \{w_0 \in W : \omega(w_0) = \infty\}.$$

The last theorem implies that $\sigma : M \times \mathbb{R}^+ \to M$ is continuous. Clearly one has $\sigma(w_0, 0) = w_0$. In order to show that (21.1) holds, we fix $s \geq 0$ and define $\phi_1(t) = \sigma(\sigma(w_0, s), t)$, for all $t \geq 0$ and $\phi_2(t) = \sigma(w_0, s + t)$, for all $t \geq 0$. Now both ϕ_1 and ϕ_2 are solutions of the Initial Value Problem:

$$\partial_t w = f(w), \qquad w(0) = \sigma(w_0, s).$$

Consequently, the uniqueness assertion in the last theorem implies that $\phi_1(t) = \phi_2(t)$, for all $t > 0$. This shows that $\sigma(w_0, s + t) = \sigma(\sigma(w_0, s), t)$, i.e., $\sigma(w_0, t)$ is a semiflow on M. In this way, every Lipschitz continuous vector field $f(w)$ on W generates a semiflow on a suitable subset $M(f)$ in W. It is possible that $M(f)$ may be empty.

Let K be a positively invariant set for the translational semiflow

$$S(\tau) : f \to f_\tau$$

on either $C(W \times \mathbb{R}, W)$, or $C(W \times \mathbb{R}^+, W)$. We will say that a set $\Omega \subset W$ is a **(positively) invariant domain** for K provided that for all $f \in K, w_0 \in \Omega$, and $t_0 \geq 0$, the maximally defined solution (ϕ, I) of (21.8), where $I = (\alpha, \omega)$, satisfies $\omega = \infty$ and $\phi(t) \in \Omega$ for all $t \geq t_0$.

Some important examples of invariant domains occur in the finite dimensional case where W is the Hilbert space $\mathbb{R}^n$. As usual, $\langle \cdot, \cdot \rangle$ is the standard inner product on $\mathbb{R}^n$. Let Ω be an open set in $\mathbb{R}^n$ with a piecewise C^2 boundary $\partial\Omega$ such that Ω lies on one side of $\partial\Omega$. Let ν denote the outward normal to Ω. We leave it as an exercise to show that Ω is an invariant domain for a positively invariant set $K \subset C_{\text{Lip}}(W \times \mathbb{R}, W)$ whenever one has

$$\langle f(w, t), \nu \rangle < 0, \qquad \text{for all } w \in \partial\Omega, f \in K, \text{ and } t \geq 0.$$

2.1.5. Flow Homomorphisms. For $i = 1, 2$, we let $S_i(t)$ denote a flow on a metric space M_i, for $t \in \mathbb{R}$, and let $\mathcal{K}_i \subset M_i$ be an invariant set for S_i. A **flow homomorphism** (from $\mathcal{K}_1$ to $\mathcal{K}_2$) is defined to be a continuous mapping $H : \mathcal{K}_1 \to \mathcal{K}_2$ with the properties that $H(\mathcal{K}_1) = \mathcal{K}_2$ and

$$H(S_1(t)m_1) = S_2(t)H(m_1), \qquad \text{for all } (m_1, t) \in \mathcal{K}_1 \times \mathbb{R}. \tag{21.9}$$

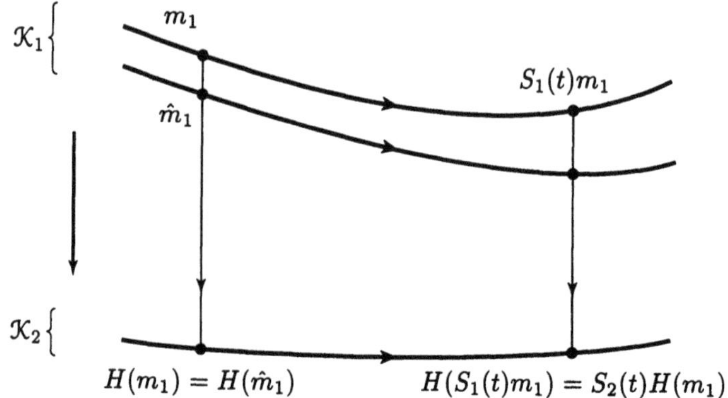

Figure 2.1. Flow Homomorphism

In this case, the invariant set $\mathcal{K}_1$ is said to be an **extension** of $\mathcal{K}_2$, see Figure 2.1.

Note that Equation (21.9) is a precise statement that a flow homomorphism "preserves" the dynamics of $\mathcal{K}_1$. A flow homomorphism $H : \mathcal{K}_1 \to \mathcal{K}_2$ need not be one-to-one. However, if $H(m_1) = H(\hat{m}_1)$, for $m_1, \hat{m}_1 \in \mathcal{K}_1$, then (21.9) implies that $H(S_1(t)m_1) = H(S_1(t)\hat{m}_1)$, for all $t \in \mathbb{R}$, i.e., H maps the entire trajectory $\gamma(m_1)$ in $\mathcal{K}_1$ onto the trajectory $\gamma(H(\hat{m}_1))$ in $\mathcal{K}_2$. Because of this fact, the dynamics on $\mathcal{K}_1$ might be "more complicated" than the dynamics on $\mathcal{K}_2$, which offers a rationale for the term "extension" defined above.

An example of a flow homomorphism is the mapping $H : T^2 \to T^1$ given by $H(\theta_1, \theta_2) = \theta_1$, where T^k is the k-dimensional torus. The dynamics on T^2 and T^1 are

$$S_1(t)(\theta_1, \theta_2) = (\theta_1 + t, \theta_2 + \omega t) \quad \text{and} \quad S_2(t)\theta_1 = \theta_1 + t,$$

respectively, where $\omega \in \mathbb{R}$. Other examples will be developed in the text.

2.2. Compact and κ-Contracting Semiflows.

Many of the applications of the theory of semiflows to partial differential equations, or to differential equations with time delays, occur in an important setting wherein the given semiflow has some smoothing property. In this section we will examine two of these properties: compactness and κ-contracting.

The semiflow $\sigma(u,t) = S(t)u$ on W is said to be **compact**, if for every bounded set B in W there is a $r = r(B)$, $0 \le r < \infty$, such that for every $t > r$, the set $S(t)B$ lies in a compact subset of W, i.e., $\mathrm{Cl}_W S(t)B$ is compact. The number $r(B)$ is referred to as a **compactification time** for $S(t)B$. If $r(B) = t_0$ can be chosen independent of B, then we will say that the semiflow σ is **compact for** $t > t_0$. The following lemma gives a sufficient condition, in terms of compact imbeddings, for a semiflow σ to be compact.

Lemma 22.1. *Let σ be a semiflow on a Banach space W. Let $V \hookrightarrow W$ be a linear subspace of W that is compactly imbedded in W. Assume that for every bounded set B in W, there is an $r = r(B) \geq 0$ such that the set $S(t)B$ is bounded in V for all $t > r$, i.e., there is a function $b = b(B,t)$ with $0 \leq b(B,t) < \infty$ such that*

$$\|S(t)u\|_V \leq b(B,t), \qquad \text{for all } t > r \text{ and } u \in B.$$

Then σ is a compact semiflow on W.

Proof. Since V is compactly imbedded in W, every bounded set in V is mapped into a compact set in W by the inclusion mapping. Hence σ is a compact semiflow. □

The **Kuratowski measure of noncompactness** is the nonnegative, real-valued function $\kappa(B)$ defined for the bounded sets $B \subset W$ that is given by

$$\kappa(B) \stackrel{\text{def}}{=} \inf\{d : B \text{ has a finite open cover of sets of diameter } < d\}.$$

If B is a nonempty, unbounded set in W, then we define $\kappa(B) = \infty$. The properties of $\kappa(B)$, which we will use herein, are given in the following lemma.

Lemma 22.2. *The Kuratowski measure of noncompactness $\kappa(B)$ on a complete metric space W satisfies the following properties:*

(1) *$\kappa(B) = 0$ if and only if $Cl_W B$ is compact.*
(2) *If $B_1 \subset B_2$, then $\kappa(B_1) \leq \kappa(B_2)$.*
(3) *$\kappa(B_1 \cup B_2) = \max(\kappa(B_1), \kappa(B_2))$.*
(4) *$\kappa(B) = \kappa(Cl_W B)$.*
(5) *If B_t is a family of nonempty, closed, bounded sets defined for $t > r$ that satisfy $B_s \supset B_t$, whenever $s \leq t$, and $\kappa(B_t) \to 0$, as $t \to \infty$, then $\bigcap_{t>r} B_t$ is a nonempty, compact set in W.*
(6) *Let B_t be given as in (5). Then for any sequences $t_n \in \mathbb{R}^+$, with $t_n \to \infty$ as $n \to \infty$, and $u_n \in B_{t_n}$, there exist subsequences, which we relabel as t_n and u_n, such that $u = \lim_{n\to\infty} u_n$ exists and $u \in \bigcap_{t>r} B_t$.*

If in addition, W is a Banach space, then the following are valid:

(7) *$\kappa(B_1 + B_2) \leq \kappa(B_1) + \kappa(B_2)$.*
(8) *$\kappa(B) = \kappa(Cl_W \operatorname{conv} B)$, where $Cl_W \operatorname{conv} B$ is the closed convex hull of B.*
(9) *If $L : W \to W$ is a bounded linear operator, then $\kappa(LB) \leq \|L\|\kappa(B)$.*

Proof. The proofs of properties (1) - (4) and (7) - (9) are rather easy, and we omit the details. In order to prove Item (5), we note that it suffices to restrict t to take on integer values, $t = n > r$, only, since one has $\cap_{t>r} B_t = \cap_{n>r} B_n$. Define B by $B \stackrel{\text{def}}{=} \cap_{n>r} B_n$. By Item (2) one has $\kappa(B) \leq \kappa(B_n)$, for all $n > r$. Hence $\kappa(B) = 0$. Since B is closed, it is compact.

In order to show that B is nonempty, we begin by selecting a point $u_n \in B_n$, for each $n > r$. Let $A_1 = \{u_n : n > r\}$ and $A_2 = \text{Cl}_W A_1$. Then $\kappa(A_1) = \kappa(A_2)$ by Items (2) and (4). We claim that $\kappa(A_1) = 0$. Indeed, since one has $A_1 \subset B_n \cup \{u_i : r < i \leq n-1\}$, it follows from Items (1), (2), and (3) and the definition of κ that $\kappa(A_1) \leq \kappa(B_n)$, for all $n \geq 1$. Hence $\kappa(A_1) = \kappa(A_2) = 0$, and A_2 is a nonempty, compact set. Now if A_1 consists of a finite number of points, then there is a $u \in A_1$ with the property that $u \in B_n$ for all $n > r$. On the other hand, if A_1 is an infinite set, then there is a point of accumulation $u \in A_2$. Since the sets B_n are closed and monotone, one has $u \in B_n$, for all $n > r$. Thus in either case, one has $u \in B$.

For Item (6) we let t_n and u_n be sequences that satisfy $t_n \to \infty$, as $n \to \infty$, and $u_n \in B_{t_n}$. Define A_n as

$$A_n \stackrel{\text{def}}{=} \text{Cl}_W\{u_m : m \geq n\}.$$

Then A_n is bounded and closed with $A_n \supset A_{n+1}$ and $\kappa(A_n) \leq \kappa(B_{t_n}) \to 0$. Hence Item (5) is applicable, and $\cap_{n \geq r} A_n$ is nonempty and compact. Also for any $u \in \cap_{n \geq r} A_n$ there is a subsequence of u_n, which we relabel as u_n, such that $u = \lim_{n\to\infty} u_n$. □

A semiflow σ on W is said to be **κ-contracting** if for every bounded set $B \subset W$, one has $\kappa(S(t)B) \to 0$, as $t \to \infty$. The semiflow σ is said to be **uniformly κ-contracting** if there is an $r \geq 0$ and a nonnegative function $k(t)$ with $k(t) \to 0$, as $t \to \infty$, such that for every bounded set B in W one has $\kappa(S(t)B) \leq k(t)\kappa(B)$, for all $t > r$. The prototype of a κ-contracting semiflow is a compact semiflow, since $\kappa(S(t)B) = 0$, for all $t > r(B)$ and for every bounded set B in W.

As we will see later, a κ-contracting semiflow on W has many nice properties. Two of the elementary properties are given in the following result.

Lemma 22.3. *Let σ be a κ-contracting semiflow on complete metric space W. Then the following properties are valid:*

(1) *Let w be a point in W, where there is a negative continuation ϕ^w and a bounded set B, such that $\phi^w(t) \in B$, for all $t \leq 0$. Then the negative hull $H^-(\phi^w)$ is a compact set in $\text{Cl}_W B$, and the α-limit set $\alpha(\phi^w)$ is a nonempty, compact invariant set for σ.*

(2) *Let B be a closed, bounded invariant set in W. Then B is a compact set.*

Proof. We will leave this as an exercise. However, it is useful to note that for $\tau \geq 0$ one has

$$\{\phi^w(t) : t \leq 0\} = S(\tau)\{\phi^w(t) : t \leq -\tau\} \subset S(\tau)B.$$

The κ-contracting property then implies that $H^-(\phi^w)$ is compact. The arguments concerning the alpha limit set and the invariant set B are straightforward. □

The following two lemmas describe sufficient conditions for a given semiflow σ to be κ-contracting. See Sell (1972) for additional properties.

Lemma 22.4. *Let $\sigma(u,t) = S(t)u$ be a semiflow on a Banach space W. Assume that one can write S is the form $S = S_1 + S_2$, i.e.,*[6]

$$S(t)u = S_1(t)u + S_2(t)u, \qquad \text{for all } t \geq 0 \text{ and } u \in W.$$

Assume further that for every $a \geq 0$ there is an $r = r(a)$, $0 \leq r < \infty$, such that the closed ball B_a, centered at the origin and of radius a, satisfies

(1) *$Cl_W S_1(t)B_a$ is compact for all $t > r(a)$, and*
(2) *there is a function $k(a,t)$ such that $k(a,t) \to 0$, as $t \to \infty$, and*

$$\|S_2(t)u\| \leq k(a,t), \qquad \text{for all } t \geq 0 \text{ and } u \in B_a.$$

Then σ is a κ-contracting semiflow on W.

Proof. Indeed, let A be any bounded set in W, and fix $a \geq 0$ so that $A \subset B_a$. Then from Lemma 22.2, we have

$$\kappa(S(t)A) \leq \kappa(S(t)B_a) = \kappa(S_1(t)B_a + S_2(t)B_a) \leq \kappa(S_2(t)B_a),$$

for all $t > r(a)$, since $\kappa(S_1(t)B_a) = 0$, for all $t > r(a)$. Next for all $\epsilon > 0$ one has

$$\kappa(S_2(t)B_a) \leq \operatorname{diam}(S_2(t)B_a) + \epsilon \leq 2k(a,t) + \epsilon, \qquad \text{for all } t \geq 0.$$

Since ϵ is arbitrary, one has $\kappa(S(t)B_a) \to 0$, as $t \to \infty$. Hence σ is a κ-contracting semiflow. If the compactification time r for $S_1(t)$ can be chosen independent of the radius a, then σ is a uniformly κ-contracting semiflow. □

A pseudometric ρ on a Banach space W is said to be **precompact** if any bounded sequence in W has a subsequence that is a Cauchy sequence with respect to ρ.

---

[6] While S does satisfy (21.2), we do not insist that either S_1 or S_2 satisfy this semigroup property.

Lemma 22.5. *Suppose that a semiflow σ on a Banach space W satisfies* (22.1)

$$\|S(t)w_1 - S(t)w_2\| \leq \phi(t)\|w_1 - w_2\| + \rho_t(w_1, w_2), \qquad w_1, w_2 \in W,\ t > \tau,$$

where $\tau \geq 0$ is a constant, ρ_t is a precompact pseudometric on W, for each $t > \tau$, and ϕ is a nonincreasing function that satisfies $\phi(t) \to 0$, as $t \to \infty$. Then the semiflow is uniformly κ-contracting.

Proof. Let B be any bounded set in W, and fix $\epsilon > 0$. Then from the definition of Kuratowski measure κ, there are open sets $B_1, \ldots, B_m$ such that $B \subset B_1 \cup \cdots \cup B_m$ and

$$\operatorname{diam} B_i \leq \kappa(B) + \epsilon, \qquad i = 1, \ldots, m.$$

Now fix $t > \tau$. Since ρ_t is a precompact pseudometric on W, there is an $\frac{\epsilon}{2}$-net for B; that is to say, there is an integer $n = n(t) \geq 1$ and a collection of sets $N_1^t, \ldots, N_{n(t)}^t$ in W such that $B \subset N_1^t \cup \cdots \cup N_{n(t)}^t$ and

$$\rho_t(x, y) \leq \epsilon, \qquad \text{for } x, y \in N_j^t,\ j = 1, \ldots, n(t).$$

As a result one has

$$B \subset \bigcup_{i=1}^{m} \bigcup_{j=1}^{n(t)} (B_i \cap N_j^t)$$

and

$$S(t)B \subset \bigcup_{i=1}^{m} \bigcup_{j=1}^{n(t)} S(t)(B_i \cap N_j^t).$$

If $w_1, w_2 \in B_i \cap N_j^t$, then (22.1) implies that

$$\|S(t)w_1 - S(t)w_2\| \leq \phi(t)(\kappa(B) + \epsilon) + \epsilon.$$

The monotonicity of ϕ then implies that

$$\operatorname{diam} S(t)(B_i \cap N_j^t) \leq \phi(t)\kappa(B) + \epsilon(\phi(\tau) + 1).$$

Since this inequality is valid for every $\epsilon > 0$, we take the limit as $\epsilon \to 0$, and use Item (3) of Lemma 22.2 to obtain $\kappa(S(t)B) \leq \phi(t)\kappa(B)$, for $t > \tau$. Since $\phi(t) \to 0$, as $t \to \infty$, we conclude that the semiflow σ is uniformly κ-contracting. □

The following commutivity relationship describes a useful property of the Kuratowski measure of noncompactness in the context of semiflows.

Lemma 22.6. *Let σ be a semiflow on W and let B be a set in M with the property that $\gamma^+(S(\tau)B)$ is bounded, for some $\tau \geq 0$. Then one has*

$$\kappa\left(H^+(S(t)B)\right) = \kappa\left(S(t)\gamma^+(B)\right), \qquad \textit{for all } t \geq \tau.$$

Proof. The equalities $\kappa(H^+(S(t)B)) = \kappa(\gamma^+(S(t)B)) = \kappa(S(t)\gamma^+(B))$, for $t \geq \tau$, follow immediately from (21.5) and Lemma 22.2, Item (4). □

2.3. Attractors and Global Attractors.

The attractors of a semiflow are very important objects. The reason for this is that much (but not all) of the longtime dynamics is represented by the dynamics on and near the attractors. Of special interest is the global attractor. Not every semiflow has a global attractor. However, many dissipative semiflows do, and when this global attractor exists, it is the depository of all the longtime dynamics of the given system.

For any two bounded sets A and B in W we define

$$\delta(B,A) \stackrel{\text{def}}{=} \sup_{u\in B}\left(\inf_{v\in A} d(u,v)\right).$$

Thus one has $\delta(B,A) \leq \epsilon$ if and only if $B \subset N_\epsilon(A)$, where $N_\epsilon(A)$ is the closed ϵ-neighborhood of A. An alternate definition is

$$\delta(B,A) = \inf\{\epsilon > 0 : B \subset N_\epsilon(A)\}.$$

The function δ is not symmetric. That is, in general one has $\delta(B,A) \neq \delta(A,B)$. If A and B are closed bounded sets in H, then the **Hausdorff distance** between A and B is given by

$$\text{dist}_{\text{Hausdorff}}(A,B) \stackrel{\text{def}}{=} \max(\delta(A,B),\delta(B,A)).$$

Note that $\text{dist}_{\text{Hausdorff}}(A,B)$ is a metric on the closed, bounded sets in W. If A is a set in W and u is a point in W, we define the distance from u to A by

$$\text{dist}_W(u,A) \stackrel{\text{def}}{=} \inf_{v\in A} d(u,v) = \delta(\{u\},A).$$

Note that $\text{dist}_W(u,A) = 0$ if and only if $u \in \text{Cl}_W A$. Also note that if w_n is any convergent sequence in W with limit w_0, then for any bounded set A in W one has $d(w_n,A) \to d(w_0,A)$, as $n \to \infty$.

We are primarily interested in the function $\delta(B,A)$ in the case where both A and B are nonempty and bounded. Nevertheless, the definition of the function δ extends in a consistent way to some other cases. For example, we can define $\delta(\emptyset,A) = 0$, for any set A, but we will leave $\delta(B,\emptyset)$ as being undefined, when B is nonempty. Let A_1 be a nonempty set. If $A_1 \subset A_2$, then $\delta(B,A_1) \geq \delta(B,A_2)$; and if $B_1 \subset B_2$, then $\delta(B_1,A_1) \leq \delta(B_2,A_1)$.

Let σ be a semiflow on $M \subset W$, and let A and B be two sets in M. We will say that A **attracts** B if

$$\delta(S(t)B,A) \to 0, \qquad \text{as } t \to \infty. \tag{23.1}$$

Notice that (23.1) is equivalent to saying that for every $\epsilon > 0$ there is a $T = T(\epsilon) \geq 0$ such that

$$\text{dist}_W(S(t)u,A) \leq \epsilon, \qquad \text{for all } t > T \text{ and } u \in B, \tag{23.2}$$

that is, $S(t)B$ lies in $N_\epsilon(A)$, for all $t \geq T$. (It is important to note the uniformity in (23.1) and (23.2). Inequality (23.2) holds for all $t \geq T$ and $u \in B$.) Since the value of $\delta(B,A)$ is unchanged if one replaces B by the closure $\text{Cl}_M B$, we see that if A attracts B, then A attracts $\text{Cl}_M B$. Also note that if A is empty, then (23.1) is valid if and only if B is empty.

2.3.1 Asymptotical Compactness. We now introduce a key concept which plays a pivotal role in the theory of attractors. Let σ be a semiflow on $M \subset W$. We will say that σ is **asymptotically compact on a set** $B \subset M$, if, for any sequences $u_n \in B$ and $t_n \to \infty$, there exist subsequences, which we relabel as u_n and t_n, with the property that the limit $v = \lim S(t_n)u_n$ exists and $v \in M$.

A semiflow σ on $M \subset W$ is said to be **ultimately bounded** if for every bounded set B in M, there is a $\tau = \tau(B) \geq 0$ such that $\gamma^+(S(\tau)B)$ is bounded. The following lemma describes some relations between the concepts of attracts, asymptotically compact, and ultimately bounded. (Also see Lemmas 23.8 and 23.9 below.)

Lemma 23.1. *Let σ be a semiflow on $M \subset W$. Then the following hold:*

(1) *Let A and B be sets in M, where A is bounded and A attracts B. Then one has $\omega(B) \subset Cl_M A$.*
(2) *Let σ be asymptotically compact on a nonempty set $B \subset M$. Then $\omega(B)$ is a nonempty, compact, invariant set in M, and $\omega(B)$ attracts B.*
(3) *Let A be a nonempty, compact set in M, and assume that A attracts a nonempty set B. Then σ is asymptotically compact on B and the conclusions of Items (1) and (2) hold.*
(4) *Let σ be asymptotically compact on a set B in M. Then there exists a $\tau \geq 0$ such that $\gamma^+(S(\tau)B)$ is bounded.*
(5) *If σ is asymptotically compact on every bounded set in M, the σ is ultimately bounded.*

Proof. Item (1): Let $\epsilon > 0$ be given. From (23.2) one has $S(t)B \subset N_\epsilon(A) \cap M$, for all $t \geq T(\epsilon)$. Since $N_\epsilon(A)$ is closed, one has $H^+(S(t)B) \subset N_\epsilon(A) \cap M$, for all $t \geq T(\epsilon)$. It follows that $\omega(B) \subset \cap_{\epsilon>0} (N_\epsilon(A) \cap M) \subset \mathrm{Cl}_M A$.

Item (2): Assume now that σ is asymptotically compact on a nonempty set $B \subset M$. It follows from the definition of asymptotic compactness and the Characterization Lemma 21.4 that $\omega(B)$ is nonempty. In order to show that $\omega(B)$ is compact, we let v_n be any sequence in $\omega(B)$. Then the Characterization Lemma implies that for each $n \geq 1$ there is a $u_n \in B$ and $t_n \geq n$ such that $d(v_n, S(t_n)u_n) \leq \frac{1}{n}$. The asymptotic compactness property allows us to choose subsequences, which we relabel as v_n, u_n and t_n, so that $v = \lim S(t_n)u_n \in \omega(B) \subset M$. Clearly one then has $v = \lim v_n$, i.e., $\omega(B)$ is a compact set. The proof of the invariance of $\omega(B)$ is identical to the argument of Lemma 21.5, and we omit the details.

In order to show that $\omega(B)$ attracts B, we proceed by contradiction, and assume that for some $\epsilon > 0$ there does not exist a time $T \geq 0$ such that

$$\mathrm{dist}_W(S(t)u, \omega(B)) \leq \epsilon, \qquad \text{for all } u \in B \text{ and } t \geq T.$$

As a result there are sequences $u_n \in B$ and $t_n \in \mathbb{R}^+$ such that $t_n \to \infty$, while

$$\mathrm{dist}_W(S(t_n)u_n, A) > \epsilon, \qquad \text{for all } n \geq 1, \tag{23.3}$$

where $A = \omega(B)$. Next we use the asymptotic compactness property to select a subsequence, which we relabel as u_n and t_n, so that the limit $v = \lim S(t_n)u_n$ exists. From the Characterization Lemma one has $v \in \omega(B)$, which contradicts (23.3).

Item (3): Now assume that A is a nonempty, compact set in M and that A attracts a nonempty set B. Let $u_n \in B$ and $t_n \to \infty$ be given sequences. One then has $\mathrm{dist}_W(S(t_n)u_n, A) \to 0$, as $n \to \infty$, because A attracts B. This implies that there are subsequences, which we relabel as u_n and t_n, and there is a sequence $v_n \in A$ such that $d_W(S(t_n)u_n, v_n) \leq \frac{1}{n}$. Since A is compact, there are further subsequences, which we relabel as u_n, t_n, and v_n, such that the limit $v = \lim_{n\to\infty} v_n$ exists, $v \in A$, and $v = \lim_{n\to\infty} S(t_n)u_n$. Hence σ is asymptotically compact on B. The remainder of the proof now follows from Items (1) and (2).

Items (4) and (5): If it were not the case that $\gamma^+(S(\tau)B)$ is bounded, for some $\tau \geq 0$, then there exist sequences $u_n \in B$ and $t_n \to \infty$ such that $d(u_0, S(t_n)u_n) \to \infty$, for any $u_0 \in M$. However, this contradicts the fact that $S(t_n)u_n$ contains a convergent subsequence. Finally, Item (5) follows immediately from the definition of ultimate boundedness and Item (4) of this lemma. □

2.3.2 Attractors and Their Properties. A set $\mathfrak{A}$ contained in M is said to be an **attractor** for the semiflow σ on M, provided that

(1) $\mathfrak{A}$ is a compact, invariant set in M, and
(2) there is a neighborhood U of $\mathfrak{A}$ in M, such that $\mathfrak{A}$ attracts every bounded set in U.

Equivalently, $\mathfrak{A}$ is an **attractor** for σ provided that $\mathfrak{A}$ is a compact, invariant set in M and there is a bounded neighborhood V of $\mathfrak{A}$ in M with the property that $\mathfrak{A}$ attracts V. The **basin of attraction** of $\mathfrak{A}$ is defined as

$$B(\mathfrak{A}) \stackrel{\text{def}}{=} \{u \in M : \mathrm{dist}_W(\sigma(u,t), \mathfrak{A}) \to 0, \ \text{as } t \to \infty\}.$$

A set $\mathfrak{A}$ is said to be a **global attractor** for σ provided that[7] $M = W$ and

(1) $\mathfrak{A}$ is a compact, invariant set in W;
(2) there is a neighborhood U of $\mathfrak{A}$ in M, such that $\mathfrak{A}$ attracts every bounded set in U; and
(3) the basin of attraction satisfies $B(\mathfrak{A}) = W$.

If $B(\mathfrak{A}) \neq W$, then $\mathfrak{A}$ is sometimes referred to as a **local attractor**. It happens that the empty set $\emptyset$ is an attractor with basin of attraction $B(\emptyset) = \emptyset$. However, a global attractor is always nonempty.

Several properties of attractors, which are needed later, are given in the following two lemmas.

[7]This definition of a global attractor differs from other definitions found in the literature. For more details on these differences, see Section 2.7.

Lemma 23.2. *Let σ be a semiflow on $M \subset W$, and let $\mathfrak{A}$ be an attractor for σ. Let U be a neighborhood of $\mathfrak{A}$ with the property that $\mathfrak{A}$ attracts every bounded set in U. Then the following are valid:*

(1) *For every nonempty, bounded set B in U, σ is asymptotically compact on B and the omega limit set $\omega(B)$ is a nonempty, compact, invariant set with $\omega(B) \subset \mathfrak{A}$, and $\omega(B)$ attracts B.*
(2) *The basin of attraction $B(\mathfrak{A})$ is an open set in M.*
(3) *For every nonempty compact set K in $B(\mathfrak{A})$ one has (a) $\mathfrak{A}$ attracts K; (b) $\omega(K)$ is a nonempty, compact, invariant set; (c) $\omega(K) \subset \mathfrak{A}$; and (d) $\omega(K)$ attracts K.*

Proof. It suffices to prove this result in the case where $\mathfrak{A}$ is nonempty. Item (1): Since $\mathfrak{A}$ is a nonempty, compact, invariant set which attracts B, Item (1) follows from Lemma 23.1.

Item (2): Let V be a bounded neighborhood of $\mathfrak{A}$ with the property that $\mathfrak{A}$ attracts V. This means that for any $\epsilon > 0$, there is a time $T = T(V, \epsilon) \geq 0$, such that $S(t)V \subset N_\epsilon(\mathfrak{A})$, for all $t \geq T$. Since we can replace V by its interior, if necessary, there is no loss in generality in assuming V to be open. Then for every $u \in B(\mathfrak{A})$, there is a time $\tau = \tau(u) \geq 0$ such that $S(\tau)u \in V$. From the continuity of σ, there is a constant $\delta > 0$ such that $S(\tau)N_\delta(u) \subset V$ as well. It then follows that $\mathfrak{A}$ attracts $N_\delta(u)$, and therefore $B(\mathfrak{A})$ is open. In particular, one has $S(t+\tau)N_\delta(u) \subset S(t)V \subset N_\epsilon(\mathfrak{A})$, for all $t \geq T = T(V, \epsilon)$. This in turn implies that

$$S(s)N_\delta(u) \subset N_\epsilon(\mathfrak{A}), \qquad \text{for all } s \geq T + \tau. \tag{23.4}$$

Item (3): Let K be a given compact set in $B(\mathfrak{A})$, and let $\epsilon > 0$ be given so that $N_\epsilon(\mathfrak{A}) \subset U$, where U is given by Item (1). By using the compactness of K, we can find a δ-net $\{u_1, \ldots, u_n\}$ in K and times $\{\tau_1, \ldots, \tau_n\}$ in $\mathbb{R}^+$ so that $S(\tau_i)N_\delta(u_i) \subset V$, for $1 \leq i \leq n$. Inequality (23.4) then implies that

$$S(s)K \subset \bigcup_{i=1}^{n} S(s)N_\delta(u_i) \subset N_\epsilon(\mathfrak{A}), \qquad \text{for all } s \geq T + \tau,$$

where $\tau = \max(\tau_1, \ldots, \tau_n)$. Hence $\mathfrak{A}$ attracts K. Finally we note that for sufficiently large τ_0 one has $S(\tau_0)K \subset U$, since $N_\epsilon(\mathfrak{A}) \subset U$, for sufficiently small $\epsilon > 0$. The remainder of the proof of Item (3) now follows from Item (1). □

Lemma 23.3 (Maximality Property). *Let σ be a semiflow on $M \subset W$, and let $U \subset M$. Let B be a closed bounded set in U, and assume that B attracts each compact set in U. Then B is maximal in the sense that every compact, invariant set K in U satisfies $K \subset B$. In particular, every compact, invariant set K in the basin of attraction of an attractor $\mathfrak{A}$ satisfies $K \subset \mathfrak{A}$.*

Proof. Let K be a compact invariant set in U. Let $u \in K$ and let ϕ^u be a negative continuation of the motion through u with the property that

$\phi^u(t) \in K$, for all $t \leq 0$, see Lemma 21.2. Let $D \overset{\text{def}}{=} \text{Cl}_W\{\phi^u(t) : t \leq 0\}$. Then $D \subset K$ and D is a compact set in U. Since K is invariant, one has $u \in S(t)D$, for all $t \geq 0$. Since B attracts D, one has

$$\text{dist}_U(u, B) \leq \delta(S(t)D, B) \to 0, \qquad \text{as } t \to \infty.$$

Since B is closed, this implies that $u \in B$, i.e., one has $K \subset B$. The remainder of the proof now follows from Lemma 23.2, Item (3). □

Lemma 23.4. *Let σ be a semiflow on $M \subset W$. Let $\mathfrak{A}$ be an attractor, and let V be any bounded neighborhood of $\mathfrak{A}$ with the property that $\mathfrak{A}$ attracts V. One then has $\mathfrak{A} = \omega(V)$.*

Proof. Since $\mathfrak{A} \subset V$, we obtain $\mathfrak{A} = \omega(\mathfrak{A}) \subset \omega(V)$. Now Lemma 23.2 implies that $\omega(V)$ is a compact, invariant set in the basin $B(\mathfrak{A})$, and the Maximality Property implies that $\omega(V) \subset \mathfrak{A}$. Hence one has $\mathfrak{A} = \omega(V)$. □

Lemma 23.5 (Minimality Property). *Let σ be a semiflow on $M \subset W$. Let $\mathfrak{A}$ be an attractor for σ, and let U be a neighborhood of $\mathfrak{A}$ in M with the property that $\mathfrak{A}$ attracts every bounded set in U. Then $\mathfrak{A}$ is minimal, in the sense that, if B is any closed bounded set in U that attracts every compact set in U, then $B \supset \mathfrak{A}$.*

Proof. Note that since $\mathfrak{A}$ is a compact set in U, B attracts $\mathfrak{A}$. From Lemma 23.1, Item (1), and the invariance of $\mathfrak{A}$, one then obtains $\mathfrak{A} = \omega(\mathfrak{A}) \subset B$. □

In the next result, we examine the issue of the connectedness of an attractor $\mathfrak{A}$. For this purpose it is convenient to note that if N is a positively invariant set for a semiflow σ on $M \subset W$, then every component of N is, itself, a positively invariant set. In particular, each component of the basin of attraction $B(\mathfrak{A})$ of an attractor $\mathfrak{A}$ is positively invariant.

Lemma 23.6 (Connectedness Lemma). *Let σ be a semiflow on $M \subset W$. Then the following hold:*

(1) *Let B be a nonempty set in M and assume that σ is asymptotically compact on B. If B is connected, then $\omega(B)$ is connected.*

Assume now that W is a Banach space and that M is an open, connected set in W. Then the following hold:

(2) *Let $\mathfrak{A}$ be a nonempty attractor for σ, and let $B(\mathfrak{A})$ denote the basin of attraction of $\mathfrak{A}$. Assume that there is an open, connected set $U_0 \subset B(\mathfrak{A})$ with $\mathfrak{A} \subset U_0$. Then the attractor $\mathfrak{A}$ is connected.*

(3) *Assume that $M = W$. Then any global attractor $\mathfrak{A}$ in W is connected.*

Proof. [8] Item (1): In order to prove the connectedness of $\omega(B)$, we proceed by contradiction. If $\omega(B)$ is not connected, then for sufficiently small $\epsilon > 0$,

[8] This proof of the Connectedness Lemma remains valid if M and W are assumed to be path-connected metric spaces.

the closed neighborhood $N_\epsilon = N_\epsilon(\omega(B))$ can be written in the form $N_\epsilon = U_\epsilon \cup V_\epsilon$, where U_ϵ and V_ϵ are disjoint, nonempty, closed sets, and both $\omega(B) \cap U_\epsilon$ and $\omega(B) \cap V_\epsilon$ are nonempty. Since $\omega(B)$ attracts B (see Lemma 23.2), there is a time $T \geq 0$ such that $S(t)B \subset N_\epsilon = U_\epsilon \cap V_\epsilon$, for all $t > T$. Since the continuous image of a connected set is connected, the set $S(t)B$ is connected for each $t \geq 0$. As a result, for each $t > T$, the set $S(t)B$ lies in either U_ϵ, or V_ϵ, but not both. Say that $S(\tau)B \subset U_\epsilon$, for some $\tau > T$. The continuity of σ implies that $S(t)B$ cannot jump to V_ϵ at any $t > \tau$. As a result one has $S(t)B \subset U_\epsilon$, for all $t \geq \tau$. Since U_ϵ is closed, one has $\omega(B) \subset U_\epsilon$, and $\omega(B) \cap V_\epsilon = \emptyset$, which is a contradiction.

Item (2): Let U_0 be an open, connected set in $B(\mathfrak{A})$ with $\mathfrak{A} \subset U_0$. If $\mathfrak{A}$ is not connected, then there exist disjoint, open sets U and V in U_0 with $\mathfrak{A} \subset U \cup V$, and there exist points $u \in \mathfrak{A} \cap U$ and $v \in \mathfrak{A} \cap V$. Since $\mathfrak{A}$ is invariant, each of the sets $\mathfrak{A} \cap U$ and $\mathfrak{A} \cap V$ are invariant as well. Therefore $S(t)u \in \mathfrak{A} \cap U$ and $S(t)v \in \mathfrak{A} \cap V$, for all $t \geq 0$. Since U_0 is open and connected, it is path-connected. Therefore there is a continuous mapping $\phi : [0,1] \to U_0$ with $\phi(0) = u$ and $\phi(1) = v$. The image $K = \phi([0,1])$ is a compact, connected set in $U_0 \subset B(\mathfrak{A})$, and by Lemma 23.2, $\mathfrak{A}$ attracts K. Since $B(\mathfrak{A})$ is positively invariant, it follows that $S(t)K$ is a compact, connected set in $B(\mathfrak{A})$, for each $t \geq 0$. Since $\mathfrak{A}$ attracts K and since $U \cup V$ is an open neighborhood of $\mathfrak{A}$, there is a time $T_0 = T_0(K) \geq 0$ such that $S(t)K \subset U \cup V$, for all $t \geq T_0$. Since U and V are disjoint, and since $S(t)K$ is connected, one finds that for each $t \geq T_0$ the set $S(t)K$ lies in U, or V, but not both. However, this contradicts the fact that $S(t)u \in U$ and $S(t)v \in V$, for all $t \geq 0$.

Item (3): This follows directly from Item (2) because one can take $U_0 = W = B(\mathfrak{A})$. □

Since each component of the basin of attraction $B(\mathfrak{A})$ is positively invariant, it follows from Lemmas 23.2 and 23.6 that, in a path-connected space M, an attractor $\mathfrak{A}$ is connected if and only if the basin $B(\mathfrak{A})$ is connected.

2.3.3. Stability, Dissipativity, and Absorbing Sets. Let σ be a semiflow on W and let $A \subset W$. The set A is said to be **Lyapunov stable**, provided that A is positively invariant and for every $\tau > 0$ and every neighborhood V of A, there is a neighborhood U of A such that

$$S(t)\,U \subset V, \qquad \text{for all } t \geq \tau. \tag{23.5}$$

The set A is said to be **uniformly asymptotically stable** if it is Lyapunov stable and there is a neighborhood U_0 of A such that A attracts U_0, i.e.,

$$\delta(S(t)U_0, A) \to 0, \qquad \text{as } t \to \infty. \tag{23.6}$$

Let σ be a semiflow on W. We will say that σ is **point dissipative**, or simply **dissipative**, if there is a nonempty bounded set A in W such that A attracts every point in W. In this case, any set B that contains an

ϵ-neighborhood $N_\epsilon(A)$, for some $\epsilon > 0$, is said to be an **absorbing set** for σ. Note that the existence of an absorbing set for σ is equivalent to stating that the semiflow is point dissipative. If B is an absorbing set, then for any $u \in W$ there is a $T = T(u, \epsilon) \geq 0$ such that

$$S(t)u \in N_\epsilon(A) \subset B, \qquad \text{for all } t \geq T.$$

In many applications the absorbing set is a closed neighborhood $N_\rho(0)$ of the origin of radius $\rho > 0$. For any closed, bounded set B in W, we define $K = K(B)$ by

$$K = K(B) \stackrel{\text{def}}{=} \{u \in B : \gamma^+(u) \subset B\}, \tag{23.7}$$

see Figure 2.2. (Note that $K(B)$ may be empty.) We will say that a continuous family of semiflows $S_\lambda(t)$ is **uniformly dissipative at** $\lambda_0 \in \Lambda$ provided that there is a neighborhood O of λ_0 in Λ and an open, bounded set A in W such that A is an absorbing set for $S_\lambda(t)$, for each $\lambda \in O$.

Lemma 23.7 (Stability Lemma). *Let σ be a semiflow on $M \subset W$ and let $\mathfrak{A}$ be an attractor in M. Then $\mathfrak{A}$ is uniformly asymptotically stable.*

Proof. Let U be a bounded neighborhood of $\mathfrak{A}$ with the property that $\mathfrak{A}$ attracts U. Then $U_0 = U$ satisfies (23.6) with $A = \mathfrak{A}$. It remains to show that $\mathfrak{A}$ is Lyapunov stable. Fix $\epsilon_0 > 0$ so that for all ϵ, $0 < \epsilon \leq \epsilon_0$, one has $N_\epsilon(\mathfrak{A}) \subset U$. It suffices to show that for every $\tau > 0$, there is a neighborhood V of $\mathfrak{A}$ such that $S(t)V \subset N_\epsilon(\mathfrak{A})$, for all $t \geq \tau$. Assume on the contrary that this is not the case. Then there is a $\tau > 0$ and there are sequences $u_n \in U$ and $t_n \in [\tau, \infty)$, such that $\text{dist}_W(u_n, \mathfrak{A}) \to 0$, as $n \to \infty$, and $S(t_n)u_n$ lies in the set $N_\epsilon(\mathfrak{A})^c$, the complement of $N_\epsilon(\mathfrak{A})$. This means that (23.3) is valid for these sequences, where $A = \mathfrak{A}$. Since $\text{dist}_W(u_n, \mathfrak{A}) \to 0$, as $n \to \infty$, and since $\mathfrak{A}$ is compact, there is a subsequence of u_n, which we will relabel as u_n, that is convergent and $u = \lim_{n\to\infty} u_n \in \mathfrak{A}$. Let $A_0 \stackrel{\text{def}}{=} \{u_m : m \geq 1\}$. Since $A_0 \subset U$ and since $\mathfrak{A}$ attracts U, we see that $\mathfrak{A}$ attracts A_0, i.e., one has $\delta(S(t)A_0, \mathfrak{A}) \to 0$, as $t \to \infty$. In other words, there is a $T = T(\epsilon) \geq \tau$ such that

$$\text{dist}_W(S(t)u_n, \mathfrak{A}) \leq \epsilon, \qquad \text{for all } t \geq T \text{ and all } n \geq 1. \tag{23.8}$$

By using the Continuity Lemma 21.1 we choose another subsequence of u_n, which we relabel again as u_n, such that

$$d(S(t)u_n, S(t)u) \leq \epsilon, \qquad \text{for } \tau \leq t \leq T \text{ and all } n \geq 1.$$

One then has,

$$\text{dist}_W(S(t)u_n, \mathfrak{A}) \leq d(S(t)u_n, S(t)u) \leq \epsilon, \tag{23.9}$$

for all $t \in [\tau, T]$ and all $n \geq 1$. Since $t_n \geq \tau$, we see that (23.8) and (23.9) contradict (23.3). □

2.3.4 Attractors for κ-Contracting Semiflows. In this section and the next, we describe some key properties of κ-contracting semiflows.

Lemma 23.8. *Let σ be a κ-contracting semiflow on a complete metric space W. Then the following holds.*

(1) *Assume that for a nonempty set B in W, there is a $\tau \geq 0$, such that $\gamma^+(S(\tau)B)$ is bounded. Then σ is asymptotically compact on B; the omega limit set $\omega(B)$ is a nonempty, compact, and invariant; and $\omega(B)$ attracts B.*

(2) *Assume that B is a bounded, open set with $H^+(S(t)B) \subset B$, for some $t \geq \tau$. Then σ is asymptotically compact on B, and $\mathfrak{A} = \omega(B)$ is an attractor with $B \subset B(\mathfrak{A})$.*

Proof. Item (1): In order to show that σ is asymptotically compact on B, we let $u_n \in B$ and $t_n \to \infty$ be given sequences. By choosing subsequences, if necessary, we can assume that $\tau \leq t_n \leq t_{n+1}$, for all n. Let $v_n = S(t_n)u_n$ and set $A_n = \mathrm{Cl}_M\{v_m : m \geq n\}$. By Lemma 22.6, one has

$$\kappa(A_n) \leq \kappa\left(H^+(S(t_n)B)\right) = \kappa(S(t_n - \tau)\gamma^+(S(\tau)B)) \to 0, \qquad \text{as } n \to \infty.$$

By Lemma 22.2, Item (6), there exist subsequences, which we relabel as u_n and t_n, such that $S(t_n)u_n$ is convergent. Hence σ is asymptotically compact on B. The remainder of Item (1) now follows from Lemma 23.1.

Item (2): Since σ is asymptotically compact on B, by Item (1), it follows from Lemma 23.1 that $\mathfrak{A} = \omega(B)$ is a compact, invariant set in W and that $\mathfrak{A}$ attracts B. Since $\omega(B) \subset H^+(S(t)B) \subset B$, it follows that B is a neighborhood of $\mathfrak{A}$. From the definitions we see that $\mathfrak{A}$ is an attractor and $B \subset B(\mathfrak{A})$. □

Lemma 23.9. *Let σ be a κ-contracting semiflow on a complete metric space W, and assume that σ is point dissipative. Then the following statements are valid:*

(1) *There is a nonempty, closed, bounded set B_0 in W that attracts all points in W.*

(2) *Let $K_0 = K(B_0)$ be given by (23.7), where B_0 satisfies Item (1). Then K_0 is a nonempty, closed, bounded, positively invariant set in W that attracts all points in W. Furthermore, σ is asymptotically compact on K_0; the omega limit set $A = \omega(K_0)$ is a nonempty, compact, invariant set that attracts all points in W; and A attracts K_0.*

Proof. Item (1): Since σ is dissipative, there is a nonempty, bounded set N in W that attracts all points in W. Then $B_0 \stackrel{\text{def}}{=} \mathrm{Cl}_W N$ is a nonempty, closed, bounded set that attracts all points in W.

Item (2): Let K_0 be given as in the statement of this lemma. Since B_0 attracts every point $u \in W$, it follows that the positive orbit $\gamma^+(u)$

is bounded. Consequently Lemma 23.8 implies that the omega limit set $\omega(u)$ is a nonempty, compact, invariant set, and Lemma 23.1 implies that $\omega(u) \subset B_0$. Since $\omega(u)$ is invariant, one has $\omega(u) \subset K_0$. Hence K_0 is nonempty. Clearly K_0 is bounded and positively invariant with $\gamma^+(K_0) \subset K_0 \subset B$. We claim that K_0 is closed. If this were not the case, then there is a convergent sequence $u_n \in K_0$ with $u = \lim_{n\to\infty} u_n \notin K_0$. Since B is closed, one has $u \in B$, and since $u \notin K_0$, there is a $\tau > 0$ such that $S(\tau)u$ lies in the open set B^c, the complement of B. From the continuity of σ one concludes that $S(\tau)u_n \in B^c$, for large n, which contradicts the fact that $\gamma^+(u_n) \subset K_0 \subset B$. Hence K_0 is closed. Now Lemma 23.8 implies that $\omega(u)$ attracts u, for each $u \in W$. Since $\omega(u) \subset K_0$, we see that K_0 attracts all points in W. Now define $A \overset{\text{def}}{=} \omega(K_0)$. Since $\gamma^+(K_0) \subset K_0$, the remainder of the proof of Item (2) now follows from Lemma 23.8, Item (1). □

The following result is an interesting characterization of an attractor for a κ-contracting semiflow in terms of uniform asymptotic stability.

Theorem 23.10 (Stability Theorem). *Let σ be a κ-contracting semiflow on $M \subset W$, where W is a complete metric space, and let $\mathfrak{A}$ be a nonempty invariant set in M. Then the following statements are equivalent:*

(1) *$\mathfrak{A}$ is an attractor for σ.*
(2) *$\mathfrak{A}$ is uniformly asymptotically stable.*
(3) *There is a neighborhood V of $\mathfrak{A}$, where $\gamma^+(S(\tau)V)$ is bounded, for some $\tau \geq 0$, and $\omega(V) = \mathfrak{A}$.*

Proof. The implication (1) ⇒ (2) follows from the Stability Lemma 23.7, and the implication (2) ⇒ (3) follows from the definitions. The implication (3) ⇒ (1) follows from Lemma 23.8. It is only the latter implication which uses the κ-contracting property. □

2.3.5 Existence of Global Attractors. The following two results describe sufficient conditions for the existence of an attractor, or a global attractor.

Theorem 23.11. *Let σ be a semiflow on $M \subset W$, where W is a complete metric space. Assume that there is a compact set K in M and a neighborhood U of K in M with the property that K attracts all bounded sets in U. Then the following are valid:*

(1) *The semiflow σ has an attractor $\mathfrak{A} \subset K$.*
(2) *The basin of attraction satisfies $B(\mathfrak{A}) \supset U$.*
(3) *The attractor satisfies $\mathfrak{A} = \omega(K)$.*
(4) *Moreover, if $M = W$ and K attracts all points in W, then $\mathfrak{A}$ is a global attractor for σ, and $\mathfrak{A}$ attracts all compact sets in W.*

Proof. It suffices to prove this when K is nonempty. Let B be a bounded neighborhood of K in U. Since K attracts B, Lemma 23.1, Item (3), implies that (i) σ is asymptotically compact on B, (ii) the omega limit set $\omega(B)$ is a nonempty, compact, invariant set with $\omega(B) \subset K$, and (iii) $\omega(B)$ attracts B. From the definition and the invariance of the omega limit set, we then obtain $\omega(\omega(B)) = \omega(B) \subset \omega(K)$. Since $K \subset B$, one finds that K attracts itself and $\omega(K) \subset \omega(B)$. Consequently one has $\omega(B) = \omega(K)$. Define $\mathfrak{A}$ by $\mathfrak{A} \stackrel{\text{def}}{=} \omega(K)$. One then has $\mathfrak{A} \subset K$, and B is a bounded neighborhood of $\mathfrak{A}$ as well. Since $\mathfrak{A} = \omega(K) = \omega(B)$ and since $\omega(B)$ attracts B, we see that $\mathfrak{A}$ is an attractor. The remainder of the proof is straightforward, and we omit the details. □

While this theorem gives some information about attractors for general semiflows, applying it may be a problem because of the difficulty of identifying an appropriate compact set K. In the case of a compact semiflow, or more generally, a κ-contracting semiflow, we are able to derive a stronger result.

Existence Theorem 23.12. *Let σ be a κ-contracting semiflow on W, where W is a complete metric space, and assume that σ is point dissipative and that for every compact set K in W, there is a $\tau \geq 0$ such that $\gamma^+(S(\tau)K)$ is bounded. Then there is a global attractor $\mathfrak{A}$ for σ, and $\mathfrak{A}$ attracts every compact set K in W.*

Assume in addition that σ satisfies one of the following properties:

(1) *σ is compact, or*
(2) *σ is ultimately bounded,*

then $\mathfrak{A}$ attracts every bounded set B in W.

Proof. Let K_0 and $A = \omega(K_0)$ be given by Lemma 23.9. We then have that A is a nonempty, compact, invariant set in W, and A attracts every point in W. We claim that there is an $\epsilon > 0$ and a $\tau \geq 0$ such that $\gamma^+(S(\tau)N_\epsilon(A))$ is a bounded set in W. We will argue this by contradiction. If this were not the case, then there are sequences w_n in W and t_n in $\mathbb{R}^+$, such that $\text{dist}(w_n, A) \to 0$, $t_n \to \infty$, and $\text{dist}(S(t_n)w_n, A) \to \infty$, as $n \to \infty$. Hence there is a sequence $u_n \in A$, such that $d(u_n, w_n) \to 0$, as $n \to \infty$. Since A is compact, there is no loss in generality in assuming that u_n is convergent, say that $u_n \to u_0$, as $n \to \infty$. One then has $w_n \to u_0$ and $K = \{w_n\} \cup \{u_0\}$ is a compact set in W. Since $\text{dist}(S(t_n)w_n, A) \to \infty$, as $n \to \infty$, this contradicts the assumption that $\gamma^+(S(\tau)K)$ be bounded for some $\tau \geq 0$.

We let $\epsilon > 0$ be chosen as above and set $B = N_\epsilon(A)$, and let $\tau \geq 0$ be fixed so that $\gamma^+(S(\tau)B)$ is bounded in W. Even though Int B is an absorbing set for σ, and, for every $w \in W$, the ω limit set $\omega(w)$ is a nonempty, compact, invariant set with $\omega(w) \subset A$, we do not claim that A is the global attractor for σ. In general, the global attractor is a larger set. For example, the invariant set A may have an "unstable manifold", which lies in B, and further, the individual solutions in the unstable manifold may leave B and remain outside B for time t in a bounded set. The global attractor $\mathfrak{A}$ must include any unstable manifold, as well as any points on

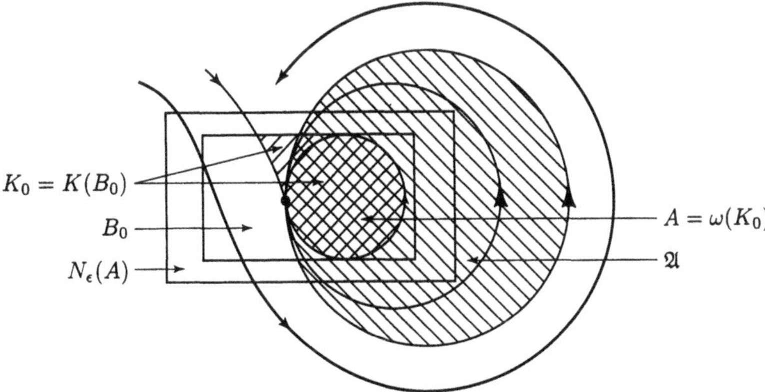

Figure 2.2. Detail for Lemma 23.9 and Theorem 23.12

the positive trajectory of motions beginning in the unstable manifold. Our proof of this theorem, which consists of several steps, is based on the theory of negative continuations of motions of the semiflow σ. The first step is to describe a set $\mathfrak{A}$, which we will later show to be the global attractor.

Step (1). The construction of $\mathfrak{A}$: In order to describe the unstable manifold for A, we define $N(B)$ to be the collection of all points $u \in B$ with the property that there is a negative continuation ϕ^u and a sequence t_n^u in $\mathbb{R}$ with $t_n^u \to -\infty$ and with $\phi^u(t_n^u) \in B$, for all n. It follows from the construction of B that for each $u \in N(B)$, the negative continuation ϕ^u satisfies $\phi^u(t) \in \gamma^+(S(\tau)B)$, for all $t \leq 0$, which in turn implies that $N(B) \subset \gamma^+(S(\tau)B)$. Furthermore, one has

$$N(B) \subset S(t)N(B) \subset S(t)\gamma^+(S(\tau)B), \qquad \text{for all } t \geq 0. \tag{23.10}$$

Since $\kappa(K) = \kappa(N(B))$, where $K = \mathrm{Cl}_W N(B)$, it follows that K is a compact set in W, by the κ-contracting property and Lemma 22.2. Since $A \subset N(B)$, it follows that $N(B)$ is nonempty and $A \subset K$, see Figure 2.2.

Next we define a special subset Γ of K by $\Gamma \stackrel{\text{def}}{=} \{u \in K : d(u, A) = \epsilon\}$, where $\epsilon > 0$ is given above with $B = N_\epsilon(A)$. Since K is compact, it follows that Γ is compact as well. Since A attracts all points in W, it follows that for each $u \in \Gamma$ there is a time $\tau(u)$ where $S(\tau(u))u \in \mathrm{Int}\, B$, the interior of B. Owing to the continuity of σ, for each $u \in \Gamma$, there is a neighborhood V_u of u with the property that $S(\tau(u))V_u \subset \mathrm{Int}\, B$. Using the compactness of Γ, we choose a finite cover $\{V_{u_1}, \dots, V_{u_m}\}$ of Γ and assume that this is ordered so that $0 \leq \tau_i \leq \tau_m$, where $\tau_i = \tau(u_i)$, for $1 \leq i \leq m$. Define $T \stackrel{\text{def}}{=} \tau(u_m)$.

Note that if Γ is empty, then for every $u \in N(B)$, the negative continuation ϕ^u satisfies $\phi^u(t) \in B$, for all $t \leq 0$. The reason for this is that $\phi^u(t)$ cannot leave B, unless there is a time $t_1 < 0$ with $d(\phi^u(t_1), A) = \epsilon$; i.e., $\phi^u(t_1) \in \Gamma$. For the same reason, one has $S(t)u \in B$, for all $t \geq 0$. In other words, if Γ is empty, then $N(B)$ is an invariant set with $N(B) \subset B$.

Since K is compact, it follows from Characterization Lemma 21.4, that K is invariant, which in turn implies that $K = N(B)$. When Γ is empty, we take $T = 0$. Whether Γ is empty or not, we define $\mathfrak{A}$ by $\mathfrak{A} \stackrel{\text{def}}{=} \gamma^+(K)$.

Now (23.10) implies that $\mathfrak{A} \subset S(t)\mathfrak{A}$, for all $t \geq 0$. Since $\mathfrak{A}$ is a positive trajectory, one has $S(t)\mathfrak{A} \subset \mathfrak{A}$, for all $t \geq 0$. Thus, $S(t)\mathfrak{A} = \mathfrak{A}$, for all $t \geq 0$, i.e., $\mathfrak{A}$ is invariant. Furthermore, the restriction of $S(t)$ to $\mathfrak{A}$ is continuous at $t = 0$. Since K is compact, it then follows from Lemma 21.1, Item (2), that the set

$$D^0 \stackrel{\text{def}}{=} \bigcup_{0 \leq t \leq T} S(t)K$$

is compact. We will next show that that $\mathfrak{A} = D^0$. Since the definitions of D^0 and $\mathfrak{A}$ imply that $D^0 \subset \mathfrak{A}$, we need to verify that for every $u \in K$ and every $t_0 \geq 0$, one has $u_0 \stackrel{\text{def}}{=} S(t_0)u \in D^0$.

Since $u \in K$, there is a sequence $u_n \in N(B)$, such that $u_n \to u$, as $n \to \infty$. From the continuity, one obtains that $S(t_0)u_n \to S(t_0)u = u_0$, as $n \to \infty$. It is convenient now to distinguish two cases. In the first case, we assume that there is a subsequence such that $S(t_0)u_n \in B$. By restricting to this subsequence, which we relabel as u_n, one has $S(t_0)u_n \in N(B)$, from the definition of $N(B)$. Hence one finds that $S(t_0)u_n \to S(t_0)u = u_0 \in K = \mathrm{Cl}_W N(B) \subset D^0$.

In the second case, we assume that $S(t_0)u_n \notin B$, for some subsequence, say, u_n. Since A attracts each point u_n, there exist two sequences t_{1n} and t_{2n} in $\mathbb{R}^+$ with the following properties: (a) one has $t_{1n} < t_0 < t_{2n}$; (b) the sequences $S(t_{1n})u_n$ and $S(t_{2n})u_n$ are in Γ; and (c) one has $S(t)u_n \notin B$, for $t_{1n} < t < t_{2n}$. It then follows from the argument above that $0 \leq t_{2n} - t_{1n} \leq T$, which implies that $t_{1n} \in [t_0 - T, t_0]$ and $t_{2n} \in [t_0, t_0 + T]$. By compactness, there exist subsequences, which we will relabel as u_n, t_{1n}, and t_{2n}, such that the following limits exist: $t_{1n} \to t_{10} \in [t_0 - T, t_0]$ and $t_{2n} \to t_{20} \in [t_0, t_0 + T]$, with $0 \leq t_{20} - t_{10} \leq T$. Since $t_{10} \leq t_0 \leq t_{20}$, this implies that $0 \leq t_0 - t_{10} \leq T$. By continuity, one then has the limit $S(t_{1n})u_n \to S(t_{10})u \in \Gamma \subset K$. Consequently, by the Continuity Lemma 21.1, one obtains

$$\lim_{n \to \infty} S(t_0)u_n = \lim_{n \to \infty} S(t_0 - t_{1n})S(t_{1n})u_n = S(t_0 - t_{10})S(t_{10})u = u_0,$$

which implies that $u_0 \in S(t_0 - t_{10})K \subset D^0$. Hence one has $\mathfrak{A} = D^0$ and we see that $\mathfrak{A}$ is a compact invariant set in W with $A \subset \mathfrak{A}$.

Step (2). Property A: *If u is any point in W with the property that there is a negative continuation ϕ^u and a sequence t_n^u in $\mathbb{R}$ with $t_n^u \to -\infty$ and with $\phi^u(t_n^u) \in B$, for all n, then one has $u \in \mathfrak{A}$.*

In order to prove Property A, we consider first the case where $u \in B$. Then one has $u \in N(B) \subset K \subset \mathfrak{A}$, from the definitions. For the case where $u \notin B$, we replace u with $v = \phi^u(t_n^u)$, for some n where $t_n^u < 0$. One then has $v \in N(B) \subset K \subset \mathfrak{A}$. Since $\mathfrak{A}$ is invariant, one obtains $u = \phi^u(0) = S(-t_n^u)v \in \mathfrak{A}$.

Step (3). Property B: *If u is any point in W with the property that there is a negative continuation ϕ^u and a bounded set B_1 such that $\phi^u(t) \in B_1$, for all $t \leq 0$, then one has $u \in \mathfrak{A}$.*

In order to prove Property B, we let ϕ^u and B_1 be given so that B_1 is bounded and $\phi^u(t) \in B_1$, for all $t \leq 0$. It then follows from Lemma 22.3 that the α-limit set $\alpha(\phi^u)$ is a nonempty, compact, invariant set with $\alpha(\phi^u) \subset \mathrm{Cl}_W B_1$. Let $u_0 \in \alpha(\phi^u)$. Since A attracts u_0, one has $\omega(u_0) \subset A$. Since $\alpha(\phi^u)$ is invariant, one has $\omega(u_0) = \omega(u_0) \cap \alpha(\phi^u)$. Hence, from the defintion of the α-limit set, there is a sequence t_n, with $t_n \to -\infty$, such that $\phi^u(t_n) \to A$; that is, there is a subsequence, which we relabel as t_n, such that $\phi^u(t_n) \in B$, for all n. It then follows from Property A that $u \in \mathfrak{A}$.

Step (4). Property C: *There is a $\tau = \tau(\epsilon) > 0$ with the following property: For any two points u and v in $N_\epsilon(\mathfrak{A})$ that satisfy:*

1. *$S(t_0)u = v$, for some $t_0 \geq 0$;*
2. *$S(t)u \in N_\epsilon(\mathfrak{A})$, for all t with $0 \leq t \leq t_0$; and*
3. *$d(v, \mathfrak{A}) = \epsilon$;*

one has $t_0 \leq \tau$.

First we define B_τ and B_∞ by

$$B_\tau \stackrel{\text{def}}{=} \bigcap_{0 \leq t \leq \tau} \mathrm{Cl}_W S(t) N_\epsilon(\mathfrak{A}), \quad \text{for } \tau \geq 0, \text{ and} \quad B_\infty \stackrel{\text{def}}{=} \bigcap_{\tau \geq 0} B_\tau.$$

One then has $\mathfrak{A} \subset B_\tau \subset N_\epsilon(\mathfrak{A}) \cap S(\tau)N_\epsilon(\mathfrak{A})$, for all $\tau \geq 0$. Since $\kappa(B_\tau) \leq \kappa(S(\tau)N_\epsilon(\mathfrak{A})) \to 0$, as $\tau \to \infty$, it follows that B_∞ is a nonempty compact set in $N_\epsilon(\mathfrak{A})$ (see Lemma 22.2).

In order to prove Property C, we need to show that for some $\tau > 0$, the set $B_\tau \cap \Gamma$ is empty. We will argue this by contradiction. Assume instead that the set $B_\tau \cap \Gamma$ is nonempty, for each $\tau \geq 0$. It then follows from Lemma 22.2 that the set $B_\infty \cap \Gamma$ is nonempty, as well. Let $v \in B_\infty \cap \Gamma$. We will now show that there is a negative continuation ϕ^v and a bounded set B_1, with $\phi^v(t) \in B_1$, for all $t \leq 0$. The existence of such a negative continuation then contradicts Property B, since one would have $v \in \mathfrak{A}$ while $d(v, \mathfrak{A}) = \epsilon > 0$

In order to construct the negative continuation ϕ^v, we begin with a sequence v_n in $B_{t_n} \cap \Gamma$, where $t_n \to \infty$ and $v_n \to v$, as $n \to \infty$. From the definition of B_{t_n}, it follows that there is a $w_n \in S(t_n)N_\epsilon(\mathfrak{A}) \subset B_{t_n}$, such that $d(v_n, w_n) < \frac{1}{n}$. Hence one has $w_n \to v$, as $n \to \infty$. Also for each w_n, there is a partial negative continuation $\phi^{w_n} : [-t_n, 0] \to W$, with $x_n \stackrel{\text{def}}{=} \phi^{w_n}(-t_n) \in N_\epsilon(\mathfrak{A})$ and $v \in B_\infty$, by Lemma 22.1, Item (6). Assume for the moment that one has a sequence $v^k \in B_\infty$, for integers $k \geq 0$, with the following properties:

$$v^0 = v \quad \text{and} \quad S(1)v^{k+1} = v^k, \qquad \text{for } k \geq 0.$$

Define $\phi^v = \phi^v(-t)$, for $t \geq 0$, by $\phi^v(-k) = v^k$, for $k \geq 0$, and $\phi^v(-k+t) = S(t)v^k$, for $0 \leq t \leq 1$. Then ϕ^v is a negative continuation of v, with

$\phi^v(-k) \in B_\infty$, for $k \geq 0$. Furthermore, one has

$$\phi^v(-t) \in B_1 \stackrel{\text{def}}{=} \bigcup_{0 \leq \tau \leq 1} S(\tau)B_\infty, \qquad \text{for all } t \geq 0. \tag{23.11}$$

We leave it as an exercise to show that B_1 is a bounded set in W, which leads to the desired contradiction (see Exercise 23.16).

How does one get the sequence v^k? One uses the partial negative continuations ϕ^{w_n} to define

$$u_n^k \stackrel{\text{def}}{=} S(t_n - k)x_n = \phi^{w_n}(-k) \in B_{t_n - k}, \qquad \text{for } 0 \leq k \leq t_n. \tag{23.12}$$

For $k \geq 0$ fixed, it follows from Lemma 22.1, Item (6) that there exist subsequences, which we relabel as u_n^k, such that $u_n^k \to v^k \in B_\infty$, as $n \to \infty$. Clearly, one has $v^0 = v$. By using the semigroup property (21.2), with (23.12) and the definition of a partial negative continuation, one obtains

$$S(1)u_n^{k+1} = S(1)\phi^{w_n}(-(k+1)) = \phi^{w_n}(-k) = u_n^k, \qquad \text{for } 0 \leq (k+1) \leq t_n.$$

By taking the limit, as $n \to \infty$, in the last equation, one obtains $S(1)v^{k+1} = v^k$, which completes the proof of Property C.

Step (5). Property D (Lyapunov Stability): *For any $t_0 > 0$, there is a $\delta > 0$ such that $\gamma^+(S(t_0)N_\delta(\mathfrak{A})) \subset \text{Int } N_\epsilon(\mathfrak{A})$, where $\epsilon > 0$ is given above.*

In order to prove Property D, we let $\tau > 0$ be given by Property C, and we fix t_0 so that $0 < t_0$. It then follows from the Continuity Lemma 21.1, Item (3), that there is a $\delta > 0$ such that if $u \in N_\delta(\mathfrak{A})$, then one has

$$d(S(t)u, \mathfrak{A}) < \epsilon, \qquad \text{for } t_0 \leq t \leq t_0 + 2\tau.$$

Property C then implies that

$$d(S(t)u, \mathfrak{A}) < \epsilon, \qquad \text{for all } t \geq t_0,$$

which in turn implies Property D.

Step (6). *The set $\mathfrak{A}$ is a global attractor for σ, and $\mathfrak{A}$ attracts each compact set in W.*

In order to prove this assertion, we first note that Property D and Lemma 23.8 imply that the omega limit set $\mathfrak{A}_0 \stackrel{\text{def}}{=} \omega(N_\delta(\mathfrak{A}))$ is a nonempty, compact, invariant set which attracts $N_\delta(\mathfrak{A})$. Since one has

$$\mathfrak{A} \subset H^+(S(t)N_\delta(\mathfrak{A})), \qquad \text{for each } t \geq t_0,$$

it follows from the definition of an omega limit set that $\mathfrak{A} \subset \mathfrak{A}_0$. Since $\mathfrak{A}_0$ is compact and invariant, it follows from Property B that $\mathfrak{A}_0 \subset \mathfrak{A}$. Hence one has $\mathfrak{A} = \omega(N_\delta(\mathfrak{A}))$, and $\mathfrak{A}$ is an attractor, since it attracts $N_\delta(\mathfrak{A})$, which is a neighborhood of $\mathfrak{A}$. Since $A = \omega(K_0) \subset K$ - see Step (1) - one has $A \subset \mathfrak{A}$. Consequently, it follows that $\mathfrak{A}$ attracts every point in W, i.e., the basin satisfies $B(\mathfrak{A}) = W$. Hence $\mathfrak{A}$ is the global attractor. The fact that $\mathfrak{A}$ attracts every compact set in $B(\mathfrak{A}) = W$ now follows from Lemma 23.2.

Step (7). The Final Step: The final step is to look at the two cases:

(1) σ is compact, or
(2) σ is ultimately bounded.

If σ is compact, then for every bounded set B_1 in W, there is a time $t_1 > 0$, such that $K_1 = \mathrm{Cl}_W S(t_1)B_1$ is compact. Since $\mathfrak{A}$ attracts K_1, it follows that it attracts B_1, as well. If instead, σ is ultimately bounded, then we use Lemma 23.8, Item (1). For each bounded set B_1 in W, the compact, invariant set $K_1 = \omega(B_1)$ attracts B_1. From the Maximality Property (Lemma 23.3), one has $K_1 \subset \mathfrak{A}$. Hence $\mathfrak{A}$ attracts B_1. □

Corollary 23.13. *Let σ be a point dissipative semiflow on W, and let V be a subspace of W that is compactly imbedded in W, i.e., one has $V \hookrightarrow W$. Assume that for every bounded set B in W there is an $r \geq 0$ such that the set $S(t)B$ is bounded in V for all $t > r$. Then σ is a compact semiflow, and it has a global attractor $\mathfrak{A}$. Furthermore, $\mathfrak{A}$ is a bounded set in V, and $\mathfrak{A}$ attracts all bounded sets in W.*

Proof. Note that under the assumptions of this corollary, one can prove that for every compact set K in W, there is a $\tau \geq 0$ such that $\gamma(S(\tau)K)$ is bounded in W. (We leave this proof as an exercise.) The facts that σ is a compact semiflow and that σ has a global attractor follow directly from Lemma 22.1 and Theorem 23.12. Since $\mathfrak{A}$ is invariant and bounded in W, one finds that $\mathfrak{A} = S(t)\mathfrak{A}$ (for $t > r$) is bounded in V. □

2.3.6 Robustness of Attractors. It is hard to overstate the importance of the role played by the Stability Lemma in the study of the longtime dynamics of infinite dimensional systems. As we now show, for example, the uniform asymptotic stability property underlies the various arguments used to prove the robustness (i.e., upper semicontinuous dependence on parameters) of attractors.

Let $\mathfrak{A}_0$ be an attractor for a given κ-contracting semiflow $S_0(t)$. Let $S_0(t)$ be imbedded into a continuous family of semiflows $S_\lambda(t)$, where $\lambda \in \Lambda$ and $S_{\lambda_0}(t) = S_0(t)$. We will say that the family $S_\lambda(t)$ is **robust at** $\mathfrak{A}_0$, or **upper semicontinuous** with respect to λ at $\lambda = \lambda_0$, provided that, for every $\epsilon > 0$, there is a neighborhood $O = O(\epsilon)$ of λ_0 in Λ such that for each $\lambda \in O$, the semiflow $S_\lambda(t)$ has an attractor $\mathfrak{A}_\lambda$ and

$$\mathfrak{A}_\lambda \subset N_\epsilon(\mathfrak{A}_0) \qquad \text{for all } \lambda \in O. \tag{23.13}$$

In other words, the family $S_\lambda(t)$ is robust at $\mathfrak{A}_0$ provided that $\mathfrak{A}_0$ is a "point" of upper semicontinuity for the imbedding of $S_0(t)$ into $S_\lambda(t)$. We now have the following result.

Theorem 23.14 (Robustness of Attractors Theorem). *Let $S_0(t)$ be a semiflow on W and let $\mathfrak{A}_0$ be an attractor for $S_0(t)$. Then the following statements are valid:*

(1) *Let U_1 be any fixed, bounded neighborhood of $\mathfrak{A}_0$ and let $S_0(t)$ be imbedded into any continuous family $S_\lambda(t)$, where each semiflow $S_\lambda(t)$, for $\lambda \in \Lambda$, is asymptotically compact on U_1. (For example, for each $\lambda \in \Lambda$, $S_\lambda(t)$ is a κ-contracting semiflow and there is a $\tau_\lambda \geq 0$ such that $\gamma_\lambda^+(S_\lambda(\tau_\lambda)U_1)$ is bounded.) Then the family $S_\lambda(t)$ is robust at $\mathfrak{A}_0$.*

(2) *Assume that $S_0(t)$ is imbedded into a continuous family $S_\lambda(t)$, where for each $\lambda \in \Lambda$, the semiflow $S_\lambda(t)$ is κ-contracting and the family $S_\lambda(t)$ is uniformly dissipative at λ_0, where $S_0(t) = S_{\lambda_0}(t)$. Then there is a neighborhood O_2 of λ_0 such that $S_\lambda(t)$ has a global attractor, for every $\lambda \in O_2$. Furthermore, for every $\epsilon > 0$, the neighborhood $O \subset O_2$ of λ_0 in Λ can be chosen so that (23.13) is satisified for any attractor $\mathfrak{A}_\lambda$ of $S_\lambda(t)$.*

Before proving this result, it should be noted that in Item (2), the semiflow $S_0(t)$ is robust at $\mathfrak{A}_0$, by Item (1). Consequently (23.13) is valid for some attractor $\mathfrak{A}_\lambda$. The new information in Item (2) is that (23.13) holds for any attractor, including the global attractor.

Proof. Item (1): Let $S_0(t)$ be imbedded into a continuous family of κ-contracting semiflows $S_\lambda(t)$, for $\lambda \in \Lambda$, where $S_{\lambda_0}(t) = S_0(t)$. Now the Stability Lemma 23.7 implies that the attractor $\mathfrak{A}_0$ is uniformly asymptotically stable. Let U_0 be a fixed neighborhood of $\mathfrak{A}_0$ that satisfies (23.6). Consequently, there is a $\tau > 0$ such that the hull $B = H^+(S_0(\tau)U_0)$ is bounded and closed, and B lies in the basin $B(\mathfrak{A}_0)$. Since $B(\mathfrak{A}_0)$ is open (see Lemma 23.2), there is a $\delta_0 > 0$, such that the bounded set $N_{\delta_0}(B)$ lies in $B(\mathfrak{A}_0)$, as well. By choosing a larger value of τ, if necessary, there is no loss in generality in assuming that $N_{\delta_0}(B) \subset U_0 \cap U_1$, where the neighborhood U_1 is given by the hypotheses. Let $\epsilon > 0$ be given so that $V = N_\epsilon(\mathfrak{A}_0) \subset U_0 \cap U_1$. It follows from (23.6) that there is a $T_0 > 2\tau$ such that $S_0(t)U_0 \subset V$, for all $t \geq T_0$. Next we use the Lyapunov stability of $\mathfrak{A}_0$ - see the Stability Lemma 23.7 - to find a δ, with $0 < \delta \leq \min(\delta_0, \epsilon)$, and a neighborhood $U = N_\delta(\mathfrak{A}_0)$ of $\mathfrak{A}_0$ such that

$$S_0(t)U \subset \operatorname{Int} N_\epsilon(\mathfrak{A}_0) \subset V, \qquad \text{for all } t \geq \tau, \tag{23.14}$$

where Int denotes the interior. One then has $U = N_\delta(\mathfrak{A}_0) \subset V = N_\epsilon(\mathfrak{A}_0) \subset U_0$. Since $\mathfrak{A}_0$ attracts $N_{\delta_0}(B)$, it follows from (23.6), there is a time $T \geq T_0 > 2\tau$, such that

$$S_0(t)(U_0 \cup N_{\delta_0}(B)) \subset \operatorname{Int} N_\delta(\mathfrak{A}_0) \subset U \qquad \text{for all } t \geq T - \tau. \tag{23.15}$$

Since $S_0(\tau)V \subset S_0(\tau)U_0 \subset B \subset \operatorname{Int} N_\delta(B) \stackrel{\text{def}}{=} U_2 \subset U_0 \cap U_1$, it follows from the continuity of $S_\lambda(t)u$ that there is a neighborhood O of λ_0 in Λ

such that one has

(23.16) $$S_\lambda(\tau)U \subset S_\lambda(\tau)V \subset U_2 \subset U_0 \cap U_1, \qquad \text{for } \lambda \in O.$$

Since $\delta \leq \delta_0$, it follows from (23.15) that by replacing O with a smaller neighborhood of λ_0, if necessary, one has

(23.17) $$S_\lambda(t)U_2 \subset U, \qquad \text{for } T-\tau \leq t \leq T \text{ and } \lambda \in O.$$

Similarly, by using (23.14), with yet a smaller neighborhood O, if necessary, one has

(23.18) $$S_\lambda(t)U \subset V, \qquad \text{for } \tau \leq t \leq T+\tau \text{ and } \lambda \in O.$$

Now we claim that for $\lambda \in O$ one has

(23.19) $$S_\lambda(nT-\tau)U_2 \subset U, \qquad \text{for } n = 1, 2, \cdots.$$

Indeed, for $n = 1$, this follows from (23.17). Now assume that (23.19) holds for some value n. Then (23.16) and (23.19) imply that

$$S_\lambda(nT)U_2 = S_\lambda(\tau)S_\lambda(nT-\tau)U_2 \subset S_\lambda(\tau)U \subset U_2.$$

By applying $S_\lambda(T-\tau)$ to the last relation and using the relationship (23.17), we obtain

$$S_\lambda((n+1)T-\tau)U_2 \subset S_\lambda(T-\tau)U_2 \subset U.$$

Hence (23.19) is valid by induction.

Next we will show that for $\lambda \in O$ one has

(23.20) $$S_\lambda(t)U_2 \subset V, \qquad \text{for all } t \geq T-\tau.$$

First note that the last inequality holds for $T-\tau \leq t \leq T$ by (23.17), since $U \subset V$. The proof for $t > T$ follows from

(23.21) $$S_\lambda(t)U_2 \subset V, \qquad \text{for } nT \leq t \leq (n+1)T, \text{ and } n = 1, 2, \cdots.$$

In order to prove (23.21), we use $T > 2\tau$, (23.18), and (23.19), with $t = nT-\tau+s$ to obtain

$$S_\lambda(t)U_2 = S_\lambda(nT-\tau+s)U_2 = S_\lambda(s)S_\lambda(nT-\tau)U_2 \subset S_\lambda(s)U \subset V,$$

since $\tau \leq s \leq T+\tau$, which in turn implies (23.21).

Now (23.20) implies that $\gamma^+(S_\lambda(T-\tau)U_2) \subset V$, and since V is closed one has $H^+(S_\lambda(T-\tau)U_2) \subset V \subset U_1$. Since $S_\lambda(t)$ is asymptotically compact on U_1, it then follows from Lemma 23.8 that $\mathfrak{A}_\lambda = \omega(U_2)$ is an attractor for $S_\lambda(t)$ with $\mathfrak{A}_\lambda \subset N_\epsilon(\mathfrak{A}_0)$

Item (2): The argument here is an adaptation of the argument used for Item (1). The main difference is that we now fix U_0 to be the common absorbing set A, for $\lambda \in O_1$, where O_1 is some neighborhood of λ_0 in Λ. Then the sets $B = H^+(S_0(\tau)A)$, $U_2 = \text{Int } N_\delta(B)$, and $V = N_\epsilon(\mathfrak{A}_0)$ are given as in the proof of Item (1). There is then a neighborhood O of λ_0 such that (23.20) holds. This in turn implies that $H^+(S_\lambda(T-\tau)U_2) \subset U_2$, and Theorem 23.12 implies that $\mathfrak{A}_\lambda = \omega(A)$ is a global attractor for $S_\lambda(t)$, since the basin satisfies $B(\mathfrak{A}_\lambda) = W$, and (23.13) is valid. □

2.3.7 Global Attractors: A Summary. For future reference, we summarize here the dynamical properties of a global attractor $\mathfrak{A}$ for a semiflow σ on W. These properties are immediate consequences of the definitions and results numbered from 23.1 to 23.14.

Theorem 23.15 (Properties of Global Attractors). *Let $\mathfrak{A}$ be a global attractor for a semiflow σ on a complete metric space W. Then the following properties are satisfied:*

(1) *$\mathfrak{A}$ is a nonempty, compact, invariant set in W.*
(2) *There is a neighborhood U of $\mathfrak{A}$ with the property that $\mathfrak{A}$ attracts every nonempty, bounded set B in U.*
(3) *If $\mathfrak{A}$ attracts a given nonempty, bounded set B in W, then the omega limit set $\omega(B)$ is a nonempty, compact, invariant set with $\omega(B) \subset \mathfrak{A}$, and $\omega(B)$ attracts B.*
(4) *The basin of attraction satisfies $B(\mathfrak{A}) = W$.*
(5) *$\mathfrak{A}$ attracts all compact sets in W, and the semiflow σ is point dissipative. In particular for every compact set $K \subset W$, the omega limit set $\omega(K)$ is a nonempty, compact, invariant set with $\omega(K) \subset \mathfrak{A}$, and $\omega(K)$ attracts K.*
(6) (Maximality Property) *$\mathfrak{A}$ is maximal with respect to (1), i.e., every compact, invariant set lies in $\mathfrak{A}$.*
(7) (Minimality Property) *$\mathfrak{A}$ is minimal in the following sense: Let U be a neighborhood of $\mathfrak{A}$ that satisfies part (2). If B is any closed set in U that attracts every compact set in U, then $B \supset \mathfrak{A}$.*
(8) *If W is a Banach space, or more generally, a path-connected metric space, then $\mathfrak{A}$ is connected.*
(9) *$\mathfrak{A}$ is uniformly asymptotically stable.*

If in addition, the semiflow σ is κ-contracting, then the following properties are valid:

(10) *There is a neighborhood U of $\mathfrak{A}$ with the property that $\gamma^+(S(\tau)U)$ is bounded, for some $\tau \geq 0$, and $\omega(U) = \mathfrak{A}$.*
(11) *$\mathfrak{A}$ attracts each compact set K in W, and $\omega(K) \subset \mathfrak{A}$.*
(12) *If in addition, the semiflow σ is ultimately bounded, then $\mathfrak{A}$ attracts each bounded set B in W, and $\omega(B) \subset \mathfrak{A}$.*

Moreover, if the semiflow σ is compact, then the following property is valid:

(13) *$\mathfrak{A}$ attracts each bounded set B in W, and $\omega(B) \subset \mathfrak{A}$.*

The next result, which summarizes Theorems 23.11 and 23.12, describes various sufficient conditions for the existence of global attractors for a given semiflow.

Theorem 23.16 (Sufficient Conditions for Existence of Global Attractors). *Let σ be a semiflow on a complete metric space W. Then each of the following conditions is a sufficient condition for the existence of a global attractor $\mathfrak{A}$ for σ.*

(1) *There is a nonempty, compact set K in W with a neighborhood U*

such that K attracts all bounded sets in U and K attracts all points in W.

(2) *The semiflow σ is κ-contracting and point dissipative, and for every compact set $K \subset W$, there is a $\tau \geq 0$ such that $\gamma^+(S(\tau)K)$ is bounded. In this case, if in addition, σ is compact, or ultimately bounded, then $\mathfrak{A}$ attracts all bounded sets in W.*

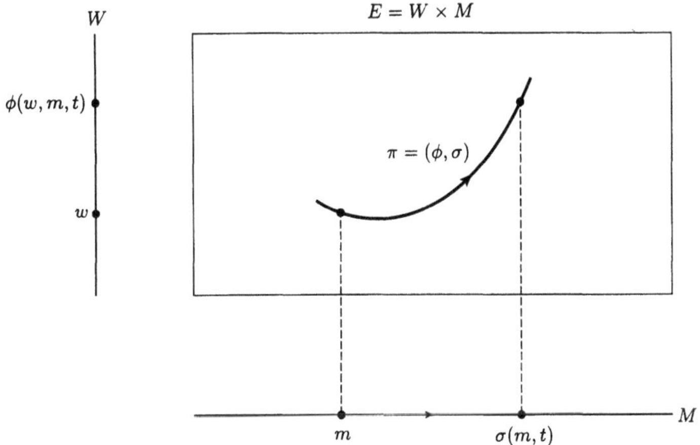

Figure 2.3. Skew Product Semiflow on $W \times M$

2.4. Skew Product Dynamics and Nonautonomous Equations.

Sometimes one encounters dynamical systems on a product space $E = W \times M$, where W and M are metric spaces. A semiflow $\pi = (\phi, \sigma)$ on $E = W \times M$ is said to be a **skew product semiflow** provided ϕ and σ have the form $\phi = \phi(w, m, t)$, and $\sigma = \sigma(m, t)$, i.e., σ does not depend on w. In other words, a semiflow $\pi = (\phi, \sigma)$ is a skew product semiflow on $E = W \times M$ if and only if σ is itself a semiflow on M. We are especially interested in skew product semiflows in the case where M, the **base space**, is a compact metric space, and W, the **fiber space**, is a Banach space, see Figure 2.3.

One place where skew product semiflows arise naturally in nonlinear dynamics occurs when one linearizes a given nonlinear vector field along a compact, invariant set. Another example is in the study of nonautonomous equations, which have time-varying coefficients.

Example 24.1 (Nonautonomous Differential Equations). Let us return to the space $C_{\text{Lip}} = C_{\text{Lip}}(W \times \mathbb{R}, W)$, where W is a given Banach space. For each $g \in C_{\text{Lip}}$ and $w_0 \in W$, we let $\phi(w_0, g, t)$ denote the maximally defined solution of the Initial Value Problem

$$\partial_t w = g(w, t), \qquad w(0) = w_0.$$

Let K be a compact, positively invariant set (with respect to translations in time) in $C_{\text{Lip}} = C_{\text{Lip}}(W \times \mathbb{R}, W)$. A subset $\Omega \subset W$ is said to be an **invariant domain** for K, provided that for each $f \in K$ and $w_0 \in \Omega$, the solution $\phi(w_0, f, t)$ of the Initial Value Problem

$$\partial_t w = f(w,t), \qquad w(0) = w_0,$$

is uniquely determined and satisfies $\phi(w_0, f, t) \in \Omega$, for all $t \geq 0$. In this case, it follows from Theorem 21.6 that the mapping $(w_0, f, t) \to \phi(w_0, f, t)$ is a continuous mapping of $\Omega \times K \times \mathbb{R}^+$ into Ω. Furthermore, the function $(w_0, f, \tau) \to \pi(w_0, f, \tau)$ is a continuous mapping of $\Omega \times K \times \mathbb{R}^+$ into $\Omega \times K$, where π is defined by

$$\pi(w_0, f, \tau) = (\phi(w_0, f, \tau), f_\tau).$$

Clearly one has $\pi(w_0, f, 0) = (w_0, f)$. In order to show that π is a semiflow on $\Omega \times K$, it suffices to show that

$$\phi(\phi(w_0, f, \tau), f_\tau, t) = \phi(w_0, f, \tau + t) \tag{24.1}$$

for all $f \in K$, $w_0 \in \Omega$, and $\tau, t \in \mathbb{R}^+$. In order to prove the validity of (24.1), we define $\phi_1(t)$ by the left side of (24.1) and $\phi_2(t)$ by the right side of (24.1). Next note that both ϕ_1 and ϕ_2 are solutions of the Initial Value Problem

$$\partial_t w = f(w, \tau + t), \qquad w(0) = \phi(w_0, f, \tau).$$

The uniqueness statement in Theorem 21.6 implies that $\phi_1(t) = \phi_2(t)$, for all $t \geq 0$.

Since $(f, \tau) \to f_\tau$ is a semiflow on K which does not depend on $w_0 \in \Omega$, the semiflow in π is a skew product semiflow on $\Omega \times K$. A typical situation where this contruction is useful occurs when $f \in C_{\text{Lip}}$ and $K \overset{\text{def}}{=} H^+(f)$ is the positive hull of f and is compact.

Example 24.2 (Partly Coupled Systems). A differential system of the form

$$\begin{aligned} \partial_t u &= f(u, v) \\ \partial_t v &= g(v), \end{aligned}$$

where $u \in U$ and $v \in V$, is called a **partly coupled system**. The solution of the v-equation does not depend on the initial condition for u. This system generates a skew product semiflow. An example of such a partly coupled system is the motion of a satellite in an n-body problem. In this case, the satellite, which is the $(n+1)$-body, is considered to have 0 gravitational mass, in which case, it does not perturb the gravitational forces acting on the other n-bodies. For this model, the v-equation represents the laws of motion for the original n-bodies, and the u-equation describes the forces acting on the satellite. See Markus and Sell (1968, 1974) for more details.

Example 24.3 (Linearized Semiflows). In this illustration we will restrict our attention to the ODE theory. The PDE theory, which is important in this book, will be treated later. Here we begin with a nonlinear differential equation

$$\partial_t v = f(v)$$

on some domain M in a Euclidean space $\mathbb{R}^n$, where the vector field $f(v)$ is smooth. Let $S(t)v_0$ denote the maximally defined solution with initial condition $v(0) = v_0$, and let K denote a compact, invariant set for this flow. Let $A(v) = \frac{\partial f}{\partial v}$ denote the Jacobian operator and let $\Phi(v_0, t)u_0$ denote the solution of the linear equation

$$\partial_t u = A(S(t)v_0)u, \qquad u(0) = u_0,$$

where $v_0 \in K$. Then the mapping

$$\pi(u_0, v_0, t) = (\Phi(v_0, t)u_0, S(t)v_0)$$

is a (linear) skew product flow on $\mathbb{R}^n \times K$.

We are especially interested in the following theorem, which describes some useful dynamical properties for a compact set K that is invariant in a given skew product semiflow $\pi = (\phi, \sigma)$ on $E = W \times M$. We will use below the subscript 2, as in K_2 and P_2, to refer to the second coordinate in $W \times M$.

Theorem 24.4. *Let $\pi = (\phi, \sigma)$ be a skew product semiflow on $E = W \times M$. Let K be a compact, invariant set for π and define K_2 by*

$$K_2 \stackrel{\text{def}}{=} \{m \in M : \text{there is an } w \in W \text{ with } (w, m) \in K\}.$$

Then the following properties are valid:

1. *The set K_2 is a compact, invariant set for the semiflow σ on the space M.*
2. *For each $m \in K_2$, the fiber of K over m, i.e., the set*

$$K(m) = \{w \in W : (w, m) \in K\}$$

is a compact set in W.

3. *One has*

$$\phi(K(m), m, t) \subset K(\sigma(m, t)), \qquad \text{for all } t \geq 0 \text{ and } m \in K_2; \tag{24.2}$$

i.e., the fiber $K(m)$ is mapped into the fiber $K(\sigma(m, t))$ by ϕ, for $m \in K_2$.

Proof. Let K_2 and $K(m)$ be defined as above. Let $P_2 : E \to M$ denote the projection $P_2(w, m) = m$. Note that P_2 is continuous and that $K_2 = P_2(K)$. Since K_2 is the continuous image of a compact set, it is compact. Next note that

$$P_2(\pi(w, m, t)) = \sigma(m, t), \qquad \text{for all } t \geq 0 \text{ and } (w, m) \in E. \tag{24.3}$$

Recall that K is invariant under σ if and only if for every $(w, m) \in K$ and every $t \geq 0$, there is a point $(\hat{w}, \hat{m}) \in K$ such that

$$\pi(\hat{w}, \hat{m}, t) = (\phi(\hat{w}, \hat{m}, t), \sigma(\hat{m}, t)) = (w, m). \tag{24.4}$$

Now for every $m \in K_2$, there is an $w \in W$ such that $(w, m) \in K$. Let $t \geq 0$ be given. It then follows from (24.4) that there is a point $(\hat{w}, \hat{m}) \in K$ such that $\sigma(\hat{m}, t) = m$. It follows that K_2 is invariant under σ.

Since $P_2^{-1}(m) = W \times \{m\}$ is a closed set and since $K(m) \times \{m\} = P_2^{-1}(m) \cap K$, it follows that each fiber $K(m)$ is a compact set in W. Finally if $w \in K(m)$, then (24.3) implies that $\phi(w, m, t) \in K(\sigma(m, t))$, which in turn implies (24.2). □

2.5. Singular Semiflows.

Let V and W be Banach spaces, where $V \mapsto W$. Assume that $\sigma(u, t) = S(t)u$ is a given semiflow on V. A mapping T is said to be a **singular semiflow on** W if the following properties are satisfied:

(1) For each $(u, t) \in W \times (0, \infty)$ one has $T(t)u \in V$.
(2) $(u, t) \to T(t)u$ is a continuous mapping of $W \times (0, \infty)$ into V.
(3) If $u \in V$, then one has $T(t)u = S(t)u$ for all $t > 0$.

In this setting, $S(t)$ is referred to as the **reduced semiflow** on V. Since $S(t)$ satisfies the semigroup property (21.2), it follows from conditions (2) and (3) that one has

$$T(t + s) = T(t)T(s), \qquad \text{for all } s, t > 0.$$

In short, a singular semiflow differs from a semiflow only in that $T(t)u$ may not be defined for $t = 0$ when $u \in W$ and $u \notin V$. As a result of condition (3), we see that for any $u \in V$, one can extend the domain of definition of $t \to T(t)u$ from $(0, \infty)$ to $[0, \infty)$ by setting $T(0)u = u$. The longtime dynamics of a singular dynamical system $T(t)$ on W are precisely the same as the longtime dynamics of the reduced dynamical system $S(t)$ on V. In particular, all the invariant sets and all the attractors of $T(t)$ lie in V. The following is an instructive example of a singular dynamical system.

Example 25.1. Let $\Omega = Q \times I$ be a product space where Q is an open, bounded set in $\mathbb{R}^n$ and I is an open, bounded interval in $\mathbb{R}$. A point in Ω will be represented in the form $(x_1, \cdots, x_n, x_{n+1})$ where $(x_1, \cdots, x_n) \in Q$ and $x_{n+1} \in I$. Let p satisfy $1 \leq p < \infty$ and set $U = L^p(\Omega) = L^p(\Omega, \mathbb{R})$. The space $V = L^p(Q) = L^p(Q, \mathbb{R})$ is the collection of those functions $v \in U$ with the property that $v = v(x_1, \cdots, x_n)$ does not depend on x_{n+1}. One has the continuous imbedding $V \mapsto U$. We define an operator M on U by $v = Mu$, where

$$v(x_1, \cdots, x_n) = \frac{1}{|I|} \int_I u(x_1, \cdots, x_n, x_{n+1})\, dx_{n+1}.$$

Note that M is a bounded, linear projection on W and the range is $\mathcal{R}(M) = V$. Each point $u \in U$ has a unique representation $u = v + w$, where $Mu = v$ and $Mw = 0$.

Next let $S(t)v$ be any dynamical system on V. For example, one might have

$$S(t)v = e^{-\lambda t} v + (1 - e^{-\lambda t}) v_0,$$

where $\lambda > 0$ and v_0 is a given point in V. A singular dynamical system $T(t)$ on U is now defined by $T(0)u = u$ and

$$T(t)u = S(t)v, \qquad t > 0,$$

where $u = v + w$ with $v \in V$ and $Mw = 0$. For this example, the range $V = \mathcal{R}(M)$ represents a slow manifold and the null space $\mathcal{N}(M)$ is a fast manifold.

Example 25.2. Here is a simple illustration of the use of singular dynamical systems in the study of singular perturbations. Consider the system of ordinary differential equations

$$\partial_t v + Av = f(\epsilon, v, w), \qquad \partial_t w + \epsilon^{-2} w = g(\epsilon, v, w) \tag{25.1}$$

on the product of two finite dimensional Banach spaces $V \times W$, where $\epsilon > 0$. We assume that A is a bounded linear operator on V, that f and g satisfy

$$f(\epsilon, \cdot) \in C_{\mathrm{Lip}}(V \times W, V), \quad g(\epsilon, \cdot) \in C_{\mathrm{Lip}}(V \times W, W),$$

and that they depend continuously on ϵ, for $\epsilon \geq 0$, with

$$\|f(\epsilon, v, w)\|_V \leq M \quad \text{and} \quad \|g(\epsilon, v, w)\|_W \leq M,$$

for all ϵ, v, and w. Furthermore, we assume that the reduced problem

$$\partial_t v + Av = f(0, v, 0), \tag{25.2}$$

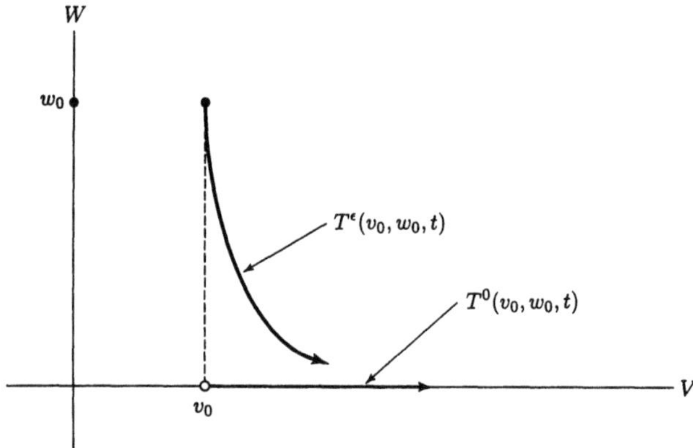

Figure 2.4. Singular Semiflow

has a global attractor $\mathfrak{A}_1$. Under these assumptions, Theorem 21.6 implies that there exist unique solutions of (25.1) and (25.2), and that these solutions are defined, for all $t \in \mathbb{R}$. We will denote the solutions of (25.1) by

$$v(t) = T_1^\epsilon(v_0, w_0, t) \quad \text{and} \quad w(t) = T_2^\epsilon(v_0, w_0, t),$$

where $v(0) = v_0 \in V$ and $w(0) = w_0 \in W$. Let $S(t)v_0$ denote the solution of (25.2) with $S(0)v_0 = v_0$, see Figure 2.4.

Now the Gronwall inequality implies that

$$\|T_2^\epsilon(v_0, w_0, t)\|_W \le e^{-\epsilon^{-2}t}\|w_0\|_W + M\epsilon^2, \qquad \text{for } t \ge 0.$$

Set $T^\epsilon = (T_1^\epsilon, T_2^\epsilon)$, for $\epsilon > 0$, and define $T^0 = (T_1^0, T_2^0)$ by

$$T_1^0(v_0, w_0, t) = S(t)v_0 \quad \text{and} \quad T_2^0(v_0, w_0, t) = \begin{cases} w_0, & \text{for } t = 0, \\ 0, & \text{for } t > 0. \end{cases}$$

Thus T^0 is a semiflow on $V \times W$ with a global attractor $\mathfrak{A}_0 = \mathfrak{A}_1 \times \{0\}$. Since T^ϵ, for $\epsilon \ge 0$, is a continuous family of semiflows, it follows from the Robustness of Attractors Theorem that for every $\eta > 0$ there is an $\epsilon_0 > 0$ such that, for $0 < \epsilon \le \epsilon_0$, the semiflow T^ϵ has an attractor $\mathfrak{A}_\epsilon$ that satisfies $\mathfrak{A}_\epsilon \subset N_\eta(\mathfrak{A}_0)$. If in addition, the spectrum of A lies entirely in the right half of the complex plane, then the inclusion $\mathfrak{A}_\epsilon \subset N_\eta(\mathfrak{A}_0)$ is valid when $\mathfrak{A}_\epsilon$ is the global attractor of T^ϵ.

2.6. Exercises.

Section 2.1

21.1. Complete the proof of Lemma 21.2, i.e., show that (21.2) implies (21.6).

21.2. Prove Lemma 21.3.

21.3. Let σ be a semiflow on $M \subset W$ and let D be a positively invariant set in M. Show that $K = (Cl_W D) \cap M$ is a positively invariant set. (Hence $H^+(B)$ is positively invariant for every set B in M.)

21.4. Let σ be a semiflow on $M \subset W$, and let B be a subset of M. Show that the omega limit set $\omega(B)$ is positively invariant.

21.5. Complete the proof of Lemma 21.5, i.e., show that $\alpha_M(B)$ and $H_M(B)$ are invariant sets.

21.6. A function $f \in C_{\text{Lip}}(W \times \mathbb{R}, W)$ is said to have **linear growth** if there exist constants a, b such that $\|f(w,t)\| \leq a\|w\| + b$, for all $w \in W$ and $t \in \mathbb{R}$. Show that if f has linear growth, then every maximally defined solution ϕ of (21.8) is defined for all $t \in \mathbb{R}$.

21.7. Complete the proof of Lemma 21.4 concerning the alpha limit set.

Section 2.2

22.1. Give a proof of Items (1 - 4) and (7 - 9) in Lemma 22.2.

22.2. Prove Lemma 22.3.

Section 2.3

23.1. A set A in W is said to be an **attracting set** for a semiflow σ on W, if A is a nonempty, bounded, positively invariant set, and if there is a bounded neighborhood U of A with the property that A attracts each point $u \in U$. Construct a semiflow on the plane $\mathbb{R}^2$, where the origin is an attracting set, but it is not an attractor.

23.2. Let σ be a semiflow on $M \subset W$. Show that if a nonempty, compact set D in M attracts itself, then the hull $H^+(D)$ is a compact set in M.

23.3. Find an example of a semiflow which is dissipative and such that there is a compact set K where $\gamma^+(K)$ is not bounded.

23.4. Let σ be a semiflow on $M \subset W$, and let B be a bounded set in M, where the omega limit set $\omega(B)$ is a nonempty, compact set that attracts B.

(1) Show that $\omega(B)$ is invariant.
(2) Assume further that $\omega(B) \subset B$. Show that $\omega(B) = \cap_{t \geq 0} S(t)B$.

23.5. Let K be a nonempty, invariant set for a semiflow on W. Show that K is Lyapunov stable if and only if for any neighborhood U of K, there is a positively invariant neighborhood $V \subset U$.

23.6. Prove Item (3) in Lemma 23.2.

23.7. Let σ be a semiflow on $M \subset W$, and assume that there is a nonempty, compact set K in M and a bounded neighborhood B of K in M with the property that K attracts B. Prove the following:

(1) For all $\tau \geq 0$, the set $S(\tau)K$ attracts B.

(2) For all $\tau \geq 0$ one has $\omega(B) \subset S(\tau)K$.
(3) For all $\tau \geq 0$ one has $\omega(B) = \bigcap_{t\geq\tau} S(t)K$.

23.8. Let K be a nonempty, compact, invariant set for the semiflow σ on M. Let $u \in K$, and let ϕ^u be a negative continuation with $\phi^u(t) \in K$, for all $t \leq 0$. Show that the sets $\alpha(\phi^u)$ and $\omega(u)$ are nonempty, compact, connected invariant sets in K.

23.9. Let σ be a κ-contracting semiflow on M, where M is a closed bounded set in a complete metric space W. Let B be a set in M with the property that there is a $u \in B$ with a globally defined solution ϕ^u passing through it with range in M.

(1) Show that $H_M^-(B)$ and $\alpha_M(B)$ are nonempty, compact sets in M.
(2) Assume, in addition, that $\gamma^+(B)$ is bounded. Show that $H^+(B)$, $H(B)$, and $\omega(B)$ are nonempty, compact sets in M.

23.10. Let σ be a κ-contracting semiflow on W.

(1) Show that every closed, bounded, invariant set in W is compact.
(2) Let B be a bounded, invariant set. Show that $K = \mathrm{Cl}_W B$ is compact and invariant.
(3) Determine whether or not the closure of an invariant set is invariant when the semiflow σ is not κ-contracting.

23.11. Construct a compact semiflow σ on a Banach space W with the property that there is an open set U in W where the positive orbit $\gamma^+(U)$ is not open.

23.12. Let K_0 and $A = \omega(K_0)$ be given by Lemma 23.8. Assume, in addition to the hypotheses of Lemma 23.8, that there is a $\tau \geq 0$ and a $\delta > 0$ such that the set $B_1 \stackrel{\text{def}}{=} \gamma^+(S(\tau)N_\delta(A))$ is bounded. Show that for every compact set K in W, there exist $\epsilon = \epsilon(K) > 0$ and $t_1 = t_1(K) > 0$, such that $S(t)N_\epsilon(K) \subset B_1$, for all $t \geq t_1$.

23.13. A semiflow σ on W is said to be **asymptotically smooth** if for any nonempty, closed, bounded, positively invariant set B in W, there is a nonempty, compact set $J \subset B$, such that J attracts B, see Hale (1988).

(1) Show that a semiflow on W is asymptotically smooth if and only if for any nonempty, closed, bounded set B in W, there is a nonempty, compact set J that attracts $\gamma^+(B)$.
(2) Show that evey κ-contracting semiflow on W that is ultimately bounded is asymptotically smooth.

23.14. Let $S_\lambda(t)$ be a continuous family of semiflows, for $\lambda \in \Lambda$. What conclusions can be drawn about the semiflows in some neighborhood of a point $\lambda_0 \in \Lambda$ when it is known that the semiflow $S_{\lambda_0}(t)$ has precisely one of the following properties: it is compact; it is κ-contracting; it is dissipative; it has a global attractor?

23.15. Prove Item (4) in Theorem 23.11.

23.16. Show that the set B_1 defined in (23.11) is bounded in W.

23.17. Show that the conclusion of Item (2) of the Robustness of Attractors Theorem 23.14 is false, even in 1-dimension, when one drops the assumption of uniform dissipativeness.

23.18 Under the assumptions of Corollary 23.13, show that for every compact set K in W, there is a $\tau \geq 0$ such that $\gamma^+(S(\tau)K)$ is bounded in W.

Section 2.4

24.1. Is it true that (24.2) can be improved to read

$$\phi(K(m), m, t) = K(\sigma(m,t)), \qquad \text{for all } t \geq 0 \text{ and } m \in K_2$$

under the hypotheses of Theorem 24.4? If so, prove it. If not, what additional hypotheses are needed to get this equality?

2.7. Commentary.

One of our objectives in this section is to give a concise road map to the vast literature on those issues of dynamical systems treated in this chapter. For additional background material we refer the reader to the excellent historical notes appearing in the monograph by Hale (1988). Also more detailed references are available in Sell (2001).

(1) Much of the theory of semiflows presented in this paper extends to sets M that are not metrizable. More generally, one can replace the metric topology on M with any weak topology and then derive a similar theory. In such cases, the topology can be described in terms of a family of seminorms on an ambient linear space W. For further information on these extensions, see Sell (1971).

Section 2.1. (2) An implicit consequence of the definition of a semiflow is the uniqueness of solutions of the Initial Value Problem. Nevertheless, the uniqueness property can be relaxed for the study of longtime dynamics, see Sell (1973, 1996) and Raugel and Sell (1993a), for example.

(3) A subtlety of the dynamics of semiflows is the difference between *invariant* and *positively invariant* sets. The consequences of the invariance condition $S(t)K = K$, for all $t \geq 0$, are much deeper than the simple relationships described in Section 2.1. Some of these consequences are developed in Section 2.3.

(4) There do exist theories of semiflows that are based on a weakening of the continuity hypothesis,

$$\sigma : M \times (0, \infty) \to M \text{ is continuous.} \tag{27.1}$$

For example, Temam (1988) requires instead that for each $t \geq 0$, the mapping $\sigma(\cdot, t) : M \to M$ be continuous, or equivalently that $\sigma : M \times [0, \infty) \to M$ be continuous, where now $[0, \infty)$ has the <u>discrete</u> topology, see Ellis (1969). Of course in using the Ellis-Temam approach, one would not be able to prove, without further assumptions, some of the results given above, including the Continuity Lemma 21.1, as well as any other results, such as the Stability Lemma 23.9, the Robustness of Attractors Theorem 23.14, and the Existence Theorem 23.12, which are based on Lemma 21.1.

The continuity property (27.1), as well as the related properties:

$$\sigma(u,\cdot) : (0,\infty) \to M \text{ is continuous for every } u \in M, \tag{27.2}$$

and

$$\sigma(u,\cdot) : [0,\infty) \to M \text{ is continuous for every } u \in M, \tag{27.3}$$

in the context of semiflows, have been studied by several researchers. In particular, Ball (1976, 1997) has a result wherein it is shown that if a solution is measurable in time, then it is continuous in time. Also, Ball (1974) and Chernoff and Marsden (1970) have shown that (27.2) implies (21.1) under reasonable conditions. Finally, Ball (1997) gives an example of a semiflow on a Hilbert space with a global attractor, where property (27.2) is satisfied, but (27.3) is not. Also see Chernoff (1975).

Section 2.2. (5) Lemma 22.4 is based on a result in Lopes and Ceron (1984). An application using Lemma 22.4 can be found in Taboada and You (1992).

Section 2.3. (6) The concept of an attractor, even in the finite dimensional setting, has evolved over the course of time. For example, this concept does not appear in the seminal books of Birkhoff (1927) or Nemytskii and Stepanov (1960). Even in the renaissance of dynamical systems theory which occured in the 1950s, it is difficult to find any papers in which the attractor concept, as described herein, played an important role. For this reason, it is difficult to identify where this concept of "attractor" first appeared. Some aspects of the concept of an attractor can be seen in the book of Bhatia and Szegö (1970), which deals with the evolution of sets in a flow. The first author recalls the early lectures of Conley in which the importance of an attractor was developed as a part of his index theory and the theory of attractor-repellor pairs. This work was subsequently published in Conley (1978).

One can find many different uses of the term "attractor" in the literature. For finite dimensional dynamics, every dynamical system is compact, and therefore, κ-contracting. As a result, the Stability Theorem 23.10 illustrates several equivalent formulations of the concept of the attractor used in the finite dimensional setting, see Conley (1978) and Hale (1988).

It should be noted that in the past some authors have attributed the term "attractor" to a much weaker concept, namely, the concept of an attracting set. A set A in W is said to be an **attracting set** for a dynamical system σ on W if A is a nonempty, bounded, positively invariant set, and if there is a bounded neighborhood U of A with the property that A attracts each point $u \in U$. An attracting set that is not Lyapunov stable is not a very rich dynamical object, since such sets behave badly under arbitrarily small, smooth perturbations of the vector field. An instructive example is given in polar coordinates in $\mathbb{R}^2$ by

$$\partial_t r = r(1-r), \qquad \partial_t \theta = 1 - a\cos\theta,$$

where a is a constant. When $a = 1$, the fixed point $(r_0 = 1, \theta_0 = 0)$ is an attracting set that is not an attractor.

(7) The concept of asymptotical compactness used herein is related to simliar concepts found in Ladyzhenskaya (1991), Hale (1988), and Ball (1997). In our usage, the concept applies to a semiflow acting on a distinguished set B in M. In particular we do not require that the semiflow itself be either κ-contracting, or asymptotically smooth.

(8) In the infinite dimensional setting, the earliest contributions to the theory of global attractors appear in Billotti and LaSalle (1971), Ladyzhenskaya (1972), and Hale, LaSalle, and Slemrod (1972). The concept of a global attractor used in this book is weaker than the concepts appearing, for example, in Hale (1988) or Temam (1988). Recall that we defined $\mathfrak{A}$ to be a global attractor for a dynamical system σ on W if the following three properties hold:

(1) $\mathfrak{A}$ is a nonempty, compact, invariant set for σ.
(2) $\mathfrak{A}$ attracts a bounded neighborhood of itself.
(3) $\mathfrak{A}$ attracts all points in W, i.e., the basin satisfies $B(\mathfrak{A}) = W$.

In Hale (1988) and Temam (1988) a set $\mathfrak{A}$ in W is defined to be a global attractor if it satisfies (1) and the following two properties hold:

(4) $\mathfrak{A}$ is maximal with respect to (1), i.e., every compact, invariant set K satisfies $K \subset \mathfrak{A}$.
(5) $\mathfrak{A}$ attracts every bounded set in W.

As seen in Lemma 23.3, the Maximality Property (4) is a consequence of (1), (2), and (3). Also properties (2) and (3) follow immediately from (1) and (5). Other authors use the Minimality Property as a part of the definition of a global attractor, see Ladyzhenskaya (1987), for example. As we have seen in Lemma 23.5, this property too is a consequence of (1), (2), and (3). See Foias and Temam (1979), Ball (1997), and Hale (1985, 1988) for additional references.

The principal advantage of using the weaker concept of a global attractor can be seen in the Existence Theorem 23.12. Under the minimal assumptions on σ, it is shown that the set $\mathfrak{A}$ satisfies Items (1), (2), and (3) above, but Item (5) is not resolved, without additional hypotheses. The main reasons for this can be traced to steps (5) and (6) in the proof of the this theorem. We believe that our concept of a global attractor is the better concept in the infinite dimensional setting. An open question is: Does there exist an example of a κ-contracting semiflow that is point dissipative on a complete metric space W in which the global attractor does not attract every bounded set in W? This issue of attracting every bounded set in W is not fully resolved by our Main Theorem, and we are unaware of any counterexample.

Semiflows that are (point) dissipative but not compact arise in the study of nonlinear wave equations, see for example, Babin and Vishik (1983a,b), Ball (1978), Feireisl and Zuazua (1993), Ghidaglia and Temam (1987), Hale (1988), Hale and Raugel (1992b), Haraux (1988), Ladyzhenskaya (1987), Marion (1989a,b), Massatt (1983), Teman (1988, Chap 2), and You (1994a). The strategy used by Hale, LaSalle, and Slemrod (1972), in addressing the dynamics of such problems, is based on a concept of asymptotic smoothness, also see Hale and Lopes (1973). Another approach, used by Ladyzhenskaya (1987) and Ball (1997), is based on the concept of asymptotic compactness. As a practical matter, a simple verification of the asymptotic smoothness or asymptotic compactness properties is based on the concept of a κ-contracting semiflow, see Lopes and Ceron (1984), Hale (1988), and Temam (1988). A basic theorem, one finds in the literature (see Hale (1988) for example) for the existence of a global attractor is the following: Assume that σ is a κ-contracting semiflow on a complete metric space W and

(1) that σ is (point) dissipative, and
(2) that the positive trajectory of every bounded set is bounded;

then there is a global attractor $\mathfrak{A}$, and $\mathfrak{A}$ attracts every bounded set B in W. As we have seen in Theorem 23.12, the hypothesis (2) can be replaced by the weaker requirement (2′) that the positive trajectory of every compact set is bounded, or ultimately bounded. An interesting application of our Theorem 23.12 occurs in the study of gradient systems, see Section 7.2.

(9) In the definition of Lyapunov stability of a set A, we require that $S(t)U \subset V$, for all $t \geq \tau$, where $\tau > 0$, see (23.5). This differs from the classical concept where one requires $S(t)U \subset V$, for all $t \geq \tau$, where $\tau = 0$, see Hale (1988, Sect. 3.3) and Bhatia and Szegö (1970). The weaker version, with $\tau > 0$, is used here in order that the Stability Lemma 23.7 be valid for dynamical systems that may fail to be continuous at $t = 0$. See Ball (1997) for some examples which furnish more information on this point.

(10) There are other properties which a given attractor may, or may not, possess. One of the most interesting of these is the property of finite dimensionality of the global attractor of a dynamical system on an infinite dimensional Banach space. The original papers addressing this issue were by Mallet-Paret (1976), who developed the theory for Hilbert spaces, and Mañé (1981), who extended the theory to Banach spaces. Subsequently, researchers in dynamical systems have made a major effort in trying to estimate the dimensions of global attractors in terms of the physical parameters of the problem. For example, in the case of the Navier-Stokes equations, the Grashof number, which is a cousin of the Reynolds number, has been used to describe upper bounds for the dimension. An excellent introduction to this rich theory can be found in Temam (1988). Also see

Sell (2001) for additional references.

(11) It is possible, in some cases, to derive more detailed information about the restriction of a semiflow to a global attractor $\mathfrak{A}$. For example, if the restriction $S_A(t)$ of $S(t)$ to a global attractor is one-to-one, then $S_A(t) : \mathfrak{A} \to \mathfrak{A}$ is a group, or a flow, on $\mathfrak{A}$, see Hale (1988, Sect. 3.10). Also in some cases, one can characterize a global attractor A as the set-theoretic union of the unstable manifolds of the stationary solutions, see Theorem 72.1, Babin and Vishik (1983a, Sect. 1 and 5), and Hale (1988, Sect. 4.3).

In this connection, researchers have been able to show the following:

(1) that some attractors have various stability and/or Morse-Smale dynamics on them: see Angenent (1986), Brunovsky and Polacik (1997), Hale and Raugel (1992c), Mallet-Paret (1988), and Matano (1979);
(2) that the attractor is the graph of a function: see Brunovsky (1990) and Jolly (1989);
(3) that the stable and unstable manifolds intersect transversally: see Henry (1985);
(4) that a Poincaré-Bendixson theory exists in some infinite dimensional problems: see Fiedler and Mallet-Paret (1989); Mallet-Paret and Smith (1990); and Mallet-Paret and Sell (1996).

What is underlying many of these results is the existence of a (Lyapunov) function which assumes only values in the nonnegative integers and which is nonincreasing along orbits: see Matano (1982) and Mallet-Paret (1988).

The application of integer-valued Lyapunov functions was recently extended to a class of differential delay equations with feedback properties, see Mallet-Paret (1988) and Mallet-Paret and Sell (1996ab). The use of these functions, as a tool for understanding the structure of the global attractor, is leading to some very interesting results in the theory of such equations, see Krisztin, Walther, and Wu (1999); and Krisztin and Walther (2001).

(12) One can find in the literature several theorems which conclude that a given global attractor is an upper semicontinuous function of the parameters of the problem, see for example Hale (1988) and Temam (1988). In most cases, these theorems are direct applications of the Robustness of Attractors Theorem 23.14, which also implies the upper semicontiuity of all attractors of a given semiflow, see Hale (1988). However, in some cases, such as partial differential equations on thin domains, one is studying the behavior of an attractor in the setting of a singular perturbation. In such cases, one is able to use the Robustness of Attractors Theorem to show that the attractor of the reduced problem is upper semicontinuous, but this is done only after finding a regularization of the singular perturbation, see Hale, Lin, and Raugel (1988), Hale and Raugel (1991, 1992a,b,c,d), for reaction diffusion equations and nonlinear wave equations, and Raugel and Sell (1993, 1994a,b) for the Navier-Stokes equations.

The related issue of the lower semicontinuity of an attractor, or more generally, of a compact invariant set $\mathcal{K}$, is a much deeper problem. In general, one cannot show that $\mathcal{K}$ admits a lower semicontinuous perturbation, unless $\mathcal{K}$ possess some additional dynamical properties. Special cases of the lower semicontinuity of compact invariant sets occur in the theory of structural stability, see Pliss and Sell (1998). It is known that, for finite dimensional systems of ordinary differential equations, if $\mathcal{K}$ is a hyperbolic set, or a Morse-Smale set, then it is structurally stable, see Pilyugin (1992) and Pliss (1977) for example. A corresponding theory for partial differential equations appears in Hale, Lin, and Raugel (1988), and Hale and Raugel (1989, 1992b). The more general aspects of lower semicontinuity, in the possible absence of structural stability, is developed in Pliss and Sell (1991, 1998). Also see Kamaev (1980).

Section 2.4. (13) The observation that the time parameter appearing in nonautonomous (i.e., time-varying) ordinary differential equations could be compactified, and thereby allowing the exploitation of skew product flow structures in analyzing the dynamics of these equations, was first made by Miller (1965), for almost periodic coefficients, and Sell (1967), for general time-varying coefficients. Also see Sell (1971), Miller and Sell (1970), and Sacker and Sell (1977) for extensions of this idea.

Some of the early users of skew product dynamics in nonlinear infinite dimensional systems include Babin and Chow (1998), Babin and Sell (2000), Chepyzhov and Vishik (1993, 1994, 1995), Chow and Yi (1994), Chow and Leiva (1995), Haraux (1988), Magalhães (1987), Sacker and Sell (1994), Shen and Yi (1995a,b,c, 1996, 1998) and Yi (1998).

(14) It is possible to use the conclusions of Theorem 25.1 to construct a good theory of longtime dynamics for problems in which some solutions need not exist for all time, i.e., the problem may have finite-time blowup, see Sell (1971).

(15) There is an alternate way of constructing solutions of the nonautonomous differential equation $\partial_t w = f(w,t)$, where f is a given element of $C_{\mathrm{Lip}} = C_{\mathrm{Lip}}(W \times \mathbb{R}, W)$. More precisely, let $\Omega \subset W$ be an invariant domain for the set K consisting of the single function f, i.e., $K = \{f\}$. For $w_0 \in \Omega$ and $t_0 \geq 0$ let $\psi(w_0, t_0, t)$ denote the solution of the Initial Value Problem $\partial_t w = f(w,t)$, $w(t_0) = w_0$. Because of Theorem 21.6 the solution $\psi(w_0, t_0, t)$ is defined for all $t \geq 0$ and satisfies $\psi(w_0, t_0, t) \in \Omega$, for all $t \geq 0$. Furthermore, ψ satisfies the following properties:

(1) ψ is continuous as a mapping of $\Omega \times \mathbb{R}^+ \times \mathbb{R}^+$ into Ω.
(2) $\psi(w_0, t_0, t_0) = w_0$
(3) $\psi(\psi(w_0, t_0, t_0 + \tau), t_0 + \tau, t) = \psi(w_0, t_0, \tau + t)$.

The proof of item (3) follows from the uniqueness statement of Theorem 21.6, along with the observation that the two functions of t defined by the left side and right side of (3) are solutions of $\partial_t w = f(w,t)$ that satisfy: $w(t_0 + \tau) = \psi(w_0, t_0, t_0 + \tau)$. Any continuous mapping of $\Omega \times \mathbb{R}^+ \times \mathbb{R}^+$ into Ω that satisfies (1), (2), and (3) is said to be a **process** on Ω, see Dafermos (1971).

While the theory of skew product flows and the theory of processes offer two ways to study the dynamics of nonautonomous problems, there are, nevertheless, important advantages in using the theory of skew product flows and skew product semiflows. The major advantage is best illustrated in the case where f is chosen in such a way that the hull $H^+(f)$ is compact and Ω is an invariant domain for $H^+(f)$. One then has

$$\psi(w_0, \tau, t) = \phi(w_0, f_\tau, t),$$

where ϕ is described in the Example 24.1. While $\phi(w_0, f_\tau, t)$ depends on w_0, τ and t, as does $\psi(w_0, \tau, t)$, the important point to note is that

> *the function $\phi(w_0, g, t)$ is defined for all g in the compact set $H^+(f)$. In other words, the "time" variable τ is now imbedded in a compact set.*

Another advantage is that, in the context of skew product semiflows, one can immediately use the full theory of semiflows. One does not have to reinvent the wheel.

It might appear that if one is given a process ψ, with no knowledge about the vector field f that generated ψ, then one may not be able to compactify the τ-variable in $\psi(w_0, \tau, t)$. This is not the case. As a matter of fact, one can use a process to construct a suitable skew product flow. See Dafermos (1975) for more details.

Section 2.5. (16) Variations of the examples in Section 2.5.2 do arise in the study of partial differential equations on thin domains and other singular perturbation problems. A thin domain problem is, in fact, a singular perturbation of a problem on the lower dimensional space. See Hale and Raugel (1991, 1992a,b,d) and Raugel and Sell (1993, 1994a,b).

(17) The fact that the Robustness of Attractors Theorem 23.14 is valid for semiflows, which may lack continuity at $t = 0$, should be useful in many problems in the theory of singular perturbations. For example, Norman (1999, 2000a,b) recently used this theorem to show that the CSTR (continuous stirred tank reactor) is the reduced problem for a chemical reactor fluid flow, where the chemical diffusivity is very large.

Additional Readings

Ball (1997); Bhatia and Szegö (1970); Billotti and LaSalle (1971); Cholewa and Dlotko (2000); Conley (1978); Hale (1988); Hale, LaSalle, and Slemrod (1972); Lorenz (1963); Nemytskii and Stepanov (1960); Sacker and Sell (1977, 1994); and Sell (1971, 1996).

3
LINEAR SEMIGROUPS

In this chapter we describe the basic notions of a linear C_0-semigroup and the related concepts of an infinitesimal generator. These concepts form infinite dimensional versions of solutions of the finite dimensional linear ordinary differential equation $\partial_t x = Ax$. In particular, the C_0-semigroup corresponds to the solution operator or the fundamental solution matrix, and the infinitesimal generator corresponds to the linear coefficient matrix A. We will see that the C_0-semigroups are linear prototypes of the semiflows described in Chapter 2.

More specifically, we will consider the Initial Value Problem on a Banach space W given by the linear equation

$$\partial_t w = Aw, \qquad w(0) = w_0 \in W, \tag{30.1}$$

for $t \geq 0$, where $\partial_t w = \frac{d}{dt} w$. The operator A is a linear operator on W. While the theory we describe here will certainly apply in the case of a bounded, linear operator, our primary interest is on an important class of unbounded linear operators with domain $\mathcal{D}(A) \subset W$. We will describe this class in detail later. For our purposes though, it is accurate to view equation (30.1) as a reformulation of a linear partial differential equation, where the linear operator A contains various spatial derivatives. This point of view is developed in some detail in Section 3.8.

We say that $w(t)$ is a **classical solution** of (30.1) on an interval $I = [0, \tau)$, where $0 < \tau \leq \infty$, provided $w : I \to W$ is strongly continuous on I, strongly differentiable on the open interval $(0, \tau)$, $w(0) = w_0$, $w(t) \in \mathcal{D}(A)$, for $0 < t < \tau$, and $\partial_t w(t) = Aw(t)$, for $0 < t < \tau$. While many problems of interest have classical solutions, there are some which do not. Nevertheless, there is an important solution concept, namely, a mild solution, which is both useful and central to the overall theory of dynamics of linear and nonlinear infinite dimensional systems. The mild solutions are generated by the C_0-semigroup, and, as we will see, every classical solution is a mild solution.

3.1. C_0-Semigroups and Infinitesimal Generators.

Let W denote a given Banach space, and let $\mathcal{L}(W)$ denote the collection of bounded linear operators on W. We will say that $T(t)$ is a **semigroup of bounded linear operators** on W if $T(t) \in \mathcal{L}(W)$, for all $t \in \mathbb{R}^+ = [0, \infty)$, and one has

$$T(0) = I \tag{31.1}$$

$$T(t)T(s) = T(t+s), \qquad s, t \in [0, \infty). \tag{31.2}$$

A semigroup of bounded linear operators on W is a **C_0-semigroup** if one has

$$\lim_{t \to 0^+} T(t)w = w, \qquad \text{for every } w \in W. \tag{31.3}$$

Equation (31.2) is referred to as the **semigroup property**, and (31.3) is a statement of strong continuity of $T(t)$ at $t = 0$, i.e, for every $w \in W$ one has $\|T(t)w - w\|_W \to 0$, as $t \to 0^+$. As we show below (see Theorem 31.3), the C_0-semigroups are prototypes of the semiflows described in Chapter 2.

Let $T(t)$, $0 \le t < \infty$, be a C_0-semigroup on W. We define its **(infinitesimal) generator** A as follows: First, the domain of A is defined as the set

$$\mathcal{D}(A) \stackrel{\text{def}}{=} \left\{ w \in W : \lim_{h \to 0^+} \frac{T(h) - I}{h} w \quad \text{exists in } W \right\}.$$

Secondly, for $w \in \mathcal{D}(A)$ we define

$$Aw \stackrel{\text{def}}{=} \lim_{h \to 0^+} \frac{T(h) - I}{h} w = \left. \frac{d^+(T(t)w)}{dt} \right|_{t=0}.$$

It is easily verified that $\mathcal{D}(A)$ is a linear subspace in W, and that A is a linear operator on W. In the sequel we will use the notation $(T(t), A)$ to denote a C_0-semigroup $T(t)$ and the associated infinitesimal generator A. On occasion we will denote the C_0-semigroup $(T(t), A)$ in the abbreviated form as e^{At}, or (e^{At}, A).

A family of bounded linear operators $T(t)$, for $t \in \mathbb{R}$, on a Banach space W is called a **C_0-group** if it satisfies (31.1), (31.3) and

$$T(t)T(s) = T(t+s), \qquad \text{for all } t, s \in \mathbb{R}.$$

A necessary and sufficient condition for a C_0-semigroup $(T(t), A)$ to be extended to a C_0-group is that for every $t \ge 0$, the operator $T(t)$ is invertible and $T(t)^{-1} \in \mathcal{L}(W)$. The extension of $T(t)$ for negative time is then given by $T(-t) = T(t)^{-1}$, for $t > 0$. For a C_0-group, the definition of its infinitesimal generator is still the same as for a C_0-semigroup. A C_0-group $T(t)$ is called a **unitary group** if it has the property that $T(t)^{-1} = T(t)^*$, for all $t \ge 0$, where L^* denotes the adjoint of the operator L.

Lemma 31.1. *Let $T(t)$ be a C_0-semigroup on W. Then there exists $M \geq 1$ and $a \in \mathbb{R}$ such that*

$$\|T(t)\| = \|T(t)\|_{\mathcal{L}} = \|T(t)\|_{\mathcal{L}(W)} \leq Me^{-at}, \qquad \textit{for } 0 \leq t < \infty.$$

Proof. We claim that there exists an $\eta > 0$ and an $M > 0$ such that

$$\|T(t)\| \leq M, \qquad \text{for } 0 \leq t \leq \eta. \tag{31.4}$$

If not, then there is a sequence $t_n \to 0^+$ such that the operator norm satisfies $\|T(t_n)\| \geq n$, while $\|T(t_n)w\|$ is bounded, for every $w \in W$. By the Uniform Boundedness Theorem (see Appendix A), the operator norms $\|T(t_n)\|$ are uniformly bounded, which is a contradiction.

Since $\|T(0)\| = 1$ one has $M \geq 1$. Define $a \stackrel{\text{def}}{=} -\eta^{-1}\log M$. Now for any $t \geq 0$ there is a unique integer n and a δ, where $0 \leq \delta < \eta$ such that $t = n\eta + \delta$. The semigroup property (31.2) then implies that

$$\|T(t)\| = \|T(n\eta + \delta)\| = \|T(\delta)T(\eta)^n\| \leq M^{n+1} \leq M^{\frac{t}{\eta}+1} = Me^{-at}. \quad \square$$

Lemma 31.2. *Let $T(t)$ be a C_0-semigroup on W. Then for every $w \in W$ the mapping $t \to T(t)w$ is a continuous mapping of $\mathbb{R}^+$ into W. Moreover, for every compact set K in W and every $t \in \mathbb{R}^+$ one has*

$$\sup_{w\in K} \|T(t+h)w - T(t)w\| \to 0, \qquad \textit{as } h \to 0, \tag{31.5}$$

where one replaces $h \to 0$ in (31.5) with $h \to 0^+$, when $t = 0$.

Proof. For each $t \geq 0$, and $h \geq 0$, one has

$$\begin{aligned}\|T(t+h)w - T(t)w\| &\leq \|T(t)[T(h)w - w]\| \\ &\leq \|T(t)\|\,\|T(h)w - w\| \leq Me^{|a|t}\|T(h)w - w\|.\end{aligned}$$

By letting $h \to 0+$ and using (31.3), we see that the mapping $t \to T(t)w$ is right continuous. Similarly if $t \geq h \geq 0$, then

$$\|T(t-h)w - T(t)w\| \leq \|T(t-h)\|\,\|w - T(h)w\| \leq Me^{|a|t}\|w - T(h)w\|,$$

which verifies the left-continuity.

In order to prove (31.5), we fix $t \in \mathbb{R}^+$, let $\epsilon > 0$ be given, and let $\{w_i : 1 \leq i \leq N\}$ be an ϵ-net for the compact set K. Next we use (31.3) to choose h_0 so that $0 < h_0 < 1$ and for $|h| \leq h_0$, one has $\|T(t+h)w_i - T(t)w_i\| \leq \epsilon$, for $i = 1, \ldots, N$. Then for any $w \in K$ there is a w_i such that $\|w - w_i\| \leq \epsilon$. From Lemma 31.1 one then has

$$\begin{aligned}\|T(t+h)w - T(t)w\| &\leq \|T(t+h)w_i - T(t)w_i\| \\ &\quad + \|T(t+h)w - T(t+h)w_i\| \\ &\quad + \|T(t)w - T(t)w_i\| \\ &\leq \epsilon + 2Me^{|a|(t+1)}\|w - w_i\| \leq \left(1 + 2Me^{|a|(t+1)}\right)\epsilon,\end{aligned}$$

which completes the proof. $\square$

Theorem 31.3. *Let $T(t)$ be a C_0-semigroup on a Banach space W. Then the mapping $(w,t) \to T(t)w$ is a continuous mapping of $W \times [0,\infty)$ into W. Consequently $T(t)w$ is a semiflow on W, and this semiflow is continuous at $t = 0$.*

Proof. Let t_n and w_n be convergent sequences in $\mathbb{R}^+$ and W with $t = \lim_{n\to\infty} t_n$ and $w = \lim_{n\to\infty} w_n$. We need to show that $\|T(t_n)w_n - T(t)w\| \to 0$, as $n \to \infty$. Without loss of generality we can assume that $t_n \leq t+1$. One then has

$$\|T(t_n)w_n - T(t)w\| \leq \|T(t_n)w_n - T(t_n)w\| + \|T(t_n)w - T(t)w\|.$$

Now Lemma 31.2 implies that $\|T(t_n)w - T(t)w\| \to 0$, as $n \to \infty$, and Lemma 31.1, in turn, implies that

$$\|T(t_n)w_n - T(t_n)w\| \leq Me^{|a|t_n}\|w_n - w\| \leq Me^{|a|(t+1)}\|w_n - w\|,$$

which converges to 0, as $n \to \infty$. Hence $T(t)w$ is a semiflow on W, and it is continuous at $t = 0$. □

A proof of the following result, which describes some elementary properties of C_0-semigroups, can be found in Pazy (1983).

Theorem 31.4. *Let $(T(t), A)$ be a C_0-semigroup on a Banach space W. Then the following statements are valid:*

(1) *For every $w_0 \in W$ one has*

$$\lim_{h\to 0} \frac{1}{h}\int_t^{t+h} T(s)w_0\, ds = T(t)w_0 \qquad 0 \leq t < \infty,$$

where for $t = 0$, the limit is taken with $h \to 0^+$.

(2) *For every $w_0 \in W$ and $t \geq 0$, one has $\int_0^t T(s)w_0\, ds \in \mathcal{D}(A)$ and*

$$A\left(\int_\tau^t T(s)w_0\, ds\right) = T(t)w_0 - T(\tau)w_0, \qquad 0 \leq \tau \leq t. \tag{31.6}$$

(3) *For every $w_0 \in \mathcal{D}(A)$ and $t \geq 0$, one has $T(t)w_0 \in \mathcal{D}(A)$ and*

$$\frac{d}{dt}T(t)w_0 = AT(t)w_0 = T(t)Aw_0, \tag{31.7}$$

where for $t = 0$, the derivative becomes a right derivative. In particular, the function $w(t) \stackrel{\text{def}}{=} T(t)w_0$ is a classical solution of (30.1), provided $w_0 \in \mathcal{D}(A)$.

(4) *For every $w_0 \in \mathcal{D}(A)$ and $0 \leq s \leq t$ one has*

$$T(t)w_0 - T(s)w_0 = \int_s^t T(\tau)Aw_0\, d\tau = \int_s^t AT(\tau)w_0\, d\tau.$$

Given a C_0-semigroup $(T(t), A)$ on W, the function $w(t) = T(t)w_0$, for any $w_0 \in W$, is said to be a **mild solution** of (30.1), see (31.7). Thus a mild solution is a classical solution whenever $w_0 \in \mathcal{D}(A)$. We will say more about the connection between mild solutions and classical solutions in Section 3.5. Next we turn to a description of several important properties shared by all infinitesimal generators of C_0-semigroups.

Corollary 31.5. *Let $(T(t), A)$ be a C_0-semigroup on a Banach space W. Then $\mathcal{D}(A)$ is a dense linear subspace in W, and A is a closed linear operator on W.*

Proof. For each $w \in W$ define $w_t \stackrel{\text{def}}{=} \frac{1}{t}\int_0^t T(s)w\,ds$, for $t > 0$. From Theorem 31.4, Item (2), we see that $w_t \in \mathcal{D}(A)$. Since the mapping $t \to T(t)w$ is a continuous mapping and $T(0) = I$, we see that $w_t \to w$, as $t \to 0^+$. Consequently, $\mathrm{Cl}_W \mathcal{D}(A) = W$, i.e., $\mathcal{D}(A)$ is dense in W. Now let $w_n \in \mathcal{D}(A)$ with $(w_n, Aw_n) \to (w, v)$ in $W \times W$, as $n \to \infty$. Then Theorem 31.4, Item (4), implies that

$$T(t)w_n - w_n = \int_0^t T(s)Aw_n ds. \tag{31.8}$$

By taking the limit as $n \to \infty$ in (31.8) and dividing by t, one obtains the identity

$$\frac{1}{t}[T(t)w - w] = \frac{1}{t}\int_0^t T(s)v\,ds, \qquad \text{for } t > 0.$$

Next by taking the limits as $t \to 0^+$, we find that the right side tends to $T(0)v = v$. As a result, the left side has a limit, which implies that $w \in \mathcal{D}(A)$ and the limit is Aw. Hence $Aw = v$, and A is a closed linear operator on W. □

The resolvent operator $R(\lambda, A) = (\lambda I - A)^{-1}$ has the following characterization in terms of the C_0-semigroup e^{At}, see Pazy (1983) for the details.

Lemma 31.6. *Let (e^{At}, A) be a C_0-semigroup on a Banach space W that satisfies $\|e^{At}\| \leq Me^{-at}$, for all $t \geq 0$. Then $\lambda \in \rho(A)$, for all λ with $\mathrm{Re}\,\lambda > -a$, the resolvent operator $R(\lambda, A) = (\lambda I - A)^{-1}$ satisfies*

$$R(\lambda, A)w = \int_0^\infty e^{-\lambda t}e^{At}w\,dt, \qquad \textit{for } \mathrm{Re}\,\lambda > -a,$$

and for $n = 1, 2, \cdots$, one has

$$\|R(\lambda, A)^n\| - \|R(\lambda, A)\|_{\mathcal{L}} \leq \frac{M}{(\mathrm{Re}\,\lambda + a)^n}, \qquad \textit{for } \mathrm{Re}\,\lambda > -a. \tag{31.9}$$

3.2. An Illustrative Example.

In this section we will give a basic construction of a ubiquitous C_0-semigroup e^{-At} with the infinitesimal generator $-A$ on a real Hilbert space H. This construction will be used extensively throughout this book. Let A be a given self-adjoint linear operator on a Hilbert space H. Assume that

(1) A is **bounded below**, i.e., there is an $a \in \mathbb{R}$ such that

$$a\|u\|^2 \le \langle Au, u\rangle, \qquad \text{for all } u \in \mathcal{D}(A), \text{and} \tag{32.1}$$

(2) A has **compact resolvent**, i.e., the resolvent operator $R(\lambda, A) = (\lambda I - A)^{-1}$ is compact for all $\lambda \in \rho(A)$, where $\rho(A)$ is the resolvent set, or regular set, of A.

The self-adjoint operator A is said to be **positive** if (32.1) holds for some $a > 0$. We will next show that $-A$ is the infinitesimal generator of a C_0-semigroup e^{-At}. Later in the chapter we will derive other properties of e^{-At}.

Since A is a self-adjoint operator, its spectrum is real. Inequality (32.1) implies that the spectrum lies in the interval $[a, \infty)$, and the fact that A has compact resolvent implies that the spectrum consists only of point spectrum, i.e., eigenvalues, each with finite multiplicity. Also the spectrum has no (finite) accumulation points. We let

$$0 < a = \lambda_1 \le \lambda_2 \le \lambda_3 \le \dots \tag{32.2}$$

denote the eigenvalues of A, repeated with the respective multiplicity, and we will let $\{e_1, e_2, e_3, \dots\}$ be the corresponding unit eigenvectors, which forms an orthonormal basis for H, see Naylor and Sell (1982). One then has

$$Ae_i = \lambda_i e_i, \qquad i = 1, 2, 3, \dots .$$

If in addition, H is infinite dimensional, then one has

$$\lambda_i \to \infty, \qquad \text{as } i \to \infty. \tag{32.3}$$

Since (32.1) remains valid when $a = \lambda_1$, the smallest eigenvalue of A, we will assume that $a = \lambda_1$ in the sequel.

Any $u \in H$ has a generalized Fourier series expansion given in terms of the orthonormal basis, i.e., one has $u = \sum_{i=1}^{\infty} c_i e_i$, where the coefficients c_i satisfy $c_i = \langle u, e_i\rangle$. Furthermore, the Parseval equality

$$\|u\|^2 = \sum_{i=1}^{\infty} |c_i|^2 = \sum_{i=1}^{\infty} |\langle u, e_i\rangle|^2 \tag{32.4}$$

is valid. The domain of A can be characterized as

$$\mathcal{D}(A) = \{u \in H : \sum_{i=1}^{\infty} |\lambda_i|^2 |\langle u, e_i\rangle|^2 < \infty\},$$

and A is given by the formula

$$Au = \sum_{i=1}^{\infty} \lambda_i \langle u, e_i \rangle e_i, \qquad \text{for } u \in \mathcal{D}(A).$$

The Spectral Mapping Theorem, as it applies to A, states that if f is any continuous, real-valued function defined on the spectrum $\sigma(A)$, then the linear operator $f(A)$ is defined by the formula

$$f(A)u = \sum_{i=1}^{\infty} f(\lambda_i) \langle u, e_i \rangle e_i, \tag{32.5}$$

where the domain of $f(A)$ is given by

$$\mathcal{D}(f(A)) = \{u \in H : \sum_{i=1}^{\infty} |f(\lambda_i)|^2 |\langle u, e_i \rangle|^2 < \infty\}.$$

We will next show that the C_0-semigroup generated by $-A$ is given by

$$e^{-At}u \stackrel{\text{def}}{=} \sum_{i=1}^{\infty} e^{-\lambda_i t} \langle u, e_i \rangle e_i. \tag{32.6}$$

The first step is to use the Parseval equality (32.4) with (32.6) to find that

$$||e^{-At}u||^2 = \sum_{i=1}^{\infty} e^{-2\lambda_i t} |\langle u, e_i \rangle|^2 \leq \sum_{i=1}^{\infty} e^{-2\lambda_1 t} |\langle u, e_i \rangle|^2 = e^{-2at} ||u||^2. \tag{32.7}$$

Hence one has $e^{-At} \in \mathcal{L}(H)$, and $||e^{-At}|| \leq e^{-at}$ for all $t \geq 0$. The identity (31.1) is trivially true for e^{-At}. The semigroup identity (31.2) is an immediate consequence of the orthonormality of $\{e_i\}$ and the Spectral Mapping Theorem (32.5) for $f(\lambda) = e^{-\lambda t}$.

Next we verify the strong continuity property (31.3). Fix $u \in H$ and let $\epsilon > 0$ be given. Choose $N \geq 1$ so that $\lambda_i > 0$ for $i \geq N+1$ and $\sum_{i=N+1}^{\infty} |\langle u, e_i \rangle|^2 \leq \epsilon/2$. Next choose $h_0 > 0$ so that for $0 < h \leq h_0$ one has

$$(e^{-\lambda_i h} - 1)^2 |\langle u, e_i \rangle|^2 \leq \frac{\epsilon}{2N}, \qquad \text{for } 1 \leq i \leq N \text{ and } 0 < h \leq h_0.$$

One then has

$$\begin{aligned} ||e^{-Ah}u - u||^2 &= \sum_{i=1}^{\infty} (e^{-\lambda_i h} - 1)^2 |\langle u, e_i \rangle|^2 \\ &\leq \sum_{i=1}^{N} (e^{-\lambda_i h} - 1)^2 |\langle u, e_i \rangle|^2 + \sum_{i=N+1}^{\infty} |\langle u, e_i \rangle|^2 \leq \epsilon, \end{aligned}$$

for $0 < h \le h_0$. Since ϵ is arbitrary, we see that (31.3) holds.

Finally we show that the infinitesimal generator of the C_0-semigroup e^{-At} is actually $-A$. For this purpose we will use the Spectral Mapping Theorem (32.5) and the fact that the orthogonal projection P_N defined by $P_N u \stackrel{\text{def}}{=} \sum_{i=1}^{N} \langle u, e_i \rangle e_i$ is continuous. Let B denote the infinitesimal generator of e^{-At}. For $u \in \mathcal{D}(B)$ one has

$$P_N \left(\lim_{h \to 0+} \frac{e^{-Ah} - I}{h} u \right) = \lim_{h \to 0+} P_N \left(\frac{e^{-Ah} - I}{h} u \right) = \sum_{i=1}^{N} -\lambda_i \langle u, e_i \rangle e_i.$$

By letting $N \to \infty$, we see that $u \in \mathcal{D}(A)$ and $Bu = -Au$. We leave it as an exercise to complete the proof by showing that $\mathcal{D}(A) \subset \mathcal{D}(B)$, and $Bu = -Au$, for all $u \in \mathcal{D}(A)$.

Let us now show that the linear operator e^{-At} is compact for each $t > 0$. If the Hilbert space H is finite dimensional, then every linear operator on H is compact. Therefore, we now assume H to be infinite dimensional. For each $u \in H$ one then has

$$\begin{aligned} \|e^{-At}u - P_N e^{-At}u\| &= \sum_{i=N+1}^{\infty} e^{-2\lambda_i t} |\langle u, e_i \rangle|^2 \\ &\le e^{-2\lambda_{N+1} t} \sum_{i=1}^{\infty} |\langle u, e_i \rangle|^2 = e^{-2\lambda_{N+1} t} \|u\|^2. \end{aligned}$$

Now (32.3) implies that for each $t > 0$ the linear operator e^{-At} is the uniform limit of linear operators with finite dimensional range. Hence e^{-At} is compact, see Naylor and Sell (1982, p. 384).

Let us summarize the results of this section. Items (1) and (2) of the following theorem are proved above, and Items (3) and (4) are proved in Section 3.6.

Theorem 32.1. *Let A be a self adjoint operator on a Hilbert space H, and assume that A is bounded below and has compact resolvent. Then the following statements are valid:*

(1) *The linear operator $-A$ generates a C_0-semigroup e^{-At}, which is given by (32.6), and one has $\|e^{-At}\| \le e^{-at}$, where $a = \lambda_1$ is the smallest eigenvalue of A.*
(2) *For each $t > 0$, the linear operator e^{-At} is compact.*
(3) *The semigroup e^{-At} is analytic, and the linear operator A is a sectorial operator.*
(4) *For any fixed $t > 0$, one has $\|e^{-A(t+h)} - e^{-At}\|_{\mathcal{L}} \to 0$, as $h \to 0$, where $\mathcal{L} = \mathcal{L}(W)$.*

3.3. Compact and κ-Contracting Semigroups.

Let $(T(t), A)$ be a C_0-semigroup on a Banach space W, and let κ be the Kuratowski measure of noncompactness on the bounded subsets of W. Now $T(t)w$, when viewed as a semiflow on W, is κ-contracting if and only if $\kappa(T(t)B) \to 0$, as $t \to \infty$, for every bounded set B in W. Likewise the semiflow $T(t)w$ is compact if and only if there is an r, $0 \leq r < \infty$, such that $\mathrm{Cl}_W T(t)B$ is compact, for every bounded set B in W and each $t > r$.

Assume that the given C_0-semigroup $T(t)$ admits a decomposition of the form $T(t) = T_1(t) + T_2(t)$, for $t \geq 0$, and that there is an r, $0 \leq r < \infty$, and a nonnegative function $k(t)$ with $k(t) \to 0$, as $t \to \infty$, and such that

(1) $\mathrm{Cl}_W T_1(t) N_{\rho_1}(0)$ is compact, for some $\rho_1 > 0$ and all $t > r$, and
(2) $\|T_2(t)w\| \leq k(t)\|w\|$, for all $t \geq 0$ and all $w \in N_{\rho_2}(0)$, for some $\rho_2 > 0$.

Then, as argued in Lemma 22.4, it follows that $T(t)w$ is a κ-contracting semiflow on W.

The following characterization of a compact C_0-semigroup is useful. A proof can be found in Pazy (1983, Theorem 3.3).

Theorem 33.1. *Let $(T(t), A)$ be a C_0-semigroup on a Banach space W. Then the following statements are equivalent.*

(1) *$T(t)$ is compact for $t > 0$.*
(2) *The infinitesimal generator A has compact resolvent, and $T(t)$ is continuous in the uniform operator topology for $t > 0$, i.e., for each $t > 0$, one has $\|T(t+h) - T(t)\|_{\mathcal{L}} \to 0$ as $h \to 0$.*

3.4. Hille-Yosida and Lumer-Phillips Theorems.

For applications of the semigroup theory of linear operators to partial differential equations and differential delay equations, one needs to know how to determine whether a given operator A on a Banach space W is eligible to be the infinitesimal generator for some C_0-semigroup $T(t)$. Oftentimes such an operator A is the Laplacian operator, or more generally, a second order elliptic differential operator, coupled with certain boundary conditions, see Section 3.8.

There are two major results in this regard. One is the Hille-Yosida Theorem, which provides necessary and sufficient conditions for A to be the generator, but the conditions are usually hard to check. The other is the Lumer-Phillips Theorem, which gives necessary and sufficient conditions for A to be the generator of a nonexpansive semigroup of linear operators and is quite useful in the Hilbert space setting. Our goal here is to present these two theorems and to sketch the proofs.

Theorem 34.1 (Hille-Yosida). *A linear operator A on a Banach space W is the infinitesimal generator of a C_0-semigroup $T(t)$ that satisfies*

$$\|T(t)\| \leq Me^{at}, \qquad \textit{for constants } M \geq 1, a \in \mathbb{R} \textit{ and all } t \geq 0,$$

if and only if both of the following conditions are satisfied:

(1) *A is a closed linear operator and the domain $\mathcal{D}(A)$ is dense in W.*
(2) *The resolvent set $\rho(A)$ contains the set $\{\lambda \in \mathbb{R} : \lambda > a\}$ and the resolvent operator $R(\lambda, A) = (\lambda I - A)^{-1}$ satisfies*

$$\|R(\lambda, A)^n\| \leq \frac{M}{(\lambda - a)^n}, \qquad \textit{for } \lambda > a \textit{ and } n = 1, 2, \ldots. \tag{34.1}$$

The proof goes through several steps, and we will use the characterization of the resolvent operator $R(\lambda, A)$ given in Lemma 31.6. The first two lemmas have easy proofs, and we omit the details.

Lemma 34.2. *Let $(T(t), A)$ be a C_0-semigroup on a Banach space W that satisfies $\|T(t)\| \leq Me^{at}$, for $t \geq 0$, and let $r \in \mathbb{R}$. Then $S(t) = e^{-rt}T(t)$ is a C_0-semigroup that satisfies $\|S(t)\| \leq Me^{(a-r)t}$. Furthermore the infinitesimal generator of $S(t)$ is $B = A - rI$, where $\mathcal{D}(B) = \mathcal{D}(A)$. Moreover the resolvent sets of A and B satisfy $\rho(B) = \rho(A) - r$ and the resolvent operators satisfy $R(\lambda, B) = R(\lambda + r, A)$.*

The last lemma is useful because it shows that by choosing $r = a$, it suffices to prove Theorem 34.1 in the case where $a = 0$. The next lemma will enable us to make one further reduction to the case where $M = 1$. This reduction is accomplished by means of the construction of a suitable equivalent norm on W.

Lemma 34.3. *Let A be a linear operator on a Banach space W with the property that every $\lambda > 0$ is in the resolvent set $\rho(A)$. Assume that there is an $M \geq 1$ such that the resolvent operator $R(\lambda, A)$ satisfies $\|\lambda^n R(\lambda, A)^n\| \leq M$, for $\lambda > 0$ and integers $n \geq 1$. Then the norm $|\cdot|$ on W that is defined by*

$$|w| = \limsup_{\lambda \to \infty} \left(\sup_{n \geq 0} \|\lambda^n R(\lambda, A)^n w\| \right),$$

satisfies $\|w\| \leq |w| \leq M\|w\|$, for $w \in W$, and

$$|R(\lambda, A)^n w| \leq \frac{1}{\lambda^n}|w|, \qquad \textit{for } \lambda > 0 \textit{ and } w \in W.$$

It then follows from Corollary 31.5 and Lemma 31.6 that conditions (1) and (2) in Theorem 34.1 are necessary conditions for A to be an infinitesimal generator. In order to prove the sufficiency, we need to study the **Yosida approximant** of A, which is defined by

$$A_\lambda \stackrel{\text{def}}{=} \lambda A R(\lambda, A) = \lambda A(\lambda I - A)^{-1}, \qquad \text{for } \lambda \in \rho(A),$$

where A is a given closed linear operator on W. Since $(\lambda I - A)R(\lambda, A) = I$, one obtains $A_\lambda = \lambda^2 R(\lambda, A) - \lambda I$, and consequently A_λ is a bounded linear operator on W. The proof of the following result can be found in Pazy (1983, Lemmas 3.2-3.4).

Lemma 34.4. *Let A be a closed linear operator on a Banach space W and the domain $\mathcal{D}(A)$ is dense in W. Assume that if $\lambda > 0$, then $\lambda \in \rho(A)$, and that the resolvent operator $R(\lambda, A) = (\lambda I - A)^{-1}$ satisfies*

$$||R(\lambda, A)|| \leq \frac{1}{\lambda}, \qquad \lambda > 0.$$

Then the following statements are valid:

(1) *The resolvent operator $R(\lambda, A)$ satisfies*

$$\lim_{\lambda \to \infty} \lambda R(\lambda, A)w = w, \qquad \text{for } w \in W.$$

(2) *The Yosida approximant A_λ satisfies*

$$\lim_{\lambda \to \infty} A_\lambda w = Aw, \qquad \text{for } w \in \mathcal{D}(A). \tag{34.2}$$

(3) *For each $\lambda > 0$, the Yosida approximant A_λ is the infinitesimal generator of the C_0-group $e^{A_\lambda t}$ on W. Moreover, if $\lambda > 0$ and $\mu > 0$, then for $t \geq 0$ one has*

$$\begin{aligned} &||e^{A_\lambda t}|| \leq 1, \\ &||(e^{A_\lambda t} - e^{A_\mu t})w|| \leq t||(A_\lambda - A_\mu)w||, \qquad \text{for all } w \in W. \end{aligned} \tag{34.3}$$

Proof of Theorem 34.1. As noted above, the necessity of conditions (1) and (2) has been shown. In order to prove the sufficiency, we first use Lemma 34.2 (with $r = a$) to reduce to the case where $a = 0$. Lemma 34.3 then allows us to construct an equivalent norm on W where $M = 1$. For simplicity, we will use $|| \cdot ||$ to denote this new norm. Finally we invoke Lemma 34.4 to complete the proof as follows. According to (34.3), one has

$$||e^{A_\lambda t}w - e^{A_\mu t}w|| \leq t||A_\lambda w - Aw|| + t||A_\mu w - Aw||, \qquad \text{for all } w \in \mathcal{D}(A).$$

Using this inequality, together with (34.2), one finds that that for each $w \in \mathcal{D}(A)$, the limit $\lim_{\lambda \to \infty} e^{A_\lambda t}w$ exists uniformly, for t in any bounded interval. Now define $T(t)$ by

$$T(t)w \stackrel{\text{def}}{=} \lim_{\lambda \to \infty} e^{A_\lambda t}w, \qquad \text{for } w \in \mathcal{D}(A) \text{ and } t \geq 0. \tag{34.4}$$

Since $||e^{tA_\lambda}|| \leq 1$, one has $||T(t)w|| \leq ||w||$, for all $w \in \mathcal{D}(A)$ and $t \geq 0$. Since $\mathcal{D}(A)$ is dense in W, there is, for each $t \geq 0$, a unique extension of

$T(t)$ to all of W and the operator norm satisfies $\|T(t)\| \leq 1$, for all $t \geq 0$. It is easy to check that $T(t)$, $t \geq 0$, is a C_0-semigroup, and we leave this as an exercise.

The remaining item to verify is that the infinitesimal generator of $T(t)$ is A. Let $B : \mathcal{D}(B) \to W$ be the generator of $T(t)$. From formulae (34.2) and (34.4) and Theorem 31.4, Item (4), we have

$$T(t)w - w = \lim_{\lambda \to \infty} \int_0^t e^{A_\lambda s} A_\lambda w \, ds = \int_0^t T(s) A w \, ds \qquad \text{for } w \in \mathcal{D}(A).$$

It follows from the definition of the infinitesimal generator and Theorem 31.4, Item (1), as these apply to $(T(t), B)$, that if $w \in \mathcal{D}(A)$, then $w \in \mathcal{D}(B)$ and $Bw = Aw$, so that $A \subseteq B$. On the other hand, since B is an infinitesimal generator for $T(t)$, conditions (1) and (2) in the Hille-Yosida Theorem apply to B, with $M = 1$ and $a = 0$. Hence $1 \in \rho(B)$. Since $A \subseteq B$, one has

$$\|Aw - w\| = \|Bw - w\| \geq b\|w\|, \qquad \text{for all } w \in \mathcal{D}(A),$$

for some constant $b > 0$. Hence, $1 \in \rho(A)$. Consequently, we find that

$$(I - B)\mathcal{D}(A) = (I - A)\mathcal{D}(A) = W.$$

Hence $\mathcal{D}(B) = (I - B)^{-1} W = (I - B)^{-1}(I - B)\mathcal{D}(A) = \mathcal{D}(A)$ and $A = B$. □

Next we have the Lumer-Phillips Theorem for a semigroup on a Hilbert space H. In order to do this, we introduce the following concepts: A linear operator $A : \mathcal{D}(A)(\subset H) \to H$ is said to be **accretive** if

$$\text{Re}\, \langle Aw, w \rangle \geq 0, \qquad \text{for all } w \in \mathcal{D}(A).$$

A is said to be **maximal accretive** if it is accretive and the range of $I + A$ satisfies $\mathcal{R}(I + A) = H$. A C_0-semigroup $T(t)$ is said to be **nonexpansive** if $\|T(t)\| \leq 1$, for all $t \geq 0$.

Theorem 34.5 (Lumer-Phillips). *Let H be a Hilbert space. Then a linear operator $-A : \mathcal{D}(A) \to H$, where $\mathcal{D}(A) \subset H$, is the infinitesimal generator of a nonexpansive C_0-semigroup e^{-At} on H if and only if both the following conditions are satisfied:*

(1) *A is a closed linear operator and $\mathcal{D}(A)$ is dense in H, and*
(2) *A is a maximal accretive operator.*

Proof. The necessity part is relatively easy and is left as an exercise. Here we show the sufficiency part. Because of the Hille-Yosida Theorem 34.1, it suffices to show that

$$\{\lambda > 0\} = \{\lambda \in \mathbb{R} : \lambda > 0\} \subset \rho(-A) \text{ and } \|R(\lambda, -A)\| \leq \frac{1}{\lambda}, \quad \text{for } \lambda > 0.$$

In order to prove this, we note that the accretive property of A implies that

$$(34.5)\qquad \|(\lambda I + A)w\| \geq \lambda\|w\|, \qquad \text{for every } \lambda > 0 \text{ and } w \in \mathcal{D}(A).$$

Consequently, $(\lambda I + A)^{-1} : \mathcal{R}(\lambda I + A) \to H$ exists and is bounded with $\|\lambda I + A\| \leq \lambda^{-1}$. It remains to show $\mathcal{R}(\lambda I + A) = H$, for all $\lambda > 0$. Define $\Lambda \stackrel{\text{def}}{=} \{\lambda > 0 : \mathcal{R}(\lambda I + A) = H\}$. By the maximal accretive property, Λ contains $\lambda = 1$. Since the resolvent set is open and $\Lambda = \{\lambda > 0\} \cap \rho(-A)$, the set Λ is an open set in $\{\lambda > 0\}$. We claim that Λ is a (relatively) closed set in $\{\lambda > 0\}$. Indeed, let λ_n be a convergent sequence in Λ with $\lambda_n \to \lambda_0 > 0$. Without loss of generality, we can assume that $\lambda_n \geq \frac{\lambda_0}{2} > 0$. Then for any $y \in H$, there is $w_n \in \mathcal{D}(A)$ such that $\lambda_n w_n + Aw_n = y$, and by (34.5), w_n satisfies

$$\|w_n\| \leq \frac{1}{\lambda_n}\|\lambda_n w_n + Aw_n\| \leq \frac{1}{\lambda_n}\|y\| \leq \frac{2\|y\|}{\lambda_0}.$$

Consequently, one has

$$\begin{aligned} \frac{\lambda_0}{2}\|w_n - w_m\| &\leq \lambda_m\|w_n - w_m\| \leq \|\lambda_m(w_n - w_m) + A(w_n - w_m)\| \\ &= |\lambda_m - \lambda_n|\,\|w_n\| \leq \frac{2\|y\|}{\lambda_0}|\lambda_m - \lambda_n|. \end{aligned}$$

Hence w_n is a Cauchy sequence in H, and therefore there is a limit $w_0 = \lim_{n\to\infty} w_n$ in H, and $\lim_{n\to\infty} Aw_n = y - \lambda_0 w_0$. Since A is a closed operator, one obtains $w_0 \in \mathcal{D}(A)$ and $Aw_0 = y - \lambda_0 w_0$, so that $y = (\lambda_0 I + A)w_0$. Consequently, $\lambda_0 \in \Lambda$, and Λ is closed in $\{\lambda > 0\}$. Therefore $\Lambda = \{\lambda > 0\}$, which completes the proof. □

3.5. Differentiable Semigroups.

As stated in the beginning of this chapter, the principal objective of introducing C_0-semigroups is to establish a theory for describing the solutions of the linear evolutionary equation

$$(35.1)\qquad \partial_t w = Aw, \qquad w(0) = w_0 \in W,$$

on a Banach space W. As shown in Theorem 31.4, Item (3), if A is the infinitesimal generator of a C_0-semigroup $T(t)$, then $T(t)w_0$ is a solution of (35.1) provided that $w_0 \in \mathcal{D}(A)$. Even though $T(t)w$ is defined for all $w \in W$, one needs a stronger concept for the semigroup of bounded linear operators, *viz.* a differentiable semigroup to solve (35.1). We introduce this concept here.

Let $(T(t), A)$ be a C_0-semigroup on W. We will say that this semigroup is **differentiable for** $t > t_0 \geq 0$, if for every $w \in W$ the mapping

$$t \to T^0(t)w \stackrel{\text{def}}{=} T(t)w$$

is strongly differentiable for $t > t_0$. In this setting, we then define $T^{(1)}(t)w$ by

$$T^{(1)}(t)w \stackrel{\text{def}}{=} \frac{d}{dt}T(t)w, \qquad t > t_0.$$

The differentiability for $t > t_0$ is equivalent to the existence of the limit

$$\left\| h^{-1}\left(T(t+h)w - T(t)w\right) - T^{(1)}(t)w \right\| \to 0, \qquad \text{as } h \to 0,$$

for $t > t_0$ and for each $w \in W$. The higher order derivatives, if they exist, are defined for $w \in W$ by induction:

$$T^{(n)}(t)w \stackrel{\text{def}}{=} \frac{d}{dt}T^{(n-1)}(t)w, \qquad \text{for } t > t_0 \text{ and } n = 2, 3, \cdots.$$

Theorem 35.1. *Let $(T(t), A)$ be a C_0-semigroup on a Banach space W. Assume that this semigroup is differentiable for $t > t_0$. Then the following are valid:*

(1) *For $t > nt_0$, $n = 1, 2, \ldots$, one has $T(t) : W \to \mathcal{D}(A^n)$ and*

$$D^nT(t) \stackrel{\text{def}}{=} T^{(n)}(t) = A^nT(t) \in \mathcal{L}(W). \tag{35.2}$$

(2) *For $t > nt_0$, $n = 1, 2, \ldots$, the operator $T^{(n-1)}(t)$ is continuous in uniform operator topology, i.e., one has*

$$\|T^{(n-1)}(t+h) - T^{(n-1)}(t)\|_{\mathcal{L}} \to 0, \qquad \text{as } h \to 0.$$

(3) *For $t > nt_0$, $n = 1, 2, \ldots$, the operator $T(t)$ is $(n-1)$-times differentiable in the uniform operator topology, i.e., one has*

$$\left\| h^{-1}\left(T^{(i-1)}(t+h) - T^{(i-1)}(t)\right) - T^{(i)}(t) \right\|_{\mathcal{L}} \to 0,$$

as $h \to 0$, for $1 \leq i \leq n-1$.

(4) *For any $w_0 \in W$, one has $T(\cdot)w_0 \in C^{0,1}_{\text{loc}}(t_0, \infty; W)$.*

Proof. First we note that since A is a closed, linear operator and $T(t)$ is bounded, then $AT(t)$ is a closed, linear operator. Assume for the moment that $AT(t)w$ is defined for all $w \in W$, and $t > t_0$. It then follows from the Closed Graph Theorem that $AT(t)$ is a bounded operator, i.e., $AT(t) \in \mathcal{L}(W)$. In order to show that $AT(t)w$ is defined for all $w \in W$ and $t > t_0$, we first fix $w \in W$. Since $T(t)w$ is differentiable for $t > t_0$, it follows that the limit

$$\lim_{h\to 0} \frac{T(t+h)w - T(t)w}{h}$$

exists. Now the semigroup property (31.2) implies that

$$\lim_{h\to 0} \frac{T(t+h)w - T(t)w}{h} = \lim_{h\to 0} \frac{T(h) - I}{h} T(t)w = AT(t)w,$$

which establishes (35.2) for $n = 1$.

From Lemma 31.1, one has $\|T(t)\| \le M_1 = Me^{|a|}$, for $0 \le t \le 1$. Now fix $t_1 > t_0$ and let t_2 satisfy $t_1 \le t_2 \le t_1 + 1$. By integrating (35.2), with $w \in W$, one obtains

$$\int_{t_1}^{t_2} T^{(1)}(s)w\,ds = \int_{t_1}^{t_2} AT(s)w\,ds = \int_{t_1}^{t_2} AT(s-t_1)T(t_1)w\,ds.$$

As was shown in the last paragraph, one has $T(t_1)w \in \mathcal{D}(A)$. Therefore, it follows from Theorem 31.4, Item (4), that

$$\int_{t_1}^{t_2} T^{(1)}(s)w\,ds = \int_{t_1}^{t_2} T(s-t_1)AT(t_1)w\,ds = [T(t_2) - T(t_1)]w,$$

for $w \in W$. Hence one has $\|(T(t_2)-T(t_1))w\| \le M_1(t_2-t_1)\|AT(t_1)\|\|w\|$ for $w \in W$, which establishes Item (2) for $n = 1$. Item (3), for $n = 1$, is vacuous, and therefore valid. The remainder of the argument is by induction. We will leave these details as an exercise.

The Hölder continuity in Item (4) now follows from Items (1), (2), and (3) of this theorem. □

We next prove the following uniqueness result.

Theorem 35.2. *Let $(T(t), A)$ be a C_0-semigroup on a Banach space W. Assume that one of the following holds:*

(1) $u_0 \in \mathcal{D}(A)$, or

(2) $T(t)$ is differentiable for $t > 0$ and $u_0 \in W$.

Then there is one and only one classical solution of (35.1), and this solution is given by $w(t) = T(t)w_0$, for $t \ge 0$.

Proof. First assume that $T(t)$ is differentiable for $t > 0$. The fact that $T(t)w_0$ is a classical solution of (35.1) follows immediately from the definition and equation (35.2). Now let $w : [0, \tau) \to W$ be any other classical solution, where $0 < \tau \le \infty$. Define $u(t,s)$ by $u(t,s) \stackrel{\text{def}}{=} T(t-s)w(s)$, for $0 \le s \le t < \tau$. Then the mapping $s \to u(t,s)$ is continuous for $0 \le s \le t$ and differentiable for $0 < s < t$. Furthermore, by the last theorem the derivative satisfies

$$\begin{aligned}\frac{\partial}{\partial s}u(t,s) &= T(t-s)\frac{d}{ds}w(s) - T^{(1)}(t-s)w(s)\\ &= T(t-s)Aw(s) - AT(t-s)w(s) = 0,\end{aligned}$$

for $0 < s < t$. One then has $u(t,t) = w(t) = T(t)w_0 = u(t,0)$, for $0 \le t < \tau$. If $T(t)$ is not differentiable for $t > 0$, but $u_0 \in \mathcal{D}(A)$, then the same argument applies by using Theorem 31.4, Item (3). □

There are some interesting and useful consequences which arise in the case that the C_0-semigroup $(T(t), A)$ is differentiable, for $t > 0$, i.e., $t_0 = 0$ in Theorem 35.1. They are summarized in the following result:

Theorem 35.3. *Let $(T(t), A)$ be a C_0-semigroup on a Banach space W. Assume that this semigroup is differentiable for $t > 0$. Then the following are valid:*

(1) *For all $w \in W$, one has*

$$T(t)w - w = \int_0^t T^{(1)}(s)\, w\, ds = \int_0^t AT(s)\, w\, ds, \qquad \textit{for all } t \ge 0. \tag{35.3}$$

(2) *For all $w \in W$, one has $AT(\cdot)w = T^{(1)}(\cdot)w \in L^1_{\mathrm{loc}}[0,\infty; W)$.*

Proof. Let $w \in W$, and let t and t_0 satisfy $0 < t_0 < t < \infty$. Then from (31.6) and (35.2) one obtains

$$T(t)w - T(t_0)w = \int_{t_0}^t T^{(1)}(s)\, w\, ds = \int_{t_0}^t AT(s)\, w\, ds, \tag{35.4}$$

which is valid in the space W. Since Theorem 31.4, Item (2) implies that the outer terms are equal at $t_0 = 0$, we can take the limit, as $t_0 \to 0^+$, in (35.4) and thereby obtain (35.3). Furthermore, (35.3) implies Item (2). □

Theorem 35.2 is important because of the connection between classical solutions of equation (35.1) and the C_0-semigroup $(T(t), A)$. If $T(t)$ is differentiable for $t > 0$, then <u>every</u> $T(t)u_0$ is a classical solution, for $u_0 \in W$. Without the property of differentiability, the best that can be said is that $T(t)u_0$ is a classical solution only when $u_0 \in \mathcal{D}(A)$. As noted above, $T(t)u_0$ is a mild solution of equation (35.1), for any C_0-semigroup $(T(t), A)$ and any $u_0 \in W$. Theorem 35.2 states that some mild solutions are classical solutions. This concept of a mild solution and the connection with classical solutions will be extended to other equations in the sequel.

3.6. Analytic Semigroups and Sectorial Operators.

A special class of C_0-semigroups, namely, the analytic semigroups, plays a fundamental role in the study of the dynamics of infinite dimensional systems. There are two principal reasons why analytic semigroups are important in the study of systems of nonlinear parabolic partial differential equations. The first reason is owing to the good information one has on the behavior of solutions as time $t \to 0^+$. In particular, we will show that, under reasonable conditions, a C_0-semigroup (e^{At}, A) on a Banach space W is an analytic semigroup if and only if there are constants $M_0 \geq 1$ and $M_1 > 0$ and an $a \in \mathbb{R}$ such that

$$\|e^{At}w\| \leq M_0 e^{-at}\|w\|, \qquad \text{and} \qquad \|Ae^{At}w\| \leq M_1 t^{-1} e^{-at}\|w\|,$$

for all $w \in W$ and $t > 0$. Since the nonlinear evolutionary equation is, in fact, a perturbation of the linear equation, by having good information about the behavior of solutions of the linear problem, as $t \to 0$, one can use this to study the Initial Value Problem for related nonlinear problems. The second reason for the importance of analytic semigroups is given in the Fundamental Theorem on Sectorial Operators, which is presented below.

A vector-valued function $f : D \to W$, where D is an open set of the complex plane $\mathbb{C}$ and W is a Banach space, is said to be **analytic** (or an **analytic mapping**) if, for any $z_0 \in D$, the strong limit

$$\lim_{z \to 0} \frac{1}{z}[f(z_0 + z) - f(z_0)]$$

exists in W. (The limit here is interpreted as $z \to 0$ in $\mathbb{C}$.) For example, if A is the infinitesimal generator of a C_0-semigroup on a Banach space W, then the resolvent operator $R(\lambda, A) = (\lambda I - A)^{-1}$ is an analytic mapping from the resolvent set $\rho(A)$ into $\mathcal{L}(W)$, see Hille and Phillips (1957).

For $\delta \in (0, \pi)$ and $\sigma \in (0, \pi)$ we define the following open sectors in the complex plane $\mathbb{C}$:

$$\begin{aligned}
\Delta_\delta &\stackrel{\text{def}}{=} \{z \in \mathbb{C} : |\arg z| < \delta, z \neq 0\}, \\
\Delta_\delta(a) &\stackrel{\text{def}}{=} a + \Delta_\delta = \{z \in \mathbb{C} : |\arg(z - a)| < \delta, z \neq a\}, \\
\Sigma_\sigma &\stackrel{\text{def}}{=} \{z \in \mathbb{C} : |\arg z| > \sigma, z \neq 0\}, \\
\Sigma_\sigma(a) &\stackrel{\text{def}}{=} a + \Sigma_\sigma = \{z \in \mathbb{C} : |\arg(z - a)| > \sigma, z \neq a\}.
\end{aligned}$$

A simple calculation shows that $\Delta_\delta(-a) = -\Sigma_\sigma(a)$, provided that $\delta = \pi - \sigma$, see Figure 3.1. As a result one has

$$\Sigma_\sigma(a) \subset \rho(A) \iff \Delta_\delta(-a) \subset \rho(-A), \qquad \text{where } \delta = \pi - \sigma, \tag{36.1}$$

for any linear operator $A : \mathcal{D}(A) \to W$, where $\mathcal{D}(A)$ is dense in W.

Let $(T(t), A)$ be a C_0-semigroup on W. We will say that $(T(t), A)$ is an **analytic semigroup** if there exists an extension of $T(t)$ to a mapping $T(z)$ defined for z in some sector $\Delta_\delta \cup \{0\}$ and satisfying the following conditions:

(1) The mapping $z \to T(z)$ is a mapping of $\Delta_\delta \cup \{0\}$ into $\mathcal{L}(W)$.
(2) $T(z_1 + z_2) = T(z_1)T(z_2)$ for all z_1 and z_2 in $\Delta_\delta \cup \{0\}$.
(3) For each $w \in W$, one has $T(z)w \to w$, as $z \to 0$ in $\Delta_\delta \cup \{0\}$.
(4) For each $w \in W$, the function $z \to T(z)w$ is an analytic mapping from Δ_δ into W.

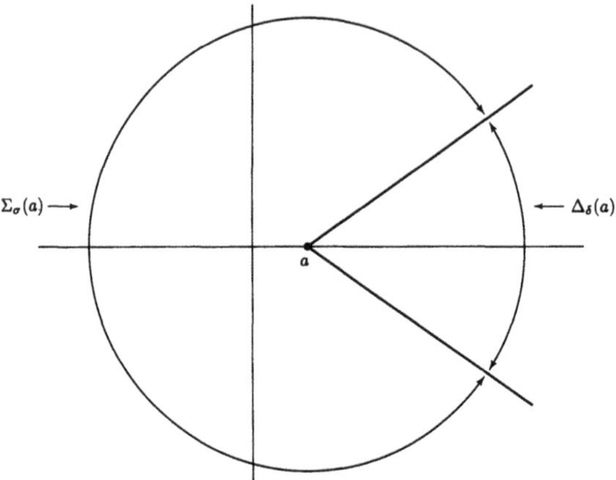

Figure 3.1. The Sectors Δ and Σ with $\delta = \pi - \sigma$

Any mapping $T(z)$ that is an extension of $T(t)$ and satisfies Items (1) - (4) above is said to be an **analytic semigroup extension** of $T(t)$. Note that $(T(t), A)$ is an analytic semigroup if and only if $(T(t)e^{rt}, A + rI)$ is an analytic semigroup, for every $r \in \mathbb{R}$.

There is, of course, the related issue of determining which properties on the infinitesimal generator A will guarantee that a given C_0-semigroup is an analytic semigroup. This prompts the following definition. A linear operator A is said to be a **sectorial operator** on W if $A : \mathcal{D}(A) \to W$, where $\mathcal{D}(A) \subset W$, and it satisfies the following two conditions:

(1) A is densely defined and closed.
(2) There exist real numbers $a \in \mathbb{R}$, $\sigma \in (0, \frac{\pi}{2})$ and $M \geq 1$, such that one has $\Sigma_\sigma(a) \subset \rho(A)$, and

$$\|R(\lambda, A)\| \leq \frac{M}{|\lambda - a|}, \qquad \text{for all } \lambda \in \Sigma_\sigma(a). \tag{36.2}$$

A sectorial operator A is said to be **positive** if it satisfies (36.2) for some $a > 0$. By using (36.1), we see that (36.2) is equivalent to
(36.3)

$$\|R(\lambda, -A)\| \leq \frac{M}{|\lambda + a|}, \qquad \text{for all } \lambda \in \Delta_\delta(-a) \subset \rho(-A), \text{ where } \delta = \pi - \sigma.$$

Also note that (36.3) is equivalent to

$$\|R(\mu, -B)\| \leq \frac{M}{|\mu|}, \qquad \text{for all } \mu \in \Delta_\delta(0), \tag{36.4}$$

where $B = A - aI$. Furthermore, A is a sectorial operator if and only if B is sectorial.

3.6.1 Characterizations of Analytic Semigroups. The first result is a technical lemma, which shows that if A is a sectorial operator, then $-A$ is

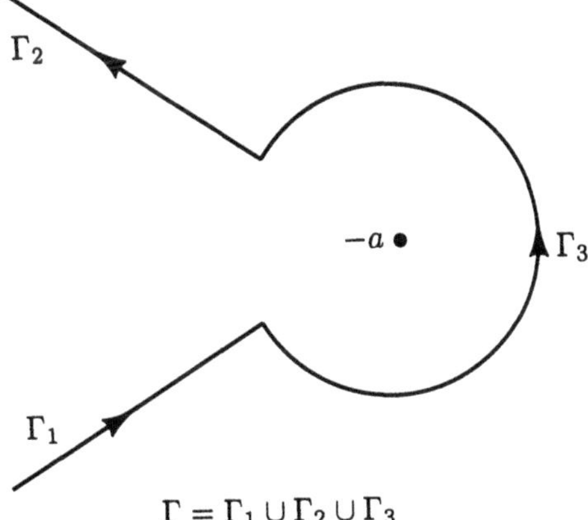

$\Gamma = \Gamma_1 \cup \Gamma_2 \cup \Gamma_3$

Figure 3.2. Detail for Lemma 36.1

the infinitesimal generator of a C_0-semigroup e^{-At}. A proof can be found in Pazy (1983), see Figure 3.2.

Lemma 36.1. *Let A be a sectorial operator on a Banach space W, where (36.2) is satisfied for appropriate constants M, a, and σ. Then $-A$ is the infinitesimal generator of a C_0-semigroup e^{-At}, and there is a constant $M_0 \geq 1$ such that $\|e^{-At}\| \leq M_0 e^{-at}$, for all $t \geq 0$. Moreover one has*

$$e^{-At} = \frac{1}{2\pi i} \int_\Gamma e^{\lambda t} R(\lambda, -A)\, d\lambda, \qquad \text{for } t > 0, \tag{36.5}$$

and $e^{-At} = I$ at $t = 0$. Here $\Gamma = \Gamma(\eta, \theta)$ is a contour in the resolvent set $\rho(-A)$ given by $\Gamma = \Gamma_1 \cup \Gamma_2 \cup \Gamma_3$, where

$$\Gamma_1 = \{\lambda = -a + re^{-i(\pi-\theta)} : r \geq \eta\}, \quad \Gamma_2 = \{\lambda = -a + \eta e^{i\varphi} : |\varphi| \leq \pi - \theta\},$$
$$\Gamma_3 = \{\lambda = -a + re^{i(\pi-\theta)} : r \geq \eta\},$$

θ satisfies $\sigma < \theta < \frac{\pi}{2}$, and η is any positive constant. The path of integration along Γ in (36.5) is counter clockwise. The integral in (36.5) exists in the uniform operator topology on $\mathcal{L}(W)$, for all $t > 0$.

The following fundamental theorem describes several characterizations of an analytic semigroup. The proof presented here is based on ideas found in Pazy (1983).

Theorem 36.2. *Let $(e^{-At}, -A)$ be a C_0-semigroup on a Banach space W, and let $M \geq 1$ and $a \in \mathbb{R}$ be chosen so that $\|e^{-At}\| \leq Me^{-at}$, for all $t \geq 0$. Then the following statements are equivalent:*

1. *e^{-At} is an analytic semigroup, and there is an analytic semigroup extension e^{-Az} of e^{-At}, defined on some sector $\Delta_\delta \cup \{0\}$, with $0 < \delta < \pi/2$, and a constant $M_1 \geq M$ such that $\|e^{-Az}\| \leq M_1 e^{-a\mathrm{Re}\, z}$, for all $z \in \Delta_\delta \cup \{0\}$.*

(2) *There is a constant M_2 such that the resolvent operator satisfies*

$$\|R(\sigma+i\tau,-A)\| \le \frac{M_2}{|\tau|}, \qquad \text{for all } \sigma > -a \text{ and } \tau \neq 0. \tag{36.6}$$

(3) *The operator A is a sectorial operator and one has*

$$\|R(\lambda,A)\| \le \frac{M_3}{|\lambda - a|}, \qquad \text{for all } \lambda \in \Sigma_\xi(a), \tag{36.7}$$

for appropriate constants $M_3 \ge 1$ and $\xi \in (0, \frac{\pi}{2})$.

(4) *The semigroup e^{-At} is differentiable for $t > 0$, and there is a constant $M_4 > 0$ such that*

$$\|Ae^{-At}\| \le \begin{cases} M_4 t^{-1} e^{-at}, & \text{for } 0 < t \le 1, \\ M_4 e^{-at}, & \text{for } t \ge 1. \end{cases} \tag{36.8}$$

Proof. (1) $\Longrightarrow$ (2): For $w \in W$ and real $\sigma > -a$ one has

$$R(\sigma+i\tau,-A)w = \int_0^\infty e^{-(\sigma+i\tau)t} e^{-At} w\, dt \tag{36.9}$$

by Lemma 31.6. Since $e^{-Az}w$ is analytic for $z \in \Delta_\delta$, it follows from the Cauchy Theorem that the path of integration in (36.9) can be shifted to any ray $z = re^{i\theta}$, where $|\theta| < \delta$ and $0 < r < \infty$. Now fix θ with $0 < \theta < \delta$. For $\tau > 0$ set $z = re^{-i\theta}$, then since $\sigma + a > 0$, (36.9) and Item (1) imply that

$$\begin{aligned} \|R(\sigma+i\tau,-A)\| &\le \int_0^\infty e^{-\text{Re}\,((\sigma+i\tau)z)} M_1 e^{-a\text{Re}\, z}\, dr \\ &\le M_1 \int_0^\infty e^{-r(\sigma\cos\theta + \tau\sin\theta)} e^{-ar\cos\theta}\, dr \le \frac{M_2}{\tau}. \end{aligned}$$

For $\tau < 0$, one uses the ray $z = re^{i\theta}$ to obtain an equivalent conclusion.

(2) $\Longrightarrow$ (3): By assumption one has $\|R(\lambda,-A)\| \le M_2|\text{Im}\,\lambda|^{-1}$, for $\text{Re}\,\lambda > -a$. Also, Lemma 31.6 implies that $\|R(\lambda,-A)\| \le M|\text{Re}\,\lambda + a|^{-1}$, for $\text{Re}\,\lambda > -a$. Consequently one has $\|R(\lambda,-A)\| \le M_3|\lambda + a|^{-1}$, for $\text{Re}\,\lambda > -a$ and an appropriate $M_3 > 0$. Hence $\rho(-A)$ contains the sector $\Delta_{\frac{\pi}{2}}(-a)$.

Using the fact that the mapping $\lambda \to R(\lambda,-A)$ is analytic for $\text{Re}\,\lambda > -a$, one can expand $R(\lambda,-A)$ in a Neuman series around $\lambda = \sigma + i\tau$, for $\sigma > -a$, to obtain

$$R(\lambda,-A) = \sum_{n=0}^\infty R(\sigma+i\tau,-A)^{n+1}(\lambda - \sigma - i\tau)^n. \tag{36.10}$$

This series converges in the uniform operator topology on $\mathcal{L}(W)$ when λ satisfies

$$\|R(\sigma + i\tau, -A)\| \, |\lambda - \sigma - i\tau| \le \nu < 1.$$

In particular, if λ is chosen with Im $\lambda = -\tau$, this series is convergent for

$$|\text{Re } \lambda - \sigma| \le \frac{\nu|\text{Im } \lambda|}{M_2} \le \frac{\nu}{\|R(\sigma + i\tau, -A)\|}$$

by (36.6). Since $\sigma > -a$ and $0 \le \nu < 1$ are arbitrary, this implies that the resolvent set $\rho(-A)$ satisfies

$$\rho(-A) \supset \left\{\lambda \in \mathbb{C} : \frac{|\text{Re } \lambda + a|}{|\text{Im } \lambda|} < \frac{1}{M_2}\right\}.$$

Combining this with the fact noted above that $\rho(-A) \supset \Delta_{\frac{\pi}{2}}(-a)$, we see that

$$\rho(-A) \supset \Delta_\eta(-a), \qquad \text{where } \eta = \frac{\pi}{2} + \arctan\left(\frac{1}{M_2}\right).$$

As a result, from (36.1) we then obtain

$$\rho(A) \supset \Sigma_\sigma(a), \qquad \text{where } \sigma = \frac{\pi}{2} - \arctan\left(\frac{1}{M_2}\right),$$

and $0 < \sigma < \frac{\pi}{2}$. Furthermore, from the above, one has

$$\|R(\lambda, A)\| = \|R(-\lambda, -A)\| \le \frac{M_3}{|\lambda - a|}, \qquad \text{for } -\lambda \in \Delta_{\frac{\pi}{2}}(-a).$$

On the other hand if $-\lambda \in \Delta_\eta(-a)\backslash\Delta_{\frac{\pi}{2}}(-a)$, then there is a ν, $0 \le \nu < 1$, with $|\text{Re } \lambda - a| = \frac{\nu}{M_2}|\text{Im } \lambda|$. For this λ one has $|\lambda - a| = C_1|\text{Im } \lambda|$, where $C_1 = (1 + (\frac{\nu}{M_2})^2)^{1/2}$. Consequently, from (36.6) and (36.10) one finds that

$$\begin{aligned}\|R(\lambda, A)\| = \|R(-\lambda, -A)\| &\le \|R(\sigma + i\tau, -A)\| \sum_{n=0}^{\infty} \nu^n \\ &\le \frac{1}{1-\nu} \frac{M_2}{|\text{Im } \lambda|} = \frac{C_2}{|\lambda - a|}.\end{aligned}$$

Hence A is a sectorial operator on W and (36.7) holds, with $\pi - \eta < \xi < \pi/2$, for a suitable M_3, with σ as given above.

(3) $\Longrightarrow$ (4): Define B by $B = A - aI$. Then (36.7) is equivalent to (36.4), and B is also a sectorial operator. Furthermore, the C_0-semigroup e^{-Bt} is differentiable for $t > 0$ if and only if e^{-At} is differentiable for $t > 0$. From Lemma 36.1, one has

$$e^{-Bt} = \frac{1}{2\pi i} \int_\Gamma e^{\lambda t} R(\lambda, -B) \, d\lambda,$$

where the contour $\Gamma = \Gamma(\eta, \theta)$ is described in the statement of that lemma. Now for fixed $t > 0$ and $w \in W$, one has

$$\frac{e^{-B(t+h)} - e^{-Bt}}{h} w = \frac{1}{2\pi i} \int_\Gamma \frac{e^{\lambda(t+h)} - e^{\lambda t}}{h} R(\lambda, -B) w \, d\lambda$$
$$= \frac{1}{2\pi i} \int_\Gamma \lambda e^{\lambda t}(1 + o(h)) R(\lambda, -B) w \, d\lambda,$$

where $o(h) \to 0$, as $h \to 0$. Assume for the moment that all of the integrals above converge in the uniform operator topology on $\mathcal{L}(W)$. Now the last of these integrals has a limit as $h \to 0$, which implies that the semigroup e^{-At} is differentiable for $t > 0$, and

$$\partial_t e^{-Bt} = \frac{1}{2\pi i} \int_\Gamma \lambda e^{\lambda t} R(\lambda, -B) \, d\lambda.$$

As an exercise, the reader should verify that the $o(h)$ part of the integral converges to 0, as $h \to 0$. It now suffices to show that the integral

$$\frac{1}{2\pi i} \int_{\Gamma(\eta,\theta)} \lambda e^{\lambda t} R(\lambda, -B) \, d\lambda$$

converges in the uniform operator topology on $\mathcal{L}(W)$. Because of the factor λ appearing in the last integral, we can let $\eta \to 0$. One then uses inequality (36.4) to get

$$\left\| \frac{1}{2\pi i} \int_{\Gamma(0,\theta)} \lambda e^{\lambda t} R(\lambda, -B) \, d\lambda \right\| \le \frac{1}{\pi} \int_0^\infty |\lambda| \, |e^{\lambda t}| \, \|R(\lambda, -B)\| \, d|\lambda|$$
$$\le \frac{M}{\pi} \int_0^\infty e^{rt\cos(\pi - \theta)} \, dr = M_5 t^{-1},$$

where $0 < \theta < \frac{\pi}{2}$. Hence the semigroup e^{-Bt} is differentiable for $t > 0$, and $\|\partial_t e^{-Bt}\| \le M_5 t^{-1}$, for $t > 0$. Now

$$\|\partial_t e^{-At}\| = \|\partial_t (e^{-at} e^{-Bt})\|$$
$$\le a e^{-at} \|e^{-Bt}\| + e^{-at} \|\partial_t e^{-Bt}\| \le M_4 t^{-1} e^{-at}, \tag{36.11}$$

for $0 < t \le 1$, where $M_4 = Ma + M_5$. From (35.2) one then has

$$A e^{-At} w = \partial_t e^{-At} w, \qquad \text{for all } w \in W \text{ and } t > 0.$$

This fact, together with (36.11), implies (36.8).

(4) $\Longrightarrow$ (1): By making M_4 larger and the constant a smaller, if necessary, one can replace (36.8) with $\|A e^{-At}\| \le M_4 t^{-1} e^{-at}$, for all $t > 0$. Now

let $T(t) = e^{-At}$. Since $T(t)$ is differentiable for $t > 0$, one can use Theorem 35.1 to verify the following, with a simple induction argument:

$$T^{(n)}(t) = \left(AT\left(\frac{t}{n}\right)\right)^n, \qquad \text{for all } t > 0 \text{ and } n = 1, 2, \ldots. \tag{36.12}$$

Combining this with (36.8) one then obtains

$$\|T^{(n)}(t)\| \le \left\|AT\left(\frac{t}{n}\right)\right\|^n \le \left(\frac{M_4 n}{t} e^{-a\frac{t}{n}}\right)^n \le \left(\frac{M_4}{t}\right)^n n^n e^{-at}. \tag{36.13}$$

Since $n!e^n \ge n^n$, by the Stirling formula, this implies that

$$\frac{1}{n!}\|T^{(n)}(t)\| \le \left(\frac{M_4 e}{t}\right)^n e^{-at}.$$

Hence the Taylor series

$$T(z) \overset{\text{def}}{=} T(t) + \sum_{n=1}^{\infty} \frac{T^{(n)}(t)}{n!}(z-t)^n$$

converges in the uniform operator topology on $\mathfrak{L}(W)$ for

$$|\text{Im } z| \le |z - t| \le \nu \left(\frac{t}{M_4 e}\right),$$

where ν is fixed and satisfies $0 < \nu < 1$. Therefore $T(z)$ is an extension of $T(t)$ and it is analytic in the sector Δ_δ, where $\delta = \arctan(\frac{\nu}{M_4 e})$.

It is easily seen that this mapping $T(z)$ satisfies properties (1), (3), and (4) in the definition of an analytic semigroup extension. In order to verify the semigroup property (2) we first fix $z_2 = t \in \mathbb{R}$. Then $S_1(z) \overset{\text{def}}{=} T(z)T(t)$ and $S_2(z) \overset{\text{def}}{=} T(z+t)$ are both analytic mappings of Δ_δ into $\mathfrak{L}(W)$. Since these two functions agree for $z = \tau \in \mathbb{R}^+$, they agree everywhere on Δ_δ, i.e., one has $T(z)T(t) = T(z+t)$, for all $z \in \Delta_\delta$ and $t \in \mathbb{R}^+$. Using the same argument and fixing $z = z_1 \in \Delta_\delta$, one finds that $T(z_1)T(z) = T(z_1 + z)$, for all $z, z_1 \in \Delta_\delta$. □

Example 36.3. Let us return to the C_0-semigroup e^{-At} constructed in Section 3.2. Recall that in this case, the linear operator A was assumed to be self-adjoint, bounded below, and with compact resolvent. By using equation (32.5) with $f(\lambda) = e^{-\lambda t}$, the semigroup $(e^{-At}, -A)$ has the Fourier series representation

$$e^{-At}u = \sum_{i=1}^{\infty} e^{-\lambda_i t}\langle u, e_i\rangle e_i, \qquad u \in H.$$

The time parameter t in this semigroup can be complexified by replacing t with the complex variable z to obtain

$$e^{-Az}u = \sum_{i=1}^{\infty} e^{-\lambda_i z}\langle u, e_i\rangle e_i, \qquad u \in H. \tag{36.14}$$

In order to verify the convergence of the infinte series in (36.14) we note that

$$\|e^{-Az}\|^2 \le \sum_{i=1}^{\infty} |e^{-\lambda_i z}|^2 |\langle u, e_i\rangle|^2 \le \sum_{i=1}^{\infty} e^{-2\lambda_i \mathrm{Re}\, z} |\langle u, e_i\rangle|^2 \le e^{-2\lambda_1 \mathrm{Re}\, z}\|u\|^2,$$

for Re $z > 0$. Hence the series is convergent for all $u \in H$, provided that Re $z > 0$, in particular e^{-Az} is well-defined for all z in the sector $\Delta_{\frac{\pi}{2}} \cup \{0\}$. Furthermore, we claim that for all $z_0 \in \Delta_{\frac{\pi}{2}}$ (Re $z_0 > 0$), the strong limit

$$\frac{1}{z}[e^{-A(z_0+z)}w - e^{-Az_0}w] \overset{s}{\to} -Ae^{-Az_0}w, \qquad \text{as } z \to 0 \text{ in } \Delta_{\pi/2}(0), \tag{36.15}$$

exists, which implies that the mapping $z \to e^{-Az}w$ is an analytic mapping from $\Delta_{\frac{\pi}{2}}$ into W. Since the proof of (36.15) is similar to the argument given in the paragraph following (32.7) in which one establishes the strong continuity of e^{-At}, we will omit the details. As a result of Theorem 36.2, we see that A is a sectorial operator.

3.6.2 Construction of Analytic Semigroups. As stated above, the principal objective of the theory of C_0-semigroups is to describe the solutions of a linear evolutionary equation on a Banach space W of the form

$$\partial_t w = Aw, \qquad w(0) = w_0 \in W. \tag{36.16}$$

Recall Theorem 35.2. When $(T(t), A)$ is a C_0-semigroup on W, then $w(t) = T(t)w_0$, is a classical solution to (36.16) provided $w_0 \in \mathcal{D}(A)$. If $(T(t), A)$ is differentiable for $t > 0$, then $T(t)w_0$ is a classical solution to the equation (36.16), for each $w_0 \in W$.

In the modern theory of partial differential equations, there is a related concept of weak solution of (36.16), a concept which is described in terms of a duality between various Hilbert spaces. Our main objective in the remainder of this section is to give an abstract framework which describes this duality and provides the basis for the construction of an analytic semigroup on a Hilbert space H. We will see later that this abstract framework, which is based on the Lax-Milgram property, arises commonly in the theory of linear partial differential equations.

In order to describe this framework, we begin with two Hilbert spaces H and V and a sesquilinear functional $a(u, v)$ on V. Our objective here is to show that under reasonable conditions a sesquilinear functional $a(u, v)$ generates a sectorial operator B, and consequently $-B$ generates an analytic semigroup e^{-Bt}.

Let V and H be two Hilbert spaces, and assume that $V \mapsto H$, i.e., $V \subset H$, and there is a constant $C > 0$ such that

$$\|v\|_H \leq C\|v\|_V, \qquad \text{for all } v \in V.$$

The **dual space** of V, which we will denote by V', is defined to be the collection of all linear functionals ℓ on V with the property that there is a constant $c \geq 0$, which depends on ℓ, such that

$$|\ell(u)| \leq c\|u\|_V, \qquad \text{for all } u \in V.$$

If $w \in V'$, we will write the value of w at the point $v \in V$ by

$$w(v) \stackrel{\text{def}}{=} \langle\langle v, w\rangle\rangle = \langle v, w\rangle_{(V,V')}, \qquad \text{for all } v \in V. \tag{36.17}$$

It is convenient to assume that the form $\langle\langle v, v'\rangle\rangle$ to be linear in v and hemilinear in v', i.e., $\langle\langle v, v'\rangle\rangle$ is a sesquilinear form. If V and H are real spaces, then so is V' and $\langle\langle \cdot, \cdot\rangle\rangle$ is a bilinear form on $V \times V'$. Note that the dual space V' becomes a Banach space with the norm given by

$$\|w\|_{V'} \stackrel{\text{def}}{=} \sup\{|\langle\langle v, w\rangle\rangle| : \|v\|_V \leq 1\}.$$

Furthermore, one has

$$|\langle\langle v, w\rangle\rangle| \leq \|v\|_V\|w\|_{V'}, \qquad \text{for all } v \in V \text{ and } w \in V'. \tag{36.18}$$

For each $w \in H$, the mapping $v \to \langle v, w\rangle_H$ is a bounded linear functional on V. This then defines a continuous imbedding $H \mapsto V'$, and one has $V \mapsto H \mapsto V'$.

We will say that a sesquilinear functional $a(u, v)$ on V is **bounded in the V-norm** if there is a constant $M \geq 0$ such that

$$|a(u, v)| \leq M\|u\|_V\|v\|_V, \qquad \text{for all } u, v \in V. \tag{36.19}$$

The sesquilinear functional $a(u, v)$ on V is said to be **coercive** if there is a $\delta > 0$ such that

$$\text{Re } a(u, u) \geq \delta\|u\|_V^2, \qquad \text{for all } u \in V. \tag{36.20}$$

A mapping $a = a(u, v)$, or more specifically, the triple $(H, V, a(\cdot, \cdot))$ is said to satisfy the **Lax-Milgram property** if the following two conditions are satisfied:

1. (1) H and V are Hilbert spaces, and one has $V \mapsto H$.
2. (2) The mapping $(u, v) \to a(u, v)$ is a sesquilinear functional on V that is bounded in the V-norm and coercive.

The following result is easily verified:

Lemma 36.4. *Let $a(u,v)$ satisfy the Lax-Milgram property. Then the following statements are valid:*

(1) *The mappings $(u,v) \to a^*(u,v)$ and $(u,v) \to s(u,v)$, where*

$$a^*(u,v) \stackrel{\text{def}}{=} \overline{a(v,u)}, \qquad s(u,v) \stackrel{\text{def}}{=} \frac{1}{2}\left[a(u,v) + a^*(u,v)\right]$$

satisfy the Lax-Milgram property and $s(u,u) = \operatorname{Re} a(u,u)$ for all $u \in V$.

(2) *The mapping*

$$(u,v) \to \langle u,v \rangle_s \stackrel{\text{def}}{=} s(u,v)$$

is a positive definite, Hermitian, sesquilinear form, i.e., an inner product, on V, and the associated norm $\|u\|_s = s(u,u)^{\frac{1}{2}}$ is equivalent to the V-norm $\|u\|_V$ on V.

A prototypical example where the Lax-Milgram property is satisfied arises in the study of the elliptic equation $-\Delta u = f$, where u, f are complex-valued functions, with Dirichlet boundary condition $u = 0$ on $\partial\Omega$, where Ω is a sufficiently smooth bounded domain in $\mathbb{R}^d$. In this case one has the compact imbedding $V \hookrightarrow H$, where $H = L^2(\Omega, \mathbb{C})$ and $V = H_0^1(\Omega, \mathbb{C})$. The sesquilinear function $a = a(u,v)$ is given by

$$a(u,v) = \int_\Omega \sum_{i=1}^d D_i u \, D_i \bar{v} \, dx = \int_\Omega \nabla u \cdot \nabla \bar{v} \, dx, \qquad \text{for } u,\, v \in V,$$

and $D_i = \frac{\partial}{\partial x_i}$. The Lax-Milgram property is satisfied, and a is positive definite and Hermitian. If the equation $a(u,v) = \langle f,v \rangle_H$ is valid for some $f \in H$ and all $v \in V$, where $u \in H^2(\Omega, \mathbb{C}) \cap V$, then one can perform an integration by parts to obtain $\langle -\Delta u, v \rangle_H = \langle f,v \rangle_H$, which implies that $-\Delta u = f$, since V is dense in H. Thus the equation $a(u,v) = \langle f,v \rangle_H$, or more generally the equation $a(u,v) = \langle\langle f,v \rangle\rangle$, when it is valid for all $v \in V$, describes a weak solution of $-\Delta u = f$. We now have the following result.

Lemma 36.5 (Lax-Milgram). *Assume that $a(u,v)$ satisfies the Lax-Milgram property. Then there is a unique bounded linear operator $B \in \mathcal{L}(V,V')$ such that*

$$a(u,v) = \langle\langle u, Bv \rangle\rangle, \qquad \text{for all } u,v \in V, \tag{36.21}$$

and the following properties are satisfied:

(1) *B is a one-to-one mapping of V onto V' with a bounded inverse, where*

$$\|B\|_{\mathcal{L}(V,V')} \le M, \qquad \|B^{-1}\|_{\mathcal{L}(V',V)} \le \delta^{-1},$$

and M and δ satisfy (36.19) and (36.20).

(2) *The restriction of operator B to $\mathcal{D}(B_H) = \{u \in V : Bu \in H\}$ is a closed linear operator B_H on H.*

(3) *If the form $a(u, v)$ is Hermitian, then B_H is a selfadjoint operator.*

(4) *If in addition, the imbedding $V \hookrightarrow H$ is compact, then the operator B_H has compact resolvent.*

Proof. Step 1. We first show that there is a bounded linear operator $B \in \mathcal{L}(V, V')$ with bounded inverse such that (36.21) is satisfied. Indeed for each $v \in V$ the mapping $u \to a(u, v)$ is a bounded linear mapping of V into $\mathbb{C}$ that satisfies (36.19). Hence there is a $v' \in V'$ such that

$$a(u, v) = \langle\langle u, v' \rangle\rangle, \qquad \text{for all } u \in V,$$

and v' satisfies $\|v'\|_{V'} \leq M\|v\|_V$, for all $v \in V$. Define B by $Bv \stackrel{\text{def}}{=} v'$. Then $B \in \mathcal{L}(V, V')$ with

$$|\langle\langle u, Bv \rangle\rangle| = |a(u, v)| \leq M\|u\|_V\|v\|_V,$$

and one has $\|B\|_{\mathcal{L}(V,V')} \stackrel{\text{def}}{=} \sup\{\|Bv\|_{V'} : \|v\|_V \leq 1\} \leq M$. Since a is coercive, one has

$$\delta\|v\|_V^2 \leq \text{Re } a(v, v) \leq |\langle\langle v, Bv \rangle\rangle| \leq \|Bv\|_{V'}\|v\|_V,$$

which implies that B is bounded below, one-to-one, and $\|B^{-1}\|_{\mathcal{L}(\mathcal{R}(B),V)} \leq \delta^{-1}$.

Step 2. We next show that there is a linear operator $G \in \mathcal{L}(V', V)$ such that

$$\langle\langle u, v' \rangle\rangle = \langle u, Gv' \rangle_V, \qquad \text{for all } u \in V \text{ and } v' \in V'. \tag{36.22}$$

Indeed, if $v' \in V'$ then from (36.18) one has

$$|\langle\langle u, v' \rangle\rangle| \leq \|v'\|_{V'}\|u\|_V \qquad \text{for all } u \in V.$$

Consequently the Riesz Representation Theorem implies that there is a unique point $v \in V$, which we will label by $v = Gv'$, such that

$$\langle\langle u, v' \rangle\rangle = \langle u, Gv' \rangle_V, \qquad \text{for all } u \in V.$$

From the argument given above, we see that

$$|\langle u, Gv' \rangle_V| \leq \|v'\|_{V'}\|u\|_V, \qquad \text{for all } u \in V,$$

which implies that $\|Gv'\|_V \leq \|v'\|_{V'}$, or equivalently, $\|G\|_{\mathcal{L}(V',V)} \leq 1$.

Step 3. We next show that there is a linear operator $L \in \mathcal{L}(V, V)$ with bounded inverse that satisfies

$$(36.23) \qquad \langle u, w \rangle_V = a(u, L^{-1}w) \qquad \text{for all } u, w \in V.$$

Indeed, by the Riesz Representation Theorem, we see that for every $v \in V$ there is a $w \in V$ such that

$$a(u, v) = \langle u, w \rangle_V, \qquad \text{for all } u \in V.$$

Now define L by $Lv \stackrel{\text{def}}{=} w$. It is easily verified that L is linear. Since a is coercive, it follows that if $a(u, v) = \langle u, Lv \rangle_V = 0$, for all $v \in V$, then $u = 0$. Hence the range $\mathcal{R}(L)$ is dense in V. By (36.19) one has

$$|\langle u, Lv \rangle_V| = |a(u, v)| \leq M\|u\|_V\|v\|_V, \qquad \text{for all } u, v \in V.$$

Consequently one has $\|L\|_{\mathcal{L}(V)} \leq M$. On the other hand,

$$\delta\|v\|_V^2 \leq \text{Re } a(v, v) \leq |\langle v, Lv \rangle_V| \leq \|v\|_V\|Lv\|_V, \qquad \text{for all } v \in V,$$

implies that L^{-1} exists with $\|L^{-1}\|_{\mathcal{L}(\mathcal{R}(L), V)} \leq \delta^{-1}$, and (36.23) is valid for $w \in \mathcal{R}(L)$. Now L^{-1} has a unique extension to V, and (36.23) holds in the limit. It follows that $\mathcal{R}(L) = V$ and L remains one-to-one.

Step 4. Finally we show that B is a mapping of V onto V', i.e., $\mathcal{R}(B) = V'$ and $B^{-1} = G$. For $v' \in V'$, we define $w \stackrel{\text{def}}{=} L^{-1}Gv' \in V$. From (36.22), (36.23), and (36.21), in that order, one finds that

$$\langle\langle u, v' \rangle\rangle = \langle u, Gv' \rangle_V = a(u, w) = \langle\langle u, Bw \rangle\rangle, \qquad \text{for all } u \in V,$$

which implies that $v' = Bw$. The proofs of Items (2) - (4) are left as an exercise for the reader. □

An Inner Product on V': The sesquilinear functional $a(u, v)$ allows one to construct an inner product $\langle \cdot, \cdot \rangle_a$ on V' by means of the formula

$$\langle w_1, w_2 \rangle_a \stackrel{\text{def}}{=} \langle B^{-1}w_1, B^{-1}w_2 \rangle_V,$$

or equivalently, with $v_i = B^{-1}w_i$, for $i = 1, 2$,

$$\langle v_1, v_2 \rangle_V = \langle Bv_1, Bv_2 \rangle_a.$$

The associated norm $\|w\|_a$ on V' then satisfies $\|v\|_V = \|Bv\|_a$. As shown in the last lemma one has

$$\delta\|v\|_V = \delta\|Bv\|_a \leq \|Bv\|_{V'} \leq M\|v\|_V = M\|Bv\|_a.$$

Thus, the Hilbert norm $\|\cdot\|_a$ is equivalent to the Banach norm $\|\cdot\|_{V'}$ on V'.

The Role of the Density of V in H: Let us summarize the situation described in the Lax-Milgram Lemma. Assume that V is dense in H. Since $V \mapsto H$, it follows that for each $v \in H$, the mapping $u \to \langle u, v\rangle_H$ is a bounded linear functional on V. Moreover, one has

$$[V \text{ is dense in } H] \Longleftrightarrow [\langle u,v\rangle_H = 0, \text{ for all } u \in V \Longrightarrow v = 0].$$

Now the imbedding $V \mapsto H$, implies $H' \mapsto V'$, which in turn implies that

$$V \mapsto H \overset{j}{\mapsto} H' \mapsto V',$$

where the imbedding $H \overset{j}{\mapsto} H'$ is the hemilinear mapping that is the inverse of the Riesz mapping from H' to H, i.e., $j(v)$ is defined by

$$j(v) : u \to \langle u, v\rangle_H.$$

This means that V is imbedded as a linear subspace of V'. Furthermore, due to the density of V in H, the mapping $B : \mathcal{D}(B) = V \to V'$ is densely defined and a closed linear operator on V'. We call this operator B the **form operator on** V' determined by the sesquilinear functional $a(u,v)$.

It is easily seen that the form operators generated by the sesquilinear functionals $a^*(u,v)$ and $s(u,v)$ are B^*, the adjoint of B, and $S = \frac{1}{2}[B+B^*]$, the selfadjoint part of B, respectively. Also note that these form operators are defined when the sesquilinear functional a has the Lax-Milgram property <u>and</u> V is dense in H. We refer to the operator B_H as the **form operator on** H determined by the sesquilinear functional $a(u,v)$. When the context is clear, sometimes one can also briefly denote B_H as B.

Theorem 36.6. *Assume that $a(u,v)$ satisfies the Lax-Milgram property, and that V is dense in H. Let B and B_H denote the form operators generated on V' and H, respectively. Then B and B_H are sectorial operators, and their negatives generate analytic semigroups $T(t)$ and $T_H(t)$ on V' and on H, respectively. Moreover, one has $T(t)|_H = T_H(t)$, for all $t \geq 0$, i.e., the restriction of $T(t)$ to H is $T_H(t)$.*

Proof. We only prove the theorem for B and leave the proof for B_H as an exercise. As a first step, we claim that the closed sector $\mathrm{Cl}_{\mathbb{C}}\Delta_{\frac{\pi}{2}}$ lies in the resolvent set $\rho(-B)$, i.e.,

$$\mathrm{Cl}_{\mathbb{C}}\Delta_{\frac{\pi}{2}} = \{\lambda \in \mathbb{C} : \mathrm{Re}\ \lambda \geq 0\} \subset \rho(-B),$$

and that

$$\|\lambda R(\lambda, -B)\|_{\mathcal{L}(V')} \leq 1 + M\delta^{-1}, \qquad \lambda \in \mathrm{Cl}_{\mathbb{C}}\Delta_{\frac{\pi}{2}}, \tag{36.24}$$

where δ is given by the coercive property (36.20). (Note that for $u \in V$ and $w \in H$, one has $\langle\langle u, w\rangle\rangle = \langle u, w\rangle_H$.) Indeed, according to (36.20), for any $u \in V$ and $\overline{\lambda} \in \mathrm{Cl}_{\mathbb{C}}\Delta_{\frac{\pi}{2}}$, one has

$$\delta\|u\|_V^2 \leq \mathrm{Re}\,\{a(u,u) + \overline{\lambda}\|u\|_H^2\} \leq |\langle\langle u, Bu + \lambda u\rangle\rangle| \leq \|(\lambda I + B)u\|_{V'}\|u\|_V.$$

Hence

$$\delta\|u\|_V \leq \|(\lambda I + B)u\|_{V'} \leq \|\lambda I + B\|_{\mathcal{L}(V,V')}\|u\|_V, \tag{36.25}$$

which implies that $\lambda I + B$ is one-to-one and the range $\mathcal{R}(\lambda I + B)$ is a closed linear subspace of V', since B is a closed operator. On the other hand, let B^* be the form operator in V' determined by the sesquilinear form $(u,v) \to a^*(u,v)$. A similar argument establishes that $\overline{\lambda}I + B^*$ is a densely defined on V', it is a one-to-one mapping of V' into V, and that $\overline{\lambda}I + B^* = (\lambda I + B)^*$. Since V' is a Hilbert space in the inner product $\langle\cdot,\cdot\rangle_a$, one has the identity

$$\mathrm{Cl}_{V'}\mathcal{R}(\lambda I + B) = \mathcal{N}((\lambda I + B)^*)^{\perp}.$$

Since $(\lambda I + B)^*$ is positive definite, we see that $\mathcal{R}(\lambda I + B)$ is a dense subspace of V'. Since $\mathcal{R}(\lambda I + B)$ is both dense and closed in V', one has $\mathcal{R}(\lambda I + B) = V'$. Therefore $\lambda \in \rho(-B)$, and (36.25) implies that $\|R(\lambda, -B)\|_{\mathcal{L}(V',V)} \leq \delta^{-1}$. Let $u = R(\lambda, -B)w$, where $w \in V'$. One then has $u \in V'$, $w = (\lambda I + B)u$ and, by the Lax-Milgram Lemma 36.5,

$$\begin{aligned}|\langle\langle v, \lambda u\rangle\rangle| &= |\langle\langle v, (\lambda I + B)u\rangle\rangle - \langle\langle v, Bu\rangle\rangle| \\ &\leq \|(\lambda I + B)u\|_{V'}\|v\|_V + \|Bu\|_{V'}\|v\|_V \\ &\leq (\|w\|_{V'} + M\|u\|_V)\|v\|_V \leq (1 + M\delta^{-1})\|w\|_{V'}\|v\|_V.\end{aligned}$$

It follows that $\|\lambda u\|_{V'} = \|\lambda R(\lambda, -B)w\|_{V'} \leq (1 + M\delta^{-1})\|w\|_{V'}$, which implies (36.24).

Next we show that there is an angle $\theta \in (0, \frac{\pi}{2})$ such that the open sector $\Delta_{\pi-\theta}$ satisfies $\Delta_{\pi-\theta} \subset \rho(-B)$. For this purpose, we fix any $b \in (0,1)$ and $\eta_0 \in \mathbb{R}^+$. For any $\lambda = \xi + i\eta$ satisfying

$$|\lambda - i\eta_0| = |\xi + i(\eta - \eta_0)| \leq \frac{|\eta_0 b|}{1 + M\delta^{-1}}, \tag{36.26}$$

it follows from (36.24) that

$$|\lambda - i\eta_0|\,\|R(i\eta_0, -B)\|_{\mathcal{L}(V')} \leq b < 1.$$

By using the Neuman expansion of resolvent operators, see Naylor and Sell (1982), one finds that any $\lambda = \xi + i\eta$ satisfying (36.26) is in the resolvent set $\rho(-B)$, and

$$R(\lambda, -B) = \sum_{k=0}^{\infty}(i\eta_0 - \lambda)^k R(i\eta_0, -B)^{k+1}.$$

Moreover, for this λ one has $\lambda R(\lambda, -B) = (\lambda - i\eta_0 + i\eta_0)R(\lambda, -B)$, and inequality (36.24) implies that
(36.27)

$$\begin{aligned}\|\lambda R(\lambda,-B)\|_{\mathcal{L}(V')} &\leq (1+|i\eta_0|\,\|R(i\eta_0,-B)\|)\sum_{k=0}^{\infty}|\lambda - i\eta_0|^k\|R(i\eta_0,-B)\|^k \\ &\leq \frac{1}{1-b}(1+|i\eta_0|\,\|R(i\eta_0,-B)\|) \\ &\leq \frac{1}{1-b}(2+M\delta^{-1}) \stackrel{\text{def}}{=} \hat{M}.\end{aligned}$$

We will denote the set of λ satisfying (36.26) as $B(i\eta_0, b)$, and define

$$\Omega(b) = \bigcup_{\eta\in\mathbb{R}+} B(i\eta, b).$$

As argued above, one has $\Omega(b) \subset \rho(-B)$, and $0 \in \rho(-B)$. Since $\rho(-B)$ is open, there is a neighborhood of the origin, denoted by B_0, such that $B_0 \subset \rho(-B)$. Therefore,

$$\Omega(b) \cup B_0 \cup \mathrm{Cl}_{\mathbb{C}}\Delta_{\frac{\pi}{2}} \subset \rho(-B). \tag{36.28}$$

Define θ by

$$\theta = \arctan\left(\frac{1+M\delta^{-1}}{b}\right) \in \left(0, \frac{\pi}{2}\right).$$

From (36.24), (36.26), (36.27), and (36.28), it follows that $-\Sigma_\theta = \Delta_{\pi-\theta} \subset \rho(-B)$, and

$$\|R(\lambda,-B)\|_{\mathcal{L}(V')} \leq \frac{\hat{M}}{|\lambda|}, \qquad \text{for } \lambda \in \Delta_{\pi-\theta}$$

From (36.2) and (36.3) we see that the last inequality is equivalent to

$$\Sigma_\theta \subset \rho(B) \qquad \text{and} \qquad \|R(\lambda,B)\|_{\mathcal{L}(V')} \leq \frac{\hat{M}}{|\lambda|}, \text{ for } \lambda \in \Sigma_\theta.$$

Thus B is a sectorial operator on V'. By Theorem 36.2, the operator $-B$ generates an analytic semigroup $T(t)$ on V'. Finally, the relation $T(t)\,|_H = T_H(t)$ follows from the proof of the Hille-Yosida Theorem and the identity $R(\lambda, -B)\,|_H = R(\lambda, -B_H)$. ☐

It will be shown in Section 3.8 that this theorem provides an answer to the question: Does a given boundary value problem associated with a differential operator generate a sectorial operator and, thereby, an analytic semigroup? If the Lax-Milgram property is satisfied and V is dense in H, the answer is yes!

3.7. Fractional Powers and Interpolation Spaces.

As we will see, sectorial operators offer many mathematical advantages in the study of both linear and nonlinear evolutionary equations. As shown in the last section, such operators lead to the theory of analytic semigroups. Another important feature, which we explore in this section, is the concept of the fractional power of an operator. It is by studying the fractional powers of certain linear operators that one is able to introduce an important class of interpolation spaces. This theory will enable us to develop a functional-analytic foundation for nonlinear equations, such as the Navier-Stokes equations. One of our principal goals in this book is to show how this functional-analytic formulation leads to valuable information concerning the long-time dynamics of the solutions of these equations.

Example 37.1. In order to motivate the general theory of fractional powers associated with certain sectorial operators, it is convenient to return to the analytic semigroup e^{-At} on a Hilbert space H, which was constructed in Section 3.2, where the operator A is now assumed to be selfadjoint, positive, and with compact resolvent. In this case the eigenvalues satisfy (32.2), and one has $\langle u, Au\rangle \geq \lambda_1\|u\|^2$, for all $u \in H$. For any $\alpha \geq 0$, the function $f(\lambda) = \lambda^\alpha$ is defined on the spectrum of A. Consequently the domain $\mathcal{D}(A^\alpha)$ of the fractional power A^α is defined by[9]

$$V^{2\alpha} \stackrel{\text{def}}{=} \mathcal{D}(A^\alpha) \stackrel{\text{def}}{=} \{u \in H : \sum_{i=1}^{\infty} \lambda_i^{2\alpha}|\langle u, e_i\rangle|^2 < \infty\}, \tag{37.1}$$

and the operator A^α is given by

$$A^\alpha u = \sum_{i=1}^{\infty} \lambda_i^\alpha \langle u, e_i\rangle e_i, \qquad \text{for all } u \in \mathcal{D}(A^\alpha).$$

For $\alpha > 0$, we let $V^{-2\alpha}$ denote the dual space of $V^{2\alpha}$, see Section 3.6.2. Thus $V^{2\alpha}$ and $V^{-2\alpha}$ are related by equation (36.17), and one has

$$\langle\langle v, w\rangle\rangle = \langle A^\alpha v, A^{-\alpha} w\rangle_H, \qquad \text{for } v \in V^{2\alpha},\ w \in V^{-2\alpha}.$$

We will expand upon this duality later in this chapter.

It is easily verified that for each $\alpha \geq 0$ the set V^α is a linear subspace in H, and $V^0 = H$. Furthermore, V^α itself becomes a Hilbert space with the V^α-inner product and the V^α-norm given by

(37.2)

$$\langle u, v\rangle_\alpha \stackrel{\text{def}}{=} \sum_{i=1}^{\infty} \lambda_i^\alpha u_i \bar{v}_i, \qquad \|v\|_\alpha^2 \stackrel{\text{def}}{=} \sum_{i=1}^{\infty} \lambda_i^\alpha |v_i|^2, \qquad u, v \in V^\alpha, \text{ and } \alpha \in \mathbb{R}.$$

[9]Here we use 2α instead of α, because it is a convenient notation for the scaling arising in the interpolation spaces.

Moreover, if $\alpha \geq \beta$, then (32.2) and (37.2) imply that $V^\alpha \mapsto V^\beta$ and

$$\|v\|_\beta^2 \leq \lambda_1^{\beta-\alpha}\|v\|_\alpha^2, \qquad \text{for all } v \in V^\alpha.$$

Notice that if $\alpha \geq \beta$, then V^α is dense in V^β. Indeed, for every $v \in V^\beta$ and $N \geq 1$ we define v^N by

$$v^N \stackrel{\text{def}}{=} (v_1, v_2, \ldots, v_N, 0, 0, \ldots) \in V^\alpha.$$

One then has $\|v - v^N\|_\beta^2 = \sum_{i=N+1}^\infty \lambda_i^\beta |v_i|^2 \to 0$, as $N \to \infty$.

Let $\alpha \geq \beta$ and set $\gamma = \theta\alpha + (1-\theta)\beta$, for $0 \leq \theta \leq 1$. One then has $V^\alpha \mapsto V^\gamma \mapsto V^\beta$, and we claim that

$$\|v\|_\gamma \leq \|v\|_\alpha^\theta \|v\|_\beta^{1-\theta}, \qquad \text{for all } v \in V^\alpha \text{ and } 0 \leq \theta \leq 1. \tag{37.3}$$

Indeed, by using the Hölder inequality one obtains

$$\|v\|_\gamma^2 = \sum_{i=1}^\infty \lambda_i^{\theta\alpha}|v_i|^{2\theta}\lambda_i^{(1-\theta)\beta}|v_i|^{2(1-\theta)} \leq \left(\sum_{i-1}^\infty \lambda_i^\alpha |v_i|^2\right)^\theta \left(\sum_{i=1}^\infty \lambda_i^\beta |v_i|^2\right)^{1-\theta},$$

which implies (37.3). Next notice that

$$\|v\|_2^2 = \sum_{i=1}^\infty \lambda_i^2 |v_i|^2 = \|Av\|_0^2, \qquad \text{for all } v \in V^2.$$

More generally for any $\alpha \in \mathbb{R}$, we define $A : V^{2+\alpha} \to V^\alpha$ by $(Av)_i = \lambda_i v_i$. One then has $\|v\|_{2+\alpha}^2 = \|Av\|_\alpha^2$, for all $v \in V^{2+\alpha}$, i.e., A is a linear isometry from $V^{2+\alpha}$ onto V^α, for every $\alpha \in \mathbb{R}$. Similarly the analytic semigroup e^{-At} on H extends to, or restricts to, an analytic semigroup on each space V^α, for $\alpha \in \mathbb{R}$, by means of the componentwise formula $(e^{-At}u)_i = e^{-\lambda_i t}u_i$, for $i = 1, 2, \ldots$. Finally, Theorem 32.1 remains valid for the analytic semigroup e^{-At} on each fractional power space V^α.

Let us summarize the above. Let I be an interval in $\mathbb{R}$. A family of Banach spaces V^α with norms $\|\cdot\|_\alpha$, defined for $\alpha \in I$, is said to be a **family of interpolation spaces on** I if the following property holds:

For $\alpha, \beta \in I$ with $\alpha \geq \beta$, one has $V^\alpha \mapsto V^\beta$, and V^α is dense in V^β. Morevover, for every θ, $0 \leq \theta \leq 1$, there is a constant $C = C(\alpha, \beta, \theta)$ such that

$$\|v\|_\gamma \leq C\|v\|_\alpha^\theta \|v\|_\beta^{1-\theta}, \qquad \textit{for all } v \in V^\alpha,$$

where $\gamma = \theta\alpha + (1-\theta)\beta$.

We have the following result.

Theorem 37.2. *Let A be a positive, selfadjoint linear operator with compact resolvent on a Hilbert space H, and define the Hilbert space V^α by (37.1) and (37.2). Let the eigenvalues of A satisfy (32.2). Then the following hold:*

(1) *V^α is a family of interpolation spaces on $\mathbb{R}$ with $V^0 = H$, see (37.3).*
(2) *A is an isometry from $V^{2+\alpha}$ to V^α, for every $\alpha \in \mathbb{R}$.*
(3) *The analytic semigroup e^{-At} on H extends to, or restricts to, an analytic semigroup on each V^α, $\alpha \in \mathbb{R}$.*
(4) *If $\alpha > \beta$, then one has a compact imbedding $V^\alpha \hookrightarrow V^\beta$.*
(5) *For every $\epsilon > 0$ there is a constant $C_\epsilon > 0$ such that*

$$\|v\|_H^2 \le \epsilon \|A^{\frac{1}{2}} v\|_H^2 + C_\epsilon \|A^{-\frac{1}{2}} v\|_H^2, \qquad \text{for all } v \in V^1 = \mathcal{D}(A^{\frac{1}{2}}). \tag{37.4}$$

Proof. The proofs of Items (1) - (4) follow from the discussion preceding the statement of the theorem. In order to prove Item (5), we argue by assuming (37.4) to be false. Then, for some $\epsilon > 0$, there is a sequence $v_n \in V^1$ such that

$$\|v_n\|_0^2 \ge \epsilon \|v_n\|_1^2 + n \|v_n\|_{-1}^2.$$

Take $u_n = \|v_n\|_1^{-1} v_n$. One then has $\|u_n\|_1 = 1$ and

$$\|u_n\|_0^2 \ge \epsilon + n\|u_n\|_{-1}^2, \qquad \text{for } n = 1, 2, \cdots. \tag{37.5}$$

Since the unit ball in the Hilbert space V^1 is weakly compact, there is a subsequence, which we relabel as u_n, that converges weakly in V^1 to a limit u. Since $V^1 \hookrightarrow V^0$, the sequence u_n converges strongly to u in V^0 and in V^{-1}. Therefore the sequence of norms $\|u_n\|_0$ and $\|u_n\|_{-1}$ are bounded. Consequently, (37.5) implies that $\|u\|_{-1} = 0$, i.e., $u = 0$. Since $\|u_n\|_0 \ge \epsilon$ implies that $\|u\|_0 \ge \epsilon$, we obtain a contradiction. □

Let P denote the orthogonal projection of H onto span$\{e_1, \cdots, e_n\}$, and set $Q = I - P$. Recall that $\{e_1, e_2, \cdots\}$ is an orthonormal basis in H of eigenvectors of A, see Section 3.2. Note that $-AQ$ generates an analytic semigroup e^{-AQt}, on QH. By using the Parseval identity and Theorem 6.10.13 of Naylor and Sell (1982), one can easily verify that, for $0 \le r < 1$, one has

$$\|A^r e^{-AQ\tau}\|_{\mathcal{L}(H)} = b(\tau) = b_r(\tau) = \begin{cases} r^r e^{-r} \tau^{-r}, & 0 < \tau \le a, \\ \lambda_{n+1}^r e^{-\lambda_{n+1}\tau}, & a < \tau < \infty, \end{cases} \tag{37.6}$$

where $a = r\lambda_{n+1}^{-1}$. Also, for $\gamma < \lambda_{n+1}$, one has

$$\begin{aligned} &\int_0^\infty b_r(\tau)\, d\tau = (1-r)^{-1} e^{-r} \lambda_{n+1}^{r-1} \\ &\int_0^\infty b_r(\tau) e^{\gamma\tau}\, d\tau \le r(1-r)^{-1}\lambda_{n+1}^{r-1} + \lambda_{n+1}^r(\lambda_{n+1} - \gamma)^{-1}. \end{aligned} \tag{37.7}$$

Now we let W denote a given Banach space and let A be a positive, sectorial operator on W. For any $\alpha > 0$, we define

$$A^{-\alpha} \stackrel{\text{def}}{=} \frac{1}{\Gamma(\alpha)} \int_0^\infty t^{\alpha-1} e^{-At}\, dt, \tag{37.8}$$

where e^{-At} is the analytic semigroup generated by $-A$, and $\Gamma(\alpha)$ is Gamma function. The reader should verify that the integral in (37.8) is well-defined and that it converges in the uniform operator topology on $\mathcal{L}(W)$. This definition is consistent with the spectral mapping definition

$$A^{-\alpha} = \int_0^\infty \lambda^{-\alpha}\, dE(\lambda),$$

where A is a positive, selfadjoint operator on a Hilbert space H with the spectral decomposition $A = \int_0^\infty \lambda\, dE(\lambda)$. Also one can check that if $\alpha = 1$ the operator defined in (37.8) is exactly the inverse of A. The proof of the following result can be found in Pazy (1983, Section 2.6).

Lemma 37.3. *For any α, $\beta > 0$, one has $A^{-\alpha} \in \mathcal{L}(W)$ and $A^{-\alpha}A^{-\beta} = A^{-(\alpha+\beta)}$. Furthermore, each $A^{-\alpha}$ is one-to-one, and one has*

$$A^{-\alpha} = \frac{\sin(\pi\alpha)}{\pi} \int_0^\infty \lambda^{-\alpha} R(\lambda, -A)\, d\lambda, \qquad \textit{for } 0 < \alpha < 1.$$

The **fractional power** A^α of the operator A is defined to be

$$A^\alpha = (A^{-\alpha})^{-1}, \qquad \text{for } \alpha > 0,$$

and the domain is given by $\mathcal{D}(A^\alpha) \stackrel{\text{def}}{=} \mathcal{R}(A^{-\alpha})$. Also define $A^0 \stackrel{\text{def}}{=} I$, the identity on W. For $\alpha > 0$, the domain $\mathcal{D}(A^\alpha)$ becomes a Banach space in the **graph norm**

$$\|u\|_\alpha \stackrel{\text{def}}{=} \|A^\alpha u\|_W, \qquad \text{for } u \in \mathcal{D}(A^\alpha). \tag{37.9}$$

Note that $\|u\|_0 = \|u\|_W$, for $u \in W$. The **interpolation space of order** 2α is defined to be $V^{2\alpha} \stackrel{\text{def}}{=} \mathcal{D}(A^\alpha)$, where $\mathcal{D}(A^\alpha)$ has the graph norm. Note that if $W = H$ is a Hilbert space, then the graph norm $\|u\|_\alpha$ on $\mathcal{D}(A^\alpha)$ is generated by the inner product

$$\langle u, v\rangle_\alpha \stackrel{\text{def}}{=} \langle A^\alpha u, (A^\alpha)^* v\rangle_H,$$

where $(A^\alpha)^*$ is the adjoint of A^α. The definition of $V^{2\alpha}$ given in (37.1) is consistent with the last equation.

Let B be a sectorial operator on a Banach space W, and let e^{-Bt} be the analytic semigroup generated by $-B$. Let $M \geq 1$ and $b \in \mathbb{R}$ be chosen so that $\|e^{-Bt}\| \le Me^{-bt}$, for all $t \ge 0$. Then it is easily seen that B is positive if and only if one can choose the term b to satisfy $b > 0$. Any sectorial operator B can be converted into a positive, sectorial operator A by setting $A = B + rI$ and choosing $r > 0$ to be sufficiently large. Since $e^{-At} = e^{-rt}e^{-Bt}$, we see that if $a = r+b$ satisfies $a > 0$, then A is a positive, sectorial operator with $\|e^{-At}\| \le Me^{-at}$, for all $t \ge 0$. We now use the positive operator A to define the fractional powers. In particular, for $\alpha \ge 0$, we define $V^{2\alpha}$ by $V^{2\alpha} \stackrel{\text{def}}{=} \mathcal{D}(A^\alpha)$. If $W = H$ is a Hilbert space, then the space $V^{-2\alpha}$ is defined as the dual to the space $V^{2\alpha}$, see Section 3.6.2. It may appear that $V^{2\alpha}$ depends on the choice of r, but this is not the case. In short, the fractional powers $V^{2\alpha}$ are uniquely determined by the given sectorial operator B. We will refer to the spaces $V^{2\alpha}$ as the **fractional power spaces generated by the sectorial operator** B. The following result describes some basic properties of the fractional power spaces.

Lemma 37.4. *Let A be a positive, sectorial operator on a Banach space W, and let e^{-At} denote the analytic semigroup on W generated by $-A$. Then for α, $\beta \ge 0$, the following properties are valid:*

(1) *The operator A^α is a densely defined, closed linear operator.*
(2) *For $\alpha \ge \beta$, one has $\mathcal{D}(A^\alpha) \mapsto \mathcal{D}(A^\beta)$, in terms of the graph norms on these spaces, and $\mathcal{D}(A^\alpha)$ is dense in $\mathcal{D}(A^\beta)$, see (37.9).*
(3) *If in addition, A has compact resolvent, then one has $\mathcal{D}(A^\alpha) \hookrightarrow \mathcal{D}(A^\beta)$, whenever $\alpha > \beta$.*
(4) *One has $A^\alpha A^\beta = A^\beta A^\alpha = A^{\alpha+\beta}$ on $\mathcal{D}(A^\gamma)$, for any $\alpha, \beta \in \mathbb{R}$, where $\gamma = \max(\alpha, \beta, \alpha+\beta)$.*
(5) *One has $A^\alpha e^{-At}u = e^{-At}A^\alpha u$, for all $u \in \mathcal{D}(A^\alpha)$ and $t \ge 0$. Furthermore, e^{-At} is an analytic semigroup on V^α, for each $\alpha \in \mathbb{R}$.*
(6) *The mapping*

$$(u,t) \to e^{-At}u : V^{2\alpha} \times [0,\infty) \to V^{2\alpha}$$

is continuous, for every $\alpha \in \mathbb{R}$.

Proof. The verification of Items (1) - (4) is left as an exercise, see Pazy (1983, Theorem 6.8). The commutivity property (5), for $\alpha \le 0$, follows immediately from the definition (37.8). In order to prove the commutivity property for $\alpha > 0$, it suffices to show that $e^{-At}A^{-\alpha}$ maps W into $\mathcal{R}(A^{-\alpha})$ and that

$$A^\alpha e^{-At}A^{-\alpha}w = e^{-At}w, \qquad \text{for } w \in W.$$

Also noted above, one has $e^{-At}A^{-\alpha} = A^{-\alpha}e^{-At}$. As a result, one finds that $\mathcal{R}(e^{-At}A^{-\alpha}) \subset \mathcal{R}(A^{-\alpha})$. Furthermore, one has

$$A^\alpha e^{-At}A^{-\alpha}w = A^\alpha A^{-\alpha}e^{-At}w = e^{-At}w, \qquad \text{for } w \in W.$$

Finally Item (6) follows from Item (5) and Theorem 31.3. □

In addition to viewing $A : V^{2+\alpha} \to V^{\alpha}$ as an isometry, in the sense described above, one also has A as an unbounded operator $A : \mathcal{D}_W(A) \to W$, where the domain of A in W is $\mathcal{D}_W(A) \stackrel{\text{def}}{=} \mathcal{D}(A)$. By using the fact that $A^{\alpha} : V^{2\alpha} \to W$ is one-to-one, we can lift the domain $\mathcal{D}_W(A)$ to any of the fractional power spaces by the identity

$$\mathcal{D}_{V^{2\alpha}}(A) \stackrel{\text{def}}{=} A^{-\alpha}(\mathcal{D}_W(A)) = V^{2\alpha+2} \subset V^{2\alpha}. \tag{37.10}$$

We refer to $\mathcal{D}_{V^{2\alpha}}(A)$ as the **domain of A in $V^{2\alpha}$**. Note that for $v_0 \in \mathcal{D}_{V^{2\alpha}}(A)$, one has $v_0 = A^{-\alpha}u_0$, for some $u_0 \in \mathcal{D}_W(A)$. We let

$$Av_0 = AA^{-\alpha}u_0 = A^{-\alpha}Au_0,$$

which is in $V^{2\alpha}$, since $Au_0 \in W$.

We now come to the Main Theorem on analytic semigroups, or sectorial operators. As we will see later, the inequalities described below in Items (2) and (3) are extremely important in the study of the dynamics of nonlinear evolutionary equations.

Theorem 37.5 (Fundamental Theorem on Sectorial Operators). *Let A be a positive, sectorial operator on a Banach space W, and let $T(t) = e^{-At}$ be the analytic semigroup generated by $-A$. Then the following statements hold:*

(1) *For any $r \geq 0$ and $t > 0$, the semigroup e^{-At} maps W into $\mathcal{D}(A^r)$, and it is strongly continuous in $t > 0$.*

(2) *For any $r \geq 0$, there is a constant $M_r > 0$ such that*

$$\|e^{-At}\|_{\mathcal{L}(W,\mathcal{D}(A^r))} = \|A^r e^{-At}\|_{\mathcal{L}(W)} \leq M_r t^{-r} e^{-at}, \qquad \text{for all } t > 0. \tag{37.11}$$

where $a > 0$ is given by (36.2).

(3) *For $0 < \alpha \leq 1$, there is a constant $K_\alpha > 0$ such that*

$$\|e^{-At}w - w\|_W \leq K_\alpha t^{\alpha}\|A^{\alpha}w\|_W, \qquad \text{for } t \geq 0 \text{ and } w \in \mathcal{D}(A^{\alpha}). \tag{37.12}$$

(4) *The functions $e^{-At}w$ are Lipschitz continuous in t, for $t > 0$ and $w \in W$. More precisely, for every $r \geq 0$, there is a constant $C_r > 0$ such that*

$$\|A^r(e^{-A(t+h)} - e^{-At})w\| \leq C_r|h|t^{-(1+r)}\|w\|, \tag{37.13}$$

for all $t > 0$ and $w \in W$.

(5) *For $w \in W$, one has*

$$e^{-At}w \subset C[0,\infty;W) \cap C^{0,1}_{\text{loc}}(0,\infty;V^{2r}), \qquad \text{for all } r \geq 0.$$

If in addition, one has $w \in V^{2\rho}$, for some $\rho > 0$, then

$$e^{-At}w \in C[0,\infty;V^{2\rho}) \cap C^{0,\rho-\alpha}_{\text{loc}}[0,\infty;V^{2\alpha}) \cap C^{0,1}_{\text{loc}}(0,\infty;V^{2r}), \tag{37.14}$$

for all $r \geq 0$, and all α with $0 \leq \alpha < \rho$.

(6) *If in addition, the sectorial operator A has compact resolvent, then the analytic semigroup e^{-At} is compact, for $t > 0$.*

As we will see, this Fundamental Theorem 37.5, along with Theorem 35.1 on differentiable C_0-semigroups, will have widespread use on the Good Ship QED as we travel in the world of nonlinear dynamics.

Proof of Theorem 37.5. (1) First note that for any positive integer m, the operator $T(t)$ maps W into $\mathcal{D}(A^m)$ for each $t > 0$ by Theorem 36.2 and Theorem 35.1, Item (1). In fact, since $T(t)$ is analytic semigroup, one has $A^m T(t) = [AT(t/m)]^m$ by (35.2) and (36.12). For any $\alpha \geq 0$, we take an integer $m > \alpha$. Then $T(t) : W \to \mathcal{D}(A^m) \subset \mathcal{D}(A^\alpha)$, for $t > 0$ by Lemma 37.2. The strong continuity in $t > 0$ follows directly from (37.11), which we prove next.

(2) Let $\alpha \geq 0$. Since A^α is closed and $T(t)$ is bounded, one has $A^\alpha T(t) \in \mathcal{L}(W)$, for $t > 0$, by Lemma 37.4 and the Closed Graph Theorem. From Theorems 35.1 and 36.2, one has $A^m T(t) = T^{(m)}(t)$, for each integer $m \geq 1$. Therefore, it follows from (36.13) that one has $\|A^m T(t)\| \leq M_m t^{-m} e^{-at}$, for $t > 0$, for an appropriate constant M_m. Now assume that α satisfies $m - 1 < \alpha < m$. Then (37.8) implies

$$
\begin{aligned}
\|A^\alpha T(t)\| = \|A^{\alpha-m} A^m T(t)\| &= \left\| \frac{1}{\Gamma(m-\alpha)} \int_0^\infty s^{m-\alpha-1} A^m T(t+s)\, ds \right\| \\
&\leq \frac{1}{\Gamma(m-\alpha)} \int_0^\infty s^{m-\alpha-1} M_m (t+s)^{-m} e^{-a(t+s)}\, ds \\
&\quad (\text{let } s = t\sigma) \\
&\leq \frac{1}{\Gamma(m-\alpha)} M_m t^{-\alpha} e^{-at} \int_0^\infty \sigma^{m-\alpha-1} (1+\sigma)^{-m} d\sigma \\
&= M_\alpha t^{-\alpha} e^{-at},
\end{aligned}
$$

for $t > 0$ and an appropriate constant M_α. Thus (37.11) is proved.

(3) Let $0 < \alpha \leq 1$ and $w \in \mathcal{D}(A^\alpha)$. Then by (35.2), Lemma 37.4, Item (5), and Item (2) of this theorem, one has

$$
\begin{aligned}
\|T(t)w - w\| = \left\| \int_0^t \frac{d}{ds} T(s) w\, ds \right\| &= \left\| \int_0^t AT(s) w\, ds \right\| \\
&= \left\| \int_0^t A^{1-\alpha} T(s) A^\alpha w\, ds \right\| \\
&\leq M_{1-\alpha} \int_0^t s^{\alpha-1} e^{-as}\, ds \|A^\alpha w\| \\
&\leq \alpha^{-1} M_{1-\alpha} t^\alpha \|A^\alpha w\|, \qquad t \geq 0.
\end{aligned}
$$

Therefore, inequality (37.12) holds with $K_\alpha = \alpha^{-1} M_{1-\alpha}$.

(4) The proof of the Lipschitz continuity uses both inequalities (37.11) and (37.12). Indeed for $t > 0$, one has

$$
\|A^r (e^{-Ah} - I) e^{-At} w\| \leq K_1 |h|\, \|A^{1+r} e^{-At} w\| \leq K_1 M_{1+r} |h| t^{-(1+r)} \|w\|.
$$

Therefore, inequality (37.13) holds with $C_r = K_1 M_{1+r}$.

(5) Since $A^\alpha e^{-At}w = e^{-At}A^\alpha w$, for $w \in V^{2\alpha}$, Item (5) follows from Theorem 31.3 and inequalities (37.12) and (37.13).

(6) Note that (37.11), with $r = 1$, implies that, for any $t > 0$, the semigroup e^{-At} maps every bounded set in W into a bounded set in $\mathcal{D}(A) = V^2$. Since A has compact resolvent, the imbedding $V^2 = \mathcal{D}(A) \hookrightarrow W$ is compact, which completes the proof of the theorem. □

The interpolation inequalities of fractional powers of sectorial operators follow. These are also called **moment inequalities.**

Theorem 37.6 (Interpolation Inequalities). *Let A be a positive, sectorial operator on a Banach space W, and let $\alpha, \beta \in \mathbb{R}$ with $\alpha \geq \beta$. Let θ satisfy $0 \leq \theta \leq 1$, and set $\gamma = \theta\alpha + (1-\theta)\beta$. Then there is a constant $C = C(\alpha,\beta,\gamma) \geq 1$ such that*

$$\|A^\gamma w\| \leq C\|A^\alpha w\|^\theta \|A^\beta w\|^{1-\theta}, \qquad \text{for } w \in \mathcal{D}(A^\alpha). \tag{37.15}$$

In particular, for each θ, $0 \leq \theta \leq 1$, there is a constant $C = C(\theta) \geq 1$ such that

$$\|A^\theta w\| \leq C\|Aw\|^\theta \|w\|^{1-\theta}, \qquad \text{for } w \in \mathcal{D}(A). \tag{37.16}$$

Consequently, the family $V^{2\alpha} = \mathcal{D}(A^\alpha)$ is a family of interpolation spaces on $\mathbb{R}$ with $V^0 = W$ and $V^1 = \mathcal{D}(A^{\frac{1}{2}})$.

Proof. Note that by setting $\gamma = \theta$, $\alpha = 1$, and $\beta = 0$, we see that (37.16) is a special case of (37.15). We will prove (37.15) in the case that $W = H$ is a Hilbert space and A is selfadjoint. The argument for the general case can be found in Krein (1971).

The spectral theory for positive, selfadjoint operators implies that

$$A^\alpha w = \int_0^\infty \lambda^\alpha \, dE(\lambda)w, \qquad \text{for } w \in \mathcal{D}(A^\alpha) \text{ and } \alpha \in \mathbb{R}.$$

Also the orthogonality of the projections $dE(\lambda)$ and the Hölder inequality imply that for any $w \in \mathcal{D}(A^\alpha)$, one has

$$\begin{aligned}
\|A^\gamma w\|_H^2 &= \left\langle \int_0^\infty \lambda^\gamma \, dE(\lambda)w, \int_0^\infty \mu^\gamma \, dE(\mu)w \right\rangle_H \\
&= \int_0^\infty \int_0^\infty (\lambda\mu)^\gamma \langle dE(\lambda)w, dE(\mu)w\rangle_H \\
&= \int_0^\infty \lambda^{2\gamma} \|dE(\lambda)w\|_H^2 \\
&= \int_0^\infty \lambda^{2\theta\alpha} \| dE(\lambda)w\|_H^{2\theta} \, \lambda^{2(1-\theta)\beta} \| dE(\lambda)w\|_H^{2(1-\theta)} \\
&\leq \left(\int_0^\infty \lambda^{2\alpha} \| dE(\lambda)w\|_H^2\right)^\theta \left(\int_0^\infty \lambda^{2\beta} \| dE(\lambda)w\|_H^2\right)^{1-\theta} \\
&= \|A^\alpha w\|_H^{2\theta} \, \|A^\beta w\|_H^{2(1-\theta)}. \qquad \square
\end{aligned}$$

Next we take a different viewpoint. Suppose that one does not have any operator like A at hand. There are only two Hilbert spaces V and H, which satisfy the conditions that $V \mapsto H$ and V is dense in H. Can one define interpolation spaces between V and H in some sense?

There are several ways to proceed; some are quite involved. In the general setting of Banach spaces, for instance, one can use the theory of trace spaces, cf. Lions and Magenes (1970). In the setting of Hilbert spaces, one can actually create an operator A by using the Lax-Milgram paradigm. Here is the key idea.

Theorem 37.7. *Let V and H be two given Hilbert spaces such that $V \mapsto H$ and V is dense in H. Then the following properties are valid:*

(1) *There exists a positive, selfadjoint operator $L : \mathcal{D}(L) = V \overset{\text{onto}}{\to} H$ such that the norms $\|w\|_V$ and $\|Lw\|_H$ are equivalent on V.*

(2) *There is a family of interpolation spaces V^α, $\alpha \in \mathbb{R}$, such that $V^1 = V$ and $V^0 = H$.*

Proof. This is an immediate application of Theorem 36.6. Observe that $s(u,v) \overset{\text{def}}{=} \langle u, v\rangle_V$ is a positive definite, Hermitian sesquilinear functional on V that satisfies the Lax-Milgram property. As a result, the form operator $A \overset{\text{def}}{=} B_H$ is a positive, selfadjoint operator mapping $V^2 = \{u \in V : Bu \in H\}$ onto H. Let A^α, $\alpha \in \mathbb{R}$, denote the fractional powers of A. Let $V^{2\alpha} = \mathcal{D}(A^\alpha)$, for $\alpha \geq 0$, and let $V^{-2\alpha}$ be the dual space to $V^{2\alpha}$. It is easily verified that the norms $\|v\|_V$ and $\|A^{\frac{1}{2}}v\|_H$ are equivalent on V. Therefore, $L = A^{\frac{1}{2}}$ satisfies Item (1) of this theorem, and V^α, $\alpha \in \mathbb{R}$, is a family of interpolation spaces satisfying Item (2) by Lemma 37.4. $\square$

The following result describes the spatial regularity of the functions $u \in \mathcal{D}(A^\alpha)$, where $0 < \alpha \leq 1$, and A is a positive, sectorial operator.

Lemma 37.8. *Let Ω be an open, bounded set in $\mathbb{R}^n$ of Lipschitz class, and let q satisfy $1 \leq q < \infty$. Assume that A is a positive, sectorial operator on the Banach space $L^q = L^q(\Omega, \mathbb{R}^M)$ with its domain $\mathcal{D}(A)$ satisfying the imbedding relationship $\mathcal{D}(A) \mapsto W^{m,q} = W^{m,q}(\Omega, \mathbb{R}^M)$, for some integer $m \geq 1$. Let $\mathcal{D}(A^\alpha)$ denote the domain of A^α, the fractional power of A, where $0 < \alpha \leq 1$. Then one has:*

(1) *$\mathcal{D}(A^\alpha) \mapsto W^{k,p}$, whenever $p \geq q$ and $k \geq 0$ is an integer with*

$$k - \frac{n}{p} < m\alpha - \frac{n}{q}. \tag{37.17}$$

(2) *$\mathcal{D}(A^\alpha) \mapsto C^{N,\lambda}(\Omega, \mathbb{R}^M)$ whenever*

$$0 < N + \lambda < m\alpha - \frac{n}{q},$$

where N is a nonnegative integer and $0 < \lambda \leq 1$.

Proof. Let $J : \mathcal{D}(A) \hookrightarrow W^{m,q}$ be the inclusion mapping, which is continuous by assumption. Since $0 < \alpha \leq 1$, it follows from inequality (37.17) and the Nirenberg-Gagliardo inequality with $\theta = 1$, see Appendix B, that one has the continuous imbedding $W^{m,q} \hookrightarrow W^{k,p}$. Hence there is a constant $C_1 > 0$ such that $\|u\|_{W^{m,q}} \leq C_1 \|Au\|_{L^q}$, for all $u \in \mathcal{D}(A)$. Now choose θ so that $k - \frac{n}{p} < m\theta - \frac{n}{q} < m\alpha - \frac{n}{q}$ and $0 < \theta < \alpha$. With $r = q$, it then follows from the Nirenberg-Gagliardo inequality, see Appendix B, that there is a constant $C_2 > 0$ such that

$$\|u\|_{W^{k,p}} \leq C_2 \|Au\|_{L^q}^{\theta} \|u\|_{L^q}^{1-\theta}, \qquad \text{for all } u \in \mathcal{D}(A). \tag{37.18}$$

Assume for the moment that there is a constant $C_5 > 0$ such that, for $0 < \alpha \leq 1$,

$$\|u\|_{W^{k,p}} \leq C_5 \|A^{\alpha} u\|_{L^q}, \qquad \text{for all } u \in \mathcal{D}(A). \tag{37.19}$$

Then (37.19) implies that the inclusion mapping $J : \mathcal{D}(A) \to W^{k,p}$ is uniformly bounded in the A^{α}-norm on $\mathcal{D}(A)$. Since $\mathcal{D}(A)$ is dense in $\mathcal{D}(A^{\alpha})$, for $\alpha \leq 1$, this implies that J has a unique, continuous extension $J : \mathcal{D}(A^{\alpha}) \to W^{k,p}$, which completes the proof of Item (1) of the lemma.

In order to verify inequality (37.19), we first note that an application of the Young inequality to (37.18) implies that there is a constant $C_3 = C_3(\theta) > 0$ such that, for all $\epsilon > 0$, one has

$$\|u\|_{W^{k,p}} \leq \epsilon^{1-\theta} \|Au\|_{L^q} + C_3 \epsilon^{-\theta} \|u\|_{L^q}, \qquad \text{for all } u \in \mathcal{D}(A). \tag{37.20}$$

Also, equation (37.8) implies that for $0 < \alpha \leq 1$, one has

$$u = \frac{1}{\Gamma(\alpha)} \int_0^{\infty} t^{\alpha-1} e^{-At} A^{\alpha} u \, dt, \qquad \text{for all } u \in \mathcal{D}(A).$$

Since $\mathcal{D}(A) \hookrightarrow W^{k,p}$, we then obtain

$$\|u\|_{W^{k,p}} \leq \frac{1}{\Gamma(\alpha)} \int_0^{\infty} t^{\alpha-1} \|e^{-At} A^{\alpha} u\|_{W^{k,p}} \, dt, \qquad \text{for all } u \in \mathcal{D}(A), \tag{37.21}$$

and all α with $0 < \alpha \leq 1$. By using (37.20) with $\epsilon = t$, one obtains

$$\int_0^1 t^{\alpha-1} \|e^{-At} A^{\alpha} u\|_{W^{k,p}} \, dt \leq \int_0^1 t^{\alpha-1} t^{1-\theta} \|A e^{-At} A^{\alpha} u\|_{L^q} \, dt + C_3 \int_0^1 t^{\alpha-1} t^{-\theta} \|e^{-At} A^{\alpha} u\|_{L^q} \, dt.$$

By applying (37.11) to the right side of the last inequality, one finds that

$$\int_0^1 t^{\alpha-1} \|e^{-At} A^{\alpha} u\|_{W^{k,p}} \, dt \leq \frac{1}{\alpha - \theta} (M_1 + C_3 M_0) \|A^{\alpha} u\|_{L^q} \tag{37.22}$$

for $u \in \mathcal{D}(A)$, since $\alpha > \theta$. Similarly, by using (37.20), with $\epsilon = 1$, $a > 0$, and (37.11), one obtains a constant $C_4 > 0$ such that

(37.23)

$$\begin{aligned}\int_1^\infty t^{\alpha-1}\|e^{-At}A^\alpha u\|_{W^{k,p}}\,dt &\leq M_1\int_1^\infty t^{\alpha-1}e^{-at}\|A^\alpha u\|_{L^q}\,dt \\ &\quad + C_3M_0\int_1^\infty t^{\alpha-1}e^{-at}\|A^\alpha u\|_{L^q}\,dt \\ &\leq C_4(M_1 + C_3M_0)\|A^\alpha u\|_{L^q}.\end{aligned}$$

By incorporating (37.22) and (37.23) into (37.21), we obtain (37.19).

The proof of Item (2) of the lemma, which uses the second Nirenberg-Gagliardo inequality in Theorem B.3, is similar, and we will omit the details. □

We will need the following result which gives a sufficient condition for a function $u \in L^2_{\text{loc}}[0,T;H)$ to satisfy $u \in C[0,T;H)$, where $0 < T \leq \infty$ and H is a Hilbert space. This will be formulated in terms of the interpolation spaces V^α generated by a given positive, selfadjoint linear operator A with compact resolvent. For a fixed $\alpha \in \mathbb{R}$, we will say that a given function $u \in L^\infty_{\text{loc}}[0,T;V^\alpha)$ has a **generalized time derivative** $\partial_t u = g$, where $g \in L^p_{\text{loc}}[0,T;V^{\alpha-1})$, for some p with $1 \leq p \leq \infty$, provided that the equation

$$A^{-\frac{1}{2}}(u(t) - u(t_0)) = A^{-\frac{1}{2}}\int_{t_0}^t g(s)\,ds, \tag{37.24}$$

holds in V^α for t_0, $t \in [0,T)$.

Lemma 37.9 (Continuity Lemma). *Let A be a positive, selfadjoint linear operator with compact resolvent on a Hilbert space H, and define the Hilbert space V^α by (37.1) and (37.2). Let $\alpha \in \mathbb{R}$ be fixed and let*

$$u \in L^\infty[0,T;V^\alpha) \cap L^2_{\text{loc}}[0,T;V^{\alpha+1})$$

be a function with a generalized time derivative $\partial_t u = g$ that satisfies

$$g \in L^2_{\text{loc}}[0,T;V^{\alpha-1}),$$

for some T with $0 < T \leq \infty$. Then the following results are valid:

(1) *One has $u \in C^{0,\frac{1}{2}}_{\text{loc}}[0,T;V^{\alpha-1}) \cap C[0,T;V^\alpha_w)$, where V^α_w denotes the weak topology on V^α.*

(2) *One has*

$$\langle\langle g,u\rangle\rangle_\alpha = \frac{1}{2}\frac{d}{dt}\|u\|^2_\alpha, \qquad \text{in } L^1_{\text{loc}}[0,T;\mathbb{R}), \tag{37.25}$$

and $\langle\langle g,u\rangle\rangle_\alpha \overset{\text{a.e.}}{=} \frac{1}{2}\frac{d}{dt}\|u\|^2_\alpha$, where $\|u\|_\alpha$ is the norm on V^α and $\langle\langle\cdot,\cdot\rangle\rangle_\alpha = \langle A^{-\frac{1}{2}}\cdot, A^{\frac{1}{2}}\cdot\rangle_\alpha$, where one has the compact imbeddings $V^{\alpha+1} \hookrightarrow V^\alpha \hookrightarrow V^{\alpha-1}$.

(3) *One has $u \in C[0,T;V^\alpha)$.*

Proof. Item (1): The Hölder continuity of u follows directly from (37.24) and the Hölder inequality. Consequently, for each $v \in V^{\alpha+1}$ one has

$$\langle A^{-\frac{1}{2}}(u(t) - u(t_0)), A^{\frac{1}{2}}v\rangle_\alpha = \langle (u(t) - u(t_0)), v\rangle_\alpha \to 0,$$

as $t \to t_0$ in $[0, T)$. This implies that $u \in C[0, T; V_w^\alpha)$, since $V^{\alpha+1}$ is dense in V^α.

Item (2): Note that by a direct calculation one has

$$\begin{aligned} &\frac{1}{h}(\|u(t+h)\|_\alpha^2 - \|u(t)\|_\alpha^2) \\ &\quad = \left\langle A^{-\frac{1}{2}}\frac{1}{h}(u(t+h) - u(t)), A^{\frac{1}{2}}(u(t+h) + u(t))\right\rangle_\alpha . \end{aligned} \tag{37.26}$$

It suffices to study the limit in (37.26) for sequences $h = h_n \to 0$. Let $h = h_n$ be such a sequence. Now the hypotheses imply that one has

$$A^{-\frac{1}{2}}\frac{1}{h}(u(t+h) - u(t)) \xrightarrow{s} A^{-\frac{1}{2}}g(t), \qquad \text{in } L^2_{\text{loc}}[0, T; V^\alpha).$$

From Appendix B, we see that

$$A^{\frac{1}{2}}(u(t+h) + u(t)) \xrightarrow{s} 2A^{\frac{1}{2}}u(t), \qquad \text{in } L^2_{\text{loc}}[0, T; V^\alpha),$$

perhaps for a subsequence of $h = h_n$. Hence the right side of equation (37.26) has the limit $\langle A^{-\frac{1}{2}}g, 2A^{\frac{1}{2}}u\rangle_\alpha$ in $L^1_{\text{loc}}[0, T; \mathbb{R})$, which implies (37.25).

Item (3): Since equation (37.24) implies that $\|u\|_\alpha^2 \in C[0, T; \mathbb{R})$, for $t \in [0, T)$ one has

$$\|u(t+h)\|_\alpha^2 - \|u(t)\|_\alpha^2 \to 0, \qquad \text{as } h \to 0. \tag{37.27}$$

From the identity

$$\|u(t+h) - u(t)\|_\alpha^2 = \|u(t+h)\|_\alpha^2 - \|u(t)\|_\alpha^2 - 2\langle u(t), u(t+h) - u(t)\rangle_\alpha,$$

we see that Item (1) and (37.27) imply Item (3). ☐

3.8. Illustrations.

In this section we will study four typical linear partial differential equations, the heat equation, the linear parabolic equations, the wave equation, and the Stokes equations. Our objective is to formulate each as a linear evolutionary equation on a Hilbert space H, so that the semigroup theory developed above is applicable.

3.8.1 Heat Equation. We begin with the 1-dimensional (1D) heat equation

$$\text{(38.1)}\qquad \begin{cases} \partial_t u - \nu \Delta u = 0, & t > 0,\ x \in \Omega \stackrel{\text{def}}{=} (0,1), \\ u(t,0) = u(t,1) = 0, & t \geq 0, \\ u(0,x) = u_0(x), & x \in \Omega, \end{cases}$$

where $u = u(t,x)$ is the temperature at time $t \geq 0$ of a bar located on the interval $[0,1]$, $\nu > 0$ is a constant, and $\Delta = d^2/dx^2$ is the one-dimensional Laplace operator. We use the Dirichlet boundary conditions in this illustration. While this is a simple equation, it has many of the features shared by more complicated parabolic-type equations, including the heat equation in higher dimensions, as we will see below.

The initial step in working on infinite dimensional dynamical systems is to formulate the original partial differential equation as an abstract evolutionary differential equation in a Banach space. The boundary conditions are incorporated into the description of the domain of the principal differential operator so that the evolutionary equation looks like an ordinary differential equation. After this set-up, one can study the solutions and further properties by using the theory of semiflows.

For equation (38.1), we fix the basic phase space to be the real Hilbert space $H = L^2(0,1)$, with norm $\|\cdot\|$ and inner product $\langle \cdot, \cdot \rangle$, and define an operator $A : \mathcal{D}(A) \to H$ by

$$A\varphi = -\nu \Delta \varphi = -\nu \frac{d^2\varphi}{dx^2}, \qquad \text{for } \varphi \in \mathcal{D}(A),$$

with

$$\mathcal{D}(A) = \{\varphi \in H : D^2\varphi \in H \text{ and } \varphi(0) = \varphi(1) = 0\},$$

where the Dirichlet boundary conditions are now incorporated into the definition of $\mathcal{D}(A)$, and D^i denotes the distribution derivatives, for $i = 1, 2$. The analysis of equation (38.1) is based on four steps.

(1) The boundary conditions, which are included in the definition of $\mathcal{D}(A)$, are well-defined: Since one has $D^2\varphi \in H = L^2(0,1) \subset L^1(0,1)$, it follows that $D\varphi \in AC[0,1]$, $\varphi \in C^1[0,1]$, and consequently $\varphi(0) = \varphi(1) = 0$ are well-defined.

(2) The domain $\mathcal{D}(A)$ can be written in the form $\mathcal{D}(A) = H^2(0,1) \cap H_0^1(0,1)$: First note that it is obvious that $\mathcal{D}(A) \subset H^2(0,1) \cap H_0^1(0,1)$. In order to prove the inverse inclusion, let $w \in H^2(0,1) \cap H_0^1(0,1)$. By the Sobolev Imbedding Theorem in Appendix B, we see that $H^2(0,1) \hookrightarrow C^1[0,1]$ so that $w \in C^1[0,1]$. Since $w \in H_0^1(0,1)$ implies that $w(0) = w(1) = 0$, one has $w \in \mathcal{D}(A)$.

(3) A is a densely defined, selfadjoint, positive operator with compact inverse: Since $C_0^\infty(0,1) \subset \mathcal{D}(A)$, and since the former is dense in H, one finds that $\mathcal{D}(A)$ is dense in H. By integrating by parts, one obtains

$$\text{(38.2)}\qquad \langle Aw, v \rangle = -\nu \int_0^1 w''(x)v(x)\,dx = \nu \int_0^1 w'(x)v'(x)\,dx = \langle w, Av \rangle,$$

for $v, w \in \mathcal{D}(A)$. Consequently, A is a symmetric operator. Moreover, we claim that $\mathcal{R}(A) = H$. Indeed, let $g \in L^2(0,1)$ be given and define f by

$$f(x) = -\frac{1}{\nu}\int_0^x dy \int_0^y g(s)\, ds + \frac{x}{\nu}\int_0^1 dy \int_0^y g(s)\, ds, \qquad \text{for } x \in [0,1].$$

It is easily seen that $f \in \mathcal{D}(A)$ and $Af = g$. By the Hellinger Lemma (see Appendix A) A is a selfadjoint operator. From (38.2) we obtain

$$\langle Au, u\rangle = \nu \int_0^1 (u'(x))^2\, dx \geq \nu\|u'\|^2 \geq \nu\|u\|^2, \qquad \text{for all } u \in \mathcal{D}(A).$$

Consequently, one has $\langle Au, u\rangle = 0$ if and only if $u = 0$. As a result, the selfadjoint operator A is positive. Moreover, $f = A^{-1}g$, i.e., A^{-1} maps $L^2(0,1)$ into $C^1(0,1)$. Therefore A^{-1} is compact.

(4) A is a sectorial operator, and $-A$ generates an analytic semigroup e^{-At}, which is compact for $t > 0$ by Theorems 32.1 and 36.2. It is easily seen that the eigenvalues of A are given by

$$\lambda_n = \nu n^2\pi^2, \qquad \text{for } n = 1, 2, \cdots,$$

and the corresponding eigenvectors $e_n(x) = \sqrt{2}\sin(n\pi x)$, $n = 1, 2, \cdots$, form an orthonormal basis for the space H. As noted in Section 3.2, the analytic semigroup e^{-At} assumes the form

$$(e^{-At}w)(x) = \sum_{n=1}^{\infty} e^{-\nu n^2\pi^2 t}\langle w, e_n\rangle e_n(x), \qquad \text{for } t \geq 0.$$

From (32.7) one has $\|e^{-At}\| \leq e^{-\nu\pi^2 t}$, for $t \geq 0$.

Now equation (38.1) can be written in the following form,

$$\partial_t u + Au = 0, \qquad t > 0,$$

with the initial condition $u(0) = u_0$ assumed to be in H. Here $u(t) = u(t,\cdot)$ is regarded as an abstract function of t with values in H, and equation (38.1), with the given boundary and initial conditions, becomes a Cauchy problem of a linear evolutionary equation in the space H. By Theorems 32.1, 35.2, and 36.2, $u(t) = e^{-At}u_0$, is the unique mild solution, for $t \geq 0$, and it a classical solution, as well.

The theory of the heat equation in higher space dimensions is included in a broader theory of linear parabolic equations, which we describe next. As we will now see, this is very similar to the 1D theory.

3.8.2 Linear Parabolic Equations. For this illustration, we let $\Omega \subset \mathbb{R}^n$ be an open, bounded domain with boundary of Lipschitz class, and let

$$A(x,D)u = \sum_{|\alpha|,|\beta|\leq m} (-1)^{|\alpha|} D^\alpha \left(a_{\alpha,\beta}(x) D^\beta\right) u, \tag{38.3}$$

be a differential operator on Ω, where $\alpha = (\alpha_1, \ldots, \alpha_n)$, α_i is a nonnegative integer, $1 \leq i \leq n$, $|\alpha| = \sum_{i=1}^{n} \alpha_i$, D^α is the partial differential operator $D^\alpha = D_1^{\alpha_1} \ldots D_n^{\alpha_n}$, $a_{\alpha,\beta}(x)$ are C^m scalar-valued functions on $\overline{\Omega} = \mathrm{Cl}_{\mathbb{R}^n}\Omega$, and $m \geq 1$ is an integer. The operator

$$A_0(x, D)u = \sum_{|\alpha|=|\beta|=m} (-1)^m D^\alpha \left(a_{\alpha,\beta}(x) D^\beta\right) u,$$

consisting of the highest order terms, is called the **principal part** of the operator $A(x, D)$. The operator $A(x, D)$ is said to be **(uniformly) elliptic of order** $2m$, if there is a constant $c > 0$, such that the coefficients of the principal part satisfy

$$\operatorname{Re} \left(\sum_{|\alpha|=|\beta|=m} a_{\alpha,\beta}(x) \zeta^\alpha \zeta^\beta \right) \geq c|\zeta|^{2m}, \tag{38.4}$$

for all $x \in \overline{\Omega}$ and $\zeta \in \mathbb{R}^n$. For example, the Laplacian operator $-\Delta$ and the biharmonic operator Δ^2 are strongly elliptic operators of order 2 and 4, respectively.

For the first result, we impose the homogeneous boundary condition

$$u \,|_\Gamma = \frac{\partial}{\partial \nu} u \,|_\Gamma = \cdots = \frac{\partial^{m-1}}{\partial \nu^{m-1}} u \,|_\Gamma = 0,$$

and we set $H = L^2(\Omega)$ and $V = H_0^m(\Omega)$. Define an operator A by $A\varphi = A(x, D)\varphi$, for $\varphi \in \mathcal{D}(A)$, where the domain is $\mathcal{D}(A) \stackrel{\text{def}}{=} H^{2m}(\Omega) \cap H_0^m(\Omega)$.

Theorem 38.1. *Assume that $A = A(x, D)$ is given as above, where A is a uniformly elliptic operator of order $2m$. Then the operator A is a sectorial operator and $-A$ is the infinitesimal generator of an analytic semigroup e^{-At}.*

Outline of the Proof. Let us restrict our treatment to the case of real coefficients, for simplicity. We define a sesquilinear form

$$b[u, v] \stackrel{\text{def}}{=} \sum_{0 \leq |\alpha|, |\beta| \leq m} \langle a_{\alpha,\beta} D^\beta u, D^\alpha v \rangle_H,$$

and then use the Garding inequality (see Friedman (1969) and Naylor and Sell (1982)) to conclude that there exist constants $k_0 \geq 0$ and $C_0 > 0$ such that

$$b[\varphi, \varphi] + k_0 \|\varphi\|_H^2 \geq C_0 \|\varphi\|_V^2, \qquad \text{for all } \varphi \in V.$$

Thus, $a(u, v) \stackrel{\text{def}}{=} b[u, v] + k_0 \langle u, v \rangle_H$ satisfies the Lax-Milgram property. One can verify that the corresponding form operator in H is exactly $A + k_0 I$. From Theorem 36.6, we see that $-B = -(A + k_0 I)$ generates an analytic semigroup e^{-Bt}. Now $e^{-At} = e^{k_0 t}e^{-Bt}$ is the analytic semigroup generated by $-A$. □

As a corollary, the initial-boundary value problem of the parabolic equation

$$\begin{cases} \partial_t u(t,x) + A(x,D)u(t,x) = 0, & t > 0, \quad x \in \Omega, \\ u|_\Gamma = \frac{\partial}{\partial \nu}u|_\Gamma = \cdots = \frac{\partial^{m-1}}{\partial \nu^{m-1}}u|_\Gamma = 0, & t \geq 0, \\ u|_{t=0} = u_0(x) \in H, & x \in \Omega, \end{cases}$$

has a unique, classical solution $u(t,x)$, which is given by

$$u(t,x) = (e^{-At}u_0)(x), \qquad \text{for } t > 0 \text{ and } x \in \Omega.$$

Note that if $u_0 \in H_0^m(\Omega) \cap H^{2m}(\Omega)$, then $e^{-At}u_0 \in C^1[0,\infty; H_0^m(\Omega) \cap H^{2m}(\Omega))$.

Besides the semigroups generated by even-order elliptic differential operators in the Hilbert space $L^2(\Omega)$ as shown above, it is also very useful to have a corresponding semigroup theory in a Banach space $W = L^p(\Omega)$, with $1 \leq p \leq \infty$. In general, such an L^p-theory will be needed in studying semilinear heat equation with polynomial nonlinearity and in seeking for optimal regularity results. Below we will address the theory of semigroups associated with uniformly elliptic differential operators and Dirichlet boundary conditions in the space $L^p(\Omega)$, $1 < p < \infty$. Some comments on the cases $p = 1$ and $p = \infty$ will be made without detailed proof.

Let $A(x,D)$ be a symmetric second-order differential operator given by (38.3), where $|\alpha| = |\beta| = 1$, i.e.,

$$A(x,D) = -\sum_{i,j=1}^{n} \frac{\partial}{\partial x_i}\left(a_{ij}(x)\frac{\partial}{\partial x_j}\right),$$

where the coefficients $a_{ij}(x) = a_{ji}(x)$ are real-valued functions in $C^1(\overline{\Omega})$. Assume that $A(x,D)$ is uniformly elliptic of order 2. Let $1 < p < \infty$, and define an operator A_p as

$$A_p u = A(x,D)u, \qquad \text{for } u \in \mathcal{D}(A_p), \tag{38.5}$$

where $\mathcal{D}(A_p) = W^{2,p}(\Omega) \cap W_0^{1,p}(\Omega)$. One can show that $A_p : \mathcal{D}(A_p) \to L^p(\Omega)$ is a closed linear operator. Note that the definition of the domain $\mathcal{D}(A_p)$ incorporates the Dirichlet boundary condition that $u|_{\partial\Omega} = 0$, if $u \in \mathcal{D}(A_p)$. We next prove the following result.

Theorem 38.2. *Let $1 < p < \infty$. Then the operator A_p given by (38.5) is a positive, sectorial operator and $-A_p$ generates an analytic semigroup $e^{-A_p t}$ on $W = L^p(\Omega)$.*

Proof. Let $q = p/(p-1)$. Denote by $\langle\langle \cdot,\cdot \rangle\rangle$ the dual product between $W = L^p(\Omega)$ and $W' = L^q(\Omega)$. We will use complex arithmetic. Note that if $u \in L^p(\Omega)$, then $u^* \in L^q(\Omega)$, where $u^* = |u|^{p-2}\overline{u}$, and $\overline{u}$ is the complex conjugate of u. Also one has

$$\langle\langle u, u^* \rangle\rangle = \int_\Omega |u|^p \, dx = \|u\|_{L^p}^p = \|u^*\|_{L^q}^q.$$

For $u \in \mathcal{D}(A_p)$, the Dirichlet boundary conditions imply that

$$\begin{aligned}
\langle\langle A_p u, u^* \rangle\rangle &= -\int_\Omega \sum_{k,j=1}^n \frac{\partial}{\partial x_k}\left(a_{kj}\frac{\partial u}{\partial x_j}\right)\overline{u}|u|^{p-2}\, dx \\
&= \int_\Omega \sum_{k,j=1}^n a_{kj}\frac{\partial u}{\partial x_j}\frac{\partial}{\partial x_k}(\overline{u}|u|^{p-2})\, dx \\
&= \int_\Omega \sum_{k,j=1}^n a_{kj}\left[|u|^{p-2}\frac{\partial u}{\partial x_j}\frac{\partial \overline{u}}{\partial x_k} + \overline{u}\frac{\partial u}{\partial x_j}\frac{\partial |u|^{p-2}}{\partial x_k}\right] dx.
\end{aligned}$$

Since one has

$$\frac{\partial}{\partial x_k}|u|^{p-2} = \frac{1}{2}(p-2)|u|^{p-4}\left(\frac{\partial u}{\partial x_k}\overline{u} + u\frac{\partial \overline{u}}{\partial x_k}\right),$$

we then obtain
(38.6)

$$\langle\langle A_p u, u^* \rangle\rangle =$$
$$\int_\Omega \sum_{k,j=1}^n a_{kj}\left[|u|^{p-2}\frac{\partial u}{\partial x_j}\frac{\partial \overline{u}}{\partial x_k} + \frac{1}{2}(p-2)|u|^{p-4}\overline{u}\frac{\partial u}{\partial x_j}\left(\frac{\partial u}{\partial x_k}\overline{u} + u\frac{\partial \overline{u}}{\partial x_k}\right)\right] dx.$$

Next define the real-valued functions $\alpha_k = \alpha_k(x)$ and $\beta_k = \beta_k(x)$ by

$$|u|^{(p-4)/2}\overline{u}\frac{\partial u}{\partial x_k} = \alpha_k + i\beta_k.$$

Then (38.6) can be written as

$$\begin{aligned}
\langle\langle A_p u, u^* \rangle\rangle &= \int_\Omega \sum_{k,j=1}^n a_{kj}\big[(\alpha_j + i\beta_j)(\alpha_k - i\beta_k) \\
&\qquad + \frac{1}{2}(p-2)(\alpha_j + i\beta_j)(\alpha_k + i\beta_k + \alpha_k - i\beta_k)\big]\, dx \\
&= \int_\Omega \sum_{k,j=1}^n a_{kj}((p-1)\alpha_k\alpha_j + \beta_k\beta_j)\, dx \\
&\qquad + i\int_\Omega \sum_{k,j=1}^n a_{kj}((p-1)\alpha_k\beta_j - \alpha_j\beta_k)\, dx.
\end{aligned} \tag{38.7}$$

Let

$$M = \max_{\substack{1 \le i,j \le n \\ x \in \overline{\Omega}}} |a_{kj}(x)|,$$

and

$$\|\alpha\|^2 = \sum_{k=1}^{n} \int_{\Omega} |\alpha_k(x)|^2 dx, \qquad \|\beta\|^2 = \sum_{k=1}^{n} \int_{\Omega} |\beta_k(x)|^2 dx.$$

Here one can justify that each $\alpha_k(x)$ and $\beta_k(x)$ is in $L^2(\Omega)$. (We leave this as an exercise.) By using (38.4), for $m = 1$, with (38.7) we then obtain

$$\text{(38.8)} \qquad \operatorname{Re} \langle\langle A_p u, u^* \rangle\rangle \ge c\left((p-1)\|\alpha\|^2 + \|\beta\|^2\right) \ge 0.$$

Furthermore, one has $\operatorname{Re} \langle A_p u, u^* \rangle = 0$ if and only if $u = 0$. Also one finds that

$$\text{(38.9)} \qquad |\operatorname{Im} \langle\langle A_p u, u^* \rangle\rangle| \le \frac{Mp}{2} \left(\varepsilon \|\alpha\|^2 + \frac{1}{\varepsilon} \|\beta\|^2 \right), \qquad \text{for any } \varepsilon > 0.$$

Next we prove that $-A_p$ generates a nonexpansive semigroup on $L^p(\Omega)$. In fact, consider any $\lambda > 0$ and any real function $u \in \mathcal{D}(A_p)$, by (38.6) and (38.4), we get

$$\begin{aligned}
\langle\langle A_p u, u^* \rangle\rangle &= (p-1) \int_{\Omega} |u|^{p-2} \left(\sum_{k,j=1}^{n} a_{kj} \frac{\partial u}{\partial x_k} \frac{\partial u}{\partial x_j} \right) dx \\
&\ge c(p-1) \int_{\Omega} |u|^{p-2} \|\nabla u\|^2 dx.
\end{aligned}$$

Furthermore, since $u^* = |u|^{p-2}\overline{u}$, we obtain

$$\text{(38.10)} \qquad \begin{aligned}
\|(\lambda I + A_p)u\|_{L^p} &\ge \frac{|\langle\langle(\lambda I + A_p)u, u^*\rangle\rangle|}{\|u^*\|_{L^q}} \ge \frac{\lambda |\langle u, u^* \rangle|}{\|u^*\|_{L^q}} \\
&= \frac{\lambda \|u\|_{L^p}^p}{\|u\|_{L^p}^{p/q}} = \lambda \|u\|_{L^p}.
\end{aligned}$$

This implies that $\lambda I + A_p$ is one-to-one. We will now show that $\lambda I + A_p$ is a mapping onto W. Since it has a closed range, we only need to show that the range $\mathcal{R}(\lambda I + A_p)$ is dense in $W = L^p(\Omega)$. If there is a $v \in L^q(\Omega)$ such that

$$\langle\langle(\lambda I + A_p)u, v\rangle\rangle = 0, \qquad \text{for all } u \in \mathcal{D}(A_p),$$

then it can be shown (and we leave it as exercise) that

$$v \in \mathcal{D}(A_q) = W^{2,q}(\Omega) \cap W_0^{1,q}(\Omega),$$

and that $\langle\langle u, (\lambda I + A_q)v\rangle\rangle = 0$, for all $u \in \mathcal{D}(A_p)$. Since $\mathcal{D}(A_p)$ is dense in W, one has $(\lambda I + A_q)v = 0$. The injection property remains valid if p is replaced by q. Hence one has $v = 0$. The denseness and closedness of $\mathcal{R}(\lambda I + A_p)$ implies that $\mathcal{R}(\lambda I + A_p) = W$. Thus we have shown that $\lambda I + A_p$ is invertible for any $\lambda > 0$, and (38.10) indicates that

$$\|(\lambda I + A_p)^{-1}\|_{\mathcal{L}(W)} \le \frac{1}{\lambda}, \qquad \text{for } \lambda > 0.$$

According to the Hille-Yosida Theorem 34.1, $-A_p$ generates a C_0-semigroup $e^{-A_p t}$ on W, with $\|e^{-A_p t}\| \le 1$, for $t \ge 0$, see inequality (34.1).

Finally we show that the semigroup $e^{-A_p t}$ is an analytic semigroup. Here we use the numerical range $S(-A_p)$ of the operator $-A_p$. From (38.8) we see that

$$S(-A_p) \subset \{\lambda \in C : |\arg \lambda| > \pi - \theta_0\},$$

where

$$\theta_0 = \arctan \frac{MP}{2c\sqrt{p-1}}, \qquad 0 < \theta_0 < \frac{\pi}{2}.$$

Note that this equality follows from (38.9) by taking $\varepsilon = \sqrt{p-1}$. It follows that there exists a constant $K_\theta > 0$ such that

$$\text{dist } (\lambda; \overline{S(-A_p)}) \ge K_\theta |\lambda|, \qquad \text{for any } \lambda \in \Delta_{\pi-\theta}.$$

Therefore $\Delta_{\pi-\theta} \subset \rho(-A_p)$ and

$$\|(\lambda I + A_p)^{-1}\|_{\mathcal{L}(W)} \le \frac{1}{K_\theta |\lambda|}, \qquad \text{for any } \lambda \in \Delta_{\pi-\theta},$$

see Pazy (1983, Theorem 1.3.9). By Theorem 36.2, $e^{-A_p t}$ is an analytic semigroup on W. □

This result can be generalized by using the Agmon estimate, see Agmon (1965). Here we only state the general results without proof.

Theorem 38.3. *Let $A(x, D)$ be a uniformly elliptic differential operator of order $2m$ with all coefficients in $C^m(\overline{\Omega})$, where Ω is a bounded domain in $\mathbb{R}^n$ with Lipschitz continuous boundary. Let $1 < p < \infty$ and define*

$$A_p u = A(x, D)u, \qquad \textit{for } u \in \mathcal{D}(A_p),$$

where $\mathcal{D}(A_p) = W^{2m,p}(\Omega) \cap W_0^{m,p}(\Omega)$. Then $-A_p$ is the infinitesimal generator of an analytic semigroup on $W = L^p(\Omega)$.

It remains to comment on the cases $p = 1$ and $p = \infty$. First for $p = \infty$, there is an essential difficulty in defining A_∞ so that the domain $\mathcal{D}(A_\infty)$ is dense in $L^\infty(\Omega)$. However, by restricting to

$$W = C_0 = \{u \in C(\overline{\Omega}) : u = 0 \text{ on } \partial\Omega\},$$

with the sup-norm, the linear operator $-\Delta$ does generate an analytic semigroup in W, provided that the region Ω is of class $C^{2m,\mu}$, for some μ with $0 < \mu \leq 1$, see Stewart (1974).

As for $p = 1$, we let $A(x, D)$ as before be a uniformly elliptic operator of order $2m$ with smooth coefficients over a bounded Lipschitz domain $\Omega \subset \mathbb{R}^n$. Set

$$A_1 u = A(x, D)u, \quad \text{for } u \in \mathcal{D}(A_1),$$

where $\mathcal{D}(A_1) = \{u \in W^{2m-1,1}(\Omega) \cap W_0^{m,1}(\Omega) : A(x, D)u \in L^1(\Omega)\}$. Then the operator $-A_1$ generates an analytic semigroup on $L^1(\Omega)$.

3.8.3 Stokes Equations. The Stokes equations represent the linear portion of the Navier-Stokes equations. In Chapter 6 we will study the latter equations in detail. Our objective here is limited to the Stokes equations alone. We wish to show that, under reasonable conditions, these equations can be reformulated as an abstract linear evolutionary equation $\partial_t u + Au = 0$ on an appropriate Hilbert space. We will show, among other things, that the associated Stokes operator A is a positive, selfadjoint operator with compact resolvent.

Let Ω be a bounded open set in $\mathbb{R}^d$, where $d = 2$ or 3, and assume that Ω is of class C^2. The **Stokes equations** on Ω are given by

$$\begin{aligned} \partial_t u - \nu \Delta u + \nabla p &= f, \\ \nabla \cdot u &= 0, \end{aligned} \tag{38.11}$$

where the velocity $u = (u_1, \ldots, u_d)$ is a d-dimensional vector field on Ω, the pressure p is a scalar field on Ω, and the forcing function $f = f(t, x)$ is a known or given d-dimensional vector field on Ω. The objective is to solve for $u = u(t, x)$ and $p = p(t, x)$ so that u and p satisfy (38.11) in Ω, and u satisfies the Initial Value Problem

$$u(0, x) = u_0(x), \qquad x \in \Omega.$$

Furthermore, we assume that the Dirichlet boundary conditions

$$u(t, x) = 0, \qquad \text{for } x \in \partial\Omega \text{ and } t > 0 \tag{38.12}$$

are satisfied.

At first sight, the Stokes equations do not appear to be a fertile land for harvesting a C_0-semigroup. There is the pressure term p and the conservation equation $\nabla \cdot u = 0$. Neither of these involve time derivatives. However, by choosing the correct state space H, one addresses both of these matters: The divergence condition $\nabla \cdot u = 0$ is automatically satisfied, and the pressure term p disappears!

We introduce the following function spaces: $L^p(\Omega) = L^p(\Omega, \mathbb{R}^d)$, for $1 \leq p \leq \infty$, $H^k(\Omega) = H^k(\Omega, \mathbb{R}^d)$, for $k = 1, 2, \ldots$, and $C^k(\Omega) = C^k(\Omega, \mathbb{R}^d)$

for $k = 1, 2, \ldots, \infty$. Because of the conservation equation $\nabla \cdot u = 0$, it is convenient to introduce the following function space

$$Z_0^\infty(\Omega) \stackrel{\text{def}}{=} \{v \in C_0^\infty(\Omega) : \nabla \cdot v = 0 \text{ in } \Omega\}.$$

The space $Z_0^\infty(\Omega)$ is a collection of divergent-free, smooth vector fields with compact support in Ω. It is a linear subspace of both $L^2 = L^2(\Omega)$ and $H^1(\Omega)$. Next we define

$$H \stackrel{\text{def}}{=} \text{Cl}_{L^2(\Omega)} Z_0^\infty(\Omega), \qquad \text{and} \qquad V \stackrel{\text{def}}{=} \text{Cl}_{H^1(\Omega)} Z_0^\infty(\Omega).$$

The inner product $\langle \cdot, \cdot \rangle_H$ on H is precisely the L^2-inner product $\langle \cdot, \cdot \rangle_{L^2}$, and the inner product $\langle \cdot, \cdot \rangle_V$ on V is the H^1-inner product, i.e.,

$$\langle u, v \rangle_V = \langle u, v \rangle_{L^2} + \sum_{i=1}^{d} \langle D_i u, D_i v \rangle_{L^2}, \qquad u, v \in V.$$

We define $\|\nabla u\|_{L^2}^2$ by

$$\|\nabla u\|_{L^2}^2 \stackrel{\text{def}}{=} \sum_{i=1}^{d} \|D_i u\|_{L^2}^2, \qquad \text{for } u \in H^1(\Omega).$$

From the Poincaré inequality, there is a constant $c > 0$ such that $\|u\|_{L^2} \leq c\|\nabla u\|_{L^2}$, for all $u \in H_0^1(\Omega)$. As a result, one has

$$\|\nabla u\|_{L^2}^2 \leq \|u\|_V^2 \leq (1 + c^2)\|\nabla u\|_{L^2}^2, \qquad \text{for all } u \in V. \tag{38.13}$$

In other words, the norm $\|\nabla u\|$ is equivalent to the V-norm $\|u\|_V$ on V.

There is a characterization of H which will be useful. In particular, one has

$$H = \text{Cl}_{L^2(\Omega)} \{v \in C^1(\overline{\Omega}) : \nabla \cdot v = 0 \text{ in } \Omega \text{ and } v \cdot n = 0 \text{ on } \partial\Omega\}.$$

Indeed, the set $Z^1(\Omega) \stackrel{\text{def}}{=} \{v \in C^1(\overline{\Omega}) : \nabla \cdot v = 0 \text{ on } \Omega \text{ and } v \cdot n = 0 \text{ on } \partial\Omega\}$ contains $Z_0^\infty(\Omega)$. However, in terms of the L^2-norm, the space $Z_0^\infty(\Omega)$ is dense in $Z^1(\Omega)$. Therefore, the closures of these two spaces in $L^2(\Omega)$ are the same.

Recall that if p and u are C^1-functions on $\overline{\Omega}$, then

$$\nabla \cdot (pu) = u \cdot \nabla p + p \nabla \cdot u, \qquad \text{in } \Omega.$$

Since Ω is of class C^2, the Gauss Divergence Theorem implies that

$$\int_\Omega u \cdot \nabla p \, dx = -\int_\Omega p \nabla \cdot u \, dx + \int_{\partial\Omega} p u \cdot n \, ds,$$

where n is the unit outward normal to the boundary $\partial\Omega$. Consequently, we see that if $\nabla \cdot u = 0$ in Ω and $u \cdot n = 0$ on $\partial\Omega$, then $\int_\Omega u \cdot \nabla p \, dx = 0$. In other words, the velocity field u is orthogonal to any gradient ∇p. Let $\mathbb{P}$ denote the orthogonal projection of $L^2(\Omega)$ onto H. $\mathbb{P}$ is sometimes referred to as the **Helmholtz**, or the **Leray**, **projection**. One then has the following result.

Lemma 38.4. *Let Ω be an open, bounded set in $\mathbb{R}^d$ of class C^2. Then $H^\perp$, the orthogonal complement of H, satisfies*

$$H^\perp = Cl_{L^2(\Omega)}\{\nabla p : p \in C^1(\overline{\Omega}, \mathbb{R})\}. \tag{38.14}$$

Proof. Let H_1 be defined by the right side of equation (38.14). As argued above, one has $\nabla p \perp H$ whenever $p \in C^1(\overline{\Omega}, \mathbb{R})$, and consequently, $H_1 \subset H^\perp$. Next we show that if $u \in C_0^\infty(\Omega, \mathbb{R}^d)$ then there is a $v \in H$ and $p \in C^\infty(\Omega, \mathbb{R}) \cap C^2(\overline{\Omega}, \mathbb{R})$ such that $u = v + \nabla p$. Indeed, define p as the solution of

$$\Delta p = \nabla \cdot u \text{ in } \Omega, \qquad \text{and} \qquad \frac{\partial p}{\partial n} = u \cdot n = 0 \text{ on } \partial\Omega. \tag{38.15}$$

Next set $v = u - \nabla p$. Then $v \in C^\infty(\Omega) \cap C^2(\overline{\Omega})$, and $\nabla \cdot v = 0$ in Ω, while $v \cdot n = 0$ on $\partial\Omega$. Hence $v \in H$ and $\nabla p \in H_1$. Finally, if $u \in L^2(\Omega)$, then u is the limit (in L^2) of a sequence $u_n \in C_0^\infty(\Omega)$. Since the projection $\mathbb{P}$ is continuous, the sequence $v_n = \mathbb{P}u_n$ converges in H to a limit v. Since H_1 is closed, the sequence $\nabla p = (I - \mathbb{P})u_n$ has its limit ∇p in H_1. Thus $H_1 = H^\perp$. □

Next we want to examine the stationary solutions of the Stokes equations, i.e., the solutions of

$$\begin{aligned} -\nu\Delta u + \nabla p &= f \\ \nabla \cdot u &= 0, \end{aligned} \tag{38.16}$$

with Dirichlet boundary conditions (38.12). The following result is proved in Constantin and Foias (1988; Theorem 3.11).

Lemma 38.5. *Let Ω be an open, bounded set in $\mathbb{R}^d$ of class C^2. Then there is a constant $c > 0$ such that, for every $f \in L^2(\Omega)$, there exist unique functions $u \in H^2(\Omega) \cap V$ and $p \in H^1(\Omega, \mathbb{R})$ that satisfy (38.16), with $\int_\Omega p\,dx = 0$, and one has*

$$||u||_{H^2(\Omega)} + ||\nabla p||_{L^2(\Omega)} \le c||f||_{L^2(\Omega)}. \tag{38.17}$$

The **Stokes operator** A is defined by $Au = \mathbb{P}(-\Delta u)$ for $u \in \mathcal{D}(A)$, where $\mathcal{D}(A) = H^2(\Omega) \cap V$. Thus, if (u,p) is the solution of (38.16) given by the last lemma, then one has $\nu Au = \mathbb{P}f$. The key, to understanding (38.11) then, is to project this equation into the space H by applying $\mathbb{P}$. We then obtain

$$\partial_t u + \nu Au = \mathbb{P}f. \tag{38.18}$$

Note that $\mathbb{P}u = u$ since $\nabla \cdot u = 0$ in Ω and, as a result of the Dirichlet boundary conditions, one has $u \cdot n = 0$ on $\partial\Omega$. Therefore $\mathbb{P}\partial_t u = \partial_t u$. Also note that the pressure term ∇p is missing in (38.18) because $\mathbb{P}\nabla p = 0$, by Lemma 38.4. While the pressure term ∇p does not appear in (38.18), it has not been lost. Indeed, if u is given by (38.18), then one uses (38.15) to find p, see Constantin and Foias (1988). We now have the following result:

Theorem 38.6. *Let Ω be an open, bounded set in $\mathbb{R}^d$ of class C^2, and let A be the Stokes operator on Ω. Then the following hold:*

(1) *The linear operator A is positive and selfadjoint.*
(2) *The inverse A^{-1} is a compact linear operator on H.*
(3) *The operator A is a positive, sectorial operator, and there exist eigenvalues satisfying (32.2), and the corresponding collection of eigenvalues $\{e_1, e_2, e_3, \dots\}$ forms an orthonormal basis for H.*

Proof. First of all, we show that the Stokes operator A is symmetric; i.e., one has

$$\langle Au, v\rangle_{L^2} = a(u,v) = \langle u, Av\rangle_{L^2}, \qquad \text{for all } u, v \in \mathcal{D}(A), \tag{38.19}$$

where $a(u,v)$ is the bilinear form

$$a(u,v) = \sum_{i=1}^{d} \langle D_i u, D_i v\rangle_{L^2}, \qquad \text{for } u, v \in V.$$

Indeed, if $u, v \in \mathcal{D}(A)$, then $\mathbb{P}u = u$ and $\mathbb{P}v = v$. Consequently, one has
(38.20)

$$\langle Au, v\rangle_{L^2} = \langle -\Delta u, v\rangle_{L^2} = -\int_\Omega \Delta u \cdot v\, dx = \sum_{i=1}^{d} \int_\Omega D_i u \cdot D_i v\, dx = a(u,v),$$

and similarly, one finds that $a(u,v) = \langle u, Av\rangle_{L^2}$.

Secondly, we show that the bilinear form $a(u,v)$ is symmetric and positive. Indeed from the definitions, one has $a(u,v) = a(v,u)$, i.e., a is symmetric. From (38.13) we obtain

$$(1 + c^2)^{-1} \|u\|_V^2 \le \sum_{i=1}^{d} \|D_i u\|_{L^2}^2 = a(u,u), \qquad \text{for all } u \in V, \tag{38.21}$$

hence $a(u,v)$ is positive.

Thirdly, we show that the Stokes operator A is selfadjoint and positive. Indeed, if $v \in \mathcal{D}(A^*)$, then there is an $f \in H$ satisfying $\mathbb{P}f = f$ and

$$\langle Au, v\rangle_{L^2} = \langle u, f\rangle_{L^2}, \qquad \text{for all } u \in \mathcal{D}(A).$$

Since $f \in H$, it follows from Lemma 38.5 that there is a $w \in \mathcal{D}(A)$ satisfying $Aw = f$. We claim that $w = v$. In order to show this, it suffices to show that

$$\langle h, w - v\rangle_{L^2} = 0, \qquad \text{for all } h \in H.$$

Let h be an arbitrary point in H. We then use Lemma 38.5 to find $\hat{w} \in \mathcal{D}(A)$ to satisfy $A\hat{w} = h$. One then obtains

$$\begin{aligned}\langle h, w - v\rangle_{L^2} &= \langle A\hat{w}, w\rangle_{L^2} - \langle A\hat{w}, v\rangle_{L^2} \\ &= \langle \hat{w}, Aw\rangle_{L^2} - \langle \hat{w}, f\rangle_{L^2} = \langle \hat{w}, f - f\rangle_{L^2} = 0.\end{aligned}$$

Hence $\mathcal{D}(A^*) \subset \mathcal{D}(A)$ and A is selfadjoint. The positivity of A follows from (38.21).

Finally, we note that the inverse A^{-1} is a compact operator. Indeed, for $f \in H$ we let $u \in \mathcal{D}(A)$ be fixed so that $Au = f$, i.e., $u = A^{-1}f$. Then (38.17) implies that

$$||A^{-1}f||_{H^2(\Omega)} \leq c||f||_{L^2(\Omega)}.$$

Since the imbedding $H^2(\Omega) \hookrightarrow L^2(\Omega)$ is compact, we see that A^{-1} is a compact operator. The remainder of the proof now follows immediately from Theorem 32.1. □

3.8.4 Wave Equation. The simplest wave equation takes on the form

$$\partial_t^2 u = \Delta u, \qquad \text{for } t > 0, x \in \Omega, \tag{38.22}$$

where Ω is an open, bounded domain in $\mathbb{R}^n$ with boundary of Lipschitz class, and $\partial_t^2 u = \frac{\partial^2 u}{\partial t^2}$. We assume here that the Dirichlet boundary condition $u(t,x) = 0$, for $x \in \partial\Omega$ and $t \geq 0$, holds and the initial condition becomes $u\mid_{t=0} = u_0(x)$ and $\partial_t u\mid_{t=0} = u_1(x)$, for $x \in \Omega$. Let $H = L^2(\Omega)$ and define an operator A on H by $A\varphi = -\Delta\varphi$, for $\varphi \in \mathcal{D}(A)$, where $\mathcal{D}(A) = H^2(\Omega) \cap H_0^1(\Omega)$, see Section 3.8.2. Then (38.22) can be written as an abstract evolutionary equation of second-order in time, i.e.,

$$\partial_t^2 u + Au = 0, \qquad \text{for } t > 0, \tag{38.23}$$

where $u = u(t) = u(t, \cdot)$, with the initial conditions $u(0) = u_0$ and $\partial_t u(0) = u_1$.

This second-order equation can be converted to a first-order system by introducing the vectors

$$w(t) = \begin{pmatrix} u(t) \\ \partial_t u(t) \end{pmatrix}, \qquad \text{and} \qquad w_0 = \begin{pmatrix} u_0 \\ u_1 \end{pmatrix}. \tag{38.24}$$

Let $V = H_0^1(\Omega) = \mathcal{D}(A^{\frac{1}{2}})$ and $E = V \times H$. Define an operator $G : \mathcal{D}(G) \to E$ by

$$G = \begin{bmatrix} 0 & -I_H \\ A & 0 \end{bmatrix},$$

where the domain is $\mathcal{D}(G) = \mathcal{D}(A) \times \mathcal{D}(A^{\frac{1}{2}}) = (H^2(\Omega) \cap H_0^1(\Omega)) \times H_0^1(\Omega)$, and I_H is the identity on H. Then the equation (38.23) can be further written as an abstract evolutionary equation of first-order in time, i.e.,

$$\partial_t w + Gw = 0, \qquad \text{for } t > 0, \tag{38.25}$$

with initial condition $w(0) = w_0 \in E$. Since the norm $||A^{\frac{1}{2}}u||_H^2$ is equivalent to the norm $||u||_V^2$ on V, we define the norm of w by

$$||w||_E^2 \stackrel{\text{def}}{=} ||A^{\frac{1}{2}}u||_H^2 + ||\partial_t u||_H^2.$$

We now show that the operator $-G$ is the infinitesimal generator of a C_0-group $T(t)$, for $t \in \mathbb{R}$. More generally we have the following result:

Lemma 38.7. *Let A be a positive, selfadjoint operator with compact resolvent on a Hilbert space H. Let $E = \mathcal{D}(A^{\frac{1}{2}}) \times H$ and define $G : \mathcal{D}(G) \to E$ by*

$$G = \begin{bmatrix} 0 & -I_H \\ A & 0 \end{bmatrix},$$

where the domain is given by $\mathcal{D}(G) = \mathcal{D}(A) \times \mathcal{D}(A^{\frac{1}{2}}) \subset E$. Then G has the following properties:

(1) *G is densely defined and closed operator.*
(2) *G is skew-adjoint, i.e., $G^* = -G$.*
(3) *G has compact resolvent.*
(4) *The spectrum $\sigma(G)$ consists of point spectrum only, and $\mu \in \sigma(G)$ if and only if $\mu = \pm i\lambda$, where $\lambda^2 \in \sigma(A)$. Also one has $Au = \lambda^2 u$ if and only if $Gw = \pm i\lambda w$, where $w = (u, \mp i\lambda u)^T$.*
(5) *$-G$ is the infinitesimal generator of a unitary group $T(t)$, where $t \in \mathbb{R}$, which is given by*

$$T(t) = \begin{bmatrix} \cos(A^{\frac{1}{2}}t) & A^{-\frac{1}{2}}\sin(A^{\frac{1}{2}}t) \\ -A^{\frac{1}{2}}\sin(A^{\frac{1}{2}}t) & \cos(A^{\frac{1}{2}}t) \end{bmatrix}, \qquad t \in \mathbb{R}, \tag{38.26}$$

where $A^{-\frac{1}{2}} = (A^{\frac{1}{2}})^{-1}$, and $\cos(A^{\frac{1}{2}}t)$ and $\sin(A^{\frac{1}{2}}t)$ are defined by (32.5) with $f(x^2) = \cos(xt)$ and $f(x^2) = \sin(xt)$, respectively.
(6) *The C_0-group $T(t)$ is differentiable, for $t \in \mathbb{R}$.*

Proof. Item (1) is elementary and is left as an exercise. Item (2): In order to prove that G is skew-adjoint, we first prove that the adjoint G^* is an extension of $-G$, i.e., $-G \subset G^*$. For any $\begin{pmatrix} \varphi \\ \psi \end{pmatrix} \in \mathcal{D}(G)$ and $\begin{pmatrix} f \\ g \end{pmatrix} \in \mathcal{D}(-G) = \mathcal{D}(G)$, we have

$$\begin{aligned}\left\langle -G\begin{pmatrix} \varphi \\ \psi \end{pmatrix}, \begin{pmatrix} f \\ g \end{pmatrix}\right\rangle_E &= \langle \psi, f\rangle_V - \langle A\varphi, g\rangle_H = \langle A^{\frac{1}{2}}\psi, A^{\frac{1}{2}}f\rangle_H - \langle A\varphi, g\rangle_H \\ &= \langle \psi, Af\rangle_H - \langle \varphi, g\rangle_V = \left\langle \begin{pmatrix} \varphi \\ \psi \end{pmatrix}, G\begin{pmatrix} f \\ g \end{pmatrix}\right\rangle_E.\end{aligned}$$

This shows that $\mathcal{D}(-G) \subset \mathcal{D}(G^*)$ and $G^* |_{\mathcal{D}(-G)} = -G$. Next we prove that $\mathcal{D}(G^*) \subset \mathcal{D}(-G) = \mathcal{D}(G)$. If $\begin{pmatrix} f \\ g \end{pmatrix} \in \mathcal{D}(G^*)$, then there is a $\begin{pmatrix} \varphi \\ \psi \end{pmatrix} \in \mathcal{D}(G)$ such that

$$\left\langle G\begin{pmatrix} u \\ v \end{pmatrix}, \begin{pmatrix} f \\ g \end{pmatrix}\right\rangle_E = \left\langle \begin{pmatrix} u \\ v \end{pmatrix}, \begin{pmatrix} \varphi \\ \psi \end{pmatrix}\right\rangle_E, \qquad \text{for all } \begin{pmatrix} u \\ v \end{pmatrix} \in \mathcal{D}(G). \tag{38.27}$$

(i) Let $u = 0$. Then (38.27) yields

$$\langle -v, f\rangle_V = \langle v, \psi\rangle_H, \qquad \text{for all } v \in V.$$

But $\langle v, f\rangle_V = \langle A^{\frac{1}{2}}v, A^{\frac{1}{2}}f\rangle_H$, which implies that $A^{\frac{1}{2}}f \in \mathcal{D}(A^{\frac{1}{2}})$, since $A^{\frac{1}{2}}$ is selfadjoint. Thus $f \in \mathcal{D}(A)$.

(ii) Let $v = 0$. Then (38.27) implies that

$$\langle Au, g\rangle_H = \langle u, \varphi\rangle_V = \langle A^{\frac{1}{2}}u, A^{\frac{1}{2}}\varphi\rangle_H, \qquad \text{for all } u \in \mathcal{D}(A).$$

Since $\mathcal{D}(A^{\frac{1}{2}}) = A^{\frac{1}{2}}\mathcal{D}(A)$ and $A^{\frac{1}{2}}$ is selfadjoint, it follows that g is in $\mathcal{D}(A^{\frac{1}{2}}) = V$.

By combining (i) and (ii), we have $\begin{pmatrix} f \\ g \end{pmatrix} \in \mathcal{D}(G)$. Consequently, one has $-G = G^*$.

Item (3). It is easy to show that $G^{-1} \in \mathcal{L}(E) \cap \mathcal{L}(E; \mathcal{D}(G))$, where $\mathcal{D}(G)$ is a Hilbert space equipped with an equivalent graph norm:

$$\left\| \begin{pmatrix} \varphi \\ \psi \end{pmatrix} \right\|_{\mathcal{D}(G)} \stackrel{\text{def}}{=} \left[\|A\varphi\|_H^2 + \|A^{\frac{1}{2}}\psi\|_H^2 \right]^{\frac{1}{2}}.$$

It suffices to prove that the imbedding $\mathcal{D}(G) \mapsto E$ is compact. However, this follows immediately from $\mathcal{D}(A) \hookrightarrow \mathcal{D}(A^{\frac{1}{2}})$ and $\mathcal{D}(A^{\frac{1}{2}}) = V \hookrightarrow H$.

Item (4). The proof of this property is straightforward, so we omit the details.

Item (5). The argument here is similar to that given in Section 3.2. Let the eigenvalues of A satisfy (32.2) and let $\{e_1, e_2, e_3, \dots\}$ denote the orthonormal basis of eigenvectors. Define $\mu_n \stackrel{\text{def}}{=} \sqrt{\lambda_n}$, for $n = 1, 2, 3, \dots$, and let $w_n^{\pm} \stackrel{\text{def}}{=} (e_n, \mp i\mu_n e_n)^T$. Then $\{w_n^{\pm} : n = 1, 2, 3, \dots\}$ forms an orthogonal basis for E. In addition, the spectral mapping formula

(38.28)
$$T(t)w \stackrel{\text{def}}{=} \sum_{n=1}^{\infty} e^{i\mu_n t}\langle w, w_n^-\rangle w_n^- + \sum_{n=1}^{\infty} e^{-i\mu_n t}\langle w, w_n^+\rangle w_n^+, \qquad \text{for } t \in \mathbb{R},$$

defines a C_0-group on E with infinitesimal generator $-G$. The formula (38.28) is a representation of the C_0-group in complex arithmetic. By direct operator calculations one can verify that the operator defined by the right side of (38.26) is a strongly continuous group of bounded linear operators and that its infinitesimal generator is exactly $-G$. The formula (38.26) is a representation of the C_0-group in real arithmetic. The details are omitted.

Item (6). By using equation (38.28) and the analyticity of e^z, one can readily show that the differentiablility of $T(t)$, for $t \in \mathbb{R}$, follows directly from the definition of a differentiable group, see Section 3.5. □

It follows from Theorem 35.2, that for all $w_0 \in E$, $T(t)w_0$ is a mild solution of equation (38.25). Also, if $w_0 \in \mathcal{D}(G)$, then $T(t)w_0$ is a classical solution of (38.25).

Next let us consider the abstract wave equation with a distributed damping term $\alpha\, \partial_t u$ given by

$$\partial_t^2 u + \alpha\, \partial_t u + Au = 0, \qquad \text{for } t \geq 0, \tag{38.29}$$

where $\alpha > 0$ is a constant, and the operator A satisfies the same conditions described in Lemma 38.7. Since A is positive, the first eigenvalue λ_1 is positive, and one has $\|A^{\frac{1}{2}}u\|^2 \geq \lambda_1\|u\|^2$, for all $u \in \mathcal{D}(A^{\frac{1}{2}})$. Let $w = w(t)$ be given by (38.24). Then equation (38.29) can be formulated as a first-order evolutionary equation in $E = \mathcal{D}(A^{\frac{1}{2}}) \times H$ as follows,

$$\partial_t w + G_\alpha w = 0, \qquad \text{for } t \geq 0, \tag{38.30}$$

where $w(0) = w_0 \in E$, and $G_\alpha : \mathcal{D}(G) \to E$ is defined by

$$G_\alpha = \begin{bmatrix} 0 & -I_H \\ A & \alpha I_H \end{bmatrix}.$$

We now have the following result:

Corollary 38.8. *For $\alpha > 0$, the operator G_α is densely defined, closed operator with compact resolvent. Furthermore, $-G_\alpha$ generates a C_0-group $e^{-G_\alpha t}$, which is differentiable for $t > 0$, and there is a constant $\mu > 0$, such that*

$$\|e^{-G_\alpha t}\|_{\mathcal{L}(E)} \leq 4e^{-\mu t}, \qquad \text{for } t \geq 0. \tag{38.31}$$

Proof. Since G_α is a bounded perturbation of the operator G defined in Lemma 38.7, most of the properties described above are easily verified, and we will omit some of the details. However, we will prove that the C_0-semigroup $e^{-G_\alpha t}$ satisfies inequality (38.31). First for any $w_0 \in \mathcal{D}(G_\alpha^\infty) = \cap_{k=1}^\infty \mathcal{D}(G_\alpha^k)$, which is dense in E, the corresponding mild solution $w(t) = w(t; w_0)$ is a classical solution to equation (38.30), and its first component function is accordingly the classical solution of the original equation (38.29).

First we take the inner product of (38.29) with u_t in H, to obtain

$$\frac{1}{2}\partial_t\|\partial_t u\|^2 + \alpha\|\partial_t u\|^2 + \frac{1}{2}\partial_t\|A^{\frac{1}{2}}u\|^2 = 0, \qquad \text{for } t \geq 0, \tag{38.32}$$

where $\|\cdot\|$ is the H-norm. Next we take the inner-product of (38.29) with εu in H, and find

$$\varepsilon\partial_t\langle\partial_t u, u\rangle - \varepsilon\|\partial_t u\|^2 + \frac{\varepsilon\alpha}{2}\partial_t\|u\|^2 + \varepsilon\|A^{\frac{1}{2}}u\|^2 = 0, \qquad \text{for } t \geq 0. \tag{38.33}$$

By adding equations (38.32) and (38.33), we get

$$\frac{1}{2}\partial_t N(t; u, \varepsilon) + \left[(\alpha - \varepsilon)\|\partial_t u\|^2 + \varepsilon\|A^{\frac{1}{2}}u\|^2\right] = 0, \qquad \text{for } t \geq 0,$$

where $N = N(t; u, \varepsilon)$ is defined by

$$N(t; u, \varepsilon) \stackrel{\text{def}}{=} \|\partial_t u\|^2 + \varepsilon\alpha\|u\|^2 + \|A^{\frac{1}{2}}u\|^2 - 2\varepsilon\langle\partial_t u, u\rangle.$$

Our next objective is to show that N is equivalent to the norm $\|w\|_E^2$ on E, provided that $\varepsilon > 0$ is sufficiently small. For this step we restrict ε to satisfy

$$0 < \varepsilon \leq \mu \stackrel{\text{def}}{=} \min\left(1, \frac{\alpha}{2}, \frac{\lambda_1}{1+\alpha}\right). \tag{38.34}$$

From the Young inequality and the fact that $\|A^{\frac{1}{2}}u\|^2 \geq \lambda_1\|u\|^2$, one obtains

$$N \leq (1+\varepsilon)\|\partial_t u\|^2 + (1+\alpha)\,\varepsilon\|u\|^2 + \|A^{\frac{1}{2}}u\|^2 \leq 2\|\partial_t u\|^2 + 2\|A^{\frac{1}{2}}u\|^2 = 2\|w\|_E^2.$$

The last inequality and (38.34) imply that

$$\frac{\varepsilon}{2}N(t;u,\varepsilon) \leq (\alpha - \varepsilon)\|\partial_t u\|^2 + \varepsilon\|A^{\frac{1}{2}}u\|^2,$$

and consequently one has

$$\partial_t N(t;u,\varepsilon) + \varepsilon N(t;u,\varepsilon) \leq 0, \qquad t \geq 0. \tag{38.35}$$

Similarly, from the Young inequality and (38.34) one obtains

$$\begin{aligned} N &\geq \|\partial_t u\|^2 + \varepsilon\alpha\|u\|^2 + \|A^{\frac{1}{2}}u\|^2 - \frac{1}{2}\|\partial_t u\|^2 - 2\varepsilon^2\|u\|^2 \\ &\geq \frac{1}{2}\|\partial_t u\|^2 + \|A^{\frac{1}{2}}u\|^2 \geq \frac{1}{2}\|w\|_E^2. \end{aligned}$$

Using the Gronwall inequality on (38.35), one obtains inequality (38.31) since, for $\varepsilon = \mu$ given by (38.34), one has the following (for all $t \geq 0$):

$$\|w(t)\|_E^2 \leq 2N(t;u,\varepsilon) \leq 2e^{-\varepsilon t}N(0;u,\varepsilon) \leq 4e^{-\varepsilon t}\|w_0\|_E^2. \quad \square$$

3.8.5 Schrödinger Equation. A problem closely related to the wave equation is the following generalization of the Schrödinger equation:

$$\partial_t u - (\lambda + i\alpha)\Delta u - (\gamma + i\beta)Vu = 0, \qquad x \in \Omega, \tag{38.36}$$

where Ω is an open, bounded set in $\mathbb{R}^d$ with boundary of Lipschitz class, and $i = \sqrt{-1}$. Here we assume that λ, γ, α, and β are real constants with $\lambda \geq 0$, and $V = V(x)$ is a potential function with $x \in \Omega$. Equation (38.36) reduces to the Schrödinger equation when one sets $\lambda = \gamma = 0$, $\alpha = h^2$ (where h is the Planck constant) and $\beta > 0$. In the study of the hydrogen atom, for example, one takes $H = L^2(\mathbb{R}^3, \mathbb{C})$, and the potential V is the Coulomb potential $V(x) = \|x\|^{-1}$, for $x \neq 0$. This problem is not of immediate interest here because the physical space $\mathbb{R}^3$ is not bounded. Instead, we consider the Ginzburg-Landau equation, which is a nonlinear perturbation of (38.36) on a bounded set Ω, see Temam (1988).

We assume that the Laplacian operator Δ in equation (38.36) has boundary conditions so that the operator A, given by $Au = -\Delta u$, is a nonnegative, selfadjoint operator with compact resolvent on the complex Hilbert space $L^2(\Omega, \mathbb{C})$. For example, one may use the Dirichlet boundary conditions, see Section 3.8.2. Let us look at the case where the potential V is constant, say $V(x) \equiv 1$. Then (38.36) becomes

$$\partial_t u + Bu = 0,$$

where $B = (\lambda + i\alpha)A - (\gamma + i\beta)I$. By using the methods of Section 3.2, one readily shows that $-B$ is the generator of a C_0-semigroup e^{-Bt}, where

$$e^{-Bt}u = \sum_{n=1}^{\infty} \exp(-(\lambda\lambda_n - \gamma)t)\exp(i(-\alpha\lambda_n + \beta)t)\langle u, e_n\rangle e_n, \qquad t \geq 0,$$

the eigenvalues of A satisfy (32.2), and $e_1, e_2, e_3, \dots$ are the corresponding eigenvectors. Notice that e^{-Bt} is a C_0-group, defined for all $t \in \mathbb{R}$, when $\lambda = \gamma = 0$.

3.8.6 The $L^1 - L^\infty$ Regularity Property of Analytic Semigroups. Let us return to the heat equation

$$\partial_t u = \nu\Delta u, \qquad (t, x) \in \mathbb{R}^+ \times \Omega, \tag{38.37}$$

where Ω is an open bounded domain in $\mathbb{R}^m$, with boundary $\partial\Omega$ of Lipschitz class, and u is a real-valued function. We assume that the Dirichlet boundary condition $u\mid_{\partial\Omega} = 0$ is satisfied, and we let $u(0, x) = u_0(x)$ denote the initial condition. As noted above (also see Exercise 38.3), $-A_1 = \nu\Delta : \mathcal{D}(A_1) \to L^1(\Omega)$ generates an analytic semigroup $T(t)$ on $L^1 = L^1(\Omega)$. For $u_0 \in L^1$, we let $u(t) = T(t)u_0$ denote the mild solution of (38.37), with $u(0) = u_0$. The following result characterizes an important $L^1 - L^\infty$ regularity property of the semigroup $T(t)$.

Theorem 38.9. *Let $T(t)$ be the analytic semigroup on $L^1(\Omega)$ described above. Then for every $t > 0$, one has $T(t) \in \mathcal{L} = \mathcal{L}(L^1(\Omega), L^\infty(\Omega))$. Morevover, there is a constant $C > 0$, which depends only on Ω, m, and ν, such that one has*

$$\|T(t)\|_{\mathcal{L}} \leq Ct^{-\frac{m}{2}}, \qquad \textit{for all } t > 0. \tag{38.38}$$

Furthermore, inequality (38.38) is valid, if one uses the Neumann boundary condition $\frac{\partial u}{\partial n} = 0$ on $\partial\Omega$ and replaces $L^1(\Omega)$ by the Banach space

$$L^1_0(\Omega) \stackrel{\text{def}}{=} \{\varphi \in L^1(\Omega) : \int_\Omega \varphi(x)\, dx = 0\}. \tag{38.39}$$

Before we prove this important result, we recall that, for Ω bounded, one has the imbedding chain

$$L^\infty(\Omega) \mapsto L^q(\Omega) \mapsto L^p(\Omega) \mapsto L^1(\Omega), \qquad \text{for } 1 \leq p \leq q \leq \infty.$$

The $L^1 - L^\infty$ regularity in Theorem 38.9 gives us the estimate of the operator norm of the semigroup $T(t)$, which maps the largest space $L^1(\Omega)$ into the smallest space $L^\infty(\Omega)$ in this chain. Consequently, all the other information about the mappings $T(t) : L^p(\Omega) \to L^q(\Omega)$, for $1 \le p \le q \le \infty$, will be available by the interpolation inequalities, as we will see.

Proof. Our approach in proving (38.38) will consist in (1) obtaining a good estimate for $\|u(t)\|_{L^p}$, for any integer p satisfying $2 \le p < \infty$, and (2) letting $p \to \infty$ while using the relationship

$$\|u(t)\|_{L^\infty} \le \limsup_{p\to\infty} \|u(t)\|_{L^p}.$$

For simplicity, we assume the diffusion coefficient satisfies $\nu = 1$, and we consider first the Dirichlet boundary conditions. The proof for $m \ge 2$ goes through four steps.[10]

Step 1. First we handle the spatial variable. Suppose that $p = 2\rho \ge 2$, with $\rho \ge 1$ being an integer. We will now show that there is an α, with $1 < \alpha < \infty$, and a $C_1 > 0$, such that

$$\partial_t \|u(t)\|^p_{L^p} + C_1 \|u(t)\|^p_{L^{\alpha p}} \le 0, \qquad \text{for } t > 0. \tag{38.40}$$

By multiplying equation (38.37) by $|u(t,x)|^{p-2}u(t,x)$, integrating over Ω, and using Green's formula, we get

$$\frac{1}{p}\partial_t \|u(t)\|^p_{L^p} + (p-1)\int_\Omega |\nabla_x u(t,x)|^2 \, |u(t,x)|^{p-2} dx = 0, \qquad \text{for } t > 0.$$

It follows that

$$\partial_t \|u(t)\|^p_{L^p} + \frac{4p(p-1)}{p^2}\int_\Omega \left|\nabla_x |u(t,x)|^{p/2}\right|^2 dx = 0, \qquad \text{for } t > 0.$$

Since $p \ge 2$, we have $2 \le \frac{4p(p-1)}{p^2} < 4$. The above equality yields

$$\partial_t \|u(t)\|^p_{L^p} + 2\int_\Omega \left|\nabla_x |u(t,x)|^{p/2}\right|^2 dx \le 0, \qquad \text{for } t > 0. \tag{38.41}$$

From the Sobolev embedding theorems (see Appendix B) one has

$$W^{j,a}(\Omega) \hookrightarrow W^{k,b}(\Omega), \qquad \text{provided that } j - \frac{m}{a} \ge k - \frac{m}{b}, \tag{38.42}$$

where $m = \dim \Omega$. Note that (38.42) implies that, for $m \ge 3$, one has

$$H^1(\Omega) \hookrightarrow L^q(\Omega), \qquad \text{where } q = \frac{2m}{m-2}.$$

[10] We omit the argument for $m = 1$. This can be found in You (1999).

Next the Poincaré inequality implies that there is a constant $C_0 > 0$, such that

$$C_0\|\varphi\|_{L^q}^q \leq \int_\Omega \|\nabla_x\varphi\|^2\,dx, \qquad \text{for all } \varphi \in H_0^1(\Omega).$$

Hence, there is a constant $C_1 > 0$, such that for $\varphi = |u(t,x)|^{\frac{p}{2}}$, one has

$$2\int_\Omega \left|\nabla_x|u(t,x)|^{p/2}\right|^2 dx \geq C_1\left(\int_\Omega |u(t,x)|^{pm/(m-2)}\,dx\right)^{\frac{m-2}{m}}, \qquad \text{for } t > 0.$$

Now for $m = 2$, one has $H^1(\Omega) \mapsto L^q(\Omega)$, for any q with $1 \leq q < \infty$, and consequently, we have

$$2\int_\Omega \left|\nabla_x|u(t,x)|^{p/2}\right|^2 dx \geq C_1\left(\int_\Omega |u(t,x)|^{pq}\,dx\right)^{\frac{1}{q}}.$$

We now define α by $\alpha = \alpha(m) = \frac{m}{m-2}$, for $m \geq 3$, and $\alpha = q \geq 2$, for $m = 2$. With this choice of α, we see that inequality (38.41) implies (38.40).

Step 2. Next one handles the time variable. Let $k \geq 1$ be any integer. Multiply (38.40) by t^{k+1} and integrate both sides over the time interval $[0,t]$ to obtain

(38.43)

$$t^{k+1}\|u(t)\|_{L^p}^p + C_1\int_0^t s^{k+1}\|u(s)\|_{L^{\alpha p}}^p\,ds \leq (k+1)\int_0^t s^k\|u(s)\|_{L^p}^p\,ds.$$

Let a constant $\beta = \beta(p)$ be given by

$$\beta = \frac{p(\alpha-1)}{p\alpha-1}, \qquad \text{or equivalently,} \qquad \alpha p = \frac{p-\beta}{1-\beta}.$$

Note that one has $0 < \beta < 1$ and $\lim_{p\to\infty}\beta(p) = \frac{\alpha-1}{\alpha}$. By using the Hölder inequality with the two exponents $1/\beta$ and $1/(1-\beta)$, for $t > 0$, one finds that

$$\begin{aligned}
\|u(t)\|_{L^p}^p &= \int_\Omega |u(t,x)|^p\,dx = \int_\Omega |u(t,x)|^\beta\,|u(t,x)|^{p-\beta}\,dx \\
&\leq \left(\int_\Omega |u|\,dx\right)^\beta \left(\int_\Omega |u|^{(p-\beta)/(1-\beta)}\,dx\right)^{1-\beta} \qquad (38.44)\\
&= \|u(t)\|_{L^1}^\beta\,\|u(t)\|_{L^{\alpha p}}^{p-\beta}.
\end{aligned}$$

We now take a particular integer $k \geq 1$ given by

$$k = \left[\frac{\alpha p}{\alpha-1}\right] = \text{integer part of } \frac{\alpha p}{\alpha-1}.$$

One then has $\limsup_{p\to\infty}(k+1)^{\frac{1}{p}} < \infty$. By a straightforward calculation, one can readily verify that $p < \beta(k+2)$.

We claim that one has

$$
\begin{aligned}
t^{k+1}\|u(t)\|_{L^p}^p + C_1 \int_0^t s^{k+1}\|u(s)\|_{L^{\alpha p}}^p \, ds \\
\le (k+1)\left(k+2-\frac{p}{\beta}\right)^{-\frac{\beta}{p}} \left(t^{\frac{(k+2)\beta}{p}-1}\right) \\
\times \left(\int_0^t s^{k+1}\|u(s)\|_{L^{\alpha p}}^p \, ds\right)^{\frac{(p-\beta)}{p}} \|u_0\|_{L^1}^\beta,
\end{aligned} \tag{38.45}
$$

Indeed, by substituting (38.44) into the right side of inequality (38.43) and noting that $\|T(t)\|_{\mathcal{L}(L^1(\Omega))} \le 1$, for $t \ge 0$, one obtains

$$
\begin{aligned}
& t^{k+1}\|u(t)\|_{L^p}^p + C_1 \int_0^t s^{k+1}\|u(s)\|_{L^{\alpha p}}^p \, ds \\
& \le (k+1)\|u_0\|_{L^1}^\beta \int_0^t s^k \|u(s)\|_{L^{\alpha p}}^{p-\beta} \, ds \\
& \le (k+1)\|u_0\|_{L^1}^\beta \int_0^t \|u(s)\|_{L^{\alpha p}}^{p-\beta} s^{\frac{(k+1)(p-\beta)}{p}} s^{k-\frac{(k+1)(p-\beta)}{p}} \, ds \\
& \le (k+1)\|u_0\|_{L^1}^\beta \left(\int_0^t s^{k+1}\|u(s)\|_{L^{\alpha p}}^p \, ds\right)^{\frac{(p-\beta)}{p}} \\
& \quad \times \left(\int_0^t s^{k+1-\frac{p}{\beta}} \, ds\right)^{\frac{\beta}{p}},
\end{aligned}
$$

which implies (38.45).

Step 3. The next step is to rewrite inequality (38.45) in terms of the parameter p. By the observation that the scalar function $h(y) = y^{\frac{1}{p}}$, for $p \ge 1$, is concave over $[0, \infty)$, we have the concavity inequality:

$$
(a+b)^{\frac{1}{p}} \ge \frac{1}{2}\left[(2a)^{\frac{1}{p}} + (2b)^{\frac{1}{p}}\right].
$$

By raising (38.45) to the power $1/p$ and by using this concavity inequality, we obtain

$$
\begin{aligned}
& 2^{(\frac{1}{p}-1)} t^{\frac{(k+1)}{p}} \|u(t)\|_{L^p} + 2^{(\frac{1}{p}-1)} (C_1)^{\frac{1}{p}} \left(\int_0^t s^{k+1}\|u(s)\|_{L^{\alpha p}}^p \, ds\right)^{\frac{1}{p}} \\
& \le (k+1)^{\frac{1}{p}} \left(k+2-\frac{p}{\beta}\right)^{-\frac{\beta}{p^2}} \left(t^{\frac{(k+2)\beta}{p^2}-\frac{1}{p}}\right) \\
& \quad \times \left(\int_0^t s^{k+1}\|u(s)\|_{L^{\alpha p}}^p \, ds\right)^{\frac{(p-\beta)}{p^2}} \|u_0\|_{L^1}^{\frac{\beta}{p}}.
\end{aligned}
$$

Next we use the Young inequality, $ab \le \epsilon a^P + C_\epsilon b^Q$, with

$$P = \frac{p}{p-\beta} \quad \text{and} \quad Q = \frac{p}{\beta}$$

and

$$a = \left(\int_0^t s^{k+1} \|u(s)\|_{L^{\alpha p}}^p \, ds\right)^{\frac{(p-\beta)}{p^2}}, \quad \text{and}$$

$$b = \left(k+2-\frac{p}{\beta}\right)^{-\frac{\beta}{p^2}} \left(t^{\frac{(k+2)\beta}{p^2}-\frac{1}{p}}\right) \|u_0\|_{L^1}^{\frac{\beta}{p}},$$

to obtain

$$\begin{aligned} 2^{\frac{1}{p}-1} t^{\frac{(k+1)}{p}} \|u(t)\|_{L^p} + 2^{\frac{1}{p}-1} (C_1)^{\frac{1}{p}} \left(\int_0^t s^{k+1} \|u(s)\|_{L^{\alpha p}}^p \, ds\right)^{\frac{1}{p}} \\ \le \epsilon (k+1)^{\frac{1}{p}} \left(\int_0^t s^{k+1} \|u(s)\|_{L^{\alpha p}}^p \, ds\right)^{\frac{1}{p}} \\ + C_\epsilon (k+1)^{\frac{1}{p}} \left(k+2-\frac{p}{\beta}\right)^{-\frac{1}{p}} \left(t^{\frac{(k+2)}{p}-\frac{1}{\beta}}\right) \|u_0\|_{L^1}. \end{aligned} \tag{38.46}$$

We claim that, due to the choice of the integer k mentioned earlier, there is a suitably small $\epsilon > 0$ such that

$$\epsilon (k+1)^{\frac{1}{p}} < 2^{\frac{1}{p}-1} (C_1)^{\frac{1}{p}}. \tag{38.47}$$

Now we move the first term on the right side of (38.46) to the left side and use (38.47) to obtain

$$\|u(t)\|_{L^p} \le 2^{1-\frac{1}{p}} C_\epsilon (k+1)^{\frac{1}{p}} \left(k+2-\frac{p}{\beta}\right)^{-\frac{1}{p}} \left(t^{\frac{1}{p}-\frac{1}{\beta}}\right) \|u_0\|_{L^1}. \tag{38.48}$$

Step 4. We now let $p \to \infty$ in (38.48). Owing to the previous choice of the integer k, we have

(38.49)

$$\begin{aligned} \left(k+2-\frac{p}{\beta}\right)^{-\frac{1}{p}} &\le \left(\frac{\alpha p}{\alpha-1}-\frac{p}{\beta}\right)^{-\frac{1}{p}} = \left[p\left(\frac{\alpha}{\alpha-1}-\frac{p\alpha-1}{p(\alpha-1)}\right)\right]^{-\frac{1}{p}} \\ &= (\alpha-1)^{\frac{1}{p}} = \begin{cases} \left(\frac{2}{m-2}\right)^{\frac{1}{p}}, & \text{if } m \ge 3; \\ (q-1)^{\frac{1}{p}}, & \text{if } m = 2. \end{cases} \end{aligned}$$

Since $q \ge 2$, (38.49) implies that

$$\left(k+2-\frac{p}{\beta}\right)^{-\frac{1}{p}} \le 2^{\frac{1}{p}}, \qquad \text{for any } p \ge 2 \text{ and any } m \ge 2.$$

Then the limit of (38.48) yields

$$\|u(t)\|_{L^\infty} \le \limsup_{p\to\infty} \|u(t)\|_{L^p} \le 2C_\epsilon M_1 \left(t^{-\lim \frac{1}{\beta}}\right) \|u_0\|_{L^1}, \tag{38.50}$$

for $t > 0$, where $M_1 = \sup\{(k+1)^{\frac{1}{p}} : k = [\alpha p/(\alpha - 1)], p \ge 2\} > 1$ and $\lim \frac{1}{\beta} = \lim_{p\to\infty} \frac{1}{\beta} = \frac{m}{2}$. We see then that (38.50) implies (38.38). This completes the proof with Dirichlet boundary conditions. Note that then constant C in (38.38) of Theorem 38.9 depends only on Ω, m, and ν. This can be seen from (38.40), (38.47), (38.48), and (38.50).

Step 5. Let us now turn to the heat equation with Neumann boundary conditions. By examining the proof through the above four steps, we can see that the Dirichlet boundary condition is crucial in only one place, and this is where the Poincaré inequality is used. While the Poincaré inequality does not generally hold for the case of the Neumann boundary conditions, it is valid when the initial condition satisfies $u_0 \in L^1_0(\Omega)$ in (38.39), see Temam (1988, p. 49). □

Based on this important theorem, we can directly derive other regularity properties for related linear semigroups. While we formulate the following result for the Dirichlet boundary conditions, there is an appropriate version for the Neumann boundary conditions.

Theorem 38.10. *Consider equation (38.37) on $\Omega \subset \mathbb{R}^m$, where $m = \dim \Omega \ge 1$. Assume that the Dirichlet boundary conditions hold, and let p and q satisfy $1 \le p < \infty$, $1 < q \le \infty$. Then the analytic semigroup $T(t)$ has the following regularity properties:*

(1) *$T(t) : L^p(\Omega) \to L^q(\Omega)$, for all $t > 0$, and there is a constant $C_1 > 0$ such that*

$$\|T(t)u_0\|_{L^q} \le C_1\, t^{-\frac{m}{2}\left(\frac{1}{p}-\frac{1}{q}\right)} \|u_0\|_{L^p}, \qquad \textit{for any } t > 0 \textit{ and } u_0 \in L^p(\Omega). \tag{38.51}$$

(2) *$T(t) : W^{r,p}(\Omega) \to W^{s,q}(\Omega)$, for all $t > 0$, and there is a constant $C_2 > 0$ such that*

$$\|T(t)u_0\|_{W^{s,q}} \le C_2 t^{-\frac{m}{2}\left(\frac{1}{p}-\frac{1}{q}\right)+\frac{1}{2}(r-s)} \|u_0\|_{W^{r,p}}, \tag{38.52}$$

for any $t > 0$ and any $u_0 \in W^{r,p}(\Omega)$.

Proof. (Sketch) If $p = 1$, then Item (1) can be shown directly from Theorem 38.9 by using the interpolation inequality

$$\|u\|_{L^q} \le \left(\|u\|_{L^1}\right)^{\frac{1}{q}} \left(\|u\|_{L^\infty}\right)^{1-\frac{1}{q}}.$$

If $p > 1$, then the rigorous proof can be made by revising the proof of Theorem 38.9, starting from (38.44). We leave this as an exercise.

As for Item (2), the mapping $T(t) : W^{r,p}(\Omega) \to W^{s,q}(\Omega)$ can be decomposed into

$$T(t) : W^{r,p}(\Omega) \to W^{s,p}(\Omega) \to W^{s,q}(\Omega), \qquad \text{for } t > 0,$$

where the first mapping contributes the factor $t^{\frac{1}{2}(r-s)}$, which can be shown by utilizing Theorems 37.5 and 38.2, while the second mapping contributes the factor $t^{-\frac{m}{2}(\frac{1}{p}-\frac{1}{q})}$, which is a corollary of Item (1). □

It should be noted that Theorem 38.10 can be generalized to other uniformly elliptic differential operators of second order.

3.9. Perturbation Theory.

In this section we will examine some basic properties of perturbations of a sectorial operator. More specifically, let A be a given sectorial operator on a Banach space W. We then seek sufficient conditions on another linear operator B in order that the perturbed operator $L \stackrel{\text{def}}{=} A + B$ be a sectorial operator on W, as well. In this section we focus on autonomous perturbations. The nonautonomous theory is treated in Chapter 4.

Throughout this section we will let A be a positive, sectorial operator on a Banach space W, and we will let V^α, for $\alpha \in \mathbb{R}$, denote the family of interpolation spaces generated by the fractional powers of A, where $\mathcal{D}(A^\alpha) = V^{2\alpha}$, for $\alpha \geq 0$. We will use the characteriztions of A given in Theorem 36.2. Note that since A is positive one has $\|e^{-At}\| \leq Me^{-at}$, for all $t \geq 0$, where $M \geq 1$ and $a > 0$. Thus inequalities (36.6), (36.7), and (36.8) are valid for this choice of a.

Note that if $\lambda \in \rho(A)$, then the resolvent $R(\lambda, A) = (\lambda I - A)^{-1}$ is bounded and satisfies (36.7). Since

$$AR(\lambda, A) = -(\lambda I - A)R(\lambda, A) + \lambda R(\lambda, A) = -I + \lambda R(\lambda, A),$$

we see that $AR(\lambda, A)$ is bounded and

$$\|AR(\lambda, A)\| \leq 1 + \frac{M_3|\lambda|}{|\lambda - a|},$$

by inequality (36.7). This implies that there is a constant $C > 0$ and $\sigma \in (0, \frac{\pi}{2})$, such that

$$\|AR(\lambda, A)\| \leq C, \qquad \text{for all } \lambda \in \Sigma_\sigma(-a) \text{ with } |\lambda - a| \geq 1. \tag{39.1}$$

Lemma 39.1. *Let B be a linear operator on W with domain satisfying $\mathcal{D}(A) \subset \mathcal{D}(B)$, where A is given above. Assume that there are constants $\mu \geq 0$ and $K \geq 0$ such that*

$$\|Bu\| \leq \mu\|Au\| + K\|u\|, \qquad \textit{for all } u \in \mathcal{D}(A).$$

If one has $\mu C < 1$, where C is given by (39.1), then $L = A+B$ is a sectorial operator on W.

Proof. Let $\lambda \in \Sigma_\sigma(-a)$ satisfy $|\lambda - a| \geq 1$. It follows from (36.7) and (39.1) that for all $w \in W$, one has

$$||BR(\lambda, A)w|| \leq \mu||AR(\lambda, A)w|| + K||R(\lambda, A)w|| \leq \left(\mu C + \frac{KM_3}{|\lambda - a|}\right)||w||.$$

Consequently, if $1 > \mu C + \frac{KM_3}{|\lambda-a|}$, then one obtains

$$\begin{aligned} ||R(\lambda, L)|| &= ||R(\lambda, A)(I - BR(\lambda, A))^{-1}|| \\ &\leq \frac{M_3}{|\lambda - a|}\left(1 - \mu C - \frac{KM_3}{|\lambda - a|}\right)^{-1}. \end{aligned}$$

Let $n \geq 2$ be chosen so that $1 - \mu C > \frac{1}{n}$, and let $\lambda \in \Sigma_\sigma(-a)$ satisfy $|\lambda - a| \geq 2nKM_3$. One then has $(1 - \mu C - \frac{KM_3}{|\lambda-a|})^{-1} \leq 2n$, and therefore $||R(\lambda, L)|| \leq \frac{2nM_3}{|\lambda-a|}$. This implies that L is a sectorial operator (see Exercise 36.1). □

Lemma 39.2. *Let $B \in \mathcal{L}(V^{2\beta}, W)$, where β satisfies $0 \leq \beta < 1$. Then $L = A+B$ is a sectorial operator on W. In particular, $A+cA^\beta$ is a sectorial operator on W, where $c \in \mathbb{R}$.*

Proof. First we note that $\mathcal{D}(A) = V^2 \subset V^{2\beta} = \mathcal{D}(B)$. Next observe that $B \in \mathcal{L}(V^{2\beta}, W)$ if and only if $BA^{-\beta} \in \mathcal{L}(W, W)$. Therefore there exist positive constants C_1 and C_2 such that for every $u \in \mathcal{D}(A)$, one has

$$||Bu|| = ||BA^{-\beta}A^\beta u|| \leq C_1||A^\beta u|| \leq C_2||Au||^\beta||u||^{1-\beta},$$

see inequality (37.16). The Young inequality then implies that for every (sufficiently small) $\mu > 0$ there is a $K = K_\mu > 0$ such that $||Bu|| \leq \mu||Au|| + K||u||$, for all $u \in \mathcal{D}(A)$. The result now follows from Lemma 39.1. □

In Theorem 44.3 we will show that the analytic semigroup e^{-Lt} satisfies the identity

$$e^{-Lt}v_0 = e^{-At}v_0 - \int_0^t e^{-A(t-s)}Be^{-Ls}v_0\, ds, \qquad \text{for } v_0 \in W,$$

and that the domain $\mathcal{D}(L)$ satisfies

$$\mathcal{D}(L) \cap V^{2\beta} \subset \mathcal{D}(A) \subset \mathcal{D}(L).$$

We will also show that, if in addition, the sectorial operator A has compact resolvent, then the analytic semigroup e^{-Lt} is compact, for $t > 0$.

The next result is a direct application of the last lemma, where the space W is replaced by $V^{2\alpha}$. The argument uses the fact that the given operator A is a sectorial operator on $V^{2\alpha}$, and the domain of A in $V^{2\alpha}$ is

$$\mathcal{D}_{V^{2\alpha}}(A) = A^{-\alpha}(\mathcal{D}_W(A)) = V^{2\alpha+2},$$

see (37.10). We leave the detailed proof as an exercise.

Lemma 39.3. *Let $B \in \mathcal{L}(V^{2\beta+2\alpha}, V^{2\alpha})$, where β satisfies $0 \le \beta < 1$. Then $L = A + B$ is a sectorial operator on $V^{2\alpha}$.*

3.10. Exercises.

Section 3.1

31.1. Let $(T(t), A)$ be a C_0-semigroup on W. Assume that there is a nonnegative function $k(t)$ with $k(t) \to 0$, as $t \to \infty$, such that $\|T(t)w\| \le k(t)\|w\|$ for all $w \in W$ and $t \ge 0$. Show that there is a constant M, $1 \le M < \infty$, and an $a > 0$ such that $\|T(t)w\| \le Me^{-at}\|w\|$, for all $w \in W$ and $t \ge 0$.

31.2. Let $A \in \mathcal{L}(W)$, and define the exponential e^{At} by

$$e^{At} = \sum_{n=0}^{\infty} \frac{t^n A^n}{n!}, \qquad \text{for } t \in \mathbb{R}.$$

Show that (e^{At}, A) is a C_0-semigroup, where the infinitesimal generator A has domain $\mathcal{D}(A) = W$. (In this case, (e^{At}, A) is a C_0-group.)

31.3. A C_0-semigroup $(T(t), A)$ on W is said to be **uniformly continuous** if

$$\lim_{h \to 0+} \|T(h) - I\|_{\mathcal{L}(W)} = 0. \tag{310.1}$$

(1) Show that if A is a bounded linear operator, then the C_0-semigroup e^{At}, which is constructed in Exercise 31.2, is uniformly continuous.
(2) Prove the converse. That is, show that if a C_0-semigroup $(T(t), A)$ is uniformly continuous, then the infinitesimal generator A is a bounded linear operator on W.
(3) The continuity described in Theorem 31.3 states that for every (t_0, w_0) in $\mathbb{R}^+ \times W$ and every $\epsilon > 0$, there is a $\delta = \delta(\epsilon, t_0, w_0) > 0$ such that $\|T(t)w - T(t_0)w_0\| \le \epsilon$ whenever $|t - t_0| \le \delta$ and $\|w - w_0\| \le \delta$. Show that (310.1) is equivalent to the claim that δ can be chosen independent of w_0 for $\|w_0\| \le 1$.

31.4. Let $(T(t), A)$ be a C_0-semigroup on W. Show that $\mathcal{D}(A^\infty) \stackrel{\text{def}}{=} \bigcap_{n \ge 1} \mathcal{D}(A^n)$ is a dense linear subspace in W. Hint: Consider all the elements

$$y = \int_0^\infty \varphi(t) T(t) w \, dt, \qquad \text{for } w \in W \text{ and } \varphi \in C_0^\infty(\mathbb{R}^+).$$

31.5. Let $(T(t), A)$ be a C_0-semigroup on W, where W is a reflexive Banach space.

(1) Prove that $(T^*(t), A^*)$ is a C_0-semigroup on the adjoint space W^*.
(2) Determine whether this remains true in a nonreflexive Banach space.

31.6. A C_0-semigroup $(T(t), A)$ on a Banach space W is said to be **exponentially stable**, if there exist constants $a > 0$ and $M \geq 1$, such that

$$||T(t)|| \leq Me^{-at}, \qquad t \geq 0.$$

It is called an L^p - **stable** semigroup, for $1 \leq p < \infty$, if for any $w \in W$, one has

$$\int_0^\infty ||T(t)w||^p dt < \infty.$$

Show that $(T(t), A)$ is exponentially stable if and only if it is L^p-stable, for some p, with $1 \leq p < \infty$, see Pazy (1983).

31.7. Let $(T(t), A)$ be a C_0-semigroup on a Banach space W. Assume that there is an a with $0 \leq a < 1$, such that for some $t > 0$ one has $||(T(t) - I)x|| \leq a||x||$, for all $x \in W$. Show that the infinitesimal generator A is in $\mathcal{L}(W, W)$ and that $T(t)$ is a C_0-group.

31.8. Let $L \in \mathcal{L}(W, W)$ with $||L|| < 1$ and set $E = I + L$. The object of this exercise is to show that there is a C_0-semigroup (e^{At}, A) on W such that $e^{At} = E$ at $t = 1$. The first step will involve the construction of a discrete semigroup $T(t)$, defined on the dyadic rational numbers, $t = k2^{-m}$ where k and m are nonnegative integers, such that $T(0) = I$ and $T(1) = E$.

(1) Show that there is a unique $B \in \mathcal{L}(W, W)$ such that $||B|| < 1$ and $(I + B)^2 = I + L$. (Define $T(k2^{-1}) \stackrel{\text{def}}{=} (I + B)^k$, for $k = 0, 1, 2, \cdots$. Thus $T(0) = I$ and $T(1) = E$.)
(2) Use induction with Item (1) to extend the definition of $T(t)$ to a semigroup defined on the dyadic rational numbers, $t = k2^{-m}$.
(3) Show that the mapping $t \to T(t)$ is uniformly continuous (strongly in W) on the set of dyadic rational numbers t with $0 \leq t \leq 1$.
(4) Show that there exists a unique continuous extension of $T(t)$ to $t \in [0, \infty)$, and that this extension, which we denote by e^{At}, is a C_0-semigroup.
(5) Show that the infinitesimal generator A of e^{At} satisfies $A \in \mathcal{L}(W, W)$ and that e^{At} is a C_0 group on W.

Section 3.4

34.1. Show that the expression $|w|$ in Lemma 34.3 can be written in the form

$$|w| = \lim_{\lambda \to \infty} \left(\sup_{n \geq 0} ||\lambda^n R(\lambda, A)^n w|| \right),$$

and that it is a norm on W.

34.2. Let $(T(t), A)$ be a C_0-semigroup on W that satisfies $||T(t)|| \leq Me^{at}$, and let A_λ be the Yosida approximant for $\lambda > a$. Show that

$$T(t)w = \lim_{\lambda \to \infty} e^{tA_\lambda} w, \qquad \text{for } w \in W.$$

34.3. Prove the necessity part of the Lumer-Phillips Theorem 34.5.

34.4. Let $(T(t), A)$ be a C_0-semigroup. Prove that

$$e^{t\sigma(A)} \stackrel{\text{def}}{=} \{e^{t\lambda} : \lambda \in \sigma(A)\} \subset \sigma(T(t)), \qquad \text{for } t \geq 0,$$

where $\sigma(A)$ and $\sigma(T(t))$ are the spectrum set of A and of $T(t)$ respectively. (Note that $e^{t\sigma(A)}$ and $\sigma(T(t))$ are not equal in general, as is demonstrated in the next exercise.)

34.5. Let H_n be an n-dimensional real Hilbert space with an orthonormal basis $\{e_{nj}$, where $j = 1, \cdots, n\}$. Let $A_n \in \mathcal{L}(H_n)$ be defined by

$$A_n = \lambda_n I + N_n = \begin{pmatrix} \lambda_n & 1 & 0 & \cdots & 0 \\ 0 & \lambda_n & 1 & \cdots & 0 \\ & & & \ddots & \\ 0 & 0 & 0 & \cdots & \lambda_n \end{pmatrix}, \qquad \text{with } \lambda_n = -\frac{1}{2} + i\omega_n,$$

and $\omega_n \to \infty$ as $n \to \infty$. Let $H = \oplus \sum_{n=1}^{\infty} H_n$ denote the Hilbert space of all $h = (h_1, h_2, \ldots)$ such that $h_n \in H_n$ and $\sum_{n=1}^{\infty} |h_n|^2 < \infty$, with the inner product

$$\langle g, h \rangle_H \stackrel{\text{def}}{=} \sum_{n=1}^{\infty} \langle g_n, h_n \rangle_{H_n}.$$

Define a linear operator A on $\mathcal{D}(A) = \{h \in H : \sum_{n=1}^{\infty} |A_n h_n|^2 < \infty\}$ by $Ah = (A_1 h_1, A_2 h_2, \ldots)$, see Huang (1985).

(1) Show that $\sigma(A) = \{\lambda_n : n = 1, 2, \cdots\}$, and $\sup_{\lambda \in \sigma(A)} \text{Re } \lambda = -\frac{1}{2} < 0$.

(2) Show that A has compact resolvent by approximating A^{-1} with linear operators of finite rank.

(3) Show that A generates a C_0-semigroup $T(t)$, $t \geq 0$, by using the Lumer-Phillips Theorem.

(4) Prove that $T(t)h = (e^{A_1 t} h_1, e^{A_2 t} h_2, \ldots)$, for $t \geq 0$, and that

$$\omega_0 = \inf_{t>0} \frac{\ln \|T(t)\|}{t} = \frac{1}{2}.$$

(5) Show that $\sigma(T(t)) \neq e^{t\sigma(A)}$. (Hint: $T(t)$ is not exponentially stable.)

34.6. Let (T(t), A) be a C_0-semigroup on a Banach space W. Let

$$B_\lambda(t)w = \int_0^t e^{\lambda(t-s)} T(s) w \, ds, \qquad \text{for } t \geq 0 \text{ and } \lambda \in C.$$

Prove that $(\lambda I - A) B_\lambda(t) w = e^{\lambda t} w - T(t) w$, for $w \in W$.

Section 3.5

35.1. Complete the proof of Theorem 35.1 on differentiability.

35.2. Show that the C_0-semigroup e^{-At} constructed in Section 3.2 is differentiable for $t > 0$. (Also see Theorem 36.2.)

Section 3.6

36.1. Let A be a densely defined, closed linear operator on a Banach space W and assume that the resolvent operator $R(\lambda, A)$ satisfies

$$||R(\lambda, A)|| \leq \frac{M}{|\lambda - a|}, \qquad \text{for all } \lambda \in \Sigma_\sigma(-a) \text{ with } |\lambda - a| \geq \rho,$$

for some $\sigma \in (0, \frac{\pi}{2})$ and some $\rho > 0$. Show that A is a sectorial operator.

36.2. Show that the Lax-Milgram Lemma 36.5 extends to the setting where H is a Hilbert space, V is a Banach space, $V \mapsto H$ with V dense in H, and where the mapping $(u, v) \to a(u, v)$ is a coercive, V-bounded sesquilinear functional on V. (Hint: Show that $s(u, v) = \frac{1}{2}(a(u, v) + \overline{a(v, u)})$ is an inner product on V and the norm $s(u, u)^{\frac{1}{2}}$ is equivalent to the V-norm on V.)

36.3. Prove Items (2) - (4) of Lemma 36.5. Show that the form operator $B : \mathcal{D}(B) = V \to V'$ is a densely defined, closed, linear operator.

36.4. Prove Theorem 36.6 for the form operator B_H.

36.5. Let A be a sectorial operator in a Banach space W. Prove that for each $\alpha \in [0, 1]$, there exists a constant $C_\alpha > 0$ such that

$$||A^\alpha(\mu I + A)^{-1}|| \leq C_\alpha \mu^{\alpha - 1}, \qquad \text{for all } \mu > 0.$$

36.6. Let $A = B_H$ be the form operator in H determined by Theorem 36.6.

(1) Show that

$$||R(\lambda, A)||_{\mathcal{L}(H,V)} \leq \left(\frac{1 + M\delta}{\delta|\lambda|}\right)^{\frac{1}{2}}, \qquad \text{for } \lambda < 0.$$

(2) Show that $||e^{-At}w||_V \leq Ct^{-\frac{1}{2}}||w||_H$, for $t > 0$ and $w \in H$, where e^{-At} is the analytic semigroup generated by $-A$, and C is a constant.

36.7. Show that the form operators B and B_H constructed in Section 3.6 are positive, sectorial operators.

Section 3.7

37.1. Prove Lemma 37.3. Hint: Use the following properties of the Gamma and Beta functions. For $0 < \alpha < 1$, one has

$$B(\alpha,\beta) = \frac{\Gamma(\alpha)\Gamma(\beta)}{\Gamma(\alpha+\beta)} \qquad \text{and} \qquad \Gamma(\alpha)\Gamma(1-\alpha) = \frac{\pi}{\sin(\pi\alpha)}.$$

37.2. Let A be a positive, sectorial operator with compact resolvent on a Banach space W. Show that the analytic semigroup e^{-At} is compact, for $t > 0$. (Compare this with Theorem 32.1.)

37.3. Let A be a maximal accretive operator in a real Hilbert space. Suppose there is a constant $\delta > 0$ such that $\langle Au, u\rangle \geq \delta\|u\|^2$ for all $u \in \mathcal{D}(A)$. Show that, for $0 < \alpha < 1$, one has

$$\langle A^\alpha u, u\rangle \geq \delta^\alpha \|u\|^2, \qquad \text{for all } u \in \mathcal{D}(A^\alpha).$$

37.4. Let A and B be positive, sectorial operators on a Banach space W with $\mathcal{D}(A) = \mathcal{D}(B)$. Assume that for some $\alpha \in [0,1)$ the linear operator $(A-B)A^{-\alpha}$ is bounded on W. Show that for every $\beta \in [0,1]$ the linear operators $A^\beta B^{-\beta}$ and $B^\beta A^{-\beta}$ are bounded on W.

37.5. Let A be a sectorial operator on a Banach space W, and define the interval J as the set $a \in \mathbb{R}$ such that $A_a = A + aI$ is a positive operator. Let A_a^α denote the fractional powers of A_a. (i) Show that, for each $\alpha \geq 0$, the domains $\mathcal{D}(A_a^\alpha)$ do not depend on $a \in J$. (ii) Show that for a_1, $a_2 \in J$, the two graph norms $\|A_{a_1}^\alpha u\|$ and $\|A_{a_2}^\alpha u\|$ are equivalent norms on $\mathcal{D}(A_a^\alpha)$. (Hint: Use Exercise 37.4.)

37.6. Let A be a sectorial operator on a space of functions defined on an open, bounded region $\Omega \subset \mathbb{R}^n$, where the boundary $\partial\Omega$ is of class C^∞. Assume that there is a sequence of integers $k_n \to \infty$ with the property that $\mathcal{D}(A^n) \subset C^{k_n}(\Omega)$. Show that every eigenvector ϕ of A corresponding to a nonzero eigenvalue λ satisfies $\phi \in C^\infty(\Omega)$.

37.7. Verify that (37.8) is well-defined and that the integral in it converges in $\mathcal{L}(W)$.

37.8. Prove Items (1) - (4) of Lemma 37.4.

37.9. Let A be a positive, selfadjoint linear operator on a Hilbert space and assume that A has compact resolvent, see Section 3.2 and Example 37.1. Let the eigenvalues satisfy (32.2) and let the fractional powers A^α be given by (37.1). For each integer $n \geq 1$, let P denote the orthogonal projection onto $PH \stackrel{\text{def}}{=} \mathrm{Span}(e_1, \cdots, e_n)$. and set $Q = I - P$.

(1) Show that e^{-AQt} is an analytic semigroup on H with infinitesimal generator AQ.
(2) Show that for every $\beta \geq 0$ and $t > 0$, one has

$$A^\beta e^{-AQt} w = e^{-AQt} A^\beta w = A^\beta Q e^{-AQt} w = e^{-AQt} A^\beta Q w,$$

for all $w \in H$. Show that the equality above holds at $t = 0$ when $w \in V^{2\beta}$.

(3) For $t > 0$ and $\beta \geq 0$, define $b(t) \stackrel{\text{def}}{=} \|A^\beta e^{-AQt}\|_{\mathcal{L}(H)}$. Show that $b(t)$ satisfies

$$b(t) = \begin{cases} \beta^\beta e^{-\beta} t^{-\beta}, & 0 < t \leq a, \\ \lambda_{n+1}^\beta e^{-\lambda_{n+1} t}, & a < t, \end{cases} \tag{310.2}$$

where $a = \beta \lambda_{n+1}^{-1}$.

(4) Show that for $0 < \nu < \lambda_{n+1}$, $0 \leq \beta < 1$, and $\alpha = 1 - \beta$, one has

$$\begin{aligned} \int_0^\infty b(t)\, dt &= \alpha^{-1} e^{-\beta} \lambda_{n+1}^{-\alpha} \\ \int_0^\infty b(t) e^{\nu t}\, dt &\leq \beta \alpha^{-1} \lambda_{n+1}^{-\alpha} + \lambda_{n+1}^\beta (\lambda_{n+1} - \nu)^{-1}. \end{aligned} \tag{310.3}$$

Section 3.8

38.1. Let $W = L^2(\mathbb{R})$ and define $Af = -f'$, where $\mathcal{D}(A) = \{f \in W : f' \in W\}$. Prove that A generates a C_0-nonexpansive semigroup $T(t)$, for $t \geq 0$. Find an explicit form of $T(t)$.

38.2. The Maximum Principle. Let $Au = -\Delta u$ be given as in Section 3.8.2, where $\mathcal{D}(A) = H^2(\Omega, \mathbb{R}) \cap H_0^1(\Omega, \mathbb{R})$, Ω is an open, bounded domain in $\mathbb{R}^m$ with a Lipschitz continuous boundary and u is a real-valued function.

(1) Show that for every $u_0 \in L^\infty = L^\infty(\Omega, \mathbb{R})$, one has

$$\lim_{t \to \infty} \|e^{-At} u_0\|_{L^\infty} = 0.$$

Hint: Show that if $u(t,x) = (e^{-At}u_0)(x)$ has a maximum in x at $x = x_0$ and $t = t_0$, then $\partial_t u(t, x_0) < 0$, when $u_0 \neq 0$.

(2) Show that Part (1) can be extended to the system of equations

$$\partial_t u_i - \Delta u_i, \qquad \text{for } 1 \leq i \leq k.$$

38.3. The next four problems involve the heat equation

$$\partial_t u - \nu \Delta u = 0, \qquad \text{on } \Omega \subset \mathbb{R}^m, \tag{310.4}$$

with Dirichlet boundary conditions on an open, bounded domain Ω with boundary of Lipschitz class. Let $A_1 u = -\nu \Delta u$ on a suitable domain $\mathcal{D}(A_1)$ in $L^1(\Omega)$. Show that $-A_1$ generates an analytic semigroup $T(t)$ on $L^1 = L^1(\Omega)$ and that

$$\|T(t)u_0\|_{L^1} \leq \|u_0\|_{L^1}, \qquad \text{for all } u_0 \in L^1 \text{ and all } t \geq 0.$$

38.4. Extend Theorem 38.9 by showing that inequality (38.38) is valid with the Neumann boundary condition $\frac{\partial u}{\partial n} = 0$ on $\partial\Omega$, provided that $u_0 \in L^1_0(\Omega)$, i.e., $\int_\Omega u_0(x)\,dx = 0$.

38.5. We return to Exercise 38.3 with the Dirichlet boundary conditions, where $m \geq 2$. Let p and q satisfy $1 \leq p < \infty$, $1 < q \leq \infty$. Verify the following:

(1) $T(t) : L^p(\Omega) \to L^q(\Omega)$, for all $t > 0$, and there is a constant $C_1 = C_1(m,p,q) > 0$, such that

$$\|T(t)u_0\|_{L^q} \leq C_1 t^{-\frac{m}{2}(\frac{1}{p}-\frac{1}{q})}\|u_0\|_{L^p}, \qquad \text{for all } u_0 \in L^p(\Omega) \text{ and all } t > 0.$$

(2) $T(t) : W^{r,p}(\Omega) \to W^{s,q}(\Omega)$, for all $t > 0$, and there is a constant $C_2 = C_2(m,p,q,r,s) > 0$, such that

$$\|T(t)u_0\|_{W^{s,q}} \leq C_2 t^{-\frac{m}{2}(\frac{1}{p}-\frac{1}{q})+\frac{1}{2}(r-s)'}\|u_0\|_{W^{r,p}},$$

for all $u_0 \in W^{r,p}(\Omega)$ and all $t > 0$.

38.6. Show that the operators ∂_t and $A^{\frac{1}{2}}$ occurring in the proof of equation (38.32) commute on the domain of $A^{\frac{1}{2}}$.

38.7. Complete the proof of Lemma 38.7, Item (5).

38.8. Extend results Exercises 38.4 and 38.5 to the case where the Laplacian $\nu\Delta$ is replaced by a strongly elliptic linear differential operator of second order.

38.9. (Hybrid System) An elastic Euler-Bernoulli beam clamped at one end and linked to a movable rigid body at the other end can be modeled as follows:

$$\begin{aligned}
\partial_t^2 u(x,t) + \partial_x^4 u(x,t) &= 0, \qquad t \geq 0, 0 < x < 1,\\
u(0,t) = \partial_x u(0,t) &= 0, \qquad t \geq 0,\\
\partial_t^2 u(1,t) - \partial_x^3 u(1,t) &= 0, \qquad t \geq 0,\\
\partial_t^2 \partial_x u(1,t) + \partial_x^2 u(1,t) &= 0, \qquad t \geq 0,
\end{aligned}$$

where the last two equations stand for the force balance and the bending moment balance at the endpoint $x = 1$. This system consists of a PDE coupled with two ODEs.

(1) Let $y(t) = \begin{bmatrix} u(\cdot,t) \\ u(1,t) \\ \partial_x u(1,t) \end{bmatrix}$, for $t \geq 0$. Let $H = L^2(0,1) \times \mathbb{R} \times \mathbb{R}$. Formulate this system as a second-order linear evolutionary equation of the form

$$\partial_t^2 y + Ay(t) = 0, \qquad t \geq 0,$$

by specifing the explicit form and domain of the linear operator $A : \mathfrak{D}(A) \to H$.

(2) Show that A is a densely defined and closed operator. Also show that it is self-adjoint and there is $\delta > 0$ such that $\langle A\varphi, \varphi\rangle_H \geq \delta\|\varphi\|_H^2$, for all $\varphi \in \mathcal{D}(A)$.
(3) Show that A has compact resolvent.
(4) Show that $\sigma(A) = \sigma_p(A) = \{\mu_n^4 : n = 1, 2, \cdots\}$, where $\mu_n > 0$, for $n = 1, 2, \cdots$, are the increasing positive roots of the transcendental equation

$$(1+\mu^4) + (1-\mu^4)\,\text{ch}\;\mu\cos\mu - (\mu^3 - \mu)\,\text{sh}\;\mu\cos\mu - (\mu^3+\mu)\,\text{ch}\;\mu\sin\mu = 0.$$

(5) Let $w(t) = \begin{bmatrix} y(t) \\ \partial_t y(t) \end{bmatrix}$ where $\partial_t y$ stands for the strong derivative in H with respect to time variable. Let $E = \mathcal{D}(A^{\frac{1}{2}}) \times H$. Reformulate this system as a first-order linear evolutionary equation of the form $\partial_t w = Gw$ for $t > 0$, where $G : \mathcal{D}(G) \to E$ generates a C_0-unitary group.

38.10. A linear Boussinesq equation is given by

$$\partial_t^2 u + \Delta^2 u + \beta\Delta u = 0, \qquad t \geq 0, \quad x \in \Omega \subset \mathbb{R}^n,$$
$$u = \frac{\partial u}{\partial \nu} = 0, \qquad t \geq 0, \quad \text{on } \Gamma = \partial\Omega,$$

where Ω is an open, bounded domain with boundary Γ of Lipschitz class and $\beta \in \mathbb{R}$.

(1) Reformulate this problem as a second-order evolutionary equation.
(2) Next reformulate it as a first-order evolutionary equation of the form $\partial_t w = Gw$, for $t \geq 0$, where the operator G generates a C_0-group in a suitable Hilbert space.

38.11. Consider the Maxwell equations

$$\epsilon\,\partial_t E = \text{curl } H, \qquad \text{div } E = 0,$$
$$-\mu\,\partial_t H = \text{curl } E, \qquad \text{div } H = 0,$$

where $\Omega \subset \mathbb{R}^3$ is an open bounded domain having piecewise smooth boundary Γ, $\epsilon > 0$ and $\mu > 0$ are electrical and magnetic permitivity constants. Let $H|_\Gamma = H_\nu^{\Gamma} + H_\tau^{\Gamma}$ be the orthogonal decomposition, with H_ν^{Γ} and H_τ^{Γ} the normal component and the tangential component to the boundary Γ. Let $W = L^2(\Omega)^6$ be the six-dimensional vector Hilbert space for the electromagnetic field, and

$$W_\alpha = \left\{ \begin{pmatrix} \varphi \\ \psi \end{pmatrix} \in W : \text{div } \varphi = 0 \,\text{and div } \psi = 0 \right\}.$$

Prove that the following operator $A : \mathcal{D}(A) \to W_\alpha$, where

$$\mathcal{D}(A) = \left\{ \begin{pmatrix} e \\ h \end{pmatrix} \in W_\alpha \cap H^1(\Omega)^6 : h_\tau^\Gamma = 0 \,\text{and } e|_\Gamma = 0 \right\},$$

and

$$A\begin{pmatrix} e \\ h \end{pmatrix} = \begin{pmatrix} \frac{1}{\epsilon}\,\mathrm{curl}\; h \\ -\frac{1}{\mu}\,\mathrm{curl}\; e \end{pmatrix},$$

is skew-symmetric and generates a C_0-group of isometries in W_α.

38.12. Consider a gaseous ignition model given by

$$\partial_t \theta = \Delta\theta + \frac{1}{C}\int_\Omega \Delta\theta\, dx, \qquad (x,t) \in \Omega \times \mathbb{R}^+,$$

where the boundary conditions are $\theta(x,t) = 0$, for $(x,t) \in \partial\Omega \times \mathbb{R}^+$, Ω is an open bounded domain in $\mathbb{R}^3$ of class C^1, and $C = C(\Omega) > 0$ is a constant. Let $H = L^2(\Omega)$. Prove that the operator A, defined by

$$Af = -(\Delta f + \frac{1}{C}\int_\Omega \Delta f dx), \qquad \text{for } f \in \mathcal{D}(A) = H^2(\Omega) \cap H_0^1(\Omega),$$

is sectorial on the space H.

38.13. Consider the linear delayed differential equation

$$\begin{aligned} &\partial_t x(t) = Ax(t) + Lx_t, \qquad t > 0, \\ &x(0) = x_0, \quad x(\theta) = \varphi(\theta), \quad \theta \in [-r, 0), \end{aligned}$$

where $x(t)$ and $x_0 \in H$ (a real Hilbert space), and $x_t(\theta) = x(t+\theta)$, for $\theta \in [-r, 0]$. Assume that $A : \mathcal{D}(A) \to H$ is the infinitesimal generator of a C_0-semigroup e^{At} on H, and

$$Lx_t = \sum_{i=0}^{k} A_i x(t - r_i) + \int_{-r}^{0} B(\theta)x(t+\theta)d\theta,$$

with $0 = r_0 < r_1 < \cdots < r_k = r < \infty$, $A_i \in \mathcal{L}(H), i = 0, 1, \cdots, k$, and $B(\cdot) \in L^\infty([-r, 0]; \mathcal{L}(H))$. Let $(x_0, \varphi) \in M^2 \overset{\text{def}}{=} H \times L^2(-r, 0; H)$.

(1) Prove that $S(t) : (x_0, \varphi) \mapsto (x(t), x_t), t \geq 0$, is a C_0-semigroup on the Hilbert space M^2, where $x(t)$ is the mild solution given by

$$x(t) = \begin{cases} T(t)x_0 + \int_0^t T(t-s)Lx_s ds, & t \geq 0, \\ \varphi(t), & -r \leq t < 0. \end{cases}$$

(2) Prove that the infinitesimal generator $\mathcal{A}$ of this semigroup $S(t)$ on M^2 is given by

$$\mathcal{D}(\mathcal{A}) = \hat{W}^2, \text{ and } \mathcal{A}\begin{pmatrix} h(0) \\ h(\cdot) \end{pmatrix} = \begin{pmatrix} Ah(0) + Lh \\ \partial_\theta h \end{pmatrix},$$

where $\hat{W}^2$ is the set of $\begin{pmatrix} f^0 \\ f^1 \end{pmatrix} \in M^2$ with the properties $f^0 = f^1(0)$, $f^1 \in AC([-r, 0); H)$, $\partial_t f^1 \in L^2(-r, 0; H)$, and $f^0 \in \mathcal{D}(A)$. Here $AC([-r, 0]; H)$ is the collection of all strongly absolutely continuous H-valued functions, and $\partial_t f^1$ denotes the strong derivative in H.

38.14. By using the Spectral Mapping Formula (32.5), show directly that the eigenvalues and eigenvectors in the proof of Lemma 38.7, Item (5), are the eigenvalues and eigenvectors of the C_0-group described in (38.26).

38.15. Complete the proof of Corollary 38.8 by showing that $e^{-G_\alpha t}$ is differentiable, for $t > 0$.

38.16. Show that the C_0-semigroup constructed in Corollary 38.8 is a C_0-group. Is this C_0 group differentiable, for all $t \in \mathbb{R}$?

38.17. Prove that each α_k and β_k, which are defined in the proof of Theorem 38.2, is in $L^2(\Omega)$.

3.11. Commentary.

Section 3.1. The strong continuity of the C_0-semigroup $(T(t), A)$, as described in (31.3), Lemma 31.2, and Theorem 31.3, is the appropriate level of continuity for studying solutions of linear partial differential equations. As shown in Section 3.8, many classes of linear partial differential equations generate semigroups with this property. If one had chosen, instead, to develop a theory based on a statement of uniform continuity, say,

$$\lim_{h \to 0+} \|T(h) - I\|_{\mathcal{L}} = 0,$$

then the generator A is necessarily a bounded linear operator, see Exercise 31.3.

Beyond the basic properties of C_0-semigroups and infinitesimal generators, there are two important classes of semigroups that have been studied in depth. One is the nonexpansive[11] semigroups with the property $\|T(t)\|_{\mathcal{L}} \leq 1$ for all $t \geq 0$. The subspace decomposition reducing the nonexpansive semigroups, and the related results in functional analysis, can be found in the classical book Sz.-Nagy and Foias (1967); also see Fuhrmann (1984).

In general, a C_0-semigroup $(T(t), A)$ on a Banach space W is said to be **exponentially stable**, if there exist constant $a > 0$ and $M \geq 1$, such that

$$\|T(t)\|_{\mathcal{L}} \leq Me^{-at}, \qquad t \geq 0.$$

It is called L^p-**stable** semigroup, $1 \leq p < \infty$, if for any $w \in W$,

$$\int_0^\infty \|T(t)w\|^p dt < \infty.$$

It can be shown that $(T(t), A)$ is exponentially stable if and only if it is L^p-stable, for some $1 \leq p < \infty$, see Pazy (1983). The class of exponentially

[11]Some authors have referred to these semigroups as *contracting* semigroups. Since one can have $\|T(t)\|_{\mathcal{L}} \equiv 1$, for all $t \geq 0$, the label *contracting* seems inappropriate.

stable semigroups is closely related to the stabilization in control theory. In the infinite dimensional case, a C_0-semigroup with the property that $\lim_{t\to\infty} \|T(t)w\| = 0$, for each $w \in W$, need not be exponentially stable.

In the Exercise 34.5, it is shown that, for infinite dimensional linear systems, one can have $\sigma(T(t)) \neq \exp(t\sigma(A))$. For an exponentially stable C_0-semigroup define

$$\sigma_0(A) \stackrel{\text{def}}{=} \sup_{\lambda\in\sigma(A)} \text{Re } \lambda \qquad \text{and} \qquad \omega_0(A) \stackrel{\text{def}}{=} \inf_{t>0} \frac{\ln \|T(t)\|}{t} = \lim_{t\to\infty} \frac{\ln \|T(t)\|}{t}.$$

Then $\sigma_0(A) \leq \omega_0(A)$ is always valid. One says that A has the **spectral determined growth** property if $\sigma_0(A) = \omega_0(A)$. It is only in this case that the exponential growth rate of $T(t)$ is determined by $\sigma_0(A)$. Then the question is which types of C_0-semigroups possess this property? Of course, the spectral determined growth property is valid in the finite dimensional case. Furthermore, we know that any analytic semigroup and any compact semigroup possess this property, see Triggiani (1975). In Hulang (1985), one finds the following necessary and sufficient condition for $\sigma_0(A) = \omega_0(A)$ in a Hilbert space:

$$\sup_{\text{Re } \lambda\geq\sigma} \|R(\lambda, A)\| < \infty, \qquad \text{for any } \sigma > \sigma_0(A).$$

A related evolutionary equation which arises in the study of the Navier-Stokes equations with moving boundaries is the problem

$$\partial_t(Bu(t)) = Au(t),$$

where A and B are appropriate linear operators on certain Banach spaces. A theory of semigroups for such problems can be found in Sauer (1981/82, 1988) and Grobbelar-van Dalsen and Sauer (1989, 1993).

Section 3.2. The illustrative example presented in this section is, in fact, a good system for testing whether a given finite dimensional concept might be generalizable to the infinite dimensional setting.

Section 3.4. Some examples in which the conditions in Hille-Yosida Theorem are directly verified can be found in Pazy (1983), Goldstein (1985), Tanabe (1979), and Reed and Simon (1975). The Lumer-Phillips Theorem admits an extension to Banach spaces, see Pazy (1983, Section 1.4).

Section 3.5. The full characterization of the generators of differentiable semigroups is provided in Pazy (1983).

An important class of C_0-semigroups is the Gevrey class, which arises in many applications such as linear elastic systems and the Navier-Stokes equations. A C_0-semigroup $T(t)$ on a Banach space W is said to be of **Gevrey class** δ, for $t > t_0$, if $T(t)$ is differentiable for $t > t_0$ and for every compact set $K \subset (t_0, \infty)$, and each $\theta > 0$, there is a constant $C = C(K, \theta)$ such that $\|T(t)\|_{\mathcal{L}} \leq C\theta^n (n!)^\delta$, for all $t \in K$ and $n = 0, 1, 2, \ldots$. The theory of C_0 semigroups of Gevrey class has been developed, and it parallels the theory of differentiable semigroups, see Taylor (1989).

Section 3.6. The use of the Lax-Milgram Property, as a framework for the study of homogeneous boundary value problems of elliptic differential operators, is developed in Lions and Magenes (1972) and Lions (1969). Theorem 36.6 describes the connection between the theory of weak solutions, based on the triplet framework $\{V, H, a(\cdot,\cdot)\}$ and the theory of mild solutions based on C_0-semigroup.

An important application arises with the fractional order Sobolev spaces

$$H^s(\Omega) = [H^m(\Omega), L^2(\Omega)]_\theta, \qquad \text{with } s = \theta m,$$

where $m > 0$ is integer and $0 < \theta < 1$. Moreover $H_0^s(\Omega)$ is the closure of $C_0^\infty(\Omega)$ in $H^s(\Omega)$, and $H^{-s}(\Omega)$ is the dual space of $H_0^s(\Omega), s > 0$. However there have been several different ways to define the fractional Sobolev spaces $W^{s,p}(\Omega)$. The detailed discussion together with the complicated imbedding properties can be found in Adams (1975) and Lions (1969).

Section 3.7. Another illustration of a singular semiflow arises naturally in the context of a positive, sectorial operator A and the associated analytic semigroup e^{-At} on a Banach space W. Let $V^{2\alpha}$ denote the fractional power spaces generated by A, where $\alpha \in \mathbb{R}$. It is shown in Lemma 37.4 that e^{-At} is an analytic semigroup on $V^{2\alpha}$, for each $\alpha \in \mathbb{R}$, and that the mapping

$$(u,t) \to e^{-At}u : V^{2\alpha} \times [0,\infty) \to V^{2\alpha}$$

is continuous, for every $\alpha \in \mathbb{R}$. Let α and β be real numbers with $\alpha > \beta$. Then one has $V = V^{2\alpha} \mapsto W = V^{2\beta}$, and that $V^{2\alpha}$ is dense in $V^{2\beta}$. Let $S(t)v = e^{-At}v$, for $v \in V$, and $T(t)w = e^{-At}w$, for $w \in W$. One then has $T(t)v = S(t)v$, for all $(v,t) \in V \times (0,\infty)$. Also one has $T(t)w \in V$, for all $(w,t) \in W \times (0,\infty)$, and the mapping

$$(w,t) \to T(t)w : W \times (0,\infty) \to V$$

is continuous. Thus, $T(t)$ is a singular dynamical system on W. A key difference here comes from the Fundamental Theorem on Sectorial Operators 37.5. The limit $\lim_{t\to 0+} \|S(t)v\|_V = \|v\|_V$ exists, while the limit $\lim_{t\to 0+} \|T(t)w\|_V$ does not, when $w \notin V$. Inequality (37.11) gives a measure of the singularity of $T(t)w$ at $t = 0$.

A full theory on the interpolation Sobolev spaces is included in Lions and Magenes (1972). A detailed treatment of fractional powers of operators can be found in Tanabe (1979) and Kato (1961, 1962). Lemma 37.8 is adapted from Friedman (1969) by using an idea of Henry (1981; Sections 1.4, 1.6).

It is shown in Sections 3.7 and 3.8 that linear evolutionary equations, of first-order and second-order in time, are linked with the standard C_0 semigroup theory. One may wonder what happens to the Cauchy problem of higher-order linear evolutionary equations, say,

$$\begin{aligned} &\partial_t^n u(t) = Au(t), && \text{for } t > 0, \\ &u^{(k)}(0) = u_k, && \text{for } k = 0, 1, \cdots, n-1, \end{aligned}$$

in some Banach space W. This problem is well-posed if the following two conditions are satisfied: (i) there is a dense subspace $\mathcal{D}$ of W such that for any $u_0, u_1, \cdots, u_k \in \mathcal{D}$, there exists a unique classical solution in W; (ii) the continuous dependence of the solution $u(t)$ on the initial data uniformly for t in any given bounded interval holds. To certain surprise, it can be shown, see Fattorini (1983), that the above Cauchy problem of higher order evolutionary equations, with $n \geq 3$, is well-posed if and only if $A \in \mathcal{L}(W)$.

Section 3.8. Sections 3.8.1 and 3.8.2 lie at the heart of the linear theory underlying the study of reaction diffusion equations. For further information, see Henry (1981) and Pazy (1983). Additional background material on the Stokes equations can be found in Constantin and Foias (1988), Doering and Gibbons (1995), Galdi (1994), Ladyzhenskaya (1963), and Temam (1977, 1983, 1988). In particular, a comprehensive theory for the Stokes equations with periodic boundary conditions is given in Temam (1983).

There are many extensions of the theory of the wave equation presented here. For example, the same conclusions as those presented in Section 3.8.4 are valid for the linear hyperbolic equation

$$\begin{aligned} &\partial_t^2 u(t,x) + A(x,D)u(t,x) = 0, \qquad t>0, x \in \Omega, \\ &u|_\Gamma = \frac{\partial u}{\partial \nu}|_\Gamma = \cdots = \frac{\partial^{m-1} u}{\partial \nu^{m-1}}|_\Gamma = 0, \qquad t \geq 0, \\ &u|_{t=0} = u_0(x), \partial_t u|_{t=0} = u_1(x), \end{aligned}$$

where $A(x,D)$ is a uniformly elliptic operator of order $2m$, see Section 3.8.2.

As we have seen, the concepts of sectorial operators and analytic semigroups arise in the study of linear parabolic evolutionary equations. In Chen and Russell (1982) it is conjectured that for a linear elastic equation

$$\partial_t^2 u - \partial_x^4 u - \rho\, \partial_x^2 \partial_t u = 0, \qquad \text{for } t > 0 \text{ and } x \in (0,1),$$

with Dirichlet boundary conditions, or the hinged boundary conditions, at both endpoints, the corresponding C_0-semigroup exhibits the features of an analytic semigroup due to such a "sufficiently strong" damping term $\rho\, \partial_x^2 \partial_t u$. This conjecture has been proved to be true by Chen and Triggiani (1989).

Additional Readings

Further information on the semigroup theory of linear operators can be found in Hille and Phillips (1957), as well as Dunford and Schwartz (1958), Engel and Nagel (2000), Friedman (1969), Goldstein (1985), Henry (1981), Kato (1966), Krein (1971), Pazy (1983), Triebel (1978), and Yosida (1980). Pazy's book is often cited and is very useful.

4
BASIC THEORY OF EVOLUTIONARY EQUATIONS

In the last chapter, we presented a theory describing solutions of a linear evolutionary equation $\partial_t u + Au = 0$, where $\partial_t u = \frac{d}{dt}u$, on a Banach space W, in terms of C_0-semigroups. As we have seen, this theory allows one to construct mild solutions of many linear partial differential equations with constant coefficients. Our objective in this chapter is to generalize this theory so that it applies first to the linear inhomogeneous equation

$$\partial_t u + Au = f(t), \tag{40.1}$$

and then the nonlinear evolutionary equation

$$\partial_t u + Au = F(u). \tag{40.2}$$

While one of our ultimate goals is to study the dynamics of the solutions of equation (40.2), there are many valuable general techniques which can be developed by focusing on the simpler problem (40.1). It is important to note that these two equations are related. In particular, a function $u = u(t)$ is a solution of equation (40.2) if and only if it is also a solution of equation (40.1), where $f(t) - F(u(t))$. As we will see, many properties of the solutions of the nonlinear problem (40.2) are derived from similar properties of the corresponding solutions of (40.1) in this way.

In the sequel, we will assume that equations (40.1) and (40.2) have the property that $-A$ is the infinitesimal generator of a C_0-semigroup. More specifically, from time to time we assume that one of the following Standing Hypotheses is satisfied .

Standing Hypothesis A. *Let A be a positive, sectorial operator on a Banach space W with associated analytic semigroup e^{-At}. Let $V^{2\alpha}$ be the family of interpolation spaces generated by the fractional powers of A, where $V^{2\alpha} = \mathcal{D}(A^\alpha)$, for $\alpha \geq 0$. Let $||A^\alpha u|| = ||A^\alpha u||_W = ||u||_{V^{2\alpha}} = ||u||_{2\alpha}$ denote the norm on $V^{2\alpha}$. See Lemma 37.4 for more information.*

Standing Hypothesis B. *The operator A is a positive, selfadjoint, linear operator, with compact resolvent, on a Hilbert space H. Consequently A satisfies the Standing Hypothesis A. Moreover, the fractional power spaces V^α are defined for all $\alpha \in \mathbb{R}$, and equation (37.2) defines the Hilbert space structure on each V^α. Also the semigroup e^{-At} is compact, for $t > 0$. See Theorem 37.2 for more information.*

Under the Standing Hypotheses A or B, the interpolation spaces V^α satisfy the continuous imbedding property $V^{2\beta} \mapsto V^{2\gamma}$, for $\gamma \le \beta$. (If in addition, A has compact resolvent, then the imbedding $V^{2\beta} \hookrightarrow V^{2\gamma}$ is compact when $\gamma < \beta$.) One consequence of the *continuous* imbeddings is a litany of related continuous imbeddings, for example,

$$C[0,T;V^{2\beta}) \mapsto C[0,T;V^{2\gamma}) \quad \text{and} \quad C^{0,\theta}_{\text{loc}}(0,T;V^{2\beta}) \mapsto C^{0,\theta}_{\text{loc}}(0,T;V^{2\gamma}),$$

for $\gamma \le \beta$. (Recall that $V^0 = W$.) One obtains similar imbeddings for $L^p_{\text{loc}}[0,T;Y)$ and $L^p_{\text{loc}}(0,T;Y)$, for $1 \le p \le \infty$ and some Banach space Y, and other function spaces, as well. In all these cases, the inclusion operator I is a bounded linear operator. An example illustrating this point is the existence of constant $C = C(\gamma,\beta,T)$, for $\gamma \le \beta$, such that for all $f \in L^p_{\text{loc}}[0,T;V^{2\beta})$, one has $f \in L^p_{\text{loc}}[0,T;V^{2\gamma})$, and $\int_0^t \|A^\gamma f\|^p\,ds \le C\int_0^t \|A^\beta f\|^p\,ds$, for all t with $0 \le t < T$.

In this chapter we have two objectives. First, we present an introduction to the existence, uniqueness, and regularity of solutions (be they mild, strong, classical, or weak solutions) for equation (40.1), under varying assumptions on A and $f(t)$. The second objective is to develop a comprehensive reference source for the applications of the above theory to the nonlinear problem (40.2), when $f(t) = F(u(t))$ and $u(t)$ is some given solution, or approximate solution, of equation (40.2). It should not be surprising that many of the issues one faces in the study of the dynamics of the solutions of equation (40.2), for example, regularity, can be reduced to related problems for equation (40.1). While the paradigm of using equation (40.1) in this capacity enables one to unify the study of many differing nonlinear problems, it has the potential of raising a number of technical issues at a rather early stage. Therefore, for the initial reading of this chapter, we make the following recommendations:

(1) Read in full the material through Lemma 42.8, including the proofs.
(2) Read the remainder of Section 4.2, skipping the proofs until they are required at a later time.
(3) Read Sections 4.3 and 4.4.
(4) Section 4.5 can be skipped for the first reading. However, it is important for Chapters 7 and 8.
(5) Read Sections 4.6 and 4.7 through the Herculean Theorem 47.6. This result is very important for the study of the dynamics of nonlinear evolutionary equations.

4.1. PDEs as Evolutionary Equations.

We begin with the study of partial differential equations. In particular, we will show next how the abstract equations (40.1) and (40.2) arise in this study. This occurs as the result of a standard imbedding of a system of (nonlinear) PDEs into an abstract (nonlinear) evolutionary equation. Let us begin with the simplest problem, a system of reaction diffusion equations of the form

$$\partial_t u - \nu \Delta u = f(u, x), \tag{41.1}$$

where $u \in \mathbb{R}^k$, $x \in \Omega$, $t \geq 0$, and Ω is a suitable smooth, open, bounded domain in $\mathbb{R}^d$, $\Delta = D_1^2 + \cdots + D_d^2$ is the Laplacian, $\nu > 0$ is a constant, and f is a smooth function. Recall that the solutions of the linear equation

$$\partial_t u - \nu \Delta u = 0 \tag{41.2}$$

are described in terms of a C_0-semigroup. More precisely, assume that (41.2) is given on Ω with suitable boundary conditions, say, $u(x) = 0$, for $x \in \partial\Omega$. In this way, (41.2) generates a linear evolutionary equation

$$\partial_t u + \nu A u = 0, \qquad u \in H, \tag{41.3}$$

where H is a Hilbert space, or sometimes a Banach space, consisting of functions $u = u(x)$ defined for $x \in \Omega$ and satisfying other conditions. For example one might have $H = L^2(\Omega, \mathbb{R}^k)$, or $H = H_0^1(\Omega, \mathbb{R}^k)$, and $-A$ is the generator of a C_0-semigroup e^{-At}.

We assume the nonlinear term $f : \mathbb{R}^k \times \Omega \to \mathbb{R}^k$ appearing in (41.1) is sufficiently smooth, and we define a nonlinear operator $F(u)$ formally by

$$F(u)(x) = f(u(x), x), \qquad \text{for } x \in \Omega. \tag{41.4}$$

A natural requirement one might impose on the nonlinear operator $F(u)$ is that it satisfy

$$u \in H \Longrightarrow F(u) \in H; \tag{41.5}$$

i.e., F maps H into itself. Naturally (41.5) imposes added conditions on the nonlinearity $f(u, x)$ appearing in (41.1), and one typically needs to find sufficient conditions on the function $f(u, x)$ so that (41.5) is valid. For example, if $f(u, x) = \sin u$, then F satisfies (41.5). While not all nonlinear PDEs will satisfy (41.5), it is, nevertheless, a good starting point for our theory.

The (abstract) nonlinear evolutionary equation generated by (41.1) is given by

$$\partial_t u + \nu A u = F(u), \qquad \text{where } u \in H,\ t \geq 0. \tag{41.6}$$

We will be seeking solutions of (41.6) that satisfy the initial condition

$$u(0) = u_0, \qquad \text{where } u_0 \in H.$$

Other systems of PDEs can also be reduced to a nonlinear evolutionary equation of the type (41.6). For example, in (41.1) the constant $\nu > 0$ can be replaced by a diagonal $k \times k$ matrix $\Gamma = \text{diag}(\gamma_1, \cdots, \gamma_k)$, where $\gamma_i > 0$, for $1 \leq i \leq k$, so that

$$\partial_t u - \Gamma \Delta u = f(u, x), \tag{41.7}$$

More generally one can study equations of the form

$$\partial_t u^i - \sum_{j=1}^{k} D_j(a_{ij}(x) D_j u^i) = f(u, x), \qquad 1 \leq i \leq k, \tag{41.8}$$

where $u = (u^1, \cdots, u^k)$, $D_j = \partial_{x_j}$, and the differential operator

$$A_0(x, D) = -\sum_{j=1}^{k} D_j(a_{ij}(x) D_j u^i), \qquad 1 \leq i \leq k$$

is a uniformly elliptic operator of order 2 over Ω (see Section 3.8.2). In either case, the linear problem ($f \equiv 0$) generates a linear evolutionary equation (41.3), where the linear operator $-A$ is the generator of a C_0-semigroup e^{-At}. For the nonlinear terms occuring in (41.7) or (41.8), one employs the same composition as used in (41.4), and thereby obtaining a similar nonlinear evolutionary equation (41.6).

Somewhat more complicated problems are illustrated by the Burgers equation

$$\partial_t u - \partial_x^2 u + u \, \partial_x u = g(x), \tag{41.9}$$

the Kuramoto-Sivashinsky equation, with $f(u, x) = \frac{1}{2}(\partial_x u)^2$, and the Cahn-Hilliard equation, with $f(u, x) = \partial_x^2(p(u))$ where $u \in \mathbb{R}^1$, Ω is an interval in $\mathbb{R}^1$, and $p(u)$ is a polynomial in u of odd degree. In these latter two equations, the (linear) Laplace operator is also replaced by the negative of the biharmonic operator and its perturbations. The most significant difference between these examples and (41.1) is that the nonlinear terms now depend on the spatial derivatives of the solution u. Equation (41.9) is a special case of

$$\partial_t u - \partial_x^2 u = f(u, \partial_x u, x). \tag{41.10}$$

As in the case of (41.1) we obtain a nonlinear evolutionary equation for (41.10) of the form (41.6), where $A = -\Delta = -\frac{\partial^2}{\partial x^2}$ (with the appropriate boundary conditions) and $F(u)$ is the nonlinearity of composition type given by $v = \partial_x u$ and

$$F(u)(x) = f(u(x), v(x), x). \tag{41.11}$$

There are several points that need to be made here.

(1) Since the evaluation of $F(u)(x)$ depends on the derivative $\partial_x u(x)$, we will require $u(x)$ to have a distributional derivative, e.g., $u \in H^1(\Omega, \mathbb{R})$.
(2) The function $F(u)$ will not be as smooth in x as is the function u. In other words, one should expect, among other things, that F satisfies $F : V \to H$, where $V \mapsto H$ and $V \neq H$. For example, one might have $V = H^1(\Omega, \mathbb{R})$ and $H = L^2(\Omega, \mathbb{R})$.
(3) As will be seen later, there is an advantage in exploiting the fact that, under appropriate boundary conditions, the Standing Hypothesis B is satisfied, and therefore e^{-At} is an analytic semigroup and the spaces V and H referred to above become $V = \mathcal{D}(A^{\frac{1}{2}})$ and $H = \mathcal{D}(A^0)$. One then has a compact imbedding $V \hookrightarrow H$.

Nonlinear partial differential equations of the type noted above are sometimes referred to as **semilinear** equations to emphasize the fact that the operator A, together with the time derivative ∂_t, appear as linear operators. Such semilinear equations are natural candidates for the nonlinear problem (40.2) and (41.11).

The partial differential equations given above are all examples of local partial differential equations because the terms, the linear and the nonlinear terms, are calculated by using local information about the function $u = u(x)$, as illustrated in equation (41.4). An example of a nonlocal equation is an equation with a convolution, such as

$$\partial_t u(x,t) - \partial_x^2 u(x,t) = \int_{-\infty}^{\infty} a(y) u(x-y,t)\, dy.$$

In order to evaluate the integral in this equation, one uses information about $u(x,t)$, for all x with $-\infty < x < \infty$.

When the partial differential equation is converted to an abstract evolutionary equation, the property of localness seems to be lost. After using the definition (41.4) to get $F = F(u)$, what is the significance of the localness in the original problem? A partial answer is given in terms of the Gâteaux, or Fréchet derivatives of F. Clearly one would use equation (41.4), or (41.11), to compute these derivatives. Each of these equations involves local terms only. For example, it is noted in Henry (1981) that the function $F(u) = \sin u$, on the one hand, is globally Lipschitz continuous, but differentiable nowhere on $L^2 = L^2(0,1;\mathbb{R})$, and on the other hand, F is analytic on $H^1 = H^1(0,1;\mathbb{R})$. For this example, F satisfies both $F : L^2 \to L^2$ and $F : H^1 \to H^1$. This feature, which is fairly common, can sometimes be exploited, as part of a bootstrap argument, to derive more information about the solutions of (40.2).

Part I: Linear Theory

4.2. Solution Concepts and the Variation of Constants Formula.

Unlike the finite dimensional theory of dynamical systems, one finds that, in the infinite dimensional setting, there are several different notions of a solution. In this setting, each solution concept generalizes a different feature of the finite dimensional theory. In order to describe these solution concepts, we will focus first on the linear inhomogeneous equation

$$\partial_t u + Au = f(t), \tag{42.1}$$

with initial condition

$$u(0) = u_0 \in W, \tag{42.2}$$

on a Banach space W, where $t \geq 0$. We assume first that $-A$ is the infinitesimal generator of a C_0-semigroup e^{-At}, and that $f \in L^1_{\text{loc}}[0,T;W)$, where $0 < T \leq \infty$. Recall that $L^1_{\text{loc}}[0,T;W)$ denotes those functions $f \in L^1_{\text{loc}}(0,T;W)$ with the property that $\int_0^\sigma \|f(t)\|\,dt < \infty$, for each σ with $0 < \sigma < T$, see Appendix C.

There are three solution concepts which are of interest here, *viz.*, mild solution, strong solution, and classical solution. A fourth concept, that of a weak solution, is presented below in Section 4.2.3. A function $u : [0,T) \to W$ is said to be a **mild solution** of (42.1)-(42.2) **in the space** W and on the interval $[0,T)$, provided that $u \in C[0,T;W)$ and u satisfies

$$u(t) = e^{-At}u_0 + \int_0^t e^{-A(t-s)}f(s)ds, \qquad t \in [0,T), \tag{42.3}$$

where the integral is in the Bochner sense and represents a point in W for each t, see Appendix C. A related formula is

$$u(t) = e^{-A(t-t_0)}u(t_0) + \int_{t_0}^t e^{-A(t-s)}f(s)ds, \qquad t \in [t_0, t_0 + T_1). \tag{42.4}$$

Indeed, by a simple change of variables one can show that if u satisfies (42.3), then it satisfies (42.4), for every $t_0 \in [0,T)$, where $T_1 = T - t_0$. Either of the formulae (42.3) or (42.4) is called the **Variation of Constants Formula**, and the respective theories are the same. We will focus on (42.3), for simplicity.

A function $u : [0,T) \to W$ is said to be a **strong solution** of (42.1) - (42.2) **in the space** W, if it satisfies the following five conditions:

(1) $u \in C[0,T;W)$ and $u(0) = u_0$;

(2) u is (strongly) differentiable in W almost everywhere (a.e.) in $(0, T)$;
(3) $\partial_t u \in L^1_{\text{loc}}[0, T; W)$ and $u(t) = u(t_0) + \int_{t_0}^t \partial_t u(s)\, ds$, for all t, $t_0 \in [0, T)$;
(4) $u(t) \in \mathcal{D}(A)$ a.e. on $(0, T)$; and
(5) u satisfies the equation

$$\partial_t u(t) + Au(t) \overset{\text{a.e.}}{=} f(t), \tag{42.5}$$

is satisfied in W, almost everywhere on $(0, T)$.

Let $f \in L^1_{\text{loc}}[0, T; W)$ and let $u = u(t)$ be a strong solution on the interval $[0, T)$. Then one has $Au \in L^1_{\text{loc}}[0, T; W)$, or equivalently, $u \in L^1_{\text{loc}}[0, T; \mathcal{D}(A))$. The issue of a solution $u(t)$ satisfying equation (42.5), or a more general equation, almost everywhere (a.e.) on an interval I will arise quite often in this book. We will denote this by $\overset{\text{a.e.}}{=}$. For example, $\partial_t u(t) + Au(t) \overset{\text{a.e.}}{=} f(t)$, on I.

Lastly, if one has $f \in C[0, T; W)$, then a solution $u : [0, T) \to W$ is said to be a **classical solution** of (42.1)-(42.2) **in the space** W, if it satisfies the following:

(1) $u \in C[0, T; W)$ and $u(0) = u_0$;
(2) u is (strongly) differentiable in W at each $t \in (0, T)$;
(3) $\partial_t u \in L^1_{\text{loc}}[0, T; W)$ and $u(t) = u(t_0) + \int_{t_0}^t \partial_t u(s)\, ds$, for all t, $t_0 \in [0, T)$;
(4) $u(t) \in \mathcal{D}(A)$ for all $t \in (0, T)$; and
(5) u satisfies the equation in (42.5) in W, everywhere on $(0, T)$.

Differences between these solution concepts can already be seen in the purely linear case where $f \equiv 0$. For example, the Variation of Constants Formula (42.3) reduces to $u(t) = e^{-At} u_0$. Hence for every $u_0 \in W$ there is a mild solution of (42.1) - (42.2), and this solution is the motion through u_0 generated by the semigroup e^{-At}. If $u_0 \in \mathcal{D}(A)$, then the mild solution $e^{-At} u_0$ is both a strong solution and a classical solution by Theorem 31.4. On the other hand, if the semigroup e^{-At} is differentiable for $t > 0$, then for every $u_0 \in W$, the mild solution $e^{-At} u_0$ is both a strong and classical solution, see Theorem 35.2.

As we now proceed to show, the general theory of these three solution concepts for the linear inhomogeneous equation (42.1) can be summarized as follows:

(1) For every $u_0 \in W$ there is a unique, mild solution of (42.1) - (42.2). (This is, in fact, an immediate consequence of the definition of a mild solution.)
(2) Every strong solution is a mild solution.
(3) Every classical solution is a strong solution.
(4) Existence implies Uniqueness: Every strong solution, as well as every classical solution, is uniquely determined by the initial condition $u_0 \in W$ and the forcing function f.

(5) A mild solution is a strong solution, or a classical solution, provided that some additional conditions are satisfied.

We begin this theory by addressing properties (2) and (3), above. Note that property (4) is an immediate consequence of properties (1), (2), and (3).

Lemma 42.1. *Let $(e^{-At}, -A)$ be a C_0-semigroup on a Banach space W and let $f \in L^1_{\text{loc}}[0,T;W)$. Then every classical solution of (42.1) in W is a strong solution, and every strong solution of (42.1) in W is a mild solution.*

Proof. The fact that every classical solution is a strong solution is an immediate consequence of the definitions. Let $u(t)$ be a strong solution of (42.1) and define $g(s) = e^{-A(t-s)}u(s)$, for $0 \le s \le t$. Then g is (strongly) differentiable and for almost all $s \in (0,t)$, the product formula for differentiation implies that

$$\begin{aligned}\partial_s g &\overset{\text{a.e.}}{=} Ae^{-A(t-s)}u(s) + e^{-A(t-s)}\partial_s u \\ &\overset{\text{a.e.}}{=} Ae^{-A(t-s)}u(s) - e^{-A(t-s)}Au(s) + e^{-A(t-s)}f(s) \overset{\text{a.e.}}{=} e^{-A(t-s)}f(s).\end{aligned}$$

Hence $g \in W^{1,1}(0,t;W)$, for any $t \in (0,T)$. By using the Newton-Leibniz Formula (see Appendix C.4, Theorem C.9 and its corollary) to integrate the above equality, we obtain (42.3). Therefore, $u(t)$ is a mild solution of equation (42.1). □

Next we will prove the uniqueness of solutions of (42.1) - (42.2).

Lemma 42.2. *Let $(e^{-At}, -A)$ be a C_0-semigroup on a Banach space W and let $f \in L^1_{\text{loc}}[0,T;W)$. Then there is at most one mild, strong, or classical solution of (42.1)-(42.2) in W.*

Proof. Assume that there are two such solutions, say $u(t)$ and $v(t)$. Then $w(t) \overset{\text{def}}{=} u(t) - v(t)$ is a mild solution of $\partial_t w + Aw = 0$, $w(0) = 0$. Hence, Lemma 31.1 implies that $\|w(t)\| \le Me^{-at}\|w(0)\|$, for $t \ge 0$, and thus $w(t) \equiv 0$. □

4.2.1 C_0-Theory. In this section we will examine the connections between these these three solution concepts for (42.1) under the assumption that $(e^{-At}, -A)$ is a C_0-semigroup on a given Banach space W. Later we will develop the theory for analytic semigroups.

Lemma 42.3. *Let e^{-At} be a C_0-semigroup and let $f \in L^1_{\text{loc}}[0,T;W)$. Let $u = u(t)$ satisfy (42.3) on $[0,T)$. Then $u \in C[0,T;W)$, and u is a mild solution of (42.1).*

Proof. As noted above, if u satisfies equation (42.3), then it satisfies equation (42.4), for every $t_0 \in [0,T)$. Next note that $e^{-A(t-t_0)}u_0 \overset{s}{\to} u(t_0)$ in W, as $t \to t_0$, by Lemma 31.2. (When $t_0 = 0$, the last limit becomes $t \to 0^+$.)

Furthermore, Lemma 31.1 implies that $\left\| \int_{t_0}^{t} e^{-A(t-s)} f(s)\, ds \right\|$ is bounded by

$$\left| \int_{t_0}^{t} \|e^{-A(t-s)}\| \, \|f(s)\| \, ds \right| \le M e^{|a|\,|t-t_0|} \left| \int_{t_0}^{t} \|f(s)\| \, ds \right| \to 0, \quad \text{as } t \to t_0,$$

since $f \in L^1_{\rm loc}[0,T;W)$. Hence, $u(t) \to u(t_0)$ in W, as $t \to t_0$. □

Theorem 42.4. *Let $(e^{-At}, -A)$ be a C_0-semigroup on a Banach space W, and let $f \in L^1_{\rm loc}[0,T;W)$. Then a mild solution $u : [0,T) \to W$ of (42.1) is a strong solution if and only if both of the following two conditions are satisfied:*

(1) *$u(t) \in \mathcal{D}(A)$ almost everywhere on $(0,T)$, and*
(2) *$Au \in L^1_{\rm loc}[0,T;W)$, i.e., $u \in L^1_{\rm loc}[0,T;\mathcal{D}(A))$.*

Proof. We will present the proof in the case that W is a Hilbert space. The general case is left as an exercise. Assume that $u(\cdot)$ is a strong solution of (42.1). Condition (1) follows from the definition of a strong solution, and Condition (2) follows from the fact that $-Au(t) \overset{\text{a.e.}}{=} \partial_t u(t) - f(t)$ and $\partial_t u \in L^1_{\rm loc}[0,T;W)$.

Next assume that a given mild solution u of (42.1) satisfies Conditions (1) and (2). Then (31.6) implies that

$$-A \int_s^t e^{-A(\sigma-s)} f(s) d\sigma \overset{\text{a.e.}}{=} e^{-A(t-s)} f(s) - f(s), \qquad \text{for } 0 < s \le t < T.$$

Let A^* denote the adjoint of the linear operator A. For any $g \in \mathcal{D}(A^*)$, the domain of A^*, we take the scalar product of the last equation with g to obtain

$$\begin{aligned} \langle e^{-A(t-s)} f(s), g \rangle &\overset{\text{a.e.}}{=} \langle f(s), g \rangle - \langle A \int_s^t e^{-A(\sigma-s)} f(s) d\sigma, g \rangle \\ &\overset{\text{a.e.}}{=} \langle f(s), g \rangle - \int_s^t \langle e^{-A(\sigma-s)} f(s), A^* g \rangle d\sigma. \end{aligned}$$

By integrating the last equality with respect to s and interchanging the order of integration in the last term, one obtains

$$\int_0^t \langle e^{-A(t-s)} f(s), g \rangle ds = \int_0^t \langle f(s), g \rangle \, ds - \int_0^t \int_0^\sigma \langle e^{-A(\sigma-s)} f(s), A^* g \rangle \, ds \, d\sigma.$$

Next by using the Variation of Constants Formula (42.3), one finds that

$$(42.6) \quad \langle u(t) - e^{-At} u_0, g \rangle = \int_0^t \langle f(\sigma), g \rangle d\sigma - \int_0^t \langle u(\sigma) - e^{-A\sigma} u_0, A^* g \rangle d\sigma.$$

Using the definition of the adjoint operator A^* and (31.6) once again, one obtains

$$\begin{aligned}\int_0^t \langle e^{-A\sigma}u_0, A^* g\rangle d\sigma &= \langle \int_0^t e^{-A\sigma}u_0\, d\sigma, A^* g\rangle \\ &= \langle A\int_0^t e^{-A\sigma}u_0 d\sigma, g\rangle = -\langle e^{-At}u_0 - u_0, g\rangle.\end{aligned}$$

Note that conditions (1) and (2) imply that $\int_0^t u(\sigma)\, d\sigma \in \mathcal{D}(A)$, for $0 \le t < T$. Now by substituting the last equation into (42.6), one gets

$$\langle u(t) - u_0 - \int_0^t f(\sigma)\, d\sigma, g\rangle = -\langle \int_0^t u(\sigma)\, d\sigma, A^* g\rangle = -\langle A\int_0^t u(\sigma)\, d\sigma, g\rangle,$$

for $0 \le t < T$. Using the fact that $\mathcal{D}(A^*)$ is dense in the Hilbert space W (see Pazy (1983)), one obtains

$$u(t) - u_0 - \int_0^t f(s) ds = -A\int_0^t u(s) ds = -\int_0^t Au(s) ds,$$

for $0 \le t < T$, or

$$u(t) = u_0 + \int_0^t [-Au(s) + f(s)] ds, \qquad \text{for } 0 \le t < T.$$

By Appendix C, this implies that $u(t)$ is differentiable almost everywhere in $(0, T)$, $u(0) = u_0$, and u satisfies (42.5). Thus u is a strong solution. □

The **Leibniz Formula** in the space W takes the form

$$\frac{d}{dt}\int_0^t e^{-A(t-s)} f(s)\, ds \overset{\text{a.e.}}{=} -A\int_0^t e^{-A(t-s)} f(s)\, ds + f(t). \tag{42.7}$$

Notice that this formula is valid in the context of a C_0-semigroup if and only if

$$v(t) \overset{\text{def}}{=} \int_0^t e^{-A(t-s)} f(s)\, ds \tag{42.8}$$

is a strong solution of equation (42.1). If the operator A is in $\mathcal{L}(W)$, then the Leibniz Formula is valid in W, for any $f \in L^1_{\text{loc}}[0, T; W)$. However, the story is much more complicated when A is not a bounded operator. We now present the first of several sufficient conditions for the Leibniz Formula to be valid.

Lemma 42.5. *Let $(e^{-At}, -A)$ be a C_0-semigroup on H, where H is a Hilbert space, or a reflexive Banach space. Let $f \in C^{0,1}_{\text{loc}}[0,T;H)$, i.e., f is locally Lipschitz continuous on $[0,T)$. Define $v = v(t)$ by (42.8). Then one has*

$$v \in C^{0,1}_{\text{loc}}[0,T;H) \cap L^1_{\text{loc}}[0,T;\mathcal{D}(A)), \tag{42.9}$$

v is a strong solution of (42.1), and the Leibniz Formula (42.7) is valid in H.

Proof. Let $t > 0$ and let $0 < h \leq \min(1, t, T-t)$. Then one has

$$\begin{aligned} v(t+h) - v(t) &= \int_0^{t+h} e^{-A(t+h-s)} f(s)\, ds - \int_0^t e^{-A(t-s)} f(s)\, ds \\ &= \int_0^t e^{-A(t-s)} \left[f(s+h) - f(s) \right] ds \\ &\quad + \int_0^h e^{-A(t+h-s)} f(s)\, ds. \end{aligned} \tag{42.10}$$

Let $M \geq 1$ and $a \in \mathbb{R}$ be given so that $\|e^{-At}\| \leq Me^{-at}$, for $t \geq 0$. One then has

$$\left\| \int_0^t e^{-A(t-s)} [f(s+h) - f(s)]\, ds \right\| \leq Mte^{|a|t} L|h|,$$

for $|h| \leq \min(1, t, T-t)$, where $L = L(t)$ satisfies

$$\|f(s+h) - f(s)\| \leq L|h|, \qquad \text{for } 0 \leq s \leq t < T,$$

and

$$\left\| \int_0^h e^{-A(t+h-s)} f(s)\, ds \right\| \leq Me^{|a|(t+1)} \sup_{0 \leq s \leq 1} \|f(s)\|\, |h|.$$

Hence v is locally Lipschitz continuous on $[0,T)$. Since H is a Hilbert space (or a reflexive Banach space), this implies that the derivative $\partial_t v(t)$ exists almost everywhere on $(0,T)$, see Pazy (1983).

Next we will show that $v(t) \in \mathcal{D}(A)$, almost everywhere on $(0,T)$. Indeed from (42.8), one has

$$\frac{e^{-Ah} - I}{h} v(t) = \frac{1}{h} \left[v(t+h) - v(t) \right] - \frac{1}{h} \int_t^{t+h} e^{-A(t+h-s)} f(s)\, ds. \tag{42.11}$$

Since $\partial_t v(t)$ exists almost everywhere, the middle term in (42.11) converges almost everywhere to $\partial_t v(t)$, as $h \to 0$. Since $e^{-A(t+h-s)} f(s) - f(t) \to 0$, uniformly for $t \leq s \leq t+h$, as $h \to 0$, the last term in (42.11) has the limit $-f(t)$. Hence the limit on the left side exists almost everywhere, and by definition it is $-Av(t)$. As a result equation (42.1) is valid almost everywhere, and v, which is a strong solution, satisfies (42.9). $\square$

Theorem 42.6. *Let $(e^{-At}, -A)$ be a C_0-semigroup on H, where H is a Hilbert space, or a reflexive Banach space. Let $f \in C^{0,1}_{\text{loc}}[0,T;H)$. Assume that one of the following two conditions hold:*

(1) $u_0 \in \mathcal{D}(A)$, *or*
(2) e^{-At} *is differentiable for* $t > 0$, *and* $u_0 \in H$.

Then the mild solution $u = u(t)$ of (42.1) is a strong solution, and the Leibniz Formula (42.7) is valid in W. If in addition $f \in C^1[0,T;H)$, then u is a classical solution.

Proof. Let u be the mild solution of (42.1). From Lemma 42.5, the function v, see (42.8), is a strong solution of (42.1). If $u_0 \in \mathcal{D}(A)$, then (31.7) implies that $e^{-At}u_0$ is differentiable, for $t > 0$. On the other hand, if the C_0-semigroup e^{-At} is differentiable for $t > 0$, then for every $u_0 \in H$, the term $e^{-At}u_0$ is differentiable, for $t > 0$, and equation (35.2) is valid, for $t > 0$, with $T(t) = e^{-At}$. In either case, we conclude that u is a strong solution of (42.1). □

4.2.2 Analytic Theory. The above results are of some use, but applying them requires that one prove something about the specified mild solution $u(t)$. This can be especially difficult in nonlinear applications, where the solution $u(t)$ is unknown, or only known implicitly. One can obtain stronger results by assuming that the C_0-semigroup e^{-At} has some additional properties, such as e^{-At} is analytic, or A is sectorial, i.e., the Standing Hypothesis A is satisfied. The key point is that, when A is a sectorial operator, then one can use the Fundamental Theorem of Sectorial Operators (Theorem 37.5).

The scale of Banach spaces $V^{2\alpha}$, which arise in the context of the Standing Hypothesis A, offers additional flexibility in the development of the theory of mild, strong, and/or classical solutions. However, along with the additional flexibility, one also encounters some additional complexity. Let us reconsider equation (42.1), where the Standing Hypothesis A is satisfied and $f \in L^p_{\text{loc}}[0,T;W)$ is given, where $1 \le p \le \infty$. However, we now fix the initial condition to satisfy $u_0 \in V^{2\rho}$, where $\rho > 0$. As we will see in the next result, the mild solution u of equation (42.1), which is defined by equation (42.3), satisfies $u \in C[0,T;V^{2\rho})$. Furthermore, equation (42.3) holds in the space $V^{2\rho}$. Because of the continuous imbeddings $V^{2\rho} \hookrightarrow W$ and $C[0,T;V^{2\rho}) \hookrightarrow C[0,T;W)$, u is also a mild solution in W and equation (42.3) also holds in W, as well. However, the story for strong solutions and/or classical solutions is different. In this case, the solution u resides in one space, while equation (42.1) holds in another! The following definitions should be helpful in keeping this straight.

We assume that $f \in L^p_{\text{loc}}[0,T;W)$ and that the Standing Hypothesis A is satisfied, where $1 \le p \le \infty$, and that $\rho \ge 0$. A function $u : [0,T) \to W$ is said to be a **mild solution** of (42.1) **in the space** $V^{2\rho}$ and on the interval $[0,T)$, provided that $u \in C[0,T;V^{2\rho})$ and equation (42.3) holds in the space $V^{2\rho}$. Note that for a mild solution, one has $u(0) = u_0 \in V^{2\rho}$, and the integral $\int_0^t e^{-A(t-s)} f(s)\,ds$ is in $V^{2\rho}$, for each $t \in [0,T)$. Specifically, the integral is a Bochner integral in W, and its value is a point in $V^{2\rho}$, for each $t \in [0,T)$. A function $u : [0,T) \to W$ is said to be a **strong solution** of (42.1) **in the space** $V^{2\rho}$, provided that the following conditions hold:

(1) $u \in C[0,T; V^{2\rho})$;
(2) u is (strongly) differentiable in W almost everywhere (a.e.) in $(0,T)$;
(3) $\partial_t u \in L^1_{\mathrm{loc}}[0,T;W)$ and $u(t) = u(t_0) + \int_{t_0}^t \partial_t u(s)\,ds$, for all t, $t_0 \in [0,T)$;
(4) $u(t) \in \mathcal{D}(A)$ a.e. on $(0,T)$; and
(5) u satisfies equation (42.5) in W and almost everywhere on $(0,T)$.

Finally, if one has $f \in C[0,T;W)$, then $u : [0,T) \to W$ is said to be a **classical solution** of (42.1) **in the space** $V^{2\rho}$, provided that the following conditions hold:

(1) $u \in C[0,T; V^{2\rho})$;
(2) u is (strongly) differentiable in W at each $t \in (0,T)$;
(3) $\partial_t u \in L^1_{\mathrm{loc}}[0,T;W)$ and $u(t) = u(t_0) + \int_{t_0}^t \partial_t u(s)\,ds$, for all t, $t_0 \in [0,T)$;
(4) $u(t) \in \mathcal{D}(A)$ for all $t \in (0,T)$; and
(5) u satisfies the equation in (42.5) in W everywhere on $(0,T)$.

Notice that if one has the continuous imbedding, $V^{2\rho} \mapsto V$, for some Banach space V, then any mild solution in $V^{2\rho}$ is a mild solution in V, as well. Similarly, the continuous imbeddings $V^{2\rho} \mapsto V \mapsto W$ (or $V^{2\rho} \mapsto W \mapsto V$), imply that any strong, or classical, solution in the space $V^{2\rho}$ is a strong, or classical, solution in V, as well. Also note that argument given in Theorem 42.4 shows that if $f \in L^1_{\mathrm{loc}}[0,T;W)$, then a mild solution u of equation (42.1) in $V^{2\rho}$, where $0 \le \rho < 1$, is a strong solution in $V^{2\rho}$ if and only if both of the following two conditions are satisfied:

(1) $u(t) \in \mathcal{D}(A) = V^2$ almost everywhere on $(0,T)$, and
(2) $Au \in L^1_{\mathrm{loc}}[0,T;W)$, i.e., $u \in L^1_{\mathrm{loc}}[0,T;\mathcal{D}(A))$.

When the Standing Hypothesis B is satisfied, then the scale of Hilbert spaces $V^{2\alpha}$ is defined for all $\alpha \in \mathbb{R}$. The definitions given above might be used for $W = V^{2\alpha_0}$, for any $\alpha_0 \in \mathbb{R}$. For example, in Chapter 6 on the Navier-Stokes equations, it is shown that any (weak) solution u of Class LH satisfies $u \in L^2_{\mathrm{loc}}[0,\infty;H)$ and it is a mild solution of the Navier-Stokes equations in the space V^{-1}, see Lemma 63.1. We now have the following result, which yields further information on the regularity of mild solutions, when $p > 1$.

Lemma 42.7. *Let $f \in L^p_{\mathrm{loc}}[0,T;W)$ and let the Standing Hypothesis A hold, where $1 \le p \le \infty$ and $0 < T \le \infty$. Then the following are valid:*

(1) *The implications*

$$\text{classical solution in } V^{2\alpha} \implies \text{strong solution in } V^{2\alpha} \implies \text{mild solution in } V^{2\alpha}, \tag{42.12}$$

as described in Lemma 42.1 for a solution u, are valid for any α *with* $\alpha \geq 0$.

(2) *Assume that* $p > 1$. *Let* $u_0 \in W$ *be fixed and let* $u = u(t)$ *satisfy (42.3) in W on* $[0, T)$. *Then u is a mild solution of equation (42.1) in W and*

$$u \in C[0, T; W) \cap C^{0,\theta}_{\text{loc}}(0, T; V^{2r}), \tag{42.13}$$

for every r with $0 \leq r < 1 - \frac{1}{p}$, *where* $0 < \theta < 1 - \frac{1}{p} - r$.

(3) *Assume that* $p > 1$ *and* $u_0 \in V^{2\rho}$, *where* $0 < \rho < 1 - \frac{1}{p}$. *Then u is a mild solution of equation (42.1) in* $V^{2\alpha}$ *and*

$$u \in C[0, T; V^{2\alpha}) \cap C^{0,\theta_1}_{\text{loc}}[0, T; V^{2r}) \cap C^{0,\theta_2}_{\text{loc}}(0, T; V^{2\sigma}), \tag{42.14}$$

for every α, r, *and* σ *with* $0 \leq \alpha \leq \rho$, $0 \leq r < \rho$, *and* $\rho \leq \sigma < 1 - \frac{1}{p}$, *where* $\theta_1 = \theta_1(r)$ *and* $\theta_2 = \theta_2(\sigma)$ *are positive.*

Proof. Item (1): The implication between classical and strong solutions in $V^{2\alpha}$ follows directly from the definitions. Let $u = u(t)$ be a strong solution of (42.1) and define $g(s) = e^{-A(t-s)}u(s)$, for $0 \leq s \leq t$. As argued in Lemma 42.1, $g(s)$ is (strongly) differentiable, for almost all $s \in (0, t)$, and one has $\partial_s g \overset{\text{a.e.}}{=} e^{-A(t-s)} f(s)$ in W on $(0, T)$. By integrating the last equation and using the Newton-Leibniz Formula one obtains

$$u(t) - e^{-At}u(0) = \int_0^t e^{-A(t-s)} f(s)\, ds, \qquad \text{for } 0 \leq t < T.$$

Since the left side of the last equation is a point in $V^{2\alpha}$, the right side is, as well. Hence, the integral, even though it is a Bochner integral in W, is in $V^{2\alpha}$ and u is a mild solution of equation (42.1) in $V^{2\alpha}$.

Items (2) and (3): Let $p > 1$ and let ρ and σ satisfy $0 \leq \rho \leq \sigma < 1 - \frac{1}{p}$. Let us first examine the continuity properties of $v = \int_0^t e^{-A(t-s)} f(s)\, ds$. From (37.11) and the Hölder inequality with $\frac{1}{p} + \frac{1}{q} = 1$, one obtains

$$\begin{aligned}\int_0^t \|A^\rho e^{-A(t-s)} f(s)\|\, ds &\leq M_\rho e^{|a|t} \int_0^t (t-s)^{-\rho} \|f(s)\|\, ds \\ &\leq M_\rho e^{|a|t} \left(\int_0^t (t-s)^{(-\rho)q}\, ds \right)^{\frac{1}{q}} \left(\int_0^t \|f\|^p\, ds \right)^{\frac{1}{p}}.\end{aligned}$$

Hence, one has $v \in C[0, T; V^{2\rho})$. The last term is finite precisely when $q = p(p-1)^{-1}$ satisfies $1 > q\sigma$, which is guaranteed by the assumption on σ. Next we note that

$$v(t+h) - v(t) = \int_0^t [e^{-Ah} - I] e^{-A(t-s)} f(s)\, ds + \int_t^{t+h} e^{-A(t+h-s)} f(s)\, ds,$$

where $0 \leq h \leq \min(1, T - t)$. Inequalities (37.11) and (37.12) and the Hölder inequality imply that there is a constant $K_1 > 0$ such that

$$\int_0^t \|A^r[e^{-Ah} - I]e^{-A(t-s)}f(s)\| \, ds \leq K_1 e^{|a|t} \int_0^t |h|^\theta (t-s)^{-(r+\theta)} \|f(s)\| \, ds$$
$$\leq K_1 e^{|a|t} |h|^\theta \left(\int_0^t (t-s)^{-q(r+\theta)} ds \right)^{\frac{1}{q}} \left(\int_0^t \|f(s)\|^p \, ds \right)^{\frac{1}{p}},$$

which is finite when $0 \leq r < 1 - \frac{1}{p}$ and $0 < \theta < 1 - \frac{1}{p} - r$. Similarly there are constants K_2 and K_3 such that

$$\int_t^{t+h} \|A^r e^{-A(t+h-s)} f(s)\| ds \leq K_2 e^{|a|h} \int_t^{t+h} (t+h-s)^{-r} \|f(s)\| \, ds$$
$$\leq K_3 |h|^\theta \left(\int_t^{t+h} \|f(s)\|^p ds \right)^{\frac{1}{p}},$$

for the same values of r and θ. (The reader should verify that similar estimates are valid for small $h < 0$.) This implies that v satisfies

$$v \in C^{0,\theta}_{\text{loc}}[0, T; V^{2r}), \qquad \text{for each } 0 \leq r < 1 - \frac{1}{p}, \tag{42.15}$$

where $0 < \theta < 1 - \frac{1}{p} - r$.

Next consider $u = u(t) = e^{-At}u_0 + v(t)$. If $u_0 \in W$, then (37.14) and (42.15) imply (42.13). Similarly, if $u_0 \in V^{2\rho}$, where $0 < \rho < 1 - \frac{1}{p}$, then (37.14) and (42.15), with the continuous imbedding $V^{2\sigma} \mapsto V^{2\rho} \mapsto V^{2\alpha}$, for $0 \leq \alpha \leq \rho \leq \sigma$, imply (42.14). □

Next we have the following result concerning the strong solutions of (42.1) and the Leibniz Formula.

Lemma 42.8. *Let $f \in L^p_{\text{loc}}[0, T; W)$, for some p with $1 \leq p \leq \infty$, and assume that the C_0-semigroup $(e^{-At}, -A)$ on W is differentiable for $t > 0$. Then the following statements are equivalent:*

(1) *The Leibniz Formula (42.7) is valid in W.*
(2) *The function $v = v(t)$ defined by (42.8) is a strong solution of (42.1) in W.*
(3) *For every $u_0 \in W$, the mild solution $u = u(t)$ of (42.1), with $u(0) = u_0$, is a strong solution of (42.1) in W.*

Proof. The equivalence of Items (1) and (2) is argued above. Since e^{-At} is a differentiable semigroup, one has $\partial_t e^{-At}u_0 = -Ae^{-At}u_0$, for all $u_0 \in W$ and $t > 0$. The identity $u(t) = e^{-At}u_0 + v(t)$ then implies the equivalence of Items (2) and (3). □

The following result gives a major improvement over Lemma 42.5.

Theorem 42.9. *Assume that the Standing Hypothesis A is satisfied and*

(42.16) $$f \in L^p_{\rm loc}[0,T;W) \cap C^{0,\theta}_{\rm loc}(0,T;W),$$

where $1 \leq p \leq \infty$, $0 < \theta \leq 1$, *and* $0 < T \leq \infty$. *Let* $v = v(t)$ *be given by equation (42.8), and for any* $u_0 \in W$, *define* $u = u(t)$ *by (42.3). Then* u *and* v *are mild solutions of (42.1) in* W *that satisfy*

(42.17)
$$v \in C[0,T;W) \cap C^{0,1-r}_{\rm loc}(0,T;V^{2r}) \cap L^1_{\rm loc}[0,T;\mathcal{D}(A)) \cap C(0,T;\mathcal{D}(A)),$$

and

(42.18) $$u \in C[0,T;W) \cap C^{0,1-r}_{\rm loc}(0,T;V^{2r}) \cap C(0,T;\mathcal{D}(A)),$$

for every r *with* $0 \leq r < 1$. *In addition, the Leibniz Formula (42.7) is valid in* W, *and* u *and* v *are strong solutions in* W, *for* $0 \leq t < T$. *Moreover, if* $u_0 \in V^{2\rho}$, *for some* ρ *with* $0 < \rho \leq 1$, *then* u *is a strong solution in* $V^{2\rho}$ *on* $[0,T)$.

Proof. It follows from Lemmas 42.3 and 42.7 that u and v are mild solutions of equation (42.1) and they belong to $C[0,T;W)$. The first part of the argument is to show that v is a strong solution of equation (42.1) and that it satisfies (42.17). Our strategy here is to approximate equation (42.1) with a family of problems

(42.19) $$\partial_t v^\nu + A v^\nu = f^\nu(t), \qquad \text{where } 0 < \nu < T,$$

and f^ν is given below. We will show that, as $\nu \to 0^+$, one has

(1) $v^\nu \to v$ in W, uniformly on compact sets in $[0,T)$;
(2) $f^\nu \to f$ in $L^1_{\rm loc}[0,T;W)$, with $f^\nu(t) \to f(t)$, uniformly on compact sets in $(0,T)$;
(3) v^ν is a strong solution of equation (42.19) in W, and $Av^\nu(t)$ is a continuous function on the interval $(0,T)$ with values in W; and
(4) $Av^\nu(t) \to Av(t)$ in W, uniformly on compact sets in $(0,T)$, which in turn implies that $\partial_t v^\nu(t) \to \partial_t v(t)$, uniformly on compact subsets in $(0,T)$.

It then follows from Lemma C.7 in Appendix C, that $\partial_t v \in L^1_{\rm loc}[0,T;W)$, and v is a strong solution of equation (42.1) in the space W.

For $\nu > 0$, we define v^ν by $v^\nu(t) = 0$, for $0 \leq t \leq \nu$, and

$$v^\nu(t) = \int_\nu^t e^{-A(t-s)} f(s)\,ds = \int_0^t e^{-A(t-s)} f^\nu(s)\,ds, \qquad \text{for } \nu < t < T,$$

where $f^\nu(s) = f(s)$, for $s \geq \nu$, and $f^\nu(s) = 0$, for $0 < s < \nu$. Clearly one has $f^\nu \in L^1_{\rm loc}[0,T;W)$, for all $\nu > 0$, and

$$\lim_{\nu \to 0^+} f^\nu(t) = f(t), \qquad \text{uniformly on compact sets in } (0,T),$$

and $f^\nu \to f$ in $L^1_{\text{loc}}[0,T;W)$. From inequality (37.11), one obtains

$$\|v^\nu(t) - v(t)\| \leq \int_0^{\min(\nu,t)} \|e^{-A(t-s)}\|_{\mathcal{L}(W)} \|f(s)\| \, ds$$
$$\leq M_0 e^{|a|t} \int_0^{\min(\nu,t)} \|f(s)\| \, ds \to 0,$$

as $\nu \to 0^+$. This implies that $v^\nu(t) \to v(t)$ in W, uniformly for t in compact sets in $[0,T)$.

The next step is to show that for ν fixed, $\nu > 0$, one has $v^\nu(t) \in \mathcal{D}(A)$, for all $t \geq 0$, and that $Av^\nu \in C[0,T;W)$. For this purpose, it is convenient to write v^ν in the form $v^\nu = v_1^\nu + v_2^\nu$, where $v_1^\nu(t) = v_2^\nu(t) = 0$, for $0 \leq t \leq \nu$, and

$$v_1^\nu(t) = \int_\nu^t e^{-A(t-s)}(f(s) - f(t)) \, ds \text{ and } v_2^\nu(t) = \int_\nu^t e^{-A(t-s)} f(t) \, ds,$$

for $\nu < t < T$. From (31.6) one obtains

$$-Av_2^\nu(t) = -A \int_\nu^t e^{-A(t-s)} f(t) \, ds = f(t) - e^{-A(t-\nu)} f(t), \qquad \text{for } t \geq \nu.$$

Notice that the last equation and (31.5) imply that $Av_2^\nu(t) \to 0$ in W, as $t \to \nu^+$. Since $v_2^\nu(t) = 0$, for $0 \leq t \leq \nu$, we conclude that $v_2^\nu(t) \in \mathcal{D}(A)$, for all $t \geq 0$, and

$$Av_2^\nu \in C[0,T;W), \qquad \text{for all } \nu \text{ with } 0 < \nu < T.$$

Turning to v_1^ν, we first note that since f is locally Hölder continuous on $(0,T)$, for any δ and τ with $0 < \delta \leq \tau < T$, there is a constant $C_{\delta,\tau}$ such that

$$\|f(t) - f(s)\| \leq C_{\delta,\tau}|t-s|^\theta, \qquad \text{for } s,t \in [\delta,\tau].$$

Then from the last inequality and (37.11), one obtains

$$\|Ae^{-A(t-s)}(f(s) - f(t))\| \leq M_1(t-s)^{-1} C_{\nu,t}(t-s)^\theta,$$

for $\nu \leq s \leq t \leq \tau < T$, where $0 < \theta \leq 1$. As a result, the integral $\int_\nu^t Ae^{-A(t-s)}[f(s) - f(t)] \, ds$ exists in W, for $t \geq \nu$, and the closedness of A implies that

$$Av_1^\nu(t) = \int_\nu^t Ae^{-A(t-s)}(f(s) - f(t)) \, ds, \qquad \text{for } t > \nu.$$

Consequently, one has

$$\|Av_1^\nu(t)\| \leq \int_\nu^t M_1 C_{\nu,\tau}(t-s)^{\theta-1} \, ds = \theta^{-1} M_1 C_{\nu,\tau}(t-\nu)^\theta,$$

for $\nu < t \leq \tau < T$. Now the last inequality implies that $Av_1^\nu(t) \to 0$, as $t \to \nu^+$. Since $v_1^\nu(t) = 0$, for $0 \leq t \leq \nu$, we conclude that $Av_1^\nu(t) \in W$, for all $t \geq 0$, and $Av_1^\nu \in C[0,T;W)$, for every ν with $0 < \nu < T$. As a result, one finds that, for each $\nu > 0$, one has $v^\nu(t) \in \mathcal{D}(A)$, for $0 \leq t < T$, and $Av^\nu \in C[0,T;W)$. Since v^ν is a mild solution of equation (42.19), in the space W, with $Av^\nu \in L^1_{\text{loc}}[0,T;W)$, it follows from Theorem 42.4 that v^ν is a strong solution in W, as well.

The next step is to show that $v(t) \in \mathcal{D}(A)$, for each $t > 0$, and

$$\lim_{\nu \to 0+} Av^\nu(t) = Av(t), \qquad \text{in } W,$$

where the last limit is valid uniformly on compact sets in $(0,T)$. Let t_1 and t_2 be fixed with $0 < t_1 < t_2 < T$, and consider ν with $0 < \nu < t_1$. Let t satisfy $t_1 \leq t \leq t_2$. From the definition of v^ν we find that

$$Av(t) - Av^\nu(t) = \int_0^\nu Ae^{-A(t-s)} f(s)\, ds, \qquad \text{for } 0 < \nu < t < T.$$

Now inequality (37.11) implies that

$$\|Ae^{-A(t-s)}\|_{\mathcal{L}(W)} \leq M_1 e^{|a|t_2}(t-s)^{-1} \leq M_1 e^{|a|t_2}(t-\nu)^{-1},$$

for $0 \leq s \leq \nu < t$. Consequently, one has

$$\|Av(t) - Av^\nu(t)\| \leq M_1 e^{|a|t_2}(t-\nu)^{-1} \int_0^\nu \|f(s)\|\, ds,$$

for $0 < \nu < t_1 \leq t \leq t_2 < T$. Hence, $Av^\nu \to Av$, as $\nu \to 0^+$, uniformly on compact sets in $(0,T)$.

We have thus shown that, for t in any compact subset of $(0,T)$, the two terms $Av^\nu(t)$ and $f^\nu(t)$ in (42.19) have uniform limits in W, as $\nu \to 0^+$. Also $\partial_t v^\nu(t) \to \partial_t v(t)$, as $\nu \to 0^+$, for each $t > 0$, since $v^\nu(t) = v(t)$, for $\nu < t$. Thus

$$\partial_t v(t) + Av(t) = f(t), \qquad \text{for all } t > 0.$$

As noted above, it then follows from Lemma C.7 in Appendix C that

$$v \in L^1_{\text{loc}}[0,T;\mathcal{D}(A)) \cap C(0,T;\mathcal{D}(A)),$$

and that v is a strong solution of (42.1) in W. From Lemma 42.8, u is a strong solution, as well.

Let us now verify directly the local Hölder continuity of v on $(0,T)$. Let t_1 and t_2 be fixed with $0 < t_1 < t_2 < T$, and set $h_0 = \frac{1}{2}\min(t_1, T - t_2)$. Let $t \in [t_1,t_2]$ and let h satisfy $0 \leq |h| \leq h_0$. We will now show that there is a constant $K = K(t_1,t_2)$, which does not depend on t or h, such that, for $0 \leq r < 1$, one has

$$\|A^r(v(t+h) - v(t))\| \leq K|h|^{1-r}, \qquad \text{for all } t \in [t_1,t_2] \text{ and } |h| \leq h_0.$$

We present here only the argument for $0 \le h \le h_0$. (The reader should verify the case where h is negative.) Since v is a mild solution on $[0, T)$, (42.4) implies that

$$v(t+h) - v(t) = (e^{-Ah} - I)v(t) + \int_t^{t+h} e^{-A(t+h-s)} f(s)\, ds,$$

for $t_1 \le t \le t_2$. Now inequality (37.12) implies that there is a constant $K_1 > 0$ such that

$$\|A^r (e^{-Ah} - I)v(t)\| \le K_1 |h|^{1-r} \sup_{t_1 \le s \le t_2} \|Av(s)\|.$$

Similarly inequality (37.11) and (42.16) imply that there is a $K_2 > 0$ such that

$$\int_t^{t+h} \|A^r e^{-A(t+h-s)} f(s)\| ds \le K_2 e^{|a|(t_2+h_0)} \sup \|f\| \int_t^{t+h} (t+h-s)^{-r} ds,$$

where the sup is taken over all s with $t_1 - h_0 \le s \le t_2 + h_0$. This implies that $v \in C^{0,1-r}_{\text{loc}}(0, T; V^{2r})$, for $0 \le r < 1$; i.e., (42.17) is valid. (42.18) then follows from (42.17) and Theorem 37.5, Item (5).

Lastly, inequality (37.11) implies that $Ae^{-At}u_0 \in L^1_{\text{loc}}[0, \infty; W)$, when $u_0 \in V^{2\rho}$ and $\rho > 0$, since

$$\begin{aligned} \int_0^t \|Ae^{-As}u_0\|\, ds &\le \int_0^t \|A^{1-\rho} e^{-As}\|_{\mathcal{L}(W)} \|A^\rho u_0\|\, ds \\ &\le M_{1-\rho} \int_0^t s^{\rho - 1}\, ds\, \|A^\rho u_0\| < \infty. \end{aligned}$$

Since $u \in C[0, T; V^{2\rho})$, it then follows from Theorem 42.4 and the definition of a strong solution, that u is a strong solution of equation (42.1) in $V^{2\rho}$ on $[0, T)$. □

The Legacy of the Data: It is useful to reflect on the possible singular behavior of mild and strong solutions of (42.1) at $t = 0$. If $u_0 \in W$ and $u_0 \notin V^{2\rho}$, for any $\rho > 0$, then (42.14) is no longer valid for any $r = \rho > 0$. It can happen that the solution u may have some singular behavior at $t = 0$, and the reason for this can be traced directly to the initial condition u_0 and the estimate (37.11). This is a common phenomenon in linear and nonlinear evolutionary equations. Problems occurring only at $t = 0$ are oftentimes traced to the initial data. For additional information on this topic, see Temam (1982).

Notice that when $p = \infty$ in Lemma 42.7, the function $v = v(t)$ defined by (42.8) satisfies $v \in L^\infty_{\text{loc}}[0, T; V^{2r})$ for $0 \le r < 1$, but $r = 1$ is not resolved; see (42.15). As a result, one is unable to conclude that $v \in L^1_{\text{loc}}[0, T; V^2)$; i.e., one cannot assert that v is a strong solution of (42.1). This means that the assumption $f \in L^\infty_{\text{loc}}[0,T;W)$ alone is not strong enough to show that the Leibniz Formula (42.7) is satisfied. However, by assuming greater regularity for the forcing function f, as in Theorem 42.9, one can obtain a stronger conclusion. In the next result we show that by assuming additional spatial regularity for the forcing function f, one obtains a significant improvement in the spatial regularity of the mild solutions.

Theorem 42.10. *In addition to the Standing Hypothesis A, let*

$$f \in L^p_{\text{loc}}[0,T;V^{2\nu}), \qquad \text{where } 1 < p \text{ and } 1 < \nu p \leq \infty. \tag{42.20}$$

Let β satisfy $0 \leq \beta < \nu - \frac{1}{p}$. Assume that $u_0 \in V^{2\mu}$, where $0 < \mu \leq 1+\beta$, and let u be the mild solution of equation (42.1) in W on $[0,T)$. Then u is a strong solution of equation (42.1) in $V^{2\mu}$ on $[0,T)$, and u satisfies

$$u \in C[0,T;V^{2\mu}) \cap C^{0,\theta_0}_{\text{loc}}[0,T;V^{2\rho}) \cap C^{0,\theta_1}_{\text{loc}}(0,T;V^{2+2\beta}), \tag{42.21}$$

where $0 \leq \rho < \mu$, $0 \leq \beta < \nu - \frac{1}{p}$, and θ_0 and θ_1 are positive. Assume instead that (42.20) is replaced by $1 < p$ and $1 < \nu p \leq \infty$ and

$$f \in L^1_{\text{loc}}[0,T;W) \cap L^p_{\text{loc}}(0,T;V^{2\nu}), \tag{42.22}$$

and $u_0 \in W$. Then for every $\tau \in (0,T)$, the translate $u_\tau = u_\tau(t)$ satsifies (42.21), where the interval $[0,T)$ is replaced by $[0,T-\tau)$. Furthermore, u_τ is a strong solution of equation (42.1) in $V^{2\mu}$ on $[0,T-\tau)$, for every μ with $0 \leq \mu \leq 1+\beta$.

Proof. Assume that (42.20) holds. Let $u = u(t) = e^{-At}u_0 + v(t)$ be a mild solution of equation (42.1) in W, where $v = v(t)$ satisfies equation (42.8). As argued in the proof of Lemma 42.7, one uses inequality (37.11) and the Hölder inequality to obtain a constant M such that

(42.23)
$$\begin{aligned}\left\| A^{1+\beta} \int_{t_0}^t e^{-A(t-s)} f(s)\, ds \right\| &\leq \int_{t_0}^t \|A^{1+\beta-\nu} e^{-A(t-s)}\|_{\mathcal{L}(W)} \|A^\nu f(s)\|\, ds \\ &\leq M \left(\int_{t_0}^t (t-s)^{-(1+\beta-\nu)q}\, ds \right)^{\frac{1}{q}} \left(\int_{t_0}^t \|A^\nu f(s)\|^p\, ds \right)^{\frac{1}{p}},\end{aligned}$$

for $0 \leq t_0 \leq t < T$, where $q = p(p-1)^{-1}$, when p is finite, and a similar but different estimate holds for $q = 1$ and $p = \infty$. Since the integrals above are finite when $0 \leq \beta < \nu - \frac{1}{p}$, one obtains $v \in L^\infty_{\text{loc}}[0,T;V^{2+2\beta})$. The argument in Lemma 42.7 for the Hölder continuity of v is applicable here, wih W replaced by $V^{2\nu}$, and hence one can conclude that

$$v \in C^{0,\theta}_{\text{loc}}[0,T;V^{2+2\beta}), \tag{42.24}$$

where $0 < \theta_1 < \nu - \beta - \frac{1}{p}$, see (42.15). Since $u_0 \in V^{2\mu}$, where $0 < \mu \leq 1+\beta$, it follows from (37.14) that

$$e^{-At}u_0 \in C[0,\infty; V^{2\mu}) \cap C^{0,\mu-\rho}_{\text{loc}}[0,\infty; V^{2\rho}) \cap C^{0,1}_{\text{loc}}(0,\infty; V^{2r}),$$

for all $r \geq 0$ and all ρ satisfying $0 \leq \rho < \mu$. Consequently, $u(t) = e^{-At}u_0 + v(t)$ satisfies (42.21).

Now (42.24) implies that $Av \in L^\infty_{\text{loc}}[0,T; W)$, and from (37.11), one finds that $Ae^{-At}u_0 \in L^1_{\text{loc}}[0,\infty; W)$, since

$$\begin{aligned}\int_0^t \|Ae^{-As}u_0\|\, ds &\leq \int_0^t \|A^{1-\mu}e^{-As}\|_{\mathcal{L}(W)}\|A^\mu u_0\|\, ds \\ &\leq M_{1-\mu}\int_0^t s^{\mu-1}\, ds\, \|A^\mu u_0\| < \infty,\end{aligned}$$

for $\mu > 0$. It then follows from Theorem 42.4 and the definition of a strong solution, that u is a strong solution of equation (42.1) in $V^{2\mu}$ on $[0,T)$.

In the case that (42.22) holds in place of (42.20), then inequality (42.23) is valid; <u>however</u> one needs to restrict t_0 to satisfy $0 < t_0 \leq t < T$. In this case, one can conclude that $v \in C^{0,\theta}_{\text{loc}}(0,T; V^{2+2\beta})$, instead of (42.24). It then follows from (37.14) that the mild solution $u(t) = e^{-At}u_0 + v(t)$ satisfies $u(\tau) \in V^{2\mu}$, for any μ with $0 \leq \mu \leq 1+\beta$ and any τ with $0 < \tau < T$. Furthermore, the translate f_τ satisfies (42.20), where the interval $[0,T)$ is replaced by $[0, T-\tau)$. The conclusions concerning the translate u_τ now follows from the arguments of the preceeding paragraphs. □

The Legacy of the Linear Semigroup: Let us reflect on the main difference between Lemma 42.7 and Theorem 42.10. By assuming more spatial regularity for the forcing function f, where $f(t) \in W$ is replaced by $f(t) \in V^{2\nu}$ (for $0 < \nu \leq 1$), one then shows that the mild solution u inherits more spatial regularity, where $u(t) \in V^{2\rho}$ (for $0 \leq \rho < 1$) is replaced by $u(t) \in V^{2+2\beta}$ (for $0 \leq \beta < \nu - \frac{1}{p}$). Since β can be chosen to be arbitrarily close to $\nu - \frac{1}{p}$, we see that Theorem 42.10 includes the implication

$$f(t) \in V^{2\nu} \Longrightarrow u(t) \in V^{2+2\nu-\frac{2}{p}-\epsilon}, \qquad \text{for } t > 0,$$

where $\epsilon > 0$ is arbitrarily small. This improvement in spatial regularity is due entirely to the action of the analytic semigroup e^{-At} in equation (42.3).

It turns out that in the case where the analytic semigroup e^{-At} is compact, for $t > 0$, one can set $\epsilon = 0$ in the last implication. However, in order to do this, we need to introduce the concept of a weak solution for equation (42.1).

4.2.3 Compact Theory and Weak Solutions. Once again, we assume that the Standing Hypothesis B is satisfied. Our objective here is to introduce a fourth solution concept, that of a weak solution, and its relationship to the other solution concepts given above. In particular, we will say that a function $u = u(t)$ is a **weak solution** of equation (42.1) **in the space** V^α, on the interval $[0, T)$, where $0 < T \leq \infty$, provided that $u(0) = u_0 \in V^\alpha$ and $f \in L^2_{\text{loc}}[0, T; V^{\alpha-1})$, and that the following properties are satisfied:

(1) one has

$$u \in L^\infty_{\text{loc}}[0, T; V^\alpha) \cap L^2_{\text{loc}}[0, T; V^{\alpha+1}); \tag{42.25}$$

(2) the function u has a time derivative $\partial_t u$ in $L^p_{\text{loc}}[0, T; V^{\alpha-1})$, for some p with $1 \leq p < \infty$, such that the equation

$$u(t) - u(t_0) = \int_{t_0}^t \partial_t u \, ds \qquad \text{holds in } V^{\alpha-1}, \text{ for } t_0,\, t \in [0, T); \tag{42.26}$$

(3) one has

$$\|u(t)\|_\alpha^2 + \int_{t_0}^t \|A^{\frac{1}{2}} u\|_\alpha^2 \, ds \leq \|u(t_0)\|_\alpha^2 + \int_{t_0}^t \|A^{-\frac{1}{2}} f\|_\alpha^2 \, ds, \tag{42.27}$$

for almost all t_0 and t with $0 \leq t_0 \leq t < T$; and

(4) the function u satisfies the weak-integrated form of equation (42.1), that is to say, one has

$$\begin{aligned} \langle A^{-\frac{1}{2}}(u(t) - u(t_0)), A^{\frac{1}{2}} v\rangle_\alpha &+ \int_{t_0}^t \langle A^{\frac{1}{2}} u(s), A^{\frac{1}{2}} v\rangle_\alpha \, ds \\ &= \int_{t_0}^t \langle A^{-\frac{1}{2}} f(s), A^{\frac{1}{2}} v\rangle_\alpha \, ds, \end{aligned} \tag{42.28}$$

for all $v \in V^{\alpha+1}$ and all t_0 and t with $0 \leq t_0 \leq t < T$.

There are several properties inherited by every weak solution of equation (42.1). In particular, one has the following result.

Lemma 42.11. *In addition to the Standing Hypotheses B, let $u = u(t)$ be a weak solution of equation (42.1) in V^α, on the interval $[0, T)$, where $0 < T \leq \infty$,*

$$u_0 \in V^\alpha \quad \textit{and} \quad f \in L^p_{\text{loc}}[0, T; V^{\alpha-1}), \qquad \textit{where } 2 \leq p \leq \infty.$$

Then the following properties are valid:

(1) *The time derivative satisfies $\partial_t u \in L^2_{\text{loc}}[0, T; V^{\alpha-1})$ and u satisfies*

$$u \in C[0, T; V^\alpha) \cap C^{0,\frac{1}{2}}_{\text{loc}}[0, T; V^{\alpha-1}).$$

(2) *One has $\partial_t u + Au \overset{\text{a.e.}}{=} f$ in the space $V^{\alpha-1}$, and u is a mild solution of equation (42.1) in V^β, for each $\beta < \alpha + (p-2)p^{-1}$.*

Proof. Item (1): Since $A^{\frac{1}{2}}V^\alpha = V^{\alpha-1}$ and $A^{\frac{1}{2}}V^{\alpha+1} = V^\alpha$, it follows from (42.25) that $Au \in L^2_{\text{loc}}[0,T;V^{\alpha-1})$. Since $f \in L^2_{\text{loc}}[0,T;V^{\alpha-1})$, one has $\partial_t u = -Au + f \in L^2_{\text{loc}}[0,T;V^{\alpha-1})$. It then follows from (42.25) and the Continuity Lemma 37.9 that $u \in C[0,T;V^\alpha)$. Moreover, equation (42.26), with $p = 2$ and the Hölder inequality implies that

$$\begin{aligned}||A^{-\frac{1}{2}}(u(t) - u(t_0))||_\alpha &\le \int_{t_0}^t ||A^{-\frac{1}{2}}\partial_t u||_\alpha \, ds \\ &\le (t-t_0)^{\frac{1}{2}} \left(\int_{t_0}^t ||A^{-\frac{1}{2}}\partial_t u||_\alpha^2 \, ds \right)^{\frac{1}{2}},\end{aligned}$$

for $0 \le t_0 \le t < T$. This, in turn, implies the Hölder continuity of u.

Item (2): We need to show that u is a mild solution. For this purpose, we set $W = V^{\alpha-1}$, $W^{2\rho} = V^{\alpha-1+2\rho}$, for $\rho \ge 0$, and $\alpha - 1 + 2\rho = \beta$. Then one has

$$\beta < \alpha + \frac{p-2}{p} \quad \Longleftrightarrow \quad \rho < 1 - \frac{1}{p}, \text{ with } p \ge 2.$$

The conclusion now follows from Lemma 42.7, Item (3). □

In order to develop the theory of weak solutions of equation (42.1) in a Hilbert space H, we begin with

$$u_0 \in V^\alpha \quad \text{and} \quad f : [0,T) \to V^{\alpha-1}.$$

This theory does have implications in the related theory of mild and strong solutions. It should be helpful to fix the theory of weak solutions in the context of Section 4.2.2, by defining $W \stackrel{\text{def}}{=} V^{\alpha-1}$ and $W^r \stackrel{\text{def}}{=} V^{\alpha-1+r}$, for $r \ge 0$. In terms of the W^r-scale of Hilbert spaces, one then has

$$u_0 \in W^1 \quad \text{and} \quad f : [0,T) \to W = W^0.$$

In the next result, we show the existence of weak solutions and the connection between these solutions and mild solutions. We will use the inequality $\lambda_1||u||_\alpha^2 \le ||A^{\frac{1}{2}}u||_\alpha^2$, where $\lambda_1 > 0$ is the first eigenvalue of the positive, selfadjoint operator A, see (32.2).

Theorem 42.12. *In addition to the Standing Hypothesis B, let*

$$u_0 \in V^\alpha \quad \textit{and} \quad f \in L^p_{\text{loc}}[0,T;V^{\alpha-1}), \quad \textit{where } 2 \le p \le \infty.$$

Then the following properties hold:

(1) *There is a unique weak solution $u = u(t)$ of equation (42.1), in the space V^α, on the interval $[0,T)$, with $u(0) = u_0$,*

$$\begin{aligned}&\partial_t u \in L^2_{\text{loc}}[0,T;V^{\alpha-1}), \quad \textit{and} \\ &u \in C[0,T;V^\beta) \cap C^{0,\theta_1}_{\text{loc}}[0,T;V^\sigma) \cap L^2_{\text{loc}}[0,T;V^{\alpha+1}),\end{aligned} \tag{42.29}$$

for each β and σ, where $\beta \leq \alpha$, $\sigma < \alpha$, and $\theta_1 = \theta_1(\sigma) > 0$. Furthermore, u is a mild solution of equation (42.1) in V^β, for each $\beta < \alpha + (p-2)p^{-1}$.

(2) *For $0 \leq t < T$, the following inequalities are valid in the space V^α:*

$$\|u(t)\|_\alpha^2 \leq e^{-\lambda_1 t}\|u_0\|_\alpha^2 + \int_0^t e^{-\lambda_1(t-s)}\|A^{-\frac{1}{2}}f\|_\alpha^2\, ds,$$

$$\int_0^t \|A^{\frac{1}{2}}u\|_\alpha^2\, ds \leq \|u_0\|_\alpha^2 + \int_0^t \|A^{-\frac{1}{2}}f\|_\alpha^2\, ds, \tag{42.30}$$

$$\int_0^t \|A^{-\frac{1}{2}}\partial_t u\|_\alpha^2\, ds \leq 2\|u_0\|_\alpha^2 + 4\int_0^t \|A^{-\frac{1}{2}}f\|_\alpha^2\, ds.$$

(3) *If $p > 2$, then u is a mild solution of equation (42.1) in V^α, and in addition to (42.29), one has*

$$u \in C_{\mathrm{loc}}^{0,\theta_2}(0,T;V^\sigma), \tag{42.31}$$

for each σ, where $\alpha \leq \sigma < \alpha + 1 - \frac{2}{p}$ and $\theta_2 = \theta_2(\sigma) > 0$.

Proof. **Existence of a Weak Solution:** The proof of this result is based on a methodology which is referred to as the Bubnov-Galerkin approximations.[12] As we will see, this methodology plays an important role in the study of solutions of evolutionary equations. However, before presenting this approach, it is useful to develop a heuristic (and nonrigorous) argument which, on the one hand, presents the main ideas, and on the other hand, motivates the appropriateness of the Bubnov-Galerkin approximations.

In order to "prove" (42.25), we apply $A^{\frac{\alpha}{2}}$ to equation (42.1), take the scalar product of equation (42.1) with $A^{\frac{\alpha}{2}}u$, and use the identity

$$\langle A^{\frac{\alpha}{2}}\partial_t u, A^{\frac{\alpha}{2}}u\rangle = \frac{1}{2}\frac{d}{dt}\|u\|_\alpha^2$$

along with the Young inequality to obtain

$$\partial_t\|u\|_\alpha^2 + \|A^{\frac{1}{2}}u\|_\alpha^2 \leq \|A^{-\frac{1}{2}}f\|_\alpha^2. \tag{42.32}$$

By using the inequality $\lambda_1\|u\|_\alpha^2 \leq \|A^{\frac{1}{2}}u\|_\alpha^2$, we see that

$$\partial_t\|u\|_\alpha^2 + \lambda_1\|u\|_\alpha^2 \leq \|A^{-\frac{1}{2}}f\|_\alpha^2.$$

The Gronwall inequality then implies the first inequality in (42.30). By integrating inequality (42.32), one obtains (42.27), which, in turn, implies

[12] See Sections 6.2, 6.8, and 6.10 for more information.

the second inequality in (42.30). In addition, by the argument of Lemma 42.11, one finds that $\partial_t u \in L^2_{\text{loc}}[0, T; V^{\alpha-1})$ and

$$\int_0^t \|A^{-\frac{1}{2}}\partial_t u\|_\alpha^2 \, ds \le 2\int_0^t \|A^{\frac{1}{2}}u\|_\alpha^2 \, ds + 2\int_0^t \|A^{-\frac{1}{2}}f\|_\alpha^2 \, ds$$
$$\le 2\|u_0\|_\alpha^2 + 4\int_0^t \|A^{-\frac{1}{2}}f\|_\alpha^2 \, ds,$$

which completes the "proof" of (42.30).

Why is this heuristic argument nonrigorous? The answer is that the reasoning is circular. One is essentially assuming the conclusion. For example, one is assuming that $\|A^{\frac{1}{2}}u\|_\alpha$ is finite in order to derive (42.32), which implies (42.30) and which, in turn, implies that $\|A^{\frac{1}{2}}u\|_\alpha$ is finite, almost everywhere. The Bubnov-Galerkin approximations offer a methodology for using the arguments given above, but do this in a rigorous manner.

In order to describe this methodology, we return to the eigenvalues of the operator A, which are repeated according to their multiplicities, as in (32.2). The associated collection of eigenvectors $\{e_1, e_2, e_3, \cdots\}$ forms an orthonormal basis for H. For each integer $n \ge 1$ we let $P = P_n$ denote the orthogonal projection of H onto Span $\{e_1, \cdots, e_n\}$, and set $Q = Q_n = I - P$. The projections P and Q are called the **spectral projections** determined by the operator A.

For each $u \in H$ we define p and q by $p = Pu$ and $q = Qu$. Notice that $p = p_n$ and $q = q_n$ depend on the n modes $\{e_1, \cdots, e_n\}$ used to define the spectral projections $P = P_n$ and $Q = Q_n$. These spectral projections commute with A, i.e., $PA = AP$ and $QA = AQ$, where the last equation is restricted to $\mathcal{D}(A)$, the domain of A. The nth order Bubnov-Galerkin approximation of (42.1) is given by the solutions of the nth order ordinary differential equation

$$\partial_t p + Ap = Pf. \tag{42.33}$$

More precisely, if $u = u(t)$ is a solution of (42.1) with initial condition $u(0) = u_0 \in H$, then the solution $p(t)$ of (42.33) that satisfies $p(0) = p_0$, where $p_0 = Pu_0$, is said to be the nth **order Bubnov-Galerkin approximation** of u. We will sometimes write $u_n(t) = p(t)$ to emphasize the dependence on n.

The strategy behind the use of the Bubnov-Galerkin approximations is to find a suitable subsequence of $u_n = p$, which is convergent and such that the relations (42.20) and (42.30) hold in the limit as $n \to \infty$, where $u = \lim u_n$ (in a suitable sense). The key to succeeding in this strategy is to derive estimates of various norms of the approximants $u_n = p$, estimates which do not depend on n. In particular, if one replaces $u = u(t)$ with $u_n = u_n(t)$, then the argument in the paragraph containing (42.32) is completely rigorous (see below). Furthermore, the right side of the estimates

(42.27) and (42.30) do not depend on n. Such estimates are carried out in detail for the Navier-Stokes equations (see Sections 6.2 and 6.8). The latter equations differ from equation (42.1) in that the Navier-Stokes equations contain a (nonlinear) inertial term $B(u,u)$. If one sets $B(u,u) \equiv 0$, then the arguments in Sections 6.2 and 6.8, with essentially no other change, apply to equation (42.1). The existence of a convergent subsequence of the Bubnov-Galerkin approximations follows from the Compactness Lemmas in Section 6.3.

The reason that the heuristic argument given above applies (rigorously) to the Bubnov-Galerkin approximations is that these approximants lie in a finite dimensional space and all norms $||\cdot||_\alpha$ on a finite dimensional space are equivalent. Consequently the norms $||A^{\frac{1}{2}}u_n||_\alpha$ and $||Au_n||_\alpha$ are finite when $u_n \in \mathcal{R}(P_n)$, since the latter space is finite dimensional.

In order to use the Compactness Lemma, we need to verify that $A^{\frac{1}{2}}u_n$ and $A^{-\frac{1}{2}}\partial_t u_n$ are bounded sequences in the space $L^2_{\rm loc}[0,T;V^\alpha)$. From the comments given above, we see that inequalities (42.30) are valid when u is replaced by u_n. Since the right sides of these inequalities do not depend on n, we see that the sequences mentioned above are bounded.

In order to verify equation (42.28), for the "limit" u, we first note that the Bubnov-Galerkin approximations u_n satisfy the identity

$$\langle u_n(t)-u_n(t_0),P_nv\rangle_\alpha+\int_{t_0}^t\langle A^{\frac{1}{2}}u_n,P_nA^{\frac{1}{2}}v\rangle_\alpha\,ds=\int_{t_0}^t\langle A^{-\frac{1}{2}}f,P_nA^{\frac{1}{2}}v\rangle_\alpha\,ds,$$

for all $v \in V^{\alpha+1}$ and for all t and t_0 with $0 \le t_0 \le t < T$. Since the Compactness I Lemma 63.2 implies that, for a suitable subsequence, one has $u_n \xrightarrow{s} u$ in V^α and $A^{\frac{1}{2}}u_n \xrightarrow{w} A^{\frac{1}{2}}u$ in $L^2_{\rm loc}[0,T;V^\alpha)$, and since $P_nA^{\frac{1}{2}}v \xrightarrow{s} A^{\frac{1}{2}}v$ in V^α, the last equation holds in the limit, which is equation (42.28).

One needs to verify that if $u_n \xrightarrow{s} u$ in $L^2_{\rm loc}[0,T;V^\alpha)$, then the weak limit of the sequence $\partial_t u_n$ in $L^2_{\rm loc}[0,T;V^{\alpha-1})$ is, in fact, $\partial_t u$. For the Navier-Stokes equations, the corresponding issue is addressed in Lemma 63.4. The reader should verify that the same argument applies here.

Uniqueness of a Weak Solution: Assume that there are two weak solutions u_1 and u_2 with $u_1(0)=u_2(0)=u_0$. Let $w=u_1-u_2$. From the definition of a weak solution and Lemma 42.11, one finds that w satisfies

$$w \in L^\infty_{\rm loc}[0,T;V^\alpha)\cap L^2_{\rm loc}[0,T;V^{\alpha+1}), \qquad \partial_t w \in L^2_{\rm loc}[0,T;V^{\alpha-1}).$$

From relations (42.25) and (42.28), one can show that[13] $\partial_t w + Aw \overset{\text{a.e.}}{=} 0$ in the space $V^{\alpha-1}$. By taking the scalar product of the last equation with w and using the Continuity Lemma 37.9, one obtains

$$\frac{1}{2}\frac{d}{dt}||w||^2_{\alpha-1}+||A^{\frac{1}{2}}w||^2_{\alpha-1}=0,$$

[13] Since w is a weak solution with $w(0)=0$, one cannot use the theory of mild solutions and C_0-semigroups to prove that $w(t)=0$, for $t \ge 0$.

which implies that

$$\partial_t \|w\|_{\alpha-1}^2 + 2\lambda_1 \|w\|_{\alpha-1}^2 \leq 0.$$

The Gronwall inequality then yields $\|w(t)\|_{\alpha-1}^2 \leq e^{-2\lambda_1 t}\|w(0)\|_{\alpha-1}^2$, for $t \geq 0$. Since $w(0) = 0$, we see that $w(t) = 0$, for all $t \geq 0$.

Other Properties: From Lemma 42.11, Item (2), we see that u is a mild solution in V^β, for each $\beta < \alpha + (p-2)p^{-1}$. In particular, u is a mild solution in V^α, when $p > 2$. Inequalities (42.30) follow from the fact that they hold for the Bubnov-Galerkin approximations u_n, and that $u_n \overset{s}{\to} u$ in V^α and $A^{\frac{1}{2}}u_n \overset{w}{\to} A^{\frac{1}{2}}u$ in $L^2_{\mathrm{loc}}[0,T;V^\alpha)$. Finally, the reader should verify that the Hölder property (42.31) follows from (42.14). □

Theorem 42.12 is sometimes used as a part of a bootstrap argument, where one has $u_0 \in V^{\alpha+1}$ and $f \in L^p_{\mathrm{loc}}[0,T;V^\alpha)$, for some p with $2 \leq p \leq \infty$. Since one has the imbeddings

$$u_0 \in V^{\alpha+1} \hookrightarrow V^\alpha \quad \text{and} \quad f \in L^p_{\mathrm{loc}}[0,T;V^\alpha) \mapsto L^p_{\mathrm{loc}}[0,T;V^{\alpha-1}),$$

Theorem 42.12 is applicable for both α and $\alpha+1$. Furthermore, the uniqueness of the weak solutions implies that the weak solution $u = u(t)$ in V^α is, in fact, a weak solution in $V^{\alpha+1}$. Hence, it is a mild solution in V^β, for $\beta < \alpha + 1 + (p-2)p^{-1}$ by Theorem 42.12. As a result, inequalities (42.30) are valid, where α is replaced by $\alpha + 1$; and from (42.29), one has

$$u \in C[0,T;V^\beta) \cap C^{0,\theta_1}_{\mathrm{loc}}[0,T;V^\sigma) \cap L^2_{\mathrm{loc}}[0,T;V^{\alpha+2}), \tag{42.34}$$

for $\beta \leq \alpha + 1$ and $\sigma < \alpha + 1$, where $\theta_1 = \theta_1(\sigma) > 0$. If in addition, $p > 2$, then (42.31) implies that for some $\theta_2 = \theta_2(\sigma) > 0$, one has

$$u \in C^{0,\theta_2}_{\mathrm{loc}}(0,T;V^\sigma), \qquad \text{for } \alpha + 1 \leq \sigma < \alpha + 2 - \frac{2}{p}. \tag{42.35}$$

Moreover, in this setting, one can also apply Theorem 42.10, which leads to valuable information about the strong solutions of equation (42.1).

Corollary 42.13. *In addition to the Standing Hypothesis B, let $u_0 \in V^\alpha$ and let $f \in L^p_{\mathrm{loc}}[0,T;V^\alpha)$, for some p with $2 \leq p \leq \infty$. Then the following statements are valid:*

(1) *The weak solution u of equation (42.1) in V^α on $[0,T)$ is a strong solution of equation (42.1) in V^α on $[0,T)$, and u satisfies*

$$u \in C[0,T;V^\alpha) \cap C^{0,\theta_0}_{\mathrm{loc}}[0,T;V^\sigma) \cap C^{0,\theta_1}_{\mathrm{loc}}(0,T;V^{\alpha+1+2\beta}), \tag{42.36}$$

for every σ and β with $\sigma < \alpha$ and $0 \leq \beta < 1 - \frac{1}{p}$, where $\theta_0 = \theta_0(\sigma)$ and $\theta_1 = \theta_1(\beta)$ are positive.

(2) *For every $\tau \in (0,T)$, one has $u(\tau) \in V^{\alpha+1}$, and the conclusions in Items (1)–(3) in Theorem 42.12 are valid on the interval $\tau \leq t < T$, with α replaced by $\alpha+1$ and u_0 replaced by $u(\tau)$. In addition, (42.34) and (42.35) are valid for u_τ, where the interval $[0,T)$ is replaced by $[0,T-\tau)$. Furthermore, the following inequalities are valid, for $0 < \tau < t < T$:*

(42.37)
$$\|u(t)\|_{\alpha+1}^2 \leq e^{-\lambda_1(t-\tau)}\|u(\tau)\|_{\alpha+1}^2 + \int_\tau^t e^{-\lambda_1(t-s)}\|f\|_\alpha^2\, ds,$$
$$\int_\tau^t \|A^{\frac{1}{2}}u\|_{\alpha+1}^2\, ds \leq \|u(\tau)\|_{\alpha+1}^2 + \int_\tau^t \|f\|_\alpha^2\, ds,$$
$$\int_\tau^t \|A^{-\frac{1}{2}}\partial_t u\|_{\alpha+1}^2\, ds \leq 2\|u(\tau)\|_{\alpha+1}^2 + 4\int_\tau^t \|f\|_\alpha^2\, ds.$$

Proof. Item(1): Define $W^{2r} = V^{\alpha-1+2r}$, for $r \geq 0$, and set $\mu = \nu = 1$. Then Theorem 42.10 implies that u is a strong solution of equation (42.1) in $W^{2\mu} = V^\alpha$ on $[0,T)$. Since $W^{2+2\beta} = V^{\alpha+1+2\beta}$, (42.21) implies (42.36).

Item (2): Since $u(\tau) \in V^{\alpha+1}$, for $\tau \in (0,T)$, the conclusion that u_τ is a strong solution of equation (42.1) in $V^{\alpha+1}$ follows from Theorem 42.10, as well. Accordingly, u_τ is a weak solution of equation (42.1) in $V^{\alpha+1}$. We now apply Theorem 42.12 once again, where α is replaced by $\alpha+1$, u_0 is replaced by $u(\tau)$, and u is replaced by the translate u_τ, to obtain (42.37). □

This leads to another question: What happens if the forcing function f has additional temporal regularity? The answers, which are given in the next two results, are rather surprising: The mild solution u acquires additional spatial regularity. Once again, it is the analytic semigroup e^{-At} that lies behind this phenomenon of converting temporal regularity into spatial regularity.

Theorem 42.14. *In addition to the Standing Hypothesis B, let*
$$f \in C[0,T;V^\alpha) \cap W^{1,p}_{\text{loc}}[0,T;V^{\alpha-1}),$$
for some p with $2 \leq p \leq \infty$, and let $g = \partial_t f$. Let $u_0 \in V^{\alpha+2}$ and define $v_0 = f(0) - Au_0 \in V^\alpha$. Let $u = u(t)$ be the weak solution of equation (42.1) in the space V^α, with $u(0) = u_0$. Then the following properties hold:

(1) *u is a mild solution of equation (42.1) in V^σ, for each $\sigma < \alpha+2$, and u is a strong solution in V^β, for each $\beta \leq \alpha$. Moreover, one has*

(42.38)
$$u \in C^1[0,T;V^\beta) \cap C[0,T;V^\nu) \cap C^{0,\theta_1}_{\text{loc}}[0,T;V^\sigma),$$

for all β, ν, and σ with $\beta \leq \alpha$, $\nu \leq \alpha+2$, and $\sigma < \alpha+2$, where $\theta_1 = \theta_1(\sigma) > 0$.

(2) *The function* $v \stackrel{\text{def}}{=} \partial_t u$ *satisfies*

$$v(t) = e^{-At}v_0 + \int_0^t e^{-A(t-r)}g(r)\,dr, \qquad \text{for } t \geq 0, \tag{42.39}$$

in any space V^β, *with* $\beta < \alpha$, *and* v *is a weak solution of* $\partial_t v + Av = g(t)$ *in the space* V^α *with*

$$v \in C[0,T;V^\beta) \cap L^2_{\text{loc}}[0,T;V^{\alpha+1}) \cap C^{0,\theta_2}_{\text{loc}}[0,T;V^\sigma), \tag{42.40}$$

for all β *and* σ *with* $\beta \leq \alpha$ *and* $\sigma < \alpha$, *where* $\theta_2 = \theta_2(\sigma) > 0$.

Proof. Since $f \in C[0,T;V^\alpha) \mapsto L^\infty_{\text{loc}}[0,T;V^\alpha)$ and $u_0 \in V^{\alpha+2} \hookrightarrow V^{\alpha+1}$, it follows from Theorem 42.12 (with $p = \infty$ and with α replaced by $\alpha+1$) that u is a mild solution of equation (42.3) in V^σ, for each $\sigma < \alpha + 2$. Also, Corollary 42.13, with the continuous imbedding $V^\alpha \mapsto V^\beta$, for $\beta \leq \alpha$, implies that u is a strong solution of (42.1) in V^β, for each $\beta \leq \alpha$. Furthermore, (42.34) is valid. Let $w = w(t)$ and $z = z(t)$ be defined by

$$w(t) = e^{-At}v_0 + \int_0^t e^{-A(t-r)}g(r)\,dr \tag{42.41}$$

and $z(t) = u_0 + \int_0^t w(s)\,ds$. Since $g \in L^2_{\text{loc}}[0,T;V^{\alpha-1})$ and $v_0 \in V^\alpha$, it follows from Theorem 42.12 that equation (42.41) is valid in V^β, for each $\beta < \alpha$, and that, with f replaced by g, w is a weak solution of equation (42.1) in V^α which satisfies (42.29), which in turn, implies (42.40). Hence there is a $\theta_3 > 0$ such that

$$z \in C^1[0,T;V^\beta) \cap C^{0,\theta_3}_{\text{loc}}[0,T;V^{\alpha+1}), \qquad \text{for } \beta \leq \alpha.$$

Assume for the moment that z is a strong solution of equation (42.1) in V^α. Since $z(0) = u_0 = u(0)$; and since every strong solution of equation (42.1) is a mild solution, see (42.12); and since the mild solution is uniquely determined by the initial condition, one has $z(t) = u(t)$, for all $t \geq 0$. As a result, one obtains $w(t) = \partial_t z(t) = \partial_t u(t) = v(t)$, for all $t > 0$. Consequently one finds that $u \in C^1[0,T;V^\beta)$, for each β with $\beta \leq \alpha$, and v satisfies (42.39) and (42.40). Furthermore, one has $Au = f - \partial_t u \in C[0,T;V^\alpha)$, which implies that $u \in C[0,T;V^{\alpha+2}) \mapsto C[0,T;V^\nu)$, for $\nu \leq \alpha+2$. As a result, the Hölder continuity in (42.38) now follows from (42.34), with $\alpha+1$ replaced by $\alpha+2$.

It remains to show that z is a strong solution of equation (42.1) in V^α.

We begin by calculating $-A\int_0^t w(s)\,ds$.

(42.42)

$$-A\int_0^t w(s)\,ds = -A\int_0^t e^{-As}v_0\,ds + (-A)\int_0^t\int_0^s e^{-A(s-r)}g(r)\,dr\,ds$$

(by the Fubini Theorem)

$$= -A\int_0^t e^{-As}v_0\,ds + (-A)\int_0^t\int_r^t e^{-A(s-r)}g(r)\,ds\,dr$$

(by equation (31.6) and the closedness of A)

$$= (e^{-At} - I)v_0 + \int_0^t (e^{-A(t-r)} - I)g(r)\,dr$$

$$= w(t) - v_0 - \int_0^t g(r)\,dr = w(t) + Au_0 - f(t).$$

Hence one has $-A(\int_0^t w(s)\,ds + u_0) = w(t) - f(t)$, or $-Az(t) = \partial_t z(t) - f(t)$ in V^α. This implies that $z \in C[0,T;V^{\alpha+2})$ and that z is a strong solution of equation (42.1) in V^α. (Recall that the domain of A in V^α is $\mathcal{D}(A)_{V^\alpha} = V^{\alpha+2}$, see (37.10).) □

When the operator A is given by $-\Delta$, with suitable boundary conditions, the last theorem states that: if the forcing function f has one derivative in time, then the solution u gains one derivative in time and two derivatives in space.

Next we return to an issue raised in Theorem 42.14. What is shown there is that if $u_0 \in V^{\alpha+2}$, then u satisfies property (42.38). One can ask what happens with other mild solutions, where $u_0 \in V^\alpha$, but $u_0 \notin V^{\alpha+2}$? An answer to this question is in the following result.

Theorem 42.15. *In addition to the Standing Hypothesis B, let*

$$f \in C[0,T;V^\alpha) \cap W^{1,2}_{\mathrm{loc}}[0,T;V^{\alpha-1}).$$

For any $u_0 \in V^\alpha$, let $u = u(t)$ denote the weak solution of equation (42.1) in V^α on $[0,T)$. Then the following hold:

(1) *u is a strong solution in V^α, and it satisfies (42.29), (42.31), and (42.36), with $p = \infty$.*
(2) *u is a mild solution in V^β, for each $\beta < \alpha + 1$.*
(3) *u satisfies*

$$u \in C^1(0,T;V^\beta) \cap C(0,T;V^\nu) \cap C^{0,\theta_4}_{\mathrm{loc}}(0,T;V^\sigma), \tag{42.43}$$

for every β, ν, and σ with $\beta \le \alpha$, $\nu \le \alpha + 2$, and $\sigma < \alpha + 2$, where $\theta_4 = \theta_4(\sigma) > 0$.

(4) *For every $\tau \in (0,T)$, the translate u_τ is a strong solution of (42.1), where f is replaced by f_τ, in $V^{\alpha+1}$ on $[0, T-\tau)$.*

Proof. Items (1) and (2) follow directly from Theorem 42.12 and Corollary 42.13. For the proof of Items (3) and (4), we first fix ϵ, where $0 < \epsilon < T$. Since $\int_0^t \|A^{\frac{1}{2}}u(s)\|_\alpha^2\, ds < \infty$, for every $t \in [0, \epsilon)$, it follows from Theorem 42.12 that there is a time $t_0 \in (0, \epsilon)$ where $u(t_0) \in V^{\alpha+1}$. Consequently, (42.34) implies that

$$u \in C[t_0, T; V^\beta) \cap C^{0,\theta_5}_{\text{loc}}[t_0, T; V^\sigma) \cap L^2_{\text{loc}}[t_0, T; V^{\alpha+2}),$$

for every β and σ with $\beta \leq \alpha + 1$ and $\sigma < \alpha + 1$, where $\theta_5 = \theta_5(\sigma) > 0$. Since $\int_{t_0}^t \|Au(s)\|_\alpha^2\, ds < \infty$, for every $t \in [t_0, T)$, it follows that there is a time $t_1 \in (t_0, \epsilon)$ where $u(t_1) \in V^{\alpha+2}$. Consequently, Theorem 42.14 is applicable, where u_0 is replaced by $u(t_1)$. As a result, one has

$$u \in C^1[t_1, T; V^\beta) \cap C[t_1, T; V^\nu) \cap C^{0,\theta_6}_{\text{loc}}[t_1, T; V^\sigma),$$

for every β, ν, and σ with $\beta \leq \alpha + 1$, $\nu \leq \alpha + 2$ and $\sigma < \alpha + 2$, where $\theta_6 = \theta_6(\sigma) > 0$. Since ϵ is arbitrary in $(0, T)$ and $0 < t_1 < \epsilon$, this implies that u satisfies (42.43), and that the translate u_τ is a strong solution in $V^{\alpha+1}$. □

There are other fruits which are yet to be gleaned in this harvest. We have not fully exploited the power of the Happy Reaper 42.12. For example, (42.31) and (42.35) were not used in Theorem 42.14. These morsels may be used at another feast.

4.2.4. The C_0-Theory, Revisited. We now return to our starting point, where $(e^{-At}, -A)$ is a C_0-semigroup on a Banach space W. Our goal here is to present some generalizations of Theorems 42.14 and 42.15, which are valid in the absence of the Standing Hypothesis A. The proofs of the following two results are left as an exercise.

Theorem 42.16. *Let $(e^{-At}, -A)$ be a C_0-semigroup on a Banach space W, and let $f \in W^{1,p}_{\text{loc}}[0, T; W)$, for some p with $1 \leq p \leq \infty$. Let $g = \partial_t f$. Let $u_0 \in \mathcal{D}(A)$ and set $y_0 = f(0) - Au_0 \in W$. Let $u = u(t)$ be the mild solution of equation (42.1) on W on $[0, T)$. Then u is a strong solution of equation (42.1) in W, and it satisfies*

$$u \in C[0, T; W) \cap C^1[0, T; W) \cap C[0, T; \mathcal{D}(A)).$$

Moreover, $y \stackrel{\text{def}}{=} \partial_t u$ is a mild solution of $\partial_t y + Ay = g(t)$ in W on $[0, T)$.

Theorem 42.17. *Let $(e^{-At}, -A)$ be a C_0-semigroup on a Banach space W that is differentiable for $t > 0$. Let $f \in W^{1,p}_{\text{loc}}[0, T; W)$, for some p with $1 \leq p \leq \infty$, and set $g = \partial_t f$. Let $u_0 \in W$. Then $u = u(t)$, which is defined by equation (42.3), is a strong solution of equation (42.1) in W on $[0, T)$ and*

$$u \in C[0, T; W) \cap C^1(0, T; W),$$

and for every $\tau \in (0,T)$, the translate u_τ is a strong solution of (42.1), with f replaced by f_τ, in W on $[0,T-\tau)$ with

$$u_\tau \in C[0,T-\tau;W) \cap C^1[0,T-\tau;W) \cap C[0,T-\tau;\mathcal{D}(A)).$$

4.3. Linear Skew Product Semiflows.

Let $\mathcal{E} = W \times M$ be given where W is a fixed Banach space (the **state space**) and M is a metric space (the **base space**). Assume that $m_t \stackrel{\text{def}}{=} \sigma(m,t)$ is a semiflow on M. A **linear skew product semiflow** on $\mathcal{E}$ is a mapping $\pi = (\Phi, \sigma)$ of the form

$$\pi(w,m,t) = (\Phi(m,t)w,\ m_t), \qquad \text{for } t \geq 0$$

with the following four properties:

(1) $\Phi(m,0) = I$, the identity operator, for all $m \in M$;
(2) $\Phi(m,t) \in \mathcal{L} = \mathcal{L}(W)$ is a bounded linear mapping from W into W that satisfies the cocycle identity

$$\Phi(m,s+t) = \Phi(m_t,s)\Phi(m,t), \qquad \text{for } m \in M \text{ and } s,t \geq 0; \tag{43.1}$$

(3) the mapping from $\mathcal{E} \times (0,\infty)$ into W given by $(w,m,t) \to \Phi(m,t)w$ is continuous; and
(4) for each $(w,m) \in \mathcal{E}$ the mapping $t \to \Phi(m,t)w$ is continuous at $t = 0$, and for each $w \in W$ the limit $\lim_{t\to 0^+} \Phi(m,t)w = w$ is uniform for m in compact sets; i.e., for every compact set $M_0 \subset M$, $w \in W$, and $\epsilon > 0$, there is a $\delta > 0$ such that $\|\Phi(m,t)w - w\| \leq \epsilon$, for all $m \in M_0$ and $t \in [0,\delta]$.

For most applications one is interested in the case where M is a compact metric space, and in many of these applications, the set M is invariant, i.e., $\sigma(M,t) = M$, for all $t \geq 0$. For such applications, the following feature is useful.

Lemma 43.1. *Let $\pi = (\Phi,\sigma)$ be a linear skew product semiflow on $\mathcal{E} = W \times M$. Then for each compact, positively invariant set $M_0 \subset M$, there exist constants $K \geq 1$ and $a \in \mathbb{R}$ such that*

$$\|\Phi(m,t)w\| \leq K\|w\|e^{at}, \qquad \textit{for all } w \in W,\ m \in M_0,\ \textit{and } t \geq 0. \tag{43.2}$$

Proof. This proof is based on Sacker and Sell (1994). Let M_0 be a given compact, positively invariant set in M. We claim that there is an $h > 0$ such that

$$K \stackrel{\text{def}}{=} \sup\{\|\Phi(m,t)\|_{\mathcal{L}} : m \in M_0,\ 0 \leq t \leq h\} < \infty,$$

where $\mathcal{L} = \mathcal{L}(W)$. If K is not finite, then there are sequences $m_n \in M_0$ and $t_n > 0$ such that $t_n \to 0$ and $\|\Phi(m_n, t_n)\|_{\mathcal{L}} \geq n$. The Uniform Boundedness Principle (see Appendix A.9) then implies that there is a $w \in W$ such that $\|\Phi(m_n, t_n)w\|$ is unbounded. However this contradicts Property (4) in the definition of a linear skew product semiflow. Hence $K < \infty$. Since $\|\Phi(m, 0)\|_{\mathcal{L}} = 1$, one has $K \geq 1$. Next we define $a \stackrel{\text{def}}{=} h^{-1} \log K$, and for each $t \geq 0$, we write $t = nh + \tau$, where n is a nonnegative integer and $0 \leq \tau < h$. Inequality (43.2) then follows from an application of the cocycle identity. (See the proof of Lemma 31.1 for a similar argument.) We will omit these details. □

There are two prototypical examples of a linear skew product semiflows in the finite dimensional setting. In the first example, $\Phi(m, t)$ is the fundamental solution operator (or solution matrix) of the linear ordinary differential equation

$$\partial_t w = A(m_t)w = A(\sigma(m, t))w$$

that satisfies $\Phi(m, 0) = I$, where $A(m)$ is a continuous $(n \times n)$ matrix-valued function defined on M. For the second example, we consider the time varying problem $m = m(t)$, where m is a $n \times n$ matrix valued function that satisfies

$$m(\cdot) \in L^\infty \cap C = L^\infty(\mathbb{R}; \mathcal{L}(\mathbb{R}^n)) \cap C(\mathbb{R}; \mathcal{L}(\mathbb{R}^n)).$$

In this case, the flow σ on $L^\infty \cap C$ is the translational flow $(m, \tau) \to m_\tau$ (see Appendix B). Also $\Phi(m, t)$ is the fundamental solution operator of the ordinary differential equation $\partial_t w = m(t)w$, see Miller and Sell (1970), Sell (1967, 1971), and Sacker and Sell (1974, 1976a,b). As we show in the next section, one can construct similar examples in the infinite dimensional setting.

Let us now restrict our treatment to the linear skew product semiflows $\pi = (\Phi, \sigma)$ on $\mathcal{E} = W \times M$, where M is a compact, metric space that is invariant under the semiflow σ. Thus one has $\sigma(M, t) = M$, for all $t \geq 0$. In the following definitions of distinguished subsets of $\mathcal{E}$, we will make use of the concept of a negative continuation $\phi^{(w,m)}(t) = (w(t), m_t)$, where $-\infty < t \leq 0$, through a point $(w, m) \in \mathcal{E}$ (see Section 2.1). (It should be noted that, unless stated to the contrary, we do not assume the negative continuation to be unique.) We define next the following subsets of $\mathcal{E}$:

(1) $\mathcal{M}$: The set of points $(w, m) \in \mathcal{E}$ such that there is a negative continuation $\phi^{(w,m)}$ through (w, m).
(2) $\mathcal{U}$: The set of points $(w, m) \in \mathcal{M}$ such that there is a negative continuation $\phi^{(w,m)}(t) = (w(t), m_t)$ that satisfies $\|w(t)\| \to 0$, as $t \to -\infty$.
(3) $\mathcal{B}^-$: The set of points $(w, m) \in \mathcal{M}$ such that there is a negative continuation $\phi^{(w,m)}(t)$ that satisfies $\sup_{t \leq 0} \|w(t)\| < \infty$.
(4) $\mathcal{B}_u^-$: The set of points $(w, m) \in \mathcal{M}$ such that there is a unique bounded negative continuation $\phi^{(w,m)}$.

(5) $\mathcal{B}^+$: The set of points $(w, m) \in \mathcal{E}$ such that $\sup_{t \geq 0} \|\Phi(m,t)w\| < \infty$.
(6) $\mathcal{S}$: The set of points $(w, m) \in \mathcal{E}$ such that $\|\Phi(m,t)w\| \to 0$, as $t \to \infty$.
(7) $\mathcal{B} \stackrel{\text{def}}{=} \mathcal{B}^- \cap \mathcal{B}^+$.

One refers to $\mathcal{U}$ as the **unstable set**, to $\mathcal{S}$ as the **stable set**, and to $\mathcal{B}$ as the **bounded set**.

A linear skew product semiflow is said to be **compact**, for $t > t_0$, where $0 \leq t_0 < \infty$, if for every bounded set $B_0 \subset W$ and each $t > t_0$, there is a compact set $B(t) \subset W$, such that $\Phi(m,t)B_0 \subset B(t)$, for all $m \in M$. Notice that the compact set $B(t)$ does not depend on $m \in M$.

There is a variation of the Kuratowski measure κ which occurs in the study of linear skew product flows. Recall that for any bounded set $N \subset W$, the function $\kappa(N)$ is defined as in Section 2.2. One says that a linear skew product semiflow $\pi = (\Phi, \sigma)$ on $\mathcal{E}$ is **uniformly κ-contracting** if for every bounded set N in W, there is a nonnegative function $\beta = \beta(t)$ with $\beta(t) \to 0$, as $t \to \infty$, and such that

$$\kappa(\Phi(m,t)N) \leq \beta(t)\kappa(N), \qquad \text{for all } m \in M.$$

The uniformity here is that the function $\beta(t)$ does not depend on $m \in M$.

A **discrete** linear skew product semiflow π on $\mathcal{E} = W \times M$ is defined by using the definition of a linear skew product semiflow, but now restricting the time t to assume values in the discrete semigroup Z^+. Thus $\sigma(m,\tau) = m \cdot \tau$ is a discrete semiflow on M, where $m \in M$, and for $\pi(w,m,\tau) = (\Phi(m,\tau)w, \sigma(m,\tau))$ one has

(1) $\Phi(m, 0) = I$, the identity operator, for all $m \in M$;
(2) $\Phi(m,\tau)$ is a bounded linear mapping from W into W that satisfies the cocycle identity:

$$\Phi(m, \sigma + \tau) = \Phi(m \cdot \tau, \sigma)\Phi(m,\tau), \qquad \text{for } m \in M \text{ and all } \sigma, \tau \in Z^+;$$

(3) for each $\tau \in Z^+$, the mapping from $\mathcal{E}$ into W given by $(w, m, \tau) \to \Phi(m,\tau)w$ is continuous.

In other words, properties (1), (2), and (3) for a linear skew product semiflow hold for the discrete times $\tau \in Z^+$. Property (4) is vacuous in this setting, and Lemma 43.1 is still valid since $K \stackrel{\text{def}}{=} \sup\{\|\Phi(m,1)\|_{\mathcal{L}} : m \in M_0\}$ is finite, for any compact set $M_0 \subset M$.

The discrete linear skew product semiflows arise in two standard ways in our study of the dynamics of evolutionary equations. First, one may begin with a linear skew product semiflow $\pi(w,m,t) = (\Phi(m,t)w, \sigma(m,t))$ defined for $t \in \mathbb{R}^+$. A discrete counterpart then arises by restricting the time t to assume values $t = n\tau$ in the discrete semigroup τZ^+, where $n = 0, 1, \cdots$ and $\tau > 0$ is fixed.

A second construction arises when $T = \{T_n\} = \{T_n : -\infty < n < \infty\}$ is a sequence with $T_n \in \mathcal{L} = \mathcal{L}(W)$, for all $n \in Z$, with

$$\|T\|_\infty \stackrel{\text{def}}{=} \sup_{n \in Z} \|T_n\|_{\mathcal{L}} < \infty,$$

that is $T \in \mathcal{W} \stackrel{\text{def}}{=} \ell_\infty(Z, \mathcal{L}(W))$, where W is a given Banach space. The base space $\mathcal{W}$ is then the collection of all such sequences, and the dynamics on $\mathcal{W}$ is the translational flow $(T, \tau) \to T \cdot \tau$, where $T \cdot \tau = \{T_{\tau+n}\}$, for $\tau \in Z$. We will use the Fréchet metric topology on $\mathcal{W}$. This topology is generated by the family of pseudonorms

$$N_n(T) = \sup\{\|T_k\|_{\mathcal{L}} : |k| \leq n\}, \qquad \text{where } n = 1, 2, \cdots;$$

and a corresponding invariant metric (see Appendices A and B) is

$$d(T) \stackrel{\text{def}}{=} \sum_{n=1}^{\infty} 2^{-n} \min(N_n(T), 1).$$

This describes the topology of uniform convergence on bounded sets in Z. The linear mapping Φ is then defined, for $k \geq 0$, by using induction and setting $\Phi(T, 0) = I$ and $\Phi(T, k)w = T_{k-1} \cdot \ldots \cdot T_1 \cdot T_0 w$, for $k \geq 1$. Then

$$\pi(w, T, k) = (\Phi(T, k)w, T \cdot k) \tag{43.3}$$

is a discrete linear skew product semiflow on $W \times \mathcal{W}$.

4.4. Perturbations of Analytic Semigroups.

Our objective in this section is to present a typical construction of a linear skew product semiflow.[14] We begin with the linear time-varying evolutionary equation

$$\partial_t u + Au = B(t)u, \qquad t \in \mathbb{R}, \tag{44.1}$$

on a Banach space W, where A satisfies the Standing Hypothesis A. From (37.11), there is an $M_0 \geq 1$ and an $a > 0$ so that

$$\|e^{-At}\|_{\mathcal{L}} \leq M_0 e^{-at}, \qquad \text{for all } t \geq 0 \tag{44.2}$$

We will consider (44.1) as a perturbation of the linear time-invariant evolutionary equation $\partial_t u + Au = 0$. In most applications, equation (44.1) arises as the linear equation one obtains by linearizing the nonlinear evolutionary equation

$$\partial_t u + Au = F(u)$$

along a globally defined motion $\phi(t)$ with range in some compact, invariant set. We will focus our attention here on equation (44.1), and postpone a detailed discussion of this linearization process until Section 4.9 and Chapter 7.

---

[14]We invite the reader to review the notation introduced with the Gronwall-Henry inequality in Appendix D.

While we develop the theory in this section based on the premise that the Standing Hypothesis A is satisfied, there is a good analogue in the case where $-A$ is the infinitesimal generator of a C_0-semigroup on a Banach space W and $B = B(t)$ is a suitable function with $B(t) \in \mathcal{L}(W)$, for $t \geq 0$. The C_0-theory is developed in the exercises.

In this section we examine the behavior of the solutions of equation (44.1) under several related scenarios. In particular, we will assume that the operator $B(t)$ in equation (44.1) satisfies $B(\cdot) \in L^\infty_{\text{loc}}[0, T; \mathcal{L}(V^{2\beta}, V^{2\alpha}))$, for various choices of β and α, where $0 \leq \beta - \alpha < 1$ and $0 < T \leq \infty$. One case of special interest is $B(t) \in \mathcal{L}(V^{2\beta}, W)$, where $0 \leq \beta < 1$. The methodology for studying any of these cases is the same.

In order to simplify the notation, we assume first that the linear operator $B(t)$ in equation (44.1) satisfies

$$B(\cdot) \in L^\infty_{\text{loc}}[0, T; \mathcal{L}(V^{2\beta}, W)), \qquad \text{where } 0 \leq \beta < 1$$

and $0 < T \leq \infty$. Let $\mathcal{L} = \mathcal{L}(V^{2\beta}, W)$ denote the Banach space of all bounded linear operators from $V^{2\beta}$ into W, with the operator norm $\|L\|_{\mathcal{L}}$, for $L \in \mathcal{L}$, where $\|L\|_{\mathcal{L}} = \sup\{\|Lv\| : \|A^\beta v\| \leq 1\}$. Note that

$$B(\cdot) \in L^\infty_{\text{loc}}[0, T; \mathcal{L}(V^{2\beta}, W)) \iff B(\cdot)A^{-\beta} \in L^\infty_{\text{loc}}[0, T; \mathcal{L}(W, W)).$$

It then follows that for each τ with $0 \leq \tau < T$, there is a constant $C_\tau \geq 0$ such that

$$\|B(t)v\| \leq C_\tau \|A^\beta v\|, \qquad \text{for } 0 \leq t \leq \tau \text{ and } v \in V^{2\beta}. \tag{44.3}$$

For any $v_0 \in V^{2\beta}$, we consider the mild solution $v = v(t)$ of equation (44.1) in $V^{2\beta}$ given by

$$v(t) = e^{-At}v_0 + \int_0^t e^{-A(t-s)} B(s)v(s)\, ds, \qquad \text{for } 0 \leq t < T, \tag{44.4}$$

and consider the mapping $v(t) \to \hat{v}(t)$ given by

$$\hat{v}(t) = e^{-At}v_0 + \int_0^t e^{-A(t-s)} B(s)v(s)\, ds, \qquad \text{for } 0 \leq t < T.$$

We now use the Contraction Mapping Theorem to show that, for small $\tau > 0$, this mapping has a unique fixed point in the space $C[0, \tau; V^{2\beta})$. First observe that $\hat{v}(t)$ is continuous in t. (Why?) Then from inequalities (37.11) and (44.3), one obtains

$$\begin{aligned}
\|A^\beta(\hat{v_1}(t) - \hat{v_2}(t))\| &\leq \int_0^t \|A^\beta e^{-A(t-s)}\|_{\mathcal{L}(W)} \|B(v_1 - v_2)\|\, ds \\
&\leq M_\beta C_\tau \int_0^t (t-s)^{-\beta} \|A^\beta(v_1(s) - v_2(s))\|\, ds \\
&\leq M_\beta (1-\beta)^{-1} C_\tau \tau^{1-\beta} \|A^\beta(v_1 - v_2)\|_{C[0,\tau;W)},
\end{aligned}$$

for all t with $0 \leq t \leq \tau$. We see then that for small τ the mapping $v(t) \to \hat{v}(t)$ is a contraction, and consequently, there is a unique fixed point for equation (44.4). The contraction property also shows that the initial value problem

$$\partial_t v + Av = B(t)v, \qquad v(0) = v_0 \tag{44.5}$$

has a unique mild solution in $V^{2\beta}$, for each $v_0 \in V^{2\beta}$, and this solution is given by equation (44.4). We leave it as an exercise to show that this solution on $0 \leq t < \tau$ has a unique maximal continuation to a solution on $0 \leq t < T$. We will denote this solution by $\Phi(B,t)v_0 = v(t)$, and we will refer to $\Phi(B,t)$ as the **solution operator** generated by B, or generated by equation (44.1).

4.4.1 A Basic Theorem. Let us now restrict our treatment to the case where

$$B(\cdot) \in L^\infty \stackrel{\text{def}}{=} L^\infty(\mathbb{R}; \mathcal{L}(V^{2\beta}, W)), \qquad \text{where } 0 \leq \beta < 1. \tag{44.6}$$

For $L \in \mathcal{L} = \mathcal{L}(V^{2\beta}, W)$, we let $\|L\|_{\mathcal{L}} = \sup\{\|Lv\| : \|A^\beta v\| \leq 1\}$ denote the operator norm. For the study of the dynamical properties of the mild solutions of equation (44.5), there are several metrics which can be used on the space L^∞, and certain subspaces of L^∞. For $B \in L^\infty$, we define the norm

$$\|B\|_\infty \stackrel{\text{def}}{=} \text{ess sup}\{\|B(t)\|_{\mathcal{L}} : t \in \mathbb{R}\},$$

and the pseudonorms

$$\|B\|_{\infty;[\tau,T]} \stackrel{\text{def}}{=} \text{ess sup}\{\|B(s)\|_{\mathcal{L}} : \tau \leq s \leq T\} = \|B_\tau\|_{\infty;[0,T-\tau]},$$

and the following (which depends on the given linear operator A)

$$\|B\|_{\{A;[\tau,T]\}} \stackrel{\text{def}}{=} \sup_{0 \leq t \leq T-\tau} \int_0^t \|A^\beta e^{-A(t-s)} B(\tau + s)\|_{\mathcal{L}}\, ds,$$

where $-\infty < \tau \leq T < \infty$. In the case of the norm $\|\cdot\|_\infty$, the space L^∞ becomes a Banach space, and we will denote the associated topology by $\mathcal{T}_\infty$. In the case of the pseudonorms $\|\cdot\|_{\infty;[\tau,T]}$ and $\|\cdot\|_{\{A;[\tau,T]\}}$, we derive two topologies $\mathcal{T}_{\text{bo}}$ and $\mathcal{T}_A$, respectively, by using the metric

$$d(B^1, B^2) = \sum_{n \in Z} 2^{-|n|} \frac{\|B^1 - B^2\|_n}{1 + \|B^1 - B^2\|_n},$$

where $\|B\|_n = \|B\|_{\infty;[n,n+1]}$ and $\|B\|_n = \|B\|_{\{A;[n,n+1]\}}$, respetively. In this way, $(L^\infty, \mathcal{T}_{\text{bo}})$ and $(L^\infty, \mathcal{T}_A)$ are Fréchet spaces (see Appendix A).

If $B \in L^\infty$, then one has $\|B\|_{\{A;[\tau,T]\}} = \|B_\tau\|_{\{A;[0,T-\tau]\}}$, with the corresponding equality valid for $\|\cdot\|_{\infty;[\tau,T]}$. Also from inequality (37.11) and the definition of the Gamma function, one has

$$\begin{aligned}
\|B\|_{\{A;[0,n]\}} &= \sup_{0\le t\le n}\int_0^t \|A^\beta e^{-A(t-s)}B(s)\|_{\mathcal{L}}\,ds\\
&\le M_\beta \sup_{0\le t\le n}\int_0^t (t-s)^{-\beta}e^{-a(t-s)}\|B\|_{\infty;[0,n]}\,ds\\
&\le M_\beta a^{\beta-1}\Gamma(1-\beta)\|B\|_{\infty;[0,n]} \le M_\beta a^{\beta-1}\Gamma(1-\beta)\|B\|_\infty.
\end{aligned}$$

This implies that the three topologies $\mathcal{T}_\infty$, $\mathcal{T}_{\rm bo}$, and $\mathcal{T}_A$ satisfy the relation

$$\mathcal{T}_A \subset \mathcal{T}_{\rm bo} \subset \mathcal{T}_\infty. \tag{44.7}$$

It is easily seen that, for $B \in L^\infty$, the mapping $v_0 \to \Phi(B,t)v_0$ is linear for each $t \ge 0$. Furthermore, one has

$$\begin{aligned}
\|A^\beta\Phi(B,t)v_0\| &\le \|A^\beta e^{-At}v_0\| + \int_0^t \|A^\beta e^{-A(t-s)}\|_{\mathcal{L}(W)}\|B(s)\Phi(B,s)v_0\|\,ds\\
&\le Me^{-at}\|A^\beta v_0\|\\
&\quad + M_\beta\|B\|_{\infty;[0,\infty)}\int_0^t e^{-a(t-s)}(t-s)^{-\beta}\|A^\beta\Phi(B,s)v_0\|\,ds,
\end{aligned}$$

where M_β is given by (37.11). With $v(t) = e^{at}\|A^\beta\Phi(B,t)v_0\|$, one can apply the Gronwall-Henry inequality (94.17) to get

$$\|A^\beta\Phi(B,t)v_0\| \le Me^{-at}E_{1-\beta,1}(\mu t)\|A^\beta v_0\|, \qquad \text{for } t \ge 0, \tag{44.8}$$

where $M > 0$ is a constant and $\mu^{1-\beta} = M_\beta\|B\|_\infty\Gamma(1-\beta)$. Thus for each $t \ge 0$, $\Phi(B,t)$ is a bounded linear operator on $V^{2\beta}$, with $\Phi(B,0)v_0 = v_0$, for all $v_0 \in V^{2\beta}$.

Let us next show that Φ satisfies the cocycle identity

$$\Phi(B,\tau+t) = \Phi(B_\tau,t)\Phi(B,\tau), \qquad \text{for } t,\ \tau \ge 0, \tag{44.9}$$

where $B_\tau(t) = B(\tau+t)$. The proof of this is based on the fact that, for each $v_0 \in V^{2\beta}$, the two functions $v_1(t) = \Phi(B,\tau+t)v_0$ and $v_2(t) = \Phi(B_\tau,t)\Phi(B,\tau)v_0$ are mild solutions in $V^{2\beta}$ of the initial value problem $\partial_t v + Av = B(\tau+t)v$, where $v(0) = \Phi(B,\tau)v_0$. Because of the uniqueness of solutions of the initial value problem, one has $v_1(t) = v_2(t)$, for all $t \ge 0$, which implies (44.9).

Next we will verify that $\Phi(B,t)v_0$ is Lipschitz continuous in $v_0 \in V^{2\beta}$ and $B \in (L^\infty, \mathcal{T}_A)$. For $i = 1,2$, we let $B_i \in L^\infty$, $v_{i0} \in V^{2\beta}$, and let $v_i = v_i(t)$ denote the mild solutions in $V^{2\beta}$ of the initial value problem

$$\partial_t v_i + Av_i = B_i(t)v_i, \qquad v_i(0) = v_{i0}.$$

Then $w = v_1 - v_2$ is a mild solution in $V^{2\beta}$ of the initial value problem

$$\partial_t w + Aw = B_1(t)w + (B_1(t) - B_2(t))v_2(t),$$

where $w(0) = w_0 = v_{10} - v_{20}$, and one has
(44.10)

$$w(t) = e^{-At}w_0 + \int_0^t e^{-A(t-s)}(B_1 - B_2)v_2(s)\,ds + \int_0^t e^{-A(t-s)}B_1 w(s)\,ds.$$

It then follows from (44.2), (44.10), and (37.11) that

$$\begin{aligned}\|A^\beta w(t)\| \le M_0 e^{-at}\|A^\beta w_0\| &+ \int_0^t \|A^\beta e^{-A(t-s)}(B_1(s) - B_2(s))v_2(s)\|\,ds \\ &+ M_\beta \|B_1\|_\infty \int_0^t (t-s)^{-\beta} e^{-a(t-s)}\|A^\beta w(s)\|\,ds.\end{aligned}$$

Let τ be fixed with $0 < \tau < \infty$. Then for $0 < t \le \tau$, one obtains

$$\begin{aligned}\|A^\beta w(t)\| \le M_0 e^{-at}\|A^\beta w_0\| &+ \|B_1 - B_2\|_{\{A;[0,\tau]\}} \sup_{0\le s\le\tau} \|A^\beta v_2(s)\| \\ &+ M_\beta \|B_1\|_\infty \int_0^t (t-s)^{-\beta} e^{-a(t-s)}\|A^\beta w(s)\|\,ds.\end{aligned}$$

By using inequality (44.8) to estimate $\sup_{0\le s\le\tau} \|A^\beta v_2(s)\|$, with $0 \le \beta < 1$, it then follows again from the Gronwall-Henry inequality, that there exist nonnegative continuous functions $a(\cdot)$ and $b(\cdot)$, such that

$$\|A^\beta w(t)\| \le a(\tau)\|A^\beta(v_{10} - v_{20})\| + b(\tau)\|B_1 - B_2\|_{\{A;[0,\tau]\}}, \tag{44.11}$$

for $0 \le t \le \tau$. Since τ is arbitrary, we see that the solution operator $\Phi(B,t)v_0$ is locally Lipschitz continuous in $B \in (L^\infty, \mathcal{T}_A)$ and $v_0 \in V^{2\beta}$, for $0 \le t < \infty$. The inclusion relation (44.7) implies that one has continuity for B in the spaces $(L^\infty, \mathcal{T}_{bo})$ and $(L^\infty, \mathcal{T}_\infty)$, as well.

The use of the Gronwall-Henry inequality arises several times in the proof of the next theorem. Various forms of the functions $E_{r,c}(\mu t)$ are used. This is a good point for the reader to review this material in Appendix D. Equation (94.20) is important in the study of the longtime dynamics of some evolutionary equations. Since this issue arises here, it is convenient to adopt the following reformulation for the functions $E_{r,c}(\mu t)$ and some of its relatives. We say that a function $E = E(t) : [0,\infty) \to \mathbb{R}$ has the **EX property** provided that E is positive, continuous, monotone nondecreasing, with $E(0) = 1$, and there is a $\mu > 0$ such that

$$\lim_{t\to\infty} \frac{1}{t}\log(E(t)) = \mu. \tag{44.12}$$

In the case that E has the EX property and equation (44.12) holds, for $\mu > 0$, we will write $E(\mu; t) \stackrel{\text{def}}{=} E(t)$. The role of μ in the notation $E(\mu; t)$ is to denote the asymptotic exponential growth rate of the function $E(t)$.

The prototype for a function satisfying the EX property is $E(\mu; t) = e^{\mu t}$. Another example is $E(\mu; t) = E_{r,c}(\mu t)$, where r and c are positive (see Appendix D). More generally, let E_1, E_2, and E_3 be functions with the EX property. Then each of the following functions also has the EX property. We assume here that the constants C, a, and ν are positive and θ satisfies $0 \leq \theta \leq 1$.

(1) $E_4(\mu; t) = 1 + Ct^a E_1(\mu; t)$.
(2) $E_5(\mu; t) = E_1(\mu; t)^\theta E_2(\mu; t)^{1-\theta}$.
(3) $E_6(\mu + \nu; t) = E_1(\mu; t) E_2(\nu; t)$.
(4) $E_7(\mu; \tau) = E_1(a\mu; t)$, where $\tau = at$. Note that in this case, one has

$$\frac{1}{\tau} \log(E_7(\mu; \tau)) = \frac{1}{at} \log(E_1(a\mu; t)) \to \mu, \qquad \text{as } t \to \infty.$$

The following theorem is rather long. Part of the reason for this is that we present here the basic theory of the continuity and regularity of mild solutions of equation (44.1). In the statement and the proof of this theorem, we let $C_0, C_1, \cdots$ denote constants which may depend on the linear operator A and the parameters β, r, and ρ, but which are independent of the term B, the time t, and the initial condition v_0.

Theorem 44.1. *Let the Standing Hypothesis A be satisfied, and let B satisfy (44.6). Then there exist functions E_1, E_2, E_3, and E_4 which satisfy the EX property, and the following statements are valid:*

(1) *For each $B \in L^\infty$ and $v_0 \in V^{2\beta}$ there is a unique mild solution $v(t) = \Phi(B, t)v_0$ of (44.5) in $V^{2\beta}$, for $0 \leq t < \infty$, satisfying the equation*

$$\Phi(B, t)v_0 = e^{-At}v_0 + \int_0^t e^{-A(t-s)} B(s)\Phi(B, s)v_0 \, ds, \tag{44.13}$$

and

$$\Phi(B, \cdot)v_0 \in C[0, \infty; V^{2\alpha}) \cap C^{0,\theta}_{\text{loc}}(0, \infty; V^{2r}), \tag{44.14}$$

for all α and r with $0 \leq \alpha \leq \beta$ and $0 \leq r < 1$, where $0 < \theta < 1 - r$.

(2) *For all $(v_0, B) \in V^{2\beta} \times L^\infty$, inequality (44.8) is valid. Furthermore, there are constants C, $\mu > 0$ such that, for $v_0 \in V^{2\beta}$ and $t > 0$, one has*

$$\|A^\beta \Phi(B, t)v_0\| \leq M_\beta t^{-\beta} e^{-at} E_1(\mu; t)\|v_0\|, \tag{44.15}$$

where $E_1(\mu; t) = C\, E_{1-\beta,1-\beta}(\mu t)$.

(3) *One has*

$$\|\Phi(B,t)v_0\| \leq M_0 e^{-at} E_2(\mu;t)\|v_0\|, \qquad \text{for } v_0 \in V^{2\beta} \text{ and } t>0, \tag{44.16}$$

where $E_2(\mu;t) = 1 + C_0\|B\|_\infty t^{1-\beta}E_1(\mu;t)$, *for* $t \geq 0$.

(4) *Furthermore, there is a unique continuous extension of* $\Phi(B,t)v_0$, *to all* $v_0 \in W$, *and (44.13) and (44.16) hold for all* $v_0 \in W$.

(5) *For each* r *with* $0 \leq r < 1$, *and for each* ρ *with* $\rho \geq 0$, *there are positive constants* C_1 *and* ν *and functions* E_3 *and* E_4, *which have the EX property, such that,*

$$\|A^r\Phi(B,t)v_0\| \leq C_1 t^{\rho-r} E_3(\nu;t)e^{-at}\|A^\rho v_0\|, \tag{44.17}$$

for $v_0 \in V^{2\rho} = \mathcal{D}(A^\rho)$ *and* $t>0$, *whenever* $0 \leq \rho \leq r < 1$. *Also, one has* $\Phi(B,\cdot)v_0 \in C[0,\infty;V^{2r})$, *as well as*

$$\|A^r\Phi(B,t)v_0\| \leq C_1 E_4(\nu;t)e^{-at}\|A^\rho v_0\|, \tag{44.18}$$

for $v_0 \in V^{2\rho}$ *and* $t>0$, *whenever* $0 \leq r \leq \rho$.

(6) *If* $v_0 \in W$, *then* $\Phi(B,t)v_0$ *satisfies*

$$\Phi(B,\cdot)v_0 \in C[0,\infty;W) \cap C^{0,\theta_1}_{\text{loc}}(0,\infty;V^{2r}), \tag{44.19}$$

for any r *with* $0 \leq r < 1-\beta$, *where* $\theta_1 = \theta_1(r) > 0$. *If* $v_0 \in V^{2\rho}$, *where* $0 < \rho < 1$, *then in addition to (44.19), one has*

$$\Phi(B,\cdot)v_0 \in C[0,\infty;V^{2\alpha}) \cap C^{0,\theta_2}_{\text{loc}}[0,\infty;V^{2r}), \tag{44.20}$$

for any α *and* r *with* $0 \leq \alpha \leq \rho$ *and* $0 \leq r < \rho$, *where* $\theta_2 > 0$.

(7) *For each* $t \in [0,\infty)$, *the mapping* $(v_0,B) \to \Phi(B,t)v_0$, *of* $V^{2\beta} \times L^\infty$ *into* $V^{2\beta}$, *is Lipschitz continuous in* v_0 *and* B, *and inequality (44.11) holds. Moreover, for each* r *with* $0 \leq r < 1$ *and each* $t \in [0,\infty)$, *the mapping* $(v_0,B) \to \Phi(B,t)v_0$, *of* $\mathcal{E}^{2r} = V^{2r} \times L^\infty$ *into* V^{2r}, *is Lipschitz continuous in* v_0 *and* B, *and there are nonnegative continuous functions* $a(\cdot)$ *and* $b(\cdot)$ *such that*

$$\|A^r w(t)\| \leq a(\tau)\|A^r(v_{10}-v_{20})\| + b(\tau)\|B_1 - B_2\|_{\infty;[0,\tau]}, \tag{44.21}$$

for $0 \leq t \leq \tau$, *where* $w(t) = \Phi(B_1,t)v_{10} - \Phi(B_2,t)v_{20}$.

Proof. Let $v_0 \in V^{2\beta}$. The proof of the existence and uniqueness of the solution $\Phi(B,t)v_0$ is given above. Since $v(t) = \Phi(B,t)v_0 \in C[0,\infty;V^{2\beta})$, it follows that

$$f(t) \stackrel{\text{def}}{=} B(t)v(t) \in L^\infty_{\text{loc}}[0,\infty;W).$$

Consequently, Lemma 42.7 implies that $\Phi(B,t)v_0$ satisfies (44.14), which completes the proof of Item (1).

Item(2): The proof of inequality (44.8) is a direct application of the Gronwall-Henry inequality. We will omit these details. In order to prove inequality (44.15), we take the norm of both sides of equation (44.13) in $V^{2\beta}$ and use inequality (37.11) to obtain

(44.22)

$$||A^{\beta}\Phi(B,t)v_0|| \leq M_{\beta}t^{-\beta}e^{-at}||v_0|| + M_{\beta}||B||_{\infty}\int_0^t (t-s)^{-\beta}e^{-a(t-s)}||A^{\beta}\Phi(B,s)v_0||\,ds,$$

for $v_0 \in V^{2\beta}$. Next with $v(t) = e^{at}||A^{\beta}\Phi(B,t)v_0||$ and $h(t) = M_{\beta}t^{-\beta}||v_0||$, inequality (44.15) follows from the Gronwall-Henry inequality, see (94.16) and (94.17), where

$$\mu^{1-\beta} = M_{\beta}||B||_{\infty}\Gamma(1-\beta).$$

Item (3): To prove inequality (44.16), we return to (44.13) and take the norm in W. For $v_0 \in V^{2\beta}$, one then has

$$||\Phi(B,t)v_0|| \leq M_0 e^{-at}||v_0|| + M_0||B||_{\infty}\int_0^t e^{-a(t-s)}||A^{\beta}\Phi(B,s)v_0||\,ds.$$

Note that from inequality (44.15) and the monotonicity of $E_1(\mu;t)$, one has

$$\int_0^t e^{-a(t-s)}||A^{\beta}\Phi(B,s)v_0||\,ds \leq M_{\beta}(1-\beta)^{-1}t^{1-\beta}E_1(\mu;t)||v_0||.$$

By using the definition of μ given above and this inequality, one readly obtains inequality (44.16), for $v_0 \in V^{2\beta}$.

Item (4): Now inequality (44.16) implies that the mapping of $V^{2\beta}$ into W given by $v_0 \to \Phi(B,t)v_0$ is continuous in the W-norm, for $t \geq 0$. Consequently, there is a unique continuous extension of $\Phi(B,t)v_0$. We will denote this extension by $\Phi(B,t)v_0$, for $v_0 \in W$. In addition, we see that the integral in equation (44.13) is valid in the space W, for $v_0 \in W$. That is, $\Phi(B,t)v_0$ is a mild solution in the space W. Furthermore, by passing to the limits in W, we see that equation (44.13), as well as inequalities (44.15) and (44.16), are valid, for all $v_0 \in W$.

Item (5): Let δ and $\bar{\delta}$ satisfy $\delta = 1 - r + \rho$, when $0 \leq \rho \leq r < 1$, and $\bar{\delta} = 1 - \beta + \rho$, when $0 \leq \rho \leq \beta < 1$. One then has $0 < \delta \leq 1$ and $0 < \bar{\delta} \leq 1$. For $0 \leq \rho \leq r < 1$, it follows from (37.11) and (44.13) that, for $v_0 \in V^{2\rho}$, one has

$$||A^{r}\Phi(B,t)v_0|| \leq M_{r-\rho}t^{-r+\rho}e^{-at}||A^{\rho}v_0|| + M_r||B||_{\infty}\int_0^t (t-s)^{-r}e^{-a(t-s)}||A^{\beta}\Phi(B,s)v_0||\,ds, \tag{44.23}$$

for $t > 0$. Now for $0 \leq \rho \leq r = \beta < 1$, inequality (44.23) implies that

$$||A^{\beta}\Phi(B,t)v_0|| \leq M_{\beta-\rho}t^{-\beta+\rho}e^{-at}||A^{\rho}v_0|| + M_{\beta}||B||_{\infty}\int_0^t (t-s)^{-\beta}e^{-a(t-s)}||A^{\beta}\Phi(B,s)v_0||\,ds.$$

The Gronwall-Henry inequality, with $h(t) = M_{\beta-\rho}t^{-\beta+\rho}\|A^{\rho}v_0\|$, then implies that

$$\|A^{\beta}\Phi(B,t)v_0\| \leq C_1 t^{-\beta+\rho}e^{-at}E_5(\nu;t)\|A^{\rho}v_0\|, \tag{44.24}$$

for $v_0 \in V^{2\rho}$ and $t > 0$, where $E_5(\nu;t) = E_{\bar{\delta},\bar{\delta}}(\nu t)$ and $\nu^{\bar{\delta}} = M_{\beta}\|B\|_{\infty}\Gamma(\bar{\delta})$. Similarly, by using inequalities (44.23) and (44.24), with $0 \leq \rho = r \leq \beta < 1$, one obtains

$$\|A^{\rho}\Phi(B,t)v_0\| \leq C_2 e^{-at}E_6(\nu;t)\|A^{\rho}v_0\|, \tag{44.25}$$

for $v_0 \in V^{2\rho}$ and $t \geq 0$, where

$$\begin{aligned} E_6(\nu;t) &= 1 + C_3\|B\|_{\infty}E_5(\nu;t)\int_0^t (t-s)^{-\rho}s^{\rho-\beta}\,ds \\ &= 1 + C_4\|B\|_{\infty}t^{1-\beta}E_5(\nu;t). \end{aligned}$$

Next for $0 \leq \rho \leq r \leq \beta < 1$, one uses the Interpolation inequality (37.15), together with inequalities (44.24) and (44.25), and $r = (1-\theta)\rho + \theta\beta$ for some θ with $0 \leq \theta \leq 1$, to obtain

$$\begin{aligned} \|A^{r}\Phi(B,t)v_0\| &\leq C_5\|A^{\beta}\Phi(B,t)v_0\|^{\theta}\|A^{\rho}\Phi(B,t)v_0\|^{1-\theta} \\ &\leq C_6 t^{\rho-r}e^{-at}E_3(\nu;t)\|A^{\rho}v_0\|, \end{aligned}$$

where E_3 is defined by $E_3(\nu;t) = E_5(\nu;t)^{\theta}E_6(\nu;t)^{1-\theta}$, for $t \geq 0$. This establishes inequality (44.17) when $0 \leq \rho \leq r \leq \beta < 1$.

Next for $0 \leq r \leq \beta < 1$ and $0 \leq r \leq \rho \leq 1$, it follows from (44.25), and the fact that $A^{r-\rho} \in \mathcal{L}(W)$ satisfies $\|A^{r}v_0\| \leq \|A^{r-\rho}\|_{\mathcal{L}(W)}\|A^{\rho}v_0\|$, that one has

$$\|A^{r}\Phi(B,t)v_0\| \leq \|A^{r-\rho}\|_{\mathcal{L}(W)}C_2e^{-at}E_6(\nu;t)\|A^{\rho}v_0\|,$$

which implies inequality (44.18), in this case. This completes the proof of inequalities (44.17) and (44.18) when $0 \leq r \leq \beta$.

Next we turn to the case where $\beta < r < 1$. First note that for $0 \leq \beta < r = \rho < 1$, $A^{\beta-r}$ is a bounded linear operator with $\|A^{\beta}\Phi v_0\| \leq \|A^{\beta-r}\|_{\mathcal{L}(W)}\|A^{r}\Phi v_0\|$, and inequality (44.23) implies that

$$\begin{aligned} \|A^{r}\Phi(B,t)v_0\| &\leq M_0e^{-at}\|A^{\rho}v_0\| \\ &\qquad + C_7\|B\|_{\infty}\int_0^t (t-s)^{-r}e^{-a(t-s)}\|A^{r}\Phi(B,s)v_0\|\,ds, \end{aligned}$$

where $C_7 = \|A^{\beta-r}\|_{\mathcal{L}(W)}M_r$. Then the Gronwall-Henry inequality yields

$$\|A^{r}\Phi(B,t)v_0\| \leq M_0e^{-at}E_7(\nu;t)\|A^{r}v_0\|, \tag{44.26}$$

for $v_0 \in V^{2r}$ and $t \geq 0$, where $E_7(\nu;t) = E_{1-r,1}(\nu t)$, and

$$\nu^{1-r} = C_7\|B\|_\infty \Gamma(1-r).$$

Next for $0 \leq \beta < r \leq \rho \leq 1$ and $r < 1$, we use $A^{r-\rho} \in \mathcal{L}(W)$ and inequality (44.26) to establish

$$\|A^r\Phi(B,t)v_0\| \leq C_8 e^{-at} E_7(\nu;t)\|A^\rho v_0\|, \qquad \text{for } v_0 \in V^{2r} \text{ and } t \geq 0,$$

which is inequality (44.18) in this case.

Finally for $0 \leq \beta < r < 1$ and $\rho \leq r < 1$, one uses the boundedness of $A^{\beta-r}$ and $A^{\rho-r}$ with inequality (44.23) to obtain

$$\|A^r\Phi(B,t)v_0\| \leq M_{r-\rho}t^{-r+\rho}e^{-at}\|A^\rho v_0\| \\ + C_7\|B\|_\infty \int_0^t (t-s)^{-r}e^{-a(t-s)}\|A^r\Phi(B,s)v_0\|\, ds.$$

The Gronwall-Henry inequality, with $h(t) = M_{r-\rho}t^{-r+\rho}\|A^\rho v_0\|$, then establishes

$$\|A^r\Phi(B,t)v_0\| \leq M_{r-\rho}t^{-r+\rho}e^{-at}E_8(\nu;t)\|A^\rho v_0\|,$$

for $v_0 \in V^{2\rho}$ and $t > 0$, where $E_8(\nu;t) = E_{1-r,\delta}(\nu t)$ and ν is given above. This completes the proof of inequalities (44.17) and (44.18).

The argument leading to the proof of inequality (44.18) also shows that the integral term in equation (44.13) exists in the space V^{2r}, when $v_0 \in V^{2\rho}$ and $0 \leq r \leq \rho < 1$. If $v_0 \in W$, then inequality (44.17) implies that $\Phi(B,\tau)v_0 \in V^{2r}$, for any $\tau > 0$. The cocycle identity (44.9) and inequality (44.18) then imply that

$$\Phi(B,\tau+\cdot)v_0 = \Phi(B_\tau,\cdot)\Phi(B,\tau)v_0 \in C[0,\infty;V^{2r}),$$

which in turn implies that $\Phi(B,\cdot)v_0 \in C(0,\infty;V^{2r})$, for all $v_0 \in W$.

Item (6): Let $f = f(t) \stackrel{\text{def}}{=} B(t)\Phi(B,t)v_0$. If $v_0 \in W$, then inequality (44.17), with $\rho = 0$ and $r = \beta$, implies that $f \in L^p_{\text{loc}}[0,\infty;W)$, where $1 \leq p < \frac{1}{\beta}$. Lemma 42.7 then implies that (44.19) holds. Now assume that $v_0 \in V^{2\rho}$, for some ρ with $0 < \rho < 1$. For $\beta \leq \rho < 1$, set $p = \infty$. For $0 < \rho < \beta < 1$, let ϵ satisfy $0 < \rho < \beta + \epsilon < 1$ and set $\frac{1}{p} = \beta + \epsilon - \rho < 1$. In either case, one finds that $f \in L^p_{\text{loc}}[0,\infty;V^{2\rho})$ - see inequalities (44.17) and (44.18) with $r = \beta$ - and $\rho < 1 - \frac{1}{p}$. Consequently, (44.20) follows from (42.14).

Item (7): For $r = \beta$ follows from inequality (44.11). The proof of inequality (44.21) for the general case, where $0 \leq r < 1$, is left as an exercise. □

We now have the following result, which provides a more complete description of the behavior of $\Phi(B,t)v_0$, as $t \to \infty$.

Corollary 44.2. *Let the Standing Hypothesis A be satisfied. Then for every $r \geq 0$ and $(v_0, B) \in W \times L^\infty$, one has*

$$\Phi(B, \cdot)v_0 \in C(0, \infty; V^{2r}).$$

Moreover, for every $r \geq 0$ and $\rho \geq 0$ there exist constants $C = C(r, \rho) > 0$ and $\nu > 0$ and functions E_1 and E_2 with the EX property such that, for $v_0 \in V^{2\rho}$ and $t > 0$, one has

$$\|A^r \Phi(B, t)v_0\| \leq C t^{\rho - r} e^{-at} E_1(\nu; t) \|A^\rho v_0\|, \tag{44.27}$$

whenever $\rho \leq r$, while

$$\|A^r \Phi(B, t)v_0\| \leq C e^{-at} E_2(\nu; t) \|A^\rho v_0\|, \tag{44.28}$$

whenever $\rho \geq r$. In particular, $\Phi(B, \cdot)v_0 \in C[0, \infty; V^{2r})$, whenever $v_0 \in V^{2\rho}$ and $\rho \geq r$.

Proof. One proves this by using mathematical induction on r. Here is the basic idea. From Theorem 44.1, we see that the conclusions are valid for $0 \leq r < 1$. Furthermore, by using the A^σ-norm, where $\|v_0\|_{2\sigma} = \|A^\sigma v_0\|$, for $v_0 \in V^{2\sigma} = \mathcal{D}(A^\sigma)$, inequality (44.17) assumes the form

$$\|A^\alpha \Phi(B, t)v_0\|_{2\sigma} \leq C t^{-\alpha} E_3(\nu; t) e^{-at} \|v_0\|_{2\sigma}, \tag{44.29}$$

which is valid for all σ and α with $0 \leq \sigma < 1$ and $0 \leq \alpha < 1$. (Note that ν and E_3 depend on α and σ.)

Now if r satisfies $1 \leq r < 2$, we write $r = \sigma + \alpha$, where $\sigma = \alpha = \frac{r}{2}$. From the cocycle identity (44.9) and the inequalities (44.17) and (44.29), one then obtains

$$\begin{aligned}
\|A^r \Phi(B, 2t)v_0\| &= \|A^\alpha \Phi(B_t, t)\Phi(B, t)v_0\|_{2\sigma} \\
&\leq C_1 t^{-\alpha} e^{-at} E_3(\mu; t) \|\Phi(B, t)v_0\|_{2\sigma} \\
&= C_1 t^{-\alpha} e^{-at} E_3(\nu, t) \|A^\sigma \Phi(B, t)v_0\| \\
&\leq C_1 C_2 t^{-\alpha} t^{-\sigma} e^{-2at} E_3(\mu; t) E_4(\nu; t) \|v_0\| \\
&\leq C_3 t^{-r} E_5(\mu + \nu; t) \|v_0\| \\
&= 2^r C_3 (2t)^{-r} E_5\left((\mu + \nu)/2; 2t\right) \|v_0\|,
\end{aligned}$$

which implies (44.27), for this value of r, when $\rho = 0$. We leave the remaining details as an exercise. □

4.4.2 Topological Issues. We see then that for each $t \geq 0$ and for each $B \in L^\infty$, one has

$$\Phi(B, t) \in \mathcal{L}(W) \cap \mathcal{L}(V^{2r}), \qquad \text{for any } r \geq 0.$$

Furthermore, as noted above, the mapping $B \to \Phi(B,t)$ is a Lipschitz continuous mapping of the space $(L^\infty, \mathcal{T}_A)$ into $\mathcal{L}(V^{2\beta})$, for $t > 0$. Moreover, one has

$$\Phi(B,t) \in \mathcal{L}(W, V^{2r}), \qquad \text{for any } r \geq 0 \text{ and any } t > 0.$$

One of the consequences of inequality (44.28) is that for every $\delta > 0$, there is a constant[15] $K_\delta > 0$, such that for all $v_0 \in V^{2\rho}$, one has

$$\|A^r \Phi(B,t)v_0\| \leq K_\delta e^{-(a-\nu-\delta)t}\|A^\rho v_0\|, \qquad \text{for } t \geq 1, \tag{44.30}$$

provided that $0 \leq r \leq \rho$. (Compare this with inequality (37.11).)

Does this prove that Φ generates a linear skew product semiflow? Not quite. There is another important issue. One needs to verify that the mapping $(B,\tau) \to B_\tau$ is continuous, for $\tau > 0$. In order to do this, we first restrict $B = B(t)$ to lie in the subspace $\mathcal{M}^\infty \subset L^\infty$, where

$$\mathcal{M}^\infty \overset{\text{def}}{=} L^\infty(\mathbb{R}; \mathcal{L}(V^{2\beta}, W)) \cap C(\mathbb{R}; \mathcal{L}(V^{2\beta}, W)). \tag{44.31}$$

For $B \in \mathcal{M}^\infty$ we define $\sigma(B,\tau) = B_\tau$ by $B_\tau(t) = B(\tau + t)$, for $\tau, t \in \mathbb{R}$. It is easily verified that $\sigma : \mathcal{M}^\infty \times \mathbb{R} \to \mathcal{M}^\infty$ is a continuous mapping, and it is a <u>flow</u> on $\mathcal{M}^\infty$, where $\mathcal{M}^\infty$ is endowed with either the topology $\mathcal{T}_A$, or $\mathcal{T}_{\text{bo}}$.[16] The key step in showing that σ is continuous is the observation that if $B \in \mathcal{M}^\infty$, then B is uniformly continuous on compact sets in $\mathbb{R}$.

From time to time we will further restrict $B = B(t)$ to be in a compact, invariant subset $\mathcal{K}$ in $\mathcal{M}^\infty$. More precisely, we fix $\mathcal{K}$ to be a compact set in $\mathcal{M}^\infty$, where $\mathcal{K}$ is invariant under the flow σ. One should note that a necessary and sufficient condition that $\mathcal{K}$ be compact in $\mathcal{M}^\infty$ is that it be bounded and uniformly equicontinuous. This follows from the Ascoli-Arzelá Theorem, see Sell (1967) and Naylor and Sell (1982, pp 148 - 149). The phase space for the linear skew product semiflow over $\mathcal{K}$ would be the restricted bundles $V^{2r} \times \mathcal{K}$.

We are also interested in the case where $B = B(t)$ lies in the space

$$\mathcal{M}^p \overset{\text{def}}{=} L^\infty(\mathbb{R}; \mathcal{L}(V^{2\beta}, W)),$$

for some β with $0 \leq \beta < 1$ and with $1 \leq p \leq \infty$. Let us now turn to the question of the dynamical properties of the mapping π, where $\pi(v_0, B, \tau) = (\Phi(B,\tau)v_0, B_\tau)$. As noted in the discussion preceeding the statement of Theorem 44.1, one of the key issues is the continuity of the mapping $(B,\tau) \to B_\tau$. If $B \in \mathcal{M}^\infty$, then this mapping is continuous in the L^∞_{loc}-topology on $\mathcal{M}^\infty$. This means that for every $t > 0$ and every $\tau \geq 0$, one has

$$\|B_{\tau+h} - B_\tau\|_{\infty;[0,t]} \to 0 \text{ and } \|B_{\tau+h} - B_\tau\|_{\{A;[0,t]\}} \to 0, \qquad \text{as } h \to 0^+.$$

---

[15] While K_δ does depend on other terms, such as r and β, its dependence on δ is most delicate, since one typically has $K_\delta \to \infty$, as $\delta \to 0^+$.

[16] Note that σ is <u>not</u> continuous in the topology $\mathcal{T}_\infty$.

If $B = B(s)$ satisfies $B \in L^\infty$, but it is not equal (almost everywhere) to a continuous function, then $(B, \tau) \to B_\tau$ is not continuous in this topology. Nevertheless, one does have continuity in an L^p_{loc}-topology for any p with $1 \le p < \infty$. For $B \in L^\infty$ and $-\infty < a \le b < \infty$, we define the pseudonorm

$$||B||_{p:[a,b)} \stackrel{\text{def}}{=} \left(\int_a^b ||B(s)||^p_{\mathcal{L}} \, ds \right)^{\frac{1}{p}}.$$

Note that $B \in L^\infty$ satisfies $||B(s)v(s)|| \le ||B(s)||_{\mathcal{L}} ||A^\beta v(s)||$, for almost all $s \ge 0$, and that

$$||B||_{p:[0,t)} \le t^{\frac{1}{p}} ||B||_{\infty:[0,t)} < \infty, \qquad \text{for each } t > 0,$$

which implies the continuous imbedding

$$L^\infty_{\text{loc}} = L^\infty_{\text{loc}}(\mathbb{R}; \mathcal{L}(V^{2\beta}, W)) \hookrightarrow L^p_{\text{loc}} = L^p_{\text{loc}}(\mathbb{R}; \mathcal{L}(V^{2\beta}, W)),$$

for $1 \le p < \infty$. This family of pseudonorms generates the L^p_{loc}**-topology** on L^∞. Now the argument leading from (44.10) to (44.11), together with the Hölder inequality with p satisfying

$$\frac{1}{1-\beta} < p \le \infty, \tag{44.32}$$

also establishes that there exist continuous functions $a(\cdot)$ and $b(\cdot)$, which depend neither on the initial conditions (v_{10} and v_{20}) nor the functions B_1 or B_2, such that one has

$$||A^\beta w(t)|| \le a(\tau) ||A^\beta (v_{10} - v_{20})|| + b(\tau) ||B_1 - B_2||_{p:[0,\tau)}, \qquad \text{for } 0 \le t \le \tau,$$

instead of (44.21). Furthermore, one has $||B_{\tau+h} - B_\tau||_{p:[0,t)} \to 0$, as $h \to 0^+$, for each $t > 0$ and each $\tau \ge 0$, as argued in Appendix B.

In the sequel we will use the notation $\mathcal{M}^p$, for $1 \le p \le \infty$, to denote the space $\mathcal{M}^p$ with the L^p_{loc}-topology. Thus one has $\mathcal{M}^p = L^\infty$, for $1 \le p < \infty$, while $\mathcal{M}^\infty$ is given by (44.31). We will also make occasional references to inequality (44.32). While some results are valid on L^∞ when $p = \infty$, any references to flows, or skew product flows, that use the continuity of the mapping $(B, \tau) \to B_\tau$, will also use the fact that $\mathcal{M}^\infty$ is given by (44.31). Without an explicit statement to the contrary, any assertion about the continuity concerning the space $\mathcal{M}^\infty$ will apply to both topologies $\mathcal{T}_A$ and $\mathcal{T}_{bo}$. We now have the following result.

Theorem 44.3. *Let the Standing Hypothesis A be satisfied, and assume that $B \in \mathcal{M}^p$, where $0 \le \beta < 1$ and p satisfies inequality (44.32). For $p < \infty$, we assume that $\mathcal{M}^p$ has the L^p_{loc}-topology, and for $p = \infty$, $\mathcal{M}^\infty$ may* *have either the $\mathcal{T}_{bo}$, or the $\mathcal{T}_A$, topology. Then the following statements are valid:*

(1) *The mapping $(B,\tau) \to B_\tau$ is a flow on the space $\mathcal{M}^p$.*

(2) *The function $\pi(v_0, B, \tau) \stackrel{\text{def}}{=} (\Phi(B,\tau)v_0, B_\tau)$ is a linear skew product semiflow on $V^{2\beta} \times \mathcal{M}^p$.*

(3) *The linear skew product semiflow π has a unique extension to a linear skew product semiflow on $W \times \mathcal{M}^p$, and equation (44.13), as well as inequalities (44.15) and (44.16), are valid for all $v_0 \in W$. In particular, the integral in equation (44.13) exists in the space W, when $v_0 \in W$.*

(4) *For each r with $r \geq 0$, the mapping $\pi = (\Phi(B,\tau), B_\tau)$ is a linear skew product semiflow on $V^{2r} \times \mathcal{M}^p$.*

Proof. First consider $B \in \mathcal{M}^\infty$, with the topology $\mathcal{T}_A$, or $\mathcal{T}_{bo}$. From the comments preceding the statement of this theorem, it follows that the mapping π satisfies properties (1), (2), an (3) in the definition of a linear skew product semiflow in the space $V^{2\beta} \times \mathcal{M}^\infty$. Also the continuity of $\Phi(B,t)w$, at $t = 0$, for each $(w, B) \in V^{2\beta} \times \mathcal{M}^\infty$, follows from (44.14). In order to show that for each $w \in W$, the limit $\lim_{t\to 0^+} \Phi(B,t)w = w$ is uniform for B in compact sets in the metric space $\mathcal{M}^\infty$, it suffices to verify this limit exists when B and t are replaced by convergent sequences B_n and t_n, see Naylor and Sell (1981). Assume that $B_n \to B_0$ in $\mathcal{M}^\infty$ and $t_n \to t_0$, as $n \to \infty$, where $0 \leq t_0 < \infty$. Now one has

$$\|A^\beta(\Phi(B_n,t_n)w - \Phi(B_0,t_0)w)\| \leq \|A^\beta(\Phi(B_n,t_n)w - \Phi(B_0,t_n)w)\| \\ + \|A^\beta(\Phi(B_0,t_n)w - \Phi(B_0,t_0)w)\|.$$

From (44.14) one obtains $\|A^\beta(\Phi(B_0,t_n)w - \Phi(B_0,t_0)w)\| \to 0$, as $n \to \infty$. Since the sequence t_n is bounded, there is a $\tau > 0$ such that $0 \leq t_n \leq \tau$, for all n. From (44.11) one obtains

$$\|A^\beta(\Phi(B_n,t)w - \Phi(B_0,t)w)\| \leq b(t)\|B_n - B_0\|_{\{A;[0,\tau]\}}, \qquad \text{for all } t \in [0,\tau].$$

This implies that $\|A^\beta(\Phi(B_n,t_n)w - \Phi(B_0,t_n)w)\| \to 0$, as $n \to \infty$. Hence π is a linear skew product semiflow on $V^{2\beta} \times \mathcal{M}^\infty$. The same argument with the use of inequality (44.18), with $r = \rho$, and inequality (44.21) establishes Items (2) and (3).

The proofs that Items (1), (2), and (3) are valid for $B \in \mathcal{M}^p$, where p satisfies inequality (44.32), follow from the comments preceding the statement of this theorem and the argument of the last paragraph. Finally, we note that Item (4) follows from Corollary 44.2. □

4.4.3 Strong Solutions. Let us now turn to the question of strong solutions of equation (44.1). As usual we assume that Hypothesis A is satisfied, that $0 \leq \beta < 1$, and that $B \in \mathcal{M}^p$, where p satisfies inequality (44.32). As in Section 4.2.2, we say that a function $v : [0,T) \to W$ is a **strong solution** of (44.1) **in** $V^{2\rho}$, where $\rho \geq 0$, provided that $v \in C[0,T;V^{2\rho})$ and

(1) v is (strongly) differentiable in W almost everywhere (a.e.) in $(0,T)$;
(2) $D_t v \in L^1_{\text{loc}}[0,T;W)$;
(3) $v(t) \in \mathcal{D}(A)$ a.e. on $(0,T)$; and
(4) v satisfies the equation $\partial_t v(t) + Av(t) \overset{\text{a.e.}}{=} B(t)v(t)$ in W, on $(0,T)$.

In light of Theorems 42.6 and 42.9, it should not be surprising that, for a strong solution, the perturbation term $B(t)$ will now be required to have additional regularity in either space, or time. We have the following result:

Theorem 44.4. *Let the Standing Hypothesis A be satisfied, and let $B \in \mathcal{M}^\infty$, where $0 \leq \beta < 1$, satisfy*

$$B \in C^{0,\theta_1}_{\text{loc}}(0,\infty;\mathcal{L}(V^{2\beta},W)), \qquad \text{for some } \theta_1 \text{ with } 0 < \theta_1 \leq 1. \tag{44.33}$$

Then the following statements are valid:

(1) *For every $w_0 \in W$, the function $w = w(t) = \Phi(B,t)w_0$ is a strong solution of equation (44.5) in W on $0 \leq t < \infty$, and one has*

$$\partial_t \Phi(B,t)w_0 \overset{\text{a.e.}}{=} [-A + B(t)]\Phi(B,t)w_0, \tag{44.34}$$

as well as

$$w \in C[0,\infty;W) \cap C^{0,1-r}_{\text{loc}}(0,\infty;V^{2r}) \cap C(0,\infty;\mathcal{D}(A)), \tag{44.35}$$

for every r and with $0 \leq r < 1$.

(2) *If $w_0 \in V^{2\rho}$, where $0 < \rho \leq 1$, then in addition to (44.35), w is a strong solution in $V^{2\rho}$ and it satisfies*

$$w \in C[0,\infty;V^{2\alpha}) \cap C^{0,\theta_1}_{\text{loc}}[0,\infty;V^{2r}),$$

for every α and r with $0 \leq \alpha \leq \rho$ and $0 \leq r < \rho$, where $\theta_1 = \theta_1(r) > 0$.

Proof. We claim that Item (1) follows from Theorems 42.9 and 44.1. Indeed, with $w_0 \in W$, (44.19) and (44.33) imply that $f(t) \overset{\text{def}}{=} B(t)\Phi(B,t)w_0$ satisfies

$$f \in L^\infty_{\text{loc}}[0,\infty;W) \cap C^{0,\theta}_{\text{loc}}(0,\infty;W),$$

for some θ with $0 < \theta \leq 1$. Hence $\Phi(B,t)w_0$ is a strong solution in W, and w satisfies (44.34) and (44.35). Item (2) now follows from Lemma 42.7 and Theorem 42.9. □

The equality (44.13) remains valid when the perturbation term B does not depend on time t, i.e., when $B \in \mathcal{L}(V^{2\beta},W)$. In this case, one has

$$\|B\|_\infty = \sup\{\|Bv\| : \|A^\beta v\| \leq 1\} = \|B\|_{\mathcal{L}(W)}$$

and equation (44.1) becomes $\partial_t u + Lu = 0$, where $L = A - B$. As noted in Lemma 39.2, the linear operator L is a sectorial operator on the space W, and consequently, e^{-Lt} is an analytic semigroup on W. Furthermore, for any $v_0 \in W$, $e^{-Lt}v_0$ is a strong solution of (44.1) in W. Consequently, by using the argument of Lemma 42.1, we see that it is a mild solution in W and, by the uniqueness of mild solutions, one has $e^{-Lt}v_0 = \Phi(B,t)v_0$. The identity (44.13) now assumes the form

$$e^{-Lt}v_0 = e^{-At}v_0 + \int_0^t e^{-A(t-s)}Be^{-Ls}v_0\,ds, \qquad \text{for } v_0 \in W. \tag{44.36}$$

Furthermore, inequality (44.17), with $\rho = 0$, now assumes the form

$$\|A^r e^{-Lt}v_0\| \le C_r t^{-r} E(\nu;t) e^{-at}\|v_0\|, \qquad \text{for } t > 0, \tag{44.37}$$

for $v_0 \in W$ and $r \ge 0$, and inequality (44.30) becomes

$$\|A^r e^{-Lt}v_0\| \le K_\delta e^{-(a-\nu-\delta)t}\|v_0\|, \qquad \text{for } t \ge 1. \tag{44.38}$$

It follows that $e^{-Lt}v_0 \in V^{2r}$, for all $r \ge 0$, $t > 0$, and $v_0 \in W$. We summarize this in the following result.

Theorem 44.5. *Let the Standing Hypothesis A be satisfied and assume that $B \in \mathcal{L}(V^{2\beta}, W)$, where β satisfies $0 \le \beta < 1$. Then the linear operator $L = A - B$ is a sectorial operator on W; the associated analytic semigroup e^{-Lt} on W satisfies equation (44.36); and inequalities (44.37) and (44.38) hold, for all $r \ge 0$. Moreover, one has $\mathcal{D}(A) = \mathcal{D}(L)$.*

4.5. Exponential Dichotomies: Existence and Robustness.

As we will see in Chapter 7, the key feature, underlying essentially all the dynamical phenomena occurring in dissipative nonlinear evolutionary equations, is the presence of certain hyperbolic structures in the semiflow generated by these equations. The primary tool for understanding these hyperbolic structures is the exponential dichotomy for a linear skew product semiflow. It is this dichotomy concept which we develop now. The material presented here is based on Pliss and Sell (1999), Sacker and Sell (1994), and Henry (1981).

Let W be a given Banach space, M be a metric space, and define the product space $\mathcal{E} = W \times M$. For any set $\mathcal{V}$ in $\mathcal{E}$, we define the **fiber** $\mathcal{V}(m_0)$ over the point $m_0 \in M$ by

$$\mathcal{V}(m_0) \stackrel{\text{def}}{=} \{(w,m) \in \mathcal{E} : (w,m) \in \mathcal{V} \text{ and } m = m_0\}.$$

Similarly for any set $S \subset M$ we define the **restriction of $\mathcal{V}$ to S** as

$$\mathcal{V}(S) \stackrel{\text{def}}{=} \{(w,m) \in \mathcal{V} \colon m \in S\} = \bigcup_{m \in S} \mathcal{V}(m).$$

Notice that $\mathcal{E}(m) = W \times \{m\}$, and $\mathcal{E}(m)$ is a Banach space, with the same structure as W. A mapping $P : \mathcal{E} \to \mathcal{E}$ is said to be a **projector** if P is continuous and has the form $P(w,m) = (P(m)w, m)$, where $P(m)$ is a (linear) projection[17] on the fiber $\mathcal{E}(m)$. This means that $P(m) : \mathcal{E}(m) \to \mathcal{E}(m)$ is a bounded linear mapping that satisfies $P(m)P(m) = P(m)^2 = P(m)$. For any projector $P : \mathcal{E} \to \mathcal{E}$ we define the **range** and **null space** by

$$\mathcal{R} = \mathcal{R}(P) = \{(w,m) \in \mathcal{E} : P(m)w = w\} \text{ and}$$
$$\mathcal{N} = \mathcal{N}(P) = \{(w,m) \in \mathcal{E} : P(m)w = 0\}.$$

Note that the fibers $\mathcal{R}(m)$ and $\mathcal{N}(m)$ are linear subspaces of $\mathcal{E}(m)$, since $P(m)$ is a linear mapping. Also since P is continuous, this means that the fibers $\mathcal{R}(m)$ and $\mathcal{N}(m)$ vary continuously in m, which implies that $P(m)$ varies continuously in the operator norm in $\mathcal{L}(W)$. The following result is easily proven.

Lemma 45.1. *Let P be a projector on $\mathcal{E}$. Then $\mathcal{R}$ and $\mathcal{N}$ are closed subsets in $\mathcal{E}$, and for all $m \in M$, one has*

$$\mathcal{R}(m) \cap \mathcal{N}(m) = \{0\} \quad \textit{and} \quad \mathcal{R}(m) + \mathcal{N}(m) = \mathcal{E}(m). \tag{45.1}$$

If P is a projector on $\mathcal{E}$, then the mapping $Q = I - P$, where $Q : \mathcal{E} \to \mathcal{E}$ is defined by $Q(w,m) = I(w,m) - P(w,m) = (w - P(m)w, m)$ and I is the identity mapping on $\mathcal{E}$, is also a projector on $\mathcal{E}$. The projector Q is called the **complementary projector** to P, and one has $\mathcal{R}(Q) = \mathcal{N}(P)$ and $\mathcal{N}(Q) = \mathcal{R}(P)$.

The range and nullspace of a projector are subbundles of $\mathcal{E}$. More precisely, a subset $\mathcal{V}$ in $\mathcal{E}$ is said to be a **subbundle** of $\mathcal{E}$ if there is a projector P on $\mathcal{E}$ with the property that $\mathcal{V} = \mathcal{R}(P)$. In this case, $\mathcal{W} = \mathcal{N}(P)$ is a **complementary** subbundle, and one has $\mathcal{E} = \mathcal{V} + \mathcal{W}$, in the sense that equations (45.1) are valid. The equation $\mathcal{E} = \mathcal{V} + \mathcal{W}$ is sometimes referred to as a **Whitney sum** of subbundles. The **trivial** subbundle $\mathcal{E}_0 = \{0\} \times M$ plays a role in the theory of exponential dichotomies.

Now let $\pi = (\Phi, \sigma)$ be a linear skew product semiflow on $\mathcal{E} = W \times M$ (see Section 4.3). For our applications, we assume that σ is a <u>flow</u> on M, where $\sigma(m,\tau) = m_\tau$. A projector P on $\mathcal{E}$ is said to be **invariant** if one has

$$P(m_t)\Phi(m,t) = \Phi(m,t)P(m), \qquad \text{for all } t \geq 0 \text{ and } m \in M. \tag{45.2}$$

Of course, if the projector P is invariant, then the projector $Q = I - P$ is invariant, as well. The invariance of a projector is equivalent to the assertion that both subbundles, $\mathcal{R}$ and $\mathcal{N}$, are positively invariant sets for the semiflow π.

[17]There is a small ambiguity in this notation since we use both $P(w,m)$ and $P(m)w$ to represent P. These two formulations really offer two ways of viewing the concept of a projector.

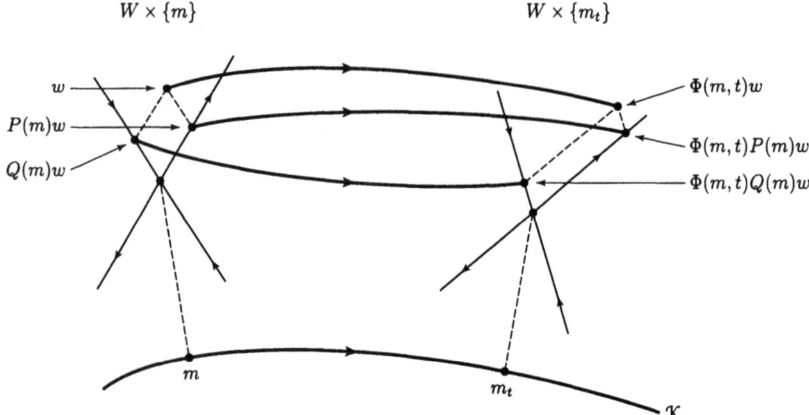

Figure 4.1. Exponential Dichotomy over $\mathcal{K}$

4.5.1 Exponential Dichotomy. In the following definition we will refer to some solutions with negative continuations. We do not assume the uniqueness of the negative continuations, but we will show that, when one has an exponential dichotomy, then in a qualified sense, some negative continuations are unique. We now consider a linear skew product semiflow $\pi = (\Phi, \sigma)$ on $\mathcal{E} = W \times M$, however, we will restrict to the case where the semiflow $\sigma(m,t) = m_t$ is a flow on M. We say that π has an **exponential dichotomy** in W and over an invariant set $\mathcal{K} \subset M$ (or π has an **exponential dichotomy on** $W \times \mathcal{K}$), see Figure 4.1, if there is a projector P on the restriction $\mathcal{E}(\mathcal{K}) = W \times \mathcal{K}$, and constants $K \geq 1$ and $\alpha > 0$ such that the following hold:

(1) The projectors P and Q are invariant on $\mathcal{E}(\mathcal{K})$, where $Q = I - P$.
(2) One has $\mathcal{R}(P(m)) \subset \mathcal{U}(m)$, for each $m \in \mathcal{K}$. For each $w \in \mathcal{R}(P(m))$, we let

$$\phi^{w,m}(t) = (\Phi(m,t)w, m_t), \qquad \text{for } t \leq 0,$$

denote <u>any</u> negative continuation with $\|\Phi(m,t)w\| \to 0$, as $t \to -\infty$.
(3) The following inequalities are valid for all $w \in W$:

$$\|\Phi(m,t)Q(m)w\| \leq K\|w\|e^{-\alpha t}, \qquad \text{for } t \geq 0 \text{ and } m \in \mathcal{K}, \tag{45.3}$$

and for any negative continuation through $(P(m)w, m)$ that remains in $\mathcal{U}$, one has

$$\|\Phi(m,t)P(m)w\| \leq K\|w\|e^{\alpha t}, \qquad \text{for } t \leq 0 \text{ and } m \in \mathcal{K}. \tag{45.4}$$

We will refer to (K, α) as the **characteristics** of the exponential dichotomy, and we will call (P, Q) the **associated projectors.** The case $P = 0$ is allowed, provided Items (1) - (3) are valid. In this case, the linear skew product semiflow is said to be **exponentially stable**.

Recall that $\phi(t) = (w(t), m_t)$ is a global motion through the point $(w_0, m) \in \mathcal{E}$ provided that $w = w(t)$ is continuous, one has

$$\Phi(m_\tau, t)w(\tau) = w(\tau + t), \qquad \text{for all } \tau \in \mathbb{R} \text{ and all } t \geq 0, \tag{45.5}$$

and $w(0) = w_0$, see (21.6). The restriction of a global motion to $\{t : t \leq 0\}$ is called a negative continuation

The concept of an exponential dichotomy for a discrete linear skew product semiflow is identical to that given above, with the single exception that the time t is now restricted to assuming values in some discrete subgroup, or discrete semisubgroup, of $\mathbb{R}$. In particular, for the discrete semiflow $\pi(w, T, \tau) = (\Phi(T, \tau)w, T \cdot \tau)$ given by (43.3) on the space $W \times \mathcal{W}$, we let $\mathcal{K}$ denote any set in the base space $\mathcal{W} = \ell_\infty(Z, \mathcal{L}(W))$. The semiflow π is said to have an **exponential dichotomy on $W \times \mathcal{K}$, with characteristics** (K, s), where $K \geq 1$ and $0 < s < 1$, provided that there is a projector P defined on $\mathcal{K}$ such that

(1) For each $T = \{T_n\} \in \mathcal{K}$, the projector sequence $P(T) = \{P_n\}$, where $P_n = P_n(T)$ depends on T, satisfies

$$P_{m+1}T_m = T_m P_m, \qquad \text{for all } m \in Z.$$

(2) One has $\mathcal{R}(P(T)) \subset \mathcal{U}(T)$, for each $T \in \mathcal{K}$. For each $w \in \mathcal{R}(P(T))$, we let

$$\phi^{w,T}(k) = (\Phi(T, k)w, T \cdot k), \qquad \text{for } k \leq 0,$$

denote <u>any</u> negative continuation with $\|\Phi(T, k)w\| \to 0$, as $k \to -\infty$.

(3) The following inequalities are valid for all $w \in W$ and $T \in \mathcal{K}$:

$$\|\Phi(T, k)Q(T)w\| = \|T_{k-1} \cdots T_1 T_0 Q_0(T)w\| \leq K\|w\|s^k, \tag{45.6}$$

for $k \geq 0$, and for any negative continuation through $(P(T)w, T)$ that remains in $\mathcal{U}$, one has

$$\|\Phi(T, k)P(T)w\| \leq K\|w\|s^{-k}, \qquad \text{for } k \leq 0. \tag{45.7}$$

It should be noted that if π has an exponential dichotomy over a set $\mathcal{K} \subset \mathcal{W}$ with characteristics (K, s), then π has an exponential dichotomy over the invariant set $\gamma(\mathcal{K})$, the trajectory of $\mathcal{K}$, with the same characteristics. In the following lemma it is shown, in addition, that π has an exponential dichotomy over the hull $H(\mathcal{K}) = \text{Cl}\ \gamma(\mathcal{K})$, with the same characteristics.

There is some additional information contained in the inequalities (45.4) and (45.7). Let $w \in \mathcal{R}(P(m))$. Then one has $w = P(m)w$. Also for $t \leq 0$ set $\tau = -t$, and let $\hat{w} = \Phi(m, t)w$. One then has $w = \Phi(m_t, \tau)\hat{w}$ and $P(m_t)\hat{w} = \hat{w}$. Now inequalities (45.4) and (45.7) can be rewritten in the form $\|P(m_t)\hat{w}\| \le K\|\Phi(m_t,\tau)\hat{w}\|e^{-\alpha\tau}$, or by dropping the hats, and by replacing $\tau \ge 0$ by $t \ge 0$ and m_t by m, one has

$$\|\Phi(m,t)P(m)w\| \ge K^{-1}\|P(m)w\|e^{\alpha t}, \tag{45.8}$$

for all $w \in W$ and all $t \ge 0$.

There are important consequences of the existence of an exponential dichotomy. It follows from (45.3) and (45.4) that

$$\mathcal{N}(P(m)) \subset \mathcal{S}(m) \quad \text{and} \quad \mathcal{R}(P(m)) \subset \mathcal{U}(m), \qquad \text{for all } m \in \mathcal{K}.$$

Consequently, Lemma 45.1 implies that

$$\mathcal{S}(m) + \mathcal{U}(m) = \mathcal{E}(m), \qquad \text{for all } m \in \mathcal{K}.$$

We now have the following result wherein, among other things, we show that certain negative continuations are unique.

Lemma 45.2. *Let $\pi = (\Phi, \sigma)$ be a linear skew product semiflow on $\mathcal{E} = W \times M$, where σ is a flow on M. Assume that π has an exponential dichotomy over an invariant set $\mathcal{K} \subset M$. Then the following statements are valid:*

1. *The bounded set satisfies $\mathcal{B}(\mathcal{K}) = \mathcal{E}_0(\mathcal{K}) = \{0\} \times \mathcal{K}$.*
2. *One has $\mathcal{S}(m) \cap \mathcal{U}(m) = \{0\}$, for all $m \in \mathcal{K}$.*
3. *One has $\mathcal{S}(m) = \mathcal{N}(P(m)) = \mathcal{R}(Q(m))$ and $\mathcal{U}(m) = \mathcal{R}(P(m)) = \mathcal{N}(Q(m))$, for all $m \in \mathcal{K}$. Thus the convergence rates in $\mathcal{S}$ and $\mathcal{U}$ are exponential over $\mathcal{K}$.*
4. *The subbundles $\mathcal{S} = \mathcal{R}(Q)$ and $\mathcal{U} = \mathcal{R}(P)$ satisfy*

$$\pi(\mathcal{S}, t) \subset \mathcal{S} \quad \textit{and} \quad \pi(\mathcal{U}, t) = \mathcal{U}, \qquad \textit{for } t \ge 0, \tag{45.9}$$

that is, $\mathcal{S}$ is positively invariant and $\mathcal{U}$ is invariant.

5. *For all $m \in \mathcal{K}$, the restriction of $\Phi(m,t)$ to $\mathcal{R}(P(m)) = \mathcal{U}(m)$ is an isomorphism of $\mathcal{R}(P(m))$ onto $\mathcal{R}(P(m_t))$, for each $t \ge 0$. Moreover, for each $w \in W$, the function $\Phi(m,t)P(m)w$ has a unique negative continuation satisfying*

$$\Phi(m,t)P(m)w \in \mathcal{R}(P(m_t)), \qquad \textit{for all } t \le 0,$$

and the strong cocycle identity

$$\Phi(m,\tau+t)P(m) = \Phi(m_\tau,t)\Phi(m,\tau)P(m), \qquad \textit{for all } \tau, t \in \mathbb{R}, \tag{45.10}$$

is valid, for all $m \in \mathcal{K}$.

6. *One has $\mathcal{U}(m) = \mathcal{B}_u^-(m)$, i.e., for each $w \in \mathcal{R}(P(m)) = \mathcal{U}(m)$, there is a unique negative continuation through (w,m), where the w-coordinate is uniformly bounded, for $t \le 0$, and this negative continuation is $\Phi(m,t)w$.*
7. *The exponential dichotomy over $\mathcal{K}$ extends to an exponential dichotomy of the closure $Cl_M\mathcal{K}$, with no change in the characteristics (K,α).*

In the case of a discrete linear skew product semiflow, all the assertions above remain valid for time t restricted to an appropriate discrete subgroup of $\mathbb{R}$.

Proof. Item (1). Let $(w, m) \in \mathcal{B}$. It then follows from the definition of $\mathcal{B}$ that there is a global motion $\phi^{(w,m)}(t) = (w(t), m_t)$ such that $w(0) = w$ and

$$\|w\|_\infty \stackrel{\text{def}}{=} \sup\{\|w(t)\| : t \in \mathbb{R}\} < \infty.$$

For $t \in \mathbb{R}$, define $u(t) \stackrel{\text{def}}{=} P(m_t)w(t)$ and $v(t) \stackrel{\text{def}}{=} Q(m_t)w(t)$. Since inequalities (45.3) and (45.4) hold at $t = 0$, one has

$$(45.11) \qquad \|u(t)\| \le K\|w\|_\infty \quad \text{and} \quad \|v(t)\| \le K\|w\|_\infty, \qquad \text{for all } t \in \mathbb{R}.$$

Furthermore, the invariance properties (45.2) and (45.5) imply that, for $\tau \in \mathbb{R}$ and $t \ge 0$, one has

$$\begin{aligned}\Phi(m_\tau, t)u(\tau) &= \Phi(m_\tau, t)P(m_\tau)w(\tau) = P(m_{\tau+t})\Phi(m_\tau, t)w(\tau)\\ &= P(m_{\tau+t})w(\tau + t) = u(\tau + t).\end{aligned}$$

Hence $(u(t), m_t)$ is a global motion with $u(0) = P(m)w(0)$. Similarly, $(v(t), m_t)$ is a global motion with $v(0) = Q(m)w(0)$. Also inequality (45.11) implies that both of the orbits $(u(t), m_t)$ and $(v(t), m_t)$ belong to the bounded set $\mathcal{B}$, for all $t \in \mathbb{R}$.

Now let $t \ge 0$ and set $\tau = -t$. Then equation (45.5) implies that $v(0) = \Phi(m_\tau, t)v(\tau)$, for all $t \ge 0$. From inequality (45.3) and the identity $Q(m_\tau)v(\tau) = v(\tau)$, we find that

$$\|v(0)\| = \|\Phi(m_\tau, t)v(\tau)\| = \|\Phi(m_\tau, t)Q(m_\tau)v(\tau)\| \le K\|v\|_\infty e^{-\alpha t},$$

for all $t \ge 0$. This implies that $v(0) = 0$ and $v(t) = 0$, for all $t \in \mathbb{R}$. Hence, $w(t) = u(t)$, for all $t \in \mathbb{R}$. Since $u(0) \in \mathcal{R}(P(m))$, it follows similarly from inequality (45.4) that

$$\|u(0)\| = \|\Phi(m_t, \tau)u(\tau)\| = \|\Phi(m_t, \tau)P(m_t)u(t)\| \le K\|u\|_\infty e^{\alpha\tau},$$

for all $\tau \le 0$, which implies that $w(0) = u(0) = 0$. Hence one has $\mathcal{B}(\mathcal{K}) = \mathcal{E}_0(\mathcal{K}) = \{0\} \times \mathcal{K}$.

Items (2) and (3). Since $\mathcal{S}(m) \cap \mathcal{U}(m) \subset \mathcal{B}(m)$, one has $\mathcal{S}(m) \cap \mathcal{U}(m) = \{0\}$, for all $m \in \mathcal{K}$. Since $\mathcal{R}(P(m)) \subset \mathcal{U}(m)$ and $\mathcal{R}(Q(m)) \subset \mathcal{S}(m)$, for all m, and $\mathcal{R}(P(m)) + \mathcal{R}(Q(m)) = W$, for all $m \in \mathcal{K}$, one obtains $\mathcal{S}(m) + \mathcal{U}(m) = W$, $\mathcal{S}(m) = \mathcal{N}(P(m))$, and $\mathcal{U}(m) = \mathcal{R}(P(m))$, for all $m \in \mathcal{K}$. Thus the convergence rates in $\mathcal{S}$ and $\mathcal{U}$ are exponential over $\mathcal{K}$, by (45.3) and (45.4).

Items (4) and (5). The inclusion $\pi(\mathcal{S}, t) \subset \mathcal{S}$ in (45.9) follows directly from the above and the definition of an exponential dichotomy. Next, the invariance of P (see equation (45.2)) implies that $\Phi(m, t)$ maps $\mathcal{R}(P(m))$

into $\mathcal{R}(P(m_t))$. From the second condition in the definition of an exponential dichotomy, we see that $\Phi(m,t)$ maps $\mathcal{R}(P(m))$ onto $\mathcal{R}(P(m_t))$, and inequality (45.8) implies that the restriction $\Phi(m,t)\mid_{\mathcal{R}(P(m))}$ is one-to-one. Thus this restriction is an isomorphism of $\mathcal{R}(P(m))$ onto $\mathcal{R}(P(m_t))$, and inequality (45.8) implies that the inverse operator is a bounded linear operator. By using this inverse operator, we see that, for each $w \in W$, the function $\Phi(m,t)P(m)w$ has a unique negative continuation with

$$\Phi(m,t)P(m)w \in \mathcal{R}(P(m_t)), \qquad \text{for all } m \in \mathcal{K} \text{ and all } t \leq 0.$$

In fact, for $t \leq 0$, we define $\Phi(m,t)P(m)$ to be the unique isomorphism

$$\Phi(m,t)P(m) : \mathcal{R}(P(m)) \to \mathcal{R}(P(m_t)) \tag{45.12}$$

with the property that $\Phi(m_t,-t)\Phi(m,t)P(m) = P(m)$. Also, for $t \geq 0$, one has

$$\Phi(m,t)\mathcal{U}(m) = \Phi(m,t)\mathcal{R}(P(m)) = \mathcal{R}(P(m_t)) = \mathcal{U}(P(m_t)).$$

This implies that $\pi(\mathcal{U},t) = \mathcal{U}$, for $t \geq 0$. The proof of equation (45.10) is left as a very important exercise.

Item (6). If there is a point $(w,m) \in \mathcal{B}^-$ for which there are two global motions, say, ϕ_1 and ϕ_2, with the w-coordinate being uniformly bounded, for $t \leq 0$, we will write $\phi_1(t) = (w_1(t), m_t)$ and $\phi_2(t) = (w_2(t), m_t)$, for $t \in \mathbb{R}$. Then w_1 and w_2 are uniformly bounded, for $t \leq 0$, and $w = w_1 - w_2$ satisfies $w(t) = 0$, for all $t \geq 0$. Hence,

$$(w(t), m_t) = (w_1(t) - w_2(t), m_t) \in \mathcal{B}(\mathcal{K}), \qquad \text{for all } t \in \mathbb{R}.$$

It follows from Item (1) that $w(t) \equiv 0$. Hence one has $\mathcal{U}(m) = \mathcal{B}_u^-(m)$. We leave the proof of Item (7) as an exercise. □

In some cases, as we now show, the unstable fiber $\mathcal{U}(m) = \mathcal{R}(P(m))$ is finite dimensional in the presence of an exponential dichotomy.

Lemma 45.3. *Let $\pi = (\Phi, \sigma)$ be a linear skew product semiflow on $\mathcal{E} = W \times M$, where σ is a flow on M. Assume that π has an exponential dichotomy over an invariant set $\mathcal{K} \subset M$. Then the following statements are equivalent:*

1. *The restriction of π to $\mathcal{E}(\mathcal{K}) = W \times \mathcal{K}$ is uniformly κ-contracting.*
2. *One has dim $\mathcal{U}(m) =$ dim $\mathcal{R}(P(m)) < \infty$, for every $m \in \mathcal{K}$.*

Proof. (1)⇒(2). For any Banach space V, we let $N_r = N_r(0)$ denote the closed ball of radius r centered at the origin 0. Now one has $\kappa(N_r) = 0$ if and only if V is finite dimensional. Let $m \in \mathcal{K}$, and let $N_1(m)$ denote the closed ball of radius 1, centered at the origin, in $\mathcal{R}(P(m))$. We have $\kappa(N_1(m)) = \kappa(J_t[\Phi(m,t)N_1(m)])$, where J_t is the inverse isomorphism in (45.12). Then by inequality (45.8), we obtain $\kappa(N_1(m)) \leq Ke^{-\alpha t}\kappa(\Phi(m,t)N_1(m))$, for all $t \geq 0$. If the linear skew product semiflow π is uniformly κ-contracting, this implies that $\kappa(N_1(m)) = 0$, or $\mathcal{R}(P(m))$ is finite dimensional.

(2)⇒(1). Assume now that $\mathcal{R}(P(m))$ is finite dimensional for each $m \in \mathcal{K}$. Let B be a bounded set in $\mathcal{E}(m)$, for some $m \in \mathcal{K}$. Then Lemma 22.2 implies that

$$\kappa(\Phi(m,t)B) \leq \kappa(\Phi(m,t)P(m)B) + \kappa(\Phi(m,t)Q(m)B),$$

and $\kappa(\Phi(m,t)P(m)B) = 0$, since $\mathcal{R}(P(m))$ is finite dimensional. However, inequality (45.3) implies that $\kappa(\Phi(m,t)Q(m)B) \leq \operatorname{diam}(B)Ke^{-\alpha t}$, for $t \geq 0$. Hence π is uniformly κ-contracting. □

A given linear skew product flow $\pi = (\Phi, \sigma)$ on $\mathcal{E} = W \times M$ can be imbedded into a one-parameter family $\pi_\lambda = (\Phi_\lambda, \sigma)$, for $\lambda \in \mathbb{R}$, by defining $\Phi_\lambda(m,t) \stackrel{\text{def}}{=} e^{-\lambda t}\Phi(m,t)$. One refers to π_λ as the **shifted semiflow**. The reason for this teminology can be appreciated by assuming that $\Phi(m,t)$ is a fundamental solution operator for the nonautonomous linear evolutionary equation $\partial_t u = m(t)u$. In this case, $\Phi_\lambda(m,t)$ is a fundamental solution operator for the shifted equation $\partial_t v = (m(t) - \lambda)v$. Note that if π is uniformly κ-contracting, then π_λ is also uniformly κ-contracting, for $\lambda > 0$. In the case of a discrete linear skew product semiflow, the shifted semiflow is defined as above. However, the interpretation in terms of a differential equation is no longer appropriate.

The set of those $\lambda \in \mathbb{R}$ for which π_λ admits an exponential dichotomy on $\mathcal{E}$ is called the **resolvent set** for π. The complement in $\mathbb{R}$ of the resolvent set is called the **dynamical spectrum** $\Sigma(\pi)$ of π. The unstable set, the stable set, the bounded set, etc, are all defined for the shifted flow π_λ and will be denoted by $\mathcal{U}_\lambda$, $\mathcal{S}_\lambda$, $\mathcal{B}_\lambda$, etc. The two sets $\mathcal{U}_\lambda$ and $\mathcal{S}_\lambda$ are monotone in λ: $\mathcal{U}_\lambda$ is nonincreasing, and $\mathcal{S}_\lambda$ is nondecreasing.

We are interested in the situation where the shifted linear skew product semiflow π_λ has an exponential dichotomy for different values of the parameter λ. More precisely, let λ and μ be given, where $\lambda < \mu$, and assume that π_λ and π_μ each has an exponential dichotomy over an invariant set $\mathcal{K}$ in M, with invariant projectors (P_λ, Q_λ) and (P_μ, Q_μ), and characteristics $(K_\lambda, \alpha_\lambda)$ and (K_μ, α_μ). By replacing these characteristics with (K, α), where $K = \max(K_\lambda, K_\mu)$ and $\alpha = \min(\alpha_\lambda, \alpha_\mu)$, we see that (K, α) serve as characteristics for both π_λ and π_μ. As noted above, one has $\mathcal{S}_\lambda \subset \mathcal{S}_\mu$ and $\mathcal{U}_\mu \subset \mathcal{U}_\lambda$, since $\lambda < \mu$.

The existence of the exponential dichotomies for the two linear skew product semiflows π_λ and π_μ has an equivalent formulation in terms of a trichotomy. We say that π has an **exponential trichotomy** over an invariant set $\mathcal{K} \subset M$, with **characteristics** λ_1, λ_2, λ_3, λ_4, and K, where $\lambda_1 < \lambda_2 \leq 0 \leq \lambda_3 < \lambda_4$ and $K \geq 1$, see Figure 4.2, if there exist three projectors P, Q, and R defined over $\mathcal{K}$ such that the following properties hold:

(1) Each of the projectors P, Q, and R is invariant on $\mathcal{E}(\mathcal{K}) = W \times \mathcal{K}$.

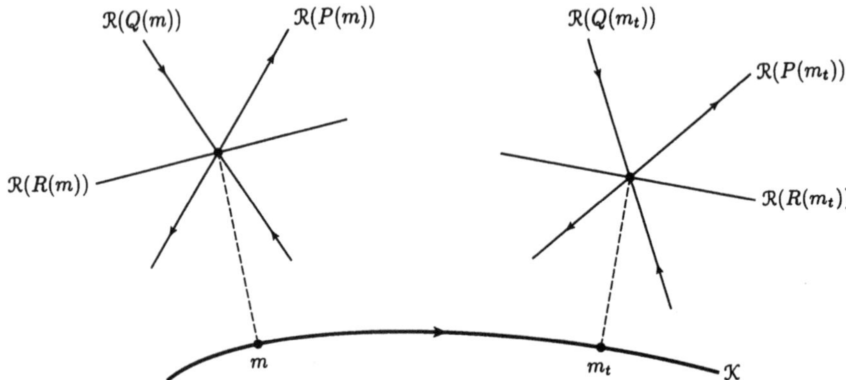

Figure 4.2. Exponential Trichotomy over $\mathcal{K}$

(2) For each $m \in \mathcal{K}$, the projections $P(m)$, $Q(m)$, and $R(m)$ commute and one has

$$I_W = P(m) + Q(m) + R(m) \text{ and}$$
$$P(m)Q(m) = P(m)R(m) = Q(m)R(m) = 0.$$

(3) For each $w \in \mathcal{R}(P(m))$ there is a negative continuation

$$\phi^{(w,m)}(t) = (\Phi(m,t)w, m_t), \qquad \text{for } t \leq 0,$$

such that $\|e^{-\lambda_4 t}\Phi(m,t)w\| \to 0$, as $t \to -\infty$. (We do not require that this negative continuation be unique. Let $(\Phi(m,t)w, m_t)$, denote any negative continuation that satisfies $\|e^{-\lambda_4 t}\Phi(m,t)w\| \to 0$, as $t \to -\infty$.)

(4) For each $w \in \mathcal{R}(R(m))$ there is a negative continuation

$$\phi^{(w,m)}(t) = (\Phi(m,t)w, m_t), \qquad \text{for } t \leq 0,$$

such that $\|e^{-\lambda_2 t}\Phi(m,t)w\| \to 0$, as $t \to -\infty$. (We do not require that this negative continuation be unique, and we let $(\Phi(m,t)w, m_t)$ denote any negative continuation that satisfies $\|e^{-\lambda_2 t}\Phi(m,t)w\| \to 0$, as $t \to -\infty$.)

(5) The following four inequalities are valid for all $w \in W$:

$$\|\Phi(m,t)Q(m)w\| \leq K\|w\|e^{\lambda_1 t}, \qquad \text{for } t \geq 0 \text{ and } m \in \mathcal{K}, \tag{45.13}$$

$$\|\Phi(m,t)P(m)w\| \leq K\|w\|e^{\lambda_4 t}, \qquad \text{for } t \leq 0 \text{ and } m \in \mathcal{K}, \tag{45.14}$$

$$\|\Phi(m,t)R(m)w\| \leq K\|w\|e^{\lambda_3 t}, \qquad \text{for } t \geq 0 \text{ and } m \in \mathcal{K}, \tag{45.15}$$

and

$$\|\Phi(m,t)R(m)w\| \leq K\|w\|e^{\lambda_2 t}, \qquad \text{for } t \leq 0 \text{ and } m \in \mathcal{K}, \tag{45.16}$$

where (45.14) and (45.16) are valid for any negative continuation, as described above.

It may be helpful to look at two finite dimensional examples of an exponential trichotomy, one with continuous-time dynamics and the other with discrete dynamics.

Continuous-Dynamics Example: Consider the linear, autonomous ordinary differential equation

$$\partial_t x = Ax, \qquad \text{where } x \in \mathbb{R}^n, \text{ or } x \in \mathbb{C}^n, \tag{45.17}$$

and A is an $n \times n$ matrix. For this example, like all autonomous problems, the base space $\mathcal{K}$ in the definition of an exponential trichotomy reduces to a set with a single point. Because the dynamics is trivial on a set with only one point, we now suppress the notation $m \in \mathcal{K}$, etc. Let e^{At} denote the solution operator for equation (45.17). Then equation (45.17) has an exponential trichotomy with characteristics $\lambda_1 < \lambda_2 \leq 0 \leq \lambda_3 < \lambda_4$ and $K \geq 1$ if and only if the eigenvalues λ of A split into three bands in the complex plane $\mathbb{C}$:

$$\text{Re } \lambda \leq \lambda_1, \quad \text{or} \quad \lambda_2 \leq \text{Re } \lambda \leq \lambda_3, \quad \text{or} \quad \lambda_4 \leq \text{Re } \lambda. \tag{45.18}$$

The range $\mathcal{R}(R)$ of the projector R is the algebraic sum of the generalized eigenspaces corresponding to eigenvalues λ with $\lambda_2 \leq \text{Re } \lambda \leq \lambda_3$, and similar characterizations apply to $\mathcal{R}(Q)$ and $\mathcal{R}(P)$. Thus $\mathcal{R}(R)$ would include any eigenvectors with eigenvalues λ, with $\text{Re } \lambda = 0$. The characteristic K, which is dependent on the norm on $\mathbb{R}^n$, or $\mathbb{C}^n$, has the property that $K \to \infty$, as the angle between any two of the spaces $\mathcal{R}(Q)$, $\mathcal{R}(R)$, or $\mathcal{R}(P)$ goes to 0.

Discrete-Dynamics Example: Next we consider a linear <u>mapping</u> $y = Ex$ on $\mathbb{R}^n$, or $\mathbb{C}^n$, where E is an $n \times n$ matrix. In the case where there exists a matrix A such that $E = e^A$, then the discrete dynamics generated by iterates of the mapping $y = Ex$ is identical to the discrete dynamics generated by restricting e^{At} to times $t \in Z$. One can show that for a given E, the equation $E = e^A$ has a solution A if and only if E has a bounded inverse, see Hartman (1964). (The derivation of A, in general, uses complex arithmetic, even when E is real.)

What happens if E is not invertible, i.e., $\mu = 0$ is an eigenvalue of E? In order to answer this question, it is useful to note that, when $E = e^A$, then $\mu \in \mathbb{C}$ is an eigenvalue of E if and only if A has an eigenvalue $\lambda \in \mathbb{C}$ with $\mu = e^\lambda$. Since $|\mu| = e^{\text{Re } \lambda}$, in this case, the three-band condition (45.18) now assumes the form

$$|\mu| \leq e^{\lambda_1}, \quad \text{or} \quad e^{\lambda_2} \leq |\mu| \leq e^{\lambda_3}, \quad \text{or} \quad e^{\lambda_4} \leq |\mu|,$$

in terms of the eigenvalues μ of E. In the case where E is not invertible, the null space $\mathcal{N}(E)$ lies in the stable manifold $\mathcal{R}(Q)$. Roughly speaking, if $x \in \mathcal{N}(E)$, then $Ex = e^{-\infty}x = 0$. See the exercises for further illustrations.

Since inequalities (45.13)–(45.16) hold at $t = 0$, for all $m \in \mathcal{K}$ and $w \in W$ one obtains

$$\|P(m)w\|,\ \|Q(m)w\|,\ \|R(m)w\| \leq K\|w\|. \tag{45.19}$$

The projector R is called the **neutral** projector. The following property is easily verified, and we leave the proof as an exercise:

Lemma 45.4. *The linear skew product semiflow π has an exponential trichotomy over an invariant set $\mathcal{K} \subset M$ with characteristics λ_1, λ_2, λ_3, λ_4, and K if and only if both of the following two properties hold:*

(1) *For any λ with $\lambda_1 < \lambda < \lambda_2$, the semiflow π_λ has an exponential dichotomy with characteristics (K, α_λ) and associated projections (P_λ, Q_λ), where*

$$2\alpha_\lambda = \min(\lambda - \lambda_1, \lambda_2 - \lambda), \quad P_\lambda = P + R, \quad \textit{and} \quad Q_\lambda = Q.$$

(2) *For any μ with $\lambda_3 < \mu < \lambda_4$, the semiflow π_μ has an exponential dichotomy with characteristics (K, α_μ) and associated projections (P_μ, Q_μ), where*

$$2\alpha_\mu = \min(\mu - \lambda_3, \lambda_4 - \mu), \quad P_\mu = P, \quad \textit{and} \quad Q_\mu = R + Q.$$

As a result, we see that Lemma 45.2 is applicable to each of the exponential dichotomies, for π_λ and for π_μ. Consequently the negative continuations $(\Phi(m,t)u, m_t)$, for $t \leq 0$, which are defined when $u \in \mathcal{R}(P(m)) + \mathcal{R}(R(m))$, are uniquely determined.

The argument leading to inequality (45.8) is applicable, in the setting of an exponential trichotomy, to both the projections $R(m)$ and $P(m)$. In particular, for all $m \in \mathcal{K}$, one has

$$\|\Phi(m,t)R(m)w\| \geq K^{-1}\|R(m)w\|e^{\lambda_2 t}, \qquad \text{for } w \in W \text{ and } t \geq 0, \tag{45.20}$$

and

$$\|\Phi(m,t)P(m)w\| \geq K^{-1}\|P(m)w\|e^{\lambda_4 t}, \qquad \text{for } w \in W \text{ and } t \geq 0. \tag{45.21}$$

Likewise, one has

$$\|\Phi(m,t)R(m)w\| \geq K^{-1}\|R(m)w\|e^{\lambda_3 t}, \qquad \text{for } w \in W \text{ and } t \leq 0. \tag{45.22}$$

The linear space $\mathcal{R}(R(m))$ is referred to as the **neutral space** for the exponential trichotomy.

In the case of a discrete linear skew product semiflow, the definition of an **exponential trichotomy** is identical to that given above with the single exception that the time t is restricted to some discrete subgroup, or discrete semisubgroup, of $\mathbb{R}$.

In the infinite dimensional setting, one is unable to convert any of the inequalities (45.3), (45.6), or (45.13) into a statement about the behavior of solutions in the stable bundle $\mathcal{S}$, or $\mathcal{S}_\lambda$, for time $t \leq 0$. The reason is that, in general, the solution $\Phi(m,t)Q(m)v$ need not exist for $t < 0$. However, a partial extension is possible in the following case: Assume that there is a $w \in W$ and a $\tau > 0$, such that $\Phi(m_{-\tau}, \tau)w = v$. Now equations (43.1) and (45.2) and inequality (45.13) imply that, for $t \geq 0$, one has

$$\|Q(m_t)\Phi(m,t)v\| = \|\Phi(m_{-\tau}, \tau + t)Q(m_{-\tau})w\| \leq K\|w\|e^{\lambda_1(\tau+t)},$$

which goes to 0, as $t \to \infty$, since $\lambda_1 < 0$. Furthermore, if $w \in \mathcal{R}(Q(m_{-\tau}))$, then one has $v \in \mathcal{R}(Q(m))$, by (45.9). We now adopt the convention of defining $\Phi(m,t)v$, for $-\tau \leq t \leq 0$, by the formula

$$\Phi(m,t)v \stackrel{\text{def}}{=} \Phi(m_{-\tau}, \tau + t)w, \qquad \text{for } -\tau \leq t \leq 0. \tag{45.23}$$

In this case, the cocycle identity (43.1) admits the following extension

$$\Phi(m, s+t)v = \Phi(m_t, s)\Phi(m,t)v, \tag{45.24}$$

for $-\tau \leq s,\ t < \infty$ with $-\tau \leq s + t$. Consequently, the argument leading to inequalities (45.20) and (45.21) now extends to give us

$$\|\Phi(m,t)Q(m)v\| \geq K^{-1}\|Q(m)v\|e^{\lambda_1 t}, \qquad \text{for } -\tau \leq t \leq 0. \tag{45.25}$$

It should be emphasized that equations (45.23) and (45.24) depend on the point w. There may be several such points that can be used. However, in every case, inequality (45.25) implies $\|\Phi(m_{-\tau}, \tau)w\| \geq K^{-1}\|Q(m)v\|e^{-\lambda_1 \tau}$.

In the definition of an exponential trichotomy, we allow for the cases where any of the three projectors P, R, or Q is the zero projector. If $R \equiv 0$, then an exponential trichotomy is an exponential dichotomy. The case where $P \equiv 0$ is of special interest. We will say that an exponential trichotomy is **stable** if $P \equiv 0$. Stable exponential trichotomies typically arise in the study of "stable" sets, such as inertial manifolds. More on this later.

Let us return to the linear skew product semiflow $\pi = (\Phi, \sigma)$ on $V^{2\beta} \times \mathcal{M}^\infty$ constructed in Section 4.4. Recall that this linear skew product semiflow consists of all mild solutions, in the space $V^{2\beta}$, generated by equation (44.1), where A satisfies Hypothesis A and $B \in \mathcal{M}^\infty$, see (44.31). By Theorem 44.1, this linear skew product semiflow has a unique extension to a linear skew product semiflow on $W \times \mathcal{M}^\infty$. Let $\mathcal{K}$ be a given compact invariant set in $\mathcal{M}^\infty$. For example, $\mathcal{K}$ might be the hull $H(B)$, where $B \in \mathcal{M}^\infty$ is uniformly continuous on $\mathbb{R}$, see Sell (1967). We then have the following result.

Lemma 45.5. *Let A satisfy the Standing Hypothesis A and let $\mathcal{K}$ be a given compact, invariant set in $\mathcal{M}^\infty$. Let $\pi = (\Phi(B,\tau), B_\tau)$ be the linear skew product semiflow generated by equation (44.1) on $W \times \mathcal{M}^\infty$, and assume that π has an exponential dichotomy on $W \times \mathcal{K}$, with characteristics $K \geq 1$ and $\alpha > 0$. Then for every $r \geq 0$ and every $\rho \geq 0$, there exist constants K_1, K_2, and K_3 such that for $x \in V^{2\rho}$ one has*

$$\|A^r \Phi(B,t)P(B)x\| \leq K_1 e^{\alpha t} \|A^\rho x\|, \qquad \text{for } t \leq 0, \tag{45.26}$$

while

$$\|A^r \Phi(B,t)Q(B)x\| \leq K_2 t^{\rho - r} e^{-\alpha t} \|A^\rho x\|, \qquad \text{for } 0 < t \leq 1, \tag{45.27}$$

whenever $r - \rho \geq 0$, and

$$\|A^r \Phi(B,t)Q(B)x\| \leq K_2 e^{-\alpha t} \|A^\rho x\|, \qquad \text{for } 0 \leq t \leq 1, \tag{45.28}$$

whenever $r - \rho \leq 0$. In addition one has

$$\|A^r \Phi(B,t)Q(B)x\| \leq K_3 e^{-\alpha t} \|A^\rho x\|, \qquad \text{for } 1 \leq t < \infty. \tag{45.29}$$

In particular, for each $r \geq 0$, there is a constant $K(r) \geq 1$, such that π has an exponential dichotomy on $V^{2r} \times \mathcal{K}$, with characteristics $K(r)$ and $\alpha > 0$.

Before doing the proof, it is important to note that the characteristic α, which describes the exponential decay rate for the dichotomy on $V^{2r} \times \mathcal{K}$, does not depend on r, for $r \geq 0$.

Proof. We will let $K_4, K_5, \cdots$ denote constants that depend on r and ρ, but are independent of time t. It follows from Corollary 44.2 that $\Phi(B,t)w \in V^{2r}$, for all $r \geq 0$, all $w \in W$, all $B \in \mathcal{K}$, and all $t > 0$. Therefore, (45.9) implies that

$$\mathcal{U} = \mathcal{U}(\mathcal{K}) \subset V^{2r} \times \mathcal{K}, \qquad \text{for all } r \geq 0.$$

Hence one has $\mathcal{R}(P(B)) \subset V^{2r}$, for all $r \geq 0$ and all $B \in \mathcal{K}$. Consequently, one has $P(B)V^{2r} \subset V^{2r}$, and the restriction $P(B)|_{V^{2r}} : V^{2r} \to V^{2r}$ is a bounded linear operator in the A^r-norm on V^r, i.e., $\|P(B)\|_{\mathcal{L}} < \infty$, for each $B \in \mathcal{K}$, where $\mathcal{L} = \mathcal{L}(V^{2r})$. We claim that

$$\sup_{B \in \mathcal{K}} \|P(B)\|_{\mathcal{L}} < \infty. \tag{45.30}$$

Indeed, if (45.30) were not true, then there is a sequence $\{B^n\}$ in $\mathcal{K}$ such that $\|P(B^n)\|_{\mathcal{L}} \to \infty$, as $n \to \infty$. Since $\mathcal{K}$ is compact, there is no loss in generality in assuming that $\{B^n\}$ is convergent, say $B^n \to B \in \mathcal{K}$. From the Uniform Boundedness Principle, there is an $x \in V^{2r}$ such that $\|A^r P(B^n)x\| \to \infty$, as $n \to \infty$. However, this contradicts the fact that

$\|A^r P(B^n)x\| \to \|A^r P(B)x\|$, as $n \to \infty$, owing to the continuity of the projector $P(B)$ in B. Hence, for each $r \geq 0$, there is a constant $K_4 = K_4(r)$ such that, one has

$$\|A^r P(B)x\| \leq K_4 \|A^r x\| \quad \text{and} \quad \|A^r Q(B)x\| \leq K_4 \|A^r x\|, \tag{45.31}$$

for all $x \in V^{2r}$ and all $B \in \mathcal{K}$. Next let ρ satisfy $\rho \geq 0$. For $t \leq 0$, $r \geq 0$, and $\tau = t - 1$, we use Corollary 44.2, (45.4), (45.10) and the continuous imbedding $V^{2\rho} \mapsto W$ to conclude that

$$\begin{aligned} \|A^r \Phi(B,t)P(B)x\| &= \|A^r \Phi(B_\tau, 1)\Phi(B, t-1)P(B)x\| \\ &\leq K_5 \|\Phi(B, t-1)P(B)x\| \\ &\leq K_6 e^{\alpha(t-1)} \|x\| \leq K_7 e^{\alpha t} \|A^\rho x\|. \end{aligned}$$

For $0 < t \leq 1$ and $0 \leq \rho \leq r$, one uses (44.27) and (45.31) to get

$$\|A^r \Phi(B,t)Q(B)x\| \leq K_8 t^{\rho - r} e^{-\alpha t} \|A^\rho Q(B)x\| \leq K_9 t^{\rho - r} e^{-\alpha t} \|A^\rho x\|,$$

and for $0 \leq r \leq \rho$, one uses (44.28), (45.31), and the continuous imbedding $V^{2\rho} \mapsto V^{2r}$ to find that

$$\begin{aligned} \|A^r \Phi(B,t)Q(B)x\| &\leq K_8 e^{-\alpha t} \|A^r Q(B)x\| \\ &\leq K_9 e^{-\alpha t} \|A^r x\| \leq K_{10} e^{-\alpha t} \|A^\rho x\|. \end{aligned}$$

Similarly for $1 < t$, $\tau = t - 1$, $r \geq 0$, and $\rho \geq 0$, one uses (44.27), (44.28), (45.2), (45.3), and (45.31) to obtain

$$\begin{aligned} \|A^r \Phi(B,t)Q(B)x\| &= \|A^r \Phi(B_\tau, 1)\Phi(B, t-1)Q(B)x\| \\ &\leq K_8 \|\Phi(B, t-1)Q(B)x\| \\ &\leq K_9 e^{-\alpha t} \|x\| \leq K_{10} e^{-\alpha t} \|A^\rho x\|. \end{aligned}$$

Finally be setting $r = \rho$ in the inequalities above, we see that π has an exponential dichotomy on $V^{2r} \times \mathcal{K}$, for each $r \geq 0$. □

4.5.2 Inhomogeneous Equations. We are interested here in the linear inhomogeneous equation

$$\partial_t v + Av = B(t)v + h(t), \tag{45.32}$$

where the Standing Hypothesis A is satisfied, $B = B(t) \in \mathcal{M}^p$, p satsifies inequality (44.32), $h \in L^1_{\text{loc}}[0, T; W)$, and $0 < T \leq \infty$. We will say that $v = v(t)$ is a **mild solution** of equation (45.32) in W on the interval $[0, T)$, if $v(0) = v_0 \in W$, $v(\cdot) \in C[0, T; W)$, and one has

(45.33)

$$v(t) = e^{-At} v_0 + \int_0^t e^{-A(t-s)} [B(s)v(s) + h(s)]\, ds, \qquad \text{for } 0 \leq t < T,$$

where the integral exists in W. By using the argument of Theorem 44.1, one can readily show that, for every $v_0 \in W$ and every $h \in L^1_{\mathrm{loc}}[0,T;W)$, there is a unique mild solution $v = v(t)$ of equation (45.32) in W on the interval $[0,T)$.

If $h \equiv 0$, then the mild solution v is given by $v(t) = \Phi(B,t)v_0$, for all $t \geq 0$. For the general case, we claim that the mild solution v of equation (45.32) satisfies

$$v(t) = \Phi(B,t)v_0 + \int_0^t \Phi(B_s, t-s)\, h(s)\, ds, \qquad \text{for } 0 \leq t < T, \tag{45.34}$$

where $v_0 \in W$, and of course B_s is the translate $B_s(t) = B(s+t)$, for $s,\, t \in \mathbb{R}$. In light of Theorem 44.1, in order to show that v satisfies equation (45.34), it suffices to show that the particular solution of (45.33), where $v_0 = 0$, satisfies

$$v(t) = \int_0^t \Phi(B_s, t-s)\, h(s)\, ds, \qquad \text{for } 0 \leq t < T.$$

For this purpose, we define $w(t) \stackrel{\text{def}}{=} \int_0^t \Phi(B_r, t-r)\, h(r)\, dr$. We will now show that

$$w(t) = \int_0^t e^{-A(t-s)}[B(s)w(s) + h(s)]\, ds, \qquad \text{for } 0 \leq t < T.$$

It follows from the uniqueness of the mild solutions that $w(t) = v(t)$, for $0 \leq t < T$. The proof that $w(t)$ satisfies the equation above involves a straightforward change of variables along with the fact that $\Phi(B_r, t-r)$ is given by

$$\Phi(B_r, t-r)w_0 = e^{-A(t-r)}w_0 + \int_0^{t-r} e^{-A(t-r-\sigma)}B_r(\sigma)\Phi(B_r,\sigma)w_0\, d\sigma,$$

for $w_0 \in W$, see equation (44.13). We will omit these details. One then has the following result.

Theorem 45.6. *Let the Standing Hypothesis A be satisfied. Let $B = B(t) \in \mathcal{M}^p$, where p satisfies inequality (44.32), and let $h \in L^1_{\mathrm{loc}}[0,T;W)$, and $0 < T \leq \infty$. Then for every $v_0 \in W$, there is a unique mild solution $v = v(t)$ of equation (45.32) in W, and v satisfies both equations (45.33) and (45.34). If in addition, one has $B \in \mathcal{M}^\infty$, h satisfies $h \in L^\infty_{\mathrm{loc}}(0,T;W)$, and $v_0 \in V^{2\rho}$, where $0 \leq \rho < 1$, then v satisfies*

$$v(\cdot) \in C[0,T;V^{2\alpha}) \cap C^{0,\theta}_{\mathrm{loc}}(0,T;V^{2(\rho+r)}),$$

for every α and r with $0 \leq \alpha \leq \rho$ and $0 \leq r < 1$, where $0 < \theta < 1 - r$.

Moreover, if B and h satisfy

(45.35)
$$B \in \mathcal{M}^p \cap C^{0,\theta_1}_{\mathrm{loc}}(0,\infty;\mathcal{L}(V^{2\beta},W)) \quad \text{and} \quad h \in L^1_{\mathrm{loc}}[0,T;W) \cap C^{0,\theta_2}_{\mathrm{loc}}(0,T;W),$$

for some θ_i with $0 < \theta_i \leq 1$, for $i = 1,2$, then v is a strong solution of equation (45.32) in W on $0 \leq t < T$, i.e., one has $v \in C[0,T;W)$, and

1. *v is (strongly) differentiable in W almost everywhere in $(0,T)$;*
2. *$\partial_t v \in L^1_{\mathrm{loc}}[0,T;W)$;*
3. *$v(t) \in \mathcal{D}(A)$ a.e. on $(0,T)$; and*
4. *v satisfies the equation*

$$\partial_t v(t) + Av(t) \overset{\text{a.e.}}{=} B(t)v(t) + h(t), \qquad \text{on } (0,T) \text{ in the space } W.$$

Furthermore, if in addition one has $v_0 \in V^{2\rho}$, with $0 < \rho \leq 1$, then

$$v \in C[0,T;V^{2\alpha}) \cap C^{0,1-r}_{\mathrm{loc}}(0,T;V^{2r}) \cap C(0,T;\mathcal{D}(A)),$$

for all α and r with $0 \leq \alpha \leq \rho$ and $0 \leq r < 1$.

Proof. The proof of the existence and uniqueness of $v(t)$ follows the paradigm of Theorem 44.1. The proof of the Hölder continuity of $v(t)$ follows from Lemma 42.7, with $p = \infty$. The fact that $v(t)$ satisfies both equations (45.33) and (45.34) is argued above. In order to show that $v = v(t)$ is a strong solution of equation (45.32) in W on the interval $0 \leq t < T - \tau$ is accomplished by showing that, when (45.35) is satisfied, then the function $f \overset{\text{def}}{=} Bv + h$ satisfies (42.16). One then can apply Theorem 42.9 to get the conclusion. We leave the details as an exercise. □

Before presenting the main theorem concerning the mild solutions of equation (45.32) in the presence of an exponential dichotomy, it is useful to note that $v = v(t)$ is a solution of equation (45.32) in W on an interval I if and only if $w = w(t)$, where $w(t) \overset{\text{def}}{=} e^{-\lambda t}v(t)$, is a solution of the shifted equation

$$\partial_t w + (A + \lambda I)w = B(t)w + g(t) \tag{45.36}$$

on I, where $g(t) = e^{-\lambda t}h(t)$. We use the notation $\Phi_\lambda(B,t) = e^{-\lambda t}\Phi(B,t)$ below.

Theorem 45.7. *Let the Standing Hypothesis A be satisfied and let*

$$B \in \mathcal{M}^p = L^\infty(\mathbb{R},\mathcal{L}(V^{2\beta},W)), \qquad \text{where } 0 \leq \beta < 1 \tag{45.37}$$

and p satisfies inequality (44.32). (See Theorem 44.3.) Assume that, for some $\lambda \in \mathbb{R}$, the shifted linear skew product semiflow $\pi_\lambda(B,x,\tau) = (\Phi_\lambda(B,\tau)x, B_\tau)$ has an exponential dichotomy in $V^{2\beta}$ over an invariant set

$\mathcal{I} \subset \mathcal{M}^p$, and let (K, α) denote the characteristics and (P, Q) the associated projectors of the dichotomy. Then the following statements are valid:

(1) *Assume that $g(t) \in L^\infty(-\infty, 0; W) \cap L^\infty_{\mathrm{loc}}(\mathbb{R}; W)$. Then for every point ξ in $\mathcal{R}(P(B))$, there is a unique mild solution $w = w(t) = w(\xi, t)$ of equation (45.36) in $V^{2\beta}$ that satisfies $P(B)w(0) = \xi$ and $w \in L^\infty(-\infty, 0; V^{2\beta})$. Moreover, for $t \le 0$, w satisfies the equation*

$$w(t) = \Phi_\lambda(B, t)\xi - \int_t^0 \Phi_\lambda(B_s, t-s)P(B_s)g(s)\,ds + \int_{-\infty}^t \Phi_\lambda(B_s, t-s)Q(B_s)g(s)\,ds. \tag{45.38}$$

(2) *Assume that $g(t) \in L^\infty(0, \infty; W) \cap L^\infty_{\mathrm{loc}}(\mathbb{R}; W)$. Then for every point ξ in $\mathcal{R}(Q(B))$, there is a unique mild solution $w = w(t) = w(\xi, t)$ of equation (45.36) in $V^{2\beta}$ that satisfies $Q(B)w(0) = \xi$ and $w \in L^\infty(0, \infty; V^{2\beta})$. Moreover, for $t \ge 0$, w satisfies the equation*

$$w(t) = \Phi_\lambda(B, t)\xi - \int_t^\infty \Phi_\lambda(B_s, t-s)P(B_s)g(s)\,ds + \int_0^t \Phi_\lambda(B_s, t-s)Q(B_s)g(s)\,ds. \tag{45.39}$$

(3) *Assume that $g(t) \in L^\infty(\mathbb{R}; W)$. Then there is a unique mild solution $w = w(t)$ of equation (45.36) in $V^{2\beta}$ with $w \in L^\infty(\mathbb{R}; V^{2\beta})$. Moreover, w satisfies the equation*

$$w(t) = -\int_t^\infty \Phi_\lambda(B_s, t-s)P(B_s)g(s)\,ds + \int_{-\infty}^t \Phi_\lambda(B_s, t-s)Q(B_s)g(s)\,ds, \tag{45.40}$$

for $t \in \mathbb{R}$. Furthermore, the time translate w_τ of equation (45.40) satisfies

$$w_\tau(t) = -\int_t^\infty \Phi_\lambda(B_{\tau+s}, t-s)P(B_{\tau+s})g_\tau(s)\,ds + \int_{-\infty}^t \Phi_\lambda(B_{\tau+s}, t-s)Q(B_{\tau+s})g_\tau(s)\,ds,$$

for all τ, $t \in \mathbb{R}$.

(4) *(Invariance Property 1.) The function $w = w(t) = w(\xi, t)$ given by equation (45.38) has a unique extension to all $t \in \mathbb{R}$, so that*

$$w(\xi(\tau), t) = w(\xi(\tau + t), 0) = w(\xi(0), \tau + t), \qquad \text{for all } \tau,\ t \in \mathbb{R}, \tag{45.41}$$

where $\xi(t)$ *is defined by* $\xi(t) \stackrel{\text{def}}{=} P(B_t)w(\xi,t)$, *for* $t \in \mathbb{R}$. *Furthermore,* $\xi(t)$ *is the unique mild solution in* $V^{2\beta}$ *of the reduced evolutionary equation*

$$\partial_t\xi + A\xi = B(t)\xi + P(B_t)g(t), \qquad \text{for } t \in \mathbb{R},$$

with $\xi(0) = \xi \in \mathcal{R}(P(B))$.

(5) *(Invariance Property 2.) The function* $w = w(t) = w(\xi,t)$ *given by equation (45.39) satisfies*

$$w(\xi(\tau),t) = w(\xi(\tau+t),0) = w(\xi(0),\tau+t), \tag{45.42}$$

for all $\tau \geq 0$ *and* $-\tau \leq t < \infty$, *where* $\xi(t)$ *is defined by* $\xi(t) \stackrel{\text{def}}{=} Q(B_t)w(\xi,t)$, *for* $t \geq 0$. *Furthermore,* $\xi(t)$ *is the unique mild solution in* $V^{2\beta}$ *of the reduced evolutionary equation*

$$\partial_t\xi + A\xi = B(t)\xi + Q(B_t)g(t), \qquad \text{for } t \geq 0,$$

with $\xi(0) = \xi \in \mathcal{R}(Q(B))$.

Proof. We will use the fact that the projectors P and Q are invariant, which implies that

$$Q(B_t)\Phi_\lambda(B_s,t-s) = \Phi_\lambda(B_s,t-s)Q(B_s), \qquad \text{for all } t \geq s, \tag{45.43}$$

and

$$P(B_t)\Phi_\lambda(B_s,t-s) = \Phi_\lambda(B_s,t-s)P(B_s), \qquad \text{for all } s,\ t \in \mathbb{R}, \tag{45.44}$$

see (45.10). We will refer to the four inequalities (45.26), (45.27), (45.28), and (45.29), with $r = \beta$ and $\rho = 0$, as the dichotomic inequalities. Note that these four inequalities hold when B is replaced by B_s, for any $s \in \mathbb{R}$.

(1) If $w = w(t)$ is defined by (45.38), then (45.43) yields $P(B)w(0) = \xi$, and

$$w(0) = \xi + \int_{-\infty}^{0} \Phi_\lambda(B_s,0-s)Q(B_s)g(s)\,ds.$$

A direct calculation with the use of the cocycle identity (44.9) shows that $(w(t),B_t)$ is a negative continuation, for $t \leq 0$, see (45.5). Furthermore, the dichotomic inequalities imply that $\|A^\beta w(t)\|$ is bounded, for $t \leq 0$. Consequently, Lemma 45.2 implies that w is uniquely determined by (45.38). It remains to show that any solution w of (45.36), with a bounded negative continuation in $V^{2\beta}$ and with $P(B)w(0) = \xi$, must satisfy (45.38). From equation (45.34) one finds that this bounded solution satisfies

$$w(t) = \Phi_\lambda(B_\tau,t-\tau)w(\tau) + \int_\tau^t \Phi_\lambda(B_s,t-s)g(s)\,ds, \tag{45.45}$$

for $\tau \le t \le 0$, see (45.5). Now we apply the projection $Q(B_t)$ to equation (45.45) and use (45.43) to find that

$$Q(B_t)w(t) = \Phi_\lambda(B_\tau, t-\tau)Q(B_\tau)w(\tau) + \int_\tau^t \Phi_\lambda(B_s, t-s)Q(B_s)g(s)\,ds,$$

for $\tau \le t \le 0$. Since $\|A^\beta w(\tau)\|$ is bounded for $\tau \le 0$, the exponential dichotomy for π_λ implies that

$$\|A^\beta \Phi_\lambda(B_\tau, t-\tau)Q(B_\tau)w(\tau)\| \le Ke^{-\alpha(t-\tau)}\|A^\beta w(\tau)\| \le K_0 e^{-\alpha(t-\tau)} \to 0,$$

as $\tau \to -\infty$. As a result, we find that

$$Q(B_t)w(t) = \int_{-\infty}^t \Phi_\lambda(B_s, t-s)Q(B_s)g(s)\,ds, \qquad \text{for all } t \le 0.$$

By applying $P(B_\tau)\Phi_\lambda(B_t, \tau - t)$ to equation (45.45) and using (45.44), we obtain

$$P(B_\tau)w(\tau) = \Phi_\lambda(B_t, \tau - t)P(B_t)w(t) - \int_\tau^t \Phi_\lambda(B_s, \tau - s)P(B_s)g(s)\,ds,$$

for $\tau \le t \le 0$. Now first set $t = 0$, and use $P(B)w(0) = \xi$, and then set $\tau = t$, to obtain

$$P(B_t)w(t) = \Phi_\lambda(B, t)\xi - \int_t^0 \Phi_\lambda(B_s, t-s)P(B_s)g(s)\,ds, \qquad \text{for all } t \le 0.$$

Since $w(t) = P(B_t)w(t) + Q(B_t)w(t)$, this implies (45.38).

(2) If $w = w(t)$ is defined by (45.39), then one has $Q(B)w(0) = \xi$,

$$w(0) = \xi - \int_0^\infty \Phi_\lambda(B_s, 0-s)P(B_s)g(s)\,ds,$$

and

$$w(t) = \Phi_\lambda(B, t)w(0) + \int_0^t \Phi_\lambda(B_s, t-s)g(s)\,ds, \qquad \text{for all } t \ge 0. \tag{45.46}$$

Thus w is a solution of (45.34). Furthermore, the dichotomic inequalities imply that $\|A^\beta w(t)\|$ is bounded for $t \ge 0$. Since the right side of equation (45.39) is uniquely determined, it remains to show that any solution w of (45.36) that is bounded in $V^{2\beta}$, for $t \ge 0$, with $Q(B)w(0) = \xi$, must satisfy (45.39). From equation (45.34) one finds that this bounded solution satisfies

$$w(\tau) = \Phi_\lambda(B_t, \tau - t)w(t) + \int_t^\tau \Phi_\lambda(B_s, \tau - s)g(s)\,ds, \tag{45.47}$$

for $0 \leq t \leq \tau$. Now we apply $P(B_t)\Phi_\lambda(B_\tau, t-\tau)$ to equation (45.47) and use (45.44) to find that

$$P(B_t)w(t) = \Phi_\lambda(B_\tau, t-\tau)P(B_\tau)w(\tau) - \int_t^\tau \Phi_\lambda(B_s, t-s)P(B_s)g(s)\,ds,$$

for $0 \leq t \leq \tau$. Since the dichotomic inequality (45.26) implies

$$\|A^\beta \Phi_\lambda(B_\tau, t-\tau)P(B_\tau)w(\tau)\| \leq Ke^{\alpha(t-\tau)}\|A^\beta w(\tau)\| \leq K_0 e^{\alpha(t-\tau)} \to 0,$$

as $\tau \to \infty$, we find that

$$P(B_t)w(t) = -\int_t^\infty P(B_t)\Phi_\lambda(B_s, t-s)g(s)\,ds, \qquad \text{for all } t \geq 0.$$

By using (45.43) and applying $Q(B_\tau)$ to equation (45.47), with $t = 0$, and by using $Q(B)w(0) = \xi$, and replacing τ with t, we obtain

$$Q(B_t)w(t) = \Phi_\lambda(B, t)\xi + \int_0^t \Phi_\lambda(B_s, t-s)Q(B_s)g(s)\,ds, \qquad \text{for all } t \geq 0.$$

Since $w(t) = P(B_t)w(t) + Q(B_t)w(t)$, this implies (45.39).

(3) Next, if $w = w(t)$ is defined by (45.40), then one has

$$w(0) = -\int_0^\infty \Phi_\lambda(B_s, -s)P(B_s)g(s)\,ds + \int_{-\infty}^0 \Phi_\lambda(B_s, -s)Q(B_s)g(s)\,ds.$$

As in Item (1), one uses (45.40) and the cocycle identity (44.9) shows that $(w(t), B_t)$ is a negative continuation, for $t \leq 0$, see (45.5). Furthermore, the dichotomic inequalities imply that $\|A^\beta w(t)\|$ is bounded for $t \leq 0$. Consequently, Lemma 45.2 implies that w is uniquely determined by (45.40), for $t \leq 0$. As in Item (2), one shows that w satisfies equation (45.46), for $t \geq 0$. Also, the dichotomic inequalities imply that $\|A^\beta w(t)\|$ is bounded for $t \geq 0$. Consequently, w is a solution for $t \geq 0$, and it is uniquely determined here as well. It remains to show that any solution w of (45.36) that is bounded for $t \in \mathbb{R}$ must satisfy (45.40). From equation (45.34) one finds that this bounded solution satisfies

$$w(t) = \Phi_\lambda(B_\tau, t-\tau)w(\tau) + \int_\tau^t \Phi_\lambda(B_s, t-s)g(s)\,ds, \qquad \text{for } \tau \leq t. \tag{45.48}$$

Now we apply the projection $Q(B_t)$ to equation (45.48) to find that

$$Q(B_t)w(t) = \Phi_\lambda(B_\tau, t-\tau)Q(B_\tau)w(\tau) + \int_\tau^t \Phi_\lambda(B_s, t-s)Q(B_s)g(s)\,ds,$$

for $\tau \le t$. Let $\tau \to -\infty$. Then inequality (45.29) implies that

$$Q(B_t)w(t) = \int_{-\infty}^{t} Q(B_t)\Phi_\lambda(B_s, t-s)g(s)\,ds.$$

Next we apply $P(B_\tau)\Phi_\lambda(B_t, \tau - t)$ to equation (45.48) and use (45.44) to find that

$$P(B_\tau)w(\tau) = \Phi_\lambda(B_t, \tau - t)P(B_t)w(t) - \int_{\tau}^{t} \Phi_\lambda(B_s, \tau - s)P(B_s)g(s)\,ds,$$

for $\tau \le t$. By first using (45.26), as $t \to \infty$, and next replacing τ with t, we find that

$$P(B_t)w(t) = -\int_{t}^{\infty} P(B_t)\Phi_\lambda(B_s, t-s)g(s)\,ds,$$

which, together with (45.43) and (45.44), imply that (45.40) is valid. The formula for the translate w_τ is proved by using a routine change of the variables of integration.

The proofs of identities (45.41) and (45.42) are based on a straightforward, but somewhat lengthy calculation, using the invariance equalities (45.43) and (45.44) along with the cocycle identities (44.9) and (45.10). The key step, in the case of the Invariance Property 1, is to show that both $w_1(t) = w(\xi(0), \tau + t)$ and $w_2(t) = w(\xi(\tau), t)$ are mild solutions that satisfy

$$w(t) = \Phi_\lambda(B_\tau, t)w(\xi(\tau), 0) + \int_{0}^{t} \Phi_\lambda(B_{\tau+s}, t-s)g_\tau(s)\,ds.$$

The uniqueness of mild solutions then implies that $w_1(t) = w_2(t)$, which in turn implies (45.41). The argument for (45.42) is similar. □

4.5.3 Discrete Inhomogeneous Equations. There is an analogue of the last theorem which applies to the study of solutions of the discrete equation

$$w_{n+1} = T_n w_n + f_n, \qquad n \in Z. \tag{45.49}$$

In this equation, we assume that $T = \{T_n\} = \{T_n : -\infty < n < \infty\}$ is a given sequence with $T_n \in \mathcal{L} = \mathcal{L}(W)$, for all $n \in Z$, and

$$\|T\|_\infty \overset{\text{def}}{=} \sup\{\|T_n\|_{\mathcal{L}} : -\infty < n < \infty\} < \infty.$$

That is, $T = \{T_n\} \in \ell_\infty(Z, \mathcal{L}(W))$. Let $\mathcal{W} \overset{\text{def}}{=} \ell_\infty(Z, W)$. We shall say that the sequence $T = \{T_n\} \in \ell_\infty(Z, \mathcal{L}(W))$ has the **Strong Boundedness Property** provided that, for every sequence $\{f_n\} \in \mathcal{W}$, there is a unique sequence $\{w_n\} \in \mathcal{W}$ that satisfies equation (45.49). The reader should observe that we are using two topologies on the spaces $\ell_\infty(Z, \mathcal{L}(W))$ and $\mathcal{W}$. For many dynamical issues it is best to use the Fréchet metric topology that is equivalent to uniform convergence on finite subsets in Z. However, for the Strong Boundedness Property we use these as Banach spaces with the sup-norm topology, which is equivalent to uniform convergence on Z.

For $T = \{T_n\} \in \ell_\infty(Z, \mathcal{L}(W))$, we define $L = L(T) = \{L_n\}$ by

$$L_n w_n = w_{n+1} - T_n w_n, \qquad \text{for any sequence } \{w_n\} \in \mathcal{W} \text{ and } n \in Z.$$

Note that $L(T) : \mathcal{W} \to \mathcal{W}$ is a bounded linear operator with $\|L(T)\|_{\mathcal{L}(\mathcal{W})} \leq 1 + \|T\|_\infty$, and equation (45.49) becomes $L_n w_n = f_n$, for $n \in Z$. Furthermore, if T satisfies the Strong Boundedness Property, then the range and null space of $L(T)$ satisfy $\mathcal{R}(L(T)) = \mathcal{W}$ and $\mathcal{N}(L(T)) = \{0\}$. Hence, $G(T) \stackrel{\text{def}}{=} L(T)^{-1}$ exists and $G(T) : \mathcal{W} \to \mathcal{W}$. It is easily verified that $G(T)$ is a closed linear operator on the Banach space $\mathcal{W}$, and by the Closed Graph Theorem G is a bounded linear operator on $\mathcal{W}$. Furthermore, one has

$$w_n = (Gf)_n = \sum_{k=-\infty}^{\infty} G_{n,k+1} f_k, \qquad \text{for } n \in Z, \tag{45.50}$$

where $G_{n,m}(T) = G_{n,m} \in \mathcal{L}(W)$, for all $m, n \in Z$, see Henry (1981, pp 229 - 232). Now equations (45.49) and (45.50) imply that

$$f_n = w_{n+1} - T_n w_n = \sum_{k=-\infty}^{\infty} (G_{n+1,k+1} - T_n G_{n,k+1}) f_k,$$

which in turn, implies that

$$G_{n+1,k+1} - T_n G_{n,k+1} = \begin{cases} I, & \text{if } n = k, \\ 0, & \text{if } n \neq k. \end{cases} \tag{45.51}$$

When $T = \{T_n\}$ has the Strong Boundedness Property, the three operators T, $L(T)$, and $G(T)$ are bounded linear operators on $\mathcal{W}$; i.e., they are elements of the Banach space $\mathcal{L}(\mathcal{W})$. We will denote the associated norms by

$$\|G(T)\|_\infty = \|G(T)\|_{\mathcal{L}(\mathcal{W})} \quad \text{and} \quad \|L(T)\|_\infty = \|L(T)\|_{\mathcal{L}(\mathcal{W})}.$$

Note that one has $\|G_{n,m}(T)\|_{\mathcal{L}(W)} \leq \|G(T)\|_\infty$, for all $m, n \in Z$. In the case of T, or more generally for any $T \in \ell_\infty(Z, \mathcal{L}(W))$, we define $T\{w_n\} = \{T_n w_n\}$, where $\{w_n\} \in \mathcal{W}$. Thus one has $T \in \mathcal{L}(\mathcal{W})$ and $\|T\|_{\mathcal{L}(\mathcal{W})} = \|T\|_\infty$.

Next assume that the sequence $T = \{T_n\} \in \ell_\infty(Z, \mathcal{L}(W))$ satisfies the Strong Boundedness Property and let $S = \{S_n\}$ be another sequence in $\ell_\infty(Z, \mathcal{L}(W))$. We will now show that if $\|T - S\|_\infty$ is sufficiently small, then the sequence $S = \{S_n\}$ satisfies the Strong Boundedness Property, as well. Indeed, one has

$$L(S) = L(T) + (T - S) = L(T)[I + G(T)(T - S)].$$

Consequently, if

$$(45.52) \qquad \|G(T)\|_\infty \sup_{n\in Z} \|T_n - S_n\|_{\mathcal{L}} = \|G(T)\|_\infty \|T - S\|_\infty < 1,$$

it follows from the Neumann Property in operator theory that $L(S)$ has a bounded linear inverse $G(S)$, and $G(S) = (I + G(T)(T - S))^{-1}G(T)$.

As we now show, there is a profound connection between the Strong Boundedness Property for $T = \{T_n\}$ and the theory of exponential dichotomies for an associated discrete linear skew product semiflow $\hat{\pi}(w, T, k) = (\Phi(T, k)w, T \cdot k)$ on the space $W \times \mathcal{Z}$, where $\mathcal{Z} = \ell_\infty(Z, \mathcal{L}(W))$, see Sections 4.3 and 4.5.1. Assume that $\hat{\pi}(w, T, k)$ has an exponential dichotomy, with characteristics (K, s), over an invariant set $\mathcal{K}$ in $\mathcal{Z}$. Let $T = \{T_n\} \in \mathcal{K}$, and let $P = \{P_n\}$ denote the associated invariant projector. Set $Q = \{Q_n\} = \{I - P_n\}$. We will now show that this sequence T has the Strong Boundedness Property by using various properties described in Lemma 45.2, as it applies to this discrete semiflow. Let $G_{n,m}$ be defined by $G_{m,m} = Q_m$, for $n = m$, and for $n > m$ set

$$(45.53) \qquad G_{n,m} \stackrel{\text{def}}{=} \Phi(T \cdot m, n - m)Q_m = T_{n-1} \cdots T_{m+1}T_m Q_m.$$

For $n < m$, we let $G_{n,m} = G_{n,m}P_m : \mathcal{R}(P_m) \to \mathcal{R}(P_n)$ denote the unique ismorphism that satisfies $-\Phi(T \cdot n, m - n)G_{n,m} = P_m$, and we write

$$(45.54) \qquad G_{n,m} = -\Phi(T \cdot m, n - m)P_m, \qquad \text{for } n < m,$$

see (45.12). Now inequalities (45.6) and (45.7) imply that $\|G_{n,m}\|_{\mathcal{L}} \leq Ks^{n-m}$, for $n \geq m$, and $\|G_{n,m}\|_{\mathcal{L}} \leq Ks^{m-n}$, for $n < m$. Given $f = \{f_n\} \in \mathcal{W}$ it is easily verified that the sequence $w = \{w_n\}$, where

$$w_n = \sum_{k\in Z} G_{n,k} f_{k-1} = \sum_{k\leq n} \Phi(T \cdot k, n - k)Q_k f_{k-1} - \sum_{n<k} \Phi(T \cdot k, n - k)P_k f_{k-1},$$

for all $n \in Z$, is a solution of equation (45.49), and one has $\|w_n\| \leq K\|f\|_\infty \frac{1+s}{1-s}$, for all $n \in Z$. Hence, the inverse $G = G(T) = L(T)^{-1}$ exists and one has

$$(45.55) \qquad \|G(T)\|_\infty \leq \sup_{n\in Z} \sum_{k\in Z} \|G_{n,k}\|_{\mathcal{L}} \leq K\frac{1+s}{1-s} \leq \frac{2K}{1-s}.$$

This leads us to a very useful and powerful characterization of an exponential dichotomy for $\hat{\pi} = (\Phi, \sigma)$. In particular, we will show that the linear skew product semiflow $\hat{\pi}$ has an exponential dichotomy over the set $\mathcal{K}_0 = \{T\}$, where $T = \{T_n\} \in \mathcal{Z}$, if and only if the given sequence T has the Strong Boundedness Property.

Theorem 45.8. *Let $\hat{\pi} = \hat{\pi}(w, T, k)$ be a discrete linear skew product semiflow on $W \times \mathcal{Z}$. Then the following assertions are valid.*

1. *If $\hat{\pi}$ has an exponential dichotomy, with characteristics (K, s), over a set $\mathcal{K}$ in $\mathcal{Z}$, then each $T \in \mathcal{K}$ has the Strong Boundedness Property, and the associated inverse operator $G(T)$ satisfies inequality (45.55).*
2. *Let $b > 0$ be fixed and let $\mathcal{K}_b$ denote the collection of all $T \in \mathcal{Z}$ that have the Strong Boundedness Property with $||G(T)||_\infty \leq b$. Then $\hat{\pi}$ has an exponential dichotomy over $\mathcal{K}_b$ with characteristics (K, s), where*

$$K \leq 2||G(T)||_\infty \quad \textit{and} \quad \frac{2b||T||_\infty}{2b||T||_\infty + 1} \leq s < 1. \tag{45.56}$$

Proof. The proof of Item (1) is given in the preceeding paragraphs. For the proof of Item (2) we let $T = \{T_n\} \in \mathcal{K}_b$, and we define $P_m \stackrel{\text{def}}{=} I - G_{m,m}$, for $m \in Z$. We will now use the following lemma, which is an easy consequence of equations (45.50) and (45.51), as is noted in the exercises.

Lemma 45.9. *Let $T \in \mathcal{Z}$. Then the following assertions are valid:*

1. *Let w_n be a sequence in W that satisfies $w_{n+1} = T_n w_n$, for $n \geq m$, and set $w_n = 0$, for $n < m$. Then the sequence $w = \{w_n\}$ is in $\mathcal{W}$ if and only if*

$$w_n = G_{n,m} w_m, \qquad \textit{for all } n \in Z, \textit{ where } w_m \in \mathcal{N}(P_m).$$

2. *Let w_n be a sequence in W that satisfies $w_{n+1} = T_n w_n$, for $n < m$, and set $w_n = 0$, for $n > m$. Then the sequence $w = \{w_n\}$ is in $\mathcal{W}$ if and only if*

$$w_n = G_{n,m+1}(-T_m w_m), \quad \textit{for all } n \in Z, \textit{ where } T_m w_m \in \mathcal{N}(G_{m+1,m+1}).$$

Continuing with the proof of Item (2) in Theorem 45.8, we define $\mathcal{N}_m$ to be the collection of all vectors $w_m \in W$ such that there exists a sequence $w = \{w_n\} \in \mathcal{W}$, with $w_{n+1} = T_n w_n$, for $n \geq m$, and $w_n = 0$, for $n < m$. It follows from Lemma 45.9 that

$$\mathcal{N}_m \subset \mathcal{N}(P_m), \qquad \text{for all } m \in Z, \tag{45.57}$$

and from the definition one has

$$T_m \mathcal{N}_m \subset \mathcal{N}_{m+1}, \qquad \text{for all } m \in Z. \tag{45.58}$$

Let $w \in W$ and set $w_n = G_{n,m} w$, for $n \geq m$, and $w_n = 0$, for $n < m$. It is easily seen that $w_{n+1} = T_n w_n$, for $n \geq m$. Hence from Lemma 45.9 and (45.57), one has $w_n = G_{n,m} w_m$, for all $n \in Z$, where $w_m \in \mathcal{N}(P_m)$. Hence

$$0 = P_m w_m = P_m G_{m,m} w = P_m(I - P_m)w, \qquad \text{for all } w \in W,$$

which implies that $P_m^2 = P_m$. Thus P_m and $G_{m,m}$ are complementary bounded linear projections on W, for each $m \in Z$. In addition, if $w \in W$ is chosen so that $w \in \mathcal{N}(P_m)$, one then has $P_m w = 0$, or $w = G_{m,m} w = w_m$, and $w = w_m \in \mathcal{N}_m$. Consequently, (45.57) implies that

$$\mathcal{N}_m = \mathcal{N}(P_m), \qquad \text{for all } m \in Z. \tag{45.59}$$

Also note that (45.58) and (45.59) imply that if $P_m w = 0$, then $P_{m+1} T_m w = 0$, since $T_m w \in \mathcal{N}(P_{m+1})$. One then has

$$T_m P_m w = P_{m+1} T_m w = 0, \qquad \text{for all } w \in \mathcal{N}(P_m). \tag{45.60}$$

Next we define $\mathcal{R}_m$ to be the collection of all vectors $w_m \in W$ such that there exists a sequence $w = \{w_n\} \in \mathcal{W}$, with $w_{n+1} = T_n w_n$, for $n < m$, and $w_n = 0$, for $n > m$. It follows from the definition that $T_m \mathcal{R}_m = \mathcal{R}_{m+1}$, for all $m \in Z$, and Lemma 45.9 implies that the restriction of T_m to $\mathcal{R}_m$ is an isomorphism of $\mathcal{R}_m$ onto $\mathcal{R}_{m+1}$. We claim that

$$\mathcal{R}_m = \mathcal{R}(P(m)), \qquad \text{for all } m \in Z.$$

Indeed, by replacing w with $\hat{w}$, where $\hat{w}_n = w_n$, for $n < m$, and $\hat{w}_n = 0$, for $n \geq m$, it follows from Lemma 45.9 that $w_m = T_{m-1} w_{m-1} = T_{m-1} \hat{w}_{m-1} \in \mathcal{N}(G_{m,m})$, for all $m \in Z$. Since $w_m \in \mathcal{R}_m$, one has $\mathcal{R}_m \subset \mathcal{N}(G_{m,m}) = \mathcal{R}(P_m)$, for all $m \in Z$. Conversely, let $w_m \in \mathcal{N}(G_{m,m})$ and set $w_n = G_{n,m} w_m$, for $n \in Z$. One then has $w_{n+1} = T_n w_n$, for $n < m$, and $w_m = G_{m,m} w_m = 0$. It then follows from equation (45.51) that $w_n = 0$, for all $n \geq m$. Hence $w_m = T_{m-1} w_{m-1} \in \mathcal{R}_m$ by Lemma 45.9. Thus one has $\mathcal{N}(G_{m,m}) \subset \mathcal{R}_m$, which implies that $\mathcal{R}_m = \mathcal{N}(G_{m,m}) = \mathcal{R}(P_m)$, for all $m \in Z$.

Next we will show that

$$T_m P_m w = P_{m+1} T_m w, \qquad \text{for all } w \in \mathcal{R}(P_m). \tag{45.61}$$

Indeed if $w \in \mathcal{R}(P_m)$, then $T_m P_m w = T_m w \in \mathcal{R}_{m+1} = \mathcal{R}(P_{m+1})$ by Lemma 45.9. This in turn implies (45.61). It then follows from (45.60) and (45.61) and the linearity of T_m and P_m that $P_{m+1} T_m = T_m P_m$, for all $m \in Z$. This implies that the projector $P = \{P_n\}$ is invariant.

Notice that equation (45.51) implies that equation (45.53) holds, for all $n > m$, with $Q_m \stackrel{\text{def}}{=} G_{m,m} = I - P_m$. Furthermore, since the restriction $T_m \mid_{\mathcal{R}(P_m)}$ is an isomorphism of $\mathcal{R}(P_m)$ onto $\mathcal{R}(P_{m+1})$, it follows that, for $n < m$, the operator $G_{n,m} = G_{n,m} P_m$ is the unique isomorphism of $\mathcal{R}(P_m)$ onto $\mathcal{R}(P_n)$ such that $-\Phi(T \cdot n, m - n) G_{n,m} = P_m$, i.e., $G_{n,m}$ satisfies (45.54).

It remains to verify inequalities (45.6), (45.7), and (45.56). For this purpose we let $S = \{S_n\}$ be defined by $S_n = \theta T_n$, where $\theta > 1$. Since $\|T - S\|_\infty = |\theta - 1| \, \|T\|_\infty$, it follows from (45.52) that, if

$$\|G(T)\|_\infty \|T - S\|_\infty \leq b|\theta - 1| \, \|T\|_\infty < 1,$$

then the sequence S has the Strong Boundedness Property. For the sequel we assume that θ satisfies

$$\text{(45.62)} \qquad \|G(t)\|_\infty \|T - S\|_\infty \le b|\theta - 1|\,\|T\|_\infty \le \frac{1}{2} \quad \text{and} \quad \theta > 1.$$

Set $s = \theta^{-1}$. We then consider the equation

$$\text{(45.63)} \qquad v_{n+1} = S_n v_n + g_n, \qquad n \in Z.$$

Let $G_{n,m}(T)$ and $G_{n,m}(S)$ denote the respective inverses for $L(T)$ and $L(S)$. Let $f = \{f_n\}$ satisfy $f_{m-1} = w_m$, where $w_m \in \mathcal{N}(P_m)$, and $f_n = 0$, for $n \neq m-1$, and let $g = f$. Now assume that $v = \{v_n\}$ is the bounded solution of equation (45.63) with this g. From Lemma 45.9, one has $v_n = G_{n,m}(S)v_m$, where $v_m \in \mathcal{N}(P_m)$. Next set $w_n = \theta^{m-n}v_n$, for $n \in Z$. Since $v_n = 0$, for $n < m$, and $\theta > 1$, it follows that $w = \{w_n\}$ is a bounded solution of equation (45.49) and $v_m = w_m$. Because of the uniqueness of the bounded solutions v and w, one has

$$\theta^{n-m} w_n = v_n = G_{n,m}(S)v_m, \qquad \text{for } n \in Z.$$

Since $v_m = w_m$, one also obtains

$$w_n = \theta^{m-n} G_{n,m}(S) w_m = G_{n,m}(T) w_m, \qquad \text{for all } m, n \in Z.$$

Since $G_{n,m}Q_m = G_{n,m}$, for all $n \ge m$, one has $\theta^{m-n}G_{n,m}(S) = G_{n,m}(T)$, for all $n \ge m$. As a result, for $n \ge m$, one obtains

$$\|\Phi(T\cdot m, n-m)Q_m\|_{\mathcal{L}} = \|G_{n,m}(T)\|_{\mathcal{L}} \le \|G(S)\|_\infty \theta^{m-n} = \|G(S)\|_\infty s^{n-m}.$$

Next one sets $f_m = -T_m w_m$, where $T_m w_m \in \mathcal{N}(G_{m+1,m+1})$, and $f_n = 0$, for $n \neq m$, and let $g = f$. Now assume that $w = \{w_n\}$ is the bounded solution of equation (45.49) with this f. From Lemma 45.9, one has $w_n = G_{n,m}(T)(-T_m w_m)$, for all $n \in Z$. Next set $v_n = \theta^{n-m}w_n$, for $n \in Z$. Since $w_n = 0$, for $n > m$, and since $\theta > 1$, it follows that $v = \{v_n\}$ is a bounded solution of equation (45.49) and $v_m = w_m$. As argued above, one then has $G_{n,m}(S) = \theta^{n-m}G_{n,m}(T)$, for all $n \le m$. Again one obtains

$$\|\Phi(T\cdot m, n-m)P_m\|_{\mathcal{L}} = \|G_{n,m}(T)\|_{\mathcal{L}} \le \|G(S)\|_\infty s^{-(n-m)}, \qquad \text{for } n < m.$$

We see then that $\hat{\pi}$ has an exponential dichotomy over $\mathcal{K}_0 = \{T\}$, with characteristics (K, s), where $K = \|G(S)\|_\infty$. Since $T \in \mathcal{K}_b$ and $s^{-1} = \theta > 1$, the second inequality in (45.56) follows from (45.62). Furthermore, since $\|(I + G(T)(T-S))w\| \ge \frac{1}{2}\|w\|$, for all $w \in W$, one has

$$\|G(S)\|_\infty = \|(I + G(T)(T-S))^{-1}G(T)\|_\infty \le 2\|G(T)\|_\infty,$$

which completes the proof of (45.56). ☐

There are numerous connections between the theory of exponential dichotomies for discrete and continuous-time linear skew product semiflows. Some of these features are developed in the Exercises and in the Commentary section. However, there is a specific connection which we require. Let $\pi = \pi(w, B, t)$ denote the linear skew product semiflow on $W \times \mathcal{M}^p$ (or on $W \times \mathcal{M}^\infty$), where inequality (44.32) holds, and $t \in \mathbb{R}^+$. Let $\hat{\pi} = \hat{\pi}(w, T, k)$ denote the discrete linear skew product semiflow on $W \times \mathcal{W}$, where $k \in Z^+$ and $\mathcal{W} = \ell_\infty(Z, W)$. The proof of the following result is left as an exercise.

Lemma 45.10. *Let π and $\hat{\pi}$ be given as above. Let $B \in \mathcal{M}^p$ (or $B \in \mathcal{M}^\infty$), where inequality (44.32) holds, and define $T = \{T_n\}$ by $T_n = \Phi(B_{n\tau}, \tau)$, for $n \in Z$, where $\tau > 0$ is fixed. Then the following statements are valid:*

1. *If π has an exponential dichotomy over the hull $H(B)$ with characteristics (K, α), then $\hat{\pi}$ has an exponential dichotomy over $\mathcal{K}_0 = \{T\}$ with characteristics (K, s), where $s = e^{-\alpha\tau}$.*
2. *If $\hat{\pi}$ has an exponential dichotomy over $\mathcal{K}_0 = \{T\}$ with characteristics (K, s), then π has an exponential dichotomy over the hull $H(B)$ with characteristics (MK, α), where $\alpha = -\frac{1}{\tau}\log(s)$ and (by Theorem 44.1, Item (3))*

$$M \stackrel{\text{def}}{=} \sup\{e^{\alpha\tau}\|\Phi(B_{n\tau}, t)\|_{\mathcal{L}} : 0 \le t \le \tau \text{ and } n \in Z\} < \infty.$$

4.5.4 Robustness Theorems. We now return to the linear skew product flow π on $W \times \mathcal{M}^p$ generated by equation (44.1), where the Standing Hypothesis A is satisfied, B satisfies (45.37), and p satisfies (44.32). Since the robustness theorems we describe here have wide applicability, it is convenient to formulate the theory in a somewhat abstract version.

In particular, let V and W be two Banach spaces where one has the continuous imbedding $V \mapsto W$. Let $\mathcal{L} = \mathcal{L}(V, W)$ denote the space of bounded linear operators from V into W. Let $\mathcal{W} = \mathcal{W}(\mathbb{R}, \mathcal{L})$ be a Fréchet space, where each $B \in \mathcal{W}$ is a function $B = B(t)$ defined on $\mathbb{R}$ and assuming values in $\mathcal{L}$. The topology on $\mathcal{W}$ is not completely arbitrary. It must satisfy two essential properties:

Hypothesis T. *The topology on $\mathcal{W}$ satisfies the following:*

1. *the mapping $(B, \tau) \to B_\tau$ is a flow on $\mathcal{W}$, where $B_\tau(t) = B(\tau + t)$, and*
2. *for each $B \in \mathcal{W}$ and $t \ge 0$, there is an operator $\Phi(B, t) \in \mathcal{L}(W)$ with the property that*

$$\pi(w, B, t) = (\Phi(B, t)w, B_t)$$

is a linear skew product semiflow on $\mathcal{E} = W \times \mathcal{W}$.

For example, it is shown in Theorem 44.3, that if the Standing Hypothesis A is satisfied and B satisfies (45.37), then the topology on $\mathcal{M}^p$ satisfies the Hypothesis T precisely when p satisfies inequality (44.32). In the case of $\mathcal{M}^\infty$, one can use either of the topologies, $\mathcal{T}_{\text{bo}}$ or $\mathcal{T}_A$.

We want to show that the concept of an exponential dichotomy over a compact invariant set $\mathcal{K} \subset \mathcal{W}$ is an open condition in the sense that if a perturbed motion B_σ remains in a prescribed neighborhood $N_\epsilon(\mathcal{K})$, for all $t \in \mathbb{R}$, then the linear skew product semiflow π has an exponential dichotomy over the hull $H(B)$, with good characteristics, and the associated projectors (P, Q) vary continuously over the neighborhood $N_\epsilon(\mathcal{K})$. The neighborhood $N_\epsilon(\mathcal{K})$ is described in terms of the metric $d(B, D) \stackrel{\text{def}}{=} d_0(B - D, 0) = d_0(B, D)$, where d_0 is an invariant metric on $\mathcal{W}$. We also present several applications of this openness condition to illustrate its importance in the study of the longtime dynamics.

Theorem 45.11 (Robustness of Dichotomies). *Assume that the topology on $\mathcal{W}$ satisfies Hypothesis T and let π be a linear skew product semiflow given by Hypothesis T on $\mathcal{E} = W \times \mathcal{W}$. Assume the π has an exponential dichotomy over a compact invariant set $\mathcal{K}$ in $\mathcal{W}$, with characteristics (K, α). Then there is an $\epsilon_0 > 0$, which depends on the characteristics, such that if $B \in \mathcal{W}$ satisfies $B_\sigma \in N_\epsilon(\mathcal{K})$, for all $\sigma \in \mathbb{R}$ and some ϵ with $0 < \epsilon \leq \epsilon_0$, then the equation (44.1), with this B, has an exponential dichotomy over the hull $H(B)$, with characteristics $(\hat{K}, \hat{\alpha})$, and the projectors (P, Q) vary continuously over $\mathcal{K} \cup H(B)$, and $H(B) \subset N_{\epsilon_0}(\mathcal{K})$.*

For the applications of this Robustness Theorem, where $\mathcal{W} = \mathcal{M}^p$ and $p < \infty$ for example, it is convenient to rewrite the basic hypothesis:

$$B_\sigma \in N_\epsilon(\mathcal{K}), \qquad \text{for all } \sigma \in \mathbb{R}, \tag{45.64}$$

in terms of the pseudo-norms and invariant metric that generates the L^p_{loc}-topology on L^∞, see Sections 4.4.1 and 4.4.2. In particular, with ϵ given with $0 < \epsilon < 8$, we fix $K = K(\epsilon) \geq 1$ so that $\sum_{|n|>K} 2^{-|n|} \leq \frac{\epsilon}{2}$. Then (45.64) is satisfied if for every $\sigma \in \mathbb{R}$, there is a $D \in \mathcal{K}$ (D will depend on σ), such that for every integer n, with $|n| \leq K$, one has

$$\int_n^{n+1} \|(B(\sigma + s) - D(s))v\|^p \, ds \leq \left(\frac{\epsilon}{8}\right)^p \|A^\beta v\|^p, \qquad \text{for all } v \in V^{2\beta}.$$

Proof of Theorem 45.11. We will prove the existence of an exponential dichotomy over the hull $H(B)$. The verification of the continuous dependence of the associated projectors (P, Q) is left as an exercise. Let $\mathcal{K}$ and B be given as in the hypotheses. As above we let $\hat{\pi} = \hat{\pi}(w, T, k)$ denote the discrete linear skew product semiflow on $W \times \mathcal{W}$, where $k \in Z^+$.

The strategy of the proof is rather simple. For a $\tau > 0$, which will be fixed later, we define the sequence of operators $S = \{S_n\}$ by $S_n \stackrel{\text{def}}{=} \Phi(B_{n\tau}, \tau)$, for $n \in Z$. Our objective is to show that S has the Strong Boundedness Property and then to use Theorem 45.8 and Lemma 45.10 to conclude that π has an exponential dichotomy over the hull $H(B)$. In order to show that S has the Strong Boundedness Property, we will use the exponential dichotomy of π over $\mathcal{K}$ to construct an auxiliary sequence $T = \{T_n\}$ with the following two properties: (1) the norm $\|T - S\|_\infty$ is sufficiently small, and (2) the discrete flow $\hat{\pi}$ has an exponential dichotomy over $\mathcal{K}_0 = \{T\}$, with good characteristics. (The linear transformation $T_n = M_n \Phi_n$ is constructed by setting $\Phi_n = \Phi(B^{(n-1)}, \tau)$ for a specific $B^{(n-1)} \in \mathcal{K}$ and then jumping to a suitable nearby point $B^{(n)} \in \mathcal{K}$, see Figure 4.3. The mapping M_n is a

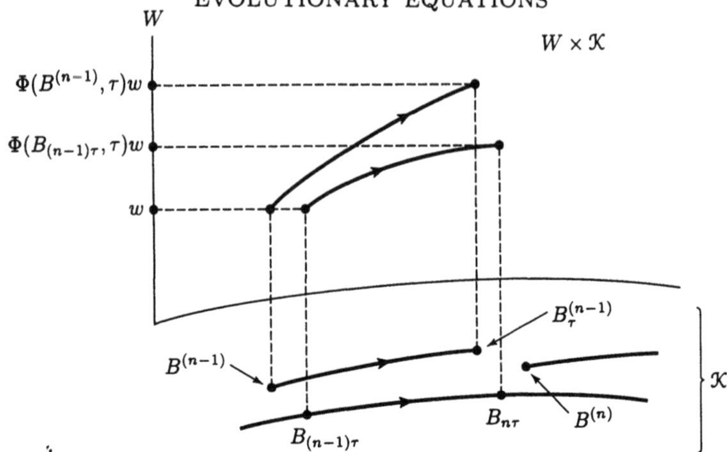

Figure 4.3. Detail for the Proof of Theorem 45.11

small impulse which moves from $B_\tau^{(n-1)}$ to $B^{(n)}$ in $\mathcal{K}$, while realigning the stable and unstable manifolds over these points.) Having done this, we then use Theorem 45.8 to conclude that T has the Strong Boundedness Property, and then we use (45.52) to conclude that S has the Strong Boundedness Property.

Now for the details. Let (K, α) be the characteristics of the exponential dichotomy, over the compact invariant set $\mathcal{K} \subset \mathcal{W}$, for the linear skew product semiflow $\pi(w, B, t) = (\Phi(B, t)w, B_t)$. Recall that the associated projectors $(P(B), Q(B))$ vary continuously for $B \in \mathcal{K}$. For B_1, $B_2 \in \mathcal{K}$, we define

$$M = M(B_1, B_2) \stackrel{\text{def}}{=} P(B_2)P(B_1) + Q(B_2)Q(B_1).$$

One then has $M\mathcal{R}(P(B_1)) \subset \mathcal{R}(P(B_2))$, and $M\mathcal{R}(Q(B_1)) \subset \mathcal{R}(Q(B_2))$. Furthermore, if

$$\|P(B_1) - P(B_2)\|_{\mathcal{L}(W)} < 1,$$

then $\|P(B_1) - P(B_2)\|_{\mathcal{L}(W)} = \|Q(B_1) - Q(B_2)\|_{\mathcal{L}(W)} < 1$, and M maps $\mathcal{R}(P(B_1))$ and $\mathcal{R}(Q(B_1))$ onto $\mathcal{R}(P(B_2))$ and $\mathcal{R}(Q(B_2))$, respectively. That is to say, one has $M\mathcal{R}(P(B_1)) = \mathcal{R}(P(B_2))$ and $M\mathcal{R}(Q(B_1)) = \mathcal{R}(Q(B_2))$. Since $M(B_1, B_1) = I$, for all $B_1 \in \mathcal{K}$, and since $P(B)$ and $Q(B)$ are unifomly continuous on the compact set $\mathcal{K}$, it follows that $M = M(B_1, B_2)$ is invertible if $d(B_1, B_2)$ is small. More precisely, for every $\delta > 0$, there is an $\eta > 0$, such that, for B_1, $B_2 \in \mathcal{K}$, one has

$$d(B_1, B_2) \le 2\eta \Longrightarrow \|I - M\|_{\mathcal{L}(W)} \le \delta \quad \text{and} \quad \|I - M^{-1}\|_{\mathcal{L}(W)} \le \delta.$$

Next we fix ρ so that $0 < \rho < \alpha$, and we fix $\tau > 0$ so that

$$K(1 + \delta)e^{-(\alpha-\rho)\tau} \le 1.$$

Define (K_1, s_1) by $K_1 = 2K$ and $s_1 = e^{-\rho\tau}$.

From the continuity property of the linear skew product semiflow, we see that for every $\mu > 0$, there is a ν, with $0 < \nu \le \eta$, such that if $D \in \mathcal{W}$

and $\hat{B} \in \mathcal{K}$ with $d(D, \hat{B}) \le \nu$, then one has $\|\Phi(D,t) - \Phi(\hat{B},t)\|_{\mathcal{L}} \le \mu$, for $0 \le t \le \tau$, and $d(D_\tau, \hat{B}_\tau) \le \eta$. In this setting, one then has

$$\begin{aligned} \|\Phi(D,\tau) - M\Phi(\hat{B},\tau)\|_{\mathcal{L}} &\le \|\Phi(D,\tau) - \Phi(\hat{B},\tau)\|_{\mathcal{L}} \\ &\quad + \|\Phi(\hat{B},\tau) - M\Phi(\hat{B},\tau)\|_{\mathcal{L}} \\ &\le \mu + \|I - M\| K_0, \end{aligned} \tag{45.65}$$

where $K_0 = \sup\{\|\Phi(\hat{B},\tau)\|_{\mathcal{L}} : \hat{B} \in \mathcal{K}\}$ is finite since $\mathcal{K}$ is compact. Now choose δ and μ so that

$$\mu \le \frac{1 - s_1}{8K_1} \quad \text{and} \quad \|I - M\|_{\mathcal{L}} \le \delta \le \min\left(\frac{1 - s_1}{8K_0K_1}, 1\right). \tag{45.66}$$

Next we let η and ν be fixed so that the corresponding implications given above are satisfied, and we set $\epsilon = \nu$. Now assume that $B_\sigma \in N_\epsilon(\mathcal{K})$, for all $\sigma \in \mathbb{R}$. Let $B^{(n)} \in \mathcal{K}$ be fixed so that $d(B_{n\tau}, B^{(n)}) \le \epsilon = \nu$, for all $n \in Z$. Let $(P, Q) = (P(B), Q(B))$ denote the invariant projectors associated with the exponential dichotomy over $\mathcal{K}$. Define $\{P_n\}$ and $\{Q_n\}$ by $P_n = P(B^{(n)})$ and $Q_n = Q(B^{(n)})$, for $n \in Z$. It then follows that

$$\|\Phi(B_{(n-1)\tau}, t) - \Phi(B^{(n-1)}, t)\|_{\mathcal{L}} \le \mu, \qquad \text{for } 0 \le t \le \tau,$$

and $d(B_{n\tau}, B_\tau^{(n-1)}) \le \eta$. Let $B_1 = B_\tau^{(n-1)}$ and $B_2 = B^{(n)}$. One then has $B_1, B_2 \in \mathcal{K}$ and

$$d(B_1, B_2) \le d(B_1, B_{n\tau}) + d(B_{n\tau}, B^{(n)}) \le \eta + \nu \le 2\eta.$$

Now define $M_n = M(B_1, B_2)$, for this choice of B_1 and B_2, and set

$$T_n = M_n \Phi(B^{(n-1)}, \tau), \qquad \text{for } n \in Z. \tag{45.67}$$

It then follows from inequalities (45.65) and (45.66) that, with $D = B_{(n-1)\tau}$ and $\hat{B} = B^{(n-1)}$, $S_n = \Phi(B_{(n-1)\tau}, \tau)$ satisfies

$$\|S_n - T_n\|_{\mathcal{L}} \le \mu + \delta K_0 \le \frac{1 - s_1}{4K_1}, \qquad \text{for all } n \in Z. \tag{45.68}$$

Furthermore, if $v_0 \in \mathcal{R}(Q_{n-1}) = \mathcal{R}(Q(B^{(n-1)}))$, then one has

$$\Phi(B^{(n-1)}, \tau)v_0 \in \mathcal{R}(Q(B_\tau^{(n-1)})) = \mathcal{R}(Q(B_1)).$$

Because of the choice of M_n, one then has

$$T_n v_0 = M_n \Phi(B^{(n-1)}, \tau) v_0 \in \mathcal{R}(Q_n) = \mathcal{R}(Q(B_2)).$$

Consequently one has $||\Phi(B^{(n-1)},t)v_0|| \le Ke^{-\alpha t}||v_0||$, for $0 \le t \le \tau$, and

$$||T_n v_0|| \le ||M_n||_{\mathcal{L}}||\Phi(B^{(n-1)},\tau)v_0|| \le (1+\delta)Ke^{-\alpha\tau}||v_0|| \le e^{-\rho\tau}||v_0||,$$

for $n \in Z$, and for some $0 < \rho < \alpha$. By using induction, one then finds that for $m > n$ one has

$$||\Phi(T\cdot n, m-n)Q_n v_0|| = ||T_{m-1}\cdots T_{n+1}T_n v_0|| \le K(1+\delta)e^{-\rho\tau(m-n)}.$$

Similarly, for $v_0 \in \mathcal{R}(P_n)$, one obtains $||\Phi(B_\tau^{(n-1)},t)M_n^{-1}P_n v_0|| \le K(1+\delta)e^{\alpha t}||v_0||$, for $-\tau \le t \le 0$, and

$$||\Phi(T\cdot n, m-n)P_n v_0|| \le K(1+\delta)e^{\rho\tau(m-n)}||v_0||, \qquad \text{for } m < n.$$

Since $\delta \le 1$, we conclude that $\hat{\pi}$ has an exponential dichotomy over $\mathcal{K}_0 = \{T\}$, with characteristics (K_1, s_1).

We are now prepared to reap the harvest! Since $\hat{\pi}$ has an exponential dichotomy over $\{T\}$ with characteristics (K_1, s_1), it follows from Theorem 45.8, Item (1) that T has the Strong Boundedness Property. Also inequality (45.55) implies that

$$||G(T)||_\infty \le \frac{2K_1}{1-s_1}.$$

By using this with inequality (45.68) we then obtain

$$||G(T)||_\infty||T-S||_\infty \le \frac{2K_1}{1-s_1}\frac{1-s_1}{4K_1} = \frac{1}{2}.$$

Consequently inequality (45.52) implies that S has the Strong Boundedness Property, and

$$||G(S)||_\infty \le ||(I+G(T)(T-S))^{-1}||_\infty||G(T)||_\infty \le 2||G(T)||_\infty \le \frac{4K_1}{1-s_1}.$$

It then follows from Theorem 45.8, Item (2) that $\hat{\pi}$ has an exponential dichotomy over $\mathcal{K}_1 = \{S\}$ with characteristics (K_2, s_2), where

$$K_2 \le 2||G(S)||_\infty \le \frac{8K_1}{1-s_1} \quad \text{and} \quad s_2 = \frac{2||G(S)||_\infty}{2||G(S)||_\infty + 1}.$$

Finally, Lemma 45.10 implies that the continuous-time semiflow π has an exponential dichotomy over the hull $H(B)$. □

Theorem 45.12 (Robustness of Trichotomies). *Assume that the linear skew product semiflow π on $W \times \mathcal{W}$ has an exponential trichotomy over a compact invariant set $\mathcal{K} \subset \mathcal{W}$. Then there is an $\epsilon_0 > 0$, which depends on the characteristics, such that if $B \in \mathcal{W}$ satisfies $B_\sigma \in N_\epsilon(\mathcal{K})$, for all $\sigma \in \mathbb{R}$*

and some ϵ with $0 < \epsilon \leq \epsilon_0$, equation (44.1), with this B, has an exponential trichotomy over $H(B)$ and the projectors (P, Q, R) vary continuously over $\mathcal{K} \cup H(B)$, where $H(B)$ is the hull of B and $H(B) \subset N_{\epsilon_0}(\mathcal{K})$.

Proof. Again, we prove only the existence of the exponential trichotomy here. The verification of the continuous dependence of the associated projectors (P, Q, R) is left as an exercise. Since an exponential trichotomy is equivalent to a pair of exponential dichotomies for the shifted flows, this result is an immediate corollary of the last theorem. However, a more instructive argument arises when one uses the proof of Theorem 45.11 with a small change in the definition of the mapping $M = M(B_1, B_2)$. In place of the two equalities

$$M\mathcal{R}(P(B_1)) = \mathcal{R}(P(B_2)), \qquad M\mathcal{R}(Q(B_1)) = \mathcal{R}(Q(B_2)),$$

one uses the trio

$$M\mathcal{R}(P(B_1)) = \mathcal{R}(P(B_2)), \quad M\mathcal{R}(Q(B_1)) = \mathcal{R}(Q(B_2)),$$
$$M\mathcal{R}(R(B_1)) = \mathcal{R}(R(B_2)),$$

where (P, Q, R) are the three projectors arising in the definition of an exponential trichotomy. □

Part II: Nonlinear Theory

We begin the study of the nonlinear theory with a brief discussion of the "nonlinear" terms which arise in the study of evolutionary equations. In particular, we describe here the spaces of the nonlinearities and the topologies on these spaces. Let V and W denote two Banach spaces, and let $C_{\mathrm{Lip}} = C_{\mathrm{Lip}}(V \times \mathbb{R}^+; W)$ denote the collection of all continuous functions $G : V \times \mathbb{R}^+ \to W$ with the property that for any bounded set B in V there exist constants $K_0 = K_0(B)$ and $K_1 = K_1(B)$, such that

$$\|G(v, t)\|_W < K_0, \qquad \text{for all } v \in B \text{ and } t \in \mathbb{R}^+, \tag{46.1}$$

and for all v_1, $v_2 \in B$ and $t \in \mathbb{R}^+$,

$$\|G(v_1, t) - G(v_2, t)\|_W \leq K_1 \|v_1 - v_2\|_V. \tag{46.2}$$

We recall here the topology of uniform convergence on bounded sets, which is described in Appendices A and B. In particular, for every bounded set B in V and every τ with $0 < \tau < \infty$, we define the pseudonorm

$$\|G\|_{\{C^0(B \times [0,\tau])\}} \stackrel{\text{def}}{=} \sup \{\|G(u, t)\|_W : u \in B \text{ and } 0 \leq t \leq \tau\}. \tag{46.3}$$

The Fréchet space topology $\mathcal{T}_{bo} = \mathcal{T}_{bo}^0$ on $C_{\rm Lip}$ is generated by a countable family of these pseudonorms $\|\cdot\|_{\{C^0(B_n\times[n-1,n])\}}$, where $n = 1, 2, \cdots$, where B_n is a sequence of bounded sets with $V = \cup_{n=1}^{\infty} B_n$, and $B_n \subset B_{n+1}$.

Of particular interest is the case where $V = W$, with $\|\cdot\| = \|\cdot\|_W$, in which case (46.1) and (46.2) are written in the form

$$\|G(w,t)\| \leq K_0, \qquad \text{for all } w \in B \text{ and } t \in \mathbb{R}^+, \tag{46.4}$$

and for all w_1, $w_2 \in B$ and $t \in \mathbb{R}^+$, one has

$$\|G(w_1,t) - G(w_2,t)\| \leq K_1\|w_1 - w_2\|, \tag{46.5}$$

Next for θ satisfying $0 < \theta \leq 1$, we define the space

$$C_{\mathrm{Lip};\theta}(V \times \mathbb{R}^+, W)$$

to be those functions G in $C_{\rm Lip}(V \times \mathbb{R}^+, W)$ with the property that for every bounded set B in V and every compact set J in $\mathbb{R}^+$, there exist a constant $K_2 = K_2(B, J) \geq 0$ such that

$$\|G(w_1,t_1) - G(w_2,t_2)\| \leq K_2(\|w_1 - w_2\| + |t_1 - t_2|^\theta), \tag{46.6}$$

for all t_1, $t_2 \in J$ and all w_1, $w_2 \in B$.

Another version of the space $C_{\rm Lip}$ arises when the Standing Hypothesis A is satisfied and one has $V = V^{2\beta}$, for some β with $0 \leq \beta < 1$. In this case, the inequalities (46.1) and (46.2) assume the form

$$\|G(v,t)\| \leq K_0, \qquad \text{for all } v \in B,\ t \in \mathbb{R}^+, \tag{46.7}$$

and for all $v_1, v_2 \in B$ and $t \in \mathbb{R}^+$, one has

$$\|G(v_1,t) - G(v_2,t)\| \leq K_1\|A^\beta(v_1 - v_2)\|, \tag{46.8}$$

where $\|\cdot\| = \|\cdot\|_W$. In this setting, there is a second Fréchet space topology $\mathcal{T}_A = \mathcal{T}_A^0$, which will sometimes be used on the space $C_{\rm Lip}$. In particular, for each bounded set B in $V^{2\beta}$ and each $\tau > 0$, we define a new pseudonorm by

$$\|G\|_{\{A;C^0(B\times[0,\tau])\}} \stackrel{\rm def}{=} \sup_{t\in[0,\tau]} \sup_{u\in B} \int_0^t \|A^\beta e^{-A(t-s)} G(u,s)\|\, ds. \tag{46.9}$$

Notice that inequality (37.11), together with the definition of Gamma function, imply that

(46.10)

$$\begin{aligned} \|G\|_{\{A;C^0(B\times[0,\tau])\}} &\leq M_\beta \sup_{t\geq 0} \int_0^t (t-s)^{-\beta} e^{-a(t-s)}\, ds\, \|G\|_{\{C^0(B\times[0,\tau])\}} \\ &\leq M_\beta a^{\beta-1}\Gamma(1-\beta)\|G\|_{\{C^0(B\times[0,\tau])\}}. \end{aligned}$$

We see then that, if G_m is a sequence in C_{Lip} with the property that $G_m \to 0$, uniformly on each bounded set $B \times [0,\tau]$ in $V^{2\beta} \times \mathbb{R}^+$, then one has $\|G_m\|_{\{C^0(B\times[0,\tau])\}} \to 0$, which in turn implies that $\|G_m\|_{\{A;C^0(B\times[0,\tau])\}} \to 0$, as $m \to \infty$.

As noted above, we use the notation $\mathcal{T}_{bo}^0$ and $\mathcal{T}_A^0$ for these two topologies. The subscript in $\mathcal{T}_{bo}^0$, refers to the topology of uniform convergence on bounded sets, or the bounded-open (bo)-topology, and the subscript in $\mathcal{T}_A^0$ is used to refer to the special role played by the linear operator A. It follows from inequality (46.10) that the topologies satisfy $\mathcal{T}_A^0 \subset \mathcal{T}_{\mathrm{bo}}^0$. (The role of the superscript ^{0} is explained shortly.)

In most applications of dynamical systems theory to nonlinear evolutionary equations, we will be interested in nonlinear terms which are Fréchet differentiable. For each integer $k \geq 1$, we define $C_F^k = C_F^k(V \times \mathbb{R}^+; W)$ to be the collection of all continuous functions $G = G(v,t)$, where $G : V \times \mathbb{R}^+ \to W$ is k-times continuously Fréchet differentiable in v. We will write the kth derivative as D^kG, for $k \geq 1$. If $G \in C_F^1(V \times \mathbb{R}^+; W)$, then for each $(v,t) \in V \times \mathbb{R}^+$, $DG(v,t)$ is a bounded linear operator, i.e., one has $DG(v,t) \in \mathcal{L} = \mathcal{L}(V,W)$. We also define

$$C_{\mathrm{Lip}}^k \stackrel{\mathrm{def}}{=} C_{\mathrm{Lip}} \cap C_F^k \stackrel{\mathrm{def}}{=} C_{\mathrm{Lip}}(V \times \mathbb{R}^+; W) \cap C_F^k(V \times \mathbb{R}^+; W).$$

Like the space $C_{\mathrm{Lip}}(V^{2\beta} \times \mathbb{R}^+; W)$, the space $C_{\mathrm{Lip}}^1(V^{2\beta} \times \mathbb{R}^+; W)$ has two topologies, $\mathcal{T}_{bo}^1$ and $\mathcal{T}_A^1$, which play an important role in the study of nonlinear dynamics. For $G \in C_{\mathrm{Lip}}^1$, we use the following pseudonorms:

(46.11)
$$\begin{aligned} \|G\|_{\{C^1(B\times[0,\tau])\}} &\stackrel{\mathrm{def}}{=} \|G\|_{\{C^0(B\times[0,\tau])\}} \\ &\qquad + \sup\{\|DG(u,t)\|_{\mathcal{L}} : (u,t) \in B \times [0,\tau]\}, \\ \|G\|_{\{A;C^1(B\times[0,\tau])\}} &\stackrel{\mathrm{def}}{=} \|G\|_{\{A;C^0(B\times[0,\tau])\}} \\ &\qquad + \sup_{t\in[0,\tau]} \sup_{u\in B} \int_0^t \|A^\beta e^{-A(t-s)} DG(u,s)\|_{\mathcal{L}}\, ds. \end{aligned}$$

By using a suitable countable family of these pseudonorms, one obtains the topologies $\mathcal{T}_{bo}^1$ and $\mathcal{T}_A^1$ on C_{Lip}^1, see Appendices A and B.

The situation where $G = G(v)$ is independent of time $t \geq 0$ deserves special consideration. For this case, we define

(46.12)
$$C_{\mathrm{Lip}} \stackrel{\mathrm{def}}{=} C_{\mathrm{Lip}}(V;W) \text{ and } C_{\mathrm{Lip}}^k \stackrel{\mathrm{def}}{=} C_{\mathrm{Lip}}^k(V;W) \stackrel{\mathrm{def}}{=} C_{\mathrm{Lip}}(V;W) \cap C_F^k(V;W),$$

for integers $k \geq 1$. If $G \in C_{\mathrm{Lip}}(V,W)$, or $G \in C_{\mathrm{Lip}}^1(V,W)$, then the two pseudonorms $\|G\|_{\{C^0(B\times[0,\tau])\}}$, or $\|G\|_{\{C^1(B\times[0,\tau])\}}$, do not depend on τ, and we will write them as

$$(46.13) \qquad \|G\|_{\{C^0(B)\}}, \quad \text{or} \quad \|G\|_{\{C^1(B)\}}, \qquad \text{respectively.}$$

In the case where the Standing Hypothesis A is satisfied and

$$G \in C_{\text{Lip}}(V^{2\beta}, W) \quad \text{or} \quad G \in C^1_{\text{Lip}}(V^{2\beta}, W),$$

we define the new pseudonorms:
(46.14)

$$\|G\|_{\{A;C^0(B)\}} \stackrel{\text{def}}{=} \sup_{t\geq 0} \sup_{u\in B} \int_0^t \|A^\beta e^{-A(t-s)} G(u)\| \, ds,$$

$$\|G\|_{\{A;C^1(B)\}} \stackrel{\text{def}}{=} \|G\|_{\{A;C^0(B)\}} + \sup_{t\geq 0} \sup_{u\in B} \int_0^t \|A^\beta e^{-A(t-s)} DG(u)\|_{\mathcal{L}} \, ds.$$

Since the analytic semigroup e^{-At} satisfies inequality (37.11), where $a > 0$, and since both G and DG are bounded on bounded sets in V, these pseudonorms are well-defined. As noted above, a suitable countable family of these pseudonorms generate Fréchet space topologies on $C_{\text{Lip}}(V^{2\beta}, W)$ and $C^1_{\text{Lip}}(V^{2\beta}, W)$. We denote these topologies by $\mathcal{T}^0_A$ and $\mathcal{T}^1_A$, respectively. If $G \in C_{\text{Lip}}(V^{2\beta}, W)$, or $G \in C^1_{\text{Lip}}(V^{2\beta}, W)$, one has

$$\|G\|_{\{A;C^0(B\times[0,\tau])\}} \leq \|G\|_{\{A;C^0(B)\}}, \text{ or } \|G\|_{\{A;C^1(B\times[0,\tau])\}} \leq \|G\|_{\{A;C^1(B)\}},$$

for all $\tau > 0$. This then implies the continuous imbeddings:

$$C_{\text{Lip}}(V^{2\beta}, W) \mapsto C_{\text{Lip}}(V^{2\beta} \times \mathbb{R}^+, W) \text{ and}$$
$$C^1_{\text{Lip}}(V^{2\beta}, W) \mapsto C^1_{\text{Lip}}(V^{2\beta} \times \mathbb{R}^+, W).$$

At this point the role of the superscript in the notation for the topologies $\mathcal{T}^0_A$ and $\mathcal{T}^1_A$, for example, should be clear. In the first case one refers to the $C^0(B)$-topology, and in the second case to the $C^1(B)$-topology. For $\mathcal{T}^1_A$ one uses the first derivative DG.

There is an important improvement in inequality (46.10) in the case where the Standing Hypothesis B is satisfied. In this situation, Exercise 37.9 and relations (310.2) and (310.3) hold. As a result, one has good information concerning the norm $\|A^\beta e^{-At}\|_{\mathcal{L}(H)}$, for $t > 0$. Therefore, when $G = G(v)$ is independent of time $t \geq 0$, one obtains

$$\begin{aligned} \|G\|_{\{A;C^0(U)\}} &\leq (1-\beta)^{-1} e^{-\beta} \lambda_1^{\beta-1} \|G\|_{\{C^0(U)\}} \\ \|G\|_{\{A;C^1(U)\}} &\leq (1-\beta)^{-1} e^{-\beta} \lambda_1^{\beta-1} \|G\|_{\{C^1(U)\}}. \end{aligned} \tag{46.15}$$

4.6. Well-Posed Problems: C_0-Theory.

We begin with the following Initial Value Problem for an abstract nonlinear evolutionary equation of the form

$$\partial_t u + Au = F(u,t), \qquad \text{for } u(t_0) = u_0 \in W \text{ and } t \geq t_0 \geq 0, \tag{46.16}$$

on a Banach space W. We now assume that the $-A$ is the infinitesimal generator of a C_0-semigroup e^{-At} and that $F \in C_{\text{Lip}} = C_{\text{Lip}}(W \times \mathbb{R}^+, W)$.

A typical occurrence of the time-dependent nonlinear evolutionary equation described here will arise when one seeks to study solutions of an autonomous problem $\partial_t v + Av = G(v)$, in the vicinity of a given bounded solution $v(t)$, for $t \geq 0$. Note that $w = u + v$ is another solution of the latter equation if and only if $u = u(t)$ is a solution of (46.16), where $F(u,t) = G(u + v(t)) - G(u)$. If $G \in C_{\text{Lip}}(W, W)$ and if $v(t)$ is bounded for $t \geq 0$, then one has $F \in C_{\text{Lip}}(W \times \mathbb{R}^+, W)$.

4.6.1 Local Existence and Uniqueness Theorems. We follow the approach used in Part I for linear inhomogeneous equations. In particular, our theory of solutions of the Initial Value Problem (46.16) begins with the Variation of Constants Formula. More specifically, let $I = [t_0, t_0 + \tau)$ be an interval in $\mathbb{R}^+$, where $\tau > 0$. A pair (u, I) is said to be a **mild solution** of (46.16) **in the space** W on I if $u : I \to W$ is a (strongly) continuous mapping and a solution of the integral equation

$$u(t) = e^{-A(t-t_0)}u_0 + \int_{t_0}^{t} e^{-A(t-s)} F(u(s), s)\, ds, \qquad t \in I. \tag{46.17}$$

A pair (u, I), where $u : I \to W$ is continuous, satisfies $u(t_0) = u_0$, is (strongly) differentiable almost everywhere with $\partial_t u$ and Au in $L^1_{\text{loc}}(I, W)$, and satisfies the differential equation

$$\partial_t u(t) + Au(t) \overset{\text{a.e.}}{=} F(u(t), t), \qquad \text{on } (t_0, t_0 + \tau), \tag{46.18}$$

is said to be a **strong solution** of (46.16) **in the space** W on I. If in addition, one has $\partial_t u \in C(I, W)$ and the differential equation in (46.18) is satisfied for $t_0 < t < t_0 + \tau$, then (u, I) is called a **classical solution** of (46.16) **in the space** W on I. Notice that (u, I) is a mild solution of (46.16) if and only if $v(t) \overset{\text{def}}{=} u(t)$ is a mild solution of the linear inhomogeneous problem

$$\partial_t v + Av = F(u(t), t), \qquad v(t_0) = u_0.$$

As a result, Lemma 42.1 implies that a classical solution, or a strong solution, if it exists, must be a mild solution. We sometimes refer to u itself as the solution in a context where the interval of definition I is clearly understood. The first result is a local existence and uniqueness theorem for the mild solutions of (46.16).

Theorem 46.1. *Let $(e^{-At}, -A)$ be a C_0-semigroup on a Banach space W, and let $F \in C_{\text{Lip}}(W \times \mathbb{R}^+, W)$. Then for every $u_0 \in W$ and $t_0 \geq 0$, the Initial Value Problem (46.16) has a unique, mild solution in W on some interval $I = [t_0, t_0 + \tau)$, for some $\tau > 0$ which is given in the proof below.*

Proof. The proof reduces to finding a fixed point of the mapping $\hat{u} = \mathcal{T}u$, where

$$\hat{u}(t) = e^{-A(t-t_0)}u_0 + \int_{t_0}^{t} e^{-A(t-s)} F(u(s), s)\, ds, \qquad t \in I = [t_0, t_0 + \tau),$$

for some $\tau > 0$. Let $M \geq 1$ and $a \in \mathbb{R}$ be chosen by Lemma 31.1 so that one has $\|e^{-At}\|_{\mathcal{L}(W)} \leq Me^{-at}$, for all $t \geq 0$. Fix b so that $\|u_0\| \leq b$ and define

$$B \stackrel{\text{def}}{=} \{u \in W : \|u\| \leq Me^{|a|}(b+1)\}. \tag{46.19}$$

Let $K_1 = K_1(B)$ be given so that (46.5) holds for $w_1, w_2 \in B$ and $t \in \mathbb{R}^+$, and let K_0 satisfy $\|F(u,t)\| \leq K_0$, for all $(u,t) \in B \times \mathbb{R}^+$. Next we define τ by

$$\tau = \min\left(1, \frac{1}{K_0}, \frac{1}{2MK_1e^{|a|}}\right), \tag{46.20}$$

and set $I = [t_0, t_0 + \tau]$. Define $\mathcal{F} \stackrel{\text{def}}{=} \{u \in C(I,W) : u(t) \in B \text{ for } t \in I\}$, where $C(I,W)$ is a Banach space with the sup-norm. Therefore $\mathcal{F}$ is a closed, convex set in $C(I,W)$. Then it is readily verified that

(1) $\mathcal{T}$ maps $\mathcal{F}$ into itself, and
(2) $\mathcal{T}$ is a contraction mapping on $\mathcal{F}$ with the contraction constant $\leq \frac{1}{2}$.

For the most part, the verification of these two properties is rather straightforward. Indeed, by using (46.20), one verifies that $\|\hat{u}(t)\| \leq Me^{|a|}(b+1)$, for $t_0 \leq t \leq t_0 + \tau$. Also, if $\hat{u}_i = \mathcal{T}u_i$, for $i = 1, 2$, one has

$$\begin{aligned}\|\hat{u}_1(t) - \hat{u}_2(t)\| &\leq \int_{t_0}^{t} Me^{|a|(t-s)}K_1\|u_1(s) - u_2(s)\|\, ds \\ &\leq \tau MK_1e^{|a|}\|u_1 - u_2\|_\infty \leq \frac{1}{2}\|u_1 - u_2\|_\infty,\end{aligned}$$

for $t_0 \leq t \leq t_0 + \tau$. In order to show that $\hat{u}$ is a continuous mapping of I into W, we note that

$$\hat{u}(t+h) - \hat{u}(t) = E_1 + E_2 + E_3,$$

where

$$\begin{cases} E_1 = e^{-A(t+h-t_0)}u_0 - e^{-A(t-t_0)}u_0, \\ E_2 = \int_{t_0}^{t} \left(e^{-A(t+h-s)}F(u(s),s) - e^{-A(t-s)}F(u(s),s)\right) ds, \\ E_3 = \int_{t}^{t+h} e^{-A(t+h-s)}F(u(s),s)\, ds. \end{cases}$$

Without loss of generality, we assume that $|h| \leq 1$. If $t = t_0 = 0$, we further restrict h to satisfy $0 \leq h \leq 1$. Now $\|E_1\| \to 0$, as $h \to 0$, by Lemma 31.2. Since the mapping of I into W given by $s \to e^{-As}F(u(s),s)$ is continuous, one has $\|E_2\| \to 0$, as $h \to 0$, by (31.5) and the Lebesgue Dominated Convergence Theorem. Since E_3 satisfies $\|E_3\| \leq MK_0e^{|a|(t+1)}|h|$, we see that $\hat{u} \in C(I,W)$. Therefore there exists a unique fixed point u of $\mathcal{T}$ in $\mathcal{F}$, and this fixed point is a mild solution of (46.16) on I. It follows from the Contraction Property that the mild solution is uniquely determined by the initial condition u_0. Clearly every mild solution must be a fixed point of $\mathcal{T}$, which completes the proof. □

By imposing additional regularity conditions on the nonlinearity $F(u,t)$, one can show that the mild solution is, in fact, a strong solution. Here is a typical result.

Theorem 46.2. *Let $(e^{-At}, -A)$ be a C_0-semigroup and let*

$$F \in C_{\text{Lip};1} = C_{\text{Lip};1}(W \times \mathbb{R}^+; W), \text{ or } F \in C_{\text{Lip}} = C_{\text{Lip}}(W; W),$$

where W is now assumed to be a Hilbert space, or a reflexive Banach space. If either $u_0 \in \mathcal{D}(A)$; or e^{-At} is differentiable, for $t > 0$, and $u_0 \in W$; then the mild solution of (46.16) in W is a strong solution in W.

Proof. Let $M \geq 1$ and $a \in \mathbb{R}$ be given by Lemma 31.1 to satisfy

$$\|e^{-At}\|_{\mathcal{L}(W)} \leq Me^{-at}, \qquad \text{for all } t \geq 0.$$

For $t_0 \geq 0$, set $J \stackrel{\text{def}}{=} [t_0, t_0 + \sigma]$ and $B \stackrel{\text{def}}{=} \{u \in W : \|u\| \leq \rho\}$. Fix K_0 so that $\|F(u,t)\| \leq K_0$, for $(u,t) \in B \times J$, and let K_2 be given by (46.6). For $t, t+h \in J$, where $h \geq 0$, one obtains

$$\begin{aligned} u(t+h) - u(t) = e^{-A(t+h-t_0)}u_0 - e^{-A(t-t_0)}u_0& \\ + \int_{t_0}^{t_0+h} e^{-A(t+h-s)} F(u(s), s)\, ds& \\ + \int_{t_0}^{t} e^{-A(t-s)} \left[F(u(s+h), s+h) - F(u(s), s)\right] ds.& \end{aligned}$$

Since $u_0 \in \mathcal{D}(A)$, equation (31.6) implies that

$$e^{-A(t+h-t_0)}u_0 - e^{-A(t-t_0)}u_0 = -\int_{t-t_0}^{t+h-t_0} e^{-As} Au_0\, ds.$$

By using the given estimate on $\|e^{-At}\|_{\mathcal{L}(W)}$, with inequalities (46.1) and (46.6), with $\theta = 1$, we see that $\|u(t+h) - u(t)\|$ is bounded by

$$Me^{|a|\sigma}\left(\|Au_0\|\,|h| + K_0|h| + K_2\int_{t_0}^{t} (|h| + \|u(s+h) - u(s)\|)\, ds\right).$$

Then with $K_3 = Me^{|a|\sigma}(\|Au_0\| + K_0 + \sigma K_2)$, the Gronwall inequality implies that

$$\|u(t+h) - u(t)\| \leq K_3\,|h| \exp(MK_2 e^{|a|\sigma}\sigma) \qquad \text{for } t,\ t+h \in J.$$

Therefore $u(t)$ is uniformly Lipschitz continuous on J, and consequently by (46.6), $F(u(t), t)$ is uniformly Lipschitz continuous on J, as well. By Theorem 42.6, the mild solution $y = y(t)$ of

$$y(t) = e^{-A(t-t_0)}u_0 + \int_{t_0}^{t} e^{-A(t-s)} F(u(s), s)\, ds, \qquad t \in J,$$

is a strong solution of the Initial Value Problem

$$\partial_t y + Ay = F(u(t), t), \qquad v(t_0) = u_0.$$

However, since u is the mild solution of (46.16), it follows from the uniqueness of mild solutions that $u(t) \equiv y(t)$, for $t \in J$. Consequently, $u(t)$ is a strong solution of (46.16) on J. Since σ is arbitrary with $0 \leq \sigma < \tau$, we see that $u(t)$ is a strong solution of (46.16) on I.

The argument for the case where $(e^{-At}, -A)$ is differentiable, for $t > 0$, proceeds along the same lines. The major difference is that, instead of using (31.6), one uses (35.2), for $t_0 > 0$, to obtain

$$\begin{aligned} e^{-A(t+h-t_0)}u_0 - e^{-A(t-t_0)}u_0 &= h\,T^{(1)}(t - t_0)u_0 + E(t - t_0, h)\,u_0 \\ &= -h\,Ae^{-A(t-t_0)}u_0 + E(t - t_0, h)\,u_0, \end{aligned}$$

where $T(t) = e^{-At}$, $T^{(1)}(t)u_0 = \partial_t(T(t)u_0)$, and $h^{-1}\|E(t - t_0, h)u_0\| \to 0$, as $h \to 0$. We leave the details as an exercise. $\square$

For the remainder of this section we will focus on mild solutions of (46.16). For the reasons given above, the theory below also applies to the strong solutions.

4.6.2 Maximally Defined Solutions. Let (u_1, I_1) and (u_2, I_2) be two mild solutions of (46.16), where $I_i = [t_0, t_0 + \tau_i)$, $i = 1, 2$, and $\tau_1 \leq \tau_2$. Owing to the uniqueness of solutions, one must have $u_1(t) = u_2(t)$, for $t \in I_1$. Hence (u_2, I_2) is an extension of (u_1, I_1). When $\tau_1 < \tau_2$, we say that (u_2, I_2) is a **proper extension** of (u_1, I_1). A solution (u, I) of (46.16) is said to be a **maximally defined solution** if (u, I) has no proper extension. We now have the following result:

Theorem 46.3. *Let $(e^{-At}, -A)$ be a C_0-semigroup on a Banach space W, and let $F \in C_{\mathrm{Lip}}(W \times \mathbb{R}^+, W)$. Then for every $u_0 \in W$ and $t_0 \in \mathbb{R}^+$, there is a unique, maximally defined, mild solution (u, I) of (46.16) in W, where $I = [t_0, t_0 + T)$. Furthermore, either $T = \infty$, or*

$$\lim_{t \to T^-} \|u(t)\| = \infty. \tag{46.21}$$

Proof. The existence and uniqueness of a maximally defined solution (u, I) is a direct consequence of Theorem 46.1. In order to prove (46.21), we proceed by contradiction. Assume that $T < \infty$ and there is a $b \in \mathbb{R}^+$ and a sequence t_n in I, with $t_n \to t_0 + T$ and $\|u(t_n)\| \leq b$. Next let B be the bounded set in W given by (46.19). Let $K_0 = K_0(B)$ and $K_1 = K_1(B)$ be given as in the proof of Theorem 46.1, and choose $M \geq 1$ and $a \in \mathbb{R}$ so that $\|e^{-At}\| \leq Me^{-at}$, for all $t \geq 0$. Let τ be given by (46.20). Since (u, I) is a unique maximally defined, mild solution, it follows from Theorem 46.1 (with t_0 replaced by t_n) that $t_0 \leq t_n < t_n + \tau \leq t_0 + T$, for all n. Since $\tau > 0$, this contradicts the fact that $t_n \to t_0 + T$. $\square$

4.6.3 Continuous Dependence of Solutions. Let $C_{\rm Lip} = C_{\rm Lip}(W \times \mathbb{R}^+, W)$. For any $F \in C_{\rm Lip}$ and $u_0 \in W$ we let $\phi(u_0, F, t)$ denote the maximally defined, mild solution of (46.16) in W, with $t_0 = 0$, that satisfies

$$\phi(u_0, F, 0) = u_0,$$

and let $I = [0, T)$ denote the interval of definition of $\phi(u_0, F, t)$, where $T = T(u_0, F)$ satisfies $0 < T \le \infty$. Next define Ξ by

$$\Xi \stackrel{\rm def}{=} \{(u_0, F, t) \in W \times C_{\rm Lip} \times \mathbb{R}^+ : 0 \le t < T(u_0, F)\}. \tag{46.22}$$

We assume that $C_{\rm Lip}$ has the topology $\mathcal{T}_{bo} = \mathcal{T}^0_{bo}$ of uniform convergence on bounded sets. In this section we will prove the following result:

Theorem 46.4. *Let $(e^{-At}, -A)$ be a C_0-semigroup on a Banach space W, and let Ξ be given by (46.22). Then the following statements are valid:*

(1) *The mapping of Ξ into W given by $(u_0, F, t) \to \phi(u_0, F, t)$ is continuous, and $\phi(u_0, F, t)$ is locally Lipschitz continuous in F and u_0.*
(2) *The set Ξ is open in $W \times C_{\rm Lip} \times \mathbb{R}^+$.*
(3) *If $\tau \in [0, T(u_0, F))$ and $t \in [0, T(\phi(u_0, F, \tau), F_\tau))$, then $\tau + t \in [0, T(u_0, F))$ and one has*

$$\phi(\phi(u_0, F, \tau), F_\tau, t) = \phi(u_0, F, \tau + t), \tag{46.23}$$

where $F_\tau(u, t) = F(u, \tau + t)$. In particular, if $F \in C_{\rm Lip}(W, W)$ is time-independent, then $F_\tau = F$, for all $\tau \ge 0$, and

$$\phi(\phi(u_0, F, \tau), F, t) = \phi(u_0, F, \tau + t). \tag{46.24}$$

Proof. First we note that each mild solution $\phi(u_0, F, t)$ is continuous in t, see Lemma 42.3. Let us now verify that $\phi(u_0, F, t)$ is Lipschitz continuous in F and u_0. Let $r > 0$ be fixed and set $B_r = \{w \in W : \|w\| \le r\}$. For $K_0 > 0$ and $K_1 > 0$, we define $C_{\rm Lip}[K_0, K_1]$ to be the collection of those $G \in C_{\rm Lip}$ with the property that G satisfies inequalities (46.4) and (46.5), for these choices of K_0 and K_1, where $B = B_{2r}$. It then follows from Theorem 46.1 that there exists a $\tau > 0$ such that if $F \in C_{\rm Lip}[K_0, K_1]$ and $u_0 \in B_r$, then one has $\|\phi(u_0, F, t)\| \le 2r$, for $0 \le t \le \tau$. Next we let $F_i \in C_{\rm Lip}[K_0, K_1]$ and let $u_i = u_i(t)$ be mild solutions of the problems

$$\partial_t u_i + Au_i = F_i(u_i), \qquad u_i(0) = u_{i0} \in B_r, \quad \text{for } i = 1, 2.$$

Let $w = u_1 - u_2$ and set $w_0 = w(0) = u_{10} - u_{20}$. Let $F_i(u_i) = F_i(u_i(t), t)$, for $i = 1, 2$. Then $w = w(t)$ is a solution of

$$\begin{aligned} w(t) &= e^{-At} w_0 + \int_0^t e^{-A(t-s)} [F_1(u_1) - F_2(u_2) + F_1(u_2)]\, ds \\ &= e^{-At} w_0 + \int_0^t e^{-A(t-s)} [F_1(u_2) - F_2(u_2)]\, ds \\ &\quad + \int_0^t e^{-A(t-s)} [F_1(u_1) - F_1(u_2)]\, ds, \end{aligned}$$

for $0 \leq t \leq \tau$. It then follows from Lemma 31.1 that there are constants $M \geq 1$ and $a \in \mathbb{R}$ such that, for $0 \leq t \leq \tau$, one has

$$\begin{aligned}\|w(t)\| \leq Me^{|a|t}\|w_0\| + M\int_0^t e^{|a|(t-s)}\|[F_1(u_2) - F_2(u_2)]\|\,ds \\ + M\int_0^t e^{|a|(t-s)}\|F_1(u_1) - F_1(u_2)\|\,ds.\end{aligned} \tag{46.25}$$

We now define $\sigma \geq 0$ by

$$\sigma = \sup\{t \geq 0 : \|u_i(s)\| \leq 2r \text{ for } i = 1, 2 \text{ and all } s \text{ with } 0 \leq s \leq t\}. \tag{46.26}$$

Since $u_{10}, u_{20} \in B_r$, one has $0 < \tau \leq \sigma$. It then follows from inequalities (46.5) and (46.25) that

$$\begin{aligned}\|w(t)\| \leq Me^{|a|t}\|w_0\| + M\tau e^{|a|t}\|F_1 - F_2\|_{\{C^0(B\times[0,\tau])\}} \\ + MK_1\int_0^t e^{|a|(t-s)}\|w(s)\|\,ds,\end{aligned}$$

for $0 \leq t \leq \tau$. It then follows from the Gronwall inequality that

$$\|w(t)\| \leq M_\tau e^{(|a|+MK_1)\sigma}\left(\|u_{10} - u_{20}\| + \|F_1 - F_2\|_{\{C^0(B\times[0,\tau])\}}\right), \tag{46.27}$$

for $0 \leq t \leq \tau$, where $M_\tau = M\max(1, \tau)$.

While the value of τ used above may be small, one can adapt this argument for large values of τ as follows. First one fixes $F_1 \in C_{\text{Lip}}$ and $u_{10} \in W$. Then let τ be any number satisfying $0 < \tau < T(u_{10}, F_1)$. Next choose $r > 0$ so that $\|\phi(u_{10}, F_1, t)\| \leq \frac{1}{2}r$, for $0 \leq t \leq \tau$. Then choose $K_0 > 0$ and $K_1 > 0$ so that $G = F_1$ satisfies (46.4) and (46.5) on B_{2r}. Then the argument above establishes that (46.27) holds for this τ. Also note that it then follows from Theorems 46.1 and 46.3 and inequality (46.27) that if $\|u_{10} - u_{20}\|$ and $\|F_1 - F_2\|_{\{C^0(B\times[0,\tau])\}}$ are sufficiently small, then $\tau \in (0, T(u_{20}, F_2))$. Hence, Ξ is an open set.

Next we will show that $\phi(u_0, F, t)$ is jointly continuous in (u_0, F, t). Let F_n, u_n and t_n be sequences with $F_n \to F$ in $(C_{\text{Lip}}, \mathcal{T}_{bo})$, $u_n \to u_0$ in W, and $t_n \to t_0$, where $t_0 \in [0, T(u_0, F))$. Pick τ so that $0 \leq t_0 < \tau < T(u_0, F)$. Then as noted in the last paragraph, for large n, one has $\tau < T(u_n, F_n)$, and inequality (46.27) implies that $\|\phi(u_n, F_n, t) - \phi(u_0, F, t)\| \to 0$, as $n \to \infty$, uniformly for $0 \leq t \leq \tau$. Since

$$\begin{aligned}\|\phi(u_n, F_n, t_n) - \phi(u_0, F, t)\| \leq \|\phi(u_n, F_n, t_n) - \phi(u_0, F, t_n)\| \\ + \|\phi(u_0, F, t_n) - \phi(u_0, F, t)\|,\end{aligned}$$

for all n, it follows from inequality (46.27) and the continuity of $\phi(u_0, F, t)$, with respect to t, that $\|\phi(u_n, F_n, t_n) - \phi(u_0, F, t_n)\| \to 0$, as $n \to \infty$. Hence the function $\phi(u_0, F, t)$ is continuous on Ξ.

Finally, we note that both

$$\phi_2(t) = \phi(u_0, F, \tau + t) \quad \text{and} \quad \phi_1(t) = \phi(\phi(u_0, F, \tau), F_\tau, t)$$

are mild solutions of the same initial value problem:

$$\partial_t u + Au = F_\tau(u, t), \qquad u(0) = \phi(u_0, F, \tau).$$

From the uniqueness of mild solutions, these two functions must be the same, which yields (46.23). Equation (46.24) now follows from equation (46.23). □

4.6.4 Construction of the Nonlinear Semiflow. In this section we describe the construction of the semiflow based on the theory of mild solutions of equation (46.16). As usual we let $(e^{-At}, -A)$ be a given C_0-semigroup on a Banach space W.

We begin with the autonomous case where $F = F(u)$ does not depend on t, i.e., one has $F \in C_{\text{Lip}}(W, W)$, and equation (46.16) assumes the form

$$\partial_t u + Au = F(u), \tag{46.28}$$

In this case, we define

$$M = M(F) \overset{\text{def}}{=} \{u_0 \in W : T(u_0, F) = \infty\},$$

and for $u_0 \in M$, which we assume to be nonempty, set

$$S(t)u_0 = \sigma(u_0, t) \overset{\text{def}}{=} \phi(u_0, F, t). \tag{46.29}$$

Clearly one has $\sigma : M \times \mathbb{R}^+ \to M$, and σ is continuous with $\sigma(u_0, 0) = u_0$. Also (46.24) implies that

$$\sigma(\sigma(u_0, t), s) = \sigma(u_0, t + s) \qquad \text{for all } u_0 \in M \text{ and } s, t \in \mathbb{R}^+.$$

σ is called the semiflow on M **generated by** the mild solutions of (46.28).

In the nonautonomous case, where $F = F(u, t)$ depends on t, one uses the theory of skew product semiflows to describe the longtime dynamics of (46.16). More specifically, if $F \in C_{\text{Lip}} = C_{\text{Lip}}(W \times \mathbb{R}^+, W)$, then for each bounded set B in W there exist $K_0 = K_0(B)$ and $K_1 = K_1(B)$ such that inequalities (46.4) and (46.5) hold for $G = F$. It is easily seen that every translate $G = F_\tau$, where $F_\tau(u, t) = F(u, \tau + t)$, also satisfies (46.4) and (46.5). Consequently, every such G in the positive hull $H^+(F)$ also satisfies these inequalities, as well, i.e., one has

$$F \in C_{\text{Lip}} \Longrightarrow H^+(F) \subset C_{\text{Lip}}.$$

Recall that $H^+(F)$ is the closure in $C = C(W \times \mathbb{R}^+, W)$ of the positive semitrajectory $\gamma^+(F) = \{F_\tau : \tau \geq 0\}$. Next we define

$$M = M(H^+(F)) \stackrel{\text{def}}{=} \{(u_0, G) \in W \times H^+(F) : T(u_0, G) = \infty\}.$$

For each $(u_0, G) \in M$ and $\tau \geq 0$, se set

$$\pi(u_0, G, \tau) = (\phi(u_0, G, \tau), G_\tau).$$

Clearly one has $\pi : M \times \mathbb{R}^+ \to M$ and π is continuous, see Theorem 46.4. Furthermore, $\pi(u_0, G, 0) = (u_0, G)$, and (46.23) implies that π is a semiflow on M. Because of the form of π, we see that π is a skew product semiflow.[18]

One cannot rule out at this point the possibility that M may be empty. However if this happens, it would be both rare and somewhat bizarre. For example, if (46.16) has a stationary solution $u(t) \equiv u_0$, $t \geq 0$, then M is nonempty. A more interesting matter is the following question of global existence:

Find sufficient conditions so that $T(u_0, G) = \infty$, for all $u_0 \in W$ and $G \in H^+(F)$. To put it another way, determine when one has $M(F) = W$, in the autonomous case, or $M(H^+(F)) = W \times H^+(F)$, in the time-varying case.

We will study this issue of global existence later in this chapter.

4.7. Well-Posed Problems: Analytic Theory.

Once again, we begin with an **Initial Value Problem** for an abstract nonlinear evolutionary equation of the form

$$\partial_t u + Au = F(u, t), \qquad u(t_0) = u_0 \in W, \tag{47.1}$$

on a Banach space W, where $t \geq t_0 \geq 0$. However, in this section we will assume that A satisfies the Standing Hypothesis A. There is no loss of generality in assuming that the sectorial operator A is positive. Indeed, since A is sectorial, there is a real number $a \in \mathbb{R}$ such that the linear operator $B = A + aI$ is a positive, sectorial operator, see Section 3.6. In this case equation (47.1) is equivalent to the equation $\partial_t u + Bu = G(u, t)$, where $G(u, t) = F(u, t) + au$. The positivity of the sectorial operator A offers a convenient way for describing the fractional power spaces V^α, for $\alpha \in \mathbb{R}$, and it permits a direct application of the Fundamental Theorem on Sectorial Operators, Theorem 37.5. If A is not positive, then one can achieve the same goals indirectly by using B and its fractional powers, since the semigroups satisfy $e^{-Bt} = e^{-at}e^{-At}$, for $t \geq 0$.

[18]Notice that if $F \in C_{\text{Lip}}(W, W)$ is independent of t, then one has $H^+(F) = \{F\}$. In this case, the skew product semiflow π reduces to $\pi(u_0, F, t) = (S(t)u_0, F)$, see (46.29).

4.7.1. Mild Solutions. We follow the approach used in the C_0-theory based on the Variation of Constants Formula. However, now we will require that the nonlinear term F satisfy $F \in C_{\text{Lip}} = C_{\text{Lip}}(V^{2\beta} \times \mathbb{R}^+; W)$, where β is fixed with $0 \leq \beta < 1$. Once again, we let $I = [t_0, t_0 + \tau)$ be an interval in $\mathbb{R}^+$, where $\tau > 0$ and let $\rho \geq 0$. A pair (u, I) is said to be a **mild solution** of (47.1) **in the space** $V^{2\rho}$ on I, provided that $u \in C(I, V^{2\rho})$ and u is a solution of the integral equation

$$u(t) = e^{-A(t-t_0)}u_0 + \int_{t_0}^{t} e^{-A(t-s)}F(u(s), s)\,ds, \qquad t \in I, \tag{47.2}$$

in $V^{2\rho}$, where the Bochner integral in W is in the space $V^{2\rho}$. Notice that the initial condition satisfies $u(0) = u_0 \in V^{2\rho}$.

There are related solution concepts. In particular, a pair (u, I) where $u : I \to V^{2\rho}$ is continuous, satisfies $u(t_0) = u_0$, is (strongly) differentiable almost everywhere with $\partial_t u$ and Au in $L^1_{\text{loc}}(I, W)$, and satisfies the differential equation

$$\partial_t u(t) + Au(t) \overset{\text{a.e.}}{=} F(u(t), t), \qquad \text{in } W \text{ and on } I \tag{47.3}$$

is said to be a **strong solution** of (47.1) **in the space** $V^{2\rho}$ on I. If in addition, one has $\partial_t u \in C(I, W)$ and the differential equation in (47.3) is satisfied everywhere in I, then (u, I) is called a **classical solution** of (47.1) **in** $V^{2\rho}$ on I.

Notice that (u, I) is a mild solution of (47.1) in $V^{2\rho}$ if and only if $v(t) \overset{\text{def}}{=} u(t)$ is a mild solution of the linear inhomogeneous problem

$$\partial_t v + Av = F(u(t), t), \qquad v(t_0) = u_0$$

in $V^{2\rho}$. As a result, Lemma 42.7 implies that a classical solution, or a strong solution, if it exists, must be a mild solution. We sometimes abbreviate the notation for a solution and write $u = (u, I)$, where the interval of definition I is understood. The first result is a local existence and uniqueness theorem for mild solutions of the Initial Value Problem (47.1).

Lemma 47.1. *Let the Standing Hypothesis A be satisfied and let*

$$F \in C_{\text{Lip}} = C_{\text{Lip}}(V^{2\beta} \times \mathbb{R}^+, W), \qquad \textit{where } 0 \leq \beta < 1. \tag{47.4}$$

Then for every $u_0 \in V^{2\beta}$ and every $t_0 \geq 0$, there is a $\tau > 0$ such that the Initial Value Problem (47.1) has a unique, mild solution $u = u(t)$ in $V^{2\beta}$ with

$$u \in C[t_0, t_0 + \tau; V^{2\beta}) \cap C^{0,\theta_1}_{\text{loc}}[t_0, t_0 + \tau; V^{2\alpha}) \cap C^{0,\theta}_{\text{loc}}(t_0, t_0 + \tau, V^{2r}), \tag{47.5}$$

for all α and r with $0 \leq \alpha < \beta$ and $0 \leq r < 1$, where $\theta_1 > 0$, and $\theta > 0$.

Proof. As in the argument of Theorem 46.1, the proof reduces to finding a fixed point of the mapping $\hat{u} = \mathcal{T}u$ defined by

$$\hat{u}(t) = e^{-A(t-t_0)}u_0 + \int_{t_0}^{t} e^{-A(t-s)}F(u(s), s)\,ds, \qquad t \in I = [t_0, t_0 + \tau],$$

for some $\tau > 0$. Let $M_0 \geq 1$ be given by (37.11). We assume now b is chosen so that $\|A^\beta u_0\| \leq b$. Define

$$B \stackrel{\text{def}}{=} \{v \in V^{2\beta} : \|A^\beta v\| \leq M_0(b+1)\}.$$

Let $K_0 = K_0(B)$ and $K_1 = K_1(B)$ be given so that (46.7) and (46.8) hold, with $G = F$, for $v, v_1, v_2 \in B$ and $t \in \mathbb{R}^+$. Next we define τ by

$$\tau^{1-\beta} = \min\left(\frac{M_0(1-\beta)}{2M_\beta K_0}, \frac{1-\beta}{2M_\beta K_1}\right), \tag{47.6}$$

and set $I = [t_0, t_0 + \tau]$. Finally define $\mathfrak{F} \stackrel{\text{def}}{=} \{u \in C(I, V^{2\beta}) : u(t) \in B \text{ for } t \in I\}$, where $\mathfrak{F}$ has the norm $\|u\|_\infty = \sup_{t \in I} \|A^\beta u(t)\|$.

We now show that $\mathcal{T}$ maps $\mathfrak{F}$ into itself. Let $\hat{u} \stackrel{\text{def}}{=} \mathcal{T}u$. Then by using (37.11), (46.7), and (47.6) we obtain

$$\begin{aligned}
\|A^\beta \hat{u}(t)\| &\leq \|e^{-A(t-t_0)} A^\beta u_0\| + \int_{t_0}^t \|A^\beta e^{-A(t-s)}\|_{\mathcal{L}(W)} K_0 \, ds \\
&\leq M_0 b + \int_{t_0}^t M_\beta K_0 (t-s)^{-\beta} \, ds \\
&\leq M_0 b + M_\beta K_0 (1-\beta)^{-1} \tau^{1-\beta} \leq M_0(b+1),
\end{aligned}$$

for $0 \leq t \leq \tau$. Hence, $\hat{u}(t) \in B$, for $0 \leq t \leq \tau$. The proof that $\hat{u}$ is a continuous mapping of I into $V^{2\beta}$, which now uses (37.12), is similar to the argument of Theorem 46.1, and we omit the details.

Next we show that for τ given by (47.6), the mapping $\mathcal{T}$ is a contraction on $\mathfrak{F}$ with contraction coefficient $\leq \frac{1}{2}$. Indeed, let $u_1, u_2 \in \mathfrak{F}$ and set $\hat{u}_i = \mathcal{T}u_i$, $i = 1, 2$. Then, for $t_0 \leq t \leq t_0 + \tau$, one has

$$\begin{aligned}
\|A^\beta (\hat{u}_1(t) - \hat{u}_2(t))\| &\leq \int_{t_0}^t \|A^\beta e^{-A(t-s)} (F(u_1(s), s) - F(u_2(s), s))\| \, ds \\
&\leq M_\beta K_1 \int_{t_0}^t (t-s)^{-\beta} \|A^\beta (u_1(s) - u_2(s))\| \, ds \\
&\leq M_\beta K_1 (1-\beta)^{-1} \tau^{1-\beta} \|u_1 - u_2\|_\infty \leq \frac{1}{2} \|u_1 - u_2\|_\infty.
\end{aligned}$$

As a result, $\mathcal{T}$ has a unique fixed point on $\mathfrak{F}$. This fixed point is the mild solution of (47.1) on I, and because of the contraction property, this solution is uniquely determined. Since $u \in C[t_0, t_0 + \tau; V^{2\beta})$, it follows that $f(t) \stackrel{\text{def}}{=} F(u(t), t)$ satisfies $f \in C[t_0, t_0 + \tau; W)$. As a result, Lemma 42.7, with $p = \infty$, implies that (47.5) holds. □

4.7.2. Strong Solutions. By imposing additional regularity conditions on the nonlinearity $F(u, t)$, one can show that the mild solution is a strong solution.

Lemma 47.2. *Let the Standing Hypothesis A be satisfied and let*

$$F \in C_{\mathrm{Lip};\theta} = C_{\mathrm{Lip};\theta}(V^{2\beta} \times \mathbb{R}^+, W),$$

where $0 \le \beta < 1$ and $0 < \theta \le 1$. Then for every $u_0 \in V^{2\beta}$, there is a $T > 0$ such that the mild solution $S(t)u_0$ of equation (47.1) in $V^{2\beta}$ is a strong solution in $V^{2\beta}$ on the interval $0 \le t < T$, and it satisfies

$$S(\cdot)u_0 \in C[0,T; V^{2\alpha}) \cap C^{0,1-r}_{\mathrm{loc}}(0,T; V^{2r}) \cap C(0,T; \mathcal{D}(A)), \tag{47.7}$$

for all α and r with $0 \le \alpha \le \beta$ and $0 \le r < 1$.

Proof. Let $u : [0,T) \to W$ be a mild solution of (47.1) in $V^{2\beta}$. Then $v(t) \overset{\mathrm{def}}{=} u(t)$ is a mild solution of the linear inhomogeneous equation

$$\partial_t v + Av = f(t), \qquad v(0) = u_0,$$

in $V^{2\beta}$, where $f = f(t) = F(u(t), t)$. Since $F \in C_{\mathrm{Lip};\theta}$, it follows from (47.5) that f is also locally Hölder continuous in W on $(0,T)$. By Lemma 42.7 and Theorem 42.9, with $p = \infty$, $v(t)$ and, hence $u(t)$, are strong solutions in the space $V^{2\beta}$ on the interval $[0,T)$, and (47.7) follows from (42.14) and (42.18). $\square$

Since $C_{\mathrm{Lip}}(V^{2\beta}, W) \subset C_{\mathrm{Lip}:\theta}$, the following result is an immediate consequence of the last lemma.

Corollary 47.3. *Let the Standing Hypothesis A be satisfied and let $F = F(u)$ satisfy $F \in C_{\mathrm{Lip}}(V^{2\beta}, W)$, where $0 \le \beta < 1$. Then for every $u_0 \in V^{2\beta}$, there is a $T > 0$ such that, the mild solution $S(t)u_0$ of the autonomous equation (46.28) is a strong solution in $V^{2\beta}$ on the interval $0 \le t < T$, and it satisfies(47.7).*

4.7.3. Maximally Defined Solutions. Let (u_1, I_1) and (u_2, I_2) be two mild solutions of (47.1), where $I_i = [t_0, t_0 + \tau_i)$, $i = 1, 2$, and $\tau_1 \le \tau_2$. Owing to the uniqueness of solutions, one must have $u_1(t) = u_2(t)$, for $t \in I_1$. Hence (u_2, I_2) is an extension of (u_1, I_1). When $\tau_1 < \tau_2$, we say that (u_2, I_2) is a **proper extension** of (u_1, I_1). A solution (u, I) of (47.1) is said to be a **maximally defined solution** if (u, I) has no proper extension. We now have the following result.

Lemma 47.4. *Let the Standing Hypothesis A be satisfied and let $F = F(u,t)$ satisfy (47.4). Then for every $u_0 \in V^{2\beta}$ and $t_0 \in \mathbb{R}^+$, there is a unique, maximally defined, mild solution (u, I) of (47.1) in $V^{2\beta}$, where $I = [t_0, t_0 + T)$. Furthermore, either $T = \infty$, or*

$$\lim_{t \to T^-} \|A^\beta u(t)\| = \infty. \tag{47.8}$$

The proof of this result, which uses Lemma 47.1, follows the paradigm of Theorem 46.3. We omit the details.

4.7.4. Continuous Dependence of Solutions. For any $F \in C_{\rm Lip}$, see (47.4), and $u_0 \in V^{2\beta}$ we let $\phi(u_0, F, t)$ denote the maximally defined, mild solution of (47.1) in $V^{2\beta}$ that satisfies $\phi(u_0, F, 0) = u_0$, and let $I = [0, T)$ denote the interval of definition of $\phi(u_0, F, t)$, where $T = T(u_0, F)$ satisfies $0 < T \leq \infty$. Next define Ξ by

$$\Xi \stackrel{\rm def}{=} \{(u_0, F, t) \in V^{2\beta} \times C_{\rm Lip} \times \mathbb{R}^+ : 0 \leq t < T(u_0, F)\}. \tag{47.9}$$

We let $(\Xi, \mathcal{T}_A^0)$ and $(\Xi, \mathcal{T}_{\rm bo}^0)$ denote the space Ξ with the respective topologies on $C_{\rm Lip}$. We now show that the mild solution mapping $S : (\Xi, \mathcal{T}_A^0) \to V^{2\beta}$ is continuous, and as a result, inequality (46.10) implies that the mapping $S : (\Xi, \mathcal{T}_{\rm bo}^0) \to V^{2\beta}$ is continuous, as well.

Theorem 47.5. *Let the Standing Hypothesis A be satisfied and let $F = F(u,t)$ satisfy (47.4). Let Ξ be given by (47.9). Then the following statements are valid:*

(1) *The mapping $(u_0, F, t) \to \phi(u_0, F, t)$ of $(\Xi, \mathcal{T}_A^0)$, or $(\Xi, \mathcal{T}_{bo}^0)$, into $V^{2\beta}$ is continuous, and $\phi(u_0, F, t)$ is locally Lipschitz continuous in F and u_0.*
(2) *The set Ξ is open in $V^{2\beta} \times C_{\rm Lip} \times \mathbb{R}^+$.*
(3) *If $\tau \in [0, T(u_0, F))$ and $t \in [0, T(\phi(u_0, F, \tau), F_\tau))$, then $\tau + t \in [0, T(u_0, F))$ and one has*

$$\phi(\phi(u_0, F, \tau), F_\tau, t) = \phi(u_0, F, \tau + t), \tag{47.10}$$

where $F_\tau(u,t) = F(u, \tau + t)$. In particular, if $F \in C_{\rm Lip}(V^{2\beta}, W)$ is autonomous, then $\phi(\phi(u_0, F, \tau), F, t) = \phi(u_0, F, \tau + t)$.

Proof. The proof of this result follows the argument used in the proof of Theorem 46.4. The only difference occurs in the verification of the Lipschitz continuity of the solution $\phi(u_0, F, t)$ with respect to $F \in C_{\rm Lip} = C_{\rm Lip}(V^{2\beta} \times \mathbb{R}^+; W)$ and $u_0 \in V^{2\beta}$. Instead of inequality (46.25), we now obtain a constant $M_1 > 0$ such that for $0 \leq t \leq \tau$, one has

$$\begin{aligned} \|A^\beta w(t)\| \leq M_1 e^{-at} \|A^\beta w_0\| &+ \int_0^t \|A^\beta e^{-A(t-s)} [F_1(u_2) - F_2(u_2)]\| \, ds \\ &+ \int_0^t \|A^\beta e^{-A(t-s)}\|_{\mathcal{L}(W)} \|F_1(u_1) - F_1(u_2)\| \, ds, \end{aligned}$$

where $a > 0$. It then follows from (37.11) and (46.9) that

$$\begin{aligned} \|A^\beta w(t)\| \leq M_1 e^{-at} \|A^\beta w_0\| &+ \|F_1 - F_2\|_{\{A; C^0(B \times [0,\tau])\}} \\ &+ M_\beta K_1 \int_0^t (t-s)^{-\beta} e^{-a(t-s)} \|A^\beta w(s)\| \, ds, \end{aligned}$$

for $0 \leq t \leq \tau$. It follows from the Gronwall-Henry inequality that, for $\mu^{1-\beta} = M_\beta K_1 \Gamma(1-\beta)$ and $E(\mu;t) = E_{1-\beta,1}(\mu t)$, there is a constant $C = C(\tau)$, such that

(47.11)

$$\|A^\beta w(t)\| \leq C\, E(\mu;\tau) \left(\|A^\beta(u_{10} - u_{20})\| + \|F_1 - F_2\|_{\{A;C^0(B\times[0,\tau])\}} \right)$$

for $0 \leq t \leq \tau$. This establishes the local Lipschitz continuity of the solution $\phi(u_0, F, t)$ in F and u_0. □

4.7.5. Construction of the Nonlinear Semiflow. In this section we describe the construction of the semiflow based on the theory of mild solutions of the nonlinear equation (47.1). We begin with the autonoumous case where $F = F(u)$ does not depend on t; i.e., one has $F \in C_{\text{Lip}}(V^{2\beta}, W)$, where $0 \leq \beta < 1$, and equation (47.1) assumes the form

$$\partial_t u + Au = F(u), \qquad u(0) = u_0 \in V^{2\beta}. \tag{47.12}$$

For $F \in C_{\text{Lip}}(V^{2\beta}, W)$ we define $M = M(F) \stackrel{\text{def}}{=} \{u_0 \in V^{2\beta} : T(u_0, F) = \infty\}$, and for $u_0 \in M$, we set

$$S(t)u_0 = \sigma(u_0, t) \stackrel{\text{def}}{=} \phi(u_0, F, t). \tag{47.13}$$

As argued in Section 4.6.4, it is easily seen that σ is a semiflow on M.

In the nonautonomous case, where $F = F(u,t)$ depends on t, one uses the theory of skew product semiflows to describe the long-time dynamics of (47.1). More specifically, let $F \in C_{\text{Lip}} = C_{\text{Lip}}(V^{2\beta} \times \mathbb{R}^+, W)$. Then $G = F$ satisfies (46.7) and (46.8). It is easily seen that every translate $G = F_\tau$, where $F_\tau(u,t) = F(u, \tau + t)$, also satisfies (46.7) and (46.8). The same is true for every $G \in H^+(F)$, and one has

$$F \in C_{\text{Lip}} \Longrightarrow H^+(F) \subset C_{\text{Lip}}.$$

Recall that $H^+(F)$ is the closure in $C(V^{2\beta} \times \mathbb{R}^+, W)$ of the positive semi-trajectory $\gamma^+(F) = \{F_\tau : \tau \geq 0\}$. Next we define

$$M = M(H^+(F)) \stackrel{\text{def}}{=} \{(u_0, G) \in W \times H^+(F) : T(u_0, G) = \infty\}.$$

For each $(u_0, G) \in M$ and $\tau \geq 0$, we set

$$\pi(u_0, G, \tau) = (\phi(u_0, G, \tau), G_\tau).$$

Clearly π is a semiflow on M. Because of the form of π, we see that π is a skew product semiflow.

The following result describes a key feature used in the study of the longtime dynamics of mild solutions of the nonlinear equation (47.12). In simple terms, we now show that "mild implies strong" - or to put it another way: the meek shall inherit the Earth!

Theorem 47.6. (The Herculean Theorem). *Let the Standing Hypothesis A be satisfied and assume that* $F \in C_{\mathrm{Lip}}(V^{2\beta}, W)$, *for some* β *with* $0 \le \beta < 1$. *Let* $\mathcal{K}$ *be a bounded set in* $V^{2\beta}$, *and assume that* $\mathcal{K}$ *is an invariant set for the semiflow* $S(t)u_0$ *given by equation (47.13). Then one has* $\mathcal{K} \subset \mathcal{D}(A) = V^2$, *and for every* $u_0 \in \mathcal{K}$, *the global mild solution* $S(t)u_0$ *is both a strong and a classical solution of equation (47.12) in* V^{2r}, *for all* $t \in \mathbb{R}$, *with*

$$S(\cdot)u_0 \in C^{0,1-r}_{\mathrm{loc}}(\mathbb{R}; V^{2r}) \cap C(\mathbb{R}; \mathcal{D}(A)), \tag{47.14}$$

and $\mathcal{K}$ *is a bounded, invariant set in* V^{2r}, *for each* r *with* $0 \le r < 1$.

Proof. Let $u_0 \in \mathcal{K}$ and let $u(t) = S(t)u_0$ be the globally defined mild solution with values in $\mathcal{K} \subset V^{2\beta}$. By Lemma 47.4, this solution is defined for all $t \ge 0$. It follows from Corollary 47.3 that $S(t)u_0$ is a strong solution in $V^{2\beta}$ on $[0, \infty)$, and it satisfies (47.7) with $T = \infty$. Since $\mathcal{K}$ is an invariant set, it follows from Lemma 21.2 that for each $\tau > 0$ there is a point $u_1 \in \mathcal{K}$ such that $S(\tau)u_1 = u_0$. Since $S(t)u_1$ satisfies (47.7) with $T = \infty$, one finds that $u_0 \in \mathcal{D}(A)$, and since τ is arbitrary, (47.7) implies (47.14). It then follows from the definitions of strong and classical solutions, that $S(t)u_0$ is a strong and classical solution in V^{2r}, for $0 \le r < 1$.

It is clear from (47.14) that $\mathcal{K}$ is an invariant set in V^{2r}, for each r with $0 \le r < 1$. Note that since $\mathcal{K}$ is a bounded set in $V^{2\beta} \hookrightarrow W$ and since F maps bounded sets in $V^{2\beta}$ into bounded sets in W, there exist $b > 0$ and $K_0 \ge 0$ so that $\|u\| \le b$ and $\|F(u)\| \le K_0$, for all $u \in \mathcal{K}$. For $u_0 \in \mathcal{K}$ and $\tau > 0$, we pick $u_1 \in \mathcal{K}$ so that $S(\tau)u_1 = u_0$. From the mild solution formula, one then has

$$u_0 = e^{-A\tau}u_1 + \int_0^\tau e^{-A(\tau-s)}F(S(s)u_1)\,ds.$$

Then for $0 \le r < 1$, one uses inequality (37.11), with parameters $M_r > 0$ and $a > 0$, and the change of variables $x = a(\tau - s)$ to obtain

$$\begin{aligned}
\|A^r u_0\| &\le \|A^r e^{-A\tau}u_1\| + \int_0^\tau \|A^r e^{-A(\tau-s)}F(S(s)u_1)\|\,ds \\
&\le M_r\tau^{-r}e^{-a\tau}\|u_1\| + M_r \int_0^\tau (\tau-s)^{-r}e^{-a(\tau-s)}\|F(S(s)u_1)\|\,ds \\
&\le M_r\tau^{-r}e^{-a\tau}b + M_r K_0 \int_0^\tau (\tau-s)^{-r}e^{-a(\tau-s)}\,ds \\
&\le M_r\tau^{-r}e^{-a\tau}b + M_r K_0\, a^{r-1}\int_0^{a\tau} x^{-r}e^{-x}\,dx.
\end{aligned}$$

Now let $\tau \to \infty$, and one obtains $\|A^r u_0\| \le M_r K_0\, a^{r-1}\Gamma(1-r)$. □

There are some standard situations whereby one can verify that for a given nonlinearity F one has $T(u_0, F) = \infty$, for all $u_0 \in V^{2\beta}$. We summarize this theory in the following statement.

Theorem 47.7. *Let the Standing Hypothesis A be satisfied and let $F = F(u,t)$ satisfy (47.4). Let $H^+(F)$ denote the hull of F. Then a sufficient condition for $T(u_0, F) = \infty$, for all $u_0 \in V^{2\beta}$, is that one of the following properties be satisfied:*

(1) *There exist nonnegative constants C_0 and C_1 such that*

$$\|F(w,t)\| \leq C_0 + C_1\|A^\beta w\|, \qquad \text{for all } (w,t) \in V^{2\beta} \times \mathbb{R}^+. \tag{47.15}$$

(2) *There is a Banach space V_0 with $V_0 \mapsto V^{2\beta}$ and one has: (i) $\phi(u_0, F, t) \in V_0$, for all $u_0 \in V^{2\beta}$ and all t with $0 < t < T(u_0, F)$, and (ii) there is a constant $C_0 \geq 1$ such that whenever $u_0 \in V_0$ and $G \in H^+(F)$, one has*

$$\|\phi(u_0, G, t)\|_{V_0} \leq C_0\|u_0\|_{V_0}, \qquad \text{for } 0 < t < T(u_0, G).$$

We will leave the proof of this result as an exercise. It is important to use equation (47.8) in the argument.

The assumption concerning the linear growth at infinity given by inequality (47.15) appears to be rather restrictive, and it is. However, it does arise in some applications. For example, in many population models, the nonlinearity F may denote the growth rates of the various species. In such models one oftentimes finds a saturation requirement, which takes on the form

$$\limsup_{\|w\| \to \infty} \frac{\|F(w,t)\|}{\|w\|} < \infty,$$

which in turn implies (47.15).

A typical application involving Part (2) of Theorem 47.7 occurs when Ω is a sufficiently smooth bounded region in $\mathbb{R}^d$, for some $d \geq 1$, and $A = -\Delta$, the Laplacian, with suitable boundary conditions. In this case one takes $V^{2\beta} = L^2(\Omega, \mathbb{R})$ and $V_0 = L^\infty(\Omega, \mathbb{R})$. The continuous imbedding $V_0 \mapsto V^{2\beta}$ is then valid, and the remaining conditions in Part (2) are derived by using a Maximum Principle argument, see Protter and Weinberger (1984). Moreover, in this case the linear semigroup e^{-At} is compact for $t > 0$, and as will be shown in the next section, this creates additional regularity in both space and time. The Sobolev Imbedding Theorems (see Appendix B) are important here. For example, when $d \leq 3$, one oftentimes obtains

$$\phi(u_0, F, t) \in H^2(\Omega, \mathbb{R}) \hookrightarrow L^\infty(\Omega, \mathbb{R}), \qquad \text{for } t > 0.$$

4.7.0. Extension of the Semiflow. The phase space for the semiflow $S(t)$ generated by the solutions of equation (47.12), see equation (47.13), is the space $V^{2\beta}$, at least in the case where $T(u_0, F) = \infty$, for all $u_0 \in V^{2\beta}$. Because of the continuous imbedding $V^{2\beta} \mapsto W$, it is natural to ask whether this semiflow admits an extension to a semiflow $\hat{S}(t)$ on W. The resolution of this issue is somewhat complicated. However, there is a practical situation where one does obtain such an extension, and this arises when the coefficients K_0 and K_1, which are used in inequalities (46.7) and (46.8), can be chosen independent of the bounded set B in $V^{2\beta}$. In order to describe this extension, we return to the nonautonomous problem (47.1).

Except for the fact that the (maximally defined) mild solutions $\phi(u_0, F, t)$ of (47.1) satisfy (47.8) and need not be defined for all time $t \geq 0$, there is a striking similarity between these solutions and the corresponding solutions $\Phi(B, t)v_0$ of the linear problem (44.5). More precisely, let $S(t)u_1 = \phi(u_1, F, t)$ and $S(t)u_2 = \phi(u_2, F, t)$ be two maximally defined solutions of (47.1) in $V^{2\beta}$ defined on intervals $[0, T_1)$ and $[0, T_2)$, respectively. Set $T = \min(T_1, T_2)$. Then for any τ with $0 \leq \tau < T$, the solutions $S(t)u_1$ and $S(t)u_2$ remain in a bounded set $B = B(\tau)$ in $V^{2\beta}$, for $0 \leq t \leq \tau$, and there is a constant $K_1 = K_1(\tau)$ such that

$$\|F(S(t)u_1, t) - F(S(t)u_2, t)\| \leq K_1(\tau)\|A^\beta(S(t)u_1 - S(t)u_2)\|,$$

for $0 \leq t \leq \tau < T$, see (46.8). It then follows from (37.11) and (47.2) that

$$\|A^\beta(S(t)u_1 - S(t)u_2)\| \leq M_\beta t^{-\beta} e^{-at}\|u_1 - u_2\|$$
$$+ M_\beta K_1 \int_0^t (t-s)^{-\beta} e^{-a(t-s)}\|A^\beta(S(s)u_1 - S(s)u_2\| \, ds,$$

for $0 \leq t \leq \tau$. (Compare with inequality (44.22).) As argued in the proof of Theorem 44.1, Item (2), the Gronwall-Henry inequality implies that

$$\|A^\beta(S(t)u_1 - S(t)u_2)\| \leq M_\beta t^{-\beta} e^{-at} E(\mu; t)\|u_1 - u_2\|, \tag{47.16}$$

for $0 \leq t \leq \tau < T$, where

$$\mu^{1-\beta} = M_\beta K_1(\tau)\Gamma(1-\beta). \tag{47.17}$$

(Compare with inequality (44.15).) Now inequality (47.16) is valid for all pairs u_1, $u_2 \in V^{2\beta}$. Of course, the times T and τ depend on u_1 and u_2. Since $V^{2\beta}$ is dense in W, inequality (47.16), which is a statement of uniform continuity of $S(t)u$ in u, suggests that there is a unique extension of $\phi(u_0, F, t)$ to initial conditions $u_0 \in W$.

In this way, one expects that much of Theorem 44.1 can be extended to the solutions of the nonlinear problem (47.1). However, one major difference is in the formula for μ. The role of the norm $\|B\|_\infty$ used in the linear problem is now replaced by $K_1(\tau)$. It can happen that $K_1(\tau) \to \infty$, as $\tau \to T^-$. It this case, equation (47.17) implies that $\mu \to \infty$, as well. Nevertheless, the analogy between the linear and the nonlinear problems is very informative. In order to describe this analogy, it is convenient to first restrict to a subset of C_{Lip}. In particular, we say define $C_{\text{Lip; Global}}$ as the collection of all functions $F \in C_{\text{Lip}}$ with the property that there is a constant $K_1 \geq 0$ such that for all (u_1, t), $(u_2, t) \in V^{2\beta} \times \mathbb{R}^+$, one has

$$\|F(u_1, t) - F(u_2, t)\| \leq K_1\|A^\beta(u_1 - u_2)\|. \tag{47.18}$$

Note that if $F \in C_{\text{Lip}}$ satisfies (47.18), then every translate F_τ satisfies (47.18), where $F_\tau(u, t) = F(u, \tau + t)$. The proofs of the following two theorems are left as exercises.

Theorem 47.8. *Let the Standing Hypothesis A be satisfied and let $F \in C_{\text{Lip; Global}}$. Then for all $u_0 \in V^{2\beta}$, the mild solution $\phi(u_0, F, t)$ exists in $V^{2\beta}$ for all $t \geq 0$, and the following statements are valid:*

1. *For each $t \in [0, \infty)$, the mapping $v_0 \to \phi(v_0, F, t)$, of $V^{2\beta}$ into $V^{2\beta}$, is Lipschitz continuous in F and v_0, where one uses an invariant metric that generates the topology of uniform convergence on bounded sets for $F \in C_{\text{Lip; Global}}$.*
2. *For $v_0 \in V^{2\beta}$ the solution $\phi(v_0, F, t)$ satisfies*

$$\phi(v_0, F, \cdot) \in C[0, \infty; V^{2\alpha}) \cap C^{0,\theta}_{\text{loc}}(0, \infty; V^{2(\beta+r)}),$$

for all α and r with $0 \leq \alpha \leq \beta$ and $0 \leq r < 1$, where $0 < \theta < 1 - r$.

3. *There is a unique continuous extension of $\phi(u_0, F, t)$ to $u_0 \in W$, for all $t \geq 0$, and one has*

$$\phi(u_0, F, \cdot) \in C[0, \infty; W) \cap C^{0,\theta_1}_{\text{loc}}(0, \infty; V^{2r}), \qquad \textit{for } u_0 \in W \textit{ and } 0 \leq r < 1,$$

where $\theta_1 = \theta_1(r) > 0$. Moreover, for such r and for each ρ with $0 \leq \rho \leq 1$, there are positive constants $C = C(r, \rho)$ and ν and functions E_1 and E_2, which have the EX property, such that, for any pair u_1, $u_2 \in V^{2\rho} = \mathcal{D}(A^\rho)$, one has

$$\|A^r(S(t)u_1 - S(t)u_2)\| \leq Ct^{\rho - r}E_1(\nu; t)e^{-at}\|A^\rho(u_1 - u_2)\|, \tag{47.19}$$

for $t > 0$, whenever $0 \leq \rho \leq r < 1$, and

$$\|A^r(S(t)u_1 - S(t)u_2)\| \leq CE_2(\nu; t)e^{-at}\|A^\rho(u_1 - u_2)\|, \tag{47.20}$$

for $t > 0$, whenever $0 \leq r \leq \rho \leq 1$.

4. *If in addition $F = F(u, t)$ satisfies $F \in C_{\text{Lip}:\theta}(V^{2\beta} \times \mathbb{R}^+, W)$, then the solution $\phi(u_0, F, t)$ satisfies*

$$\phi(v_0, F, \cdot) \in C[0, \infty; V^{2\alpha}) \cap C(0, \infty; \mathcal{D}(A)) \cap C^{0,1-r}_{\text{loc}}(0, \infty; V^{2r}), \tag{47.21}$$

for all $v_0 \in V^{2\beta}$ and all α and r with $0 \leq \alpha \leq \beta$ and $0 \leq r < 1$.

In the following theorem we will study the skew product semiflow π generated by equation (47.1), where $F \in C_{\text{Lip; Global}}$. This semiflow is given by the equation

$$\pi(u_0, F, \tau) = (\phi(u_0, F, \tau), F_\tau), \tag{47.22}$$

for $u_0 \in W$ and $F \in C_{\text{Lip; Global}}$, where $F_\tau(u, t) = F(u, \tau + t)$.

Theorem 47.9. *Let the Standing Hypothesis A be satisfied and let* $F \in C_{\text{Lip; Global}}$. *Then the following statements are valid:*

(1) *The mapping* π *given by equation (47.22) on* $W \times C_{\text{Lip; Global}}$ *is a skew product semiflow on* $W \times C_{\text{Lip;Global}}$, *where* $C_{\text{Lip;Global}}$ *has the Fréchet space topology of uniform convergence on bounded sets.*

(2) *For each* r *with* $0 \leq r < 1$, *the mapping* π *is also a skew product semiflow on* $V^{2r} \times C_{\text{Lip; Global}}$. *Moreover, for each* $u_0 \in W$, *one has* $\phi(u_0, F, t) \in V^{2r}$, *for* $t > 0$, *and inequalities (47.19) and (47.20) are valid.*

What happens in the general case, where $F \in C_{\text{Lip}}$, but (47.18) is not satisfied? In the case of the study of bounded solutions, the story is pretty much the same. Assume that one is interested in the study of a class of solutions $\phi(u_0, F, t)$, with the property that there is a constant $a_0 > 0$ such that each solution in this class satisfies $\|A^\beta \phi(u_0, F, t)\| \leq a_0$, for all $t \geq 0$. In this case we claim that one can replace F with another function F^a, where $F^a \in C_{\text{Lip; Global}}$ and $\phi(u_0, F, t) = \phi(u_0, F^a, t)$, for all $t \geq 0$ and all solutions in this class. The idea here is to fix a constant $a > a_0$ and to consider a family of functions $F = F(\lambda, u, t)$, which depend on a parameter λ, where $\lambda \in \Lambda$ and Λ is a compact metric space. We define

$$C^\Lambda_{\text{Lip}} \stackrel{\text{def}}{=} \{F(\lambda, u, t) \in C(\Lambda \times V^{2\beta} \times \mathbb{R}^+; W)\}$$

such that for every bounded set $B \subset V^{2\beta}$, there exist constants $K_0 = K_0(B)$ and $K_1 = K_1(B)$, such that

$$\|F(\lambda, u, t)\| \leq K_0, \qquad \text{for all } (\lambda, u, t) \in \Lambda \times B \times \mathbb{R}^+ \tag{47.23}$$

and for all (λ, u_1, t), $(\lambda, u_2, t) \in \Lambda \times B \times \mathbb{R}^+$, one has

$$\|F(\lambda, u_1, t) - F(\lambda, u_2, t)\| \leq K_1 \|A^\beta(u_1 - u_2)\|. \tag{47.24}$$

Similarly, we define $C^\Lambda_{\text{Lip; Global}}$ as the collection of all $F \in C^\Lambda_{\text{Lip}}$ such that K_0 and K_1 can be chosen independent of the bounded set B in $V^{2\beta}$. We now have the following result.

Lemma 47.10 (Preparation Lemma). *Let* $F = F(\lambda, u, t) \in C^\Lambda_{\text{Lip}}$. *Then for every* $u_0 \in V^{2\beta}$ *and every* $a > 0$, *there is a function* $F^a \in C^\Lambda_{\text{Lip}}$ *that satisfies the following properties:*

(1) *One has* $F^a(\lambda, u, t) = F(\lambda, u, t)$, *for all* $u \in V^{2\beta}$ *with* $\|A^\beta(u - u_0)\| \leq a$ *and all* $\lambda \in \Lambda$, *and all* $t \geq 0$.

(2) *One has* $F^a(\lambda, u, t) = 0$, *for all* $u \in V^{2\beta}$ *with* $\|A^\beta(u - u_0)\| \geq 2a$ *and all* $\lambda \in \Lambda$, *and all* $t \geq 0$.

(3) *There are positive constants* L_0 *and* L_1, *such that* $F^a \in C^\Lambda_{\text{Lip; Global}}$,

$$\|F^a(\lambda, u, t)\| \leq L_0, \qquad \textit{for all } (\lambda, u, t) \in \Lambda \times V^{2\beta} \times \mathbb{R}^+, \tag{47.25}$$

and for all (λ, u_1, t), $(\lambda, u_2, t) \in \Lambda \times V^{2\beta} \times \mathbb{R}^+$. *one has*

$$\|F^a(\lambda, u_1, t) - F^a(\lambda, u_2, t)\| \le L_1 \|A^\beta(u_1 - u_2)\|. \tag{47.26}$$

(4) *If* W *is a Hilbert space, and one has* $F \in C^k_F(V^{2\beta} \times \mathbb{R}^+; W)$, *for each* $\lambda \in \Lambda$ *and some* $k \ge 1$, *then* $F^a \in C^k_F(V^{2\beta} \times \mathbb{R}^+; W)$.

Proof. Let $u_0 \in V^{2\beta}$ and $a > 0$ be given. We define F^a by the formula $F^a(\lambda, u, t) = \theta_a(u) F(u, t)$, where $\theta_a : V^{2\beta} \to [0, 1]$ is a Lipschitz continuous function that satisfies $\theta_a(u) = 1$, for $\|A^\beta(u - u_0)\| \le a$, $\theta_a(u) = 0$, for $\|A^\beta(u - u_0)\| \ge 2a$, and

$$|\theta_a(u_1) - \theta_a(u_2)| \le \frac{2}{a}\|A^\beta(u_1 - u_2)\|, \qquad \text{for all } u_1,\ u_2 \in V^{2\beta}.$$

For example, one might use $\theta_a(u) = 2 - \frac{1}{a}\|A^\beta(u - u_0)\|$, for $a \le \|A^\beta(u - u_0)\| \le 2a$.

Let B be the bounded set $B = \{u \in V^{2\beta} : \|A^\beta(u - u_0)\| \le 2a\}$ and let $K_0 = K_0(B)$ and $K_1 = K_1(B)$ be given by inequalities (47.23) and (47.24). It is easily verified that inequality (47.25) holds with $L_0 = K_0$. For inequality (47.26), we first note that it is obvious when $\|A^\beta(u_i - u_0)\| > 2a$, for $i = 1, 2$, because $\theta_a(u_i) = 0$. If one has $\|A^\beta(u_i - u_0)\| \le 2a$, for $i = 1, 2$, then

$$\begin{aligned}
&\|F^a(\lambda, u_1, t) - F^a(\lambda, u_2, t)\| \\
&\quad = \|\theta_a(u_1)(F(\lambda, u_1, t) - F(\lambda, u_2, t)) + (\theta_a(u_1) - \theta_a(u_2))F(\lambda, u_2, t)\| \\
&\quad \le \theta_a(u_1)\|F(\lambda, u_1, t) - F(\lambda, u_2, t)\| + |\theta_a(u_1) - \theta_a(u_2)|\,\|F(\lambda, u_2, t)\|,
\end{aligned}$$

which implies (47.26) with $L_1 = K_1 + 2K_0 a^{-1}$. Finally, if one has $\|A^\beta(u_1 - u_0)\| \le 2a < \|A^\beta(u_2 - u_0)\|$ (or vice versa), then we choose $w \in V^{2\beta}$ on the line segment joining u_1 and u_2 in $V^{2\beta}$ so that $\|A^\beta(w - u_0)\| = 2a$. From the argument above we see that inequality (47.26) is valid when w replaces u_2. Since $\|A^\beta(u_1 - w)\| \le \|A^\beta(u_1 - u_2)\|$, this implies that (47.26) is valid as stated. If W is a Hilbert space, the θ_a can be chosen to be a C^∞-function. Item (4) follows from this fact. □

In addition to equation (47.1), we are interested in the solutions of the modified equation

$$\partial_t u + Au = F^a(u, t), \qquad u(0) = u_0 \in V^{2\beta}.$$

Since F^a satisfies inequalities (47.25) and (47.26), it follows from Theorem 47.7 that the mild solutions of the last equation are defined for all $t \ge 0$. Furthermore, since $F(u, t) = F^a(u, t)$ when $\|A^\beta(u - u_0)\| \le a$, it follows that the solutions satisfy

$$\phi(u_1, F^a, t) = \phi(u_1, F, t), \qquad \text{on any interval } I = [0, \tau],$$

where $\|A^\beta(\phi(u_1, F^a, t) - u_0)\| \le a$, for $0 \le t \le \tau$. In an application of this lemma one would choose the parameter $a > 0$ large enough so that the bounded solutions of (47.1) of interest would satisfy $\|A^\beta(\phi(u_1, F, t) - u_0)\| < a$, for all $t \ge 0$. For example, where it is appropriate, a may be chosen so that the global attractor $\mathfrak{A}$ satisfies $\mathfrak{A} \subset \{u \in V^{2\beta} : \|A^\beta(u - u_0)\| < a\}$.

A nice application of the last result arises when one is studying the dynamics of the solutions of the autonomous equation (47.12) in the neighborhood of a closed, bounded, invariant set $\mathcal{K}$ in $V^{2\beta}$. In doing this, there is freedom in the choice of the two parameters $u_0 \in V^{2\beta}$ and $a > 0$ used in the Preparation Lemma 47.10. For example, u_0 might be chosen to be a fixed point (i.e., a stationary solution of (47.12)) in, or near to, the set $\mathcal{K}$. The parameter $a > 0$ is chosen so that the set $\{u \in V^{2\beta} : \|A^\beta(u - u_0)\| < a\}$ is a suitable neighborhood of the closed set $\mathcal{K}$.

4.8. Regularity and Compactness Properties.

In this section we show that, under reasonable assumptions, the solutions of the Initial Value Problem (46.16) have additional regularity or smoothness properties. In particular, we examine these issues, which in the case of applications to PDEs, amount to increased smoothness of solutions in the temporal and/or the spatial variables. In addition, we seek conditions under which the semiflow generated by (46.16) is compact.

4.8.1 Compactness Properties. As noted in Theorem 23.12, if a semiflow σ on a Banach space W is point dissipative and compact, for $t > t_0$, then there exists a global attractor $\mathfrak{A}$. In this section we will establish sufficient conditions under which the semiflow generated by the mild soluitons of the autonomous evolutionary equation

$$(48.1) \qquad \partial_t u + Au = F(u), \qquad u(0) = u_0 \in W,$$

be compact, or κ-contracting.

Lemma 48.1. *Let the Standing Hypothesis A be satisfied and assume that A has compact resolvent. Let* $F = F(u)$ *satisfy* $F \in C_{Lip}(V^{2\beta}, W)$, *where* $0 \le \beta < 1$, *and define* $S(t)u_0 = \sigma(u_0, t) \stackrel{\text{def}}{=} \phi(u_0, F, t)$. *Then for every bounded set* $B \subset V^{2\beta}$, *there is a time* $T = T(B)$, *with* $0 < T \le \infty$, *and such that for each* t, *with* $0 < t < T$, *the set* $S(t)B$ *lies in a compact subset of* $V^{2\beta}$.

Proof. From Item (6) of Theorem 37.5, we see that the analytic semigroup e^{-At} is compact, for $t > 0$. Let B_0 be a given bounded set in $V^{2\beta}$, and fix $b > 0$ so that $\|A^\beta v\| \le b$, for all $v \in B_0$. We will now use the notation given in the proof of Lemma 47.1. Let B be given by $B = \{v \in V^{2\beta} : \|A^\beta v\| \le M_0(b + 1)\}$. One then has $B_0 \subset B$, since $M_0 \ge 1$. Let τ is given by (47.6). As shown in Lemma 47.1, one then obtains $S(t)B_0 \subset B$, for $0 \le t \le \tau$.

Furthermore, one has

$$S(t)B_0 = \{e^{-At}u_0 + \int_0^t e^{-A(t-\eta)}F(S(\eta)u_0)\,d\eta : u_0 \in B_0\}$$
$$\subset N_1 + N_2 + N_3,$$

for $0 < t \le \tau$, where the sets N_1, N_2, and N_3 are defined by

$$N_1 = \{e^{-At}u_0 : u_0 \in B_0\},$$
$$N_2 = \left\{\int_{t-\delta}^t e^{-A(t-\eta)}F(S(\eta)u_0)\,d\eta : u_0 \in B_0\right\}, \quad \text{and}$$
$$N_3 = e^{-A\delta}\left\{\int_0^{t-\delta} e^{-A(t-\delta-\eta)}F(S(\eta)u_0)\,d\eta : u_0 \in B_0\right\},$$

and δ is arbitrary and satisfies $0 < \delta < t$. Now inequality (37.11) implies that, for $u_0 \in B_0$, one has

$$\left\|\int_{t-\delta}^t A^\beta e^{-A(t-s)}F(S(s)u_0)\,ds\right\| \le K_0 M_\beta \int_{t-\delta}^t (t-s)^{-\beta}\,ds$$
$$= K_0 M_\beta (1-\beta)^{-1}\delta^{1-\beta}.$$

The Kuratowski measure of noncompactness κ on the Banach space $V^{2\beta}$ satisfies

$$\kappa(N_2) \le \operatorname{diam}_{V^{2\beta}}(N_2) \le C\delta^{1-\beta}, \text{ and } \kappa(S(t)B_0) \le \kappa(N_1) + \kappa(N_2) + \kappa(N_3),$$

for $C = 2K_0M_\beta(1-\beta)^{-1}$, see Lemma 22.2. Since $N_1 = e^{-At}B_0$ and $N_3 = e^{-A\delta}B_1$ are precompact sets, one has $\kappa(N_1) = 0$ and $\kappa(N_3) = 0$. By letting $\delta \to 0$, we conclude that $\kappa(S(t)B_0) = 0$, for $t > 0$, which completes the proof. □

The argument given above readily extends to the interval $[0, T)$, where

$$T = T(B_0) \stackrel{\text{def}}{=} \sup\{\tau : S(t)B_0 \text{ is bounded in } V^{2\beta} \text{ for all } t \text{ with } 0 \le t \le \tau\}.$$

We have seen here that $T \ge \tau$, where τ is given by (47.6). We then have the following results concerning global attractors for the nonlinear equation (48.1).

Theorem 48.2. *Let the Standing Hypothesis A be satisfied and assume that A has compact resolvent. Assume that the nonlinearity satisfies $F \in C_{Lip}(V^{2\beta}, W)$, and define $S(t)u_0 = \sigma(u_0, t) \stackrel{\text{def}}{=} \phi(u_0, F, t)$, for $u_0 \in V^{2\beta}$. Assume that the maximally defined solutions $\phi(u_0, F, t)$ are defined for all $t \ge 0$, and that there is a $\rho > 0$ such that*

$$\limsup_{t\to\infty} \|A^\beta S(t)u_0\| \le \rho, \qquad \text{for all } u_0 \in V^{2\beta}. \tag{48.2}$$

Then the semiflow $S(t)$ has a global attractor $\mathfrak{A}$ in $V^{2\beta}$, and $\mathfrak{A}$ attracts all bounded sets in $V^{2\beta}$.

The argument is simple. Since inequality (48.2) establishes that the semiflow $S(t)$ is point dissipative in $V^{2\beta}$, the conclusion follows from Lemma 48.1 and the Existence Theorem 23.12. The proofs of the following two results are left as exercises.

Lemma 48.3. *Let the Standing Hypothesis A be satisfied and assume that A has compact resolvent. Let $B = B(t)$ satisfy $B \in \mathcal{M}^\infty$, for some β with $0 \le \beta < 1$. Then, for each $t > 0$, the solution operator $\Phi(B,t)$ for equation (44.1) maps bounded sets in V^{2r} into compact sets in V^{2r}, for each r with $0 \le r < 1$.*

Theorem 48.4. *Let A be a sectorial operator, where the analytic semigroup e^{-At} is κ-contracting. For $\alpha \ge 0$ define $V^{2\alpha}$ by $V^{2\alpha} = \mathcal{D}((A+aI)^\alpha)$, where $a \in \mathbb{R}$ is chosen so that $A + aI$ is positive. Assume that $F \in C_{\text{Lip}}(V^{2\beta}, W)$, for some β with $0 \le \beta < 1$, and let $S(t)u_0 = \phi(u_0, F, t)$ denote the maximally defined mild solution of equation (48.1) in $V^{2\beta}$, for $t \ge 0$. Then the following are valid:*

(1) *The operator $S(t)$ is κ-contracting on $V^{2\beta}$.*
(2) *If in addition, there is an absorbing set for equation (48.1) in $V^{2\beta}$, and for every compact set $K \subset V^{2\beta}$, there is a time $\tau \ge 0$, such that $\gamma^+(S(\tau)K)$ is bounded in $V^{2\beta}$, then the semiflow $S(t)$ has a global attractor $\mathfrak{A}$ in $V^{2\beta}$.*

4.8.2 Regularity in Space and Time. Lemma 48.1 and Theorem 48.2 show that in general the solution $\phi(u_0, F, t)$ has greater spatial regularity than the initial condition u_0. These two results are the nonlinear analogues of Theorem 42.10 for the linear inhomogeneous equation (42.1). By using a theory of weak solutions, with the Standing Hypothesis B, as in Section 4.2.3, one can sometimes obtain weak solutions with higher spatial regularity. However, as noted in Theorem 42.12, these weak solutions need not be mild solutions of the problem in the spaces of higher regularity.[19]

There is an analogue of Theorem 42.14 in the nonlinear setting, when the nonlinearity $F = F(u,t)$ has greater temporal regularity. As we now show, the mild solutions have additional temporal and spatial regularity.

Theorem 48.5. *Let the Standing Hypothesis B hold on a Hilbert space $W = H$ and let $F = F(u,t)$ satisfy*

$$F \in C_{\text{Lip}}(V^{2\beta} \times \mathbb{R}^+, H) \cap C^1_F(V^{2\beta} \times \mathbb{R}^+, H), \qquad \text{with } 0 \le \beta < 1.$$

Assume further that there is a p with $2 \le p \le \infty$ such that if $w : [0,T) \to V^{2\beta}$ is continuous and strongly differentiable for $0 < t < T$, then $f(t) = F(w(t),t)$ satisfies

$$f \in C[0,T; H) \cap W^{1,p}_{\text{loc}}[0,T; V^{-1}).$$

[19]See Chapter 6 on the weak solutions of the Navier-Stokes equations, for example.

Let $\partial_t f = \partial_u F\, \partial_t u + \partial_t F$, where $\partial_u F$ and $\partial_t F$ are the (partial) Fréchet derivatives of F. Let $u_0 \in \mathcal{D}(A) = V^2$ and define $v_0 = F(u_0, 0) - Au_0 \in H$. Let $u = u(t) = \phi(u_0, F, t)$ denote the maximally defined mild solution of equation (47.1) in $V^{2\beta}$ on the interval $[0, T)$, where $0 < T \le \infty$. Then u satisfies the following properties:

(1) *The function u is a strong solution of equation (47.1) in $V^{2\beta}$. Moreover, for each r with $0 \le r < 1$, there is a $\theta = \theta(r) > 0$ such that one has*

$$u \in C^1[0, T; H) \cap C[0, T; V^2) \cap C^{0,\theta}_{\mathrm{loc}}[0, T; V^{2r}).$$

(2) *Set $g = \partial_t f = \partial_u F\, \partial_t u + \partial_t F$. Then, for each $\alpha < 0$, the function $v \stackrel{\mathrm{def}}{=} \partial_t u$ satisfies*

$$v(t) = e^{-At} v_0 + \int_0^t e^{-A(t-r)} g(r)\, dr, \qquad \text{for } t \ge 0.$$

Also v is a mild solution of $\partial_t v + Av = g(t)$ in V^α, and it is a weak solution in H with

$$v \in C[0, T; H) \cap L^2_{\mathrm{loc}}[0, T; V^1) \cap C^{0,\theta_1}_{\mathrm{loc}}[0, T; V^{2\alpha}),$$

where $\theta_1 = \theta_1(\alpha) > 0$.

(3) *If instead, one has $u_0 \in V^{2\beta}$, then $u = u(t) = \phi(u_0, F, t)$ satisfies*

$$u \in C^1(0, T; V^{2\beta}) \cap C(0, T; V^{2(1+\beta)}) \cap C^{0,\theta_2}_{\mathrm{loc}}(0, T; V^{2r}), \tag{48.3}$$

where $\theta_2 = \theta_2(r) > 0$, for $0 \le r < 1 + \beta$. Furthermore, for each such r, u is a strong solution of (47.1) in V^{2r} on $[t_1, T)$, for any t_1 with $0 < t_1 < T$.

Proof. This is a direct application of Theorem 42.14 and Theorem 42.15 with $V^\alpha = H$ and $f(t) = F(\phi(u_0, F, t), t)$. □

4.9. The Linearized Equation.

In this section we address an issue which arises when one has a given solution $u = u(t)$ of the nonlinear evolutionary equation

$$\partial_t u + Au = F(u), \qquad u(0) = u_0, \tag{49.1}$$

and one seeks to linearize this equation along this solution $u(t)$. For this purpose we assume that the Standing Hypothesis A is satisfied and that the nonlinearity F satisfies

$$F \in C^1_{\mathrm{Lip}} \stackrel{\mathrm{def}}{=} C_{\mathrm{Lip}}(V^{2\beta}, W) \cap C^1_F(V^{2\beta}, W), \qquad \text{where } 0 \le \beta < 1.$$

Recall that the derivative DF is said to satisfy the **Hölder property** if DF is Hölder continuous on every compact set $\mathcal{K}$ in $V^{2\beta}$.

Let $\mathcal{K}$ be a given compact, invariant set in $V^{2\beta}$ for the mild solutions of equation (49.1). For each $u_0 \in \mathcal{K}$, we let $S(t)u_0$, for $t \in \mathbb{R}$, denote the global solution in $\mathcal{K}$ passing through u_0, and define $B = B(t)$ by $B(t) = DF(S(t)u_0)$, for $t \in \mathbb{R}$. Since the derivative DF is continuous, it is bounded on $\mathcal{K}$, and consequently one has

$$B(\cdot) \in \mathcal{M}^\infty = L^\infty(\mathbb{R}, \mathcal{L}(V^{2\beta}, W)) \cap C(\mathbb{R}, \mathcal{L}(V^{2\beta}, W)).$$

As a result, the theory of Section 4.4 is immediately applicable to

$$\partial_t v + Av = DF(u(t))v, \tag{49.2}$$

which is referred to as the **linearized equation** associated with (49.1).

However, before turning to the linearized equation, *per se*, we should make note of additional information which can be brought to bear on the problem. Since $F \in C_{\text{Lip}}(V^{2\beta}, W)$, it follows from the Herculean Theorem 47.6 that $\mathcal{K} \subset \mathcal{D}(A)$; the mild solution $u(t) = S(t)u_0$ is a strong solution of equation (49.1); and (47.14) is valid. Furthermore, if the derivative DF satisfies the Hölder property, then $B(t) = DF(u(t))$ is also Hölder continuous in t since one has

$$\begin{aligned} \|DF(u(t_1)) - DF(u(t_2))\|_{\mathcal{L}} &\leq K_1 \|A^\beta(u(t_1) - u(t_2)\|^\theta \\ &\leq K_2 |t_1 - t_2|^{\theta(1-\beta)}, \end{aligned} \tag{49.3}$$

for some θ with $0 < \theta \leq 1$. The following theorem is now an immediate consequence of these remarks and Theorems 44.1.

Theorem 49.1. *Let the Standing Hypothesis A be satisfied and assume that the nonlinearity $F = F(u)$ is in $C^1_{\text{Lip}}(V^{2\beta}, W)$. Define $S(t)u_0 = \sigma(u_0, t) \stackrel{\text{def}}{=} \phi(u_0, F, t)$, and assume that the maximally defined mild solutions $S(t)u_0$ are defined for all $t \geq 0$ and that there is a $\rho > 0$ such that*

$$\limsup_{t \to \infty} \|A^\beta S(t)u_0\| \leq \rho, \qquad \textit{for all } u_0 \in V^{2\beta}.$$

For each $u_0 \in \mathcal{K}$, where $\mathcal{K}$ is a given compact, invariant set for equation (49.1), we define $\Phi(u_0, t) = \Phi(B, t)$, where $B(t) = DF(S(t)u_0)$, to be the solution operator generated by equation (49.2). Then

$$\pi(v_0, u_0, t) \stackrel{\text{def}}{=} (\Phi(u_0, t)v_0, S(t)u_0) \tag{49.4}$$

is a linear skew product semiflow on $V^{2\beta} \times \mathcal{K}$. Furthermore, if in addition, the derivative DF has the Hölder property, then for $u_0 \in \mathcal{K}$ and $v_0 \in V^{2\beta}$, *the solutions $S(t)u_0$ and $\Phi(u_0, t)v_0$ are strong solutions of equations (49.1) and (49.2), respectively.*

There are some situations in the study of the dynamics of evolutionary equations where one requires the greater uniformity afforded by the Fréchet derivative instead of the Gâteaux derivative. However, it is noteworthy that the construction of a linear skew product semiflow requires only the Gâteaux derivative. The major role to be played by the Fréchet derivative is in the theater of nonlinear equations. The uniform error estimates occuring with the Fréchet derivative are used in an essential way in the study of the dynamics of nonlinear evolutionary equations. We will use this feature, for example, in the study of the saddle property, the center manifold, and the inertial manifold in Chapters 7 and 8.

There is one useful situation where the Gâteaux derivative does afford some interesting information. Let $u_i = u_i(t)$, for $i = 1, 2$, be two solutions of equation (49.1), and set $w = u_1 - u_2$. Then w is a solution of the equation

$$\partial_t w + Aw = F(u_1) - F(u_2).$$

If F is continuously Gâteaux differentiable, i.e., $F \in C^1_G(V^{2\beta}, W)$, then one can write

$$F(u_1) - F(u_2) = \int_0^1 DF(u_2 + \theta(u_1 - u_2))\, d\theta\, (u_1 - u_2), \tag{49.5}$$

where now D stands for the Gâteaux derivative. As a result we see that w is a solution of the linear equation

$$\partial_t w + Aw = B(t)w,$$

where $B(t) = \int_0^1 DF(u_2(t) + \theta(u_1(t) - u_2(t))\, d\theta$. Since $F \in C^1_G(V^{2\beta}, W)$, one has $B(\cdot) \in C[0,T; \mathcal{L}(V^{2\beta}, W))$, where $[0, T)$ is the intersection of the two intervals of existence for the solutions u_1 and u_2. Hence, one can get some qualitative information about w in this way.

4.9.1. Differentiability of Mild Solutions. In addition to the Standing Hypothesis A, let us now assume that

$$F \in C_{\mathrm{Lip}}(V^{2\beta}, W) \cap C^1_F(V^{2\beta}, W). \tag{49.6}$$

We let $S(v, t)$ denote the maximally defined mild solution of

$$S(v,t) = e^{-At}v + \int_0^t e^{-A(t-s)} F(S(v,s))\, ds, \qquad \text{for } v \in V^{2\beta}, \tag{49.7}$$

as constructed in Section 4.7. It follows from Lemma 47.2 that $S(v, t)$ is a strong solution in W of equation (49.1), for every $v \in V^{2\beta}$, and it satisfies (47.7). For such a solution we set $B(v)(t) = DF(S(v,t))$ and let $\Phi(v,t) = \Phi(B(v),t)$ be the solution operator generated by $B(v)$, see Section 4.4. Note that one has

$$B(v)(\cdot) \in L^{\infty}_{\text{loc}}[0,T;\mathcal{L}(V^{2\beta},W)) \cap C[0,T;\mathcal{L}(V^{2\beta},W)),$$

for some $T = T(v)$ with $0 < T \leq \infty$. If in addition, the Fréchet derivative DF has the Hölder property, then $B(v)(\cdot)$ satisfies (49.3), for t_1, $t_2 \in [0,T)$, and $\Phi(B(v),t)$ is a strong solution of (44.1) for $0 \leq t < T$, by Theorem 44.4. For each $w \in W$, we define the derivative DS by

$$DS(v_0,t)w = \frac{\partial}{\partial v} S(v,t)w \,|_{v=v_0}, \qquad \text{for } w,\ v_0 \in V^{2\beta},$$

whenever the limit for the derivative exists in W; i.e., $DS(v_0,t)$ is a strong derivative. We now have the following result.

Theorem 49.2. *Let the Standing Hypothesis A be satisfied and assume that the nonlinear term $F = F(u)$ satisfies (49.6) and that the Fréchet derivative DF has the Hölder property. Then the following properties are valid:*

(1) *For each $t \geq 0$ and each $v \in V^{2\beta}$, the mild solution $S(v,t)$ is Fréchet differentiable in v and the derivative $DS(v,t)$ satisfies*

$$DS(v,t)w = \Phi(B(v),t)w, \qquad \textit{for all } w \in V^{2\beta},$$

in the sense that

$$DS(v,t)w = e^{-At}w + \int_0^t e^{-A(t-s)} DF(S(v,s))\, DS(v,s)w\, ds, \tag{49.8}$$

for v, $w \in V^{2\beta}$.

(2) *Let $S(v,t)$ be a solution of equation (49.1). Then one has*

$$\Phi(B(v),t)G(v) = DS(v,t)G(v) = G(S(v,t)), \qquad \textit{for } t \geq 0, \tag{49.9}$$

where $G(v) = -Av + F(v)$, for $v \in \mathcal{D}(A)$.

Comments on the Proof. In order to prove that $S = S(v,t)$ is differentiable in v, one rederives the existence of S by using a pair of successive approximations given by

$$S_{n+1}(v,t) = e^{-At}v + \int_0^t e^{-A(t-s)} F(S_n(v,s))\, ds$$

$$\Psi_{n+1}(v,t)w = e^{-At}w + \int_0^t e^{-A(t-s)} DF(S_n(v,s))\Psi_n(v,s)w\, ds,$$

where $\Psi_n(v,t) \in \mathcal{L}(V^{2\beta}, V^{2\beta})$, for each n and (v,t). One shows that, for t small, this pair of sequences has a limit (S, Ψ), see Section 4.7. Also one shows that if S_n is differentiable with $\Psi_n(v,t)w = DS_n(v,t)w$, then S_{n+1} is differentiable and one has $DS_{n+1}(v,t)w = \Psi_{n+1}(v,t)w$. Finally one shows that the limiting equations

$$S(v,t) = e^{-At}v + \int_0^t e^{-A(t-s)}F(S(v,s))\,ds,$$

$$\Psi(v,t)w = e^{-At}w + \int_0^t e^{-A(t-s)}DF(S(v,s))\Psi(v,s)w\,ds,$$

hold, and that the identity $DS_{n+1}(v,t)w = \Psi_{n+1}(v,t)w$ holds in the limit. Hence S is differentiable. The fact that S is Fréchet differentiable and that equation (49.8) holds follows from a direct calculation based on the Chain Rule. Also, the error term arising from the Fréchet differentiability of F allows one to calculate and estimate the error term for S.

For Item (2) one uses the Euler method for integration to show that

$$DS(v,t)G(v) = \lim_{h\to 0}\frac{1}{h}[S(v+hG(v),t) - S(v,t)] = \partial_t S(v,t) - G(S(v,t)).$$

We leave the verification of the first equality in (49.9) as an exercise.

4.9.2. The Linear Skew Product Semiflow, Revisited. Once again we assume that the Standing Hypothesis A is satisfied. Our objective here is to study how the linear skew product semiflow π, see equation (49.4), depends on the nonlinear term $F \in C^1_{\mathrm{Lip}} = C^1_{\mathrm{Lip}}(V^{2\beta}, W)$, where $0 \le \beta < 1$. For this purpose, we will require some additional notation. For $G \in C^1_{\mathrm{Lip}}$ and $u_0 \in V^{2\beta}$, we let $S^G(t)u_0$ denote the maximally defined mild solution of

$$\partial_t u + Au = G(u), \qquad u(0) = u_0, \tag{49.10}$$

and we let $I = [0, \omega(u_0, G))$ denote the interval of definition of this solution (see Section 4.7). Next we define

$$B^{(u_0,G)}(t) \stackrel{\text{def}}{=} DG(S^G(t)u_0), \qquad \text{for } 0 < t < \omega(u_0, G),$$

and we consider the linear problem

$$\partial_t v + Av = B^{(u_0,G)}(t)\,v, \quad v(0) = v_0 \qquad \text{for } 0 \le t < \omega(u_0, G), \tag{49.11}$$

where $v_0 \in V^{2\beta}$. Let $v(t) = \Phi^G(u_0,t)v_0$ denote the mild solution of (49.11), for $0 \le t < \omega(u_0, G)$ (see Section 4.4). Next we define a mapping π^G formally by the equation

$$\pi^G(v_0, u_0, t) \stackrel{\text{def}}{=} \left(\Phi^G(u_0,t)v_0, S^G(t)u_0\right), \qquad \text{for } 0 \le t < \omega(u_0, G). \tag{49.12}$$

Notice that the range of the mapping π^G is in $V^{2\beta} \times V^{2\beta}$, while the domain is

$$\mathcal{D}(\pi^G) = V^{2\beta} \times \Xi = \left\{(v_0, u_0, G, t) : v_0 \in V^{2\beta} \text{ and } (u_0, G, t) \in \Xi\right\},$$

where Ξ is given by equation (47.9), with F replaced by G. The proof of the following result is left as an exercise.

Theorem 49.3. *Let the Standing Hypothesis A be satisfied and let π^G be given by (49.12). Assume that C^1_{Lip} has the topology $\mathcal{T}^1_A$. Then $\pi^G : V^{2\beta} \times \Xi \to V^{2\beta} \times V^{2\beta}$ is continuous.*

4.10. Exercises.

Section 4.2

42.1. Let $(e^{-At}, -A)$ be a C_0-semigroup on a Banach space W, and let $f \in C[0,T;W)$. Show that a mild solution u of (42.1) is a classical solution if and only if both of the following two conditions are satisfied:

(1) $u(t) \in \mathcal{D}(A)$, for all $t \in (0,T)$, and
(2) $Au \in L^1_{\mathrm{loc}}[0,T;W)$.

42.2. Let A be a bounded linear operator on W and let $f \in C[0,T;W)$. Show that a mild solution of (42.1) is a classical solution.

42.3. Let $(e^{-At}, -A)$ be a C_0-semigroup on the Banach space W. Show that if $u_0 \in \mathcal{D}(A)$ and $f \in C^1[0,\infty;W)$, then the mild solution (42.3) is a classical solution to (42.1).

42.4. Show that a C_0-semigroup $(e^{-At}, -A)$ is differentiable for $t > 0$ if and only if for every $f \in C[0,\infty;W)$ the Initial Value Problem $\partial_t u + Au = f(t)$, $u(0) = 0$, has a classical solution.

42.5. Show that Theorem 42.9 admits the following extension, where W is a Hilbert space, and the Standing Hypothesis B is satisfied. Assume that

$$A^{-\epsilon} f \in L^1_{\mathrm{loc}}[0,T;W) \cap C^{0,\theta}_{\mathrm{loc}}(0,T;W),$$

where $0 \le \epsilon < 1$, $0 < \theta \le 1$, and $0 < T \le \infty$. For any $u_0 \in W$, define $u = u(t)$ by (42.3). Then u is a mild solution of (42.1) in the space $V^{2\gamma}$, for each $\gamma \le 0$. Also u satisfies $u \in C[0,T;V^{2\gamma})$, for each $\gamma \le 0$, and

$$u \in C^{0,\theta}_{\mathrm{loc}}(0,T;V^{2\gamma}), \qquad \text{for each } \gamma \le -\epsilon.$$

In addition, for each $\gamma \le -\epsilon$, the Leibniz Formula (42.7) is valid in $V^{2\gamma}$, for all $t > 0$, and u is a strong solution in $V^{2\gamma}$.

42.6. Assume that in addition to the hypotheses of Theorem 42.10, one has $u_0 \in V^{2r}$, for some r with $0 \le r \le 1 + \beta$; then the solution $u = u(t)$ of (42.1) satisfies $u \in C^{0,\theta}_{\mathrm{loc}}[0,T;V^{2r})$, for an appropriate $\theta = \theta(r)$ with $0 < \theta \le 1$.

42.7. Let A satisfy the Standing Hypothesis B. Let $f = f_1 + f_2$, where

$$f_1 \in W^{1,2}_{\mathrm{loc}}[0,T;H) \qquad \text{and} \qquad f_2 \in L^1_{\mathrm{loc}}[0,T;H) \cap L^p_{\mathrm{loc}}(0,T;V^{2\alpha})$$

for some α with $0 < \alpha \le 1$ and p with $1 < \alpha p \le \infty$. Show that for any $u_0 \in H$, the mild solution $u = u(t)$ of equations (42.1) - (42.2) satisfies (42.43) and u is a strong solution of (42.1) on $[t_1, T)$, for any $t_1 > 0$.

42.8. The following exercise is to generalize the conclusions of Theorem 42.14 to a C_0-semigroup $(e^{-At}, -A)$ on a Hilbert space H. Let $f \in W^{1,1}_{\text{loc}}[0,T;H)$, and let $g = \partial_t f$ denote the time derivative of f. Let $u_0 \in \mathcal{D}(A)$ and define $u = u(t)$ by equation (42.3). Show that if $v = \partial_t u$, and $v_0 = f(0) - Au_0$, then (1) the function u is a strong solution of (42.1) with $u \in C^1[0,T;H) \cap C[0,T;\mathcal{D}(A))$, and (2) the function v satisfies $v(t) = e^{-At}v_0 + \int_0^t e^{-A(t-r)}g(r)\,dr$, for all $t \geq 0$.

42.9. (1) Derive sharp upper bounds for the Hölder exponents appearing in Theorems 42.10, 42.12, 42.14, and in Theorem 42.15. (2) Do the same for (42.14) in Lemma 42.7.

42.10. In some situations, one may encounter functions f that are bounded in $[\epsilon, T]$, for every ϵ with $0 < \epsilon < T$, and that satisfy $f(t) \to \infty$, as $t \to 0^+$. To deal with such functions, one can consider certain **weighted spaces**. An example is the following: Let X denote a real Banach space, $[a,b]$ a bounded interval, and $\mu \in \mathbb{R}$. Define

$$B_\mu(a,b;X) = \{f : (a,b] \to X : \|f\|_{B_\mu} \stackrel{\text{def}}{=} \sup_{a<t\leq b} (t-a)^\mu \|f(t)\| < \infty\}$$

and

$$C^{0,\alpha}_\mu(a,b;X) = C^{0,\alpha}_{\text{loc}}((a,b];X) \cap B_\mu(a,b;X).$$

Assume that $f \in L^\infty_{\text{loc}}((0,T];W) \cap B_\theta(0,T;W)$, for some θ with $0 \leq \theta \leq 1$, and let v be given by equation (42.8). Show that $v \in C^{0,1-\theta}([0,T];W)$ and that there is a constant $C > 0$ such that $\|v\|_{C^{0,1-\theta}} \leq C\|f\|_{B_\theta}$, for all f.

42.11. Prove Theorems 42.16 and 42.17. (Hint: Show that the arguments used for Theorems 42.14 and 42.15 can be adapted to this situation.)

Section 4.4

44.1. (1) Show that the functions E_1, $E_2, \cdots$ which have the EX property and which arise in Theorem 44.1 and Corollary 44.2, satisfy $E_i(t) \to 1$, as $B \to 0$ in the Fréchet spaces $(L^\infty, \mathcal{T}_A)$, or $(L^\infty, \mathcal{T}_{bu})$, see (44.7).

(2) Show that one has

$$E_i \to 1 \text{ in } L^\infty_{\text{loc}}[0,\infty;\mathbb{R}), \qquad \text{as } B \to 0 \text{ in } L^\infty_{\text{loc}}[0,\infty;\mathcal{L}(V^{2\beta},W)),$$

where L^∞_{loc} refers to the Fréchet space topologies (see Appendix B).

44.2. Show that the term μ appearing in inequalities (44.15)-(44.16) can be replaced by $\mu = \mu(t)$, where $\mu(t)^{1-\beta} = M_\beta \|B\|_{L^\infty(0,t;\mathbb{R})}\Gamma(1-\beta)$.

44.3. In addition to the hypotheses of Theorem 44.1, assume that

$$B(\cdot) \in \mathcal{M}^\infty \cap C^{0,\theta}_{\text{loc}}(\mathbb{R}, \mathcal{L}(V^{2\beta}, W)),$$

for some θ with $0 < \theta \leq 1$. Show that (44.14) can be strengthened to read

$$\Phi(B,\cdot)v_0 \in C^{0,1-\alpha}_{\text{loc}}[0,\infty;V^{2\alpha}) \cap C^{0,\theta}_{\text{loc}}(0,\infty;V^{2\gamma}),$$

for all α and γ with $0 \le \alpha \le \beta < \gamma < \beta + 1$, where $\theta = \theta(\gamma) > 0$.

44.4. Let the Standing Hypotheses A be satisfied, and let

$$B(\cdot) \in L^p_{\text{loc}} = L^p_{\text{loc}}[0, \infty; \mathcal{L}(V^{2\beta}, W)), \qquad \text{for some } 0 \le \beta < 1,$$

where p satisfies (44.32) and $p < \infty$.

(1) Show that there exists a unique solution $w(t) = \Phi(B, t)w_0$ in $V^{2\beta}$ of the equation (44.1) that satisfies $w(0) = w_0$, for each $w_0 \in V^{2\beta}$.
(2) Show that the cocycle identity (44.9) holds and that the mapping $\pi(v, B, \tau) = (\Phi(B, \tau)v, B_\tau)$ is a linear skew product semiflow on $V^{2\beta} \times L^p_{\text{loc}}$.
(3) Determine which of the conclusions of Theorem 44.1 extend to this case.

44.5. Use the methodology of Theorem 42.14 to derive a theory of solutions of equation (44.1) under the added assumption that

$$B(\cdot) \in C^\infty[0, T; \mathcal{L}(W, V^{-2\epsilon})), \qquad \text{where } \epsilon > 0.$$

Try to derive analogues of the Fundamental Theorem on Sectorial Operators in this setting.

44.6. Show that there are points $B \in \mathcal{M}^\infty$ with the property that the translation mapping $(B, \tau) \to B_\tau$ is <u>not</u> continuous when $\mathcal{M}^\infty$ has the Banach space sup-norm topology of uniform convergence on $\mathbb{R}$. (This shows the importance of the Fréchet space topology on $\mathcal{M}^\infty$, for our theory.)

44.7. Complete the proofs of Corollary 44.2 and Theorem 44.3.

44.8. Let A be a sectorial operator with compact resolvent on a Banach space W. Show that for each $B \in \mathcal{M}^p$, where p satisfies inequality (44.32), the mapping $\Phi(B, t)$ is uniformly κ-contracting. Is this mapping compact for $t > 0$?

44.9. Let the Standing Hypothesis A be satisfied, and let $\Phi(B, t)$ be the solution operator generated by $B \in \mathcal{M}^p$ for some β with $0 \le \beta < 1$, where p satisfies inequality (44.32). Show that the following hold:

(1) For each $w_0 \in W$, the mapping $(s, t) \to \Phi(B_s, t - s)w_0$ is a continuous mapping of (s, t) into W, for $s \le t$.
(2) For each $T \ge 0$ and each $w_0 \in W$, one has

$$\lim_{h \to 0+} \frac{1}{h} \int_t^{t+h} \Phi(B_s, T - s)w_0 \, ds = \Phi(B_t, T - t)w_0, \qquad \text{for } 0 \le t < T < \infty,$$

and

$$\lim_{h \to 0+} \frac{1}{h} \int_t^{t+h} \Phi(B_s, t + h - s)w_0 \, ds = w_0,$$

where the integral and the limits hold in W.

44.10. Under the hypotheses of Theorem 44.6, show that for every $w_0 \in W$, the function $w(t) = \Phi(B_s, t-s)w_0$, for $t > 0$ and $0 \le s < t$, is a strong solution of equation (44.5) in W, for $0 \le s < t < \infty$, and it satisfies $w(s) = w_0$, as well as

$$\partial_t \Phi(B_s, t-s)w_0 \overset{\text{a.e.}}{=} [-A + B(t)]\Phi(B_s, t-s)w_0, \qquad \text{for } 0 \le s < t < \infty.$$

44.11. Derive sharp upper bounds for the Hölder exponents in Theorem 44.1.

44.12. The objective of this exercise is to develop an analogue of the theory of Section 4.4 to the case where $-A$ is the infinitesimal generator of a C_0-semigroup, which we will denote as e^{-At}, on a Banach space W. Let us note at the outset that, for the most part, the proofs needed in the analogue theory can be obtained from the arguments in Section 4.4 by setting $\beta = 0$. (Indeed, with $\beta = 0$, one has $V^{2\beta} = W$, and inequality (37.11) reduces to Lemma 31.1.)

Let L^∞ be given by (44.6), where $\beta = 0$, $V^{2\beta} = W$, and $\mathcal{L} = \mathcal{L}(W)$. Let $\mathcal{M}^\infty$ be given by (44.31), with $V^{2\beta} = W$, and set $\mathcal{M}^p = L^\infty$, for $1 \le p < \infty$. We will use the topologies described in Section 4.4 on these spaces (see Theorem 44.3).

(1) Show that for each $B \in L^\infty$ and each $v_0 \in W$, there is a unique mild solution $v(t) = \Phi(B,t)v_0$ of (44.5), and it satisfies (44.13) and $\Phi(B,\cdot)v_0 \in C[0,\infty,W)$.
(2) Use the Gronwall inequality to derive the analogue of (44.8).
(3) Show that for each $t \ge 0$, the mapping $(B, v_0) \to \Phi(B,t)v_0$ is Lipschitz continuous in B and v_0, as a mapping of $L^\infty \times W$ into W, see (44.21) with $r = 0$.
(4) Let $B \in \mathcal{M}^p$, where $1 \le p \le \infty$. Show that the mapping $(B,\tau) \to B_\tau$ is a flow on $\mathcal{M}^p$.
(5) Let $B \in \mathcal{M}^p$, where $1 < p \le \infty$. Show that the function $\pi(v_0, B, \tau) = (\Phi(B,\tau)v_0, B_\tau)$ is a linear skew product semiflow on $W \times \mathcal{M}^p$.
(6) Let $B \in \mathcal{L}(W)$ and set $L = A - B$. Use the properties proved above to show that $-L$ is the infinitesimal generator of a C_0-semigroup on W and $\mathcal{D}(L) = \mathcal{D}(A)$.

44.13. The constant ν appearing in Theorem 44.1 and Corollary 44.2 depends on r and ρ. Find an estimate of this dependence for $\beta < r < 1$. What happens as $r \to \infty$ in Corollary 44.2?

44.14. In reference to Theorem 44.1, Item (6), assume that $v_0 \in V^{2\rho}$, with $\rho = 1$. Is the relation (44.20) valid in this case?

Section 4.5

45.1. Show that the projector P arising in the definition of an exponential dichotomy is uniquely determined.

45.2. Let $\pi = (\Phi, \sigma)$ be a linear skew product <u>flow</u> on $W \times M$. Thus $\sigma(m,t)$, $\Phi(m,t)$, and the inverse $\Phi^{-1}(m,t)$ are well defined for all $m \in M$

and all $t \in \mathbb{R}$. Let $m_0 \in M$. Show that π has an exponential dichotomy over the orbit $\hat{M} = \gamma(m_0)$ if and only if there is a bounded linear projection P_0 on W and constants $K \geq 1$ and $\alpha > 0$ such that

$$\|\Phi(m_0, t) P_0 \Phi^{-1}(m_0, s)\| \leq K e^{\alpha(t-s)}, \qquad \text{for all } t \leq s,$$

and

$$\|\Phi(m_0, t)(I - P_0)\Phi^{-1}(m_0, s)\| \leq K e^{-\alpha(t-s)}, \qquad \text{for all } s \leq t.$$

45.3. Prove Item (7) of Lemma 45.2.

45.4. Verify the strong cocycle identity (45.10).

45.5. Assume that π has an exponential dichotomy over an invariant set $\mathcal{J} \subset \mathcal{M}^\infty$. Show that π has an exponential dichotomy, with the same characteristics, over $\mathcal{K} = \mathrm{Cl}(\mathcal{J})$, where the closure is taken in the topology of uniform convergence on bounded sets in $\mathbb{R}$.

45.6. Assume that π admits an exponential dichotomy in $\mathcal{E}(\mathcal{K})$, where $\mathcal{K}$ is a compact, invariant set in $\mathcal{M}^\infty$ that satisfies

$$\mathcal{K} \subset C^{0,\theta_1}_{\mathrm{loc}}(\mathbb{R}; \mathcal{L}(V^{2\beta}, W)),$$

for some θ_1 with $0 < \theta_1 \leq 1$.

(1) Show that the projectors $P(B)$ and $P(B_\tau)$ satisfy

$$P(B_\tau) = \Phi(B, \tau) P(B) \left[\Phi(B, \tau)\right]^{-1}, \qquad \text{for } \tau \in \mathbb{R}.$$

(2) Show that $P(B_\tau)$ is locally Lipschitz continuous in τ, see Henry (1981, Lemma 7.6.2).

45.7. Consider the restriction of π to $\mathcal{E}(\mathcal{K})$, where $\mathcal{K}$ is some compact, invariant set in $\mathcal{M}^\infty$. Assume that, for some r with $0 < r < 1$, the linear skew product semiflow has an exponential dichotomy on $\mathcal{E}^{2r}(\mathcal{K})$ with characteristics $K \geq 1$ and $\alpha > 0$. Does it necessarily follow that π has an exponential dichotomy on $\mathcal{E}(\mathcal{K})$ with the same characteristic α? (Compare with Lemma 45.4.)

45.8. Determine whether the Lemma 45.5 is valid in the case that $\mathcal{K}$ is a closed, invariant set, but not necessarily compact. What happens if $\mathcal{K}$ is invariant, but not necessarily closed?

45.9. The object of this exercise is to show that the alternate Variation of Constants Formula (45.34) admits a partial extension to $t \leq 0$ in the presence of an exponential trichotomy, or an exponential dichotomy. Let $\mathcal{K}$ be a closed invariant set in $\mathcal{M}^p$, where p satisfies (44.32). Assume that the linear skew product semiflow π has an exponential trichotomy over $\mathcal{K}$, and let $\{P, Q, R\} = \{P^u, P^s, P^o\}$ denote the associated projectors over $\mathcal{K}$. Assume that $g = g(t) \in L^1_{\mathrm{loc}}(\mathbb{R}; W)$. Show that for each $(B, v_0) \in \mathcal{K} \times W$ and for $i = u, o$, the function

$$v(t) \stackrel{\text{def}}{=} \Phi(B, t) P^i(B) v_0 + \int_0^t \Phi(B_s, t - s) P^i(B_s) g(s)\, ds, \qquad \text{for } t \leq 0,$$

is a negative continuation through $P^i(B)v_0$ for the mild solution of the evolutionary equation

$$\partial_t v + Av = B(t)v + P^i(B_t)g(t), \qquad \text{for } t \le 0,$$

and it satisfies $P^i(B_t)v(t) = v(t)$, for all $t \le 0$.

45.10. Let $\mathcal{K}_0$ be a set in $\mathcal{W} = \ell_\infty(Z, \mathcal{L}(W))$ with the property that there is a $b > 0$ such that every $T \in \mathcal{K}_0$ has the Strong Boundedness Property with $\|G(T)\|_\infty \le b$. Define $\mathcal{K}$

$$\mathcal{K} = \text{Closure} \bigcup_{\tau \in Z} \{T \cdot \tau : T \in \mathcal{K}_0\},$$

and the closure is taken in the Fréchet metric topology on $\mathcal{W}$. Show that each $T \in \mathcal{K}$ has the Strong Boundedness Property and $\|G(T)\|_\infty \le b$.

45.11. Show that the set $\mathcal{K}_b$ given in Theorem 45.8 is a closed, invariant set in the Fréchet metric topology on $\mathcal{W} = \ell_\infty(Z, \mathcal{L}(W))$.

45.12. The process used in the proof of Theorem 45.8 to construct the operator sequence $T = \{T_n\}$ is an example of a linear evolutionary equation with impulses, where discrete jumps occur at specified times $t_n = n\tau$, for $n \in Z$. This can be reformulated without the impulses by changing the given evolutionary equation over a specific time intervals. In particular assume that, for a small $\sigma > 0$, one has

$$\partial_t u + Au = B(t)u, \qquad \text{for } t_{n-1} < t < t_n - \sigma, \tag{410.1}$$

where A satisfies the Standing Hypothesis A, $B \in \mathcal{M}^p$, and p satisfies (44.32). On the remaining intervals one has

$$\partial_t u = E_n u, \qquad \text{for } t_n - \sigma < t < t_n, \tag{410.2}$$

where $E_n \in \mathcal{L}(V^{2\beta}, V^{2\beta}) \mapsto \mathcal{L}(V^{2\beta}, W)$ is chosen so that $M = M(B_1, B_2)$ satisfies

$$M(\Phi(B_{t_{n-1}}, t_n - \sigma), B_{t_n}) = e^{E_n \sigma},$$

see (45.67). Problem (410.1) and (410.2) can be written in the more succinct form $\partial_t u = D(t)u$, for $t \in \mathbb{R}$, where $D(t) = -A(t) + B(t)$, or $D = (A, B)$, and

$$D \in \mathcal{F} \stackrel{\text{def}}{=} L^\infty(\mathbb{R}, \mathcal{L}(V^2, W)) \times L^\infty(\mathbb{R}, \mathcal{L}(V^{2\beta}, W)).$$

Note that $A(t) = A$, for $t_{n-1} < t < t_n - \sigma$, and $A(t) = 0$, for $t_n - \sigma < t < t_n$. Let

$$\|D(t)\|_{\mathcal{L}} \stackrel{\text{def}}{=} \|A(t)\|_{\mathcal{L}_1} + \|B(t)\|_{\mathcal{L}_2} \qquad \text{for } t \in \mathbb{R},$$

where $\mathcal{L}_1 = \mathcal{L}(V^2, W)$ and $\mathcal{L}_2 = \mathcal{L}(V^{2\beta}, W)$. Assume that $\mathcal{F}$ has the Fréchet metric topology generated by the pseudonorms $\|\cdot\|_{p:[0,\tau)}$, for $1 \le p < \infty$ and $0 < \tau < \infty$, where

$$\|D\|_{p:[0,\tau)} \stackrel{\text{def}}{=} \left(\int_0^\tau \|D(t)\|_{\mathcal{L}}^p \, dt \right)^{\frac{1}{p}}.$$

(1) Let $w_0 \in W$ and let $D = D(t)$ be given as above. Use the operator $\Phi(B,t)$ to construct a solution operator $\Phi(D,t)w_0$, for $t \geq 0$.

(2) Let D_τ be defined by $D_\tau(t) = D(\tau + t)$. for $\tau, t \in \mathbb{R}$. Show that the cocycle identity $\Phi(D_\tau, t)\Phi(D,\tau) = \Phi(D, \tau + t)$, for all $\tau \in \mathbb{R}$ and $t \geq 0$ is valid.

(3) Show that, for $1 \leq p < \infty$, the mapping $(D,\tau) \to D_\tau$ is a flow on $\mathfrak{F}$.

(4) Let π given by $\pi(v_0, D, \tau) = (\Phi(D,\tau)v_0, D_\tau)$, where $v_0 \in V^{2\beta}$ and $D = (A,B) \in \mathfrak{F}$. Show that for $1 < p(1-\beta) < \infty$, the mapping π is a linear skew product semiflow on $V^{2\beta} \times \mathfrak{F}$.

(5) Show that there exist continuous functions $a(\cdot)$ and $b(\cdot)$, which depend on p for $1 \leq p < \infty$, such that if $D_1 = (A_1, B_1)$ and $D_2 = (A_2, B_2)$ are two points in $\mathfrak{F}$, then one has

$$||A^\beta w(t)|| \leq a(\tau)||A^\beta(v_{10} - v_{20})|| + b(\tau)||D_1 - D_2||_{p;[0,\tau]}, \qquad \text{for } 0 \leq t \leq \tau,$$

where $w(t) = \Phi(D_1,t)v_{10} - \Phi(D_2,t)v_{20}$.

(6) Show that for the proper choice of $B \in \mathcal{M}^p$, one has $T_n = \Phi(D_n, \tau)$, for $n \in Z$.

45.13. Prove Lemma 45.9.

45.14. Derive sharp upper bounds for the Hölder exponents appearing in Theorem 45.6.

45.15. Extend the theory of the Hölder continuity in Theorem 45.6 to the case where p satisfies (44.32) with $p < \infty$.

45.16. (This is a continuation of Exercise 44.12.) Derive an analogue of Theorem 45.7, in the setting of Exercise 44.12, where $-A$ is the infinitesimal generator of a C_0-semigroup on a Banach space W.

45.17. The objective of this exercise is to examine the meaning of an exponential trichotomy in the setting of the partially coupled linear systems

$$\begin{aligned} \partial_t u &= Au + Bv \\ \partial_t v &= Zu + Dv \end{aligned} \tag{410.3}$$

and

$$\begin{aligned} \partial_t u &= Au + Bv \\ \partial_t v &= Zu + \epsilon^{-2}Ev \end{aligned} \tag{410.4}$$

on the product $W = U \times V$ of two Banach spaces U and V. In these equations $Z = 0$ is always the 0-operator, and ϵ will be a small positive parameter. In this exercise, which has several parts, we will make various assumptions about the spaces U and V and the operators A, B, D, and E. In each case, three operators A, D, and E are assumed to be infinitesimal generators of C_0-semigroups, which we will denote by e^{At}, e^{Dt}, and e^{Et}. We also assume the C_0-semigroup e^{Et} to be exponentially stable, i.e., there are constants $K_1 \geq 1$ and $\mu > 0$ such that

$$||e^{Et}||_{\mathcal{L}(V)} \leq K_1 e^{-\mu t}, \qquad \text{for all } t \geq 0.$$

(1) Let $\mathbb{A}$ and $\mathbb{A}_\epsilon$ be operators defined by

$$\mathbb{A} = \begin{pmatrix} A & B \\ 0 & D \end{pmatrix} \quad \text{and} \quad \mathbb{A}_\epsilon = \begin{pmatrix} A & B \\ 0 & \epsilon^{-2}E \end{pmatrix}.$$

Show that $\mathbb{A}$ and $\mathbb{A}_\epsilon$ are infinitesimal generators of C_0-semigroups $e^{\mathbb{A}t}$ and $e^{\mathbb{A}_\epsilon t}$ on W.

(2) Find formulae for the semigroups $e^{\mathbb{A}t}$ and $e^{\mathbb{A}_\epsilon t}$ in terms of the semigroups e^{At}, e^{Dt}, and e^{Et} and the operator B.

(3) Show that $e^{\mathbb{A}t}$ and $e^{\mathbb{A}_\epsilon t}$ are analytic semigroups if and only if e^{At}, e^{Dt} and e^{Et} are analytic semigroups. In the case of analytic semigroups:
   (3a) Show that $e^{\mathbb{A}t}$ and $e^{\mathbb{A}_\epsilon t}$ are compact for $t > 0$ if and only if e^{At}, e^{Dt} and e^{Et} are compact for $t > 0$.
   (3b) Show that, for ϵ small, $e^{\mathbb{A}t}$ and $e^{\mathbb{A}_\epsilon t}$ are κ-contracting if and only if e^{At} and e^{Dt} are κ-contracting. (Note that E is absent here.)

(4) Assume that W is finite dimensional. Show that both $e^{\mathbb{A}t}$ and $e^{\mathbb{A}_\epsilon t}$ have exponential trichotomies. Describe the projectors P, R, and Q on W in terms of the spectral properties of A, D, and E, where ϵ is small.

(5) Assume that $e^{\mathbb{A}t}$ and $e^{\mathbb{A}_\epsilon t}$ are analytic semigroups on W and that they are compact for $t > 0$. Show that the conclusions in the last Item are applicable in this case. Show that the dimensions of the neutral and unstable manifolds are finite in this setting.

(6) To what extent do the conclusions of Item (5) extend to the case where $e^{\mathbb{A}t}$ and $e^{\mathbb{A}_\epsilon t}$ are κ-contracting?

(7) To what extent do the conclusions of the last item extend to the case where $e^{\mathbb{A}t}$ and $e^{\mathbb{A}_\epsilon t}$ are simply C_0-semigroups?

Section 4.6

46.1. Complete the proof of Theorem 46.2.

46.2. Let $(e^{-At}, -A)$ be a C_0-semigroup on a Banach space W and let $F \in C_{\text{Lip}}(W \times \mathbb{R}^+, W)$ be such that there exist nonnegative constants C_0 and C_1 such that

$$||F(w,t)|| \le C_0 + C_1||w||, \qquad \text{for all } (w,t) \in W \times \mathbb{R}^+.$$

Show that $T(u_0, F) = \infty$, for all $u_0 \in W$.

46.3. Determine whether the two families of pseudonorms $||G||_{\{A;C^0(B)\}}$ and $||G||_{\{A;C^0(B\times[0,\tau])\}}$ generate the same topologies on $C_{\text{Lip}}(V^{2\beta}, W)$. What happens with the analogous pseudonorms on $C^1_{\text{Lip}}(V^{2\beta}, W)$?

Section 4.7

47.1. In terms of the notation of Lemma 47.1, where $\epsilon = 1 - \beta$,

(1) show that for $t \in [t_0, t_0 + \tau)$, where τ is given by (47.6), one has

$$\|(S(t)u_0 - S(t)v_0)\| \leq M_\beta t^{-\beta} E(\mu; t) e^{-at} \|u_0 - v_0\|,$$

for all $u_0, v_0 \in V^{2\beta}$ with $\|A^\beta u_0\| \leq b$ and $\|A^\beta v_0\| \leq b$; and

(2) show that one has

$$\|A^\beta(S(t)u_0 - S(t)v_0)\| \leq M E(\mu; t) e^{-at} \|u_0 - v_0\|,$$

for all $u_0, v_0 \in V^{2\beta}$ satisfying the conditions above.

Compare these inequalities with those in Theorem 44.1 on the linear problem.

47.2. Show that on each compact subinterval of the interval of definition $[0, T(u_0, F))$, the mild solution $\phi(u_0, F, t)$ is locally Lipschitz continuous in F and u_0, where one uses the metrizable topology of uniform convergence on bounded sets in C_{Lip} and the norm topology on $V^{2\beta}$. (Part of the problem here is the construction of a good metric for C_{Lip}.)

47.3. Prove Lemma 47.4.

47.4. In the proof of the Herculean Theorem 47.6, one obtains $u \in C^{0,1-r}_{\text{loc}}(\mathbb{R}, V^{2r})$, for $0 \leq r < 1$. Show that one has in this case $u \in C^{0,1-r}(\mathbb{R}, V^{2r})$, i.e., the Hölder coefficient can be chosen to be independent of $t \in \mathbb{R}$.

47.5. In the notation of the Herculean Theorem 47.6, show that the mild solution $S(t)u_0$ satisfies $S(\cdot)u_0 \in C^1(\mathbb{R}; W)$, for each $u_0 \in \mathcal{K}$.

47.6. In the context of the Herculean Theorem 47.6, explain why, without further assumptions, one cannot conclude that u is a mild solution in V^2, even though one has $u \in C(\mathbb{R}, V^2)$.

47.7. Prove Theorem 47.7.

47.8. Prove Theorem 47.8.

47.9. Prove Theorem 47.9.

47.9. Review the theory of Section 4.7 and determine which additional properties are valid when the Standing Hypothesis B holds instead of the Standing Hypothesis A.

Section 4.8

48.1. Let the Standing Hypothesis A be satisfied and let the nonlinearity $F = F(u)$ be in $C_{\text{Lip}}(V^{2\beta}, W)$, for some β with $0 \leq \beta < 1$. Let $u(t) = \phi(u_0, F, t)$ be the maximally defined, mild solution of (47.12) on the interval $I = [0, T)$, where $T = T(u_0, F)$ satisfies $0 < T \leq \infty$. Show that for every γ, with $0 \leq \gamma < 1$, the mapping $t \to \partial_t u(t)$ maps $(0, T)$ into $V^{2\gamma}$ and is locally Hölder continuous on $(0, T)$.

48.2. Prove Lemma 48.3.

48.3. Prove Theorem 48.4.

Section 4.9

49.1. Prove Theorem 49.2. In particular, derive a formula for the error term in the first order Taylor expansion of $S(v,t)$ in terms of the given function F.

49.2. Determine to what extent that Theorem 49.2 is valid when the assumption that $F \in C_F^1(V^{2\beta}, W)$ is replaced by $F \in C_G^1(V^{2\beta}, W)$.

49.3. Prove Theorem 49.3.

49.4. In the spirit of Exercise 44.12, derive the analogue of the theory of Section 4.9 to treat the case where $-A$ is the infinitesimal generator of a C_0-semigroup on a Banach space W.

4.11. Commentary.

Section 4.2. The reason we cannot conclude that the weak solution $u = u(t)$ given in Theorem 42.12 is a mild solution in the space V^α is due to the fact that that the integral $\int_0^t e^{-A(t-s)} f(s)\,ds$ need not exist in V^α for $f \in L^2_{\mathrm{loc}}[0,T;V^{\alpha-1})$. Indeed, inequality (37.11) implies that

$$\begin{aligned}\int_0^t \|e^{-A(t-s)} f(s)\|_\alpha\,ds &\le \int_0^t \|A^{\frac{1}{2}} e^{-A(t-s)}\|_\alpha \,\|A^{-\frac{1}{2}} f(s)\|_\alpha\,ds \\ &\le \int_0^t K(t-s)^{-\frac{1}{2}} \|A^{-\frac{1}{2}} f(s)\|_\alpha\,ds,\end{aligned}$$

and the later integral is generally not finite for $t > 0$, since $(t-s)^{-\frac{1}{2}} \notin L^2$.

In order to deal with some regularity questions related to solutions of parabolic partial differential equations with a sectorial operator A on a Banach space W and a variety of nonlinearities which behave badly, as $t \to 0^+$, one can use certain **weighted function spaces**, which also give rise to a family of interpolation spaces and a family of 'temporary" functions spaces. For example, in Lunardi (1995), the family of "intermediate" spaces, where $\alpha \ge 0$ and

$$\begin{aligned}D_A(\alpha, p) &\stackrel{\text{def}}{=} \{w \in W : \|t^{1-\alpha-\frac{1}{p}} A e^{-At} w\| \in L^p(0,1)\}, \\ D_A(\alpha) &\stackrel{\text{def}}{=} \{w \in D_A(\alpha,\infty) : \lim_{t\to 0^+} t^{1-\alpha} A e^{-At} w = 0\},\end{aligned}$$

for some p with $1 \le p \le \infty$, and the family of temporary function spaces

$$B_\mu(a,b;W) \stackrel{\text{def}}{=} \{f : (a,b] \to W : \|f\|_{B_\mu} \stackrel{\text{def}}{=} \sup_{a<t\le b} (t-a)^\mu \|f(t)\| < \infty\},$$

$$C_\mu^{0,\alpha}(a,b;W) \stackrel{\text{def}}{=} C_{\mathrm{loc}}^{0,\alpha}((a,b];W) \cap B_\mu(a,b;W), \text{ etc}$$

are used in the analysis of mild solutions of both linear and semilinaer equations.

Section 4.3. The use of skew product dynamics to study the behavior of solutions of linear ordinary differential equations with time-varying coefficients originated in Sacker and Sell (1974, 1976a,b), and Selgrade (1975). Also see Sell (1978), Coppel (1978), Sacker (1978), Johnson (1978a,b, 1979, 1980a, 1980b), Johnson and Moser (1982), and Johnson, Palmer, and Sell (1987). Several authors have contributed to the extension of this theory to infinite dimensional systems of linear evolutionary equations, see for example, Sacker and Sell (1994), Shen and Yi (1993a,b, 1994, 1996), and Yi (1996).

Section 4.5. The theory of exponential dichotomies, and its connections with nonlinear dynamics (see Chapter 7), goes back to the work of Lyapunov (1892) on the stable manifold theorem. This relationship was certainly known to, and used by, Poincaré (1890, 1892) in his study of transversal intersections of stable and unstable manifolds. The formulation used in this volume, with applications to finite dimensional systems of linear equations with time-varying coefficients, appears in Perron (1928, 1930a,b), Coppel (1965), and Sacker and Sell (1974, 1976a,b). Also see Hartman (1964); Hirsch, Pugh, and Shub (1977); and Pliss (1964, 1966, 1969, 1977). Several authors have contributed to the extension of this theory to infinite dimensional systems of linear evolutionary equations, see for example, Shen and Yi (1993a,b, 1994, 1996), Henry (1981), Massera and Schäfer (1966), Sacker and Sell (1994), and Yi (1996). Some of the theory presented in Section 4.5 is adapted from Henry (1981; Chap 7) and Pliss and Sell (1999). For example, Lemma 45.5 is an extension of a result of Henry (1981, p 226), who proves this result for $0 \leq r < 1$. The extension for all $r \geq 0$ is a consequence of Corollary 44.2. The characterization of an exponential dichotomy, as given in Theorem 45.8, appears in Henry (1981). However, the proof given here is from Pliss and Sell (1999). We feel that the newer proof offers better insight into the geometric properties of an exponential dichotomy. The two Robustness Theorems 45.11 and 45.12 are taken from Pliss and Sell (1999), as well. As noted in this reference, these theorems provide a solid foundation that various hyperbolic structures vary continuously in terms of the parameters of the problem. We will return to this issue again in Chapter 7.

One rather interesting application of exponential dichotomies is the connection with the Floquet theory for time periodic (or almost periodic) linear systems in both finite and infinite dimensions; see, for example, Johnson (1978a,b, 1979, 1980a,b); Johnson and Moser (1982); Johnson and Sell (1981); and Chow. Lu, and Mallet-Paret (1994, 1995).

In the case of nonautonomous linear ordinary differential equations there is a commonly used dichotomy concept, which seemingly applies to a single equation

$$\partial_t u = A(t)u, \qquad u \in \mathbb{R}^n, \tag{411.1}$$

where one has $A(\cdot) \in C(\mathbb{R}, \mathcal{L}(\mathbb{R}^n, \mathbb{R}^n) \cap L^\infty(\mathbb{R}, \mathcal{L}(\mathbb{R}^n, \mathbb{R}^n))$. Let $\Phi(t)$ denote the fundamental solution of (411.1), i.e., $\Phi(0) = I$, and $\Phi(t) \in \mathcal{L}(\mathbb{R}^n, \mathbb{R}^n)$ with $\partial_t \Phi(t) = A(t)\Phi(t)$, for all $t \in \mathbb{R}$. We say that equation (411.1) has an **exponential dichotomy** if there exists a linear projection P_0 on $\mathbb{R}^n$ and constants $K \geq 1$ and $\alpha > 0$ such that

$$\|\Phi(t)(I - P_0)\Phi^{-1}(s)u_0\| \leq K\|u_0\|e^{-\alpha(t-s)}, \qquad \text{for all } s \leq t,$$

and

$$\|\Phi(t)P_0\Phi^{-1}(s)u_0\| \leq K\|u_0\|e^{\alpha(t-s)}, \qquad \text{for all } s \geq t.$$

It should be noted that this concept of an exponential dichotomy depends heavily on the existence of the inverse $\Phi^{-1}(s)$, for $s \in \mathbb{R}$. An extension of this concept to a nonautonomous evolutionary equation on a Banach space is possible, but it would have limited applicability because of this requirement on the existence of inverses. (For example, it would not apply to equation (44.1) when W is infinite dimensional and A has compact resolvent.) In the concept of an exponential dichotomy used in this chapter, we are able to bypass this issue because we replace the requirement of the existence of inverses with the invariance property (45.2). While it appears that the definition of an exponential dichotomy used for equation (411.1) applies only to that equation, this is not the case. In fact, when equation (411.1) admits such an exponential dichotomy, then every translated equation $\partial_t u = A_\tau(t)u$ also has an exponential dichotomy, with the same choice of K and α. As a result, one then obtains an exponential dichotomy over the hull $H(A)$ with characteristics (K, α) (see Exercise 45.5). Also see Perron (1928, 1930a,b), Massera and Schäffer (1958, 1959a,b), and Coppel (1965, 1967, 1968, 1978, 1984).

The restriction of the concept of an exponential dichotomy for a linear skew product semiflow $\pi = (\Phi, \sigma)$ on $\mathcal{E} = W \times M$, to the case where σ is a flow, as opposed to a semiflow, on the base space M, is not a serious restriction. If instead, the set M is not invariant under σ, then one can replace M by

$$\hat{M} \stackrel{\text{def}}{=} \{\phi^m(t) : \phi^m \text{ is a global solution in } M \text{ through } m\},$$

where $\hat{M}$ has the topology of uniform convergence on bounded sets. One then uses the linear skew product semiflow $\hat{\pi}(w, \phi^m, \tau) = (\Phi(m, \tau), \phi^m_\tau)$ in the new space $\hat{\mathcal{E}} = W \times \hat{M}$ in place of π.

The concept of the dynamical spectrum arises naturally in the study of the autonomous linear equation $\partial_t u + Au = 0$, when A is a sectorial operator with compact resolvent on a Banach space W. (This includes, of course, all autonomous finite dimensional systems of ordinary differential equations.) In this setting, the base space $M = \{A\}$ reduces to a single point, and $\pi(v, A, t) = (e^{-At}v, A)$, for all $v \in W$ and $t \geq 0$. Also $\Sigma(\pi)$ consists of precisely those real numbers Re λ, where λ is an eigenvalue of A.

For a linear skew product flow on $\mathcal{E} = W \times M$, where the Banach space W is finite dimensional, say, $\dim W = n$, it is shown in Sacker and Sell (1978) that the dynamical spectrum $\Sigma(\pi)$ is the union of k compact, nonoverlapping intervals

$$\Sigma(\pi) = \bigcup_{i=1}^{k} [a_i, b_i],$$

where $1 \leq k \leq n$ and $a_k \leq b_k < \cdots < a_2 \leq b_2 < a_1 \leq b_1$. If λ_i, for $i = 0, 1, \cdots, k$, are chosen so that

$$\lambda_k < a_k \leq b_k < \lambda_{k-1} < \cdots < \lambda_2 < a_2 \leq b_2 < \lambda_1 < a_1 \leq b_1 < \lambda_0,$$

then for each $\lambda = \lambda_i$, the shifted linear skew product flow π_λ has an exponential dichotomy over M with projector P_λ. Furthermore, each of the sets

$$\mathcal{V}_i \stackrel{\text{def}}{=} \mathcal{S}_{\lambda_{i-1}} \cap \mathcal{U}_{\lambda_i}, \qquad \text{for } i = 1, \cdots, k,$$

is a subbundle of $\mathcal{E}$, and there is an integer n_i, with $1 \leq n_i \leq n$ and $n_1 + \cdots + n_k = n$, such that $\dim \mathcal{V}_i(m) = n_i$, for each $m \in M$. In addition these subbundles are disjoint, in the sense that

$$\mathcal{V}_i(m) \cap \left(\sum_{j \neq i} \mathcal{V}_j(m) \right) = \{0\}, \qquad \text{for all } i = 1, \cdots, k$$

and one has $\mathcal{V}_1(m) + \cdots + \mathcal{V}_k(m) = \mathcal{E}(m) = W \times \{m\}$, for all $m \in M$. These subbundles are called **spectral subbundles**, and the intervals $[a_i, b_i]$ are called **spectral intervals**.

To a limited extent, this spectral theory can be extended to a general linear skew product semiflow π, at least in the case where π is uniformly κ-contracting, see Chow and Leiva (1995), Magalhaes (1987), and Sacker and Sell (1994). In this setting, the dynamical spectrum $\Sigma(\pi)$ is nonempty, and it may consist of a countably many nonoverlapping closed intervals, each of finite length, or it may consist of a finite number of closed nonoverlapping intervals. If $\Sigma(\pi)$ consists of a finite number of closed nonoverlapping intervals and if W is infinite dimensional, then one of these intervals is unbounded, i.e., it is a ray $(-\infty, b_0]$, where $b_0 \in \mathbb{R}$.

There are important connections between the dynamical spectrum and the Lyapunov exponents, see[20] Johnson, Palmer, and Sell (1987). First of all, the largest Lyapunov exponent must lie in the rightmost spectral interval $[a_1, b_1]$. In an appropriate sense, the endpoints of the spectral intervals are all Lyapunov exponents, sometimes with differing ergodic measures on the base space M. In the special case where M is the hull of an almost

[20] In this paper, the authors address the finite dimensional problem only. However, many of these ideas extend to uniformly κ-contracting linear skew product semiflows on infinite dimensional spaces.

periodic motion, then there is a unique ergodic measure on M, and all of the endpoints are Lyapunov exponents with respect to this measure. Since there are at most n Lyapunov exponents (owing to the Millionščikov-Oseledec Theorem), see Millionščikov (1968) and Oseledec (1968), it follows that when the dynamical spectrum contains n intervals (i.e., the maximal numbers of intervals), in the case of almost periodic coefficients, all the spectral intervals reduce to points, see Johnson and Sell (1981).

As noted above, in the finite dimensional setting, the dynamical spectrum $\Sigma(\pi)$ is a compact set in $\mathbb{R}$, see Sacker and Sell (1978). For the infinite dimensional setting, one can show, in some cases, that $\Sigma(\pi)$ is a closed set in $\mathbb{R}$, see Magalhaes (1987). Since the characteristics $\Lambda = (\lambda_1, \lambda_2, \lambda_3, \lambda_4)$ for an exponential trichotomy must lie in the complement of $\Sigma(\pi)$, an open set in $\mathbb{R}$, it is clear that these characteristics are not uniquely determined. Nevertheless, it is possible to show that all the characteristics, including $K \geq 1$, have suitable semicontinuity properties, under small perturbations of the linear dynamics. For example, the Robustness Theorem 45.12 admits an extension whereby the charateristics $(K(B), \Lambda(B))$ can be chosen so that $(K(B), \Lambda(B)) \to (K(0), \Lambda(0))$, as $B \to 0$.

In Lemma 45.2 it is shown that if a linear skew product semiflow $\pi = (\Phi, \sigma)$ on $\mathcal{E}(M) = W \times M$ has an exponential dichotomy over an invariant set M, then the bounded set $\mathcal{B}(M)$ is trivial, i.e., one has $\mathcal{B}(m) = \{0\}$, for all $m \in M$. The converse question is considered in Sacker and Sell (1974, 1976a,b), for the finite dimensional case, and in Sacker and Sell (1994), in the Banach space setting. In both settings one has an Alternative Theorem, which reads roughly as follows: Assume that $\mathcal{B}(M)$ is trivial, that M is compact and invariant, and that Φ is uniformly κ-contracting on W. Then either π has an exponential dichotomy over M, or the flow σ has a specific gradient-like structure on M. Some applications of this Alternative Theorem are given in Sacker and Sell (1976a,b, 1994).

Let us now look at a few applications for the Robustness Theorem for Dichotomies. We will focus here on several descriptions of the set $\mathcal{K}$ in $\mathcal{M}^\infty$ and what it means for B_σ to satisfy (45.64). We let N_ϵ^∞ denote an ϵ-neighborhood in $\mathcal{M}^\infty$ and N_ϵ^p an ϵ-neighborhood in $\mathcal{M}^p$. Let $\mathcal{K} = H(D)$ be the hull in $\mathcal{M}^\infty$ of $D \in \mathcal{M}^\infty$ and let $B \in \mathcal{M}^\infty$ satisfy

$$||B(t) - D(t)|| \leq \epsilon, \qquad \text{for all } t \in \mathbb{R}.$$

It is easily verified that one then has $B_\sigma \in N_\epsilon^\infty(\mathcal{K})$. More generally, if $B \in \mathcal{M}^p$ satisfies

$$\int_t^{t+1} ||B - D||^p \, ds \leq \epsilon^p, \qquad \text{for all } t \in \mathbb{R},$$

then one has $B_\sigma \in N_\epsilon^p(\mathcal{K})$. In the infinite dimensional case, the latter version of the Robustness Theorem, with $\beta = 0$, can be found in Chow and Leiva (1995) and Leiva (1998). Also see Henry (1981). The finite dimensional version can be found in Coppel (1978), for example.

The next application arises in the context of linear equations with time-varying, quasi periodic coefficients. Let T^k denote the k-dimensional torus, and let $\theta = (\theta_1, \cdots, \theta_k) \in T^k$ and $\omega \in \mathbb{R}^k$. Let $\sigma(\omega; \theta, t) = \theta + \omega t$ denote a twist flow on T^k, where $\omega \in \mathbb{R}^k$ will be treated as a parameter. Let $E = E(\theta) : T^k \to \mathcal{L}(V^{2\beta}, W)$ be a continuous function. Let $B(t) = E(\sigma(\omega; \theta_0, t))$, for some $\theta_0 \in T^k$, and $D(\theta, t) = D(t) = E(\sigma(\omega_0; \theta, t))$, for $t \in \mathbb{R}$ and some $\omega_0 \in \mathbb{R}^k$. The mapping $J : \theta \to D(\theta, \cdot)$ is a continuous mapping of T^k into $\mathcal{M}^\infty$, and the image $\mathcal{K} = J(T^k)$ is a compact invariant set in the translational flow on $\mathcal{M}^\infty$. As a matter of fact, $\mathcal{K}$ is itself a torus with dimension satisfying $\dim \mathcal{K} \leq k$. Since E is a continuous mapping, it follows that for every $\epsilon > 0$ there is a $\delta > 0$ with the property that $B_\sigma \in N_\epsilon^\infty(\mathcal{K})$ whenever $\|\omega - \omega_0\|_{\mathbb{R}^k} \leq \delta$. On the other hand, if the only vectors n, m in Z^k, that satisfy $m\omega + n\omega_0 = 0$, are $m = n = 0$, then one has

$$\sup_{t \in \mathbb{R}} \|B(t) - D(t)\| = \text{Total Variation}(E) = \sup_{\theta_1, \theta_2 \in T^k} \|E(\theta_1) - E(\theta_2)\|,$$

which in general can be large. Other applications are given in Pliss and Sell (1999).

Sections 4.6 and 4.7. In the subsequent chapters in this volume, we will present many applications to nonlinear dynamics of the theory developed in the first four chapters. In some cases, such as the nonlinear wave equation, the dynamical theory builds on the C_0-theory developed in Section 4.6. In other cases, such as reaction diffusion equations or the Navier-Stokes equations, we use the analytic theory of Section 4.7. While the C_0-theory is also applicable to the study of differential delay equations and functional differential equations, we focus more on the theory of nonlinear partial differential equations in this volume. (Also see Section 7.7.) For a good introduction to the dynamics of functional differential equations, the reader should consult: Bellman and Cooke (1963); Hale (1977); Hale and Verduyn Lunel (1993); Mallet-Paret (1988); Mallet-Paret and Sell (1996a,b); Krisztin, Walther, and Wu (1999); and Krisztin and Walther (2000).

Additional Readings

Hale and Verduyn Lunel (1993), Henry (1981), Lunardi (1995), Pazy (1983), Pliss and Sell (1991, 1997, 1999, 2000), Sacker and Sell (1994), Shen and Yi (1996, 1998), Tanabe (1979), Temam (1982, 1988), and Yi (1996, 1998).

5 NONLINEAR PARTIAL DIFFERENTIAL EQUATIONS

In this chapter we turn our attention to the study of the dynamical properties of solutions of nonlinear partial differential equations. We are especially interested here in those nonlinear evolutionary equations which arise in the analysis of two broad classes of partial differential equations: parabolic evolutionary equations and hyperbolic evolutionary equations. While our usage of the terms "parabolic" and "hyperbolic" in this context is motivated by related concepts arising in the basic classification of partial differential equations, we will attribute these terms, instead, to certain dynamical features of the linear ancestry of the underlying nonlinear problems. More precisely, the linear prototypes of the partial differential equations of interest here include the heat equation and the wave equation:

$$\partial_t u - \nu \Delta u = 0 \quad \text{and} \quad \partial_t^2 u - \nu \Delta u = 0,$$

on a suitable domain Ω in $\mathbb{R}^d$, with various boundary conditions and initial conditions

The dynamical properties of these linear equations, and others, are developed in Section 3.8. For the heat equation, and its relatives, one obtains a linear evolutionary equation $\partial_t u + Au = 0$, where $-A$ is the infinitesimal generator of a C_0-semigroup e^{-At}, on a Banach space W. In many cases of interest, e^{-At} is an analytic semigroup. For the wave equation, and its kin, one obtains a first-order system $\partial_t w + Gw = 0$, where $w = (u, \partial_t u)$, and $-G$ is the infinitesimal generator of a C_0-group e^{-Gt}. In the case of parabolic evolutionary equations, the underlying partial differential equations include: reaction diffusion equations, equations of convection, the Navier-Stokes equations, the Kuramoto-Sivashinsky equation, and the Cahn-Hilliard equation. While hyperbolic evolutionary equations include: the nonlinear wave equation, the Petrovsky equation, the Korteweg-de Vries equation, and the nonlinear Schrödinger equation.

The basic paradigm we use in the study of each of these nonlinear problems is to derive an alternate, but equivalent, formulation of the problem as a nonlinear evolutionary equation

$$\partial_t u + Au = F(u), \tag{50.1}$$

on a Banach space, or a Hilbert space. We will assume that the associated linearized problem $\partial_t u + Au = 0$ gives rise to either a C_0-semigroup or a C_0-group e^{-At} with infintesimal generator $-A$. The basic idea is to use e^{-At} to find solutions of equation (50.1) satisfying appropriate initial conditions, as is described in Sections 4.6 and 4.7, for example. As noted in Chapter 4, there are several types of solutions: mild, weak, strong, and classical. Our first objective is to show that the Initial Value Problem for equation (50.1) is well-posed and that the solutions generate a nonlinear semiflow or flow. Of course, we are interested in the longtime dynamics of the semiflow or flow. For example, one of goals in this chapter is to show that, in many cases, the semiflow has a global attractor. Other very important dynamical issues will be treated in Chapters 7 and 8.

The following Standing Hypothesis, which is used in Chapter 4, will have an encore here in the dynamical theory of nonlinear partial differential equations.

Standing Hypothesis A. *Let A be a positive, sectorial operator on a Banach space W with associated analytic semigroup e^{-At}. Let $V^{2\alpha}$ be the family of interpolation spaces generated by the fractional powers of A, where $V^{2\alpha} = \mathcal{D}(A^\alpha)$, for $\alpha \geq 0$. Let $\|A^\alpha u\| = \|A^\alpha u\|_W = \|u\|_{V^{2\alpha}} = \|u\|_{2\alpha}$ denote the norm on $V^{2\alpha}$. See Lemma 37.4 for more information.*

In some applications, for example the Navier-Stokes equations, the ambient space W is a Hilbert space, which we will denote by H, and the following property is satisfied:

Standing Hypothesis B. *The operator A is a positive, selfadjoint, linear operator, with compact resolvent, on a Hilbert space H. Consequently A satisfies the Standing Hypothesis A. Moreover, the fractional power spaces V^α are defined for all $\alpha \in \mathbb{R}$, and equation (37.2) defines the Hilbert space structure on each V^α. Also the semigroup e^{-At} is compact, for $t > 0$. See Theorem 37.2 for more information.*

In Section 5.1, we examine the basic theory of reaction diffusion equations in space dimension $d \leq 3$. Here we will encounter the first examples of parabolic evolutionary equations. Section 5.2 is devoted to an investigation of various nonlinear wave equations, and it serves as an introduction to hyperbolic evolutionary equations. Equations of convection are the subject of Section 5.3. While this is a return to the area of parabolic evolutionary equations, it is of special interest because of a new nonlinear term involving spatial derivatives of the solutions. This feature gives rise to a new portfolio of analytic techniques. Finally in Section 5.4 (the Kuramoto-Sivashinsky equation) and in Section 5.5 (the Cahn-Hilliard equation), we encounter parabolic evolutionary equations, where the leading linear term is a higher order elliptic operator.

5.1. Reaction Diffusion Equations.

In this section we study the basic dynamical properties for solutions of reaction diffusion equations (**RDEs**). In general, this type of problem includes nonlinear heat-transfer equations (such as the Chafee-Infante equation), nerve equations (such as the Hodgkin-Huxley equations), ecology equations (such as predator-prey equations), and chemical reaction equations (such as Belousov-Zhabotinsky reactor). In a broad sense, a reaction diffusion equation can be viewed as any semilinear parabolic equation with a linear diffusion term and an nonlinearity which does not involve any spatial derivatives of the solution.

The role of the Laplacián Δ in the RDE is to model the dissipation in the underlying physical system. When the RDE is reformulated as a nonlinear evolutionary equation as in (50.1), the resulting linear operator A is oftentimes $A = -\nu\Delta$, with associated homogeneous boundary conditions, where ν is a positive constant, which measures the strength of the dissipation. Thus the interaction between the dissipation, within the physical region $\Omega \subset \mathbb{R}^d$, and the behavior at the boundary $\partial\Omega$ is contained in the operator A and its domain $\mathcal{D}(A)$. Furthermore, all this information is incorporated into the dynamical properties of the C_0-semigroup e^{-At}.

The nonlinear term $F = F(u)$ models all the remaining features of the underlying physical system, for example, the stabilizing and/or destabilizing properties of the medium; the role of any inhomogeneous boundary conditions, which are typical features of real-world problems; and the very important role of the dimension $m = \dim \Omega$ of the physical region. The **complexity** of the model problem (50.1) depends on all these factors. As we will see, the most significant factor affecting the complexity is the dimension m, followed by the growth rate of the function F.

5.1.1. The Chafee-Infante Equation. A typical problem, with many very rich dynamical properties, was proposed in Chafee and Infante (1974) and Chafee (1974). Their work is one of the earliest studies of stability and bifurcation problems in infinite dimensions, and the semilinear heat equation

$$\partial_t u = \nu\Delta u + f(u), \qquad \text{where } (t,x) \in \mathbb{R}^1 \times \Omega \text{ and } u \in \mathbb{R}, \tag{51.1}$$

which spawned this work, is called the Chafee-Infante equation. We assume here that the Dirichlet[21] boundary conditions $u(t,x) = 0$ hold on the

[21] Other boundary conditions include the Neumann boundary condition where $\frac{\partial u}{\partial n} = 0$ on $\partial\Omega$ and n denotes the unit outward normal. Also one sometimes encounters "mixed" boundary conditions of the from $\theta(x)u(x) = (1 - \theta(x))\frac{\partial u}{\partial n}(x) = 0$ on $\partial\Omega$, where $0 \leq \theta(x) \leq 1$.

boundary $\partial\Omega$ of an open, bounded set Ω in $\mathbb{R}^m$, and that the boundary is of Lipschitz class. The initial condition for (51.1) is $u(0,x) = u_0(x)$, for $x \in \Omega$. Also we assume that $\nu > 0$ is a constant and that $f : \mathbb{R} \to \mathbb{R}$ is a C^2-function satisfying

$$\limsup_{|u|\to\infty} \frac{f(u)}{u} \leq 0. \tag{51.2}$$

Note that the polynomial function $f(u) = \sum_{i=0}^{2k-1} b_i u^i$ of odd degree, where b_i are constants with $b_{2k-1} < 0$ and $k \geq 1$, satisfies inequality (51.2). For example, $f(u) = (u-a)(1-u^2)$, where $-1 < a < 1$. Inequality (51.2) is sometimes called an *asymptotically monotone condition*, or a *sign condition.*

As noted above, our approach to the study of the Problem (51.1) is to use the paradigm of Chapter 4 and to convert this problem to an abstract evolutionary equation

$$\partial_t u + Au = F(u), \qquad u \in H, \tag{51.3}$$

where the solution $u = u(t)$ and the initial condition $u(0) = u_0$ assume values in some Hilbert space H, which is the ambient space, or the phase space, for the problem. Possible choices for H include: $H = L^2(\Omega) = L^2(\Omega, \mathbb{R})$, $H = H^1(\Omega)$, or $H = H_0^1(\Omega)$.

The first step in reducing Problem (51.1) to the evolutionary equation (51.3) is to identify the linear operator A. In this case we fix

$$A\varphi(x) = -\nu\Delta\varphi(x), \tag{51.4}$$

with the Dirichlet boundary conditions. As noted in Section 3.8, the domain of A is $\mathcal{D}(A) = H^2(\Omega) \cap H_0^1(\Omega)$. Furthermore, the Standing Hypotheses B is satisfied on $H = L^2(\Omega)$. Let $V = V^1$ and $H = V^0$. For the nonlinear term, we define the Nemytskii operator $F : \varphi \to F(\varphi)$ by

$$F(\varphi)(x) = f(\varphi(x)), \qquad \text{for } x \in \Omega.$$

Recall that $V^1 = H_0^1(\Omega)$ and $V^2 = \mathcal{D}(A) = H_0^1(\Omega) \cap H^2(\Omega)$, see Section 3.8. We let $\langle \cdot, \cdot \rangle = \langle \cdot, \cdot \rangle_H$ and $\|\cdot\| = \|\cdot\|_H$ denote the inner product and norm on H. For $\varphi \in V = V^1 = \mathcal{D}(A^{\frac{1}{2}})$, we define $\|\varphi\|_V^2 = \|\varphi\|^2 + \|A^{\frac{1}{2}}\varphi\|^2$, and set $\|\varphi\|_{V^2}^2 = \|\varphi\|^2 + \|A\varphi\|^2$, for $\phi \in V^2$. It follows from the Poincaré inequality that there is a constant $C_0 > 0$ such that

$$\|A^{\frac{1}{2}}\varphi\| \leq \|\varphi\|_V \leq C_0 \|A^{\frac{1}{2}}\varphi\|, \qquad \text{for all } \varphi \in V. \tag{51.5}$$

Let us first examine the case where $m = \dim \Omega = 1$ and Ω is a bounded interval of length $|\Omega|$. By the Sobolev imbedding theorems, there are continuous imbeddings

$$V \mapsto H^1(\Omega) \hookrightarrow C^{0,\frac{1}{2}}(\Omega) \hookrightarrow C = C(\Omega) \mapsto L^\infty(\Omega) \mapsto H = L^2(\Omega), \tag{51.6}$$

see Appendix B. Hence, there is a constant C_1 such that $\|\varphi\|_\infty \le C_1\|\varphi\|_V$, where $\|\varphi\|_\infty$ is the L^∞-norm of φ. Since $f : \mathbb{R} \to \mathbb{R}$ is continuous, this implies that $f(\varphi(x))$ is continuous, for each $\varphi \in V$. Therefore, $F : V \to C \mapsto H$. For $f \in C^2(\mathbb{R};\mathbb{R})$ and $r > 0$, we define $K_i = K_i(r)$ by

$$K_i = \max\left\{\left|D^i f(\xi)\right| : |\xi| \le C_1 r\right\}, \qquad \text{for } i = 1, 2.$$

Assume now that $\varphi_i \in V$, for $i = 1, 2$, satisfy $\varphi_i \in B(0, r)$, i.e., $\|\varphi_i\|_V \le r$. Then one has

$$|Df(\varphi_2(x) + \lambda(\varphi_1(x) - \varphi_2(x)))| \le K_1 \qquad \text{for } 0 \le \lambda \le 1.$$

Consequently, from the Mean Value Theorem, one has

$$\begin{aligned}\|F(\varphi_1) - F(\varphi_2)\|^2 &= \int_\Omega |f(\varphi_1(x)) - f(\varphi_2(x))|^2\, dx \\ &= \int_\Omega \left|\int_0^1 Df(\varphi_2(x) + \lambda(\varphi_1(x) - \varphi_2(x)))\, d\lambda\, (\varphi_1(x) - \varphi_2(x))\right|^2 dx \\ &\le \int_\Omega K_1^2 |\varphi_1(x) - \varphi_2(x)|^2 dx = K_1^2\|\varphi_1 - \varphi_2\|^2 \le K_1^2\|\varphi_1 - \varphi_2\|_V^2.\end{aligned}$$

It follows that $F \in C_{\mathrm{Lip}}(V, H)$. Moreover, we claim that $F \in C_{\mathrm{Lip}}(V, V)$. In order to prove this, one needs to estimate the L^2-norm of

$$Df(\varphi_1)\partial\varphi_1 - DF(\varphi_2)\partial\varphi_2 = Df(\varphi_1)\partial\varphi_1 - Df(\varphi_2)\partial\varphi_2 \pm Df(\varphi_1)\partial\varphi_2$$

in terms of $\|\varphi_1 - \varphi_2\|_V$, where $\partial\varphi = \partial_x\varphi(x)$. From the argument above, one readily verifies that

$$\|Df(\varphi_1)\left[\partial\varphi_1 - \partial\varphi_2\right]\|^2 \le K_1^2\|\varphi_1 - \varphi_2\|_V^2,$$

for $\varphi_1, \varphi_2 \in B(0, r)$. Similarly, one obtains

$$\|(Df(\varphi_1) - Df(\varphi_2))\partial\varphi_2\|^2 \le K_2^2\|\varphi_1 - \varphi_2\|_\infty^2\|\partial\varphi_2\|^2.$$

These two inequalities then imply that

$$\|Df(\varphi_1)\partial\varphi_1 - D(\varphi_2)\partial\varphi_2\|^2 \le (K_1^2 + K_2^2 r^2)\|\varphi_1 - \varphi_2\|_V^2,$$

for φ_1, $\varphi_2 \in B(0, r)$. The reader should verify that the Nemytskii function $F(u)$ is twice Fréchet differentiable on V and one has $F \in C_F^2(V, H) \cap C_F^1(V, V)$, with the derivatives given by

$$\begin{aligned}(DF(u)v)(x) &= f'(u(x))\, v(x), \quad \text{and} \\ (D^2F(u)\langle v, w\rangle)(x) &= f''(u(x))\, v(x)\, w(x),\end{aligned}$$

for $u, v, w \in V$.

Thus the Chafee-Infante Problem (51.1) can be reformulated in terms of the evolutionary equation (51.3) on V. With this formulation, the original unknown function $u(t, x)$ in (51.1) is viewed as an abstract function $u(t)$ of time t with $u(t) = u(t, \cdot)$ assuming values in V. The first issue after the above set-up is the existence, uniqueness, and the regularity of the mild solution $u(t) = S(t)u_0$ of equation (51.3) with $S(0)u_0 = u_0$.

Theorem 51.1. *Let (51.3) be the evolutionary equation generated by the Chafee-Infante Problem in 1D, where $f \in C^2(\mathbb{R}, \mathbb{R})$ satisfies inequality (51.2). Then for every $u_0 \in V = V^1$, there is a unique globally defined mild solution*

$$S(t)u_0 \stackrel{\text{def}}{=} u(t) = e^{-At}u_0 + \int_0^t e^{-A(t-s)} f(u(s))\, ds$$

of the initial value Problem (51.3) in V, where $u = u(t)$ exists for all $t \geq 0$,

$$u \in C[0,\infty; V^1) \cap C^{0,1-r}_{\text{loc}}(0,\infty; V^{1+2r}) \cap C(0,\infty; V^3), \tag{51.7}$$

for all r with $0 \leq r < 1$, and

$$u \in C^1(0,\infty; V). \tag{51.8}$$

Also, $S(t)$ is a semiflow on V^1 which is locally Lipschitz continuous in $u_0 \in V^1$, and it is point dissipative and compact, for $t > 0$, in V^1. Moreover, there is a global attractor $\mathfrak{A}$ for this semiflow with $\mathfrak{A} \subset V^3$. Also for each r with $0 \leq r < 1$, one has

$$S(\cdot)u_0 \in C^{0,1-r}_{\text{loc}}(\mathbb{R}, V^{1+2r}) \cap C(\mathbb{R}, V^3), \qquad \textit{for each } u_0 \in \mathfrak{A}. \tag{51.9}$$

In addition, the global attractor $\mathfrak{A}$ attracts all bounded sets in V^1. Finally, for each $u_0 \in V^1$, the mild solution $S(t)u_0$ in V^1 is a strong solution in V^1, and for each $u_0 \in \mathfrak{A}$, the mild solution $S(t)u_0$ is a classical solution in V^{2r}, for each r with $0 \leq r < 1$.

Proof. Since the nonlinearity F satisfies $F \in C_{\text{Lip}}(V,V) \cap C_{\text{Lip}}(V,H)$, we can use the theory developed in Section 4.7 to study the mild solutions of equation (51.3). In fact we can, and we will, do this twice!

Since $F \in C_{\text{Lip}}(V,H)$, we apply the theory of Section 4.7, with $W = H$ and $V^{2\beta} = V^1$ ($\beta = \frac{1}{2}$). One concludes that for every $u_0 \in V$, there is a unique, globally defined mild solution $S(t)u_0$, for $0 \leq t < T$, where $T = T(u_0)$ satisfies $0 < T \leq \infty$. Furthermore, $S(t)u_0$ is a strong solution in V, it satisfies (47.7), and $S(t)u_0$ is jointly continuous in (t, u_0), as described in Theorem 47.5. We will show shortly that $T(u_0) = \infty$, for all $u_0 \in V$, which implies that $S(t)$ is a semiflow on V, and (47.7) implies that $u = u(t) = S(t)u_0$ satisfies

$$u \in C[0,\infty; V^{2\alpha}) \cap C^{0,1-r}_{\text{loc}}(0,\infty; V^{2r}) \cap C(0,\infty; V^2),$$

for all α and r with $0 \leq \alpha \leq \frac{1}{2}$ and $0 \leq r < 1$.

For the greater regularity of the mild solutions of equation (51.3), we will take advantage of the fact that $F \in C^1_{\text{F}}(V,V) \cap C_{\text{Lip}}(V,V)$. Since the given solution $u(t) = S(t)u_0$ satisfies (47.7), it follows that $f = f(t) \stackrel{\text{def}}{=} F(S(t)u_0)$ satisfies

$$f \in C[0,T;V) \cap C^{0,1}_{\text{loc}}(0,T;V). \tag{51.10}$$

As a result, Theorem 42.9 - with $p = \infty$ and V replaced by W, i.e., $W = V^1$ and $W^{2r} = V^{1+2r}$ - implies that (51.7) holds. Recall that the domain of A in the space $V^1 = W$ is $\mathcal{D}_{V^1}(A) = V^3 = W^2$, see (37.10). In this case, Lemma 47.2 leads to the stonger conclusion (51.7), and one finds that $u(t)$ is a strong solution on $\epsilon \leq t < \infty$ in the space $W = V$, for every $\epsilon > 0$. Furthermore, we are able to apply relation (48.3) in Theorem 48.5 and conclude that u satisfies (51.8), assuming of course that $T(u_0) = \infty$, for all $u_0 \in V$. In order to show that $T(u_0) = \infty$, we use Lemma 47.4 and show that

$$\lim_{t\to\tau^-} \|S(t)u_0\|_V = \infty$$

does not occur for any $u_0 \in V$ and any τ with $0 < \tau < \infty$.

The Energy Method. First we take the inner-product in H of both sides of equation (51.1) with the multiplier $u = u(t) \in V$ to get

$$\int_\Omega (\partial_t u)u\,dx - \nu \int_\Omega (\Delta u)u\,dx = \int_\Omega f(u)\,u\,dx. \tag{51.11}$$

By the sign condition (51.2), for any $\epsilon > 0$, there is a constant $N_\epsilon > 0$ such that $|f(s)| \leq \epsilon|s|$, when $|s| > N_\epsilon$. This implies that

$$|f(s)| \leq C_\epsilon + \epsilon|s|, \qquad \text{for } s \in \mathbb{R}, \tag{51.12}$$

where $C_\epsilon = \max\{|f(s)| : |s| \leq N_\epsilon\}$. Let $D_\epsilon = C_\epsilon N_\epsilon |\Omega|$. By integrating the second term of (51.11) by parts and using the Schwarz inequality, we obtain

$$\frac{1}{2}\partial_t\|u\|^2 + \nu\|\partial_x u\|^2 \leq D_\epsilon + \epsilon\|u\|^2, \qquad \text{for any } \epsilon > 0, \tag{51.13}$$

since $u \in C[0,T;H) \mapsto L^2_{\text{loc}}[0,T;H)$ (see the Continuity Lemma 37.9). Now set $\epsilon = \mu/2$ and $\mu = \nu\lambda_1$, where λ_1 is the first eigenvalue of A. From the Gronwall inequality we obtain

$$\|u(t)\|^2 \leq e^{-\mu t}\|u_0\|^2 + 2\mu^{-1}D_\epsilon, \qquad \text{for } t \geq 0. \tag{51.14}$$

Hence, there is no blowup in H, and the mild solution $S(t)u_0$ exists in H, for all $t \geq 0$, by Lemma 47.4.

While this establishes the lack of blowup in H of mild solutions of equation (51.3), we need to go further and examine this issue in the space V. As noted above, for each $u_0 \in V$, the mild solution u satisfies

$$u \in C[0,T;V) \cap C(0,T;V^3) \cap C^1(0,T;V), \tag{51.15}$$

where $T = T(u_0)$. Next we take the scalar product of equation (51.3) with Au and obtain

$$\langle \partial_t u, Au \rangle + \nu ||Au||^2 = \langle F(u), Au \rangle.$$

Since $u \in C[0,T;V) \mapsto L^2_{\text{loc}}[0,T;V)$, it follows from the Continuity Lemma 37.9 and the Young inequality that

$$\frac{1}{2}\partial_t ||A^{\frac{1}{2}}u||^2 + \nu ||Au||^2 \le ||F(u)||\,||Au|| \le \nu ||Au||^2 + \frac{1}{4\nu}||F(u)||^2.$$

Since inequality (51.12) implies that $|f(s)|^2 \le 2C_\epsilon^2 + 2\epsilon^2|s|^2$, this implies that

$$||F(u)||^2 \le 2C_\epsilon^2|\Omega| + 2\epsilon^2||u||^2 \le 2C_\epsilon^2|\Omega| + 2\epsilon^2\lambda_1^{-1}||A^{\frac{1}{2}}u||^2.$$

Hence, one finds that

$$\partial_t ||A^{\frac{1}{2}}u||^2 \le \nu^{-1}C_\epsilon^2|\Omega| + \nu^{-1}\epsilon^2\lambda_1^{-1}||A^{\frac{1}{2}}u||^2.$$

The Gronwall inequality and (51.5) then imply that

$$C_0^{-2}||u||_V^2 = ||A^{\frac{1}{2}}u||^2 \le e^{\epsilon^2\nu^{-1}\lambda_1^{-1}t}||u_0||_V^2 + C, \qquad \text{for } t \ge 0, \tag{51.16}$$

for some constant $C > 0$, where C does not depend on the initial condition. This implies that the mild solution $S(t)u_0$ does not blow up in finite time in the space V. Hence $T(u_0) = \infty$, for all $u_0 \in V$. As argued in Section 4.7, $S(t)$ is then a semiflow on V, and it is locally Lipschitz continuous in $u_0 \in V$.

Notice that inequality (51.14) was not used in showing that $T(u_0) = \infty$. While inequality (51.14) does show that $S(t)$ is point dissipative in H, inequality (51.16) does not show the point dissipativity in V. *This is actually a common situation one encounters with energy method arguments. One needs different methods in differing spaces.* Point dissipativity in V is valid, but to prove this fact we need an additional tool. In this case, the new tool is a Lyapunov function.

The Lyapunov Function. Consider the function $E(\varphi) : V \to \mathbb{R}$, defined by

$$E(\varphi) = \int_\Omega \left[\frac{\nu}{2}\left(\varphi'(x)\right)^2 - G(\varphi(x))\right] dx, \tag{51.17}$$

where $G(s) = \int_0^s f(r)\,dr$. As noted above, $u = u(t) = S(t)u_0$ is a strong solution in V. We now use the fact that the Dirichlet boundary condition $u(t,0) = 0$, for all $x \in \partial\Omega$, implies that $\partial_t u(t,0) = 0$, for all $x \in \partial\Omega$, as well. Then by using (51.7) and (51.8) and integrating by parts, where n denotes the unit outward normal to $\partial\Omega$, we have

$$\begin{aligned}\partial_t E(S(t)u_0) &= \int_\Omega [\nu\partial_x u \cdot \partial_t\partial_x u - F(u)\partial_t u]\,dx \\ &= \int_{\partial\Omega} \nu\partial_t u \cdot \frac{\partial u}{\partial n}\,dS - \int_\Omega |\partial_t u(t,x)|^2\,dx = -||\partial_t u||^2.\end{aligned} \tag{51.18}$$

Let $u_0 \in V^2$. Since $f(t) = F(S(t)u_0)$ satisfies (51.10), it follows that $f(t)$ is differentiable almost everywhere on $[0, \infty)$ and

$$f \in C[0, \infty; V^1) \cap W^{1,\infty}_{\text{loc}}[0, \infty; V^1).$$

It then follows from Theorem 42.14 that (42.38) holds, where $T = \infty$. Hence $u \in C^1[0, \infty; V)$. By integrating equation (51.18) over $[0, t]$ and using the Newton-Leibniz formula, we obtain

$$E(S(t)u_0) \leq E(u_0), \qquad \text{for } t \geq 0. \tag{51.19}$$

For any initial data $u_0 \in V$, there is a sequence $\{u_n\} \subset D(A)$ such that $u_n \to u_0$ in V. By the continuous dependence of the mild solution on initial data and the continuity of $E(\varphi)$ in $\varphi \in V$, we conclude that the inequality (51.19) actually holds for any $u_0 \in V$.

We now show that for small $\epsilon > 0$ there is a constant $D_\epsilon > 0$, such that

$$C_0^{-2}\frac{\nu}{4}\|\varphi\|_V^2 \leq E(\varphi) + D_\epsilon, \qquad \text{for any } \varphi \in V. \tag{51.20}$$

Note that $\|\varphi\|^2 \leq \lambda_1^{-1}\|A^{\frac{1}{2}}\varphi\|^2 \leq \lambda_1^{-1}\|\varphi\|_V^2$, for $\varphi \in V$. From inequality (51.12), we find that $|G(s)| \leq C_\epsilon N_\epsilon + \epsilon|s|^2$, and consequently, for $D_\epsilon = C_\epsilon N_\epsilon|\Omega|$, one has

$$\left|\int_\Omega G(u(t,x))\,dx\right| \leq D_\epsilon + \epsilon\lambda_1^{-1}\|u\|_V^2.$$

As a result, (51.5) and (51.17) imply that

$$C_0^{-2}\frac{\nu}{2}\|\varphi\|_V^2 \leq \frac{\nu}{2}\|A^{\frac{1}{2}}\varphi\|^2 = \frac{\nu}{2}\|\partial_x\varphi\|^2 \leq E(\varphi) + D_\epsilon + \epsilon\lambda_1^{-1}\|\varphi\|_V^2.$$

In addition, one has

$$E(\varphi) \leq D_\epsilon + \left(\frac{\nu}{2} + \frac{\epsilon}{\lambda_1}\right)\|\varphi\|_V^2, \qquad \text{for all } \varphi \in V. \tag{51.21}$$

By taking $\epsilon \leq C_0^{-2}\frac{1}{4}\nu\lambda_1$, one obtains inequality (51.20), as well as

$$C_0^{-2}\frac{\nu}{4}\|S(t)u_0\|_V^2 \leq E(S(t)u_0) + D_\epsilon \leq E(u_0) + D_\epsilon, \tag{51.22}$$

for any $t \geq 0$.

Let us return now to the properties of the semiflow on V. Inequalities (51.21) and (51.22) show that, for every bounded set B in V, the positive orbit $\gamma^+(B)$ is bounded in V. Furthermore, Lemma 48.1 implies that $S(t)$ is compact, for $t > 0$. Assume for the moment that $S(t)$ is point dissipative in V. Then $S(t)$ is ultimately bounded in V, and it follows from the Existence

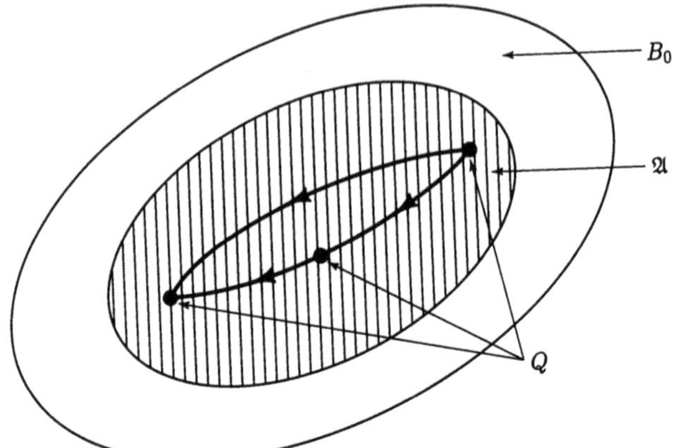

Figure 5.1. Detail for the Proof of Theorem 51.1

Theorem 23.12 that there is a global attractor $\mathfrak{A}$ in V and $\mathfrak{A}$ attracts all bounded sets in V. Furthermore the Herculean Theorem 47.6 implies that relation (51.9) and the remaining conclusions of Theorem 51.1 are valid. It remains to show that $S(t)$ is point dissipative in V.

We define a set Q in V as the set of all stationary solutions of equation (51.3), i.e, $u_0 \in Q$ if and only if $\partial_t S(t)u_0 \equiv 0$ and one has $Au_0 = F(u_0)$. Owing to (51.18) and (51.19), we see that $u_0 \in Q$ if and only if

$$Q \stackrel{\text{def}}{=} \{u_0 \in V : E(S(t)u_0) = E(u_0), \text{ for all } t \geq 0\}.$$

It follows that Q is a closed, invariant set in V. Also for any $u_0 \in Q$, one has $\|u_0\|_V^2 = \langle Au_0, u_0 \rangle = \langle F(u_0), u_0 \rangle$, which implies that

$$\|u_0\|_V^2 = \int_\Omega f(u_0)u_0 \, dx \leq D_\epsilon + \epsilon\lambda_1^{-1}\|u_0\|_V^2,$$

by inequality (51.12). For $0 < \epsilon \leq \min(\frac{1}{4}\nu\lambda_1, \frac{1}{2}\lambda_1)$, this implies that $\|u_0\|_V^2 \leq 2D_\epsilon$. Hence Q is a closed, bounded, invariant set in V. Since $S(t)$ is compact for $t > 0$, this implies that Q is a compact set in V.

Since E is bounded below in V (inequality (51.20)) and nonincreasing (inequality (51.19)), this implies that for each $u_0 \in V$, the limit $e(u_0) \stackrel{\text{def}}{=} \lim_{t\to\infty} E(S(t)u_0)$ exists, and that the omega limit set $\omega(u_0)$ satisfies $\omega(u_0) \subset \{\varphi \in V : E(\varphi) = e(u_0)\}$. It follows that $\omega(u_0)$ is a nonempty compact, invariant set with $\omega(u_0) \subset Q$, see Lemma 23.8, and Q is a nonempty set in V. Furthermore, by Lemma 23.6, $\omega(u_0)$ is a connected set in V.

We now define the real number E_0 and the bounded set B_0 by

$$E_0 \stackrel{\text{def}}{=} \max\{E(\varphi) : \varphi \in Q\} \quad \text{and} \quad B_0 \stackrel{\text{def}}{=} \{\varphi \in V : E(\varphi) < 2E_0\}.$$

By inequality (51.20), we see that B_0 is a bounded set in V. Also, for every $u_0 \in V$, there is a time $\tau = \tau(u_0) \geq 0$, such that $E(S(t)u_0) < 2E_0$, for all $t \geq \tau$. Hence $S(t)$ is point dissipative in V and B_0 is an absorbing set for for $S(t)$, see Figure 5.1. □

There is additional information concerning the dynamics of $S(t)$ on V contained in the proof of the last theorem. We summarize these features in the following result, where we will use the definitions given in the last proof.

Corollary 51.2. *Under the hypotheses of Theorem 51.1, the following statements hold.*

(1) *For any bounded set B in V and any $\tau > 0$, the positive orbit $\gamma^+(S(\tau)B)$ is precompact in V.*
(2) *For each $u_0 \in V$, the ω-limit set $\omega(u_0)$ is a nonempty, compact, connected, invariant set in V with $\omega(u_0) \subset Q \cap \{\varphi \in V : E(\varphi) = e(u_0)\} \subset \mathfrak{A}$.*
(3) *The set Q is a nonempty, compact, invariant set in V.*
(4) *For every point $u_0 \in \mathfrak{A}$, one has $E(u_0) \le E_0$.*

Proof. Item (1) is proved in Theorem 51.1. For each $u_0 \in V$ and $\tau > 0$, the positive hull $H^+(S(\tau)u_0)$ is compact in V, and therefore by Lemmas 21.4 and 23.6, the omega limit set $\omega(u_0)$ is nonempty, compact, connected and invariant. This invariance and the monotonicity (51.19) imply that for any $\varphi \in \omega(u_0)$, one has $E(\varphi) = E(S(t)\varphi) = e(u_0)$, for all $t \ge 0$. It follows that

$$\partial_t E(S(t)\varphi) \equiv 0,\ t \ge 0, \qquad \text{so that } \partial_t(S(t)\varphi) \equiv 0, \qquad \text{for } t \ge 0,$$

by (51.18). Therefore, $S(t)\varphi = \varphi$, for $t \ge 0$, and $\varphi \in Q$, which completes the proof of Item (2). Item (3) is proved in Theorem 51.1, and Item (4) follows from Item (2) and $E(u_0) = e(u_0) = E(\omega(u_0)) \le E_0$ for any $u_0 \in \mathfrak{A}$. □

Let us now turn to the question: What happens to the dynamics of equation (51.1) when $m = \dim \Omega \ge 2$? One can get some valuable insight into the basic issues involved by restricting to the case where $2 \le m \le 3$, which we now do. In this case, one no longer has the Sobolev imbedding $H^1(\Omega) \mapsto L^\infty(\Omega)$, an imbedding which plays a central role in the analysis of the solutions of equation (51.1) presented above, see (51.6). However, when $m \le 3$, one does have the imbeddings

$$H^2(\Omega) \hookrightarrow C^{0,\frac{1}{2}}(\Omega) \hookrightarrow C - C(\Omega) \mapsto L^\infty(\Omega) \mapsto L^2(\Omega).$$

Furthermore, the Sobolev space $H^1(\Omega)$ satisfies

$$H^1(\Omega) \mapsto L^p(\Omega), \qquad \text{for every } p \text{ with } 1 \le p < \infty, \tag{51.23}$$

when $m = 2$, and when $m = 3$, one has (see Appendix B)

$$H^1(\Omega) \mapsto L^p(\Omega), \qquad \text{for every } p \text{ with } 1 \le p < 6. \tag{51.24}$$

The key ideas of analyzing the dynamics of equation (51.1), when $2 \le m \le 3$, can best be described in the special case where

$$f(u) = \sum_{i=0}^{2k-1} b_i u^i \tag{51.25}$$

is a polynomial in u, with odd degree $(2k - 1)$, where $b_{2k-1} < 0$. We claim that the following assertions are valid:

The case $m = 2$. *For any polynomial $f(u)$ given by (51.25), the Nemytskii function $F = F(u)(x) = f(u(x))$ satisfies*

$$F \in C_{\mathrm{Lip}}(V, H) \cap C_F^1(V, H) \cap C_{\mathrm{Lip}}(V, V). \tag{51.26}$$

The case $m = 3$. *For any polynomial $f(u)$ given by (51.25), where the degree of f is 3, the Nemytskii function F satisfies (51.26).*

The verification that F satisfies (51.26), either in the $m = 2$, or the $m = 3$ case, involves a variation in the argument used above in the 1D case, $m = 1$. Let us examine the $m = 3$ case more carefully, where we now assume that degree$(f) = 3$. In order to estimate $\|F(\varphi_1) - F(\varphi_2)\|^2$, for $\varphi_1, \varphi_2 \in V$, one cannot use the 1D argument, because one no longer has the imbedding $V \mapsto L^\infty$. However, owing to (51.24), one can use an $L^p - L^q$ Hölder inequality, with $1 < p, q < \infty$, and $p^{-1} + q^{-1} = 1$. As a result, one can verify that there is a polynomial $P = P(r)$, with nonnegative coefficients, such that if $\varphi_1, \varphi_2 \in V$ with $\|\varphi_i\|_V \leq r$, for $i = 1, 2$, and $\|\varphi_1 - \varphi_2\|_V \leq 1$, then one has

$$\|F(\varphi_1) - F(\varphi_2)\|^2 \leq P(r^2)\|\varphi_1 - \varphi_2\|_V^2. \tag{51.27}$$

This implies that $F \in C_{\mathrm{Lip}}(V, H)$. Similarly, one can verify the validity of (51.26), and we invite the reader to check the details.

Note that the sign condition $b_{2k-1} < 0$ is not used in the verification of (51.26). Since F satisfies (51.26) in the $m = 2$ and the $m = 3$ cases, it follows, as argued in Theorem 51.1, that for each $u_0 \in V$, there is a unique mild solution $u = u(t)$ of the Initial Value Problem (51.3) on some interval $0 \leq t < T = T(u_0)$, where $0 < T(u_0) \leq \infty$, and that u is regular in the sense that (51.15) is valid. The proof that $T(u_0) = \infty$, for every $u_0 \in V$, follows the methodology used in Theorem 51.1. In the $m = 2$ and $m = 3$ cases, one verifies that

$$E(\varphi) = \int_\Omega \left[\frac{\nu}{2}|\nabla\varphi(x)|^2 - G(\varphi(x))\right] dx$$

is well-defined for every $\varphi \in V$, and it is a Lyapunov function on V. In particular, the three inequalities (51.19)–(51.21) are valid. (As in the 1D case, the sign condition $b_{2k-1} < 0$ plays an essential role in the verification of these inequalities.) As a result, we obtain the following extension of Theorem 51.1.

Theorem 51.3. *Consider the Chafee-Infante Problem, where $2 \leq m \leq 3$. Assume that f satisfies (51.25), where $b_{2k-1} < 0$, and that degree$(f) = 3$ when $m = 3$. Then for every $u_0 \in V$, there is a unique globally defined mild solution $u = u(t) = S(t)u_0$ of the Initial Value Problem (51.3) in V, for all $t \geq 0$, and (51.7) and (51.8) are satisfied. Also, $S(t)$ is a semiflow on V which is locally Lipschitz continuous in $u_0 \in V$, point dissipative, and* *compact for $t > 0$ in V. Moreover, there is a global attractor $\mathfrak{A}$ for this semiflow with $\mathfrak{A} \subset V^3$ and, for each r with $0 \le r < 1$, (51.9) holds. In addition, the global attractor $\mathfrak{A}$ attracts all bounded sets in V. Finally, for each $u_0 \in V$, $S(t)u_0$ is a strong solution in V, and for each $u_0 \in \mathfrak{A}$, $S(t)u_0$ is a classical solution in V, and the conclusions of Corollary 51.2 are valid.*

Since the semiflow $S(t)$ on V generated by the Chafee-Infante Problem is a gradient system, in the sense used in Section 7.2, the global attractor has additional properties. See Theorems 72.1, 72.2, and 72.3, for details.

Our analysis of the Chafee-Infante Problem (51.1) can be extended to spatially dependent reaction diffusion equations, such as

$$\partial_t u = \nu \Delta u + f(x, u), \tag{51.28}$$

with homogeneous Dirichlet or Neumann boundary condition. For example, $f(x, u)$ may be a polynomial function of odd degree given by

$$f(x, u) = \sum_{i=0}^{2k-1} b_i(x) u^i,$$

where each $b_i \in L^\infty(\Omega)$ and $b_{2k-1}(x) \le -\beta < 0$, for some constant β. In general, one assumes that $f : \Omega \times \mathbb{R} \to \mathbb{R}$ satisfies the **Carathéodory conditions**:

(C1) For each $s \in \mathbb{R}$, the function $f(\cdot, s)$ is Lebesgue measurable in Ω;
(C2) For almost every $x \in \Omega$, the function $f(x, \cdot)$ is continuous in $\mathbb{R}$;
(C3) For almost every $x \in \Omega$, the function $f(x, \cdot)$ is differentiable in $\mathbb{R}$ and the partial derivative $\partial_s f$ satisfies (C1) and (C2).

We also assume that there are positive constants c_i, for $1 \le i \le 4$, and an integer $p \ge 2$ such that

$$\begin{aligned} |f(x, s)| &\le c_1 |s|^{p-1} + c_2, \quad \text{for } (x, s) \in \Omega \times \mathbb{R}, \\ |\partial_s f(x, s)| &\le c_3 |s|^{p-2} + c_4, \quad \text{for } (x, s) \in \Omega \times \mathbb{R}. \end{aligned} \tag{51.29}$$

Under these conditions, the **Nemytskii mapping**

$$F : g(x) \mapsto f(x, g(x)), \quad x \in \Omega,$$

has the properties shown in Appendix B. Let $q \in \mathbb{R}$ satisfy $\frac{1}{p} + \frac{1}{q} = 1$. Then one has $F \in C^1_{\mathrm{Lip}}(L^p(\Omega); L^q(\Omega))$. In addition, we replace (51.2) by the asymptotical monotone condition

$$\limsup_{|s| \to \infty} \sup_{x \in \Omega} \frac{f(x, s)}{s} \le 0. \tag{51.30}$$

In view of the combined effect of the spatial dependence in f and a higher dimensional domain Ω, the choice of an appropriate phase space has to be done with care. In order that the Volterra integral

$$J(t; u_0) = \int_0^t e^{-A(t-\tau)} F(S(\tau)u_0)\, d\tau, \qquad t \geq 0,$$

make sense, we require that the phase space U satisfy the followng conditions:

(a) F maps U into $H = L^2(\Omega)$, and
(b) the Volterra integral converges in the space U for almost all $t \in [0, T(u_0))$, where $[0, T(u_0))$ is the interval of definition for the maximally defined mild solution.

The choice for $H = L^2(\Omega)$ is appropriate, because the analytic semigroup is defined on this space. In addition, we require the validity of the following:

(c) the imbedding properties of Sobolev spaces, as shown in Appendix B,
(d) the properties of the Nemytskii mapping based on the assumptions, and
(e) the $L^1 - L^\infty$ regularity and the extensions of the analytic semigroup e^{-At}, as shown in Sections 3.8.6 and Theorem 38.10.

Three examples of proper phase spaces are the following:

1) $U = V \cap L^p(\Omega)$, where $2 \leq p < \infty$ and $V = V^{2\beta}$, for some $0 \leq \beta < 1$, based on the Standing Hypothesis A or the Standing Hypothesis B. In particular, for (51.28), V can be $H_0^1(\Omega)$ or $H^1(\Omega)$, depending on the boundary conditions.
2) $U = V \cap L^\infty(\Omega)$, where V can be as above.
3) $U = W^{1,q}(\Omega)$, for some $q \geq 1$, as a Sobolev space defined in Appendix B.

With respect to choices 1) and 2), the norm on U is

$$\|\varphi\|_U = \|\varphi\|_V + \|\varphi\|_{L^p(\Omega)}, \qquad 1 \leq p \leq \infty,$$

and U is a separable Banach space (see Appendix A). In the exercises, more details are developed based on the degree of polynomial f, as well as the dimension of Ω.

We present the following result on the global dynamics of the spatially dependent Problem (51.28) over a higher dimensional domain $\Omega \subset \mathbb{R}^m$.

Theorem 51.4. *Assume that the nonlinear term f in (51.28) satisfies the conditions of Carathéodory and that (51.29) and (51.30) hold. Let $m =$ dim Ω and assume that*

$$2 \leq p < \infty, \quad \text{if } m = 1 \text{ or } 2, \quad \text{or} \quad 2 \leq p < \frac{2m}{m-2}, \quad \text{if } m \geq 3.$$

Then for any given $u_0 \in U = V \cap L^p(\Omega)$, where $V = V^1 = H_0^1(\Omega)$ for Dirichlet boundary condition and $V = \dot{H}^1(\Omega)$ (with zero average) for Neumann boundary condition, there exists a unique globally defined, mild solution $u(t)$ in V of equation (51.28), such that

$$u \in C[0,\infty;U) \cap L^2_{\text{loc}}[0,\infty;V) \cap L^p_{\text{loc}}[0,\infty;L^p(\Omega)).$$

Moreover, there exists a global attractor $\mathfrak{A}$ in space U, for the associated semiflow $S(t)$.

Sketch of the Proof. The theory described in Sections 4.7.1 to 4.7.3 can be adapted to show the local existence of the solution to (51.28) in the intersection space U with the aid of Theorem 38.10. We leave it as an exercise. The global existence follows from the *a priori* estimates:

$$\frac{1}{2}\partial_t \|u\|^2 + b(u,u) + c_5 \int_\Omega |u(t,x)|^p \, dx \le c_6 |\Omega|, \tag{51.31}$$

where c_i, for $5 \le i \le 8$, are positive constants, $b(u,u) = \nu \|\nabla u\|^2$, and

$$\|\partial_t u\|^2 + \partial_t \left[\frac{1}{2} b(u,u) + \int_\Omega \Phi(x,u) \, dx \right] = 0, \tag{51.32}$$

with $\Phi(x,s) = \int_0^s f(x,\sigma)\,d\sigma$ satisfying $c_7|s|^p - c_8 \le \Phi(x,s) \le c_7|s|^p + c_8$. Then one applies the Uniform Gronwall inequality (see Appendix D) to show the existence of an absorbing set in space U, based on the relations (51.31) and (51.32). Finally, by the compact imbedding of $U \hookrightarrow H = L^2(\Omega)$, the semiflow $S(t)$ is compact in H, for $t > 0$. By a bootstrap argument, this semiflow is also compact in U. Therefore, a global attractor $\mathfrak{A}$ exists by Theorem 23.12. □

5.1.2. Systems of Equations and Inhomogeneous Boundary Conditions. In Section 5.1.1 we examined the basic dynamics of a single scalar-valued reaction diffusion equation (RDE), where the solution $u(t,x)$ assumes values in $\mathbb{R}$. In many applications, however, one encounters coupled systems of RDEs, such as given by the two-dimensional vector (u,v) where

$$\begin{cases} \partial_t u = \nu \Delta u + f(u,v) \\ \partial_t v = \nu \Delta v + g(u,v) \end{cases}$$

on a smooth, bounded, open domain Ω in $\mathbb{R}^m$. For example, in the study of chemical kinetics, one oftentimes has multi-dimensional vectors $u = (u_1.u_2,\cdots,u_N)$ and systems of the form

$$\partial_t u_i = \nu_i \Delta u_i + f_i(u_1,u_2,\cdots,u_N), \qquad 1 \le i \le N,$$

or more complicated problems, such as

$$\partial_t u = D\Delta u + f(u), \tag{51.33}$$

where u and f are now N-dimensional vectors and D is an $N \times N$ matrix.

While it is not surprising that such systems inevitably lead to greater algebraic complexity in their formulation, the basic strategy for their study follows closely the paradigm used above. For example, in the case of equation (51.33), one first studies the linear problem

$$\partial_t u = D\Delta u, \tag{51.34}$$

with suitable boundary conditions. For example, one may have homogeneous boundary conditions of Dirichlet, or Neumann type, or more generally Robin - or mixed - boundary conditions of the form

$$\theta(x)u(x) + (1 - \theta(x))\frac{\partial u}{\partial n}(x) = 0, \qquad \text{for } x \in \partial\Omega, \tag{51.35}$$

where $0 \leq \theta(x) \leq 1$, for $x \in \partial\Omega$. In order to follow the paradigm of Chapter 4, the boundary conditions should insure that the associated linear operator $Au = -D\Delta u$, with these boundary conditions, is a sectorial operator on a suitable Banach space or Hilbert space W, such as $L^2(\Omega, \mathbb{R}^N)$, and thus the Standing Hypothesis A is satisfied. This requirement already imposes a restriction on the matrix D, for example, that D be a **positive (definite) matrix** in the sense that there is a $c_0 > 0$ such that[22]

$$\sum_{i,j=1}^{N} D_{ij}\xi_i\xi_j \geq c_0\|\xi\|^2 = c_0 \sum_{i=1}^{N} \xi_i^2, \qquad \text{for all } \xi = (\xi_1, \xi_2, \cdots, \xi_N) \in \mathbb{R}^N.$$

If D is a diagonal matrix with positive entries on the diagonal, then it is positive.

The next step is to consider the Nemytskii operator $F(u)(x) = f(u(x))$. The goal here is to show that F is sufficiently well behaved so that $F \in C_{\text{Lip}}(V^{2\beta}, W)$, for some β with $0 \leq \beta < 1$. At this point, the theory of Section 4.7 becomes fully operational.

In the mathematical modeling of real-world problems, one sometimes encounters homogeneous boundary conditions, such as (51.35). Inhomogeneous boundary conditions of various types, such as $u(x) = \phi(x)$, or $\frac{\partial u}{\partial n} = \psi(x)$, for $x \in \partial\Omega$, are far more common. The dependence of the longtime dynamics of the problem on the boundary conditions is a problem of great practical interest. (In this illustration, one treats $\phi = \phi(x)$, or $\psi = \psi(x)$, as a known function and assumes that ϕ has sufficient smoothness so that the theory we describe next is valid.)

What is typically done in such a case is to "homogenize" the boundary conditions. This is accomplished by finding a good function $w = w(x)$, which is defined for $x \in \bar{\Omega} = \Omega \cup \partial\Omega$, such that $w(x) = \phi(x)$, or $\partial_n w(x) = \psi(x)$, for all $x \in \partial\Omega$. After doing this, one introduces a change of variables

[22]Compare this with the condition of uniform ellipticity presented in Chapter 3.

$u = v + w$ into the basic equation, say equation (51.33). Since $\partial_t w \equiv 0$, the equation for v is

$$\partial_t v = D\Delta v + g(x, v) \tag{51.36}$$

where $g(x,v) = f(v+w(x)) + D\Delta w(x)$ and v now satisfies the homogeneous boundary conditions $v(x) = 0$, or $\frac{\partial v}{\partial n} = 0$, for $x \in \partial\Omega$. Thus equation (51.33), with the inhomogeneous boundary conditions, is replaced by the related equation (51.36), with homogeneous boundary conditions.

5.1.3. Partly Dissipative Systems. The longtime dynamics of partly dissipative reaction diffusion systems are of interest, see (Marion 1989) and (Shao 1994). The following is a typical system:

$$\begin{aligned} \partial_t u &= D\Delta u + f(x, u, v), \\ \partial_t v &= G(x, u)v + h(x, u), \end{aligned} \tag{51.37}$$

where $u \in \mathbb{R}^{m_1}$ and $v \in \mathbb{R}^{m_2}$, $m_1 + m_2 = N$, and $G(x, u)$ is an $m_2 \times m_2$ matrix, with entries being C^2-functions of x and u. The feature is that the v-equation has vanishing diffusion coefficients. Thus, the v-equation in (51.37) is essentially an "ordinary" differential equation, and there is no spatial smoothing or dissipation in the v-coordinate of the solution. An example of such a system is the system of FitzHugh-Nagumo equations from neurobiology:

$$\begin{aligned} \partial_t u &= d\Delta u + f(u) + \gamma v, \\ \partial_t v &= -\sigma v + \delta u, \end{aligned}$$

where the coefficients $d, \delta, \gamma, \sigma$ are in $\mathbb{R}$, and d and σ are positive. The function f is a polynomial of odd degree with a negative leading coefficient. Other examples can be found in Keener and Tyson (1992) and Sell and Varghese (1994).

The "usual suspects" of Dirichlet, Neumann, or periodic are typical boundary conditions for the system (51.37). A pivot concept arising in this study is a **positively invariant region** $\mathcal{D}$ in $\mathbb{R}^N$, see Chueh, Conley and Smoller (1977) and Smoller (1983), for example. Such a region is a closed, bounded, convex domain that is positively invariant for the semiflow generated by the mild solutions of (51.37). The functions f, G and h are assumed to be C^2 mappings on $\bar{\Omega} \times \mathcal{D}$, where G is uniformly positive definite. One can show that the induced semiflow $S(t)$ associated with (51.37) has an attractor $\mathfrak{A}$ in $V = V_1 \times L^2(\Omega)^{m_2}$ that attracts any bounded sets in $H = L^2(\Omega, \mathcal{D})$, where

$$V_1 = \begin{cases} H_0^1(\Omega)^{m_1}, & \text{for Dirichlet boundary condition,} \\ H^1(\Omega)^{m_1}, & \text{for Neumann boundary condition,} \\ H_{\text{per}}^1(\Omega)^{m_1}, & \text{for periodic boundary condition,} \\ & \text{with } \Omega = (0, L)^m. \end{cases}$$

Note that the attractor has finite Hausdorff and fractal dimensions, see Marion (1989).

5.2. Nonlinear Wave Equations.

The nonlinear wave equations (**NWEs**), which are analyzed in this section, arise as perturbations of various linear wave equations, such as

$$\partial_t^2 u - \nu \Delta u = h, \quad \text{or} \quad \partial_t^2 u + \nu \Delta^2 u = h,$$

where $\partial_t^2 u = \frac{\partial^2 u}{\partial t^2}$. Like their linear cousins, the NWEs are used to model wave propagation, such as water waves, or electromagnetic waves, or elastic vibrations, but now in nonlinear media. For example, the Gordon family

$$\partial_t^2 u + \alpha\, \partial_t u - \Delta u + \beta\, f(u) = h, \tag{52.1}$$

which has several branches, including the Sine's ($f(u) = \sin u$) and the Klein's ($f(u) = |u|^k u$), are used for models of electrical dynamics and relativistic quantum mehanics.

One oftentimes encounters a damping effect owing to friction in the underlying physical system. In the NWE, this is the source of dissipativity, and it is modeled by a linear term $\alpha\, \partial_t u$, where $\alpha > 0$, as in (52.1), or by a nonlinear term $g(\partial_t u)$. The nonlinear version is treated in Section 5.2.3.

In a broad sense, a **nonlinear wave equation** can be any semilinear hyperbolic evolutionary equation featuring a second-order time derivative. However, there is a technical difference between the hyperbolic evolutionary equations here and the traditional hyperbolic partial differential equations. For example, Petrovsky equation $\partial_t^2 u + \nu \Delta^2 u = h$ generates a hyperbolic evolutionary equation, but it is not a hyperbolic partial differential equation since this equation is of second order in time variable and of fourth order in spatial variables.

Our main objective in this section is to develop the basic dynamical theory of NWEs. However, before embarking on this journey, it may be helpful to recall the differences between the linear theories of parabolic and hyperbolic partial differential equations presented in Section 3.8. While the linear evolutionary equation $\partial_t u + Au = 0$ derived in the parabolic case typically came with a sectorial operator A and an analytic semigroup e^{-At}, the corresponding equation $\partial_t u + Gu = 0$ in the hyperbolic case gave birth to a C_0-group e^{-Gt}. The spectra of A and G march to different drummers!

What are the consequences of this difference? Unlike the parabolic evolutionary equations for which the principal (spatial) differential operator is typically a uniformly elliptic operator with good dissipativity properties, the principal (spatial) differential operator in hyperbolic cases will not provide an impact on dissipativity. The driving component for dissipativity in the hyperbolic evolutionary equations is typically a friction term in the form of an external or internal damping. If such a damping happens to be locally distributed, or if the damping acts only on the boundary, as is often the case in control theory, then one encounters new difficulties in determining the asymptotic and global dynamics in these "weakly damped" scenarios. While we examine here one such problem, many issues of nonlinear dynamics for weakly damped nonlinear wave equations remain open.

Furthermore, even if there exists a uniformly distributed damping, such as the linear damping $\alpha\,\partial_t u$, the proof of the "asymptotical compactness" of the semiflow generated by the mild solutions is demanding. The reason is that the linear semigroup $T(t)$ generated by the principal (spatial) differential operator will not be compact, for $t > t_0$. Hence, the burden of proving asymptotical compactness will depend on a suitable κ-contracting property, as we will see below.

5.2.1 Abstract Nonlinear Wave Equations. Consider a hyperbolic evolutionary equation on a Hilbert space H of the following form,

$$\partial_t^2 u + \alpha\,\partial_t u + Au + f(u) = h, \qquad \text{for } t > 0, \tag{52.2}$$

with boundary conditions $u(0) = u_0$ and $\partial_t u(0) = u_1$, where $\alpha > 0$ is a constant. We assume that $h = h(t) \in L^\infty(0,\infty; H)$, and we let

$$\|h\|_\infty = \operatorname{ess\,sup}\{\|h(t)\|_H : t \in (0,\infty)\}.$$

Also we assume that A is a linear operator on H that satisfies the Standing Hypotheses B.

As in the case of linear wave equations addressed early in Section 3.8.4, we take the product Hilbert space $W = V^1 \times H = V \times H$ to be the phase space for this abstract nonlinear wave equation (52.2). The W-norm will be denoted by $\|\cdot\|_W$. More generally we define $W^s = V^{s+1} \times V^s$, for $s \in \mathbb{R}$. Thus $W^0 = W$.

We assume here that $f : V \to H$ has the **compactness property**. That is to say, f maps bounded sets in V into compact sets in H. We assume further that

$$f \in C_{\text{Lip}}(V, H), \qquad \text{with } f(0) = 0; \tag{52.3}$$

that f is a **gradient operator**, that is, there exists a functional $F : V \to \mathbb{R}$, such that $F(0) = 0$,

$$F(\varphi) - F(\psi) = \int_0^1 \langle f(\psi + \theta(\varphi - \psi)), \varphi - \psi\rangle_H \, d\theta, \tag{52.4}$$

for all $\varphi, \psi \in V$; and that for every $\epsilon > 0$ there exist positive constants C_ϵ, D_ϵ, and Q_ϵ such that

$$-\epsilon\|u\|_H^2 - C_\epsilon \le F(u) \le D_\epsilon + \epsilon\|u\|_V^2, \qquad \text{for any } u \in V, \tag{52.5}$$

and

$$-\epsilon\|u\|_V^2 - Q_\epsilon \le \langle f(u), u\rangle, \qquad \text{for any } u \in V. \tag{52.6}$$

Then the second-order evolutionary equation (52.2) can be formulated into a first-order evolutionary equation in space W,

$$\partial_t w + Gw = R(w) + g, \qquad \text{for } t > 0, \tag{52.7}$$

with initial condition $w(0) = w_0 \in W$, see (38.30), where

$$G = \begin{pmatrix} 0 & -I \\ A & \alpha I \end{pmatrix}, \tag{52.8}$$

the domain is $\mathcal{D}(G) = V^2 \times V^1 = \mathcal{D}(A) \times V$, I is the identity on H, and

$$R(w) = \begin{pmatrix} 0 \\ -f(u) \end{pmatrix}, \quad g = \begin{pmatrix} 0 \\ h \end{pmatrix}, \quad w(t) = \begin{pmatrix} u(t) \\ \partial_t u(t) \end{pmatrix}, \quad w_0 = \begin{pmatrix} u_0 \\ u_1 \end{pmatrix}.$$

Let us turn next to two typical examples of this general construction.

Example 52.1 The Sine-Gordon equation is given by

$$\partial_t^2 u + \alpha\, \partial_t u - d\, \partial_x^2 u + \beta \sin u = h, \qquad \text{for } t > 0 \text{ and } x \in (0,1), \tag{52.9}$$

where $\alpha > 0$, $d > 0$, and $\beta \in \mathbb{R}$ are constants. We assume that the Dirichlet boundary conditions $u(0) = u(1) = 0$ hold. In this problem, the nonlinearity is $f(\varphi(x)) = \beta \sin \varphi(x)$. The linear operator $A = -d\, \partial_x^2$, with Dirichlet boundary conditions, leads to the spaces

$$H = L^2(0,1), \; V = \mathcal{D}(A^{\frac{1}{2}}) = H_0^1(0,1), \text{ and } W = H_0^1(0,1) \times L^2(0,1).$$

As noted in Corollary 38.8, the corresponding linear operator $-G$, see equation (52.8) in this case, has compact resolvent and generates an exponentially stable C_0-semigroup e^{-Gt}, which is differentiable for $t > 0$.

We claim that the nonlinearity $f(u)$ satisfies the compactness property and conditions (52.3)–(52.6). First of all, for any u_1, $u_2 \in V$ and for $x \in \Omega = (0,1)$, we have

$$f(u_1(x)) - f(u_2(x)) = \beta \int_0^1 \cos(u_2(x) + \theta(u_1(x) - u_2(x)))\, d\theta (u_1(x) - u_2(x)).$$

Owing to the continuous imbedding $V \mapsto H$, one has

$$\|f(u_1) - f(u_2)\|_H \le |\beta|\, \|u_1 - u_2\|_H \le |\beta| c_0 \|u_1 - u_2\|_V, \tag{52.10}$$

for some $c_0 > 0$. Moreover, for any $u \in V$, the distribution derivative of $f(u)$ is given by

$$\partial_x f(u) = \partial_x (f(u(x))) = \beta \cos(u(x)) \partial_x u,$$

so that

$$\|f(u)\|_V \le \|f(u)\|_H + \|\partial_x f(u)\|_H \le |\beta|(1 + \|\partial_x u\|_H) \le |\beta|(1 + c_0\|u\|_V).$$

Therefore, f maps any bounded set in V into a bounded set in V. Owing to the compact imbedding $V \hookrightarrow H$, we see that f has the compactness property. Also, inequality (52.10) shows that (52.3) is satisfied.

Next we define $F : V \to \mathbb{R}$ as

$$F(u) = \int_0^1 \langle f(tu), u \rangle \, dt = \int_0^1 \int_\Omega \beta \sin(tu(x))u(x)\, dx\, dt, \tag{52.11}$$

for $u \in V$. Using equation (52.11), we have

$$F(u) \le |F(u)| \le |\beta| \int_\Omega |u(x)|\, dx \le |\beta|\, \|u\|_H \le D_\epsilon + \epsilon\|u\|_V^2, \tag{52.12}$$

for $u \in V$, where $D_\epsilon = \epsilon^{-1}|\beta|^2\lambda_1^{-1}$, and λ_1 is the first eigenvalue of A. On the other hand, for any $\epsilon > 0$, from (52.11) we can get

$$F(u) \ge -|F(u)| \ge -|\beta||\Omega|^{1/2}\|u\|_H = -|\beta|\, \|u\|_H \ge -\epsilon\|u\|_H^2 - C_\epsilon, \tag{52.13}$$

for $u \in V$, where $C_\epsilon = |\beta|^2(4\epsilon)^{-1}$. As a result, inequalities (52.12) and (52.13) imply (52.5) and (52.6) with $Q_\epsilon = (\epsilon\lambda_1)^{-1}|\beta|^2$. Thus we have shown that the nonlinearity of the Sine-Gordon equation (52.9) in 1D satisfies the compactness property and conditions (52.3)–(52.6). Actually, the verification given here holds in higher space dimensions, as well.

Example 52.2. The Klein-Gordon equation is given by

$$\partial_t^2 u + \alpha\, \partial_t u - \Delta u + \gamma\, |u|^k u = h, \qquad \text{for } t > 0 \text{ and } x \in \Omega, \tag{52.14}$$

with Dirichlet boundary conditions $u = 0$ on $\Gamma = \partial\Omega$. Here we assume that α and γ are positive constants, $k \ge 0$ and Ω is a bounded domain with smooth boundary Γ in $\mathbb{R}^m$, where $m \le 3$, such that Ω is located locally on one side of Γ. The nonlinearity is $f : \varphi(x) \to \gamma|\varphi(x)|^k\varphi(x)$, and the linear term is $A = -\Delta$, with the Dirichlet boundary conditions. As noted in Section 3.8.4, the corresponding choice of spaces in this case are

$$H = L^2(\Omega), \quad V^1 = V = \mathcal{D}(A^{\frac{1}{2}}) = H_0^1(\Omega), \quad \text{and} \quad W = V \times H.$$

We leave it as an exercise to verify that if k satisfies

$$0 \le k < \infty \quad (\text{when } m = 1 \text{ or } 2), \quad \text{or} \quad 0 \le k \le 2 \quad (\text{when } m = 3),$$

then the nonlinearity in equation (52.14) satisfies the compactness property as well as the conditions (52.3)–(52.6).

Now we return to the general nonlinear wave equation (52.2) and the reduced first-order evolutionary equation (52.7) in the space W. The first result is the global existence of mild solutions.

Theorem 52.3. *Let $h \in L^\infty(0,\infty;H) \cap C^{0,1}_{\rm loc}(0,\infty;H)$, and assume that the hypotheses on f, including (52.3)-(52.6), are valid. Then for every $w_0 \in W$, equation (52.7) has a unique globally defined, mild solution $w = S(t)w_0 = w(t;w_0)$ with $w(0) = w_0$ and*

$$w \in C[0,\infty;W) \cap C^1(0,\infty;W^{-1}). \tag{52.15}$$

Moreover, if h is autonomous, then $w(t;w_0)$ is a strong solution in W and $S(t)$ is a semiflow on W, which is point dissipative and ultimately bounded in W.

Proof. In Corollary 38.8, we have shown that the linear operator $-G$ in equation (52.7) is the infinitesimal generator of an exponentially stable C_0-semigroup e^{-Gt} in the space W, and this semigroup is differentiable, for $t > 0$. Since the nonlinearity f satisfies (52.3), it follows from Theorems 46.1 and 46.3 that for every $w_0 \in W$ there is a $T = T(w_0) > 0$ such that there exists a maximally defined mild solution $w(t)$ in W on the interval $[0,T)$ of equation (52.7), which is uniquely determined by the initial condition w_0. In particular, $w \in C[0,T;W)$. Furthermore, $S(t)w_0$ is locally Lipschitz continuous in $w_0 \in W$ (see Theorem 46.4). Also, Theorem 46.2 implies that $S(t)w_0$ is a strong solution in W. If we are able to show that $T(w_0) = \infty$, for every $w_0 \in W$, then it follows from Section 4.6.4 that $S(t)w_0$ is a semiflow on W, when h is autonomous.

In order to show that $T(w_0) = \infty$, we use the functional $E : W \to \mathbb{R}$ given by

$$E(\zeta) = \frac{1}{2}\|\varphi\|_V^2 + \frac{1}{2}\|\psi\|_H^2 + F(\varphi) - \langle h,\varphi\rangle, \qquad \text{with } \zeta = \begin{pmatrix}\varphi\\ \psi\end{pmatrix} \in W, \tag{52.16}$$

where $F(\varphi)$ satisfies (52.4) - (52.5). By choosing $\epsilon = 1/8$, letting $C = C_{1/8}$ in inequality (52.5), and using the Young inequality, we obtain

(52.17)
$$\begin{aligned} E(\zeta) &\geq \frac{1}{2}\|\varphi\|_V^2 + \frac{1}{2}\|\psi\|_H^2 - \left(\frac{1}{8}\|\varphi\|_H^2 + C\right) - \left(\frac{1}{8}\|\varphi\|_H^2 + 2\|h\|_\infty^2\right) \\ &\geq \frac{1}{4}\left\|\begin{pmatrix}\varphi\\ \psi\end{pmatrix}\right\|_W^2 - (C + 2\|h\|_\infty^2) = \frac{1}{4}\|\zeta\|_W^2 - (C + 2\|h\|_\infty^2). \end{aligned}$$

By Theorem 46.2, for any $w_0 \in \mathcal{D}(G) = V^2 \times V^1$, the mild solution $w(t) = S(t)w_0$ of equation (52.7) in W is a strong solution in W. Consequently, $E(w(t))$ is differentiable with respect to time t almost everywhere, and from (52.2) we obtain

$$\begin{aligned} \partial_t E(w(t)) &= \frac{1}{2}\,\partial_t\langle A^{1/2}u(t), A^{1/2}u(t)\rangle + \frac{1}{2}\,\partial_t\langle \partial_t u(t), \partial_t u(t)\rangle \\ &\quad + \langle f(u(t)), \partial_t u(t)\rangle - \langle h, \partial_t u(t)\rangle \\ &= \langle Au(t), \partial_t u(t)\rangle + \langle \partial_t^2 u(t), \partial_t u(t)\rangle \\ &\quad + \langle f(u(t)), \partial_t u(t)\rangle - \langle h, \partial_t u(t)\rangle \\ &= -\alpha\|\partial_t u(t)\|_H^2 \leq 0, \end{aligned}$$

for almost every $t \in (0,T)$. Hence one has $E(w(t)) \leq E(w_0)$, whenever $w_0 \in \mathcal{D}(G)$. From inequality (52.17), it follows that for any $w_0 \in \mathcal{D}(G)$, one has

$$\tag{52.18} \frac{1}{4}||w(t;w_0)||_W^2 \leq (C+2||h||_\infty^2)+E(w(t;w_0)) \leq C+2||h||_\infty^2+E(w_0),$$

for any $t \in [0,T(w_0))$. The fact that the right side of this inequality does not depend on time, together with equation (46.21), implies that $T(w_0) = \infty$, for all $w_0 \in \mathcal{D}(G)$. Since $\mathcal{D}(G)$ is dense in W, the continuous dependence of the mild solution $S(t)w_0$ on $w_0 \in W$ and the continuity of $E(\zeta)$ imply that inequality (52.18) is valid for all $w_0 \in W$. Therefore, $T(w_0) = \infty$, for all $w_0 \in W$, and $S(t)w_0$ is a semiflow on W.

Next we show that the semiflow $S(t)w_0$ is point dissipative by the *multiplier method*, a method which is commonly used in the theory of nonlinear partial differential equations. In particular, we will use the multiplier $u_t + \epsilon u$, where a new constant $\epsilon > 0$, will be specified later. We use here the abbreviated notation $u_t = \partial_t u$ and $u_{tt} = \partial_t^2 u$.

By taking the scalar product in H of equation (52.2) with $u_t + \epsilon u$, we have

$$\tag{52.19} \langle u_t, u_{tt}+\alpha u_t+Au+f(u)\rangle+\langle \epsilon u, u_{tt}+\alpha u_t+Au+f(u)\rangle = \langle u_t+\epsilon u, h\rangle.$$

Next we define the functional $L : W \to \mathbb{R}$ by

$$L(\zeta) = \frac{1}{2}||\psi||_H^2 + \frac{1}{2}||\varphi||_V^2 + F(\varphi) + \frac{\epsilon\alpha}{2}||\varphi||_H^2 + \epsilon\langle\varphi,\psi\rangle,$$

where $\zeta = (\varphi,\psi)^T$. With $K_\epsilon = 1+\epsilon^{-1}$, and $w_0 \in \mathcal{D}(G)$, we see that (52.19) implies that

$$\begin{aligned}\partial_t\Big[\frac{1}{2}||u_t||_H^2 + \frac{1}{2}||A^{1/2}u||_H^2 + F(u) + \frac{\varepsilon\alpha}{2}\,||u||_H^2 + \varepsilon\,\langle u,u_t\rangle\Big]& \\ + [(\alpha-\varepsilon)||u_t||_H^2 + \varepsilon||A^{\frac{1}{2}}u|| + \varepsilon\langle u, f(u)\rangle]& \\ < (\varepsilon||u_t||_H^2 + \varepsilon^{-1}||h||_H^2) + (\varepsilon^2\lambda_1^{-1}||A^{\frac{1}{2}}u||_H^2 + ||h||_H^2)&.\end{aligned}$$

Thus one has

$$\begin{aligned}\partial_t L(w(t)) + \Big[(\alpha-\epsilon)||u_t||_H^2 + \epsilon||A^{\frac{1}{2}}u||_H^2 + \epsilon\langle u, f(u)\rangle\Big]& \\ \leq (\epsilon||u_t||_H^2 + \epsilon^{-1}||h||_\infty^2) + \Big(\epsilon^2\lambda_1^{-1}||A^{\frac{1}{2}}u||_H^2 + ||h||_\infty^2\Big)&,\end{aligned}$$

or

$$\tag{52.20} \begin{aligned}\partial_t L(w(t)) + \Big[(\alpha-2\epsilon)||u_t||_H^2 + \epsilon(1-\epsilon\lambda_1^{-1})||A^{\frac{1}{2}}u||_H^2 + \epsilon\langle u, f(u)\rangle\Big]& \\ \leq K_\epsilon||h||_\infty^2&,\end{aligned}$$

for $t > 0$. Now (52.6) implies

$$(52.21)\qquad \begin{aligned}(\alpha - 2\epsilon)\|u_t\|_H^2 + \epsilon(1 - \epsilon\lambda_1^{-1})\|A^{1/2}u\|_H^2 + \epsilon\langle u, f(u)\rangle \\ \geq (\alpha - 2\epsilon)\|u_t\|_H^2 + \epsilon(1 - \epsilon\lambda_1^{-1} - \epsilon)\|u\|_V^2 - \epsilon Q_\epsilon.\end{aligned}$$

Also (52.5) implies
(52.22)

$$\begin{aligned} L(w(t)) &= \frac{1}{2}\|u_t\|_H^2 + \frac{1}{2}\|u\|_V^2 + F(u) + \frac{\epsilon\alpha}{2}\|u\|_H^2 + \epsilon\langle u, u_t\rangle \\ &\leq \frac{1}{2}\|u_t\|_H^2 + \frac{1}{2}\|u\|_V^2 + (D_\epsilon + \epsilon\|u\|_V^2) \\ &\quad + \frac{\epsilon\alpha\lambda_1^{-1}}{2}\|u\|_V^2 + \epsilon(\lambda_1^{-1}\|u\|_V^2 + \|u_t\|_H^2) \\ &= \left(\frac{1}{2} + \epsilon\right)\|u_t\|_H^2 + \left[\frac{1}{2} + \epsilon + \epsilon\lambda_1^{-1}\left(\frac{\alpha}{2} + 1\right)\right]\|u\|_V^2 + D_\epsilon. \end{aligned}$$

We now fix $\epsilon > 0$ so that
(52.23)

$$\alpha - 2\epsilon \geq \epsilon\left(\frac{1}{2} + \epsilon\right) \quad \text{and} \quad \epsilon(1 - \epsilon\lambda_1^{-1} - \epsilon) \geq \epsilon\left[\frac{1}{2} + \epsilon + \epsilon\lambda_1^{-1}\left(\frac{\alpha}{2} + 1\right)\right].$$

Then from (52.20), (52.21) and (52.22), for any $w_0 \in \mathcal{D}(G)$ and any $t > 0$, we have

$$\partial_t L(w(t)) + \epsilon L(w(t)) \leq K_\epsilon\|h\|_\infty^2 + M_\epsilon,$$

where $M_\epsilon = \epsilon(Q_\epsilon + D_\epsilon)$. From the Gronwall inequality, we then obtain

$$(52.24)\qquad L(w(t)) \leq e^{-\epsilon t}L(w_0) + \epsilon^{-1}(K_\epsilon\|h\|_\infty^2 + M_\epsilon), \qquad \text{for } t \geq 0,$$

for $w_0 \in \mathcal{D}(G)$. By the continuity of the mild solution with respect to initial data and the continuity of $L(w)$, as noted above, inequality (52.26) holds for all $w_0 \in W$.

In addition to inequality (52.23), by choosing a smaller value of $\epsilon > 0$ if necessary, we can assume that $0 < 6\epsilon < \min(3, \lambda_1)$. In this case we find that

$$(52.25)\qquad \begin{aligned} L(\zeta) &\geq \frac{1}{2}\|\psi\|_H^2 + \frac{1}{2}\|\varphi\|_V^2 + F(\varphi) + \epsilon\langle\varphi, \psi\rangle \\ &\geq \frac{1}{2}\|\psi\|_H^2 + \frac{1}{2}\|\varphi\|_H^2 - \epsilon\|\varphi\|_H^2 - C_\epsilon - \frac{\epsilon}{2}(\|\varphi\|_H^2 + \|\psi\|_H^2) \\ &\geq \frac{1}{2}(1 - \epsilon)\|\psi\|_H^2 + \frac{1}{2}(1 - 3\epsilon\lambda_1^{-1})\|\varphi\|_V^2 - C_\epsilon \\ &\geq \frac{1}{4}(\|\psi\|_H^2 + \|\phi\|_V^2) - C_\epsilon = \frac{1}{4}\|\zeta\|_W^2 - C_\epsilon, \end{aligned}$$

where the constant C_ϵ is given in inequality (52.5). Finally, we combine (52.24) and (52.25) to obtain

$$(52.26)\qquad \frac{1}{4}\|w(t; w_0)\|_W^2 \leq e^{-\epsilon t}L(w_0) + C_\epsilon + \frac{1}{\epsilon}(K_\epsilon\|h\|_\infty^2 + M_\epsilon),$$

for $t \geq 0$, which implies that

$$\limsup_{t\to\infty} \|w(t;w_0)\|_W^2 \leq 4\left[C_\epsilon + \epsilon^{-1}(K_\epsilon\|h\|_\infty^2 + M_\epsilon)\right]. \tag{52.27}$$

Since the right side of inequality (52.27) does not depend on the initial condition w_0, the semiflow $S(t)$ is point dissipative and ultimately bounded in W. □

Next we show that there exists a global attractor $\mathfrak{A}$ for the semiflow $S(t)$ generated by the mild solutions of equation (52.2).

Theorem 52.4. *In addition to the hypotheses of Theorem 52.3, assume that $h \in H$ is independent of time t. Then there exists a global attractor $\mathfrak{A}$ for the semiflow $S(t)w_0 = w(t;w_0)$ on W given by Theorem 52.3. Moreover, $\mathfrak{A}$ attracts every bounded set in W.*

Proof. According to Theorem 52.3 and the Existence Theorem 23.12, it suffices to show here that $S(t)$ is κ-contracting. We will use Lemma 22.4 for this purpose. From the Variation of Constants Formula for mild solutions, we see that

$$S(t)w_0 = e^{-Gt}w_0 + S_1(t)w_0, \qquad \text{for all } t \geq 0 \text{ and } w_0 \in W, \tag{52.28}$$

where e^{-Gt} is the C_0-group generated by $-G$, and

$$\begin{aligned} S_1(t)w_0 &= \int_0^t e^{-G(t-s)}[R(w(s;w_0)) + g]\,ds \\ &= \int_0^t e^{-Gs}[R(w(t-s;w_0)) + g]\,ds, \end{aligned}$$

since g is independent of time t. Furthermore, from Corollary 38.8, there are constants $M \geq 1$ and $\mu > 0$ such that

$$\|e^{-Gt}w_0\|_W \leq Me^{-\mu t}\|w_0\|_W, \qquad \text{for all } w_0 \in W \text{ and } t \geq 0. \tag{52.29}$$

We will now show that for every $a > 0$ there is a $r = r(a)$, with $0 \leq r < \infty$, such that the closed ball $B_a(0)$ centered at the origin and with radius a satisfies

$$Cl_W\, S_1(t)B_a(0) \text{ is compact in } W, \qquad \text{for all } t > r(a).$$

Since (52.16) and (52.5) imply that there are constants $K^1, K^2(h) \geq 0$ such that

$$E(w) \leq K^1\|w\|_W^2 + K^2(h), \qquad \text{for all } w \in W,$$

it follows from from inequality (52.18) that for every $a > 0$ and every given $t \geq 0$, the set

$$U_a(t) \stackrel{\text{def}}{=} \{w(t-s;w_0) : w_0 \in B_a(0) \text{ and } 0 \leq s \leq t\}$$

is bounded in W. Recall that

$$R(w(t-s;w_0))+g=\begin{pmatrix}0\\ h-f(u(t-s;u_0,u_1))\end{pmatrix},$$

where $u(t-s;u_0,u_1)$ is the V-component of the vector function $w(t-s;w_0)$.

Since $f:V\to H$ has the compactness property, it follows that $R(U_a(t))+g$ is a precompact set in W, for every $a\geq 0$ and every $t\geq 0$. Furthermore, we claim that, for every $a\geq 0$ and every $t\geq 0$, the set

$$V_a(t)\stackrel{\text{def}}{=}\{e^{-Gs}[R(U_a(t))+g]:0\leq s\leq t\}$$

is a precompact set in W. Indeed, it suffices to show that $V_a(t)$ is totally bounded in W. Let $M\geq 1$ be the constant that appeared in (52.29). Let $\eta>0$ be arbitrary. Then there is a finite set $\{w_1,\cdots,w_m\}\subset R(U_a(t))+g$ such that for every $w\in R(U_a(t))+g$, there is an $i\in\{1,\cdots,m\}$ with

$$\|w-w_i\|_W<\frac{\eta}{2M}.$$

Fix this $i=i(w)$. Since $e^{-Gs}:[0,\infty)\to W$ is strongly continuous, there exist a finite set $\{s_1^i,\cdots,s_{n_i}^i\}\subset[0,t]$ such that for any $s\in[0,t]$, there is an integer $j\in\{1,\cdots,n_i\}$ with

$$\|e^{-Gs}w_i-e^{-Gs_j^i}w_i\|_W<\frac{\eta}{2}.$$

Without loss of generality, we can assume that s_j^i is nondecreasing in j. Then for any $w\in R(U_a(t))+g$ and any $s\in[0,t]$, there must exist indices i and j such that

$$\begin{aligned}\|e^{-Gs}w-e^{-Gs_j^i}w_i\|_W&\leq\|e^{-Gs}w-e^{-Gs}w_i\|_W+\|e^{-Gs}w_i-e^{-Gs_j^i}w_i\|_W\\&<M\frac{\eta}{2M}+\frac{\eta}{2}=\eta.\end{aligned}$$

This shows that the set $V_a(t)$ is totally bounded with the η-net given by

$$N=\left\{e^{-Gs_j^i}w_i:j=1,\cdots,n_i;\ i=1,\cdots,m\right\}.$$

Thus we have proved that for every $a\geq 0$ and every given $t\geq 0$, the set $V_a(t)$ is precompact in W. Then by the Mazur Theorem (see Appendix A) the closed convex hull $C\ell_W\mathrm{Conv}(V_a(t))$, must be compact in W.

Finally, we note that for every given $w_0\in B_a(0)$ and $t>0$, one has

$$\frac{1}{t}\int_0^t e^{-Gs}[R(w(t-s;w_0))+g]\,ds\in C\ell_W\mathrm{Conv}(V_a(t)),$$

so that

$$\|f(u)\|_V \leq \|f(u)\|_H + \|\partial_x f(u)\|_H \leq |\beta|(1 + \|\partial_x u\|_H) \leq |\beta|(1 + c_0\|u\|_V).$$

Therefore, f maps any bounded set in V into a bounded set in V. Owing to the compact imbedding $V \hookrightarrow H$, we see that f has the compactness property. Also, inequality (52.10) shows that (52.3) is satisfied.

Next we define $F : V \to \mathbb{R}$ as

$$F(u) = \int_0^1 \langle f(tu), u\rangle \, dt = \int_0^1 \int_\Omega \beta \sin(tu(x))u(x) \, dx \, dt, \tag{52.11}$$

for $u \in V$. Using equation (52.11), we have

$$F(u) \leq |F(u)| \leq |\beta| \int_\Omega |u(x)| \, dx \leq |\beta| \, \|u\|_H \leq D_\epsilon + \epsilon\|u\|_V^2, \tag{52.12}$$

for $u \in V$, where $D_\epsilon = \epsilon^{-1}|\beta|^2\lambda_1^{-1}$, and λ_1 is the first eigenvalue of A. On the other hand, for any $\epsilon > 0$, from (52.11) we can get

$$F(u) \geq -|F(u)| \geq -|\beta||\Omega|^{1/2}\|u\|_H = -|\beta| \, \|u\|_H \geq -\epsilon\|u\|_H^2 - C_\epsilon, \tag{52.13}$$

for $u \in V$, where $C_\epsilon = |\beta|^2(4\epsilon)^{-1}$. As a result, inequalities (52.12) and (52.13) imply (52.5) and (52.6) with $Q_\epsilon = (\epsilon\lambda_1)^{-1}|\beta|^2$. Thus we have shown that the nonlinearity of the Sine-Gordon equation (52.9) in 1D satisfies the compactness property and conditions (52.3)–(52.6). Actually, the verification given here holds in higher space dimensions, as well.

Example 52.2. The Klein-Gordon equation is given by

$$\partial_t^2 u + \alpha \, \partial_t u - \Delta u + \gamma \, |u|^k u = h, \qquad \text{for } t > 0 \text{ and } x \in \Omega, \tag{52.14}$$

with Dirichlet boundary conditions $u = 0$ on $\Gamma = \partial\Omega$. Here we assume that α and γ are positive constants, $k \geq 0$ and Ω is a bounded domain with smooth boundary Γ in $\mathbb{R}^m$, where $m \leq 3$, such that Ω is located locally on one side of Γ. The nonlinearity is $f : \varphi(x) \to \gamma|\varphi(x)|^k\varphi(x)$, and the linear term is $A = -\Delta$, with the Dirichlet boundary conditions. As noted in Section 3.8.4, the corresponding choice of spaces in this case are

$$H = L^2(\Omega), \quad V^1 = V = \mathcal{D}(A^{\frac{1}{2}}) = H_0^1(\Omega), \quad \text{and} \quad W = V \times H.$$

We leave it as an exercise to verify that if k satisfies

$$0 \leq k < \infty \quad (\text{when } m = 1 \text{ or } 2), \quad \text{or} \quad 0 \leq k \leq 2 \quad (\text{when } m = 3),$$

then the nonlinearity in equation (52.14) satisfies the compactness property as well as the conditions (52.3)–(52.6).

Now we return to the general nonlinear wave equation (52.2) and the reduced first-order evolutionary equation (52.7) in the space W. The first result is the global existence of mild solutions.

Then the original problem (52.30) is formulated as an abstract nonlinear wave equation in H:

$$\partial_t^2 u + \delta\, \partial_t u + \alpha\, Au + (\beta + \|A^{\frac{1}{4}}u\|_H^2)A^{\frac{1}{2}}u = 0, \qquad \text{for } t > 0. \tag{52.31}$$

with initial conditions $u(0) = u_0$ and $\partial_t u(0) = u_1$, where $u(t) = u(\cdot, t)$ assumes values in H. As usual, we let $V = V^1 = \mathcal{D}(A^{\frac{1}{2}})$ with the norm $\|u\|_V = \|A^{\frac{1}{2}}u\|_H$. Note that this norm is equivalent to the norm of $H^2(0,1)$. As before, we let $\|\cdot\|_W$ denote the norm on the product space $W = V \times H$.

We define the linear operator G by

$$G = \begin{pmatrix} 0 & -I_H \\ \alpha A & \delta I_H \end{pmatrix},$$

where the domain is $\mathcal{D}(G) = \mathcal{D}(A) \times V = V^2 \times V^1 \subset W$. Let $v = \partial_t u$ and set $w = \begin{pmatrix} u \\ v \end{pmatrix}$ and $w_0 = \begin{pmatrix} u_0 \\ u_1 \end{pmatrix}$. The nonlinear term is

$$f(w) = \begin{pmatrix} 0 \\ -(\beta + \|A^{\frac{1}{4}}u\|_H^2)A^{\frac{1}{2}}u \end{pmatrix}. \tag{52.32}$$

One can check that $f \in C_{\mathrm{Lip}}(W, W)$, and $f : \mathcal{D}(A) \to \mathcal{D}(A)$. Then the initial value problem (52.31) can be written as a first-order evolutionary equation in W

$$\partial_t w + Gw = f(w), \qquad \text{for } t > 0 \text{ and } w(0) = w_0 \in W. \tag{52.33}$$

By Corollary 38.8, the operator $-G$ generates a C_0-semigroup e^{-Gt} on W. Furthermore, e^{-Gt} is differentiable, for $t > 0$, exponentially stable, and $-G$ has compact resolvent.

It follows from the theory of Section 4.6 that, for $w_0 \in W$, there exists a mild solution

$$w(t; w_0) = S(t)w_0 = e^{-Gt}w_0 + \int_0^t e^{-G(t-s)} f(S(s)w_0)\, ds$$

of the equation (52.33) in W on $[0, T)$, for some $T = T(w_0) > 0$. We claim that for this problem, the mild solutions exist for all $t \geq 0$ and that the associated semiflow $S(t)w_0$ is (point) dissipative and ultimately bounded on W.

Theorem 52.5. *Consider the equation (52.33) on W, with the assumptions made above, where f is given by (52.32). For any $w_0 \in W$, there exists a unique globally defined mild solution $w = w(t) = S(t)w_0$ of equation (52.33) with $w \in C[0, \infty; W)$. Furthermore, the mapping $S(t)$ generated by* *the mild solutions of (52.33) is a point dissipative semiflow on W and it is ultimately bounded.*

Proof. Since the nonlinearity f satisfies $f \in C_{\text{Lip}}(W, W)$, it follows from Section 4.6 (see Theorems 46.1 - 46.3) that for every $w_0 \in W$, there is a unique, maximally defined, mild solution $w(t) = S(t)w_0$ of equation (52.33), defined on an interval $0 \leq t < T = T(w_0)$, where $0 < T(w_0) \leq \infty$. Furthermore, since the semigroup e^{-Gt} is differentiable, for $t > 0$, the mild solution is also a strong solution in W on $[0, T)$ (see Theorem 46.2).

The idea of proving the global existence of solutions and the pointwise dissipativity of the associated semiflow is simple: find suitable *a priori* estimates. Let $\epsilon > 0$ be an arbitrary constant. Taking the inner-product of equation (52.31) with the multiplier $2\partial_t u + \epsilon u$ in H, we get

$$\partial_t L(t) + N(t) = \frac{1}{4}\epsilon\,\beta^2, \qquad t \in [0, T). \tag{52.34}$$

where $L(t)$ and $N(t)$ are two time-dependent functionals from W to $\mathbb{R}$ given by

$$L(t) = \|\partial_t u\|_H^2 + \alpha\|A^{\frac{1}{2}}u\|_H^2 + \epsilon\langle\partial_t u, u\rangle + (\beta/2 + \|A^{\frac{1}{4}}u\|_H^2)^2, \tag{52.35}$$

and
(52.36)

$$N(t) = (2\delta - \epsilon)\|\partial_t u\|_H^2 + \epsilon\,\alpha\|A^{\frac{1}{2}}u\|_H^2 + \epsilon\,\delta\langle\partial_t u, u\rangle + \epsilon\left(\beta/2 + \|A^{\frac{1}{4}}u\|_H^2\right)^2.$$

Now choose $\epsilon > 0$ sufficiently small, such that

$$0 < \epsilon < \min\left\{1,\ \alpha\lambda_1,\ 4\delta\left(3 + \frac{\delta^2}{\alpha\lambda_1}\right)\right\}, \tag{52.37}$$

where $\lambda_1 > 0$ is the smallest eigenvalue of A. Then we deduce that

$$\begin{aligned}
N(t) = \epsilon\Bigg[&\left(\frac{2\delta - \epsilon}{\epsilon}\right)\|\partial_t u\|_H^2 + \alpha\|A^{\frac{1}{2}}u\|_H^2 + \left(\delta - \frac{\epsilon}{2}\right)\langle\partial_t u, u\rangle \\
&+ \frac{\epsilon}{2}\langle\partial_t u, u\rangle + \left(\frac{\beta}{2} + \|A^{\frac{1}{4}}u\|_H^2\right)^2\Bigg] \\
\geq \epsilon\Bigg[&\left(\frac{2\delta - \epsilon}{\epsilon} - \frac{\delta - \epsilon/2}{2\eta}\right)\|\partial_t u\|_H^2 + \frac{\alpha}{2}\|A^{\frac{1}{2}}u\|_H^2 \\
&+ \left(\frac{\alpha}{2}\|A^{\frac{1}{2}}u\|_H^2 - \frac{\eta}{2}\left(\delta - \frac{\epsilon}{2}\right)\|u\|_H^2\right) + \frac{\epsilon}{2}\langle\partial_t u, u\rangle + \left(\frac{\beta}{2} + \|A^{\frac{1}{4}}u\|_H^2\right)^2\Bigg]
\end{aligned}$$

(now fix $\eta = \alpha\lambda_1\delta^{-1}$)

$$\begin{aligned}
\geq \epsilon\Bigg[&\left(\frac{2\delta - \epsilon}{\epsilon} - \frac{2\delta - \epsilon}{4\alpha\lambda_1\delta^{-1}}\right)\|\partial_t u\|_H^2 + \frac{\alpha}{2}\|A^{\frac{1}{2}}u\|_H^2 \\
&+ \frac{\epsilon}{2}\langle\partial_t u, u\rangle + \left(\frac{\beta}{2} + \|A^{\frac{1}{4}}u\|_H^2\right)^2\Bigg].
\end{aligned}$$

Since (52.37) implies that

$$\frac{2\delta-\epsilon}{\epsilon}-\frac{2\delta-\epsilon}{4\alpha\lambda_1\delta^{-1}}=\frac{2\delta}{\epsilon}-1-\frac{\delta^2}{2\alpha\lambda_1}+\frac{\epsilon}{4\alpha\lambda_1\delta^{-1}}\geq\frac{1}{2},$$

we then obtain
(52.38)

$$N(t)\geq\frac{\epsilon}{2}\left[||\partial_t u||_H^2+\alpha||A^{\frac{1}{2}}u||_H^2+\epsilon\langle\partial_t u,u\rangle+2\left(\frac{\beta}{2}+||A^{\frac{1}{4}}u||_H^2\right)^2\right].$$

Since

$$2\left(\frac{\beta}{2}+||A^{\frac{1}{4}}u||_H^2\right)^2\geq\left(\beta+||A^{\frac{1}{4}}u||_H^2\right)^2-\frac{\beta^2}{2},$$

relations (52.38) and (52.35) imply that

$$N(t)\geq\frac{\epsilon}{2}L(t)-\frac{1}{4}\epsilon\beta^2. \tag{52.39}$$

Substituting (52.39) into (52.34), we then get

$$\partial_t L(t)+\frac{\epsilon}{2}L(t)\leq\frac{\epsilon}{2}\beta^2,\qquad\text{for } t\in[0,T).$$

From the Gronwall inequality we obtain

$$L(t)\leq e^{-\frac{\epsilon}{2}t}L(0)+\beta^2,\qquad\text{for } t\in[0,T).$$

Now inequality (52.37) also implies that

$$2L(t)\geq||\partial_t u||_H^2+\alpha||A^{\frac{1}{2}}u||_H^2+(\beta+||A^{\frac{1}{4}}u||_H^2)^2,$$

which in turn implies that

$$\min(1,\alpha)||S(t)w_0||_W^2\leq2\left(e^{-\frac{\epsilon}{2}t}L(0)+\beta^2\right),\qquad\text{for } t\in[0,T), \tag{52.40}$$

where $L(0)$ depends on initial datum w_0. Inequality (52.40) shows that for any $w_0\in W$, the solution $S(t)w_0$ exists in $C[0,\infty;W)$, so that $T=\infty$, and that

$$\limsup_{t\to\infty}||S(t)w_0||_W^2\leq2\beta^2(\min(1,\alpha))^{-1}.$$

This indicates that the semiflow $S(t)$ generated by the mild solutions of equation (52.33) is point dissipative and ultimately bounded in W. □

The next step is to show the existence of a global attractor, by proving that the semiflow is κ-contracting, with the aid of Lemma 22.5, instead of Lemma 22.4.

Theorem 52.6. *Under the assumptions made above, there exists a global attractor $\mathfrak{A}$ for the semiflow $S(t)$ generated by the mild solutions of equation (52.33) in W and $\mathfrak{A}$ attracts every bounded set in W.*

Proof. According to the Existence Theorem 23.12, since the semiflow $S(t)$ is point dissipative and ultimately bounded, it suffices to show that $S(t)$ is κ-contracting. Let us consider two trajectories of equation (52.33), $S(t)w_{0i} = w_i(t)$, for $i = 1, 2$, with

$$w_1(t) = \begin{pmatrix} u_1(t) \\ v_1(t) \end{pmatrix} = \begin{pmatrix} u_1(\cdot,t) \\ \partial_t u_1(\cdot,t) \end{pmatrix} \text{ and } w_2(t) = \begin{pmatrix} u_2(t) \\ v_2(t) \end{pmatrix} = \begin{pmatrix} u_2(\cdot,t) \\ \partial_t u_2(\cdot,t) \end{pmatrix},$$

where the initial data $w_i(0) = w_{0i}$ are in a given bounded set B in W. Let $z(t) = w_1(t) - w_2(t) = \begin{pmatrix} y(t) \\ \partial_t y(t) \end{pmatrix}$, where $y(t) = u_1(\cdot,t) - u_2(\cdot,t)$. Consider next the equation satisfied by the first component $y(\cdot)$ of $z(\cdot)$. By using an integration by parts with equation (52.30), one obtains

$$\begin{aligned} \partial_t^2 y + \delta\,\partial_t y + \alpha\,\partial_x^4 y &= \left(\beta + \int_\Omega |\partial_x u_1(\xi,t)|^2\,d\xi\right)\partial_x^2 y \\ &\quad - \left(\int_\Omega y(\xi,t)\,(\partial_x^2 u_1 + \partial_x^2 u_2)(\xi,t)\,d\xi\right)\partial_x^2 u_2, \end{aligned} \tag{52.41}$$

for $t > 0$. Taking the inner-product of equation(52.41) with $2\partial_t y + \epsilon\, y$, we obtain

$$\frac{d}{dt}P(t) + Q(t) = h\,(t; u_1, u_2, y)\,, \tag{52.42}$$

where

$$\begin{aligned} P(t) &= ||\partial_t y||_H^2 + \alpha||\partial_x^2 y||_H^2 + \epsilon\langle y, \partial_t y\rangle_H + \frac{1}{2}\epsilon\delta||y||_H^2 \\ &\quad + \left(\beta + \int_\Omega |\partial_x u_1(\xi,t)|^2\,d\xi\right)||\partial_x y||_H^2, \\ Q(t) &= (2\delta - \epsilon)||\partial_t y||_H^2 + \epsilon\alpha||\partial_x^2 y||_H^2 \\ &\quad + \epsilon\left(\beta + \int_\Omega |\partial_x u_1(\xi,t)|^2\,d\xi\right)||\partial_x y||_H^2, \end{aligned}$$

and

$$h(t; u_1, u_2, y) = -\left\langle 2\partial_t y + \epsilon\, y, \left(\int_\Omega y(\xi,t)(\partial_x^2 u_1 + \partial_x^2 u_2)(\xi,t)\,d\xi\right)\partial_x^2 u_2\right\rangle.$$

Similarly, as argued in Theorem 52.5, one can assert that, for small $\epsilon > 0$,

$$Q(t) - \frac{1}{2}\epsilon P(t) \geq \frac{1}{2}\epsilon\left(\beta + \int_\Omega |\partial_x u_1(\xi,t)|^2\,d\xi\right)||\partial_x y||_H^2. \tag{52.43}$$

With this $\epsilon > 0$ fixed so that (52.43) holds, we let

$$g(t; u_1, u_2, y) = \frac{1}{2}\epsilon\left(\beta + \int_\Omega |\partial_x u_1(\xi, t)|^2\, d\xi\right) ||\partial_x y||_H^2$$
$$= -\frac{1}{2}\epsilon\left(\beta + \int_\Omega |\partial_x u_1(\xi, t)|^2\, d\xi\right) \langle y, \partial_x^2 y\rangle_H.$$

Then from equations (52.42) and (52.43) it follows that

$$\text{(52.44)} \quad \partial_t P(t) + \frac{1}{2}\epsilon\, P(t) \leq h(t; u_1, u_2, y) - g(t; u_1, u_2, y), \qquad \text{for } t > 0.$$

Now inequality (52.40) implies that the two trajectories $S(t)w_{0i}$, for $i = 1, 2$, are uniformly bounded in W, for all $t \geq 0$, i.e., there is a constant $M(B) > 0$ such that

$$||S(t)w_{0i}||_W \leq M(B), \qquad \text{for } t \geq 0, \text{ and } i = 1, 2.$$

Since $W \subset H^2(\Omega) \times L^2(\Omega)$, this implies that

$$||u_i(t)||_H,\ ||\partial_t u_i(t)||_H,\ ||\partial_x u_i(t)||_H,\ ||\partial_x^2 u_i(t)||_H\ \leq M(B),$$

for $t \geq 0$ and $i = 1, 2$. Consequently, from the definitions of h and g and the Schwarz inequality, one can show that there exists a constant $L(B)$ such that

$$\text{(52.45)} \qquad \left\{ \begin{array}{l} |h(t; u_1, u_2, y)| \\ |g(t; u_1, u_2, y)| \end{array} \right. \leq L(B)||y(t)||_H^2, \qquad \text{for any } t > 0.$$

Now by integrating inequality (52.44) and substituting the inequalities (52.45) into the Volterra integral in the integrated solution of the differential inequality (52.44), we get

$$\text{(52.46)} \qquad P(t) \leq e^{-\frac{\epsilon}{2}t}P(0) + 4\epsilon^{-1}L(B)||y||_{\infty:[0,t]}^2,$$

where $||y||_{\infty:[0,t]} = ||y||_{L^\infty(0,t;H)}$. By choosing a smaller value for $\epsilon > 0$, if necessary, it follows that there exist positive constants $b = b(\alpha, \epsilon, \delta)$ and $d = d(\alpha, \epsilon, \delta)$ such that

$$P(t) \geq b||z(t)||_W^2 + 2\epsilon^{-1}g(t; u_1, u_2, y)$$

and

$$P(0) \leq d||z(0)||_W^2 + 2\epsilon^{-1}g(0; u_1, u_2, y).$$

From inequality (52.46), it follows that for $t > 0$, and w_{01}, $w_{02} \in B$, one has

(52.47)

$$||S(t)w_{01} - S(t)w_{02}||_W^2 \leq db^{-1}e^{-\frac{\epsilon}{2}t}||w_{01} - w_{02}||_W^2 + CL(B)||y||_{\infty;[0,t]}^2,$$

where $C = 8b^{-1}\epsilon^{-1}$. Then it is easy to conclude that for $t > 0$ and $w_{0i} \in B$, for $i = 1, 2$, one has

$$\|S(t)w_{01} - S(t)w_{02}\|_W^2 \leq \phi(t)\|w_{01} - w_{02}\|_W^2 + \rho_t(w_{01}, w_{02})^2 \tag{52.48}$$

where $\phi(t) = db^{-1}e^{-\frac{\epsilon}{2}t}$, and $\rho_t(w_{01}, w_{02})^2 = CL(B)\|u_1 - u_2\|^2_{\infty;[0,t]}$.

We claim that the term $\rho_t(w_{01}, w_{02})$ in (52.48) is a precompact pseudometric on the space W. This can be proved by applying the Ascoli-Arzelá Theorem, see Naylor and Sell (1982). Indeed for any given bounded sequence $\{w_0^n\}$ in $W \subset H^2(\Omega) \times L^2(\Omega)$, let

$$S(t)w_0^n = w^n(t) = \begin{pmatrix} u^n(\cdot, t) \\ \partial_t u^n(\cdot, t) \end{pmatrix}, \qquad \text{for } n = 1, 2, \cdots,$$

be the mild solutions in W. From inequality (52.40), we see that, for all $t \geq 0$, the sequence $S(t)w_0^n$ is uniformly bounded in $H^2(\Omega) \times L^2(\Omega)$. Hence, one has

(1) $u^n(\cdot, t)$ is bounded in $H^2(\Omega)$, and consequently, it is precompact in $L^2(\Omega)$, for each $t > 0$, and
(2) $\partial_t u^n(\cdot, t)$ is bounded in $L^\infty(0, \infty; L^2(\Omega))$. Hence, there is a constant $K \geq 0$ with the property that $\|u^n(t_1) - u^n(t_2)\|_{L^2(\Omega)} \leq K|t_1 - t_2|$, for any n and any t_1, $t_2 \geq 0$.

Consequently, the sequence $u^n(\cdot, t)$, as an H-valued functions, is equicontinuous in t, for $t \geq 0$. Therefore, by the Ascoli-Arzelá Theorem there is a subsequence of $u^n(\cdot, t)$ that is convergent in the Fréchet space $L^\infty_{loc}[0, \infty; H)$, and this is a Cauchy sequence with respect to $\rho_t(w_{01}, w_{02})$.

Finally, with inequality (52.48) and the properties satisfied by ϕ_t and ρ_t, it follows from Lemma 22.5 that the semiflow $S(t)$ on W is κ-contracting, since the bounded set B is arbitrary. By the Existence Theorem 23.12, it follows that there is a global attractor $\mathfrak{A}$, and $\mathfrak{A}$ attracts every bounded set B in W. □

5.2.3 Nonlinear Wave Equations With Local Damping. As noted earlier, a significant feature of dissipative hyperbolic evolutionary equations is that the dissipativity is caused mainly by the damping term, be it linear or nonlinear. In this section, we shall explore the existence of global attractors for nonlinear wave equations with a *local or weak damping* in the form of $d(x)g(u_t)$, where $d(x)$ is a function of spatial variable with its support Υ being a proper (maybe very small) subset of the entire domain Ω.

The main objective here is to prove the existence of global attractors for a NWE with this type of locally distributed damping. Since our goal is to describe the key ideas, we will focus on the 1D (one spatial dimension) case here, with a few observations on the corresponding 2D and 3D results added later.

Let $\Omega = (a, b) \subset \mathbb{R}$ be a bounded, open interval on the real line. Consider the problem:

$$\partial_t^2 u + d(x)g(\partial_t u) - \Delta u + f(u) = 0, \qquad x \in \Omega, \quad t \in \mathbb{R}^+, \tag{52.49}$$

with boundary conditions $u(a,t) = u(b,t) = 0$, for $t \in \mathbb{R}^+$, and initial conditions $u(x,0) = u_0(x)$ and $u_t(x,0) = u_1(x)$, for $x \in \overline{\Omega} = [a,b]$. We will use the abbreviated notation $u_t = \partial_t u$, $u_{tt} = \partial_t^2 u$, and $u_x = \partial_x u$, and we make the following assumptions concerning the functions d, g, and f appearing in equation (52.49).

(Ad) $d \in C^1(\overline{\Omega})$, and one has $0 \leq d(x) \leq d_1$, for any $x \in \overline{\Omega}$, and $0 < d_0 \leq d(x)$, for any $x \in \Upsilon \stackrel{\text{def}}{=} [a,\alpha) \cup (\beta, b]$, with $a < \alpha < \beta < b$, where d_0, d_1 are positive constants.

(Ag) $g \in C^1(\mathbb{R})$ is strictly increasing, $g(0) = 0$, and there are two constants g_0 and g_1 such that

$$0 < g_0 \leq \liminf_{|z|\to\infty} g'(z) \leq \limsup_{|z|\to\infty} g'(z) \leq g_1 < \infty.$$

(Af) $f \in C^2(\mathbb{R})$ satisfies the polynomial growth condition $|f''(z)| \leq c(1+|z|^{2p-1})$, with an integer $p \geq 1$, and the asymptotical sign condition

$$\liminf_{|z|\to\infty} \frac{f(z)}{z} \geq 0.$$

Assumption (Ad) allows for $d(x) = 0$ in $\Omega \backslash \Upsilon = [\alpha, \beta]$. Thus the damping $d(x)\,g(u_t)$ is assumed to be effective only in Υ, which may be very small and with support near the boundary $\partial\Omega = \{a, b\}$. The reader should verify that owing to Assumption (Ag), one has:

(1) the function g is uniformly Lipschitz continuous on $\mathbb{R}$;

(2) there exists a constant $\rho_1 > 0$ such that

$$g(z)^2 \leq \rho_1 g(z) z, \qquad \text{for all } z \in \mathbb{R}; \quad \text{and} \tag{52.50}$$

(3) for any $\epsilon > 0$ there exists a constant $\rho_2 = \rho_2(\epsilon) > 0$ such that

$$z^2 \leq \epsilon + \rho_2(\epsilon) g(z) z, \qquad \text{for all } z \in \mathbb{R}. \tag{52.51}$$

As usual, we will let $\langle \cdot, \cdot \rangle$ and $\|\cdot\|$ denote the inner product and the norm on $H = L^2(\Omega)$. Let $A : \mathcal{D}(A) \to H$ be given by $A\varphi = -\Delta\varphi$, with $\mathcal{D}(A) = H^2(\Omega) \cap H_0^1(\Omega)$. As noted in Chapter 3, this operator A satisfies the Standing Hypothesis B, and the fractional powers of A satisfy: $A^{\frac{1}{2}}\varphi = \partial_x \varphi$, with $\mathcal{D}(A^{\frac{1}{2}}) = H_0^1(\Omega)$ and one has $H_0^{\frac{1}{4}}(\Omega) \subset \mathcal{D}(A^{\frac{1}{4}}) \subset H^{\frac{1}{4}}(\Omega)$. Also, the Poincaré inequality implies that there is a constant $\lambda_0 > 0$ such that

$$\|\varphi\|^2 \leq \lambda_0 \|\partial_x \varphi\|^2, \qquad \text{for all } \varphi \in H_0^1(\Omega). \tag{52.52}$$

We consider the Hilbert space $W = H_0^1(\Omega) \times L^2(\Omega)$, with the norm $\|w\|_W^2 = \|\partial_x\varphi\|^2 + \|\psi\|^2$, where $w = (\varphi, \psi)^T \in W$. Hence, Problem (52.49) reduces to an abstract evolutionary equation:

$$\partial_t w + Gw = R(w), \qquad \text{where } w = \begin{pmatrix} u \\ \partial_t u \end{pmatrix} \in W, \tag{52.53}$$

and

$$G = \begin{pmatrix} 0 & -I \\ A & 0 \end{pmatrix}, \; R(w)(x) = \begin{pmatrix} 0 \\ -d(x)g(\partial_t u) - f(u) \end{pmatrix}, \text{ and } w_0 = \begin{pmatrix} u_0 \\ u_1 \end{pmatrix}.$$

Since the nonlinear term $R = R(w)$ satsifies $R \in C_{\mathrm{Lip}}(W, W)$, it follows from Section 4.6 (See Theorems 46.1 - 46.3) that for each $w_0 \in W$, there is a unique maximally defined mild solution $w(t) = S(t)w_0$ in W on $0 \leq t < \tau(w_0)$, where $0 < \tau(w_0) \leq \infty$. Furthermore, $w(t)$ is a strong solution on this interval. Below we will show that $\tau(w_0) = \infty$, for every $w_0 \in W$. It will then follow that $S(t)w_0$ is a semiflow on W. The main result of interest here is the following:

Theorem 52.7. *Under Assumptions (Ad), (Ag) and (Af), there exists a global attractor $\mathfrak{A} \subset W$ for the semiflow $S(t)$ generated by the mild solutions of equation (52.53) on W, and $\mathfrak{A}$ attracts all bounded sets in W.*

The proof of Theorem 52.9 will be provided through a series of lemmas. Before doing that though, it is useful to make some observations on the difficulties one encounters in this proof. Since the damping is effective only locally in a neighborhood Υ of the boundary, and the nonlinearity is distributed over the entire region Ω, one needs to exercise special care in deriving the *a priori* estimates in order to establish the dissipative property of the semiflow. It will be seen that one new feature in the proof of dissipativity lies in utilizing multipliers which involve functions of the spatial variable x. Another novelty occurs in invoking a unique continuation property for the linear evolutionary equation with a spatially varying perturbation. Also, the proof of the κ-contracting property of the semiflow is somewhat different from the cases studied above, since we encounter a delicate interaction between the nonlinear terms $g(\partial_t u)$ and $f(u)$.

The first lemma establishes a general inequality for the energy functional

$$E(y)(t) = E(y; f)(t) \overset{\text{def}}{=} \frac{1}{2}\|z(t)\|_W^2 + \int_\Omega F(y(x,t))\, dx, \tag{52.54}$$

where $F(r) = \int_0^r f(s)\, ds$, for $r \in \mathbb{R}$,

$$\|z(t)\|_W^2 = \|\partial_x y(t)\|^2 + \|\partial_t y(t)\|^2,$$

and $y(t) = y(x,t)$ is a solution of equation (52.49) with $y(a,t) = y(b,t) = 0$, for $t \geq 0$. It is convenient in this regard to treat a somewhat more general problem in place of (52.49). Let us consider

$$\partial_t^2 y + D(x, \partial_t y) - \Delta u + f(u) = 0, \qquad x \in \Omega, \quad t \in (0,T), \tag{52.55}$$

with boundary condition $y(a,t) = y(b,t) = 0$, for $t \in [0,T)$, where $D = D(x,v)$ satisfies $D \in C^1(\overline{\Omega} \times \mathbb{R}; \mathbb{R})$ and $D(x,v)v \geq 0$, for any $v \in \mathbb{R}$. Note that if D is defined by $D(x,v) \stackrel{\text{def}}{=} d(x)g(v)$, then the Problem (52.49) is a special case of (52.55) under Assumptions (Ad) and (Ag).

Equation (52.55), like equation (52.49), also has the formulation as a first-order evolutionary equation (52.53). Indeed, one obtains the same linear term Gw, but now the nonlinear term assumes the form

$$R(w)(x) = \begin{pmatrix} 0 \\ -D(x, \partial_t y) - f(u) \end{pmatrix}, \quad \text{and} \quad w = \begin{pmatrix} y \\ \partial_t y \end{pmatrix}.$$

We will also use the functional $E(y)$ when referring to the solutions of (52.55). Let $y = y(x,t)$ be any strong solution of equation (52.55) in W on $(0,T)$, where $0 < T < \infty$. Then the following are valid:

i) Since y is a strong solution of equation (52.55), we have

$$\partial_t E(y)(t) = -\int_\Omega D(x, \partial_t y)\partial_t y \, dx \leq 0. \tag{52.56}$$

As a result, for any $t \in [0,T]$, one has

$$E(y)(T) - E(y)(t) = -\int_t^T \int_\Omega D(x, \partial_t y)\partial_t y \, dx \, dt \leq 0. \tag{52.57}$$

ii) By the asymptotic sign condition satisfied by f, it follows from the argument of Section 5.1.1 that, for any $\delta > 0$, there exists a constant $C_\delta > 0$ such that

$$F(r) = \int_0^r f(s)\, ds \geq -\delta r^2 - C_\delta, \qquad \text{for any } r \in \mathbb{R}.$$

In particular, with $\delta = (4\lambda_0)^{-1}$, where λ_0 is given in (52.52), we see that $E(y)(t)$ is bounded below in W, i.e., one has

$$E(y)(t) \geq \frac{1}{2}\|z(t)\|_W^2 - \frac{1}{4\lambda_0}\|y(\cdot,t)\|^2 - c_0 \geq \frac{1}{4}\|z(t)\|_W^2 - c_0, \tag{52.58}$$

where $c_0 = (b-a)C_\delta$.

There are several terms that arise in the subsequent analysis of the solutions of equation (52.49) and which we will write in the abbreviated form $I_i = I_i(y)$, for $i \geq 1$. In particular, we fix $x_0 \in (a,b)$ for the duration of the analysis. Next define

$$I_1 = \int_0^T \int_\Omega (|y|^2 + |y|^{2+2p})\, dx\, dt, \quad I_2 = \int_0^T \int_\Omega D(x, \partial_t y)^2 \, dx\, dt,$$

$$I_3 = \int_0^T \int_\Omega D(x, \partial_t y)\, y \, dx\, dt, \quad I_4 = \int_0^T \int_\Omega D(x, \partial_t y)\, \partial_t y \, dx\, dt,$$

$$I_5 = \int_0^T \int_\Upsilon |\partial_t y|^2 \, dx\, dt, \quad I_{10} = I_1 + I_2, \quad I_{11} = I_{10} + I_4,$$

$$I_6 = I_6(x_0) = \int_0^T [(b - x_0)|\partial_x y(b,t)|^2 + (x_0 - a)|\partial_x y(a,t)|^2]\, dt.$$

Lemma 52.8. *Let the Assumptions (Ad), (Ag), and (Af) hold. Then there is a constant $C_1 > 0$ such that, for every strong solution $y = y(x,t)$ of equation (52.55) in W on $0 \le t \le T$, one has*
(52.59)
$$\int_0^T E(y)(t)\,dt \le -\left[\int_\Omega \partial_t y[(x-x_0)\partial_x y + y]dx\right]_0^T + C_1 I_{10} + \frac{3}{2} I_6(x_0).$$

Proof. First multiply equation (52.55) by $q(x)\partial_x y$, where $q(x) = 3(x - x_0)$, then integrate the two sides of the equality over the region $\Omega \times (0,T)$ to obtain

$$\left[\int_\Omega (\partial_t y\, q\, \partial_x y)dx\right]_0^T + I_{20} + I_{21} = \frac{3}{2} I_6(x_0), \tag{52.60}$$

where

$$I_{20} = \frac{1}{2}\int_0^T \int_\Omega \partial_x q(|\partial_t y|^2 + |\partial_x y|^2 - 2F(y))\,dx\,dt,$$
$$I_{21} = \int_0^T \int_\Omega D(x, \partial_t y)\, q\, \partial_x y\, dx\, dt.$$

Next, multiply equation (52.55) by y and integrate the two sides of the equality over $\Omega \times (0,T)$ to get

$$\left[\int_\Omega (\partial_t y\, y)\, dx\right]_0^T + I_3 = I_{22}, \tag{52.61}$$

where

$$I_{22} = \int_0^T \int_\Omega (|\partial_t y|^2 - |\partial_x y|^2 - f(y)y)\,dx\,dt.$$

By adding the two equations (52.60) and (52.61), we get
(52.62)
$$\begin{aligned}\int_0^T E(y)(t)\,dt < &-\left[\int_\Omega \partial_t y(q\,\partial_x y + y)\,dx\right]_0^T \\ &- \int_0^T \int_\Omega D(x, \partial_t y)(q\,\partial_x y + y)\,dx\,dt \\ &+ \int_0^T \int_\Omega \left(4F(y) - |\partial_x y|^2 - f(y)\,y\right)\,dx\,dt + \frac{3}{2} I_6(x_0).\end{aligned}$$

By using the Young and the Poincaré inequalities, it follows that, for any small $\epsilon > 0$, there exists a constant $c(\epsilon) > 0$ such that

$$\begin{aligned}
\Big|\int_0^T \int_\Omega & D(x, \partial_t y)(q\,\partial_x y + y)\,dx\,dt\Big| \\
&\leq \int_0^T \int_\Omega |D(x, \partial_t y)|(|q\,\partial_x y| + |y|)\,dx\,dt \\
&\leq \int_0^T (\epsilon \int_\Omega (|q|^2|\partial_x y|^2 + |y|^2)\,dx + c(\epsilon)\int_\Omega D(x, \partial_t y)^2\,dx)\,dt \\
&\leq \epsilon\,(\|q\|^2_{L^\infty(\Omega)} + \lambda_0)\int_0^T \|\partial_x y(t)\|^2\,dt \\
&\qquad + c(\epsilon)\int_0^T \int_\Omega D(x, \partial_t y)^2\,dx\,dt.
\end{aligned}$$

Now (Af) implies that $|\int_0^T \int_\Omega (4F(y) - f(y)y)\,dx\,dt| \leq c_0 I_1$. By substituting the last two inequalities into the right-hand side of (52.62), with $\epsilon > 0$ fixed so that $\epsilon(\|q\|^2_{L^\infty(\Omega)} + \lambda_0) < 1$, we get inequality (52.59) with $C_1 = \max(c(\epsilon), c_0)$. □

The next lemma is to estimate the term $\|y_t(\cdot, T)\|^2 + \|y_x(\cdot, T)\|^2$. In this regard, we have the following result:

Lemma 52.9. *Let the Assumptions (Ad), (Ag), and (Af) hold. Then there is a constant $C_2 > 0$ such that for any strong solution y of (52.55) and for a sufficiently large $T > 0$, one has*

$$\|z(T)\|^2_W \leq C_2(I_1 + I_2 + I_4 + I_5 + 1). \tag{52.63}$$

Proof. Let ν be the unit outward normal vector on the boundary $\partial\Omega$, i.e., $\nu(x) = +1$ if $x = b$, and $\nu(x) = -1$ if $x = a$. Let $G(x_0) = \{x \in \partial\Omega : (x - x_0)\cdot\nu(x) > 0\}$, for $x_0 \in (a, b)$. Then $G(x_0) = \partial\Omega = \{a, b\}$, and there are open neighborhoods Υ_1 of a and Υ_2 of b, such that $C\ell(\Upsilon_1 \cup \Upsilon_2) \cap \Omega \subset \Upsilon$. Let $q_0 \in W^{1,\infty}(\overline{\Omega})$ satisfy the following conditions:

$$q_0(x) = \begin{cases} \nu(x), & \text{for } x \in G(x_0), \\ 0, & \text{for } x \in \Omega\backslash(\Upsilon_1 \cup \Upsilon_2). \end{cases}$$

From (52.60) with $q = q_0$, it follows that there is a constant $c(q_0) > 0$ such that

$$\begin{aligned}
\int_0^T \int_{G(x_0)} |\partial_x y|^2\,ds\,dt &\leq \int_0^T \int_{\partial\Omega} (q_0\cdot\nu)|\partial_x y|^2\,ds\,dt \\
&= \int_0^T [|\partial_x y(b,t)|^2 + |\partial_x y(a,t)|^2]\,dt \leq 2I_{31} + I_{32},
\end{aligned} \tag{52.64}$$

where

$$I_{31} = \left[\int_\Omega (\partial_t y\, q_0\, \partial_x y)\, dx\right]_0^T$$

$$I_{32} = c(q_0)\int_0^T \int_{(\Upsilon_1\cup\Upsilon_2)\cap\Omega} (|D|^2 + |\partial_t y|^2 + |\partial_x y|^2 + |y|^2 + |y|^{2p+2})\, dx\, dt.$$

Next we wish to eliminate the term $\int_0^T \int_{(\Upsilon_1\cup\Upsilon_2)\cap\Omega} |\partial_x y|^2\, dx\, dt$ in inequality (52.64). First we revise the argument in Lemma 52.8 leading to (52.61) and multiply equation (52.55) by $\xi_0\, y$, in place of y alone, where $\xi = \xi_0(x) \in W^{2,\infty}(\overline{\Omega})$ satisfies

$$\xi_0(x) \in [0,1], \text{ for } x \in \overline{\Omega}, \quad \xi_0(x) = 1, \text{ for } x \in \Upsilon_1 \cup \Upsilon_2,$$
$$\xi_0(x) = 0 \text{ for } x \in \Omega\backslash\Upsilon.$$

In this case we find that there is a constant $c(\xi_0) > 0$ such that

$$\int_0^T \int_\Omega \xi_0\, |\partial_x y|^2\, dx\, dt \le c(\xi_0)(I_{10} + I_{33}) - \left[\int_\Omega (\partial_t y\, \xi_0\, y)\, dx\right]_0^T,$$

where

$$I_{33} = \int_0^T \int_\Upsilon |\partial_t y|^2\, dx\, dt.$$

By using (52.59) with (52.64) and the last inequality, one obtains

$$\begin{aligned}\int_0^T E[y;f](t)\, dt \le k_1 \int_0^T \int_\Omega (D(x,\partial_t y)^2 + |y|^2 + |y|^{2p+2}) dx\, dt\\ + k_1 \int_0^T \int_\Upsilon |y_t|^2 dx\, dt + J\end{aligned} \tag{52.65}$$

where

$$J = \left[\int_\Omega \partial_t y[(-(x - x_0) + 2c(x_0)q_0)\partial_x y - (1 + c(x_0)c(q_0)\xi_0)y]\, dx\right]_0^T,$$

$c_0 = c(x_0) = (3/2)\max(b - x_0, x_0 - a)$, and $k_1 = k_1(x_0, q_0, \xi_0) > 0$ is a constant. Similarly, with a suitable $k_2 = k_2(x_0, q_0, \xi_0) > 0$, the Schwarz inequality and (52.58), one

$$J \le k_2 \left[E[y;f](T) + \frac{1}{2}\int_0^T \int_\Omega D(x,\partial_t y)\partial_t y\, dx\, dt + c_0\right] \tag{52.66}$$

Observe that (52.57) implies that $TE(y)(T) \le \int_0^T E(y)(t)\, dt$. Hence from inequalities (52.65) and (52.66), it follows that

$$TE(y)(T) \le \int_0^T E(y)(t)\, dt \le k_1(I_{10}+I_{33})+k_2[E(y)(T)+\frac{1}{2}I_4+c_0]. \tag{52.67}$$

Consequently, by (52.58), and after an algebraic arrangement, we get

$$\|z(T)\|_W^2 \leq \frac{4k_1 + 2k_2}{T - k_2}(I_{11} + I_{33}) + \frac{k_2\, c_0}{T - k_2}.$$

For $T \geq 2\,k_2$, inequality (52.63) holds with $C_2 = 4 + \max(2c_0, \frac{k_1}{k_2})$. □

In order to show that the semiflow generated by the mild solutions of (52.53) is point dissipative, we need the next technical lemma.

Lemma 52.10. *Let the Assumptions (Ad), (Ag), and (Af) hold. Then there is a bounded set B_1 in W such that for any given bounded set $B \subset W$ there exists a constant $K(B) > 0$ with the following property: if u is a strong solution of the Problem (52.49) in W, with the initial datum $w_0 = (u_0, u_1)^T \in B$, then*

$$E(u)(T) - E(u)(0) < -K(B),$$

as long as $S(T)w_0 \in \gamma^+(B)\backslash B_1$.

Proof. We will argue by contradiction. If the conclusion is false, then for any $M > 0$, there is a bounded set $B^M \subset W$ and a sequence of solutions $S(t)w_0^n$ of the Problem (52.53) with

$$w_0^n \in B^M \text{ and } S(T)w_0^n \in \gamma^+(B^M), \qquad \text{for all } n \geq 1, \tag{52.68}$$

and such that for all $n \geq 1$, one has

$$M \leq \|S(T)w_0^n\|_W^2 = \|\partial_t u^n(\cdot, T)\|^2 + \|\partial_x u^n(\cdot, T)\|^2, \tag{52.69}$$

while

$$E(u^n)(T) - E(u^n)(0) \to 0, \qquad \text{as } n \to \infty. \tag{52.70}$$

According to (52.57), with $D(x, z) = d(x)g(z)$, the limit (52.70) implies that

$$\int_0^T \int_\Omega d(x)g(\partial_t u^n)\partial_t u^n \, dx\, dt \to 0, \qquad \text{as } n \to \infty. \tag{52.71}$$

The two properties described earlier in (52.56) and (52.58), along with (52.68), imply that the sequence of orbit segments $\{S(t)w_0^n : 0 \leq t \leq T\}$ lies in a bounded set in $L^\infty(0, T; W)$. Hence, there exists a subsequence, which we relabel as $S(t)w_0^n$, and a limit function $U(x, t)$, such that

$$S(t)w_0^n \to \begin{pmatrix} U \\ \partial_t U \end{pmatrix} \quad \text{weak}^* \text{ in } L^\infty(0, T; W), \qquad \text{as } n \to \infty. \tag{52.72}$$

Note that in the 1D case, one has the imbeddings $H_0^1(\Omega) \hookrightarrow C(\overline{\Omega}) \mapsto L^r(\Omega)$, for any $r \in [1, \infty]$, by the Sobolev imbedding theorem. Hence, the polynomial growth condition in (Af) implies that

$$f(u^n) \to f(U) \text{ weakly in } L^2(0,T; L^2(\Omega)), \qquad \text{as } n \to \infty.$$

Moreover, by utilizing the Schwarz inequality, (52.50) and (52.71), we have

$$d(x)g(\partial_t u^n) \to 0 \text{ weakly in } L^2(0,T; L^2(\Omega)), \qquad \text{as } n \to \infty.$$

Therefore, by taking the limit, as $n \to \infty$, in equation (52.49) with $u = u^n$, one finds that the limit function U is a weak solution (see Lions (1969)) of the boundary value problem:

$$\partial_t^2 U - \Delta U + f(U) = 0, \qquad \text{in } \Omega \times (0,T), \tag{52.73}$$

with the boundary condition $U\,|_{\partial\Omega \times [0,T]} = 0$. Besides, (52.71) and (52.72), together with Assumption (Ad), imply that

$$\partial_t U|_{\Upsilon \times [0,T]} = 0. \tag{52.74}$$

Let $V = \partial_t U$. Then V is a weak solution of the boundary value problem:

$$\partial_t^2 V - \Delta V + f'(U)V = 0, \qquad \text{in } \Omega \times (0,T), \tag{52.75}$$

with boundary conditions $V\,|_{\partial\Omega \times [0,T]} = 0$ and $V_t|_{\Upsilon \times [0,T]} = 0$, where $f'(U) \in L^\infty(\Omega \times [0,T])$. Equation (52.75) turns out to be a linear wave equation with a perturbation of potential term whose coefficient function is essentially bounded. By the unique continuation property in the potential theory, cf. Ruiz (1992), or in this 1D case one can show directly, the boundary condition for (52.75) implies that

$$V = \partial_t U = 0 \text{ in } \Omega \times (0,T),$$

which in turn, implies that $U(x,t) \equiv U(x)$ is a stationary weak solution of the elliptic problem:

$$-\Delta U + f(U) = 0, \qquad \text{in } \Omega,$$

with boundary condition $U\,|_{\partial\Omega} = 0$, see Adams (1965).

On the other hand, by combining (52.69) and (52.63) (in which one replaces y by u^n), one obtains

$$\begin{aligned}
M &\le \|S(T)w_0^n\|_W^2 \\
&\le C_2\Big(\int_0^T \int_\Omega (d(x)^2 g(\partial_t u^n)^2 + d(x)g(\partial_t u^n)\partial_t u^n + |u^n|^2 + |u^n|^{2p+2})\, dx\, dt \\
&\quad + \int_0^T \int_\Upsilon |\partial_t u^n|^2\, dx\, dt + 1\Big).
\end{aligned}$$

Then (52.50), (52.51), and (52.71) imply that

$$\int_0^T \int_\Omega (d(x)^2 g(\partial_t u^n)^2 + d(x) g(\partial_t u^n)\partial_t u^n)\, dx\, dt \to 0, \qquad \text{as } n \to \infty$$

and

$$\int_0^T \int_\Upsilon |\partial_t u^n|^2\, dx\, dt \le \epsilon(b-a)T + \frac{\rho_2(\epsilon)}{d_0} \int_0^T \int_\Upsilon d(x) g(\partial_t u^n)\partial_t u^n\, dx\, dt,$$

which tends to $\epsilon(b-a)T$, as $n \to \infty$. Since $\epsilon > 0$ can be arbitrarily small, it follows that

$$0 < M \le C_2 \left[\limsup_{n\to\infty} \int_0^T \int_\Omega (|u^n|^2 + |u^n|^{2p+2})\, dx\, dt + 1 \right].$$

Finally, from the compact imbedding $H_0^1(\Omega) \hookrightarrow L^r(\Omega)$, for $1 \le r \le \infty$, and by using (52.72) and the Lebesgue Dominated Convergence Theorem, we conclude that

$$0 < M \le C_2 \left(T \int_\Omega (|U|^2 + |U|^4)\, dx + 1 \right),$$

where $M > 0$ can be arbitrarily large, but $U(x)$ is a specific function in $C(\overline{\Omega})$. This is a contradiction, which completes the proof. □

We next show that there exists an absorbing set for the semiflow in W.

Corollary 52.11. *The semiflow $S(t)w_0$ generated by the mild solutions of equation (52.53) in W is point dissipative, i.e., there exists an absorbing set B_0 in W.*

Proof. First of all, the two properties of the energy functional $E(u)(t)$ described by (52.57) and (52.58) show that when time evolves, the W-norm of any solution of equation (52.49) is bounded. Thus for every initial datum w_0 in W, the mild/strong solution $S(t)w_0$ exists for all $t \ge 0$, and this generates a semiflow on W (see Section 4.6).

Moreover, as argued above, for any given $t \ge 0$, $S(t)$ maps bounded sets in W into bounded sets in W. We now claim that the set $B_0 = \gamma^+(B_1)$ is an absorbing set in W, where B_1 is the bounded set given in Lemma 52.10. We leave it as an exercise to prove that B_0 itself is a bounded set in W. Next we claim that, for any bounded set $Z \subset W$, there is a finite time $t_0 = t_0(Z)$ such that

$$S(t)(Z) \subset B_0, \qquad \text{for all } t > t_0. \tag{52.76}$$

In fact, if this is not true, then there is a bounded set Z_0, an initial point $w_0 \in Z_0$, and an increasing time sequence $\{t_n\}$, such that $t_n \to \infty$, and

$$S(t_n)w_0 \in \gamma^+(Z_0)\backslash B_0 \subset \gamma^+(Z_0)\backslash B_1, \qquad \text{for } n = 1, 2, \cdots.$$

By Lemma 52.10, then there is a constant $K = K(Z_0) > 0$, with

$$E(u)(t_n) - E(u)(t_{n-1}) < -K, \qquad \text{for } n = 1, 2, \cdots .$$

As a result, one finds that

$$0 \leq \frac{1}{4}\|S(t_n)w_0\|_W^2 \leq c_0 + E(u)(t_n) \leq c_0 + E(u)(0) - nK \to -\infty,$$

see (52.58). This is a contradiction, and consequently, the absorbing property (52.76) holds. □

In order to prove Theorem 52.7, the next step is to show that the semiflow $S(t)w_0$ generated by the mild solutions of equation (52.53) in W is κ-contracting. As before, we seek to split $S(t)$ into $S(t)w_0 = S_1(t)w_0 + S_2(t)w_0$, where $S_1(t)w_0$ is "asymptotically stable" and $S_2(t)w_0$ is "ultimately compact", see (52.28) and (52.48), for example. However, in the present problem, the damping term $d(x)\,g(\partial_t u)$ is locally supported, and consequently, the technique for the construction of S_1 and S_2 is different.

We begin with the nonlinear term f in equation (52.49). Let $f = f_1 + f_2$, where $f_1(z) = z$ and $f_2(z) = f(z) - z$. For $w_0 \in B_0$, where B_0 is the absorbing set, we let $S(t)w_0 = S_1(t)w_0 + S_2(t)w_0$, with $w(t) = S(t)w_0$, $w^{(i)}(t) = S_i(t)w_0$, for $i = 1, 2$,

$$w = \begin{pmatrix} u \\ \partial_t u \end{pmatrix}, \quad w^{(1)} = \begin{pmatrix} v \\ \partial_t v \end{pmatrix}, \quad w^{(2)} = \begin{pmatrix} \psi \\ \partial_t \psi \end{pmatrix}, \quad \text{and } w_0 = \begin{pmatrix} u_0 \\ u_1 \end{pmatrix}.$$

One then has $u = v + \psi$. We require that $v = v(t) = v(x, t)$ be a solution of the nonautonomous problem:

$$\text{(52.77)} \quad \partial_t^2 v + d(x)[g(\partial_t u) - g(\partial_t u - \partial_t v)] - \Delta v + f_1(v) = 0, \qquad \text{in } \Omega \times \mathbb{R}^+,$$

with the Dirichlet boundary condition $v(x,t) = 0$, for $x \in \partial\Omega$, and the initial conditions $v(x, 0) = u_0(x)$ and $\partial_t v(x, 0) = u_1(x)$, for $x \in \overline{\Omega}$, while $\psi = \psi(t) = \psi(x, t)$ satisfies

$$\text{(52.78)} \quad \partial_t^2 \psi + d(x)g(\partial_t \psi) - \Delta\psi + f_1(v + \psi) - f_1(v) = -f_2(u), \qquad \text{in } \Omega \times \mathbb{R}^+,$$

with the same boundary condition and initial conditions $\psi(x, 0) = 0$ and $\partial_t \psi(x, 0) = 0$, for $x \in \overline{\Omega}$. In equations (52.77) and (52.78), $u = u(x, t)$ represents the solution of the original Problem (52.49).

In the sequel, whenever we mention a solution v of equation (52.77), or a solution ψ of equation (52.78), we mean that $w^{(1)}(t)$, or $w^{(2)}(t)$, is a mild solution of the corresponding first-order evolutionary equation formulated from equation (52.77), or equation (52.78), respectively. The first result concerns the longtime behavior of v.

Lemma 52.12. *Let the assumptions (Ad), (Ag), and (Af) hold, and let B_0 be the absorbing set given above. Then there exists a function $\beta(t): \mathbb{R}^+ \to \mathbb{R}^+$, such that β is independent of u and v, with $\lim_{t\to\infty}\beta(t) = 0$, and $\|w^{(1)}(t)\|_W^2 \le \beta(t)$, for $t \ge 0$ and for any solution v of equation (52.77) with $(v(0), \partial_t v(0))^T \in B_0$.*

Proof. We return to the definition of $E(y) = E(y; f)$ in equation (52.54). Similar to the energy functional $E(y; f)(t)$ treated in Lemmas 52.8 and 52.9, we will use $E(v; f_1)$ here. Since g is strictly increasing by Assumption (Ag), one obtains

$$\begin{aligned}\partial_t E(v; f_1)(t) &= -\int_\Omega d(x)[g(\partial_t u) - g(\partial_t u - \partial_t v)]\partial_t v\, dx \\ &= -\int_\Omega d(x)[g(\partial_t u) - g(\partial_t u - \partial_t v)][\partial_t u - (\partial_t u - \partial_t v)]\, dx \le 0.\end{aligned}$$

In order to prove this lemma, it suffices to establish the following statement: (ST) With B_0 given as above, there is a $T_0 \ge 0$, such that for any $N > 0$, there is a constant $K_0 = K_0(N) > 0$ such that for any $w_0 \in B_0$, one has

$$E(v; f_1)(T) - E(v; f_1)(0) \le -K_0 < 0, \qquad \text{for } T > T_0,$$

whenever $0 < N \le \|w^{(1)}(T)\|_W^2$.

In order to prove (ST), we set $F_1(r) = \int_0^r f_1(s)\, ds$. Since $\int_\Omega F_1(v(x,t))\, dx \ge 0$, one has $E(v; f_1)(t) \ge \frac{1}{2}\|w^{(1)}(t)\|_W^2$, for $t \ge 0$. Then an adaption of the inequality (52.67), with $\{y, f\}$ being replaced by $\{v, f_1\}$, and the underlying equation (52.49) replaced by equation (52.77), respectively, leads to

$$\begin{aligned}TE(v;f_1)(T) &\le \int_0^T E(v; f_1)(t)\, dt \\ &\le K_1\left\{\int_0^T\int_\Omega (D^2 + |v|^2 + |v|^{2p+2})\, dx\, dt + \int_0^T\int_\Upsilon |\partial_t v|^2\, dx\, dt\right\} \\ &\quad + K_2\left(E(v; f_1)(T) + \frac{1}{2}\int_0^T\int_\Omega D\,\partial_t v\, dx\, dt\right) \\ &\le K\left(J_{11}(v) + J_{33}(v) + E(v; f_1)(T)\right),\end{aligned}$$

where

$$J_{11}(v) = \int_0^T\int_\Omega (D^2 + D\,\partial_t v + |v|^2 + |v|^{2p+2})\, dx\, dt$$

$$J_{33}(v) = \int_0^T\int_\Upsilon |\partial_t v|^2\, dx\, dt,$$

$D = D(v) = D(x, \partial_t u, \partial_t v) = d(x)[g(\partial_t u) - g(\partial_t u - \partial_t v)]$, and K_1 and K_2 are positive constants. It follows that for $T > K = \max(K_1, K_2)$, one has

$$E(v; f_1)(T) \leq \frac{K}{T-K}(J_{11}(v) + J_{33}(v)). \tag{52.79}$$

We now prove the Statement (ST) by contradiction. If it is false, there is constant $N_0 > 0$ and a sequence of solutions $\{w_n^{(1)}\}$, with initial conditions in B_0, such that for any given $T > K$, one has $0 < N_0 \leq \|w_n^{(1)}(T)\|_W^2$, for all $n \geq 1$. However,

$$E(v^n; f_1)(T) - E(v^n; f_1)(0) = -\int_0^T \int_\Omega D(v^n)\partial_t v^n \, dx \, dt \to 0, \tag{52.80}$$

as $n \to \infty$. By repeating the argument presented in the proof of Lemma 52.10, we find that there is a subsequence of $\{w_n^{(1)}\}$, which we relabel as $\{w_n^{(1)}\}$, such that

$$w_n^{(1)} \to 0 \text{ weak* in } L^\infty(0, T; W), \qquad \text{as } n \to \infty. \tag{52.81}$$

Notice that the only stationary solution to equations (52.73) - (52.74), with f replaced by f_1, is $U \equiv 0$. Next letting $v = v^n$ in (52.79) and then passing to the limit, as $n \to \infty$, and arguing as in the proof of Lemma 52.10, we obtain

$$\begin{aligned} 0 < N_0 &\leq \|w_n^{(1)}(T)\|_W^2 \leq E(v^n; f_1)(T) \\ &\leq \frac{K}{T-K} \limsup_{n\to\infty} \left[J_{11}(v^n) + J_{33}(v^n)\right]. \end{aligned} \tag{52.82}$$

Finally, by using (52.50), (52.51), (52.80), and (52.81), one can veryify that all the terms in the lim sup bracket of inequality (52.82) converge to zero, as $n \to \infty$. It leads to a contradiction that $0 < N_0 \leq 0$. Hence, Statement (ST) holds. □

The next result deals with the longtime behavior of ψ.

Lemma 52.13. *Let the assumptions (Ad), (Ag), and (Af) hold. Then for each $t > 0$, the set $C\ell_W S_2(t) B_0$ is compact in W. Moreover, for any given bounded set B, there is a $\tau(B) \in [0, \infty)$, such that the set $C\ell_W S_2(t)B$ is compact in W, for each $t > \tau(B)$.*

Proof. Since B_0 is an absorbing set, the second statement is an obvious consequence of the first. Hence, we focus on the first statement. In particular, it suffices to show that for each $t > 0$, the set $S_2(t)B_0$ lies in a bounded set in the product space $Y = H^{\frac{3}{4}}(\Omega) \times H^{\frac{1}{4}}(\Omega)$, since one has the compact imbedding $Y \hookrightarrow W$. From the interpolation properties of sectorial operators discussed in Chapter 3, one has

$$H_0^s(\Omega) \subset \mathcal{D}(A^{\frac{s}{2}}) \subset H^s(\Omega), \qquad \text{for any } s \geq 0.$$

Since A is a positive, self-adjoint operator, the norm $\|A^{\frac{s}{2}}\varphi\|$ and the norm $\|\varphi\|_{H^s(\Omega)}$ are equivalent on the space $\mathcal{D}(A^{\frac{s}{2}})$. By taking the inner-product in $L^2(\Omega)$ of equation (52.78) with the multiplier $2A^{\frac{1}{2}}\partial_t\psi + \epsilon A^{\frac{1}{2}}\psi$, for arbitrary $\epsilon > 0$, we get

$$\begin{aligned}\partial_t(\|A^{\frac{1}{4}}\partial_t\psi\|^2 + \epsilon\langle A^{\frac{1}{4}}\partial_t\psi, A^{\frac{1}{4}}\psi\rangle + \|A^{\frac{3}{4}}\psi\|^2 + \|A^{\frac{1}{4}}\psi\|^2) \\ + \epsilon\left(-\|A^{\frac{1}{4}}\partial_t\psi\|^2 + \|A^{\frac{3}{4}}\psi\|^2 + \|A^{\frac{1}{4}}\psi\|^2\right) \\ = -\langle 2A^{\frac{1}{2}}\partial_t\psi + \epsilon A^{\frac{1}{2}}\psi, d(\cdot)g(\partial_t\psi) + f_2(u)\rangle.\end{aligned} \tag{52.83}$$

Hereafter we will use $c(B_0)$ to denote a local constant that depends on the absorbing set B_0, but is independent of time and the initial data.

From Corollary 52.11 and Lemma 52.12, we see that, as long as $w_0 \in B_0$, the solution $S(t)w_0$ of equation (52.53) and the solution $S_1(t)w_0$ of equation (52.77) are uniformly bounded in W, for $t \geq 0$. Hence, for $w_0 \in B_0$, the solution $S_2(t)w_0$ of equation (52.78) is uniformly bounded in W, as well. Next observe that, since g is strictly increasing, one has

(52.84)

$$\begin{aligned}|-\langle 2A^{\frac{1}{2}}\partial_t\psi, d(\cdot)g(\partial_t\psi)\rangle| &= 2\left|\int_\Omega d(x)\,\partial_x\int_0^{\partial_t\psi(x,t)} g(s)\,ds\right| \\ &= 2\left|\left[d(x)\int_0^{\partial_t\psi(x,t)} g(s)\,ds\right]_{x=a}^{x=b} - \int_\Omega d'(x)\int_0^{\partial_t\psi(x,t)} g(s)\,ds\right| \\ &= 2\left|\int_\Omega d'(x)\int_0^{\partial_t\psi(x,t)} g(s)\,ds\right| \leq 2\|d\|_{C^1}\|\partial_t\psi\|^2 \leq c(B_0).\end{aligned}$$

Furthermore, since f and f_2 are C^1 functions and $\|u(t)\|_{C(\Omega)}$ is uniformly bounded, one has

(52.85)

$$\begin{aligned}\left|-\langle 2A^{\frac{1}{2}}\partial_t\psi, f_2(u)\rangle\right| &\leq 2\|\partial_t\psi\|^2\|A^{\frac{1}{2}}f_2(u)\|^2 = 2\|\partial_t\psi\|^2\|f_2'(u)\partial_x u\|^2 \\ &\leq c(B_0)\|\partial_t\psi\|^2\|u\|^2_{H^1(\Omega)} \leq c(B_0).\end{aligned}$$

Moreover, since g is uniformly Lipschitz continuous by Assumption (Ag), we have

(52.86)

$$\begin{aligned}|\langle\epsilon A^{\frac{1}{2}}\psi, d(\cdot)g(\partial_t\psi) + f_2(u)\rangle| &\leq \epsilon\|\psi\|_{H^1(\Omega)}(\|d(\cdot)g(\partial_t\psi)\| + \|f_2(u)\|) \\ &\leq \epsilon\|\psi\|_{H^1(\Omega)}(\|d\|_{C(\overline{\Omega})}L_g\|\partial_t\psi\| + \|f_2(u)\|) \leq c(B_0),\end{aligned}$$

for $0 < \epsilon < 1$, where $L_g > 0$ satisfies $|g(\partial_t\psi)| = |g(\partial_t\psi) - g(0)| \leq L_g|\partial_t\psi|$. By incorporating (52.84), (52.85), and (52.86) in (52.83) and defining

$$\Pi[\psi](t) \stackrel{\text{def}}{=} \|A^{\frac{1}{4}}\partial_t\psi\|^2 + \epsilon\langle A^{\frac{1}{4}}\partial_t\psi, A^{\frac{1}{4}}\psi\rangle + \|A^{\frac{3}{4}}\psi\|^2 + \|A^{\frac{1}{4}}\psi\|^2, \quad \text{for } t > 0,$$

we obtain

$$\begin{aligned}\partial_t \Pi[\psi](t) + \epsilon \Pi[\psi] &\leq 2\epsilon \|A^{\frac{1}{4}} \partial_t \psi\|^2 + \epsilon^2 \langle A^{\frac{1}{4}} \partial_t \psi, A^{\frac{1}{4}} \psi \rangle + c(B_0) \\ &\leq 2\epsilon \|A^{\frac{1}{4}} \partial_t \psi\|^2 + \epsilon^2 \|\partial_t \psi\| \|A^{\frac{1}{2}} \psi\| + c(B_0) \\ &\leq 2\epsilon \|A^{\frac{1}{4}} \partial_t \psi\|^2 + c(B_0), \quad \text{for } t > 0.\end{aligned}$$

Since $\psi(x, 0) = \partial_t \psi(x, 0) \equiv 0$, the Gronwall inequality then implies that

$$\Pi[\psi](t) \leq c(B_0)\epsilon^{-1} + 2\epsilon \int_0^t e^{-\epsilon(t-s)} \|A^{\frac{1}{4}} \partial_t \psi(s)\|^2 \, ds, \qquad \text{for } t \geq 0.$$

Fix ϵ sufficiently small so that $2\Pi[\psi](t) \geq \|A^{\frac{1}{4}} S_2(t) w_0\|_W^2$ and $0 < \epsilon < 1$. It follows that, for $t > 0$, one has

$$\begin{aligned}e^{\epsilon t} \|A^{\frac{1}{4}} \partial_t \psi\|^2 &\leq e^{\epsilon t} \left(\|A^{\frac{1}{4}} \partial_t \psi\|^2 + \|A^{\frac{3}{4}} \psi\|^2 \right) \\ &\leq e^{\epsilon t} 2c(B_0)\epsilon^{-1} + 4\epsilon \int_0^t e^{\epsilon s} \|A^{\frac{1}{4}} \partial_t \psi(s)\|^2 \, ds.\end{aligned} \tag{52.87}$$

By applying the Gronwall inequality to the outer two terms in inequality (52.87), we get

$$\|A^{\frac{1}{4}} \partial_t \psi\|^2 \leq 2c(B_0)\epsilon^{-1} + 8c(B_0)(3\epsilon)^{-1} e^{3\epsilon t} \leq 5c(B_0)\epsilon^{-1} e^{3\epsilon t}, \qquad \text{for } t \geq 0.$$

By substituting the last inequality into (52.87), we obtain

$$\|A^{\frac{1}{4}} \partial_t \psi(\cdot, t)\|^2 + \|A^{\frac{3}{4}} \psi(\cdot, t)\|^2 \leq 2c(B_0)\epsilon^{-1} + 5c(B_0)\epsilon^{-1} e^{3\epsilon t}, \qquad \text{for } t \geq 0.$$

This shows that the set $S_2(t)B_0$ lies in a bounded set of space $Y = H^{\frac{3}{4}}(\Omega) \times H^{\frac{1}{4}}(\Omega)$, which completes the proof. □

Proof of Theorem 52.7. We note that: (1) Corollary 52.11 shows that the semiflow generated by the mild solutions of equation (52.53) is (point) dissipative and ultimately bounded in W, and (2) Lemmas 22.3, 52.12, and 52.13 imply that the semiflow is κ-contracting. Consequently, the Existence Theorem 23.12 applies, and there exists a global attractor $\mathfrak{A} \subset W$ for this semiflow and that $\mathfrak{A}$ attracts all bounded sets in W. □

5.3. Equations of Convection.

In many chemical processes, geographical phenomena, and industrial applications, convection is a significant factor. In this section, we will focus on the asymptotic dynamics of a one-dimensional, scalar, convective reaction diffusion equation of the form

$$\partial_t u = a\, \partial_x^2 u + \partial_x (f(u)) + g(u) + h(x), \qquad x \in (0, L), \quad t > 0, \tag{53.1}$$

with homogeneous Dirichlet boundary condition, $u(0,t) = u(L,t) = 0$. In general, an important feature of an equation with convection is the occurrence of a term, sometimes linear and sometimes nonlinear, with a first-order spatial derivative. In equation (53.1), $\partial_x(f(u))$ is such a term. The Burgers equation

$$\partial_t u = a\,\partial_x^2 u + \partial_x(u^2) = a\,\partial_x^2 u + 2u\,\partial_x u$$

is an example of an equation with convection.

From the viewpoint of physics, at the onset of convection there is a chaotic tendency that the state loses stability, but as a result of convection concurred with diffusion mechanism the whole process eventually tends to be dissipative. Because of these two interacting aspects of a convective diffusion process, the mathematical treatment of the corresponding diffusion-convection equations is usually more delicate than the simpler reaction diffusion equations.

Here we will demonstrate that under certain growth conditions of functions f and g, equation (53.1) generates a dissipative dynamical system. More precisely, we assume that $a > 0$ is a constant, that $f : \mathbb{R} \to \mathbb{R}$ and $g : \mathbb{R} \to \mathbb{R}$ are polynomial functions with $f(0) = g(0) = 0$, and that $h \in H_0^1(0, L)$ is a given function. Moreover, we require that $g(u) = \sum_{i=0}^{2k-1} b_i u^i$ be of odd degree, where where b_i are constants with $b_{2k-1} < 0$ and $k \geq 1$, As a result, there exist constants $c_0 \geq 1$, $r_0 > 0$, and an integer $m \geq 1$, such that

$$\begin{aligned} |f'(s)| &\leq c_0\,(1 + |s|^m)\,, \\ |f''(s)| &\leq c_0\,(1 + |s|^{m-1})\,, \\ |g'(s)| &\leq c_0\,(1 + |s|^m)\,, \end{aligned} \tag{53.2}$$

and

$$g(s)s \leq 0, \qquad \text{for } |s| \geq r_0. \tag{53.3}$$

The inequalities (53.2) are called growth conditions, and (53.3) is a sign condition. (See (51.2).)

5.3.1 Construction of the Semiflow. Let $\Omega = (0, L)$. For simplicity we will let $a = 1$. To set up the mathematical framework, we consider $H = L^2(\Omega)$ with the usual norm and inner-product, which are denoted by $\|\cdot\|$ and $\langle\cdot,\cdot\rangle$, respectively. Define the operator A by $A\varphi = -\partial_x^2\varphi$, for $\varphi \in \mathcal{D}(A) \stackrel{\text{def}}{=} H^2(\Omega) \cap H_0^1(\Omega)$. As shown in Chapter 3, the Standing Hypothesis B is satisfied, and one has $A^{\frac{1}{2}}\varphi = \partial_x\varphi$, where $\mathcal{D}(A^{\frac{1}{2}}) = V^1 = H_0^1(\Omega) \stackrel{\text{def}}{=} V$, see (51.6).

We next define the nonlinear operator $B : V \to H$ by

$$B(u)(x) = \partial_x(f(u(x))), \qquad \text{for } x \in \Omega. \tag{53.4}$$

The following lemma confirms that the operator B is well-defined and has specified properties. We will denote the norm of Banach space $L^p(\Omega)$ by $\|\cdot\|_p$, for $1 \leq p \leq \infty$. Note that $\|\cdot\|_2 = \|\cdot\|$.

Lemma 53.1. *Under the given assumptions on f, there exist constants $C_0 > 0$ and $C_1 > 0$ such that $B(u) \in H$, $\|B(u)\| \le C_0 (1 + \|u\|_\infty^m) \|A^{\frac{1}{2}} u\|$, for all $u \in V$,*

$$(53.5) \qquad \langle B(u), u^k \rangle = 0, \qquad \textit{for all integers } k \ge 1 \textit{ and } u \in V,$$

and for all $u \in V^2 = \mathcal{D}(A)$, one has

$$(53.6) \qquad \|B(u)\|^2 \le C_1 (L + \|u\|_{2m}^{2m}) \|A^{\frac{1}{2}} u\| \, \|Au\|.$$

Proof. From (51.6) and (53.2), one obtains $B(u) \in H$, for each $u \in V$, since

$$|B(u)(x)|^2 = |f'(u)\partial_x u|^2 \le C_0^2 (1 + \|u\|_\infty^m)^2 |\partial_x u|^2,$$

which is Lebesgue measurable and integrable. Since $\|\partial_x u\| = \|A^{\frac{1}{2}} u\|$, for $u \in \mathcal{D}(A^{\frac{1}{2}})$, we obtain

$$(53.7) \qquad \|B(u)\|^2 = \int_\Omega |f'(u)\partial_x u|^2 \, dx \le C_0^2 (1 + \|u\|_\infty^m)^2 \|A^{\frac{1}{2}} u\|^2.$$

By an integration-by-parts, we find $\langle B(u), u^k \rangle = -k \int_\Omega \partial_x \varphi(u) \, dx = 0$, where $\varphi(s) = \int_0^s f(t) t^{k-1} dt$, for $k \ge 1$. This implies that (53.5) is valid. In order to prove (53.6), we note that $\mathcal{D}(A) = V^2 \hookrightarrow W^{1,\infty}(\Omega)$. So if $u \in \mathcal{D}(A)$, then $\partial_x u \in L^\infty(\Omega)$. As a result, one has

(53.8)

$$\|B(u)\|^2 \le \int_\Omega C_0^2 (1 + |u(x)|^m)^2 |\partial_x u|^2 \, dx \le C_0^2 \|\partial_x u\|_\infty^2 \int_\Omega (1 + |u(x)|^m)^2 \, dx.$$

According to the Nirenberg-Gagliardo inequality (see Appendix B), there exists a constant $C > 0$, such that

$$(53.9) \qquad \|u\|_\infty \le C \|u\|^{\frac{1}{2}} \|A^{\frac{1}{2}} u\|^{\frac{1}{2}}, \qquad \text{for all } u \in V.$$

By replacing u by $A^{\frac{1}{2}} u$ in (53.9) and using $\|A^{\frac{1}{2}} u\| = \|\partial_x u\|$, we get

$$(53.10) \qquad \|\partial_x u\|_\infty^2 \le C^2 \|A^{\frac{1}{2}} u\| \, \|Au\|, \qquad \text{for } u \in V^2.$$

By substituting (53.10) and $(1 + |u|^m)^2 \le 2(1 + |u|^{2m})$ into (53.8) and using the Schwarz inequality, we obtain (53.6), for $C_1 = \sqrt{2} C_0 C$. □

The problem generated by equation (53.1) with homogeneous Dirichlet boundary conditions is now reformulated as the evolutionary equation:

$$(53.11) \qquad \partial_t u + Au = F(u), \qquad \text{where } u(0) = u_0 \in H,$$

and $F(u) = B(u) + g(u) + h$. Since f and g are polynomials and $h \in V$, one can readily show that

$$F \in C_{\mathrm{F}}^2(V, H) \cap C_{\mathrm{F}}^1(V, V).$$

It then follows from the theory of Section 4.7 that, for every $u_0 \in V$, there is a unique, maximally defined, mild solution $u = u(t)$ of (53.11) in V on $[0, T)$, where $0 < T = T(u_0) \le \infty$, and u is a strong solution in V which satisfies (47.7). It follows that $f(t) = F(u(t))$ satisfies

$$f \in C[0, T; V) \cap C_{\mathrm{loc}}^{0,1}(0, T; V).$$

Since V is a Hilbert space, one has $C_{\mathrm{loc}}^{0,1}(0, T; V) \subset W_{\mathrm{loc}}^{1,\infty}(0, T; V)$. We now have two lemmas.

Lemma 53.2. *Under the standing assumptions on f and g, let $u_0 \in V$ and let $u(t) = S(t)u_0$ denote the maximally defined mild solution of equation (53.11) in V on $[0, T(u_0))$, and let $\lambda_1 > 0$ be the smallest eigenvalue of A. Then there exists a positive constant M_1, which does not depend on the initial data, such that for all $u_0 \in V$ and $0 \le t < T(u_0)$, one has*

$$\|S(t)u_0\|^2 \le \|u_0\|^2 e^{-\lambda_1 t} + M_1 \lambda_1^{-1}. \tag{53.12}$$

Proof. Let $u = u(t)$ be the (local) solution of equation (53.11) with $u(0) = u_0$. The reader should verify that the following steps are justified, owing to the stated properties of the function $u = u(t)$ and the imbedding relations in (51.6). By taking the inner product in H of equation (53.11) with u^{2p-1}, where $p \ge 1$, we obtain

$$\begin{aligned} (2p)^{-1}\partial_t \int_\Omega u^{2p}\,dx &= \langle \partial_t u, u^{2p-1}\rangle \\ &= \langle -Au + B(u) + g(u) + h, u^{2p-1}\rangle \\ &= -\langle Au, u^{2p-1}\rangle + \langle g(u) + h, u^{2p-1}\rangle \\ &= -(2p-1)\int_\Omega u^{2p-2}(\partial_x u)^2\,dx \\ &\quad + \int_\Omega g(u)u^{2p-1}dx + \int_\Omega h(x)u^{2p-1}dx. \end{aligned} \tag{53.13}$$

Using (53.3), we get $\int_\Omega g(u)u^{2p-1}dx \le K_p \stackrel{\text{def}}{=} L \sup_{|u|\le r_0} |g(u)u^{2p-1}|$. Recall that

$$\|A^{\frac{1}{2}}u\|^2 \ge \lambda_1 \|u\|^2, \qquad \text{for all } u \in V. \tag{53.14}$$

From the Young inequality, for every $\epsilon > 0$ and for all α and β with $\frac{1}{\alpha} + \frac{1}{\beta} = 1$ and $1 < \alpha < \infty$, there is a constant $C_0 = C_0(\epsilon, \alpha, \beta) > 0$ such that $\varphi\psi \le \epsilon\varphi^\alpha + C_0\psi^\beta$, whenever $\varphi, \psi \ge 0$. Next we set

$$\epsilon_1 = \frac{1}{2p}\left(\frac{\lambda_1}{p}\right)^{1-2p}, \quad \alpha = 2p, \quad \text{and } \beta = \frac{2p}{2p-1},$$

with $\varphi = h(x)$ and $\psi = u(x)^{2p-1}$. We then obtain a constant $C_1 > 0$ such that

$$\int_\Omega h(x)u^{2p-1}dx \le \epsilon_1 \int_\Omega h^{2p}dx + C_1 \int_\Omega u^{2p}dx. \tag{53.15}$$

Since inequality (53.14) implies that

$$\lambda_1 \int_\Omega u^{2p}\,dx \le p^2 \int_\Omega u^{2(p-1)}(\partial_x u)^2\,dx, \tag{53.16}$$

by substituting first (53.16) into (53.15), and next (53.15) into (53.13), we find

$$\frac{1}{2p}\partial_t \int_\Omega u^{2p}\,dx \leq -\frac{2p-1}{2}\int_\Omega u^{2p-2}(\partial_x u)^2\,dx + K_p + \epsilon_1 \|h\|_{2p}^{2p}. \tag{53.17}$$

With $M_p = 2pK_p + (\lambda_1 p^{-1})^{1-2p}\|h\|_{2p}^{2p}$, we then obtain

$$\partial_t \int_\Omega u^{2p} dx + p(2p-1)\int_\Omega u^{2(p-1)}(\partial_x u)^2 dx \leq M_p. \tag{53.18}$$

By using inequality (53.16) in (53.18), we obtain

$$\partial_t \|u\|_{2p}^{2p} = \partial_t \int_\Omega u^{2p}\,dx \leq -\frac{2p-1}{p}\lambda_1 \int_\Omega u^{2p}\,dx + M_p,$$

and the Gronwall inequality implies that

$$\|u(t)\|_{2p}^{2p} \leq \|u_0\|_{2p}^{2p} e^{-\frac{(2p-1)\lambda_1}{p}t} + \frac{p}{(2p-1)\lambda_1}M_p, \tag{53.19}$$

for $0 \leq t < T(u_0)$. With $p = 1$, this implies (53.12). □

Lemma 53.3. *Under the given assumptions on f and g, one has $T(u_0) = \infty$, for every $u_0 \in V$, and the mild solutions of equation (53.11) generate a semiflow on V.*

Proof. By integrating inequality (53.18), for $p = 1$, with respect to t, we obtain

$$\frac{1}{t_2 - t_1}\int_{t_1}^{t_2} \|A^{\frac{1}{2}}u(s)\|^2 ds \leq \frac{1}{t_2 - t_1}\|u(t_1)\|^2 + M_1, \tag{53.20}$$

for $0 \leq t_1 < t_2 < T(u_0)$. By taking the inner product in H of equation (53.11) with Au, and using (53.6), we find

$$\begin{aligned}\frac{1}{2}\partial_t\|A^{\frac{1}{2}}u\|^2 &= \langle \partial_t u, Au\rangle = \langle -Au + B(u) + g(u) + h, Au\rangle \\ &\leq -\|Au\|^2 + C_1(L + \|u\|_{2m}^{2m})^{\frac{1}{2}}\|A^{\frac{1}{2}}u\|^{\frac{1}{2}}\|Au\|^{\frac{3}{2}} \\ &\quad + |\langle g(u) + h, Au\rangle|.\end{aligned} \tag{53.21}$$

From (53.2) and (53.10), we obtain

$$\begin{aligned}|\langle g(u), Au\rangle| &= |\langle A^{\frac{1}{2}}g(u), A^{\frac{1}{2}}u\rangle| - \left|\int_\Omega g'(u)(\partial_x u)^2\,dx\right| \\ &\leq C_0 \int_\Omega (1 + |u|^m)(\partial_x u)^2\,dx \\ &\leq C_0\left(\|A^{\frac{1}{2}}u\|^2 + \|\partial_x u\|_\infty^2 \int_\Omega |u|^m\,dx\right) \\ &\leq C_0\left(\|A^{\frac{1}{2}}u\|^2 + C^2\|A^{\frac{1}{2}}u\|\,\|Au\|\,\|u\|_m^m\right).\end{aligned} \tag{53.22}$$

From the Young inequality, we find constants C_2, C_3, and C_4, such that

$$C_1^{\frac{1}{2}}(L+\|u\|_{2m}^{2m})^{\frac{1}{2}}\|A^{\frac{1}{2}}u\|^{\frac{1}{2}}\|Au\|^{\frac{3}{2}} \le \frac{1}{4}\|Au\|^2 + C_2(L+\|u\|_{2m}^{2m})^2\|A^{\frac{1}{2}}u\|^2,$$
$$C_0C^2\|u\|_m^m\|A^{\frac{1}{2}}u\|\,\|Au\| \le \frac{1}{4}\|Au\|^2 + C_3\|u\|_m^{2m}\|A^{\frac{1}{2}}u\|^2,$$
$$|\langle h, Au\rangle| \le \frac{1}{4}\|Au\|^2 + C_4\|h\|^2.$$

Now substituting these inequalities into inequality (53.21), we obtain (53.23)

$$\partial_t\|A^{\frac{1}{2}}u\|^2 \le -\|Au\|^2 + b(u)\|A^{\frac{1}{2}}u\|^2 + 2\|h\|^2 \le b(u)\|A^{\frac{1}{2}}u\|^2 + 2\|h\|^2,$$

where $b(u) = 2C_0 + 2C_2L^2 + 4C_2L\|u\|_{2m}^{2m} + 2C_2\|u\|_{2m}^{4m} + 2C_3\|u\|_m^{2m}$. By means of a lengthy calculation, which uses (53.19) alternately, with $2p = m$ and $2p = 2m$, one finds a function $\Psi(u_0)$ of the initial data $v_0 \in V$ and a constant D_m, which does not depend on u_0. such that

$$|b(u(t))| \le \Psi(u_0)e^{-\lambda_1 t} + D_m, \qquad \text{for } 0 \le t < T(u_0). \tag{53.24}$$

The reader should verify that, it follows from inequalities (53.23), (53.24) and the Gronwall inequality that $\|A^{\frac{1}{2}}u(t)\|^2$ cannot blow up in finite time. (A good estimate of the norm in V is given below.) Hence, Lemma 47.4 implies that $T(u_0) = \infty$, for every $u_0 \in V$. The construction in Section 4.7 then implies that the mild solutions on V generate a semiflow $S(t)$ on V. □

5.3.2. Global Attractor in V. In this section, we will introduce constants L_i and N_i, with $0 \le i \le 3$, for various *a priori* estimates of the mild solutions of equation (53.11) in V. It is important to note that the constants N_i do not depend on the initial data u_0 of the mild solution $S(t)u_0$. For example, (53.12) implies that

$$\|S(t)u_0\|^2 \le L_0e^{-\lambda_1 t} + N_0, \qquad \text{for all } t \ge 0, \tag{53.25}$$

where $L_0 = \|u_0\|^2$ and $N_0 = M_1\lambda_1^{-1}$. Likewise, inequalities (53.20) and (53.25) imply that

$$\int_{t-1}^t \|A^{\frac{1}{2}}S(s)u_0\|^2\,ds \le L_1e^{-\lambda_1 t} + N_1, \qquad \text{for all } t \ge 1, \tag{53.26}$$

and (53.24) yields

$$\int_{t-1}^t |b(u(s))|\,ds \le L_2e^{-\lambda_1 t} + N_2, \qquad \text{for all } t \ge 1.$$

We then apply the Uniform Gronwall inequality to (53.23) and use the inequality

$$e^{xD} \le 1 + xDe^{D}, \qquad \text{for } 0 \le x \le 1 \text{ and } D \ge 0, \tag{53.27}$$

to obtain

$$\|A^{\frac{1}{2}} S(t)u_0\|^2 \le L_3 e^{-\lambda_1 t} + N_3, \qquad \text{for all } t \ge 1. \tag{53.28}$$

By choosing a larger value for L_3, if necessary, one can extend inequality (53.28) to hold, for all $t \ge 0$. Inequality (53.28) and Lemma 47.4 show that $T(u_0) = \infty$, for all $u_0 \in V$, since N_3 does not depend on u_0. We are now prepared for the main result.

Theorem 53.4. *Under the given assumptions on f and g, the semiflow $S(t)$ generated by the mild solutions of equation (53.11) in V is point dissipative in V. Furthermore, $S(t)$ is compact in V, for $t > 0$. Hence, there is a global attractor $\mathfrak{A}$ in V, and $\mathfrak{A}$ attracts all bounded sets in V. Moreover, one has $\mathfrak{A} \subset V^2 = \mathfrak{D}(A)$, and for each $u_0 \in \mathfrak{A}$, (47.14) is valid.*

Proof. The point dissipativity of $S(t)$ follows from inequality (53.28), since N_3 does not depend in the initial data $u_0 \in V$. The fact that $S(t)$ is compact, for $t > 0$, follows from Lemma 48.1, and the existence of a global attractor is a consequence of the Existence Theorem 23.12, also see Theorem 48.2. The remaining conclusions follow from the Herculean Theorem 47.6. □

Up to this point in the argument, we have only used the assumption that $h \in L^2(\Omega)$. Under the stronger assumption that $h \in V$, one can show the following:

(1) $V^{\frac{3}{2}} = \mathfrak{D}(A^{\frac{3}{4}})$ is a positively invariant set for $S(t)$.
(2) $V^2 = \mathfrak{D}(A)$ is a positively invariant set for $S(t)$.
(3) $\mathfrak{A}$ is compact in $V^{\frac{3}{2}}$.
(4) $\mathfrak{A}$ is point dissipative in V^2.

5.4. Kuramoto-Sivashinsky Equation.

In this section, and the next, we examine two fundamental illustrations of nonlinear partial differential equations wherein the leading linear operator is of higher order. While in each case these equations can be reformulated as a parabolic evolutionary equation, the methodology for analyzing the dynamical properties of these equations differs significantly from those developed in Sections 5.1 and 5.3.

The first of these involves the equation

$$\partial_t v + \nu \Delta^2 v + b\Delta v + a\,v + v\partial_x v = 0,$$

which is called the Kuramoto-Sivashinsky equation (KSE). Here a, b, and ν are real constants and b and ν are positive. This equation arises in the study of various pattern formation phenomena involving some kind of phase turbulence or phase transition. See for example, Kuramoto (1978) in connection with reaction diffusion systems and Sivashinsky (1977) in modeling flame front propagation of mild combusion. This type equation also arises in the study of: convective hydrodynamics, see Swift and Hohenberg (1977); plasma confinement in toroidal devices, see La Quey, et al (1975); viscous film flow, see Aimar and Denar (1982); and bifurcating solutions of the Navier-Stokes equations, see Shang and Sivashinsky (1983).

Let us first address the mathematical setting for the 1D problem. By using various rescalings in x, t, and v spaces, the KSE assumes several equivalents forms. It is convenient to use the version

$$\partial_t u + \partial_x^4 u + b^2 L^2\, \partial_x^2 u + a\, u + u\, \partial_x u = 0, \tag{54.1}$$

for $x \in \Omega = \Omega_\pi = (-\pi, \pi)$, where b and L are positive constants and $a \in \mathbb{R}$. With the scaling we have chosen, the viscosity ν has been fixed with $\nu = 1$. We will describe the boundary conditions later, and this will lead to a parabolic evolutionary equation. The leading term $\partial_x^4 v$, being a uniformly elliptic operator, is dissipative, while the lower order term $b^2 L^2 \partial_x^2 u$ is antidissipative. the parameter L assumes the role of a bifurcation parameter in this problem. By means of another rescaling, a variation of (54.1) arises on the interval $\Omega_L = (-L/2, L/2)$. Thus in equation (54.1), one should view the parameter L as representing a cell size.

With periodic boundary conditions, we obtain two different problems for the KSE: (1) the antisymmetric case and (2) the general case, both of which are treated below. The main technical obstacle we face in our goal to show the existence of a global attractor, is the matter of showing that the induced nonlinear semiflow is (point) dissipative. As it turns out, this is a nontrivial and novel issue.

Let $L^2_{\text{per}} = L^2_{\text{per}}(\Omega)$ denote the Fourier space of L^2-periodic functions of period 2π on Ω. Let $\langle \cdot, \cdot \rangle$ and $\|\cdot\|$ denote the usual inner product and norm on L^2_{per}. We let $H^k_{\text{per}} = H^k_{\text{per}}(\Omega)$ denote the associated Sobolev spaces of periodic functions, for $k = 1, 2, \cdots$. These function spaces split into two complementary spaces consisting of the **even** and the **odd** periodic functions, respectively. Of special interest here is the space of odd periodic functions

$$L^2_{\text{op}} = L^2_{\text{odd per}} = \{\varphi \in L^2_{\text{per}} : \varphi(-x) = -\varphi(x), \text{ for } x \in \Omega\}.$$

We define $H^k_{\text{op}} = H^k_{\text{per}} \cap L^2_{\text{op}}$, for $k = 1, 2, \cdots$.

The nonlinear term $v = u\, \partial_x u$ in equation (54.1) has the property that $v \in L^2_{\text{op}}$, whenever $u \in H^1_{\text{op}}$. We will use this observation to show that the KSE generates a semiflow on a suitable subspace of L^2_{op}. (This is the antisymmetric case.) More generally, this semiflow on the space of odd periodic functions extends to a semiflow on a suitable subspace of L^2_{per}. (This is the general case.)

5.4.1. The Antisymmetric Case. In this section, we consider the KSE (54.1), with spatially periodic boundary conditions:

$$\partial_x^j u(-\pi, t) = \partial_x^j u(\pi, t), \qquad \text{for } j = 0, 1, 2, 3, \text{ and } t > 0, \tag{54.2}$$

and with initial condition

$$u(x, 0) = u_0(x), \qquad x \in \Omega = \Omega_\pi. \tag{54.3}$$

Let $H = L^2_{\text{op}}$, and note that

$$\int_\Omega u(x)\, dx = 0, \qquad \text{for all } u \in H = L^2_{\text{op}}. \tag{54.4}$$

Let $V = H^2_{\text{op}} = H^2_{\text{per}} \cap H$. On the space V we define the inner product

$$\langle \varphi, \psi \rangle_V = \int_\Omega \varphi''(x)\psi''(x) dx$$

and norm $\|\varphi\|_V^2 = \langle \varphi, \varphi \rangle_V$. Note that, on the space V, this V-norm is equivalent to the inherent norm $\|\cdot\|_{H^2}$ of Sobolev space $H^2(\Omega)$, because Condition (54.4) implies that there is a constant $C_0 > 0$ such that

$$\|\varphi\| \leq C_0 \|\varphi'\| \leq C_0^2 \|\varphi''\|.$$

We define the linear differential operator A on $\mathcal{D}(A) \stackrel{\text{def}}{=} H^4_{\text{op}} \stackrel{\text{def}}{=} V^2$ by

$$A\varphi = \partial_x^4 \varphi, \qquad \text{for } \varphi \in \mathcal{D}(A).$$

Since A is uniformly elliptic, the Standing Hypothesis B is satisfied (see Section 3.8). We let e^{-At} denote the associated analytic semigroup on H.

Next we define a nonlinear operator $F : V \to H$ by

$$F(\varphi) = -\left(b^2 L^2 \varphi'' + a\varphi + \varphi\varphi'\right), \qquad \text{for } \varphi \in V.$$

The Problem (54.1)–(54.2) can then be written as an parabolic evolutionary equation

$$\partial_t u + Au = F(u), \qquad \text{for } t > 0. \tag{54.5}$$

It can be verified that $F \in C_{\text{Lip}}(V, H)$. Therefore, it follows from the theory of Section 4.7 that for every $u_0 \in V$, there is a unique, maximally defined, mild solution $S(t)u_0$ in V of equation (54.5) on the interval $[0, T)$, where $0 < T = T(u_0) \leq \infty$.

In order to show that $T(u_0) = \infty$, for every $u_0 \in V$. we will show that there is an absorbing set B_0 in V. That is, we will show that for every $u_0 \in V$, there is a time $\tau = \tau(u_0) \in [0, T(u_0))$ such that $S(t)u_0 \in B_0$, for $\tau \leq t < T$. Since the absorbing set B_0 is bounded in V, it will then follow from Lemma 47.4 that: (1) $T(u_0) = \infty$, for all $u_0 \in V$, (2) $S(t)$ is a semiflow on V, and (3) $S(t)$ is point dissipative on V.

While this outline should be familiar, we encounter a new obstacle for the KSE. None of the methods used earlier seem to be applicable, either to the KSE on H^2_{op}, or to the KSE on H^2_{per}. What we will show is that there is a function $\Psi = \Psi(x) \in H^4_{\text{op}}$ such that a variation of an earlier argument is applicable to the KSE written in terms of a new variable w, where

$$u(x,t) = w(x,t) + \Psi(x).$$

The initial condition then satisfies $u_0 = w_0 + \Psi$. In terms of the w variable, the KSE becomes

$$\begin{aligned}\partial_t w + \partial_x^4 w + b^2L^2\partial_x^2 w + aw + w\,\partial_x w + \Psi\partial_x w + \Psi' w\\ = -\Psi^{(4)} - b^2L^2\Psi^{(2)} - \Psi\Psi' - a\Psi.\end{aligned} \tag{54.6}$$

Next we multiply equation (54.6) by w and then integrate over Ω to obtain

$$\begin{aligned}\frac{1}{2}\partial_t\|w\|^2 + \|\partial_x^2 w\|^2 - b^2L^2\|\partial_x w\|^2 + \int_\Omega w^2\partial_x w\,dx\\ + \int_\Omega \Psi\,w\,\partial_x w\,dx + \int_\Omega \Psi' w^2\,dx + a\|w\|^2\\ = \int_\Omega (-\Psi^{(4)} - b^2L^2\Psi^{(2)} - a\Psi - \Psi\Psi')w\,dx.\end{aligned}$$

Using the periodicity of w and Ψ and integration-by-parts, one obtains

$$\int_\Omega w^2\partial_x w\,dx = 0 \quad\text{and}\quad \int_\Omega \Psi\,w\,\partial_x w\,dx = -\frac{1}{2}\int_\Omega \Psi' w^2\,dx.$$

Hence we find

$$\begin{aligned}\frac{1}{2}\partial_t\|w\|^2 + \|\partial_x^2 w\|^2 - b^2L^2\|\partial_x w\|^2 + a\|w\|^2 + \frac{1}{2}\int_\Omega \Psi' w^2\,dx\\ = -\int_\Omega \left(\Psi^{(4)} + b^2L^2\Psi^{(2)} + a\Psi + \Psi\Psi'\right) w\,dx.\end{aligned} \tag{54.7}$$

Now define a bilinear form $(\cdot,\cdot)_{\beta\Psi}$ as follows:

(54.8)

$$(g_1,g_2)_{\beta\Psi} = \int_\Omega g_1''g_2''\,dx - b^2L^2\int_\Omega g_1'g_2'\,dx + a\int_\Omega g_1g_2\,dx + \beta\int_\Omega g_1g_2\Psi'\,dx,$$

where $g_1(x)$ and $g_2(x)$ are sufficiently smooth functions defined on $\overline{\Omega}$, and β is a real constant. Next define $\mathcal{L}$ to be the associated linear differential operator:

$$\mathcal{L} = \frac{\partial^4}{\partial x^4} + b^2L^2\frac{\partial^2}{\partial x^2} + aI = \partial_x^4 + b^2L^2\partial_x^2 + aI,$$

where $\mathcal{D}(\mathcal{L}) = \mathcal{D}(A)$. By means of integration-by-parts, we see that the bilinear form defined by (54.8) can be written as

$$(g_1, g_2)_{\beta\Psi} = \int_\Omega g_1 (\mathcal{L} + \beta\Psi') g_2 \, dx = \langle g_1, \mathcal{L} g_2 \rangle + \beta \langle g_1, \Psi' g_2 \rangle. \tag{54.9}$$

The parameter $\beta \in \mathbb{R}$ and the function $\Psi(x)$ will be chosen later. It is interesting that in terms of this bilinear form, equation (54.7) can be written as

$$\frac{1}{2}\partial_t \|w\|^2 + (w, w)_{\frac{1}{2}\Psi} = -(w, \Psi)_\Psi. \tag{54.10}$$

Let $Q(g)$ be the quadratic form defined by

$$Q(g) = \int_\Omega \left[g''(x)^2 + g(x)^2 \right] dx = \|g\|^2 + \|\partial_x^2 g\|^2, \qquad \text{for } g \in V. \tag{54.11}$$

We are now prepared to answer the question: How should one choose Ψ? The answer is in the following result.

Lemma 54.1. *There exists a function $\Psi \in H^2_{\mathrm{op}}$ such that for any $\beta \in [1/4, 1]$, one has*

$$(g, g)_{\beta\Psi} \geq \pi \, Q(g), \qquad \text{for all } g \in V = H^2_{\mathrm{op}}.$$

We will postpone the proof of this technical lemma and turn our attention directly to the main result. Note that, owing to the periodicity of Ψ, we have $(\Psi, \Psi)_{\beta\Psi} = (\Psi, \Psi)_{0\Psi} = \langle \Psi, \mathcal{L}\Psi \rangle \stackrel{\mathrm{def}}{=} R_0(\Psi)$, for any $\beta \geq 0$.

Theorem 54.2. *For any $u_0 \in V = H^2_{\mathrm{op}}$, there exists a unique, globally defined, mild solution $S(t)u_0$ in V of equation (54.5), and $S(t)$ is a semiflow on V. Moreover, the semiflow is point dissipative in V and compact in V, for $t > 0$. Hence the KSE has a global attractor $\mathfrak{A}$ in V, and $\mathfrak{A}$ attracts all bounded sets in V. Furthermore, one has $\mathfrak{A} \subset H^4_{\mathrm{op}}$, and for each $u_0 \subset \mathfrak{A}$, one has*

$$S(\cdot)u_0 \in C^{0,1-r}_{\mathrm{loc}}(\mathbb{R}; V^{2r}) \cap C(\mathbb{R}; \mathcal{D}(A)),$$

for each r with $0 \leq r < 1$, and $V^{2r} = \mathcal{D}(A^r)$.

Proof. From equation (54.9), (54.10), (54.11), and Lemma 54.1, we have

$$\begin{aligned} \partial_t \|w\|^2 &= -2(w, w)_{\frac{1}{2}\Psi} - 2(w, \Psi)_\Psi \\ &\leq -2(w, w)_{\frac{1}{2}\Psi} + \frac{1}{2}(w, w)_\Psi + 2(\Psi, \Psi)_\Psi \\ &= -\frac{3}{2}(w, w)_{\frac{1}{3}\Psi} + 2R_0(\Psi) \leq -\frac{3\pi}{2} Q(w) + 2R_0(\Psi) \\ &= -\frac{3\pi}{2} \left(\|w\|^2 + \|\partial_x^2 w\|^2 \right) + 2R_0(\Psi). \end{aligned} \tag{54.12}$$

Now (54.12) implies that $\partial_t\|w\|^2 \le -4\|w\|^2 + 2|R_0(\Psi)|$, since $\frac{3\pi}{2} \ge 4$. From the Gronwall inequality, we then obtain

$$\|w(t)\|^2 \le e^{-4t}\|w_0\|^2 + \frac{1}{2}|R_0(\Psi)|, \qquad \text{for } 0 \le t < T(u_0). \tag{54.13}$$

Inequality (54.13) then implies that there constants L_0 and N_0, where N_0 depends on Ψ, but does not depend on u_0, such that $u(t) = S(t)u_0$ satisfies

$$\|u(t)\|^2 \le L_0\, e^{-4t} + N_0, \qquad \text{for } 0 \le t < T(u_0). \tag{54.14}$$

Also, (54.12) implies that $\partial_t\|w\|^2 + 4\|\partial_x^2 w\|^2 \le 2|R_0(\Psi)|$. By integrating this inequality, one obtains

$$4\int_t^{t+\tau} \|\partial_x^2 w(s)\|^2\, ds \le \|w(t)\|^2 + 2|R_0(\Psi)|, \tag{54.15}$$

for $0 \le t < t+\tau < \min(t+1, T(u_0))$. Since $\|\partial_x^2 u\|^2 \le 2\|\partial_x^2 w\|^2 + 2\|\Psi''\|^2$, inequalities (54.13) and (54.15) imply that there exist constants L_1 and N_1, where N_1 does not depend on u_0, such that

$$\int_t^{t+\tau} \|\partial_x^2 u(s)\|^2\, ds \le L_1 e^{-4t} + N_1, \tag{54.16}$$

for $0 \le t < t+\tau < \min(t+1, T(u_0))$.

Next we multiply equation (54.1) by $\partial_x^4 u$ and integrate over Ω, to obtain

$$\frac{1}{2}\partial_t\|\partial_x^2 u\|^2 + \|\partial_x^4 u\|^2 + b^2L^2\int_\Omega \partial_x^2 u\, \partial_x^4 u\, dx + a\|\partial_x^2 u\|^2 + \int_\Omega u\, \partial_x u\, \partial_x^4 u\, dx = 0. \tag{54.17}$$

From the Young inequality, the Nirenberg-Gagliardo inequalities, and the interpolation inequalities, one obtains

$$\left|b^2L^2\int_\Omega \partial_x^2 u\, \partial_x^4 u\, dx\right| \le \frac{1}{4}\|\partial_x^4 u\|^2 + b^4L^4\|\partial_x^2 u\|^2,$$
$$\left|\int_\Omega u\, \partial_x u\, \partial_x^4 u\, dx\right| \le \frac{1}{4}\|\partial_x^4 u\|^2 + C_0^2\|u(t)\|_\infty^2\|\partial_x^2 u\|^2,$$

where $\|u(t)\|_\infty = \sup\{\|u(s)\| : 0 \le s \le t\} \le L_0 e^{-4t} + N_0$, for $0 \le t < T(u_0)$. By combining the last two inequalities with (54.17), we obtain

$$\frac{1}{2}\partial_t\|\partial_x^2 u\|^2 + \frac{1}{2}\|\partial_x^4 u\|^2 \le \left(|a| + b^2L^2 + C_0^2[L_0 e^{-4t} + N_0]\right)\|\partial_x^2 u\|^2, \tag{54.18}$$

which implies that

$$\partial_t\|\partial_x^2 u\|^2 \le 2\left(|a| + b^2L^2 + C_0^2[L_0 + N_0]\right)\|\partial_x^2 u\|^2.$$

By applying the Gronwall inequality to the last differential inequality, it follows from Lemma 47.4 that $T(u_0) = \infty$, for all $u_0 \in V$. As argued in Section 4.7, $S(t)$ is a semiflow on V.

We now set $\tau = 1$ in inequality (54.16). It then follows from inequality (54.16) that by applying the Uniform Gronwall Inequality to (54.18) one obtains constants L_2 and N_2, where N_2 does not depend on u_0, such that

$$\|\partial_x^2 u(t)\|^2 \, ds \leq L_2 e^{-4t} + N_2, \qquad \text{for } t \geq 0.$$

The last inequality implies that $S(t)$ is point dissipative in V. From Lemma 48.1, we see that $S(t)$ is compact, for $t > 0$. Hence, the Existence Theorem 23.12 implies that there is a global attractor $\mathfrak{A}$ in V, and $\mathfrak{A}$ attracts all bounded sets in V. The remaining conclusions follow from the Herculean Theorem 47.6. □

Proof of Lemma 54.1. Recall that any real-valued function $v \in L^2_{\text{op}}$, has the Fourier series expansion

$$v(x) = i \sum_{n \in \mathbb{Z}} v_n e^{inx}, \qquad \text{with } v_n = -v_n \in \mathbb{R} \text{ and } v_0 = 0.$$

Similarly, if $\Psi \in H^4_{\text{op}}$, then Ψ' has the expansion

$$\Psi'(x) = -\sum_{n \in \mathbb{Z}} \psi_n e^{inx}, \qquad \text{with } \psi_n = \psi_{-n} \in \mathbb{R} \text{ and } \psi_0 = 0.$$

Since the functions $\{i(e^{inx} - e^{-inx}) : n \in \mathbb{Z}\}$ are eigenfunctions of $\mathcal{L}$, with eigenvalues $\lambda_n = n^4 - b^2 L^2 n^2 + a$, the Parseval equality implies that

$$(2\pi)^{-1} \langle v, \mathcal{L} v \rangle = \sum_{n \in \mathbb{Z}} \lambda_n v_n^2 = 2 \sum_{n > 0} \lambda_n v_n^2, \qquad \text{for } v \in H^4_{\text{op}}.$$

Similarly, one has

$$\begin{aligned}
(2\pi)^{-1} \int_\Omega v(x)^2 \Psi'(x) \, dx &= (2\pi)^{-1} \langle v, \Psi' v \rangle \\
&= (2\pi)^{-1} \sum_{k,\ell,m} \int_\Omega v_k \, v_m \, \psi_\ell \, e^{i(k+\ell+m)x} \, dx \\
&= \sum_{k+\ell+m=0} v_k \, v_m \, \psi_\ell = \sum_{k,m} v_k \, v_m \, \psi_{-k-m} - \sum_{k,m} v_k \, v_m \, \psi_{|k+m|}.
\end{aligned}$$

As a result, it follows from (54.9) that

$$\begin{aligned}
&(2\pi)^{-1} (v, v)_{\beta\Psi} \\
&\quad = 2 \sum_{n>0} \lambda_n v_n^2 \\
&\qquad + \beta \sum_{k,m>0} v_k \, v_m \left(\psi_{|k+m|} - \psi_{|k-m|} + \psi_{|-k-m|} - \psi_{|-k+m|} \right) \\
&\quad = 2 \left[\sum_{n>0} (\lambda_n + \beta \psi_{2n}) \, v_n^2 + 2\beta \sum_{k>m>0} v_k \, v_m \left(\psi_{|k+m|} - \psi_{|k-m|} \right) \right].
\end{aligned}$$

Let $\tau_n^2 = \frac{1}{2}(n^4+1)$ and $p_n = \tau_n v_n$. We will require that, for $\beta \geq \frac{1}{4}$, the coordinate ψ_{2n} satisfy $\psi_{2n} \geq 0$ and

$$\lambda_n + \beta\psi_{2n} \geq \tau_n^2. \tag{54.19}$$

Then the form $(v,v)_{\beta\Psi}$ is bounded below by

$$\begin{aligned}(v,v)_{\beta\Psi} &\geq 4\pi\left(\sum_{n>0} p_n^2 + 2\beta\sum_{k>m>0} p_k \frac{\psi_{|k+m|}-\psi_{|k-m|}}{\tau_k\tau_m} p_m\right)\\ &= 4\pi\langle p,(I+2\beta\Gamma)p\rangle_{\ell^2},\end{aligned} \tag{54.20}$$

where $\langle\cdot,\cdot\rangle_{\ell^2}$ stands for the inner product of Hilbert space ℓ^2, $p=\{p_n\}\in\ell^2$, and $\Gamma:\ell^2\to\ell^2$ is the linear operator whose matrix representation has the entries

$$\Gamma_{km} = \frac{\psi_{|k+m|}-\psi_{|k-m|}}{\tau_k\tau_m}, \qquad \text{for } k>m$$

and $\Gamma_{km}=0$, if $k\leq m$.

Next we choose a sequence $\psi = \{\psi_n\}\in\ell^2$, such that (54.19) holds and the Hilbert-Schmidt norm of the operator $4\beta\Gamma$ is less than 1. For this purpose, we set $f(x)=e^{-x}$. Then we define $\psi=\{\psi_n\}$ as follows: $\psi_{2n+1}=0$, for any integer $n\geq 0$, $\psi_{-2n}=-\psi_{2n}$ and, for $n\geq 1$, we set

$$\psi_{2n} = \begin{cases}\delta, & \text{if } 1\leq n\leq N,\\ \delta f\left(\frac{n}{N}\right), & \text{if } N<n,\end{cases} \tag{54.21}$$

where δ is a positive constant, and N is a positive integer, which we now define. Inequality (54.19) is not difficult to satisfy with $\psi_{2n}\geq 0$. Indeed, for $n\geq N_0$, where

$$N_0 = \min\{m\in\mathbb{Z}^+ : \lambda_n\geq\tau_n^2 \text{ for all } n\geq m\},$$

one can use any value for $\psi_{2n}\geq 0$. If $N_0\geq 1$, we set $\psi_{2n}=\delta$, for $1\leq n\leq N_0$, where $\delta\geq 4\max\{\tau_n^2-\lambda_n : 1\leq n\leq N_0\}>0$. (A simple calculation shows that $\delta = 2(b^4L^4+2|a|+1)$ satisfies this inequality.) We now fix the value of N by $N=\max\left(N_0,\left[16\delta^2\right]\right)+1$. Note that $|\psi_{k+m}-\psi_{k-m}|=0$, whenever $0<k\leq m$, $0<k+m\leq 2N$, and

$$|\psi_{k+m}-\psi_{k-m}| \leq \delta\sup_{x\geq 0}|f'(x)|\cdot\frac{k+m-(k-m)}{2N}\leq\frac{\delta m}{N},$$

for all $k>m>0$. Hence we have the estimate of the Hilbert-Schmidt norm of Γ as follows

$$
\begin{aligned}
\|\Gamma\|_{H.S.}^2 &= \sum_{k>m>0} \left| \frac{\psi_{k+m} - \psi_{k-m}}{\tau_k \tau_m} \right|^2 \\
&\leq \frac{\delta^2}{N^2} \sum_{m=1}^{N} \sum_{k=2N-m+1}^{\infty} \frac{m^2}{\tau_m^2 \tau_k^2} + \frac{\delta^2}{N^2} \sum_{m=N+1}^{\infty} \sum_{k=m+1}^{\infty} \frac{m^2}{\tau_m^2 \tau_k^2} \\
&\leq \frac{\delta^2}{N^2} \sum_{m=1}^{N} \frac{m^2}{\tau_m^2} \int_{2N-m}^{\infty} \frac{dx}{\frac{1}{2}(x^4+1)} + \frac{\delta^2}{N^2} \sum_{m=N+1}^{\infty} \frac{m^2}{\tau_m^2} \int_{m}^{\infty} \frac{dx}{\frac{1}{2}(x^4+1)} \\
&\leq \frac{2\delta^2}{3N^2} \sum_{m=1}^{N} \frac{1}{(2N-m)^3} \frac{m^2}{\tau_m^2} + \frac{2\delta^2}{3N^2} \sum_{m=N+1}^{\infty} \frac{1}{m\tau_m^2} \\
&\leq \frac{2\delta^2}{3} \left(\frac{1}{N^5} \sum_{m=1}^{N} \frac{m^2}{\tau_m^2} + \frac{1}{N^2} \sum_{m=N+1}^{\infty} \frac{1}{m\tau_m^2} \right) \\
&\leq \frac{2\delta^2}{3} \left(\frac{1}{N^5} \int_0^N \frac{x^2}{\frac{1}{2}(x^4+1)} \, dx + \frac{1}{N^2} \int_N^{\infty} \frac{dx}{\frac{x}{2}(x^4+1)} \right) \\
&\leq \frac{2\delta^2}{3} \left(\frac{1}{N^5} \int_0^N dx + \frac{1}{N^2} \int_N^{\infty} \frac{dx}{x^3} \right) \leq \frac{\delta^2}{N}.
\end{aligned}
$$

Since $N \geq \left[16\delta^2\right] + 1$, one has

$$
\|4\,\beta\,\Gamma\|_{H.S.} \leq 4\,\|\Gamma\|_{H.S.} \leq 1, \qquad \text{for any } \beta \in [1/4, 1]\,. \tag{54.22}
$$

Substituting (54.22) into (54.20), for any $\beta \in [1/4, 1]$, we obtain

$$
\begin{aligned}
(v, v)_{\beta\Psi} &\geq 4\pi \langle p, (I + 2\beta\Gamma)p \rangle_{\ell^2} \geq 2\pi \|p\|_{\ell^2}^2 \\
&= 2\pi \sum_{n>0} v_n^2 \tau_n^2 = \pi \sum_{n \in \mathbb{Z}} \tau_n^2 v_n^2 = \pi\, Q(v),
\end{aligned}
$$

for any $v \in H^2_{\mathrm{op}}$. □

5.4.2. The General Case. From the proof of Theorem 54.2, it can be seen that the assumption that $u_0 \in H^2_{\mathrm{op}}$, instead of $u_0 \in H^2_{\mathrm{per}}$, is used only in the proof of Lemma 54.1. The KSE is well formulated in the full space L^2_{per} of all 2π-periodic functions. In particular, for each $u_0 \in H^2_{\mathrm{per}}$, there is a unique, maximally defined mild solution in H^2_{per} of equation (54.5). In equation (54.5), the disturbances of low wave numbers are amplified, especially worse if $a < 0$, while the disturbances of high wave numbers are damped. The $u\,\partial_x u$ term is regarded as a nonlinear energy-transfer mechanism that can transfer energy from low to high wave numbers and thus prevent an unbounded growth of modes of low wave numbers. If this heuristic idea can be made rigorous, then the hypothesis of oddness can be removed. This work was successfully completed by Il'yashenko (1992), Collet, etal (1993), and Goodman (1994). Here is the main result.

Theorem 54.3. *For any $u_0 \in V = H^2_{\mathrm{per}}$, there exists a unique, globally defined mild solution $S(t)u_0$ in V of equation (54.5), and $S(t)$ is a semiflow on V. Moreover, the semiflow is point dissipative in V and compact, for $t > 0$. Hence the KSE has a global attractor $\mathfrak{A}$ in V, and $\mathfrak{A}$ attracts all bounded sets in V. Furthermore, one has $\mathfrak{A} \subset H^4_{\mathrm{per}}$, and for each $u_0 \in \mathfrak{A}$, one has*

$$S(\cdot)u_0 \in C^{0,1-r}_{\mathrm{loc}}(\mathbb{R}; V^{2r}) \cap C(\mathbb{R}; \mathcal{D}(A)),$$

for each r with $0 \leq r < 1$.

5.5. Cahn-Hilliard Equation.

One of the most often and extensively studied examples of dissipative systems is the Cahn-Hilliard equation. It was initially introduced by Cahn and Hilliard (1958) as a model equation for describing the dynamics of pattern formation via phase transition, which was phenomenologically observed in phase separation of a binary solution under sufficient cooling. This kind of pattern formation occurs in alloys, polymer solutions, and liquid mixtures, cf. Novick-Cohen and Segel (1984) and references therein. In this section, we will prove the existence of various attractors for the Cahn-Hilliard equation in the space $H^2(\Omega)$.

Let Ω be an open, bounded domain in $\mathbb{R}^m$, where $1 \leq m \leq 3$, with a smooth boundary denoted by $\Gamma = \partial\Omega$. The Cahn-Hilliard equation is

$$\partial_t u + \nu\Delta^2 u = \Delta(f(u)), \qquad \text{for } x \in \Omega \text{ and } t \geq 0, \tag{55.1}$$

where $\nu > 0$ is a constant, Δ stands for the Laplacian operator, and f is a polynomial function of the degree $2p-1$, namely,

$$f(u) = \sum_{j=1}^{2p-1} a_j u^j, \qquad \text{for an integer } p \geq 2, \tag{55.2}$$

with its leading coefficient $a_{2p-1} > 0$. Thus, we have $f(0) = 0$. We assume that the space dimension m and the integer p used in equation (55.2) satisfy

$$p = \begin{cases} \text{any positive integer}, & \text{if } m = 1 \text{ or } 2, \\ 2, & \text{if } m = 3. \end{cases} \tag{55.3}$$

The original 1D Cahn-Hilliard equation has the nonlinear term $f(u) = -\gamma_1 u + \gamma_2 u^3$, with constants γ_1 and $\gamma_2 > 0$. In terms of the function

$$K(u) \stackrel{\text{def}}{=} -\nu\Delta\, u + f(u), \tag{55.4}$$

the Cahn-Hilliard equation can be written as

$$\partial_t u = \Delta\, K(u). \tag{55.5}$$

We will also use the Landau-Ginzburg free energy functional

$$J(u) \stackrel{\text{def}}{=} \int_\Omega \left[\frac{\nu}{2}|\nabla u|^2 + F(u)\right] dx, \qquad \text{where } F(z) = \int_0^z f(s)\, ds. \tag{55.6}$$

We assume that the solutions of equation (55.1) satisfy the Neumann boundary condition

$$\partial_n u = \partial_n(\Delta u) = 0, \qquad \text{for } x \in \Gamma \text{ and } t \geq 0, \tag{55.7}$$

where $\partial_n = \frac{\partial}{\partial n}$ stands for the outward normal derivative along the boundary Γ. This is sometimes called a **non-flux** boundary condition. Given an initial condition

$$u(x,0) = u_0(x), \qquad \text{for } x \in \overline{\Omega}, \tag{55.8}$$

we will investigate the initial-boundary value problem (55.1), (55.7), and (55.8).

Notice that for every number $\rho \in \mathbb{R}$, the function $u(x) \equiv \rho$, for $x \in \Omega$, is a stationary solution of equation (55.1), with boundary condition (55.7). This implies that the Cahn-Hilliard equation cannot have a global attractor, since the set Q of stationary solutions is unbounded. Nevertheless, this equation does have a host of attractors with many interesting dynamical properties. For example, the property

$$\overline{u}_0 \stackrel{\text{def}}{=} \frac{1}{|\Omega|}\int_\Omega u_0(x)\, dx = \frac{1}{|\Omega|}\int_\Omega u(x,t)\, dx, \qquad \text{for } t \geq 0 \tag{55.9}$$

is referred to as a preservation of mass. With the boundary conditions (55.7), one can readily check that every strong solution of (55.1) satisfies equation (55.9).

5.5.1 Construction of the Semiflow. In order to reformulate the Cahn-Hilliard equation as a parabolic evolutionary equation, we use the Sobolev spaces $H^j = H^j(\Omega)$, for integers $j \geq 0$, where $H^0 = L^2(\Omega)$. As usual the inner product and norm on H^0 is denoted by $\langle\cdot,\cdot\rangle = \langle\cdot,\cdot\rangle_0$ and $\|\cdot\| = \|\cdot\|_0$. For $j \geq 0$ one has $\|u\|_{j+1}^2 = \|u\|_j^2 + \sum_{|\alpha|=j+1}\|D^\alpha u\|_0^2$, with a similar construction for the inner product. For $u \in H^j$, with $j \geq 1$, we define $\|\nabla u\|_{j-1}^2 \stackrel{\text{def}}{=} \sum_{1\leq|\alpha|\leq j}\|D^\alpha u\|_0^2$, and we define v by $u = v + \overline{u}$, where $\overline{u}$ is given by (55.9). Since v satisfies the Poincaré inequality, there is a constant $C > 0$, such that $\|v\|_0^2 \leq C^2\|\nabla v\|_0^2 = C^2\|\nabla u\|_0^2$. It then follows from the Minkowski inequality that

$$N_j(u) \stackrel{\text{def}}{=} \left(\|\nabla u\|_{j-1}^2 + |\overline{u}|^2\right)^{\frac{1}{2}}, \qquad \text{for } j = 1, 2, \cdots,$$

is a norm on H^j, and it is equivalent to the standard norm on H^j, that is, there are positive constants $C_5 = C_5(j)$ and $C_6 = C_6(j)$, such that

$$C_5^2\|u\|_j^2 \leq N_j(u)^2 \leq C_6^2\|u\|_j^2, \qquad \text{for all } u \in H^j \text{ and } j = 1, 2, \cdots. \tag{55.10}$$

According to the Sobolev Imbedding Theorem (see Appendix B) one has $H^1(\Omega) \mapsto L^r(\Omega)$, for any r with $1 \leq r < \infty$, when $m = 1$ or 2; and $H^1(\Omega) \mapsto L^r(\Omega)$, for any r with $1 \leq r \leq 6$, when $m = 3$. Because of (55.3), we see that for every integer $k \geq 0$, there is a constant $c_k \geq 0$, such that for $u \in H^1$, one has

$$\|f^{(k)}(u)\|^2 \leq c_k^2 \left(1 + \|u\|_1^{2(2p-1-k)}\right), \qquad \text{for } 0 \leq k \leq 2p-1, \tag{55.11}$$

where $f^{(k)}(u) = D^k f(u)$ is the kth derivative of the polynomial function $f(u)$.

Next we reformulate the Cahn-Hilliard Problem as an evolutionary equation. We begin with the linear operator A, which is the biharmonic operator and is defined as $A = \Delta^2$ on the domain

$$\mathcal{D}(A) = V^4 \stackrel{\text{def}}{=} \{\varphi \in H^4 : \partial_n \varphi = \partial_n \Delta\varphi = 0 \text{ on } \Gamma\}. \tag{55.12}$$

For the nonlinear term, we define $G = G(u)$ by $G(u) = \Delta(f(u))$. The Cahn-Hillard Problem with the boundary conditons (55.7) is now reformulated as the evolutionary equation

$$\partial_t u + \nu A u = G(u). \tag{55.13}$$

We begin by analyzing the linear operator A.

Lemma 55.1. *Let A be given as above and define $B = A + aI$, where $a > 0$. Then B is a positive, selfadjoint, linear operator on H^0 with compact resolvent, i.e., the Standing Hypothesis B is satisfied. The negative operators $-A$ and $-B$ generate an analytic semigroups that satisfy $e^{-At} = e^{at}e^{-Bt}$, for $t \geq 0$. Moreover, the domain $\mathcal{D}(B^{\frac{1}{2}}) = \mathcal{D}(A^{\frac{1}{2}})$ satisfies $\mathcal{D}(A^{\frac{1}{2}}) = V^2 \stackrel{\text{def}}{=} \{\varphi \in H^2 : \partial_n\varphi = 0 \text{ on } \Gamma\}$, with $A^{\frac{1}{2}}\varphi = -\Delta\varphi$, for $\varphi \in \mathcal{D}(A^{\frac{1}{2}})$.*

Proof. Since $C_0^\infty(\Omega)$, the space of C^∞ functions with compact support in Ω, is dense in H^0 and $C_0^\infty(\Omega) \subset \mathcal{D}(A)$, we see that $\mathcal{D}(A)$ is dense in H^0. The reader should verify that $A : \mathcal{D}(A) \to H^0$ is a symmetric operator with $\langle Au, u\rangle \geq 0$, for all $u \in \mathcal{D}(A)$, i.e., A is nonnegative. Since the null space $\mathcal{N}(A)$ contains the the spatially constant functions, we see that A is not positive.

We claim that the range satisfies $\mathcal{R}(A) = H^0$. Indeed, one can solve the problem $Au = h$, for any $h \in H^0$, by solving two successive Neumann problems, first

$$-\Delta v = h, \qquad \text{and then} \qquad -\Delta u = v, \qquad \text{with } \partial_n v = \partial_n u = 0.$$

Even though the solutions v and u are not unique (since $\mathcal{N}(A) \neq \{0\}$), the solutions do exist. Furthermore, from the regularity theory for solutions of elliptic operators, one finds that $v \in H^2$ and $u \in H^4$, see Lions and Magenes (1968, Theorem 5.1) and Edmunds and Evans (1987, Chapter VII). Furthermore, the solution u satisfies the boundary conditions (55.7). Since $\mathcal{R}(A) = H^0$ and A is symmetric, it follows from Schechter (1981, Theorem 2.6.3) that A is a self-adjoint operator. Also, for $a > 0$, the operator $B = A + aI$ satisfies the Standing Hypothesis B. Let L be defined by $L\varphi = -\Delta\varphi$, where $\varphi \in \{\varphi \in H^2 : \partial_n\varphi = 0 \text{ on } \Gamma\}$. The two successive Neumann problems used to solve $Au = h$ show that $L(v) = h$ and $L(u) = v$, or $Au = L^2 u = h$. Hence, $L = A^{\frac{1}{2}}$. The remaining properties of the lemma now follow from this observation. □

The fractional powers V^α are defined in Section 3.2 in reference to the operator B. We will rescale these fractional powers so that $V^4 = \mathcal{D}(A) = \mathcal{D}(B)$. One then has $V^{4\alpha} \subset H^{4\alpha}$, for $0 \le \alpha \le 1$. As a result, one obtains $\mathcal{D}(A^{\frac{1}{2}}) = \mathcal{D}(B^{\frac{1}{2}}) = V^2$, or more generally, $V^{4\alpha} = \mathcal{D}(A^\alpha) = \mathcal{D}(B^\alpha)$, for $\alpha \ge 0$.

Let us now return to the nonlinear term $G = G(u)$ and the polynomial $f = f(u)$. For smooth functions u one has

$$\nabla f(u) = f^{(1)}(u)\nabla u, \quad \Delta f(u) = f^{(2)}(u)|\nabla u|^2 + f^{(1)}(u)\Delta u,$$
$$\nabla\Delta f(u) = f^{(3)}(u)|\nabla u|^2\nabla u + f^{(2)}(u)\left(2\Delta u\nabla u + \nabla|\nabla u|^2\right) + f^{(1)}(u)\nabla\Delta u,$$

where $|\nabla u|^2 = \nabla u \cdot \nabla u$. As a result of (55.11) and the imbeddings $H^2 \mapsto L^\infty$ and $H^2 \mapsto W^{1,6}$, for $1 \le m \le 3$, one can readily verify that

$$G \in C_{\text{Lip}}(H^2, H^0) \cap C_{\text{Lip}}(H^3, H^1). \tag{55.14}$$

The proof of the following result is left as an exercise.

Lemma 55.2. *Let $G = G(u)$, $K = K(u)$, and $J = J(u)$ be given as above. Then in addition to (55.14), one has*

$$K \in C_{\text{Lip}}(H^2, H^0), \quad \nabla K \in C_{\text{Lip}}(H^3, H^0), \quad \textit{and} \quad J \in C^1_{\text{Lip}}(H^2, \mathbb{R}),$$

where the Fréchet derivative $DJ(u)$ satisfies

$$DJ(u)v = \nu\langle\nabla u, \nabla v\rangle + \langle f(u), v\rangle, \qquad \textit{for } u, v \in H^2.$$

We are now in position to invoke the theory of Section 4.7 to obtain solutions of the Cahn-Hilliard equation.

Lemma 55.3. *Let the Cahn-Hilliard equation be given as above. Then for each $u_0 \in V^2$, there is a unique, maximally defined, mild solution $S(t)u_0$ of equation (55.13) in V^2 on $[0,T)$, where $0 < T = T(u_0) \le \infty$. Furthermore, $S(t)u_0$ is a strong solution in V^2, and it satisfies*

$$S(\cdot)u_0 \in C[0,T;V^2) \cap C^{0,1-r}_{\text{loc}}(0,T;V^{4r}) \cap C(0,T;\mathcal{D}(A)), \tag{55.15}$$

for all r with $0 \leq r < 1$. *In addition, one has* $S(\cdot)u_0 \in C[0, T; V^3)$, *whenever* $u_0 \in V^3 = \mathcal{D}(A^{\frac{3}{4}}) \subset H^3$.

Proof. Since the Standing Hypothesis B is satisfied and $G \in C_{\text{Lip}}(H^2, H^0)$, this follows directly from the theory of Section 4.7 (see Lemma 47.2). By the same token, since $G \in C_{\text{Lip}}(H^3, H^1) \mapsto C_{\text{Lip}}(H^3, H^0)$, Lemma 47.2 is applicable once again, and one has $S(\cdot)u_0 \in C[0, T_3; V^3)$, whenever $u_0 \in V^3$. It follows from (47.8) that $T_3 = T(u_0)$. □

Lemma 55.4. *Let J and K be given by equations (55.4) and (55.6) and let p satisfy (55.2) and (55.3). Then the following properties hold:*

(1) *There exist positive constants* f_0, C_0, *and* C_1 *such that*

$$-f_0 \leq F(s) \leq C_0 s^{2p} + f_0, \qquad \text{for } s \in \mathbb{R}.$$

and for all $u \in H^1$, *one has*

$$0 \leq J(u) - \frac{\nu}{2}\|\nabla u\|_0^2 + f_0\,|\Omega| \leq C_1\left(\|\nabla u\|_0^2 + |\overline{u}|^2\right)^p + 2f_0\,|\Omega|. \tag{55.16}$$

(2) *For any* $u_0 \in V^2$, *the solution* $S(t)u_0$ *of (55.1) in* V^2 *satisfies*

$$\partial_t J(S(t)u_0) = -\|\nabla K(S(t)u_0)\|^2 \quad \text{and}$$
$$\frac{\nu}{2}\|\nabla S(t)u_0\|_0^2 - f_0|\Omega| \leq J(S(t)u_0) \leq J(u_0),$$

for $0 \leq t < T(u_0)$.

(3) *There exists a continuous function* $L = L(r, s)$ *such that for any* $u_0 \in V^2$, *the solution* $S(t)u_0$ *of (55.1) in* V^2 *satisfies*

$$\|\nabla S(t)u_0\|^2 \leq C_6^2\|S(t)u_0\|_1^2 \leq L = L(\|\nabla u_0\|, |\overline{u}_0|), \qquad \text{for } 0 \leq t < T(u_0).$$

Proof. The first inequality in Item (1) follows directly from the definition of F. For the second inequality, we recall that the continuous imbedding $H^1 \mapsto L^{2p}$ implies that there is a constant $C > 0$ such that $\|u\|_{L^{2p}}^{2p} \leq C^{2p}\|u\|_1^{2p}$, for all $u \in H^1$. The result now follows from the definition of J and inequality (55.10).

For Item (2), we let $u_0 \in V^2$. It then follows from Lemmas 55.2 and 55.3 that $\nabla K(S(\cdot)u_0) \in C(0, T; H^0)$ and $J(S(\cdot)u_0) \in C^1(0, T; \mathbb{R})$. It then follows from Lemma 55.2, formulae (55.4), (55.5), (55.7), and our perennial friend, integration-by-parts, that for $0 < t < T$, one has

$$\begin{aligned}
\partial_t J(S(t)u_0) &= DJ(S(t)u_0)\partial_t S(t)u_0 \\
&= \nu\langle \nabla S(t)u_0, \nabla \partial_t S(t)u_0\rangle + \langle f(S(t)u_0), \partial_t S(t)u_0\rangle \\
&= -\nu\langle \Delta S(t)u_0, \partial_t S(t)u_0\rangle + \langle f(S(t)u_0), \partial_t S(t)u_0\rangle \\
&= \langle K(S(t)u_0), \partial_t S(t)u_0\rangle = \langle K(S(t)u_0), \Delta K(S(t)u_0)\rangle \\
&= -\langle \nabla K(S(t)u_0), \nabla K(S(t)u_0)\rangle = -\|\nabla K(S(t)u_0)\|^2.
\end{aligned}$$

Now if $u_0 \in V^3$, then $\nabla K(S(\cdot)u_0) \in C[0,T;H^0)$, by Lemmas 55.2 and 55.3. In this case, one obtains

$$J(S(t)u_0) = J(u_0) - \int_0^t \|\nabla K(S(s)u_0)\|^2\, ds \le J(u_0), \qquad \text{for } 0 \le t < T.$$

Since $J : H^2 \to \mathbb{R}$ is continuous and V^3 is dense in V^2, and since $T(u_0) \le \liminf_{n\to\infty} T(u_n)$, for any sequence u_n in V^2, with $u_n \to u_0$, (as is shown in the proof of Theorem 46.4), the last inequality is valid for all $u_0 \in V^2$. Hence Item (2) is valid. The conclusion in Item (3) now follows from inequality (55.16). □

In order to study the longtime dynamics of the Cahn-Hilliard semiflow $S(t)$, we write H^2 in the form

$$H^2 = \bigcup_{a\in\mathbb{R}} H_a^2, \qquad \text{where } H_a^2 \stackrel{\text{def}}{=} \{\varphi \in H^2 : \overline{\varphi} = a\}, \text{ for } a \in \mathbb{R}.$$

The set H_0^2 is a closed linear subspace of H^2, and for each $a \in \mathbb{R}$, with $a \neq 0$, one has $H_a^2 = H_0^2 + a$, i.e., H_a^2 is a hyperplane in H^2. (We treat a as a spatially constant function here.) Owing to the preservation of the mean value in (55.9), we see that each hyperplane H_a^2 is a closed, positively invariant set for the semiflow $S(t)$. For each compact set K in $\mathbb{R}$, we define

$$H_K^2 \stackrel{\text{def}}{=} \bigcup_{a\in K} H_a^2. \tag{55.17}$$

Similarly, one has $V^2 = \cup_{a\in\mathbb{R}} V_a^2$, where $V_a^2 = V^2 \cap H_a^2$, and we set $V_K^2 = \cup_{a\in K} V_a^2$, for any compact set K in $\mathbb{R}$. The proof of the following technical lemma is deferred until the end of this section.

Lemma 55.5. *Let the Cahn-Hilliard equation (55.1) be given with the boundary condition (55.7). For any compact set K in $\mathbb{R}$, let H_K^2 be given by (55.17). Then there exist positive constants C_7 and C_8 and a number $\sigma \in [0,1)$, such that for any $u_0 \in V_K^2$ and $t \in [0, T(u_0))$, one has $\|\Delta S(t)u_0\|^2 \le C_7\|\Delta^2 S(t)u_0\|^2$, and*

$$\partial_t\|\Delta S(t)u_0\|^2 + \nu\|\Delta^2 S(t)u_0\|^2 \le C_8(1 + \|\Delta^2 S(t)u_0\|^{2\sigma}). \tag{55.18}$$

Lemma 55.6. *Let $z \in C(0,T;\mathbb{R}^+)$ and $y \in C[0,T;\mathbb{R}^+)$, where y is absolutely continuous and $0 < T \le \infty$. Assume that*

$$\begin{aligned} y'(t) &\le -az(t) + b(1 + z(t)^\sigma), \\ y(t) &\le cz(t), \end{aligned} \tag{55.19}$$

for $t > 0$, where $\sigma \in [0,1)$ and a, b, c are positive constants. Then $y'(t) < 0$, whenever $y(t) > cz_0$, and for $0 \le t < T$, one has

$$\limsup_{t\to T^-} y(t) \le cz_0 \quad \textit{and} \quad y(t) \le \max(y(0), cz_0), \tag{55.20}$$

where $r = z_0$ is the unique positive root of the equation $-ar + b(1+r^\sigma) = 0$.

Proof. Let $\delta(s) = -as + b(1+s^\sigma)$, for $s \geq 0$, and define

$$\eta(s) \stackrel{\text{def}}{=} \begin{cases} \delta(s_0) & \text{for } 0 \leq s \leq s_0, \\ \delta(s) & \text{for } s_0 < s, \end{cases}$$

where $s_0 > 0$ is the unique maximum point of $\delta(s)$. One then has $\delta(s) \leq \eta(s)$, for $s \geq 0$, and η is nonincreasing. From the two inequalities of (55.19), we get

$$y'(t) \leq \delta(z(t)) \leq \eta(z(t)) \leq \eta(c^{-1}y(t)), \qquad \text{for } t > 0.$$

It follows that $y'(t) < 0$, whenever $c^{-1}y(t) > z_0$, which implies (55.20). □

This brings us to the first theorem concerning the semiflow generated by the mild solutions of the Cahn-Hilliard Problem.

Theorem 55.7. *Let the Cahn-Hilliard Problem (55.1) be given with the boundary conditions (55.7). Assume that (55.2) and (55.3) hold. For $u_0 \in V^2$, the maximally defined, mild solution $S(t)u_0$ of (55.13) exists for all $t \geq 0$. Furthermore, $S(t)$ is a semiflow on V^2 and the following properties are satisfied:*

(1) *For each $u_0 \in V^2$, $S(t)u_0$ is a strong solution of (55.13) and it satisfies*

$$S(\cdot)u_0 \in C[0,\infty;V^2) \cap C^{0,1-r}_{\text{loc}}(0,\infty;V^{4r}) \cap C(0,\infty;\mathcal{D}(A)),$$

for all r with $0 \leq r < 1$.

(2) *The subspace V^3 is a positively invariant subspace for this semiflow, and one has $S(\cdot)u_0 \in C[0,T;V^3)$, whenever $u_0 \in V^3$.*

(3) *$S(t)$ is point dissipative on V^2_K, for each compact set K in $\mathbb{R}$.*

(4) *The semiflow is compact on V^2, for $t > 0$.*

Proof. Let $u_0 \in V^2$ and set $T = T(u_0)$. Let $K = \{\overline{u}_0\}$, $y = y(t) = \|\Delta S(t)u_0\|^2$, $z = z(t) = \|\Delta^2 S(t)u_0\|^2$, and $(a,b,c) = (\nu, C_8, C_7)$. From (55.15), we see that $y \in C[0,T;\mathbb{R}^+)$ and $z \in C(0,T;\mathbb{R}^+)$. Since $\partial_t S(t)u_0 = -AS(t)u_0 + G(S(t)u_0)$, for $t > 0$, one has $\partial_t S(\cdot)u_0 \in C(0,T;V^0)$, by (55.15) and the continuity of $G : H^2 \to H^0$. For each $\tau \in (0,T)$, we set $u_\tau = S(\tau)u_0$. Since $S(\tau + t)u_0 = S(t)u_\tau$, one has

$$S(\cdot)u_\tau \in C[0,T-\tau;\mathcal{D}(A)) \quad \text{and}$$
$$\partial_t S(\cdot)u_\tau \in C[0,T-\tau;V^0) \subset L^2_{\text{loc}}[0,T-\tau;V^0).$$

It then follows from the argument of the Continuity Lemma 37.9, that

$$\partial_t \|A^{\frac{1}{2}} S(t)u_\tau\|^2 = \partial_t \|\Delta S(t)u_\tau\|^2 = 2\langle \partial_t S(t)u_\tau, AS(t)u_\tau \rangle.$$

Hence one has $y \in C^1(0,T;\mathbb{R}^+)$. As a result, Lemma 53.6 is applicable. Since $y(t) \leq \max(y(0), cz_0)$, it follows from Lemma 47.4 that $T = \infty$. As a result, Item (1) now follows from Lemma 55.3. Since $G \in C_{\mathrm{Lip}}(H^3, H^1) \mapsto C_{\mathrm{Lip}}(H^3, H^0)$, Item (2) follows from (55.15) and Theorem 46.2. For Item (3), we let K be any compact set in $\mathbb{R}$. Since $T = \infty$, Item (3) follows from (55.20). Indeed, the set $B = \{u_0 \in V^2 : \|\Delta u_0\|^2 < 2C_7 z_0\}$ is an absorbing set. Finally, Item (4) follows from Lemma 48.1. $\square$

5.5.2 Attractors for the Cahn-Hillard Equation. As noted above, the semiflow $S(t)$ generated by the mild solutions of the Cahn-Hilliard equation cannot have a global attractor in V^2, because the set

$$Q \overset{\text{def}}{=} \{u_0 \in H^2 : S(t)u_0 = u_0, \text{ for all } t \geq 0\}$$

of stationary solutions of equation (55.13), is an unbounded, invariant set in V^2. For a compact set K in $\mathbb{R}$, set $Q_K \overset{\text{def}}{=} Q \cap H^2_K = Q \cap V^2_K$. We now have the following result.

Theorem 55.8. *Let the Cahn-Hilliard Problem (55.1) be given with the boundary conditions (55.7). For any nonempty compact set K in $\mathbb{R}$, let V^2_K be given by (55.17). Then the following statements hold:*

(1) *The semiflow $S(t)$ on V^2_K has a nonempty, compact attractor $\mathfrak{A}_K$ in V^2_K, and $\mathfrak{A}_K$ attracts all bounded sets in V^2_K.*
(2) *The attractor $\mathfrak{A}_K$ has a lamination*

$$\mathfrak{A}_K = \bigcup_{a \in K} \mathfrak{A}_a,$$

where each $\mathfrak{A}_a$ is the nonempty, compact attractor in V^2_a given by Item (1).
(3) *The attractor $\mathfrak{A}_K$ in Item (2) is a bounded, invariant set in V^{4r}, for each r with $0 \leq r < 1$ and $\mathfrak{A}_K \subset V^4$.*
(4) *For each r with $0 \leq r < 1$, and for every $u_0 \in \mathfrak{A}_K$, the global mild solution $S(t)u_0$ is a classical solution of the Cahn Hilliard equation, for all $t \in \mathbb{R}$, and one has*

$$S(\cdot)u_0 \in C^{0,1-r}_{\mathrm{loc}}(\mathbb{R}; V^{4r}) \cap C(\mathbb{R}; V^4).$$

(5) *The set Q_K is nonempty, compact, and invariant, and $Q_K \subset \mathfrak{A}_K$.*
(6) *For each $u_0 \subset V^2$, the omega limit set $\omega(u_0)$, of the solution $S(t)u_0$, is a nonempty, compact, connected, invariant set in Q_a, where $a = \overline{u}_0$.*
(7) *For each $u_0 \in \mathfrak{A}_K$, the alpha limit set $\alpha(u_0)$, of the solution $S(t)u_0$, is a nonempty, compact, connected, invariant set in Q_a, where $a = \overline{u}_0$.*
(8) *The attractor $\mathfrak{A}_K$ satisfies $\mathfrak{A}_K = W^u(Q_K)$, where $W^u(Q_K)$ is the unstable set associated with Q_K (see Section 2.1.1).*

Proof. Items (1) and (2). Since the semiflow $S(t)$ is point dissipative on V_K^2 and compact, for $t > 0$, it follows from the Existence Theorem 23.12 that there is an global attractor $\mathfrak{A}_K$ for the restricted semiflow on V_K^2, and $\mathfrak{A}_K$ attracts all bounded sets in V_K^2.

Items (3) and (4). This follows from the Herculean Theorem 47.6.

Item (5). Let $\rho \in K$. Then $v(x) \equiv \rho$ is in Q_K. Hence Q_K is a nonempty, invariant set. It follows from Lemmas 55.5 and 55.6 that for $v \in Q_K$ one has $S(t)v = v$, for all $t \geq 0$. Therefore, one has $\partial_t \|\Delta S(t)v\|^2 = 0$, which implies that $\|\Delta v\|^2 \leq C_7 z_0$. Hence Q_K is a bounded set in V_K^2. Since $S(t)Q_K = Q_K$, for $t > 0$, we see that Q_K is compact, as well.

Item (6). Let $u_0 \in V^2$ and set $a = \overline{u}_0$. From Lemma 55.4, Item (2), we obtain the existence of the limits

$$\lim_{t\to\infty} J(S(t)u_0) = \ell^+(u_0) \geq -f_0|\Omega| \quad \text{and} \quad \lim_{t\to\infty} \partial_t J(S(t)u_0) = 0.$$

Since V_a^2 is positively invariant and closed, Theorem 23.15, implies that $\omega(u_0)$ is a nonempty, compact, invariant set in $\mathfrak{A}_a$. By Lemma 23.6, Item (1), $\omega(u_0)$ is connected. From the continuity of $J : H^2 \to \mathbb{R}$, it follows that $J(S(t)v) = \ell^+(u_0)$, for all $v \in \omega(u_0)$ and all $t \in \mathbb{R}$. Hence, for each $v \in \omega(u_0)$, one has

$$0 = \partial_t J(S(t)v) = -\|\nabla K(S(t)v)\|^2, \qquad \text{for all } t \in \mathbb{R}.$$

Thus $\nabla K(S(t)v) = 0$, which in turn implies that $\partial_t S(t)v = \Delta K(S(t)v) = 0$, for all $t \in \mathbb{R}$. Hence $v \in Q \cap \mathfrak{A}_a = Q_a$.

Item (7). Let $u_0 \in \mathfrak{A}_K$ and set $= \overline{u}_0$. Since $\mathfrak{A}_K$ is nonempty and compact, and since $J : \mathfrak{A}_K \to \mathbb{R}$ is continuous, there is a $J_0 \in \mathbb{R}$ with $J(u) \leq J_0$, for all $u \in \mathfrak{A}_K$. Since $J(S(t)u_0)$ is monotone and bounded above, the following limits exist: $\lim_{t\to-\infty} J(S(t)u_0) = \ell^-(u_0) \leq J_0$ and $\lim_{t\to-\infty} \partial_t J(S(t)u_0) = 0$. The remainder of the argument follows the paradigm of Item (6). We omit the details.

Item (8). The proof of this result is an adaptation of the argument used in Theorem 72.1, Item (6). We leave this as an exercise. □

Proof of Lemma 55.5. We will give the argument for the case where $m = 3$. A minor adaptation of this methodology is used for the easier cases, $m = 1$ or 2, see Temem (1988). Let $\alpha = \max\{|\rho| : \rho \in K\}$. We let $u = S(t)u_0$, for $u_0 \in V_K^2$ and $0 \leq t < T(u_0)$. By the Nirenberg-Gagliardo inequality (see Appendix B) there exist positive constants C_1, C_2, and C_3 such that

$$\begin{aligned} \|u\|_{L^\infty} &\leq C_1 \|\Delta^2 u\|^{1/6} \|u\|_{L^6}^{5/6}, \\ \|\nabla u\|_{L^4} &\leq C_2 \|\Delta^2 u\|^{1/4} \|\nabla u\|^{3/4}, \\ \|\Delta u\| &\leq C_3 \|\Delta^2 u\|^{1/2} \|\nabla u\|^{1/2}. \end{aligned} \tag{55.21}$$

We will let L_j, for $1 \leq j \leq 4$ denote positive constants which depend on the compact set $K \subset \mathbb{R}$ and the initial datum $u_0 \in V_K^2$. These constants do not depend on $t \in [0,T)$. Since f is a polynomial of degree $2p-1$, with $p=2$, there exist positive constants β_1 and β_2 such that

$$|f'(s)| \leq \beta_1 \left(1+|s|^2\right) \quad \text{and} \quad |f''(s)| \leq \beta_2 \left(1+|s|\right), \quad \text{for all } s \in \mathbb{R}.$$

Hence there is a constant $C_4 > 0$ such that

$$\begin{aligned}||\Delta f(u)|| &\leq ||f'(u)||_{L^\infty}||\Delta u|| + ||f''(u)||_{L^\infty}||\nabla u||_{L^4}^2 \\ &\leq C_4\beta_1 \left(1+||u||_{L^\infty}^2\right) ||\nabla u||^{1/2}||\Delta^2 u||^{1/2} \\ &\quad + C_4\beta_2 \left(1+||u||_{L^\infty}\right) ||\nabla u||^{3/4}||\Delta^2 u||^{1/4}.\end{aligned}$$

From Lemma 55.4, Item (3), there exist constants L_1 and L_2 such that

$$||\Delta f(u)|| \leq L_1 \left(1+||u||_{L^\infty}^2\right) ||\Delta^2 u||^{1/2} + L_2 \left(1+||u||_{L^\infty}\right) ||\Delta^2 u||^{1/4}.$$

Similarly, we use (55.21), the continuous imbedding $H^1 \mapsto L^6$, and Lemma 55.4, Item (3) to obtain a constant L_3 such that

$$||u||_{L^\infty} \leq L_3 ||\Delta^2 u||^{1/6}.$$

By combining the last two inequalities, we find a constant L_4 such that

$$\begin{aligned}||\Delta f(u)|| &\leq L_1 \left(1+L_3^2||\Delta^2 u||^{\frac{1}{3}}\right) ||\Delta^2 u||^{\frac{1}{2}} \\ &\quad + L_2 \left(1+L_3||\Delta^2 u||^{\frac{1}{6}}\right) ||\Delta^2 u||^{\frac{1}{4}} \qquad (55.22) \\ &\leq L_4 \left(1+||\Delta^2 u||^{5/6}\right).\end{aligned}$$

Next we form the L^2 inner-product of equation (55.1) with the multiplier $\Delta^2 u$ and obtain

$$\begin{aligned}\frac{1}{2}\partial_t||\Delta u||^2 + \nu \left\|\Delta^2 u\right\|^2 &\leq \left|\int_\Omega \Delta f(u)\Delta^2 u \, dx\right| \\ &\leq \frac{\nu}{2}||\Delta^2 u||^2 + \frac{1}{2\nu}\int_\Omega |\Delta f(u)|^2 \, dx. \qquad (55.23)\end{aligned}$$

Inequality (55.18) now follows from inequalities (55.22) and (55.23) with $\sigma = \frac{5}{6}$.

As noted in Smoller (1983, p 112), for any $g \in H^2(\Omega)$ with $\int_\Omega g\, dx = 0$ and $\partial g/\partial n = 0$ on Γ, one has $||g||^2 \leq c_0||\nabla g||^2 \leq c_1||\Delta g||^2$, where c_0 and c_1 are positive constants depending only on Ω. With $u(t) \in \mathcal{D}(A)$, for $t > 0$, we have $\int_\Omega \Delta u\, dx = 0$ and $\partial_n \Delta u = 0$ on Γ. Hence, one has $||\Delta u||^2 \leq c_1||\Delta^2 u||^2$, which completes the proof. □

Remarks. (1) As noted above, the space V^3 is a positive invariant set for the semiflow $S(t)$, and $\mathfrak{A}_K$ is a bounded, invariant set in V^3. One can adapt the methodology presented here to show that (a) $\mathfrak{A}_K$ is a compact set in V^3 and (b) it is an attractor in the H^3-metric.

(2) If it happens that for some $a \in \mathbb{R}$, the set $J(Q_a)$ is finite; for example, Q_a may be finite, then there is a Morse decomposition on $\mathfrak{A}_a$ (see Theorem 72.3).

(3) There is an alternate proof that Q_K is bounded, and we present the main ideas here. Note that $v \in Q$ if and only if there is a $\rho \in \mathbb{R}$, such that v is a solution of the elliptic problem

$$-\nu\Delta v + f(v) = \rho, \qquad x \in \Omega, \tag{55.24}$$

with Neumann boundary condition $\partial_n v = 0$ on Γ. The constant ρ is not arbitrary. It must satisfy $\rho = \frac{1}{|\Omega|}\int_\Omega f(v)\,dx$. That is to say, the latter equation is a constraint on v and/or ρ. Let $\alpha = \max\{|\rho| : \rho \in K\}$. By taking the L^2 inner-product of (55.24) with v itself, we get

$$\nu\|\nabla v\|^2 = -\int_\Omega f(v)v\,dx + \rho\int_\Omega v\,dx = -\int_\Omega f(v)v\,dx + \overline{v}\int_\Omega f(v)\,dx. \tag{55.25}$$

Note that f satisfies $-f(s)s \le -\frac{1}{2}a_{2p-1}s^{2p} + c$, for all $s \in \mathbb{R}$, where $c > 0$ is a constant. Also, for any $\epsilon > 0$, there is a constant $c_\epsilon > 0$ such that $|f(s)| \le \epsilon\, a_{2p-1}s^{2p} + c_\epsilon$, for all $s \in \mathbb{R}$. As a result, equation (55.25) implies

$$\begin{aligned}\nu\|\nabla v\|^2 &\le \left(-\frac{1}{2} + \epsilon|\overline{v}|\right) a_{2p-1}\int_\Omega v^{2p}\,dx + (c + c_\epsilon|\overline{v}|)|\Omega| \\ &\qquad \{\text{for } \epsilon < (2\alpha)^{-1}\} \\ &\le (c + c_\epsilon|\overline{v}|)|\Omega| \le (c + c_\epsilon\alpha)|\Omega|.\end{aligned} \tag{55.26}$$

Next we take the L^2 inner-product of equation (55.24) with Δv to obtain

$$-\nu\|\Delta v\|^2 + \int_\Omega f(v)\Delta v\,dx = \rho\int_\Omega \Delta v\,dx = 0. \tag{55.27}$$

Now

$$\int_\Omega f(v)\Delta v\,dx = -\int_\Omega f'(v)|\nabla v|^2\,dx \le m_0\|\nabla v\|^2, \tag{55.28}$$

where $m_0 = \max(0, \max\{-f'(s) : s \in \mathbb{R}\})$. By using (55.26) with equations (55.28) and (55.27), we get

$$\nu\|\Delta v\|^2 \le \frac{m_0}{\nu}(c + c_\epsilon\alpha)|\Omega|.$$

Since there is a constant $C > 0$ such that $\|u\|_2^2 \le C(\|\Delta u\|^2 + \alpha^2)$, for all $u \in \mathfrak{D}(A)$, we see that $Q \cap V_K^2$ is a bounded set in V^2.

5.6. Exercises.

Section 5.1

51.1. In the 1D Chafee-Infante Problem, let f be a polynomial that satisfies (51.2). Show that the Nemytskii mapping $F(u)(x) = f(u(x))$ satisfies

$$F \in C^1_F(V, V) \cap C^2_F(V, H).$$

51.2. In the 1D Chafee-Infante Problem, let $f \in C^3(\mathbb{R}, \mathbb{R})$ and (51.2) be satisfied. Show that the corresponding Nemytskii mapping F satisfies

$$F \in C^1_F(V^2, V^2) \cap C^2_F(V^2, V^1) \cap C^3_F(V^2, H).$$

Can it be generalized to the 2D and 3D domains?

51.3. Complete the proof of (51.27) for $m = 2$ and $m = 3$ cases.

51.4. Verify the inequality (51.28).

51.5. Prove Theorem 51.3 in detail.

51.6. (1) Show that the conclusions of Theorem 51.1 and Corollary 51.2 remain valid when the Dirichlet boundary conditions are replaced by the Neumann boundary conditions $\partial_n u = 0$ on $\partial\Omega$. (2) What happens if one has the mixed boundary condition: $\tau u(x) + (1 - \tau)\partial u(x)/\partial n = 0$ on $\partial\Omega$, where $0 < \tau < 1$?

51.7. Prove that, for $u_0 \in H$ and under the same assumptions as in the paragraph containing equation (51.1), there exists a unique mild solution $u(t) = S(t)u_0$ of equation (51.3), such that

$$u \in C[0, \infty; H) \cap C(0, \infty; V^1) \cap C^{0,1-r}_{\text{loc}}(0, \infty; V^{2r}) \cap C(0, \infty; V^2),$$

for any r with $0 \leq r < 1$.

51.8. Prove that the semiflow $S(t)$ on H defined in Exercise 51.7 is compact, for $t > 0$.

51.9. As in Theorem 51.4, define $\Pi : \Psi \to C([0, \tau]; U)$ by

$$(\Pi u)(t) = e^{-At}u_0 + \int_0^t e^{-A(t-s)}F(u(s))\,ds, \qquad t \in [0, \tau],$$

where $\Psi = \{\varphi \in C([0, \tau]; U) \mid \|\varphi(t)\|_U \leq \rho\}$ and $u_0 \in U$ is fixed. Use Theorem 38.10 to show that

$$\|(\Pi u)(t)\|_U \leq M_0 \|u_0\| + K_0 \int_0^t \left(\frac{C}{(t-s)^\alpha} + \frac{C}{(t-s)^\gamma} \right) ds,$$

where M_0, K_0 and C are constants and $\alpha, \gamma < 1$.

51.10. Under the assumptions of Theorem 51.4, use inequalities (51.31) and (51.32), as well as the Uniform Gronwall inequality to show that the solution semiflow of (51.28) has an absorbing set in space U.

51.11. In addition to the assumptions of Theorem 51.4, let us assume that there is a constant $0 < \tau \le 1$ and a positive continuous function $g_0(r)$ such that

$$\left| \frac{\partial f}{\partial s}(x, s_1) - \frac{\partial f}{\partial s}(x, s_2) \right| \le g_0(r) \, |s_1 - s_2|^\tau$$

for all $x \in \Omega$ and any s_1, s_2 in $[-r, r]$. Prove that the Fréchet derivative of the solution semigroup $DS(t)$ is locally Hölder continuous uniformly in $t \in [0, T]$ for any given $T > 0$.

51.12. Let Ω be any open, bounded domain with locally Lipschitz continuous boundary in $\mathbb{R}^d$. Let $f(s)$ be a polynomial of degree $p \ge 1$. In order that f maps $V = H^1(\Omega) \cap L^r(\Omega)$ into V itself, what is the sharpest choice of r? Answer this question for $d = 3$, 4, 5 and 6. What is the sharpest choice of q if $V = W^{1,q}(\Omega)$?

51.13. For the partly dissipative evolutionary equation

$$\partial_t \begin{pmatrix} u \\ v \end{pmatrix} + \begin{pmatrix} A & 0 \\ 0 & \gamma I \end{pmatrix} \begin{pmatrix} u \\ v \end{pmatrix} = \begin{pmatrix} f(u) - v \\ \delta u \end{pmatrix},$$

where $A = -\partial_x^2$ with $\mathcal{D}(A) = H^2(0, L) \cap H_0^1(0, L)$, $\gamma > 0$ and $\delta \in \mathbb{R}$ are constants, prove that the solution semigroup $S(t)$ on $W = H_0^1(\Omega) \times L^2(\Omega)$ has a decomposition $S(t) = S_1(t) + S_2(t)$ such that $S_1(t) \begin{pmatrix} u_0 \\ v_0 \end{pmatrix} \to 0$ at a uniform rate as $t \to \infty$, and $\kappa(S_2(t)B) \to 0$, as $t \to \infty$, where κ is the Kuratowski measure and B is any given bounded set in W.

51.14. Consider the initial boundary value problem

$$\begin{aligned} \partial_t u &= \partial_x^2 u + au - bu^2, && x \in \Omega = (0, 1), \ t > 0, \\ u(0, t) &= u(1, t) = 0, && t \ge 0, \\ u(x, 0) &= u_0(x) \ge 0, && 0 < x < 1, \end{aligned}$$

where a and b are positive constants. Assume that $u_0 \in H_0^1(0, 1)$. Prove that the mild solution $u(x, t)$ is nonnegative on $\overline{\Omega} = [0, 1]$ and that solution exists for $t \in [0, \infty)$.

51.15. Consider a Chafee-Infante Problem with a forcing term,

$$\begin{aligned} \partial_t u - \partial_x^2 u + f(x, u) &= a \sin x, && x \in \Omega = (0, \pi), \ t > 0, \\ u(0, t) = u(\pi, t) &= 0, && t > 0, \\ u(x, 0) &= u_0(x), && x \in \overline{\Omega}, \end{aligned}$$

where $f(x, u) = \lambda u + \beta(x) u^3$, with $\beta \in C(\overline{\Omega})$. Determine conditions satisfied by the constant λ and function β under which there exists a unique global mild solution for each $u_0 \in H_0^1(\Omega)$ and a global attractor exists for the semiflow on $H_0^1(\Omega)$.

51.16. Verify that the semiflow $S(t)$ on V^1 described in Theorem 51.1 is a gradient system (see Section 7.2).

51.17. Consider a Hodgkin-Huxley equation

$$\partial_t u - \nu \partial_x^2 u + f_1(w)u - f_2(w) = 0, \quad (x,t) \in (0,1) \times \mathbb{R}^+,$$
$$\partial_t w = -h_1(u)w + h_2(u),$$

with the Neumann boundary conditions $\partial_x u(0,t) = \partial_x u(1,t) = 0$, $t \geq 0$, and initial conditions $u(x,0) = u_0(x)$ and $w(x,0) = w_0(x)$, $x \in [0,1]$. Assume that $\nu > 0$, f_1 and f_2 are polynomials such that $f_1(w) \geq a > 0$, h_1 and h_2 are uniformly bounded real functions on $\mathbb{R}$ such that h_1' and h_2' are locally bounded, $h_1(u) \geq b > 0$ and $h_2(u) \geq 0$. Also assume that there is a fixed constant $w^* > 0$ such that $0 \leq w_0(x) \leq w^*$. Prove the following:

(a) For any $\begin{pmatrix} u_0 \\ w_0 \end{pmatrix} \in H^1(0,1) \times L^2(0,1) \stackrel{\text{def}}{=} Y$, the mild solution exists uniquely in Y for $t \geq 0$.

(b) The generated solution semiflow $S(t)$ on Y is point dissipative.

(c) There exists a global attractor $\mathfrak{A}$ in Y for $S(t)$.

(d) Prove that there exists a positively invariant region for this problem.

51.18. For the solution semiflow $S(t)$ of equation (51.28) described in Theorem 51.4, show that the equilibrium set Q is bounded in U and bounded in $H^2(\Omega)$.

51.19. Prove that the global attractor $\mathfrak{A}$ in Theorem 51.1 is a bounded set in space $L^\infty(\Omega)$.

51.20. Show that the set B_0 constructed in the proof of Theorem 51.1 is connected.

51.21. Sketch the analogue of Figure 5.1, where Q is an infinite set with isolated points.

Section 5.2

52.1. Check that under the assumptions of Example 52.2, the nonlinear term in equation (52.14) satisfies all the conditions (52.3)–(52.5).

52.2. Consider the following nonlinear wave equation,

$$\varepsilon \partial_t^2 u + \partial_t u + Au + g(u) = f,$$

where the Standing Hypothesis B is satisfied by the linear operator $A : \mathcal{D}(A) \to H$ and H is a real Hilbert space. Let $V = \mathcal{D}(A^{1/2})$, $f \in L^\infty(0,\infty; H) \cap C[0,\infty; H)$, and $g \in C^1_{\text{Lip}}(V, H)$. Define $E_0 = V \times H$ and $E_1 = \mathcal{D}(A) \times V$. Prove the following:

(a) For any initial datum $w_0 = \begin{pmatrix} u(0) \\ \partial_t u(0) \end{pmatrix} = \begin{pmatrix} u_0 \\ u_1 \end{pmatrix} \in E_0$, there exists a unique mild solution $S(t)w_0$ in E_0 for $t \in [0,\infty)$.

(b) Define two functionals on E_0 as follows,

$$\Gamma_\varepsilon(z) = \frac{1}{2}|u|_H^2 + \epsilon\langle u, v\rangle_H + \varepsilon|v|_H^2 + \|u\|_V^2,$$
$$K_\varepsilon(z) = \|u\|_V^2 + \langle u, v\rangle_H + \varepsilon|v|_H^2,$$

where $z = (u, v)^T \in E_0$. Verify that if $\varepsilon \leq 1$, then $\Gamma_\varepsilon(z)$ and $K_\varepsilon(z)$ are equivalent to $\|z\|_{E_0}^2$ and to $\|z\|_{Q_N E_0}^2$, respectively, when N is sufficiently large, see (d) below.

(c) Using multiplier $2\partial_t y + y$ to show that there is a constant $\alpha > 0$ such that

$$\partial_t \Gamma_\varepsilon(Y(t)) \leq \alpha \Gamma_\varepsilon(Y(t)), \quad t > 0,$$

if $\varepsilon \in (0, 1)$. Here u_1, u_2 are any two solutions of this equation, $y = u_1 - u_2$, and $Y(t) = (y(t), \partial_t y)^T$.

(d) Let $\sigma(A) = \{\lambda_n\}^\infty$, where λ_n is increasing and repeated to its multiplicity. Let $P_n : H \to \text{Span}\,\{e_1, \ldots, e_n\}$ and $Q_n = I - P_n$ be orthogonal projections. Define $\varphi(t) = Q_N y(t)$ and $\Phi(t) = (\varphi(t), \partial_t \varphi)^T$. Use the multiplier $2\partial_t \varphi + \varphi/\varepsilon$ to show that $\Phi(t)$ satisfies the differential inequality

$$\partial_t K_\varepsilon(\Phi(t)) + \frac{1}{2\varepsilon} K_\varepsilon(\Phi(t)) \leq \beta(\lambda_{N+1})\|y(t)\|_V^2, \quad t > 0,$$

where $\beta(\lambda_{N+1})$ is a constant depending on λ_{N+1}.

52.3. Consider 1D damped Boussinesq equation with the homogeneous Dirichlet boundary conditions and the initial conditions,

$$\partial_t^2 u + \delta\partial_t u + \alpha\partial_x^4 u + \beta\partial_x^2 u - k\,|\partial_x u|^2\,\partial_x^2 u + \gamma\partial_x^2(u^2) = 0, \; t > 0, \; x \in \Omega,$$
$$u|_{\partial\Omega} = \left.\frac{\partial u}{\partial n}\right|_{\partial\Omega} = 0, \quad t > 0$$
$$u(x, 0) = u_0(x), \quad \partial_t u(x, 0) = u_1(x), \quad x \in \overline{\Omega},$$

where $\Omega = (0, 1)$, δ, α and k are positive constants, β and γ are real constants. Formulate this problem to a nonlinear evolutionary equation in the form

$$\partial_t w + Aw = F(w),$$

with $w(t) = (u(t), \partial_t u)^T$. Then prove that its mild solution for any $w_0 = (u_0, u_1)^T \in \mathcal{D}(A^{1/2}) \times L^2(\Omega) \stackrel{\text{def}}{=} W$ exists globally and that its solution semiflow is dissipative.

52.4. (a) Discuss the effect if the term $-k|\partial_x u|^2\partial_x^2 u$ in the equation of Exercise 52.3 is changed to $k\,\partial_x u\,\partial_x^2 u$ with any real constant k.

(b) Can the results in Exercise 52.3 be generalized to 2D domain and 3D domain? Prove your claim.

52.5. Study the nonlinear beam equation with a structural damping,

$$\begin{aligned} &u_{tt} + \alpha u_{xxxx} - \delta u_{xxt} \\ &\quad - [a + b\int_0^1 |u_x(\xi,t)|^2\,d\xi + q\int_0^1 (u_x u_{xt})(\xi,t)\,d\xi]u_{xx} = f, \end{aligned}$$

with $u(0,t) = u_{xx}(0,t) = u(1,t) = u_{xx}(1,t) = 0$, where α, δ, b and q are positive constants and a is a real constant. Define $A = \partial_x^4$ with domain

$$\mathcal{D}(A) = \{\varphi \in H^4(0,1) : \varphi(0) = \varphi''(0) = \varphi(1) = \varphi''(1) = 0\}.$$

Use the similar approach shown in Section 5.2.2 to investigate the mild solutions of the evolutionary equation in space $E = \mathcal{D}(A^{1/2}) \times L^2(0,1)$, formulated from this nonlinear beam equation. Try to prove that the induced semiflow is dissipative in E and has a global attractor in $E^1 = \mathcal{D}(A) \times \mathcal{D}(A^{1/2})$ which attracts all bounded sets in E.

52.6. Consider the following nonlinear wave equation over a bounded domain $\Omega \subset \mathbb{R}^d$ ($d = 2$ or 3) with a smooth boundary,

$$\partial_t^2 v + g(x, \partial_t v) - \Delta v + f(v) = 0, \quad (x,t) \in \Omega \times \mathbb{R}^+,$$

with the homogeneous Dirichlet boundary condition, where $g \in C^1(\Omega, \mathbb{R})$, $f \in C^2(\mathbb{R})$ and $g(x,\varphi)\varphi \geq 0$. For any given functions $h \in W^{1,\infty}(\Omega)^3$ and $\eta \in W^{1,\infty}(\Omega)$, prove two identities hold for any strong solution v of this problem on $[0,T]$,

$$\begin{aligned} &\int_\Omega \partial_t v (h \cdot \nabla v)\,dx\Big|_{t=0}^{t=T} + \frac{1}{2}\int_0^T \int_\Omega (\operatorname{div} h)\Big[|\partial_t v|^2 - |\nabla v|^2 - 2F(v) \\ &\qquad + \sum_{k,j} \frac{\partial h_k}{\partial x_j}\frac{\partial v}{\partial x_k}\frac{\partial v}{\partial x_j} + g(x,\partial_t v)(h\cdot\nabla v)\Big]\,dx\,dt \\ &\qquad = \frac{1}{2}\int_0^T \int_{\partial\Omega} (h\cdot n)\left|\frac{\partial v}{\partial n}\right|^2 ds\,dt, \end{aligned} \tag{56.1}$$

and

$$\begin{aligned} &\int_\Omega v\partial_t v\,dx\Big|_{t=0}^{t=T} + \int_0^T\int_\Omega g(x,\partial_t v)v\,dx\,dt \\ &= \int_0^T \int_\Omega \{[|\partial_t v|^2 - |\nabla v|^2 - f(v)v] + v\nabla\eta\cdot\nabla v\}\,dx\,dt, \end{aligned} \tag{56.2}$$

where $F(z) = \int_0^z f(\sigma)\,d\sigma$ and n is the outward normal vector to the boundary $\partial\Omega$.

52.7. Verify (52.60), (52.61) and (52.65).

52.8. Try to generalize the results shown in Lemma 51.8 and Lemma 51.9 to the higher dimensional cases, $\Omega \subset \mathbb{R}^d$, $d = 2$ or 3.

52.9. Consider the singularly perturbed Hodgkin-Huxley equations,

$$\varepsilon \partial_t^2 u + (N + \varepsilon f_3(w))\partial_t u - \partial_x^2 u + f_1(w)u - f_2(w) = 0, \quad (x,t) \in (0,1) \times \mathbb{R}^+,$$
$$\partial_t w = -h_1(u)w + h_2(w),$$

with all the assumptions in Exercise 51.17 satisfied and, in addition, assume that $N > 0$ is a fixed constant, $\varepsilon > 0$ is a constant that can be arbitrarily small, and f_3 is a polynomial.

(1) First reformulate this problem to an evolutionary equation in the following form,

$$\partial_t U = G_\varepsilon U + F_\varepsilon(U), \quad t > 0,$$

on a Hilbert space $E = H^1(\Omega) \times L^2(\Omega) \times L^2(\Omega)$, $\Omega = (0,1)$.

(2) Show that there is a constant $\varepsilon_0 > 0$, such that for any $0 < \varepsilon \leq \varepsilon_0$, the solution semiflow $S_\varepsilon(t)$ on E is well-defined and dissipative.

(3) Prove that for any $0 < \varepsilon \leq \varepsilon_0$ there exists a global attractor $\mathfrak{A}_\varepsilon$ for the semiflow $S_\varepsilon(t)$ in E.

(4) Is it true that the global attractor $\mathfrak{A}_\varepsilon$ sits in space $H^2(\Omega) \times H^1(\Omega) \times H^1(\Omega)$?

(5) Is there any nesting property or relationship among the global attractors $\mathfrak{A}_\varepsilon$ for all ε in $(0, \varepsilon_0]$?

52.10. Prove that the set $B_0 = \gamma^+(B_1)$ in the proof of Corollary 52.11 is a bounded set in W.

Section 5.3

53.1. Show that the nonlinear part $F(u)$ in (53.11) satisfies

$$F \in C_F^2(V, H) \cap C_F^1(V, V).$$

53.2. Suppose $\Omega \subset \mathbb{R}^2$ or $\mathbb{R}^3$ is a bounded domain with smooth boundary and let

$$B(u)(x) = \text{div}\,_x f(u(x)).$$

Generalize the results in Lemma 53.1 to this operator B.

53.3. Work on the convective diffusion equation (53.1) with Neumann boundary condition $\partial_x u(0,t) = \partial_x u(L,t) = 0$. Show that with the appropriate choice of phase space, the same conclusion in Theorem 53.4 holds.

53.4. Prove that for any $u_0 \in V$ and $h \in H_0^1(\Omega)$, the solution $u(t) = S(t)u_0$ of the Initial Value Problem (53.11) satisfies

$$u \in C(0, \infty; V^2) \cap L^2_{\text{loc}}[0, \infty; V^3).$$

53.5. If $h = h(x,t)$, instead of $h(x)$, is a time-dependent function in $L^\infty[0,\infty;H)$, or $L^2(0,\infty;H)$, respectively, then what portion of the theory presented in Section 5.3 remains valid?

Section 5.4

54.1. Fill the detailed proof of equality preceding (54.19).

54.2. Verify that the two inequalities preceding (54.22) are valid.

54.3. Let $J(v,\varphi)$ be a functional defined by

$$J(v,\varphi) = \left\|\partial_x^2 v\right\|^2 - \beta \left\|\partial_x v\right\|^2 + \frac{1}{2}\int_\Omega \varphi'(x)v^2(x)\,dx, \qquad \text{for } v \in V,$$

where $\beta > 0$ is a given constant, $V = H^2(\Omega) \cap H_0^1(\Omega)$ and $\Omega = (-L/2, L/2)$. Prove that for any given $\mu > 0$, there exist a constant $\varepsilon > 0$ and a function $\varphi \in C_0^\infty(\Omega)$ such that the following coercive property is satisfied:

$$J(v,\varphi) \geq \varepsilon \left\|\partial_x^2 v\right\|^2 + \mu\|v\|^2, \quad \text{for any } v \in V,$$

where $\|\cdot\|$ stands for the $L^2(\Omega)$ norm.

54.4. Consider the Swift-Hohenberg equation

$$\partial_t u + \partial_x^4 u + \partial_x^2 u + u\partial_x u + \alpha u = 0, \quad x \in \Omega = (-L/2, L/2), \quad t > 0,$$

with the "rigid" boundary conditions $u(\pm L/2,t) = \partial_x u(\pm L/2,t) = 0$, $t \geq 0$. Let $u(x,t) = v(x,t) + \varphi(x)$, where $\varphi(x)$ is a gauge function that satisfies the coercive property in Exercise 54.3. Then prove that if u is the mild solution of the evolutionary equation formulated from the Swift-Hohenberg equation with $u(x,0) = u_0(x) \in L^2(\Omega)$, then

$$\|v(\cdot,t)\|^2 \leq \|u_0 - \varphi\|^2 e^{-\mu t} + \mu^{-2}\|F\|^2, \quad \text{for } t \in [0, T(u_0)),$$

where $F = -(\partial_x^4\varphi + \partial_x^2\varphi + \varphi\partial_x\varphi + \alpha\varphi)$, and $[0,T(u_0))$ is the maximal existence interval.

54.5. Prove that the semiflow $S(t)$ generated in $H - L^2(\Omega)$ from Swift-Hohenberg equation has a global attractor in $V = H_0^2(\Omega)$ which attracts all bounded sets in H.

54.6. Prove that the same conclusion holds in Exercise 54.5 if the "rigid" boundary conditions are replaced by the "hinged" boundary conditions

$$u(\pm L/2,t) = \partial_x^2 u(\pm L/2,t) = 0, \quad t \geq 0.$$

54.7. Consider the Burgers-Sivashinsky equation

$$\partial_t u + u\partial_x u = \partial_x^2 u + u, \quad x \in \Omega, \quad t > 0,$$

where $\Omega = (-L/2, L/2)$, u satisfies the L-periodic boundary condition, and the condition $\int_\Omega u(x,t)\,dx = 0$. Prove that if $\|u(\cdot,0)\| = \|u_0\| \le C_1(L)$ (which is a constant), then there is a constant $C_2(L)$ such that

$$\|u(\cdot,t)\| = \|S(t)u_0\| \le C_2(L) \text{ for all } t \ge 0,$$

here the norm is of $L^2(\Omega)$.

54.8. (a) Rescale the KSE to hold on the interval $\Omega_L = (-L/2, L/2)$.

(b) Show that there is a constant C_0, independent of the period L, such that if the initial datum satisfies $u_0(x) \in H^2_{\text{op}}(\Omega_L)$, for the KSE on Ω_L, then

$$\limsup_{t\to\infty} \|S(t)u_0\|_{L^2(\Omega)} \le C_0 L^{8/5}.$$

54.9. Prove that the same conclusion in Exercise 54.8 holds with a possibly different constant $C_0 > 0$ if $u_0 \in \mathcal{P}^0_L$ for the KSE, i.e., u_0 is an L-periodic function with mean value zero on Ω.

54.10. (a) Prove that for any function $\psi \in H^2_{\text{op}}(\Omega_L)$, one has

$$\|\psi'\|_{L^\infty} \le \frac{\sqrt{L}}{2\sqrt{3}}\|\psi\|_{L^2}.$$

(b) Prove that for any periodic function $\varphi \in H^1(\Omega_L)$, with $\int_{\Omega_L} \varphi\,dx = 0$, one has

$$\|\varphi\|_{L^\infty} \le \frac{\sqrt{2}}{2}\|\varphi\|^{1/2}_{L^2}\|\varphi\|^{1/2}_{H^1}.$$

54.11. Study the generalized Kuramoto-Sivashinsky equation,

$$\partial_t u + \nu\partial_x^4 u + b\partial_x^2 u + \partial_x(f(u)) + g(u) = 0,$$

where f and g are polynomials, and $f(s)$ has even degree with a positive leading coefficient. Try to generalize the results in Sections 5.4.1 and 5.4.2 to this equation with certain assumptions.

Section 5.5

55.1. Verify inequality (55.11).

55.2. Prove that for any bounded set S in H^3, there is a constant $C(f, s, \Omega)$ such that

$$\begin{aligned}
&\|\Delta f(u) - \Delta f(v)\|_{H^1} \\
&\le C(f, S, \Omega)(\|u - v\|_{L^\infty} + \|\nabla u - \nabla v\|_{L^2} \\
&+\|\Delta u - \Delta v\|_{L^2} + \|\nabla(\Delta u) - \nabla(\Delta v)\|_{L^2}),
\end{aligned}$$

for any u, $v \in S$.

55.3. Prove (55.14) in detail.

55.4. Prove Lemma 55.2.

55.5. Use the Nirenberg-Gagliardo inequality (see Appendix B) to show that

$$\|u\|_{L^\infty} \le \begin{cases} C\|\nabla u\| + |\bar{u}|, & \text{if } n = 1, \\ C\left\|\Delta^2 u\right\|^{\lambda} \|u\|_{L^q}^{1-\lambda}, & \text{with } \lambda = (1 + 3q/2)^{-1}, \text{ if } n = 2, \\ C\left\|\Delta^2 u\right\|^{1/6} \|u\|_{L^6}^{5/6}, & \text{if } n = 3; \end{cases}$$

$$\|\nabla u\|_{L^4} \le \begin{cases} C\left\|\Delta^2 u\right\|^{1/6} \|\nabla u\|^{5/6}, & \text{if } n = 1 \text{ or } n = 2, \\ C\left\|\Delta^2 u\right\|^{1/4} \|\nabla u\|^{3/4}, & \text{if } n = 3, \end{cases}$$

and

$$\|\Delta u\| < \begin{cases} C\left\|\Delta^2 u\right\|^{1/3} \|\nabla u\|^{2/3}, & \text{if } n = 1 \text{ or } n = 2, \\ C\left\|\Delta^2 u\right\|^{1/2} \|\nabla u\|^{1/2}, & \text{if } n = 3, \end{cases}$$

where $\|\cdot\|$ is the L^2 norm and $\bar{u}$ stands for the mean value of u, and C represents different positive constants.

55.6. Complete the proof of Items (7) and (8) in Theorem 55.8 in detail.

55.7. In Theorem 55.8, show that $\mathfrak{A}_K \subset V^4$.

55.8. Under the assumptions of Lemma 55.5, prove that there exist positive constants $\beta_1(K)$ and $\beta_2(K)$ which depend on the compact set $K \subset \mathbb{R}$, such that

$$\|\Delta f(u)\| \le \beta_1(K)(1 + \left\|\Delta^2 u\right\|^{1/3}), \quad \text{if } n = 1,$$

and

$$\|\Delta f(u)\| \le \beta_2(K)(1 + \left\|\Delta^2 u\right\|^{3/4}), \quad \text{if } n = 2,$$

where $n = \dim(\Omega)$ and the norm is of $L^2(\Omega)$. This implies (55.18) for $n = 1$ and $n = 2$.

55.9. Prove that $\mathfrak{A}_K$ is a compact set in V^3 and it is an attractor in the H^3 metric.

55.10. Show that if $u_0 \in V^3$, then the mild solution of (55.13) with this initial status satisfies $u \in L^\infty(0, \infty; W^{1,6}(\Omega))$ and

$$\|\nabla(\Delta f(u(t)))\|^2 \le h(r)(1 + \left\|\nabla(\Delta^2 u(t))\right\|), \quad t > 0,$$

for any u_0 in the ball $B_{V^3}(0; r)$, where $h(r) > 0$ is a positive constant depending on r.

55.11. Consider the viscous Cahn-Hilliard equation,

$$(1-\alpha)\partial_t u - \alpha\Delta(\partial_t u) + \Delta^2 u - \Delta f(u) = 0, \quad x \in \Omega, \quad t > 0$$
$$u = \Delta u = 0, \quad x \in \partial\Omega, \quad t \geq 0,$$
$$u(x,0) = u_0(x), \quad x \in \Omega,$$

where $\alpha \in [0,1]$ is a parameter, $\Omega \subset \mathbb{R}^d$ $(1 \leq d \leq 3)$ is a bounded domain with smooth boundary $\partial\Omega$, and $f \in C^3(\mathbb{R})$. Define $A = -\Delta$ with domain $\mathcal{D}(A) = H^2(\Omega) \cap H_0^1(\Omega)$ and $F(s) = \int_0^s f(\sigma)\, d\sigma$. Let

$$K(u) = Au + f(u), \quad J(u) = \int_\Omega [\frac{1}{2}|\nabla u|^2 + F(u)]\, dx,$$
$$K_1(u) = K(u) + \alpha\partial_t u, \quad K_2(u) = K(u) + (1-\alpha)A^{-1}(\partial_t u).$$

Show that if $u(t)$ is a strong solution of this problem, then it satisfies

$$(1-\alpha)\partial_t J(u) + \alpha(1-\alpha)\, \|\partial_t u\|^2_{L^2(\Omega)} + \|K_1(u)\|^2_{H^1(\Omega)} = 0,$$

and

$$\alpha\partial_t J(u) + \alpha(1-\alpha)\, \|\partial_t u\|^2_{H^{-1}(\Omega)} + \|K_2(u)\|^2_{L^2(\Omega)} = 0.$$

55.12. In the viscous Cahn-Hilliard equation described in Exercise 55.11, assume that f is a polynomial of odd degree and with a positively leading coefficient. Prove that if u_0 is in the ball of radius r centered at the origin of $H = L^2(\Omega)$, then there exist positive constants M (independent of r) and t_1 (depending on r) such that the solution $S_\alpha(t)u_0$ satisfies

$$\|S_\alpha(t)u_0\|_{H^2(\Omega)} \leq M, \quad \text{for } t \geq t_1.$$

55.13. Show that for every $\alpha \in (0,1)$, there exists a global attractor $\mathfrak{A}_\alpha$ in $H^2(\Omega)$ for the semiflow $S_\alpha(t)$ of the viscous Cahn-Hilliard equation described in Exercise 55.11. Determine whether $\mathfrak{A}_\alpha$ is continuous in $\alpha \in (0,1)$ with respect to the Hausdorff distance in $H^2(\Omega)$.

55.14. Discuss the behavior of global attractors $\mathfrak{A}_\alpha$, $0 < \alpha < 1$, when $\alpha \to 0$ and $\alpha \to 1$, respectively.

5.7. Commentary.

Section 5.1. Since the same equation (51.1) has been extensively used as a model for heat transfer processes, it is also called a nonlinear *heat equation.* Some authors refer to equation (51.1) as the *Allen-Cahn equation*, see Allen and Cahn (1979). This citation seems to be inaccurate because the work of Chafee-Infante (1974) on this problem appeared several years earlier.

In general, there are two approaches for proving the local existence and uniqueness of solutions to an initial-boundary value problem of an evolutionary PDE. One is the **(operator) semigroup** approach that we take in Section 5.1. By this approach, one has to set up a functional framework and formulate the original problem as an initial value problem in a Banach space, then check that the conditions on the linear part and the nonlinear part of the equation required by an abstract existence theorem (as we have presented in Chapter 4) are satisfied, and finally apply that theorem.

The other approach is called the **Bubnov-Galerkin method** (see Sections 6.2, 6.8, and 6.10 for details). This method features a sequence of finite dimensional approximations of solutions usually with respect to the variational form (or called weak form) of the original problem. Then one can use either compactness techniques, or monotonicity techniques, or penalization techniques to reach the limit that gives a (weak) solution. The best reference for this method is Lions (1969).

While the semigroup approach seems simpler, the Bubnov-Galerkin method has its own merits. The latter is more flexible in studying the regularity of solutions in various function spaces and more convenient in searching and analyzing numerical approximate solutions. Furthermore, the Bubnov-Galerkin approach leads to a concept of a weak solution even when the semigroup approach is not applicable, as is illustrated with the weak solutions of Leray-Hopf class in Chapter 6.

It is difficult to pinpoint the first paper(s) on the existence of global attractors for reaction diffusion equations/systems in bounded domains. To some extent, this was known in the 1970s, and several major papers appeared in the 1980s. See for example, Chafee and Infante (1974); Mallet-Paret (1976); Mañé (1977); Chueh, Conley, and Smoller (1977); Conway, Hopf, and Smoller (1978); Henry (1981); Babin and Vishik (1983a,b); and Marion (1987). The corresponding results for unbounded domains in weighted Sobolev spaces have been established in Babin and Vishik (1990). Applications of the longtime dynamics of RDE to population dynamics, such as predator-prey models, appear in Murray (1993), Redlinger (1995) and Smith (1995).

Structure of Global Attractors. Although in general the issue of the structure of a global attractor remains quite open, many insightful results have already been obtained. We have mentioned the structure of gradient systems, see Section 7.2. Besides, it is proved in Brunovsky (1990) that for scalar reaction diffusion equations with Dirichlet boundary conditions the global attractor is the graph of a C^1 mapping from a subset with a nonempty interior of a subspace, whose dimension equals the maximal Morse index of all the equilibrium solutions. Also see Fusco (1987), Brunovsky and Fiedler (1988, 1989), and Jolly (1989) for related results.

Blowup. If we do not impose an asymptotical sign condition and adequate growth conditions on the nonlinearity of equation (51.1), then the mild solution and/or other types of solutions may blowup or, in other words, global solutions may not exist for some initial data. Two simple examples can be shown in terms of (51.1) with the same boundary conditions, where

(i) $f(u) = |u|^{p-1}u$;
(ii) $f(u) = e^u$.

Many articles have contributed results on the blowup sets, blowup times, blowup solutions, and blowup rates. Chen and Matano (1989) has proved that a solution to a 1D semilinear heat equation with Dirichlet or Neumann boundary conditions will blowup at most at finite points if the initial value is continuous. A typical blowup rate for (51.1) in case (i) is

$$(T-t)^{1/(p-1)}u(x,t) \to \text{const} \ , \ \text{as } t \to T,$$

and in case (ii) is

$$u(x,t) - \log \frac{1}{T-t} \to 0, \text{ as } t \to T,$$

where $|x - x_0| \leq C|T-t|^{1/2}$ and x_0 is a blowup point. See Chen (1990), Giga and Kohn (1987), and Herrero and Velazquez (1992).

It might appear that blowup (in finite time) is the antithesis of longtime dynamics, and perhaps that is the case. This makes the paper Chen and Matano (1989) all the more interesting, because these authors study the blowup issue for the semilinear heat equation by using longtime dynamics. The key idea they propose is to rescale time t near the blowup time and to rewrite the original equation in these new variables. By doing this they are able to use the longtime dynamics of the new equation to show that the solutions of the original equation have an "asymptotic profile", as they approach blowup.

Lap Number. An intrinsic notion associated with 1D semilinear parabolic equations

$$\partial_t u = a(x,t)\partial_x^2 u + b(x,t)\partial_x u + f(t,u,\partial_x u)$$

is the lap number of u for each time t, denoted by $\gamma(u(t))$, that is an integer-valued function of t and roughly evaluates the complexity of the piecewise monotone behavior of the solution $u(x,t)$ when time evolves. More precisely, the lap number characterizes the number of sign-change times in the x-direction. This concept is introduced in Matano (1982), in which it is shown that, under certain assumptions, the lap number of any solution to the above equation with either of the three types of boundary conditions is *nonincreasing* with time t.

In Henry (1985) and Angenent (1986), Matano's idea has been exploited to investigate the Fréchet derivative of the solution semigroup and the transversality between stable and unstable manifolds with respect to hyperbolic equilibrium points. In the 1D case, along this direction more profound studies also reveal the connection of the dimensions of unstable manifolds and the number of zeros of the associated stationary solutions. Some other results on the structural dynamics of infinite dimensional Morse-Smale systems generated by reaction diffusion or heat equations can be found in Henry (1985).

Linearization. The theory of linearization and diffeomorphism near a fixed point for ODEs has been extensively and deeply studied. For the reaction diffusion problem (51.1), suppose $u = \varphi(x)$ is a hyperbolic stationary solution of this problem, then there is a fundamental question: Is the semiflow nearby this equilibrium topologically similar to that of the associated linear problem? In other words, can the nonlinear problem be linearized by at least a homeomorphism in a neighborhood of φ? The results answering these questions are usually referred to as the *Hartman-Grobman Theorem*. In Lu (1991), such a theorem and the existence of invariant foliations have been established for 1D reaction diffusion equations.

Quasilinear Parabolic Equations. As a related area, the general theory of the existence, uniqueness, and dynamics of solutions to quasilinear parabolic equations can be found in Amann (1986, 1989, 1990). The existence of global attractors for quasilinear equations have been addressed in Babin and Vishik (1989). In Dung (1997), for a class of weakly coupled quasilinear parabolic systems, the existence of global attractors is proved via the L^∞ dissipativity of the semiflow based on the invariance principle, or through the weak L^p dissipativity with the aid of a nonlinear Gronwall inequality.

Section 5.2. The abstract nonlinear wave equation (52.2) is a model for many concrete nonlinear wave equations in the form of

$$\partial_t^2 u + \alpha \partial_t u - \Delta u + f(u) = h.$$

The conditions on the nonlinearity can be directly imposed on the scalar function f and on its anti-derivative, instead of regarding it as a Nemytskii mapping on function spaces, see Temam (1988) and many other articles. The theory of the existence of global attractors for damped nonlinear wave equations was first established in Ghidaglia and Temam (1987) and Ladyzhenskaya (1987).

In Babin and Vishik (1983b, 1989), a concept of an (E_1, E) global attractor is used for nonlinear wave equations, where $E_1 \hookrightarrow E$. In this setting, a global attractor $\mathfrak{A}$ is located in the space E_1. However, $\mathfrak{A}$ only attracts bounded sets in E, in terms of the E-metric. Related results appear in Haraux (1985).

The nonlinear damped beam equation (52.30) is from Woinowsky (1950), and it has been studied, along with its variations, by many authors. The results in Section 5.2.2 have been generalized to more sophisticated nonlinear beam models, especially the large space structure model with structural damping and Balakrishnan-Taylor damping of full exponent, see You (1993b, 1996a).

The technique of decomposition of the nonlinear term $f = f_1 + f_2$ in equation (52.49) and the corresponding decomposition of $u = v + \psi$ is a particular case of the more delicate Arrieta-Carvalho-Hale decompositions, see Arrieta, et al (1992).

Theorem 52.7 in Section 5.2.3 can be generalized to 2D and 3D cases with some restrictions. We state the theorem here and make a few remarks concerning the modification of the 1D proof.

Theorem 57.1. *Consider the initial boundary value problem (52.49) over a 2D or 3D domain Ω and let the assumptions* (Ad), (Ag), *and* (Af) *be satisfied with the following changes:*

(a) $\Upsilon = Q \cap \Omega$, *and Q is an open connected neighborhood of the boundary $\partial\Omega$;*

(b) Ω *and* $\Omega \backslash C\ell(\Upsilon)$ *are nonempty, convex sets with piecewise smooth boundaries;*

(c) *the power in the growth condition of* (Af) *satisfies $p \le 1$ for the 3D case.*

Then there exists a global attractor $\mathfrak{A} \subset W$ for the solution semiflow $S(t)$ of this problem, and attracts all bounded sets in W.

Comments on the Proof. In conducting similar *a priori* estimates as in Lemmas 52.8 and 52.9, one should use the multiplier $q(x)\nabla y + \xi(x) y$ and use the Greens formula instead of the integration-by-parts. Then, in constructing two weighting functions q_0 and ξ_0, the convexity condition in (b) will be used. With the restriction (c) in the 3D case, all the technical steps can get through.

Unbounded Domains. Nonlinear wave equations defined over an unbounded domain or an exterior domain have many significant applications in aviation, hydraulic, and hydronautic dynamics. Many results on the asymptotic behaviors (including global attractors) for these problems have been known. See, for example, Babin and Vishik (1990); Mochizuki and Motai (1995, 1996); and Feireisl (1995, 1997).

For the study of global solutions and dynamics of nonlinear wave equations over an exterior domain, there are two available approaches. One uses the Nash-Moser-Hörmander type iteration scheme, and the other is to acquire *a priori* bounds on the solutions by combining the energy estimates of the local solutions with Klainerman decay estimates, see Datti (1990) and references therein.

Weak Damping. As we have said in this section, for nonlinear wave equations the dissipativity is primarily caused by a damping term, linear or nonlinear, structural $\partial_x^2 \partial_t u$ or external $\partial_t u$. Recently, the longtime behavior of nonlinear wave equations with weak dampings has aroused many research interests. What kind of damping is weak damping? There is no official answer to this question. One usually recognizes the following types of damping as a weak damping:

(i) A locally distributed damping in the form of $d(x)g(\partial_t u)$, where the function $d(x)$ has a local or small support that is a proper subset of the domain Ω, as is in Section 5.2.3.

(ii) A nonlinear and asymptotically monotone damping in the form of $g(\partial_t u)$, where g is asymptotically monotone and g' may not be bounded.

(iii) A linear or nonlinear boundary damping, which takes place only on the boundary or a part of it. In a 1D wave equation (as a vibrating string equation), the velocity damping may only occur at one of the endpoints.

Some interesting results in cases (i) and (ii) have been established. See Lopes (1986), Nakao (1990), Feireisl and Zuazua (1993), You (1993b, 1996b, 2000), and references therein. However, as far as we are aware, the globally dissipative dynamics of nonlinear wave equations or hyperbolic evolutionary equations with the type (iii) damping remains an open problem.

Quasilinear Wave Equations. For quasilinear wave equations such as

$$\partial_t^2 u - \Delta \partial_t u + f(|\nabla u|^2)\Delta u + g(u) = 0,$$

or degenerated wave equations such as

$$K(x,t)\partial_t^2 u - \alpha \partial_t u + \Delta u + f(u, \nabla u) = 0,$$

the global existence and the longtime dynamics of solutions require special techniques and scrutiny. Some known results can be found in Assila (1998) and Cavalcanti, et al. (1998).

Section 5.3. The model equation (53.1) is a typical convective reaction diffusion equation and it can also be called a heat convection equation. There are many other forms of convective diffusion equations. In analyzing the solution behavior and conducting *a priori* estimates, one can see from this section that no Lyapunov functional is involved. In general, the semiflow generated by a convective diffusion equation may not be a gradient system.

For higher-dimensional convective equations with dissipative phenomena, see Escobedo and Zuazua (1991); Escobedo, et al. (1994); Lukaszewicz and Krzyzanowski (1997); Ueda and Matsuda (1998); and Inoue (1999).

Section 5.4. The Fourier series method used in Section 5.4.1 for constructing a gauge function is a useful tool in the investigations of KSE. This method stems from Nicolaenko, Sheurer and Temam (1985). Recently in Jolly, Rosa and Temam (2000) it has been used to recalculate the radius of an absorbing ball and estimate the dimension of an inertial manifold for the KSE.

In Duan and Ervin (1998), the effect of a nonlocal term (the periodic Hilbert transform) on the global dynamics of a generalized Kuramoto-Sivashinsky equation is studied. The result turns out that there exists a family of maximal attractors parametrized by the mean values of the initial data. A comprehensive generalized KSE in the form of

$$\partial_t u + \alpha \Delta u + \beta \Delta^2 u + \nabla \cdot f(u) + \Delta g(u) + h(u) = \varphi(x)$$

with the periodic boundary conditions over a higher dimensional domain has been investigated in Guo and Su (1993). This equation even includes the Cahn-Hilliard equation as a particular case. Under certain assumptions, it is shown that there exists a global attractor in space $H^2(\Omega)$ of finite fractal dimension for this problem. The rich dynamics related to the complicated set of stationary solutions of KSE has been thoroughly examined and evidenced in Michelson (1996).

In Il'yashenko (1992), for the 1D Kuramoto-Sivashinsky equation (54.1), with the coefficient $a = 0$ and with the periodic boundary conditions and without the antisymmetric restriction, it is proved by means of the analysis of the **transition of energy** that a global attractor exists for the semiflow. This theory provides an intrinsic explanation of the dissipative behavior that KSE exhibits. Namely, a theorem therein shows that the dynamical system of KSE transfers the energy from low modes to high modes when time evolves. While this transition of energy proceeds, the high modes eventually dominate and the dissipative fourth-order linear term makes the entire dynamics absorbed. Also see Collet, et al (1993) and Goodman (1994).

Section 5.5. There are two types of phase separation: nucleation or spinodal decomposition. The dynamics of nucleation for the Cahn-Hilliard equation have been analyzed by Cahn and Hilliard (1959); Pego (1989); and Bates and Fife (1993), for example. The dynamics of spinodal decomposition have been studied by Cahn (1961), Hilliard (1970), Hoyt (1989), and Eyre (1993).

The Cahn-Hilliard equation has attracted wideapread attention by researchers, see Elliott and Zheng (1986); Temam (1988); Nicolaenko, Scheurer and Temam (1989); Novick-Cohen (1990); Alikakos, Bates and Fusco (1991); Dlotko (1994); Caffarelli and Muler (1995); Chen and Kowalczyk (1996); and Fusco and Alikakos (1998). A comprehensive overview of the Cahn-Hilliard equation is presented in Novick-Cohen (1998).

In the proof of Lemma 55.5, we only present the argument for the case where the space dimension $m = 3$. Then three inequalities in (?) play a key role in the estimates. For cases $m = 1$ or 2, the same argument is valid based on the corresponding inequalities that have been put in Exercise 55.5.

Theorem 55.8, on the existence of the attractor in V_K^2, can be generalized. The semiflow $S(t)$, restricted on V_K^3, has a nonempty compact attractor in V_K^3.

Viscous Cahn-Hilliard Equations. Recently, the viscous Cahn-Hilliard equations in the following form stimulates many interests,

$$(1-\alpha)\partial_t u = \Delta w,$$
$$\alpha\partial_t u = \Delta u + f(u) + w,$$

or equivalently,

$$(1-\alpha)\partial_t u - \alpha\Delta\partial_t u + \Delta^2 u + \Delta f(u) = 0,$$

where $0 \le \alpha \le 1$ is a parameter, over a bounded regular domain Ω in R^m, $1 \le m \le 3$. When $\alpha = 0$, it is a standard Cahn-Hilliard equation. When $\alpha = 1$, it becomes a reaction diffusion equation. Numerical evidence shows the insensitivity of the global attractor for this equation to change with α. In Elliot and Stuart (1996), it is shown that for each $\alpha \in [0,1]$, the solution semiflow $S_\alpha(t)$ is a gradient system and is asymptotically compact, so that $S_\alpha(t)$ has a global attractor $\mathfrak{A}_\alpha$ that is continuous in α. More recent works including the structure of the global attractors can be found in Grinfeld and Novick-Cohen (1999).

Generalizations. A system of Cahn-Hilliard equations such as

$$\partial_t u = \Delta K(u), \quad K(u) = -\Gamma\Delta u + \nabla_u\Phi(u),$$

with the vector Neumann boundary conditions, where Γ is a positive definite matrix, $\Phi : R^N \to R$ is sufficiently smooth, and the underlying domain Ω is in R^m, $m \le 3$, has been studied in Cholewa and Dlotko (1994) and Li and Zhong (1998). This can be used as a model for the pattern formation of an alloy consisting of $N+1$ components. It is proved that this system has global attractors on some affine subspaces of $H^2(\Omega)^N$ under certain reasonable assumptions on Φ. The Lyapunov functional used in this scenario is

$$J(u) = \int_\Omega \mathrm{tr}\,(\nabla u^T\Gamma\nabla u)\,dx + \int_\Omega \Phi(u)\,dx.$$

Other Dissipative Equations.

In addition to the nonlinear evolutionary partial differential equations presented above, there are many other nonlinear PDEs which exhibit some sort of dissipativity, and the longtime behavior of these equations is of interest. In the remainder of this section, we shall briefly comment on some of these dissipative PDEs and touch on some of the known results in the vast sea of articles.

Fully Nonlinear Parabolic Equations. The fully nonlinear equations

$$\partial_t u = \Delta f(u) + g(u),$$

where $f \in C^2(\mathbb{R})$ and $g \in C^1(\mathbb{R})$ that can be power functions $f = u^m$ and $g = u^p$ for instance, arise in many models such as for gas/fluid flows in porous media and for biological population dynamics. There has been an abundant set of articles addressing various topics of the solutions to this type of equations. In Eden, Michaux and Rakotoson (1991) and Andreu, etal (1998), the existence of global attractors in space $L^\infty(\Omega)$ for the semiflow of weak solutions is proved by the methodology provided in DiBenedetto (1983). The same equation on $\Omega = R^N$ is studied in Feireisl, Laurençot and Simondon (1996), in which it is proved that a global attractor exists in a weighted Banach space. Also, see Peletier and Zhao (1993) and You (1996d).

Korteweg - de Vries Equations. Apart from the well-known soliton solutions and the inverse scattering transforms for KdV equations featuring the interactions of dispersion and nonlinearity, one can explore its longtime dynamics if a strongly damped term $\alpha\Delta u$ and/or weakly damped term βu are/is imposed to the equation. The existence of absorbing sets, global attractor and inertial manifolds have been proved in several settings; see Ghidaglia (1988), You (1996c), Moise and Rosa (1997), and Hayashi and Naumkin (1998).

Ginzburg-Landau Equations and Nonlinear Schrödinger Equations. The general form of these two equations is

$$\partial_t u - (\nu + i\alpha)\Delta u + (\kappa + i\beta)|u|^2 u + \gamma u = 0,$$

where ν, α, κ and β are real parameters, ν and $\kappa \geq 0$. Sometimes, the names of the two equations are referred to interchangeably. This equation, over a bounded domain of dimension $m = 1$ or 2 with Dirichlet, Neumann, or periodic boundary conditions, can be used to describe the evolution of instability waves in a variety of dissipative systems such as in fluid dynamics or in superconductivity phenomena. In Temam (1988), it is shown that the solution semiflow has a global attractor in $L^2(\Omega)$. In Ghidaglia and Heron (1987), it is proved that the attractor attracts all trajectories with $u_0 \in H^k(\Omega)$ in H^k norm.

The weakly damped nonlinear Schrödinger equation on an unbounded domain has been studied in Goubet (1998). Despite the fact that the zero-order weak dissipation term has no smoothing effect, it is proved that a global attractor exists in the energy space because it can be shown that all strong solutions converge as $t \to \infty$ to a set constituted by smoother solutions. A recent work, Abounouh and Goubet (2000), proves that in the case of 2D thin domain, the global attractor turns out to be more regular by using the technique of splitting the Fourier series and the anisotropic Sobolev inequalities.

Boussinesq Equations. An abstract theorem on the existence of a weak global attractor is proved with the criterion of asymptotical compactness in You (1997a,b). The result is stated as follows.

Theorem 57.2. *Let W be a reflexive, separable Banach space. Suppose that a semiflow $S(t)$ is weakly continuous on W for each fixed $t > 0$, and there exists a constant $\mu > 0$ and weakly continuous functional $H(\cdot)$ and $K(\cdot)$ on W, such that for any $w_0 \in W$,*

$$\partial_t \left(\|S(t)w_0\|^2 + H(S(t)w_0)\right) + \mu\|S(t)w_0\|^2 = K(S(t)w_0),$$

for almost all $t > 0$, then $S(t)$ is asymptotically compact.

Based on this result in You (1997b), it is shown that for the 2D weakly damped Boussinesq equations

$$\partial_t^2 u + \delta\partial_t u + \alpha\Delta^2 u + \beta\Delta u - \kappa|\nabla u|^2\Delta u + \gamma\Delta(u^2) = 0,$$

over a bounded smooth domain Ω in $\mathbb{R}^2$, with boundary conditions $u = \partial_n u = 0$ on $\partial\Omega$, where δ, α, and κ are positive constants and β and γ are real constants, the global mild solution defines a dissipative semiflow $S(t)$ that possesses a global attractor in the energy space. It is interesting to note that the relationship of Boussinesq equations versus nonlinear wave equations is somehow similar to the relationship of Cahn-Hilliard equations or Kuramoto-Sivashinsky equations versus reaction diffusion equations.

Von Kármán Equations. The global dynamics of von Kármán equations describing the nonlinear vibration of a thin plate or a shell with or without thermal effects are of interest both theoretically and in applications. The typical von Kármán equation is in the following form,

$$\begin{aligned} &\partial_t^2 u + \alpha\partial_t u + \Delta^2 u - [u + f, v + \tau] + g(\nabla u) = h(x), \\ &\Delta^2 v = -[u + 2f, u], \quad x \in \Omega \text{ and } t > 0, \end{aligned}$$

with boundary conditions $u = \partial_n u = v = \partial_n v = 0$ on $\partial\Omega$, where Ω is a smooth bounded domain in $\mathbb{R}^2$. Here u is the deflection of the plate, v is the Airy stress function, and $[u, v] = \partial_{x_1}^2 u\partial_{x_2}^2 v - 2\partial_{x_1x_2}^2 u\partial_{x_1x_2}^2 v + \partial_{x_2}^2 u\partial_{x_1}^2 v$. One can work on the energy functional to show that, with a damping term such as above, the strong solution semiflow is dissipative; see Menzala and Zuazua (1998). The existence of a global attractor has been shown under specific conditions by Chueshov (1991) and Lasiecka (1999).

Coupled Parabolic-Hyperbolic Equations. Oftentimes a mathematical model involves two coupled PDEs of different nature. Let us just mention two examples. One is the dissipative Zakharov system in laser and plasma physics that consists of a complex field described by a nonlinear Schrödinger equation and a real field governed by a nonlinear wave equation or a Boussinesq equation. Another is the singularly perturbed Hodgkin-Huxley system where a damped nonlinear wave equation with a small second-order coefficient is coupled with several degenerate parabolic equations. Under certain conditions, one can use combined methodology to deal with these systems and succeed in proving the existence of a global attractor and some other asymptotic properties; see Flahaut (1991) and Fitzgibbon, Parrott and You (1995). In the latter work, the upper semicontinuity of the global attractor with respect to the coefficients of the perturbed leading term is also shown.

Additional Readings

Babin and Vishik (1992), Friedman (1969), Hale (1988), Henry (1981), Ladyzhenskaya (1991), and Temam (1988).

6
NAVIER-STOKES DYNAMICS

Longtime dynamical issues arise in many areas in the world about us. For example, one finds them in various fluid flows as illustrated by 1) heat transfer and its effects on global climate modeling and weather prediction; 2) flows of multiphased fluids and oil recovery; 3) behavior of chemical solutes in lakes, harbors and river basins; 4) geothermal consequences of magma flows; and 5) fluid flows in thin films. In order to understand the modeling of such phenomena, one needs good mathematical tools coupled with sound insight into the physics and the chemistry of the problems.

The physics and chemistry of the problem depend, of course, on many factors. For example, in the study of the motion of a fluid contained in a moving capsule, one need only examine the internal forces acting on the fluid coupled with the external forces acting on the capsule. In a thermal convection problem, such as a magma flow, one finds couplings between the internal forces acting on the fluid with the force of gravity and a buoyancy force due to the thermal gradients. Likewise, in an oceanic model one has couplings with the thermal and saline gradient forces. In more complex problems, such as the behavior of chemical solutes and pollutants in water beds, one then encounters chemical reactions, which in turn create a thermal gradient and thereby affect the ambient fluid flow.

What underlies all these models is the description of the internal forces acting on the fluid itself, see Navier (1827), Poisson (1831), and Stokes (1845). These internal forces are modeled by the Navier-Stokes equations. In this book we will be interested primarily in the dynamics of incompressible, viscous fluid motion.

Even though the primary physical applications of Navier-Stokes dynamics are to fluid flows in 3D, the Navier-Stokes equations themselves make good sense in 2D as well. Furthermore the analysis of the 2D problem is meaningful for certain physical appplications, because one can give a rigorous proof that a suitable 2D problem is a good approximation to a 3D problem on thin 3D domains, see Raugel and Sell (1993a,b, 1994). As we will show below, the 2D problem does have a global attractor for both the weak solutions and the strong solutions.

In the 3D problem, one quickly encounters several difficulties in trying to build a theory of global attractors. While it is the case in 3D that for given data (u_0, f), there is a weak solution $u = u(t)$ of Class LH (Leray-Hopf) and that $u(t)$ remains a weak solution for all time $t \geq 0$, it is not known whether this solution is uniquely determined by the data (see Section 6.3). As a result, one cannot conclude that the mapping $S(t) : u_0 \to S(t)u_0 = u(t)$ satisfies the semigroup property required for a semiflow. On the other hand, for *good* data (u_0, f), where $u_0 \in V$, the initial value problem does have a unique strong solution on some interval $[0, T)$. However, it is not known whether this strong solution continues to exist for *all* time $t \geq 0$, i.e., it is not known whether $T = \infty$ (see Section 6.4.3). These two issues appear to be possible obstructions to the development of a theory of global attractors for the 3D Navier-Stokes equations. In the case of weak solutions, one faces the possibility of nonuniqueness, and in the case of strong solutions, it is the Global Regularity Problem (see Section 6.4.3). However, with an alternate point of view, one bypasses both of these issues. In particular, we will show the following:

(1) for every smooth bounded domain Ω in $\mathbb{R}^3$ and for all suitable forcing functions f, the class of weak solutions of the Navier-Stokes equations on Ω generates a semiflow on a suitable phase space, and
(2) under further regularity assumptions on f (for example, f is in H and is independent of time), this semiflow has a global attractor.

Since this theory is restricted to weak solutions, one avoids the Global Regularity Problem. The nonuniqueness issue is resolved by replacing H with another phase space W, in which each point is a weak solution. Thus if a given datum $u_0 \in H$ admits several weak solutions, then each of these solutions corresponds to different points in the space W.

We will follow the *modus operandi* used above and assume that the following Standing Hypothesis is satisfied:

Standing Hypothesis A. *Let A be a positive, sectorial operator on a Banach space W with associated analytic semigroup e^{-At}. Let $V^{2\alpha}$ be the family of interpolation spaces generated by the fractional powers of A, where $V^{2\alpha} = \mathcal{D}(A^\alpha)$, for $\alpha \geq 0$. Let $\|A^\alpha u\| = \|A^\alpha u\|_W = \|u\|_{V^{2\alpha}} = \|u\|_{2\alpha}$ denote the norm on $V^{2\alpha}$. See Lemma 37.4 for more information.*

As a matter of fact, in the context of the Navier-Stokes equations, the linear operator A is the Stokes operator. As noted in Section 3.8.3, the Stokes operator satisfies the stronger condition:

Standing Hypothesis B. *The operator A is a positive, selfadjoint, linear operator, with compact resolvent, on a Hilbert space H. Consequently A satisfies the Standing Hypothesis A. Moreover, the fractional power spaces*

V^α are defined for all $\alpha \in \mathbb{R}$, and equation (37.2) defines the Hilbert space structure on each V^α. Also the semigroup e^{-At} is compact, for $t > 0$. See Theorem 37.2 for more information.

6.1. Formulation as a Nonlinear Evolutionary Equation.

The Navier-Stokes equations for an incompressible, viscous fluid motion assume the form

$$\begin{aligned} \partial_t u - \nu \Delta u + (u \cdot \nabla)u + \nabla p &= f \\ \nabla \cdot u &= 0 \end{aligned} \tag{61.1}$$

on an open, bounded domain Ω in $\mathbb{R}^d$ of class C^2, where $d = 2$ or 3. In (61.1) the term u is a d-dimensional vector field on Ω, and $u = u(x,t)$ represents the velocity of the fluid at the point $x \in \Omega$ and time t. The pressure term $p = p(x,t)$ is a scalar field on Ω. The function $f = f(x,t)$ is a d-dimensional vector field on Ω. One should view $f(x,t)$ as representing an internal force on the fluid, for example, a gravity or buoyancy force. The coefficient of viscosity ν is a positive constant. The first equation in (61.1) is sometimes called the **momentum equation**, and the second equation $\nabla \cdot u = 0$ is referred to as the **conservation equation**. The nonlinear term $(u \cdot \nabla)u$ in (61.1) is referred to as the **inertial term**. In the sequel, we will develop several solution theories for (61.1), where the solution u also satisfies the initial value condition

$$u(x,0) = u_0(x), \tag{61.2}$$

and certain boundary conditions. The ordered pair (u_0, f) is referred to as the **data** of the Navier-Stokes equations. The term u_0 is the **initial condition** and f is the **forcing function**.

In order for the problem (61.1) to be well-posed we require that appropriate boundary conditions be satisfied. For many physical problems, the boundary conditions describe a force acting on the boundary, such as fluid entering or leaving a cavity, or a non-slip condition caused by the motion of a segment of the boundary of Ω. However, in the body[23] of this chapter we will focus primarily on homogeneous boundary conditions. In particular we will assume that the Dirichlet boundary conditions (DBC)

$$u(x,t) = 0, \qquad \text{for } x \in \partial\Omega \text{ and } t \geq 0 \tag{61.3}$$

hold. Other than the obvious requirement that the pressure $p = p(x,t)$ lie in a space where the gradient ∇p makes sense, there are no other side conditions imposed upon p. In the case that Ω is a rectangle (in $\mathbb{R}^2$) or a

[23] A brief introduction to inhomogeneous boundary conditions is given in Section 6.7.

parallelepiped (in $\mathbb{R}^3$), one can also study (61.1) where the periodic boundary conditions (PBC)

$$u(x + L_i\hat{e}_i, t) = u(x,t), \qquad x \in \mathbb{R}^d, \quad t \geq 0 \tag{61.4}$$

hold with $\Omega = \Pi_{i=1}^d (0, L_i)$ and $\{\hat{e}_1, \cdots, \hat{e}_d\}$ is the standard basis in $\mathbb{R}^d$. In the case of the (PBC), one can readily show that if the data (u_0, f) satisfy

$$\int_\Omega u_0 \, dx = \int_\Omega f \, dx = 0, \tag{61.5}$$

then the solution $u = u(x,t)$ satisfies $\int_\Omega u(x,t)\, dx = 0$, for all $t \geq 0$. Therefore in the (PBC) case, we also require that (61.5) holds.

6.1.1 The Stokes Operator, Revisited. As noted in Section 3.8.3, the Stokes Operator A arises from the study of the linear problem, where the inertial term $(u \cdot \nabla)u$ is set equal to 0 to obtain the Stokes equations:

$$\begin{aligned} \partial_t u - \nu \Delta u + \nabla p &= f, \\ \nabla \cdot u &= 0. \end{aligned} \tag{61.6}$$

Recall that the Stokes operator arises when one projects (61.6) into the Hilbert space

$$H = \text{Cl}_{\,L^2(\Omega)} \{u \in C_0^\infty(\Omega) : \nabla \cdot u = 0\},$$

or equivalently, into

$$H = \{u \in L^2(\Omega, \mathbb{R}^d) : \nabla \cdot u = 0 \text{ in } \Omega \text{ and } u \cdot n = 0 \text{ on } \partial\Omega\},$$

for Dirichlet boundary conditions. For periodic boundary conditions, one uses

$$H = \text{Cl}_{\,L^2_{\text{per}}(\Omega)} \{u \in C^\infty_{\text{per}}(\Omega)) : \nabla \cdot u = 0 \text{ and } \int_\Omega u \, dx = 0\},$$

see Temam (1983). Throughout this chapter, the inner product and norm on H will be written in the form $\langle \cdot, \cdot \rangle$ and $\|\cdot\|$, respectively.

Let $\mathbb{P}$ denote the Helmholtz, or Leray, projection, i.e., the orthogonal projection of $L^2(\Omega)$, or $L^2_{\text{per}}(\Omega)$, onto H. Recall that $H^\perp$, the orthogonal complement of H, is

$$H^\perp = \text{Cl}_{L^2(\Omega)} \{\nabla\phi : \phi \in C^1(\overline{\Omega}, \mathbb{R})\},$$

thus the pressure staisfies $\mathbb{P}\nabla p = 0$ in Ω. We assume that the forcing function $f = f(t)$ satisfies $f \in L^\infty(0, \infty; H)$. In this case one has $\mathbb{P}f = f$. By applying $\mathbb{P}$ to (61.6) one obtains

$$\partial_t u + \nu A u = f, \tag{61.7}$$

where $Au = -\mathbb{P}\Delta u$ and the appropriate boundary conditions are satisfied.

As shown in Section 3.8.4, the Stokes operator A is a positive, self-adjoint, linear operator with compact resolvent. Therefore the Standing Hypothesis B is satisfied, and the eigenvalues of A satisfy (32.2) and (32.3). The three spaces $V = V^1$, $H = V^0$, and V^{-1} have special significance in the theory of the Navier-Stokes equations. Recall that the Stokes operator A has the property that $\mathcal{D}(A^{\frac{1}{2}}) = V$ and the imbeddings $V \hookrightarrow H \hookrightarrow V^{-1}$ are compact. The inner products and the norms on V^1 and V^{-1} will be described in terms of the A^α-inner products and norms. In particular one has

$$\langle u, v\rangle_V = \langle A^{\frac{1}{2}}u, A^{\frac{1}{2}}v\rangle = \langle A^{\frac{1}{2}}u, A^{\frac{1}{2}}v\rangle_{L^2}, \qquad \text{for } u, v \in V,$$

and

$$\langle u, v\rangle_{V^{-1}} = \langle A^{-\frac{1}{2}}u, A^{-\frac{1}{2}}v\rangle = \langle A^{-\frac{1}{2}}u, A^{-\frac{1}{2}}v\rangle_{L^2}, \qquad \text{for } u, v \in V^{-1}.$$

Furthermore the imbeddings $V \hookrightarrow H \hookrightarrow V^{-1}$ gives rise to a duality and a bilinear form $\langle\langle\cdot,\cdot\rangle\rangle = \langle\langle\cdot,\cdot\rangle\rangle_{(V^{-1},V)}$, where

$$\langle\langle u, v\rangle\rangle \stackrel{\text{def}}{=} \langle A^{-\frac{1}{2}}u, A^{\frac{1}{2}}v\rangle = \langle A^{-\frac{1}{2}}u, A^{\frac{1}{2}}v\rangle_{L^2}, \qquad \text{for } u \in V^{-1} \text{ and } v \in V.$$

In particular, $\langle\langle u, v\rangle\rangle = \langle u, v\rangle$, for $u \in H$ and $v \in V$.

The positivity of A yields $\|Au\|^2 \geq \lambda_1^2\|u\|^2$, for all $u \in \mathcal{D}(A)$. In terms of the fractional powers, one then has

$$\|A^\alpha u\|^2 \geq \lambda_1^{2\alpha}\|u\|^2, \qquad \text{for } u \in \mathcal{D}(A^\alpha) \text{ and } \alpha \in \mathbb{R}. \tag{61.8}$$

We will also use the interpolation inequality (37.15), which states that if $\gamma = \theta\alpha + (1-\theta)\beta$, where $\alpha, \beta, \gamma \in \mathbb{R}$, $\alpha \geq \beta$, and $0 \leq \theta \leq 1$, then there is a constant $C > 0$ such that

$$\|A^\gamma u\| \leq C\|A^\alpha u\|^\theta \|A^\beta u\|^{1-\theta}, \qquad \text{for all } u \in \mathcal{D}(A^\alpha). \tag{61.9}$$

The Helmholtz projection $\mathbb{P}$ can be applied to the Navier-Stokes equations (61.1), as well, and one obtains

$$\partial_t u + \nu Au + B(u, u) = f, \tag{61.10}$$

which is referred to as the **Navier-Stokes (evolutionary) equation**, where $B(u, v)$ is the bilinear form $B(u, v) \stackrel{\text{def}}{=} \mathbb{P}(u \cdot \nabla)v$.

As noted in Section 3.8.3, the pressure term p does not appear in (61.10) because $\mathbb{P}\nabla p = 0$. The pressure can be recovered by applying the complementary projection $I - \mathbb{P}$ to (61.1) to obtain

$$\nabla p = (I - \mathbb{P})(\nu\Delta u - (u \cdot \nabla)u) + (I - \mathbb{P})f, \tag{61.11}$$

where by assumption we have $(I - \mathbb{P})f = 0$. As shown in Lemma 38.5, if one finds a solution $u = u(t)$ of (61.10) on some interval I, then one can use (61.11) to solve for the pressure p. Since the pressure field p is completely determined by the velocity field u, one sometimes refers to p as a *slave variable.* To put it another way, the dynamics of (61.1) is completely determined by the dynamics of (61.10), the Navier-Stokes evolutionary equation.

6.1.2 The Nonlinearity. In addition to the bilinear term $B(u,v)$, we want to study the trilinear form

$$b(u,v,w) \stackrel{\text{def}}{=} \sum_{i,j=1}^{d} \int_\Omega u_i(D_i v_j) w_j dx = \langle (u\cdot\nabla)v, w\rangle, \tag{61.12}$$

where u, v, and w lie in appropriate subspaces of $L^2(\Omega, \mathbb{R}^d)$, and $D_i = \frac{\partial}{\partial x_i}$. Note that in the case where $w \in H$ and $\mathbb{P}w = w$, one has

(61.13)

$$\langle B(u,v), w\rangle = \langle \mathbb{P}(u\cdot\nabla)v, w\rangle = \langle (u\cdot\nabla)v, \mathbb{P}w\rangle = \langle (u\cdot\nabla)v, w\rangle = b(u,v,w).$$

For $u, v, w \in H$ we claim that $b(u,v,w) = -b(u,w,v)$. Indeed by integrating by parts and by using the boundary conditions and the divergence free condition $\nabla\cdot u = \sum_i D_i u_i = 0$, one obtains

$$\begin{aligned} b(u,v,w) &= \sum_{i,j=1}^{d} \int_\Omega u_i(D_i v_j) w_j \, dx \\ &= -\sum_{i,j=1}^{d} \int_\Omega (D_i u_i) v_j w_j \, dx - \sum_{i,j=1}^{d} \int_\Omega u_i v_j (D_i w_j) \, dx \\ &= 0 - b(u,w,v), \end{aligned} \tag{61.14}$$

for sufficiently smooth functions $u, v, w \in H$. Consequently one finds that

$$b(u,v,v) = 0, \qquad \text{for smooth functions } u, v \in H. \tag{61.15}$$

The trilinear form $b(u,v,w)$ satisfies several basic Sobolev inequalities. We will formulate these inequalities in terms of the norms $\|\cdot\|_\alpha$ on the fractional power spaces V^α. We note that the Sobolov spaces $H^{2\alpha}(\Omega)$ satisfy $V^{2\alpha} = \mathcal{D}(A^\alpha) \subset H^{2\alpha}(\Omega)$, for $0 \le \alpha \le 1$, and there exist constants d_1 and d_2, which depend on α, with $0 < d_1 \le d_2$, such that

$$d_1 \|u\|_{H^{2\alpha}(\Omega)} \le \|A^\alpha u\|_{L^2(\Omega)} = \|u\|_{2\alpha} \le d_2 \|u\|_{H^{2\alpha}(\Omega)}, \tag{61.16}$$

for all $u \in V^{2\alpha} = \mathcal{D}(A^\alpha)$. In other words, the norms $\|u\|_{H^{2\alpha}}$ and $\|A^\alpha u\|$ are equivalent on $\mathcal{D}(A^\alpha)$. If $v \in H^{\alpha+1}(\Omega)$, where $\alpha \ge 0$, we define $\|\nabla v\|_{H^\alpha}$ by

$$\|\nabla v\|_{H^\alpha}^2 \stackrel{\text{def}}{=} \sum_{i=1}^{d} \|D_i v\|_{H^\alpha}^2.$$

Note that one has

$$\|\nabla v\|_{H^\alpha}^2 \le \|v\|_{H^{\alpha+1}}^2, \qquad \text{for all } v \in H^{\alpha+1}. \tag{61.17}$$

The following lemma is basic to deriving all of the auxiliary inequalities for the trilinear form b.

Lemma 61.1. *Let Ω be an open, bounded set in $\mathbb{R}^d$ of class C^ℓ, where $d = 2$ or 3 and ℓ is an integer with $\ell \geq 2$. Let α_i, $i = 1,2,3$, satisfy $0 \leq \alpha_1 \leq \ell$, $0 \leq \alpha_2 \leq \ell - 1$, $0 \leq \alpha_3 \leq \ell$,*

$$\alpha_1 + \alpha_2 + \alpha_3 \geq \frac{d}{2} \quad \text{and}$$
$$(\alpha_1, \alpha_2, \alpha_3) \notin \{(\frac{d}{2}, 0, 0),\ (0, \frac{d}{2}, 0),\ (0, 0, \frac{d}{2})\}.$$

Then there is a positive constant $C = C(d, \alpha_1, \alpha_2, \alpha_3, \Omega)$ such that

$$|b(u,v,w)| \leq \begin{cases} C\|u\|_{\alpha_1}\|v\|_{\alpha_2+1}\|w\|_{\alpha_3}, \\ C\|A^{\alpha_1/2}u\|\,\|A^{(1+\alpha_2)/2}v\|\,\|A^{\alpha_3/2}w\|, \end{cases} \tag{61.18}$$

for all $u \in V^{\alpha_1}$, $v \in V^{\alpha_2+1}$, and $w \in V^{\alpha_3}$.

Proof. We present the argument for $\ell = 2$. The general proof can be found in Constantin and Foias (1988). Let us first consider the case where $0 \leq \alpha_i < \frac{d}{2}$, for $i = 1,2,3$. In this case, we define q_i by

$$\frac{1}{q_i} = \frac{1}{2} - \frac{\alpha_i}{d}, \qquad i = 1,2,3.$$

One then has $\frac{1}{q_1} + \frac{1}{q_2} + \frac{1}{q_3} \leq 1$, and $2 \leq q_i < \infty$, for $i = 1,2,3$. We now apply the Hölder inequality to a typical term in (61.12) to obtain

$$\left| \int_\Omega u_i D_i v_j w_j \, dx \right| \leq c_1 \|u\|_{L^{q_1}} \|\nabla v\|_{L^{q_2}} \|w\|_{L^{q_3}}.$$

From the Sobolev Imbedding Theorem (see Appendix B), one has $H^{\alpha_i} \mapsto L^{q_i}$, for $i = 1,2,3$, and consequently

$$|b(u,v,w)| \leq c_2 \|u\|_{H^{\alpha_1}} \|\nabla v\|_{H^{\alpha_2}} \| \, \|w\|_{H^{\alpha_3}}.$$

As a result, (61.16) and (61.17) imply that $b(u,v,w)$ satisfies (61.18).

If $\alpha_i > \frac{d}{2}$, for some i, but $\alpha_j \neq \frac{d}{2}$, for any j, then the above argument is modified by setting $q_i = \infty$ and using the Sobolev imbedding $H^{\alpha_i} \mapsto L^\infty$.

If $\alpha_i = \frac{d}{2}$, for some i, and $\alpha_1 + \alpha_2 + \alpha_3 > \frac{d}{2}$, then one replaces the α_j with β_j, where $0 \leq \beta_j \leq \alpha_j$ and $\beta_j \neq \frac{d}{2}$, for all j, while $\beta_1 + \beta_2 + \beta_3 > \frac{d}{2}$. The argument above then yields

$$|b(u,v,w)| \leq c_3 \|u\|_{H^{\beta_1}} \|\nabla v\|_{H^{\beta_2}} \| \, \|w\|_{H^{\beta_3}}.$$

Since $H^\alpha \mapsto H^\beta$, whenever $0 \leq \beta \leq \alpha$, this completes the proof. □

Note that if $(\alpha_1, \alpha_2, \alpha_3)$ satisfies the hypotheses of Lemma 61.1 for $d = 3$, then the hypotheses are automatically satisfied, for $d = 2$. In other words, any $\mathbb{R}^3$ estimate (61.18) is valid in $\mathbb{R}^2$.

One immediate application of the last lemma is that, for $d = 2$, or 3, there is a constant $C_0 > 0$ such that

$$\|B(u,v)\|,\ \|B(v,u)\| \le C_0\|A^{\frac{5}{8}}u\|\,\|A^{\frac{5}{8}}v\|. \tag{61.19}$$

Indeed, one uses (61.18) with $\alpha_3 = 0$ and $(\alpha_1, \alpha_2) = (\frac{5}{4}, \frac{1}{4})$. Similarly, by using the identity $\langle A^{-\frac{1}{4}}B(u,v), w\rangle = b(u,v,A^{-\frac{1}{4}}w)$ with $\alpha_3 = \frac{1}{2}$ and $(\alpha_1, \alpha_2) = (1, 0)$, one finds that, for $d = 2$, or 3, there is a constant $C_1 > 0$ such that

$$\|A^{-\frac{1}{4}}B(u,v)\|,\ \|A^{-\frac{1}{4}}B(v,u)\| \le C_1\|A^{\frac{1}{2}}u\|\,\|A^{\frac{1}{2}}v\|. \tag{61.20}$$

By using (61.19), together with the interpolation inequality (61.9), one finds that there is a constant $c_1 > 0$ such that

$$\|B(u,v)\| \le c_1\|A^{\frac{1}{2}}u\|^{\frac{3}{4}}\|Au\|^{\frac{1}{4}}\|A^{\frac{1}{2}}v\|^{\frac{3}{4}}\|Av\|^{\frac{1}{4}}, \tag{61.21}$$

for $d = 2$, or 3, provided that $u, v \in \mathcal{D}(A)$. Consequently, for $d = 2$, or 3, and for all $u \in \mathcal{D}(A)$, one has

$$\|B(u,u)\| \le c_1\|A^{\frac{1}{2}}u\|^{\frac{3}{2}}\|Au\|^{\frac{1}{2}}. \tag{61.22}$$

If we combine (61.18) with the interpolation inequalities (61.9), we obtain various auxiliary estimates. These estimates bring out subtle, albeit important, differences depending on the dimension d of the physical space Ω. For example, for $d = 2$, there is a constant $C_2 > 0$ such that

$$|b(u,v,w)| \le \begin{cases} C_2\|u\|^{\frac{1}{2}}\|A^{\frac{1}{2}}u\|^{\frac{1}{2}}\|A^{\frac{1}{2}}v\|^{\frac{1}{2}}\|Av\|^{\frac{1}{2}}\|w\|, \\ \qquad \text{for } u \in \mathcal{D}(A^{\frac{1}{2}}), v \in \mathcal{D}(A), \text{ and } w \in H, \\ C_2\|u\|^{\frac{1}{2}}\|A^{\frac{1}{2}}u\|^{\frac{1}{2}}\|A^{\frac{1}{2}}v\|\,\|w\|^{\frac{1}{2}}\|A^{\frac{1}{2}}w\|^{\frac{1}{2}}, \\ \qquad \text{for } u, v, w \in \mathcal{D}(A^{\frac{1}{2}}). \end{cases} \tag{61.23}$$

The first inequality in (61.23) comes about by applying (61.9) to (61.18) with $\alpha_1 = \frac{1}{2}$, $\alpha_2 = \frac{1}{2}$ and $\alpha_3 = 0$, while in the second inequality one uses instead $\alpha_1 = \frac{1}{2}$, $\alpha_2 = 0$ and $\alpha_3 = \frac{1}{2}$.

For $d = 2$ or 3, there is a constant $C_3 > 0$ such that

$$|b(u,v,w)| \le \begin{cases} C_3\|A^{\frac{1}{2}}u\|\,\|A^{\frac{1}{2}}v\|^{\frac{1}{2}}\|Av\|^{\frac{1}{2}}\|w\|, \\ \qquad \text{for } u \in \mathcal{D}(A^{\frac{1}{2}}), v \in \mathcal{D}(A), \text{ and } w \in H, \\ C_3\|u\|^{\frac{1}{2}}\|A^{\frac{1}{2}}u\|^{\frac{1}{2}}\|A^{\frac{1}{2}}v\|^{\frac{1}{2}}\|Av\|^{\frac{1}{2}}\|w\|^{\frac{1}{2}}\|A^{\frac{1}{2}}w\|^{\frac{1}{2}}, \\ \qquad \text{for } u, w \in \mathcal{D}(A^{\frac{1}{2}}), \text{ and } v \in \mathcal{D}(A) \\ C_3\|u\|^{\frac{1}{4}}\|A^{\frac{1}{2}}u\|^{\frac{3}{4}}\|A^{\frac{1}{2}}v\|\,\|w\|^{\frac{1}{4}}\|A^{\frac{1}{2}}w\|^{\frac{3}{4}}, \\ \qquad \text{for } u, v, w \in \mathcal{D}(A^{\frac{1}{2}}). \end{cases} \tag{61.24}$$

The first inequality in (61.24) comes about by using the interpolation inequality (61.9) and (61.18) with $\alpha_1 = 1$, $\alpha_2 = \frac{1}{2}$ and $\alpha_3 = 0$, while the second follows with $\alpha_i = \frac{1}{2}$, $i = 1,2,3$. For the third inequality one uses $\alpha_1 = \alpha_3 = \frac{3}{4}$ and $\alpha_2 = 0$. The difference between (61.23) and (61.24) seems small at first sight, but as we will see, this will lead to profound differences between the dynamics of the Navier-Stokes equations in 2D and 3D.

Next we seek estimates of the quantity $A^{-\frac{1}{2}}B(u,v)$. Note that for sufficiently smooth functions $u, v, w \in H$ one has

$$|\langle A^{-\frac{1}{2}}B(u,v), w\rangle| = |\langle B(u,v), A^{-\frac{1}{2}}w\rangle| = b(u,v,A^{-\frac{1}{2}}w) = -b(u, A^{-\frac{1}{2}}w, v).$$

By using the interpolation inequality (61.9) and (61.18) with $b(u, A^{-\frac{1}{2}}w, v)$, where $d = 2$, $\alpha_1 = \frac{1}{2}$, $\alpha_2 = 0$, and $\alpha_3 = \frac{1}{2}$, we find that there is a constant C_4 such that

$$\|A^{-\frac{1}{2}}B(u,v)\| \leq C_4\|u\|^{\frac{1}{2}}\|A^{\frac{1}{2}}u\|^{\frac{1}{2}}\|v\|^{\frac{1}{2}}\|A^{\frac{1}{2}}v\|^{\frac{1}{2}}, \tag{61.25}$$

for $d = 2$ and $u, v \in V$. By using the interpolation inequality (61.9) and (61.18) with $b(u, A^{-\frac{1}{2}}w, v)$, where $d = 3$, $\alpha_1 = \alpha_3 = \frac{3}{4}$ (or $\alpha_1 = 1$, $\alpha_3 = \frac{1}{2}$) and $\alpha_2 = 0$, we find that there is a constant C_5 such that, for $d = 2$ or 3, one has
(61.26)

$$\|A^{-\frac{1}{2}}B(u,v)\|,\ \|A^{-\frac{1}{2}}B(v,u)\| \leq \begin{cases} C_5\|u\|^{\frac{1}{4}}\|A^{\frac{1}{2}}u\|^{\frac{3}{4}}\|v\|^{\frac{1}{4}}\|A^{\frac{1}{2}}v\|^{\frac{3}{4}}, \\ C_5\|A^{\frac{1}{4}}v\|\,\|A^{\frac{1}{2}}u\|, \end{cases}$$

for $u, v \in V$ (or $v \in V^{\frac{1}{2}}$ and $u \in V$). Similarly, by using a larger value of C_5 above, if necessary, one shows that

$$\|A^{-\frac{3}{4}}B(u,v)\|,\ \|A^{-\frac{3}{4}}B(v,u)\| \leq C_5\|v\|\,\|A^{\frac{1}{2}}u\|, \tag{61.27}$$

for $v \in H$ and $u \in V$ and for $d = 2$ or 3.

Next we derive the following auxiliary estimate: There exist constants $C_6 > 0$ and $C_7 > 0$ such that

$$\|A^{\frac{1}{2}}B(u,v)\| \leq \begin{cases} C_6\|A^{\frac{1}{2}}u\|^{\frac{3}{4}}\|Au\|^{\frac{1}{4}}\,\|Av\|, & \text{if } d = 2, \\ C_7\|A^{\frac{1}{2}}u\|^{\frac{1}{4}}\|Au\|^{\frac{3}{4}}\,\|Av\|, & \text{if } d = 2 \text{ or } 3, \end{cases} \tag{61.28}$$

provided that $u, v \in \mathcal{D}(A)$. In order to prove (61.28), we note that for any w in $C_0^3(\Omega, \mathbb{R}^d)$ that satisfies $\nabla\cdot w = 0$ in Ω, one has $-\nabla\cdot\Delta w = -\Delta(\nabla\cdot w) = 0$ in Ω and $-\Delta w\,|_{\partial\Omega} = 0$. Hence one has $-\Delta w \in H$ and $Aw = -\Delta w$. As a result, one obtains $b(u,v,Aw) = b(u,v,-\Delta w)$, for all $u, v \in V$. In this case, by using integration by parts, one obtains

$$b(u,v,Aw) = b(u,v,-\Delta w) = -\sum_{i,j,k=1}^{d}\int_\Omega u_i D_i v_j D_k^2 w_j\, dx = S_1 + S_2$$

where

$$S_1 = \sum_{i,j,k=1}^{d} \int_\Omega D_k u_i D_i v_j D_k w_j \, dx \text{ and } S_2 = \sum_{i,j,k=1}^{d} \int_\Omega u_i D_k D_i v_j D_k w_j \, dx.$$

By applying the argument of Lemma 61.1, and inequalities (61.9) and (61.16) with $\alpha_1 = \frac{1}{4}$, $\alpha_2 = 1$, and $\alpha_3 = 0$ (in 2D) and with $\alpha_1 = \frac{1}{2}$, $\alpha_2 = 1$, and $\alpha_3 = 0$ (in 3D), one finds that there are constant c_2 and c_3 such that

$$|S_1| \leq \begin{cases} c_2 \|u\|_{\frac{5}{4}} \|v\|_2 \|w\|_1, & \text{if } d = 2 \\ c_3 \|u\|_{\frac{3}{2}} \|v\|_2 \|w\|_1, & \text{if } d = 3. \end{cases}$$

On the other hand, by using $\alpha_1 = \frac{5}{4}$ and $\alpha_2 = \alpha_3 = 0$ (in 2D) and $\alpha_1 = \frac{7}{4}$ and $\alpha_2 = \alpha_3 = 0$ (in 3D), one finds that there are constants c_4 and c_5 such that

$$|S_2| \leq \begin{cases} c_4 \|u\|_{\frac{5}{4}} \|v\|_2 \|w\|_1, & \text{if } d = 2 \\ c_5 \|u\|_{\frac{7}{4}} \|v\|_2 \|w\|_1, & \text{if } d = 3. \end{cases}$$

By using the continuous imbedding $H^{\frac{7}{4}} \mapsto H^{\frac{3}{2}}$, we see that there are constants c_6 and c_7 such that[24]

$$(61.29) \qquad |b(u, v, -\Delta w)| \leq \begin{cases} c_6 \|A^{\frac{5}{8}} u\| \, \|Av\| \, \|A^{\frac{1}{2}} w\|, & \text{if } d = 2 \\ c_7 \|A^{\frac{7}{8}} u\| \, \|Av\| \, \|A^{\frac{1}{2}} w\|, & \text{if } d = 2 \text{ or } 3. \end{cases}$$

Since the set $\{w \in C_0^3(\Omega, \mathbb{R}^d) : \nabla \cdot w = 0 \text{ in } \Omega\}$ is dense in $\mathcal{D}(A)$, and since $b(u, v, A^{\frac{1}{2}} w) = \langle A^{\frac{1}{2}} B(u, v), w \rangle$, by (61.13), inequalities (61.9) and (61.29) imply (61.28).

Inequality (61.28) in turn implies that

$$(61.30) \qquad \|A^{\frac{1}{2}} B(u, u)\| \leq \begin{cases} C_6 \|A^{\frac{1}{2}} u\|^{\frac{3}{4}} \|Au\|^{\frac{5}{4}}, & \text{if } d = 2 \\ C_7 \|A^{\frac{1}{2}} u\|^{\frac{1}{4}} \|Au\|^{\frac{7}{4}}, & \text{if } d = 3. \end{cases}$$

Furthermore, inequalities (61.8), (61.9) and (61.28) imply that there is a constant $C_8 > 0$ such that

$$(61.31) \qquad \lambda_1^{\frac{1}{2}} \|B(u, v)\| \leq \|A^{\frac{1}{2}} B(u, v)\| \leq C_8 \|Au\| \, \|Av\|,$$

for $u, v \in \mathcal{D}(A)$ and for $d = 2$ or 3. It then follows from inequalities (61.31), (61.19), and (61.20) that the bilinear term $B = B(u, v)$ satisfies

$$B : V^2 \times V^2 \to V^1, \quad B : V^{\frac{5}{4}} \times V^{\frac{5}{4}} \to H, \quad \text{and} \quad B : V^1 \times V^1 \to V^{-\frac{1}{2}}.$$

[24] It should be noted that the argument given here implies a somewhat stronger conclusion in place of this inequality. In particular, in the factor $\|A^\alpha u\|$, where $\alpha = \frac{5}{8}$ (in 2D) and $\alpha = \frac{7}{8}$ (in 3D), one can use any α satisfying $\frac{1}{2} < \alpha$ (in 2D) and $\frac{3}{4} < \alpha$ (in 3D). However, the coefficients c_6 and c_7 may change with α.

This implies that the $B = B(u,u)$ satisfies

$$B \in C^2_{\text{Lip}}(V^2, V^1) \cap C^2_{\text{Lip}}(V^{\frac{5}{4}}, H) \cap C^2_{\text{Lip}}(V^1, V^{-\frac{1}{2}}). \tag{61.32}$$

Recall that $C^k_{\text{Lip}}(V,W) = C_{\text{Lip}}(V,W) \cap C^k_F(V,W)$, see (46.12). If the forcing term f is sufficiently smooth, e.g., $f \in V^1$, then the evolutionary equations generated by the nonlinear, or the linearized, Navier-Stokes equations are evolutionary equations on any of the spaces: $V^{-\frac{1}{2}}$, H, and V^1.

We will use the coeffcients C_0, C_1, C_2, $C_3, \cdots$, in the sequel, sometimes without explicit reference. Also, we will introduce several nonnegative functions which depend on the data (u_0, f). These functions will be labeled as L_i, M_i, and K_i, for $i = 0, 1, \cdots$. What is important is to notice that the functions labeled L_i, $i = 0, 1, \cdots$, do not depend on the initial data u_0.

6.2. Bubnov-Galerkin Approximations.

We return now to the eigenvalues of the Stokes operator A, which are enumerated (with multiplicities) as in (32.2). The associated collection of eigenvectors $\{e_1, e_2, e_3, \cdots\}$ forms an orthonormal basis for H. For each integer $n \geq 1$ we let $P = P_n$ denote the orthogonal projection of H onto $\text{Span}\{e_1, \cdots, e_n\}$, and set $Q = Q_n = I - P$. The projections P and Q are called the **spectral projections** determined by the Stokes operator A.

For each $u \in H$ we define[25] $p = p_n$ and $q = q_n$ by $p = Pu = P_n u$ and $q = Qu = Q_n u$. These spectral projections commute with A, i.e., $PA = AP$ and $QA = AQ$, where the last equation is restricted to $\mathcal{D}(A)$, the domain of A. Next we apply P and Q to the Navier-Stokes evolutionary equation (61.10) to obtain the equivalent system

$$\begin{cases} \partial_t p + \nu Ap + PB(p+q, p+q) = Pf, \\ \partial_t q + \nu Aq + QB(p+q, p+q) = Qf, \end{cases} \tag{62.1}$$

where the p-equation in (62.1) is an n-dimensional equation, while the q-equation is infinite dimensional.

The nth order Bubnov-Galerkin approximation of (61.10) is given by the solutions of the nth order ordinary differential equation

$$\partial_t p + \nu Ap + PB(p,p) = Pf, \tag{62.2}$$

which is obtained from (62.1) by setting $q = 0$ in the p-equation and ignoring the q-equation. More precisely, if $u = u(t)$ is a given solution of (61.10) with initial condition $u(0) = u_0$, then the solution $p(t)$ of (62.2) that satisfies

[25]The symbol p has two roles to play in this chapter: a pressure term and a Bubnov-Galerkin approximant. The context should make it clear which mask this actor is wearing at any given time.

$p(0) = p_0$, where $p_0 = Pu_0$, is said to be the nth **order Bubnov-Galerkin approximation** of u. We will sometimes write $u_n(t) = p(t)$ to emphasize the dependence on n.

There are several solution concepts for the Navier-Stokes equations. In this section we will derive some basic properties of the Bubnov-Galerkin approximations which are used to develop the theory of these solutions. For this purpose we let $u_n = p$ be a solution of the nth order Bubnov-Galerkin system (62.2) with initial condition

$$p(0) = p_0 = Pu_0, \tag{62.3}$$

where u_0 is a given element in H. Later in this development we will take a suitable limit as $n \to \infty$. In order to study the limit, we will require various estimates of the solution $u_n = p$ of (62.2), estimates which are independent of n so that they hold in the limit. These estimates are described in the next three lemmas given below. The proofs of these lemmas will be presented in Section 6.8.

The first result concerning the Bubnov-Galerkin approximations is applicable in either the 2D or 3D cases.

Lemma 62.1. *For $d = 2$, or 3, we let $u_n = p = p(t)$ be a maximally defined solution of the Bubnov-Galerkin initial value problem (62.2) - (62.3) on the interval $I = (\alpha, \omega)$. Then the following properties hold:*

(1) *One has $[0, \infty) \subset I$ and*

$$\|p(t)\|^2 \le e^{-\nu\lambda_1 t}\|u_0\|^2 + \frac{\|f\|_\infty^2}{\nu^2\lambda_1^2}(1 - e^{-\nu\lambda_1 t}), \qquad \text{for } t \ge 0, \tag{62.4}$$

as well as

$$\nu \int_{t_1}^{t} \|A^{\frac{1}{2}}p(s)\|^2\, ds \le \|p(t_1)\|^2 + \frac{t - t_1}{\nu\lambda_1}\|f\|_\infty^2, \qquad \text{for } t \ge t_1 \ge 0. \tag{62.5}$$

(2) *The sequence u_n is in a bounded set in $L^2(0,T;V) \cap L^\infty(0,T;H)$, the sequence Au_n lies in a bounded set in $L^2(0,T;V^{-1})$, and the sequence $D_t u_n$ lies in a bounded set in $L^p(0,T;V^{-1})$, for every $T > 0$, where*

$$p = \frac{4}{3}\,(\text{when } d = 3) \text{ and } p = 2\,(\text{when } d = 2), \tag{62.6}$$

Thus, there is a constant $b = b(T) = b(T; d, \|u_0\|^2, \|f\|_\infty^2)$ such that

$$\int_0^T \|A^{-\frac{1}{2}}D_t u_n\|^p\, ds \le b(T), \qquad \text{for all } n \ge 1. \tag{62.7}$$

(3) *For every $t > t_1 \geq 0$, there is a $t_0 \in (t_1, t)$ such that,*

$$\|A^{\frac{1}{2}}p(t_0)\|^2 < \frac{2}{t-t_1}\left(\frac{\|u_0\|^2}{\nu}e^{-\nu\lambda_1 t_1} + \frac{\|f\|_\infty^2}{\nu^2\lambda_1}\left(t - t_1 + \frac{1}{\nu\lambda_1}\right)\right). \tag{62.8}$$

(4) *For $t \geq t_0 \geq 0$, the following energy inequality holds:*

$$\frac{1}{2}\|p(t)\|^2 + \nu\int_{t_0}^t \|A^{\frac{1}{2}}p(s)\|^2 ds \leq \frac{1}{2}\|p(t_0)\|^2 + \int_{t_0}^t \langle f(s), p(s)\rangle\, ds. \tag{62.9}$$

(5) *The Variation of Constants Formula is valid, i.e., for $t \geq t_0 \geq 0$, one has*

$$p(t) = e^{-\nu A(t-t_0)}p(t_0) + \int_{t_0}^t e^{-\nu A(t-s)}P[f(s) - B(p(s), p(s))]\, ds. \tag{62.10}$$

There is a consequence of the last lemma which adds to the story of global attractors.

Corollary 62.2. *Let $f \in H$ be a time-independent forcing function. For every integer $n \geq 1$ and for $d = 2$ or 3, the nth order Bubnov-Galerkin system (62.2) has a global attractor $\mathfrak{A}_n \subset P_nH$.*

Proof. It follows from the basic theory of ordinary differential equations on finite dimensional spaces and Lemma 62.1, Item (1), that the mapping $(p_0, t) \to p(t) = S(t)p_0$ defines a compact semiflow on P_nH and that this semiflow is dissipative. The conclusion then follows from the Existence Theorem 23.12. □

Next we turn to additional properties of the Bubnov-Galerkin approximations for the 2D Navier-Stokes equations.

Lemma 62.3. *Let $d = 2$, and let $u_n = p = p(t)$ be a maximally defined solution of the Bubnov-Galerkin problem (62.2) - (62.3) on the interval $[0, \infty)$. Then the following properties hold:*

(1) *For $t > 0$, one has*

$$\begin{aligned} \partial_t\|A^{\frac{1}{2}}p\|^2 + \nu\lambda_1\|A^{\frac{1}{2}}p\|^2 &\leq \partial_t\|A^{\frac{1}{2}}p\|^2 + \nu\|Ap\|^2 \\ &\leq \frac{2}{\nu}\|f\|_\infty^2 + \frac{27C_2^4}{2\nu^3}\|p\|^2\|A^{\frac{1}{2}}p\|^4. \end{aligned} \tag{62.11}$$

(2) *There exist constants K_j, for $j = 0, 1$, M_2 and $L_2 = L_2(\|f\|_\infty)$, where L_2 is independent of u_0, such that*

$$\|A^{\frac{1}{2}}p(t)\|^2 \leq \begin{cases} K_0 + K_1 t^{-1}, & \text{for } 0 < t \leq 1, \\ M_2 e^{-\nu\lambda_1 t} + L_2, & \text{for } t \geq 1. \end{cases} \tag{62.12}$$

(3) *If in addition, one has* $u_0 \in V$, *then (by replacing* M_2 *with a larger value, if necessary)*

$$\|A^{\frac{1}{2}}p(t)\|^2 \leq M_2(1 + \|A^{\frac{1}{2}}u_0\|^2)e^{-\nu\lambda_1 t} + L_2, \qquad \text{for } t \geq 0. \tag{62.13}$$

Furthermore, there exist constants $L_3 = L_3(\|f\|_\infty)$, *where* L_3 *is independent of* u_0, K_j, *for* $j = 2, 3, 4$, *and* M_3 *such that*

$$\int_{t_0}^{t} \|Ap(s)\|^2 ds \leq K_2 + K_3 t_0^{-1}, \qquad \text{for } 0 < t_0 \leq t \leq 1, \tag{62.14}$$

and

$$\int_{t-1}^{t} \|Ap(s)\|^2\, ds \leq M_3 e^{-\nu\lambda_1 t} + L_3, \qquad \text{for } t \geq 1, \tag{62.15}$$

as well as

$$\int_{0}^{t} \|Ap(s)\|^2\, ds \leq \nu^{-1}(\|A^{\frac{1}{2}}u_0\|^2 + tK_4), \qquad \text{for } t \geq 0. \tag{62.16}$$

(4) *For each* $t \geq 0$, *the integral* $\int_0^t \|D_t p\|^2\, ds$ *has a bound, which depends on* t, *but which is independent of* n.

For the 3D Navier-Stokes equations one has the following result.

Lemma 62.4. *Let* $d = 3$, *and let* $u_n = p = p(t)$ *be a maximally defined solution of the Bubnov-Galerkin problem (62.2) - (62.3) on the interval* $[0, \infty)$. *Then the following properties hold:*

(1) *For* $t > 0$, *one has*

$$\begin{aligned} \partial_t\|A^{\frac{1}{2}}p\|^2 + \nu\lambda_1\|A^{\frac{1}{2}}p\|^2 &\leq \partial_t\|A^{\frac{1}{2}}p\|^2 + \nu\|Ap\|^2 \\ &\leq \frac{2}{\nu}\|f\|_\infty^2 + \frac{27C_3^4}{2\nu^3}\|A^{\frac{1}{2}}p\|^6. \end{aligned} \tag{62.17}$$

(2) *For every* $B \geq 0$ *and* $b \geq 0$ *there exists a time* $T_0 = T_0(B, b)$, *and for every* $t_0 \geq 0$ *there is a function* $\rho(t) = \rho(B, b, t_0; t)$, *defined for* $t_0 \leq t < t_0 + T_0$, *such that*

$$\|A^{\frac{1}{2}}p(t)\|^2 \leq \rho(t)^2, \qquad t_0 \leq t < t_0 + T_0, \tag{62.18}$$

provided that $\|f\|_\infty^2 \leq B$ *and* $\|A^{\frac{1}{2}}p(t_0)\|^2 \leq b$.

(3) *In addition, one has*

$$\int_{t_1}^{t} \|Ap(s)\|^2\, ds \leq \frac{1}{\nu}\|A^{\frac{1}{2}}p(t_1)\|^2 + \frac{2(t - t_1)}{\nu^2}\|f\|_\infty^2 + \frac{27C_3^4}{2\nu^2}\int_{t_1}^{t} \rho(s)^6\, ds, \tag{62.19}$$

for $t_0 \leq t_1 \leq t < t_0 + T_0$.

(4) *For each* t *with* $0 \leq t < T_0$, *the integral* $\int_0^t \|D_t p\|^2\, ds$ *has a bound, which depends on* t, *but which is independent of* n.

6.3. Weak Solutions.

There are at least five solution concepts for the Navier-Stokes equations: classical solutions, strong solutions, mild solutions, and two forms of weak solutions: weak solutions (of Class LH) and generalized weak solutions. In this section we will derive the basic theory of weak solutions, as developed in the pioneering works of Leray (1933, 1934a,b) and Hopf (1951). The basic idea behind this theory is rather simple. We let $u_n = p$ be a solution of the nth order Bubnov-Galerkin system

$$\partial_t p + \nu Ap + PB(p,p) = Pf, \tag{63.1}$$

with an appropriate initial condition,

$$p(0) = p_0 = Pu_0, \tag{63.2}$$

where u_0 is a given element in H, and we take a suitable limit as $n \to \infty$. In addition to the perennial issue of the existence of this limit, which is in fact a weak solution, we will need to know the basic properties of this solution. Those properties of the Bubnov-Galerkin approximations that are used below are presented in Section 6.2.

As stated above, we assume that Ω is an open, bounded region in $\mathbb{R}^d$ of class C^2, where $d = 2$ or 3, and that the forcing function f satisfies $f \in L^\infty(0,\infty;H)$. For example, one may have $f \in H$ independent of t. We define the norms $\|\cdot\|_\infty$ and $\|\cdot\|_{\infty,[0,T]}$ by

$$\|f\|_\infty = \text{ess sup}\{\|f(t)\|_H : 0 < t < \infty\},$$

and

$$\|f\|_{\infty,[0,T]} = \text{ess sup}\{\|f(t)\|_H : 0 < t < T\}. \tag{63.3}$$

We will make use of the Fréchet spaces

$$L^2_{\text{loc}}[0,\infty;H), \quad L^2_{\text{loc}}[0,\infty;V), \quad \text{and} \quad L^p_{\text{loc}}[0,\infty;V^{-1})$$

in this theory. The reader may wish to review the properties of these spaces (see Appendices A and B). The space $C[0,T;H_w)$ will denote the collection of all functions $u : [0,T) \to H$, that are continuous in the weak topology on H, i.e., for each $v \in H$, the mapping $t \to \langle u(t), v\rangle_H$ is a continuous mapping from $[0,T)$ into the scalar field, $\mathbb{R}$ or $\mathbb{C}$. We will use the symbols $\xrightarrow{s}$ and $\xrightarrow{w}$ to denote, respectively, strong and weak convergence in a Hilbert space.

In the following paragraph, we give the definition of a weak solution $u = u(t)$ of the Navier-Stokes equations. In order to formulate this definition, we need a new concept of the time-derivative $D_t u$. In particular, a function

$u : (0, T) \to H$ is said to have a **time derivative** g **in the space** V^{-1}, provided that for some p with $1 \le p < \infty$ there is a function $g \in L^p_{\text{loc}}[0, T; V^{-1})$ such that one has

$$u(t) - u(t_0) = \int_{t_0}^{t} g(s)\, ds, \qquad \text{for } 0 \le t_0 \le t < T, \tag{63.4}$$

where the integral exists in the space V^{-1}. An equivalent formulation of (63.4) is that $A^{-\frac{1}{2}}(u(t) - u(t_0)) = \int_{t_0}^{t} A^{-\frac{1}{2}} g\, ds$, where the integral exists in the space H. Note that (63.4) implies that for all $v \in V$ one has

$$\langle u(t) - u(t_0), v\rangle = \int_{t_0}^{t} \langle\langle g(s), v\rangle\rangle\, ds = \int_{t_0}^{t} \langle A^{-\frac{1}{2}} g(s), A^{\frac{1}{2}} v\rangle\, ds. \tag{63.5}$$

By differentiating (63.5) with respect to t, one obtains

$$D_t\langle u(t), v\rangle \overset{\text{a.e.}}{=} \langle\langle g(t), v\rangle\rangle, \qquad \text{for each } v \in V. \tag{63.6}$$

We will use the integrated form (63.5), instead of (63.6), in the sequel. If (63.4) or (63.5) holds, then we will write g in the form $g = D_t u$. Notice that $D_t u$ refers to a weak derivative.

6.3.1 The Leray-Hopf Theory. Let $f \in L^\infty(0, \infty; H)$ be given. We will say that a function u on $[0, \infty)$ is a **weak solution** of the Navier-Stokes equations of **Class LH** provided that $u(0) = u_0$, where $u_0 \in H$, and the following four properties hold:

(1) One has $u \in L^\infty(0, \infty; H) \cap L^2_{\text{loc}}[0, \infty; V)$.

(2) The function u has a time-derivative g in the space V^{-1}, so that (63.4) is valid with $g = D_t u$, where $g = D_t u \in L^p_{\text{loc}}[0, \infty; V^{-1})$, for some p with $1 \le p < \infty$.

(3) For almost all t and almost all t_0 with $0 < t_0 < t$, one has

$$\|u(t)\|^2 \le e^{-\nu\lambda_1(t - t_0)}\|u(t_0)\|^2 + c_0^2\|f\|_\infty^2, \tag{63.7}$$

where $c_0^2 = (\nu\lambda_1)^{-2}$, and

$$\|u(t)\|^2 + 2\nu\int_{t_0}^{t} \|A^{\frac{1}{2}} u(s)\|^2 ds \le \|u(t_0)\|^2 + 2\int_{t_0}^{t} \langle f(s), u(s)\rangle\, ds. \tag{63.8}$$

(4) The function u satisfies

(63.9)

$$\langle u(t) - u(t_0), v\rangle + \nu\int_{t_0}^{t} \langle A^{\frac{1}{2}} u, A^{\frac{1}{2}} v\rangle\, ds + \int_{t_0}^{t} b(u, u, v)\, ds = \int_{t_0}^{t} \langle f, v\rangle\, ds,$$

for all $v \in V$, and for all t and t_0 with $t \ge t_0 \ge 0$.

Notice that (63.9) is the integrated weak form of the Navier-Stokes Equations (61.10), where $\langle Au, v\rangle$ has been replaced by $\langle A^{\frac{1}{2}}u, A^{\frac{1}{2}}v\rangle$.

There are several properties of weak solutions of Class LH that follow directly from the definition. We summarize these properites in the following lemma. We will use (61.25) and (61.26) which imply that for $u \in V$ one has

$$\|A^{-\frac{1}{2}}B(u,u)\| \leq \begin{cases} C_4\|u\|\,\|A^{\frac{1}{2}}u\|, & \text{for } d=2,\\ C_5\|u\|^{\frac{1}{2}}\|A^{\frac{1}{2}}u\|^{\frac{3}{2}}, & \text{for } d=2 \text{ or } 3.\end{cases}$$

Consequently, Item (1) in the definition of a weak solution of Class LH implies that

$$B(u,u) \in L^p_{\text{loc}}[0,\infty; V^{-1}), \tag{63.10}$$

where p satisfies (62.6). Similarly one finds that $b(u,u,v) \in L^1_{\text{loc}}[0,\infty;\mathbb{R})$, for each $v \in V$.

Lemma 63.1. *Let Ω be an open, bounded domain in $\mathbb{R}^d$ of class C^2, where $d = 2$ or 3, and let $f \in L^\infty(0,\infty;H)$. Let $u = u(t)$ be any weak solution of Class LH on $[0,\infty)$ and let $c_0 = (\nu\lambda_1)^{-1}$. Then the following properties are valid.*

(1) *The solution u satisfies $u \in L^2_{\text{loc}}[0,\infty;H)$.*
(2) *The solution u satisfies $u \in C[0,\infty;H_w)$. In particular, $u = u(t)$ is uniquely determined, for all $t \geq 0$, by equation (63.9).*
(3) *Inequality (63.7), as well as*

$$\nu\int_{t_0}^t \|A^{\frac{1}{2}}u\|^2\,ds \leq \|u(t_0)\|^2 + (t-t_0)\,c_0\|f\|^2_\infty, \tag{63.11}$$

are valid for almost all $t_0 \in (0,\infty)$ and for all $t \geq t_0$.
(4) *The solution u satisfies the inequality*

$$\|u(t)\|^2 \leq e^{-\nu\lambda_1 t}\|u\|^2_\infty + c_0^2\|f\|^2_\infty, \qquad \textit{for all } t \geq 0, \tag{63.12}$$

where $\|u\|_\infty = \|u\|_{L^\infty(0,\infty;H)}$.
(5) *The function u satisfies (63.9) in the space $L^1_{\text{loc}}[0,\infty;V^{-1})$, for all $t > 0$. Consequently, one has*

$$D_t u + \nu A u + B(u,u) \overset{\text{a.e.}}{=} f, \qquad \textit{in the space } V^{-1}. \tag{63.13}$$

(6) *Let p satisfy (62.6). Then $D_t u$ is in the space $L^p_{\text{loc}}[0,\infty;V^{-1})$, and one has $u \in C^{0,\theta}_{\text{loc}}[0,\infty;V^{-1})$, where $\theta = \frac{p-1}{p}$.*
(7) *The function u is a mild solution in the space V^{-1}, i.e., the Variation of Constants Formula*

$$u(t) = e^{-\nu A(t-t_0)}u(t_0) + \int_{t_0}^t e^{-\nu A(t-s)}[f(s) - B(u(s),u(s))]\,ds, \tag{63.14}$$

for $0 \leq t_0 < t < \infty$, is valid in the space V^{-1}.

(8) *Let $\{v_n\}$ be any sequence in $L^2_{\text{loc}}[0, \infty; H)$ that satisfies the following two properties: (a) there is a function $a \in C[0, \infty; \mathbb{R}^+)$, such that, for each n, one has*

$$\|v_n(t)\| \leq a(t), \qquad \textit{for almost all } t \geq 0,$$

(b) and one has $v_n \xrightarrow{s} u$ in $L^2_{\text{loc}}[0, \infty; H)$, as $n \to \infty$. Then

$$\|u(t)\| \leq a(t), \qquad \textit{for all } t \geq 0.$$

Proof. Item (1) is a consequence of the definition and the continuous imbedding

$$L^\infty(0, \infty; H) \mapsto L^2_{\text{loc}}[0, \infty; H).$$

To verify Item (2) we let $v \in V$. As noted above, one has $b(u, u, v) \in L^1_{\text{loc}}[0, \infty; \mathbb{R})$. Similarly one has $\langle A^{\frac{1}{2}} u, A^{\frac{1}{2}} v\rangle$ and $\langle \mathbb{P} f, v\rangle$ in $L^1_{\text{loc}}[0, \infty; \mathbb{R})$. As a result, it follows from (63.9) that for $t_0 > 0$ one has

$$\lim_{t \to t_0} \langle u(t) - u(t_0), v\rangle = 0, \qquad \text{for all } v \in V,$$

and for $t_0 = 0$ one obtains

$$\lim_{t \to 0+} \langle u(t) - u(0), v\rangle = 0, \qquad \text{for all } v \in V.$$

Item (2) now follows from the fact that V is dense in H.

The fact that (63.7) is valid for all $t \geq t_0 > 0$ follows from the definition and the lower semicontinuity relationship $\|u(t)\|^2 \leq \liminf_{t_n \to t} \|u(t_n)\|^2$, which is valid for any $u \in C[0, \infty; H_w)$. By using the Schwarz and the Young inequalities, we find

$$\begin{aligned}
\left| 2 \int_{t_0}^{t} \langle f, u\rangle \, ds \right| &\leq 2\|A^{-\frac{1}{2}} f\|_\infty \int_{t_0}^{t} \|A^{\frac{1}{2}} u\| \, ds \\
&\leq 2\lambda_1^{-\frac{1}{2}} \|f\|_\infty (t - t_0)^{\frac{1}{2}} \left(\int_{t_0}^{t} \|A^{\frac{1}{2}} u\|^2 \, ds \right)^{\frac{1}{2}} \\
&\leq \nu \int_{t_0}^{t} \|A^{\frac{1}{2}} u\|^2 \, ds + \frac{1}{\nu \lambda_1} (t - t_0) \|f\|_\infty^2.
\end{aligned}$$

By combining this with (63.8) we obtain (63.11). Inequality (63.12) follows from (63.7). Indeed, from (63.7) and Item (3) in this lemma, we see that

$$\|u(t)\|^2 \leq e^{-\nu \lambda_1 (t - t_0)} \|u\|_\infty^2 + c_0^2 \|f\|_\infty^2, \qquad \text{for all } t \geq t_0,$$

and for a dense set of $t_0 > 0$. By taking limit as $t_0 \to 0^+$, we obtain (63.12).

Since $u \in L^2_{\mathrm{loc}}[0,\infty;V)$, one has

$$\int_{t_0}^{t} \langle A^{\frac{1}{2}}u(s), A^{\frac{1}{2}}v\rangle \, ds = \langle \int_{t_0}^{t} A^{\frac{1}{2}}u(s)\, ds, A^{\frac{1}{2}}v\rangle, \qquad \text{for all } v \in V.$$

Similarly one obtains

$$\int_{t_0}^{t} \langle f(s), v\rangle \, ds = \langle \int_{t_0}^{t} A^{-\frac{1}{2}}f(s)\, ds, A^{\frac{1}{2}}v\rangle, \qquad \text{for all } v \in V.$$

Next we have $B(u,u) \in L^p_{\mathrm{loc}}[0,\infty;V^{-1})$, where $p = \frac{4}{3}$, which implies that

$$\int_{t_0}^{t} b(u(s), u(s), v)\, ds = \langle \int_{t_0}^{t} A^{-\frac{1}{2}}B(u(s),u(s))\, ds, A^{\frac{1}{2}}v\rangle, \qquad \text{for all } v \in V.$$

Lastly we have

$$\langle u(t) - u(t_0), v\rangle = \langle A^{-\frac{1}{2}}(u(t) - u(t_0))\, ds, A^{\frac{1}{2}}v\rangle, \qquad \text{for all } v \in V.$$

By combining the last four equalities with (63.9), we conclude that

$$A^{-\frac{1}{2}}(u(t) - u(t_0)) = \int_{t_0}^{t} A^{-\frac{1}{2}}G(u)\, ds,$$

where $G = G(u) = -\nu Au - B(u,u) + f$. From the definition of the time derivative in the space V^{-1}, we then see that u satisfies (63.13) in the space V^{-1}, for almost all $t > 0$.

Now (63.10) implies that $A^{-\frac{1}{2}}B(u,u) \in L^p_{\mathrm{loc}}[0,\infty;H)$, where p satisfies (62.6). Similarly one has

$$A^{-\frac{1}{2}}Au \in L^2_{\mathrm{loc}}[0,\infty;H) \mapsto L^p_{\mathrm{loc}}[0,\infty;H) \quad \text{and}$$
$$A^{-\frac{1}{2}}f \in L^{\infty}[0,\infty;H) \mapsto L^p_{\mathrm{loc}}[0,\infty;H).$$

Therefore, one obtains $D_t u \in L^p_{\mathrm{loc}}[0,T;V^{-1})$. Item (6) then follows from (63.13) and the Hölder inequality.

For the proof of Item (7) set $\phi = f - B(u,u)$ and consider the linear inhomogeneous equation $\partial_t u + \nu Au = \phi$. Owing to (63.10) one has $\phi \in L^p_{\mathrm{loc}}[0,T;V^{-1})$, where p satisfies (62.6). Now Item (2) implies that $u(t_0) \in H \hookrightarrow V^{-1}$, for any $t_0 \geq 0$. Hence, Item (7) follows from observation that, for a constant $C > 0$, one has

$$\int_0^t \|A^{-\frac{1}{2}}\phi(s)\|\, ds \leq C\, t^{\frac{1}{4}} \left(\int_0^t \|A^{-\frac{1}{2}}\phi(s)\|^{\frac{4}{3}} ds\right)^{\frac{3}{4}} < \infty, \quad \text{for } 0 \leq t < T.$$

In order to prove Item (8), we first note that $v_n(t) \stackrel{s}{\to} u(t)$, in H, for almost all $t \geq 0$. Therefore, the set $\mathcal{A} = \{t \in \mathbb{R}^+ : \|u(t)\| \leq a(t)\}$ is a set of full measure in $\mathbb{R}^+ = [0, \infty)$. Let $t_0 \in \mathbb{R}^+$. Then there is a sequence $\{t_m\}$ in $\mathcal{A}$ with $t_m \to t_0$, as $m \to \infty$. Since $u \in C[0, \infty; H_w)$, one has

$$\|u(t_0)\| \leq \liminf_{m\to\infty} \|u(t_m)\| \leq \lim_{m\to\infty} a(t_m) = a(t_0). \quad \square$$

We will let $W_{LH}(f)$ denote the collection of all $u \in L^2_{\text{loc}}[0, \infty; H)$ such that u is a weak solution of Class LH for a given $f \in L^\infty(0, \infty; H)$. Because of Item (6) in the last lemma, in the sequel we will restrict the quantity p, which appears in the condition $D_t u \in L^p_{\text{loc}}[0, \infty; V^{-1})$ in the definition of a weak solution of Class LH, to satisfy (62.6).

The first step in obtaining weak solutions is to show that the sequence $u_n = p$ of Bubnov-Galerkin approximations has a subsequence with a suitable limit. In order to do this we will need an important compactness lemma. In this lemma we will be making assumptions about a sequence of functions in the Lebesgue space $L^2(a, b; V)$. Note that if $u \in L^2(a, b; V)$, then $\int_a^b \|u\|^2 \, ds \leq \lambda_1^{-1} \int_a^b \|A^{\frac{1}{2}} u\|^2 \, ds$, see (61.8). Since the forcing function f plays no direct role in the next lemma, we make no assumptions on f.

The first of two compactness lemmas is based on the compact imbeddings

$$V^{\alpha+1} \hookrightarrow V^\alpha \hookrightarrow V^{\alpha-1}, \qquad \text{where } \alpha \in \mathbb{R}.$$

As usual we let $\langle \cdot, \cdot \rangle_\alpha = \langle \cdot, \cdot, \rangle_{V^\alpha}$ and $\| \cdot \|_\alpha = \| \cdot \|_{V^\alpha}$ denote the inner product and the norm on V^α.

Lemma 63.2 (Compactness I). *Let u_n be a sequence in $L^2(a, b; V^{\alpha+1})$, for some $\alpha \in \mathbb{R}$, where $-\infty < a < b < \infty$. Assume that the following properties are satisfied: (1) u_n is bounded in $L^2(a, b; V^{\alpha+1})$, and (2) each u_n has a time derivative in the space $V^{\alpha-1}$ and the sequence $D_t u_n$ is bounded in $L^p(a, b; V^{\alpha-1})$, for some p satisfying $1 < p < \infty$. Then there exists a subsequence of u_n, which we relabel as u_n, and functions $u \in L^2(a, b; V^{\alpha+1})$ and $g \in L^p(a, b; V^{\alpha-1})$ such that the following properties hold:*

1. *One has $u_n \xrightarrow{\text{w}} u$ in $L^2(a, b; V^{\alpha+1})$.*
2. *One has $D_t u_n \xrightarrow{\text{w}} g$ in $L^p(a, b; V^{\alpha-1})$.*
3. *One has $u_n \xrightarrow{\text{s}} u$ in $L^2(a, b; V^\alpha)$ and in $L^2(a, b; V^{\alpha-1})$.*
4. *For almost every $t \in (a, b)$, one has $u_n(t) \xrightarrow{\text{s}} u(t)$ in $V^{\alpha-1}$.*
5. *For almost every $t \in (a, b)$, one has $u_n(t) \xrightarrow{\text{s}} u(t)$ in V^α.*

Proof. Since the sequences u_n and $D_t u_n$ are bounded in $L^2(a, b; V^{\alpha+1})$ and $L^p(a, b; V^{\alpha-1})$, respectively, there exists a common subsequence of u_n and $D_t u_n$, which we relabel as u_n and $D_t u_n$, where u_n is weakly convergent in the space $L^2(a, b; V^{\alpha+1})$, and $D_t u_n$ converges weakly in $L^p(a, b; V^{\alpha-1})$, with weak limits u and g, respectively. This establishes Items (1) and (2). Consequently, for every $w \in L^2(a, b; V^{\alpha+1})$, one has

$$\int_a^b \langle A^{\frac{1}{2}}(u_n(s) - u(s)), A^{\frac{1}{2}} w(s) \rangle_\alpha \, ds \to 0, \qquad \text{as } n \to \infty, \tag{63.15}$$

and

$$\int_a^b \langle A^{-\frac{1}{2}}(D_t u_n - g), A^{-\frac{1}{2}}\hat{w}\rangle_\alpha \, ds \to 0, \qquad \text{as } n \to \infty,$$

for every $\hat{w} \in L^q(a,b;V^{\alpha-1})$, where q satisfies $q^{-1}+p^{-1}=1$. (Note that $1 < q < \infty$, since $1 < p < \infty$.)

It would be nice to know that $D_t u = g$, which would shorten the argument below because one could then use the indentity $u(t) - u(t_1) = \int_{t_1}^t D_t u \, ds$ to argue that there is a constant $K > 0$ such that

$$\text{(63.16)} \quad \|A^{-\frac{1}{2}}(u(t)-u(t_1))\|_\alpha \le K|t-t_1|^{\frac{1}{q}}, \qquad \text{for almost all } t_1,\, t \in (a,b).$$

Since we will need this Hölder continuity property, we proceed in an alternate way.

For Item (3), we want to show that an appropriate subsequence of u_n, which we relabel as u_n, converges strongly to u in $L^2(a,b;V^\alpha)$, i.e., one has $\int_a^b \|v_n\|_\alpha^2 \, ds \to 0$, as $n \to \infty$, where $v_n = u_n - u$. From inequality (37.4), with H replaced by V^α, we find that for every $\epsilon > 0$ there is a $C_\epsilon > 0$ such that

$$\|v_n(t)\|_\alpha^2 \le \epsilon \|A^{\frac{1}{2}} v_n(t)\|_\alpha^2 + C_\epsilon \|A^{-\frac{1}{2}} v_n(t)\|_\alpha^2,$$

for almost all $t \in (a,b)$. Since $\int_a^b \|A^{\frac{1}{2}} v_n\|_\alpha^2 \, dt$ is bounded uniformly in n, it follows that

$$\int_a^b \|v_n\|_\alpha^2 \, dt \le \epsilon \sup_{n \ge 1} \left(\int_a^b \|A^{\frac{1}{2}} v_n\|_\alpha^2 \, dt \right) + C_\epsilon \int_a^b \|A^{-\frac{1}{2}} v_n\|_\alpha^2 \, dt.$$

Therefore it suffices to show that for an appropriate subsequence, which we relabel as v_n, one has $\int_a^b \|A^{-\frac{1}{2}} v_n\|_\alpha^2 \, ds \to 0$, as $n \to \infty$, i.e., v_n converges strongly to 0 in $L^2(a,b;V^{\alpha-1})$.

Let I be any subinterval in $(a.b)$, and let χ_I denote the characteristic function of I. We claim the subsequence u_n satisfies $\int_I u_n \, ds \overset{w}{\to} \int_I u \, ds$ in $V^{\alpha+1}$. Indeed, for any $w \in V^{\alpha+1}$ one has

$$\begin{aligned}\left\langle A^{\frac{1}{2}} \int_I v_n(s)\, ds, A^{\frac{1}{2}} w \right\rangle_\alpha - \int_I \langle A^{\frac{1}{2}} v_n(s), A^{\frac{1}{2}} w\rangle_\alpha \, ds \\ = \int_a^b \langle A^{\frac{1}{2}} v_n(s), A^{\frac{1}{2}} \chi_I(s) w\rangle_\alpha \, ds,\end{aligned}$$

which goes to 0, as $n \to \infty$, by (63.15). As a result of the compact imbeddings $V^{\alpha+1} \hookrightarrow V^\alpha \hookrightarrow V^{\alpha-1}$, we see that there is a subsequence, which we relabel as u_n, that satisfies

$$\text{(63.17)} \quad \begin{aligned} &\lim_{n\to\infty} \| \int_I u_n(s)\, ds - \int_I u \, ds\|_\alpha = 0, \quad \text{and} \\ &\lim_{n\to\infty} \|A^{-\frac{1}{2}}(\int_I u_n(s)\, ds - \int_I u\, ds)\|_\alpha = 0. \end{aligned}$$

Let ϵ satisfy $0 < \epsilon < b - a$. Next define the functions $w_{n,\epsilon}$ and w_ϵ by

$$w_{n,\epsilon}(t) \stackrel{\text{def}}{=} \frac{1}{\epsilon}\int_{t-\epsilon}^{t} u_n(s)\,ds, \qquad w_\epsilon(t) \stackrel{\text{def}}{=} \frac{1}{\epsilon}\int_{t-\epsilon}^{t} u(s)\,ds,$$

for $a + \epsilon \leq t \leq b$, and

$$w_{n,\epsilon}(t) \stackrel{\text{def}}{=} w_{n,\epsilon}(a+\epsilon), \qquad w_\epsilon(t) \stackrel{\text{def}}{=} w_\epsilon(a+\epsilon),$$

for $a \leq t < a+\epsilon$. By using the facts that both u_n and u are in $L^2(a,b;V^{\alpha-1})$, one can readily verify that $w_{n,\epsilon}$ and w_ϵ are in $L^2(a,b;V^{\alpha-1})$, for all $0 < \epsilon < b - a$.

As shown in (63.17), for each t with $a < t \leq b$ and each ϵ with $0 < \epsilon < b - a$, one has

$$w_{n,\epsilon}(t) \stackrel{\text{s}}{\to} w_\epsilon(t) \text{ in } V^\alpha \text{ and } V^{\alpha-1}, \qquad \text{as } n \to \infty. \tag{63.18}$$

Furthermore, it follows from the hypotheses of this lemma, the Hölder inequality, and the identity $u_n(t) - u_n(t_1) = \int_{t_1}^{t} D_t u_n\,ds$ that there is a constant $K > 0$ such that

$$\|A^{-\frac{1}{2}}(u_n(t) - u_n(t_1))\|_\alpha \leq \left|\int_{t_1}^{t} \|A^{-\frac{1}{2}} D_t u_n\|_\alpha\,ds\right| \leq K|t - t_1|^{\frac{1}{q}}, \tag{63.19}$$

for all $n \geq 1$, and for $t_1,\, t \in (a,b)$. Next we show that for this K one has

$$\|A^{-\frac{1}{2}}(w_{n,\epsilon}(t) - w_{n,\epsilon}(t_1))\|_\alpha \leq K|t - t_1|^{\frac{1}{q}}, \qquad \text{for all } n \geq 1, \tag{63.20}$$

and for $t_1,\, t \in (a,b)$. Indeed for $a + \epsilon \leq t_1 \leq t \leq b$ we have

$$\begin{aligned}\|A^{-\frac{1}{2}}(w_{n,\epsilon}(t) - w_{n,\epsilon}(t_1))\|_\alpha &= \frac{1}{\epsilon}\left\|\int_{t-\epsilon}^{t} A^{-\frac{1}{2}}(u_n(s) - u_n(s + t_1 - t))\,ds\right\|_\alpha \\ &\leq \frac{1}{\epsilon}\int_{t-\epsilon}^{t} \|A^{-\frac{1}{2}}(u_n(s) - u_n(s + t_1 - t))\|_\alpha\,ds \leq K|t - t_1|^{\frac{1}{q}},\end{aligned}$$

by (63.19). The remaining cases, where $a \leq t_1 < a + \epsilon \leq t \leq b$ and $a \leq t_1 \leq t < a + \epsilon$, are easily verified and we will omit the details.

Next we claim that

$$\lim_{n\to\infty} \int_a^b \|A^{-\frac{1}{2}}(w_{n,\epsilon} - w_\epsilon)\|_\alpha^2\,dt = 0, \tag{63.21}$$

and that

$$\lim_{\epsilon\to 0^+} \int_a^b \|A^{-\frac{1}{2}}(w_\epsilon - u)\|_\alpha^2\,dt = 0 \tag{63.22}$$

Now (63.21) follows from the hypotheses of this lemma, (63.18), and the Lebesgue Dominated Convergence Theorem, since

$$\begin{aligned}\|A^{\frac{1}{2}}(w_{n,\epsilon}(t) - w_\epsilon(t))\|_\alpha^2 &\le 2\|A^{-\frac{1}{2}}w_{n,\epsilon}(t)\|_\alpha^2 + 2\|A^{-\frac{1}{2}}w_\epsilon(t)\|_\alpha^2 \\ &\le 2\epsilon^{-1}\lambda_1^{-2}\int_a^b \|A^{\frac{1}{2}}u_n\|_\alpha^2\, ds + 2\|A^{-\frac{1}{2}}w_\epsilon(t)\|_\alpha^2,\end{aligned}$$

for almost all $t \in (a,b)$. It then follows from (63.18) that inequality (63.20) holds in the limit, i.e.,

$$\|A^{-\frac{1}{2}}(w_\epsilon(t) - w_\epsilon(t_1))\|_\alpha \le K|t - t_1|^{\frac{1}{q}}, \tag{63.23}$$

for almost all t_1, $t \in (a,b)$ and all ϵ with $0 < \epsilon < b - a$.

For each $\tau \in \mathbb{R}$, we define $u_\tau(t) = u(\tau + t)$, for $\tau + t \in (a,b)$ and $u_\tau(t) = 0$, for $\tau + t \notin (a,b)$. Note that the time translation mapping $\tau \to u_\tau$ is a continuous mapping of $\mathbb{R}$ into any Lebesgue space $L^r(a,b;X)$, where X is a Banach space and r satisfies $1 \le r < \infty$, see the Translation Lemma in Appendix B. This means that for any $\eta > 0$ there is an $\epsilon_0 > 0$ such that

$$\int_a^b \|u_\tau - u\|_X^r\, dt \le \eta, \qquad \text{whenever } |\tau| \le \epsilon_0. \tag{63.24}$$

Next observe that by a change of variables, one has

$$w_\epsilon(t) = \frac{1}{\epsilon}\int_0^\epsilon u(\sigma + t - \epsilon)\, d\sigma = \frac{1}{\epsilon}\int_0^\epsilon u_{\sigma-\epsilon}(t)\, d\sigma.$$

Then from the Fubini theorem for the interchange of the order of integration, we obtain

$$\int_a^b \|A^{-\frac{1}{2}}(w_\epsilon - u)\|_\alpha^2\, dt \le \frac{1}{\epsilon}\int_0^\epsilon\int_a^b \|A^{-\frac{1}{2}}(u_{\sigma-\epsilon} - u)\|_\alpha^2\, dt\, d\sigma.$$

Note that $\int_a^b \|A^{-\frac{1}{2}}(w_\epsilon - u)\|_\alpha^2\, dt \le \eta$, whenever $0 < \epsilon \le \epsilon_0$, due to (63.24) with $X = V^{\alpha-1}$ and $r = 2$. This completes the proof of (63.22). As a result, inequality (63.23) holds in the limit, i.e., (63.16) is valid.

Since $v_n = u_n - u$, one has

$$v_n(t) - v_n(t_1) = \int_{t_1}^t D_t u_n(s)\, ds - [u(t) - u(t_1)]. \tag{63.25}$$

By integrating (63.25) with respect to t_1 over the interval $I = (t - \epsilon, t)$ one obtains

$$\begin{aligned}v_n(t) = \frac{1}{\epsilon}\int_I v_n(t_1)\, dt_1 &+ \frac{1}{\epsilon}\int_{t-\epsilon}^t (s - t + \epsilon) D_t u_n(s)\, ds \\ &- \frac{1}{\epsilon}\int_{t-\epsilon}^t [u(t) - u(t_1)]\, dt_1.\end{aligned} \tag{63.26}$$

Now the hypotheses of this lemma and the Hölder inequality imply that there is a constant $C > 0$ such that

$$\begin{aligned}
&\frac{1}{\epsilon}\int_{t-\epsilon}^{t}(s-t+\epsilon)\|A^{-\frac{1}{2}}D_t u_n\|_\alpha\, ds \\
&\qquad \le \frac{1}{\epsilon}\left(\int_{t-\epsilon}^{t}(s-t+\epsilon)^q\, ds\right)^{\frac{1}{q}}\left(\int_{t-\epsilon}^{t}\|A^{-\frac{1}{2}}D_t u_n\|_\alpha^p\, ds\right)^{\frac{1}{p}} \\
&\qquad \le \left(\frac{1}{q+1}\right)^{\frac{1}{q}}\epsilon^{\frac{1}{q}}\left(\int_a^b \|A^{-\frac{1}{2}}D_t u_n\|_\alpha^p\, ds\right)^{\frac{1}{p}} \le C\epsilon^{\frac{1}{q}},
\end{aligned} \tag{63.27}$$

for all n. Similarly, from (63.16) one finds that

$$\frac{1}{\epsilon}\int_{t-\epsilon}^{t}\|A^{-\frac{1}{2}}(u(t)-u(t_1))\|_\alpha\, dt_1 \le K\epsilon^{\frac{1}{q}}. \tag{63.28}$$

For any $\epsilon_0 > 0$ we fix ϵ so that $0 < (K+C)\epsilon^{\frac{1}{q}} < \epsilon_0$, where K and C are given by (63.27) and (63.28). It then follows from (63.26), (63.27), and (63.28) that

$$\|A^{-\frac{1}{2}}v_n(t)\|_\alpha \le \frac{1}{\epsilon}\int_I \|A^{-\frac{1}{2}}v_n(t_1)\|_\alpha\, dt_1 + \epsilon_0.$$

From (63.17) we see that $\int_I \|A^{-\frac{1}{2}}v_n(t_1)\|_\alpha\, dt_1 \to 0$, as $n \to \infty$. Since $\epsilon > 0$, it follows that there is an $N \ge 1$ such that

$$\frac{1}{\epsilon}\int_{t-\epsilon}^{t}\|A^{-\frac{1}{2}}v_n(t_1)\|_\alpha\, dt_1 \le \epsilon_0, \qquad \text{for all } n \ge N.$$

Since ϵ_0 is arbitrary and almost every $t \in (a,b)$ is a Lebesgue point, it follows that $\|A^{-\frac{1}{2}}v_n(t)\|_\alpha \to 0$, as $n \to \infty$, for almost every $t \in (a,b)$, which establishes Item (4). Furthermore, one obtains

$$\int_0^T \|A^{-\frac{1}{2}}v_n\|_\alpha^r\, dt \to 0, \qquad \text{as } n \to \infty \text{ and } 1 \le r < \infty,$$

by the Lebesgue Dominated Convergence Theorem, which implies Item (3).

Finally Item (3) implies that there is a set E in (a,b) of Lebesgue measure 0 and a subsequence of u_n, which we will relabel as u_n, such that $u_n(t_0)$ converges strongly in V^α to $u(t_0)$, for $t_0 \in \mathbb{R}^+ \backslash E$, which is Item (5). □

In the application of this compactness lemma to the theory of solutions of the Navier-Stokes equations, we will be studying solutions that lie in the Fréchet spaces $L^2_{\text{loc}}[0,\infty;V)$ and $L^2_{\text{loc}}(0,\infty;V)$, as well as other related spaces.[26] In particular, we recall that a set B is bounded in $L^2_{\text{loc}}(0,\infty;V)$ if and only if one has

$$\sup\{\int_{1/m}^{m} \|A^{\frac{1}{2}}\varphi\|^2\,dt : \varphi \in B\} < \infty, \qquad \text{for each } m \geq 1,$$

and B is a bounded set in $L^2_{\text{loc}}[0,\infty;V)$ if and only if one has

$$\sup\{\int_{m-1}^{m} \|A^{\frac{1}{2}}\varphi\|^2\,dt : \varphi \in B\} < \infty, \qquad \text{for each } m \geq 1.$$

Lemma 63.3 (Compactness II). *Let φ^n be a sequence in $L^2_{\text{loc}}(0,\infty;V)$, with the properties that: (1) φ^n is bounded in $L^2_{\text{loc}}(0,\infty;V)$, and (2) each φ^n has a time derivative in the space V^{-1} and the sequence $D_t\varphi^n$ is bounded in $L^p_{\text{loc}}(0,\infty;V^{-1})$, for some p satisfying $1 < p < \infty$. Then there exists a subsequence of φ^n, which we relabel as φ^n, and functions $\varphi \in L^2_{\text{loc}}(0,\infty;V)$ and $g \in L^p_{\text{loc}}(0,\infty;V^{-1})$ such that the following properties hold:*

1. *One has $\varphi^n \overset{\text{w}}{\to} \varphi$ in $L^2_{\text{loc}}(0,\infty;V)$.*
2. *One has $D_t\varphi^n \overset{\text{w}}{\to} g$ in $L^p_{\text{loc}}(0,\infty;V^{-1})$.*
3. *One has $\varphi^n \overset{\text{s}}{\to} \varphi$ in $L^2_{\text{loc}}(0,\infty;H)$ and in $L^2_{\text{loc}}(0,\infty;V^{-1})$.*
4. *For almost every $t \in (0,\infty)$, one has $\varphi^n(t) \overset{\text{s}}{\to} \varphi(t)$ in V^{-1}.*
5. *For almost every $t \in (0,\infty)$, one has $\varphi^n(t) \overset{\text{s}}{\to} \varphi(t)$ in H.*

If in addition, φ^n is a bounded set in $L^2_{\text{loc}}[0,\infty;H)$ and D_tu_n is a bounded set in $L^p_{\text{loc}}[0,\infty;H)$, then the limit functions φ and g are in $L^2_{\text{loc}}[0,\infty;H)$ and $L^p_{\text{loc}}[0,\infty;H)$, respectively, and the limits in Items (1), (2), and (3) are valid in $L^r_{\text{loc}}[0,\infty;X)$, where $(r,X) = (2,V)$, (p,V^{-1}), and $(2,H)$ or $(2,V^{-1})$, respectively.

Proof. It follows from the hypotheses that for each integer $m \geq 1$, the sequence φ^n is a bounded sequence in the Lebesgue space $L^2(I_m;V)$, where I_m is the interval $((m+1)^{-1}, m+1)$, and the sequence of derivatives $D_t\varphi^n$ is a bounded sequence in $L^p(I_m;V^{-1})$. Therefore the hypotheses of Compactness I (Lemma 63.2), with $\alpha = 0$, are satisfied on I_m. As a result, one can find subsequences of the sequence φ^n so that the conclusions of Compactness I are valid on I_m. We will do this in a systematic way.

For $m = 1,2,3,\ldots$, we construct subsequences φ^n_m and functions φ_m and g_m with the following properties:

1. One takes $\varphi_1 = 0$ and $\varphi^n_1 = \varphi^n$, for all n, where φ^n is the sequence described in the hypotheses of this lemma.
2. For $m = 1,2,3,\ldots$, each sequence φ^n_{m+1} is a subsequence of φ^n_m and is chosen so that $\varphi^n_{m+1} = \varphi^n_m$, for $n = 1,\ldots,m$ and $m = 1,2,3,\cdots$. Furthermore, this subsequence and the functions φ_{m+1} and g_{m+1} satisfy the conclusions of Compactness I on the interval I_m.

[26]Note the difference: [0 and (0. This is a good point to review the notational conventions and the related theory of Fréchet spaces described in Appendices A and B.

Finally, the diagonal subsequence φ_n^n and the functions φ and g defined by $\varphi(t) = \lim_{m\to\infty} \varphi_m(t)$ and $g(t) = \lim_{m\to\infty} g_m(t)$, for $t \in \mathbb{R}^+$, satisfy the conclusions of this lemma. The verification that these limits exist and that of the "if in addition" clause is valid is left as an exercise. □

In the next lemma, we will apply Compactness II (Lemma 63.3) to a sequence of Bubnov-Galerkin approximations. Among other things, we will show that $D_t\varphi = g$, where φ and g are the limit functions in Compactness II.

Lemma 63.4. *Let Ω be an open, bounded domain in $\mathbb{R}^d$ of class C^2 and let $f \in L^\infty(0,\infty;H)$. For $n = 1,2,\ldots$, let $u_n = p$ be solutions of (63.1) with the property that the initial conditions $u_n(0) = p(0)$ lie in a bounded set in H, i.e., there is a constant $C > 0$ such that $\|u_n(0)\| \le C$, for all $n \ge 1$. Then the sequence $u_n = u_n(t)$ lies in a bounded set in $L^2_{\text{loc}}[0,\infty;V)$, and the sequence of derivatives $D_t u_n$ lies in a bounded set in $L^p_{\text{loc}}[0,\infty;V^{-1})$, where p satisfies (62.6). (In particular, the hypotheses of Compactness II are satisfied).*

Choose any subsequence of u_n so that the initial conditions $u_n(0) \xrightarrow{w} u_0 \in H$, and choose any further subsequence, which we relabel as u_n, such that the conclusions in Compactness II are valid, with limit functions $u \in L^2_{\text{loc}}[0,\infty;V)$ and $g \in L^p_{\text{loc}}[0,\infty;V^{-1})$. Then, after a possible change of $u(t)$ on a set of measure 0, the following hold:

1. *one has $u \in L^\infty(0,\infty;H) \cap L^2_{\text{loc}}[0,\infty;V)$;*
2. *Equation (63.9) is valid, for all $t \ge t_0 \ge 0$;*
3. *one has $D_t u = g$ in V^{-1}; and*
4. *the function u satisfies (63.7) and (63.8).*

Proof. Since $\|u_n(0)\| \le C$, for all $n \ge 1$, it follows from (62.4) that u_n is a bounded set in the Lebesgue space $L^\infty(0,\infty;H)$. Furthermore, from Lemma 62.1, we see that u_n is a bounded set in the Fréchet space $L^2_{\text{loc}}[0,\infty;V)$ and $D_t u_n$ is a bounded set in $L^p_{\text{loc}}[0,\infty;V^{-1})$, where p satisfies (62.6). Therefore the hypotheses of Compactness II are satisfied. Next let u_n now denote the subsequence and let u and g denote the two limit functions satisfying the conclusions of Compactness II.

Let $v \in V$ be given. By taking the scalar product of (63.1) in H with v and integrating in time, for $t \ge t_0 \ge 0$, we obtain

$$\begin{aligned} &\langle u_n(t) - u_n(t_0), v\rangle + \nu \int_{t_0}^t \langle A^{\frac{1}{2}} u_n(s), A^{\frac{1}{2}} v\rangle\, ds \\ &\quad + \int_{t_0}^t b(u_n(s), u_n(s), P_n v)\, ds = \int_{t_0}^t \langle f(s), P_n v\rangle\, ds, \end{aligned} \tag{63.29}$$

As argued in Lemma 63.1, the last equation can be written in the form

$$\langle A^{-\frac{1}{2}}(u_n(t) - u_n(t_0)), A^{\frac{1}{2}} v\rangle = \langle \int_{t_0}^t A^{-\frac{1}{2}} G(u_n)\, ds, A^{\frac{1}{2}} v\rangle,$$

where $G(u_n) = -\nu A u_n + P_n[-B(u_n, u_n) + f]$. We will now show that $G(u_n)$ has an appropriate weak limit in the space $L^p_{\mathrm{loc}}[0, \infty; V^{-1})$. Since $P_n v \xrightarrow{s} v$ in V, one has

$$\int_{t_0}^{t} \langle f(s), P_n v\rangle \, ds \to \int_{t_0}^{t} \langle f(s), v\rangle \, ds, \qquad \text{as } n \to \infty.$$

Since $u_n \xrightarrow{w} u$ in $L^2_{\mathrm{loc}}[0, \infty; V)$, one has

$$\lim_{n\to\infty} \int_{t_0}^{t} \langle A^{\frac{1}{2}} u_n(s), A^{\frac{1}{2}} v\rangle \, ds = \int_{t_0}^{t} \langle A^{\frac{1}{2}} u(s), A^{\frac{1}{2}} v\rangle \, ds, \quad \text{for } 0 \le t_0 \le t < \infty.$$

Assume for the moment that the trilinear term $b(u_n, u_n, P_n v)$ satisfies

$$\lim_{n\to\infty} \int_{t_0}^{t} b(u_n(s), u_n(s), P_n v) \, ds = \int_{t_0}^{t} b(u(s), u(s), v) \, ds, \tag{63.30}$$

for $0 \le t_0 \le t < \infty$. Then by using Item (5) of Compactness II, one can take the limit as $n \to \infty$ in (63.29) to obtain

$$\begin{aligned} \langle u(t) - u(t_0), v\rangle + \nu \int_{t_0}^{t} \langle A^{\frac{1}{2}} u(s), A^{\frac{1}{2}} v\rangle \, ds + \int_{t_0}^{t} b(u(s), u(s), v) \, ds \\ \overset{\text{a.e.}}{=} \int_{t_0}^{t} \langle f(s), v\rangle \, ds \end{aligned} \tag{63.31}$$

for $0 \le t_0 \le t < \infty$ and $t_0, t \in (0, \infty)\backslash E$. Now each of the terms with integrals in (63.31) is continuous in t and t_0. Therefore, by changing $u(t)$ on a set of measure 0, if necessary, we can assume, and we do assume, that (63.31) is valid for all t and t_0 with $0 \le t_0 \le t < \infty$, i.e., u satisfies (63.9).

We now use (63.31) to show that $D_t u = g$. The argument leading up to (63.31) implies that $G(u_n) \xrightarrow{w} G(u)$ in the space $L^p_{\mathrm{loc}}[0, \infty; H)$, where $G(u) = -\nu A u - B(u, u) + f$. However, from Compactness II one has $D_t u_n = G(u_n) \xrightarrow{w} g$ in $L^p_{\mathrm{loc}}[0, \infty; V^{-1})$. From the uniqueness of the limits one then obtains $G(u) = g$, i.e., $D_t u = g$.

Because of the fact that $\{u_n : n \ge 1\}$ is a bounded set in the Lebesgue space $L^\infty(0, \infty; H)$, it follows from Item (5) of Compactness II that the limit function u is in the space $L^\infty(0, \infty; H)$, as well. Moreover, inequality (62.4) holds in the limit, which shows that u satisfies (63.7).

In order to prove the energy inequality (63.8), we take the inner product of equation (62.2) with $u_n = p$ to obtain

$$\frac{1}{2}\partial_t \|u_n\|^2 + \nu \|A^{\frac{1}{2}} u_n\|^2 \overset{\text{a.e.}}{=} \langle Pf, u_n\rangle \overset{\text{a.e.}}{=} \langle f, u_n\rangle,$$

since (61.15) holds and $Pu_n = u_n$. By integrating this in time, we obtain

$$\|u_n(t)\|^2 + 2\nu \int_{t_0}^{t} \|A^{\frac{1}{2}} u_n(s)\|^2 ds \le \|u_n(t_0)\|^2 + 2 \int_{t_0}^{t} \langle f(s), u_n(s) \rangle \, ds. \tag{63.32}$$

Now restrict t_0 to satisfy $t_0 \in (0,T) \backslash E$, see Compactness II, Item (5). Then

$$\lim_{n\to\infty} \left(\|u_n(t_0)\|^2 + 2 \int_{t_0}^{t} \langle f(s), u_n(s) \rangle \, ds \right) = \|u(t_0)\|^2 + 2 \int_{t_0}^{t} \langle f(s), u(s) \rangle \, ds.$$

In order to take limits on the left side of (63.32), we use the fact that, since $u_n \overset{w}{\to} u$ in $L^2(0,T;V)$, the norms satisfy the semicontinuity relationship

$$\int_{t_0}^{t} \|A^{\frac{1}{2}} u\|^2 \, ds \le \liminf_{n\to\infty} \int_{t_0}^{t} \|A^{\frac{1}{2}} u_n\|^2 \, ds, \qquad 0 \le t_0 \le t \le T, \tag{63.33}$$

see Appendix A.9. By using: $\limsup(a_n + b_n) \ge \limsup a_n + \liminf b_n$, we see that $\frac{1}{2}\|u(t)\|^2 + \nu \int_{t_0}^{t} \|A^{\frac{1}{2}} u(s)\|^2 \, ds$ is bounded above by

$$\begin{aligned} \limsup_{n\to\infty} \|u_n(t)\|^2 &+ \liminf_{n\to\infty} 2\nu \int_{t_0}^{t} \|A^{\frac{1}{2}} u_n(s)\|^2 \, ds \\ &\le \limsup_{n\to\infty} \left(\|u_n(t)\|^2 + 2\nu \int_{t_0}^{t} \|A^{\frac{1}{2}} u_n(s)\|^2 \, ds \right) \\ &\le \|u(t_0)\|^2 + 2 \int_{t_0}^{t} \langle f(s), u(s) \rangle \, ds, \end{aligned}$$

which gives (63.8).

It remains to verify (63.30). Note that the trilinear form b satisfies

$$\begin{aligned} b(u_n, u_n, P_n v) - b(u, u, v) = b(u_n - u, u_n, P_n v) &+ b(u, u_n - u, P_n v) \\ &+ b(u, u, P_n v - v). \end{aligned}$$

Since $b(u_n - u, u_n, P_n v) = -b(u_n - u, P_n v, u_n)$, it follows from (61.24) that

$$|b(u_n - u, u_n, P_n v)| \le C_3 \|u_n - u\|^{\frac{1}{4}} \|A^{\frac{1}{2}}(u_n - u)\|^{\frac{3}{4}} \|u_n\|^{\frac{1}{4}} \|A^{\frac{1}{2}} u_n\|^{\frac{3}{4}} \|A^{\frac{1}{2}} P_n v\|.$$

Similarly, one finds that

$$|b(u, u_n - u, P_n v)| \le C_3 \|u\|^{\frac{1}{4}} \|A^{\frac{1}{2}} u\|^{\frac{3}{4}} \|u_n - u\|^{\frac{1}{4}} \|A^{\frac{1}{2}}(u_n - u)\|^{\frac{3}{4}} \|A^{\frac{1}{2}} P_n v\|.$$

Now the Hölder inequality implies that $\int_{t_0}^t |b(u_n - u, u_n, P_n v)|\,ds$ is bounded above by the product

$$C_3\|A^{\frac{1}{2}}P_n v\| \left(\int_{t_0}^t \|u_n - u\|^2 ds\right)^{\frac{1}{8}} \left(\int_{t_0}^t \|u_n\|^2 ds\right)^{\frac{1}{8}} \times$$
$$\left(\int_{t_0}^t \|A^{\frac{1}{2}}(u_n - u)\|^2 ds\right)^{\frac{3}{8}} \left(\int_{t_0}^t \|A^{\frac{1}{2}}u_n\|^2 ds\right)^{\frac{3}{8}}.$$

Notice that the first integral above converges to 0, as $n \to \infty$, by Item (3) of Compactness II. On the other hand, the other three integrals are uniformly bounded in n by inequalities (62.4), (62.5), and (63.32). Since $\|A^{\frac{1}{2}}P_n v\| = \|P_n A^{\frac{1}{2}}v\| \le \|A^{\frac{1}{2}}v\|$, for $v \in V$, we see that

$$\int_{t_0}^t |b(u_n - u, u_n, P_n v)|\,ds \to 0, \qquad \text{as } n \to \infty.$$

A similar argument shows that

$$\int_{t_0}^t |b(u, u_n - u, P_n v)|\,ds \to 0, \qquad \text{as } n \to \infty.$$

Let s be chosen so that $t_0 < s < t$, $u(s) \in V$, and $B(u(s), u(s)) \in V^{-1}$, by (61.26). Since $v \in V$, one has $(I - P_n)v \in V$, and therefore

$$\begin{aligned}|b(u(s), u(s), P_n v - v)| &= |\langle B(u(s), u(s)), (I - P_n)v\rangle| \\ &\le \|A^{-\frac{1}{2}}B(u(s), u(s))\|\,\|(I - P_n)A^{\frac{1}{2}}v\| \to 0,\end{aligned}$$

as $n \to \infty$. It follows then that

$$\lim_{n\to\infty} b(u, u, P_n v) \overset{\text{a.e.}}{=} b(u, u, v).$$

From (61.26) and the fact that

$$b(u, u, P_n v) - \langle \mathbb{P}(u \cdot \nabla)u, P_n v\rangle = \langle A^{-\frac{1}{2}}B(u, u), P_n A^{\frac{1}{2}}v\rangle,$$

we obtain

$$|b(u, u, P_n v)| \le C_5\|A^{\frac{1}{2}}v\|\,\|u\|^{\frac{1}{2}}\|A^{\frac{1}{2}}u\|^{\frac{3}{2}}.$$

From the Hölder inequality we infer that $\|u\|^{\frac{1}{2}}\|A^{\frac{1}{2}}u\|^{\frac{3}{2}} \in L^1_{\text{loc}}[0, \infty; \mathbb{R})$. Therefore one has

$$\int_{t_0}^t |b(u, u, P_n v - v)|\,ds \to 0, \qquad \text{as } n \to \infty$$

by the Lebesgue Dominated Convergence Theorem. □

The main result on weak solutions of Class LH is the following theorem.

Theorem 63.5 (Leray-Hopf). *Let Ω be an open, bounded domain in $\mathbb{R}^d$ of class C^2, where $d = 2$ or 3, and let $f \in L^\infty(0,\infty;H)$. Then for every $u_0 \in H$ there is a weak solution (of Class LH) $u = u(t)$ of (61.10) satisfying $u(0) = u_0$, and one has*

$$D_t u \in L^p_{\text{loc}}[0,\infty;V^{-1}), \tag{63.34}$$

where p satisfies (62.6).

Proof. Let u_n be the sequence of Bubnov-Galerkin approximations that satisfy $u_n(0) = P_n u_0$, see (62.2)-(62.3). As a result, the hypotheses of Lemma 63.4 are satisfied, and $u_n(0) \overset{s}{\to} u_0$ in H and in V^{-1}. Let $u \in L^2_{\text{loc}}[0,\infty;V)$ and $g \in L^p_{\text{loc}}[0,\infty;V^{-1})$ be given by Lemma 63.4, where p satisfies (62.6). Then Items (1) - (4) in Lemma 63.4 imply that u is a weak solution of Class LH with $u(0) = u_0$, that $D_t u = g$, and that (63.34) is valid. □

In the case of the 2D Navier-Stokes equations we will now show that the weak solutions are uniquely determined by the initial data $u_0 \in H$. The corresponding issue of uniqueness for the 3D equations will be studied in the following section.

Corollary 63.6 (2D Uniqueness). *Let Ω be an open, bounded domain in $\mathbb{R}^2$ of class C^2 and let $f \in L^\infty(0,\infty;H)$. Then for every $u_0 \in H$ there is precisely one weak solution (of Class LH) $u = u(t)$ of (61.10) satisfying $u(0) = u_0$. In addition one has $u \in C[0,\infty;H)$. Moreover, every weak solution (of Class LH) is the limit of a sequence of Bubnov-Galerkin approximations.*

Proof. Assume on the contrary that there are two weak solutions, u_1 and u_2. Let $w = u_1 - u_2$. For $\hat{u} = u_1, u_2, w$, one has

$$\hat{u} \in L^2_{\text{loc}}[0,\infty;V) \cap C[0,\infty;H_w) \quad \text{and} \quad D_t\hat{u} \in L^2_{\text{loc}}[0,T;V^{-1}). \tag{63.35}$$

Furthermore w satisfies the differential equation

$$\langle\langle w_t, v\rangle\rangle + \nu\langle A^{\frac{1}{2}}w, A^{\frac{1}{2}}v\rangle + b(w,u_2,v) + b(u_1,w,v) \overset{\text{a.e.}}{=} 0, \tag{63.36}$$

for each $v \in V$, and $\langle\langle w_t, w\rangle\rangle \overset{\text{a.e.}}{=} \frac{1}{2}\frac{d}{dt}\|w\|^2$, by the Continuity Lemma 37.9. By setting $v = w(t)$ in (63.36), we obtain

$$\frac{1}{2}\partial_t\|w\|^2 + \nu\|A^{\frac{1}{2}}w\|^2 + b(w,u_2,w) \overset{\text{a.e.}}{=} 0.$$

Now (61.23) implies that

$$\begin{aligned}\frac{1}{2}\partial_t\|w\|^2 + \nu\|A^{\frac{1}{2}}w\|^2 &\le |b(w,u_2,w)| \le C_2\|A^{\frac{1}{2}}u_2\|\,\|A^{\frac{1}{2}}w\|\,\|w\| \\ &\le \nu\|A^{\frac{1}{2}}w\|^2 + \frac{1}{4\nu}C_2^2\|A^{\frac{1}{2}}u_2\|^2\|w\|^2,\end{aligned}$$

almost everywhere. The Gronwall inequality then implies that

$$(63.37)\quad \|w(t)\|^2 \le \|w(0)\|^2 \exp\left(\frac{C_2^2}{2\nu}\int_0^t \|A^{\frac{1}{2}}u_2\|^2\,ds\right), \qquad \text{for all } t \ge 0.$$

Since $u_2 \in L^2_{\text{loc}}[0,\infty;V)$, the integral in (63.37) is finite. Since $w(0) = 0$, we obtain $w(t) = 0$, for all $t \ge 0$.

It also follows from (63.35) and the Continuity Lemma 37.9, Item (3), that every weak solution u (of Class LH) satsifies $u \in C[0,\infty;H)$. The fact that every weak solution (of Class LH) is the limit of a sequence of Bubnov-Galerkin approximations is now a direct consequence of the uniqueness of weak solutions and the proof of the Leray-Hopf Theorem. □

6.3.2 Generalized Weak Solutions. As we have seen in the existence theory of global attractors, it helps when the phase space for the semiflow is a complete metric space (see Section 2.3.6). Since the Fréchet space $L^2_{\text{loc}} = L^2_{\text{loc}}[0,\infty;H)$ is complete, it would follow that the space $W_{LH} = W_{LH}(f)$ of weak solutions of Class LH is complete in terms of the invariant metric given on L^2_{loc} (see Appendices A and B), provided that W_{LH} is a closed set in L^2_{loc}. Unfortunately, we are unable to prove that W_{LH} is closed. Instead, we will imbed W_{LH} into a larger class W of generalized weak solutions and we will show that W is closed. The generalized weak solutions, like the weak solutions of Class LH, will reside in the space $L^2_{\text{loc}}[0,\infty;H)$, but we will relax some conditions in the definition to allow for the possibility of a singularity at $t = 0$, see Figure 6.1.

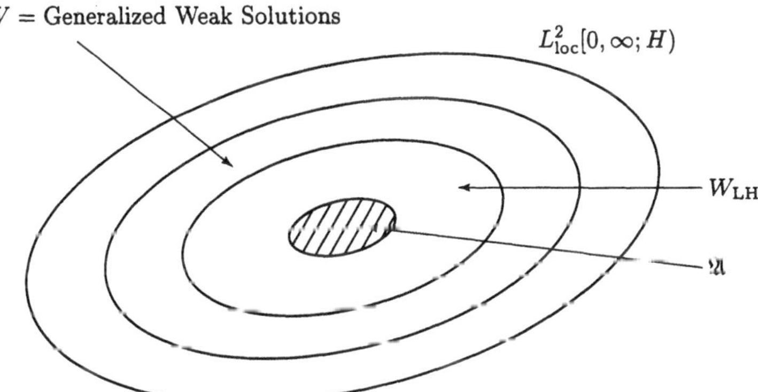

Figure 6.1. Generalized Weak Solutions, Class LH and Global Attractor $\mathfrak{A}$

We will say that a function $\varphi \in L^2_{\text{loc}}[0,\infty;H)$ is a **generalized weak solution**, and write $\varphi \in W(f)$, provided that one has

(1) $\varphi \in L^2_{\text{loc}}[0,\infty;H) \cap L^\infty_{\text{loc}}(0,\infty;H) \cap L^2_{\text{loc}}(0,\infty;V)$;
(2) $D_t\varphi \in L^p_{\text{loc}}(0,\infty;V^{-1})$, for $p = \frac{4}{3}$;

(3) for almost all t and almost all t_0 with $t > t_0$ inequalities (63.7) and (63.8) are valid;
(4) and for all $t \geq t_0 > 0$, equation (63.9) is valid, for all $v \in V$.

Some spaces arise quite often in this theory, and it is convenient to adopt some special notation for this situation. In particular, we define

$$Y \stackrel{\text{def}}{=} L^\infty(0,\infty;L^2(\Omega)) \quad \text{and} \quad Z \stackrel{\text{def}}{=} L^2_{\text{loc}}[0,\infty;H) \times Y,$$

and, for any $N_0 \geq 0$, we let

$$Y(N_0) \stackrel{\text{def}}{=} \{f \in Y : \|f\|_\infty \leq N_0\} \text{ and } Z(N_0) \stackrel{\text{def}}{=} L^2_{\text{loc}}[0,\infty;H) \times Y(N_0).$$

The space R is the product space

$$R \stackrel{\text{def}}{=} L^2_{\text{loc}}[0,\infty;H \times L^2(\Omega)) = L^2_{\text{loc}}[0,\infty;H) \times L^2_{\text{loc}}[0,\infty;L^2(\Omega)). \tag{63.38}$$

It follows from the definitions that one has $W_{LH}(f) \subset W(f)$, for every $f \in Y$. As argued in Lemma 63.1, we now have the following result:

Generalized Solution Proposition. *Let $\varphi \in W(f)$, for some $f \in Y$. Then the following hold:*

(1) *φ satisfies $\varphi \in C(0,\infty;H_w)$.*
(2) *Inequalities (63.7) and (63.11) are valid, for almost all $t_0 \in (0,\infty)$ and for all $t \geq t_0$, for $u = \varphi$.*
(3) *$u = \varphi$ satisfies the inequality (63.12), for all $t \geq 0$.*
(4) *The function $u = \varphi$ satisfies (63.9), for almost all $t_0 > 0$ and for all $t > 0$. Consequently, equation (63.13) holds in the space V^{-1}.*
(5) *Let p satisfy (62.6). Then $D_t u = D_t\varphi$ is in $L^p_{\text{loc}}(0,\infty;V^{-1})$, and $u = \varphi$ is in $C^{0,\theta}_{\text{loc}}(0,\infty;V^{-1})$, where $\theta = \frac{p-1}{p}$.*
(6) *The function $u = \varphi$ is a mild solution in the space V^{-1}, i.e., the Variation of Constants Formula (63.14) is valid for almost all $t_0 > 0$ and for all $t \geq t_0$.*
(7) *Let $\{v_n\}$ be any sequence in $L^2_{\text{loc}}[0,\infty;H)$ that satisfies the following two properties: (a) there is a function $a \in C(0,\infty;\mathbb{R}^+)$, such that, for each n, one has*

$$\|v_n(t)\| \leq a(t), \qquad \textit{for almost all } t > 0,$$

(b) and one has $v_n \xrightarrow{s} u$ in $L^2_{\text{loc}}[0,\infty;H)$, as $n \to \infty$. Then

$$\|u(t)\| \leq a(t), \qquad \textit{for all } t > 0.$$

Next we define $\mathcal{W}$ to be the collection of all ordered pairs (φ, f) in Z such that $\varphi \in W(f)$, and $\mathcal{W}_{LH}$ is defined to be the collection of all such ordered pairs satisfying $\varphi \in W_{LH}(f)$. Let $\mathcal{W}(N_0) = \mathcal{W} \cap Z(N_0)$ and $\mathcal{W}_{LH}(N_0) = \mathcal{W}_{LH} \cap Z(N_0)$. We will show below that $\mathcal{W}(N_0)$ is a closed set in $Z(N_0)$, for each $N_0 \geq 0$. This will then imply that for each $f \in Y$ the fiber $W(f)$ is a closed set in $L^2_{\text{loc}}[0, \infty; H)$.

We will require some properties of the generalized weak solutions. These properties are described in a series of lemmas. First note that it follows from the definition that a set $\hat{B} \subset W(f)$ is bounded if and only if

$$\sup\{\int_n^{n+1} \|\varphi\|^2 \, ds : \varphi \in \hat{B}\} < \infty, \qquad \text{for each } n = 0, 1, 2, \ldots. \tag{63.39}$$

Consequently, if $\hat{B}$ is a bounded set in $W(f)$, then

$$\sup\{\int_0^1 \|\varphi\|^2 \, ds : \varphi \in \hat{B}\} < \infty.$$

Among other things, the first lemma establishes the converse of the last implication. We define the set

$$\hat{B}(M_0, N_0) \stackrel{\text{def}}{=} \{(\varphi, f) \in \mathcal{W} : \int_0^1 \|\varphi\|^2 \, ds \leq M_0^2 \text{ and } \|f\|_\infty^2 \leq N_0^2\} \tag{63.40}$$

for nonnegative numbers M_0 and N_0.

Lemma 63.7. *Let $c_0 = (\nu\lambda_1)^{-1}$. Then the following statements are valid:*

(1) *For each $(\varphi, f) \in \hat{B}(M_0, N_0)$ one has*

$$\|\varphi(\tau)\|^2 \leq \tau^{-1} M_0^2 + c_0^2 N_0^2, \qquad \text{for } 0 < \tau \leq 1, \tag{63.41}$$

and

$$\|\varphi(\tau)\|^2 \leq e^{-\nu\lambda_1(\tau-1)}(M_0^2 + c_0^2 N_0^2) + c_0^2 N_0^2, \qquad \text{for } \tau \geq 1. \tag{63.42}$$

(2) *For each $(\varphi, f) \in \hat{B}(M_0, N_0)$ one has*

$$\int_T^{T+1} \|\varphi\|^2 \, ds \leq e^{-\nu\lambda_1(T-1)}(M_0^2 + c_0^2 N_0^2) + c_0^2 N_0^2, \qquad \text{for } T \geq 1. \tag{63.43}$$

(3) *The set $\hat{B}(M_0, N_0)$ is bounded in $\mathcal{W}$.*

(4) *For each τ and T with $0 < \tau \leq T < \infty$, there exist positive constants $M_i = M_i(\tau, T)$, for $i = 1, 2$, such that the following two inequalities are valid:*

$$\sup\{\int_\tau^T \|A^{\frac{1}{2}}\varphi\|^2 \, ds : (\varphi, f) \in \hat{B}(M_0, N_0)\} \leq M_1^2 \tag{63.44}$$

and for $p = 4/3$, one has

$$\sup\{\int_\tau^T \|A^{-\frac{1}{2}} D_t\varphi\|^p \, ds : (\varphi, f) \in \hat{B}(M_0, N_0)\} \leq M_2^p. \tag{63.45}$$

(5) *If in addition, one has $\varphi \in L^2_{\rm loc}[0, \infty; V) \cap L^\infty(0, \infty; H)$, then the constants M_1 and M_2 in Item (4) can be chosen to be independent of τ, for $0 < \tau \leq 1$, i.e., the limits $M_i(0, T) = \lim_{\tau \to 0+} M_i(\tau, T)$ exist, for $i = 1, 2$.*

Proof. We will use the Generalized Solution Proposition here, and we begin with $(\varphi, f) \in \hat{B}(M_0, N_0)$. For $0 < \tau \leq 1$, the set $\{t \in (0, \tau) : \|\varphi(t)\|^2 \leq \tau^{-1} M_0^2\}$ must have positive measure, since $\int_0^\tau \|\varphi\|^2 \, ds \leq \int_0^1 \|\varphi\|^2 \, ds \leq M_0^2$. By using inequality (63.7) one then finds that there is a time t_φ, with $0 < t_\varphi < \tau$ such that $\|\varphi(t_\varphi)\|^2 \leq \tau^{-1} M_0^2$ and

$$\|\varphi(t)\|^2 \leq e^{-\nu\lambda_1(t - t_\varphi)} \|\varphi(t_\varphi)\|^2 + c_0^2 N_0^2, \qquad \text{for all } t \geq t_\varphi. \tag{63.46}$$

As a result we obtain (63.41) and (63.42). Also inequality (63.43) is a direct consequence of (63.42). The fact that $\hat{B}(M_0, N_0)$ is a bounded set in $\mathcal{W}$ now follows from (63.43).

In order to derive (63.44) and (63.45), there is no loss of generality in restricting to the case where $0 < \tau \leq 1 \leq T < \infty$. For inequality (63.44) with $t = T$, we let $t_0 = t_\varphi$ be chosen so that one has $0 < t_0 < \tau$ and $\|\varphi(t_0)\|^2 \leq \tau^{-1} M_0^2$. Then (63.11) implies that

$$\nu \int_\tau^T \|A^{\frac{1}{2}} \varphi\|^2 \, ds \leq \tau^{-1} M_0^2 + T c_0 N_0^2, \tag{63.47}$$

which implies (63.44).

In order to prove (63.45), we note that the equality

$$D_t\varphi = -\nu A\varphi - B(\varphi, \varphi) + \mathbb{P} f$$

is valid in the space $L^p_{\rm loc}(0, \infty; V^{-1})$, for $p = \frac{4}{3}$. Now the Minkowski inequality implies that

$$\left(\int_\tau^T \|A^{-\frac{1}{2}} D_t\varphi\|^p \, ds\right)^{\frac{1}{p}}$$

is bounded above by

(63.48)

$$\nu \left(\int_\tau^T \|A^{\frac{1}{2}} \varphi\|^p \, ds\right)^{\frac{1}{p}} + \left(\int_\tau^T \|A^{-\frac{1}{2}} B(\varphi, \varphi)\|^p \, ds\right)^{\frac{1}{p}} + (T - \tau)^{\frac{1}{p}} \lambda_1^{-1} N_0,$$

and the Hölder inequality implies that for $p = \frac{4}{3}$ one has

$$\text{(63.49)} \qquad \int_\tau^T \|A^{\frac{1}{2}}\varphi\|^p \, ds \le (T-\tau)^{\frac{1}{3}} \left(\int_\tau^T \|A^{\frac{1}{2}}\varphi\|^2 \, ds \right)^{\frac{2}{3}}.$$

Next we note that the first inequality of (61.26) implies that
(63.50)

$$\int_\tau^T \|A^{-\frac{1}{2}}B(\varphi,\varphi)\|^p \, ds \le C_5^p \|\varphi\|_{\infty;[\tau,t]}^{\frac{2}{3}} \left(\int_\tau^T \|A^{\frac{1}{2}}\varphi\|^2 \, ds \right)^{\frac{3}{4}} (T-\tau)^{\frac{1}{4}}.$$

By combining (63.50) with (63.47), (63.48), and (63.49), we obtain (63.45).

Item (5) follows from the argument of the last paragraph along with the two observations: (i) that inequality (63.49) remains valid with $\tau = 0$, and (ii) that one can replace the term $\|\varphi\|_{\infty;[\tau,T]}^{\frac{2}{3}}$ in (63.50) with $\|\varphi\|_\infty^{\frac{2}{3}}$. □

As a corollary to the last lemma, we show that if $\hat{B}$ is a bounded set in $\mathcal{W}$, then $\hat{B}$ is bounded in some other spaces as well.

Lemma 63.8. *Let $\hat{B}$ be a bounded set in $\mathcal{W}(N_0)$, for some $N_0 \ge 0$. Then the following hold:*

(1) *$\hat{B}$ is a bounded set in $L^2_{\text{loc}}(0,\infty;V) \times Y(N_0)$.*
(2) *The set $(\{D_t\varphi, f) : (\varphi, f) \in \hat{B}\}$ is bounded in $L^p_{\text{loc}}(0,\infty;V^{-1}) \times Y(N_0)$, for $p = 4/3$.*

Proof. Since $\hat{B}$ is bounded, it is contained in $\hat{B}(M_0, N_0)$ for some nonnegative number M_0, see (63.40). Lemma 63.8 then follows from Lemma 63.7, Item (4), and the definition of boundedness. □

The next lemma gives a sufficient condition for a generalized weak solution to be a solution of Class LH.

Lemma 63.9. *Let $(\varphi, f) \in \mathcal{W}$ with $\varphi \in L^2_{\text{loc}}[0,\infty;V) \cap L^\infty(0,\infty;H)$. Then one has $(\varphi, f) \in \mathcal{W}_{LH}$.*

Proof. In reference to the definitions, we need only to verify that if (φ, f) satisfies the hypotheses of this lemma, then the following hold:

(1) Equation (63.9) is valid at $t_0 = 0$, and
(2) $D_t\varphi \in L^p_{\text{loc}}[0,\infty, V^{-1})$, for $p = \frac{4}{3}$.

Since one has $\varphi \in L^2_{\text{loc}}[0,\infty;V) \cap L^\infty(0,\infty;H)$, it follows from (61.24) that $b(\varphi,\varphi,v)$ and $\langle A^{\frac{1}{2}}\varphi, A^{\frac{1}{2}}v\rangle$ lie in $L^1_{\text{loc}}[0,\infty;\mathbb{R})$, for every $v \in V$. Since $\langle f, v\rangle \in L^1_{\text{loc}}[0,\infty;\mathbb{R})$, as well, one can let $t_0 \to 0^+$ in equation (63.9) to obtain Item (1). Item (2) follows from Lemma 63.7, Item (5). □

Finally we show that for each $N_0 \ge 0$, the space $\mathcal{W}(N_0)$ is a closed set in Z.

Lemma 63.10. *Let (φ^n, f^n) be a convergent sequence in Z with limit (φ_0, f_0) in Z. Assume that $\varphi^n \in W(f^n)$ and that $f^n \in Y(N_0)$, for some $N_0 \geq 0$ and for all n. Then there is a function $\varphi \in W(f_0)$ such that $\varphi \overset{\text{a.e.}}{=} \varphi_0$. Moreover, for each $N_0 \geq 0$, the space $\mathcal{W}(N_0)$ generated by the weak solutions of the Navier-Stokes equations is a closed set in Z.*

Proof. Let (φ^n, f^n) and (φ_0, f_0) be given as in the hypotheses, where $\varphi^n \in W(f^n)$ and $\|f^n\|_\infty \leq N_0$, for all n. Note that the function φ_0 satisfies

$$\varphi_0 \overset{\text{a.e.}}{=} \lim_{n\to\infty} \varphi^n, \qquad \text{on } (0,\infty). \tag{63.51}$$

Since (φ^n, f^n) is a convergent sequence, it is bounded in Z. Therefore, there exists a nonnegative number M_0 such that $(\varphi^n, f^n) \in \hat{B}(M_0, N_0)$, for all n, see (63.40). It then follows from Lemma 63.8 that the hypotheses of the Compactness II (Lemma 63.3) are satisfied. After a relabeling, we let φ^n denote the subsequence and we let $\varphi \in L^2_{\text{loc}}(0,\infty; V)$ denote the limiting function given by Compactness II. It then follows from Compactness II and (63.51) that $\varphi \overset{\text{a.e.}}{=} \varphi_0$. We will now show that $\varphi \in W(f_0)$.

Since one has $\varphi_0 \in L^2_{\text{loc}}[0,\infty; H)$, it follows that $\varphi \in L^2_{\text{loc}}[0,\infty; H)$. By using Compactness II again with the inequalities (63.41) and (63.42) (as applied to the sequence (φ^n, f^n)), we see that $\varphi \in L^\infty_{\text{loc}}(0,\infty; H)$. Also one has $\varphi \in L^2_{\text{loc}}(0,\infty; V)$ and $D_t\varphi \in L^p_{\text{loc}}(0,\infty; V^{-1})$ by Items (1) and (2) of Compactness II. Thus the first two conditions in the definition of a generalized weak solution are satisfied. It remains to verify (63.7), (63.8), and (63.9).

Because of Compactness II, we claim that equation (63.9), which is valid for each φ^n, is valid (almost everywhere) in the limit, as $n \to \infty$. Indeed, one has

$$\int_{t_0}^{t} \langle (A^{\frac{1}{2}}\varphi^n - A^{\frac{1}{2}}\varphi), A^{\frac{1}{2}}v\rangle\, ds = \langle \int_{t_0}^{t} (A^{\frac{1}{2}}\varphi^n - A^{\frac{1}{2}}\varphi)\, ds, A^{\frac{1}{2}}v\rangle \to 0,$$

as $n \to \infty$, by Item (1) of Compactness II. In order to show that

$$\int_{t_0}^{t} (b(\varphi^n, \varphi^n, v) - b(\varphi, \varphi, v))\, ds \to 0, \qquad \text{as } n \to \infty, \tag{63.52}$$

we note that the trilinearity of b implies that

$$b(\varphi^n, \varphi^n, v) - b(\varphi, \varphi, v) = b(\varphi^n - \varphi, \varphi^n, v) + b(\varphi, \varphi^n - \varphi, v).$$

Since the solutions φ^n satisfy (63.41) and (63.42), it follows from Compactness II that φ satisfies (63.41) and (63.42), almost everywhere. By using this fact, inequality (61.24), and Item (1) of Compactness II, we see that both

$$\int_{t_0}^{t} b(\varphi^n - \varphi, \varphi^n, v)\, ds \qquad \text{and} \qquad \int_{t_0}^{t} b(\varphi, \varphi^n - \varphi, v)\, ds$$

converge to 0, as $n \to \infty$, which implies (63.52). In order to show that

$$\int_{t_0}^{t} (\langle f^n, \varphi^n \rangle - \langle f_0, \varphi \rangle)\, ds \to 0, \qquad \text{as } n \to \infty, \tag{63.53}$$

we note that

$$\langle f^n, \varphi^n \rangle - \langle f_0, \varphi \rangle = \langle (f^n - f_0), \varphi^n \rangle + \langle f_0, \varphi^n - \varphi \rangle.$$

As a result, (63.53) follows from the convergence of the sequence f^n, the Schwarz inequality, and Item (1) of Compactness II. By restricting t and t_0 to be in the set $(0, \infty) \setminus E$ and using Item (5) of Compactness II, we see that (63.9) holds almost everywhere in the limit, as $n \to \infty$. By changing $\varphi(t)$ and $\varphi(t_0)$ on a set of measure zero, if necessary, we see that (63.9) is valid for all $t \geq t_0 > 0$.

Inequality (63.7) follows from Item (5) of Compactness II. Also inequality (63.8) is valid for the sequence φ^n. As $n \to \infty$, the right side of (63.8) has the limit

$$\|\varphi(t_0)\|^2 + 2 \int_{t_0}^{t} \langle f_0, \varphi \rangle\, ds, \qquad \text{for } t_0 \in (0, \infty) \setminus E.$$

From Item (1) of Compactness II and the lower semicontinuity property for weak convergence, one obtains

$$\int_{t_0}^{t} \|A^{\frac{1}{2}} \varphi\|^2\, ds \leq \liminf_{n \to \infty} \int_{t_0}^{t} \|A^{\frac{1}{2}} \varphi^n\|^2\, ds.$$

Therefore, for $t, t_0 \in (0, \infty) \setminus E$, we see that inequality (63.8) is valid for φ. Consequently, we have $\varphi \in W(f_0)$. □

6.3.3 The Uniqueness Problem. We have seen in Corollary 63.6 that the weak solutions (of Class LH) of the 2D Navier-Stokes equations are uniquely determined by the data (u_0, f), when $f \in L^\infty(0, \infty; H)$ and $u_0 \in H$. While there are some exceptions, as we will see below, the weak solutions (of Class LH) of the 3D Navier-Stokes equations are not known to be uniquely determined by the data.

These facts have implications on the possible use of the Bubnov-Galerkin approximations in the study of weak solutions of the 3D Navier-Stokes equations. As seen in the proof of the Leray-Hopf Theorem 63.5, for every choice of data, there is a sequence of Bubnov-Galerkin approximations which converge to a corresponding weak solution of the Navier-Stokes equations. However, if the weak solution is not uniquely determined by the data, then it can happen that there exists weak solutions which are not the limit of a sequence of Bubnov-Galerkin approximations. In such a case, one cannot use the Bubnov-Galerkin method to derive properties (other than existence) of the weak solutions for the 3D Navier-Stokes equations.

For the 2D and the 3D problems, we will show below that if the data (u_0, f) satisfy $u_0 \in V$ and $f \in L^\infty(0,\infty; H)$, then there is a time T, with $0 < T \leq \infty$, such that the weak solution with this data is uniquely determined on any subinterval of $[0, T)$. (In this case, the solution happens to be a strong solution, see Section 6.4.) The uniqueness property in this case opens the possibility of using the Bubnov-Galerkin approximations to study the weak solution on this interval. We will take advantage of this feature below.

6.4. Strong Solutions.

In this section we turn to the theory of strong solutions of the Navier-Stokes evolutionary equation (61.10). As usual, we assume that Ω is an open, bounded domain in $\mathbb{R}^d$ of class C^2, where $d = 2$, or 3, and that $f \in L^\infty(0,\infty; H)$. A key feature that differentiates the theory of weak and strong solutions concerns the sense in which the solution satisfies the Navier-Stokes equations. We have seen in Lemma 63.1, that a weak solution satisfies the Navier-Stokes equations (almost everywhere) in the space V^{-1}. It will be a consequence of our definition that a strong solution will satisfy the Navier-Stokes equations (almost everywhere) in the space H.

The initial condition u_0 for a strong solution is required to be in the space V. Since a weak solution satisfies $u \in L^2_{\text{loc}}(0,\infty; V)$, it follows that one has $u(t_0) \in V$, for almost all $t_0 > 0$. In the existence theorem proved below, we will show that for every such t_0, there is a time T_0, with $0 < T_0 \leq \infty$, such that the restriction of the weak solution u to the interval $[t_0, t_0 + T_0)$ is a strong solution.

Let I denote a right-open interval in $\mathbb{R}^+$, i.e., either $I = [t_0, \hat{T})$, where $0 \leq t_0 < \hat{T} \leq \infty$, or $I = (\tau, \hat{T})$, where $0 \leq \tau < \hat{T} \leq \infty$ A function $u = u(t)$ is said to be a **strong solution** of the Navier-Stokes equations on the interval I, provided that for every $t_0 \in I$ the following hold.

(1) The function is the restriction of a weak solution of Class LH to the interval $[t_0, \hat{T})$.
(2) One has $u(t_0) \in V$.
(3) The solution satisfies

$$u \in L^\infty_{\text{loc}}[t_0, \hat{T}; V) \cap L^2_{\text{loc}}[t_0, \hat{T}; \mathcal{D}(A)). \tag{64.1}$$

For any interval I in $\mathbb{R}^+$ and any Banach space W, we let $L^p_{\text{loc}}(I; W)$ denote the collection of strongly measurable functions $u : I \to W$, such that for every compact set $K \subset I$, one has $\int_K \|u(t)\|^p_W \, dt < \infty$, where $1 \leq p < \infty$, see Appendix C. Note that this space is a Fréchet space whenever I is a right-open interval, as given above.

One sometimes refers to a strong solution of the Navier-Stokes equations as a **regular solution**. The term *regular* refers to the fact that the vorticity

$v = \nabla \times u$ is in the space $L^\infty_{\rm loc}[t_0, \hat{T}; L^2(\Omega))$. A strong solution $u(t)$ of (61.10), on an interval $[t_0, T_0)$, is said to be **maximally defined** if either $T_0 = \infty$, or $u(t)$ has no proper extension, as a strong solution, to an interval $[t_0, \hat{T})$, where $\hat{T} > T_0$.

The relation (64.1) seems like a modest beginning for the theory of strong solutions, and it is. However, many good features follow from it. In particular, there are several properties of strong solutions which follow directly from the definition. First of all, all the properties given in Lemma 63.1 are valid on the interval $[t_0, \infty)$, since a strong solution is also a weak solution of Class LH. Additional properties are given in the following lemma. Note that it follows from (61.21), (61.30), and (64.1) that if u is a strong solution on $[t_0, \hat{T})$, then one has

$$B(u,u) \in \begin{cases} L^4_{\rm loc}[t_0, \hat{T}; H) \cap L^{\frac{8}{5}}_{\rm loc}[t_0, \hat{T}; V), & \text{for } d = 2, \\ L^4_{\rm loc}[t_0, \hat{T}; H) \cap L^{\frac{8}{7}}_{\rm loc}[t_0, \hat{T}; V), & \text{for } d = 2 \text{ or } 3. \end{cases} \tag{64.2}$$

It should be noted that, since the nonlinear term B satisfies (61.32), this opens the possibility of using a boot strap method for studying the regularity of the strong solutions of the Navier-Stokes equations. We will have more to write on this issue later.

Lemma 64.1. *Let Ω be an open, bounded domain in $\mathbb{R}^d$ of class C^2, where $d = 2$ or 3, and let $f \in L^\infty(0, \infty; H)$. Let $u = u(t)$ be any strong solution of (61.10) on the interval $I = [t_0, \hat{T})$, or $I = (\tau, \hat{T})$. Then the following properties are valid.*

(1) *The properties given in Lemma 63.1 are valid on I.*
(2) *The time derivative $\partial_t u$ is in the space $L^2_{\rm loc}(I; H)$, and the solution u is in the space $C(I; V)$. Also, one has*

$$\partial_t u + \nu A u + B(u,u) \overset{\rm a.e.}{=} f, \qquad \textit{in the space } H. \tag{64.3}$$

(3) *One has*

$$\partial_t B(u,u) \in L^2_{\rm loc}(I; V^{-\frac{3}{2}}), \tag{64.4}$$

and $\partial_t B(u,u) = B(u, \partial_t u) + B(\partial_t u, u)$.
(4) *For any $\hat{t}_0 \in I$, the strong solution u is a mild solution of equation (61.10) in both the spaces H and V on $[\hat{t}_0, \hat{T})$, and one has*

$$u \in C^{0,\theta_0}_{\rm loc}[t_0, \hat{T}; H) \cap C^{0,\theta_1}_{\rm loc}(\hat{t}_0, \hat{T}; V)$$

and $B(u,u) \in C^{0,\theta_1}_{\rm loc}(\hat{t}_0, \hat{T}; V^{-\frac{1}{2}})$, for some $\theta_0 > 0$ and $\theta_1 > 0$.
(5) *Let $\{v_n\}$ be any sequence in $L^2_{\rm loc}(I, V)$ that satisfies the following two properties: (1) there is a function $a \in C(I; \mathbb{R}^+)$ such that*

$$\|A^{\frac{1}{2}} v_n(t)\|^2 \le a(t), \qquad \textit{for almost all } t \in I,$$

and (2) one has $v_n \overset{\rm s}{\to} u$ in $L^2_{\rm loc}(I; V)$. Then one obtains

$$\|A^{\frac{1}{2}} u(t)\|^2 \le a(t), \qquad \textit{for all } t \in I.$$

Proof. Item (1) is argued above. We will prove Items (2) - (5) for the case $I = [t_0, \hat{T})$. Since u is a strong solution, one has $u \in L^2_{\text{loc}}[I; \mathcal{D}(A))$, and consequently,

$$\int_{t_0}^{t} \langle A^{\frac{1}{2}} u(s), A^{\frac{1}{2}} v \rangle \, ds = \langle \int_{t_0}^{t} Au(s) \, ds, v \rangle, \qquad \text{for all } v \in V \text{ and } t_0 < t < \hat{T}.$$

Similarly, since $f \in L^\infty(0, \infty; H)$, one obtains

$$\int_{t_0}^{t} \langle f(s), v \rangle \, ds = \langle \int_{t_0}^{t} f(s) \, ds, v \rangle, \qquad \text{for all } v \in V.$$

Also from (64.2), one has $B(u, u) \in L^4_{\text{loc}}[I; H)$, which implies that

$$\int_{t_0}^{t} b(u(s), u(s), v) \, ds = \langle \int_{t_0}^{t} B(u(s), u(s)) \, ds, v \rangle, \qquad \text{for all } v \in V.$$

As a result, (63.9) implies that

$$\langle u(t) - u(t_0) - \int_{t_0}^{t} G(u) \, ds, v \rangle = 0, \qquad \text{for all } v \in V, \tag{64.5}$$

where $G(u) = -\nu Au - B(u, u) + f$. Since V is dense in H, it follows that (64.5) holds for all $v \in H$. Hence one has

$$u(t) - u(t_0) = \int_{t_0}^{t} G(u) \, ds, \qquad \text{for all } t \geq t_0 \geq 0. \tag{64.6}$$

By differentiating equation (64.6) with respect to t, one finds that (64.3) is valid, in the space H. Since the three terms Au, f, and $B(u, u)$ are in $L^2_{\text{loc}}[I; H)$, it follows from (64.3) that $\partial_t u \in L^2_{\text{loc}}[I; H)$. The fact that $u \in C(I; V)$ then follows from the Continuity Lemma 37.9.

In order to prove Item (3), we use the Vitali Convergence Theorem, see Appendix A.10, to assert that, since $\partial_t u \in L^2_{\text{loc}}[t_0, \hat{T}; H)$, the difference quotient $\delta_h u(t) = h^{-1}(u(t + h) - u(t))$, where $h \neq 0$, satisfies

$$\int_K \|\delta_h u - \partial_t u\|^2 ds \to 0, \qquad \text{as } h \to 0,$$

for every compact subinterval K in $I = [t_0, \hat{T})$. Since

$$\delta_h B(u(t), u(t)) = B(u(t + h), \delta_h u(t)) + B(\delta_h u(t), u(t)),$$

it follows from (61.27), the convergence result just proven, and the bilinearity of B, that

$$\int_K \|A^{-\frac{3}{4}}(\delta_h B(u, u) - B(u, \partial_t u) - B(\partial_t u, u))\|^2 \, ds \to 0, \qquad \text{as } h \to 0.$$

In order to prove Item (4) we set $\phi = f - B(u,u)$. It then follows from (64.2) that $\phi \in L^4_{\text{loc}}[I;H)$. By applying Theorem 42.12 (with $p = 4$ and $\alpha = 1$), we see that u is a mild solution of $\partial_t u + \nu Au = \phi$ in the space V^β, for each β with $\beta < \frac{3}{2}$. Consequently, (42.29), with $\sigma = 0$, implies that $u \in C^{0,\theta_0}_{\text{loc}}[I;H)$, where $\theta_0 > 0$. By using (42.31), with $\sigma = 1$, one concludes that $u \in C^{0,\theta_1}_{\text{loc}}(t_0,\hat{T};V)$, for some $\theta_1 > 0$.

Next we note that $B(u,u) \in C^{0,\theta_1}_{\text{loc}}(t_0,\hat{T};V^{-\frac{1}{2}})$, for some $\theta_1 > 0$, owing to the Hölder continuity of u and the fact that (61.20) implies

$$\begin{aligned}&\|A^{-\frac{1}{4}}(B(u(t+h),u(t+h)) - B(u(t),u(t)))\| \\ &\qquad \le C_1\|A^{\frac{1}{2}}(u(t+h)-u(t))\|\,(\|A^{\frac{1}{2}}u(t+h)\| + \|A^{\frac{1}{2}}u(t)\|).\end{aligned}$$

From Item (2) one has $u \in C[I;V)$. The proof of Item (5) now follows by using the argument for Item (8) in Lemma 63.1. The reader should check the details of the proof for the case where $I = (\tau,\hat{T})$ is an open interval in $\mathbb{R}^+$. □

It will be shown shortly that the strong solutions of the Navier-Stokes equations are uniquely determined by the data (u_0, f). (For the 2D case, see Corollary 63.6.) We will let $S(f,t)u_0$ denote the maximally defined strong solution u of (61.10) with $u(0) = u_0$, where $u_0 \in V$, and we let $[0,T)$ denote the time interval of definition of this solution, where $T = T(u_0, f)$ satisfies $0 < T \le \infty$.

This brings us to an interesting and profound feature of the Navier-Stokes equations. In the theory of strong solutions presented below, we will exploit a Gronwall (differential) inequality. This inequality depends on the space dimension d. In particular, we will show that the function $r(t) = \|A^{\frac{1}{2}}u(t)\|^2$ satisfies

$$\text{(64.7)} \qquad \frac{dr}{dt} \le a + br^d, \qquad \text{on an interval,}$$

where a and b are appropriate positive constants and $d = 2$ or 3. This nonlinear differential inequality leads to an estimate of the form $r(t) \le s(t)$, where $s(t)$ is a solution of the differential equation

$$\text{(64.8)} \qquad \frac{ds}{dt} = a + bs^d, \qquad s(0) = r(0).$$

It happens that for $d \ge 2$, the solution $s(t)$ of equation (64.8) always blows up in finite time when $s(0) \ge 0$. While it does not follow from this fact that the function $r(t)$ blows up, one begins to wonder! However, we do have additional information about the solutions. In particular, since every strong solution is a weak solution of Class LH, we know that

$$\int_{t_0}^{t_1} r(t)\,dt = \int_{t_0}^{t_1} \|A^{\frac{1}{2}}u(t)\|^2\,dt < \infty, \qquad \text{for } 0 \le t_0 < t_1 < \infty.$$

As a consequence, in the 2D case one can use a Uniform Gronwall Inequality, Lemma D.3, to find an upper bound for $r(t)$ for all $t \geq 0$. It is the use of the Uniform Gronwall Inequality that leads to a better theory in 2D. As we explain in Section 6.4.3, the Uniform Gronwall Inequality does not appear to be applicable in the 3D theory.

The construction of strong solutions for the Navier-Stokes equations is based on the Bubnov-Galerkin method. In this case we will use the Compactness I Lemma 63.2, with $\alpha = 1$, which differs from the theory of weak solutions presented in Section 6.3. We begin the analysis for strong solutions with the sequence of Bubnov-Galerkin approximations u_n used in the proof of the Leray-Hopf Theorem 63.5. One then has $u_n \overset{s}{\to} u$ in $L^2_{\text{loc}}[0,\infty;H)$, where u is a weak solution (of Class LH) satisfying $u(0) = u_0 \in H$. We will then use the lemmas in Section 6.2 to argue that u_n satisfies the hypotheses of Compactness I for $\alpha = 1$, when $u_0 \in V$. It then follows from this lemma that there is a subsequence, which we will relabel as u_n, such that $u_n \overset{s}{\to} u$ in $L^2_{\text{loc}}[0,T;V)$ and $u_n \overset{w}{\to} u$ in $L^2_{\text{loc}}[0,T;V^2)$, for a suitable T with $0 < T \leq \infty$. Now for the details.

6.4.1 Two-Dimensional Theory. The first result, which is an existence theorem for strong solutions in 2D, implies that every weak solution $u(t)$ immediately becomes a strong solution at time $t = t_0$, for any $t_0 > 0$. We also derive several estimates on various norms of $u(t)$. These estimates will be important for describing the longtime dynamics of the 2D Navier-Stokes equations. The next theorem relies heavily on Lemma 62.3.

Theorem 64.2. *Let Ω be an open, bounded domain in $\mathbb{R}^2$ of class C^2 and let $f \in L^\infty(0,\infty;H)$. Let $u = u(t)$ be any weak solution of the 2D Navier-Stokes evolutionary equation (61.10) on $[0,\infty)$ with initial condition $u(0) = u_0 \in H$. Then for each $t_0 > 0$, $v(t) = u(t+t_0)$ is a strong solution of (61.10) on $[0,\infty)$ with initial condition $v(0) = u(t_0)$, and $\partial_t u \in L^2_{\text{loc}}(0,\infty;H)$. Furthermore, for $i = 2,3$ there exists constants $M_i = M_i(\|u_0\|, \|f\|_\infty)$ and $L_i = L_i(\|f\|_\infty)$, where L_i is independent of u_0, and there exist $K_j = K_j(\|u_0\|, \|f\|_\infty)$, for $j = 0,1,2,3$, such that*

$$(64.9) \qquad \|A^{\frac{1}{2}}u(t)\|^2 \leq \begin{cases} K_0 + K_1 t^{-1}, & \text{for } 0 < t \leq 1, \\ M_2 e^{-\nu\lambda_1 t} + L_2, & \text{for } t \geq 1. \end{cases}$$

Also one has

$$(64.10) \qquad \int_{t_0}^{t} \|Au(s)\|^2 ds \leq K_2 + K_3 t_0^{-1}, \qquad \text{for } 0 < t_0 \leq t \leq 1,$$

and

$$(64.11) \qquad \int_{t-1}^{t} \|Au(s)\|^2\, ds \leq M_3 e^{-\nu\lambda_1 t} + L_3, \qquad \text{for } t \geq 1.$$

In addition, if $u_0 \in V$*, then there is a* $K_4 = K_4(\|A^{\frac{1}{2}}u_0\|, \|f\|_\infty)$*, such that*

$$\|A^{\frac{1}{2}}u(t)\|^2 \leq M_2(1+\|A^{\frac{1}{2}}u_0\|^2)e^{-\nu\lambda_1 t} + L_2, \qquad \text{for } t \geq 0, \tag{64.12}$$

and

$$\int_0^t \|Au(s)\|^2\, ds \leq \nu^{-1}(\|A^{\frac{1}{2}}u_0\|^2 + tK_4), \qquad \text{for } t \geq 0. \tag{64.13}$$

Proof. Let $u = \lim u_n$ be any weak solution of (61.10) as given in Lemma 63.4 and Theorem 63.5. It follows from Lemma 62.3 that the hypotheses of the Compactness I Lemma 63.2 are satisfied, for $u_0 \in H$ and $\alpha = 1$. Therefore, there exists a subsequence of u_n, which we relabel as u_n, such that $u_n \overset{s}{\to} u$ in $L^2_{\mathrm{loc}}(0,\infty;V)$, $u_n \overset{w}{\to} u$ in $L^2_{\mathrm{loc}}(0,\infty;\mathcal{D}(A))$, and $\partial_t u_n \overset{w}{\to} \partial_t u$ in $L^2_{\mathrm{loc}}(0,\infty;H)$. Moreover, if $u_0 \in V$, then these three limits are valid in $L^2_{\mathrm{loc}}[0,\infty;W)$, for the corresponding choices of W. From Lemma 64.1, Item (5), we see that (62.12) holds in the limit, where $I = (0,\infty)$. Consequently, one has $u(t) \in \mathcal{D}(A^{\frac{1}{2}})$, for all $t > 0$, and (64.9) holds. Finally, inequalities (62.13) - (62.16) hold in the limit, which implies that the four inequalities (64.10) - (64.13) are valid. □

6.4.2 Three-Dimensional Theory. Like the 2D theory, the theory of strong solutions for the 3D Navier-Stokes equations begins with the Bubnov-Galerkin system (62.2). However, instead of starting with the 2D auxiliary estimate (61.23), one uses the 3D version (61.24), which leads to inequality (62.17), see Section 6.8. We now have the following result.

Lemma 64.3. *Let* Ω *be an open, bounded domain in* $\mathbb{R}^3$ *of class* C^2*. For each* $b > 0$ *and* $B > 0$*, there is a time* $T_1 = T_1(b,B) > 0$*, such that if* $f \in L^\infty(0,\infty;H)$*, with* $\|f\|_\infty \leq b$*, and* $u_0 \in V$*, with* $\|A^{\frac{1}{2}}u_0\| \leq B$*, then for any* $t_0 \geq 0$*, there is a strong solution* $u = u(t)$ *of (61.10) on* $[t_0, t_0+T_1)$ *that satisfies* $u(t_0) = u_0$*. Moreover, this strong solution* $u(t)$ *satisfies*

$$\|A^{\frac{1}{2}}u(t)\|^2 \leq \rho(t)^2, \qquad t_0 \leq t < t_0 + T_1, \tag{64.14}$$

where $\rho(t)^2$ *is given in the proof. In addition, one has*
(64.15)

$$\int_{t_1}^t \|Au(s)\|^2\, ds \leq \frac{1}{\nu}\|A^{\frac{1}{2}}u(t_1)\|^2 + \frac{2(t-t_1)}{\nu^2}\|f\|_\infty^2 + \frac{27C_3^4}{2\nu^2}\int_{t_1}^t \rho(s)^6\, ds,$$

for $t_0 \leq t_1 \leq t < t_0 + T_1$*, where* C_3 *is given in inequality (62.17).*

Proof. In order to show the existence of a strong solution on the interval $t_0 \leq t < t_0 + T_1$, we will use Lemma 62.4, which asserts that the Bubnov-Galerkin approximations satisfy (62.17), which yields

$$\partial_t\|A^{\frac{1}{2}}p\|^2 \leq \frac{2}{\nu}\|f\|_\infty^2 + \frac{27C_3^4}{2\nu^3}\|A^{\frac{1}{2}}p\|^6, \qquad \text{for } t \geq t_0 \geq 0.$$

It then follows that there is a $T_1 > 0$ such that inequality (62.18) is valid, for $t_0 \le t < t_0+T_1$, where $\rho(t)^2 = s(r_0, a, b, t)$ is the solution of the equation $\frac{ds}{dt} = a + bs^3$, $s(t_0) = r_0$, $\|A^{\frac{1}{2}}p(t_0)\|^2 \le \|A^{\frac{1}{2}}u_0\|^2 \le r_0$, $a = 2\nu^{-1}\|f\|_\infty^2$, and $b = 27C_3^4(2\nu^3)^{-1}$ (see Lemma D.5). It follows from Lemma 62.4 that the hypotheses of the Compactness I Lemma 63.2 are satisfied, for $u_0 \in V$ and $\alpha = 1$. Therefore, there exists a subsequence of u_n, which we relabel as u_n, such that $u_n \overset{s}{\to} u$ in $L^2_{\text{loc}}[t_0, t_0 + T_1; V)$, $u_n \overset{w}{\to} u$ in $L^2_{\text{loc}}[t_0, t_0 + T_1; \mathcal{D}(A))$, and $\partial_t u_n \overset{w}{\to} \partial_t u$ in $L^{\frac{4}{3}}_{\text{loc}}[t_0, t_0 + T_1; H)$. Let u and g denote the limit functions given by Lemma 63.4 and Theorem 63.5. Since the Bubnov-Galerkin approximants satisfy inequality (64.14), it follows from Lemma 64.1, Item (5), that the strong solution $u = u(t)$ satisfies (64.14), as well. □

The main result for the existence of strong solutions for the 3D Navier-Stokes equations is the following.

Theorem 64.4. *Let Ω be an open, bounded domain in $\mathbb{R}^3$ of class C^2 and let $f \in L^\infty(0, \infty; H)$. Then for every $u_0 \in V$ there is a time $T = T(u_0, f)$ with $0 < T \le \infty$ and such that the following hold:*

(1) *There is precisely one strong solution $u(t)$ of (61.10) on $[0, T_0)$ with $u(0) = u_0$, where T_0 is given by Lemma 62.4. Furthermore, this solution is the limit of a subsequence of Bubnov-Galerkin approximations.*
(2) *There is a maximally defined, strong solution $u(t)$ of (61.10) on $[0, T)$, where $T_0 \le T \le \infty$, and $u(t)$ is uniquely determined on any subinterval of $[0, T)$.*
(3) *One has $u(t) \in \mathcal{D}(A^{\frac{1}{2}})$, for $0 \le t < T$, and $\|A^{\frac{1}{2}}u(t)\|^2 \le \rho(t)^2$, for $0 \le t < T_0$, where T_0 and $\rho(t)$ are given by Lemma 62.4.*
(4) *Moreover, for $0 \le t_1 \le t < T_0$, one has*

$$\int_{t_1}^{t} \|Au(s)\|^2 \, ds \le \frac{1}{\nu}\|A^{\frac{1}{2}}u(t_1)\|^2 + \frac{2(t-t_1)}{\nu^2}\|f\|_\infty^2 + \frac{27C_3^4}{2\nu^2}\int_{t_1}^{t} \rho(s)^6 \, ds.$$

(5) *If $T < \infty$, then one has*

$$\lim_{t \to T^-} \|A^{\frac{1}{2}}u(t)\|^2 = \infty. \tag{64.16}$$

Proof. Items (3) and (4) follow from Lemma 64.3. The argument used to prove the uniqueness of the stong solution is similar to that used in Corollary 63.6 for the 2D theory. For the 3D argument one uses the fact that $\partial_t u \in L^2_{\text{loc}}[0, T_0; H)$. Let u_i, $i = 1, 2$, be two strong solutions of (61.10) with $u_i(0) = u_0 \in V$, for $i = 1, 2$. Let $\tau > 0$ be given by Lemma 64.3, so that both u_1 and u_2 are strong solutions on the $[0, \tau)$, and set $w = u_1 - u_2$. It then follows from Lemma 64.3 that

$$w \in L^\infty_{\text{loc}}[0, \tau; V) \cap L^2_{\text{loc}}[0, \tau; \mathcal{D}(A)), \text{ and } \partial_t w \in L^2_{\text{loc}}[0, \tau; H).$$

Furthermore w satisfies the differential equation

$$\langle\langle \partial_t w, v\rangle\rangle + \nu\langle A^{\frac{1}{2}}w, A^{\frac{1}{2}}v\rangle + b(w, u_2, v) + b(u_1, w, v) \overset{\text{a.e.}}{=} 0,$$

for each $v \in V$. From the Continuity Lemma 37.9, one has

$$\frac{1}{2}\frac{d}{dt}||w||^2 + \nu||A^{\frac{1}{2}}w||^2 + b(w, u_2, w) \overset{\text{a.e.}}{=} 0.$$

Now (61.24) and the Young inequality imply that

$$\begin{aligned}\frac{1}{2}\partial_t||w||^2 + \nu||A^{\frac{1}{2}}w||^2 &\le |b(w, u_2, w)| \le C_3||A^{\frac{1}{2}}u_2||^{\frac{1}{2}}||Au_2||^{\frac{1}{2}}||A^{\frac{1}{2}}w||\,||w|| \\ &\le \nu||A^{\frac{1}{2}}w||^2 + \frac{C_3^2}{4\nu}||A^{\frac{1}{2}}u_2||\,||Au_2||\,||w||^2,\end{aligned}$$

almost everywhere. The Gronwall inequality, Lemma D.1, then implies that, for all $t \in [0, \tau)$, one has

$$||w(t)||^2 \le ||w(0)||^2 \exp\left(\frac{C_3^2}{2\nu}\int_0^t ||A^{\frac{1}{2}}u_2||\,||Au_2||\,ds\right). \tag{64.17}$$

Since $u_2 \in L^\infty_{\text{loc}}[0, \tau; V) \cap L^2_{\text{loc}}[0, \tau; \mathcal{D}(A))$, it follows from the Schwarz inequality that the integral in (64.17) is finite. Since $w(0) = 0$, we obtain $w(t) = 0$, for all $t \in [0, T)$. Thus completes the proof of Items (1) and (2).

We will prove (64.16) by contradiction. Assume that

$$\liminf_{t\to T^-} ||A^{\frac{1}{2}}u(t)||^2 < \infty.$$

Then there is a sequence $t_n \to T^-$ and a number $r_0 > 0$ such that $||A^{\frac{1}{2}}u(t_n)||^2 \le r_0$. Without any loss of generality we can assume that t_n satisfies $T - t_n < T_0$, where T_0 is given by Lemma 64.3. Let $v(t)$ be the strong solution of (61.10) on the interval $[t_n, t_n + T_0)$ that satisfies $v(t_n) = u(t_n)$. By uniqueness, one has $v(t) = u(t)$, for $t_n \le t < T$. The function $w(t)$ defined by $w(t) = u(t)$, for $0 \le t < T$, and $w(t) = v(t)$, for $T \le t < t_n + T_0$, is a strong solution. It is also a proper extension of $u(t)$, which contradicts the fact that $u(t)$ is a maximally defined strong solution. □

6.4.3 The Global Regularity Problem. It is useful to compare and contrast the conclusions and arguments of the last theorem for the 3D Navier-Stokes equations with those of Theorem 64.2 for the 2D problem. While in the case of the 2D Navier-Stokes equations one has strong solutions defined for all time t, $0 < t < \infty$, together with good information on the $A^{\frac{1}{2}}$-norm and the A-norm (as stated in Theorem 64.2), a similar theory of strong solutions for the 3D problem is valid only on an interval $[0, T)$, where T is positive, but *it may be finite*! The reason behind this is due to the differences in the basic inequality (62.11), arising in the 2D theory, and the corresponding inequality (62.17), occurring in the 3D theory. In the 2D case, one is able to obtain results which are better than the corresponding results in the 3D theory, because one is able to use the Uniform Gronwall Inequality to exploit (62.11). Basically this allows one to reduce the nonlinear problem (62.11) to a linear problem. (See the proof of Lemma 62.3 in Section 6.8.) Without further assumptions, the same method does not work for the 3D theory because of the higher order nonlinearity $||A^{\frac{1}{2}}p||^6$ appearing in (62.17).

In both the 2D and 3D theories, the Bubnov-Galerkin approximations $u_n(t) = p(t)$ satisfy $p(t) \in \mathcal{D}(A^{\frac{1}{2}})$ and $p(t) \in \mathcal{D}(A)$, for all $t \geq 0$, because the linear subspace P_nH lies in $\mathcal{D}(A) \subset \mathcal{D}(A^{\frac{1}{2}})$, for all $n \geq 1$. The main point of inequality (62.18) used in the 3D argument is that the bound of $||A^{\frac{1}{2}}p(t)||^2$ given by $\rho(t)^2$ is independent of n only on a finite time interval $[t_0, t_0 + T_0)$. What this suggests is that a given weak solution $u(t)$ may be a strong solution on a finite interval $[t_0, T_1)$ and then loose its regularity in the sense that (64.16) holds. In this connection we will say that a maximally defined, strong solution $u(t)$ of (61.10) on the interval $[t_0, T)$ is **globally regular** if $T = \infty$, i.e., one has

$$u \in L^{\infty}_{\text{loc}}[t_0, \infty; V) \cap L^2_{\text{loc}}[t_0, \infty; \mathcal{D}(A)).$$

It follows from the theory of weak solutions that $u \in L^2_{\text{loc}}[0, \infty; V)$. The lack of global regularity for a given maximally defined, strong solution $u(t)$ means that one may have $u(0) \in V$, while $u(t) \in \{v \in H : v \notin V\}$, for some $t > 0$.

We define the **globally regular set** GR to be the collection of all data (u_0, f), where $u_0 \in V$ and $f \in L^{\infty}(0, \infty; H)$, with the property that the maximally defined strong solution of (61.10) satisfying $u(0) = u_0$ is a strong solution on the infinte interval $[0, \infty)$. This leads us to THE major unresolved issue concerning the Navier-Stokes equations, namely,

> *Is every maximally defined, strong solution of the 3D Navier-Stokes equations globally regular?*

In terms of the globally regular set GR, the last question can be reformulated as the **Global Regularity Problem**, which reads,

> *Does one always have* $GR = V \times L^{\infty}(0, \infty; H)$?

A related problem is: Fix $f \in L^{\infty}(0, \infty; H)$ and ask whether one has $V \times \{f\} \subset GR$. As we have seen, the answer to all these questions, in the case of the 2D Navier-Stokes equations, is Yes.

What is known about the Global Regularity Problem in 3D? First of all, let us note what inequality (64.14) in Lemma 64.3 does <u>not</u> say. While we do have $||A^{\frac{1}{2}}u(t)||^2 \leq \rho(t)^2$ on a finite interval $[t_0, t_0 + T_0)$, and while the term on the right side of (64.14) does blow up in finite time, this does not say that $\|A^{\frac{1}{2}}u(t)\|^2$ blows up in finite time. To the best of our knowledge, there does not exist any example[27] of the 3D Navier-Stokes equations on a bounded domain in $\mathbb{R}^3$ in which there is a strong solution that blows up in finite time. There is no known counterexample to a positive solution of the Global Regularity Problem. On the other hand, there is no known example of a system of Navier-Stokes equations in 3D and a datum $f \in L^\infty(0,\infty;H)$ for which one has $V \times \{f\} \subset GR$. In short, no counterexamples; no theorems. The Global Regularity Problem is completely open. However, the recent result in Nečas, Ružička, and Šverak (1996) is very interesting and relevant.

Next we will show that the 3D Navier-Stokes equations does have globally regular solutions provided the data (u_0, f) are small. In particular, we now show that GR contains an small open neighborhood of the origin in $V \times L^\infty(0,\infty;H)$.

Corollary 64.5. *Let Ω be an open, bounded domain in $\mathbb{R}^3$ of class C^2. If the data satisfy $(u_0, f) \in V \times L^\infty(0,\infty;H)$ with*

$$\|A^{\frac{1}{2}}u_0\|^2 \le B_0 \qquad \text{and} \qquad \|f\|_\infty^2 \le \frac{\nu^2\lambda_1}{4}B_0, \tag{64.18}$$

where $B_0 = 8^{-\frac{1}{2}}\nu^{\frac{3}{2}}\lambda_1^{\frac{1}{2}}(3\sqrt{3}C_3^2)^{-1}$, then the strong solution of (61.10) with $u(0) = u_0$ is globally regular, i.e., $T = T(u_0, f) = \infty$. In other words, the globally regular set GR contains a small neighborhood of the origin in $V \times L^\infty(0,\infty;H)$.

Proof. Let the data (u_0, f) satisfy (64.18) and define τ by

$$\tau \stackrel{\text{def}}{=} \sup\{t : 0 \le t < T \text{ and } \|A^{\frac{1}{2}}u(s)\|^2 \le 2B_0 \text{ for all } 0 \le s \le t\}.$$

If $T < \infty$, then (64.16) is valid, and by Lemma 64.1, Item (3), one has $0 < \tau < T$, with $\|A^{\frac{1}{2}}u(\tau)\|^2 = 2B_0$. Also, from the definition of B_0, one has

$$\frac{27C_3^4}{\nu^2}\|A^{\frac{1}{2}}u(t)\|^4 \le \frac{\nu\lambda_1}{2}, \qquad \text{for } 0 \le t \le \tau. \tag{64.19}$$

Now Theorem 64.4 implies that $\partial_t u \in L^2_{\text{loc}}[0,T;H)$. Let $w \stackrel{\text{def}}{=} A^{\frac{1}{2}}u$. Then one has $\partial_t w \in L^2_{\text{loc}}[0,T;V^{-1})$. It then follows from Continuity Lemma 37.9 that $((\partial_t w, w)) = \frac{1}{2}\frac{d}{dt}\|A^{\frac{1}{2}}u\|^2$. The argument used in Corollary 63.6, for the 2D problem, applies to the strong solution u on $[0,T)$. Instead of using (61.23), one uses (61.24) to obtain

$$\frac{1}{2}\partial_t\|A^{\frac{1}{2}}u\|^2 + \nu\|Au\|^2 \le \frac{\nu}{2}\|Au\|^2 + \frac{1}{\nu}\|f\|_\infty^2 + \frac{27C_3^4}{2\nu^3}\|A^{\frac{1}{2}}u\|^6. \tag{64.20}$$

[27] As usual, we assume that the viscosity ν is fixed, finite, and positive, but otherwise arbitrary.

Since $\lambda_1\|A^{\frac{1}{2}}u\|^2 \le \|Au\|^2$, we obtain

$$\partial_t\|A^{\frac{1}{2}}u\|^2 + \left(\nu\lambda_1 - \frac{27C_3^4}{\nu^3}\|A^{\frac{1}{2}}u\|^4\right)\|A^{\frac{1}{2}}u\|^2 \le \frac{2}{\nu}\|f\|_\infty^2.$$

By using (64.18), (64.19), and (64.20), together with the Gronwall inequality we obtain

$$\|A^{\frac{1}{2}}u(\tau)\|^2 \le \|A^{\frac{1}{2}}u_0\|^2 \exp\left(-\frac{\nu\lambda_1}{2}\tau\right) + \frac{4}{\nu^2\lambda_1}\|f\|_\infty^2 < 2B_0,$$

which is a contradiction. Hence $T(u_0, f) = \infty$. □

6.4.4. The Linearized Equation. In this section we want to study solutions of the linear time-varying equation

$$\partial_t v + \nu A v + B(v, u_1) + B(u_2, v) = g \quad \text{and} \quad \nabla \cdot v = 0, \tag{64.21}$$

where A is the Stokes operator and B is our friend $B(u,v) = \mathbb{P}((u \cdot \nabla)v)$. For equation (64.21) we treat u_i, $i = 1, 2$, and g as known functions defined for t in an interval $0 \le t < T$, where $0 < T \le \infty$, and we seek a solution $v = v(t)$ of equation (64.21) on the interval $[0, T)$, subject to the Initial Value Problem

$$v(0) = v_0, \qquad \text{where } v_0 \in H. \tag{64.22}$$

We assume here that the functions u_1 and u_2 satisfy

$$u_1, u_2 \in L^\infty_{\text{loc}}[0, T; V) \cap L^2_{\text{loc}}[0, T; \mathcal{D}(A)). \tag{64.23}$$

Note that property condition (64.23) is shared by all strong solutions of the Navier-Stokes equations on some interval $[0, T)$, which depends on the data of the problem. Furthermore, we will require that the forcing term g satisfy $g \in L^\infty(0, \infty; H)$. We define $\rho = \rho(t)$ and $\sigma = \sigma(t)$ by

$$\begin{aligned} \rho = \rho(t) &\stackrel{\text{def}}{=} \max\left(\|A^{\frac{1}{2}}u_1\|^2_{\infty;[0,t]}, \|A^{\frac{1}{2}}u_2\|^2_{\infty;[0,t]}\right), \\ \sigma = \sigma(t) &\stackrel{\text{def}}{=} \max\left(\int_0^t \|Au_1\|^2\, ds, \|\int_0^t \|Au_2\|^2\, ds\right), \end{aligned} \tag{64.24}$$

for $0 \le t < T$. Note that from (64.23), one has $\rho \in L^\infty_{\text{loc}}[0, T; \mathbb{R})$. Also, one has $\int_0^t \rho(s)^2\, ds \le t\rho(t)^2$, for $t \ge 0$, since ρ is monotone nondecreasing.

We will say that a function $v = v(t)$ is a **weak solution** (of Class LH) of (64.21) on $[0, T)$ provided that $v(0) = v_0 \in H$; one has

$$v \in L^\infty_{\text{loc}}[0, T; H) \cap L^2_{\text{loc}}[0, T; V); \tag{64.25}$$

the function v has a time-derivative $\partial_t v$ in the space $L^p_{\rm loc}[0,T;V^{-1})$, so that

$$v(t) - v(t_0) = \int_{t_0}^{t} \partial_t v\, ds, \qquad \text{for } 0 \le t < T,$$

for some p with $1 \le p < \infty$; and the function v satisfies

$$\begin{aligned}\langle v(t) - v(t_0), w\rangle + \nu \int_{t_0}^{t} \langle A^{\frac{1}{2}} v, A^{\frac{1}{2}} w\rangle\, ds + \int_{t_0}^{t} b(v, u_1, w)\, ds \\ + \int_{t_0}^{t} b(u_2, v, w)\, ds = \int_{t_0}^{t} \langle g, w\rangle\, ds,\end{aligned} \tag{64.26}$$

for all $w \in V$, and for all t and t_0 with $0 \le t_0 \le t < T$.

Lemma 64.6. *Let Ω be an open, bounded domain in $\mathbb{R}^d$ of class C^2, where $d = 2$ or 3, and let $g \in L^\infty(0,\infty;V^{-1})$. Let u_1 and u_2 satisfy (64.23) on an interval $[0,T)$, where $0 < T \le \infty$. Then, for every $v_0 \in H$, the following are valid:*

(1) *There is a unique weak solution $v = v(t)$ of equation (64.21) in the space H, on the interval $[0,T)$.*
(2) *One has*

$$\partial_t v \in L^2_{\rm loc}[0,T;V^{-1}) \qquad \text{and} \qquad v \in C[0,T;H). \tag{64.27}$$

(3) *There exists a function $\rho_4 = \rho_4(t)$, which depends on u_1 and u_2, and there exist constants D_1 and D_2, which do not depend on the data (v_0, g, u_1, u_2), such that, for $0 \le t < T$, this solution satisfies*

$$\|v(t)\|^2 + \int_0^t \|A^{\frac{1}{2}} v\|^2\, ds \le \rho_4(t)\left(D_1 \|v_0\|^2 + D_2\, t \|A^{-\frac{1}{2}} g\|^2_\infty\right). \tag{64.28}$$

Proof. The first step in the proof is to use the Bubnov-Galerkin method and the compactness lemmas in Section 6.3 to find a solution $v = v(t)$ of the equation

$$Jv \stackrel{\rm def}{=} \partial_t v + \nu A v = \phi, \qquad \text{where } \phi = g - B(v, u_1) - B(u_2, v), \tag{64.29}$$

in the space $V^{-\frac{1}{2}}$. The second step, which uses a bootstrap argument, is to show that this solution v has the properties described in the lemma. In the course of the proof, we will use various positive constants $c_0, c_1, \cdots$, which arise in the argument. It is important to note that these constants are independent of the data (v_0, g, u_1, u_2) and the rank n of the spectral projections $P = P_n$ associated with the Bubnov-Galerkin method.

From inequality (61.27) and the interpolation inequality (61.9), one obtains

$$\|A^{-\frac{3}{4}} B(u,v)\|^2,\ \|A^{-\frac{3}{4}} B(v,u)\|^2 \le c_1 \|A^{-\frac{1}{4}} v\|\, \|A^{\frac{1}{4}} v\|\, \|A^{\frac{1}{2}} u\|^2, \tag{64.30}$$

for all $v \in V^{\frac{1}{2}}$ and $u \in V$. By applying the spectral projection $P = P_n$ to equation (64.21), one obtains the Bubnov-Galerkin approximation equation

$$\partial_t v^n + \nu A v^n = P\phi^n = P\,Jv^n, \tag{64.31}$$

where $v^n = Pv = Pv(t)$ and $v^n(0) = Pv_0$. Since P is both an orthogonal projection and a spectral projection, it commutes with fractional powers of A, and one has

$$\|A^{-r}v^n(0)\|^2 \leq \lambda_1^{-2s}\|A^{s-r}v_0\|^2, \qquad \text{for } 0 \leq s \leq r. \tag{64.32}$$

Next we note that inequalities (64.24), (64.30), and (64.32), with the Young inequality, imply that

$$\begin{aligned} \|A^{-\frac{3}{4}}Jv^n\|^2 &\leq 3\lambda_1^{-\frac{1}{2}}\|A^{-\frac{1}{2}}g\|_\infty^2 + 6c_1\rho\|A^{-\frac{1}{4}}v^n\|\,\|A^{\frac{1}{4}}v^n\| \\ &\leq \frac{\nu^2}{4}\|A^{\frac{1}{4}}v^n\|^2 + c_3\rho^2\|A^{-\frac{1}{4}}v^n\|^2 + 3\lambda_1^{-\frac{1}{2}}\|A^{-\frac{1}{2}}g\|_\infty^2. \end{aligned} \tag{64.33}$$

By taking the inner product of equation (64.31) with $A^{-\frac{1}{2}}v^n$, one obtains

$$\begin{aligned} \frac{1}{2}\partial_t\|A^{-\frac{1}{4}}v^n\|^2 + \nu\|A^{\frac{1}{4}}v^n\|^2 &\leq |(A^{-\frac{3}{4}}Jv^n, A^{\frac{1}{4}}v^n)| \\ &\leq \frac{\nu}{4}\|A^{\frac{1}{4}}v^n\|^2 + \frac{1}{\nu}\|A^{-\frac{3}{4}}Jv^n\|^2. \end{aligned}$$

The last two inequalities then imply that

$$\partial_t\|A^{-\frac{1}{4}}v^n\|^2 + \nu\|A^{\frac{1}{4}}v^n\|^2 \leq c_0\rho^2\|A^{-\frac{1}{4}}v^n\|^2 + c_2\|A^{-\frac{1}{2}}g\|_\infty^2. \tag{64.34}$$

From (64.32) and (64.34), the Gronwall inequality implies that

$$\|A^{-\frac{1}{4}}v^n(t)\|^2 \leq \rho_1(t)(\lambda_1^{-\frac{1}{2}}\|v_0\|^2 + c_2t\|A^{-\frac{1}{2}}g\|_\infty^2), \tag{64.35}$$

where $\rho_1 = \rho_1(t) = e^{c_0t\rho(t)^2}$, for $t \geq 0$. By integrating (64.34) and using (64.35), we find that

$$\nu\int_0^t \|A^{\frac{1}{4}}v^n\|^2\,ds \leq \rho_2(t)\left(\lambda_1^{-\frac{1}{2}}\|v_0\|^2 + c_2t\|A^{-\frac{1}{2}}g\|_\infty^2\right), \tag{64.36}$$

for $t \geq 0$, where $\rho_2 = 1 + c_0t\rho^2\rho_1$. This shows that the sequence v^n forms a bounded set in $L^2_{\text{loc}}[0,T;V^{\frac{1}{2}})$. It follows from (64.33) and (64.36) that Jv^n and Av^n are in $L^2_{\text{loc}}[0,T;V^{-\frac{3}{2}})$. Also, (64.31) and (64.36) imply that the sequence $\partial_t v^n = -\nu Av^n + Jv^n$ forms a bounded set in $L^2_{\text{loc}}[0,T;V^{-\frac{3}{2}})$.

It then follows from the Compactness Lemmas (see Section 6.3), that there is a subsequence of v^n, which we relabel as v^n, and there is a function v such that, $v^n \overset{s}{\to} v$ in $L^2_{\text{loc}}[0,T;V^{-\frac{1}{2}})$, $v^n \overset{w}{\to} v$ in $L^2_{\text{loc}}[0,T;V^{\frac{1}{2}})$, and $\partial_t v^n \overset{w}{\to}$

$\partial_t v$ in $L^2_{\text{loc}}[0,T;V^{-\frac{3}{2}})$. Furthermore, inequalities (64.35) and (64.36) hold in the limit. Thus for $t \geq 0$, one has

$$\nu \int_0^t \|A^{\frac{1}{4}}v\|^2\, ds \leq \rho_2(t) \left(\lambda_1^{-\frac{1}{2}}\|v_0\|^2 + c_2 t \|A^{-\frac{1}{2}}g\|^2_\infty\right). \tag{64.37}$$

Since inequality (64.35) holds in the limit, one has $v \in L^\infty_{\text{loc}}[0,T;V^{-\frac{1}{2}})$. Consequently, $\partial_t v \in L^2_{\text{loc}}[0,T;V^{-\frac{3}{2}})$, and one has $v \in C[0,T;V^{-\frac{1}{2}})$ by the Continuity Lemma 37.9.

The proof of the uniqueness of v, which also uses some facets of the Continuity Lemma, is similar to the argument of the uniqueness of weak solutions of the 2D Navier-Stokes equations (see Corollary 63.6). Therefore, we will omit the details.

Now we can employ a bootstrap argument to show that v has greater regularity. Since $v \in L^2_{\text{loc}}[0,T;V^{\frac{1}{2}})$, it follows from (61.26) and (64.23) that both $B(v,u_1)$ and $B(u_2,v)$, and therefore ϕ, belong to $L^2_{\text{loc}}[0,T;V^{-1})$. Applying Theorem 42.12, with $\alpha = 0$, to equation (64.29), we conclude that v is a weak solution of (64.29) in the space H. Consequently, v is a weak solution of equation (64.21), which completes the proof of Item (1) in the lemma. Since v satisfies (64.25), the argument used above, along with the Continuity Lemma 37.9, imply Item (2).

By replacing A with νA, the inequalities (42.30), with $\alpha = 0$, imply that

$$\max\left(\|v(t)\|^2,\ \nu\int_0^t \|A^{\frac{1}{2}}v\|^2\, ds\right) \leq \|v_0\|^2 + \nu^{-1}\int_0^t \|A^{-\frac{1}{2}}\phi\|^2\, ds. \tag{64.38}$$

Since $g \in L^\infty(0,\infty;V^{-1})$, inequalities (61.26) and (64.24), along with the Young inequality, imply that $\|A^{-\frac{1}{2}}\phi\|^2 \leq 3\|A^{-\frac{1}{2}}g\|^2_\infty + 6C_5^2\rho\|A^{\frac{1}{4}}v\|^2$. It then follows from inequality (64.37) that

$$\int_0^t \|A^{-\frac{1}{2}}\phi\|^2\, ds \leq \rho_3(t)\left(\lambda_1^{-\frac{1}{2}}\|v_0\|^2 + D_2\, t\|A^{-\frac{1}{2}}g\|^2_\infty\right), \tag{64.39}$$

where $\rho_3 = 1 + 6C_5^2\nu^{-1}\rho\rho_2$ and $D_2 = 3 + c_2$. Hence, inequalities (64.28) follow from inequalities (64.38) and (64.39), with $\rho_4 = \nu^{-1} + \nu^{-2}\rho_3$ and $D_1 = \max(1, \lambda_1^{-\frac{1}{2}})$. $\square$

Strong Solutions: A function $v = v(t)$ is said to be a **strong solution** of (64.21) on the interval $[0,T)$, where T satisfies $0 < T \leq \infty$, provided that $v(0) = v_0 \in V$; the function v is the restriction of a weak solution, as defined in the paragraph containing (64.25), on the interval $[0,T)$; and v satisfies

$$v \in L^\infty_{\text{loc}}[0,T;V) \cap L^2_{\text{loc}}[0,T;\mathcal{D}(A)). \tag{64.40}$$

We have the following result on strong solutions.

Theorem 64.7. *Let Ω be an open, bounded domain in $\mathbb{R}^3$ of class C^2, where $d = 2$ or 3, and let $g \in L^\infty(0,\infty;H)$. Let u_1 and u_2 satisfy (64.23) on an interval $[0,T)$, where $0 < T \leq \infty$. Then, for every $v_0 \in V$, the following properties hold:*

1. *There is a unique strong solution $v = v(t)$ of equation (64.21) in the space V, on the interval $[0,T)$.*
2. *One has $\partial_t v \in L^2_{\text{loc}}[0,T;H)$ and $v \in C(0,T;V)$.*
3. *The function v is a mild solution of equation (64.21), in both the spaces H and V, and there is $\theta_0 > 0$, such that $v \in C^{0,\theta_0}_{\text{loc}}(0,T;H)$.*
4. *There exists a positive, continuous function $\rho_8 = \rho_8(t)$, which depends on u_1 and u_2, and there exist constants D_3 and D_4, which do not depend on the data (v_0, g, u_1, u_2), such that, for $0 \leq t < T$, this solution satisfies*

$$(64.41)\qquad \|A^{\frac{1}{2}}v(t)\|^2 + \nu\int_0^t \|Av\|^2\,ds \leq \rho_8(t)\left(D_3\|A^{\frac{1}{2}}v_0\|^2 + D_4\,t\|g\|^2_\infty\right).$$

Proof. The argument uses a bootstrap argument with Theorem 42.12, for various choices of α, to analyze the solutions of the equation (64.29). In particular, the inequalities (42.30) are valid, where f is replaced by ϕ and A by νA. We will use the conditions (64.23) and (64.24), for $u = u_1$, or u_2. Also we will make repeated use of the inequality $\|A^\alpha w\|^2 \leq \lambda_1^{2(\alpha-\beta)}\|A^\beta w\|^2$, for $\alpha \leq \beta$. As usual, we let $c_0, c_1, \cdots$ denote positive constants, which do not depend on the data (v_0, g, u_1, u_2) nor on the time t. The proof is presented in two major steps.

Step 1. Since $v_0 \in V \hookrightarrow H$ and $g \in L^\infty(0,\infty;H) \mapsto L^\infty(0,\infty;V^{-1})$, the hypotheses of Lemma 64.6 are satisfied. By using (61.20), (64.24), $\|A^{-\frac{1}{4}}g\|^2 \leq \lambda_1^{-\frac{1}{2}}\|g\|^2$, and the Young inequality, one obtains

$$\|A^{-\frac{1}{4}}\phi\|^2 \leq 6C_1^2\rho^2\|A^{\frac{1}{2}}v\|^2 + 3\lambda_1^{-\frac{1}{2}}\|g\|^2_\infty,$$

see (64.29). It then follows from inequalities (64.28) that

$$(64.42)\qquad \int_0^t \|A^{-\frac{1}{4}}\phi\|^2\,ds \leq \rho_5(t)\left(D_1\|A^{\frac{1}{2}}v_0\|^2 + (D_2 + 3\lambda_1^{\frac{1}{2}})t\|g\|^2_\infty\right),$$

where $\rho_5 = \lambda_1^{-1}(1+6C_1^2\rho^2\rho_4)$. Since $v_0 \in V \hookrightarrow V^{\frac{1}{2}}$, it follows from Theorem 42.12, with $\alpha = \frac{1}{2}$, $f = \phi$, and A replaced by νA, that

$$(64.43)\qquad v \in C[0,T;V^{\frac{1}{2}}) \cap L^2_{\text{loc}}[0,T;V^{\frac{3}{2}})$$

and that inequalities (42.30) hold for this choice of α. Thus one has

$$\|A^{\frac{1}{4}}v(t)\|^2 \leq \|A^{\frac{1}{4}}v_0\|^2 + \nu^{-\frac{1}{2}}\int_0^t \|A^{-\frac{1}{4}}\phi\|^2\,ds,$$

$$\nu^{\frac{3}{4}}\int_0^t \|A^{\frac{3}{4}}v\|^2\,ds \leq \|A^{\frac{1}{4}}v_0\|^2 + \nu^{-\frac{1}{2}}\int_0^t \|A^{-\frac{1}{4}}\phi\|^2\,ds.$$

From these inequalities and inequality (64.42) one finds that

$$\max\left(\|A^{\frac{1}{4}}v(t)\|^2,\ \nu^{\frac{3}{4}}\int_0^t \|A^{\frac{3}{4}}v\|^2\,ds\right) \le \rho_6(t)\left((D_1+\lambda_1^{-\frac{1}{2}})\|A^{\frac{1}{2}}v_0\|^2 + (D_2+3\lambda_1^{\frac{1}{2}})t\|g\|_\infty^2\right), \tag{64.44}$$

for $0 \le t < T$, where $\rho_6 = 1+\nu^{-\frac{1}{2}}\rho_5$.

Step 2. Next it follows from inequality (61.19) and the interpolation inequality (61.9) that one has

$$\|B(u,v)\|,\ \|B(v,u)\| \le c_{10}\|A^{\frac{1}{2}}u\|^{\frac{3}{4}}\|Au\|^{\frac{1}{4}}\|A^{\frac{1}{4}}v\|^{\frac{1}{4}}\|A^{\frac{3}{4}}v\|^{\frac{3}{4}}, \tag{64.45}$$

for $u = u_1$ and $u = u_2$. By using (64.23), (64.43), and (64.45), one can readily show that $B(v,u_1)$ and $B(u_2,v)$, and consequently ϕ, belong to the space $L^2_{\mathrm{loc}}[0,T;H)$.

From inequalities (64.24), (64.44), and (64.45) and the Hölder inequality, we get

$$\begin{aligned}\int_0^t \|B(u,v)\|^2\,ds &\le c_{11}\int_0^t \|A^{\frac{1}{2}}u\|^{\frac{3}{2}}\|Au\|^{\frac{1}{2}}\|A^{\frac{1}{4}}v\|^{\frac{1}{2}}\|A^{\frac{3}{4}}v\|^{\frac{3}{2}}ds\\ &\le c_{11}\left(\int_0^t \|A^{\frac{1}{2}}u\|^6\|Au\|^2\,ds\right)^{\frac{1}{4}}\left(\int_0^t \|A^{\frac{1}{4}}v\|^{\frac{2}{3}}\|A^{\frac{3}{4}}v\|^2\,ds\right)^{\frac{3}{4}}\\ &\le c_{11}\rho^{\frac{3}{4}}\sigma^{\frac{1}{4}}\rho_6(t)\left((D_1+\lambda_1^{-\frac{1}{2}})\|A^{\frac{1}{2}}v_0\|^2 + (D_2+3\lambda_1^{\frac{1}{2}})t\|g\|_\infty^2\right).\end{aligned} \tag{64.46}$$

From the Young inequality, we see that $\int_0^t \|\phi\|^2 ds$ is bounded by

$$t\|g\|_\infty^2 + 2c_{11}\rho^{\frac{3}{4}}\sigma^{\frac{1}{4}}\rho_6(t)\left((D_1+\lambda_1^{-\frac{1}{2}})\|A^{\frac{1}{2}}v_0\|^2 + (D_2+3\lambda_1^{\frac{1}{2}})t\|g\|_\infty^2\right).$$

We then obtain

$$\int_0^t \|\phi\|^2\,ds \le \rho_7(t)\left((D_1+\lambda_1^{-\frac{1}{2}})\|A^{\frac{1}{2}}v_0\|^2 + D_4\,t\|g\|_\infty^2\right), \tag{64.47}$$

where $\rho_7 = 1+2c_{11}\rho^{\frac{3}{4}}\sigma^{\frac{1}{4}}\rho_6$ and $D_4 = D_2+3(1+\lambda_1^{\frac{1}{2}})$.

Since $v_0 \in V$, it follows from Theorem 42.12, with $\alpha = 1$, and inequality (64.47) that v satisfies (64.40), i.e., v is a strong solution of equation (64.21). Furthermore, inequalities (42.30) are valid for $\alpha = 1$. Thus one has

$$\|A^{\frac{1}{2}}v(t)\|^2,\ \nu\int_0^t \|Av\|^2\,ds \le \|A^{\frac{1}{2}}v_0\|^2 + \nu^{-1}\int_0^t \|\phi\|^2\,ds.$$

As a result, inequalities (64.41) follow from (42.30) and (64.47), with $D_3 = D_1+\lambda_1^{-\frac{1}{2}}+1$ and $\rho_8 = 1+\nu^{-1}\rho_7$. In addition, Items (2) and (3) of this theorem also follow from Theorem 42.12. Finally, since every strong solution is a weak solution, and since the weak solutions are uniquely determined by the data (v_0, g), it follows from Lemma 64.6, that the strong solution v is uniquely determined, as well. □

In the proofs of Lemma 64.6 and Theorem 64.7, we placed a special effort in deriving formulae for the coefficients $\rho_4(t)$ and $\rho_8(t)$ appearing in the results. It is important to note that, given two strong solutions u_1 and u_2 of the Navier-Stokes equations, ρ_4 and ρ_8 depend only on: (1) the physical parameters (ν and λ_1), (2) some coefficients arising in the auxiliary inequalities in Section 6.1, and (3) the functions ρ and σ given by (64.24). This observation will be used in the next result, where we show that the strong solutions of the Navier-Stokes equations are locally Lipschitz continuous functions of the data (u_0, f).

Theorem 64.8. *Let Ω be an open, bounded domain in $\mathbb{R}^d$ of class C^2, where $d = 2$ or 3. For $i = 1, 2$, let $u_i = u_i(t)$ denote two maximally defined strong solutions of the Navier-Stokes equations defined on the intervals $[0, T_i)$, with data (u_{i0}, f_i), where $u_{i0} \in V$ and $f_i \in L^\infty(0, \infty; H)$. Let $T = \min(T_1, T_2)$ and define ρ and σ by (64.24). Let $\rho_4(t)$, $\rho_8(t)$, and D_i, for $1 \le i \le 4$, be given by Lemma 64.6 and Theorem 64.7. Then the following hold:*

(64.48)
$$\|u_1(t) - u_2(t)\|^2 \le \rho_4(t) \left(D_1 \|u_i(0) - u_2(0)\|^2 + D_2\, t \|A^{-\frac{1}{2}}(f_1 - f_2)\|_\infty^2 \right),$$

and

(64.49)
$$\|A^{\frac{1}{2}}(u_1(t) - u_2(t))\|^2$$
$$\le \rho_8(t) \left(D_3 \|A^{\frac{1}{2}}(u_1(0) - u_2(0))\|^2 + D_4\, t \|f_1 - f_2\|_\infty^2 \right).$$

If in addition u_1, $u_2 \in L^\infty(0, \infty; V)$, then one has $T = \infty$, and instead of (64.24), ρ can be chosen to be independent of time t with

$$\rho = \max(\|A^{\frac{1}{2}} u_1\|_\infty^2, \|A^{\frac{1}{2}} u_2\|_\infty^2). \tag{64.50}$$

Proof. The proof is a direct application of Lemma 64.6 and Theorem 64.7. Let $v = u_1 - u_2$ and $g = f_1 - f_2$. Then one has $v \in L^\infty_{\text{loc}}[0, T; V) \cap L^2_{\text{loc}}[0, T; V^2)$, and the initial condition $v(0) = v_0 = u_1(0) - u_2(0) \in V$. Furthermore, a direct calculation shows that v is a strong solution of

$$\partial_t v + \nu A v + B(v, u_1) + B(u_2, v) = g$$

on the interval $0 \le t < T$. Then (64.48) and (64.49) follow directly from Lemma 64.6 and Theorem 64.7. Finally, if u_1, $u_2 \in L^\infty(0, \infty; V)$, then it follows from Theorem 64.2 and (64.16) that $T = \infty$. In this case one can replace ρ by the larger value given by (64.50). □

The inequalities (64.48) and (64.49) are valid for both the 2D and the 3D problems. In the 2D case, however, one always has $T = \infty$.

6.4.5. Higher Regularity of Solutions. By making additional assumptions on the forcing term f, one can derive stronger regularity properties for certain solutions of the Navier-Stokes equations. We will illustrate this feature in two cases. In the first case we will assume that the forcing function is differentiable in time t, and in the second case, we will assume that f has additional smoothness in the spatial variable x.

Theorem 64.9. *Let Ω be an open, bounded domain in $\mathbb{R}^d$ of class C^2, where $d = 2$ or 3. Assume that the data (u_0, f) satisfy $u_0 \in V^2 = \mathcal{D}(A)$ and*

$$f \in C[0,\infty; H) \cap L^\infty(0,\infty; H) \cap W^{1,2}_{\mathrm{loc}}[0,\infty; V^{-1}).$$

Let $u = u(t)$ be the maximally defined strong solution of equation (64.3) on the interval $[0,T)$, where $0 < T \le \infty$, and set $v \overset{\mathrm{def}}{=} \partial_t u$. Then one has

$$u \in C^1[0,T;H) \cap C[0,T;V^2), \qquad v \in C[0,T;H) \cap L^2_{\mathrm{loc}}[0,T;V),$$

and v is a weak solution of the equation

$$\partial_t v + \nu A v + B(u,v) + B(v,u) = g \tag{64.51}$$

in the space H, where $g = \partial_t f$ and $v(0) = -\nu A u_0 - B(u_0,u_0) + f(0)$. Also, one has $B(u,u) \in C[0,T;H)$. In addition, v is a mild solution of equation (64.51), and it satisfies the Variation of Constants Formula

$$v(t) = e^{-\nu A t} v_0 + \int_0^t e^{-\nu A(t-s)} [g - B(u,v) - B(v,u)]\, ds$$

in the space V^β, for each $\beta < 0$.

Proof. The argument follows the bootstrap argument used in Lemma 64.6. We define $\phi \overset{\mathrm{def}}{=} f - B(u,u)$ and $\psi \overset{\mathrm{def}}{=} \partial_t \phi$. The functions ϕ and ψ are well-defined by Lemma 64.1 and the assumptions on f. From hypotheses of this theorem, the imbedding $V^\alpha \hookrightarrow V^\beta$, for $\alpha > \beta$, and Items (4) and (5) of Lemma 64.1, one finds that

$$\psi \in L^2_{\mathrm{loc}}[0,T;V^{-\frac{3}{2}}). \qquad \text{and} \qquad \phi \in C(0,T;V^{-\frac{1}{2}}) \cap W^{1,2}_{\mathrm{loc}}[0,T;V^{-\frac{3}{2}}).$$

Therefore from Theorem 42.14, with $\alpha = -\frac{1}{2}$, one finds that

$$u \in C^1[0,T;V^{-\frac{1}{2}}) \cap C[0,T;V^{\frac{3}{2}}) \quad \text{and} \quad v \in C[0,T;V^{-\frac{1}{2}}) \cap L^2_{\mathrm{loc}}[0,T;V^{\frac{1}{2}}). \tag{64.52}$$

Next we will to show that $\partial_t B(u,u) \in L^2_{\mathrm{loc}}[0,T;V^{-1})$ and

$$\partial_t B(u,u) = B(v,u) + B(u,v), \qquad \text{in } L^2_{\mathrm{loc}}[0,T;V^{-1}).$$

The proof of this identity follows the argument used for Item (4) in Lemma 64.1. Since $v = \partial_t u \in L^2_{\mathrm{loc}}[0,T;V^{\frac{1}{2}})$, by the Vitali Convergence Theorem, see Appendix A.10, the difference quotient $\delta_h u(t) = h^{-1}(u(t+h) - u(t))$ satisfies $\int_I \|A^{\frac{1}{4}}(\delta_h u - v)\|^2 \, ds \to 0$, as $h \to 0$, for every compact subinterval I in $[0,T)$. Since $u \in C[0,T;V^{\frac{3}{2}})$, it then follows from (61.19) that $B(u,u) \in C[0,T;H)$. Furthermore, one has

$$\int_I \|A^{-\frac{1}{2}}(\delta_h B(u,u) - B(v,u) - B(u,v))\|^2 \, ds \to 0, \qquad \text{as } h \to 0.$$

Therefore one obtains ϕ, $B(u,u) \in C[0,T;H) \cap W^{1,2}_{\text{loc}}[0,T;V^{-1})$. By applying Theorem 42.14 once again, this time with $\alpha = 0$, the conclusions of this lemma follow. □

In the following corollary,[28] we examine the case where the forcing function is smooth in time, but the initial condition u_0 is in V, and not in V^2.

Corollary 64.10. *Let Ω be an open, bounded domain in $\mathbb{R}^d$ of class C^2, where $d = 2$ or 3. Assume that the data (u_0, f) satisfy $u_0 \in V$ and*

$$f \in C[0,\infty;H) \cap L^\infty(0,\infty;H) \cap W^{1,2}_{\text{loc}}[0,\infty;V^{-1}).$$

Let $u = u(t)$ be the maximally defined strong solution of equation (61.10) on the interval $[0,T)$, where $0 < T \le \infty$, and set $v \overset{\text{def}}{=} \partial_t u$. Then one has

$$u \in C^1(0,T;H) \cap C(0,T;V^2), \qquad v \in C(0,T;H) \cap L^2_{\text{loc}}(0,T;V),$$

and v is a weak solution of the equation

$$\partial_t v + \nu A v + B(u,v) + B(v,u) = g \tag{64.53}$$

in the space H on the interval $[\epsilon, T)$, for every ϵ with $0 < \epsilon < T$.

Proof. Since $u \in L^2_{\text{loc}}[0,T;V^2)$, one has $\int_0^t \|Au\|^2 \, ds < \infty$, for each t with $0 < t < T$. Therefore, there is a set of times t_0 with full measure in $(0,T)$, such that $u(t_0) \in V^2$. As a result Theorem 64.9 applies on the interval $[t_0, T)$. By taking the union of these intervals $[t_0,T)$, over all such t_0, one obtains the conclusions of this corollary. □

Theorem 64.11. *Let Ω be an open, bounded domain in $\mathbb{R}^d$ of class C^∞, for $d = 2$, or 3, and let $f \in L^\infty(0,\infty;V)$. Let $u = u(t)$ be the maximally defined strong solution of the Navier-Stokes evolutionary equation (61.10) on $[0,T)$, with $u(0) = u_0$, where $0 < T \le \infty$. Then the following are valid:*

(1) *If $u_0 \in V^2 = \mathcal{D}(A)$, then one has*

$$u \in C[0,T;V^2) \cap L^2_{\text{loc}}[0,T;V^3), \qquad \textit{and} \qquad \partial_t u \in L^2_{\text{loc}}[0,T;V). \tag{64.54}$$

[28] Compare this result with Theorem 42.15.

In addition one has $u \in C^{0,\theta_1}_{\text{loc}}[0,T;V)$, *for some* $\theta_1 > 0$, *and the following inequalitites are valid:*

(64.55)

$$\|Au(t)\|^2 \leq \left(\frac{1}{t}\int_0^t \|Au\|^2\,ds + t\,c_2\|A^{\frac{1}{2}}f\|^2_\infty\right)\exp(c_1\int_0^t \|Au\|^2\,ds),$$

for $0 < t \leq 1$, *and*

(64.56)

$$\|Au(t)\|^2 \leq \left(\int_{t-1}^t \|Au\|^2\,ds + c_2\|A^{\frac{1}{2}}f\|^2_\infty\right)\exp(c_1\int_{t-1}^t \|Au\|^2\,ds),$$

for $1 \leq t < T$, *as well as*

$$\nu\int_0^t \|A^{\frac{3}{2}}u\|^2\,ds \leq \|Au_0\|^2 + c_1\int_0^t \|Au\|^4\,ds + t\,c_2\|A^{\frac{1}{2}}f\|^2_\infty. \tag{64.57}$$

(2) *If one has* $u_0 \in V$, *then* u *and* $\partial_t u$ *satisfy*

$$u \in C(0,T;V^2)\cap L^2_{\text{loc}}(0,T;V^3) \quad \textit{and} \quad \partial_t u \in L^2_{\text{loc}}(0,T;V)$$

and the inequalities given above are valid on the interval $[\tau,T)$, *for every* τ *with* $0 < \tau < T$.

Proof. It is possible to prove this theorem by use of a multistep bootstrap argument, as done above (see Theorem 64.9). This approach is outlined in the exercises. There is however, an alternate approach, which is based on the Bubnov-Galerkin method, and which is of independent interest because it illustrates how the Uniform Gronwall Inequality can be used for 3D Navier-Stokes equations. Up to this point the Uniform Gronwall Inequality has only been used for 2D problems.

As usual we let $p = u_n$ denote the Bubnov-Galerkin approximations as given in Sections 6.2, where $u_n(0) = Pu_0$. If one takes the scalar product of equation (62.2) with A^2p, while using the Young inequality and inequality (61.31), one obtains

$$\partial_t\|Ap\|^2 \leq \partial_t\|Ap\|^2 + \nu\|A^{\frac{3}{2}}p\|^2 \leq c_1\|Ap\|^4 + c_2\|A^{\frac{1}{2}}f\|^2. \tag{64.58}$$

Let us now look at the case where $u_0 \in V^2$. By using the Uniform Gronwall Inequality on (64.58), one then obtains

(64.59)

$$\|Au_n(t)\|^2 \leq \left(\frac{1}{t}\int_0^t \|Au_n\|^2\,ds + t\,c_2\|A^{\frac{1}{2}}f\|^2_\infty\right)\exp(c_1\int_0^t \|Au_n\|^2\,ds),$$

for $0 < t \leq 1$, and

(64.60)

$$\|Au_n(t)\|^2 \leq \left(\int_{t-1}^t \|Au_n\|^2\,ds + c_2\|A^{\frac{1}{2}}f\|^2_\infty\right)\exp(c_1\int_{t-1}^t \|Au_n\|^2\,ds),$$

for $1 \le t < T$. By integrating (64.58) one has

$$\nu \int_0^t \|A^{\frac{3}{2}} u_n\|^2 \, ds \le \|Au_0\|^2 + c_1 \int_0^t \|Au_n\|^4 \, ds + t\, c_2 \|A^{\frac{1}{2}} f\|_\infty^2,$$

for $t \ge 0$. One then shows that the hypotheses of the Compactness I Lemma 63.2 are valid for $\alpha = 2$. As a result, there exists a subsequence of u_n, which we relabel as u_n, that converges to u in the sense described in the Compactness I and II Lemmas. Furthermore, the above inequalities hold in the limit, which completes the proof of the theorem. □

6.5. Navier-Stokes Dynamics: Global Attractors.

In this section we examine the issue of global attractors for the Navier-Stokes equations. We are interested here in both the 2D and the 3D problems, and we begin with the 2D theory.

6.5.1 Two-Dimensional Theory. There are two semiflows for the 2D Navier-Stokes equations: the semiflow on V generated by the strong solutions and the semiflow on H generated by the weak solutions. We will let $S_s(t)u_0 = S_s(t,f)u_0$ denote the strong solution in V generated by the data (u_0, f), where $u_0 \in V$ and $f \in L^\infty(0,\infty;H)$. Likewise let $S_w(t)u_0 = S_w(t,f)u_0$ denote the weak solution in H generated by the data (u_0, f), where now $u_0 \in H$.

For the theory of global attractors, we will examine the weak solutions first. In particular, we now return to the weak solutions (of Class LH) for the 2D Navier-Stokes equations.

Theorem 65.1. *Let Ω be an open, bounded set in $\mathbb{R}^2$ of class C^2 and let $f \in H$ be a time-independent forcing function, Then*

$$\hat{\sigma}(t, u_0) = S_w(t)u_0 = S_w(t,f)u_0$$

is a semiflow on H which is point dissipative and compact for $t > 0$. Furthermore, there exists a global attractor $\mathfrak{A}_w = \mathfrak{A}_w(f)$ for $S_w(t) = S_w(t,f)$, with $\mathfrak{A}_w \subset V$, and $\mathfrak{A}_w$ attracts all bounded sets in H. Moreover, the semiflow $S_w(t,f)$ is robust (i.e., upper semicontinuous) at $\mathfrak{A}_w(f)$, for every $f \in H$.

Proof. The identity $S_w(0)u_0 = u_0$ is immediate, and the uniqueness of the weak solutions (Corollary 63.6) implies that $S_w(t)u_0$ satisfies the semigroup property. It follows from Theorem 64.2 that for each $t_0 > 0$, the function $u(t + t_0)$ is a strong solution of equation (61.10) on the interval $[0, \infty)$. For $i = 1, 2$, let $u_i = u_i(t)$ denote two weak solutions of equation (61.10) with data $(u_{i0}, f_i) \in H \times H$, where f_i is independent of time t. Notice that inequalities (64.9) and (64.10) imply that

$$u_1, u_2 \in L^\infty(t_0, \infty; V) \cap L^2_{\mathrm{loc}}[t_0, \infty; \mathcal{D}(A)),$$

for every $t_0 > 0$. It then follows from (64.48) that, for $t \geq t_0 > 0$, one has
(65.1)

$$\|u_1(t) - u_2(t)\|^2 \leq \rho_4(t)\left(D_1\|u_1(t_0) - u_2(t_0)\|^2 + D_2\, t\|A^{-\frac{1}{2}}(f_1 - f_2)\|^2\right).$$

We claim that the mapping of $(0,\infty) \times H \times H$ given by $(t, f, u_0) \to S_w(t, f)u_0$ is continuous on its domain. Indeed, let t_n, f_n and u_n be convergent sequences where $t_n \to t_0$ in $(0,\infty)$, $t_0 > 0$, $f_n \overset{s}{\to} f_0$ in H, and $u_n \overset{s}{\to} u_0$ in H, as $n \to \infty$. Since the sequence t_n belongs to a compact set in $(0,\infty)$, say $t_n \leq T$, it follows from inequality (65.1) that

$$\begin{aligned}&\|S_w(t_n, f_n)u_n - S_w(t_n, f_0)u_0\|^2\\ &\qquad\leq k(T) \times (D_1\|u_n - u_0\|^2 + D_2\, T\lambda_1^{-1}\|f_n - f_0\|^2),\end{aligned}$$

which converges to 0, as $n \to \infty$. Since $S_w(t)u_0 \in C[t_0, \infty; H)$, by Corollary 63.6, one has

$$\|S_w(t_n)u_0 - S_w(t_0)u_0\| \to 0, \qquad \text{as } n \to \infty.$$

It follows that for f fixed, $f \in H$, $S_w(t, f)u_0$ is a semiflow on H.

Now inequality (63.12) implies that $S_w(t)$ is point dissipative on H. Also (64.9) implies that for each $t > 0$, $S_w(t)$ maps a bounded set in H into a bounded set in V. It then follows from Corollary 23.13 that the semiflow $S_w(t)$ is compact for $t > 0$. By the Existence Theorem 23.12, this semiflow has a global attractor $\mathfrak{A}_w = \mathfrak{A}_w(f)$. Since $S_w(t, f)$ depends continously on $f \in H$, the final statement of this theorem follows from the Robustness Theorem 23.14. □

Let us turn to the issue of global attractors for the strong solutions $S_s(t, f)u_0$ of the 2D Navier-Stokes equations, where $u_0 \in V$.

Theorem 65.2. *Let Ω be an open, bounded set in $\mathbb{R}^2$ of class C^2 and let $f \in H$ be a time-independent forcing function, Then*

$$\sigma(t, u_0) = S_s(t)u_0 = S_s(t, f)u_0$$

is a semiflow on V which is point dissipative and compact for $t > 0$. Furthermore, there exists a global attractor $\mathfrak{A}_s = \mathfrak{A}_s(f)$ for σ, and one has $\mathfrak{A}_s = \mathfrak{A}_w$, where $\mathfrak{A}_w$ is the global attractor given by Theorem 65.1. Also $\mathfrak{A}_s = \mathfrak{A}_w$ attracts all bounded set in V, and the attractor $\mathfrak{A}_w$ is compact in the space V. Moreover, the semiflow $S_s(t, f)$ is robust at $\mathfrak{A}_s(f)$, for every $f \in H$. In addition, one has $\mathfrak{A}_s = \mathfrak{A}_w \subset V^2$, and for every $u_0 \in \mathfrak{A}_s$, the globally defined strong solution $S(t)u_0$ satisfies

$$S(\cdot)u_0 \in C^{0,1-r}_{\text{loc}}(\mathbb{R}; V^{2r}) \cap C(\mathbb{R}, V^2), \tag{65.2}$$

and $\mathfrak{A}_s$ is a bounded, invariant set in V^{2r}, for each r with $0 \leq r < 1$.

Proof. The identity $S_s(0)u_0 = u_0$ is immediate, and the uniqueness of the strong solutions implies that $S_s(t)u_0$ satisfies the semigroup property. It follows from inequalities (64.12) and (64.13) in Theorem 64.2 that $u(t) = S_s(t,f)u_0$ satisfies

$$u \in L^\infty(0,\infty;V) \cap L^2_{\text{loc}}[0,\infty;\mathcal{D}(A)).$$

For $i = 1, 2$, let $u_i = u_i(t)$ denote two strong solutions of equation (61.10) with data (u_{i0}, f_i), where $u_{i0} \in V$ and $f_i \in H$. It then follows from Corollary 64.8 that, for $t \geq 0$, one has

(65.3)

$$\|A^{\frac{1}{2}}(u_1(t) - u_2(t))\|^2 \leq \rho_8(t)\left(D_3\|A^{\frac{1}{2}}(u_{10} - u_{20})\|^2 + D_4\, t\|f_1 - f_2\|^2\right).$$

By repeating the argument used in the last theorem for the weak solutions, one can easily show that the mapping of $[0,\infty) \times H \times V$ into V given by $(t, f, u_0) \to S_s(t,f)u_0$ is continuous on its domain. It follows that for f fixed, $f \in H$, $S_s(t,f)u_0$ is a semiflow[29] on V.

Now inequality (64.12) implies that $S_s(t)$ is point dissipative on V since the term L_2 is independent of u_0. Assume for the moment that this semiflow $S_s(t)$ is compact for $t > 0$. The existence of a global attractor $\mathfrak{A}_s = \mathfrak{A}_s(f)$ for this semiflow then follows from the Existence Theorem 23.12. From Theorem 64.2 we see that the weak solutions $S_w(t)u_0$, for $u_0 \in H$, satisfy $S_w(t)u_0 \in V$, for all $t > 0$. Consequently, one has

$$S_w(t+\tau)u_0 = S_s(t)S_w(\tau)u_0, \qquad \text{for } u_0 \in H,\ \tau > 0, \text{ and } t > 0.$$

It then follows that $\mathfrak{A}_s(f) = \mathfrak{A}_s = \mathfrak{A}_w$ and that $\mathfrak{A}_w$ is compact in V. Since $S_s(t,f)$ depends continously on $f \in H$, the robustness of $\mathfrak{A}_s(f)$ follows from the Robustness Theorem 23.14.

In order to show that the semiflow $S_s(t)$ is compact, for $t > 0$, we have two arguments, both of which are informative. Define $F = F(u)$ by $F(u) = f - B(u,u)$. Then the Navier-Stokes equation (61.10) assumes the form (47.12), see Section 4.7.5. Since $f \in H$, it follows from (61.32) that $F \in C^2_{\text{Lip}}(V^1, V^{-\frac{1}{2}})$. Since A has compact resolvent, it follows from Lemma 48.1 that $S_s(t)$ is compact, for $t > 0$.

For the second argument, which uses additonal information about the strong solutions of the Navier-Stokes equations, we let $N_V(0,r_0)$ denote the set of $u_0 \in V$ such that $\|A^{\frac{1}{2}}u_0\| \leq r_0$, where $r_0 > 0$. From inequality (64.13) one then has

$$\int_0^t \|AS_s(\tau)u_0\|^2\, d\tau \leq \nu^{-1}(r_0^2 + t\, K_4) \overset{\text{def}}{=} K(t), \qquad \text{for all } t \geq 0,$$

for each $u_0 \in N_V(0,r_0)$. Consequently, for each $u_0 \in N_V(0,r_0)$ and each $t > 0$, the set

$$\{t_0 \in (0,t) : \|AS_s(t_0)u_0\|^2 \leq 2t^{-1}K(t)\}$$

[29] Notice that this semiflow is jointly continuous at $t = 0$ (see Section 2.1).

has Lebesgue measure $\geq \frac{1}{2}t$. What we need to show is that for each $t > 0$ and for each sequence $u_n \in N_V(0, r_0)$, there is a subsequence of $S_s(t)u_n$ that is strongly convergent in V. Let u_n be a sequence in $N_V(0, r_0)$, and let t_n be chosen so that $t_n \in (0, t)$ and $\|AS_s(t_n)u_n\|^2 \leq 2t^{-1}K(t)$. Because of the compact imbedding $V^2 \hookrightarrow V^1$, it then follows that the sequence $S_s(t_n)u_n$ lies in a compact set in V. Therefore there exist subsequences of t_n and $S_s(t_n)u_n$, which we relabel as t_n and $S_s(t_n)u_n$, such that $t_n \to t_0 \in [0, t]$ and $S_s(t_n)u_n \xrightarrow{s} v_0 \in V$. By using the semigroup property for $S_s(\cdot)$ and the joint continuity property proved above, one has

$$S_s(t)u_n = S_s(t - t_n)S_s(t_n)u_n \xrightarrow{s} S_s(t - t_0)v_0, \qquad \text{in } V, \text{ as } n \to \infty,$$

which completes the proof of the compactness property.

For the remaining conclusions, which concern the regularity of the global attractor $\mathfrak{A} = \mathfrak{A}_s = \mathfrak{A}_w$, we note that $\mathfrak{A}$ is a bounded, invariant set in V^1. ($\mathfrak{A}$ is of course, compact in V^1.) By the Herculean Theorem 47.6, with $\beta = \frac{3}{4}$, one finds that $\mathfrak{A} \subset V^{\frac{3}{2}}$ and $\mathfrak{A}$ is a bounded, invariant subset of $V^{\frac{5}{4}}$. Since $F \in C^2_{\text{Lip}}(V^{\frac{5}{4}}, H)$ by (61.32), we are in the position to apply the Herculean Theorem once again. As a result, we see that (65.2) holds and that $\mathfrak{A}$ is a bounded, invariant set in V^{2r}, for each r with $0 \leq r < 1$. □

We have seen here the Herculean Theorem 47.6 and (61.32) used as a part of a bootstrap argument. Moreover, if $f \in V^1$, then (61.32) implies that $F \in C^2_{\text{Lip}}(V^2, V^1)$. In principle, one should expect that, in this case, one has $\mathfrak{A}_s \subset V^3$ and that for any $u_0 \in \mathfrak{A}_s$ one has $S(\cdot)u_0 \in C^{0,1-r}_{\text{loc}}(\mathbb{R}; V^{1+2r}) \cap C(\mathbb{R}, V^3)$, for all r with $0 \leq r < 1$.

6.5.2 Three-Dimensional Theory. In order to describe the theory of global attractors for the 3D Navier-Stokes equations, we return to the theory of generalized weak solutions developed in Section 6.3.2. A special situation arises in connection with the topology on the spaces

$$Y(N_0) \overset{\text{def}}{=} \{f \in Y : \|f\|_\infty \leq N_0\} \text{ and } Z(N_0) \overset{\text{def}}{=} L^2_{\text{loc}}[0, \infty; H) \times Y(N_0),$$

where $N_0 \geq 0$ is fixed. Because of the continuous imbedding

$$L^\infty(0, \infty; L^2(\Omega)) \mapsto L^2_{\text{loc}}[0, \infty; L^2(\Omega)),$$

we see that $Y(N_0)$ is a subset of $L^2_{\text{loc}} = L^2_{\text{loc}}[0, \infty; L^2(\Omega))$. Recall that $R = L^2_{\text{loc}}[0, \infty; H \times L^2(\Omega))$, see (63.38). For each $N_0 \geq 0$, the space $Y(N_0)$ is a closed, bounded set in L^2_{loc} in the L^2_{loc}-topology. Similarly, for each $N_0 \geq 0$, the space $Z(N_0)$ is a closed set in Z in the R-topology.

For a given function $f : (0, \infty) \to X$, where X is a set, we define the **time translate** f_τ by

$$f_\tau(t) = f(\tau + t), \qquad t \geq 0,$$

where $\tau \geq 0$. Note that if $f \in L^\infty(0,\infty; L^2(\Omega))$, then $\|f_\tau\|_\infty^2 \leq \|f\|_\infty^2$, for all $\tau \geq 0$.

Let $\mathfrak{F}$ be a bounded set in Y. Thus $\mathfrak{F} \subset Y(N_0)$, for some $N_0 \geq 0$. Define $\gamma^+(\mathfrak{F}) = \{f_\tau : f \in \mathfrak{F} \text{ and } \tau \geq 0\}$. Next define the **hull** to be

$$H^+(\mathfrak{F}) \stackrel{\text{def}}{=} \text{Cl}_{L^2_{\text{loc}}}(\gamma^+(\mathfrak{F})).$$

One then has $H^+(\mathfrak{F}) \subset Y(N_0)$, as well, i.e., $H^+(\mathfrak{F})$ is a Y-bounded set and it is closed in L^2_{loc}. We are especially interested in the case where the hull is a compact set in L^2_{loc}. Recall that the hull $H^+(\mathfrak{F})$ is a compact set in L^2_{loc} if and only if for every T with $0 < T < \infty$, one has

$$\sup\{\int_0^T \|f(s+\tau+h) - f(s+\tau)\|^2\, ds : f \in \mathfrak{F}, \tau \geq 0\} \to 0, \tag{65.4}$$

as $h \to 0$, see Dunford and Schwartz (1958, Part 1, pp 298-301). Condition (65.4) will be satisfied if $\mathfrak{F}$ is Y-bounded, $\mathfrak{F} \subset C[0,\infty; L^2(\Omega))$, and $\mathfrak{F}$ is a uniformly equicontinuous family of mappings from $[0,\infty)$ into $L^2(\Omega)$. For example, the hull is compact if $\mathfrak{F}$ is a bounded set in the Hölder space $C^\alpha[0,\infty; L^2(\Omega))$, for some $\alpha > 0$. For the remainder of this section we will assume that the hull $H^+(\mathfrak{F})$ is a compact set.

For $f \in L^\infty(0,\infty; H)$ we will use the spaces $W_{LH}(f)$ and $W(f)$ of weak solutions defined in Section 6.3. Let p satisfy $1 \leq p < \infty$. Then for each $\phi \in L^p_{\text{loc}}(0,\infty; X)$, the time translate ϕ_τ satisfies $\phi_\tau \in L^p_{\text{loc}}(0,\infty; X)$, for all $\tau \geq 0$. As an application of Lemma 63.9, we now prove the following result.

Lemma 65.3. *Let $f \in Y$ and $\varphi \in W(f)$. Then for every $\tau > 0$, the time translate φ_τ is in the space $W_{LH}(f_\tau)$.*

Proof. Let $\varphi \in W(f)$, and fix $\tau > 0$. Note that the time translate φ_τ is a solution of (61.10), where f is replaced by f_τ, i.e., one has $\varphi_\tau \in W(f_\tau)$. It then follows from (63.41) and (63.42) that the time translate φ_τ is in $L^\infty(0,\infty; H)$. Also (63.43) implies that $\varphi_\tau \in L^2_{\text{loc}}[0,\infty; V)$. The result now follows from Lemma 63.9. ☐

For the remainder of this section we will assume that $f \in H$. As a result, f does not depend on time, and one has

$$W \stackrel{\text{def}}{=} W(f) = W(f_\tau) \quad \text{and} \quad W_{LH} \stackrel{\text{def}}{=} W_{LH}(f) = W_{LH}(f_\tau),$$

for all $\tau \geq 0$. We let $S(t)$ be the mapping given by

$$S : (\tau, \varphi) \to S(\tau)\varphi = \varphi_\tau. \tag{65.5}$$

As shown in Lemma 63.10, W is a closed set in $L^2_{\text{loc}}[0,\infty; H)$. Hence W is a complete metric space. In order to apply the Existence Theorem 23.12 to this situation, we will verify the following properties for $S(t)$:

(1) The mapping $S(t)$ is a semiflow on $L^2_{\text{loc}} = L^2_{\text{loc}}[0,\infty; L^2(\Omega))$, and $L^2_{\text{loc}}[0,\infty; H)$ and W are positively invariant subsets of L^2_{loc}.

(2) The restriction of the semiflow $S(t)$ to W is compact for $t > 0$.
(3) The restriction of the semiflow $S(t)$ to W is point dissipative.

We will do this next. But in addition to these properties, which imply that there is a global attractor $\mathfrak{A}$ in W, we will prove the following:

(4) The global attractor $\mathfrak{A}$ lies in W_{LH}, and $\mathfrak{A}$ attracts all bounded sets in W_{LH}.
(5) For every bounded set B in W and for every $t > 0$, the compact set $\mathrm{Cl}_W S(t)B$ lies in W_{LH}.

Lemma 65.4. *Let $f \in H$ and set $W = W(f)$ and $W_{LH} = W_{LH}(f)$. Then the mapping $S(t)$ given by (65.5) is a semiflow on $L^2_{\mathrm{loc}} = L^2_{\mathrm{loc}}[0,\infty; L^2(\Omega))$. Furthermore, the sets $L^2_{\mathrm{loc}}[0,\infty; H)$, W, and W_{LH} are positively invariant sets in L^2_{loc}.*

Proof. The main issue in showing that (65.5) defines a semiflow on L^2_{loc}, is to verify the continuity of the mapping $(\tau,\varphi) \to S(\tau)\varphi = \varphi_\tau$, for $(\tau,\varphi) \in (0,\infty)\times L^2_{\mathrm{loc}}$. Let τ_n and φ^n be convergent sequences where $\varphi^n \to \varphi$ in L^2_{loc} and $\tau_n \to \tau$ in $(0,\infty)$. Since φ^n is convergent in L^2_{loc}, it is bounded. Since $\tau > 0$, there is no loss in generality in assuming that $0 < \frac{1}{2}\tau \le \tau_n \le 2\tau$.

Now we will show that

$$d(\varphi^n_{\tau_n} - \varphi_{\tau_n}) \to 0, \qquad \text{as } n \to \infty, \tag{65.6}$$

where d is the invariant metric on L^2_{loc} given in Appendices A and B. Let a and b be given where $0 \le a < b < \infty$. It suffices to show that

$$\int_a^b \|\varphi^n_{\tau_n} - \varphi_{\tau_n}\|^2\, ds \to 0, \qquad \text{as } n \to \infty.$$

Now for $\frac{1}{2}\tau \le \sigma \le 2\tau$ one has

$$\int_a^b \|\varphi^n_\sigma - \varphi_\sigma\|^2\, ds = \int_{a+\sigma}^{b+\sigma} \|\varphi^n - \varphi\|^2\, ds \le \int_{a+\frac{1}{2}\tau}^{b+2\tau} \|\varphi^n - \varphi\|^2\, ds.$$

Since $\varphi^n \to \varphi$ in L^2_{loc}, it follows that $\varphi^n_\sigma \to \varphi_\sigma$ in L^2_{loc}, uniformly for $\frac{1}{2}\tau \le \sigma \le 2\tau$. This implies (65.6). From (92.5) in Appendix B, one has

$$d(\varphi_{\tau_n} - \varphi_\tau) \to 0, \qquad \text{as } n \to \infty. \tag{65.7}$$

Therefore one gets $d(\varphi^n_{\tau_n} - \varphi_\tau) \to 0$, as $n \to \infty$. The positive invariance of the space $L^2_{\mathrm{loc}}[0,\infty; H)$ is obvious, and the positive invariance of W and W_{LH} follow from Lemma 65.3. □

Lemma 65.5. *Let $f \in L^2(\Omega)$ and set $W = W(f)$ and $W_{LH} = W_{LH}(f)$. Then the restriction of the semiflow $S(t)$ to W is compact for $t > 0$; i.e., for each bounded set $\hat{B}$ in W and for each $\tau > 0$, the set $S(\tau)\hat{B}$ lies in a compact subset in W. Moreover, one has $Cl_W S(\tau)\hat{B} \subset W_{LH}$.*

Proof. If $S(\tau)\hat{B}$ lies in a compact set in W for some $\tau > 0$, then by the semigroup property (21.2), $S(\tau + t)\hat{B}$ lies in a compact set in W, for each $t > 0$. Let τ be fixed where $0 < \tau \leq 1$. Since W is a metric space, it suffices to verify that $S(\tau)\hat{B}$ is sequentially compact in W. Let φ^n be a bounded sequence in W. Then one has $(\varphi^n, f) \in \hat{B}(M_0, \|f\|_\infty)$, for all n, and for some $M_0 > 0$. As a result, it follows from (63.41) that

$$\|S(\tau)\varphi^n\|_\infty^2 \leq \tau^{-1}M_0^2 + 2c_0^2\|f\|_\infty^2, \qquad \text{for all } n. \tag{65.8}$$

From (63.44) one finds that

$$\int_m^{m+1} \|A^{\frac{1}{2}}S(\tau)\varphi^n\|^2\, ds \leq M_1(\tau + m, \tau + m + 1)^2, \qquad \text{for all } n, \tag{65.9}$$

for $m = 0, 1, 2, \cdots$. From Lemma 63.7 we see that the hypotheses of the Compactness II Lemma 63.3 are satisfied, for $\alpha = 0$. After a relabelling, we let φ^n and $\varphi \in L^2_{\text{loc}}(0, \infty; V)$ denoted the subsequence and the limit function given by the conclusions of Lemma 63.3. Since (65.8) implies that $\|S(\tau + t)\varphi\|^2 \leq \tau^{-1}M_0^2 + 2c_0^2\|f\|_\infty^2$, for almost all $t > 0$, it follows from the Generalized Solution Proposition, Item (7), that

$$\|S(\tau)\varphi\|_\infty^2 \leq \tau^{-1}M_0^2 + 2c_0^2\|f\|_\infty^2. \tag{65.10}$$

From Lemma 63.2, Item (1), and (65.9) one obtains the semicontinuity property for weak limits:

$$\begin{aligned} \int_m^{m+1} \|A^{\frac{1}{2}}S(\tau)\varphi\|^2\, ds &\leq \liminf_{n\to\infty} \int_m^{m+1} \|A^{\frac{1}{2}}S(\tau)\varphi^n\|^2\, ds \\ &\leq M_1(\tau + m, \tau + m + 1)^2, \end{aligned} \tag{65.11}$$

for $m = 0, 1, 2, \cdots$. Now (65.10) and (65.11) imply that

$$S(\tau)\varphi \in L^\infty(0, \infty; H) \cap L^2_{\text{loc}}[0, \infty; V), \tag{65.12}$$

and from Lemma 65.3 we conclude that $S(\tau)\varphi \in W_{LH} \subset W$. Hence $S(\tau)\hat{B}$ lies in a compact set in W.

In order to show that the set $C\ell_W S(\tau)\hat{B}$ lies in W_{LH}, we let $\varphi_0 \in C\ell_W S(\tau)\hat{B}$ be given. Then there is a sequence $\varphi^n \in \hat{B}$ with the property that $S(\tau)\varphi^n \to \varphi_0$ in W, i.e., $S(\tau)\varphi^n \xrightarrow{s} \varphi_0$ in $L^2_{\text{loc}}[0, \infty; H)$. Now each of the functions $S(\tau)\varphi^n$ satisfies (65.10) and (65.11). Furthermore, it follows from Lemma 63.7 that the sequence $S(\tau)\varphi^n$ satisfies the hypotheses of

Lemma 63.3, for $\alpha = 0$. Hence, there is a subsequence, which we will relabel as $S(\tau)\varphi^n$ and which satisfies the conclusions of the same lemma. Since $S(\tau)\varphi^n \overset{s}{\to} \varphi_0$ in $L^2_{\text{loc}}[0,\infty;H)$, it follows from Lemma 63.3 that $S(\tau)\varphi^n \overset{w}{\to} \varphi_0$ in $L^2_{\text{loc}}(0,\infty;V)$. Similarly, by using Lemma 63.2, with $\alpha = 0$ and $(a,b) = (0,1)$, one has $S(\tau)\varphi^n \overset{w}{\to} \varphi_0$ in $L^2(0,1;V)$, perhaps by taking yet another subsequence. It follows then that inequality (65.11), with φ_0 replacing $S(\tau)\varphi$, is satisfied. Hence, $\varphi_0 \in L^2_{\text{loc}}[0,\infty;V)$. As argued in the previous paragraph, by using the Generalized Solutions Proposition once again, we see that inequality (65.10) is satisfied, with φ_0 replacing $S(\tau)\varphi$. Hence $\varphi_0 \in L^\infty(0,\infty;H)$. It then follows from Lemma 63.9 that $\varphi_0 \in W_{LH}$. □

Lemma 65.6. *Let $f \in L^2(\Omega)$ and set $W = W(f)$ and $W_{LH} = W_{LH}(f)$. Then the restriction of the semiflow $S(t)$ to W is point dissipative.*

Proof. In order to verify the point dissipative property, we define U to be the set of all $\varphi \in W$ such that $\int_m^{m+1} \|\varphi\|^2\, ds \le 2c_0^2\|f\|^2$, for all integers $m \ge 0$. It follows from (63.39) that U is a bounded set in W. Let $(\varphi, f) \in \hat{B}(M_0, \|f\|_\infty)$. From inequality (63.43) one finds that

$$\int_m^{m+1} \|S(\tau)\varphi\|^2\, ds \le e^{-\nu\lambda_1\tau}K^2 + c_0^2\|f\|^2, \qquad \text{for all } \tau \ge 1 \text{ and all } m \ge 0,$$

where $K^2 = e^{\nu\lambda_1}(M_0^2 + c_0^2\|f\|^2)$. Let $\tau_0 \ge 0$ satisfy $e^{-\nu\lambda_1\tau_0}K^2 \le c_0^2\|f\|^2$. One then has

$$\int_m^{m+1} \|S(\tau)\varphi\|^2\, ds \le 2c_0^2\|f\|^2, \qquad \text{for all } \tau \ge \tau_0 \text{ and all } m \ge 0.$$

This implies that $S(\tau)\varphi \in U$, for all $\tau \ge \tau_0$, i.e., the semiflow $S(\tau)$ is point dissipative. □

Theorem 65.7. *Let Ω be an open, bounded domain in $\mathbb{R}^d$ of class C^2, where $d = 2$ or 3, and let $f \in H$. Set $W = W(f)$ and $W_{LH} = W_{LH}(f)$. Then there exists a global attractor $\mathfrak{A}$ for both the generalized weak solutions, and the weak solutions of Class LH, of the Navier-Stokes equations on Ω, and one has $\mathfrak{A} \subset W_{LH}$. Furthermore, $\mathfrak{A}$ attracts all bounded sets in W.*

Proof. The existence of the global attractor $\mathfrak{A}$ in W and the fact that it attracts all bounded sets in W now follows from the Existence Theorem 23.12 and four Lemmas 65.3–65.6. Since $\mathfrak{A}$ is invariant, it follows from Lemma 65.3 that $\mathfrak{A} \subset W_{LH}$, see Figure 6.1. □

As is usual, the global attractor $\mathfrak{A}$ consists entirely of solutions of the Navier-Stokes equations which are defined for all $t \in \mathbb{R}$. The reader should verify that it follows from the arguments in Lemma 63.7 that if $\varphi \in \mathfrak{A}$, then there are constants c_1 and c_2 such that one has

(1) $\|\varphi(t)\|^2 \le c_0^2\|f\|^2$, for all $t \in \mathbb{R}$.
(2) $\int_t^{t+1} \|A^{\frac{1}{2}}\varphi\|^2\, ds \le \nu^{-1}c_0\|f\|$, for all $t \in \mathbb{R}$.
(3) $\int_t^{t+1} \|A^{-\frac{1}{2}}D_t\varphi\|^p\, ds \le (c_1\|f\| + c_2\|f\|^{\frac{5}{2}})^p$, for all $t \in \mathbb{R}$, where $p = \frac{4}{3}$.

6.5.3 Nonautonomous Problems. In studying the time-independent problem in Theorem 65.7 (where the forcing function $f \in H$ does not depend on time t), we did use the fact that if φ is a weak solution of the Navier-Stokes equations, then for every $\tau \geq 0$, the time translate φ_τ is also a weak solution. When the forcing function depends on time, this conclusion is no longer valid. Instead, φ_τ is a solution of the translated problem

$$\partial_t u + \nu A u + B(u,u) = \mathbb{P} f_\tau.$$

In order to develop a dynamical theory to handle this situation, we use the traditional approach of skew-product flows, see Sell (1967a,b, 1973); Sacker and Sell (1977, 1994); Vishik (1992); Raugel and Sell (1993a,b, 1994); Shen and Yi (1998); and Babin and Sell (2000).

Let us look first at the 2D problem, where $f \in L^\infty(0,\infty;H)$ now depends on time t. The (positive) hull of f is defined by

$$H^+(f) = \mathrm{Cl}\{f_\tau : \tau \geq 0\}, \tag{65.13}$$

where $f_\tau(x,t) = f(x,\tau+t)$. In order to complete this defintion, one needs to specify the topology in which one takes the closure in (65.13). Since $f \in L^\infty(0,\infty;H)$, there are many possible choices for this topology; see Sell (1967, 1971), Miller and Sell (1970), and Sacker and Sell (1977). In particular, one need only require that the following three properties hold:

1. The mapping $(\tau, g) \to g_\tau$ defines a semiflow on $H^+(f)$.
2. The solution operators $S_w(t,g)u_0$ (or $S_s(t,g)u_0$) depend continuously on the parameters t, g, and u_0, for $t > 0$, $g \in H^+(f)$, and $u_0 \in H$ (or $u_0 \in V$).
3. The hull $H^+(f)$ is a compact set in this topology.

The two continuity properties in (1) and (2) insure that the skew product flow

$$\pi(g, u_0, \tau) = (g_\tau, S(\tau, g)u_0)$$

is a semiflow on $H^+(f) \times H$, when $S = S_w$, and on $H^+(f) \times V$, when $S = S_s$. The compactness of $H^+(f)$ and the methodology used above for the autonomous case imply that the skew product flow satisfies the hypotheses of the Existence Theorem 23.12, and therefore π has a global attractor $\mathfrak{A}$ in $H^+(f) \times H$, or in $H^+(f) \times V$, and $\mathfrak{A}$ attracts all bounded sets in $H^+(f) \times H$, or $H^+(f) \times V$. By using Theorem 24.4 on compact invariant sets for skew product flows, one readily verifies that $\mathfrak{A}$ satisfies the following.

Theorem 65.8. *For the 2D problem, let $H^+(f)$ be a compact set, as given above. Let $\mathfrak{A}_2$ denote the global attractor generated by the semiflow $(\tau, g) \to g_\tau$ on $H^+(f)$. Then there is a global attractor $\mathfrak{A}$ in $H^+(f) \times H$, or in $H^+(f) \times V$, and the following hold:*

1. $\mathfrak{A} \subset \mathfrak{A}_2 \times H$, *or* $\mathfrak{A} \subset \mathfrak{A}_2 \times V$.

(2) *If* $(g, u_0) \in \mathfrak{A}$, *then one has* $g \in \mathfrak{A}_2$.
(3) $\mathfrak{A}$ *attracts all bounded sets in* $H^+(f) \times H$, *or in* $H^+(f) \times V$.
(4) *For every* $g \in \mathfrak{A}_2$, *the set* $\{u_0 \in H : (g, u_0) \in \mathfrak{A}\}$, *or* $\{u_0 \in V : (g, u_0) \in \mathfrak{A}\}$, *is a nonempty, compact set in* H, *or* V.

The construction of a suitable skew product for the weak solutions of the 3D Navier-Stokes equations proceeds along a similar path. In this case we let $\mathfrak{F}$ be a set in $Y(N_0)$, for some $N_0 \geq 0$, and let $H^+(\mathfrak{F})$ be the hull of $\mathfrak{F}$. Then $H^+(\mathfrak{F}) \subset Y(N_0)$, and it follows from the general theory of semiflows that the hull is a positively invariant set for the semiflow on $L^2_{\text{loc}} = L^2_{\text{loc}}[0, \infty; L^2(\Omega))$ given by (65.5). For the remainder of this section we will assume that the hull $H^+(\mathfrak{F})$ is a compact set, see (65.4). Since the time-translation mapping $S(\tau)$ given by (65.5) is a semiflow on the hull, and since the hull is compact, it follows from the Existence Theorem 23.12 that there is a global attractor $\mathfrak{A}_2 \subset H^+(\mathfrak{F})$, and $\mathfrak{A}_2$ attracts all bounded sets in $H^+(\mathfrak{F})$.

Define $\mathcal{W}(H^+(\mathfrak{F}))$, and $\mathcal{W}_{LH}(H^+(\mathfrak{F}))$, to be the collection of all (φ, f) in $\mathcal{W}$, or in $\mathcal{W}_{LH}$, respectively, with the property that $f \in H^+(\mathfrak{F})$. Define the mapping

$$S(\tau)(\varphi, f) = (\varphi_\tau, f_\tau), \qquad \text{for } \tau \geq 0, \tag{65.14}$$

where $(\varphi, f) \in R = L^2_{\text{loc}}[0, \infty; H \times L^2(\Omega))$, see (63.38). From the comments made above, we see that S maps $[0, \infty) \times R$ into R. This semiflow is related to the skew-product dynamics used in Sacker and Sell (1977, 1994) and Sell (1967a,b, 1973). From Lemma 65.3 one has $S(\tau)(\varphi, f) \in \mathcal{W}_{LH}$, whenever $(\varphi, f) \in \mathcal{W}$ and $\tau > 0$. The argument of Lemma 65.4 now extends in a straightforward manner to establish the following result.

Lemma 65.9. *For the 3D problem, let* $H^+(\mathfrak{F})$ *be a* Y*-bounded set that is compact in the Fréchet space* $L^2_{\text{loc}}[0, \infty; L^2(\Omega))$. *Then the mapping* S *given by (65.14) is a semiflow on* R. *Furthermore, the sets* $\mathcal{W}(H^+(\mathfrak{F}))$ *and* $\mathcal{W}_{LH}(H^+(\mathfrak{F}))$ *are positively invariant subsets in this semiflow.*

Likewise the argument of Lemma 65.5 establishes the following result.

Lemma 65.10. *For the 3D problem, let* $H^+(\mathfrak{F})$ *be a* Y*-bounded set that is compact in the Fréchet space* $L^2_{\text{loc}}[0, \infty; L^2(\Omega))$. *Then the restriction of the semiflow* $S(t)$ *to* $\mathcal{W}(H^+(\mathfrak{F}))$ *is compact for* $t > 0$, *i.e., for each bounded set* B *in* $\mathcal{W}(H^+(\mathfrak{F}))$ *and for each* $\tau > 0$, *the set* $S(\tau)B$ *lies in a compact subset in* $\mathcal{W}$. *Moreover one has* $C\ell_R(S(\tau)B) \subset \mathcal{W}_{LH}(H^+(\mathfrak{F}))$, *for* $\tau > 0$.

Lastly the argument of Lemma 65.6 now establishes the following fact.

Lemma 65.11. *Let* $H^+(\mathfrak{F})$ *be a* Y*-bounded set that is compact in the Fréchet space* $L^2_{\text{loc}}[0, \infty; L^2(\Omega))$. *Then the restriction of the semiflow* $S(t)$ *to* $\mathcal{W}(H^+(\mathfrak{F}))$ *is point dissipative.*

Using these lemmas, with the Existence Theorem 23.12 and Theorem 24.4 on compact invariant sets for skew product flows, one can prove the following result.

Theorem 65.12. *For the 3D problem, let $H^+(\mathfrak{F})$ be a Y-bounded set that is compact in the Fréchet space $L^2_{\mathrm{loc}}[0,\infty; L^2(\Omega))$. Let $\mathfrak{A}_2$ denote the global attractor generated by the semiflow $(\tau, f) \to f_\tau$ on $H^+(\mathfrak{F})$. Then there is a global attractor $\mathfrak{A}$ in $\mathcal{W}(H^+(\mathfrak{F}))$, and the following hold:*

(1) $\mathfrak{A} \subset \mathcal{W}_{LH}(H^+(\mathfrak{F}))$.
(2) $\mathfrak{A}$ *attracts all bounded sets in* $\mathcal{W}(H^+(\mathfrak{F}))$.
(3) *If* $(\varphi, f) \in \mathfrak{A}$, *then one has* $f \in \mathfrak{A}_2$.
(4) *For every* $f \in \mathfrak{A}_2$, *the set* $\{\varphi \in W(f) : (\varphi, f) \in \mathfrak{A}\}$ *is a nonempty, compact set in* $W(f)$.

6.6. The Kwak Transformation.

In this section we describe a nonlinear transformation introduced in Kwak (1992b) to imbed the Navier-Stokes equations into a system of reaction diffusion equations. This transformation has the effect of changing the (nonlinear) inertial term $(u \cdot \nabla)u$ into a linear part plus some *algebraic* nonlinearities. We will denote the 2D and the 3D Navier-Stokes equations by 2DNS and 3DNS here.

As above we consider the Navier-Stokes equation (61.10) with periodic boundary conditions on $Q_2 = (0, \ell_1) \times (0, \ell_2)$ and $Q_3 = Q_2 \times (0, \ell_3)$, where $\ell_i > 0$, for $i = 1, 2, 3$. In this section we will impose an additional spatial smoothness assumption on the forcing term f and assume that $f \in D(A) \subset H^2(Q_2)$. One of the reasons for this added smoothness assumption is to insure that the global attractor for the 2DNS lies in the space $V^3 = V \cap H^3(Q_2, \mathbb{R}^2)$. See Kwak (1992b) for more details.

The Kwak Transformation is a nonlinear change of variables in either the 2DNS or the 3DNS, which has the property that it imbeds the Navier-Stokes equations on Ω into a system of reactions diffusion equations on Ω of the form

$$\frac{\partial}{\partial t} U + \mathbb{A} U = F(U), \tag{66.1}$$

where F is Lipschitz continuous and does not contain any terms with spatial derivatives. The significance of this transformation, as we will see, is that the nonlinear term $(u \cdot \nabla)u$ is decomposed into simpler pieces which are easier to analyze.

The first step is to handle the nonlinear term $B(u, u) = \mathbb{P}_d(u \cdot \nabla)u$, where $d = 2$, or 3, in a different form. We will assume that the vector field u is a column vector and that the gradient ∇ is also a column vector. In the 3D case, one has

$$u = \begin{pmatrix} u_1 \\ u_2 \\ u_3 \end{pmatrix}, \qquad \nabla = \begin{pmatrix} D_1 \\ D_2 \\ D_3 \end{pmatrix},$$

and we define

$$u \otimes u \overset{\text{def}}{=} uu^T = \begin{pmatrix} u_1^2 & u_1u_2 & u_1u_3 \\ u_1u_2 & u_2^2 & u_2u_3 \\ u_1u_3 & u_2u_3 & u_3^2 \end{pmatrix},$$

where ^{T} denotes the matrix transpose operation. Let Σ^d denote the collection of all $(d \times d)$ real symmetric matrices. Then one has $u \otimes u \in \Sigma^d$, for $d = 2$, or 3, and

$$(u \cdot \nabla)u = [\nabla^T(uu^T)]^T = [\nabla^T(u \otimes u)]^T.$$

Next define r by $r \overset{\text{def}}{=} u \otimes u = uu^T$. Then equation (61.10) can be rewritten as

$$\partial_t u + \nu A u + \mathbb{P}_d(\nabla^T r)^T = \mathbb{P}_d f. \tag{66.2}$$

The next step is to calculate the equation of motion for the new variable $r \in \Sigma^d$, given that u satisfies (61.10), or (66.2). As an intermediate step in this calculation, we define

$$z = (z^{(1)}, z^{(2)}), \qquad \text{or} \qquad z = (z^{(1)}, z^{(2)}, z^{(3)}),$$

for $d = 2$, or 3, where $z^{(i)} = D_i u = \partial_{x_i} u$, for $1 \le i \le d$. It is convenient to define the operation $\nabla * u$ by

$$\nabla * u \overset{\text{def}}{=} z.$$

Notice that for smooth u, one has $\nabla \cdot z^{(i)} = 0$, for $1 \le i \le d$, since $\nabla \cdot u = 0$. Next we apply D_i to (66.2) to obtain

$$\partial_t z^{(i)} + \nu A z^{(i)} + \mathbb{P}_d D_i(\nabla^T r)^T = \mathbb{P}_d D_i f, \qquad 1 \le i \le d.$$

Note that with the periodic boundary conditions one has $D_i \mathbb{P}_d u = \mathbb{P}_d D_i u$, for all $u \in V$. The equation for z then becomes

$$\partial_t z + \nu A z + \mathbb{P}_d \nabla * (\nabla^T r)^T = \mathbb{P}_d \nabla * f. \tag{66.3}$$

By using the identity $z_1^{(1)} + z_2^{(2)} + z_3^{(3)} = \nabla \cdot u = 0$, it is easily verified that $(\nabla^T r)^T = M(u, z)$, where

$$M(u,z) = \begin{pmatrix} u_1 z_1^{(1)} + u_2 z_1^{(2)} + u_3 z_1^{(3)} \\ u_1 z_2^{(1)} + u_2 z_2^{(2)} + u_3 z_2^{(3)} \\ u_1 z_3^{(1)} + u_2 z_3^{(2)} + u_3 z_3^{(3)} \end{pmatrix}.$$

Finally the equation for r becomes

$$\partial_t r - \nu \Delta r = N(u,z) + (\mathbb{P}_d f \otimes u + u \otimes \mathbb{P}_d f), \tag{66.4}$$

where

$$N(u,z) = -2\nu \sum_{i=1}^{d} z^{(i)} \otimes z^{(i)} - (\mathbb{P}_d M(u,z) \otimes u + u \otimes \mathbb{P}_d M(u,z)).$$

By combining (66.2), (66.3), and (66.4), we obtain the following system:

$$\begin{cases} \partial_t z + \nu A z + \mathbb{P}_d \nabla * (\nabla^T r)^T & = \mathbb{P}_d \nabla * f \\ \partial_t u + \nu A u + \mathbb{P}_d (\nabla^T r)^T & = \mathbb{P}_d f \\ \partial_t r - \nu \Delta r & = N(u,z) + (\mathbb{P}_d f \otimes u + u \otimes \mathbb{P}_d f). \end{cases} \tag{66.5}$$

The **Kwak Transformation** is then defined by

$$U = G(u) = (G_1, G_2, G_3), \qquad u \in V, \tag{66.6}$$

where $G_1(u) = \nabla * u$, $G_2(u) = u$, and $G_3(u) = u \otimes u = uu^T$.

The system of equations (66.5) for u, z, and r describes a candidate for the system of reaction diffusion equations (66.1), where $U = (z, u, r)^T$. In fact, the left side of (66.5) defines a candidate for the linear operator $\frac{\partial}{\partial t} U + \mathbb{A} U$, while the right side of this system is a candidate for the nonlinearity $F(U)$. Note that if equation (66.1) is defined in this way, it remains meaningful even if the vector U is not in the range of the Kwak Transformation G. As a result, one has considerable latitude in the construction of the reaction diffusion equation (66.1). For example, any nonlinearity $F(U)$ that agrees with the right side of (66.5) in a neighborhood of the range of G is another candidate for realizing the Kwak imbedding of the Navier-Stokes equations into a system of reaction diffusion equations.

This observation opens up the possibility that one might be able to select the nonlinearity $F(U)$ so that equation (66.1) agrees with (66.5) in a neighborhood of the range of G and other desirable properties hold. At the same time, one can add linear stabilizing terms of the form $k_1^2(z - G_1(u))$ to the left side of the z-equation in (66.5) without changing the dynamics of (66.5) on the range of G. Similarly one can add $k_1^2 u$ and $k_1^2 r$ to both sides of the u- and r-equations without altering these dynamics. Since we are primarily interested in the longtime dynamics, it suffices to insure that all such changes do not alter equation (66.5) on $G(\mathfrak{A}_{\mathrm{NS}})$, where $\mathfrak{A}_{\mathrm{NS}}$ is the global attractor of the 2DNS on Q_2. One objective then is to make such alterations so that the new system is dissipative, and as a result, equation (66.1) will have a global attractor $\mathfrak{A}_{\mathrm{RD}}$. Moreover, $\mathfrak{A}_{\mathrm{RD}}$ will necessarily satisfy the relationship $\mathfrak{A}_{\mathrm{RD}} \supset G(\mathfrak{A}_{\mathrm{NS}})$. The revised system we seek is

$$\begin{cases} \partial_t z + \nu A z + \mathbb{P}_d \nabla * (\nabla^T r)^T + k_1^2 (z - G_1(u)) & = F_1(U) \\ \partial_t u + \nu A u + \mathbb{P}_d (\nabla^T r)^T + k_1^2 u & = F_0(U) + k_1^2 u \\ \partial_t r - \nu \Delta r + k_1^2 r & = F_2(U) + k_1^2 r, \end{cases} \tag{66.7}$$

where $F = (F_1, F_0, F_2)$ agrees with the right side of (66.5) in a neighborhood of $G(\mathfrak{A}_{\text{NS}})$.

Let us restrict our attention for the moment to the 2DNS on Q_2. In this case we consider (66.7) on the function space

$$\mathcal{H} = \mathcal{H}(Q_2) \stackrel{\text{def}}{=} H(Q_2)^3 \times L^2(Q_2; \Sigma^2).$$

The following result, for the 2DNS on Q_2, is proved in Kwak (1992b).

Theorem 66.1. *Let the 2DNS be given on Q_2 with periodic boundary conditions. Then there is a system of reaction diffusion equations of the form (66.7) on $\mathcal{H} = \mathcal{H}(Q_2)$ with the following properties:*

(1) *Equation (66.7) agrees with (66.5) in a neighborhood of $G(\mathfrak{A}_{\text{NS}})$, where G is the Kwak Transformation (66.6), and $\mathfrak{A}_{\text{NS}}$ is the global attractor of the 2DNS.*
(2) *The linear operator $\mathbb{A}$, in equation (66.1), is a sectorial operator with compact resolvent on $\mathcal{H}$. Furthermore, the fractional powers of $\mathbb{A}$ are well-defined, and we set $\mathcal{V} = D(\mathbb{A}^{\frac{1}{2}})$.*
(3) *The nonlinearity $F = F(U)$ is a locally Lipschitz continuous mapping of $\mathcal{V}$ into $\mathcal{H}$. Furthermore F is a Lipschitz continuous mapping of the domain $D(\mathbb{A})$ into itself, and there exist constants K_0 and K_1 such that one has $\|\mathbb{A}F(U)\| \leq K_0$, for all $U \in D(\mathbb{A})$ and*

$$\|\mathbb{A}(F(U_1) - F(U_2))\| \leq K_1 \|\mathbb{A}(U_1 - U_2)\|, \qquad \text{for all } U_1, U_2 \in D(\mathbb{A}).$$

(4) *Equation (66.7) is dissipative on the space $\mathcal{V}$, and there is a global attractor $\mathfrak{A}_{\text{RD}}$, which is a bounded set in $D(\mathbb{A})$ that satisfies*

$$\mathfrak{A}_{\text{RD}} = G(\mathfrak{A}_{\text{NS}}).$$

Moreover, the restriction $G : \mathfrak{A}_{\text{NS}} \to \mathfrak{A}_{\text{RD}}$ is a homeomorphism which preserves all the longtime dynamics of the 2DNS.

Notice that, in both the 2D and 3D problems, the linear operator defined by the left side of (66.7) can be written in the upper triangular block form as

$$\mathbb{A} = \begin{pmatrix} \nu A + k_1^2 & * & * \\ 0 & \nu A + k_1^2 & * \\ 0 & 0 & -\nu\Delta + k_1^2 \end{pmatrix},$$

where * denotes various partial differential operators. This operator $\mathbb{A}$ is not self-adjoint. However, because of the block triangular form for $\mathbb{A}$, it is not difficult to show, in the periodic case, that $\mathbb{A}$ has exactly the same spectrum as $(\nu A + k_1^2 I)$, where A is the Stokes operator. (Of course, the multiplicities and the dimensions of the generalized eigenspaces, for these two operators will differ.) Hence, the spectrum $\sigma(\mathbb{A})$ consists of real eigenvalues only.

One of the steps used in the proof of Theorem 66.1 in Kwak (1992b) is a modification of the 2DNS given by (61.10), prior to the application of the Kwak Transformation G. This modification has the effect of leaving the original equation (61.10) unchanged in a bounded set B_0 and forcing the nonlinear terms to vanish outside another bounded set B_1, where $B_0 \subset B_1$, and both of these sets are open sets in $D(A)$. In Kwak (1992b) it is shown that the sets B_0 and B_1 can be chosen so that

(1) $\mathfrak{A}_{\mathrm{NS}} \subset B_0$, and
(2) the semiflow generated by the modified Navier-Stokes equations is dissipative and has the same global attractor $\mathfrak{A}_{\mathrm{NS}}$.

6.7. Related Nonlinear Systems.

Up to this point, the emphasis in this chapter has been based on the mathematical theory of incompressible, viscous fluid flows with homogeneous boundary conditions. While this theory is applicable to the study of such a fluid in a container subject to external forces, such as a gravity or a Coriolos force, one needs to incorporate additional equations into the model in order to study the effect of buoyancy forces resulting from heat or chemical differentials within the fluid. For example, in oceanic models one would augment the Navier-Stokes equations with two partial differential equations describing the evolution of heat and salinity within the ocean. In a model of a chemical plant, where the heat differential is caused by chemically reacting species within the fluid, one would add a family of partial differential equations describing the evolution of the concentrations of the various chemical species.

In this section we will describe two basic model problems: the Bénard convection model and a model of chemically reacting fluids. We will not develop the full mathematical theory of these models here. Instead we will focus on a brief description of the theory with references to the literature for further study. It should be noted that these models are the subject of ongoing research, and one can expect to see many important developments in the dynamical theory in the years ahead.

The models which we present here use the Boussinesq theory of incompressible fluid flows. This theory is based upon a physical argument, first proposed by Boussinesq, which concludes that, in the presence of small heat differentials, a good approximation to the physics of the underlying problem is formed by assuming the density of the fluid to be constant, both in time and in space. This in turn leads to the conservation equation $\nabla \cdot u = 0$ for the fluid. As we will explain below, the Boussinesq models are good mathematical models, even when the heat differentials are large. Of course, the connection between such a model and the underlying physical problem, in the context of large heat differentials, is an important scientific and technical issue. In order to resolve this issue fully, it may be important to use the theory of compressible fluid flows. However, that is another story. The Boussinesq models form the first step in trying to resolve this issue, and they are the focus of this section.

6.7.1 Inhomogeneous Boundary Conditions. There is another issue of concern in the theory of fluid flows which has serious and substantive dynamical consequences. In real physical problems, one practically never encounters the periodic boundary conditions (61.4), and one oftentimes has **inhomogeneous** boundary conditions in place of the homogeneous Dirichlet boundary conditions (61.3). In order to develop a theory for the inhomogeneous problems, a theory which parallels the theory of homogeneous problems given above, one introduces the **homogenization technique** for the boundary conditions. However, the homogenization of the boundary conditions for incompressible fluid flows, with $\nabla \cdot u = 0$, introduces new mathematical issues which do not occur in the case of reaction diffusion equations.

Consider an incompressible fluid flow on a smooth bounded domain Ω in $\mathbb{R}^d$, where $d = 2$ or 3. Assume further that the boundary of Ω consists of a finite number of components, say, $\partial\Omega = \Gamma_1 \cup \cdots \cup \Gamma_N$, where $N = 1$ is allowed. Next we assume that the velocity term u satisfies the boundary condition

$$u(x,t) = \phi(x), \qquad \text{for } t \geq 0,\ x \in \partial\Omega, \tag{67.1}$$

where ϕ is a sufficiently smooth function. In a typical application, one might have $\partial\Omega = \partial\Omega_i \cup \partial\Omega_w \cup \partial\Omega_e$, where i, w, e stand for *ingress, wall, egress* - that is $\partial\Omega_i$ contains the ingress points, where $\phi(x) \cdot n(x) < 0$, for $x \in \partial\Omega_i$, and $n = n(x)$ is the unit outward normal to $\partial\Omega$; $\partial\Omega_e$ contains the egress points, where $\phi(x) \cdot n(x) > 0$, for $x \in \partial\Omega_e$; and $\partial\Omega_w$ is the wall with $\phi(x) \cdot n(x) = 0$, for $x \in \partial\Omega_w$, see Norman (1999). Because of the conservation equation $\nabla \cdot u = 0$, one has the consistency condition

$$0 = \int_{\partial\Omega} u \cdot n\, dS = \int_{\partial\Omega} \phi \cdot n\, dS, \tag{67.2}$$

which is another condition that ϕ must satisfy. For example, one might have

$$\int_{\partial\Omega_i} \phi \cdot n\, dS \leq 0, \quad \int_{\partial\Omega_w} \phi \cdot n\, dS = 0, \quad \text{and} \quad \int_{\partial\Omega_e} \phi \cdot n\, dS \geq 0,$$

while

$$\int_{\partial\Omega_i} \phi \cdot n\, dS = -\int_{\partial\Omega_e} \phi \cdot n\, dS.$$

As usual, the fluid flow satisfies the Navier-Stokes equations (61.1) in Ω. However, the boundary conditions (61.3) or (61.4) are now replaced by (67.1)-(67.2). In order to analyze the Navier-Stokes equations in this case, one makes a change of variables $u = v + w$, where $w = w(x)$ is to be chosen so that

(1) $\partial_t w \equiv 0$, i.e., w does not depend on time t;
(2) $w = \phi$ on $\partial\Omega$; and
(3) $\nabla \cdot w = 0$ in Ω.

If such a function w exists, it is called a **homogenization term**. If w is sufficiently smooth, then $v = v(x,t)$ is a solution of the equations

$$\begin{aligned} \partial_t v - \nu\Delta v + (v\cdot\nabla)v + (w\cdot\nabla)v + (v\cdot\nabla)w + \nabla p = g \\ \nabla\cdot v = 0, \end{aligned} \tag{67.3}$$

where $g = f - (w\cdot\nabla)w + \nu\Delta w$, and

$$v(x,t) = 0, \qquad \text{for } t \geq 0,\ x \in \partial\Omega. \tag{67.4}$$

Thus the equation for v differs from (61.1) and (61.3) in that it contains the two linear terms $(v\cdot\nabla)w$ and $(w\cdot\nabla)v$ and the forcing term f is replaced by g. In other words, the problem (67.3) - (67.4) is a perturbation of the homogeneous problem (61.1) and (61.3).

The mathematical problem we now face is to determine what effect the homogenization term w has on the longtime dynamics of equation (67.3). However, before addressing this issue of dynamics, there is a more basic matter concerning the existence and uniqueness of w.

The basic theory on the existence of homogenization terms w can be found in Solonnikov and Ščadilov (1973), and Galdi (1994). Suffice it to say here that if the boundary of Ω and the function ϕ are sufficiently smooth, then a sufficiently smooth homogenization term always exists with $(w\cdot\nabla)w$ and Δw in H, but it need not be unique. As is shown in the exercises, the lack of uniqueness of the homogenization term w has no effect on the dynamics of (61.1) and (67.1) - (67.2). As a matter of fact, since w is not uniquely determined, this opens the possiblity of requiring w to satisfy additional conditions which are desirable for the study of the longtime dynamics of (61.1) and (67.1) - (67.2).

The theory of the existence and uniqueness of (weak and strong) solutions of the inhomogeneous problem (67.3) - (67.4) is based upon the Bubnov-Galerkin method, and it follows very closely the theory presented in Sections 6.3 and 6.4 for the homogeneous problem. However, equation (67.3) now presents a new challenge concerning the matter of the dissipativity of the weak solutions. To illustrate this, consider the following abbreviated and heuristic proof of inequality (63.7), which describes the dissipation of energy for the weak solutions of equation (61.1). By taking the scalar product of equation (61.1) with u and using the orthogonality relationship $u \perp \nabla p$, together with the Identity (61.15), the inequality (61.8) and the Young inequality, one obtains

$$\partial_t\|u\|^2 + \nu\lambda_1\|u\|^2 \leq \partial_t\|u\|^2 + \nu\|A^{\frac{1}{2}}u\|^2 \leq \frac{1}{\nu\lambda_1}\|f\|_\infty^2.$$

Inequality (63.7) now follows by applying the Gronwall inequality to the above.

In the case of equation (67.3), one uses the same approach and takes the scalar product of equation (67.3) with v. One then obtains the inequalitites

$$\partial_t \|v\|^2 + \nu\lambda_1 \|v|^2 \leq \partial_t \|v\|^2 + \nu \|A^{\frac{1}{2}} v\|^2 \leq \frac{1}{\nu\lambda_1} \|g\|_\infty^2 + 2|b(v, w, v)|.$$

In general one does not have $b(v, w, v) = 0$, and as a result, one finds that the homogenization term w and/or the boundary term ϕ can affect the longtime dynamics of the inhomogeneous problem.

6.7.2 Bénard Convection. In the next model we study the behavior of a 2D or 3D fluid in which the fluid is affected by two external forces, a gravitational force which pulls the fluid particles in one direction and a counteracting buoyancy force which is caused by maintaining a fixed thermal gradient across the physical space. We present a brief description of this model here. More details can be found in Foias, Manley, and Temam (1987) and Doering and Gibbon (1995).

Let Q be a given smooth bounded domain in the space $\mathbb{R}^{d-1}$, where $d = 2$ or 3. Set $\Omega = Q \times (0, 1)$. Let $x = (x_1, \cdots, x_d)$ denote the coordinates in Ω. We let u denote the velocity field of the fluid and we let $T = T(x, t)$ denote the temperature of the fluid. The equations of evolution in Ω are then given by

$$\begin{aligned} \partial_t u - \nu\Delta u + (u \cdot \nabla)u + \nabla p &= -g[1 + \alpha(T_1 - T)]e_d, \\ \partial_t T - \kappa\Delta T + (u \cdot \nabla)T &= 0, \\ \nabla \cdot u &= 0. \end{aligned}$$

In these equations ν, κ, α, and g are positive constants, and e_d is the unit vector in the x_d-direction. The quantity ν represents the kinematic viscosity of the fluid, κ is the thermometric conductivity, α is the volume expansion coefficient of the fluid, and g is the gravitational force. Notice that the term $-g\, e_d$ can be dropped by replacing the pressure p by $p + gx_d$. (The term gx_d is called the **hydrostatic pressure**.) One then has

$$\begin{aligned} \partial_t u - \nu\Delta u + (u \cdot \nabla)u + \nabla p &= g\alpha(T - T_1)e_d, \\ \partial_t T - \kappa\Delta T + (u \cdot \nabla)T &= 0, \\ \nabla \cdot u &= 0. \end{aligned} \tag{67.5}$$

The boundary conditions for the model are given by the following formulae:

$$\begin{cases} u = 0 & \text{at } x_d = 0 \text{ and } x_d = 1, \\ T = T_0 & \text{at } x_d = 0, \\ T = T_1 & \text{at } x_d = 1, \end{cases} \tag{67.6}$$

at the top and bottom, where $\delta T \stackrel{\text{def}}{=} T_0 - T_1$ is assumed to satisfy $\delta T \geq 0$, and on the sides one has

$$u = 0 \text{ and } \frac{\partial T}{\partial n} = 0 \qquad \text{on } \partial Q \times (0,1). \tag{67.7}$$

The identity $\frac{\partial T}{\partial n} = 0$ in (67.7) corresponds to the physical assumption that the walls of the container form a perfect insulator.

The boundary conditions (67.6) - (67.7) for the equations (67.5) can be homogenized by first introducing the new variable $\theta = g\alpha[(T - T_0) + x_d \delta T]$, and then replacing p by $p + \frac{1}{2}\tau x_d^2$, where $\tau \stackrel{\text{def}}{=} g\,\alpha\,\delta T$ is treated as as (bifurcation) parameter. (Notice that θ and τ represent changes in a new temperature scale.) One then obtains the new system

$$\begin{aligned} \partial_t u - \nu \Delta u + (u \cdot \nabla) u + \nabla p &= \theta e_d, \\ \partial_t \theta - \kappa \Delta \theta + (u \cdot \nabla)\theta &= \tau u_d, \\ \nabla \cdot u &= 0. \end{aligned}$$

The boundary conditions for the model are given by the following formulae:

$$\begin{cases} u = 0 & \text{at } x_d = 0 \text{ and } x_d = 1, \\ \theta = 0 & \text{at } x_d = 0 \text{ and } x_d = 1, \end{cases}$$

at the top and bottom, while (67.7) holds on the sides.

One sometimes finds Bénard convection modeled with other boundary conditions. For example, in Foias, Manley, and Temam (1987), the domain Q is assumed to be either an interval or a rectangle, and the side boundary conditions (67.7) are replaced by requiring that u and T be periodic in space. On the other hand, in Doering and Gibbon (1995) the side conditions (67.7) are used, while the top and bottom condtions for u are replaced by the "stress free" conditions:

$$u_3 = \partial_{x_3} u_1 = \partial_{x_3} u_2 = 0 \qquad \text{at } x_3 = 0 \text{ and } x_3 = 1,$$

in the 3D problem.

6.7.3 Chemically Reacting Flows. For the theory of chemically reacting fluid flows, we will borrow heavily from the presentation in Norman (1999, 2000a,b). In this case one adds different physics to the Bénard problem in that the thermal differential is now caused by the heat produced by chemically reacting agents within the fluid. The equations of evolution, which describe the coupling between u, T, and N chemical species with respective mass fractions Y_j, for $j = 1, \cdots, N$, are given by

$$\begin{aligned} \partial_t u - \nu \Delta u + (u \cdot \nabla) u + \nabla p &= f(T), \\ \partial_t T - \kappa \Delta T + (u \cdot \nabla) T &= -\sum_{j=1}^{N} h_j W_j(Y, T), \\ \partial_t Y_j - d \Delta Y_j + (u \cdot \nabla) Y_j &= W_j(Y, T), \\ \nabla \cdot u &= 0 \end{aligned}$$

in Ω, where $f(T)$ is the effective buoyancy force acting on the fluid, $W_j(Y,T) = W_j(Y_1, \cdots, Y_N, T)$ is the change in the mass fractions Y_j owing to the reaction, h_j is the enthalpy of the jth species, and d is the diffusivity of the chemical species. The term $\text{Le} = d^{-1}$ is called the Lewis number, and the viscosity $\nu = \text{Pr}$ is the Prandl number. Since Y_j is the mass fraction of the jth species it satisfies $0 \leq Y_j \leq 1$, for $j = 1, \cdots, N$, and $\sum_{j=1}^N Y_j = 1$. Thus the vector $Y = (Y_1, \cdots, Y_N)$ assumes values the set $[0,1]^N \subset \mathbb{R}^N$.

The inequality $W_j(Y,T) > 0$ corresponds to an increase in the jth species owing to the chemical reaction, and $W_j < 0$ yields a decrease. One assumes that there is a temperature $T_{\min}$ such that the terms W_j satisfy the following relationships:

(1) Each W_j is Lipschitz continuous in Y and T, and the mass conservation identity holds, i.e.,

$$\sum_{j=1}^{N} W_j(Y,T) = 0, \qquad \text{for all } (Y,T) \in [0,1]^N \times [T_{\min}, \infty).$$

(2) The implication $W_j < 0 \Longrightarrow Y_j > 0$ is valid. (For example, one has

$$W_j(Y,T) = \alpha_j - \beta_j Y_j, \qquad \text{for } j = 1, \cdots, N,$$

where $\alpha_j = \alpha_j(Y,T) \geq 0$ and $\beta_j = \beta_j(Y,T) > 0$.)

(3) There is a real number $B \in [0, \infty)$ such that

$$|W_j(Y,T)| \leq B, \qquad \text{for all } (Y,T) \in [0,1]^N \times [T_{\min}, \infty),$$

and $-\sum_{j=1}^N h_j W_j(Y, T_{\min})) \geq 0$, for all $Y \in [0,1]^N$.

A commonly used model for chemically reacting fluid flows is the Arrhenius model in which W_j assumes the form

$$W_j(Y,T) = \sum_{k=1}^{m_j} A_k e^{-E_k/R_0 T} \prod_{m=1}^{N} C_m^{\nu_{k,m}},$$

where the A_k are the frequency factors, the E_k are the activation energies, R_0 is the universal gas constant, $C_m = Y_m M_m$, the $\nu_{k,m}$ are nonnegative integers, and $T_{\min} = 0$, see Norman (1999).

For this model, these equations are satisfied on a smooth, bounded domain Ω in $\mathbb{R}^3$. The boundary conditions for the problem involve inhomogeneous terms. One begins by decomposing $\partial\Omega$ into three parts - $\partial\Omega = \partial\Omega_i \cup \partial\Omega_e \cup \partial\Omega_w$ - where $\partial\Omega_i$ is that portion of the boundary where the fluid (with the chemical species) flows into Ω, $\partial\Omega_e$ is that portion where the fluid flows out of Ω, and $\partial\Omega_w$ is a wall where there is no flux across the wall. In particular, one assumes that

$$\begin{aligned} u(x,t) &= \phi(x), \qquad \text{where } \phi \cdot n < 0, \\ \frac{\partial T}{\partial n} &- (\phi(x) \cdot n) T(x,t) = T_f(x), \\ d\frac{\partial Y_j}{\partial n} &- (\phi(x) \cdot n) Y_j(x,t) = Y_{j,f}(x), \end{aligned}$$

on $\partial\Omega_i$, for $j = 1, \cdots, N$. On $\partial\Omega_e$ one assumes that

$$u(x,t) = \phi(x), \qquad \text{where } \phi \cdot n > 0,$$
$$\frac{\partial T}{\partial n}(x,t) = \frac{\partial Y_j}{\partial n}(x,t) = 0,$$

and on $\partial\Omega_w$ one assumes that

$$u(x,t) = 0 \quad \text{and} \quad \frac{\partial T}{\partial n}(x,t) = \frac{\partial Y_j}{\partial n}(x,t) = 0,$$

for $j = 1, \cdots, N$. Finally one requires that ϕ satisfy the consistency Identity (67.2), or

$$\int_{\partial\Omega_i} \phi \cdot n\, dS + \int_{\partial\Omega_e} \phi \cdot n\, dS = 0.$$

We assume that the boundary data ϕ, T_f, and $Y_{j,f}$ are sufficiently smooth so that there are functions in $H^2(\Omega)$ that satisfy the boundary conditions. A physical interpretation of these boundary conditions can be found in Norman (1999).

A comprehensive theory of the longtime dynamics of the chemically reacting fluid flows, as described here, can be found the in three papers Norman (1999, 2000a,b). Since one is working in the class of weak solutions, it is not at all clear that there exist physically reasonable solutions, for appropriate choice of the data. Nevertheless, it is shown that there exists a good supply of physically reasonable weak solutions, and the semiflow generated by these solutions has a global attractor, for the 3D problem. The description of the global attractor for this problem is similar to that derived for the 3D Navier-Stokes equations alone in Section 6.5.2. A related theory of global attractors for 2D chemically reacting fluid flows, with different boundary conditions, can be found in Marion (1991), Manley and Marion (1992), and Manley, Marion, and Temam (1993).

6.8. Proofs for the Bubnov-Galerkin Approximations.

We now turn to the proofs of the three basic lemmas given in Section 6.2.

Proof of Lemma 62.1. By taking the scalar product of (62.2) with p and using the Young inequality and (61.15), one obtains

$$\frac{1}{2}\partial_t\|p\|^2 + \nu\|A^{\frac{1}{2}}p\|^2 \overset{\text{a.e.}}{=} \langle Pf, p\rangle \overset{\text{a.e.}}{=} \langle A^{-\frac{1}{2}}Pf, A^{\frac{1}{2}}p\rangle$$
$$\le \frac{\nu}{2}\|A^{\frac{1}{2}}p\|^2 + \frac{1}{2\nu\lambda_1}\|f\|_\infty^2. \tag{68.1}$$

This in turn implies that

$$\partial_t\|p\|^2 + \nu\lambda_1\|p\|^2 \le \partial_t\|p\|^2 + \nu\|A^{\frac{1}{2}}p\|^2 \le \frac{1}{\nu\lambda_1}\|f\|_\infty^2. \tag{68.2}$$

Since P is an orthogonal projection one has $\|p_0\|^2 \le \|u_0\|^2$. As a result, the Gronwall inequality then implies (62.4). By integrating the second inequality in (68.2) from t_1 to t, one obtains (62.5). It is a consequence of (62.5) that

$$\int_0^t \|A^{\frac{1}{2}}p(s)\|^2 ds \le \frac{1}{\nu}\|u_0\|^2 + \frac{t}{\nu^2\lambda_1}\|f\|_\infty^2, \qquad \text{for } t \ge 0, \tag{68.3}$$

since $\|p(0)\| \le \|u_0\|$. Furthermore (62.4) and (62.5) imply that

$$\int_{t-1}^t \|A^{\frac{1}{2}}p(s)\|^2\, ds \le M_0 e^{-\nu\lambda_1 t} + L_0, \qquad \text{for } t \ge 1, \tag{68.4}$$

where

$$M_0 = \nu^{-1}e^{\nu\lambda_1}\|u_0\|^2 \quad \text{and} \quad L_0 = \frac{\|f\|_\infty^2}{\nu^2\lambda_1}\left(1 + \frac{1}{\nu\lambda_1}\right).$$

It follows from (68.3) that for any $T > 0$ the sequence $u_n = p$ lies in a bounded set in $L^2(0,T;V)$. Recall that A is an isometry from $V^1 = V$ onto the dual space $V^{-1} = V'$, i.e., $\|Av\|_{-1} = \|v\|_1$ for all $v \in V$ (see Section 3.7). Since $\|A^{\frac{1}{2}}p\|_0 = \|p\|_1 = \|Ap\|_{-1}$, we see that $\|Ap\|$ satisfies

$$\int_0^t \|Ap(s)\|_{-1}^2 ds = \int_0^t \|A^{\frac{1}{2}}p(s)\|^2 ds \le \frac{1}{\nu}\|u_0\|^2 + \frac{t}{\nu^2\lambda_1}\|f\|_\infty^2, \qquad \text{for } t \ge 0,$$

that is, $Au_n = Ap$ lies in a bounded set in $L^2(0,T;V^{-1})$, for every $T > 0$.

Next let us examine the nonlinear term $PB(p,p)$ in (62.2). Since the orthogonal projection P commutes with A^α, for any $\alpha \in \mathbb{R}$, one has

$$\|PB(p,p)\|_{-1} = \|A^{-\frac{1}{2}}PB(p,p)\| \le \|A^{-\frac{1}{2}}B(p,p)\|.$$

From (61.9) and (61.26), we then obtain

$$\|PB(p,p)\|_{-1} \le C_5\|p\|^{\frac{1}{2}}\|A^{\frac{1}{2}}p\|^{\frac{3}{2}}. \tag{68.5}$$

From (62.4), (68.3) and (68.5) we see that there is a constant c_6 such that, for all $t \ge 0$, one obtains

(68.6)

$$\int_0^t \|P_nB(p,p)\|_{-1}^{\frac{4}{3}}\, ds \le c_0 \int_0^t \|A^{\frac{1}{2}}p\|^2 ds \le c_6\left(\frac{1}{\nu}\|u_0\|^2 + \frac{t}{\nu^2\lambda_1}\|f\|_\infty^2\right),$$

Let us now turn to the derivative $\partial_t u_n = \partial_t p$, which satisfies

$$\partial_t u_n \overset{\text{a.e.}}{=} -\nu A u_n - P_nB(u_n,u_n) + P_n f.$$

Since $L^2(0,T;V^{-1}) \mapsto L^{\frac{4}{3}}(0,T;V^{-1})$ and $f \in L^{\frac{4}{3}}(0,T;V^{-1})$, for every $T > 0$, it follows from (68.3) and (68.6) that the sequence $\partial_t u_n$ lies in a bounded set in $L^{\frac{4}{3}}(0,T;V^{-1})$, when $d=3$. In 2D we use (61.25) instead of (61.26) to obtain

$$\|PB(p,p)\|_{-1} \leq \|A^{-\frac{1}{2}}B(p,p)\| \leq C_4\|p\|\,\|A^{\frac{1}{2}}p\|. \tag{68.7}$$

From (62.4), (68.3) and (68.7) we see that there is a constant c_7 such that, for $n \geq 1$, one then obtains

$$\int_0^t \|P_nB(p,p)\|_{-1}^2\,ds \leq c_7 \int_0^t \|A^{\frac{1}{2}}p\|^2\,ds \leq c_7\left(\frac{1}{\nu}\|u_0\|^2 + \frac{t}{\nu^2\lambda_1}\|f\|_\infty^2\right).$$

It then follows that the sequence $\partial_t u_n$ lies in a bounded set in $L^2(0,T;V^{-1})$, for every $T>0$, when $d=2$. Moreover, (62.7) holds.

If (62.8) were false, then from (62.4) one has

$$\begin{aligned}\int_{t_1}^t \|A^{\frac{1}{2}}p\|^2\,ds &\geq 2\left(\frac{\|u_0\|^2}{\nu}e^{-\nu\lambda_1 t_1} + \frac{\|f\|_\infty^2}{\nu^2\lambda_1}\left(t-t_1+\frac{1}{\nu\lambda_1}\right)\right)\\ &\geq \frac{2}{\nu}\left(\|p(t_1)\|^2 + \frac{t-t_1}{\nu\lambda_1}\|f\|_\infty^2\right),\end{aligned}$$

which contradicts (62.5). In order to prove the energy inequality (62.9), we return to (68.1) to obtain

$$\frac{1}{2}\partial_t\|p\|^2 + \nu\|A^{\frac{1}{2}}p\|^2 \overset{\text{a.e.}}{=} \langle f,p\rangle.$$

By integrating this from t_0 to t, we obtain (62.9). Finally the Variation of Constants Formula is a standard result from the finite dimensional theory of ordinary differential equations. □

Proof of Lemma 62.3. (1) We return to the Bubnov-Galerkin system (62.2) with initial condition (62.3). By taking the scalar product of (62.2) with Ap, and by using the 2D auxiliary estimate (61.23) together with the Young inequality, we obtain

$$\begin{aligned}\frac{1}{2}\partial_t\|A^{\frac{1}{2}}p\|^2 + \nu\|Ap\|^2 &\leq |\langle f,Ap\rangle| + |b(p,p,Ap)|\\ &\leq \frac{\nu}{4}\|Ap\|^2 + \frac{1}{\nu}\|f\|_\infty^2 + C_2\|p\|^{\frac{1}{2}}\|A^{\frac{1}{2}}p\|\,\|Ap\|^{\frac{3}{2}}\\ &\leq \frac{\nu}{2}\|Ap\|^2 + \frac{1}{\nu}\|f\|_\infty^2 + \frac{27C_2^4}{4\nu^3}\|p\|^2\|A^{\frac{1}{2}}p\|^4,\end{aligned}$$

which implies (62.11).

(2) As a result of (62.11), one obtains

$$\partial_t\|A^{\frac{1}{2}}p\|^2 \leq \left(\frac{27C_2^4}{2\nu^3}\|p\|^2\|A^{\frac{1}{2}}p\|^2\right)\|A^{\frac{1}{2}}p\|^2 + \frac{2}{\nu}\|f\|_\infty^2. \tag{68.8}$$

By applying the Gronwall inequality to (68.8), one finds that

$$\|A^{\frac{1}{2}}p(t)\|^2 \leq \|A^{\frac{1}{2}}p(t_0)\|^2 \exp\left(\int_{t_0}^{t} g(s)\,ds\right) + \frac{2}{\nu}\|f\|_\infty^2 \int_{t_0}^{t} \exp\left(\int_{s}^{t} g(r)\,dr\right) ds,$$

for $0 \leq t_0 < t < \infty$, where $g = \frac{27C_2^4}{2\nu^3}\|p\|^2\|A^{\frac{1}{2}}p\|^2$. Now (62.4) and (62.5) imply that

$$\begin{aligned} \int_s^t g(s)\,ds &\leq \frac{27C_2^4}{2\nu^3} \sup_{s\leq r\leq t} \|p(r)\|^2 \left(\|p(s)\|^2 + \frac{(t-s)}{\nu\lambda_1}\|f\|_\infty^2\right) \\ &\leq \frac{27C_2^4}{2\nu^3}\left(e^{-\nu\lambda_1 s}\|u_0\|^2 + \frac{\|f\|_\infty^2}{\nu^2\lambda_1^2}\right) \times \\ &\quad \left(e^{-\nu\lambda_1 s}\|u_0\|^2 + \frac{\|f\|_\infty^2}{\nu\lambda_1}\left(t-s+\frac{1}{\nu\lambda_1}\right)\right) \\ &\leq K_6 + K_7(t-s), \end{aligned} \tag{68.9}$$

for $0 \leq s \leq t$, where

$$K_6 = \frac{27C_2^4}{2\nu^3}\left(\|u_0\|^2 + \frac{\|f\|_\infty^2}{\nu^2\lambda_1^2}\right)^2 \quad \text{and} \quad K_7 = \frac{27C_2^4}{2\nu^3}\left(\|u_0\|^2 + \frac{\|f\|_\infty^2}{\nu^2\lambda_1^2}\right)\frac{\|f\|_\infty^2}{\nu\lambda_1}.$$

Consequently one obtains

$$\|A^{\frac{1}{2}}p(t)\|^2 \leq \|A^{\frac{1}{2}}p(t_0)\|^2 e^{K_6} e^{K_7(t-t_0)} + \frac{2}{\nu K_7}\|f\|_\infty^2 e^{K_6} e^{K_7(t-t_0)}, \tag{68.10}$$

for $0 \leq t_0 < t < \infty$. Let us now restrict t to $0 < t \leq 1$ and set $t_1 = 0$. Then choose t_0 with $0 < t_0 < t \leq 1$ so that (62.8) holds. By combining this with (68.10), we then obtain $\|A^{\frac{1}{2}}p(t)\|^2 \leq K_0 + K_1 t^{-1}$, for $0 < t \leq 1$, where

$$K_0 = \|f\|_\infty^2\left(\frac{1}{\nu^2\lambda_1} + \frac{2}{\nu K_7}\right)e^{(K_6+K_7)} \quad \text{and}$$

$$K_1 = \left(\frac{2\|u_0\|^2}{\nu} + \frac{2\|f\|_\infty^2}{\nu^3\lambda_1^2}\right)e^{(K_6+K_7)}.$$

This is the first inequality in (62.12).

In order to obtain the second inequality in (62.12), we will need an improved version of (68.10). This improvement, which is based on the Uniform Gronwall inequality (see Lemma D.3), is argued as follows. By integrating (62.11), one has

$$\nu\int_{t-1}^{t}\|Ap\|^2\,ds \leq \|A^{\frac{1}{2}}p(t-1)\|^2 + \frac{2\|f\|_\infty^2}{\nu} + \frac{27C_2^4}{2\nu^3}\int_{t-1}^{t}\|p\|^2\|A^{\frac{1}{2}}p\|^4\,ds < \infty, \tag{68.11}$$

for $t \geq 1$, and

$$\text{(68.12)} \quad \nu \int_0^t \|Ap\|^2\, ds \leq \|A^{\frac{1}{2}}p_0\|^2 + t\frac{2\|f\|_\infty^2}{\nu} + \frac{27C_2^4}{2\nu^3}\int_0^t \|p\|^2\|A^{\frac{1}{2}}p\|^4\, ds,$$

for $t \geq 0$. From (62.4) and (68.4) one obtains
(68.13)

$$\begin{aligned}
\int_{t-1}^t g(s)\, ds &= \frac{27C_2^4}{2\nu^3}\int_{t-1}^t \|p\|^2\|A^{\frac{1}{2}}p\|^2\, ds \\
&\leq \frac{27C_2^4}{2\nu^3}\sup_{t-1\leq s\leq t}\|p(s)\|^2\int_{t-1}^t \|A^{\frac{1}{2}}p\|^2\, ds \\
&\leq \frac{27C_2^4}{2\nu^3}\left(e^{-\nu\lambda_1 t}e^{\nu\lambda_1}\|u_0\|^2 + \frac{\|f\|_\infty^2}{\nu^2\lambda_1^2}\right)(M_0e^{-\nu\lambda_1 t} + L_0) \\
&\leq M_1e^{-\nu\lambda_1 t} + L_1
\end{aligned}$$

for $t \geq 1$, where

$$M_1 = \frac{27C_2^4}{2\nu^3}\left(e^{\nu\lambda_1}\|u_0\|^2(M_0 + L_0) + M_0\frac{\|f\|_\infty^2}{\nu^2\lambda_1^2}\right) \text{ and } L_1 = \frac{27C_2^4}{2\nu^3}\frac{\|f\|_\infty^2}{\nu^2\lambda_1^2}L_0.$$

From the Uniform Gronwall inequality applied to (62.11), one finds that

$$\text{(68.14)} \quad \|A^{\frac{1}{2}}p(t)\|^2 \leq \left(M_0e^{-\nu\lambda_1 t} + L_0 + \frac{2}{\nu}\|f\|_\infty^2\right)\exp\left(M_1e^{-\nu\lambda_1 t} + L_1\right),$$

for $t \geq 1$. By using the inequality

$$\text{(68.15)} \qquad e^{xD} \leq 1 + xDe^D, \qquad \text{for } 0 \leq x \leq 1, \text{ and } D \geq 0,$$

with $x = e^{-\nu\lambda_1 t}$ and $D = M_1$ in (68.14), one then obtains the second inequality in (62.12), where

$$M_2 = M_0e^{L_1}(1 + M_1e^{M_1}) + (L_0 + 2\nu^{-1}\|f\|_\infty^2)e^{L_1}M_1e^{M_1}$$

and $L_2 = (e^{L_1} + 1)(L_0 + 2\nu^{-1}\|f\|_\infty^2)$.

In order to prove (62.13), when $u_0 \in V$, we return to (68.10) and set $t_0 = 0$. Since $\|A^{\frac{1}{2}}p_0\| \leq \|A^{\frac{1}{2}}u_0\|$, for $u_0 \in V$, we find that

$$\text{(68.16)} \quad \|A^{\frac{1}{2}}p(t)\|^2 \leq \left(\|A^{\frac{1}{2}}u_0\|^2 + \frac{2}{\nu K_7}\|f\|_\infty^2\right)e^{K_6+K_7}, \qquad 0 \leq t \leq 1.$$

By enlarging M_2 and L_2, if necessary, we obtain (62.13), for $0 \leq t \leq 1$, and for $t \geq 1$, (62.13) follows from (62.12). By integrating (62.11), we obtain

$$\text{(68.17)} \quad \nu\int_{t_0}^t \|Ap\|^2\, ds \leq \|A^{\frac{1}{2}}p(t_0)\|^2 + \frac{2\|f\|_\infty^2}{\nu}(t - t_0) + \int_{t_0}^t g\|A^{\frac{1}{2}}p\|^2\, ds,$$

for $0 < t_0 \leq t$, where g is given above. For $0 < t_0 \leq t \leq 1$, one then has

$$\int_{t_0}^{t} g\|A^{\frac{1}{2}}p\|^2\,ds \leq \sup_{t_0\leq s\leq t} \|A^{\frac{1}{2}}p(s)\|^2 \int_{t_0}^{t} g\,ds \leq (K_0 + K_1 t_0^{-1})(K_6 + K_7)$$

by (62.12) and (68.9). Consequently, we obtain (62.14) where

$$K_2 = \frac{K_6 + K_7 + 1}{\nu} K_0 + \frac{2\|f\|_\infty^2}{\nu^2} \qquad \text{and} \qquad K_3 = \frac{K_6 + K_7 + 1}{\nu} K_1.$$

Inequality (62.15) now follows from (62.12) and (68.11), where

$$M_3 = M_2 \nu^{-1} e^{\nu\lambda_1}(1 + M_1 + L_1) + M_1 \nu^{-1} L_2 \quad \text{and}$$
$$L_3 = \nu^{-1}(L_2 + 2\nu^{-1}\|f\|_\infty^2 + L_2 L_1).$$

In order to prove (62.16) we use (68.17) with $t_0 = 0$. Now (62.4) and (68.16) imply that

$$\|p(s)\|^2\|A^{\frac{1}{2}}p(s)\|^4 \leq \left(\|p_0\|^2 + \frac{\|f\|_\infty^2}{\nu^2\lambda_1^2}\right)\left(\|A^{\frac{1}{2}}u_0\|^2 + \frac{2\|f\|_\infty^2}{\nu K_7}\right)^2 e^{2(K_6+K_7)},$$

for $0 \leq s \leq 1$, and (62.4) and (62.12) imply that

$$\|p(s)\|^2\|A^{\frac{1}{2}}p(s)\|^4 \leq \left(\|p_0\|^2 + \frac{\|f\|_\infty^2}{\nu^2\lambda_1^2}\right)(M_2 + L_2)^2 \qquad 1 \leq s < \infty.$$

Since $\|p_0\| \leq \|u_0\|^2 \leq \lambda_1^{-1}\|A^{\frac{1}{2}}u_0\|^2$ and $\|A^{\frac{1}{2}}p_0\|^2 \leq \|A^{\frac{1}{2}}u_0\|^2$, inequality (62.16) follows from (68.17) and the last two inequalities.

Finally, it follows from inequalities (61.22), (62.13), and (62.16) that the integral $\int_0^t \|PB(p,p)\|^2\,ds$ has a bound that does not depend of the order n of the Bubnov-Galerkin approximation. Since $\partial_t p = Pf - \nu Ap - PB(p,p)$, it follows from this fact, inequality (62.16), and the fact that $f \in L^\infty(0,\infty;H)$ that $\int_0^t \|\partial_t p\|^2\,ds$ also has a bound that is independent of n. □

Before turning to Lemma 62.4 and the 3D theory, it is appropriate to reflect on why the Uniform Gronwall inequality can be used on (62.11). The reason for this is that g, which is the coefficient of $\|A^{\frac{1}{2}}p\|^2$ in (68.8), satisfies (68.13), i.e.,

$$\text{(68.18)} \qquad \int_{t-1}^{t} g(s)\,ds \leq M_1 e^{\nu\lambda_1 t} + L_1 \leq (M_1 + L_1), \qquad t \geq 1.$$

As we will now see, in the 3D theory we are unable to derive the appropriate analog of (68.18). This fact lies at the heart of the profound differences between the 2D and 3D theories. The reason for this, as we will see, is that in the 3D case one obtains the differential inequality (62.17). This differs from the 2D version (62.11) in an essential way, viz., the last term in (62.11) contains the factor $\|p\|^2\|A^{\frac{1}{2}}p\|^4$, as opposed to $\|A^{\frac{1}{2}}p\|^6$, which occurs in (62.17). The coefficient of $\|A^{\frac{1}{2}}p\|^2$ now becomes

$$g(t) = \text{const } \times \|A^{\frac{1}{2}}p(t)\|^4.$$

While g is integrable, it is not known whether the integral $\int_{t-1}^{t} g(s)\,ds$ has an upper bound that is independent of n. In other words, one does not know whether the weak solution $u(t)$ is in $L^4_{\text{loc}}[0,\infty;V)$ in the 3D case.

Proof of Lemma 62.4. By taking the scalar product of the Bubnov-Galerkin approximate equation (62.2) with Ap, and by using (61.24) together with the Young inequality, one obtains

$$\begin{aligned}\frac{1}{2}\partial_t\|A^{\frac{1}{2}}p\|^2 + \nu\|Ap\|^2 &\le |\langle f, Ap\rangle| + |b(p,p,Ap)| \\ &\le \frac{\nu}{4}\|Ap\|^2 + \frac{1}{\nu}\|f\|_\infty^2 + C_3\|A^{\frac{1}{2}}p\|^{\frac{3}{2}}Ap\|^{\frac{3}{2}} \\ &\le \frac{\nu}{2}\|Ap\|^2 + \frac{1}{\nu}\|f\|_\infty^2 + \frac{27C_3^4}{4\nu^3}\|A^{\frac{1}{2}}p\|^6.\end{aligned} \tag{68.19}$$

As a result, one obtains (62.17), which in turn yields

$$\partial_t\|A^{\frac{1}{2}}p\|^2 \le \frac{2}{\nu}\|f\|_\infty^2 + \frac{27C_3^4}{2\nu^3}\|A^{\frac{1}{2}}p\|^6. \tag{68.20}$$

We now use Lemma D.4 to conclude that

$$\|A^{\frac{1}{2}}p(t)\|^2 \le r(t), \qquad t_0 \le t < t_0 + T_1, \tag{68.21}$$

where r and T_1 are defined in Lemma D.4, $\|A^{\frac{1}{2}}p(t_0)\|^2 \le r_0$, $a = 2\nu^{-1}\|f\|_\infty^2$, and $b = 27C_3^4(2\nu^3)^{-1}$. So we set $T_0 = T_1$ and $\rho(t)^2 = r(t)$. By returning to (62.17) and integrating the second inequality, we obtain (62.19).

Finally, the proof that, for each $t \in [0,T)$, the integral $\int_0^t \|\partial_t p\|^2\,ds$ has a bound which depends on t, but which is independent of the order n of the Bubnov-Galerkin approximation, is identical to the argument used in the proof of Lemma 62.3. The reader should verify the details. □

6.9. Exercises.

Section 6.1

61.1. The **vorticity** v of a velocity field u on an open region Ω in $\mathbb{R}^3$ is defined as $v = \nabla \times u$.

(1) Show that if $u = u(x,t)$ is a solution of (61.10) on an interval I, then $\nabla \cdot v = 0$ and v is a solution of

$$\partial_t v - \nu\Delta v + (u\cdot\nabla)v - (v\cdot\nabla)u = \nabla \times f. \tag{69.1}$$

(2) For an open region Ω in $\mathbb{R}^2$, the vorticity is a scalar quantity $v = D_1u_2 - D_2u_1$. Show that equation (69.1) now assume the form

$$\partial_t v - \nu\Delta v + (u\cdot\nabla)v = D_2 f_1 - D_1 f_2.$$

Let $u = u(x,t)$ be a solution of (61.10) on a simply connected region Ω in $\mathbb{R}^2$. Then the Green Theorem implies that there is a scalar field φ on Ω that satisfies $u = (D_2\varphi, -D_1\varphi)$. The function φ is referred to as the **stream function** for u.

(3) Show that the vorticity v satisfies $v = -\Delta\varphi$.

(4) Show that φ is a solution of the equation

$$\partial_t\Delta\varphi - \nu\Delta^2\varphi + D_2\varphi D_1\Delta\varphi - D_1\varphi D_2\Delta\varphi = D_2 f_1 - D_1 f_2.$$

61.2 (Other Coordinate Systems). Let $\vec{u} = (u_1, u_2) = u_1\vec{i} + u_2\vec{j}$ be a vector field on an open region Ω in $\mathbb{R}^2$, where $0 \notin \Omega$. In polar coordinates, set

$$\vec{r} = \frac{1}{r}(x,y) = (\cos\theta, \sin\theta) \quad \text{and} \quad \vec{\theta} = \frac{1}{r}(-y,x) = (-\sin\theta, \cos\theta).$$

The polar coordinates of the vector field $\vec{u}$ are defined by

$$u_r = \vec{u}\cdot\vec{r} = u_1\cos\theta + u_2\sin\theta, \quad \text{and} \quad u_\theta = \vec{u}\cdot\vec{\theta} = -u_1\sin\theta + u_2\cos\theta.$$

For the vector field $\vec{u}$, we define $\Delta\vec{u}$ by $\Delta\vec{u} = (\Delta u_1, \Delta u_2)$.

(1) Show that for any scalar field φ on Ω, one has

$$\text{grad}\,\varphi = \nabla\varphi = \partial_r\varphi + \frac{1}{r}\partial_\theta\varphi \quad \text{and} \quad \Delta\varphi = \partial_r^2\varphi + \frac{1}{r}\partial\varphi + \frac{1}{r^2}\partial_\theta\varphi.$$

(2) Show that the divergence is

$$\text{div}\,\vec{u} = \nabla\cdot\vec{u} = \partial_r u_r + \frac{1}{r}u_r + \frac{1}{r}\partial_\theta u_\theta.$$

(3) Show that in polar coordinates one has

$$(\Delta\vec{u})_r = \Delta u_r - \frac{1}{r^2}u_r - \frac{2}{r}\partial_\theta u_\theta$$
$$(\Delta\vec{u})_\theta = \Delta u_\theta - \frac{1}{r^2}u_\theta + \frac{2}{r}\partial_\theta u_r.$$

(4) Let $\vec{u}$ and $\vec{v}$ be two vector fields on Ω. Show that

$$((\vec{v}\cdot\nabla)\vec{u})_r = v_r\,\partial_r u_r + \frac{1}{r}v_\theta\,\partial_\theta u_r - \frac{1}{r}v_\theta u_\theta$$
$$((\vec{v}\cdot\nabla)\vec{u})_\theta = v_r\,\partial_r u_\theta + \frac{1}{r}v_\theta\,\partial_\theta u_\theta - \frac{1}{r}v_\theta u_r.$$

(5) Derive the Navier-Stokes equations in polar coordinates.

(6) Derive the Navier-Stokes equations in cylindrical coordinates.

(7) Derive the Navier-Stokes equations in spherical coordinates.

61.3. Show that the following are valid:

(1) One has $B(u,v) \in H$, whenever $u, v \in \mathcal{D}(A)$, and
(2) there is a constant $C > 0$ such that

$$\|B(u,v)\| \leq C\|A^{\frac{1}{2}}u\|^{\frac{1}{2}}\|Au\|^{\frac{1}{2}}\|A^{\frac{1}{2}}v\|^{\frac{1}{2}}\|Av\|^{\frac{1}{2}}$$

for all $u, v \in \mathcal{D}(A)$. (Hint: Try to reduce this question to an interpolation inequality.)

61.4. Show that there is a constant $C > 0$ such that

$$\|AB(u,u)\| \leq C\|Au\|\,\|A^{\frac{3}{2}}u\|, \qquad \text{for all } u \in \mathcal{D}(A^{\frac{3}{2}}).$$

What can be derived about $\|A^m B(u,u)\|$, when $m > 1$?

61.5 Let Ω be a smooth bounded region in $\mathbb{R}^3$ and let W^α denote the collection of all functions $w = w(x_1, x_2)$ in V^α such that w does not depend on x_3.

(1) Show that if α_i, for $i = 1, 2, 3$, are nonnegative real numbers that satisfy

$$1 \leq \alpha_1 + \frac{2}{3}(\alpha_2 + \alpha_3),$$

and $(\alpha_1, \alpha_2, \alpha_3)$ is not equal to one of the following $(1,0,0)$, $(0, \frac{3}{2}, 0)$, $(0, 0, \frac{3}{2})$, then there is a constant $C = C(\alpha_1, \alpha_2, \alpha_3, \Omega)$ such that

$$|b(v, u_2, u_3)| \leq C\|v\|_{\alpha_1}\|u_2\|_{\alpha_2+1}\|u_3\|_{\alpha_3},$$
$$|b(u_2, v, u_3)| \leq C\|u_2\|_{\alpha_2}\|v\|_{\alpha_1+1}\|u_3\|_{\alpha_3},$$
$$|b(u_2, u_3, v)| \leq C\|u_2\|_{\alpha_2}\|u_3\|_{\alpha_3+1}\|v\|_{\alpha_1},$$

for all $v \in W^{\alpha_1+1}$, $u_2 \in V^{\alpha_2+1}$, and $u_3 \in V^{\alpha_3+1}$.

(2) Show that if α_i, for $i = 1, 2, 3$, are nonnegative real numbers that satisfy

$$1 \leq (\alpha_1 + \alpha_2) + \frac{2}{3}\alpha_3,$$

and $(\alpha_1, \alpha_2, \alpha_3)$ is not equal to one of the following $(1,0,0)$, $(0,1,0)$, $(0, 0, \frac{3}{2})$, then there is a constant $C = C(\alpha_1, \alpha_2, \alpha_3, \Omega)$ such that

$$|b(v_1, v_2, u)| \leq C\|v_1\|_{\alpha_1}\|v_2\|_{\alpha_2+1}\|u\|_{\alpha_3},$$
$$|b(v_1, u, v_2)| \leq C\|v_1\|_{\alpha_1}\|u\|_{\alpha_3+1}\|v_2\|_{\alpha_2},$$
$$|b(u, v_1, v_2)| \leq C\|u\|_{\alpha_3}\|v_1\|_{\alpha_1+1}\|v_2\|_{\alpha_2},$$

for all $v_1 \in W^{\alpha_1+1}$, $v_2 \in W^{\alpha_2+1}$, and $u \in V^{\alpha_3+1}$.

61.6. Show that with $d = 2$ inequality (61.19) can be replaced by

$$\|B(u,v)\|,\ \|B(v,u)\| \leq C_r\|A^{\frac{1}{2}+r}u\|\,\|A^{\frac{1}{2}+r}v\|, \qquad \text{for any } r \text{ with } 0 < r.$$

Section 6.3

63.1. Verify the limit in (63.30). (Hint: Use $\alpha_1 = \alpha_3 = 1$ and $\alpha_2 = 0$ with (61.18) to get appropriate bounds on $b(u_n - u, u, v)$, $b(u_n, u_n - u, v)$ and $b(u_n, u_n, P_n v - v)$. Then apply Lemma 62.1.)

63.2. Extend the theory of weak solutions of the Navier-Stokes equations to the case where $f \in L^\infty(0,\infty; H)$ is replaced by $f \in L^\infty(0,\infty; L^2(\Omega))$.

63.3. For the 2D Navier-Stokes equations prove the following:

(1) The class of weak solutions $W_{LH}(f)$ is homeomorphic to the Hilbert space H, when $f \in L^\infty(0,\infty; H)$.

(2) For $N_0 \geq 0$, the space $\mathcal{W}(N_0)$ is homeomorphic to

$$H \times \{f \in L^\infty(0,\infty; H) : \|f\|_\infty \leq N_0\}.$$

63.4. Complete the proof of Lemma 63.3.

63.5. Under the assumptions of Lemma 63.3, show that the limiting functions φ and g satisfy $D_t\varphi = g$. (Compare with Lemma 63.4.)

63.6. Let u^1 and u^2 be two weak solutions of Class LH for the Navier-Stokes equations and let $t_0 \in (0,\infty)$. Define $v = v(t)$ by $v(t) = u^1(t)$, for $0 \leq t < t_0$, and $v(t) = u^2(t)$, for $t_0 \leq t < \infty$. Show that v is a weak solution of Class LH if and only if $u^1(t_0) = u^2(t_0)$.

63.7. Prove that inequality (63.46) holds for all $t \geq t_\phi$, as opposed to almost all such t.

Section 6.4

64.1. Let $[0,T)$ denote the interval of existence for the maximally defined, strong solution of (61.10) with $u(0) = u_0$, where $u_0 \in V$, $f \in L^\infty = L^\infty(0,\infty; H)$ and $T = T(u_0, f)$. Define Σ to be the collection of (u_0, f, t) such that $u_0 \in V$, $f \in L^\infty$ and $t \in [0,T)$. Show that Σ is an open set in $V \times L^\infty \times \mathbb{R}^+$, where L^∞ has the topology of uniform convergence on compact sets, i.e., $f_n \to f$ if and only if for every τ, $0 < \tau < \infty$, one has

$$\operatorname*{ess\,sup}_{0 \leq s \leq \tau} \|f_n(s) - f(s)\| \to 0, \qquad \text{as } n \to \infty.$$

64.2. Let $u(t)$ be a weak solution of the 3DNS on $[0,\infty)$, where $u(0) \in V$ and $f \in L^\infty(0,\infty; H)$. Show that $u(t)$ is a globally regular solution if and only if $u \in L^4_{\mathrm{loc}}[0,\infty; V)$.

64.3. Extend Theorem 64.4 to the case where the forcing function f satisfies $f \in L^p_{\mathrm{loc}}[0,T; H)$, for some p with $2 < p \leq \infty$. Show that the strong solution u satisfies $u \in C^{0,\theta}_{\mathrm{loc}}[0,T; V)$, for an appropriate $\theta > 0$.

64.4. Let $f \in L^\infty(0,\infty; H)$, $u_0 \in V$, and let u denote the strong solution of the 2D or 3D Navier-Stokes equations on the interval $[0,T)$, where $0 <$

$T \leq \infty$. Show that for all β with $0 \leq \beta < 2$, one has $u \in C^{0,\theta}_{\text{loc}}[0,T;V^\beta)$, where $\theta = \frac{2-\beta}{2}$.

64.5. Let $u \in L^\infty_{\text{loc}}[0,T;V)$ and $t_0 \in (0,T)$. Assume that u is a strong solution of the Navier-Stokes equations on $[0,t_0)$ and on $[t_0,T)$. Show that $u \in C[0,T;V)$ and that u is a strong solution on $[0,T)$.

64.6. Show that there is a generalized weak solution $u = \varphi$ of the Navier-Stokes equations with the following properties: (1) φ is not a weak solution of Class LH, and (2) φ is a strong solution of the Navier-Stokes equations on $(0,\infty)$.

64.7. This is a generalization of Item (1) of Theorem 64.7 for the 3D Navier-Stokes equations. Let $u_0 \in V$ and let $v = v(t)$ denote the strong solution of the Navier-Stokes equations on $[0,T)$. Show that for every weak solution $u = u(t)$ of Class LH satisfying $u(0) = u_0$ one has $u(t) = v(t)$, for $0 \leq t < T$. (Hint: Try to adapt the 2D argument in Corollary 63.6 to this case.)

64.8. (The Leray Property.) Assume that $f \equiv 0$ for the 3D Navier-Stokes equations. Show that for every $u_0 \in H$ there is a $T = T(u_0) \geq 0$, such that any weak solution $u = u(t)$ of Class LH with $u(0) = u_0$ satisfies: (1) $u(t) \in \mathcal{D}(A^{\frac{1}{2}})$, for all $t \geq T$; and (2) $\|A^{\frac{1}{2}}u(t)\| \to 0$, as $t \to \infty$.

64.9. (Continuation of Exercise 64.8.) For the 2D and the 3D Navier-Stokes equations assume that $f \in H$. Show that there is an $R_0 = R_0(\nu, \lambda_1) > 0$ such that if $\|f\| < R_0$, then the following hold:

(1) For every weak solution $u = u(t)$ of Class LH with $u(0) = u_0 \in H$, there is a $T = T(u_0) \geq 0$ such that $u(t) \in \mathcal{D}(A^{\frac{1}{2}})$, for all $t \geq T$.
(2) There is a unique equilibrium point (or stationary solution) v of the Navier-Stokes equations, i.e., $\nu Av + B(v,v) = f$.
(3) The set $\mathfrak{A} = \{v\}$ is a global attractor for the weak solutions of the Navier-Stokes equations.

64.10. Show that if v is any weak solution (of Class LH) of equation (64.21), then the following three properties hold:

(1) The solution v satisfies $v \in C[0,T;H_w)$.
(2) The function v satisfies

$$\partial_t v + \nu Av + B(u_2,v) + B(v,u_2) \overset{\text{a.e.}}{=} g, \qquad \text{in the space } V^{-1}.$$

(3) The time derivative $\partial_t v$ is in the space $L^2_{\text{loc}}[0,T;V^{-1})$.

64.11. The theory of the linear problem in Section 6.4.4 does not include the case where u_1 and u_2 are weak solutions; i.e., one has

$$u_1,\, u_2 \in L^\infty(0,\infty;H) \cap L^2_{\text{loc}}[0,\infty;V) \tag{69.2}$$

in place of the stronger requirement given by (64.23). What are the difficulties in extending the linear theory to the case where (69.2) is satisfied in 3D? What happens in 2D?

64.12. For the 2D Navier-Stokes equations where the forcing function f satisfies $f \in L^\infty(0,\infty;V)$ and $u_0 \in V$, show that for $i = 4,5$ there exist constants $L_i = L_i(\|f\|_\infty)$, where L_i is independent of u_0, and $M_i = M_i(\|u_0\|, \|f\|_\infty)$ such that

$$\|Au(t)\|^2 \leq M_4 e^{-\nu\lambda_1 t} + L_4, \qquad \text{for } t \geq 2,$$

and

$$\int_{t-1}^{t} \|A^{\frac{3}{2}}u(s)\|^2\,ds \leq M_5 e^{-\nu\lambda_1 t} + L_5, \qquad \text{for } t \geq 2.$$

Show that, if in addition, one has $u_0 \in \mathcal{D}(A)$, then

$$\|Au(t)\|^2 \leq M_4(1 + \|Au_0\|^2)e^{-\nu\lambda_1 t} + L_4, \qquad \text{for } t \geq 0,$$

and

$$\int_0^t \|A^{\frac{3}{2}}u(s)\|^2\,ds \leq \nu^{-1}(\|Au_0\|^2 + tK_4), \qquad \text{for } t \geq 0.$$

where $K_4 = K_4(\|Au_0\|, \|f\|_\infty)$.

64.13. Is it the case that, if $u_0 \in \mathcal{D}(A)$, then there always is a globally regular strong solution $u = u(t)$ of the 3D Navier-Stokes equations with $u(0) = u_0$?

64.14. Let $(u_0, f) \in GR$, where GR is the globally regular set. Determine whether or not there is an $\epsilon = \epsilon(u_0, f) > 0$ such that for all $v_0 \in V$ with $\|A^{\frac{1}{2}}(u_0 - v_0)\| \leq \epsilon$, one has $(v_0, f) \in GR$.

64.15. The following steps develop the bootstrap argument for Theorem 64.11. The argument uses inequalities (61.19) and (61.28), together with the interpolation inequality (61.9).

(1) Show that there exist constants $\beta_1 < \beta_2 < \beta_3 < \beta_4$, where $\frac{2}{5} \leq \beta_1$, $\frac{3}{5} \leq \beta_2$, $\frac{2}{3} \leq \beta_3$, and $\frac{4}{5} \leq \beta_4$, such that

$$\|A^{\frac{\beta_i}{2}} B(u(t), u(t))\| \leq \rho_i(t)\|Au(t)\|, \qquad \text{for } 0 \leq t < T,$$

where $\rho_i = \rho_i(t)$ is a nondecreasing function defined for $0 \leq t < T$, for $i = 1, 2, 3, 4$.

(2) Show that

$$u \in C[0,T;V^{1+\beta_i}) \cap L^2_{\text{loc}}[0,T;V^{2+\beta_i}),$$

for $i = 1, 2, 3, 4$.

(3) Show that $B(u,u) \in L^2_{\text{loc}}[0,T;V)$.

(4) Use these properties to show that (64.54) and the other conclusions in Theorem 64.11 are valid.

Section 6.5

65.1. In this exercise we focus on the 3D Navier-Stokes equations. The purpose here is to show that, in a qualified sense, the strong solution operator $S_s(t, f)$ on V, is compact for $t > 0$. Let $\mathcal{F}$ denote a bounded subset of $L^\infty(0, \infty; H)$ with the property that $\mathcal{F}$ is a compact set in $L^2_{\text{loc}}[0, \infty; H)$. Let $\mathcal{B}$ denote a bounded subset of V.

(1) Show that there is a time $T_1 = T_1(\mathcal{F}, \mathcal{B}) > 0$ with the property that for every $(u_0, f) \in \mathcal{B} \times \mathcal{F}$, there is a strong solution of equation (61.10) with data (u_0, f) on the interval $[0, T_1)$.

(2) Show that for every t with $0 < t < T_1$, the set

$$\{S_s(t, f)u_0 : (u_0, f) \in \mathcal{B} \times \mathcal{F}\}$$

lies in a compact subset of V.

65.2. Let $f \in H$ be fixed and set $W = W(f)$ and $W_{LH} = W_{LH}(f)$, for the 3D Navier-Stokes equations. For each $\varphi \in W$ we define

$$L(\varphi) \stackrel{\text{def}}{=} \operatorname*{ess\,sup}_{0<t<\infty} \|A^{\frac{1}{2}}\varphi(t)\|.$$

(1) Show that the function L assumes values in the extended interval $[0, \infty]$ and that it is monotone nonincreasing; i.e., one has $L(\varphi_\tau) \leq L(\varphi_\sigma) \leq L(\varphi)$, whenever $0 \leq \sigma \leq \tau$.

(2) Show that the function L is lower semicontinuous on the space W. As a result, if $\varphi^n \to \varphi$ in W, then one has

$$L(\varphi) \leq \liminf_{n\to\infty} L(\varphi^n).$$

We will say that a weak solution φ is **ultimately regular** if there is a time $t = t_\varphi > 0$ such that

$$L(\varphi_\tau) < \infty, \qquad \text{for all } \tau \geq t_\varphi. \tag{69.3}$$

Show that if a given weak solution φ is ultimately regular, then the following properties are valid:

(3) Each point ϕ in the omega limit set $\omega(\varphi)$ lies in the global attractor $\mathfrak{A}$ and one has

$$L(\phi) \leq \liminf_{\tau\to\infty} L(\varphi_\tau) < \infty.$$

(4) The set $w(\varphi)$ defined by

$$w(\varphi) \stackrel{\text{def}}{=} \{\phi(0) : \phi \in \omega(\varphi)\}$$

is a bounded set in V and it consists entirely of strong solutions. In particular, if $u_0 \in w(\varphi)$, then there is one and only one weak solution ϕ with $\phi(0) = u_0$, and this solution is a strong solution for all $t \in \mathbb{R}$. Moreover, one has $\phi \in \mathfrak{A}$.

(5) There are positive constants K_1 and K_2 such that

$$\|A^{\frac{1}{2}}\phi(t)\|^2 \leq K_1, \qquad \text{for } t \in \mathbb{R} \text{ and } \phi \in \omega(\varphi), \tag{69.4}$$

and

$$\int_t^{t+1} \|A\phi(t)\|^2 \leq K_2, \qquad \text{for } t \in \mathbb{R} \text{ and } \phi \in \omega(\varphi),$$

Define L_0 by $L_0 \stackrel{\text{def}}{=} \inf_{\phi\in\mathfrak{A}} L(\phi)$. Assume that L_0 is finite, or equivalently, that there is at least one weak solution $\varphi \in W$ that satisfies (69.3). For any L_1 with $L_0 \leq L_1 < \infty$, define

$$\mathfrak{A}_1 \stackrel{\text{def}}{=} \{\phi \in \mathfrak{A} : L(\phi) \leq L_1\}.$$

Show that the set $\mathfrak{A}_1$ satisfies the following:

(6) The set $\mathfrak{A}_1$ is a nonempty, closed, positively invariant set in $\mathfrak{A}$.

(7) The omega limit set $\omega(\mathfrak{A}_1)$ is a nonempty, compact, invariant set with

$$\mathfrak{A}_2 \stackrel{\text{def}}{=} \omega(\mathfrak{A}_1) \subset \mathfrak{A}_1 \subset \mathfrak{A}.$$

(8) The set $\Gamma_2 \stackrel{\text{def}}{=} \{\phi(0) : \phi \in \mathfrak{A}_2\}$ is a bounded set in V, and (69.4) is valid for all $\phi \in \mathfrak{A}_2$ with $K_1 = L_1$.

65.3. One says that the weak solutions are **ultimately regular on a set** U in W if for each $\varphi \in U$, the weak solution φ is ultimately regular.[30] Assume now that the weak solutions are ultimately regular on a set U, where U is a neighborhood of the global attractor $\mathfrak{A}$ in W. Show that for every point $\varphi \in U$, the omega limit set $\omega(\varphi)$ is nonempty, compact and consists entirely of strong solutions. Furthermore, one has $\omega(\varphi) \subset \mathfrak{A}$ by the maximality property (see Section 2.3.7). (We caution the reader that, in general, the attractor $\mathfrak{A}$ is larger than the union of the omega limit sets in it; i.e., there may be a point $\phi \in \mathfrak{A}$, where ϕ does not lie in any omega limit set.)

65.4. Let $\mathfrak{A}$ and $\mathfrak{A}_2$ be given as in Theorem 65.8. For $g \subset \mathfrak{A}_2$, define the fiber

$$\mathfrak{A}(g) = \{u_0 \subset H(\text{or } V) : (g, u_0) \in \mathfrak{A}\}.$$

Show that $\pi(g, \mathfrak{A}(g), \tau) = \mathfrak{A}(g_\tau)$, for all $g \in \mathfrak{A}_2$ and all $\tau \geq 0$. (Compare this with Exercise 24.1.) What happens in the case of Theorem 65.12?

[30] Notice that in the case of an infinite set U, the concept of being ultimately regular on U does not imply any uniformity in the time t_φ appearing in (69.3), even when the set U is compact, or when U is the global attractor $\mathfrak{A}$, or when U is the entire space W.

Section 6.7

67.1. Let w_1 and w_2 be two homogenization terms for the inhomogeneous equation (67.1) - (67.2) on Ω. Assume that w_1 and w_2 are sufficiently smooth. Show that if v_1 is a weak solution (or strong solution) of equation (67.3) with $w = w_1$, then $v_2 = v_1 + w_1 - w_2$ is a weak solution (or strong solution) of equation (67.3) with $w = w_2$.

67.2. Let w_1 and w_2 be given as in Exercise 67.1, and assume that $d = 2$; i.e., Ω is a 2D domain. Let $\mathfrak{A}_1$ and $\mathfrak{A}_2$ denote the global attractors associated with the dynamics of equation (67.3) for $w = w_1$ and $w = w_2$, respectively. Show that $\mathfrak{A}_1$ and $\mathfrak{A}_2$ are homeomorphic sets and that the homeomorphism maps the solutions in $\mathfrak{A}_1$ onto the solutions in $\mathfrak{A}_2$.

6.10. Commentary.

The modern mathematical theory of the Navier-Stokes equations began with the pioneering work of Leray (1933, 1934a,b) and Hopf (1951), and the presentation we make here is based on these works. The theory presented here, which is based on linear and nonlinear functional analysis, can also be found in Ladyzhenskaya (1963, 1972, 1991), Foias and Prodi (1967), Temam (1977, 1983, 1988), von Wahl (1985), Constantin and Foias (1988), Galdi (1994), and Doering and Gibbon (1995). A recent comprehensive survey of the theory of the Navier-Stokes equations appears in Temam (1999). An alternate, but equivalent, formulation of the Navier-Stokes equations is given in terms of the stress tensor. For more information, see Solonnikov and Ščadilov (1973)

Section 6.2. The Bubnov-Galerkin method dates back to the period 1911 - 1915. The basic issue is to find an approximate solution of a nonlinear equation of the form $f(x) = h$ on a Hilbert space H. In this problem h is given and one seeks an approximation for x. Let $\{\phi_1, \phi_2, \phi_3, \dots\}$ be a given orthonormal basis in H. Then for each integer $n \geq 1$, the approximate solution x_n is given by $x_n = c_1\phi_1 + \cdots + c_n\phi_n$, where $(c_1, \dots, c_n)$ is a solution of the finite nonlinear system

$$\langle f(c_1\phi_1 + \cdots + c_n\phi_n), \phi_i\rangle = \langle h, \phi_i\rangle, \qquad i = 1, ..., n.$$

This method was used first[31] by I G Bubnov in 1911, and it appears in Bubnov (1913), also see Bubnov (1912, 1914). Subsequently, Galerkin (1915) showed that the method can be extended to the case where x and h reside in different Hilbert spaces, as well as the case where x and h lie in the

[31] A detailed description of the history of the Bubnov-Galerkin method is given in Grigolyuk (1971). Also see the Soviet Mathematical Encyclopedia (1977), vol. 1, p 842. (We are grateful to S Pilyugin and V A Pliss for their help in researching the Russian literature for historical references on the Bubnov-Galerkin method.)

same space, but one uses different Fourier series expansions in the domain and range of f. To the best of our knowledge, the first application of the Bubnov-Galerkin method to the study of evolutionary problem is in Faedo (1949).

The quantity

$$G = \frac{||f||_{\infty}^2}{\nu^2 \lambda_1} \text{ in 2D} \quad \text{or} \quad G = \frac{||f||_{\infty}^2}{\nu^2 \lambda_1^{\frac{3}{4}}} \text{ in 3D}$$

appears quite often in the theory of the Navier-Stokes equations. This quantity is referred to as the **Grashof number**, see Foias, et al (1983). The Grashof number is similar to the Reynolds number in the sense that the dynamical complexity of the solutions increases as G increases.

Section 6.3. In many places in this chapter, the norm $||f||_{\infty}$ can be replaced by the smaller quantity $||f||_{\infty,[0,t]}$. The reason for this is that the solutions of (61.10), or (62.2), on an interval $[0, \tau)$ do not depend on values of the forcing function f for $t > \tau$. This property is referred to as **causality**, and it is a common feature of parabolic partial differential equations, see Naylor and Sell (1982).

In this volume we present the theory of the Navier-Stokes equations on the assumption that the boundary of the physical domain Ω is of class C^2, or that that the equations satisfy periodic boundary conditions, in which case Ω is a torus. For recent theory with boundaries of Lipschitz class, see Deuring and von Wahl (1995).

It should be noted that the weak solutions of the 3D Navier-Stokes equations do not offer a fertile ground for the applications of the methods of nonlinear dynamics, as presented, for example, in Chapter 7. The reason for this is that there is no known theory of linearization along weak solutions that leads to the important feature of Fréchet differentiability as illustrated in (61.32). Nevertheless, it is still possible to find a global attractor for the 3D Navier-Stokes equations, as is shown in Section 6.5.2.

Section 6.4. As noted in Section 6.4, the Global Regularity Problem is completely open. The special result described in Corollary 64.5, where it is shown that small data give birth to globally regular solutions, has many alternate formulations. Perhaps the most interesting variation is a result of Kato and Fujita (1966), which is presented in Temam (1999). In this case, it is shown that if f is small and $u_0 \in V^1$, with $||A^{\frac{1}{8}} u_0||^2$ small, say, $||A^{\frac{1}{8}} u_0||^2 \leq B_1$, then there is a globally regular solution. Since for every $N > 1$, there is a $u_0^N \in V^1$ with $||A^{\frac{1}{2}} u_0^N||^2 > N$ and $||A^{\frac{1}{8}} u_0^N||^2 \leq B_1$, it follows that the set of data (u_0, f) that give birth to globally regular solutions is <u>unbounded</u> in the V^1-component.

While Corollary 64.5 is applicable to every smooth bounded domain in $\mathbb{R}^3$, one does get stronger results in the case of a <u>thin</u> 3D domain, as noted in Raugel and Sell (1993a,b, 1994). The basic idea is that in a thin 3D domain , one can prove that the longtime dynamics of the Navier-Stokes equations is well-approximated by the longtime dynamics of a suitable 2D problem. Also see Bondarevsky (1996), Temam and Ziane (1996, 1997), and Raugel and Sell (2000).

The concepts of weak and strong solutions of the Navier-Stokes equations (61.10) are related to the same-named concepts for the the linear inhomogeneous equation (40.1) (see Sections 4.2.2 and 4.2.3). In order to see this relationship, it is convenient to replace the term $f(t)$ in (40.1) with $\phi(t)$, where $\phi = f - B(u,u)$ comes from (61.10). In these variables, the theory of weak solutions developed in Section 4.2.3 is applicable in this chapter when $\phi \in L^2_{\mathrm{loc}}[0,T;V^{-1})$, as is the case for the 2D Navier-Stokes equations (see Lemma 63.1). Since one has $\phi \in L^{\frac{4}{3}}_{\mathrm{loc}}[0,T;V^{-1})$ for the 3D Navier-Stokes equations, the theory of weak solutions developed in Section 6.3 is more general than that of Section 4.2.3. Of course, there is no serious mathematical difficulty in extending the theory of Section 4.2.3 to the case where $f(t) \in L^p_{\mathrm{loc}}[0,T;V^{-1})$, for $1 \leq p < 2$. However, in such an extension, one is not able to invoke the Continuity Lemma 37.9 to conclude that the weak solution u satisfies $u \in C[0,T;V^{\alpha})$.

For the strong solutions there are many remarkable similarities in the two theories, in spite of the fact that one begins with definitions from opposite ends of the theoretic highway. The similarities are due to Lemma 64.1 and equation (64.3). Because of this result, one can show that, if $\phi = f - B(u,u)$, where $f \in L^{\infty}(0,\infty;H)$ and u satisfies (64.1) with $u(t_0) = u_0 \in V$, then u is a strong solution of (61.10) if and only if u is a strong solution of (40.1), where $f(t)$ is replaced by $\phi(t)$.

For a comparison of the classical solutions and the strong solutions of the Navier-Stokes equations, the reader should consult Constantin and Foias (1988).

Section 6.5. Even though it has been known for over 20 years that there is a global attractor for these equations in two-dimensions (2D), see Ladyzhenskaya (1972), the 3D problem proved to be a different matter, see Navier (1827), Poisson (1831), Stokes (1845), Leray (1933, 1934a, 1934b), and Hopf (1951). The theory presented in this volume is due to Sell (1996), and is based, in part, on earlier work for ordinary differential equations, see Sell (1973). For an extension of this theory to chemically reacting fluid flows with inhomogeneous boundary conditions, see Norman (1999, 2000a,b).

In another direction, for the Navier-Stokes equations on suitable thin 3D domains. it was shown in Raugel and Sell (1993a,b, 1994, 2001), Bondarevsky (1996), and Temam and Zaine (1996, 1997) that the weak solutions do have a global attractor, and that this attractor consists entirely of strong solutions. However, the current theory is limited to thin domains.

It is instructive to compare the results presented in Section 6.5.2 for the 3D Navier-Stokes equations with those of Foias and Temam (1987). The latter authors show that there is a **universal attracting set** in H; that is, there is a set Γ in H with the following properties:

(1) The set Γ is defined as the collection of all $u_0 \in H$ for which there

exists a globally defined weak solution $\varphi \in L^\infty(-\infty, \infty; H)$ with $\varphi(0) = u_0$. It is shown that Γ is nonempty and bounded in H.

(2) Every weak solution $\varphi \in W_{LH}$ satisfies $\varphi(t) \overset{w}{\to} \Gamma$ in H_w, as $t \to \infty$.

(3) The set Γ is compact in H_w.

(4) One has $\Gamma \subset V$ if and only if Γ is a bounded set in V.

(5) The set $\Gamma \cap V$ is weakly dense in Γ, i.e., it is dense in the weak topology H_w.

(6) The set Γ contains a set Γ_{reg}, where Γ_{reg} is weakly open and weakly dense in Γ, and for every $u_0 \in \Gamma_{\text{reg}}$ there is an $a > 0$ such that, for any weak solution φ with $\varphi(t) \in \Gamma$, for all $t \in \mathbb{R}$ and $\varphi(0) = u_0$, the restriction $\varphi|_{(-a,a)}$ is uniquely determined and $\varphi(t) \in \Gamma \cap V$, for all $t \in (-a, a)$.

It should be noted that this theory does not address the issue of whether Property (3) in the definition of a global attractor is valid.

In our notation, we note that if $\varphi \in \mathfrak{A}$, then one has $\varphi \in C(-\infty, \infty; H_w)$. Furthermore, the set $\{\varphi \in \mathfrak{A}\}$ is compact in $C(-\infty, \infty; H_w)$ in the topology of uniform convergence on compact sets, see Constantin, Foias, and Temam (1985, chap 1). As a result, the evaluation mapping of $\mathfrak{A}$ into H_w given by $\varphi \to \varphi(0)$ is continuous, and the set $\Gamma_0 = \{\varphi(0) : \varphi \in \mathfrak{A}\}$ satisfies the six properties listed in the last paragraph.

We now have the global attractor $\mathfrak{A}$ with two topologies. First one has

$$\mathfrak{A} \subset L^2_{\text{loc}} = L^2_{\text{loc}}[0, \infty; H)$$

with the L^2_{loc}-topology, and secondly one has

$$\mathfrak{A} \subset C(-\infty, \infty; H_w)$$

with the metrizable topology of uniform convergence on compact sets in $(-\infty, \infty)$. As noted in Sell (1996), the two topologies agree on the space $\mathfrak{A}$. By using the fact that Γ_0 is the continuous image of the global attractor $\mathfrak{A}$, which is compact, one can show that the two sets Γ and Γ_0 are the same.

Let us summarize the four principal advantages of the point of view developed in Section 6.5.2.

(1) This approach includes an overall framework for the study of the weak solutions of the Navier-Stokes equations in the context of dynamical systems, or semiflows.

(2) In this framework we are able to apply the Existence Theorem 23.12 to the study of weak solutions, and thereby show the existence of a global attractor $\mathfrak{A}$.

(3) The global attractor $\mathfrak{A}$ attracts all bounded sets of weak solutions.

(4) The global attractor $\mathfrak{A}$ satisfies the 13 properties listed in Theorem 23.15.

The importance of the fact that the global attractor attracts all bounded sets in W cannot be overemphasized. It is this feature, along with the

Lyapunov stability of the attractor, that is the source of various robustness theories of global attractors, see the Robustness Theorem 23.14.

Section 6.6. The Kwak transformation has a counterpart in the theory of equations with convection, see Kwak (1992a).

Section 6.7. The topic of the longtime dynamics of a fluid flow caused by the influence of the boundary conditions is a classical area of analysis. Perhaps the most studied problem is that of the Couette-Taylor flow, which is described in some detail in Section 7.5. The homogenization process described in Section 6.7 uses, in an essential way, the fact that the time derivative $\partial_t w$ of the homogenization term is known. In our case, one has $\partial_t w \equiv 0$. Similar theories could be developed if the boundary term ϕ were time dependent, but known and independent of the solutions of the Navier-Stokes equations. This happens, for example, in the Couette-Taylor flow and in a tidal basin where the motion of the ocean is affected by the motion of the moon around the earth. A far more complicated situation is encountered when there is a time-dependent dynamical coupling between the fluid motion in Ω and its behavior on the boundary $\partial\Omega$. For example, one might have a rotating solid object within a fluid, where the speed of rotation is dependent on the speed of the fluid flow around the object. An interesting illustration of such dynamical boundary conditions is treated in Sauer (1988) and Grobbelar-van Dalsen and Sauer (1989, 1993). Also see Salvi (1988, 1990). This is clearly an area deserving further study.

The connection between turbulence seen in various fluid flows and the Navier-Stokes equations used to model these fluids has attracted the attention of researchers for many decades. Hundreds, if not thousands, of papers have been published addressing this issue. In some of these works, one can find suggestions that the full picture of turbulence can be explained in terms of the longtime dynamics of the solutions of the Navier-Stokes equations. However, the application of the techniques of dynamical systems to the study of turbulence is still in its naissance. A very important development in the use of dynamical methods for the study of turbulence is a recent treatise on this subject, see Foias, et al (2001). With the appearance of this major work, Foias and his colleagues have opened a new door for the applications of dynamical systems.

As noted earlier, the Navier-Stokes equations are widely accepted as good model equations for the study of fluid flows. Nevertheless, these equations, especially in 3D, are not user-friendly. The search for an alternate model, a model which is good from the point of view of physics and more tractable from the mathematical point of view, has attracted the efforts of many researchers. Of special note is a new model based on the Camassa-Holm equations, see S Chen, et al (1999a,b). The Camassa-Holm model for 3D fluids has a global attractor consisting entirely of globally regular solutions, see Foias, Holm, and Titi (2002). For this reason alone, it is more tractable than the Navier-Stokes model. The Camassa-Holm models certainly warrant further study.

Additional Readings

Ball (1996); Constantin and Foias (1988); Doering and Gibbon (1995); Foias, et al (2001); Foias, Holm, and Titi (2002); Foias, Manley, and Temam (1987); Foias and Prodi (1967); Foias and Temam (1980, 1987); Galdi (1994); Grobbelar-van Dalsen and Sauer (1989, 1993); Hale (1988); Hopf (1951); Kirchgässner (1975); Kwak (1992a,b); Ladyzhenskaya (1963, 1972, 1991); Leray (1933, 1934a,b); Lions (1969); Lions and Magenes (1972); Raugel and Sell (1993a,b, 1994, 2002); Salvi (1988, 1990); Sauer (1988); Sell (1981b, 1996); Serrin (1959, 1962, 1963, 1972); Solonnikov and Ščadilov (1973); Temam (1977, 1983, 1988, 1999); and von Wahl (1985).

7 MAJOR FEATURES OF DYNAMICAL SYSTEMS

In this chapter we will examine a number of the classical issues arising in the study of the dynamics of differential equations on Banach spaces. The corresponding theories for the finite dimensional problems appear in a number of sources, as noted in the Commentary Section. Our emphasis here will be on the infinite dimensional theory, in the context of general nonlinear evolutionary equations. Most of the applications will be to the theory of solutions of partial differential equations. In particular, we will examine various features of the longtime dynamics of the evolutionary equation

$$\partial_t u + Au = F(u), \tag{70.1}$$

as well as the perturbed equation

$$\partial_t y + Ay = F(y) + G(y) \tag{70.2}$$

on a Banach space W. More specifically, from time to time we assume that one of the following Standing Hypotheses is satisfied .

Standing Hypothesis A. *Let A be a positive, sectorial operator on a Banach space W with associated analytic semigroup e^{-At}. Let $V^{2\alpha}$ be the family of interpolation spaces generated by the fractional powers of A, where $V^{2\alpha} = \mathcal{D}(A^{\alpha})$, for $\alpha \geq 0$. Let $||A^{\alpha}u|| = ||A^{\alpha}u||_W = ||u||_{V^{2\alpha}} = ||u||_{2\alpha}$ denote the norm on $V^{2\alpha}$. See Lemma 37.4 for more information.*

Standing Hypothesis B. *The operator A is a positive, selfadjoint, linear operator, with compact resolvent, on a Hilbert space H. Consequently A satisfies the Standing Hypothesis A. Moreover, the fractional power spaces V^{α} are defined for all $\alpha \in \mathbb{R}$, and equation (37.2) defines the Hilbert space structure on each V^{α}. Also the semigroup e^{-At} is compact, for $t > 0$. See Theorem 37.2 for more information.*

As noted in Section 4.7, there is no loss of generality in assuming that the sectorial operator A is positive. The more general problem can easily be reduced to this case.

Under the assumption that the Standing Hypothesis A is satisfied, we will make use of the spaces $C_{\mathrm{Lip}} \overset{\mathrm{def}}{=} C_{\mathrm{Lip}}(V^{2\beta}, W)$, and for each integer $k \geq 1$ we let

$$(70.3) \qquad C^k_{\mathrm{Lip}} \overset{\mathrm{def}}{=} C_{\mathrm{Lip}}(V^{2\beta}, W) \cap C^k_F(V^{2\beta}, W), \qquad \text{where } 0 \leq \beta < 1,$$

see Section 4.6. Recall that if $F \in C^1_{\mathrm{Lip}}$, then F is bounded and Lipschitz continuous on each bounded set B in $V^{2\beta}$, and the Fréchet derivative DF is bounded on B, as well. In particular, there are constants $K_0 = K_0(B)$, and $K_1 = K_1(B)$, such that $\|F(u)\| \leq K_0$, for all $u \in B$, and

$$(70.4) \qquad \|F(u_1) - F(u_2)\| \leq K_1 \|A^\beta(u_1 - u_2)\|, \qquad \text{for all } u_1, u_1 \in B.$$

The space $C^k_F(V^{2\beta}, W)$ denotes the space of functions $F : V^{2\beta} \to W$ that are k-times continuously Fréchet differentiable. Whenever B is an open, bounded, convex set, inequality (70.4) implies that $\|DF(u)\|_{\mathcal{L}} \leq K_1$, for $u \in B$, where $\mathcal{L} = \mathcal{L}(V^{2\beta}, W)$. We say that DF has the **Hölder (or the Lipschitz) property**, when DF is Hölder (or Lipschitz) continuous on each compact set $\mathcal{K}$ in $V^{2\beta}$.

For each $u_0 \in V^{2\beta}$ we let $u(t) = S(t)u_0$ denote the unique maximally defined mild solution of equation (70.1), as constructed in Section 4.7. Thus one has

$$u(t) = e^{-At}u_0 + \int_0^t e^{-A(t-s)} F(u(s))\, ds, \qquad \text{for } 0 \leq t < T,$$

where $[0, T)$ is the maximal interval of definition of $u(t)$. Since $F = F(u)$ is autonomous, Corollary 47.3 implies that $S(t)u_0$ is always a strong solution (in W) and (47.7) is valid. Any negative continuation of u_0, that is bounded in $V^{2\beta}$, for $t \leq 0$, will also be denoted by $S(t)u_0$. For any mild solution $S(t)u_0$, we set $B(t) = DF(S(t)u_0)$. Then one has $B \in C[0, T; \mathcal{L})$. Of special interest here is the case where $u_0 \in \mathcal{K}$, and $\mathcal{K}$ is a given compact invariant set for the semiflow generated by the mild solutions of equation (70.1). In this case, as noted in the Herculean Theorem 47.6, one has $\mathcal{K} \subset V^2 = \mathcal{D}(A)$, and

$$B \in \mathcal{M}^\infty = \mathcal{M}^\infty(\mathbb{R}; \mathcal{L}) \overset{\mathrm{def}}{=} C(\mathbb{R}; \mathcal{L}) \cap L^\infty(\mathbb{R}; \mathcal{L}).$$

As noted in Section 4.4, the space $\mathcal{M}^\infty$ has two topologies, $\mathcal{T}_{\mathrm{bo}}$ and $\mathcal{T}_A$, where $\mathcal{T}_A \subset \mathcal{T}_{\mathrm{bo}}$, and the translational flow mapping $(B, \tau) \to B_\tau$, where $B_\tau(t) = B(\tau + t)$, is continuous in either of these topologies. The mapping of $\mathcal{K}$ into $\mathcal{M}^\infty$ given by

$$(70.5) \qquad u_0 \to B(t) = DF(S(t)u_0)$$

is a **(flow) homomorphism** of $\mathcal{K}$ into $\mathcal{M}^\infty$. This means that (1) this mapping is a continuous mapping of $\mathcal{K}$ into $(\mathcal{M}^\infty, \mathcal{T}_{\text{bo}})$, or into $(\mathcal{M}^\infty, \mathcal{T}_A)$, and (2) this mapping commutes with the flow, i.e., the mapping (70.5) satisfies

$$S(\tau)u_0 \to B_\tau(t) = DF(S(\tau+t)u_0) = DF(S(t)S(\tau)u_0),$$

for all τ, $t \in \mathbb{R}$ and $u_0 \in \mathcal{K}$. We denote the image of this mapping by $\hat{\mathcal{K}}$. When one has a flow homomorphism, then the compact, invariant set $\mathcal{K}$ is said to be an **extension** of $\hat{\mathcal{K}}$ (see Section 2.1.5). Roughly speaking, this allows for the dynamics on $\mathcal{K}$ to have greater complexity than that on $\hat{\mathcal{K}}$.

In this chapter, we will be studying the longtime dynamics of the semiflow generated by the mild solutions of equation (70.1) in the vicinity of a distinguished compact invariant set $\mathcal{K}$. Such a set $\mathcal{K}$ might be, for example, a fixed point of (70.1), a time-periodic orbit, or a compact manifold. (It could also be a more complicated set, even with variations in the local dimension.) We will also be studying the theory of exponential dichotomies, or exponential trichotomies (ED-ET, for short) for the linearization of equation (70.1) along solutions lying in $\mathcal{K}$. In the context of the scale of spaces V^{2r}, $r \geq 0$, given by the Standing Hypothesis A, it is appropriate to ask which of these spaces, if any, is the best choice for the phase space for the nonlinear semiflow generated by (70.1) (see Section 4.7). As one might expect, there is no best choice that works in all situations. However, since we assume that the nonlinear term F is in $C^1_{\text{Lip}}(V^{2\beta}, W)$, the space $V^{2\beta}$, itself, seems to be the most natural choice for developing our theory. Our reasons for choosing $V^{2\beta}$ are summarized as follows:

(1) The space $V^{2\beta}$ is the cornerstone for both the linear and nonlinear theories growing out of Sections 4.4 and 4.7.
(2) If $\mathcal{K}$ is a compact invariant set in $V^{2\beta}$, then $\mathcal{K}$ is compact in $V^{2\alpha}$, for each α with $0 \leq \alpha \leq \beta$.
(3) If the linear skew product semiflow π has an ED-ET on $V^{2\alpha} \times \mathcal{K}$, for some α with $0 \leq \alpha \leq \beta$, then Lemma 45.5 implies that π has an ED-ET on $V^{2r} \times \mathcal{K}$, for every $r \geq \alpha$.
(4) The techniques developed in this chapter readily extend to the case where π has an ED-ET on $V^{2\alpha} \times \mathcal{K}$, for some α with $0 < \beta < \alpha < 1$. (We invite the reader to verify this assertion.)
(5) In our setting, one automatically finds that $\mathcal{K} \subset V^2 = \mathcal{D}(A)$, by the Herculean Theorem 47.6, and $\mathcal{D}(A) \subset V^{2r}$, for every r with $0 \leq r \leq 1$.

The dynamical theory developed in this chapter applies to many of the evolutionary equations studied above. For example, the Chafee-Infante problem in space dimension m, where $1 \leq m \leq 3$, falls into this context (see Section 5.1). Another example is the Navier-Stokes equations, where the nonlinearity F assumes the form $F(u) = -B(u,u) + f$, A is the Stokes operator, and the bilinear term $B(u,v)$ satisfies

$$B : V^2 \times V^2 \to V^1,\ B : V^{\frac{5}{4}} \times V^{\frac{5}{4}} \to H, \text{ and } B : V^1 \times V^1 \to V^{-\frac{1}{2}},$$

see Section 6.1. In this case, the inertial term $G(u) = B(u,u)$ satisfies (61.32), i.e.,

$$B \in C^2_{\mathrm{Lip}}(V^2, V^1) \cap C^2_{\mathrm{Lip}}(V^{\frac{5}{4}}, H) \cap C^2_{\mathrm{Lip}}(V^1, V^{-\frac{1}{2}}).$$

If the forcing term f is sufficiently smooth, say $f \in V^1$, then the evolutionary equations generated by the Navier-Stokes equations are evolutionary equations on either of the spaces, $V^{-\frac{1}{2}}$, H, and V^1. Furthermore, in each of these cases, the Fréchet derivative DF is in $C^1_{\mathrm{Lip}}(V^{2\beta}; \mathcal{L}(V^{2\beta}; W))$, for a suitable $\beta \geq 0$, and it has the Lipschitz property.

We are now ready to embark. The present journey will take us pass some of the familiar islands and placid harbors of the finite dimensional dynamics. However, in the open seas, we need to traverse carefully the rough waters and eddies of the infinite dimensional world, lest they capsize the Good Ship QED.

7.1. Local Dynamics Near an Equilibrium.

An **equilibrium point**[32] (or **stationary solution**) for the evolutionary equation (70.1) is a point u_0 in $\mathcal{D}(A)$, with the property that $Au_0 = F(u_0)$, i.e., $\partial_t u_0 \equiv 0$. Our interest in this section is to study the dynamics of equation (70.1) in the vicinity of the equilibrium point u_0. In particular we will examine the unstable manifold, the stable manifold, and the center manifold for this equilibrium point. Without additional structure, these manifolds need not exist. However, as we will see, in many noteworthy cases they do. We assume here that $F = F(u) \in C^1_{\mathrm{Lip}}$, see (70.3).

Before turning to these basic concepts, we define the **stable** and **unstable sets**, $W^s(u_0)$ and $W^u(u_0)$, for an equilibrium point u_0, as follows:

$$W^s(u_0) = \{y \in W : \|S(t)y - u_0\| \to 0, \text{ as } t \to \infty\}$$

and $W^u(u_0)$ is the collection of all $y \in W$ such that there exists a negative continuation ϕ^y through y that satisfies $\|\phi^y(t) - u_0\| \to 0$, as $t \to -\infty$.

Consider next the linearized equation

$$\partial_t v + Av - DF(u_0)v = \partial_t v + Lv = 0 \tag{71.1}$$

on W, where $L = A - DF(u_0)$ and $DF(u_0)$ is the Fréchet derivative of F at u_0. Then $-L$ is the infinitesimal generator of an analytic semigroup e^{-Lt} on W (see Theorem 44.5). We say that the equilibrium point u_0 is **hyperbolic in** W if the semigroup e^{-Lt} has an exponential dichotomy on W (see Section 4.5). In this case there is a bounded projection P on the

[32] See Exercise 71.1, for a connection with mild solutions.

phase space W and **characteristics (of hyperbolicity)** $K \geq 1$ and $\alpha > 0$ such that the following three properties hold:

(1) One has $Pe^{-Lt} = e^{-Lt}P$, for all $t \geq 0$.

(2) If $v_0 \in \mathcal{N}(P) = \mathcal{R}(Q)$, where $Q = I - P$ is the complementary projection, then one has

$$\|e^{-Lt}v_0\| \leq K\|v_0\|e^{-\alpha t}, \qquad \text{for } t \geq 0. \tag{71.2}$$

(3) If $v_0 \in \mathcal{R}(P) = \mathcal{N}(Q)$, then there is a negative continuation through v_0. This negative continuation, which we will denote by $e^{-Lt}v_0$, satisfies

$$\|e^{-Lt}v_0\| \leq K\|v_0\|e^{\alpha t}, \qquad \text{for } t \leq 0. \tag{71.3}$$

Notice that Q also satisfies $Qe^{-Lt} = e^{-Lt}Q$. As noted in Lemma 45.2, the linear subspace $\mathcal{S}(-L) = \mathcal{N}(P)$ is positively invariant under e^{-Lt}, while $\mathcal{U}(-L) = \mathcal{R}(P)$ is invariant. Furthermore, for each $v_0 \in \mathcal{R}(P)$, there is a unique negative continuation that satisfies (71.3). Moreover, the exponential estimates above imply that the spectrum $\sigma(L)$ does not have any points in the strip $\{x + iy : -\alpha < x < \alpha\}$ in the complex plane. The linear spaces $\mathcal{R}(P)$ and $\mathcal{R}(Q)$ are called, respectively, the **unstable** and **stable linear manifolds (subspaces)**. The **index**, $\text{ind}(u_0)$ of a hyperbolic equilibrium point u_0 is the dimension of the unstable linear manifold, i.e., $\text{ind}(u_0) = \dim \mathcal{R}(P) = \text{rank } P$. When $P = 0$, the equilibrium point u_0 is said to be **exponentially stable**.

As noted in Lemma 39.2 and Theorem 44.5, the operator L is a sectorial operator. Assume for the moment that L has compact resolvent. In this case, the spectrum, $\sigma(-L)$, contains only eigenvalues with no finite point of accummulation. Consequently, the equilibrium point u_0 is hyperbolic if and only if $\sigma(-L)$ does not meet the imaginary axis; i.e., $\sigma(-L) \cap \{i\omega : \omega \in \mathbb{R}\} = \emptyset$. In this case, the projection P will have finite rank, and it can be computed with the Dunford-Taylor integral

$$P = \frac{1}{2\pi i} \oint_{\Gamma} (\lambda I + L)^{-1}\, d\lambda,$$

where Γ is a simple closed curve in the complex plane containing in its interior precisely those eigenvalues λ of L, with Re $\lambda > 0$, see Sell and You (1992).

The shifted flow is given by $e^{-\lambda t}e^{-Lt}$, where $\lambda \in \mathbb{R}$ (see Section 4.5). It can happen that for some $\lambda \in \mathbb{R}$, the shifted flow has an exponential dichotomy. In this case, the inequalities (71.2) and (71.3) are valid where e^{-Lt} is replaced by $e^{-\lambda t}e^{-Lt}$. Of course the characteristics $K = K_\lambda \geq 1$ and $\alpha = \alpha_\lambda > 0$, as well as the associated projections (P_λ, Q_λ), will depend on λ. Lemma 45.5 is valid in this case, where $\mathcal{K} = \{u_0\}$, and B is given by (70.5). If $e^{-\lambda t}e^{-Lt}$ has an exponential dichotomy on W, then it has an exponential dichotomy on V^{2r}, for each $r \geq 0$. Of particular interest to us is the case $r = \beta$. By replacing the characteristic $K = K_\lambda$ with a larger value, if necessary, and leaving $\alpha = \alpha_\lambda$ unchanged, the following inequalities are valid for $x \in W$:

$$(71.4) \qquad \|A^\beta e^{-\lambda t} e^{-Lt} P_\lambda x\| \leq K e^{\alpha t} \|x\|, \qquad \text{for } t \leq 0;$$

while

$$(71.5) \qquad \|A^\beta e^{-\lambda t} e^{-Lt} Q_\lambda x\| \leq K t^{-\beta} e^{-\alpha t} \|x\|, \qquad \text{for } 0 < t \leq 1;$$

and

$$(71.6) \qquad \|A^\beta e^{-\lambda t} e^{-Lt} Q_\lambda x\| \leq K e^{-\alpha t} \|x\|, \qquad \text{for } 1 \leq t < \infty.$$

Let u_0 be an equilibrium point for equation (70.1), and define v by $v = u - u_0$. Then $u = u(t)$ is a solution of equation (70.1) if and only if $v = v(t)$ is a solution of

$$(71.7) \qquad \partial_t v + Lv = E(u_0, v),$$

where the error term satisfies

$$(71.8) \qquad E(u_0, v) = F(u_0 + v) - F(u_0) - DF(u_0)v, \qquad \text{for } v \in V^{2\beta}.$$

It follows that, as a function of v, one has

$$(71.9) \qquad E(u_0, \cdot) \in C_{\text{Lip}}(V^{2\beta}, W) \cap C^1_F(V^{2\beta}, W),$$

with $E(u_0, 0) = 0$. Note that the derivative $DE(u_0, v) = \frac{\partial}{\partial v} E(u_0, v)$ satisfies $DE(u_0, 0) = 0$. Since $DE(u_0, v) \in \mathfrak{L} = \mathfrak{L}(V^{2\beta}, W)$, for each $v \in V^{2\beta}$, and since it is continuous in v, there is a nonnegative, monotone nondecreasing function $\gamma = \gamma(\rho)$, defined for $0 \leq \rho < \rho_1$, for some $\rho_1 > 0$, such that $\gamma(\rho) \to 0$, as $\rho \to 0$ and $\|DE(u_0, v)\|_{\mathfrak{L}} \leq \gamma(\rho)$, for all $v \in V^{2\beta}$ with $\|A^\beta v\| \leq \rho$. Since one has

$$E(u_0, v_1) - E(u_0, v_2) = \int_0^1 DE(u_0, v_2 + s(v_1 - v_2))\, ds\, (v_1 - v_2),$$

and $\|A^\beta(v_2 + s(v_1 - v_2))\| \leq \rho$, when $\|A^\beta v_1\|, \|A^\beta v_2\| \leq \rho$ and $0 \leq s \leq 1$, one finds that

$$(71.10) \qquad \|E(u_0, v_1) - E(u_0, v_2)\| \leq \gamma(\rho) \|A^\beta(v_1 - v_2)\|,$$

whenever $\|A^\beta v_1\|, \|A^\beta v_2\| \leq \rho$. Since $E(u_0, 0) = 0$, it follows from inequality (71.10) that for $\|A^\beta v\| = \|A^\beta(u - u_0)\| \leq \rho$, one has

$$(71.11) \qquad \|E(u_0, v)\| \leq \gamma(\rho) \|A^\beta v\| \leq \rho\gamma(\rho).$$

We will apply the Preparation Lemma 47.10, with $\Lambda = \{p_0\}$ being a set with a single point p_0, to the error function $E(u,v)$, where u assumes values in a given compact set $\mathcal{K}$ in $V^{2\beta}$. For the application at hand, we set $\mathcal{K} = \{u_0\}$, where u_0 is the given equilibrium point for (70.1). Recall that for each $a > 0$, we obtain a new cutoff function $E^a = E^a(u_0, v)$ that satisfies $E^a(u_0, v) = E(u_0, v)$, for $\|A^\beta v\| = \|A^\beta(u - u_0)\| \leq a$,

$$\|E^a(u_0, v)\| \leq \gamma(\rho)\|A^\beta v\| \leq K_0(\rho), \tag{71.12}$$

for all $v \in V^{2\beta}$ with $\|A^\beta v\| \leq \rho$, where $K_0(\rho) = \rho\gamma(\rho)$, $0 \leq \rho \leq \rho_0 = 2a$, and $K_0(\rho) = \rho_0\gamma(\rho_0)$, for $\rho \geq \rho_0$. Also

$$\|E^a(u_0, v_1) - E^a(u_0, v_2)\| \leq K_1(\rho)\|A^\beta(v_1 - v_2)\|, \tag{71.13}$$

for all v_1, $v_2 \in V^{2\beta}$, where

$$K_1 = K_1(\rho) = \begin{cases} \gamma(\rho), & \text{for } 0 < \rho \leq \rho_0, \\ \gamma(\rho_0), & \text{for } \rho \geq \rho_0. \end{cases} \tag{71.14}$$

Thus one has $K_0(\rho) \to 0$ and $K_1(\rho) \to 0$, as $\rho \to 0^+$.

In addition to equation (71.7), we are also interested in the modified equation

$$\partial_t v + Av - DF(u_0)v = \partial_t v + Lv = E^a(u_0, v), \tag{71.15}$$

where $E^a(u_0, v)$ is given above. For $v_0 \in V^{2\beta}$, we let $S(t)v_0$ denote the maximally defined mild solution in the space $V^{2\beta}$ of equation (71.7), with $S(0)v_0 = v_0$ (see Section 4.7). Similarly, $S^a(t)v_0$ will denote the maximally defined mild solutions of equation (71.15). As noted in Sections 4.7.5 and 4.7.6, $S(t)v_0$ is a semiflow on $M(F) = \{v_0 \in V^{2\beta} : T(F, v_0) = \infty\}$, while $S^a(t)v_0$ is a semiflow on $V^{2\beta}$, since $E^a \in C_{\text{Lip; Global}}$. Furthermore if v_0 satisfies $\|A^\beta v_0\| < a$, then there is an interval $I = (t_1, t_2)$, with $0 \in I$, such that $S(t)v_0 = S^a(t)v_0$, for all $t \in I$. In addition I is maximal in the sense that both of the following properties are satisfied:

(1) Either $t_2 = \infty$, or one has $\|A^\beta S(t)v_0\| = \|A^\beta S^a(t)v_0\| \to a$, as $t \to t_2^-$.
(2) There exists a (partial) negative continuation of equation (71.7) through v_0 if and only if equation (71.15) has a similar negative continuation, and in that case, either (i) $t_1 = -\infty$, or (ii) the negative continuations terminate at t_1, or (iii) one has $\|A^\beta S(t)v_0\| = \|A^\beta S^a(t)v_0\| \to a$, as $t \to t_1^+$, where $S(t)v_0$ and $S^a(t)v_0$ denote the negative continuations.

In terms of the u-variable, where $u = u_0 + v$, equation (70.1) can be rewritten as

$$\partial_t u + L(u - u_0) = E(u_0, u - u_0), \tag{71.16}$$

while the modified equation (71.15) assumes the form

$$\partial_t u + L(u - u_0) = E^a(u_0, u - u_0), \tag{71.17}$$

Of course, these two equations agree in the set where $\|A^\beta(u - u_0)\| \leq a$.

7.1.1. Unstable and Stable Manifold Theorems. Let A satisfy the Standing Hypothesis A, and assume that the nonlinearity F is in C^1_{Lip}. Let u_0 be an equilibrium point for equation (70.1). A subset M in a space W_0, where W_0 is a linear subspace of W, is said to be a **local Lipschitz continuous manifold of radius $\rho > 0$ at a point u_0 in W_0** provided that there is a closed linear subspace V of W and there is a one-to-one mapping

$$\varphi : \mathcal{D}_V(\rho) = \{v \in V : \|v\| < \rho\} \to M \subset W_0 \subset W$$

such that the following properties are valid:

(1) One has $u_0 \in M$ and $\varphi(0) = u_0$.
(2) The mapping $\varphi : \mathcal{D}_V(\rho) \to M$ is Lipschitz continuous with a Lipschitz continuous inverse.

The manifold M is said to be **smooth (of Class C^1)** if, in addition,

(3) the mapping φ satisfies $\varphi \in C^1_F(\mathcal{D}_V(\rho), W_0)$.

In this case, the hyperspace given by the mapping $v \to \varphi(v_0) + D\varphi(v_0)(v - v_0)$ is said to be the **tangent space to M at $\varphi(v_0)$**, for each $v_0 \in \mathcal{D}_V(\rho)$, provided that

$$\lim_{v \to v_0} \frac{\|\varphi(v) - \varphi(v_0) - D\varphi(v_0)(v - v_0)\|_{W_0}}{\|v - v_0\|_V} = 0.$$

In the case where the linear subspace V has either finite dimension n, or finite codimension n, we say that the manifold M has, respectively, **dimension** n, or **codimension** n, at the point u_0. We do not exclude the boring possibility that the dimension can be $n = 0$.

A manifold M in $V^{2\beta}$ is said to be **locally positively invariant** for a semiflow $S(t)$, if for each point $v_0 \in M$ there is a time $\omega > 0$ such that the following two properties hold:

(1) One has $S(t)v_0 \in M$, for $0 \leq t < \omega$;
(2) Either one has $\omega = \infty$, or one of the following holds:
    (a) the maximally defined solution $S(t)v_0$ is not defined for $t > \omega$, or
    (b) one has $S(t_n)v_0 \notin M$, for some sequence $t_n > \omega$, with $t_n \to \omega$.

A manifold M in $V^{2\beta}$ is said to be **locally negatively invariant** for a semiflow $S(t)$, if for each point $v_0 \in M$ there exists a (partial) negative continuation of v_0, which we will denote by $S(t)v_0$, and a time $\alpha < 0$ such that the following two properties hold:

(1) One has $S(t)v_0 \in M$, for $\alpha < t \leq 0$;
(2) Either one has $\alpha = -\infty$, or one of the following holds:
    (a) there is no (partial) negative continuation $S(t)v_0$ defined for $t < \alpha$, or
    (b) any negative continuation satisfies $S(t_n)v_0 \notin M$, for some sequence $t_n < \alpha$, with $t_n \to \alpha$.

If a manifold M is both locally positively invariant and locally negatively invariant, then we will say that M is **locally invariant**.

Let $u_0 \in V^{2\beta}$ be an equilibrium point for equation (70.1). From the Herculean Theorem 47.6, we see that $u_0 \in \mathcal{D}(A) = V^2$. We will say that a subset $M = W^u_{\text{loc}}(u_0)$ of $V^{2\beta}$ is a **(local) unstable manifold** at u_0, provided that the following three properties hold:

(1) The set M is a local, Lipschitz continuous[33] manifold of radius $\rho > 0$ at u_0.

(2) For all $u_1 \in M$, there is a unique negative continuation $\phi^{u_1}(t)$ such that
$$\sup_{t\leq 0} \|A^\beta(\phi^{u_1}(t) - u_0)\| < \infty,$$
and $\|A^\beta(\phi^{u_1}(t) - u_0)\| \to 0$, as $t \to -\infty$, with $\phi^{u_1}(t) \in M$, for all $t \leq 0$.

(3) If $u_1 \in V^{2\beta}$ has the property that there is a nonnegative continuation $\phi^{u_1}(t)$ where $\|A^\beta(\phi^{u_1}(t) - u_0)\| < \rho$, for all $t \leq 0$, and $\|A^\beta(\phi^{u_1}(t) - u_0)\| \to 0$, as $t \to -\infty$, then $u_1 \in M$.

Similarly, we will say that a subset $M = W^s_{\text{loc}}(u_0)$ of $V^{2\beta}$ is a **(local) stable manifold** at u_0, provided that the following three properties hold:

(1) The set M is a local, Lipschitz continuous manifold of radius $\rho > 0$ at u_0.

(2) For all $u_1 \in M$ with $\|A^\beta(u_1 - u_0)\| < \rho$, one has $\|A^\beta(S(t)u_1 - u_0)\| \to 0$, as $t \to \infty$, and $S(t)u_1 \in M$, for all $t \geq 0$.

(3) If $u_1 \in W$ has the property that $\|A^\beta(S(t)u_1 - u_0)\| < \rho$, for all $t \geq 0$, and $\lim_{t\to\infty} \|A^\beta(S(t)u_1 - u_0)\| = 0$, then $u_1 \in M$.

Hence, a local unstable manifold is locally negatively invariant, while a local stable manifold is locally positively invariant. Next we have the Saddle Point Property for the nonlinear equation (70.1), see Figure 7.1.

Theorem 71.1. (Saddle Point Property). *Let the Standing Hypothesis A be satisfied, and let $F \in C^1_{\text{Lip}}$. Let u_0 be a hyperbolic equilibrium point for equation (70.1) in the space $V^{2\beta}$, and let $K \geq 1$ and $\alpha > 0$ be the characteristics of the hyperbolicity. Let ν satisfy $0 < \nu < \alpha$. Then the following are valid:*

(1) *There exists a local Lipschitz continuous unstable manifold $M^u = W^u_{\text{loc}}(u_0)$ in $V^{2\beta}$. Furthermore, the unstable linear manifold is the tangent space to M^u at u_0, and for every $u_1 \in M^u$ there is a negative continuation, which we denote by $S(t)u_1$, for $t \leq 0$, such that*
$$\|A^\beta(S(t)u_1 - u_0)\| \leq 2K\|A^\beta(u_1 - u_0)\|e^{\nu t}, \qquad \textit{for all } t \leq 0.$$
In particular, for $\rho > 0$, but small, the unstable set $W^u(u_0)$ satisfies
$$\{y \in W^u(u_0) : \|y - u_0\| < \rho\} = \{y \in W^u_{\text{loc}}(u_0) : \|y - u_0\| < \rho\}.$$

[33]The Lipschitz continuity here is in reference to the A^β-norm on the space $V^{2\beta}$.

(2) *There exists a local Lipschitz continuous stable manifold* $M^s = W^s_{\text{loc}}(u_0)$ *in* $V^{2\beta}$. *Furthermore, the stable linear manifold is the tangent space to* M^s *at* u_0, *and for every* $u_1 \in M^s$ *one has*

$$\|A^\beta(S(t)u_1 - u_0)\| \leq 2K\|A^\beta(u_1 - u_0)\|e^{-\nu t}, \qquad \text{for all } t \geq 0.$$

In particular, for $\rho > 0$, *but small, the stable set* $W^s(u_0)$ *satisfies*

$$\{y \in W^s(u_0) : \|y - u_0\| < \rho\} = \{y \in W^s_{\text{loc}}(u_0) : \|y - u_0\| < \rho\}.$$

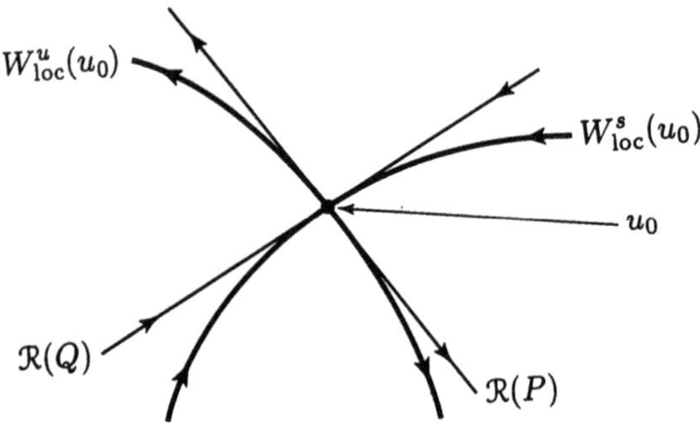

Figure 7.1. Saddle Point Property

In the proof of this theorem, it will be seen that ν can be chosen arbitrarily close to the characteristic α by making $\|A^\beta(u_1 - u_0)\|$ small. Also notice that we will show that, near the point u_0, the stable and unstable sets $W^s(u_0)$ and $W^u(u_0)$ are (locally) homeomorphic to the linear stable and unstable subspaces at u_0, respectively.

Since we assume the equilibrium point u_0 to be hyperbolic, this means that the Saddle Point Property gives information about the nonlinear dynamics near u_0 in a situation where the spectrum of the linear operator $L = A - DF(u_0)$ does not intersect some band about the imaginary axis in the complex plane. However, the stable and unstable manifolds can exist even when L has purely imaginary spectral values. In order to derive this broader result, while also proving the Saddle Point Property, we will need the following lemma.

Lemma 71.2. *Let the Standing Hypothesis A be satisfied, and let* $F \in C^1_{\text{Lip}}$. *Let* u_0 *be an equilibrium point for equation (70.1) and set* $L = A - DF(u_0)$. *Let* e^{-Lt} *denote the analytic semigroup on* W *generated by* $-L$. *Assume that for some* $\lambda \in \mathbb{R}$, *the shifted semigroup* $e^{-\lambda t}e^{-Lt}$ *has an exponential dichotomy on* W *with characteristics* $K \geq 1$ *and* $\alpha > 0$ *and with associated projections* (P_λ, Q_λ).

Then there exists a local Lipschitz continuous manifold $M^u_\lambda = W^u_{\lambda,\mathrm{loc}}(u_0)$ in $V^{2\beta}$ of radius $\rho > 0$ at u_0, and the following properties are valid:

(1) *For all $u_1 \in M^u_\lambda$, there is a unique negative continuation, which we will denote by $S(t)u_1$, such that $\sup_{t\le 0} \|e^{-\lambda t}A^\beta(S(t)u_1 - u_0)\| < \infty$. Moreover, there is a constant ν with $0 < \nu < \alpha$, such*

$$\|e^{-\lambda t}A^\beta(S(t)u_1 - u_0)\| \le 2K\|A^\beta(u_1 - u_0)\|e^{\nu t}, \qquad \text{for all } t \le 0. \tag{71.18}$$

(2) *If $u_1 \in V^{2\beta}$ has the property that there is a nonnegative continuation $S(t)u_1$ such that $\|e^{-\lambda t}A^\beta(S(t)u_1 - u_0)\| < \rho$, for all $t < 0$, then one has $u_1 \in M^u_\lambda$ and inequality (71.18) holds.*

(3) *The manifold M^u_λ is tangent to $u_0 + \mathcal{R}(P_\lambda)$ at the point u_0.*

In addition, there exists a manifold $M^s_\lambda = W^s_{\lambda,\mathrm{loc}}(u_0)$ in $V^{2\beta}$ of radius $\rho > 0$ at u_0, which is locally Lipschitz continuous, and the following properties are valid:

(4) *There is a constant ν with $0 < \nu < \alpha$, such that, for all $u_1 \in M^s_\lambda$ with $\|A^\beta(u_1 - u_0)\| < \rho$, the mild solution $S(t)u_1$ of (70.1) satisfies*

$$\|e^{-\lambda t}A^\beta(S(t)u_1 - u_0)\| \le 2K\|A^\beta(u_1 - u_0)\|e^{-\nu t}, \qquad \text{for all } t \ge 0. \tag{71.19}$$

(5) *If $u_1 \in V^{2\beta}$ has the property that the mild solution $S(t)u_1$ of (70.1) satisfies $\|e^{-\lambda t}A^\beta(S(t)u_1 - u_0)\| < \rho$, for all $t \ge 0$, then one has $u_1 \in M$ and inequality (71.19) holds.*

(6) *The manifold M^s_λ is tangent to $u_0 + \mathcal{R}(Q_\lambda)$ at the point u_0.*

Proof. Our argument uses the Lyapunov-Perron method. We study first the modified equations (71.15) and (71.17), where the parameter a will be fixed later. Let $E^a(v) = E^a(u_0, v)$. We begin with the unstable manifold. Let ν be a parameter satisfying $0 < \nu < \alpha$. (We will use this parameter to estimate the rate of exponential decay of the solutions.) For each $\xi \in \mathcal{R}(P_\lambda) \subset V^{2\beta}$, we consider the mapping $z \to \hat{z}$ defined formally, for $t \le 0$, by $\hat{z} = \mathcal{T}z$, where

(71.20)
$$\begin{aligned}\hat{z}(t) &= e^{-(\lambda+\nu)t}e^{-Lt}\xi \\ &\quad + \int_\infty^t e^{-(\lambda+\nu)(t-s)}e^{-L(t-s)}Q_\lambda e^{-(\lambda+\nu)s}E^a(e^{(\lambda+\nu)s}z(s))\,ds \\ &\quad - \int_t^0 e^{-(\lambda+\nu)(t-s)}e^{-L(t-s)}P_\lambda e^{-(\lambda+\nu)s}E^a(e^{(\lambda+\nu)s}z(s))\,ds.\end{aligned}$$

We define $\mathcal{F}_\rho$ to be those $z \in \mathcal{F} = C((-\infty, 0]; V^{2\beta}) \cap L^\infty(-\infty, 0; V^{2\beta})$ with

$$\|A^\beta z\|_\infty \overset{\text{def}}{=} \sup_{t\le 0}\|A^\beta z(t)\| \le \rho.$$

Notice that owing to (71.12), the function

$$g(s) = e^{-(\lambda+\nu)s}E^a(e^{(\lambda+\nu)s}z(s)), \qquad \text{for } s \le 0,$$

is in $L^\infty(-\infty, 0; W)$, with $\|g\|_\infty \le K_a\|A^\beta z\|_\infty$, where K_a depends on a. Owing to Lemma 45.2 and the fact that $e^{-Lt}\xi = e^{-Lt}P_\lambda\xi$, we see that the two terms $e^{-Lt}\xi$ and $P_\lambda e^{-L(t-s)}g(s)$ in (71.20) are well defined for $t \le s \le 0$.

The next step is to show that if $z \in \mathcal{F}_\rho$, then the integrals in equation (71.20) exist in the space $V^{2\beta}$. From (71.4) and (71.12) one has

$$\begin{aligned}\int_t^0 \|A^\beta e^{-(\lambda+\nu)(t-s)}e^{-L(t-s)}P_\lambda e^{-(\lambda+\nu)s}E^a(e^{(\lambda+\nu)s}z(s))\|\,ds \\ \le K\gamma(\rho)\|A^\beta z\|_\infty \int_t^0 e^{\alpha(t-s)}e^{-\nu(t-s)}\,ds \\ \le \frac{K}{\alpha-\nu}\gamma(\rho)\|A^\beta z\|_\infty,\end{aligned}$$

for all $t \le 0$. Similarly, inequalities (71.5), (71.6), and (71.12) imply that

$$\begin{aligned}\int_{-\infty}^t \|A^\beta e^{-(\lambda+\nu)(t-s)}e^{-L(t-s)}Q_\lambda e^{-(\lambda+\nu)s}E^a(e^{(\lambda+\nu)s}z(s))\|\,ds \\ \le K\gamma(\rho)\|A^\beta z\|_\infty\Big(\int_{-\infty}^{t-1} e^{-\alpha(t-s)}e^{-\nu(t-s)}\,ds \\ + \int_{t-1}^t (t-s)^{-\beta}e^{-\alpha(t-s)}e^{-\nu(t-s)}\,ds\Big) \\ \le \left(\frac{K}{\alpha+\nu}+\frac{K}{1-\beta}\right)\gamma(\rho)\|A^\beta z\|_\infty,\end{aligned}$$

for all $t \le 0$. Since $\xi \in \mathcal{R}(P_\lambda)$, inequality (71.4) implies that

$$\|A^\beta e^{-\lambda t}e^{-Lt}e^{-\nu t}\xi\| \le K\|A^\beta\xi\|e^{(\alpha-\nu)t}, \qquad \text{for } t \le 0.$$

Since $\alpha - \nu > 0$, it then follows that

$$\|A^\beta \hat{z}(t)\| \le K\|A^\beta\xi\| + \left(\frac{2K\alpha}{\alpha^2-\nu^2}+\frac{K}{1-\beta}\right)\gamma(\rho)\|A^\beta z\|_\infty, \tag{71.21}$$

for all $t \le 0$. Let $\rho_0 = \rho_0(\nu)$ be chosen so that $(\frac{2K\alpha}{\alpha^2-\nu^2}+\frac{K}{1-\beta})\gamma(\rho_0) \le \frac{1}{4}$, and $K_1(\rho_0) \le K$, see (71.14). If $\xi \in \mathcal{R}(P_\lambda)$ satisfies $K\|A^\beta\xi\| \le \frac{1}{4}\rho$, then

$$\|A^\beta \hat{z}(t)\| \le K\|A^\beta\xi\| + \frac{1}{4}\|A^\beta z\|_\infty \le \rho, \tag{71.22}$$

for all $t \leq 0$, whenever $0 < \rho \leq \rho_0(\nu)$. Hence, $\hat{z} \in L^\infty(-\infty, 0; V^{2\beta})$. By applying Lemma 42.7 and Theorem 45.7 to $\partial_t v + Lv = g(t)$, we see that $\hat{z} \in C((-\infty, 0]; V^{2\beta})$. Consequently, $\mathcal{T}$ maps $\mathfrak{F}_\rho$ into $\mathfrak{F}_\rho$, whenever $0 < \rho \leq \rho_0$.

The next step is to show that $\mathcal{T}$ is a strict contraction on $\mathfrak{F}_\rho$, under the given assumptions on ξ. Indeed if $\hat{z}_i = \mathcal{T} z_i$, for $i = 1, 2$, then for $t \leq 0$, one obtains

$$\begin{aligned}\hat{z}_1(t) - \hat{z}_2(t) = \int_{-\infty}^{t} & e^{-(\lambda+\nu)(t-s)} e^{-L(t-s)} Q_\lambda e^{-(\lambda+\nu)s} \\ & \times (E^a(e^{(\lambda+\nu)s} z_1(s)) - E^a(e^{(\lambda+\nu)s} z_2(s)))\, ds \\ - \int_{t}^{0} & e^{-(\lambda+\nu)(t-s)} e^{-L(t-s)} P_\lambda e^{-(\lambda+\nu)s} \\ & \times (E^a(e^{(\lambda+\nu)s} z_1(s)) - E^a(e^{(\lambda+\nu)s} z_2(s)))\, ds.\end{aligned}$$

By using inequalities (71.4), (71.5), (71.6), (71.13), and (71.14), and the arguments given above, with $K_1(\rho) \leq K$, for $0 < \rho \leq \rho_0$, one obtains

$$\begin{aligned}\int_{t}^{0} & \|A^\beta e^{-(\lambda+\nu)(t-s)} e^{-L(t-s)} P_\lambda e^{-(\lambda+\nu)s} \\ & \times (E^a(e^{(\lambda+\nu)s} z_1(s)) - E^a(e^{(\lambda+\nu)s} z_2(s)))\|\, ds \\ & \leq \frac{K}{\alpha - \nu} \gamma(\rho)\, \|A^\beta(z_1 - z_2)\|_\infty\end{aligned}$$

and

$$\begin{aligned}\int_{-\infty}^{t} & \|A^\beta e^{-(\lambda+\nu)(t-s)} e^{-L(t-s)} P_\lambda e^{-(\lambda+\nu)s} \\ & \times \left[E^a(e^{(\lambda+\nu)s} z_1(s)) - E^a(e^{(\lambda+\nu)s} z_2(s))\right] \|\, ds \\ & \leq \left(\frac{K}{\alpha+\nu} + \frac{K}{1-\beta}\right) \gamma(\rho)\, \|A^\beta(z_1 - z_2)\|_\infty,\end{aligned}$$

for all $t \leq 0$ and all ρ with $0 < \rho \leq \rho_0$. Since $(\frac{2K\alpha}{\alpha^2-\nu^2} + \frac{K}{1-\beta})\gamma(\rho_0) \leq \frac{1}{4}$, this implies that

$$\|A^\beta(\hat{z}_1 - \hat{z}_2)\|_\infty \leq \frac{1}{2}\|A^\beta(z_1 - z_2)\|_\infty, \qquad \text{for } 0 < \rho \leq \rho_0,$$

which shows that $\mathcal{T}$ is a strict contraction on $\mathfrak{F}_\rho$. Since $\mathfrak{F}_\rho$ is a complete metric space, for each ρ with $0 < \rho \leq \rho_0(\nu)$, $\mathcal{T}$ has a unique fixed point $z(t) = z(\xi, t)$, and

$$\|A^\beta z(t)\| \leq 2K\|A^\beta \xi\|, \qquad \text{for all } t \leq 0, \tag{71.23}$$

by inequality (71.22). Furthermore, it follows from the Contraction Mapping Theorem that $z(\xi, t)$ is locally Lipschitz continuous in ξ, i.e., there is a constant k such that

$$\|A^\beta(z(\xi_1, t) - z(\xi_2, t))\| \le k\|A^\beta(\xi_1 - \xi_2)\|, \qquad \text{for all } t \le 0;$$

and for all $\xi_i \in \mathcal{R}(P_\lambda) \cap B_{V^{2\beta}}(0, \rho/(4K))$, for $i = 1, 2$, see Naylor and Sell (1982), for example. Furthermore, one has

$$z(\xi, 0) = \xi + \int_{-\infty}^{0} Q_\lambda e^{Ls} E^a(e^{(\lambda+\nu)s} z(s))\, ds \stackrel{\text{def}}{=} \xi + \varphi(\xi).$$

Since $P_\lambda Q_\lambda = 0$, one has $P_\lambda z(\xi, 0) = P_\lambda \xi = \xi$. Consequently the mapping given by

$$\xi \to u_0 + z(\xi, 0) = u_0 + \xi + \varphi(\xi)$$

maps $\mathcal{R}(P_\lambda) \cap B_{V^{2\beta}}(0, \rho/(4K))$ into $V^{2\beta}$ and φ is Lipschitz continuous. The inverse mapping is $P_\lambda(z(\xi, 0) + u_0) = \xi$, and it too is Lipschitz continuous.

Next we define $w(t) = w(\xi, t)$, $v(t) = v(\xi, t)$, and $u(t) = u(\xi, t)$ by $w(t) = e^{\nu t} z(t)$, $v(t) = e^{\lambda t} w(t) = e^{(\lambda+\nu)t} z(t)$, and $u(t) = u_0 + v(t)$. One then has

$$v(\xi, 0) = w(\xi, 0) = z(\xi, 0) = \xi + \varphi(\xi) \quad \text{and} \quad u(\xi, 0) = u_0 + \xi + \varphi(\xi),$$

for all $\xi \in \mathcal{R}(P_\lambda) \cap B_{V^{2\beta}}(0, \rho/(4K))$. A direct calculation shows that $v(t)$ and $u(t)$ are, respectively, mild solutions of modified equations (71.15) and (71.17). Also inequality (71.23) implies that

$$\|A^\beta w(\xi, t)\| \le 2K\|A^\beta \xi\| e^{\nu t}, \qquad \text{for } t \le 0,$$

and

$$\|e^{-\lambda t} A^\beta(u(\xi, t) - u_0)\| \le 2K\|A^\beta \xi\| e^{\nu t}, \qquad \text{for } t \le 0. \tag{71.24}$$

This proves that

$$M^u \stackrel{\text{def}}{=} \text{Graph}(\xi \to u_0 + \xi + \varphi(\xi)), \qquad \text{for } \xi \in B_{V^{2\beta}}(0, \rho/(4K)), \tag{71.25}$$

is a local Lipschitz continuous manifold in $V^{2\beta}$ and that it satisfies properties (1) and (2) in the definition of an unstable manifold. Also Item (1) in the statement of the lemma is proved.

For Property (3) in the definition of an unstable manifold, we note that if $\phi(t)$ is any such negative continuation of equation (71.17), then Theorem 45.7, Item (1) implies that $w(t) = e^{-\lambda t}(\phi(t) - u_0)$ must satisfy equation (45.38). However, this in turn implies that $z(t) = w(t)e^{-\nu t}$ is a fixed point of $\mathcal{T}$. Hence one has $z(t) = z(\xi, t)$, where ξ satisfies $\xi = P_\lambda(\phi(0) - u_0)$.

This completes the proof of Property (3) in the definition, as well as Item (2) in the statement of the lemma.

For Item (3) in the statement of the lemma, we note that

$$\begin{aligned}\|A^\beta(z(\xi,0)-\xi)\| &\le \int_{-\infty}^{0} \|A^\beta e^{(\lambda+\nu)s} e^{Ls} Q_\lambda e^{-(\lambda+\nu)s} E^a(e^{(\lambda+\nu)s} z(\xi,s))\|\, ds \\ &\le \left(\frac{K}{\alpha+\nu} + \frac{K}{1-\beta}\right)\gamma(\rho)\|A^\beta z\|_\infty,\end{aligned}$$

where $\gamma(\rho)$ can be replaced by $\gamma(\|A^\beta z\|_\infty)$. Since $\|A^\beta z\|_\infty \le 2K\|A^\beta \xi\|$, this implies that

$$\|A^\beta \xi\|^{-1}\|A^\beta(z(\xi,0)-\xi)\| \to 0, \qquad \text{as } \|A^\beta \xi\| \to 0^+,$$

which completes the proof of Item (3).

The solution $u(\xi,t)$ constructed above is given by the nonlinear dynamics of the perturbed equation (71.17). That means that $v(\xi,t) = S^a(t)v(\xi,0)$, for $t \le 0$. In addition, it follows from Theorem 45.7 that $v(\xi,t)$ satisfies

$$v(\xi(\tau),t) = v(\xi(0),\tau+t), \qquad \text{for all } \tau,\ t \in \mathbb{R}, \tag{71.26}$$

where $\xi(t) = P_\lambda v(\xi,t)$, for all $t \in \mathbb{R}$. Moreover, $\xi(t)$ is the unique mild solution of the reduced equation

$$\partial_t \xi(t) + L\xi(t) = P_\lambda E^a(v(\xi(t),0)), \qquad \text{where } \xi(0) = \xi \tag{71.27}$$

and $\xi \in \mathcal{R}(P_\lambda) \cap B_{V^{2\beta}}(0,\rho/(4K))$. Now (71.26) implies that $v(\xi,t) = v(\xi(t),0)$, for all $t \in \mathbb{R}$. Thus $v(\xi,t)$ is a slave variable, and its dynamics are completely determined by the solutions of (71.27).

For the unmodified equation (70.1), or (71.16), we define

$$M_\lambda^u = W_{\lambda,\mathrm{loc}}^u(u_0) \stackrel{\mathrm{def}}{=} \{u \in M^u : \|A^\beta(u-u_0)\| \le a\},$$

where M^u is given by (71.25). Since the equations (71.16) and (71.17) agree for $\|A^\beta v\| \le a$, it follows that M_λ^u satisfies all the properties required for the original equation (70.1), or equation (71.16), in the definition of the unstable manifold, and all the related items in the lemma.

In the case of the stable manifold, one begins with $\xi \in \mathcal{R}(Q_\lambda)$ and defines a mapping formally by $\hat{z} = \mathcal{T}z$, where now equation (71.20) is replaced by

$$\begin{aligned}\hat{z}(t) =& e^{-(\lambda-\nu)t} e^{-Lt}\xi \\ &+ \int_0^t e^{-(\lambda-\nu)(t-s)} e^{-L(t-s)} Q_\lambda e^{-(\lambda-\nu)s} E^a(e^{(\lambda-\nu)s} z(s))\, ds \\ &- \int_t^\infty e^{-(\lambda-\nu)(t-s)} e^{-L(t-s)} P_\lambda e^{-(\lambda-\nu)s} E^a(e^{(\lambda-\nu)s} z(s))\, ds,\end{aligned}$$

for $t \geq 0$. We let $\mathcal{F}_\rho$ denote those $z \in \mathcal{F} = C[0,\infty;V^{2\beta}) \cap L^\infty(0,\infty;V^{2\beta})$ with

$$\|A^\beta z\|_\infty \stackrel{\text{def}}{=} \sup_{t\geq 0} \|A^\beta z(t)\| \leq \rho.$$

As argued in the case of the unstable manifold, one shows that

$$\|A^\beta \hat{z}(t)\| \leq K\|A^\beta \xi\| + \left(\frac{2K\alpha}{\alpha^2 - \nu^2} + \frac{K}{1-\beta}\right)\gamma(\rho)\|A^\beta z\|_\infty,$$

for all $t \geq 0$. With $\rho_0 = \rho_0(\nu)$ satisfying $(\frac{2K\alpha}{\alpha^2-\nu^2} + \frac{K}{1-\beta})\gamma(\rho_0) \leq \frac{1}{4}$ and $K\|A^\beta\xi\| \leq \frac{\rho}{4}$, then $\|A^\beta \hat{z}(t)\|$ satisfies (71.22), for all $t \geq 0$, whenever $0 < \rho \leq \rho_0$. Similarly, the argument for the contraction mapping leads to

$$\|A^\beta(\hat{z}_1 - \hat{z}_2)\|_\infty \leq \frac{1}{2}\|A^\beta(z_1 - z_2)\|_\infty, \qquad \text{for } 0 < \rho \leq \rho_0.$$

Hence $\mathcal{T}$ is a strict contraction on $\mathcal{F}_\rho$ in this case as well. The remainder of the argument now proceeds in a predictable fashion. We will omit the details. In this case, we note that inequality (71.24) is replaced by

$$\|e^{-\lambda t} A^\beta(u(\xi,t) - u_0)\| \leq 2K\|A^\beta\xi\| e^{-\nu t}, \qquad \text{for } t \geq 0.$$

Also it follows from Theorem 45.7, that $v(\xi,t)$ satisfies

$$v(\xi(\tau),t) = v(\xi(0),\tau + t),$$

for all $\tau \geq 0$ and $-\tau \leq t < \infty$, where $\xi(t) = Q_\lambda v(\xi,t)$, for all $t \geq 0$. Moreover, $\xi(t)$ is the unique mild solution of the reduced equation

$$\partial_t \xi(t) + L\xi(t) = Q_\lambda E^a(v(\xi(t),0)), \qquad \text{where } \xi(0) = \xi \tag{71.28}$$

and $\xi \in \mathcal{R}(Q_\lambda) \cap B_{V^{2\beta}}(0,\rho/(4K))$. Now (71.26) implies that $v(\xi,t) = v(\xi(t),0)$, for all $t \in \mathbb{R}$. Thus $v(\xi,t)$ is a slave variable, and its dynamics are completely determined by the solutions of (71.28). □

Proof of Theorem 71.1. Theorem 71.1 now follows directly from Lemma 71.2 by setting $\lambda = 0$. □

There is an important consequence of the Saddle Point Property which will be used later. In particular, we now show that, in the set of all stationary solutions of equation (70.1), the hyperbolic equilibrium points are isolated.

Corollary 71.3. *Let the Standing Hypothesis A be satisfied, and let $F \in C^1_{\text{Lip}}$. Let u_0 be a hyperbolic equilibrium point for equation (70.1) in the space $V^{2\beta}$. Then there is an $\epsilon = \epsilon(u_0) > 0$ such that, the only stationary* *solution of equation (70.1) in the set* $\{u \in V^{2\beta} : \|A^\beta(u - u_0)\| < \epsilon\}$ *is the point* u_0*, itself.*

Proof. Assume that there is a sequence of stationary solutions u_n of equation (70.1) with $\|A^\beta(u_n - u_0)\| \to 0$, as $n \to \infty$. It then follows from the Saddle Point Property that, for n sufficiently large, one has $u_n \in W^s_{\text{loc}}(u_0) \cap W^u_{\text{loc}}(u_0)$. Since the spaces $W^s_{\text{loc}}(u_0)$ and $W^u_{\text{loc}}(u_0)$ are, respectively, tangent to the stable and unstable linear manifolds at u_0, this implies that for $\rho > 0$, but small, the only point in $W^s_{\text{loc}}(u_0) \cap W^u_{\text{loc}}(u_0)$ is u_0, itself. Hence, $u_n = u_0$, for n sufficiently large. □

7.1.2. Center Manifold Theorem. As before, we assume here that the Standing Hypothesis A is satisfied, and that $F = F(u) \in C^1_{\text{Lip}}$. Let u_0 be an equilibrium point for equation (70.1) and set $L = A - DF(u_0)$, as above. Our objective now is to study the behavior of solutions of equation (70.1) in the vicinity of u_0 in the case that the spectrum $\sigma(L)$ contains points on, or near to, the imaginary axis in the complex plane. For this purpose, we will assume that the analytic semigroup e^{-Lt} has an exponential trichotomy.

More precisely, we assume now that the linear semigroup e^{-Lt} has an exponential trichotomy on W with characteristics λ_1, λ_2, λ_3, λ_4, and K, where $\lambda_1 < \lambda_2 \leq 0 \leq \lambda_3 < \lambda_4$ and $K \geq 1$, see Section 4.5. As a result there are three projections P, Q, and R on W, each of which is invariant under e^{-Lt}, and these projections commute with one another and satisfy $PQ = PR = QR = 0$ and $I = P + Q + R$. Furthermore, for each $u = Pw \in \mathcal{R}(P)$ there is a unique negative continuation $e^{-Lt}u$, such that for any $w \in W$, one has

$$\|e^{-Lt}Pw\| \leq K\|w\|e^{\lambda_4 t}, \qquad \text{for } t \leq 0 \text{ and } w \in W.$$

Also, for each $u = Rw \in \mathcal{R}(R)$ there is a unique negative continuation $e^{-Lt}u$, such that for any $w \in W$, one has

$$\|e^{-Lt}Rw\| \leq K\|w\|e^{\lambda_2 t}, \qquad \text{for } t \leq 0 \text{ and } w \in W.$$

In addition, the following two inequalities are valid:

$$\|e^{-Lt}Qw\| \leq K\|w\|e^{\lambda_1 t}, \qquad \text{for } t \geq 0 \text{ and } w \in W,$$

and

$$\|e^{-Lt}Rw\| \leq K\|w\|e^{\lambda_3 t}, \qquad \text{for } t \geq 0 \text{ and } w \in W.$$

It follows from (45.10) - see Lemma 45.2 - that

$$Pe^{-Lt} = e^{-Lt}P \quad \text{and} \quad Re^{-Lt} = e^{-Lt}R, \qquad \text{for all } t \in \mathbb{R}.$$

The inequalities (71.4) - (71.6) have counterparts in the case of an exponential trichotomy as well. In particular, by replacing the characteristic

K with a larger value, if necessary, the following inequalities are valid for $w \in W$:

$$\|A^\beta e^{-Lt} P w\| \leq K e^{\lambda_4 t} \|w\|, \qquad \text{for } t \leq 0; \tag{71.29}$$

while

$$\|A^\beta e^{-Lt} Q w\| \leq K t^{-\beta} e^{\lambda_1 t} \|w\|, \qquad \text{for } 0 < t \leq 1; \tag{71.30}$$

and

$$\|A^\beta e^{-Lt} Q w\| \leq K e^{\lambda_1 t} \|w\|, \qquad \text{for } 1 \leq t < \infty. \tag{71.31}$$

In addition, one has

$$\|A^\beta e^{-Lt} R w\| \leq K e^{\lambda_2 t} \|w\|, \qquad \text{for } t \leq 0; \tag{71.32}$$

while

$$\|A^\beta e^{-Lt} R w\| \leq K t^{-\beta} e^{\lambda_3 t} \|w\|, \qquad \text{for } 0 < t \leq 1; \tag{71.33}$$

and

$$\|A^\beta e^{-Lt} R w\| \leq K e^{\lambda_3 t} \|w\|, \qquad \text{for } 1 \leq t < \infty. \tag{71.34}$$

It is the portion of the spectrum $\sigma(-L)$ that lies in the band $\lambda_2 < \text{Re}\,\lambda < \lambda_3$ that gives rise to the projection R. By making different choices for λ_2 and λ_3, one may obtain different projections R. For example, in order to study various bifurcations, it is oftentimes useful to fix λ_2 and λ_3, as we have, and to study the behavior of the solutions of equation (70.1) as the nonlinearity $F(u)$ changes with respect to some of the parameters of the model.[34] Note that the following implications are valid: One has $R = 0$ if and only if $e^{-\mu t} e^{-Lt}$ has an exponential dichotomy, for all μ with $\lambda_2 < \mu < \lambda_3$, which in turn is valid, if and only if

$$\{\lambda \in \sigma(-L) : \lambda_2 < \text{Re}\,\lambda < \lambda_3\} = \emptyset.$$

The construction of the center manifold will be accomplished by expressing the manifold as the graph of a suitable function φ, where the domain of definition of φ is a neighborhood of the origin in the space $\mathcal{R}(R)$. It should be noted in advance that, except for the trivial case where the center manifold is 0-dimensional, the center manifold is never uniquely determined.

[34]See Marsden and McCracken (1976) for an application of this idea to the study of Hopf bifurcations.

A subset $M = W^o_{\text{loc}}(u_0)$ is said to be a **center manifold** at u_0, provided that the following three properties hold:

(1) The set M is a local Lipschitz continuous manifold of radius $\rho > 0$ at u_0.

(2) (Local Invariance) For every $u \in M$, there is a (partial) negative continuation $\phi^u(t)$ of u, defined for t satisfying $t_1 < t \leq 0$, where $-\infty \leq t_1 < 0$, and there is a time t_2 with $0 < t_2 \leq \infty$, such that one has $S(t)u \in M$, for $t_1 < t < t_2$, where $S(t)u = \phi^u(t)$, for $t \leq 0$, Moreover, the following hold:
   (a) either $t_2 = \infty$, or $\|A^\beta(S(t)u - u_0)\| \to \rho$, as $t \to t_2^-$, and
   (b) either $t_1 = -\infty$, or $\|A^\beta(S(t)u - u_0)\| \to \rho$, as $t \to t_1^+$.

(3) If $u \in V^{2\beta}$ is such that there is a negative continuation $S(t)u$, for $t \leq 0$, with $\|A^\beta(S(t)u - u_0)\| < \rho$, for all $t \in \mathbb{R}$, then $u \in M$.

Theorem 71.4. (Center Manifold). *Let the Standing Hypothesis A be satisfied, and let $F \in C^1_{\text{Lip}}$. Let u_0 be an equilibrium point for equation (70.1) and set $L = A - DF(u_0)$. Assume that the linear semigroup e^{-Lt} has an exponential trichotomy on W with characteristics λ_1, λ_2, λ_3, λ_4, and K, where $\lambda_1 < \lambda_2 \leq 0 \leq \lambda_3 < \lambda_4$ and $K \geq 1$. Let P, Q, and R be the associated projections on W as given above. Then the following are valid:*

(1) *There is a center manifold $M^o_{\text{loc}} = W^o_{\text{loc}}(u_0)$ at u_0, and M^o_{loc} is given as the graph of a Lipschitz continuous mapping $\xi \to u_0 + \xi + \varphi(\xi)$, where*

$$\varphi : \{\xi \in \mathcal{R}(R) : \|A^\beta \xi\| < \rho\} \to \mathcal{R}(P) + \mathcal{R}(Q) \subset V^{2\beta},$$

for some $\rho > 0$.

(2) *The manifold M^o_{loc} is tangent to $u_0 + \mathcal{R}(R)$ at u_0.*

(3) ***(Reduction Principle)** The manifold M^o_{loc} is locally invariant under the nonlinear semiflow generated by (70.1), and the solution $u(t) = u(u_1, t)$ of equation (70.1) with $u_1 = u_0 + \xi_0 + \varphi(\xi_0) \in M$ satisfies*

$$u(t) = u_0 + \xi(t) + \varphi(\xi(t)), \qquad \text{for } t \in I,$$

where $\xi = \xi(t)$ is the mild solution of the reduced evolutionary equation

$$\partial_t \xi + L\xi = RE(u_0, \xi + \varphi(\xi)), \qquad \text{with } \xi(0) = \xi_0 \in \mathcal{R}(R), \tag{71.35}$$

and I is the maximal interval with $0 \in I$ and $u(t) \in M$, for $t \in I$.

Before presenting the proof of this theorem, it is useful to reflect upon the linear and nonlinear geometric significance of the center manifold, see Figure 7.2. As noted in Lemma 45.4, the existence of an exponential trichotomy for e^{-Lt} is equivalent to the existence of a pair of exponential dichotomies, one for $e^{-\lambda t}e^{-Lt}$, and the other for $e^{-\mu t}e^{-Lt}$, where $\lambda_1 < \lambda < \lambda_2$

and $\lambda_3 < \mu < \lambda_4$. Let $\mathcal{S}_\lambda$, $\mathcal{U}_\lambda$, $\mathcal{S}_\mu$, and $\mathcal{U}_\mu$ denote the respective stable and unstable linear subspaces in W. Then one has

$$W = \mathcal{S}_\lambda + \mathcal{U}_\lambda = \mathcal{S}_\mu + \mathcal{U}_\mu = \mathcal{S}_\lambda + (\mathcal{U}_\lambda \cap \mathcal{S}_\mu) + \mathcal{U}_\mu,$$

and

$$\mathcal{S}_\lambda = \mathcal{R}(Q), \quad \mathcal{U}_\lambda \cap \mathcal{S}_\mu = \mathcal{R}(R), \quad \mathcal{U}_\mu = \mathcal{R}(P).$$

Thus the linear spaces $\mathcal{U}_\lambda$ and $\mathcal{S}_\mu$ have transversal intersection (see Appendix A). For the modified nonlinear problem given by equations (71.15) and (71.17), it is shown in the argument of Lemma 71.2, that there exist four nonlinear manifolds M^s_λ, M^u_λ, M^s_μ, and M^u_μ, where each of these manifolds is, respectively, homeomorphic to $\mathcal{S}_\lambda$, $\mathcal{U}_\lambda$, $\mathcal{S}_\mu$, and $\mathcal{U}_\mu$.

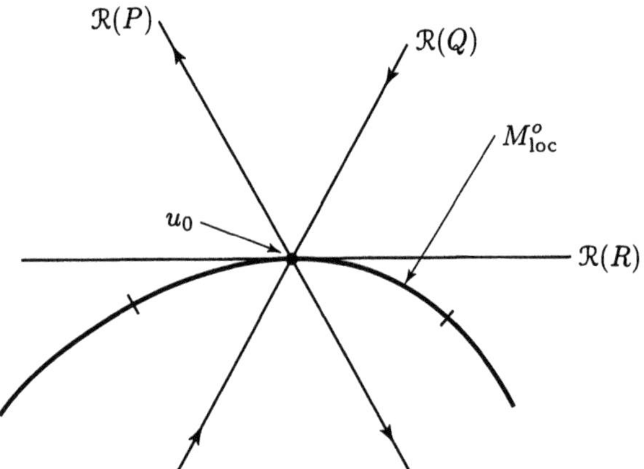

Figure 7.2. Center Manifold Theorem

What will be proved in the Center Manifold Theorem is the existence of an invariant manifold M^o_{loc} for the modified equation (71.17), where M^o_{loc} is homeomorphic to $\mathcal{U}_\lambda \cap \mathcal{S}_\mu$. This suggests (but does not prove) that the tangent spaces of the two manifolds M^u_λ and M^s_μ have a transversal intersection and that $M^o_{\text{loc}} = M^u_\lambda \cap M^s_\mu$. While the last equation may, in fact, be correct, our proof of the existence of M^o_{loc} does not establish this equality, since we do not have a direct analysis of the behavior of the solutions on the manifolds M^u_λ and M^s_μ. Instead we will present an *ab initio* construction of M^o_{loc} by finding suitable functions $w = w(t)$ which satisfy a shifted equation and which are bounded, for all $t \in \mathbb{R}$.

Specifically we use the Lyapunov-Perron method once again to seek bounded, mild solutions of the hybrid problem,

$$\begin{aligned} &\partial_t w + (A + \lambda I)w = DF(u_0)w + e^{-\lambda t} E^a(e^{\lambda t} w), \qquad t \le 0, \\ &\partial_t w + (A + \mu I)w = DF(u_0)w + e^{-\mu t} E^a(e^{\mu t} w), \qquad t \ge 0, \end{aligned} \tag{71.36}$$

for suitable parameters λ and μ, where $\lambda < 0 < \mu$ and $E^a(v) = E^a(u_0, v)$. Since $E^a(0) = 0$, inequality (71.13) implies that if $w \in L^\infty(\mathbb{R}, V^{2\beta})$, then

$g = g(t) \in L^\infty(\mathbb{R}, W)$, where

$$g(t) = \begin{cases} e^{-\lambda t} E^a(e^{\lambda t} w(t)), & t \leq 0, \\ e^{-\mu t} E^a(e^{\mu t} w(t)), & t \geq 0. \end{cases}$$

If $w = w(t)$ is a solution of (71.36) that is bounded in $V^{2\beta}$, for $t \in \mathbb{R}$, then the function

$$v(t) = \begin{cases} e^{\lambda t} w(t), & t \leq 0, \\ e^{\mu t} w(t), & t \geq 0, \end{cases} \tag{71.37}$$

is a mild solution of the modified equation (71.15), for $t \in \mathbb{R}$, with the property that $e^{-\lambda t} v(t)$ is bounded, for $t \leq 0$, and $e^{-\mu t} v(t)$ is bounded, for $t \geq 0$. We will show that for each $\xi \in \mathcal{R}(R)$, there is a unique solution $w = w(\xi, t)$ of (71.36) that is bounded for $t \in \mathbb{R}$ and $Rw(0) = \xi$. As we will see, the center manifold for the modified equation (71.17) is the graph of the mapping $\xi \to u_0 + \xi + \varphi(\xi)$, where $w(\xi, 0) = \xi + \varphi(\xi)$.

Proof of Theorem 71.4. As in the proof of Lemma 71.2, we study first the modified equations (71.15) and (71.17), where the parameter a will be fixed later. We will use a modification of the Lyapunov-Perron argument given in Lemma 71.2. In this case, the mapping $\mathcal{T}$, where $\hat{w} = \mathcal{T} w$, is now defined for $w = w(t)$ satisfying $w \in \mathcal{F} \stackrel{\text{def}}{=} C(\mathbb{R}, V^{2\beta}) \cap L^\infty(\mathbb{R}, V^{2\beta})$. We will use the norm

$$\|A^\beta w\|_\infty = \sup_{t \in \mathbb{R}} \|A^\beta w(t)\|.$$

Let $\mathcal{F}_\rho = \{w \in \mathcal{F} : \|A^\beta w\|_\infty \leq \rho\}$. The parameters λ and μ used below are assumed to satisfy $\lambda_1 < \lambda < \lambda_2$ and $\lambda_3 < \mu < \lambda_4$ (see Lemma 45.4). Let $\xi \in \mathcal{R}(R)$. The image $\hat{w}(t)$, for $t \leq 0$, is defined by

$$\begin{aligned} \hat{w}(t) = e^{-\lambda t} e^{-Lt} &\left(\xi - \int_0^\infty P e^{\mu s} e^{Ls} e^{-\mu s} E^a(e^{\mu s} w(s))\, ds \right) \\ &- e^{-\lambda t} e^{-Lt} \int_t^0 (P + R) e^{\lambda s} e^{Ls} e^{-\lambda s} E^a(e^{\lambda s} w(s))\, ds \\ &+ e^{-\lambda t} e^{-Lt} \int_{-\infty}^t Q e^{\lambda s} e^{Ls} e^{-\lambda s} E^a(e^{\lambda s} w(s))\, ds, \end{aligned}$$

and for $t \geq 0$, one uses

$$\begin{aligned} \hat{w}(t) = e^{-\mu t} e^{-Lt} &\left(\xi + \int_{-\infty}^0 Q e^{\lambda s} e^{Ls} e^{-\lambda s} E^a(e^{\lambda s} w(s))\, ds \right) \\ &+ e^{-\mu t} e^{-Lt} \int_0^t (R + Q) e^{\mu s} e^{Ls} e^{-\mu s} E^a(e^{\mu s} w(s))\, ds \\ &- e^{-\mu t} e^{-Lt} \int_t^\infty P e^{\mu s} e^{Ls} e^{-\mu s} E^a(e^{\mu s} w(s))\, ds, \end{aligned}$$

From inequalities (71.30) and (71.31) and Lemma 45.4 and the Preparation Lemma 47.10, for $t \le 0$ we get

$$\int_{-\infty}^{t} \|A^\beta e^{-\lambda(t-s)} e^{-L(t-s)} Q e^{-\lambda s} E^a(e^{\lambda s} w(s))\| \, ds$$
$$\le K\gamma(\rho)\|A^\beta w\|_\infty \left(\int_{-\infty}^{t} e^{-(\lambda-\lambda_1)(t-s)} \, ds + \int_{t-1}^{t} \cdots \, ds \right)$$
$$\le \left(\frac{K}{\lambda - \lambda_1} + \frac{K}{1-\beta} \right) \gamma(\rho)\|A^\beta w\|_\infty.$$

Similarly, for $t \ge 0$ one finds that

$$\int_{0}^{t} \|A^\beta e^{-\mu(t-s)} e^{-L(t-s)} Q e^{-\mu s} E^a(e^{\mu s} w(s))\| \, ds$$
$$\le \left(\frac{K}{\lambda - \lambda_1} + \frac{K}{1-\beta} \right) \gamma(\rho)\|A^\beta w\|_\infty,$$

since $\mu > \lambda$. Likewise, inequalities (71.10) and (71.29) and Lemma 45.4 yield

$$\int_{t}^{\infty} \|A^\beta e^{-\mu(t-s)} e^{-L(t-s)} P e^{-\mu s} E^a(e^{\mu s} w(s))\| \, ds \le \frac{K}{\lambda_4 - \mu} \gamma(\rho)\|A^\beta w\|_\infty,$$

for $t \ge 0$, and

$$\int_{t}^{0} \|A^\beta e^{-\lambda(t-s)} e^{-L(t-s)} P e^{-\mu s} E^a(e^{\mu s} w(s))\| \, ds \le \frac{K}{\lambda_4 - \mu} \gamma(\rho)\|A^\beta w\|_\infty,$$

for $t \le 0$. Also inequalities (71.10) and (71.32) and Lemma 45.4 imply that

$$\int_{t}^{0} \|A^\beta e^{-\lambda(t-s)} e^{-L(t-s)} R e^{-\lambda s} E^a(e^{\lambda s} w(s))\| \, ds \le \frac{K}{\lambda_2 - \lambda} \gamma(\rho)\|A^\beta w\|_\infty,$$

for $t \le 0$, while inequalities (71.10), (71.33), and (71.34) and Lemmas 45.4 yield

$$\int_{0}^{t} \|A^\beta e^{-\mu(t-s)} e^{-L(t-s)} R e^{-\mu s} E^a(e^{\mu s} w(s))\| \, ds$$
$$\le \left(\frac{K}{\mu - \lambda_3} + \frac{K}{1-\beta} \right) \gamma(\rho)\|A^\beta w\|_\infty, \qquad \text{for } t \ge 0.$$

Now fix λ and μ by setting $\lambda = \frac{1}{2}(\lambda_1 + \lambda_2)$ and $\mu = \frac{1}{2}(\lambda_3 + \lambda_4)$. Define α by $\alpha = \frac{1}{2}\min(\lambda_2 - \lambda_1, \lambda_4 - \lambda_3)$. By using the fact that $K \ge 1$, together with the inequalities above, one then finds that

$$\|A^\beta \hat{w}(t)\| \le K\|A^\beta \xi\| + \left(\frac{4K}{\alpha} + \frac{3K}{1-\beta} \right) \gamma(\rho)\|A^\beta w\|_\infty, \tag{71.38}$$

for all $t \in \mathbb{R}$. We see then that $\hat{w} \in L^\infty(\mathbb{R}, V^{2\beta})$. By using the argument of Lemma 42.7, one can show that $\hat{w} \in C(\mathbb{R}, V^{2\beta})$. Consequently, for every $\xi \in \mathcal{R}(R)$, we see that $\mathcal{T}$ maps $\mathcal{F}$ into itself.

The next step is to show that, for each $\xi \in \mathcal{R}(R)$, the mapping $\mathcal{T}$ is a strict contraction on $\mathcal{F}$. The series of estimates leading up to inequality (71.38) are also applicable here, where $E^a(e^{\lambda s}w(s))$ is now replaced by $(E^a(e^{\lambda s}w_1(s)) - E^a(e^{\lambda s}w_2(s)))$, etc. As a result, one finds that

$$\|A^\beta(\hat{w}_1(t) - \hat{w}_2(t))\| \leq \left(\frac{4K}{\alpha} + \frac{3K}{1-\beta}\right)\gamma(\rho)\|A^\beta(w_1 - w_2)\|_\infty, \tag{71.39}$$

for all $t \in \mathbb{R}$. Let $\rho_0 > 0$ be chosen so that $(\frac{4K}{\alpha} + \frac{3K}{1-\beta})\gamma(\rho_0) \leq \frac{1}{2}$, and set $2a = \rho_0$. It then follows from (71.39) that

$$\|A^\beta(\hat{w}_1 - \hat{w}_2)\|_\infty \leq \frac{1}{2}\|A^\beta(w_1 - w_2)\|_\infty \tag{71.40}$$

and $\mathcal{T}$ is a strict contraction on the complete metric space $\mathcal{F}_\rho$, for each ρ with $0 < \rho \leq \rho_0$. Furthermore, it follows from the Contraction Mapping Theorem that the unique fixed point $w(t) = w(\xi, t)$ is locally Lipschitz continuous in $\xi \in V^{2\beta}$, and inequalities (71.38) and (71.40) imply that $\|A^\beta w(\xi, t)\| \leq 2K\|A^\beta \xi\|$, for all $t \in \mathbb{R}$ and $0 < \rho \leq \rho_0$. Note that the function $w = w(t) = w(\xi, t)$ is a mild solution of the hybrid equation (71.36). Conversely, Theorem 45.7 implies that if $w = w(t)$ is any function in $C(\mathbb{R}, V^{2\beta}) \cap L^\infty(\mathbb{R}, V^{2\beta})$ that is a solution of (71.36), then w is a fixed point of $\mathcal{T}$. One then has $w(0) = \xi + P\int_0^\infty \cdots + Q\int_{-\infty}^0 \cdots$, from the definition of $\mathcal{T}$, and hence $Rw(0) = \xi \in \mathcal{R}(R)$ and $w(t) = w(\xi, t)$.

From the definition of $\mathcal{T}$, we see that for any $\xi \in \mathcal{R}(R)$, one has $w(\xi, 0) = \xi + \varphi(\xi)$, where

$$\varphi(\xi) = \int_{-\infty}^0 Qe^{Ls}E^a(e^{\lambda s}w(s))\,ds - \int_0^\infty Pe^{Ls}E^a(e^{\mu s}w(s))\,ds,$$

and

$$\|A^\beta\varphi(\xi)\| \leq \left(\frac{4K}{\alpha} + \frac{3K}{1-\beta}\right)\gamma(\rho)\|A^\beta w(\xi, \cdot)\|_\infty \leq \gamma(\rho_0)^{-1}\gamma(\rho)K\|A^\beta\xi\|, \tag{71.41}$$

for $\|A^\beta w(\xi, t)\| \leq 2K\|A^\beta\xi\| \leq \rho$ and $0 < \rho \leq \rho_0$. Since $\varphi(\xi) \in \mathcal{R}(P) + \mathcal{R}(Q)$, it follows that $R\varphi(\xi) = 0$ and $Rw(\xi, 0) = \xi$, for all $\xi \in \mathcal{R}(R)$. Hence the mapping $\xi \to w(\xi, 0)$ is locally Lipschitz continuous with a Lipschitz continuous inverse. We define M^o by

$$M^o \stackrel{\text{def}}{=} \text{Graph}(\xi \to u_0 + \xi + \varphi(\xi)), \quad \text{where } \xi \in \mathcal{R}(R) \cap B_{V^{2\beta}}(0; \rho/(4K)).$$

The mapping $\xi \to u_0 + \xi + \varphi(\xi)$ is a Lipschitz continuous homeomorphism of $\mathcal{R}(R) \cap B_{V^{2\beta}}(0, \rho/(4K))$ onto M^o with a Lipschitz continuous inverse. Next we show that M^o is a center manifold for the modified equation (71.17).

A direct calculation, with $v(t) = v(\xi, t)$ defined by (71.37), yields

$$(71.42)\quad v(\xi,t) = e^{-Lt}v(\xi,0) + \int_0^t e^{-L(t-s)}E^a(v(\xi,s))\,ds, \qquad \text{for all } t \in \mathbb{R}.$$

That is, $v(\xi, t)$ is the unique mild solution of (71.15), with $v(\xi, 0) = w(\xi, 0) = \xi + \varphi(\xi)$, and $u(\xi, t) = u_0 + v(\xi, t)$ is the unique mild solution of (71.17) satisfying $u(\xi, 0) = u_0 + v(\xi, 0)$. In addition, one has

$$\|A^\beta v(\xi,t)\| \le \begin{cases} 2K\|A^\beta\xi\|e^{\lambda t}, & \text{for } t \le 0 \\ 2K\|A^\beta\xi\|e^{\mu t}, & \text{for } t \ge 0. \end{cases}$$

Moreover, the solution $v(\xi, t)$ has a negative continuation, for $t \le 0$, and this negative continuation satisfies (71.42), as well.

Now Theorem 45.7 and Lemma 71.2 imply that

$$v(\xi(\tau), t) = v(\xi, \tau + t), \qquad \text{for all } \tau,\, t \in \mathbb{R},$$

where $\xi(t) = Rv(\xi, t)$, for all $t \in \mathbb{R}$. Thus $\xi(t)$ is a mild solution of the reduced equation

$$(71.43)\qquad \partial_t\xi(t) + L\xi(t) = RE^a(v(\xi(t), 0)), \qquad \xi(0) = \xi \in \mathcal{R}(R).$$

Hence $v(\xi, t)$ is a slave variable and its dynamics are completely determined by the dynamics of the reduced problem, which completes the proof of the invariance of M^o and the Reduction Principle for the modified equation.

Next let $v(t)$ be any global solution of the modified equation (71.15) that satisfies $\|A^\beta v(t)\| \le \rho$, for all $t \in \mathbb{R}$, for some ρ with $0 \le \rho < \infty$. Let $w = w(t)$ be determined by equation (71.37). Since $\lambda < 0 < \mu$, it follows that w is a bounded solution in $V^{2\beta}$ of the hybrid equation (71.36). Consequently, Theorem 45.7 implies that w is a fixed point of $\mathcal{T}$. Therefore, one necessarily has $w(t) = w(\xi, t)$, for all $t \in \mathbb{R}$, where $\xi = Rw(0)$. Hence $w(0) \in M^o$. Since (71.41) implies that $\|A^\beta\xi\|^{-1}\|A^\beta\varphi(\xi)\| \to 0$, as $\|A^\beta\xi\| \to 0$, it follows that M^o is tangent to $u_0 + \mathcal{R}(R)$ at u_0.

This completes the proof of the center manifold theorem for the modified equation (71.17). In the case of the original equation (71.16), one replaces M^o by the local manifold

$$M^o_{\text{loc}} = \{u \in M^o : \|A^\beta(u - u_0)\| \le a\}.$$

As noted above, equations (71.16) and (71.17) agree when $\|A^\beta(u - u_0)\| \le a$, which completes the proof. □

The assumption of an exponential trichotomy for the linear semigroup e^{-Lt} contains more information than the conclusions stated above. The following result is an immediate corollary of the Center Manifold Theorem 71.4 and Lemma 71.2.

Theorem 71.5. *Assume that the hypotheses of the Center Manifold Theorem are satisfied. Then there exists a center manifold* M^o_{loc} *such that the conclusions of Theorem 71.4 hold. In addition, for* λ *and* μ *satisfying* $\lambda_1 < \lambda < \lambda_2$ *and* $\lambda_3 < \mu < \lambda_4$, *there exist a stable manifold* $M^s_\lambda = W^s_{\lambda,\mathrm{loc}}(u_0)$ *and an unstable manifold* $M^u_\mu = W^u_{\mu,\mathrm{loc}}(u_0)$, *as described in Lemma 71.2.*

Since the center manifold M^o_{loc} can contain stable and/or unstable modes, with small decay rates, one sometimes refers to M^s_λ as the **strongly stable manifold** and to M^u_μ as the **strongly unstable manifold**. Unlike the situation described in the Saddle Point Property, in the presence of a nontrivial center manifold M^o_{loc}, where $\dim M^o_{\mathrm{loc}} \geq 1$, the stable and unstable sets $W^s(u_0)$ and $W^u(u_0)$ can have a rather complicated structure. However, one always has

$$W^s_{\mathrm{loc}}(u_0) \subset W^s(u_0) \quad \text{and} \quad W^u_{\mathrm{loc}}(u_0) \subset W^u(u_0).$$

7.1.3. Perturbation Theory for Equilibria. We turn now to the study of the perturbed equation (70.2), where the perturbation term G is "small" in the vicinity of a given equilibrium point u_0 for equation (70.1). We require that both F and G are in the space C^1_{Lip}, and we set

$$\Omega = \Omega_r = \{u \in V^{2\beta} : \|A^\beta(u - u_0)\| \leq r\}.$$

We will use the pseudonorm $\|G\|_{\{C^1(\Omega_r)\}}$, see (46.13), in the following result.

Theorem 71.6. *Let the Standing Hypothesis A be satisfied, and let the nonlinearities* F *and* G *be in the space* C^1_{Lip}. *Let* u_0 *be a hyperbolic equilibrium point for equation (70.1) in the space* $V^{2\beta}$. *Then for every* $r > 0$, *there is a* $\delta > 0$, *such that if* G *satisfies*

$$\|G\|_{\{C^1(\Omega_r)\}} \leq \delta,$$

then the perturbed equation (70.2) has a unique fixed point $y^G \in \Omega_r$, *and* y^G *is hyperbolic for equation (70.2). Furthermore, the associated projections* (P^G, Q^G) *vary continuously in* G, *and the characteristics* K^G *and* α^G *can be chosen to be continuous at* $G = 0$.

Proof. We will prove the existence, uniqueness, and hyperbolicity of the fixed point y^G The proofs of the continuous dependence of the characteristics and the projections are left as exercises.

It is convenient to use the change of variables $y = u_0 + w$ in equation (70.2), where u_0 is a given equilibrium point of equation (70.1). Observe that $y = y(t)$ is a mild solution of (70.2) if and only if $w = w(t)$ is a mild solution of

$$\partial_t w + Lw = E(u_0, w) + G(u_0 + w) = H(w),$$

where $L = A - DF(u_0)$ and E satisfies (71.8) - (71.11), for $\gamma = \gamma(\rho)$ with $\gamma(\rho) \to 0$, as $\rho \to 0$. We will use the Lyapunov-Perron argument to construct the fixed point y^G. In particular, the mapping $\mathcal{T}$, where $\hat{w} = \mathcal{T}w$, is now defined, for $t \in \mathbb{R}$, by

$$\hat{w}(t) = \int_{-\infty}^{t} Qe^{-L(t-s)} H(w(s))\, ds - \int_{t}^{\infty} Pe^{L(t-s)} H(w(s))\, ds,$$

where $w \in \mathfrak{F}_\rho$, and $\mathfrak{F}_\rho$ is the set of $w \in \mathfrak{F} = C(\mathbb{R}, V^{2\beta}) \cap L^\infty(\mathbb{R}, V^{2\beta})$ with

$$\|A^\beta w\|_\infty = \sup_{t \in \mathbb{R}} \|A^\beta w(t)\| \leq \rho.$$

Once again, we use the methodology developed for the Center Manifold Theorem, with $R = 0$, and inequalities (71.4), (71.5), and (71.6), for $\lambda = 0$, together with inequalities (71.10) and (71.11), to show that if $\|A^\beta w\|_\infty \leq \rho$, then

$$\|A^\beta \hat{w}(t)\| \leq \left(\frac{2K}{\alpha} + \frac{K}{1-\beta} \right) (\gamma(\rho)\|A^\beta w\|_\infty + \delta),$$

and

$$\|A^\beta(\hat{w}_1(t) - \hat{w}_2(t))\| \leq \left(\frac{2K}{\alpha} + \frac{K}{1-\beta} \right) (\gamma(\rho) + \delta)\|A^\beta(w_1 - w_2)\|_\infty$$

for all $t \in \mathbb{R}$. We now fix $\rho_0 > 0$ so that $(\frac{2K}{\alpha} + \frac{K}{1-\beta})\gamma(\rho_0) \leq \frac{1}{3}$ and we choose δ so that $(\frac{2K}{\alpha} + \frac{K}{1-\beta})\delta \leq \min(\frac{1}{3}, \frac{2}{3}\rho)$. A straightforward calculation then shows that $\mathcal{T}$ has a unique fixed point $w \in \mathfrak{F}_\rho$ and that $\|A^\beta w\|_\infty \leq \rho$. At this point $w = w(t)$ satisfies the integral equation

$$w(t) = \int_{-\infty}^{t} Qe^{-L(t-s)} H(w(s))\, ds - \int_{t}^{\infty} Pe^{L(t-s)} H(w(s))\, ds,$$

and it appears to depend on time $t \in \mathbb{R}$. However, by means of a simple change of variables, one can show that the translate $w_\tau(t) = w(\tau + t)$ is also a solution of the same integral equation. The uniqueness of the fixed points of $\mathcal{T}$ implies that $w_\tau = w$, for all $\tau \in \mathbb{R}$. It follows that $w^G \stackrel{\text{def}}{=} w(t)$ is independent of time t, and w^G satisfies

$$w^G = \int_{-\infty}^{0} Qe^{-L(-s)} H(w^G)\, ds - \int_{0}^{\infty} Pe^{L(-s)} H(w^G)\, ds.$$

Also $y^G \stackrel{\text{def}}{=} u_0 + w^G$ is a fixed point for the perturbed equation (70.2). The hyperbolicity of y^G follows from an immediate application of the Robustness of Dichotomies Theorem 45.11, perhaps with a smaller value for δ. $\square$

7.2. Dynamics of Gradient Systems.

As noted in Chapter 5, the Chafee-Infante equation is an example of a significant class of dynamical systems called gradient systems. In this section we will study such systems on an abstract level. The objective here is to develop a framework in which one may obtain applications across a broad range of dynamical systems, in both finite and infinite dimensions.

We begin with your favorite Banach space W. We say that $S = S(t)$, or more precisely, the pair $(S, L) = (S(t), L(w))$ is a **gradient system** on W if $S(t)$ is a κ-contracting semiflow on W and there is a functional $L : W \to \mathbb{R}$ for which the following five properties hold:

(G1) The functional $L : W \to \mathbb{R}$ is strongly continuous;
(G2) one has $L(w) \to \infty$, as $\|w\| \to \infty$;
(G3) the functional L is bounded below; i.e., there is a constant $C_L \geq 0$ such that $0 \leq L(w) + C_L$, for all $w \in W$;
(G4) for all $w \in W$, one has $L(S(t)w) \leq L(w)$, for all $t \geq 0$; and
(G5) the set

$$W_0 \stackrel{\text{def}}{=} \{w \in W : L(S(t)w) = L(w), \text{ for all } t \geq 0\} \tag{72.1}$$

is a bounded set in W.

If (S, L) is a gradient system on a Banach space W, then it is easily verified that the set W_0 given by (72.1) is a nonempty, closed, bounded set in W which satisfies

$$S(t_2)W_0 \subset S(t_1)W_0 \subset W_0, \qquad \text{for all } t_2 \geq t_1 \geq 0. \tag{72.2}$$

It then follows from the κ-contracting property that the omega limit set satisfies

$$K_0 \stackrel{\text{def}}{=} \omega(W_0) = \cap_{\tau \geq 0} \text{Cl}_W S(\tau) W_0 \subset W_0, \tag{72.3}$$

and it is a nonempty, compact, invariant set in W. Furthermore, K_0 is the largest compact invariant set in[35] W_0. Note that, while W_0 is positively invariant, it need not be invariant. When W_0 is not invariant, one has $K_0 \subsetneqq W_0$. Also note that K_0 can be characterized as the collection of all $w \in W_0$ such that there is a global solution φ^w passing through w with $\varphi^w(t) \in W_0$ and $L(\varphi^w(t)) = L(w)$, for all $t \in \mathbb{R}$. We next define the equilibrium set

$$Q \stackrel{\text{def}}{=} \{w_0 \in W : S(t)w_0 = w_0, \text{ for all } t \geq 0\}. \tag{72.4}$$

Notice that Q is a closed, invariant set for $S(t)$, and one has $Q \subset W_0$. Consequently, one has $Q \subset K_0$. Since K_0 is compact, it follows that Q is

[35] We do not claim here that K_0 is the largest compact invariant set in W.

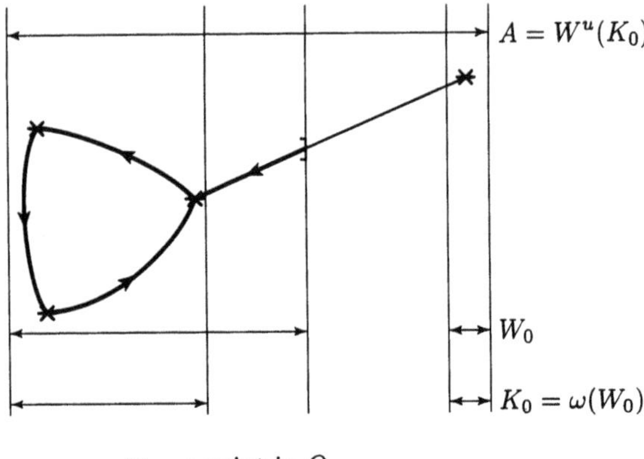

Figure 7.3. Detail for Gradient System

compact, as well, see Figure 7.3, where $Q \subsetneqq K_0 = \omega(W_0) \subsetneqq W_0 \subsetneqq A$ and A is a compact invariant set consisting of 4 equilibria and 4 transit orbits.

For each $r \in \mathbb{R}$, we define $B_r^L = \{w \in W : L(w) \leq r\}$. It then follows from the continuity of L and Property (G2) that B_r^L is a closed, bounded set in W, for each $r \in \mathbb{R}$. From Property (G3), we see that B_r^L is empty, for $r < -C_L$. Furthermore, Property (G4) implies that for every $\tau \geq 0$, one has $S(\tau)B_r^L \subset B_r^L$, for each $r \in \mathbb{R}$. Also Property (G2) and the continuity of L imply that B_s^L is a bounded neighborhood of B_r^L, whenever, $r < s$.

Next we let K denote any nonempty, compact set in W. Since L is continuous, one has $-C_L \leq r(K) < \infty$, where $r(K) = \max\{L(w) : w \in K\}$. Since $L(w) \leq r(K)$, for all $w \in K$, and $K \subset B_{r(K)}^L$, one has

$$S(\tau)K \subset S(\tau)B_{r(K)}^L \subset B_{r(K)}^L, \qquad \text{for all } \tau \geq 0.$$

It follows then that $\gamma^+(K) \subset B_{r(K)}^L$, i.e., $\gamma^+(K)$ is bounded for every nonempty, compact set K in W.

As noted in (72.2), the set W_0 is positively invariant under the semiflow $S(t)$, and $K_0 = \omega(W_0)$ is the largest compact, invariant set in W_0. Furthermore, Properties (G1), (G3), and (G4) imply that for every $w_0 \in W$, the omega limit set is nonempty and satisfies $\omega(w_0) \subset W_0$. Since $\mathrm{Cl}_W\gamma^+(w_0) \subset B_{L(w_0)}^L$, it follows that $\omega(w_0)$ is a nonempty, compact, connected, invariant set in W_0, for every $w_0 \in W$, see Lemmas 23.6 and 23.8. We now define r_0 by

$$r_0 = r(K_0) \stackrel{\text{def}}{=} \max\{L(w) : w \in K_0\}. \tag{72.5}$$

One then has $\omega(w_0) \subset B_{r_0}^L$, for every $w_0 \in W$. Furthermore, for every $w_0 \in W$, there is a time $\tau = \tau(w_0) \geq 0$, such that $L(S(t)w_0) \leq r_0 + 1$, for all $t \geq \tau$. This implies that the set $B_{r_0+1}^L$, which is a bounded neighborhood of $B_{r_0}^L$, is an absorbing set for $S(t)$, and the semiflow $S(t)$ is point dissipative on W. It then follows from the Existence Theorem 23.12 that there is a global attractor $\mathfrak{A}$ for $S(t)$ and $\mathfrak{A} \subset \omega(B^L_{r_0+1}) \subset B^L_{r_0}$ (see Lemma 23.9). In the following result, we make reference to the unstable set $W^u(K_0)$ (see Section 2.1.1).

Theorem 72.1. *Let (S, L) be a gradient system on a Banach space W, and let W_0, K_0, Q, B^L_r, and $\mathfrak{A}$ be given as above. Then the following statements are valid:*

(1) *For every $r \in \mathbb{R}$, the set B^L_r is a closed bounded set in W that satisfies $S(t)B^L_r \subset B^L_r$, for all $t \geq 0$.*
(2) *The semiflow $S(t)$ is κ-contracting, point dissipative, and for every compact set K in W, the positive orbit $\gamma^+(K)$ is a bounded set in W.*
(3) *The semiflow $S(t)$ has a global attractor $\mathfrak{A}$, which satisfies $Q \subset K_0 \subset \mathfrak{A} \subset B_{r_0}$, where r_0 is given by (72.5), and K_0 is given by (72.3).*
(4) *For every $w_0 \in W$, the omega limit set $\omega(w_0)$ is a nonempty, compact, connected, invariant set in K_0.*
(5) *For every $w_0 \in \mathfrak{A}$, the alpha limit set $\alpha(w_0)$ is a nonempty, compact, connected, invariant set in K_0.*
(6) *The global attractor $\mathfrak{A}$ satisfies $\mathfrak{A} = W^u(K_0)$, where $W^u(K_0)$ is the unstable set associated with K_0.*

Proof. Items (1), (2), and (3) are essentially proved in the three paragraphs preceding the statement of the theorem. Items (4) and (5) follow from the properties of the alpha and omega limit sets (see Lemmas 23.6 and 23.8 and Exercise 23.7). For Item (6), we first note that $K_0 \subset \mathfrak{A}$. Since both K_0 and $\mathfrak{A}$ are invariant, one has $W^u(K_0) \subset W^u(\mathfrak{A}) = \mathfrak{A}$. In order to show that $\mathfrak{A} \subset W^u(K_0)$, we let $w_0 \in \mathfrak{A}$. Then the global orbit $\{S(t)w_0 : t \in \mathbb{R}\}$ lies in $\mathfrak{A}$, and by Item (5) one has $\alpha(w_0) \subset K_0$. Thus one has $w_0 \in W^u(K_0)$. □

There is an interesting and delicate issue that arises in the study of gradient systems, an issue which lies at the heart of the definition of a global attractor for a semiflow. This issue arises only in the infinite dimensional setting. In particular, if (S, L) is a gradient system on an infinite dimensional Banach space W, then it can be the case that there is a sequence $w_n \in W$, where $L(w_n) \to \infty$, while $\limsup_{n\to\infty} \|w_n\| < \infty$. In other words, there may be a bounded set B in W with the property that $B \not\subset B_r$, for any $r \in \mathbb{R}$. As a result, it is possible then, that there is a bounded set B in W with the property that $\gamma^+(S(\tau)B)$ is unbounded, for each $\tau \geq 0$. Of course, none of this can happen if the functional L has the additional property of mapping each bounded set in W into a bounded set in $\mathbb{R}$. Or to put it another way, if Property (G2) is replaced by the stronger condition

(G2′) *one has $L(w) \to \infty$, if and only if, $\|w\| \to \infty$,*

which in turn implies that $S(t)$ is ultimately bounded, and the following holds.

Theorem 72.2. *In addition to the hypotheses of Theorem 72.1, assume that the functional L maps bounded sets in W into bounded sets in $\mathbb{R}$; i.e., (G2′) holds. Then the global attractor $\mathfrak{A}$ attracts all bounded sets in W.*

Proof. Under these hypotheses, the semiflow is ultimately bounded, and the conclusion follows from the last theorem and the Existence Theorem 23.12. □

As noted in Section 5.1, the pair (S, E), where E is given by equation (51.17), and $S(t)$ denotes the semiflow, generated by the mild solutions of the Chafee-Infante problem (51.1) on $W = V = V^1$, is a gradient system when the physical space dimension m, satisfies $1 \leq m \leq 3$. In this case one has $Q = K_0 = W_0$ (see Section 5.1). Furthermore, inequality (51.22) implies that E maps bounded sets in V into bounded sets in $\mathbb{R}$. As a result, both Theorems 72.1 and 72.2 are valid for the Chafee-Infante problem when $1 \leq m \leq 3$.

The situation where $Q = K_0 = W_0$, which arises in the Chafee-Infante problem, is a typical occurrence in the study of gradient systems. It then follows from Theorem 72.1, Item (6), that $\mathfrak{A} = W^u(Q)$. If in addition, Q is a finite set, then one obtains the identity

$$\mathfrak{A} = \bigcup_{z \in Q} W^u(z). \tag{72.6}$$

The verification of (72.6) is based on the fact that for any $w \in W^u(Q) = \mathfrak{A}$, the α-limit set $\alpha(w)$ is a nonempty, compact, connected set in Q (see Exercise 23.7). Since Q is a finite set, this implies that $\alpha(w) = \{z\}$, for some $z \in Q$, and consequently $w \in W^u(z)$, i.e., (72.6) is valid.

In the literature, one sometimes finds a proof of (72.6) under the stronger assumption that Q is a finite set, and each equilibrium point $z \in Q$ is hyperbolic. As we have seen, the assumption of hyperbolicity is not needed in the proof of (72.6). Nevertheless, there is some added information one obtains when each equilibrium point is hyperbolic, and this is the Saddle Point Property, Theorem 71.1.

A semiflow $S(t)$ on a metric space W is said to have a **Morse decomposition** on W if the following hold:

(1) There exist a finite number of disjoint invariant sets $\mathcal{K}_1, \cdots, \mathcal{K}_k$ in W, each of which is nonempty and compact;
(2) for each $w \in W$, there is an i with $1 \leq i \leq k$ and such that $\omega(w) \subset \mathcal{K}_i$; and
(3) for any compact invariant set $\mathcal{K}$ in W and each $w \in \mathcal{K}$, one of the following holds: either there is an integer i, with $1 \leq i \leq k$, such that $S(t)w \in \mathcal{K}_i$, for all $t \in \mathbb{R}$, or there exist integers $i = i(w)$ and $j = j(w)$, with $1 \leq i < j \leq k$, such that $\alpha(w) \subset \mathcal{K}_j$ and $\omega(w) \subset \mathcal{K}_i$.

The sets $\mathcal{K}_1, \cdots, \mathcal{K}_k$ are referred to as the **Morse sets**, see Figure 7.4.

Let us now return to a gradient system (S, L) on W that has the property that $L(S(t)w) = L(w)$, for all $t \geq 0$, if and only if $S(t)w = w$, for all $t \geq 0$.

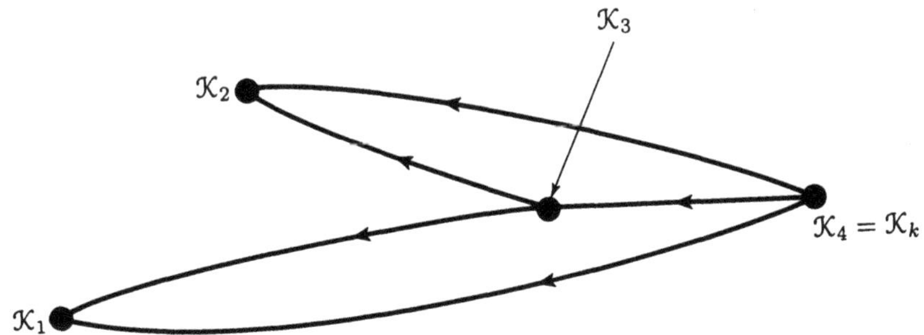

Figure 7.4. Typical Morse Decomposition

Recall that the Chafee-Infante problem has this property. Next assume that the set Q, of equilibrium points of S, is a finite set. In this case one has the following property, which the reader should verify.

Theorem 72.3. *Let (S, L) be a gradient system on W. Then the following properties hold:*

(1) *Assume that $L(S(t)w) = L(w)$, for all $t \geq 0$, if and only if $S(t)w = w$, for all $t \geq 0$. Assume further that $S(t)$ has at most a finite number of stationary solutions. i.e., the set Q is finite. Then there is a Morse decomposition on W and the Morse sets $\mathcal{K}_1, \cdots, \mathcal{K}_k$ satisfy $Q = \cup_{i=1}^k \mathcal{K}_i$; that is to say, each Morse set $\mathcal{K}_i$ consists of a single equilibrium point.*

(2) *Let W_0 be given by (72.1). Assume that $L(W_0)$ is a finite set, say $L(W_0) = \{\ell_1, \cdots, \ell_k\}$, where $\ell_1 < \cdots < \ell_k$. Then there is a Morse decomposition on W and the Morse sets are precisely the sets $\mathcal{K}_i \stackrel{\text{def}}{=} \{w \in \mathcal{K} : L(w) = \ell_i\}$, for $1 \leq i \leq k$.*

(3) *Under the conditions of either Item (1) or Item (2), there is a global attractor $\mathfrak{A}$ for the semiflow $S(t)$ on W. Moreover, for each point w in $\mathfrak{A}$, either one has $w \in \mathcal{K}_i$, for some i with $1 \leq i \leq k$, or there exist integers $1 \leq i < j \leq k$ such that the global solution $S(t)w$ satisfies*

$$\ell_i = \lim_{t\to\infty} L(S(t)w) < \lim_{t\to-\infty} L(S(t)w) = \ell_j,$$

and the α and ω-limit sets satisfy $\alpha(w) \subset \mathcal{K}_j$ and $\omega(w) \subset \mathcal{K}_i$, i.e., $\gamma(w)$ is a transit orbit.

7.3. Behavior Near a Periodic Orbit.

Let us return now to the nonlinear evolutionary equation (70.1). Instead of studying the flow in the vicinity of a given equilibrium point, we now study the flow generated by (70.1) in the vicinity of a given periodic solution. Once again we assume that A satisfies the Standing Hypothesis A, and that the nonlinearity $F = F(u)$ is in C^1_{Lip}. We begin with the mild solutions of equation (70.1) (see Section 4.7).

A mild solution $u(t) = \phi(t)$ of equation (70.1) is said to be **periodic** (in time) if ϕ satisfies

$$\phi(t) = e^{-At}\phi(0) + \int_0^t e^{-A(t-s)} F(\phi(s))\, ds, \qquad \text{for all } t \in \mathbb{R},$$

in the space $V^{2\beta}$, and one has

$$\phi(0) = \phi(T), \qquad \text{for some } T > 0. \tag{73.1}$$

Any time $T > 0$ satisfying equation (73.1) is said to be a period of the solution ϕ. The **minimum period** τ is defined to be the infimum of all $T > 0$ that satisfy (73.1). Since $\phi(t)$ is continuous in t, one has $\phi(0) = \phi(\tau)$ and $\tau \geq 0$. It follows that any equilibrium point is a periodic solution (with $\tau = 0$). However, we are specifically interested here in the case where the minimal period is positive. In the sequel, we will denote this minimal period by T. A periodic solution that is not an equilibrium point is referred to as a **nontrivial** periodic solution.

If $\phi(t) = S(t)u_0$ is a periodic solution, then $\Gamma \overset{\text{def}}{=} \gamma^+(u_0)$ satisfies $\Gamma = \{S(\tau)u_0 \in V^{2\beta} : \text{ for } 0 \leq \tau \leq T\}$. Since Γ is the continuous image of a compact set, $[0, T]$, it is a compact set in $V^{2\beta}$. It is also an invariant set for the semiflow generated by the mild solutions. Since $F \in C_{\text{Lip}}(V^{2\beta}, W)$, it follows from the Herculean Theorem 47.6, that $\Gamma \subset \mathcal{D}(A) = V^2$, and the global solution $\phi(t)$ is a classical solution of equation (70.1), for all $t \in \mathbb{R}$.

Let $DF(u)$ denote the Fréchet derivative of F at the point $u \in V^{2\beta}$. For the given periodic solution $\phi(t)$ and $\tau \in [0, T]$, define

$$B_\tau(t) = B(\tau + t) \overset{\text{def}}{=} DF(\phi(\tau + t)) = DF(S(t)\phi(\tau)), \qquad \text{for } t \in \mathbb{R}.$$

Since $F \in C^1_{\text{Lip}}$, it follows that $B \in \mathcal{M}^\infty$, see (44.31). Also the hull $H(B)$ satisfies

$$H(B) = \text{Closure}_{\mathcal{M}^\infty}\{B_\tau : \tau \in \mathbb{R}\} = \{B_\tau : 0 \leq \tau \leq T\},$$

with closure taken in the topology of uniform convergence on bounded sets, and $H(B)$ is homeomorphic to a circle. The mapping $\phi(\tau) \to B_\tau$ is a flow homomorphism from Γ into $\mathcal{M}^\infty$. The linearized equation associated with the periodic solution ϕ is $\partial_t v + Av = B(t)v$. For each τ with $0 \le \tau \le T$, we will let $\Phi(B_\tau, t)$ denote the solution operator generated by B_τ (see Theorem 44.1). Since B_τ is periodic in time τ with period T, one has $\Phi(B_{\tau+T}, t) = \Phi(B_\tau, t)$, for all $\tau \in \mathbb{R}$ and all $t \ge 0$. We will use the linear skew product semiflow

$$(73.2) \qquad \pi(v, \phi(\tau), t) = (\Phi(\phi(\tau), t)v, S(t)\phi(\tau)) = (\Phi(\phi(\tau), t)v, \phi(\tau + t)),$$

where $\Phi(\phi(\tau), t) = \Phi(B_\tau, t)$ is the solution operator generated by B_τ.

Next we introduce the time-dependent change of variables $u = \phi + v$ into equation (70.1). Observe that $u = u(t)$ is a mild solution of (70.1) if and only if $v = v(t)$ is a mild solution of

$$\partial_t v + Av = B(t)v + E(\phi, v),$$

where $E(\phi, v)$ satisfies (71.8) and (71.9), with $E(\phi, 0) = 0$, and the Fréchet derivative satisfies $DE(\phi, 0) = \frac{\partial}{\partial v} E \mid_{(\phi,0)} = 0$. Furthermore, $E(\phi, v)$ satisfies inequalities (71.10) and (71.11), for all t with $0 \le t \le T$.

Next we study the linear dynamics on $\mathcal{E}(H(B)) = V^{2\beta} \times H(B)$ generated by

$$\pi_\lambda(v, \phi(\tau), t) = (\Phi_\lambda(\phi(\tau), t)v, S(t)\phi(\tau)),$$

where $\lambda \in \mathbb{R}$ and $\pi_0 = \pi$. We claim that for a nontrivial periodic solution, where Γ is homeomorphic to a circle and not merely a point, the linear skew product semiflow π_0 never has an exponential dichotomy over $H(B)$. Equivalently, we claim that the number $\lambda = 0$ lies in the dynamical spectrum of π (see Section 4.5). Then equation (49.9) implies that for all $t \in \mathbb{R}$ and all $\tau \in [0, T]$, one has

$$(73.3) \qquad \Phi(\phi(\tau), t)G(\phi(\tau)) = G(S(t)\phi(\tau)) = G(\phi(\tau + t)),$$

where $G(u) = -Au + F(u)$. We use here Theorem 44.1, Item (4), in which it is shown that $\Phi(\phi(\tau), t)w$ is well-defined, for all $w \in W$. Since $\Gamma \subset \mathcal{D}(A)$ and $\phi(t) = S(t)u_0$ is a nontrivial periodic solution of equation (70.1), one has $G(S(t)u_0) \neq 0$, for all $t \in \mathbb{R}$. Since $G(u_0) \in W$, it follows from Theorem 44.1, Item (4), that $\Phi(\phi(\tau), t)G(\phi(\tau)) \in V^{2\beta}$, for all $t \in \mathbb{R}$. Since the mapping $t \to \Phi(\phi(\tau), t)G(\phi(\tau))$ is a continuous mapping of $\mathbb{R}$ into $V^{2\beta}$, we see that

$$G(\Gamma) = \{G(S(t)\phi(\tau)) = \Phi(\phi(\tau), t)G(\phi(\tau)) : t \in \mathbb{R}\}$$

is a compact set in $V^{2\beta}$, and $\Phi(\phi(\tau), t)G(\phi(\tau))$ is bounded in $V^{2\beta}$, for $t \in \mathbb{R}$. In other words, the fibre $\mathcal{B}(\phi(\tau))$, of the bounded subbundle $\mathcal{B}(\Gamma)$ over Γ, contains the 1-dimensional space Span $(G(\phi(\tau))) \subset V^{2\beta}$, and by Lemma 45.2, Item (1), the linear skew product semiflow π_0 does not have an exponential dichotomy over Γ.

With the comments of the last paragraph in mind, we are now prepared to introduce a concept of hyperbolicity for a periodic orbit Γ for equation (70.1). We do this in the context of exponential trichotomies (see Section 4.5) for the linear skew product semiflow π given by (73.2). Specifically, we say that a periodic orbit Γ is **hyperbolic** if the associated linear skew product semiflow π has an exponential trichotomy over Γ, with characteristics K and λ_1, λ_2, λ_3, and λ_4, and associated projectors $\{P, Q, R\}$, where one has $\lambda_1 < \lambda_2 \leq 0 \leq \lambda_3 < \lambda_4$ and

$$\dim \mathcal{R}(R(u_1)) = \dim \Gamma, \qquad \text{for all } u_1 \in \Gamma.$$

For a nontrivial periodic orbit, one has $\dim \Gamma = 1$. However, if $\dim \Gamma = 0$ (i.e., if $\Gamma = \{u_0\}$ is a trivial periodic orbit), then Γ is hyperbolic if and only if the stationary solution u_0 is hyperbolic.

Let Γ be a given hyperbolic, periodic orbit of equation (70.1) and consider the perturbed equation (70.2), where the terms F and G are in C^1_{Lip}. For $r > 0$, we let $\Omega_r = N(\Gamma, r)$ be the set consisting of all $u \in V^{2\beta}$ such that $\|A^\beta(u - v)\| \leq r$, for some $v \in \Gamma$. We will use the pseudonorm $\|G\|_{\{C^1(\Omega_r)\}}$ in the following result:

Theorem 73.1. *Let the Standing Hypothesis A be satisfied and assume that both F and G are in C^1_{Lip}. Let Γ be a hyperbolic, periodic orbit for the unperturbed problem (70.1), and let K and λ_1, λ_2, λ_3, and λ_4 denote the characteristics of the associated exponential trichotomy generated by the linear skew product semiflow π over Γ. Then for every $r > 0$, there is a $\delta = \delta(r) > 0$, such that if*

$$\|G\|_{\{C^1(\Omega_r)\}} \leq \delta, \tag{73.4}$$

then there is a homeomorphism $h = h^G : \Gamma \to V^{2\beta}$, where $\|A^\beta(h(v) - v)\| \leq r$, for all $v \in \Gamma$, such that the perturbed equation (70.2) has an invariant periodic orbit $\Gamma^G = h(\Gamma)$. Furthermore, the periodic orbit Γ^G is hyperbolic for the semiflow generated by equation (70.2), and the characteristics of the exponential trichotomy over Γ^G can be chosen to be continuous at $G = 0$.

We will not give a proof of Theorem 73.1 here, since the argument in the next section will include Theorem 73.1 as a special case. Among other things, we will show that, in the infinite dimensional setting, the restriction (73.4) can be weakened (see Theorems 74.14 and 74.15, as well as Section 7.5.2).

7.4. Invariant Manifolds.

We continue our analysis of the longtime dynamics of the nonlinear evolutionary equation (70.1), where the Standing Hypothesis A holds and $F \in C^1_{\text{Lip}}$. Instead of studying the flow in the vicinity of a given equilibrium point, or a periodic orbit, we now will study the flow generated by the mild solutions of equation (70.1) in the vicinity of a smooth, compact, connected, invariant manifold M of class C^2 in $V^{2\beta}$. The relation $M \subset \mathcal{D}(A) = V^2$ follows from the Herculean Theorem 47.6. Let $\Omega \stackrel{\text{def}}{=} N(M, \sigma_2)$ denote the neighborhood (in $V^{2\beta}$) of M of fixed radius $\sigma_2 > 0$. Since $F \in C^1_{\text{Lip}}$, it follows that both F and DF are bounded on bounded sets in $V^{2\beta}$. In particular, F and DF are bounded on Ω. In the theory presented below we will assume that the perturbation term G is in C^1_{Lip}, and it satisfies

$$\|G\|_{\{A;C^1(\Omega)\}} \leq \delta \tag{74.1}$$

(see (46.14)), where δ will be assumed to be small.[36]

Let $\mathcal{K} \subset V^{2\beta}$ denote a given compact, invariant set generated by the mild solutions of equation (70.1), where $F \in C^1_{\text{Lip}}$. Let $\pi^F = \pi$, $\Phi^F = \Phi$, and $S^F = S$ satisfy

$$\pi(v, u_0, t) = (\Phi(u_0, t)v, S(t)u_0),$$

where π is the associated linear skew product semiflow on $V^{2\beta} \times \mathcal{K}$ (see Section 4.9). Recall that both $\Phi^F(u_0, t)$ and $S^F(t)u_0$ depend continuously on F in the two topologies $(C^1_{\text{Lip}}, \mathcal{T}^1_{\text{A}})$ and $(C^1_{\text{Lip}}, \mathcal{T}^1_{\text{bo}})$ (see Theorem 49.3). From Theorem 49.2, one has $DS(u_0, t)v = \Phi(u_0, t)v$, in the sense that equation (49.8) holds. If in addition, one has $DF \in C_{\text{Lip}}(V^{2\beta}, \mathcal{L}(V^{2\beta}, W))$, then $B(t) = DF(S(t)u_0)$ is Hölder continuous in t, since $S(t)u_0$ itself is Hölder continuous in t, see (47.7). In this case, $\Phi(u_0, t)v$ is a strong solution of equation (44.5) in W (see Theorem 44.4).

Some important dynamical properties of compact, invariant manifolds for equation (70.1) are presented in the following theorem. In particular, we examine here the induced linear skew product semiflow on the tangent bundle TM, see Kahn (1980).

Theorem 74.1. *Let the Standing Hypothesis A be satisfied for equation (70.1), and assume that $F \in C^1_{\text{Lip}}$. Let M be a compact, connected, invariant manifold of class C^1 in $V^{2\beta}$, and let $\pi(w, u; t) = (\Phi(u, t)w, S(t)u)$ denote the linear skew product semiflow on $V^{2\beta} \times M$. Then the following properties are valid:*

(1) *The tangent bundle TM is an invariant set for π; i.e., $\pi(TM; t) = TM$, for all $t \geq 0$.*
(2) *There is a $K \geq 1$ and an $a \geq 0$ such that for any $(w, u) \in TM$, the globally defined solution $\Phi(u, t)w$ satisfies*

$$K^{-1}e^{-a|t|}\|A^\beta w\| \leq \|A^\beta \Phi(u, t)w\| \leq Ke^{a|t|}\|A^\beta w\|, \qquad \textit{for all } t \in \mathbb{R}.$$

[36]The theory presented in this section is basically from the paper by Pliss and Sell (2001a).

(3) *For each $T > 0$, the mapping $u \to S(t)u$ is uniformly, locally Lipschitz continuous in $u \in M$, on the interval $[-T, T]$. In particular, there exist a $\rho = \rho(T) > 0$ and a $L_1 = L_1(T) > 0$ such that*

$$\|A^\beta(S(t)u_1 - S(t)u_2)\| \leq L_1 \|A^\beta(u_1 - u_2)\|,$$

for $t \in [0, T]$, or $t \in [-T, 0]$, provided that $\|A^\beta(S(t)u_1 - S(t)u_2)\| \leq \rho$, for all $t \in [0, T]$, or $t \in [-T, 0]$.

Proof. Let $T_u M$ denote the tangent space to the manifold M at the point $u \in M$. We claim that for each $t \geq 0$ and each $w \in T_u M$, one has $\Phi(u,t)w \in T_{S(t)u}M$. For comparisons with Section 4.9, we set $S(t)u = S(u,t)$. Indeed from Theorem 49.2, one can calculate $\Phi(u,t)w$ in terms of a directional derivative, i.e.,

$$\Phi(u,t)w = \lim_{h \to 0} \frac{1}{h} \left(S(u + hw, t) - S(u,t) \right).$$

Now equation (49.7) implies that as $h \to 0^+$, one has

$$\begin{aligned}
&\frac{1}{h}(S(u+hw,t) - S(u,t)) \\
&\qquad = e^{-At}w + \int_0^t e^{-A(t-s)} \frac{1}{h}[F(S(u+hw,s)) - F(S(u,s))]\, ds \\
&\qquad \to e^{-At}w + \int_0^t e^{-A(t-s)} DF(S(u,s))\Phi(u,s)w\, ds \\
&\qquad = \Phi(u,t)w,
\end{aligned}$$

which implies that $\Phi(u,t)w$ is tangent to the trajectory through u at the point $S(t)u$, i.e., $\Phi(u,t)w \in T_{S(t)u}M$, for each $t \geq 0$.

Since $\Phi(u,0) = I$, it follows from the continuity in time t that for small $t > 0$, $\Phi(u,t)$ maps a basis of $T_u M$ onto a linearly independent set in $T_{S(t)u}M$. Since M is connected, the dimension $\dim T_u M = \dim M$ is independent of $u \in M$, it follows that $\Phi(u,t)$ maps a basis of $T_u M$ onto a basis in $T_{S(t)u}M$. Hence one has $\pi(TM; t) = TM$, for small $t \geq 0$. It follows from the semigroup property that $\pi(TM; t) = TM$, for all $t \geq 0$. Items (2)and (3) are now easily verfied, and we leave the details as an exercise. □

In addition to the given equation (70.1), we will also study the behavior of solutions of the perturbed equation (70.2) in the vicinity of M, where both F and G are in C^1_{Lip}. We will use two norms to measure the size of the perturbation term G, namely, $\|G\|_{\{C^1(\Omega)\}}$ and $\|G\|_{\{A;C^1(\Omega)\}}$, where $\Omega = N(M, \sigma_2)$ is the neighborhood of M that is fixed above. Also see (46.13) - (46.14). For $0 < \rho \leq \operatorname{diam}_{V^{2\beta}}(M)$ and $0 \leq \sigma \leq \sigma_2$, we define $b^F = b^F(\rho, \sigma)$ by

(74.2)

$$b_1^F(\rho,\sigma) \stackrel{\text{def}}{=} \sup_{y_1, y_2 \in N(M,\sigma)} \left\{ \|DF(y_1) - DF(y_2)\|_{\mathcal{L}} : \|A^\beta(y_1 - y_2)\| \leq \rho \right\},$$

where $N(M,0) = M$ and $\mathcal{L} = \mathcal{L}(V^{2\beta}, W)$. Since DF is bounded on Ω, it follows that $b_1^F(\rho,\sigma)$ is finite-valued, and since DF is continuous and M is compact, one has $b_1^F(\rho,\sigma) \to 0$, as $(\rho,\sigma) \to 0$.

The Class Σ: In the sequel we will use the class Σ consisting of all positive, real-valued functions $\beta_1 = \beta_1(r) = \beta_1(r_1, r_2)$, defined for $r = (r_1, r_2)$ with $0 < r_i < r_{i0}$, where $r_{i0} = r_{i0}(\beta_1) > 0$, for $i = 1, 2$, and satisfying $\beta_1(r) \to 0$, as $r \to 0$. For example, if some real-valued function $\xi(\epsilon)$ is of order $o(\epsilon)$, as $\epsilon \to 0$, then one can write this in the form $|\xi(\epsilon)| = \epsilon\beta_1(\epsilon)$, where $\beta_1 \in \Sigma$. The function $b_1^F(\rho,\sigma)$ is an element of Σ. Other examples which arise below are $\beta_1(\epsilon,\delta)$, $\beta_2(\epsilon,\delta)$, and $b_2^F(\rho)$. We will let $\beta_1, \beta_2, \cdots$, and $b_0, b_1^F, \cdots$ denote various elements of Σ.[37] We will use the terms $\beta_1, \beta_2, \cdots$ as local variables, which may be redefined from time to time. Global variables, which have a unique definition in this work, will be denoted by $b_0, b_1^F, b_2^F, \cdots$. The superscript F will be used to denote elements of Σ which depend on F, but which are independent of the perturbation term G.

In order to study the mild solutions of equations (70.1) and (70.2), we introduce the time-dependent change of variables $y = S(t)u_0 + w$ into equation (70.2), where $S(t)u_0$ is a given mild solution of equation (70.1). Observe that $y = y(t)$ is a mild solution of (70.2) if and only if $w = w(t)$ is a mild solution of

$$\begin{aligned} \partial_t w + Aw &= F(S(t)u_0 + w) - F(S(t)u_0) + G(S(t)u_0 + w) \\ &= B(t)w + H(S(t)u_0, w), \end{aligned} \tag{74.3}$$

where $B(t) = DF(S(t)u_0)$, $H(S(t)u_0, w) = E(S(t)u_0, w) + G(S(t)u_0 + w)$, and E satisfies (71.8) - (71.11), for some $\gamma \in \Sigma$. Also $E(S(t)u_0, 0) = 0$, and the Fréchet derivative satisfies $DE(S(t)u_0, 0) = \frac{\partial}{\partial w} E \mid_{(S(t)u_0, 0)} = 0$.

For each $u_0 \in M$, we let $\Phi(u_0, t)$ denote the fundamental solution operator of the linear system

$$\partial_t w + Aw = DF(S(t)u_0)\, w, \qquad \text{for } t > 0,$$

that satisfies $\Phi(u_0, 0) = I$, where I is the identity operator on $V^{2\beta}$. Since $F \in C_{\text{Lip}}(V^{2\beta}, W)$, it follows from the Herculean Theorem 47.6 that the manifold M satisfies $M \subset \mathcal{D}(A) = V^2$, and for every $u_0 \in M$, the global solution $S(t)u_0$ is a classical solution in M, for all $t \in \mathbb{R}$. Furthermore, one can show that equation (73.3) holds in this case, where Γ is replaced by M. As argued in the case of the periodic orbit, Lemma 45.2 implies that the linear skew product semiflow π <u>never</u> has an exponential dichotomy over M, when the manifold M is a set containing more than one point. An exponential trichotomy is possible, and we will use this approach to define the important concept of normal hyperbolicity for the manifold M.

---

[37] Note that the subscript in β_i will distinguish an element $\beta_i \in \Sigma$ from the special parameter β used in the Standing Hypothesis A and the A^β-norm on $V^{2\beta} = \mathcal{D}(A^\beta)$.

However, before doing this, it is convenient to recall some properties of the nonlinear flow $S(t)$ on $V^{2\beta}$ and the linear skew product semiflow $\pi = \pi(t)$ on M.

Since M is invariant, one has $S(t)M = M$, for all $t \geq 0$, and M is the union of the trajectories of all global solutions that lie in M. As noted in equations (45.33) and (45.34), if $w = w(t)$ is a mild solution of equation (74.3) with $w(0) = w_0 = y_0 - u_0$ and $u_0 \in M$, then w satisfies two Variation of Constants formulae:

$$(74.4) \qquad w(t) = e^{-At}w_0 + \int_0^t e^{-A(t-s)}[B(s)w(s) + H(S(s)u_0, w(s))]\, ds$$

and

$$(74.5) \qquad w(t) = \Phi(u_0, t)w_0 + \int_0^t \Phi(S(s)u_0, t-s)H(S(s)u_0, w(s))\, ds.$$

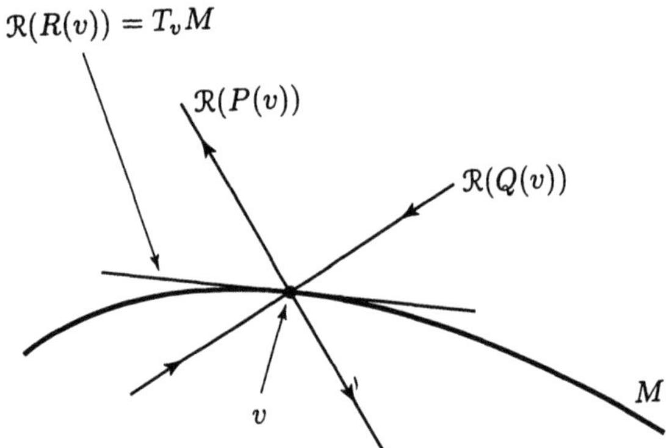

Figure 7.5. Normally Hyperbolic Invariant Manifold M

7.4.1. Statement of Theorems. We say that the compact, connected, invariant manifold M is **normally hyperbolic** if the linear skew product semiflow π has an exponential trichotomy in $V^{2\beta} \times M$, with characteristics $\lambda_1 < \lambda_2 \leq 0 \leq \lambda_3 < \lambda_4$ and $K \geq 1$ and associated invariant projectors $\{P, Q, R\}$, such that the neutral space $\mathcal{R}(R(v))$ satisfies

$$(74.6) \qquad \mathcal{R}(R(v)) = T_vM, \qquad \text{for all } v \in M,$$

see Figure 7.5. We allow here for the projection P to satisfy $P \equiv 0$. In this case, the exponential trichotomy is stable, and we say that the manifold M is normally hyperbolic and stable. In the case of a manifold that is normally hyperbolic and stable, only the characteristics K, λ_1, and λ_2 play any role in the perturbation theory described below.

In the sequel we will let $k = \dim M$ denote the dimension of M. The exponential trichotomy over M is said to be of **Lipschitz class** if the projections $P(v)$, $Q(v)$, and $R(v)$ are locally Lipschitz continuous functions on M. It is convenient to use the notation $P^s = Q$, $P^u = P$, $P^o = R$, $U^s(v) = \mathcal{R}(P^s(v))$, and $U^u(v) = \mathcal{R}(P^u(v))$, for $v \in M$, in the sequel. Since we assume M to be of class C^2, the projections $R(v) = P^o(v)$ and $Q^o(v) = I - P^o(v) = P^s(v) + P^u(v)$ are (Fréchet) differentiable mappings with respect to $v \in M$. We will denote the derivative of P^o at $v \in M$ by $DP^o(v)$. Since one has $P^o(v)P^o(v) = P^o(v)$, it follows from a simple calculation that $P^o(v)DP^o(v)P^o(v) = 0$, for all $v \in M$.

We are now prepared to describe the main results of this section. In the first result, we require that M be normally hyperbolic, where the associated exponential trichotomy is of Lipschitz class. We then argue that if the perturbation term G satisfies inequality (74.1), where $\delta > 0$ is sufficiently small, then the perturbed equation (70.2) has a normally hyperbolic invariant manifold M^G. Furthermore, M^G is homeomorphic to M, and the homeomorphism $h : M \to M^G$ is close to the identity mapping. The second result is a Shadow Theorem which compares the nonlinear dynamics on the two manifolds M and M^G.

Main Theorem. *Let the Standing Hypothesis A be satisfied. Let F and G be in C^1_{Lip} and let M be a compact, connected, invariant C^2-manifold in $V^{2\beta}$ for the unperturbed equation (70.1). Assume that M is normally hyperbolic and that the associated exponential trichotomy is of Lipschitz class.*

Then for every $\epsilon > 0$ there is a $\delta = \delta(\epsilon) > 0$ such that if $\|G\|_{\{A;C^1(\Omega)\}} \le \delta$, then there is a Lipschitz homeomorphism $h : M \to V^{2\beta}$ with the following properties:

1. *The manifold $M^G \overset{\text{def}}{=} h(M)$ is an invariant manifold for the perturbed equation (70.2).*
2. *Both manifolds M and M^G lie in $\mathcal{D}(A) = V^2$. Furthermore, M^G is of class C^1, it is normally hyperbolic for equation (70.2), and the characteristics for the associated exponential trichotomy on M^G can be chosen to vary continuously at $G = 0$.*
3. *One has $\|A^\beta(h(v) - v)\| \le 2\epsilon$, for all $v \in M$.*

Moreover, for each $u_0 \in M$, $y_0 \in M^G$, and $0 \le r < 1$, the mild solutions $S(t)u_0$ and $y(t, y_0)$ are classical solutions with $S(\cdot)u_0, y(\cdot, y_0) \in C^{0,1-r}_{\text{loc}}(\mathbb{R}; V^{2r}) \cap C(\mathbb{R}; \mathcal{D}(A))$.

For the next result, we introduce the concept of a shadow semiflow $S_1^G(t)$, which depends on the perturbations term G, but which acts on the unperturbed manifold M. The terminology arises because this semiflow acts as a "shadow" to the nonlinear dynamics on the perturbed manifold M^G. In particular, we let $S_1(t)$ and $S_2(t) = S_2^G(t)$ denote the semiflows in $V^{2\beta}$ generated by the maximally defined mild solutions of equations (70.1) and

(70.2), respectively. Let $h : M \to V^{2\beta}$ be a continuous mapping, where $M^G \stackrel{\text{def}}{=} h(M)$ is an invariant set for the perturbed equation (70.2). We say that a continuous mapping $S_1^G(t)u_0 : M \times [0, \infty) \to M$ is a **shadow semifow** for the nonlinear dynamics $S_2^G(t)$ on M^G, if it satisfies

$$(74.7) \qquad S_2^G(t)h(u_0) = h(S_1^G(t)u_0), \qquad \text{for all } (u_0, t) \in M \times [0, \infty).$$

As noted in the Main Theorem, we will show the existence of such a mapping $h = h^G : M \to V^{2\beta}$, for each G satisfying inequality (74.1), with δ sufficiently small. In this case, we say that the shadow semiflow $S_1^G(t)$ is G-**continuous** if it is continuous in the $\mathcal{T}_A^1$ topology generated by $\|G\|_{\{A;C^1(\Omega)\}}$. This means that if G_n and u_n are convergent sequences in $C_{\text{Lip}}(V^{2\beta}, W)$ and $V^{2\beta}$, respectively, with $\|G_n - G\|_{\{A;C^1(\Omega)\}} \to 0$ and $\|A^\beta(u_n - u)\| \to 0$, as $n \to \infty$, then

$$\|A^\beta(S_1^{G_n}(t)u_n - S_1^G(t)u)\| \to 0, \qquad \text{as } n \to \infty,$$

uniformly for t in compact sets in $[0, \infty)$. We will prove the following result:

Shadow Theorem. *Let the hypotheses of the Main Theorem be satisfied, and let $h = h^G : M \to V^{2\beta}$ and $M^G = h(M)$ satisfy the conclusions. Then there exists a G-continuous shadow semiflow $S_1^G(t)$ on M, for every G satisfying inequality (74.1), where δ is given in the Main Theorem. Furthermore, when $G \equiv 0$, then $S_1^0(t) = S_1(t)$ on M.*

7.4.2. Local Coordinates near M. The next step in our analysis is to derive a good local coordinate system in the vicinity of the manifold M. We present these details in the next three lemmas. First note that by using a larger value of the characteristic K, if necessary, we can assume that, for some $a_0 \geq 0$, one has

$$(74.8) \qquad \|A^\beta \Phi(u_0, t)w\| \leq K e^{a_0 t} \|A^\beta w\|, \qquad \text{for all } t \geq 0 \text{ and } w \in V^{2\beta},$$

where $u_0 \in M$, as well as

$$(74.9) \qquad \|A^\beta \Phi(u_0, t)w\| \leq K t^{-\beta} e^{a_0 t} \|w\|, \qquad \text{for all } t > 0 \text{ and } w \in W.$$

For $u_0 \in M$, we let $D(u_0, \rho)$ denote the closed, k-dimensional disk in $\mathcal{R}(P^o(u_0))$, centered at the origin, of radius ρ; i.e.,

$$D(u_0, \rho) = \{p \in \mathcal{R}(P^o(u_0)) : \|A^\beta p\| \leq \rho\}.$$

We will use $D(u_0, \rho)$ to define local coordinates in the vicinity of a point $u_0 \in M$. Since $u_0 + D(u_0, \rho)$ is tangent to M at u_0, there is a ρ small enough, say, $0 < \rho \leq \rho_2$, where ρ_2 does not depend on $u_0 \in M$, and there is a function $f : D(u_0, \rho) \to \mathcal{R}(Q^o(u_0)) \subset V^{2\beta}$ such that the image

$$(74.10) \qquad \mathcal{D}_\rho(u_0) = \{u_0 + p + f(p) : p \in D(u_0, \rho)\} = u_0 + \text{Graph } f$$

contains an open neighborhood of u_0 in M. ($\mathcal{D}_\rho(u_0)$ is the Pliss disk at u_0.) The Lipschitz property is equivalent to saying that there is an $L_0 > 0$ such that one has

$$\|P^i(u_1) - P^i(u_2)\|_{\mathcal{L}} \leq L_0\|A^\beta(u_1 - u_2)\|, \tag{74.11}$$

for all u_1, $u_2 \in \mathcal{D}_\rho(u_0)$, $u_0 \in M$, and $i = s, o, u$, where $\mathcal{L} = \mathcal{L}(V^{2\beta}, V^{2\beta})$. The complementary projector $Q^o \stackrel{\text{def}}{=} I - P^o$ is invariant and $Q^o(u) = P^s(u) + P^u(u)$, for all $u \in M$. Since the inequalities (45.13) - (45.15) hold at $t = 0$, one has

$$\|P^i(u)\|_{\mathcal{L}} \leq K, \qquad \text{for } u \in M \text{ and } i = s, o, u. \tag{74.12}$$

Because of the Lipschitz property for M, the tangent space to the curve $r = f(p)$ is Lipschitz continuous, which implies that the function f, itself, is of class $C^{1,1}$. Since M is of class C^2, the function f is of class C^2, as well. Since the space corresponding to $r = 0$ coincides with the neutral space $U^o(u_0) = \mathcal{R}(R(u_0))$, it follows that $f(0) = 0$ and the derivative $D_p f = \frac{\partial f}{\partial p}$ satisfies $D_p f(0) = 0$. Furthermore, the second derivative $D_p^2 f = \frac{\partial^2 f}{\partial p^2}$ satisfies

$$\|D^2 f(p)\|_{\mathcal{BL}} \leq \hat{L}, \qquad \text{for } \|A^\beta p\| \leq \rho \text{ and all } u_0 \in M,$$

where $\|D^2 f(p)\|_{\mathcal{BL}}$ denotes the bilinear operator norm, i.e.,

$$\|D^2 f(p)\|_{\mathcal{BL}} = \max\{\|A^\beta D^2 f(p)(u, v)\| : \|A^\beta u\|, \|A^\beta v\| \leq 1\}.$$

(The constant $\hat{L}$ depends on the Lipschitz coefficient for the mapping $v \to P^o(v)$, for $v \in M$, and is independent of the base point v.) By using a larger value for L_0, if necessary, one then has the validity of

$$\|A^\beta(f(p_1) - f(p_2))\| \leq L_0\rho\|A^\beta(p_1 - p_2)\|, \tag{74.13}$$

for $\|A^\beta p_1\|$, $\|A^\beta p_2\| \leq \rho$, and

$$\|A^\beta f(p)\| \leq L_0\|A^\beta p\|^2 \leq L_0\rho\|A^\beta p\| < L_0\rho^2, \qquad \text{for } \|A^\beta p\| \leq \rho, \tag{74.14}$$

as well as inequality (74.11). Consequently, one has the following result, which treats the radii ρ of the disks $\mathcal{D}_\rho(u_0)$ as a parameter.

Lemma 74.2. *Let the hypotheses of the Main Theorem be satisfied. Then there exists a $\rho_2 > 0$ with $4K^2L_0\rho_2 \leq 1$, such that for all $u_0 \in M$ and all u_1, $u_2 \in \mathcal{D}_\rho(u_0)$, where $0 < \rho \leq \rho_2$, inequalities (74.11), (74.13), and (74.14) are valid, and one has*

$$\begin{aligned}
\frac{3}{4}\|A^\beta P^o(u_0)(u_1 - u_0)\| &\leq (1 - L_0\rho)\|A^\beta P^o(u_0)(u_1 - u_0)\| \\
&\leq \|A^\beta(u_1 - u_0)\| \\
&\leq (1 + L_0\rho)\|A^\beta P^o(u_0)(u_1 - u_0)\| \\
&\leq \frac{5}{4}\|A^\beta P^o(u_0)(u_1 - u_0)\|.
\end{aligned}$$

In addition, one obtains

(74.15)
$$\left\{\begin{array}{l} \|A^\beta P^s(u_0)(u_1-u_0)\| \\ \|A^\beta P^u(u_0)(u_1-u_0)\| \\ \|A^\beta Q^o(u_0)(u_1-u_0)\| \end{array}\right. \leq K^2 L_0 \|A^\beta(u_1-u_0)\|^2 \leq K^2 L_0 \rho \|A^\beta(u_1-u_0)\|.$$

Proof. The argument uses $4K^2L_0\rho_2 \leq 1$ and $K \geq 1$. Let $u_1 = u_0+p+f(p)$, where $P^o(u_0)p = p$ and $Q^o(u_0)f(p) = f(p)$. The first batch of inequalities follow from

$$\|A^\beta p\| - \|A^\beta f(p)\| \leq \|A^\beta(u_1-u_0)\| \leq \|A^\beta p\| + \|A^\beta f(p)\|$$

and (74.14). Since $Q^o(u_0)(u_1-u_0) = f(p)$, (74.15) follows from (74.12) and (74.14), as well. □

We will denote a typical point $v \in \mathcal{D}_\rho(u_0)$ in the form $v = u_0+p+f(p)$, where $\|A^\beta p\| < \rho \leq \rho_2$. One then has

(74.16)
$$\begin{gathered} \|A^\beta(v-u_0)\| \leq (1+L_0\rho)\rho, \\ \|A^\beta(v_1-v_2)\| \leq (1+L_0\rho)\|A^\beta(p_1-p_2)\| \leq \frac{5}{4}\|A^\beta(p_1-p_2)\|. \end{gathered}$$

where $v_i = u_0+p_i+f(p_i)$ and $\|A^\beta p_i\| \leq \rho$, for $i = 1, 2$. We now have the following result:

Lemma 74.3. *Let the hypotheses of the Main Theorem be satisfied. Then for all $u_0 \in M$, all $u_1 \in \mathcal{D}_\rho(u_0)$, all ρ with $0 < \rho \leq \rho_2$, and all $t \geq 0$, the following are valid:*

$$\begin{aligned} \|A^\beta\Phi(u_0,t)P^o(u_0)(u_1-u_0)\| &\leq K^2\|A^\beta(u_1-u_0)\|e^{\lambda_3 t}, \\ \|A^\beta\Phi(u_0,t)P^s(u_0)(u_1-u_0)\| &\leq K^3L_0\rho\|A^\beta(u_1-u_0)\|e^{\lambda_1 t}, \\ \|A^\beta\Phi(u_0,t)P^u(u_0)(u_1-u_0)\| &\leq K^3L_0\rho\|A^\beta(u_1-u_0)\|e^{a_0 t}, \\ \|A^\beta\Phi(u_0,t)(u_1-u_0)\| &\leq K^2\|A^\beta(u_1-u_0)\|e^{\lambda_3 t} \\ &\quad + K^3L_0\rho\|A^\beta(u_1-u_0)\|e^{a_0 t}, \\ \|A^\beta\Phi(u_0,t)P^o(u_0)(u_1-u_0)\| &\geq K^{-1}\|A^\beta P^o(u_0)(u_1-u_0)\|e^{\lambda_2 t}, \\ \|A^\beta\Phi(u_0,t)(u_1-u_0)\| &\geq (4(5K)^{-1}e^{\lambda_2 t} - K^3L_0\rho e^{a_0 t}) \\ &\quad \times \|A^\beta(u_1-u_0)\|, \end{aligned}$$

where ρ_2 is given by Lemma 74.2.

Proof. The proofs of these inequalities follow from Lemma 74.2; inequality (74.14); the exponential trichotomy inequalities (45.13), (45.15), (45.20);

and inequality (74.8). For example,

$$\begin{aligned}
&\|A^\beta\Phi(u_0,t)(u_1-u_0)\| \\
&\quad\le \|A^\beta\Phi(u_0,t)P^o(u_0)(u_1-u_0)\| + \|A^\beta\Phi(u_0,t)Q^o(u_0)(u_1-u_0)\| \\
&\quad\le K\|A^\beta(u_1-u_0)\|e^{\lambda_3 t} + K\|A^\beta Q^o(u_0)(u_1-u_2)\|e^{a_0 t} \\
&\quad\le K\|A^\beta(u_1-u_0)\|e^{\lambda_3 t} + K^3 L_0\rho\|A^\beta(u_1-u_2)\|e^{a_0 t}.
\end{aligned}$$

where we used here the projector identity $Q^o(u_0) = [Q^o(u_0)]^2$. □

While the inequalities in this lemma are valid for all $t \ge 0$, we will be using them when t is restricted to a finite interval $0 \le t \le 2T$, where $T > 0$ is fixed as follows: With the characteristics K, λ_1, λ_2, λ_3, and λ_4 of the compact, invariant set M given by the exponential trichotomy on M, we seek a real number $T > 0$ such that

$$(74.17)\qquad \begin{cases} 4K^2 e^{\lambda_1\tau} & <1, \\ 96K^2 e^{(\lambda_1-\lambda_2)\tau} & <1, \\ 16K^2 e^{-\lambda_4\tau} & <1, \\ 48K^3 e^{(\lambda_3-\lambda_4)\tau} & <1, \end{cases} \qquad \text{for } T \le \tau \le 2T.$$

Note that each of the exponents in the inequalities (74.17) is negative. Consequently, there does exist a time $T > 0$ such that, for all $\tau \ge T$, these inequalities are satisfied. We fix one such T for the sequel. We will use the fact that (74.17) is valid for all τ with $T \le \tau \le 2T$.

The next step is to construct a local coordinate system near M by restricting this coordinate system to a suitable neighborhood of each disk $\mathcal{D}_\rho(u_0)$. We begin by choosing ρ_1 and σ_1 so that $0 < \rho_1 \le \rho_2$, $0 < \sigma_1 \le \sigma_2$, and

$$(74.18)\qquad \text{Convex Hull}\,(N(\mathcal{D}_{\rho_1}(u_0),\sigma_1)) \subset \Omega = N(M,\sigma_2),$$

and $M \cap N(\mathcal{D}_{\rho_1}(u_0),\sigma_1) \subset \mathcal{D}_{\rho_2}(u_0)$, for each $u_0 \in M$. Since M is compact, the parameters ρ_1 and σ_1 can be chosen to be uniform for $u_0 \in M$. The relationship (74.18) is important because it enables us to get a good estimate for the effective Lipschitz coefficient for the nonlinear perturbation term G in terms of $\|G\|_{\{A;C^1(\Omega)\}}$, where $G \in C^1_{\mathrm{Lip}}$. In particular, let $w_i = w_i(t)$ denote two continuous functions with $w_i(t) \in N(\mathcal{D}_{\rho_1}(u_0),\sigma_1)$, for $0 \le t < t_0$ and $i = 1,2$, where $0 < t_0 \le \infty$. Then (74.18) implies that $\lambda w_1(s) + (1-\lambda)w_2(s) \in \Omega$, for $0 \le s < t_0$ and $0 \le \lambda \le 1$. As a result, one has

$$G(w_1) - G(w_2) = \int_0^1 DG(w_2 + \theta(w_1 - w_2))\,d\theta(w_1 - w_2),$$

for $0 \le s < t_0$, which implies that

(74.19)

$$e^{-A(t-s)}[G(w_1) - G(w_2)] = \int_0^1 e^{-A(t-s)} DG(w_2 + \theta(w_1 - w_2))\,d\theta(w_1 - w_2),$$

for $0 \le s \le t < t_0$. We claim that if G satisfies inequality (74.1), then

(74.20)
$$\int_0^t \|A^\beta e^{-A(t-s)}[G(w_1) - G(w_2)]\| \, ds \le \delta \sup_{0 \le s \le t} \|A^\beta(w_1(s) - w_2(s))\|.$$

Indeed from equation (74.19) one has

$$\int_0^t \|A^\beta e^{-A(t-s)}[G(w_1) - G(w_2)]\| \, ds$$
$$\le \int_0^t \int_0^1 \|A^\beta e^{-A(t-s)} DG\|_{\mathcal{L}} \, d\theta \, \|A^\beta(w_1(s) - w_2(s))\| \, ds$$
$$\le \|G\|_{\{A;C^1(\Omega)\}} \sup_{0 \le s \le t} \|A^\beta(w_1(s) - w_2(s))\|.$$

We will require that σ_1 and (especially) ρ_1 satisfy a few auxiliary properties. In particular, by using the Lipschitz property, one can show that if the radius ρ_1 of the disks $\mathcal{D}_{\rho_1}(u_0)$ and the number σ_1 are replaced by smaller values, if necessary, then the coordinate system we next describe is valid in the vicinity of each disk $\mathcal{D}_{\rho_1}(u_0)$.

Lemma 74.4. *Let M be a compact manifold of class C^2 in $V^{2\beta}$. Then there exist $\rho_1 > 0$ and $\sigma_1 > 0$, such that $0 < \rho_1 \le \frac{3}{10}\rho_2$, $0 < \sigma_1 \le \sigma_2$, relation (74.18) is valid and, for every $u_0 \in M$ and every $y \in N(\mathcal{D}_{\rho_1}(u_0), \sigma_1)$, the following hold:*

1. *There is one and only one point $v \in \mathcal{D}_{\rho_2}(u_0)$ such that $y - v \in U^s(v) \oplus U^u(v)$, where $U^s(v) = \mathcal{R}(P^s(v))$ and $U^u(v) = \mathcal{R}(P^u(v))$. Furthermore, the mapping $\psi : y \to v \overset{\text{def}}{=} \psi(y) = \psi(u_0, y)$ is of class C^2 on $N(\mathcal{D}_{\rho_1}(u_0), \sigma_1)$ with $\psi(u_0, u_1) = \psi(u_1) = u_1$, for all $u_1 \in \mathcal{D}_{\rho_1}(u_0)$.*
2. *If in addition, one has $\|A^\beta(y - u_0)\| < 2\sigma_1$, then $v = \psi(y)$ satisfies $v \in \mathcal{D}_{\rho_1}(u_0)$.*
3. *Moreover, the Fréchet derivative $D\psi(y)$ of $\psi(y)$ with respect to y, where $y \in N(\mathcal{D}_{\rho_1}(u_0), \sigma_1)$, satisfies*

(74.21)
$$D\psi(y) = R(v) = P^o(v) = P^o(\psi(y)).$$

4. *The mapping ψ satisfies $\psi(y) = \psi(u_0, y) = y - \phi(u_0, y) = y - \phi(y)$, where $D\phi(y) = Q^o(v) = Q^o(\psi(y))$. The mapping ϕ has the property that for all $v \in M$, one has*

(74.22)
$$\phi(v + n) = n, \qquad \text{for } n \in \mathcal{R}(Q^o(v)).$$

5. *Let $y_i \in N(\mathcal{D}_{\rho_1}(v_0), \sigma_1)$, for some $v_0 \in M$ and set $v_i = \psi(y_i)$, for $i = 1, 2$. Then one has*

(74.23)
$$v_1 - v_2 = \psi(y_1) - \psi(y_2) = P^o(v_2)(y_1 - y_2) + e_3,$$

and there is a $b_2^F \in \Sigma$ *such that* $e_3 = e_3(y_2, y_1 - y_2)$ *satisfies*

$$\|A^\beta e_3\| \leq b_2^F(\rho)\|A^\beta(y_1 - y_2)\|, \tag{74.24}$$

whenever $\|A^\beta(y_1 - y_2)\| \leq \rho \leq \rho_1$.

(6) *In the sequel, we will require that*

$$C_2\, b_2^F(\rho_1) \leq \frac{1}{144K^2} e^{2\lambda_2 T} \leq K, \tag{74.25}$$

in which case, one has $\|v_1 - v_2\| \leq 2K\|y_1 - y_2\|$. *The constant* C_2 *is defined in Lemma 74.7, and it satisfies* $C_2 \geq 1$.

Proof. The proofs of Items (3) and (4) follow directly from Items (1) and (2), and the proofs of Items (1) and (2), as we now show, are an application of a Collared, or Tubular, Neighborhood Theorem for M.

For $0 < \rho \leq \rho_1 \leq \rho_2$, we let $y \in N(\mathcal{D}_\rho(u_0), \sigma_1)$. The defining relationship for the point $v \in \mathcal{D}_{\rho_2}(u_0)$ is that $v = u_0 + p + f(p)$, for some point $p \in \mathcal{R}(P^o(u_0))$ with $\|A^\beta p\| \leq \rho_2$, and that $y - v \in \mathcal{R}(Q^o(v))$. We now define $P_0 = P^o(u_0)$, $Q_0 = Q^o(u_0)$, $P = P^o(v)$, and $Q = Q^o(v)$. Our goal first is to find a point $p \in \mathcal{R}(P_0)$ so that

$$y - u_0 - p - f(p) = y - v = Q(y - v) = Q(y - u_0) - Qp - Qf(p).$$

In other words, p must satisfy

$$p = J(p, y) \stackrel{\text{def}}{=} P(y - u_0) + QP_0 p - Pf(P_0 p). \tag{74.26}$$

We will now show that equation (74.26) has a unique fixed point p in a suitable space. Since $p = P_0 p$, $Q_0 P_0 = 0$, and $P_0 f(P_0 p) = 0$, one has

$$J(p, y) = P(y - u_0) + (Q - Q_0)p - (P - P_0)f(p).$$

Assume for the moment that $\|A^\beta(y - u_0)\| < 2\sigma_1$. In this case, we will show that there is such a p, where $\|A^\beta p\| \leq \rho_1$, and $v = u_0 + p + f(p) \in \mathcal{D}_{\rho_1}(u_0)$. A direct calculation, using inequalities (74.11) and (74.14), leads to

$$\|A^\beta J(p, y)\| \leq 2K\sigma_1 + L_0\left(\|A^\beta p\| + \|A^\beta f(p)\|\right)^2 \leq 2K\sigma_1 + L_0\rho_1^2(1 + L_0\rho_1)^2.$$

Since $p = P_0 p$, the fixed point $p = J(p, y)$ will satisfy $\|A^\beta p\| \leq \rho_1$, provided that

$$L_0\rho_1(1 + L_0\rho_1)^2 \leq \frac{1}{2} \quad \text{and} \quad 2K\sigma_1 \leq \frac{\rho_1}{2}. \tag{74.27}$$

After a lengthy calculation which uses $v_i = u_0 + p_i + f(p_i)$, for $i = 1, 2$, and inequalities (74.11) and (74.13), one finds that

$$\|A^\beta(J(p_1, y) - J(p_2, y))\| \leq K_1\|A^\beta(p_1 - p_2)\|,$$

where $K_1 = 2(1+L_0\rho_1)\sigma_1 + 2(1+L_0\rho_1)^2\rho_1$. When ρ_1 and σ_1 are chosen so that (74.18) and (74.27) hold, as well as,

$$2(1+L_0\rho_1)\sigma_1 \leq \frac{1}{3} \quad \text{and} \quad 2(1+L_0\rho_1)^2\rho_1 \leq \frac{1}{3}, \tag{74.28}$$

then $J(p,y)$ is a strict contraction on $\mathcal{D}_{\rho_1}$ and there is a unique fixed point for $p = J(p,y)$. Since P_0p is also a fixed point of equation (74.26), where $\|A^\beta p\| \leq \rho_1$ and $\|A^\beta P_0 p\| \leq \rho_1$, it follows that $p = P_0p$, i.e., $p \in \mathcal{R}(P_0)$. Thus p also satisfies the equation $p = P_0J(p,y)$, and the point $v = u_0 + p + f(p)$ satisfies $v \in \mathcal{D}_{\rho_1}(u_0)$. This in turn implies that $y - v = Q(y-v)$.

It remains to verify the conclusion when y does not satisfy the added requirement that $\|A^\beta(y-u_0)\| < 2\sigma_1$. Since $y \in N(\mathcal{D}_\rho(u_0),\sigma_1)$, there is a point $u_1 \in \mathcal{D}_\rho(u_0)$ such that $\|A^\beta(y-u_1)\| < 2\sigma_1$ and $y \in N(\mathcal{D}_\rho(u_1),\sigma_1)$. From the argument of the last paragraph, there is then a point $v \in \mathcal{D}_{\rho_1}(u_1)$ with $y - v \in \mathcal{R}(Q^o(v))$. Now inequality (74.16) implies that

$$\|A^\beta(v-u_0)\| \leq \|A^\beta(v-u_1)\| + \|A^\beta(u_1-u_0)\| \leq 2(1+L_0\rho_1)\rho_1.$$

With $p = v - u_0 - f(p)$, it follows from (74.14) that

$$\|A^\beta p\| \leq 2(1+L_0\rho_1)\rho_1 + L_0\rho_1\|A^\beta p\|.$$

One finds that $(1+L_0\rho_1) \leq \frac{5}{4}$ and $(1-L_0\rho_1)^{-1} \leq \frac{4}{3}$, since Lemma 74.2 implies $4L_0\rho_1 \leq 4L_0\rho_2 \leq 4K^2L_0\rho_2 \leq 1$. As a result, one obtains $\|A^\beta p\| \leq \frac{10}{3}\rho_1$. Since $\rho_1 \leq \frac{3}{10}\rho_2$, we see that $v \in \mathcal{D}_{\rho_2}(u_0)$, as desired.

If $u_0 \in M$ and $y_0 \in N(\mathcal{D}_\rho(u_0),\sigma_1)$ are chosen so that $Q_0(y_0-u_0) = y_0 - u_0$, then one has $\psi(y_0) = u_0$. In other words, one has $J(p,y_0) = 0$ whenever $P_0(y_0-u_0) = 0$. Next let w be given where $y = y_0 + w \in N(\mathcal{D}_\rho(u_0),\sigma_1)$ and $\|A^\beta w\|$ is small. Let $e_3 = e_3(y_0,w)$ be defined by

$$e_3(y_0,w) \stackrel{\text{def}}{=} \psi(y_0+w) - \psi(y_0) - P_0w = v - v_0 - P_0w.$$

By means of a straightforward calculation, which uses $P_0(DP_0)P_0 = 0$, where $DP_0 = DP^o(v_0)$, one can show that $\lim_{\|A^\beta w\|\to 0}\|A^\beta w\|^{-1}e_3(y_0,w) = 0$. This proves that $\psi(y)$ is Fréchet differentiable, and equations (74.21) and (74.23) hold, that inequality (74.24) is valid, and that $b_2^F \in \Sigma$. The second inequality in (74.25) follows from the facts that $K \geq 1$, $\lambda_2 \leq 0$, and $T > 0$. □

Notation: We will denote the new (nonlinear) coordinates of the point y by

$$y = v + s + u = v + n, \tag{74.29}$$

where $\|y - v\| < \sigma_1$, $v \in M$, $s \in U^s(v)$, $u \in U^u(v)$, and $n = s + u$. By (74.22) one then has $\phi(y) = n = s + u$. In the sequel we fix ρ_1 and σ_1

so that relations (74.18), (74.25), (74.27), (74.28), and $\rho_1 \le \frac{3}{10}\rho_2$ hold. This new coordinate system for the point y depends on the base point $u_0 \in M$. If two disks $\mathcal{D}_{\rho_1}(u_0)$ and $\mathcal{D}_{\rho_1}(u_1)$ have a nontrivial intersection and $y \in N(\mathcal{D}_{\rho_1}(u_0), \sigma_1) \cap N(\mathcal{D}_{\rho_1}(u_1), \sigma_1)$, then $\psi(u_0, y) = \psi(u_1, y)$, i.e., the coordinate representations agree. More generally, let $y = y(t) = y(t, y_0)$ be a solution of the perturbed equation (70.2), where $y_0 = v_0 + n_0$, $v_0 = u_0 \in M$, and $n_0 \in \mathcal{R}(Q^o(v_0))$. Assume that one has $y(t) \in N(\mathcal{D}_{\rho_1}(S(t)u_0), \sigma_1)$ for t in some interval I. Then the local coordinate representation

$$y(t) = v(t) + n(t), \qquad \text{for all } t \in I, \tag{74.30}$$

where $v(t) \in \mathcal{D}_{\rho_2}(S(t)u_0)$ and $P^o(v(t))n(t) = 0$, is well defined.

In addition to (74.18), (74.25), (74.27), (74.28), and $\rho_1 \le \frac{3}{10}\rho_2$, we require that ρ_1 satisfy

$$3K^3 L_0 \rho_1 e^{a_0 t} \le K^2 e^{\lambda_3 t} \text{ and } 10K^3 L_0 \rho_1 e^{a_0 t} \le e^{\lambda_2 t}, \tag{74.31}$$

for $0 \le t \le 2T$, and that ρ_1 and σ_1 satisfy

$$C_3\, b_1^F(\rho_1, \sigma_1) \le \frac{1}{144K^2} e^{2\lambda_2 T}, \tag{74.32}$$

where $C_3 = C_3(2T) > 0$ is defined in Lemma 74.7 and b_1^F is given by equation (74.2). Since $\lambda_2 \le \lambda_3 \le a_0$, it then follows from inequality (74.31) and Lemma 74.3 that for $v_1 \in \mathcal{D}_{\rho_1}(v_0)$, for $0 \le t \le 2T$, one has

$$\frac{1}{2K}\|A^\beta(v_1 - v_0)\| e^{\lambda_2 t} \le \|A^\beta \Phi(v_0, t)(v_1 - v_0)\| \le 2K^2 \|A^\beta(v_1 - v_0)\| e^{\lambda_3 t}. \tag{74.33}$$

7.4.3. The Dynamics on M. The results of the last three lemmas give valuable information about the geometry of compact, connected, invariant manifolds M of class C^2. We will also be interested in such manifolds which may be lacking in smoothness, but which have the property that the induced linear skew product semiflow π has an exponential trichotomy. This is developed in the following lemma wherein we assume M to be a Lipschitz manifold, and not necessarily of class C^2. These are manifolds where the representation $f(p)$ given by equation (74.10) is a Lipschitz continuous function. While we do assume here that π has an exponential trichotomy on M, we do not require that the tangency condition (74.6) be satisfied.

Lemma 74.5. *Let the Standing Hypothesis A be satisfied and let $F \in C^1_{\text{Lip}}$. Let M be a compact, connected, invariant manifold of Lipschitz class in $V^{2\beta}$ for the unperturbed equation (70.1). Assume that the linear skew product semiflow π has an exponential trichotomy on M. Then there is a $b_3^F \in \Sigma$ such that for any two points u_1, $u_2 \in M$ with $\|A^\beta(u_1 - u_2)\| \le \rho$, where $0 < \rho \le \rho_1$, there is a function $H_2(t)$ with the property that*

$$S(t)u_1 - S(t)u_2 = \Phi(u_2, t)(u_1 - u_2) + H_2(t), \qquad \text{for all } t \in [0, 2T], \tag{74.34}$$

and

$$\|A^\beta H_2(t)\| \le b_3^F(\rho)\|A^\beta(u_1 - u_2)\|, \qquad \text{for all } t \in [0, 2T]. \tag{74.35}$$

Proof. Let u_1, $u_2 \in M$ satisfy $\|A^\beta(u_1 - u_2)\| \le \rho$, where $0 < \rho \le \rho_1$, and set $w = w(t) = S(t)u_1 - S(t)u_2$, for $0 \le t \le 2T$. Then w is a mild solution of equation (74.3) with $G \equiv 0$ and $H \equiv E$. It follows from equation (74.5) that

$$w(t) = \Phi(u_2, t)(u_1 - u_2) + \int_0^t \Phi(S(s)u_2, t - s)E(S(s)u_2, w(s))\, ds,$$

for $0 \le t \le 2T$. Hence equation (74.34) is valid, for $0 \le t \le 2T$, with

$$H_2(t) = \int_0^t \Phi(S(s)u_2, t - s)E(S(s)u_2, w(s))\, ds.$$

In order to verify that inequality (74.35) is valid, we note that inequalities (71.11) and (74.9) imply that there is a $\gamma \in \Sigma$ such that

$$\|A^\beta H_2(t)\| \le \gamma(\sigma) \int_0^t (t - s)^{-\beta} e^{a_0(t-s)} \|A^\beta w(s)\|\, ds,$$

provided that $\|A^\beta w(s)\| \le \sigma$, for $0 \le s \le t$. Now inequality (47.11), with $F_1 = F_2 = F$, implies that there is a constant $C_2 = C_2(2T) \ge 1$ such that

$$\|A^\beta w(t)\| = \|A^\beta(S(t)u_1 - S(t)u_2))\| \le C_2\|A^\beta(u_1 - u_2)\|, \tag{74.36}$$

for $0 \le t \le 2T$. If $\|A^\beta(u_1 - u_2)\| \le \rho$ and $C_2\rho = \sigma$, then $\|A^\beta w(t)\| \le \sigma$, for $0 \le t \le 2T$. With $b_3^F(\rho) = C_2(1 - \beta)^{-1}(2T)^{1-\beta}e^{2a_0T}\gamma(C_2\rho)$, one obtains (74.35). □

Lemma 74.6. *Let the Standing Hypothesis A be satisfied and let $F \in C^1_{\text{Lip}}$. Let M be a compact, connected, invariant C^2-manifold in $V^{2\beta}$ for the unperturbed equation (70.1). Assume that M is normally hyperbolic and that the associated exponential trichotomy is of Lipschitz class. Then there is a ρ_0, with $0 < \rho_0 \le \rho_1$, such that for any points v_1, $v_2 \in M$ with $\|A^\beta(v_1 - v_2)\| \le \rho_0$, one has $\|A^\beta(S(t)v_1 - S(t)v_2)\| \le \rho_1$, for $-2T \le t \le 2T$; and*

$$\frac{1}{4K}\|A^\beta(v_1 - v_2)\|e^{\lambda_2 t} \le \|A^\beta(S(t)v_1 - S(t)v_2)\| \le 4K^2\|A^\beta(v_1 - v_2)\|e^{\lambda_3 t}, \tag{74.37}$$

for $0 \le t \le 2T$; while, for $-2T \le t \le 0$, one has

$$\frac{1}{4K^2}\|A^\beta(v_1 - v_2)\|e^{\lambda_3 t} \le \|A^\beta(S(t)v_1 - S(t)v_2)\| \le 4K\|A^\beta(v_1 - v_2)\|e^{\lambda_2 t}. \tag{74.38}$$

Proof. Let $b_3^F \in \Sigma$ be given by Lemma 74.5 and fix $\rho_0 > 0$ so that $C_2\rho_0 \leq \rho_1$, where $C_2 \geq 1$ is given by (74.36), and ρ_1 satisfies the conditions stated above, as well as

$$b_3^F(\rho_1) \leq \frac{1}{48K} e^{2\lambda_2 T}. \tag{74.39}$$

Since $\lambda_2 \leq 0 \leq \lambda_3$ and $K \geq 1$, one has

$$b_3^F(\rho) \leq \frac{1}{48K} e^{\lambda_2 t} \leq 2K^2 e^{\lambda_3 t}, \qquad \text{for } 0 \leq t \leq 2T \text{ and } 0 < \rho \leq \rho_1.$$

Now equation (74.34) implies that

$$\begin{aligned} \|A^\beta \Phi(v_2,t)(u_1 - v_2)\| - \|A^\beta H_2(t)\| &\leq \|A^\beta (S(t)u_1 - S(t)v_2)\| \\ &\leq \|A^\beta \Phi(v_2,t)(u_1 - v_2)\| + \|A^\beta H_2(t)\|, \end{aligned}$$

for $0 \leq t \leq 2T$. This fact, together with (74.33), imply inequality (74.37). Inequality (74.38) follows directly from (74.37). We will omit these details. □

In addition to the functions b_1^F, b_2^F, and b_3^F introduced above, we define b_4^F and b_5^F by

$$\begin{aligned} b_4^F &= b_4^F(\rho, 2\epsilon) = K(C_3\, b_1^F(\rho, 2\epsilon) + C_2\, b_2^F(\rho)), \\ b_5^F &= b_5^F(\rho, 2\epsilon) = 3b_4^F(\rho, 2\epsilon) + b_3^F(\rho), \end{aligned} \tag{74.40}$$

where the coefficients C_2 and C_3 are defined below in Lemma 74.7.

7.4.4. Perturbed Dynamics Near M. Let $C(M, V^{2\beta})$ denote the Banach space of continuous functions $f : M \to V^{2\beta}$ with the sup-norm

$$\|f\|_\infty = \sup\{\|A^\beta f(v)\| : v \in M\}.$$

Next we define two function classes which are subsets of $C(M, V^{2\beta})$: $\mathcal{F} = \mathcal{F}(\epsilon, \ell)$ and $\mathcal{G} = \mathcal{G}(\epsilon, \ell)$, where the parameters $\epsilon > 0$ and $\ell > 0$ will be chosen later. A vector-valued function f is said to belong to $\mathcal{F}(\epsilon, \ell)$, if $f \in C(M, V^{2\beta})$ and, for each $v \in M$, one has $f(v) \in U^s(v) = \mathcal{R}(P^s(v))$ with $\|A^\beta f(v)\| < \epsilon$, and the restriction of f to each disk $\mathcal{D}_\rho(u_0)$ in M, where $0 < \rho \leq \rho_1$, is Lipschitz continuous with Lipschitz coefficient ℓ. Similarly, a vector-valued function g is said to belong to $\mathcal{G}(\epsilon, \ell)$, if $g \in C(M, V^{2\beta})$ and, for each $v \in M$, one has $g(v) \in U^u(v) = \mathcal{R}(P^u(v))$ with $\|A^\beta g(v)\| \leq \epsilon$, and the restriction of g to each disk $\mathcal{D}_\rho(v)$ in M is Lipschitz continuous with Lipschitz coefficient ℓ. Since $\mathcal{U}^s = \{(v, n) : v \in M,\, n \in U^s(v)\}$ and $\mathcal{U}^u = \{(v, n) : v \in M,\, n \in U^u(v)\}$ are closed subsets of $M \times V^{2\beta}$, see Sacker and Sell (1974, 1976a,b), it follows that for every $\epsilon > 0$ and $\ell > 0$ the spaces

$\mathcal{F}(\epsilon, \ell)$ and $\mathcal{G}(\epsilon, \ell)$ are closed sets in $C(M, V^{2\beta})$. Consequently, the product space $\mathcal{F}(\epsilon, \ell) \times \mathcal{G}(\epsilon, \ell)$ is a complete metric space with the metric

$$\|(f_1, g_1) - (f_2, g_2)\|_\infty \stackrel{\text{def}}{=} \|f_1 - f_2\|_\infty + \|g_1 - g_2\|_\infty,$$

where $(f_i, g_i) \in \mathcal{F} \times \mathcal{G}$, for $i = 1, 2$.

In the argument given below, our objective will be to find $(f, g) \in \mathcal{F} \times \mathcal{G}$ so that the mapping h, which is defined by $h(u) = u + f(u) + g(u)$, for $u \in M$, satisfies the conclusions of the Main Theorem. The pair (f, g) will be found as a fixed point of a suitable mapping A_T.

Let $u_0 \in M$ and $y_0 \in \Omega$ be given, and set $w(t) = y(t, y_0) - S(t)u_0$, with $w_0 = y_0 - u_0$. It follows that $w(t)$ satisfies equation (74.4). Let $\epsilon > 0$ satisfy $\|A^\beta w_0\| < \epsilon$ and G satisfy inequality (74.1). Then with $F_1 = F$ and $F_2 = F + G$, inequality (47.11) implies that there is a constant $K_1 = K_1(t_0) > 0$ such that

$$\|A^\beta w(t)\| \leq K_1(t_0)(\epsilon + \delta), \qquad \text{for } 0 \leq t \leq t_0.$$

If ϵ and δ satisfy $K_2(\epsilon + \delta) \leq \sigma_1 \leq \sigma_2$, where $K_2 = K_1(2T)$, then one can choose t_0 so that $t_0 \geq 2T$ and

$$\|A^\beta w(t)\| \leq K_2(\epsilon + \delta), \qquad \text{for } 0 \leq t \leq 2T. \tag{74.41}$$

Next we recall that $B(t) = DF(S(t)u_0)$, and we define

$$e = e(t) = e(t, y_0) = w(t) - \Phi(u_0, t)w_0, \tag{74.42}$$

where $w(0) = w_0 = y_0 - u_0$. It follows from equation (74.3) that $e(0) = 0$, and $e(t)$ is the mild solution of

$$\partial_t e(t) + Ae(t) = B(t)e(t) + E(S(t)u_0, w(t)) + G(S(t)u_0 + w(t)).$$

As a result of equations (44.13) and (74.4), we see that e satisfies

$$\begin{aligned} e(t) = &\int_0^t e^{-A(t-s)} B(s)e(s)\, ds + \int_0^t e^{-A(t-s)} E(S(s)u_0, w(s))\, ds \\ &+ \int_0^t e^{-A(t-s)} G(S(s)u_0 + w(s))\, ds. \end{aligned} \tag{74.43}$$

Now inequalities (74.1), (71.11), and (74.9) imply that there is a $\gamma \in \Sigma$ such that, for $0 \leq t \leq 2T$, one has

$$\begin{aligned} \|A^\beta e(t)\| \leq M_\beta \|B\|_\infty &\int_0^t (t-s)^{-\beta} e^{-a(t-s)} \|A^\beta e(s)\|\, ds \\ &+ \int_0^t (t-s)^{-\beta} e^{-a(t-s)} \|A^\beta w(s)\| \gamma(\|A^\beta w(s)\|)\, ds + \delta. \end{aligned}$$

From inequality (74.41), one obtains a $\beta_1 \in \Sigma$, where

$$\|A^\beta e(t)\| \le (\epsilon+\delta)\beta_1(\epsilon,\delta) + \delta + M_\beta \|B\|_\infty \int_0^t (t-s)^{-\beta} e^{-a(t-s)} \|A^\beta e(s)\|\, ds,$$

for $0 \le t \le 2T$. It then follows from the Gronwall-Henry inequality that there is a constant $C_1 > 0$ and a $b_0 \in \Sigma$ such that

$$\|A^\beta e(t)\| \le (\epsilon+\delta) b_0(\epsilon,\delta) + C_1\delta, \qquad \text{for } 0 \le t \le 2T. \tag{74.44}$$

Special Notation: For $i = 1, 2$, we define $y_i = y_i(t) = y(t, y_{i0})$, $v_i = v_i(t) = v(t, y_{i0})$, $n_i = n_i(t) = n(t, y_{i0})$, $s_i = s_i(t) = s(t, y_{i0})$, $u_i = u_i(t) = u(t, y_{i0})$, and $S_i = S_i(t) = S(t)v_{i0}$, where $\psi(y_{i0}) = v_{i0}$ with $v_{i0} \in \mathcal{D}_{\rho_1}(v_0)$, for some $v_0 \in M$, and $n_{i0} = y_{i0} - v_{i0} = s_{i0} + u_{i0}$. Also we define

$$\begin{aligned}
\Delta y = \Delta y(t) = y_1 - y_2, &\quad \Delta v = \Delta v(t) = v_1 - v_2,\\
\Delta n = \Delta n(t) = \Delta y - \Delta v, &\quad \Delta S = \Delta S(t) = S_1 - S_2,\\
\Delta w = \Delta w(t) = \Delta y - \Delta S, &\quad \Delta z = \Delta z(t) = \Delta v - \Delta S,\\
\Delta s = \Delta s(t) = s_1 - s_2, \quad \text{and} &\quad \Delta u = \Delta u(t) = u_1 - u_2.
\end{aligned}$$

We let the functions $E_S = E_S(t)$, $E_y = E_y(t)$, $E_v = E_v(t)$, and $E_s = E_s(t)$ be defined, for $0 \le t \le 2T$, by

$$\begin{aligned}
\Delta S(t) &= \Phi(v_{20}, t)\Delta S(0) + E_S(t),\\
\Delta y(t) &= \Phi(v_{20}, t)\Delta y(0) + E_y(t),\\
\Delta v(t) &= \Phi(v_{20}, t)\Delta v(0) + E_v(t),\\
\Delta s(t) &= \Phi(v_{20}, t)P^s(v_{20})\Delta n(0) + E_s(t).
\end{aligned} \tag{74.45}$$

This Special Notation is used in the following result and in the sequel.

Lemma 74.7. *Let the hypotheses of Lemma 74.6 be satisfied, and let $G \in C^1_{\mathrm{Lip}}$. Then there exist σ_0, with $0 < \sigma_0 \le \sigma_1$, $\epsilon_0 > 0$, $\delta_0 = \delta_0(\epsilon) > 0$, and nonnegative constants C_0, C_1, C_2, and C_3, which depend on the characteristics of the exponential trichotomy on M and the time T, such that $C_2 \ge 1$, $2\epsilon_0 \le \sigma_0$, $C_0(\epsilon_0 + \delta_0) \le \min(\rho_1, \sigma_1)$, and whenever $0 < \epsilon \le \epsilon_0$, $0 < \delta \le \delta_0$, and $\|G\|_{\{A;C^1(\Omega)\}} \le \delta$, then the conclusions of Lemma 74.6 hold and the following are valid:*

(1) *For any $v_0 \in M$ and $y_0 = v_0 + n_0$, where $n_0 \subset U^s(v_0) + U^u(v_0)$ and $\|A^\beta n_0\| \le 2\epsilon \le \sigma_0$, one has*

$$\left.\begin{cases}
\|A^\beta(y(t, y_0) - S(t)v_0)\|\\
\|A^\beta(v(t, y_0) - S(t)v_0)\|\\
\frac{1}{2}\|A^\beta n(t, y_0)\|
\end{cases}\right\} \le C_0(\epsilon+\delta), \quad \textit{for } 0 \le t \le 2T. \tag{74.46}$$

Furthermore, $w(t) = y(t, y_0) - S(t)v_0$ *satisfies equation (74.3), with* $w(0) = n_0 = y_0 - v_0$. *Also* $y(t) = y(t, y_0) = v(t) + n(t)$ *satisfies (74.30), and inequality (74.44) holds.*

(2) *Assume that* $v_{i0} \in M$ *with* $\|A^\beta \Delta v(0)\| \le \rho_0$ *and* $\|A^\beta (y_{i0} - v_{i0})\| \le 2\epsilon \le \sigma_0$, *for* $i = 1, 2$. *Then* $y_i(t) = y(t, y_{i0}) \in N(\mathfrak{D}_{\rho_1}(S(t)v_{10}), \sigma_1)$, *and*

$$\left.\begin{cases} \|A^\beta \Delta y(t)\| \\ \|A^\beta \Delta v(t)\| \\ \|A^\beta \Delta n(t)\| \end{cases}\right. \le C_2 \|A^\beta \Delta y(0)\|, \qquad \text{for } 0 \le t \le 2T. \tag{74.47}$$

(3) *Whenever* $\|A^\beta (y_{i0} - v_{i0})\| \le 2\epsilon \le \sigma_0$, *with* $v_{i0} \in M$, *for* $i = 1, 2$, *and* $\|A^\beta \Delta y(0)\| \le C_2^{-1} \rho$, *with* $0 < \rho \le \rho_0$, *then*

(74.48)
$$\|A^\beta \int_0^t \Phi(S(s)v_{20}, t - s) \int_0^1 [DF(y_2 + \theta(y_1 - y_2)) - DF(y_2)]\, d\theta\, \Delta y(s)\, ds\| \le C_3\, b_1^F(\rho, 2\epsilon) \|A^\beta \Delta y(0)\|, \qquad \text{for } 0 \le t \le 2T,$$

where b_1^F *is given by (74.2).*

(4) *There exist* $\beta_1, \beta_2 \in \Sigma$ *such that, for* $0 \le t \le 2T$, *one has*

$$\begin{aligned} &\|A^\beta E_S(t)\| \le b_3^F(\rho) \|A^\beta \Delta v(0)\|, \\ &\|A^\beta E_y(t)\| \le \left(C_3 b_1^F(\rho, 2\epsilon) + \beta_1(\epsilon, \delta)\right) \|A^\beta \Delta y(0)\|, \\ &\|A^\beta E_v(t)\| \le \left(b_4^F(\rho, 2\epsilon) + \beta_2(\epsilon, \delta)\right) \|A^\beta \Delta y(0)\|. \end{aligned} \tag{74.49}$$

(5) *Let* $y_i = y_{i0}$ *and* $v_i = v_{i0}$, $i = 1, 2$, *be given as in Item (2), and that* $\|A^\beta \Delta y(0)\| \le 3 \|A^\beta \Delta v(0)\|$. *Then for* $0 < \epsilon \le \epsilon_0$ *and* $0 < \delta \le \delta_0$, *one has*

$$\frac{1}{2} \|A^\beta \Delta S(t)\| \le \|A^\beta \Delta v(t)\| \le \frac{3}{2} \|A^\beta \Delta S(t)\|, \qquad \text{for } 0 \le t \le 2T. \tag{74.50}$$

Define $H_5(t)$ *by* $\Delta v(t) = \Delta S(t) + H_5(t)$, *for* $t \in [0, 2T]$. *Then there is a* $\beta_3 \in \Sigma$ *such that, for* $0 \le t \le 2T$, *one has*

$$\|A^\beta H_5(t)\| = \|A^\beta \Delta z(t)\| \le \left(b_5^F(\rho, 2\epsilon) + \beta_3(\epsilon, \delta)\right) \|A^\beta \Delta v(0)\|. \tag{74.51}$$

(6) *Under the conditions stated in Item (5), for* $0 \le t \le 2T$, *one has*

$$\frac{1}{8K} e^{\lambda_2 t} \|A^\beta \Delta v(0)\| \le \|A^\beta \Delta v(t)\| \le 6K^2 e^{\lambda_3 t} \|A^\beta \Delta v(0)\|. \tag{74.52}$$

Proof. The specifications of the parameters ϵ_0, δ_0, etc, will be made in the course of the proof. First note that with the exception of the estimate on $z(t) = v(t, y_0) - S(t)u_0$, the proof of Item (1) is given in the argument preceding the statement of the lemma. Indeed, inequality (74.41) implies that $w(t) = y(t, y_0) - S(t)v_0$ satisfies (74.46), with $C_0 \geq K_2$, $2\epsilon_0 \leq \sigma_0$, and $C_0(\epsilon_0 + \delta_0) \leq \min(\rho_1, \sigma_1)$. In order to prove the second inequality in (74.46), we note that $\psi(y(t, y_0)) = v(t, y_0)$ and $\psi(S(t)u_0) = S(t)u_0$, by Lemma 74.4. Hence one has

$$\begin{aligned} ||A^\beta(v(t, y_0) - S(t)u_0)|| &= ||A^\beta(\psi(y(t, y_0)) - \psi(S(t)u_0))|| \\ &\leq L||A^\beta(y(t, y_0) - S(t)u_0)||, \end{aligned}$$

for $0 \leq t \leq 2T$, since ψ is Lipschitz continuous on $N(\mathfrak{D}_{\rho_1}(u_0), \sigma_1)$. By replacing C_0 with $\max(1, L)C_0$, where L is the Lipschitz coefficient of ψ, we see that $z(t)$ satisfies (74.46), as well. This completes the proof of Item (1).

For Item (2), we first note that, since $C_0(\epsilon_0 + \delta_0) \leq \sigma_1$, it follows from Item (1) that $||A^\beta(y(t, y_{i0}) - S(t)v_{i0})|| \leq \sigma_1$, for $0 \leq t \leq 2T$. Since $||A^\beta \Delta S(0)|| \leq \rho_0$, one has $||A^\beta \Delta S(t)|| \leq \rho_1$, by Lemma 74.6. Hence one has $y(t, y_{i0}) \in N(\mathfrak{D}_{\rho_1}(S(t)v_{10}), \sigma_1)$, for $0 \leq t \leq 2T$. The fact that $\Delta y(t)$ satisfies (74.47), for some $C_2 \geq 1$, follows directly from the Lipschitz continuity of mild solutions of equation (70.2), see inequality (47.11), with $F_1 = F + G$ and $F_2 = F$. Since $v(t, y_{i0}) = \psi(y(t, y_{i0}))$, for $i = 1, 2$, the Lipschitz continuity of ψ implies that $\Delta v(t)$ and $\Delta n(t)$ satisfy (74.47), as well, with C_2 replaced by $\max(1, L)C_2$. Since $0 \leq \beta < 1$, the proof of Item (3) then follows from inequalities (74.2), (74.9), (74.32), and (74.47). (Note that $||A^\beta \Delta y(t)|| \leq C_2||A^\beta \Delta y(0)|| \leq \rho$, for $0 \leq t \leq 2T$.)

For Item (4) we note that, since $E_S(t) = H_2(t)$ (see equation (74.34)), the inequality for $||A^\beta E_S(t)||$ in (74.49) follows from inequality (74.35). For the term E_y, we note that $\Delta y = \Delta y(t)$ is a mild solution of the equation

$$\partial_t \Delta y + A\Delta y = F(y_1) - F(y_2) + G(y_1) - G(y_2).$$

We next define $\hat{R} = \hat{R}(t)$ so that the last equation becomes

$$\partial_t \Delta y + A\Delta y = DF(S_2)\Delta y + \hat{R},$$

which is a variation of equation (45.32). Thus one has

$$\begin{aligned} \hat{R} = [DF(y_2) - DF(S_2)]\Delta y + G(y_1) - G(y_2) \\ + \int_0^1 [DF(y_2 + \theta\Delta y) - DF(y_2)]\, d\theta\, \Delta y \end{aligned}$$

and

$$E_y(t) = \int_0^t \Phi(S_2(s), t - s)\hat{R}(s)\, ds.$$

From inequalities (74.9), (74.46), and (74.47) one obtains a $\beta_4 \in \Sigma$ such that for $0 \leq t \leq 2T$, one has

$$\|A^\beta \int_0^t \Phi(S_2(s), t-s)[DF(y_2(s)) - DF(S_2(s))]\Delta y(s)\, ds\| \\ \leq \beta_4(\epsilon, \delta)\|A^\beta \Delta y(0)\|.$$

Next we claim that there is a constant $c_6 = c_6(2T) > 0$ such that

(74.53)
$$\left\| A^\beta \int_0^t \Phi(S_2(s), t-s)[G(y_1(s)) - G(y_2(s))] \right\| ds \leq c_6\, \delta \|A^\beta \Delta y(0)\|,$$

for $0 \leq t \leq 2T$. Indeed, let $r = r(t)$ be defined by

$$r(t) = \int_0^t \Phi(S_2(s), t-s)[G(y_1(s)) - G(y_2(s))]\, ds.$$

From the alternate Variation of Constants Formula (74.4), one finds that

$$r(t) = \int_0^t e^{-A(t-s)}[B(s)r(s) + G(y_1(s)) - G(y_2(s))]\, ds,$$

where $B(s) = DF(S_2(s))$. Now inequalities (74.20) and (74.47) imply that

$$\int_0^t \|A^\beta e^{-A(t-s)}[G(y_1(s)) - G(y_2(s))]\|\, ds \leq C_2\delta\|A^\beta \Delta y(0)\|,$$

for $0 \leq t \leq 2T$. By using inequality (37.11), one has

$$\|A^\beta r(t)\| \leq C_2\delta\|A^\beta \Delta y(0)\| + M_\beta K_1 \int_0^t (t-s)^{-\beta}\|A^\beta r(s)\|\, ds,$$

where $\|DF(u)\|_{\mathcal{L}(V^{2\beta}, W)} \leq K_1$, for all $u \in M$. Inequality (74.53) now follows from the Gronwall-Henry inequality (see Appendix D). By combining (74.53) with the preceeding inequalities given in this proof and with inequality (74.48), we see that the inequality for $\|A^\beta E_y(t)\|$ in (74.49) is valid.

For the term Δv, we will use equations (74.23) and (74.45) and inequality (74.11). It then follows from equation (74.45) that E_v satisfies

$$E_v = P^o(S_2)\Phi(v_{20}, t)\Delta n(0) + P^o(S_2)E_y + [P^o(v_2) - P^o(S_2)]\Delta y + e_3,$$

see Lemma 74.4, Item (5). Since P^o is invariant, one has

$$P^o(S_2)\Phi(v_{20}, t)\Delta n(0) = \Phi(v_{20}, t)P^o(v_{20})\Delta n(0).$$

Since $n = Q^o(v)\,n$, one has

$$\Delta n(0) = Q^o(v_{10})n_{10} - Q^o(v_{20})n_{20} = Q^o(v_{20})\Delta n(0) + [Q^o(v_{10}) - Q^o(v_{20})]n_{10}.$$

Hence, $P^o(v_{20})\Delta n(0) = P^o(v_{20})[Q^o(v_{10}) - Q^o(v_{20})]n_{10}$, and it follows from the Lipschitz continuity of Q^o and inequalities (74.9), (74.12), and Item (1) that there is a constant $c_7 = c_7(2T) > 0$ such that

$$\|A^\beta P^o(S_2)\Phi(v_{20}, t)\Delta n(0)\| \le c_7(\epsilon + \delta)\|A^\beta \Delta y(0)\|, \qquad \text{for } 0 \le t \le 2T.$$

Recall that $\|A^\beta P^o(S_2)E_y(t)\| \le K\|A^\beta E_y(t)\|$, for $t \ge 0$. The continuity of P^o and inequalities (74.46) and (74.47) imply that there is a $\beta_5 \in \Sigma$ such that

$$\|A^\beta [P^o(v_2) - P^o(S_2)]\Delta y(t)\| \le \beta_5(\epsilon, \delta)\|A^\beta \Delta y(0)\|, \qquad \text{for } 0 \le t \le 2T.$$

Using these last three inequalities with inequalities (74.12) and (74.24), and the fact that $\lambda_2 \le 0$, we conclude that the inequality for $\|A^\beta E_v(t)\|$ in (74.49) is valid, as well.

In order to prove Item (5), we note that, since the solutions of (70.2) depend continuously on G in the topology $\mathcal{T}_A^1$, one has

$$\lim_{(\epsilon,\delta)\to(0,0)} y(t, y_0) = S(t)v_0, \qquad \text{in } V^{2\beta},$$

where the limit is uniform for $(v_0, t) \in M \times [0, 2T]$. In addition, one has

$$\begin{aligned} \lim_{(\epsilon,\delta)\to(0,0)} (y(t, y_{10}) - y(t, y_{20})) &= \lim_{(\epsilon,\delta)\to(0,0)} (v(t, y_{10}) - v(t, y_{20})) \\ &= S(t)v_{10} - S(t)v_{20}, \end{aligned} \tag{74.54}$$

in the space $V^{2\beta}$, uniformly for $0 \le t \le 2T$. We claim that, for every $r > 0$, there exist $\epsilon_1 = \epsilon_1(r) > 0$ and $\delta_1 = \delta_1(r) > 0$ such that

$$\frac{1}{2} < \frac{\|A^\beta \Delta v(t)\|}{\|A^\beta \Delta S(t)\|} \le \frac{3}{2}, \qquad \text{for } 0 < \epsilon \le \epsilon_1 \text{ and } 0 < \delta \le \delta_1, \tag{74.55}$$

whenever, $\|A^\beta \Delta v(0)\| \ge r$. Indeed, inequalities (74.37) and (74.47) imply that the middle term in (74.55) is bounded and uniformly continuous, for $0 \le t \le 2T$, on the set of v_{10}, $v_{20} \in M$ with $\|A^\beta \Delta v(0)\| \ge r$. It then follows from equation (74.54) that there exist $\epsilon_1 = \epsilon_1(r) > 0$ and $\delta_1 = \delta_1(r) > 0$ such that inequality (74.55) is valid, which in turn implies that inequality (74.50) holds under the same conditions.

For the sequel we will fix $r = C_2^{-1}\rho_0$, where ρ_0 is given by Lemma 74.6. It remains to verify inequality (74.50), for $\|A^\beta \Delta v(0)\| \le C_2^{-1}\rho$, where $0 < \rho \le \rho_0$. Note that, if $\Delta z = \Delta v - \Delta S$ satisfies

$$\|A^\beta \Delta z(t)\| \le \frac{1}{2}\|A^\beta \Delta S(t)\|, \qquad \text{for } 0 \le t \le 2T, \tag{74.56}$$

then inequality (74.50) is valid. In order to prove inequality (74.56), we note that $\Delta S(0) = \Delta v(0)$ and $z(t) = E_v(t) - E_S(t)$. Let v_{i0} and y_{i0}, for $i = 1, 2$, satisfy

$$\|A^\beta \Delta y(0)\| \leq 3\|A^\beta \Delta v(0)\|. \tag{74.57}$$

It then follows from inequalities (74.40), (74.49), and (74.57) that inequality (74.51) holds with $\beta_3 = 3\beta_2$. Next one chooses $\epsilon_2 = \delta_2 > 0$ so that $\beta_3(\epsilon_2, \epsilon_2) \leq \frac{1}{16K} e^{2\lambda_2 T}$. Since $\lambda_2 \leq 0$, it follows from inequalities (74.25), (74.32), and (74.39) that one has

$$b_4^F(\rho, 2\epsilon) \leq b_5^F(\rho, 2\epsilon) \leq \frac{1}{16K} e^{2\lambda_2 T} \leq \frac{1}{16}, \tag{74.58}$$

for $0 < \rho \leq \rho_0$ and $0 \leq 2\epsilon \leq \min(\sigma_0, 2\epsilon_2) \leq \sigma_1$. Now set ϵ_0 and $\delta_0(\epsilon)$ so that

$$2\epsilon_0 = \min(\sigma_0, C_0^{-1} \min(\rho_1, \sigma_1), 2\epsilon_1(r), 2\epsilon_2) \quad \text{and} \quad \delta_0(\epsilon) = \min(\epsilon, \delta_1(r), \delta_2),$$

for $0 < \epsilon \leq \epsilon_0$. One then has $2\epsilon_0 \leq \sigma_0$ and $C_0(\epsilon_0 + \delta_0) \leq \min(\rho_1, \sigma_1)$. It then follows from inequality (74.37) that inequality (74.56) is valid, for $0 < \epsilon \leq \epsilon_0$ and $0 < \delta \leq \delta_0(\epsilon)$. Finally, inequality (74.52) follows directly from inequalities (74.37) and (74.51). □

For any pair $(f, g) \in \mathcal{F} \times \mathcal{G} = \mathcal{F}(\epsilon, \ell) \times \mathcal{G}(\epsilon, \ell)$, we define $h : M \to V^{2\beta}$ by

$$y_0 = h(u_0) = u_0 + f(u_0) + g(u_0), \qquad \text{for } u_0 \in M.$$

We assume that $0 < \epsilon \leq \min(\rho_0, \frac{1}{3}\sigma_0)$. Since $\|A^\beta(y_0 - u_0)\| \leq 2\epsilon < \sigma_0$, the solution $y(t, y_0)$ of the perturbed equation (70.2) will remain in $N(\mathcal{D}_{\rho_1}(S(t)u_0), \sigma_1)$, for $0 \leq t \leq 2T$. The mapping $v(t, y_0) = \psi(y(t, y_0)) = \psi(u_0, y(t, y_0))$, which is defined by the local coordinate representation and which is valid for $0 \leq t \leq 2T$, admits a well-defined extension

$$v(t, y_0) = \psi^e(y(t, y_0)) = \psi(S(t)u_0, y(t, y_0)),$$

on a larger time interval, as long as $\|A^\beta(y(t, y_0) - S(t)u_0)\| \leq \rho_1$. Furthermore, if t_i satisfies $0 \leq t_i \leq 2T$, for $i = 1, 2$, and $0 \leq t_1 + t_2 \leq 2T$, then one has

$$v(t_1, y(t_2, y_0)) = v(t_1 + t_2, y_0) = v(t_2, y(t_1, y_0)).$$

Indeed, we define $S_2(t)y_0 = y(t, y_0)$ to be the unique mild solution of the perturbed equation (70.2) with $S_2(0)y_0 = y_0$. Since one has $S_2(t_1)S_2(t_2) = S_2(t_1 + t_2) = S_2(t_2)S_2(t_1)$, one obtains

$$\begin{aligned} v(t_1, y(t_2, y_0)) &= \psi^e(S_2(t_1)y(t_2, y_0)) = \psi^e(S_2(t_1)S_2(t_2)y_0) \\ &= \psi^e(S_2(t_1 + t_2)y_0) = v(t_1 + t_2, y_0) \\ &= \psi^e(S_2(t_2)S_2(t_1)y_0) = v(t_2, y(t_1, y_0)). \end{aligned} \tag{74.59}$$

Let $(f, g) \in \mathcal{F} \times \mathcal{G}$ be given and consider the collection of all solutions $y(t, y_0)$ of the perturbed equation (70.2) with $y_0 = h(u_0) \stackrel{\text{def}}{=} u_0 + f(u_0) + g(u_0)$, where $u_0 \in M$. For $0 < \epsilon \leq \epsilon_0$ and $0 \leq \delta \leq \delta_0$, one has $\|A^\beta(y(t, y_0) - S(t)u_0)\| \leq \sigma_1$, for $0 \leq t \leq 2T$. We claim that

$$M_t \stackrel{\text{def}}{=} \{v(t, h(u_0)) : u_0 \in M\} = M, \qquad \text{for each } t \in [0, 2T]. \tag{74.60}$$

Since $v(0, h(u_0)) = u_0$, it follows that $M_0 = M$. For $t > 0$ we note that the mapping $(v_0, t) \to v(t, h(v_0))$ is a continuous mapping of $M \times [0, 2T]$ into M. Since M is compact, it follows that M_t is compact. Furthermore, inequality (74.52) implies that the mapping $v_0 \to v(t, h(v_0))$ is an open mapping. Thus M_t is open. Since M is connected, this implies that $M_t = M$, for all $t \in [0, 2T]$. Furthermore, owing to the choice of ρ_1 (see (74.18)), we see that the mapping $v_0 \to v(t, h(v_0))$ is a homeomorphism of M onto M_t, for each $t \in [0, 2T]$.

7.4.5. Proofs of Theorems. The basic idea in the proofs of the Main Theorem and the Shadow Theorem is to construct a mapping $(f, g; \tau) \to (\bar{f}_\tau, \bar{g}_\tau)$, which is defined on $\mathcal{F} \times \mathcal{G}$, for $T \leq \tau \leq 2T$. We will show that for ϵ and δ small, and for $T \leq \tau \leq 2T$, the following hold:

(1) One has $\|A^\beta \bar{f}_\tau(v)\| \leq \frac{3}{4}\epsilon$. (Lemma 74.8.)
(2) The function $\bar{f}_\tau$ is Lipschitz continuous and in $\mathcal{F}$. (Lemma 74.9.)
(3) The mapping $(f, g) \to \bar{f}_\tau$ is contracting on $\mathcal{F}$. (Lemma 74.10.)
(4) The function $\bar{g}_\tau$ satisfies the same three properties. (Lemmas 74.11, 74.12, and 74.13).
(5) Let (f, g) denote the fixed point of the mapping $A_\tau : (f, g) \to (\bar{f}_\tau, \bar{g}_\tau)$, and set $h(v) = v + f(v) + g(v)$, for $v \in M$. Then $M^G = h(M)$ is an invariant set for the perturbed equation (70.2), and the other properties of the Main Theorem and the Shadow Theorem are valid. (Theorems 74.14 and 74.15.)

For each pair $(f, g) \in \mathcal{F}(\epsilon, \ell) \times \mathcal{G}(\epsilon, \ell)$, where $0 < \ell \leq 1$ we define a new function $\bar{f}_\tau$ by

$$\bar{f}_\tau(v(\tau, y_0)) = P^s(v(\tau, y_0))(y(\tau, y_0) - v(\tau, y_0)), \tag{74.61}$$

for $T \leq \tau \leq 2T$, where $y_0 = u_0 + f(u_0) + g(u_0)$. Note that for each $u_0 \in M$, it follows from (74.29) that

$$\bar{f}_\tau(v(\tau, y_0)) = s(\tau, y_0) \in U^s(v(\tau, y_0)), \qquad \text{for } T \leq \tau \leq 2T,$$

since $U^s(u)$ is the range of the projection $P^s(u)$. equation (74.61) gives the value of $\bar{f}_\tau$ at the point $v(\tau, y_0) \in M$, see Figure 7.6. According to Lemma 74.4 and (74.60), we see that $\bar{f}_\tau$ is well-defined everywhere on M, and the mapping $(u_0, \tau) \to \bar{f}_\tau(u_0)$ is a continuous mapping of $M \times [T, 2T]$ into $V^{2\beta}$.

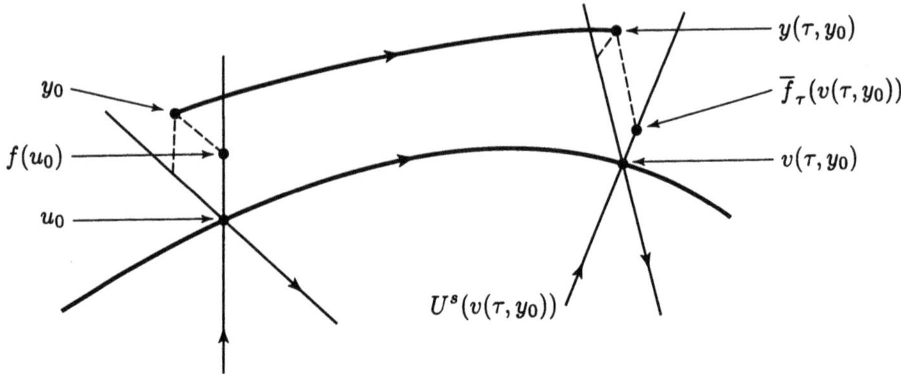

Figure 7.6. The Mapping $f \to \bar{f}_\tau$

Lemma 74.8. *Let the hypotheses of Lemma 74.7 be satisfied, and let ϵ_0 and δ_0 be given by Lemma 74.7. Then there is an ϵ_4, with $0 < \epsilon_4 \leq \epsilon_0$, such that for all ϵ with $0 < \epsilon \leq \epsilon_4$, there is a $\delta_4 = \delta_4(\epsilon)$, with $0 < \delta_4 \leq \delta_0$, such that if $\|G\|_{\{A;C^1(\Omega)\}} \leq \delta = \delta_4$, and if $f \in \mathcal{F}(\epsilon, \ell)$ and $g \in \mathcal{G}(\epsilon, \ell)$, where $0 < \ell \leq 1$, then one has*

$$\|A^\beta \bar{f}_\tau(v)\| \leq \frac{3}{4}\epsilon, \qquad \textit{for all } v \in M \textit{ and } T \leq \tau \leq 2T. \tag{74.62}$$

Proof. According to (74.60), it suffices to verify (74.62) when $v = v(\tau, y_0)$ and $y_0 = u_0 + f(u_0) + g(u_0)$, for some $u_0 \in M$ and $T \leq \tau \leq 2T$. Let $w(t) = y(t, v_0) - S(t)u_0$. From (74.42) one has $w(t) = \Phi(u_0, t)w(0) + e(t)$, for $0 \leq t \leq 2T$, where e satisfies (74.44). Also one has $w(0) = y_0 - u_0 = f(u_0) + g(u_0)$. From the definition of $\bar{f}_\tau$ in (74.61) we have

$$\|A^\beta \bar{f}_\tau(v(\tau, y_0))\| = \|A^\beta P^s(v(\tau, y_0))\,(w(\tau) + S(\tau)u_0 - v(\tau, y_0))\,\|,$$

and from (74.42) we obtain $\|A^\beta \bar{f}_\tau(v(\tau, y_0))\| \leq I_1 + I_2 + I_3 + I_4$, where

$$\begin{aligned}
I_1 &= \|A^\beta P^s(v(\tau, v_0))\Phi(u_0, \tau)f(u_0)\| \\
I_2 &= \|A^\beta P^s(v(\tau, v_0))\Phi(u_0, \tau)g(u_0)\| \\
I_3 &= \|A^\beta P^s(v(\tau, v_0))e(\tau)\| \\
I_4 &= \|A^\beta P^s(v(\tau, y_0))(S(\tau)u_0 - v(\tau, y_0))\|.
\end{aligned}$$

From the invariance property (45.2) we obtain

$$P^s(S(\tau)u_0)\Phi(u_0, \tau)f(u_0) = \Phi(u_0, \tau)f(u_0),$$

and inequalities (45.13) and (74.17) imply that

$$\|A^\beta \Phi(u_0, t)f(u_0)\| \leq Ke^{\lambda_1 \tau}\epsilon < \frac{1}{4}\epsilon, \qquad \text{for } T \leq \tau \leq 2T.$$

Using this fact, the continuity of P^s, and inequality (74.46), we find that there is a $\beta_3 \in \Sigma$ such that

$$I_1 \leq \left\|A^\beta \left[P^s(v(\tau,y_0)) - P^s(S(\tau)u_0)\right]\Phi(u_0,\tau)f(u_0)\right\| + \left\|A^\beta\Phi(u_0,\tau)f(u_0)\right\|$$
$$\leq (4\beta_3(\epsilon,\delta)+1)\|A^\beta\Phi(u_0,\tau)f(u_0)\| \leq \beta_3(\epsilon,\delta)\epsilon + \frac{1}{4}\epsilon.$$

Since the projectors are invariant, $P^s(v)P^u(v) = 0$, and $g(u_0) \in U^u(u_0)$, one obtains $P^s(S(\tau)u_0)\Phi(u_0,\tau)g(u_0) = 0$. Therefore from (74.46) and the continuity of P^s, we find that there is a $\beta_4 \in \Sigma$ such that

$$I_2 = \left\|A^\beta \left(P^s(v(\tau,y_0)) - P^s(S(\tau)u_0)\right)\Phi(u_0,\tau)g(u_0)\right\|$$
$$\leq \beta_4(\epsilon,\delta)\|A^\beta g(u_0)\| \leq \beta_4(\epsilon,\delta)\epsilon.$$

From (74.12) and (74.44) one has $I_3 \leq K((\epsilon+\delta)b_0(\epsilon,\delta) + C_1\delta)$. Lastly, since $v(\tau,y_0) \in \mathcal{D}_{\rho_1}(S(\tau)u_0)$, it follows from (74.15) and (74.46) that

$$I_4 \leq K^2L_0\|A^\beta(S(\tau)u_0 - v(\tau,y_0))\|^2 \leq C_0^2K^2L_0(\epsilon+\delta)^2.$$

From these estimates on I_1, I_2, I_3, and I_4, we find that

$$\|A^\beta \bar{f}_\tau(v(\tau,y_0))\| \leq (\epsilon+\delta)\beta_5(\epsilon,\delta) + \frac{1}{4}\epsilon + KC_1\delta,$$

for some $\beta_5 \in \Sigma$. Next choose $\epsilon_4 > 0$ so that $\beta_5(\epsilon_4,\epsilon_4) \leq \frac{1}{8}$ and $0 < \epsilon_4 \leq \epsilon_0$. Then with

$$\delta_4 = \delta_4(\epsilon) = \min\left(\delta_0, \epsilon_4, \frac{1}{4}\left(KC_1 + \frac{1}{4}\right)^{-1}\epsilon\right), \qquad \text{for } 0 < \epsilon \leq \epsilon_4,$$

inequality (74.62) holds whenever $v \in M$ is of the form $v = v(\tau,y_0)$. □

In the following result, we argue that $\bar{f}_\tau(v)$ is (locally) Lipschitz continuous in v, on each disk $\mathcal{D}_{\rho_0}(u_0) \subset M$.

Lemma 74.9. *Let the hypotheses of Lemma 74.7 be satisfied. Then there is an ϵ_5 with $0 < \epsilon_5 \leq \epsilon_4$ such that for all ϵ with $0 < \epsilon \leq \epsilon_5$, there exist $\delta_5 = \delta_5(\epsilon)$ with $0 < \delta_5 \leq \delta_4$, where ϵ_4 and δ_4 are given in Lemma 74.8, and for all ρ with $0 < \rho \leq \rho_0$, there is an $\ell_5 = \ell_5(\rho,\epsilon)$ with $8KC_2\,\ell_5(\rho_0,\sigma_0) \leq 1$ and $0 \leq \ell_5(\rho,\epsilon) < 1$, such that if $\|G\|_{\{A;C^1(\Omega)\}} \leq \delta \leq \delta_5$ and if $f \in \mathcal{F} = \mathcal{F}(\epsilon,\ell)$ and $g \in \mathcal{G} = \mathcal{G}(\epsilon,\ell)$ with $\ell \leq 1$ and $\epsilon \leq \epsilon_5$, then the restriction of $\bar{f}_\tau$ to the disk $\mathcal{D}_\rho(v_0)$, for $0 < \rho \leq \rho_0$, is Lipschitz continuous with Lipschitz coefficient ℓ_5, for every $v_0 \in M$ and for $T \leq \tau \leq 2T$. Moreover, one has $\bar{f}_\tau \in \mathcal{F}$, and $\ell_5(\rho,\epsilon) \in \Sigma$.*

Proof. Let $(f,g) \in \mathcal{F}\times\mathcal{G}$ be given. We assume that ϵ and δ satisfy $0 < \epsilon \leq \epsilon_4$ and $0 < \delta \leq \delta_4$ so that Lemma 74.8 holds. We will use the Special Notation, where $v_{10} \in \mathcal{D}_\rho(v_{20})$, for $0 < \rho \le \rho_0$, and $y_{i0} = h(v_{i0}) = v_{i0} + f(v_{i0}) + g(v_{i0})$, for $i = 1, 2$. Since $\|A^\beta \Delta y(0)\| \le (1 + 2\ell)\|A^\beta \Delta v(0)\|$, and $\ell \le 1$, we see that inequality (74.57) is valid.

According to (74.60), it will suffice to show that, under the hypotheses stated in this lemma, there is a suitable $\ell_5 = \ell_5(\rho, \epsilon)$ such that

$$\frac{\|A^\beta(\bar{f}_\tau(v_1(\tau)) - \bar{f}_\tau(v_2(\tau)))\|}{\|A^\beta(v_1(\tau) - v_2(\tau))\|} \le \ell_5, \qquad \text{for } T \le \tau \le 2T.$$

Let $\Delta s = \Delta s(t) = \Phi(v_{20}, t)P^s(v_{20})\Delta n(0) + E_s(t)$ be given as in (74.45), where

$$\Delta s = \bar{f}_\tau(v_1) - \bar{f}_\tau(v_2) = P^s(v_1)n_1 - P^s(v_2)n_2, \qquad \text{at } t = \tau.$$

By a straightforward calculation, which makes use of the Special Notation in (74.45) and equation (74.23), one finds that $E_s = E_s(t)$ satisfies

$$\begin{aligned} E_s = {} & [P^s(v_1) - P^s(v_2)]n_1 + [P^s(v_2) - P^s(S_2)]\Delta n \\ & + P^s(S_2)[Q^o(v_2) - Q^o(S_2)]\Delta y + P^s(S_2)E_y \\ & - P^s(S_2)e_3(y_2, \Delta y) + P^s(S_2)\Phi(v_{20}, t)\Delta v(0), \end{aligned}$$

for $0 \le t \le 2T$, see Lemma 74.4, Item (5). By using inequalities (74.11), (74.12), (74.46), (74.47), (74.57), and $K \ge 1$, one finds that each of the terms

$$\|A^\beta(P^s(v_1) - P^s(v_2))n_1\|, \quad \|A^\beta(P^s(v_2) - P^s(S_2))\Delta n\| \\ \text{and} \quad \|A^\beta P^s(S_2)[Q^o(v_2) - Q^o(S_2)]\Delta y\|$$

is bounded by $6C_0C_2KL_0(\epsilon + \delta)\|A^\beta \Delta v(0)\|$. Also from the invariance of the projectors and inequality (74.15), one finds that there is a constant $C_4 = C_4(2T) > 0$ such that

$$\|A^\beta P^s(S_2)\Phi(v_{20}, t)\Delta v(0)\| \le C_4\rho\|A^\beta \Delta v(0)\|.$$

Moreover, Lemma 74.7, (74.12), and (74.24) yield

$$\|A^\beta P^s(S_2)e_3(y_2, \Delta y)\| \le KC_2\, b_2^F(\rho)\|A^\beta \Delta y(0)\|.$$

From inequalities (74.12) and (74.49), we obtain

$$\|A^\beta P^s(S_2)E_y(t)\| \le K\left(C_3\, b_1^F(\rho, 2\epsilon) + \beta_1(\epsilon, \delta)\right)\|A^\beta \Delta y(0)\|,$$

for $0 \le t \le 2T$. Consequently, these estimates and (74.47) imply that there are functions $\beta_4, \beta_5, \beta_6 \in \Sigma$ such that $\beta_5 \le \beta_6$ and

$$\begin{aligned} \|A^\beta E_s\| & \le (C_4\rho + \beta_4(\epsilon, \delta))\|A^\beta \Delta v(0)\| \\ & \quad + \left(b_4^F(\rho, 2\epsilon) + \beta_5(\epsilon, \delta)\right)\|A^\beta \Delta y(0)\| \\ & \le \left(C_4\rho + 3b_4^F(\rho, 2\epsilon) + \beta_6(\epsilon, \delta)\right)\|A^\beta \Delta v(0)\|, \end{aligned} \tag{74.63}$$

for $0 \leq t \leq 2T$. It follows from inequality (45.13) that, for $t \geq 0$, one has

$$
\|A^\beta \Phi(v_{20}, t) P^s(v_{20}) \Delta n(0)\| \leq \begin{cases} Ke^{\lambda_1 t} \|A^\beta P^s(v_{20}) \Delta n(0)\|, \\ Ke^{\lambda_1 t} \|A^\beta \Delta n(0)\|. \end{cases} \tag{74.64}
$$

Since $\|A^\beta \Delta n(0)\| \leq 2\ell \|A^\beta \Delta v(0)\|$, inequalities (74.63) and (74.64) imply that

$$
\|A^\beta \Delta s(t)\| \leq \left(2Ke^{\lambda_1 t}\ell + C_4\rho + 3b_4^F(\rho, 2\epsilon) + \beta_6(\epsilon, \delta)\right) \|A^\beta \Delta v(0)\|,
$$

for $0 \leq t \leq 2T$ and $0 < \rho \leq \rho_0$. By replacing ρ_0 and σ_0 with a smaller values, if necessary, we may assume that

$$
\begin{Bmatrix} 16K^3 C_2 L_0 \, \rho_0 e^{-\lambda_4 T} \\ 48K C_4 e^{-\lambda_2 T} \rho_0 \\ 144K^2 e^{-\lambda_2 T} b_4^F(\rho_0, \sigma_0) \end{Bmatrix} \leq 1. \tag{74.65}
$$

Next we choose ϵ_5 so that $0 < \epsilon_5 \leq \epsilon_4$ and

$$
48Ke^{-\lambda_2 T} \beta_6(\epsilon_5, \epsilon_5) \leq 1. \tag{74.66}
$$

Set $\delta_5(\epsilon) = \min(\epsilon, \delta_4(\epsilon))$, for $0 < \epsilon \leq \epsilon_5$. Since $\ell \leq 1$, it then follows from the last three inequalities and inequalities (74.17) and (74.52) that

$$
\frac{\|A^\beta \Delta s(\tau)\|}{\|A^\beta \Delta v(\tau)\|} \leq \ell_5(\rho, \epsilon) \leq \frac{2}{3}, \qquad \text{for } T \leq \tau \leq 2T,
$$

$0 < \epsilon \leq \epsilon_5$ and $0 < \rho \leq \rho_0$, where

$$
\ell_5 = \ell_5(\rho, \epsilon) = \frac{1}{6}\ell + \left(C_4\rho + 3b_4^F(\rho, 2\epsilon) + \beta_6(\epsilon, \epsilon)\right) 8Ke^{-\lambda_2 T}. \tag{74.67}
$$

Equation (74.67) defines a mapping $\ell \to \ell_5$, where $0 \leq \ell = \ell(\rho, \epsilon) \leq 1$. Since this mapping is a strict contraction in the L^∞-norm, there is a unique fixed point $\ell_5 = \ell$, where $\ell_5 = \frac{5}{6}(C_4\rho + 3b_4^F(\rho, 2\epsilon) + \beta_6(\epsilon, \epsilon)) 8Ke^{-\lambda_2 T}$, i.e., $\ell_5 = \ell_5(\rho, \epsilon) \in \Sigma$. Since $\ell_5 \in \Sigma$, one can choose smaller values of ρ_0 and σ_0, if necessary, to insure that $8KC_2 \, \ell_5(\rho_0, \sigma_0) \leq 1$. □

Lemma 74.10. *Let the hypotheses of Lemma 74.7 be satisfied. Then for every ϵ with $0 < \epsilon \leq \epsilon_5$ and $\delta \leq \delta_5(\epsilon)$, where ϵ_5 and δ_5 are given in Lemma 74.9, if $\|G\|_{\{A; C^1(\Omega)\}} \leq \delta$, then for all $(f_i, g_i) \in \mathcal{F}(\epsilon, \ell) \times \mathcal{G}(\epsilon, \ell)$, where $\ell \leq 1$, one has*

$$
\|A^\beta (\bar{f}_{\tau,1}(u) - \bar{f}_{\tau,2}(u))\| \leq \frac{1}{2} \|(f_1, g_1) - (f_2, g_2)\|_\infty, \qquad \textit{for all } u \in M,
$$

where $\bar{f}_{\tau,i}$ *are given by (74.61), for* $i = 1, 2$.

Proof. Let τ be fixed where $T \leq \tau \leq 2T$. Once again we will use the Special Notation, where now one has

$$v_{10} = v_{20} = v_0 \in M \quad \text{and} \quad y_{i0} = v_0 + f_i(v_0) + g_i(v_0), \qquad \text{for } i = 1, 2.$$

Recall that $\bar{f}_i(v_i(\tau)) = s_i(\tau) = P^s(v_i(\tau))n_i(\tau)$, where $\bar{f}_i = \bar{f}_{\tau,i}$, for $i = 1, 2$.
By Lemma 74.7, the functions $y_i(t)$, $v_i(t)$ and $n_i(t)$, $s_i(t)$, and $u_i(t)$ are Lipschitz continuous functions of the initial data y_{i0}. From the construction of these solutions one has $\Delta v(0) = 0$, $\Delta y(0) = \Delta n(0)$, and

$$\text{(74.68)} \quad \|A^\beta \Delta y(0)\| \leq \|A^\beta \Delta f(v_0)\| + \|A^\beta \Delta g(v_0)\| \leq \|(f_1, g_1) - (f_2, g_2)\|_\infty,$$

where $\Delta f(v_0) = f_1(v_0) - f_2(v_0)$ and $\Delta g(v_0) = g_1(v_0) - g_2(v_0)$. Also note that

$$\text{(74.69)} \qquad P^s(v_2)\Delta s = \Delta \bar{f}(v_2) + P^s(v_2)[\bar{f}_1(v_1) - \bar{f}_1(v_2)], \qquad \text{at } t = \tau,$$

where $\Delta \bar{f}(v_0) = \bar{f}_1(v_0) - \bar{f}_2(v_0)$. From the first inequality in (74.63), one finds that

$$\text{(74.70)} \quad \|A^\beta E_s\| \leq \left(b_4^F(\rho, 2\epsilon) + \beta_5(\epsilon, \epsilon)\right) \|A^\beta \Delta y(0)\|, \qquad \text{for } 0 \leq t \leq 2T.$$

Since $P^s(v_0)\Delta n(0) = \Delta f(v_0)$, inequality (74.64) becomes

$$\|A^\beta \Phi(v_{20}, t) P^s(v_{20}) \Delta n(0)\| \leq K e^{\lambda_1 t} \|A^\beta \Delta f(v_0)\|, \qquad \text{for } t \geq 0.$$

Therefore, from (74.70), one obtains

$$\|A^\beta \Delta s(t)\| \leq K e^{\lambda_1 t} \|A^\beta \Delta f(v_0)\| + \left(b_4^F(\rho, 2\epsilon) + \beta_5(\epsilon, \delta)\right) \|A^\beta \Delta y(0)\|,$$

for $0 \leq t \leq 2T$. Since $K \geq 1$, $\lambda_2 \leq 0$, and $\beta_5 \leq \beta_6$, it then follows from inequalities (74.65), (74.66), (74.68), and (74.69) that, for $t = \tau$, one has:

$$\begin{aligned}
\|A^\beta \Delta \bar{f}(v_2)\| &\leq \|A^\beta P^s(v_2)\Delta s\| + \|A^\beta P^s(v_2)[\bar{f}_1(v_1) - \bar{f}_1(v_2)]\| \\
&\leq K\|A^\beta \Delta s\| + K C_2\, \ell_5(\rho, \epsilon)\|A^\beta \Delta y(0)\| \\
&\leq K^2 e^{\lambda_1 \tau} \|A^\beta \Delta f(v_0)\| \\
&\qquad + \left(K C_2 \ell_5(\rho, \epsilon) + K b_4^F(\rho, 2\epsilon) + K\beta_5(\epsilon, \epsilon)\right) \|A^\beta \Delta y(0)\| \\
&\leq \left(K^2 e^{\lambda_1 \tau} + K C_2\, \ell_5(\rho, \epsilon) + K b_4^F(\rho, 2\epsilon) + K\beta_5(\epsilon, \delta)\right) \\
&\qquad \times \|(f_1, g_1) - (f_2, g_2)\|_\infty.
\end{aligned}$$

From Lemma 74.9 and inequalities (74.17), (74.65), and (74.66), we see that each term: $K^2 e^{\lambda_1 \tau}$, $K C_2\, \ell_5(\rho, \epsilon)$, $K b_4^F(\rho, 2\epsilon)$, and $K\beta_5(\epsilon, \delta)$ is $\leq \frac{1}{8}$. □

In the case where the manifold M is normally hyperbolic and stable, that is, when the projector P^u satisfies $P^u = 0$, then we have finished with the preliminary lemmas. The function $g \in \mathcal{G}$ is fixed to satisfy $g \equiv 0$. In this case, one can proceed directly to Theorem 74.14 below. However, in the general case, we need to extend our journey and to develop analogues of the last three lemmas in order to find a suitable function $g \in \mathcal{G}$. On this portion of the journey we will experience a serious encounter with the rough waters of the infinite dimensional world. Special care is needed for the Good Ship QED in order to overcome some of the difficulties which arise when one tries to reverse time in a semiflow.

Let $(f, g) \in \mathcal{F} \times \mathcal{G}$. We now seek to define a new function $\bar{g}_\tau$, which is a companion to the function $\bar{f}_\tau$ given by (74.61). Among other things, we want $\bar{g}_\tau(u_0)$ to be in $U^u(u_0)$, for every $u_0 \in M$. Let $u_0 \in M$ be given, and define $y_0 = y_0(V) = u_0 + f(u_0) + V$, where $V \in U^u(u_0)$ will be treated as a parameter. Consider the equation

$$g(v(\tau, y_0(V))) = P^u(v(\tau, y_0(V)))(y(\tau, y_0(V)) - v(\tau, y_0(V))). \tag{74.71}$$

Our objective is to show that if ϵ and δ are sufficiently small, then equation (74.71) has a unique solution $V \in U^u(u_0)$. In this case we will denote this solution by $V = \bar{g}_\tau(u_0)$, see Figure 7.7. Before proving this, it is convenient to write equation (74.71) in the abbreviated form $g(v) = P^u(v)(y - v) = P^u(v)n$, where $y = y(\tau, y_0(V))$, $v = v(\tau, y_0(V))$ and $n = y - v$. Note that equation (74.71) holds in the subspace $\mathcal{R}(P^u(v))$. By adding and subtracting the terms $P^u(S)y$, $P^u(S)S$, and $P^u(S)v$ in equation (74.71), where $S = S(\tau)u_0$, we see that equation (74.71) takes on the equivalent form

$$g(v) = P^u(S)(y - S) - P^u(S)(v - S) + (P^u(v) - P^u(S))(y - v). \tag{74.72}$$

Notice that each of the terms y, v, g, and $P^u(v)$ are Lipschitz continuous functions of the parameter V, while the term S does not depend on V.

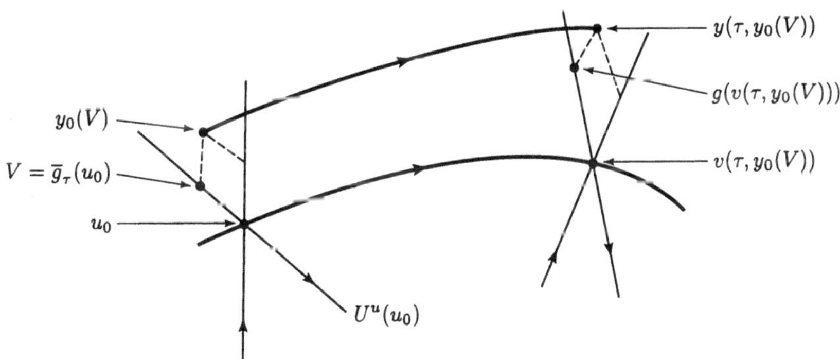

Figure 7.7. The Mapping $g \to \bar{g}_\tau$

Lemma 74.11. *Let the hypotheses of Lemma 74.7 be satisfied. Then there is an* $\epsilon_7 > 0$ *such that* $\epsilon_7 \leq \epsilon_5$ *and for all* ϵ *with* $0 < \epsilon \leq \epsilon_7$, *there is a* $\delta_7 = \delta_7(\epsilon)$ *with* $0 < \delta_7 \leq \delta_5$, *where* ϵ_5 *and* δ_5 *are given in Lemma 74.10, such that if* $\|G\|_{\{A;C^1(\Omega)\}} \leq \delta_7$, *and if* $f \in \mathcal{F}(\epsilon, \ell)$ *and* $g \in \mathcal{G}(\epsilon, \ell)$ *for any* ℓ *with* $0 < \ell \leq 1$, *then for each* $u_0 \in M$ *and* $\tau \in [T, 2T]$, *there is a unique solution* $V = \bar{g}_\tau(u_0)$ *of (74.71), where* $V \in U^u(u_0)$, *with* $\|A^\beta V\| \leq \frac{3}{4}\epsilon$. *Moreover,* $\bar{g}_\tau(u_0)$ *is continuous for* $u_0 \in M$ *and* $T \leq \tau \leq 2T$.

Proof. Let $w = w(t) = y(t, y_0) - S(t)v_0$, where $y_0 = v_0 + f(v_0) + V$, $V \in U^u(v_0)$, and $\|A^\beta V\| \leq \epsilon$. From (74.42) one has $w(t) = \Phi(v_0, t)w(0) + e(t, y_0)$, for $0 \leq t \leq 2T$, where $w(0) = w_0 = f(v_0) + V$. Since the dependence of $e(t, y_0)$ on V is especially important, we will write this as $e(t, y_0) = \hat{e}(V, t)$. Since $f(v_0) \in U^s(v_0)$ and $V \in U^u(v_0)$, one has $P^u(v_0)f(v_0) = 0$ and $P^u(v_0)V = V$. It then follows from the invariance of the projectors that

$$P^u(S)\Phi(v_0, \tau)f(v_0) = \Phi(v_0, \tau)P^u(v_0)f(v_0) = 0,$$
$$P^u(S)\Phi(v_0, \tau)w(0) = \Phi(v_0, \tau)P^u(v_0)w(0) = \Phi(v_0, \tau)V,$$

where $S = S(\tau)v_0$. Thus by applying $P^u(S)$ to (74.42), one obtains

$$P^u(S)(y - S) = \Phi(v_0, \tau)V + P^u(S)\hat{e}(V, \tau), \qquad \text{for } T \leq \tau \leq 2T, \tag{74.73}$$

where $\Phi(v_0, \tau)V = \Phi(v_0, \tau)P^u(v_0)V = P^u(S)\Phi(v_0, \tau)V$, by the invariance of the projectors. Now Lemma 45.2, Item (5), implies that $\Phi(v_0, \tau)P^u(v_0)$ has a unique extension, for all $\tau \in \mathbb{R}$, and the stronger cocycle condition:

$$\Phi(S(\tau)v_0, s)\Phi(v_0, \tau)P^u(v_0) = \Phi(v_0, s + \tau)P^u(v_0), \qquad \text{for all } s,\ \tau \in \mathbb{R},$$

is valid. In particular, with $s = -\tau$, one has

$$\Phi(S, -\tau)P^u(S)\Phi(v_0, \tau) = \Phi(S, -\tau)\Phi(v_0, \tau)P^u(v_0) = P^u(v_0).$$

Consequently one has $\Phi(S, -\tau)P^u(S)\Phi(v_0, \tau)V = P^u(v_0)V = V$. Next we multiply equation (74.72) on the left with $\Phi(S, -\tau)P^u(S)$, use the last equality and equation (74.73), and thereby rewrite equation (74.72) in the form

$$V = \Gamma(V) = \Gamma(V, v_0), \tag{74.74}$$

where $v_0 \in M$, $V \in U^u(v_0)$, $(f, g) \in \mathcal{F} \times \mathcal{G}$, and
(74.75)

$$\Gamma(V, v_0) = \Gamma(V, v_0; f, g)$$
$$\stackrel{\text{def}}{=} \Phi(S, -\tau)P^u(S)\left[-\hat{e}(V, \tau) + (v - S) + [P^u(S) - P^u(v)]n + g(v)\right].$$

Notice that the function $\Gamma(V, v_0)$ defined by equation (74.75) is well-defined on

$$\mathcal{D}(\Gamma) \stackrel{\text{def}}{=} \{(V, v_0) \in V^{2\beta} \times M : V \in \mathcal{R}(P^u(v_0)) \text{ and } \|A^\beta V\| \leq \epsilon\}.$$

Also note that the range of Γ lies in $U^u(v_0)$, for each $v_0 \in M$. Indeed, the invariance of the projectors implies that $P^u(v_0)\Phi(S, -\tau) = \Phi(S, -\tau)P^u(S)$, which in turn, implies that $P^u(v_0)\Gamma(V, v_0) = \Gamma(V, v_0)$, for all $(V, v_0) \in \mathcal{D}(\Gamma)$. (The reader should verify that one has $V = \Gamma(V, v_0)$ if and only if $g(v) = P^u(v)n$.)

Since $w(t) = w(t, y_0(V)) = y(t, y_0(V)) - S(t)v_0$ is a solution of equation (74.5), it follows from (74.42) that

$$\hat{e}(V, t) = \int_0^t \Phi(S(s)v_0, t - s)H(S(s)v_0, w(s))\, ds.$$

Inequality (74.44) implies that

$$\|A^\beta \hat{e}(V, t)\| \leq (\epsilon + \delta)b_0(\epsilon, \delta) + C_1\delta, \qquad \text{for all } (V, v_0) \in \mathcal{D}(\Gamma), \tag{74.76}$$

and $0 \leq t \leq \tau$. Since $w(s)$ is Lipschitz continuous in V, it follows from inequalities (71.10) and (74.20) that $\hat{e}(V, t)$ is Lipschitz continuous in V, as well. We claim that there is a function $\beta_3 \in \Sigma$ such that for $0 \leq t \leq \tau$, one has

$$\|A^\beta(\hat{e}(V_1, t) - \hat{e}(V_2, t))\| \leq \beta_3(\epsilon, \delta)\|A^\beta(V_1 - V_2)\|, \tag{74.77}$$

where (V_i, v_0) lie in $\mathcal{D}(\Gamma)$, for $i = 1, 2$. In order to prove inequality (74.77), we return to the formulation of $e(t)$ given in equation (74.42). For $(V_i, v_0) \in \mathcal{D}(\Gamma)$ we set $y_{i0} = v_0 + f(v_0) + V_i$, $w_{i0} = f(v_0) + V_i$, $w_i(t) = w(t, y_{i0})$, and

$$\hat{e}(V_i, t) = \hat{e}_i(t) = w_i(t) - \Phi(v_0, t)w_i(0), \qquad \text{for } i = 1, 2.$$

Set $m = m(t) = \hat{e}_1(t) - \hat{e}_2(t)$. Then (74.43) implies that m satisfies

$$m(t) = \int_0^t e^{-A(t-s)}B(s)m(s)\, ds + \int_0^t e^{-A(t-s)}[E_1(s) - E_2(s)]\, ds + \int_0^t e^{-A(t-s)}[G_1(s) - G_2(s)]\, ds,$$

where $E_i(s) = E(S(s)v_0, w_i(s))$ and $G_i(s) = G(S(s)v_0 + w_i(s))$, for $i = 1, 2$. Next it follows from inequalities (37.11), (71.10), (74.20), and (74.76) that there is a constant $K_7 > 0$ and a $\beta_4 \in \Sigma$ such that

$$\|A^\beta m(t)\| \leq K_7 \int_0^t (t - s)^{-\beta} e^{-a(t-s)}\|A^\beta m(s)\|\, ds + \beta_4(\epsilon, \delta)\|A^\beta(V_1 - V_2)\|.$$

This fact, with the Gronwall-Henry inequality, implies inequality (74.77).

The next step in the argument is to show that $\Gamma(V, v_0)$, the right side of equation (74.74), is a contraction in the V-variable under the conditions stated in this lemma. In particular, we will now show that for small ϵ and δ, the following two properties are valid:

(1) For any $(V, v_0) \in \mathcal{D}(\Gamma)$ one has $\|A^\beta\Gamma(V, v_0)\| \leq \frac{3}{4}\epsilon$.
(2) One has $\|A^\beta(\Gamma(V_1, v_0) - \Gamma(V_2, v_0))\| \leq \frac{1}{2}\|A^\beta(V_1 - V_2)\|$, for any $(V_i, v_0) \in \mathcal{D}(\Gamma)$, for $i = 1, 2$.

Since one has $g(v) \in U^u(v)$, it follows that $P^u(v)g(v) = g(v)$, and therefore

$$g(v) = P^u(S)g(v) + [P^u(v) - P^u(S)]g(v).$$

By applying $\Phi(S, -\tau)P^u(S)$ to the last equation and using the continuity of P^u and inequalities (45.14), (74.17), and (74.46), we find that there exists a $\beta_5 \in \Sigma$ such that

$$\begin{aligned} \|A^\beta \Phi(S, -\tau)P^u(S)g(v)\| &\leq Ke^{-\lambda_4\tau}\|A^\beta g(v)\| \\ &\quad + \|A^\beta \Phi(S, -\tau)P^u(S)[P^u(v) - P^u(S)]g(v)\| \\ &\leq \left(\frac{1}{4} + \beta_5(\epsilon, \delta)\right)\epsilon. \end{aligned}$$

Since $\|A^\beta V\| \leq \epsilon$, it follows from inequalities (45.14), (74.15), (74.17), and (74.46) that

$$\begin{aligned} \|A^\beta \Phi(S, -\tau)P^u(S)(v - S)\| &\leq Ke^{-\lambda_4\tau}\|A^\beta P^u(S)(v - S)\| \\ &\leq \frac{1}{4}K^2 L_0\|A^\beta(v - S)\|^2 \leq \frac{1}{4}K^2 L_0 C_0^2(\epsilon + \delta)^2. \end{aligned}$$

The continuity of P^u, Lemma 74.7 and inequality (74.76) imply that there is a constant $C_5 > 0$ and a $\beta_6 \in \Sigma$, such that

$$\begin{aligned} \|A^\beta \Phi(S, -\tau)P^u(S)[P^u(v) - P^u(S)]n\| &\leq (\epsilon + \delta)\beta_6(\epsilon, \delta), \\ \|A^\beta \Phi(S, -\tau)P^u(S)\hat{e}(V, \tau)\| &\leq (\epsilon + \delta)\beta_6(\epsilon, \delta) + C_5\delta. \end{aligned}$$

By putting these estimates together, one finds a $\beta_7 \in \Sigma$ such that

$$\|A^\beta \Gamma(V, v_0)\| \leq \frac{1}{4}\epsilon + C_5\delta + (\epsilon + \delta)\beta_7(\epsilon, \delta), \qquad \text{for all } (V, v_0) \in \mathfrak{D}(\Gamma).$$

Now choose $\bar{\epsilon}$ so that $0 < \bar{\epsilon} \leq \epsilon_5$, where ϵ_5 is given in Lemma 74.9, so that $\beta_7(\bar{\epsilon}, \bar{\epsilon}) \leq \frac{1}{8}$, and set $\bar{\delta}(\epsilon) = \min(\epsilon, \frac{1}{4C_5}\epsilon, \delta_5(\epsilon))$, where δ_5 is given in Lemma 74.9. Then with $\delta \leq \bar{\delta}$ and $0 < \epsilon \leq \bar{\epsilon}$, one has $\|A^\beta \Gamma(V, v_0)\| \leq \frac{3}{4}\epsilon$.

In order to prove the Lipschitz property for $\Gamma(V)$, we let $(V_i, v_0) \in \mathfrak{D}(\Gamma)$, for $i = 1, 2$. Next we define $S = S(\tau)v_0$, $y_i = y(\tau, y_0(V_i))$, $v_i = v(\tau, y_0(V_i))$, $\Gamma_i = \Gamma(V_i, v_0)$, and $g_i = g(v_i)$, for $i = 1, 2$. In this case one has $\Delta y(0) = \Delta n(0) = V_1 - V_2$, $\Delta v(0) = 0$, and $\Delta S(t) = 0$. In order to estimate $\|A^\beta(\Gamma(V_1, v_0) - \Gamma(V_2, v_0))\|$, we note that (74.75) implies that

$$\begin{aligned} \Gamma_1 - \Gamma_2 = \Phi(S, -\tau)P^u(S)[&-(\hat{e}_1 - \hat{e}_1) + \Delta v + \Delta g \\ &+ [P^u(S) - P^u(v_2)]\Delta n + [P^u(v_2) - P^u(v_1)]n_1], \end{aligned}$$

where $\hat{e}_i = \hat{e}(V_i, \tau)$, for $i = 1, 2$, $\Delta n = n_1 - n_2$, $\Delta v = v_1 - v_2$, and $\Delta g = g_1 - g_2$. From (74.77) there is a constant $c_5 > 0$ such that

$$\|A^\beta \Phi(S, -\tau)P^u(S)(\hat{e}_1 - \hat{e}_2)\| \leq c_5\beta_3(\epsilon, \delta)\|A^\beta(V_1 - V_2)\|.$$

Also, there is a $\beta_8 \in \Sigma$ such that

$$\left\{ \begin{array}{l} \|A^\beta \Phi(S,-\tau)P^u(S)[P^u(S) - P^u(v_2)]\Delta n\| \\ \|A^\beta \Phi(S,-\tau)P^u(S)[P^u(v_2) - P^u(v_1)]n_1\| \end{array} \right. \leq \beta_8(\epsilon,\delta)\|A^\beta(V_1 - V_2)\|.$$

Since $\ell \leq 1$, one has $\|A^\beta \Delta g\| \leq \|A^\beta \Delta v\|$. From inequalities (45.14) and (74.17), one has

$$\|A^\beta \Phi(S,-\tau)P^u(S)[\Delta v + \Delta g]\|| \leq 2Ke^{-\lambda_4 \tau}\|A^\beta \Delta v(\tau)\| \leq \|A^\beta \Delta v(\tau)\|.$$

From (74.45) one has $\Delta v(t) = E_v(t)$, since $\Delta v(0) = 0$. It then follows from inequality (74.49) that

$$\|A^\beta \Delta v(\tau)\| = \|A^\beta E_v(\tau)\| \leq (b_4^F(\rho, 2\epsilon) + \beta_2(\epsilon,\delta))\|A^\beta(V_1 - V_2)\|.$$

As a result, we find that there is a $\beta_9 \in \Sigma$ such that

$$\|A^\beta(\Gamma(V_1,v_0) - \Gamma(V_2,v_0))\| \leq (b_4^F(\rho,2\epsilon) + \beta_9(\epsilon,\delta))\|A^\beta(V_1 - V_2)\|.$$

Next we let ϵ_7 be chosen so that $0 < \epsilon_7 \leq \bar{\epsilon}$ and $\beta_9(\epsilon_7,\epsilon_7) \leq \frac{3}{8}$. Then set $\delta_7(\epsilon) = \bar{\delta}(\epsilon))$, for $0 < \epsilon \leq \epsilon_7$. Since inequality (74.65) implies that $b_4^F(\rho, 2\epsilon) \leq \frac{1}{144}$, one obtains

$$\|A^\beta(\Gamma(V_1,v_0) - \Gamma(V_2,v_0))\| \leq \frac{1}{2}\|A^\beta(V_1 - V_2)\|,$$

for all $(V_i, v_0) \in \mathcal{D}(\Gamma)$, for $i = 1,2$.

Finally the function $\bar{g}_\tau(v_0)$ is a continuous function of $v_0 \in M$, because: (1) the mapping $(V,v_0) \to \Gamma(V,v_0)$ given by equation (74.75) is continuous on $\mathcal{D}(\Gamma)$; and (2) the value $\bar{g}_\tau(v_0)$ is the unique fixed point of a contraction mapping. Similarly, one can show that $\bar{g}_\tau(v_0)$ is jointly continuous in v_0 and τ. □

In the next two lemmas, we show that $\bar{g}_\tau(v_0)$ is locally Lipschitz continuous in v_0 and that the mapping $(f,g) \to \bar{g}_\tau$ is contracting.

Lemma 74.12. *Let the hypotheses of the Main Theorem be satisfied. Then there is an ϵ_8 with $0 < \epsilon_8 \leq \epsilon_7$ such that for all ϵ with $0 < \epsilon \leq \epsilon_8$, there exists a $\delta_8 = \delta_8(\epsilon)$ with $0 < \delta_8 \leq \delta_7$, where ϵ_7 and δ_7 are given in Lemma 74.11, and, for $0 < \rho \leq \rho_0$, there exists an $\ell_8 = \ell_8(\rho,\epsilon) > 0$ with $16KC_2\ell_8(\rho_0,\sigma_0) \leq 1$ and $\ell_8(\rho,\epsilon) < 1$, such that if $G \in C^1_{\text{Lip}}$ satisfies $\|G\|_{\{A;C^1(\Omega)\}} \leq \delta_8$ and if $f \in \mathcal{F} = \mathcal{F}(\epsilon,\ell)$ and $g \in \mathcal{G} = \mathcal{G}(\epsilon,\ell)$, with $\ell \leq 1$ and $\epsilon \leq \epsilon_8$, then the restriction of g_τ to the disk $\mathcal{D}_\rho(v_0)$ is Lipschitz continuous with Lipschitz coefficient ℓ_8, for every $v_0 \in M$. Moreover, one has $\bar{g}_\tau \in \mathcal{G}$ and $\ell_8(\rho,\epsilon) \in \Sigma$.*

Proof. Let $(f,g) \in \mathcal{F} \times \mathcal{G}$ be given as in the statement of the lemma, and let τ be fixed, where $T \leq \tau \leq 2T$. While the proof of the Lipschitz continuity of $\bar{g}_\tau(v_0)$ with respect to v_0 has some similarities to the argument of the Lipschitz continuity of $\bar{f}_\tau(v_0)$ used in Lemma 74.9, one encounters some new issues which arise when one needs to study the behavior of a semiflow for time $t < 0$.

For $i = 1, 2$, we let $\hat{v}_{i0} \in M$ be given, where $\|A^\beta \Delta\hat{v}(0)\| \le \rho$, for some ρ with $0 < \rho \le \rho_0$, and $\hat{v}(0) = \hat{v}_{10} - \hat{v}_{20}$. We define $\hat{y}_{i0}$ by $\hat{y}_{i0} = \hat{v}_{i0} + f(\hat{v}_{i0}) + V_i$, where $V_i = \bar{g}_\tau(\hat{v}_{i0})$ is the solution of equation (74.71) given by Lemma 74.11. We define $\Delta\hat{y}(0) = \hat{y}_{10} - \hat{y}_{20}$, where $y(t, \hat{y}_{i0})$ is the solution of equation (70.2) through $\hat{y}_{i0}$, and we set $y_{i0} = y(\tau, \hat{y}_{i0})$ and $y_{i0} = v_{i0} + n_{i0}$ (see equation (74.30)), for $i = 1, 2$. Like the situation described in (45.23), the solution of equation (70.2) passing through y_{i0} admits a (partial) negative continuation, for $-\tau \le t \le 0$. We will denote this negative continuation by $y_i = y_i(t) = y(t, y_{i0})$, for $-\tau \le t \le 0$, and one has

$$y_i(t) = y(t, y_{i0}) = y(\tau + t, \hat{y}_{i0}), \qquad \text{for } -\tau \le t \le 0,$$

and $y_i(-\tau) = y(-\tau, y_{i0}) = \hat{y}_{i0}$. The decomposition $y(t, y_{i0}) = v(t, y_{i0}) + n(t, y_{i0})$ given by equation (74.30) is valid for $-\tau \le t \le 0$, and one has

$$v_i(t) = v(t, y_{i0}) = v(\tau + t, \hat{y}_{i0}), \quad \text{and} \quad n_i(t) = n(t, y_{i0}) = n(\tau + t, \hat{y}_{i0}),$$

for $-\tau \le t \le 0$. Similarly, we consider the solution $S_i(t) \stackrel{\text{def}}{=} S(\tau + t)\hat{v}_{i0}$ of equation (70.1), for $-\tau \le t \le 0$, and we set $\Delta S(t) = S_1(t) - S_2(t)$. It follows from inequality (74.37) that, at $t = 0$, one has

$$\|A^\beta \Delta S(0)\| = \|A^\beta (S(\tau)\hat{v}_{10} - S(\tau)\hat{v}_{20})\| \le 4K^2 e^{\lambda_3 \tau} \|A^\beta \Delta\hat{v}(0)\|. \tag{74.78}$$

We will use the Special Notation (see (74.45)), where $\Delta n(t) = \Delta u(t) + \Delta s(t)$, $\Delta z(t) = \Delta v(t) - \Delta S(t)$, and $z_i = v_i - S_i$ are thus defined for $-\tau \le t \le 0$. Note that

$$\begin{aligned} \Delta u(-\tau) = V_1 - V_2,\ \Delta u(0) = g(v_{10}) - g(v_{20}),\ \text{and} \\ \Delta s(0) = \bar{f}(v_{10}) - \bar{f}(v_{20}), \end{aligned} \tag{74.79}$$

where $\bar{f}(v_{i0}) = \bar{f}_\tau(v_{i0})$, for $i = 1, 2$, is given by Lemma 74.8. According to Lemma 74.9, one has

$$\|A^\beta \Delta n(0)\| \le (\ell + \ell_5)\|A^\beta \Delta v(0)\|, \qquad \text{where } \ell_5 = \ell_5(\rho, \epsilon) < 1. \tag{74.80}$$

The remainder of the argument now follows the pattern used in the proof of Lemma 74.9. We define $E_u = E_u(t)$ by the equation

$$\Delta u = \Delta u(t) = \Phi(S(\tau)\hat{v}_{20}, t) P^u(S(\tau)\hat{v}_{20}) \Delta n(0) + E_u(t), \tag{74.81}$$

for $-\tau \le t \le 0$. As argued in Lemma 74.9, in the case of Δs and E_s, one has

$$\Delta u(t) = P^u(v_1(t))n_1(t) - P^u(v_2(t))n_2(t), \qquad \text{for } -\tau \le t \le 0,$$

and

$$E_u(t) = E_{1u}(t) + P^u(S_2)\Phi(S(\tau)\hat{v}_{20}, t)\Delta v(0),$$

where $E_{1u} = E_{1u}(t)$ satisfies

$$E_{1u} = [P^u(v_1) - P^u(v_2)]n_1 + [P^u(v_2) - P^u(S_2)]\Delta n + P^u(S_2)E_y + P^u(S_2)[Q^o(v_2) - Q^o(S_2)]\Delta y - P^u(S_2)e_3(y_2, \Delta y),$$

for $-\tau \le t \le 0$, see Lemma 74.4, Item (5). Also, there is a function $\beta_3 \in \Sigma$ such that

$$\|A^\beta E_{1u}(t)\| \le \left(b_4^F(\rho, 2\epsilon) + \beta_3(\epsilon, \delta)\right) \|A^\beta \Delta\hat{y}(0)\|, \qquad \text{for } -\tau \le t \le 0.$$

From the invariance of the projectors and inequalities (45.14) and (74.15), one obtains

$$\begin{aligned}\|A^\beta P^u(S_2)\Phi(S(\tau)\hat{v}_{20}, t)\Delta v(0)\| &\le K e^{\lambda_4 t}\|A^\beta P^u(S(\tau)\hat{v}_{20})\Delta v(0)\| \\ &\le K^3 L_0 \rho\, e^{\lambda_4 t}\|A^\beta \Delta v(0)\|,\end{aligned}$$

for $-\tau \le t \le 0$. It follows from inequality (45.14) and equation (74.80) that

$$\|A^\beta \Phi(S(\tau)\hat{v}_{20}, t)P^u(S(\tau)\hat{v}_{20})\Delta n(0)\| \le K e^{\lambda_4 t}(\ell + \ell_5)\|A^\beta \Delta v(0)\|,$$

for $t \le 0$. Thus, for $-\tau \le t \le 0$, one has

$$\begin{aligned}\|A^\beta E_u\| \le\, & K^3 L_0\, \rho\, e^{\lambda_4 t}\|A^\beta \Delta v(0)\| \\ & + \left(b_4^F(\rho, 2\epsilon) + \beta_3(\epsilon, \delta)\right) \|A^\beta \Delta\hat{y}(0)\|,\end{aligned} \tag{74.82}$$

and it follows from equations (74.79) and (74.81) and the last three inequalities that at $t = -\tau$, one has

$$\begin{aligned}\|A^\beta (V_1 - V_2)\| \le\, & K e^{-\lambda_4 \tau}(\ell + \ell_5 + K^2 L_0 \rho)\|A^\beta \Delta v(0)\| \\ & + \left(b_4^F(\rho, 2\epsilon) + \beta_3(\epsilon, \delta)\right) \|A^\beta \Delta\hat{y}(0)\|.\end{aligned} \tag{74.83}$$

Note that, since $\ell \le 1$, one has

$$\|A^\beta \Delta\hat{y}(0)\| \le 2\|A^\beta \Delta\hat{v}(0)\| + \|A^\beta (V_1 - V_2)\|. \tag{74.84}$$

In Lemma 74.7 we derived inequality (74.56) under the assumption that inequality (74.57) is valid. Our next goal is to study the analogue of inequality

(74.56), for $-\tau \le t \le 0$, where (74.84) now holds in place of (74.57). Since $\Delta z(-\tau) = \Delta v(-\tau) - \Delta S(-\tau) = 0$, one has $\Delta z(t) = E_v(t) - E_S(t)$, for $-\tau \le t \le 0$. It then follows from inequality (74.49) that

$$\begin{aligned}\|A^\beta \Delta z(t)\| &\le \|A^\beta E_v(t)\| + \|A^\beta E_S(t)\| \\ &\le b_3^F(\rho)\|A^\beta \Delta \hat{v}(0)\| + \left(b_4^F(\rho, 2\epsilon) + \beta_2(\epsilon, \delta)\right) \|A^\beta \Delta \hat{y}(0)\|,\end{aligned}$$

for $-\tau \le t \le 0$. It then follows from (74.40), (74.84), and the last inequality that

$$\begin{aligned}\|A^\beta \Delta z(t)\| \le & \left(b_5^F(\rho, 2\epsilon) + 2\beta_2(\epsilon, \delta)\right) \|A^\beta \Delta \hat{v}(0)\| \\ & + \left(b_4^F(\rho, 2\epsilon) + \beta_2(\epsilon, \delta)\right) \|A^\beta (V_1 - V_2)\|,\end{aligned}$$

for $-\tau \le t \le 0$. By using the last inequality with inequality (74.78) and the definition of $\Delta z(t)$ at $t = 0$, one finds that

$$\begin{aligned}\|A^\beta \Delta v(0)\| &\le \|A^\beta \Delta S(0)\| + \|A^\beta \Delta z(0)\| \\ &\le \left(4K^2 e^{\lambda_3 \tau} + b_5^F(\rho, 2\epsilon) + 2\beta_2(\epsilon, \delta)\right) \|A^\beta \Delta \hat{v}(0)\| \\ &\quad + \left(b_4^F(\rho, 2\epsilon) + \beta_2(\epsilon, \delta)\right) \|A^\beta (V_1 - V_2)\|.\end{aligned} \tag{74.85}$$

Next we note that $Ke^{-\lambda_4 \tau}(\ell + \ell_5 + K^2 L_0 \rho) \le \frac{9}{64}$, owing to inequalities (74.17), $\ell + \ell_5 < 2$, and $4K^2 L_0 \rho \le 1$ (see Lemma 74.2). As a result, inequalities (74.83), (74.84), and (74.85) imply that, for some $\beta_4 \in \Sigma$, one has

$$\begin{aligned}\|A^\beta (V_1 - V_2)\| \le & Ke^{-\lambda_4 \tau}(\ell + \ell_5 + K^2 L_0 \rho) \\ & \times \left(4K^2 e^{\lambda_3 \tau} + b_5^F(\rho, 2\epsilon) + 2\beta_2(\epsilon, \delta)\right) \|A^\beta \Delta \hat{v}(0)\| \\ & + \left(2b_4^F(\rho, 2\epsilon) + 2\beta_3(\epsilon, \delta)\right) \|A^\beta \Delta \hat{v}(0)\| \\ & + \left(2b_4^F(\rho, 2\epsilon) + \beta_4(\epsilon, \delta)\right) \|A^\beta (V_1 - V_2)\|.\end{aligned}$$

Next we impose the first of two conditions on ϵ_8 and $\delta_8(\epsilon)$ by requiring that $0 < \epsilon_8 \le \epsilon_7$, $\beta_4(\epsilon_8, \epsilon_8) \le \frac{1}{8}$, and $\delta_8(\epsilon) = \min(\epsilon, \delta_7(\epsilon))$, for $0 < \epsilon \le \epsilon_8$. From inequality (74.58), one then has $2b_4^F(\rho, 2\epsilon) + \beta_4(\epsilon, \delta) \le \frac{1}{2}$, for $0 < \delta \le \delta_8$, which in turn implies that, for some $\beta_5 \in \Sigma$, one has

$$\|A^\beta (V_1 - V_2)\| \le \ell_8(\rho, \epsilon)\|A^\beta \Delta \hat{v}(0)\|,$$

where

$$\ell_8(\rho, \epsilon) = 8K^3 e^{-(\lambda_4 - \lambda_3)\tau}(\ell + \ell_5(\rho, \epsilon) + K^2 L_0 \rho) + 2b_5^F(\rho, 2\epsilon) + \beta_5(\epsilon, \delta), \tag{74.86}$$

see equation (74.40). From inequality (74.58), one has $2b_5^F(\rho, 2\epsilon) \le \frac{1}{8}$. Since $(\ell + \ell_5 + K^2 L_0 \rho) < 3$, inequality (74.17) yields

$$8K^3 e^{-(\lambda_4 - \lambda_3)\tau}(\ell + \ell_5 + K^2 L_0 \rho) < \frac{1}{2}.$$

The second condition on ϵ_8 and $\delta_8(\epsilon)$ is that we require that $\beta_5(\epsilon_8, \epsilon_8) \leq \frac{3}{8}$. One then obtains $\ell_8(\rho, \epsilon) < 1$. Finally, one can treat $\ell = \ell_8$ as the (unique) fixed point for equation (74.86), in which case, we conclude that $\ell_8 \in \Sigma$. Since $\ell_8 \in \Sigma$, we obtain $16KC_2\,\ell_8(\rho_0, \sigma_0) \leq 1$, by choosing smaller values of ρ_0 and σ_0, if necessary. □

In the following result we derive the contracting property for the mapping $(f, g) \to \bar{g}_\tau$.

Lemma 74.13. *Let the hypotheses of the Main Theorem be satisfied. Then there is an ϵ_9 with $0 < \epsilon_9 \leq \epsilon_8$ and for every ϵ with $0 < \epsilon \leq \epsilon_9$ there is a $\delta_9 = \delta_9(\epsilon) \leq \delta_8(\epsilon)$, where ϵ_8 and δ_8 are given in Lemma 74.12, such that if $G \in C^1_{\mathrm{Lip}}$ satisfies $\|G\|_{\{A;C^1(\Omega)\}} \leq \delta_9$, then for all $(f_i, g_i) \in \mathcal{F}(\epsilon, \ell) \times \mathcal{G}(\epsilon, \ell)$, where $\ell \leq 1$, and $\bar{g}_{\tau,i}$ is given by Lemma 74.11, with (f, g) replaced by (f_i, g_i), for $i = 1, 2$. One then has*

$$\|A^\beta(\bar{g}_{\tau,1}(u) - \bar{g}_{\tau,2}(u))\| \leq \frac{1}{2}\|(f_1, g_1) - (f_2, g_2)\|_\infty \qquad \textit{for all } u \in M.$$

Proof. Once again we will use the Special Notation (74.45), as well as the notation of Lemma 74.12. The reader should notice that, with some minor modifications, which arise owing to the perennial issue of the time-reversibility of semiflows, the argument follows the methodology of Lemma 74.10.

Let $(f_i, g_i) \in \mathcal{F} \times \mathcal{G}$ be given, for $i = 1, 2$, and fix τ so that $T \leq \tau \leq 2T$. Let $v_0 \in M$ be fixed and set $\hat{v}_{i0} = v_0$, $\hat{y}_{i0} = v_0 + f_i(v_0) + V_i$, where $V_i = \bar{g}_{\tau,i}(v_0)$ is the solution of equation (74.71), with (f, g) replaced by (f_i, g_i), given by Lemma 74.11, so that one has

$$g(v(\tau, \hat{y}_{i0})) = P^u(v(\tau, \hat{y}_{i0}))n(\tau, \hat{y}_{i0}), \qquad \text{for } i = 1, 2.$$

In the notation of Lemma 74.12, one now has

$$\Delta\hat{v}(0) = 0 \quad \text{and} \quad \Delta\hat{y}(0) = \Delta\hat{n}(0) - f_1(v_0) - f_2(v_0) + V_1 - V_2.$$

As in Lemma 74.12, we study the various functions $y_i(t), v_i(t), \cdots$, for $-\tau \leq t \leq 0$ and $i = 1, 2$. Since $\Delta\hat{v}(0) = 0$, one has $S(\tau)\hat{v}_{10} = S(\tau)\hat{v}_{20}$. Thus we set $S_1(t) = S_2(t) = S(\tau + t)v_0$, for $-\tau \leq t \leq 0$. We note that $E_S(t) = 0$, for $-\tau \leq t \leq 0$. From Lemmas 74.8 and 74.11, we see that, at $t = 0$, $\Delta n(0) = \Delta u(0) + \Delta s(0)$ satisfies

$$\text{(74.87)} \qquad \Delta u(0) = g_1(v_{10}) - g_2(v_{20}) \quad \text{and} \quad \Delta s(0) = \bar{f}_1(v_{10}) - \bar{f}_2(v_{20}).$$

Since $g_i(v_{i0}) = P^u(v_{i0})n_{i0}$ and $\bar{f}_i(v_{i0}) = P^s(v_{i0})n_{i0}$, where $y_{i0} = y(\tau, \hat{y}_{i0}) = v_{i0} + n_{i0}$ and $n_{i0} = s_{i0} + u_{i0}$, one has $g_i(v_{i0}) = P^u(v_{i0})g_i(v_{i0})$ and $\bar{f}_i(v_{i0}) =$

$P^s(v_{i0})\bar{f}_i(v_{i0})$, for $i=1,2$. After a lengthy calculation, which uses (74.87), along with (74.11), (74.12), and Lemmas 74.7 and 74.8, one finds that (74.88)

$$\|A^\beta P^u(S(\tau)\hat{v}_{20})\Delta n(0)\| \le \|A^\beta \Delta g(v_{20})\| + K(\ell + L_0\epsilon)\|A^\beta \Delta v(0)\| + \beta_4(\epsilon,\delta)\|A^\beta \Delta\hat{y}(0)\|,$$

for some $\beta_4 \in \Sigma$. We also note that inequality (45.14) implies that (74.89)

$$\|A^\beta \Phi(S(\tau)\hat{v}_{20}, -\tau)P^u(S(\tau)\hat{v}_{20})\Delta n(0)\| \le Ke^{-\lambda_4\tau}\|A^\beta P^u(s(\tau)\hat{v}_{20})\Delta n(0)\|.$$

Since $\Delta\hat{v}(0) = 0$, it follows from (74.45) that $\Delta v(t) = E_v(t)$, for $-\tau \le t \le 0$, and inequality (74.49) implies that, at $t=0$, one has

$$\|A^\beta \Delta v(0)\| = \|A^\beta E_v(0)\| \le \left(b_4^F(\rho, 2\epsilon) + \beta_2(\epsilon,\delta)\right)\|A^\beta \Delta\hat{y}(0)\|. \tag{74.90}$$

Now $\|A^\beta E_u(t)\|$ satisfies (74.82), for $-\tau \le t \le 0$. By using inequality (74.90), with the fact that $K^3 L_0\,\rho\, e^{-\lambda_4\tau} \le \frac{1}{16}$ (see Lemma 74.2 and (74.17)), one obtains

$$\|A^\beta E_u(-\tau)\| \le \left(2b_4^F(\rho, 2\epsilon) + \beta_5(\epsilon,\delta)\right)\|A^\beta \Delta\hat{y}(0)\|, \tag{74.91}$$

for some $\beta_5 \in \Sigma$. It then follows from inequalities (74.17), (74.81), (74.88), (74.89), (74.90), and (74.91) that, at $t=-\tau$, where $\Delta u(-\tau) = V_1 - V_2 = \Delta\bar{g}(v_0)$, one finds a $\beta_6 \in \Sigma$ such that

$$\begin{aligned}\|A^\beta \Delta\bar{g}(v_0)\| \le Ke^{-\lambda_4\tau}\left(\|A^\beta \Delta g(v_0)\| + Kb_4^F(\rho,2\epsilon)\|A^\beta \Delta\hat{y}(0)\|\right)\\ + \left(2b_4^F(\rho,2\epsilon) + \beta_6(\epsilon,\delta)\right)\|A^\beta \Delta\hat{y}(0)\|.\end{aligned} \tag{74.92}$$

Note that $(2+K^2e^{-\lambda_4\tau})b_4^F(\rho,2\epsilon) \le \frac{1}{36}$, owing to inequalities (74.65), $K \ge 1$ and $\lambda_2 \le 0 < \lambda_4$. By substituting $\Delta\hat{y}(0) = \Delta f(v_0) + \Delta\bar{g}(v_0)$ into inequality (74.92) and using (74.17), one then obtains

$$\begin{aligned}\|A^\beta \Delta\bar{g}(v_0)\| \le \left(\frac{1}{36} + \beta_6(\epsilon,\delta)\right)\|A^\beta \Delta f(v_0)\| + \frac{1}{4}\|A^\beta \Delta g(v_0)\|\\ + \left(\frac{1}{36} + \beta_6(\epsilon,\delta)\right)\|A^\beta \Delta\bar{g}(v_0)\|.\end{aligned} \tag{74.93}$$

Finally, we fix ϵ_9 so that $0 < \epsilon_9 \le \epsilon_8$ and $\beta_6(\epsilon_9,\epsilon_9) \le \frac{1}{8}$, and we set $\delta_9(\epsilon) = \min(\epsilon, \delta_8(\epsilon))$, for $0 < \epsilon \le \epsilon_9$. One then has $\beta_6(\epsilon,\delta) \le \frac{1}{8}$, for $0 < \delta \le \delta_9(\epsilon)$, and this lemma now follows from inequality (74.93). □

The storms of the infinite dimensional world have passed! We are now ready to reboard the Good Ship QED for its placid journey to the next port of call, the proofs of the Main Theorem and the Shadow Theorem. As a first step, we give a more precise formulation of a portion of the Main Theorem.

Theorem 74.14. *Let the hypotheses of the Main Theorem be satisfied. Then for each $\epsilon > 0$ there exists a $\delta = \delta(\epsilon) > 0$ and, for $0 < \rho \leq \rho_0$, there is an $\ell = \ell(\rho, \epsilon) < 1$, where $\delta(\epsilon)$, $\ell(\rho, \epsilon) \in \Sigma$, such that if $\|G\|_{\{A;C^1(\Omega)\}} \leq \delta$, then there is a continuous mapping $h : M \to V^{2\beta}$ such that the following properties hold:*

1. *the image $M^G = h(M)$ is a compact invariant set for (70.2);*
2. *for each $v \in M$, one has $\phi(v) \in U^s(v) \oplus U^u(v)$, where $h(v) = v + \phi(v)$;*
3. *the restriction of h to any disk $\mathcal{D}_{\rho_0}(u_0)$, where $u_0 \in M$, is Lipschitz continuous with*

$$\|A^\beta(\phi(v_1) - \phi(v_2))\| \leq 2\ell \|A^\beta(v_1 - v_2)\|, \tag{74.94}$$

   for all $v_1, v_2 \in \mathcal{D}_{\rho_0}(u_0)$, and the perturbed manifold M^G is of Lipschitz class;
4. *one has $\|A^\beta \phi(v)\| \leq 2\epsilon$, for all $v \in M$; and*
5. *the mapping $h : M \to M^G$ is a (local) Lipschitz continuous homeomorphism with a (local) Lipschitz continuous inverse, and consequently, M^G is a Lipschitz manifold.*

Proof of Theorem 74.14 and the Shadow Theorem. Let T be given by (74.17), and for $T \leq \tau \leq 2T$, let A_τ be the mapping on $\mathcal{F} \times \mathcal{G} = \mathcal{F}(\epsilon, \ell_1) \times \mathcal{G}(\epsilon, \ell_1)$ defined by

$$A_\tau : (f, g) \to (\bar{f}_\tau, \bar{g}_\tau),$$

where $\ell_1 = 1$, $\bar{f}_\tau$ is given by equation (74.61), and $V = \bar{g}_\tau(u_0)$ is given by Lemma 74.11, see equations (74.71) and (74.74). Let ϵ_9 and δ_9 be given by Lemma 74.13, and let ℓ_5 and ℓ_8 be given by Lemmas 74.9 and 74.12. Define $\ell = \ell_9(\rho, \epsilon)$ by

$$\ell = \ell_9(\rho, \epsilon) = \max(\ell_5(\rho, \epsilon), \ell_8(\rho, \epsilon)), \qquad \text{for } 0 < \epsilon \leq \epsilon_9 \text{ and } 0 < \rho \leq \rho_0.$$

From the construction of ℓ_5 and ℓ_8 one has $\ell_9(\rho, \epsilon) < 1$ and $\ell_9(\rho, \epsilon) \in \Sigma$. Let ϵ be fixed with $0 < \epsilon \leq \epsilon_0$, and set $\delta = \delta_9(\epsilon)$.

Assume that $\|G\|_{\{A;C^1(\Omega)\}} \leq \delta$. From Lemmas 74.9 and 74.12, we see that A_τ maps $\mathcal{F} \times \mathcal{G}$ into itself, for each τ with $T \leq \tau \leq 2T$. Also Lemmas 74.10 and 74.13 imply that A_τ is a strict contraction on $\mathcal{F} \times \mathcal{G}$. Since $\mathcal{F} \times \mathcal{G}$ is a complete metric space, the mapping A_τ has a unique fixed point. Note that, owing to equations (74.61) and (74.71), the pair (f, g) is a fixed point of A_τ, where $T \leq \tau \leq 2T$, if and only if the mapping h given by

$$h(u) = u + f(u) + g(u), \qquad \text{for } u \in M, \tag{74.95}$$

satisfies

(74.96)

$$f(v(\tau, h(u))) = P^s(v(\tau, h(u)))(y(\tau, h(u)) - v(\tau, h(u))) = s(\tau, h(u))$$
$$g(v(\tau, h(u))) = P^u(v(\tau, h(u)))(y(\tau, h(u)) - v(\tau, h(u))) = u(\tau, h(u)),$$

for $u \in M$. We now fix $\tau = T$ and we let (f_0, g_0) denote the fixed point of A_T. Let $h = h_0$ satisfy (74.95), where (f, g) are replaced by (f_0, g_0), and define $M^G = h_0(M)$. Note that equations (74.96), with $\tau = T$, can be rewritten in the form

$$y(T, h_0(u)) = h_0(v(T, h_0(u))), \qquad \text{for } u \in M,$$

or equivalently, the solution $S_2^G(t)h_0(u) = y(t, h_0(u))$ of equation (70.2) satisfies

$$S_2^G(T)h_0(u) = h_0(v(T, h_0(u))), \qquad \text{for } u \in M. \tag{74.97}$$

It follows that $S_2^G(nT)M^G = M^G$, for all integers $n \geq 0$.

We next define the Shadow Flow $S_1^G(t)$ on M by

$$S_1^G(t)u_0 \stackrel{\text{def}}{=} \psi^e(S_2^G(t)h_0(u_0)) = v(t, h_0(u_0)), \qquad \text{for } u_0 \in M, \tag{74.98}$$

and $t \in [0, \infty)$. Note that ψ^e, which is an extension of ψ, see equation (74.59), is also the inverse of h_0. Since $S_2^G(nT)M^G = M^G$, for all integers $n \geq 0$, it follows that for each $y_0 \in M^G$, there is a global solution, which we denote by $S_2^G(t)y_0$, for $t \in \mathbb{R}$, with $S_2^G(nT)y_0 \in M^G$, for all $n \in Z$. Consequently, equation (74.98) defines $S_1^G(t)u_0$, for all $t \in \mathbb{R}$. Since ψ^e and S_2^G are continuous, equation (74.98) implies that $S_1^G(t)$ depends continuously on G, as $G \to 0$ in the topology $\mathcal{T}_A^1$.

For $0 \leq t < T$ we define a mapping $h_t : M \to V^{2\beta}$ by

$$h_t(S_1^G(t)u_0) = S_2^G(t)h_0(u_0) = y(t, h_0(u_0)), \qquad \text{for } u_0 \in M. \tag{74.99}$$

Because of (74.60), we see that h_t is well-defined for all $u_0 \in M$, and the definition extends to $0 \leq t \leq T$. From (74.97) we see that $h_T = h_0$. Also from (74.29) we see that one has the local coordinate representation

$$h_t(v(t, h(u_0))) = y(t, h(u_0)) = v(t, h(u_0)) + s(t, h(u_0)) + u(t, h(u_0)),$$

for $0 \leq t \leq T$, where $s(t, h(u_0)) \in U^s(v(t, h(u_0)))$ and $u(t, h(u_0)) \in U^u(v(t, h(u_0)))$. Now define (f_t, g_t), for $0 \leq t \leq T$, by

$$f_t(v(t, h(u_0))) = s(t, h(u_0)) \quad \text{and} \quad g_t(v(t, h(u_0))) = u(t, h(u_0)). \tag{74.100}$$

Note that one has $h_T = h = h_0$, $f_T = f_0$, and $g_T = g_0$. From (74.59) one obtains the commutivity relation $S_1^G(T)S_1^G(t) = S_1^G(t)S_1^G(T)$, for $0 \leq t \leq T$. In addition, the semiflow $S_2^G(t)$ satisfies $S_2^G(T)S_2^G(t) = S_2^G(t)S_2^G(T)$, as well. By using these commutivity relations and (74.99), one then obtains

$$\begin{aligned} S_2^G(T)h_t(S_1^G(t)u_0) &= S_2^G(T)S_2^G(t)h(u_0) = S_2^G(t)S_2^G(T)h(u_0) \\ &= S_2^G(t)h_T(S_1^G(T)u_0) = S_2^G(t)h(S_1^G(T)u_0) \\ &= h_t(S_1^G(t)S_1^G(T)u_0) = h_t(S_1^G(T)S_1^G(t)u_0), \end{aligned} \tag{74.101}$$

for all $u_0 \in M$ and $0 \le t \le T$. Now (74.97) and (74.101) imply that (f_t, g_t), where (f_t, g_t) is given by (74.100), is a fixed point of A_T. The next question is: which space does (f_t, g_t) reside in? Is (f_t, g_t) in the space $\mathcal{F} \times \mathcal{G}$? If so, then the uniqueness of the fixed point for A_T implies that $h = h_T = h_t$ and $(f_t, g_t) = (f_T, g_T)$. We show next that (f_t, g_t) does indeed lie in $\mathcal{F} \times \mathcal{G}$.

We claim that there is an ϵ_{10}, with $0 < \epsilon_{10} \le \epsilon_9$, such that, for $0 < \epsilon \le \epsilon_{10}$, there is a $t_0 > 0$ where $(f_t, g_t) \in \mathcal{F} \times \mathcal{G}$, for $0 \le t \le t_0$. We will use Lemmas 74.8, 74.9, 74.11, and 74.12. Since the fixed point $(f_0, g_0) = (f_T, g_T)$ is in the range of A_T, one has $\|A^\beta f_0(u)\| \le \frac{3}{4}\epsilon$ and $\|A^\beta g_0(u)\| \le \frac{3}{4}\epsilon$, for all $u \in M$. By continuity, there is a $t_1 > 0$ such that $\|A^\beta f_t(u)\| \le \epsilon$ and $\|A^\beta g_t(u)\| \le \epsilon$, for all $u \in M$ and $0 \le t \le t_1$. Since $0 \le \ell < 1$, it follows from Lemmas 74.9 and 74.12 that there is a $t_2 > 0$ such that both f_t and g_t are Lipschitz continuous on each disk $\mathcal{D}_\rho(v_0)$, for $0 < \rho \le \rho_0$, with Lipschitz coefficient 1, for $0 \le t \le t_2$. By setting $t_0 = \min(t_1, t_2)$, we conclude that $(f_t, g_t) \in \mathcal{F} \times \mathcal{G}$, for $0 \le t \le t_0$.

Since the fixed point of A_T is unique, we have $h_t = h$, for all t, with $0 \le t \le t_0$. By iteration of this argument, we conclude that $h_t = h$, for all $t \ge 0$. This implies that equation (74.7) holds, i.e., $S_1^G(t)$ is a G-continuous shadow semiflow on M. From (74.98) and (74.7) we see that ψ^e is the inverse of h. Since h and ψ^e are both (locally) Lipschitz continuous, we see that $h : M \to M^G$ is a Lipschitz continuous homeomorphism and that M^G is a Lipschitz manifold. Also, $S_1^G(t) : M \to M$ is a semiflow in the sense that $S_1^G(t)u_0$ is jointly continuous in $(u_0, t) \in M \times [0, \infty)$; $S_1^G(0)u_0 = u_0$; and the semigroup property $S_1^G(t)S_1^G(s) = S_1^G(t+s)$, for all s, $t \in \mathbb{R}$, is valid. In addition, one has $S_1^G(t)M = M$, for all $t \ge 0$, by (74.60). This completes the proof of Theorem 74.14 and the Shadow Theorem. □

One can readily verify that $v(t) = v(t, h(u_0)) = S_1^G(t)u_0$ is the unique mild solution of the differential equation

$$\partial_t v + Av = P^o(v)[F(v + \phi(v)) + G(v + \phi(v))] = P^o(v)[F(h(v)) + G(h(v))].$$

with $v(0) = u_0 \in M$. The Herculean Theorem 47.6 implies that $v(t)$ is a strong solution of this equation and that $M^G \subset \mathcal{D}(A)$. Since $h_t = h = h_T$, for all $t \ge 0$, the equalities in (74.101) remain valid when T is replaced by an arbitrary $\sigma \ge 0$. As a result, for all $(u_0, \sigma, t) \in M \times \mathbb{R}^+ \times \mathbb{R}^+$, one then obtains

$$S_2^G(t+\sigma)h(u_0) = S_2^G(t)h(S_1^G(\sigma)u_0). \tag{74.102}$$

Theorem 74.15. *Let the hypotheses of the Main Theorem be satisfied. Then for each $\epsilon > 0$ there exists a $\delta = \delta(\epsilon) > 0$ and, for $0 < \rho \le \rho_0$, there is an $\ell = \ell(\rho, \epsilon) < 1$, where $\delta(\epsilon)$, $\ell(\rho, \epsilon) \in \Sigma$, such that if $\|G\|_{\{A, C^1(\Omega)\}} \le \delta$, then the conclusions of Theorem 74.14 are valid and the following properties hold:*

(6) *The linear skew product semiflow π^G, generated by the linearization of equation (70.2) on the invariant manifold M^G, has an exponential trichotomy over M^G.*

(7) *The manifold M^G is of class C^1, and it is normally hyperbolic for the nonlinear dynamics generated by the mild solutions of equation (70.2).*

Proof of Theorem 74.15 and the Main Theorem. Since the Main Theorem is included in Theorems 74.14 and 74.15, it suffices to verify Items (6) and (7). We let $S_1(t)$, $S_1^G(t)$, $S_2^G(t)$, and h be given as in the proof of Theorem 74.14. In order to show that the compact manifold M^G is normally hyperbolic for the solutions of the perturbed equation (70.2), when δ is sufficiently small, we will use the Robustness of Trichotomies Theorem 45.12. For this purpose, we set $\mathcal{W} = \mathcal{M}^\infty(\mathbb{R};\mathcal{L})$, and we let d denote the translation invariant metric on the Fréchet space $\mathcal{W}$ described in Appendix A. By hypotheses, the linear skew product semiflow $\pi(v_0, B; \tau) = (\Phi(B,\tau)v_0, B_\tau)$ on $V^{2\beta} \times \mathcal{M}^\infty$ has an exponential trichotomy over $\mathcal{K} \subset \mathcal{W} = \mathcal{M}^\infty(\mathbb{R};\mathcal{L})$, where $\mathcal{K} = \{DF(S(\cdot)u_0) : u_0 \in M\} \subset \mathcal{M}^\infty(\mathbb{R};\mathcal{L})$. With G given and satisfying inequality (74.1), we now consider the collection of those $B \in \mathcal{M}^\infty$ given by

$$B(t) = B(u_0, t) \stackrel{\text{def}}{=} \left[DF(S_2^G(t)h(u_0)) + DG(S_2^G(t)h(u_0))\right],$$

for $(u_0, t) \in M \times \mathbb{R}$, It then follows from equation (74.102) and the definition of B that for all $(u_0, \sigma, t) \in M \times \mathbb{R} \times \mathbb{R}$, one has

$$B_\sigma(u_0, t) = B(u_0, \sigma + t) = B(S_1^G(\sigma)u_0, t), \tag{74.103}$$

Since both $S_2^G(t)$ and $S_1^G(t)$ are continuous in G, it follows that

$$S_1^G(t)u_0 \to S_1(t)u_0 \quad \text{and} \quad S_2^G(t)h(u_0) \to S_1(t)u_0, \qquad \text{as } \delta \to 0,$$

(and as $G \to 0$ in the space $(C^1_{\text{Lip}}, \mathcal{T}^1_A)$), uniformly for (u_0, t) in compact subsets of $M \times [0, \infty)$. Since $h(v) \to v$, as $\delta \to 0$, it follows from the continuity of the Fréchet derivatives DF and DG (see Theorem 44.4) that

$$DF(S_2^G(t)u_0) \to DF(S_1(t)u_0) \quad \text{and} \quad DG(S_2^G(t)h(u_0)) \to 0,$$

as $\delta \to 0$, in the space $\mathcal{L} = \mathcal{L}(V^{2\beta}, W)$, uniformly for (u_0, t) in compact subsets of $M \times [0, \infty)$.

Now let $\epsilon_0 > 0$ be given by the Robustness of Trichotomies Theorem 45.12 for the exponential trichotomy for π over the unperturbed manifold M. Let $\delta(\epsilon)$ and $\ell(\rho, \epsilon)$ be given by Theorem 74.14, for $0 < \epsilon \leq \epsilon_0$. Let $N = N(\epsilon_0)$ be an integer with $\sum_{n=N+1}^{\infty} 2^{-n} \leq \frac{1}{2}\epsilon_0$. Next we fix $\delta > 0$ so that $0 < \delta \leq \delta(\epsilon_0)$ and

$$\|B(v_0, t) - DF(S_1(t)v_0)\|_{\mathcal{L}} \leq \frac{1}{2}\epsilon_0, \qquad \text{for all } (v_0, t) \in M \times [-N, N].$$

As a result, it follows from equation (74.103) that

$$\left\| B_\sigma(u_0,t) - DF(S_1(t)S_1^G(\sigma)u_0) \right\|_{\mathcal{L}} \leq \frac{1}{2}\epsilon_0,$$

for all $(u_0,\sigma,t) \in M \times \mathbb{R} \times [-N,N]$. It follows then that $B_\sigma \in N_{\epsilon_0}(\mathcal{K})$, for all $\sigma \in \mathbb{R}$. The Robustness of Trichotomies Theorem 45.12 then implies that Item (6) is valid.

In order to prove Item (7) it suffices to verify equation (74.6) for the linear skew product semiflow generated by the linearization of equation (70.2) over the manifold M^G. Indeed, since the projections $R_G(v) = P_G^o(v)$ vary continuously with $v \in M^G$, equation (74.6) implies that the manifold M^G is of class C^1.

We denote the linear skew product semiflow π^G over M^G in the form

$$\pi^G(v_0,y_0,t) = (\Phi^G(y_0,t)v_0, y(t,y_0)),$$

for $(v_0,y_0,t) \in V^{2\beta} \times M^G \times [0,\infty)$, where $y(t,y_0) = S_2^G(t)y_0$ and $y_0 = h(u_0)$, for some $u_0 \in M$, see Section 4.9. Since the manifold M^G is invariant under the nonlinear dynamics $S_2^G(t)$, for each $y_0 \in M^G$, there is a global solution, which we denote by $y(t) = y(t,y_0)$, through y_0, and one has $y(t,y_0) \in M^G$, for all $t \in \mathbb{R}$. The linear operator $\Phi^G(y_0,t)$ is the solution operator for the linear variational equation

$$\partial_t v + Av = [DF(y(t,y_0)) + DG(y(t,y_0))]\, v.$$

Since F and G are in C^1_{Lip}, one has

$$F(y+w) + G(y+w) = F(y) + G(y) + [DF(y) + DG(y)]w + H(y,w),$$

where the last equation redefines the term $H = H(y,w)$. We note that there is a $\gamma \in \Sigma$, such that for any $y \in M^G$, the error term H satisfies

$$\|H(y,w_1) - H(y,w_2)\| < \gamma(\sigma)\|A^\beta(w_1 - w_2)\|,$$

whenever $\|A^\beta w_1\|$, $\|A^\beta w_2\| \leq \sigma$, and

$$\|H(y,w)\| \leq \gamma(\sigma)\|A^\beta w\| \leq \sigma\,\gamma(\sigma), \qquad \text{for } \|A^\beta w\| \leq \sigma.$$

Let $y_i = y_i(t) = y(t,y_{i0})$, for $i = 1,2$, denote two solutions of the perturbed equation (70.2), where $y_{20} \in M^G$, and set $w = w(t) = y_1 - y_2$. From the alternate Variation of Constants Formula (74.5), we see that w satisfies

$$w(t) = \Phi^G(y_{20},t)w(0) + \int_0^t \Phi^G(y(s,y_{20}),t-s)\, H(y(s,y_{20}),w(s))\, ds,$$

where $w(0) = y_{10} - y_{20}$. Let (P_G^s, P_G^o, P_G^u) denote the invariant projectors on M^G, where G satisfies (74.1). Set $Q_G^o = P_G^s + P_G^u$. Next let K and

λ_i, for $1 \le i \le 4$, denote the charactersitics of the exponential trichotomy over M^G. While these characteristics depend on G, we do not include this dependence explicitly in this notation.

The proof of the validity of the relation (74.6) for the perturbed dynamics on M^G is via an argument by contradiction involving two steps: (1) a basic inequality for the dynamics on M^G, and (2) the connection between this inequality and the Lipschitz property for the mapping $h : M \to M^G$.

Step 1: We let $\epsilon > 0$ and $\delta = \delta(\epsilon) > 0$ be given so that Item (5) and Theorem 74.14 hold. Set $\delta_1 = \delta$. We will now show that there is a δ_2 with $0 < \delta_2 \le \delta_1$ such that if $\|G\|_{\{A;C^1(\Omega)\}} \le \delta_2$, then for every $u_0 \in M$ the space $U_G^o(y_0)$ is tangent to the set M^G at the point y_0, where $y_0 = h(u_0)$. Observe that $U_G^o(y_0)$ is tangent at $y_0 \in M^G$ if and only if, for every convergent sequence $y_i \in M^G$, with $y_i \to y_0$ in $V^{2\beta}$, one has

$$\lim_{i\to\infty} \|A^\beta Q_G^o(y_0)(y_i - y_0)\| \, \|A^\beta(y_i - y_0)\|^{-1} = 0.$$

If on the contrary, the space $U_G^o(y_0)$ is not tangent to M^G at the point y_0, where $y_0 = h(u_0)$ and $u_0 \in M$, then there exist a constant $\eta > 0$ and a sequence of points $u_i \in \mathcal{D}_{\rho_0}(u_0)$, for $i = 1, 2, \ldots$, such that $u_i \to u_0$ in $M \subset V^{2\beta}$, as $i \to \infty$, and $y_i = h(u_i)$ satisfies

$$(74.104) \quad \|A^\beta Q_G^o(y_0)(y_i - y_0)\| \ge \eta \|A^\beta(y_i - y_0)\| > 0, \qquad \text{for all } i \ge 1.$$

By taking a subsequence, if necessary, we can assume that either

$$(74.105) \quad \|A^\beta P_G^u(y_0)(y_i - y_0)\| \ge \frac{1}{2}\|A^\beta Q_G^o(y_0)(y_i - y_0)\|, \qquad \text{for all } i \ge 1,$$

or

$$(74.106) \quad \|A^\beta P_G^s(y_0)(y_i - y_0)\| \ge \frac{1}{2}\|A^\beta Q_G^o(y_0)(y_i - y_0)\|, \qquad \text{for all } i \ge 1.$$

Next define the two cone bundles $\mathcal{V}_1$ and $\mathcal{V}_2$ over M^G by

$$\mathcal{V}_1 \stackrel{\text{def}}{=} \{(w, y) \in V^{2\beta} \times M^G : \|A^\beta Q_G^o(y)w\| \le 2\|A^\beta w\|\},$$

and for $C > 0$,

$$\mathcal{V}_2(C) \stackrel{\text{def}}{=} \{(w, y) \in V^{2\beta} \times M^G : C\|A^\beta P_G^o(y)w\| \le \|A^\beta Q_G^o(y)w\|\}.$$

Let $(w, y) \in \mathcal{V}_2(C)$, where $C \ge 2$. Then one has

$$\|A^\beta P_G^o(y)w\| \le C\|A^\beta P_G^o(y)w\| \le \|A^\beta Q_G^o(y)w\| \le \|A^\beta w\| + \|A^\beta P_G^o(y)w\|,$$

which implies that $(C - 1)\|A^\beta P_G^o(y)w\| \le \|A^\beta w\|$. If $(w, y) \notin \mathcal{V}_1$, then

$$2\|A^\beta w\| < \|A^\beta Q_G^o(y)w\| \le \|A^\beta w\| + \|A^\beta P_G^o(y)w\|,$$

which implies that $\|A^\beta w\| < \|A^\beta P^o_G(y)w\|$. Since $C \geq 2$, one finds that

$$\|A^\beta P^o_G(y)w\| \leq (C-1)\|A^\beta P^o_G(y)w\| \leq \|A^\beta w\| < \|A^\beta P^o_G(y)w\|,$$

which is a contradiction. We have thus shown that $\mathcal{V}_2(C) \subset \mathcal{V}_1$, whenever $C \geq 2$. We now fix $C = C_4 \geq 20$ and set $\mathcal{V}_2 = \mathcal{V}_2(C_4) \subset \mathcal{V}_1$. To put it another way, the implication
(74.107)

$$C_4\|A^\beta P^o_G(y)w\| \leq \|A^\beta Q^o_G(y)w\| \quad \Longrightarrow \quad \|A^\beta Q^o_G(y)w\| \leq 2\|A^\beta w\|$$

is valid, with $C_4 \geq 20$.

Since M^G is invariant under $S^G_2(t)$, it follows that, for any $y_1 \in M^G$, the solution through y_1 has a negative continuation in M^G, which we will denote by $y(t, y_1)$, for $t \in \mathbb{R}$. Let y_0 and y_i be points in M^G, where $y_i \to y_0$, as $i \to \infty$. Since the manifold M^G is of Lipschitz class, we can use Lemma 74.5 and equation (74.34) to define $H_2(t)$ by

$$y(t, y_i) - y(t, y_0) = \Phi^G(y_0, t)(y_i - y_0) + H_2(t), \qquad \text{for } t \geq 0.$$

It follows from Lemma 74.5, as applied here,[38] that there is a $b_3 \in \Sigma$ such that, if y_0 and y_i satisfy $\|A^\beta(y_i - y_0)\| \leq \rho$, then one has

$$\|A^\beta H_2(t)\| \leq b_3(\rho)\|A^\beta(y_i - y_0)\|, \qquad \text{for all } t \in [0, 2T]. \tag{74.108}$$

Our next objective is to show that in either case (74.105) or (74.106), there is a time $\tau \in \mathbb{R}$ and an integer $i_0 \geq 1$, with the property that

$$\begin{aligned}\|A^\beta Q^o_G(y(\tau, y_0))(y(\tau, y_i) - y(\tau, y_0))\| \\ \geq C_4\|A^\beta P^o_G(y(\tau, y_0))(y(\tau, y_i) - y(\tau, y_0))\|\end{aligned} \tag{74.109}$$

for $i \geq i_0$, i.e., $(y(\tau, y_i) - y(\tau, y_0), y(\tau, y_0)) \in \mathcal{V}_2$. By combining (74.107) and (74.109), we find that, for $i > i_0$, one has

$$\|A^\beta(y(\tau, y_i) - y(\tau, y_0))\| \geq \frac{1}{2}\|A^\beta Q^o_G(y(\tau, y_0))(y(\tau, y_i) - y(\tau, y_0))\|. \tag{74.110}$$

In order to prove inequality (74.109), we first treat the case where inequality (74.105) holds. Choose $i_0 \geq 1$ so that $\|A^\beta(y_i - y_0)\| < \rho$, for $i \geq I_0$, and let H_2 be given as above. Hence $H_2(t)$ satisfies inequality (74.108), for $i \geq i_0$. We now fix $\tau > 0$ and after that fix $\rho > 0$ so that

$$(C_4 + 1)K\, b_3(\rho) \leq (C_4 + 1)Ke^{\lambda_3\tau} \leq \frac{\eta}{4K}e^{\lambda_4\tau}, \tag{74.111}$$

[38]We drop the superscript F here, because b_3 now depends on both F and G.

where C_4 is given above. For an estimate of $\|A^\beta \Phi^G(y_0,t)(y_i - y_0)\|$, for $0 \le t \le T$, we will use the identity

$$y_i - y_0 = P_G^s(y_0)(y_i - y_0) + P_G^o(y_0)(y_i - y_0) + P_G^u(y_0)(y_i - y_0).$$

By using inequalities (74.104) and (74.105), one obtains

$$\|A^\beta P_G^u(y_0)(y_i - y_0)\| \ge \frac{\eta}{2}\|A^\beta(y_i - y_0)\|,$$

and by combining this with inequality (45.21), we find that, for all $t \ge 0$, one has

$$\begin{aligned} \|A^\beta \Phi^G(y_0,t)P_G^u(y_0)(y_i - y_0)\| &\ge K^{-1}e^{\lambda_4 t}\|A^\beta P_G^u(y_0)(y_i - y_0)\| \\ &\ge \frac{\eta}{2K}e^{\lambda_4 t}\|A^\beta(y_i - y_0)\|. \end{aligned} \tag{74.112}$$

In addition, inequalities (45.15) and (45.13) imply that

$$\begin{aligned} \|A^\beta \Phi^G(y_0,t)P_G^s(y_0)(y_i - y_0)\| &\le Ke^{\lambda_1 t}\|A^\beta(y_i - y_0)\| \\ \|A^\beta \Phi^G(y_0,t)P_G^o(y_0)(y_i - y_0)\| &\le Ke^{\lambda_3 t}\|A^\beta(y_i - y_0)\|. \end{aligned}$$

Next we use Lemma 74.5 to estimate both sides of inequality (74.109). Now for $i \ge i_0$ and $0 \le t \le \tau$, and since $\lambda_1 < \lambda_3$, one has

$$\begin{aligned} &\|A^\beta Q_G^o(y(t,y_0))(y(t,y_i) - y(t,y_0))\| \\ &\quad \ge \|A^\beta Q_G^o(y(t,y_0))\Phi^G(y_0,t)(y_i - y_0)\| - \|A^\beta Q_G^o(y(t,y_0))H_2(t)\| \\ &\qquad \text{by the invariance of } Q_G^o \\ &\quad \ge \|A^\beta \Phi^G(y_0,t)P_G^u(y_0)(y_i - y_0)\| - \|A^\beta \Phi^G(y_0,t)P_G^s(y_0)(y_i - y_0)\| \\ &\qquad - \|A^\beta Q_G^o(y(t,y_0))H_2(t)\| \\ &\qquad \text{from the inequalities above} \\ &\quad \ge \left(\frac{\eta}{2K}e^{\lambda_4 t} - Ke^{\lambda_3 t} - Kb_3(\rho)\right)\|A^\beta(y_i - y_0)\|. \end{aligned}$$

Similarly, one obtains

$$\begin{aligned} &\|A^\beta P_G^o(y_0(t))(y(t,y_i) - y(t,y_0))\| \\ &\qquad \le \|A^\beta P_G^o(y_0(t))\Phi^G(y_0,t)(y_i - y_0)\| + \|A^\beta P_G^o(y_0(t))H_2(t)\| \\ &\qquad \le \|A^\beta \Phi^G(y_0,t)P_G^o(y_0)(y_i - y_0)\| + \|A^\beta P_G^o(y_0(t))H_2(t)\| \\ &\qquad \le (Ke^{\lambda_3 t} + Kb_3(\rho))\|A^\beta(y_i - y_0)\|. \end{aligned}$$

Inequality (74.109) is now a consequence of inequality (74.111).

We now assume that inequality (74.106) is valid. While this case appears to be similar to the case where inequality (74.105) holds, there are some rapids ahead! In particular, we encounter the issue of the negative continuations of solutions of the linear problem, $\Phi^G(y_2,t)w$, for $t \leq 0$ and $w \in U_G^s(y_2)$. When w is in $U_G^u(y_2)$, or in $U_G^o(y_2)$, then one can use the exponential trichotomy property to define $\Phi^G(y_2,t)w$, for $t \leq 0$, and inequalities (45.20) - (45.21) lead to good information about the growth of $\Phi^G(y_2,t)w$, for $t \geq 0$ (see inequality (74.112), for example). On the other hand, if $w \in U_G^s(y_2)$, then in general, one cannot extend the linear solution for $t < 0$. This is a problem even when $w = P_G^s(y_2)(y_1 - y_2)$, where $y_1, y_2 \in M^G$ and the nonlinear solutions $y(t,y_1)$ and $y(t,y_2)$ are defined for all $t \in \mathbb{R}$. However, this is no problem for the Good Ship QED, as we shoot the rapids. We need to use a different approach, an approach which uses the partial extension given in (45.23) and (45.25). Enjoy the ride!

Assume now that inequality (74.106) holds and let $\tau < 0$ and $\nu > 0$ be given. The values of τ and ν will be fixed later. Recall that $S_2^G(t)y_i = y(t,y_i)$ and $S_2^G(t)y_0 = y(t,y_0)$ are the global solutions of equation (70.2) passing through y_i and y_0, respectively. We define $\hat{y}_i = y(\tau,y_i)$, $\hat{y}_0 = y(\tau,y_0)$, $\hat{z}_i = \hat{y}_i - \hat{y}_0$, and $z_i = \Phi^G(\hat{y}_0,-\tau)\hat{z}_i$, for $i \geq 1$. For the purpose of using Lemma 74.5 in this case, we define $H_2(t)$ by

$$S_2^G(t)\hat{y}_i - S_2^G(t)\hat{y}_0 = \Phi^G(\hat{y}_0,t)(\hat{y}_i - \hat{y}_0) + H_2(t), \qquad \text{for } t \geq 0.$$

Let i_0 be chosen so that $\|A^\beta \hat{z}_i\| \leq \rho$, for $i \geq i_0$. From inequality (74.108) one obtains $\|A^\beta H_2(t)\| \leq b_3(\rho)\|A^\beta \hat{z}_i\|$, for $0 \leq t \leq -\tau$. By setting $t = -\tau$ in the previous equation, we obtain

$$y_i - y_0 = \Phi^G(\hat{y}_0,-\tau)\hat{z}_i + H_2(-\tau) = z_i + H_2(-\tau). \tag{74.113}$$

From the last equation and inequality (74.108), we obtain

$$\|A^\beta z_i\| \leq (1 + b_3(\rho))\|A^\beta(y_i - y_0)\|, \tag{74.114}$$

and inequalities (74.35), (74.104), and (74.106) yield

$$\begin{aligned}\|A^\beta P_G^s(y_0)z_i\| &\geq \|A^\beta P_G^s(y_0)(y_i - y_0)\| - \|A^\beta P_G^s(y_0)H_2(-\tau)\| \\ &\geq \left(\frac{\eta}{2} - Kb_3(\rho)\right)\|A^\beta(y_i - y_0)\|.\end{aligned}$$

Next we require that $\tau < 0$ and $\rho > 0$ satisfy $Kb_3(\rho) \leq \frac{\eta}{4}$, which implies that

$$\|A^\beta P_G^s(y_0)z_i\| \geq \frac{\eta}{4}\|A^\beta(y_i - y_0)\|. \tag{74.115}$$

We now use equation (45.23) to define $\Phi^G(y_0,t)z_i = \Phi^G(\hat{y}_0,-\tau+t)\hat{z}_i$, for $\tau \leq t \leq 0$. With $t = \tau$, this implies that $\hat{y}_i - \hat{y}_0 = \hat{z}_i = \Phi^G(y_0,\tau)z_i$. Now inequalities (45.25) and (74.115) imply that

$$\begin{aligned}\|A^\beta \Phi(y_0,\tau)P_G^s(y_0)z_i\| &\geq K^{-1}e^{\lambda_1\tau}\|A^\beta P_G^s(y_0)z_i\| \\ &\geq \frac{\eta}{4K}e^{\lambda_1\tau}\|A^\beta(y_i - y_0)\|.\end{aligned} \tag{74.116}$$

Next we observe that inequalties (45.14), (45.16), and (74.114) imply that

$$
\begin{aligned}
\|A^{\beta}\Phi^{G}(y_0,\tau)P_G^{u}(y_0)z_i\| &\le Ke^{\lambda_2\tau}\|A^{\beta}z_i\| \\
&\le K(1+b_3(\rho))e^{\lambda_4\tau}\|A^{\beta}(y_i-y_0)\|,
\end{aligned}
\tag{74.117}
$$

since $\lambda_2 < 0 < \lambda_4$ and $\tau < 0$, and

$$
\|A^{\beta}\Phi^{G}(y_0,\tau)P_G^{o}(y_0)z_i\| \le Ke^{\lambda_2\tau}\|A^{\beta}z_i\| \le K(1+b_3(\rho))e^{\lambda_2\tau}\|A^{\beta}(y_i-y_0)\|.
$$

From inequalities (74.116), (74.117), and $\lambda_2 < \lambda_4$, we get

$$
\begin{aligned}
&\|A^{\beta}\Phi^{G}(y_0,\tau)Q_G^{o}(y_0)z_i\| \\
&\qquad \ge \|A^{\beta}\Phi^{G}(y_0,\tau)P_G^{s}(y_0)z_i\| - \|A^{\beta}\Phi^{G}(y_0,\tau)P_G^{u}(y_0)z_i\| \\
&\qquad \ge \left(\frac{\eta}{4K}e^{\lambda_1\tau} - K(1+b_3(\rho))e^{\lambda_2\tau}\right)\|A^{\beta}(y_1-y_0)\|.
\end{aligned}
$$

We now fix $\tau < 0$ and afterwards fix $\rho > 0$ so that

$$
(C_4+1)e^{\lambda_2\tau} \le \frac{\eta}{8K^2}e^{\lambda_1\tau}, \quad \text{and} \quad b_3(\rho) \le \min\left(\frac{\eta}{4K}, 1\right).
$$

With this choice of τ and ρ, we see that inequality (74.115) holds and

$$
C_4K(1+b_3(\rho))e^{\lambda_2\tau} \le \frac{\eta}{4K}e^{\lambda_1\tau} - K(1+b_3(\rho))e^{\lambda_2\tau},
$$

which implies that inequality (74.109) is valid for $i \ge i_0$.

Step 2: Let us next return to the shadow semiflow $S_1^G(t)v_i = v(t,y_i)$ on M. Recall that $y(t,y_i) = h(v(t,y_i))$ and $y(t,y_0) = h(v(t,y_0))$, for all $t \in \mathbb{R}$. Since $y_i \to y_0$ in $V^{2\beta}$, it follows from inequality (74.47) that there is an $i_1 \ge i_0$ such that

$$
\|A^{\beta}(y(t,y_i) - y(t,y_0))\| \le \epsilon, \quad \text{and} \quad \|A^{\beta}(v(t,y_i) - v(t,y_0))\| \le \epsilon,
$$

for $i \ge i_1$, and $0 \le t \le 2T$. Let $\bar{v}_i = v(\tau,y_i)$, and $\bar{y}_i = y(\tau,y_i) = h(\bar{v}_i)$, for $i \ge 1$, and set $\bar{v}_0 = v(\tau,y_0)$ and $\bar{y}_0 = y(\tau,y_0) = h(\bar{v}_0)$. By choosing a subsequence, which we relabel as $\bar{v}_i$, one has $\bar{v}_i \in \mathcal{D}_{\rho_0}(\bar{v}_0)$, $\bar{v}_i \to \bar{v}_0$, and $\bar{y}_i \to \bar{y}_0$, as $i \to \infty$. Since $\ell \le 1$, it follows from inequality (74.94) that

$$
\|A^{\beta}(\bar{y}_i - \bar{y}_0)\| = \|A^{\beta}(h(\bar{v}_i) - h(\bar{v}_0))\| \le 3\|A^{\beta}(\bar{v}_i - \bar{v}_0)\|.
\tag{74.118}
$$

By applying $P^o(\bar{v}_0)$ to $\bar{y}_i - \bar{y}_0 = (\bar{y}_i - \bar{v}_i) - (\bar{y}_0 - \bar{v}_0) + (\bar{v}_i - \bar{v}_0)$, we obtain

$$
P^o(\bar{v}_0)(\bar{y}_i - \bar{y}_0) = P^o(\bar{v}_0)(\bar{y}_i - \bar{v}_i) - P^o(\bar{v}_0)(\bar{y}_0 - \bar{v}_0) + P^o(\bar{v}_0)(\bar{v}_i - \bar{v}_0).
\tag{74.119}
$$

Since $(\bar{y}_0 - \bar{v}_0) \in U^s(\bar{v}_0) \oplus U^u(\bar{v}_0)$ and $(\bar{y}_i - \bar{v}_i) \in U^s(\bar{v}_i) \oplus U^u(\bar{v}_i)$, one has

$$
P^o(\bar{v}_0)(\bar{y}_0 - \bar{v}_0) = 0 \quad \text{and} \quad P^o(\bar{v}_i)(\bar{y}_i - \bar{v}_i) = 0, \qquad \text{for } i \ge 1.
\tag{74.120}
$$

Consequently, the term $P^o(\bar{v}_0)(\bar{y}_i - \bar{v}_i)$ assumes the form

$$P^o(\bar{v}_0)(\bar{y}_i - \bar{v}_i) = (P^o(\bar{v}_0) - P^o(\bar{v}_i))(\bar{y}_i - \bar{v}_i). \tag{74.121}$$

From inequality (74.11) we obtain

$$\|A^\beta(P^o(\bar{v}_i) - P^o(\bar{v}_0))(\bar{y}_i - \bar{v}_i)\| \leq L_0\|A^\beta(\bar{v}_i - \bar{v}_0)\|\,\|A^\beta(\bar{y}_i - \bar{v}_i)\|.$$

By combining this inequality with (74.121) and using Item (4) of Theorem 74.14, one obtains $\|A^\beta P^o(\bar{v}_0)(\bar{y}_i - \bar{v}_i)\| \leq 2L_0\epsilon\|A^\beta(\bar{v}_i - \bar{v}_0)\|$. From Lemma 74.2 one finds that $\|A^\beta P^o(\bar{v}_0)(\bar{v}_i - \bar{v}_0)\| \geq \frac{4}{5}\|A^\beta(\bar{v}_i - \bar{v}_0)\|$. By combining the last two inequalities with (74.119) and (74.120), it follows that

$$\|A^\beta P^o(\bar{v}_0)(\bar{y}_i - \bar{y}_0)\| \geq \frac{1}{5}(4 - 10L_0\epsilon)\|A^\beta(\bar{v}_i - \bar{v}_0)\|. \tag{74.122}$$

Let us next estimate the norm $\|A^\beta P^o_G(\bar{y}_0)(\bar{y}_i - \bar{y}_0)\|$. First note that

$$P^o_G(\bar{y}_0)(\bar{y}_i - \bar{y}_0) = P^o(\bar{v}_0)(\bar{y}_i - \bar{y}_0) + (P^o_G(\bar{y}_0) - P^o(\bar{v}_0))(\bar{y}_i - \bar{y}_0). \tag{74.123}$$

Since one has

$$\lim_{(\epsilon,\delta)\to(0,0)} P^o_G(y_0) = P^o(v_0), \qquad \text{in } \mathcal{L}(V^{2\beta}, W),$$

uniformly for $v_0 \in M$, there is a $\beta_3 \in \Sigma$ such that

$$\|A^\beta(P^o_G(\bar{y}_0) - P^o(\bar{v}_0))(\bar{y}_i - \bar{y}_0)\| \leq \beta_3(\epsilon,\delta)\|A^\beta(\bar{y}_i - \bar{y}_0)\|,$$

for small ϵ and δ. As a result, (74.122) and (74.123) imply that

$$\|A^\beta P^o_G(\bar{y}_0)(\bar{y}_i - \bar{y}_0)\| \geq \frac{1}{5}(4 - 10L_0\epsilon)\|A^\beta(\bar{v}_i - \bar{v}_0)\| - \beta_3(\epsilon,\delta)\|A^\beta(\bar{y}_i - \bar{y}_0)\|. \tag{74.124}$$

Now inequalities (74.100) and (74.110) imply that

$$\|A^\beta(\bar{y}_i - \bar{y}_0)\| \geq \frac{C_4}{2}\|A^\beta P^o_G(\bar{y}_0)(\bar{y}_i - \bar{y}_0)\|.$$

By combining the last inequality with (74.124), we conclude that

$$\|A^\beta(\bar{y}_i - \bar{y}_0)\| \geq \frac{C_4(2 - 5L_0\epsilon)}{5}\|A^\beta(\bar{v}_i - \bar{v}_0)\| - \frac{C_4\beta_3(\epsilon,\delta)}{2}\|A^\beta(\bar{y}_i - \bar{y}_0)\|,$$

which implies that

$$\|A^\beta(\bar{y}_i - \bar{y}_0)\| \geq \frac{C_4(4 - 10L_0\epsilon)}{(10 + 5C_4\beta_3(\epsilon,\delta))}\|A^\beta(\bar{v}_i - \bar{v}_0)\|.$$

Let ϵ_9 and δ_9 be given by Lemma 74.13. Now choose ϵ_{10} so that $0 < \epsilon_{10} \leq \epsilon_9$, $L_0\epsilon_{10} \leq \frac{1}{10}$, and $C_4\beta_3(\epsilon_{10}, \epsilon_{10}) \leq 1$. Set $\delta_{10}(\epsilon) = \min(\epsilon, \delta_9(\epsilon))$, for $0 < \epsilon \leq \epsilon_{10}$. Since $C_4 \geq 20$, one finds that

$$||A^\beta(\bar{y}_i - \bar{y}_0)|| \geq 4||A^\beta(\bar{v}_i - \bar{v}_0)||,$$

which contradicts inequality (74.118). By setting $\delta_2 = \delta_{10}(\epsilon_{10})$, we thereby complete the proof of Theorem 74.15 and the Main Theorem. □

The norm $||G||_{\{A;C^1(\Omega)\}}$ used here has a role to play on other settings. For example, the Perturbation Theorem 73.1 is valid when the inequality (73.4) is replaced by (74.1). Indeed, if Γ is a hyperbolic, periodic orbit for equation (70.1), as described in Theorem 73.1, then the hypotheses on $M = \Gamma$, as formulated in Theorems 74.14 and 74.15, hold. The reader should verify that, in this case, Γ is a C^2-manifold of Lipschitz class.

7.5. Applications.

In this section we will examine two applications of the dynamical theories presented in the preceding four sections. While both of these applications are taken from the theory of the Navier-Stokes equations, the methodology developed here is widely applicable. We first seek to develop a rigorous foundation for the dynamical features seen in the classical Couette-Taylor flow. The second application arises in the numerical analysis of fluid flows. What we want to show here is that numerical methods based on the Bubnov-Galerkin approximations can lead to good information about the longtime dynamics of the underlying problem.[39]

These two applications, both of which are connected with the Couette-Taylor flow, address a very deep and important scientific issue: namely,

What is really known about the longtime dynamics of fluid dynamics?

For our purposes, we interpret "really known" as "rigorously known". After all, this is what mathematics is all about!

The Couette-Taylor flow refers to two patterns which arise in one of the classical fluid dynamics experiments, see Taylor (1923), and Chossat and Iooss (1994). With the modern tool of laser optics for making measurements of observables in the fluid flows and the related spectral analysis of the resulting data, one now has a good picture of some of the longtime dynamical features which occur before the onset of turbulence in real systems, see Gollub and Swinney (1975), and Andereck, Liu, and Swinney (1986). While good information is coming in at the experimental level, what can be said about the model itself? The model, which we turn to next, is described in terms of the Navier-Stokes equations.

[39]The theory presented in this section is basically from the paper by Pliss and Sell (2001a).

7.5.1. The Couette-Taylor Flow. We now consider a fluid flow between two concentric right circular cylinders, and the two cylinders are allowed to rotate with independent, but fixed, angular velocities, see Andereck, Liu, and Swinney (1986) and Chossat and Iooss (1994). While the two angular velocities lead to a two-parameter bifurcation problem, our purpose here is adequately served by considering the outer cylinder to be held fixed while the inner cylinder is made to rotate with a constant angular velocity ω, see Taylor (1923), Gollub and Swinney (1975), and Golubitsky and Stewart (1986). We will study the longtime dynamics of this problem for various choices of the parameter ω.

The fluid motion in the region Ω between the two concentric cylinders is given by the Navier-Stokes equations:

$$\begin{aligned} &\partial_t u - \nu\Delta u + (u \cdot \nabla)u + \nabla p = f_g \\ &\nabla \cdot u = 0. \end{aligned} \tag{75.1}$$

Both (u_1, u_2, u_3), cartesian coordinates, and (u_r, u_θ, u_z), cylindrical coordinates, are used to describe a vector field u. Recall that in these coordinate systems, one has

$$u = (u_1, u_2, u_3) = u_1 \overrightarrow{i} + u_2 \overrightarrow{j} + u_3 \overrightarrow{k} = u_r \overrightarrow{r} + u_\theta \overrightarrow{\theta} + u_z \overrightarrow{k},$$

where $\overrightarrow{i}, \overrightarrow{j}, \overrightarrow{k}$ forms the standard cartesian basis, $u_3 = u_z$,

$$\begin{aligned} \overrightarrow{r} &= \frac{1}{r}(x, y, 0) = (\cos\theta, \sin\theta, 0), \quad \text{and} \\ \overrightarrow{\theta} &= \frac{1}{r}(-y, x, 0) = (-\sin\theta, \cos\theta, 0). \end{aligned}$$

Recall that $u_r = u \cdot \overrightarrow{r}$ and $u_\theta = u \cdot \overrightarrow{\theta}$. We fix the region $\Omega = An \times [0, L]$, so that An is the annular region $0 \leq \theta \leq 2\pi$, $\eta \leq r \leq 1$, where $0 < \eta < 1$, and $0 \leq z \leq L$.

For the Couette-Taylor flow, the external force f_g is a gravitational force, which we assume to satisfy $f_g = -g\overrightarrow{k}$, where g is the gravitational constant. It is convenient to replace the pressure p in (75.1) with $p + p_0$, where $p_0(x, y, z) = -gz$ is the hydrostatic pressure. In this case, one has $\nabla p_0 = f_g$, and these two terms can be cancelled. As a result, (75.1) reduces to

$$\begin{aligned} &\partial_t u - \nu\Delta u + (u \cdot \nabla)u + \nabla p = 0 \\ &\nabla \cdot u = 0. \end{aligned} \tag{75.2}$$

On the boundary $\partial An \times [0, L]$, we require that

$$\begin{aligned} \text{angular velocity} &= u_\theta = \omega, \quad \text{at } r = \eta \\ \text{radial velocity} &= u_r = 0, \quad \text{at } r = \eta \\ u_\theta &= u_r = 0, \quad \text{at } r = 1, \end{aligned} \tag{75.3}$$

and we assume that all terms are spatially periodic in z. We will refer to equation (75.2), with the boundary conditions (75.3) as the **spatially periodic CT model.**

The first step in reducing this problem to the theory considered above is to "homogenize" the boundary conditions. This is accomplished by making a change of variables $u = v + w^c$ and $p = q + p^c$ in equation (75.2), where w^c and p^c will be chosen so that the following conditions hold: (1) $\nabla \cdot w^c = 0$ and $\partial_t w^c = 0$ in Ω, (2) w^c satisfies the boundary conditions (75.3), and (3) it is periodic in z. In addition, we require that $\partial_t p^c = 0$ and that (w^c, p^c) be a stationary solution of equation (75.2), i.e., one has

$$-\nu \Delta w^c + (w^c \cdot \nabla) w^c + \nabla p^c = 0. \tag{75.4}$$

It can happen that equation (75.4) has many solutions satisfying the boundary conditions (75.3), however, there is one solution of special interest because it exists for all values of the viscosity $\nu > 0$. This is the Couette solution, or the Couette flow. In particular, we define $w^c = w_\theta(r)\overrightarrow{\theta}$, where $w_\theta = C(r^{-1} - r)$, with $C = \omega\eta(1-\eta^2)^{-1}$ and

$$p^c = \frac{1}{2} C^2 \left(-\frac{1}{r^2} - 4 \log r + r^2 \right) = C^2 \int \frac{1}{r} w_\theta^2 \, dr.$$

Note that (w^c, p^c) is a solution of (75.4), for every $\nu > 0$.

In terms of the variable $v = u - w^c$, equation (75.2) can be rewritten as

$$\begin{aligned} \partial_t v - \nu \Delta v + (v \cdot \nabla) v + (v \cdot \nabla) w^c + (w^c \cdot \nabla) v + \nabla q = 0 \\ \nabla \cdot v = 0. \end{aligned} \tag{75.5}$$

Also the vector field v should satisfy the homogeneous Dirichlet boundary conditions $v(x, y, z, t) = 0$, for $(x, y) \in \partial An \times [0, L]$, and v is periodic in z. The system of equations (75.5), with the Dirichlet boundary conditions, is typical of what arises as the result of homogenization of the boundary conditions for the Navier-Stokes equations. In short, one has the basic Navier-Stokes equations, as described in Chapter 6, but now augmented by adding the linear terms $(v \cdot \nabla) w^c$ and $(w^c \cdot \nabla) v$.

As in Chapter 6, we now use the Helmholtz projection $\mathbb{P}$ on (75.5) to obtain

$$\partial_t v + \nu A v + B(v, v) + \omega [B(v, \hat{w}) + B(\hat{w}, v)] = 0, \tag{75.6}$$

where $\hat{w}$ is given by $\omega \hat{w} = w_\theta(r)\overrightarrow{\theta}$. Owing to the special boundary conditions on (75.5), the range H of the Helmholtz projection differs slightly from the construction given in Section 3.8. In particular, H is now the closure in $L^2(\Omega, \mathbb{R}^3)$ of the collection of all functions $u = u(x, y, z)$ in $C^\infty(\bar{\Omega}, \mathbb{R}^3)$ with the property that u is periodic in z and the set $\{(x, y, z) \in \bar{\Omega} : u(x, y, z) = 0\}$ contains a neighborhood of $\partial An \times [0, L]$, see Chossat and Iooss (1994).

$$\begin{pmatrix}\text{Couette}\\ \text{flow}\end{pmatrix} \longrightarrow \begin{pmatrix}\text{Taylor}\\ \text{vortex}\end{pmatrix} \longrightarrow \begin{pmatrix}\text{Wavy}\\ \text{vortex}\end{pmatrix} \longrightarrow \begin{pmatrix}\text{Modulated}\\ \text{vortex}\end{pmatrix} \longrightarrow ???$$

$$T_c^0 \longrightarrow T_t^0 \longrightarrow T^1 \longrightarrow T^2 \longrightarrow ???$$

Figure 7.8. Bifurcations in the Couette-Taylor Flow

Since the Stokes operator A satisfies the Standing Hypothesis B, we see that at $\omega = 0$, the stationary solution $v \equiv 0$ of equation (75.6) is stable and hyperbolic. It then follows from Theorem 71.6 that there is an $\omega_1 > 0$ such that the zero solution $v \equiv 0$ of equation (75.6) is stable and hyperbolic for $0 \leq \omega < \omega_1$. This stability and hyperbolicity will persist for $\omega > 0$ until an eigenvalue of the linear operator

$$L^c v = \nu A v + \omega[B(v, \hat{w}) + B(\hat{w}, v)]$$

crosses the imaginary axis in the complex plane. We will denote this value of ω as $\omega = \omega_1$. For $0 \leq \omega < \omega_1$, the experimental results and the rigorous analytical results are essentially in full agreement. The laminar flow pattern associated with the Couette flow is denoted by T_c^0 in Figure 7.8.

It may be helpful at this time to recall some of the experimental results appearing in the literature, see Gollub and Swinney (1975); Lvov, Predtechensky, and Chernykh (1983); and Andereck, Liu, and Swinney (1986). However, before doing this, it is important to emphasize a basic difference between the CT model described above and the real world situation one encounters in the experimental setup. In particular, the assumption of spatial periodicity in the z-variable is not a realistic assumption for the experimental setup. In the experiments, one encounters different boundary conditions on the top and bottom, and consequently, the CT model does not directly apply to this situation. The conventional wisdom, which is used in this case, is that if the aspect ratio $\frac{L}{1-\eta}$ is large enough, then the spatially periodic CT model should be a "reasonable" approximation to the real world phenomena seen in the experiments. It would be nice to have a theorem and proof which might substantiate a part of this conventional wisdom, but that remains an open problem.

A natural issue which then arises is, Why bother with the spatially periodic CT model in the first place? Why not work with a more realistic model with different boundary conditions on the top and bottom? These are very good questions, and perhaps that is what is ultimately needed. However, it should be noted that the analysis of such a model is far from a trivial matter. For example, if one replaces the spatial periodicity in z in (75.3) with a nonslip boundary condition on the top and bottom, then the Couette flow, as described above, is no longer a solution of the problem for $\omega \neq 0$. In fact, any nontrivial laminar flow satisfying the latter boundary conditions is not continuous at the top and bottom. Consequently, this flow is not in $C(\bar{\Omega}, \mathbb{R}^3)$. Owing to the compact imbedding $H^2(\Omega, \mathbb{R}^3) \hookrightarrow C(\bar{\Omega}, \mathbb{R}^3)$, it is not in $H^2(\Omega, \mathbb{R}^3)$ either. Needless to say, in order to get a good theory for the experimental setup, one needs to be very careful in the modeling of the fluid flow at the top and bottom of the container.

So much for our role as Devil's Advocate. Let us now accept conversion to the religion of conventional wisdom and put our worries aside, as we examine the experimental results. Based on the experiments, one sees the following scenarios: As ω crosses ω_1, a bifurcation of the Couette flow occurs, and for $\omega > \omega_1$, one observes a new pattern T_t^0, the Taylor vortex flow, see Kirchgässner and Sorger (1969) and Chossat and Iooss (1994). The latter flow is a new stationary solution $u = w^t$ for the original problem (75.2) and (75.3). (We use the superscript t here to refer to Taylor, who shares this symbol with Father Time.)

Next for ω in the range $\omega_1 < \omega < \omega_2$, the Taylor vortex flow is stable and hyperbolic. In this case, we consider the linearization of the Navier-Stokes equation (75.5) along the Taylor flow w^t, where the linear operator L^c is now replaced by

$$L^t v = \nu A v + [B(v, w^t) + B(w^t, v)], \qquad \text{for } \omega > \omega_1.$$

Since the Taylor solution w^t depends on the parameter ω, the linear operator L^t depends on ω, as well. As ω crosses the value ω_2, where an eigenvalue of L^t crosses into the unstable zone, another bifurcation occurs and a new pattern, the "wavy vortex flow" appears (see Figure 7.8). In this case, the bifurcation appears to be a Hopf bifurcation, and for $\omega > \omega_2$, one obtains a time-periodic solution of the Navier-Stokes equations. The orbit of this time-periodic solution is a 1D manifold T^1, i.e., a circle. The T^1-pattern resulting from this periodic solution persists for $\omega_2 < \omega < \omega_3$.

As ω crosses ω_3 a secondary bifurcation occurs resulting in another pattern, called the "modulated waves". What is generally believed is that the time-periodic orbit T^1 bifurcates into a 2D torus T^2, as ω crosses ω_3. For this reason, we will denote the modulated wave pattern by T^2. The pattern T^2 persists over a parameter range $\omega_3 < \omega < \omega_4$.

As ω crosses ω_4, it is not clear what occurs. It may be a Hopf-Landau bifurcation where the 2D torus T^2 bifurcates into a 3D torus T^3, see Chenciner and Iooss (1979), Sell (1979), and Haken (1981, 1983). Or it may be a more complicated behavior, including strange attractors, see Ruelle and Takens (1971) and Sell (1981). Or it may be something quite different. No one really knows. Each opinion is, at best, an educated guess, and each comes with its pros and cons.

What is "really known" about these scenarios? On the assumption that the three bifurcations, (where ω crosses ω_1, ω_2, and ω_3) are as described, then the Theorems 71.6, 74.14, and 74.15 guarantee the persistence of the patterns depicted in Figure 7.8, for ω in the ranges $\omega_1 < \omega < \omega_2$, $\omega_2 < \omega < \omega_3$, and $\omega_3 < \omega < \omega_4$. In order to apply the theorems from Section 7.4, one begins with the linearized equation along the manifold T^1 or T^2, and then one studies the perturbed problem which arises after making a small change in the parameter ω. The reason that Theorems 74.14 and 74.15 are applicable is that the terms in equation (75.6) depend continuously in ω, even in the C^1-topology. (We assume here that the manifolds have the required smoothness, and that the exponential trichotomy is of Lipschitz class.) Consequently inequality (74.1) is valid in each of these ranges, for small values of δ, provided that one makes a very small change in the parameter ω. It should be noted that the basic Hopf bifurcation theory assures the normal hyperbolicity of the periodic orbit T^1 and the 2D torus T^2 immediately after the bifurcation, see Sacker (1964, 1969), Ruelle and Takens (1971), Marsden and McCracken (1976), and Chow and Hale (1982).

What then is "really known" about the bifurcations themselves? This has been a subject of serious study over the last 30 years, see Kirchgässner (1975) and Chossat and Iooss (1994), for example. Since one has an explicit formula for the Couette solution w^c, it should be not surprising that much of the analytical work has focused on bifurcations in the vicinity of w^c. An indepth analysis of such bifurcations, which is based on (1) a smooth center manifold theorem and the associated normal forms and (2) the spatial symmetries enjoyed by the solutions of the Navier-Stokes equations in the region Ω, appears in Chossat and Iooss (1994). Some of this analysis also applies to the Taylor vortex solution w^t. While the existence of the bifurcation values ω_1 and ω_2 can be argued successfully based on such analyses, the computation of these values, in a specific case, is best left to the computer.

The analysis of the CT model for ω in the region $\omega > \omega_2$ is much more complicated. Even at the computational level, this offers great challenges. In principle, one can develop a center manifold for the study of bifurcations near a periodic orbit, and to some extent, for the study of the dynamics near a compact, invariant manifold, see Sell (1978) and Chow, Liu, and Yi (1999). For example, in the case of a periodic orbit, one can mimic the finite dimensional setting and construct a Poincaré mapping for a suitable normal cross section to the periodic orbit. However, whatever method one chooses, for applications to the CT model, one will need to use sophisticated numerical methods to locate the bifurcation values ω_3 and ω_4, see for example, Heywood and Rannacher (1982 - 1990). We turn to the numerical issues next.

7.5.2. The Bubnov-Galerkin Approximations. A common concern arising in the study of fluid flows is the connection between the theory of the Navier-Stokes equations, and the numerical study of these equations. While a complete study of this connection lies in the domain of approximation dynamics,[40] we can give some insight into the basic theory by focusing on the Bubnov-Galerkin approximations.

For this purpose, let us return to equation (75.6) and assume that, for

[40]See the forthcoming monograph Approximation Dynamics by G R Sell.

some $\omega > 0$, $M = M_\omega$ is a compact, invariant manifold of class C^2 in V^1. Since the Stokes operator A satisfies the Standing Hypothesis B, we let A^α denote the fractional powers of A with $V^{2\alpha} = \mathcal{D}(A^\alpha)$, for $\alpha \geq 0$. We note that equation (75.6) is of the form (70.1), where

$$F = F(v) = -[B(v,v) + \omega B(v,\hat{w}) + \omega B(\hat{w},v)]. \tag{75.7}$$

Furthermore, $F \in C^2_{\text{Lip}}$, see (61.32). Consequently, if M is normally hyperbolic in V^1, and the associated trichotomy is of Lipschitz class, then Theorems 74.14 and 74.15 are applicable in the study of the strong solutions of (75.6).

As in Section 6.2, we let $\{e_1, e_2, e_3, \cdots\}$ denote the orthonormal basis in H of eigenvectors of the Stokes operator A. For each integer $n \geq 1$, we let $P = P_n$ denote the orthogonal projection of H onto Span $\{e_1, \cdots, e_n\}$, and set $Q = Q_n = I - P$. For each $u \in H$ we define $p = p_n$ and $q = q_n$ by $p = Pu$ and $q = Qu$. Notice that these variables depend on the n modes $\{e_1, \cdots, e_n\}$ used to define the spectral projections $P = P_n$ and $Q = Q_n$. When we apply P and Q to the evolutionary equation (75.6), we obtain the equivalent system

$$\begin{cases} \partial_t p + \nu A p = PF(p+q), \\ \partial_t q + \nu A q = QF(p+q), \end{cases} \tag{75.8}$$

where F is given by equation (75.7). Notice that the p-equation in (75.8) is an n-dimensional equation, while the q-equation is infinite dimensional. The nth order Bubnov-Galerkin approximation of (75.6) is given by the solutions of the nth order ordinary differential equation

$$\partial_t p + \nu A p = PF(p), \tag{75.9}$$

which is obtained from (75.8) by setting $q = 0$ in the p-equation and ignoring the q-equation (see Section 6.2).

Since equation (75.9) is an ordinary differential equation on a finite dimensional space PH, and since equation (75.8) is an infinite dimensional system on H, it may not appear initially that there is any hope in comparing the longtime dynamics of the solutions of these two equations. Nevertheless, a good comparison can be made by imbedding equation (75.9) into a suitable system on the full space H. This imbedding is accomplished by appending a q-equation to the p-equation in (75.9) in such a way that the longtime dynamics of the new (p,q)-system is identical to the longtime dynamics of the ordinary differential equation (75.9). Such an imbedding is not unique. For example, the system

$$\partial_t p + \nu A p = PF(p), \quad \partial_t q + \nu A q = 0$$

has this property, provided that λ_{n+1}, the $(n+1)^{\text{st}}$ eigenvalue of the Stokes operator A, is positive. Another system, which seems better for our purposes, is

$$\begin{cases} \partial_t p + \nu A p = PF(p+q), \\ \partial_t q + \nu A q = 0. \end{cases} \tag{75.10}$$

Let $q(t) = e^{-\nu AQt} q_0$ be any solution of the q-equation in (75.10), where $q_0 \in QH \cap V^1$. One then has $\|A^{\frac{1}{2}} q(t)\|^2 \leq e^{-\lambda_{n+1}t} \|A^{\frac{1}{2}} q_0\|^2$, for $t \geq 0$. Consequently, it follows that equations (75.9) and (75.10) have the same longtime dynamics in the space V^1.

One can rewrite equations (75.8) and (75.10) in the form (70.1) and (70.2), where $F = F(u)$ is given by equation (75.7), and $G(u) = -QF(u)$. Thus one has $PF(u) = F(u) + G(u)$. In this way we see that equation (75.10) is a perturbation of equation (75.8), or equivalently, equation (75.6). This leads us to the following question:

Is the Main Theorem applicable to this perturbation problem (70.1)-(70.2)? As we now show, the answer is yes, provided that the eigenvalue λ_{n+1} is sufficiently large. In order to verify this, let us begin with a compact, invariant manifold M of class C^2 in V^1 for the original Navier-Stokes equations (75.6) and assume that M is normally hyperbolic and the associated exponential trichotomy is of Lipschitz class. For example, M may be the torus T^2 seen in the Couette-Taylor flow. Next let $U = N_R(0)$ be an open, neighborhood of the origin in V^1, of radius $R > 0$, where $M \subset U$ and $w^c \in U$.

As noted above, the term $F = F(u)$ satisfies $F \in C^2_{\text{Lip}}(V^1, V^{-\frac{1}{2}})$, see (61.32). Thus one has $2\beta = \frac{3}{2}$. Consequently, inequalities (46.7) and (46.8) imply that there exist $K_0 = K_0(\omega) > 0$ and $K_1 = K_1(\omega) > 0$ such that

$$\|A^{-\frac{1}{4}} F(u)\| \leq K_0 \quad \text{and} \quad \|A^{-\frac{1}{4}} DF(u)\|_{\mathcal{L}(V^1,H)} \leq K_1, \tag{75.11}$$

for all $u \in U$. Since Q is an orthogonal projection on V^α, for every $\alpha \in \mathbb{R}$, it follows that the perturbation term $G(u) = -QF(u)$ satisfies (75.11), as well. This implies that $\|G\|_{\{C^1(U)\}} \leq K_0 + K_1$, which need not be small!! However, since

$$A^\beta e^{-At} QF = (AQ)^\beta e^{-(AQ)t} QF \quad \text{and}$$
$$A^\beta e^{-At} QDF = (AQ)^\beta e^{-(AQ)t} QDF,$$

it follows from Exercise 37.9 and inequality (46.15) that

$$\|G\|_{\{A;C^1(U)\}} \leq (1-\beta)^{-1} e^{-\beta} \lambda_{n+1}^{\beta-1} \|G\|_{\{C^1(U)\}}, \qquad \text{where } \beta = 3/4.$$

We see that for λ_{n+1} sufficiently large, or equivalently, for n sufficiently large (say $n \geq N_\omega$), one has $\|G\|_{\{A;C^1(U)\}} \leq \delta$, where δ is given by the Main Theorem. As a result, we see that for any $n \geq N_\omega$, where $\lambda_n < \lambda_{n+1}$, the ordinary differential equation (75.9) has a compact, invariant, normally hyperbolic manifold M_n with $\dim M_n = \dim M$. Furthermore, the manifold M_n, as imbedded in the problem (75.10), is "close to" M.

Remarks. (1) Since the nonlinearity F satisfies (61.32), one can use this to generate a bootstrap argument to conclude that the solutions on the manifold M have greater regularity than $M \subset \mathcal{D}(A) = V^2$, as noted in the Herculean Theorem 47.6. In order to prove this, one needs to note that the exponential trichotomy on the space V^1 induces an exponential trichotomy on V^2 with good characteristics, see Lemma 45.5. Then $f(t) = F(S(t)u_0)$ has greater regularity. As a result one can use the theory of Section 4.7 to conclude that $M \subset V^3 \subset H^3(\Omega_0)$, with a corresponding improvement in (47.14).

(2) The fact that the pseudonorm $||G||_{\{A;C^1(U)\}}$ may be small, even when the pseudonorm $||G||_{\{C^1(U)\}}$ is large, plays an important role even in the study of some ordinary differential equations. For example, this issue arises when comparing the longtime dynamics generated by two multigrid methods used for the calculation of approximate solutions of a partial differential equation. Under the appropriate setup, a normally hyperbolic, invariant manifold for a fine grid calculation can be well approximated by a nearby normally hyperbolic, invariant manifold for a coarse grid calculation. This theory is developed further in the forthcoming monograph, Sell (2001).

7.6. Nonautonomous Problems.

The theory of dynamical systems extends readily to nonautonomous evolutionary equations of the form

$$\partial_t u + Au = F(u, t)$$

by using the construction based on the theory of skew product semiflows. This construction, which is widely used in the finite dimensional theory of ordinary differential equations, plays a similar role in the infinite dimensional setting. The main advantage of using skew product semiflows can be illustrated by studying the time-dependent forcing of equation (70.1) given by

$$\partial_t u + Au = F(u) + f(t), \tag{76.1}$$

where $f \in \mathcal{G} \stackrel{\text{def}}{=} L^\infty(\mathbb{R}, W) \cap C(\mathbb{R}, W)$ and $F \in C^1_{\text{Lip}}(V^{2\beta}, W)$, with $0 \leq \beta < 1$. Recall that the mapping given by $(f, \tau) \to f_\tau$, where $f_\tau(t) = f(\tau + t)$, for τ, $t \in \mathbb{R}$, generates a flow on $\mathcal{G}$, when $\mathcal{G}$ has the topology of uniform convergence on bounded sets, see Appendices A and B. For $\delta > 0$, we define $\mathcal{G}_\delta$ to be the collection of $f \in \mathcal{G}$ that satisfy $||f||_\infty \leq \delta$, where

$$||f||_\infty = ||f||_{L^\infty} = \sup\{||f(t)|| : t \in \mathbb{R}\}.$$

Similarly, we define $\mathcal{F} \stackrel{\text{def}}{=} L^\infty(\mathbb{R}, V^{2\beta}) \cap C(\mathbb{R}, V^{2\beta})$, and we will use the norm $||A^\beta \phi||_\infty$, for $\phi \in \mathcal{F}$. For $\rho > 0$, we define $\mathcal{F}_\rho$ to be the collection of $\phi \in \mathcal{F}$ that satisfy $||A^\beta \phi||_\infty \leq \rho$. We are now prepared for the following basic theorem.

Theorem 76.1. *Let the Standing Hypothesis A be satisfied and let $F = F(u)$ be in $C^1_{\text{Lip}}(V^{2\beta}; W)$. Assume that the unperturbed problem (70.1) has a hyperbolic stationary solution $u_0 \in V^{2\beta}$, and let (K, α) be the characteristics of hyperbolicity. Then for every $\rho > 0$, there is a $\delta > 0$ such that if $f \in \mathcal{G}_\delta$, then the perturbed equation (76.1) has a unique solution $u = \phi \stackrel{\text{def}}{=} S(f) \in \mathcal{F}_\rho$. Moreover, the mapping $S : \mathcal{G}_\delta \to \mathcal{F}_\rho$ given by $\phi = S(f)$ satisfies $S(f_\tau) = \phi_\tau = S(f)_\tau$, for all $f \in \mathcal{G}_\delta$ and $\tau \in \mathbb{R}$. Furthermore, for all f, $g \in \mathcal{G}_\delta$, one has*

$$\|A^\beta(S(f) - S(g))\|_\infty \leq 2\left(\frac{2K}{\alpha} + \frac{K}{1-\beta}\right)\|f - g\|_\infty. \tag{76.2}$$

Hence, S is a flow homomorphism from $\mathcal{G}_\delta$ onto $S(\mathcal{G}_\delta) \subset \mathcal{F}_\rho$.

Proof. As in Section 7.1, we let $L = A - DF(u_0)$. Let (K, α) denote the characteristics for the exponential dichotomy for equation (71.1) and let $\{P, Q\}$ be the associated projectors. Let $E(v) = E(u_0, v)$ be given by (71.8). Note that $E(v)$ satisfies (71.10) and (71.11), where $\gamma(\rho) \in \Sigma$. By using the change of variables $u = v + u_0$, one can rewrite equation (76.1) in the form

$$\partial_t v + Lv = E(v) + f(t), \qquad \text{for } f \in \mathcal{G}. \tag{76.3}$$

We now use the Lyapunov-Perron method and seek a fixed point of the mapping $\phi \to \hat{\phi} = \mathcal{T}(f)\phi$ given by

$$\begin{aligned}\hat{\phi}(t) = &\int_{-\infty}^{t} Qe^{-L(t-s)}[E(\phi(s)) + f(s)]\,ds \\ &- \int_{t}^{\infty} Pe^{-L(t-s)}[E(\phi(s)) + f(s)]\,ds.\end{aligned} \tag{76.4}$$

We now show that, for $\delta > 0$ and $\rho > 0$ small enough and for each $f \in \mathcal{G}_\delta$, the mapping $\hat{\phi} = \mathcal{T}(f)\phi$ given by (76.4) maps $\mathcal{F}_\rho$ into itself, and it is a strict contraction on $\mathcal{F}_\rho$. Indeed, since $\|f\|_\infty \leq \delta$ and $\|A^\beta\phi\|_\infty \leq \rho$, it follows from inequalities (71.4) - (71.6), with $\lambda = 0$, and (71.10) and (71.11), that

$$\begin{aligned}\|A^\beta\hat{\phi}(t)\| \leq\ & K(\rho\gamma(\rho) + \delta))\times \\ &\left(\int_{-\infty}^{t-1} e^{-\alpha(t-s)} + \int_{t-1}^{t} (t-s)^{-\beta}e^{-\alpha(t-s)} + \int_{t}^{\infty} e^{\alpha(t-s)}\right) ds \\ \leq\ & \left(\frac{2K}{\alpha} + \frac{K}{1-\beta}\right)(\rho\gamma(\rho) + \delta),\end{aligned}$$

for all $t \in \mathbb{R}$. Similarly, if ϕ_1, $\phi_2 \in \mathcal{F}_\rho$ and $f \in \mathcal{G}_\delta$ is fixed, then one has

$$\|A^\beta(\hat{\phi}_1(t) - \hat{\phi}(t))\| \leq \left(\frac{2K}{\alpha} + \frac{K}{1-\beta}\right)\gamma(\rho)\|A^\beta(\phi_1 - \phi_2)\|_\infty,$$

for all $t \in \mathbb{R}$. Now fix $\rho > 0$ so that $(\frac{2K}{\alpha} + \frac{K}{1-\beta})\gamma(\rho) \leq \frac{1}{2}$ and let $\delta > 0$ satisfy $(\frac{2K}{\alpha} + \frac{K}{1-\beta})\delta \leq \frac{\rho}{2}$. We then obtain $\|A^\beta \hat{\phi}\| \leq \rho$, whenever, $\phi \in \mathcal{F}_\rho$, and

$$(76.5) \quad \|A^\beta(\mathcal{T}(f)\phi_1 - \mathcal{T}(f)\phi_2)\|_\infty = \|A^\beta(\hat{\phi}_1 - \hat{\phi}_2)\|_\infty \leq \frac{1}{2}\|A^\beta(\phi_1 - \phi_2)\|_\infty,$$

for all ϕ_1, $\phi_2 \in \mathcal{F}_\rho$ and all $f \in \mathcal{G}_\delta$. Hence, for a fixed f in $\mathcal{G}_\delta$, the mapping $\phi \to \mathcal{T}(f)\phi$ is a strict contraction on $\mathcal{F}_\rho$. We will denote the unique fixed point of this mapping by $\phi = S(f)$. The argument used to prove Theorem 45.7, Item (3), applies here to conclude that ϕ is a mild solution of equation (76.3). The proof that $S(f_\tau) = \phi_\tau = (Sf)_\tau$, for all $f \in \mathcal{G}_\delta$ and $\tau \in \mathbb{R}$, follows by making a simple change of variables in equation (76.4), after replacing $\hat{\phi}$ by the fixed point ϕ.

Now for ϕ fixed in $\mathcal{F}_\rho$, one can use the argument above to show that

$$(76.6) \qquad \|A^\beta(\mathcal{T}(f)\phi - \mathcal{T}(g)\phi)\|_\infty \leq \left(\frac{2K}{\alpha} + \frac{K}{1-\beta}\right)\|f - g\|_\infty,$$

for all f, $g \in \mathcal{G}_\delta$. Let $f_i \in \mathcal{G}_\delta$ and set $\phi_i = Sf_i = \mathcal{T}(f_i)\phi_i$, for $i = 1, 2$. Since

$$\begin{aligned}\|A^\beta(Sf_1 - Sf_2)\|_\infty &\leq \|A^\beta(\mathcal{T}(f_1)\phi_1 - \mathcal{T}(f_2)\phi_1)\|_\infty \\ &\quad + \|A^\beta(\mathcal{T}(f_2)\phi_1 - \mathcal{T}(f_2)\phi_2)\|_\infty,\end{aligned}$$

inequality (76.2) now follows from inequalities (76.5) and (76.6), with $f = f_1$ and $g = f_2$. □

What are the implications of this theorem on the study of the longtime dynamics of the mild solutions of equation (76.1)? First we note that, since S is a flow homomorphism, it preserves the dynamics of the translational flows $(f, \tau) \to f_\tau$ on $\mathcal{G}_\delta$ and $(\phi, \tau) \to \phi_\tau$ on $S(\mathcal{G}_\delta) \subset \mathcal{F}_\rho$. In order to compare these dynamics with the dynamics generated by the mild solutions of equation (76.1), one should use the skew product semiflow π given by $\pi(f, u_0; \tau) = (f_\tau, S(f, \tau)u_0)$, where $S(f, t)u_0$ is the maximally defined mild solution of equation (76.1) that satisfies $S(f, 0)u_0 = u_0$ (see Theorem 47.5). Recall that the domain of definition of π is

$$\mathcal{D}(\pi) = \Xi = \{(f, u_0, t) \in \mathcal{G}_\delta \times V^{2\beta} \times \mathbb{R}^+ : 0 \leq t < \omega(f, u_0))\},$$

where $I = [0, \omega(f, u_0))$ is the interval of definition of $S(f, \cdot)u_0$. The cocycle property (47.10) implies that if $\tau \in [0, \omega(f, u_0))$ and $t \in [0, \omega(f_\tau, S(f, \tau)u_0))$, then $\tau + t \in [0, \omega(f, u_0))$ and

$$S(f_\tau, t)S(f, \tau)u_0 = S(f, \tau + t)u_0.$$

Since $\phi = S(f)$ is a mild solution of (76.1), it satisfies

$$S(f, t)\phi(0) = \phi(t) = e^{-At}u_0 + \int_0^t e^{-A(t-s)}[F(\phi(s)) + f(s)]\, ds, \qquad \text{for } t \geq 0.$$

Note that $\omega(f, \phi(0)) = \infty$ and that $\phi(t)$ is a global solution of equation (76.1) passing through $\phi(0)$. Now Theorem 45.7, Item (3), implies that

$$S(f_\tau, t)\phi(\tau) = \phi(\tau + t), \qquad \text{for all } \tau \in \mathbb{R} \text{ and } t \geq 0. \tag{76.7}$$

Hence the translational flow $(\phi, \tau) \to \phi_\tau$ and the second component of the semiflow defined by π agree along the trajectory through $(f, \phi(0)) \in \mathcal{G}_\delta \times V^{2\beta}$. Furthermore, equation (76.7) has a negative continuation $\phi(t)$, for $t \leq 0$.

We now define the mapping $J : \mathcal{G}_\delta \to \mathcal{G}_\delta \times V^{2\beta}$ by

$$J(f) = (f, \phi(0)), \qquad \text{where } \phi = S(f).$$

Since $\phi_\tau = (S(f))_\tau = S(f_\tau)$, one has $J(f_\tau) = (f_\tau, \phi(\tau))$, for all $f \in \mathcal{G}_\delta$ and all $\tau \in \mathbb{R}$. Thus J is a flow homomorphism from $\mathcal{G}_\delta$ onto $J(\mathcal{G}_\delta) \subset \mathcal{G}_\delta \times V^{2\beta}$. For each $f \in \mathcal{G}_\delta$, we define the fiber $\mathcal{E}(f)$ of $J(\mathcal{G}_\delta)$ by

$$\mathcal{E}(f) = \{u_0 \in V^{2\beta} : \phi(0) = u_0, \text{ where } \phi = S(f)\}.$$

It follows from Theorem 76.1 that $\mathcal{E}(f)$ contains exactly one point, for each $f \in \mathcal{G}_\delta$. In the language of covering spaces, $J(\mathcal{G}_\delta)$ is a 1-cover of $\mathcal{G}_\delta$, see Sacker and Sell (1977). While each fiber $\mathcal{E}(f)$ contains a single point, it does not follow that $\mathcal{E}(f) \neq \mathcal{E}(g)$, whenever $f \neq g$. However, the flow homomorphism property for S implies that, for f, $g \in \mathcal{G}_\delta$, one has

$$\mathcal{E}(f) = \mathcal{E}(g) \Longrightarrow \mathcal{E}(f_\tau) = \mathcal{E}(g_\tau), \qquad \text{for all } \tau \in \mathbb{R}.$$

It follows from (76.2) that J is Lipschitz continuous. Also the inverse mapping $J^{-1} : J(\mathcal{G}_\delta) \to \mathcal{G}_\delta$, given by $J^{-1}(f, \phi(0)) = f$, is Lipschitz continuous, as well. Hence $\mathcal{G}_\delta$ and $J(\mathcal{G}_\delta)$ are homeomorphic.

Next let $\mathcal{K}$ be a compact invariant set in $\mathcal{G}_\delta$. Then $S(\mathcal{K})$ is a compact invariant set in $\mathcal{F}_\rho$. If $\mathcal{K}$ is a minimal set (see Section 7.7) then $S(\mathcal{K})$ is minimal as well. Since $\mathcal{K}$ is an extension of $S(\mathcal{K})$, this means that the dynamics on $\mathcal{K}$ may have "greater complexity" than the dynamics on $S(\mathcal{K})$ (see Section 2.1.5). Let us illustrate the significance of the last statement in the setting where $f(t)$ is a quasi periodic function of t.

There is a convenient and useful characterization of quasi periodic functions, and we will use it as a definition. We begin with an arbitrary torus T^m of dimension $m \geq 1$. Let $\sigma(\theta, t) = \theta + \omega t$ denote a **twist flow** on T^m, where $\theta = (\theta_1, \cdots, \theta_m) \in T^m$ and $\omega = (\omega_1, \cdots, \omega_m) \in \mathbb{R}^m$. We assume that the vector ω is chosen so that it is **independent** over the integers, i.e., one has $n \cdot \omega \neq 0$, whenever $n = (n_1, \cdots, n_m) \in Z^m$ with $n \neq 0$. The latter assumption on ω implies that each trajectory $\gamma(\theta) = \{\theta + \omega t : t \in \mathbb{R}\}$ is dense in T^m, and therefore, T^m is a minimal set. T^m is, in fact, the prototypical example of a quasi periodic minimal set.

Let X be a Banach space. A function $g : \mathbb{R} \to X$ is said to be a **quasi periodic function** if there exist

(1) an integer $m \geq 1$ and a torus T^m,
(2) a vector $\omega \in \mathbb{R}^m$ that is independent over the integers, and
(3) a continuous function $G : T^m \to X$, such that, for some $\theta_0 \in T^m$, one has

$$g(t) = G(\theta_0 + \omega t), \qquad \text{for all } t \in \mathbb{R}.$$

Thus for each $\theta \in T^m$, the function $\mathfrak{G}(\theta) = \mathfrak{G}(\theta)(t)$ defined by $\mathfrak{G}(\theta)(t) \stackrel{\text{def}}{=} G(\theta + \omega t)$, for $t \in \mathbb{R}$, is a quasi periodic function and $g(t) = \mathfrak{G}(\theta_0)(t)$. It follows from the definition that, every quasi periodic function $g = g(t)$ is an element of the space $L^\infty(\mathbb{R}, X) \cap C(\mathbb{R}, X)$. As is usual, we will use the topology of uniform convergence on bounded sets (in $\mathbb{R}$) to construct the hull $H(g) = C\ell\{g_\tau : \tau \in \mathbb{R}\}$, which is a quasi periodic minimal set. Since g_τ satisfies

$$g_\tau(t) = g(\tau + t) = G(\theta + \omega t) = \mathfrak{G}(\theta)(t), \qquad \text{where } \theta = \theta_0 + \omega\tau,$$

we see that $\mathfrak{G}(\theta) \in H(g)$, for all θ of the form $\theta = \theta_0 + \omega\tau$, where $\tau \in \mathbb{R}$. Since the trajectory $\gamma(\theta_0)$ is dense in T^m, it follows that $H(g) = \{\mathfrak{G}(\theta) : \theta \in T^m\}$ and the flow on $H(g)$ is

$$\sigma(\mathfrak{G}(\theta), \tau)(t) = \mathfrak{G}(\theta + \omega\tau)(t) = G((\theta + \omega\tau) + \omega t), \qquad \text{for } \theta \in T^m.$$

In other words, $\mathfrak{G} : T^m \to \mathfrak{G}(T^m) \subset L^\infty(\mathbb{R}, X) \cap C(\mathbb{R}, X)$ is a flow homomorphism, and T^m, with the given twist flow, is an extension of $\mathfrak{G}(T^m)$, with the translational flow. One can show that $\mathfrak{G}(T^m)$ is a torus of dimension k, where[41] $0 \leq k \leq m$. See Cartwright (1967, 1969), Sacker and Sell (1977), and Shen and Yi (1998), for example.

The Quasi Periodic Case. An application of Theorem 76.1 arises when $f = f(t)$ is a quasi periodic function. For example, equation (76.1) might represent the equations of motion for an oceanic fluid flow problem, where $f = f(t)$ is the quasi periodic variation in the forcing term owing to the gravitational forces exerted by the Sun and the Moon. (We assume here, for simplicity, that the Sun-Earth-Moon system is a quasi periodic solution of the 3-body problem of celestial mechanics.) The proof of the following result is left as an exercise.

Corollary 76.2. *Let the Hypotheses of Theorem 76.1 be satisfied, and let δ, ρ, and S be given by that result. Assume now that $f \in \mathcal{G}_\delta$ is a quasi periodic function and let $\mathcal{K}$ denote the hull of f in $\mathcal{G}_\delta$. Then $S(\mathcal{K})$ is a torus and a quasi periodic minimal set in $\mathcal{F}_\rho$ and $\dim(S(\mathcal{K})) \leq \dim(\mathcal{K})$.*

---

[41]One has $k = 0 \Leftrightarrow T^0$ is a set with a single point $\Leftrightarrow G$ is a constant function.

7.7. Other Topics of Dynamical Systems.

The topics presented in this chapter fall into the general area of perturbation theories in dynamical systems. There are legions of other applications which we are unable to cover. Many of them would require their own treatise, and some are developed elsewhere. Instead we offer the reader a brief commentary and selected references as an entrée to the vast literature of dynamical systems. Additional references can be found on the World Wide Web at Sell (2001).

7.7.1. Differential Delay Equations. Functional Differential Equations. The C_0-theory for nonlinear problems, which is developed in Chapter 4, is applicable to the study of differential delay equations and functional differential equations. For a good introduction to the dynamics of functional differential equations, the reader should consult: Bellman and Cooke (1963); Hale (1977); Hale and Verduyn Lunel (1993); Mallet-Paret (1988); Mallet-Paret and Sell (1996a,b); Krisztin, Walther, and Wu (1999); and Krisztin and Walther (2000).

7.7.2. Bifurcation Theory. In the perturbation theorems described above, one finds that some dynamical features in a semiflow are preserved under small perturbations of the underlying evolutionary equation. In particular, under the presence of a suitable hyperbolic structure, as represented by an exponential dichotomy or an exponential trichotomy, one finds that equilibrium points, periodic orbits, or invariant manifolds vary continuously with respect to the parameters of the problem. On the other hand, if the hyperbolic structure fails at a given value of the parameter, then the continuous variation of the dynamical feature can fail, as well. It is at this point that bifurcation theory enters the dynamics arena.

The classical theory of Hopf bifurcation at an equilibrium point can be found in Andronov (1929) and Hopf (1942). In the case of the bifurcation of a periodic orbit, see Sacker (1964) and Ruelle and Takens (1971). For the bifurcation of higher-dimensional tori, see Sell (1979) and Chenciner and Iooss (1979). Recent aspects of the Melnikov (1963) theory involving time-periodic perturbations of homoclinic orbits appear in Palmer (1984), and for quasi periodic perturbations, see Meyer and Sell (1989). General features of perturbation theory and bifurcation theory with higher codimension appear in Takens (1974), Bogdanov (1974), Chow and Hale (1982), V I Arnold (1983), Guckenheimer and Holmes (1983), Lin (1990), and Hale and Kocak (1991). Global bifurcation and generic bifurcations can be found in Fiedler (1988) and Chow, Hale, and Mallet-Paret (1975, 1976). Extensions to the infinite dimensional setting can be found in Marsden and McCracken (1976), Chow and Mallet-Paret (1977, 1978), and Chossat and Iooss (1994).

7.7.3. Ergodic Theory. Ergodic theory is a subject which lies at the intersection of dynamical systems and operator theory. More specifically, let $S(t)$ be a semiflow on a compact metric space $\mathcal{K}$. A measure μ is said to be a **probability measure** on $\mathcal{K}$ if $\mu(\mathcal{K}) = 1$ and $\mu(M) \geq 0$, for each measurable set M. A measure μ on $\mathcal{K}$ is said to be **invariant**, with respect to the semiflow $S(t)$, if it is a probability measure and $\mu(S(t)^{-1}(M)) = \mu(M)$, for every measurable set M and every $t \geq 0$. An invariant measure μ on $\mathcal{K}$ is said to be an **ergodic measure** for $S(t)$ if, for every measurable, invariant set M in $\mathcal{K}$, one has $\mu(M) = 0$, or 1.

The ergodic theorem for dynamical systems provides sufficient conditions for an invariant measure μ on the set $\mathcal{K}$ to satisfy

$$\lim_{T\to\infty} \frac{1}{T} \int_0^T f(S(t)u_0)\,dt \overset{\text{a.e.}}{=} \int_{\mathcal{K}} f\,d\mu, \tag{77.1}$$

for all continuous functions f, where the equality $\overset{\text{a.e.}}{=}$ refers to the invariant measure μ, see Birkhoff (1931a,b). Related work of Krylov and Bogoliubov (1937) examines the general of invariant measure and ergodic measures for a semiflow on a compact, metric space. Notice that equation (77.1) is a statement equating a "time" average with a "space" average. The flow on the unit circle in the plane, generated by the ordinary differential equation $\partial_t\theta = \sin(\theta/2)$, shows that equation (77.1) holds only for $\theta_0 = 0$. This flow has a unique invariant measure, the atomic measure $\mu(0) = 1$. Thus equation (77.1) is valid μ-almost everywhere. For the general case, it is known that equation (77.1) is valid, for any ergodic measure μ.

There are hundreds of papers written on the connections between ergodic theory and the theory of dynamical systems generated by evolutionary equations. Here is a small list of references. For differential equations with almost periodic and almost automorphic coefficients, see Johnson (1980a, 1981), Sell (1981a), Johnson and Moser (1982), Shen and Yi (1998), and Yi (1998). Applications to the statistical theory of fluid flows and turbulence can be found in Foias and Temam (1980, 1987) and Foias, et al (2001). The Multiplicative Ergodic Theorem and related topics appear in Liao (1966, 1973), Oseledec (1968), Millionščikov (1968), Pesin (1977), Ruelle (1979), Katok (1980), and Johnson, Palmer, and Sell (1987). The following volumes contain some very useful material: Nemytskii and Stepanov (1960), Ellis (1969), and Sinai (1994).

A recent and significant development in the general area of dynamical systems is the application of the basic techniques to the theory of **random** dynamical systems. In this new setting, a random process is imbedded into the skew product flow/semiflow format. In order to accommodate the randomness of the system, it becomes appropriate to replace the continuity of the mapping π (see Section 2.4) with a suitable form of measurability. The probability measure associated with the random processes then becomes an invariant measure on the base space of the skew product flow/semiflow. See L Arnold (1995, 1998) and the references contained therein for more information.

7.7.4. Dimension Theory. Rather early in the study of the dynamics of infinite dimensional systems, it was noted that the attractors and global

There once was a Square, such a square little Square,
And he loved a trim Triangle;
But she was a flirt and around her skirt
Vainly she made him dangle.
Oh he wanted to wed, and he had no dread
Of domestic woes and wrangles;
For he thought that his fate was to procreate
Cute little squares and triangles.

Now it happened one day on that geometric way
There swaggered a big bold Cube,
With a haughty stare he made that Square
Have the air of a perfect boob;
To his solid spell the Triangle fell,
And she thrilled with love's sweet sickness,
For she took delight in his breadth and height -
But how she adored his thickness!

So that poor little Square just died of despair,
For his love he could not strangle;
While the bold Cube led to the bridal bed
That cute and acute Triangle.
The Square's sad lot she has long forgot,
And his passionate pretensions ...
For she dotes on her kids - Oh such cute Pyramids -
In a world of three dimensions.

Robert W Service

attractors oftentimes have finite dimension, see Mallet-Paret (1976) and Mañé (1981). More recently, new techniques, which are based on the theory of Lyapunov exponents, have enabled researchers to derive good estimates for upper bounds on the dimension of the global attractor, for a wide class of problems, see for example, Douady and Oesterlé (1980), Constantin and Foias (1985), and Eden, Foias and Temam (1991). An excellent treatment of the dimension theory can be found in Temam (1988), and we refer the reader to this source for more details.

The theory of Lyapunov exponents is applicable to the study of solutions of the linearized equation evaluated along trajectories in a compact invariant set for a nonlinear semiflow. The basic idea is to study the asymptotic exponential growth rate of the norm of the solutions $||\Phi(m, t)v||$ in the linear skew product semiflow generated by the mild solutions of the linearized equation (see Section 4.9). The Lyapunov exponent is defined by

$$\lambda = \lambda(m, v) \stackrel{\text{def}}{=} \limsup_{t \to \infty} \frac{1}{t} \|\Phi(m, t)v\|,$$

where m is a point in a compact invariant set $\mathcal{K}$. We let μ denote any ergodic measure on $\mathcal{K}$. We assume now that the operator $\Phi(m, t)$ is compact for all $t > t_0$, where $t_0 \geq 0$. In a finite dimensional space, or in a Hilbert space, the Multiplicative Ergodic Theorem states that there is an invariant measurable set $M \subset \mathcal{K}$, with $\mu(M) = 1$, such that the following hold:

(1) there is a sequence of exponents

$$\cdots < \lambda_3 < \lambda_2 < \lambda_1 < \infty \quad \text{and integers} \quad n_1, n_2, n_3, \cdots,$$

such that $n_i \geq 1$;

(2) for all $(m, v) \in M \times V^{2\beta}$ one has $\lambda(m, v) = \lambda_i$, for some integer $i \geq 1$;

(3) for almost all $m \in M$, the subspace

$$V_i(m) = \text{Span}\{v \in V^{2\beta} : \lambda(m, v) = \lambda_i\}$$

is well-defined with $\dim(V_i(m)) = n_i$, for $i \geq 1$; and

(4) the linear spaces $V_i(m)$ are "measurable" functions of $m \in M$.

See Oseledec (1968), Millionščikov (1968), Ruelle (1979), and Johnson, Palmer, and Sell (1987). For integers $m \geq 1$, we define

$$E_m \stackrel{\text{def}}{=} n_1\lambda_1 + n_2\lambda_2 + \cdots + n_m\lambda_m \quad \text{and} \quad N_m \stackrel{\text{def}}{=} n_1 + n_2 + \cdots + n_m.$$

The advantage of the Lyapunov exponents is that, by using the wedge product of the linear skew product semiflow, one can show that E_m is a good measure of the asymptotic growth rate of the infinitesimal N_m-dimensional volume elements in the nonlinear semiflow on $\mathcal{K}$. If m is large enough so that $E_m < 0$, then N_m is a upper bound for the dimension of $\mathcal{K}$, see Temam (1988).

7.7.5. Minimal Set: The Basic Building Block. Let M be a compact, nonempty, invariant set for a flow σ. M is said to be a **minimal set** if M does not contain a proper subset which is also a nonempty, compact, invariant set for σ. Examples of minimal sets include the equilibrium points, the periodic orbits, and the hull of an almost periodic motion, see Sell (1971). Birkhoff (1927) characterized minimal sets in terms of a recurrence property of the flow σ. Also see Bohr (1925 - 1926), Favard (1933), Furstenberg (1963, 1967), and Ellis (1969).

The study of minimal sets in dynamical systems has developed into a broad tapestry with many interesting and profound features. The connection with ergodic theory is especially noteworthy. In the case where $\mathcal{K}$ is an almost periodic minimal set - i.e., the hull of an almost periodic motion, - it is known that there exists a unique ergodic measure on $\mathcal{K}$. The proof of this fact is based on the group-theoretic methods lying behind the construction of the Haar measure on $\mathcal{K}$. By using the concept of the enveloping semigroup, Ellis (1969) has extended the group theoretic methods to the study of minimal sets in general.

While the existence of minimal sets with more than one ergodic measure had been known for some time, the fact that this can occur in the dynamics of the solutions of a 2-dimensional linear ordinary differential equation with almost periodic coefficents, came as a surprise, see Johnson (1980a, 1981). What underlies this development is the concept of an almost automorphic, minimal set, see Veech (1965), Shen and Yi (1998), and Yi (1998).

7.7.6. Singular Perturbations. A rapidly growing area of perturbation theory in dynamical systems centers around various theories of singular perturbations. While these theories come in differing guises, they share a common dynamical heritage. First there are problems with multiple time-scales, as is illustrated by the system

$$\epsilon^{-1}\partial_t u = f(u,v), \quad \partial_t v = g(u,v), \tag{77.2}$$

where $\epsilon > 0$ is a small parameter. One way to study the behavior of the solutions of equation (77.2) is to introduce a new time scale τ, where $\tau = \epsilon\, t$. In this case, one obtains the equivalent system

$$\partial_\tau u = f(u,v), \quad \partial_\tau v = \epsilon\, g(u,v). \tag{77.3}$$

One can then set $\epsilon = 0$ to obtain a reduced problem

$$\partial_\tau u = f(u,v), \quad \partial_\tau v = 0.$$

While the latter system of equations is simpler than equation (77.3), it becomes difficult now to relate this to the original problem because the connection between the time scales, $\tau = \epsilon\, t$, looses its meaning at $\epsilon = 0$.

Another approach for problem (77.2) is to focus on those solutions that remain within some bounded set, as $\epsilon \to 0$. In this case, the functions $f(u,v)$ and $g(u,v)$ remian bounded, at least in the finite dimensional case. This implies that $\partial_t u$ must tend to 0, as $\epsilon \to 0$. An alternate reduced problem for (77.2) is

$$f(u,v) = 0, \quad \partial_t v = g(u,v).$$

In many applications of singular perturbations, one is in the fortunate situation where the Implicit Function Theorem is applicable, and one can solve the equation $f(u,v) = 0$ for $u = \Phi(v)$, which typically represents a manifold in the uv-space. The dynamics on this manifold M_0 is then given by the solutions of $\partial_t v = g(\Phi(v), v)$.

A connection with dynamical systems arises when one poses the problem: does there exist an invariant manifold M_ϵ for problem (77.2), for small $\epsilon > 0$, where M_ϵ is "close to" M_0? In order to address this question, one seeks to make a change of variables in the vicinity of the manifold M_0, so that in the new variables equation (77.2) reduces to a **regular** perturbation. For example, one might obtain a system

$$\partial_t y + Ay = F(y) + G(y, \epsilon), \tag{77.4}$$

where $G(\cdot, \epsilon) \to 0$ in a suitable space, as $\epsilon \to 0$, and M_0 is as invariant manifold for equation (77.4) at $\epsilon = 0$.

We use the equation (77.4) here simply to illustrate that some standard perturbation theories, as described in this chapter, for example, may be applicable in the study of singular perturbations. For more information on this paradigm, the reader should consult Fenichel (1979), Fusco and Hale (1989), Hale (1989), and Jones (1995).

The process that converts a singular perturbation problem like (77.2) into a regular problem like (77.4) is called a **regularization** of the singular perturbation. This is a key feature in any singular perturbation problem. We now present two illustrations of the regularization process in shadow systems and thin domains.

Shadow Systems. Consider the following system of reaction diffusion equations with differing diffusion rates.

$$\begin{aligned} \partial_t u &= \epsilon^{-2} \Delta u + f(u, v) \\ \partial_t v &= \Delta v + g(u, v), \end{aligned} \tag{77.5}$$

where the Laplacian operator satisfies the Neumann boundary conditions $\partial_n u = \partial_n v = 0$ on $\partial\Omega$ and Ω is a open, bounded domain in $\mathbb{R}^m$ with smooth boundary, and u and v are vector quantities, say $u \in \mathbb{R}^r$ and $v \in \mathbb{R}^s$. We also assume, for simplicity, that f and g are smooth, bounded functions with

$$(f, g) \in L^\infty(\mathbb{R}^r \times \mathbb{R}^s, \mathbb{R}^r \times \mathbb{R}^s).$$

Because of the Neumann boundary conditions, the solutions of the ordinary differential equation

$$\partial_t u = f(u, v), \quad \partial_t v = g(u, v) \tag{77.6}$$

on $\mathbb{R}^r \times \mathbb{R}^s$, are also solutions of system (77.5). The solutions of (77.6) do <u>not</u> depend on the physical parameter $x \in \Omega$. These solutions are referred to as **homogeneous** solutions of system (77.5).

The mathematical role of small $\epsilon > 0$ in problem (77.5) is to introduce a strong stabilizing effect on the u-coordinate of the solutions. As ϵ gets "smaller", u becomes "more homogeneous". The reduced problem (at $\epsilon = 0$) is called the **shadow system**, and it is a coupled ODE-PDE:

$$\begin{aligned} \partial_t u &= \int_\Omega f(u,v)\,dx \\ \partial_t v &= \Delta v + g(u,v). \end{aligned} \tag{77.7}$$

See Nishiura (1982), Fusco and Hale (1989), and Hale and Sakamoto (1989). When this problem is converted to an evolutionary equation, one obtains

$$\partial_t w + A_\epsilon w = F(w)$$

for system (77.5), where

$$w = \begin{pmatrix} u \\ v \end{pmatrix}, \quad A_\epsilon = \begin{pmatrix} -\epsilon^{-2}\Delta & 0 \\ 0 & -\Delta \end{pmatrix}, \quad F(w) = \begin{pmatrix} f(u,v) \\ g(u,v) \end{pmatrix},$$

with the given boundary conditions. When f and g are smooth functions, this problem generates a semiflow on the space $H^1 = H^1(\Omega, \mathbb{R}^r \times \mathbb{R}^s)$, for example. The reduced problem (77.7) generates a singular semiflow on H^1, see Section 2.5.

Thin Domains. Another illustration of singular perturbations is seen in the analysis of evolutionary equations arising when the physical space Ω is a "thin" domain: for example, the 3D Navier-Stokes equations on $\Omega_\epsilon = (0,1)^2 \times (0,\epsilon)$, where ϵ is positive, but small. If one rescales this problem to the dilated domain $Q_3 = (0,1)^3$ by means of a change of variables $y_1 = x_1$, $y_2 = x_2$, and $y_3 = \epsilon\, x_3$, one obtains the problem

$$\begin{aligned} \partial_t u - \nu\Delta_\epsilon u + (u \cdot \nabla_\epsilon)u + \nabla_\epsilon p &= f, \\ \nabla_\epsilon \cdot u &= 0, \end{aligned} \tag{77.8}$$

on Q_3, where $\nabla_\epsilon = (D_1, D_2, \epsilon^{-1}D_3)$, $\Delta_\epsilon = D_1^2 + D_2^2 + \epsilon^{-2}D_3^2$, and $D_i = \frac{\partial}{\partial x_i}$, for $i = 1,2,3$. We assume here that (77.8) satisfies spatially periodic boundary conditions on Q_3. Because of the differential terms $\epsilon^{-1}D_3$ and $\epsilon^{-2}D_3^2$ in (77.8), we see that the Navier-Stokes equations on a thin 3D domain Ω_ϵ is singular perturbation of the 2D problem.

The regularization of the problem (77.8) is facilitated by using the orthogonal projection M, where $v = Mu$ is defined by

$$v(x_1, x_2) = \int_0^1 u(x_1,, x_2, s)\,ds$$

and $w = (I - M)u$. By using M, with the Helmholtz projection onto the vector fields with $\nabla_\epsilon \cdot u = 0$ in Q_3, one obtains the following system of equations, which is equivalent to (77.8):

$$\begin{aligned} \partial_t v + \nu A_\epsilon v + MB_\epsilon(v+w, v+w) &= Mg \\ \partial_t w + \nu A_\epsilon w + (I - M)B_\epsilon(v+w, v+w) &= (I-M)g, \end{aligned} \tag{77.9}$$

where $A_\epsilon u = -\mathbb{P}_\epsilon \Delta_\epsilon u$, $B_\epsilon(u^1, u^2) = \mathbb{P}_\epsilon((u^1 \cdot \nabla_\epsilon)u^2)$, and $g = \mathbb{P}_\epsilon f$.

When the parameter ϵ is small, then the w-equation is "superstable". As ϵ gets "smaller" $w(t)$ becomes "smaller" at a "faster" rate. The reduced problem (at $\epsilon = 0$) is the system

$$\begin{aligned} \partial_t \overline{v} + \nu A_0 \overline{v} + M B_0(\overline{v}, \overline{v}) &= Mg \\ w(t) = 0, \quad \text{for } t > 0. \end{aligned} \tag{77.10}$$

The reduced problem (77.10) generates a singular semiflow given by

$$T(0)(v_0, w_0) = (v_0, w_0) \text{ and } T(t)(v_0, w_0) = (S_2(t)v_0, 0), \quad \text{for } t > 0,$$

where $S_2(t)v_0$ is the solution of the reduced 2D Navier-Stokes equations, see Hale and Raugel (1992a,b, 1995), Raugel and Sell (1993a,b, 1994), Raugel (1995), Temam and Ziane (1996, 1997).

In this thin domain problem, the semiflow generated by the strong solutions of the reduced problem (77.10) is a singular semiflow (see Section 2.5). Note that the solution $u(t) = v(t) + w(t)$ of (77.10) satisfies $w(t) = 0$, for $t > 0$, even though $w(0) = (I - M)u_0$ need not be zero. Singular semiflows are typical features of the regularization process for a singular perturbation.

Not all singular perturbation problems appear in the form given above. In some cases the vector field itself may have singularities, as in the n-body problem in celestial mechanics. An interesting regularization of the triple collison problem occurs in replacing the singularity of the vector field with the McGehee manifold, see McGehee (1974), Moeckel (1981, 1985), and Xia (1992), for example.

7.7.7. Approximation Dynamics. In the theory of approximation dynamics, which is a relatively new theory, one seeks to answer the question: how well does the longtime dynamics of a given system (S) of differential equations approximate the longtime dynamics induced by a small perturbation of the system (S)? The issue of approximation dynamics is a fundamental issue one faces, for example, when trying to study the relationships between <u>longtime</u> dynamics of the solutions of a given partial differential equation and that of a very high dimensional system of ordinary differential equations formed by the standard spatial discretizations of the given partial differential equation.

Before turning to the historical issues underlying the theory of approximation dynamics, it is important to note that there are two major differences between "approximation dynamics" and "classical approximation" theory. These differences, in the context of equations (70.1) and (70.2), are the following:

(1) While in classical approximation theory, one begins with a given solution of equation (70.1) and one seeks an approximate solution which will be "close to" the given solution on some <u>finite</u> interval

$0 \leq t \leq T < \infty$, in approximation dynamics one seeks approximations over infinite time intervals, $0 \leq t < \infty$, or $-\infty < t < \infty$. One may view the approximate solution as a real solution, for some approximate equation, say (70.2). In classical approximation theory, the error between the approximate solution and the given solution grows as $T \to \infty$, oftentimes at an exponential rate.

(2) While in classical approximation theory, one usually begins with a single solution of equation (70.1), in approximation dynamics, one begins with an ensemble of solutions $\mathcal{K}$, say a compact invariant set for equation (70.1). One of the goals of approximation dynamics is to show that if the perturbation term G is "small enough", then the perturbed equation (70.2) has a compact invariant set $\mathcal{K}^G$ and that $\mathcal{K}^G$ is "close to" $\mathcal{K}$.

This paradigm should look familiar. We have seen it:

(1) in Section 4.5 in the Robustness Theorems for Exponential Dichotomies and Exponential Trichotomies,
(2) in Section 7.1 in the Center Manifold Theorem, and the Saddle Point Property, and
(3) in Sections 7.4 and 7.5 in the perturbation theorems for normally hyperbolic invariant manifolds.

In each of these cases, we find that a dynamical feature of the unperturbed equation is mapped onto a similar dynamical feature of the perturbed equation, and that the mapping converges to the identity as the perturbation term converges to 0. Thus these dynamical features are "points of continuity" of the longtime dynamics of the various semiflows.

The origins of approximation theory can be found, for example, in the works of Krylov and Bogoliubov (1934); Bogoliubov and Mitropolsky (1955); Hale (1961); Sacker (1969); Fenichel (1971); Hirsch, Pugh, and Shub (1977); and Wiggins (1994) on the robustness of normally hyperbolic invariant manifolds for ordinary differential equations. The material presented in Section 7.4 is based on Pliss and Sell (2000). An alternate approach to the infinite dimensional theory is in Bates, Lu, and Zeng (1999). A related development, which led to the proof that hyperbolic sets (see below) are also "points of continuity" of smooth flows, can be found, for example, in Levinson (1950), Peixoto (1959), Markus (1961), Pliss (1966, 1969, 1977), Anosov (1967), Smale (1967), Moser (1969), Franks (1972), Robinson (1974), V I Arnold (1983), Mañé (1988), and Pilyugin (1992, 1999).

A **hyperbolic set** in a semiflow is a compact invariant set $\mathcal{K}$ with the following properties: (1) there are no equilibrium points in $\mathcal{K}$, (2) there is an exponential trichotomy on $\mathcal{K}$, where the rank of the neutral projection $R(m)$ is 1 (see Section 4.5), and (3) the periodic orbits are dense in $\mathcal{K}$, see V I Arnold (1983).

More recently, the theory of approximation dynamics was extended to cover a class of "foliated" compact invariant sets in Pliss and Sell (1991, 1998, 2001b). While these sets do have an exponential trichotomy, there is no restriction on the equilibrium points or the periodic orbits. The prototypes of such a foliated set include the product of two hyperbolic sets and the product of a hyperbolic set with a compact connected smooth manifold.

In one sense, the dynamical issues related to numerical studies of the solutions of evolutionary equations fit into the theory of discrete semiflows (see Chapter 2). Thus one restricts time $t = nh$ to be in the discrete group (or semigroup) Zh (or Z^+h), where $h > 0$ is the time step. Many of the theories described above extend readily to this case, oftentimes with only minor modifications. However, in order to use results arising in such discrete semiflows to study an underlying continuous time evolutionary equation, one needs to develop theories which have good behavior as $h \to 0^+$. Such a study appears in Hagen (1996), for example. With the dependence on the time step, the theory of time discretizations of evolutionary equations fits into the broader area of approximation dynamics.

For other connections between dynamical systems and numerical analysis, see Stuart and Humphries (1996), as well as Jolly (1989); Foias, Jolly, Kevrekidis, and Titi (1991); M J Friedman and Doedel (1993); Stuart (1994); Foias and Jolly (1995); and Van Vleck (1995).

7.7.8. Hamiltonian Systems. As noted in the opening lines of this volume, it was the problem of the stability of the solar system which gave birth to the modern theory of dynamical systems. The N-body problem, which is the traditional model for the solar system, is an example of a Hamiltonian system. More generally, any system of equations of the form

$$\partial_t x = \partial_y H(x, y), \quad \partial_t y = -\partial_x H(x, y) \tag{77.11}$$

is a Hamiltonian system, where $x, y \in \mathbb{R}^n$ and $H : \mathbb{R}^n \times \mathbb{R}^n \to \mathbb{R}$ is a sufficiently smooth function, see Birkhoff (1927), Siegel and Moser (1971), and Meyer and Hall (1992). Notice that $H(x(t), y(t)) = \text{const}$ is invariant under the solutions of (77.11). In mechanical systems, H represents the total (kinetic plus potential) energy of the system, and the model typically arises in frictionless problems. While this feature differs significantly from the dissipative problem treated in this volume, we have looked at the dynamical properties which arise by adding small friction, or some kind of damping, to the nonlinear wave equations, see Chapters 3 and 5.

In the infinite dimensional arena, there have been some significant developments in the study of the dynamics of Hamiltonian systems. For example, it was discovered that the Korteweg-de Vries equation satisfies a countably infinite number of conservation laws, and the dynamics are reduced to a completely integrable Hamiltonian system, see Lax (1976) and the references contained therein. For more information on this rapidly growing area, see Chernoff and Marsden (1974); Markus and Meyer (1974, 1980); Mielke (1991); Marsden, Ratiu, and Raugel (1991); Y Li et al (1996); McLaughlin and Shatah (1996); and Bourgain (1998, 1999); for example.

7.8. Exercises.

Section 7.1

71.1. Assume that $u_0 = e^{-At}u_0 + \int_0^t e^{-A(t-s)}F(u_0)\,ds$, for all $t \geq 0$, where $F \in C^1_{\text{Lip}}$. Show that $u_0 \in \mathcal{D}(A)$ and that $Au_0 = F(u_0)$.

71.2. Complete the proof of Theorem 71.1 in the case of the stable manifold.

71.3. Let A be a sectorial operator on a Banach space W and assume that u_0 is a hyperbolic point for equation (70.1), where F is in $C_{\text{Lip}}(V^{2\beta}, W)$. For the linearized equation (71.1), determine the set $\mathcal{B}$ of all points $v_0 \in W$ that satisfiy the following two properties:

(1) there is a nonnegative continuation ϕ through v_0 with the property that $\sup_{t\leq 0} \|\phi(t)\| < \infty$, and
(2) one has $\sup_{t\geq 0} \|e^{-Lt}v_0\| < \infty$, where $L = A - DF(u_0)$.

71.4. Modify the proof of the Saddle Point Property, Theorem 71.4, so as to use only the original nonlinearity $E(v)$ in place of $E^a(v)$, see Henry (1981).

71.5. Complete the proof of the Perturbation Theorem 71.6 by showing that the associated projections (P^G, Q^G) vary continuously in G and that the characteristics K^G and α^G can be chosen to vary continuously in G, at $G = 0$.

Section 7.2

72.1. Show that the set W_0 defined in (72.1) is nonempty and closed.

72.2. Let $S(t)$ be a κ-contracting semiflow on a Banach space W, and assume that the set Q of stationary solutions is a bounded set in W.

(1) Show that Q is compact.
(2) Assume further that each point $u_0 \in Q$ is hyperbolic. Show that Q is a finite set.

72.3. In Theorem 72.1, show that $K_0 \subset \mathfrak{A}$.

72.4. Extend the theory of gradient systems to semiflows on Fréchet spaces.

72.5. Find example(s) of gradient systems where (1) $K_0 \neq W_0$, and/or (2) $Q \neq K_0$.

72.6. Let the Standing Hypothesis A be satisfied and assume that the associated semiflow e^{-At} is compact, for $t > 0$. Let $F \in C^1_{\text{Lip}}$, and let Q denote the set of all stationary solutions of equation (70.1), i.e., $u_0 \in Q$ if and only if $Au_0 = F(u_0)$.

(1) Let $u_0 \in Q$. Show that the spectrum $\sigma(-A + DF(u_0))$ does not intersect the imaginary axis, Re $\lambda = 0$ if and only if u_0 is hyperbolic.
(2) Let $u_0 \in Q$. Show that the spaces $W^u_{\text{loc}}(u_0)$ and $W^o_{\text{loc}}(u_0)$, which are given by the Center Manifold Theorem, are finite dimensional.

(3) Show that Q is a compact set in $V^{2\beta}$ if and only if it is a bounded set in $V^{2\beta}$.

(4) Assume that Q is a bounded set in $V^{2\beta}$, and that each $u_0 \in Q$ is hyperbolic. Show that Q is a finite set.

72.7. (1) Prove Theorem 72.3. In the case of Theorem 72.3, Item (2), determine whether Formula (72.6) can be replaced with

$$\mathfrak{A} = \bigcup_{i=1}^{k} \bigcup_{z \in \mathcal{K}_i} W^u(z).$$

Section 7.4

74.1. Complete the proof of Items (2) and (3) in Theorem 74.1.

74.2. Give detailed proofs of Lemmas 74.2 and 74.3

74.3. Assume that $\rho > 0$ satisfies $4KL_0\rho < 1$. The following steps will lead to a proof that there is a bounded linear transformation

$$L = L(v) \in \mathcal{L}_0 = \mathcal{L}(\mathcal{R}(Q^o(u_0)), \mathcal{R}(P^o(u_0))), \qquad \text{for } v \in \mathcal{D}_\rho(u_0),$$

such that $p_0 = L(v)q_0 \in \mathcal{R}(P^o(u_0))$ and $q = q_0 + L(v)q_0 \in \mathcal{R}(Q^o(v))$, for all $q_0 \in \mathcal{R}(Q^o(u_0))$.

(1) Show that L must satisfy $L = F(Q, L)$, where

$$F(Q, L) \stackrel{\text{def}}{=} QP_0L + QQ_0 - Q_0 = (Q - Q_0)P_0L + QQ_0 - Q_0,$$

where $Q_0 = Q^o(u_0)$, $Q = Q^o(v)$, $P_0 = P^o(u_0)$, and $P = P^o(v)$.

(2) Show that $\|F(Q, L_1) - F(Q, L_2)\|_{\mathcal{L}_0} \le KL_0\rho\|L_1 - L_2\|_{\mathcal{L}_0}$.

(3) Show that the fixed point $L = L(v)$ satisfies

$$\|L(v)\|_{\mathcal{L}_0} \le KL_0\rho(1 - KL_0\rho)^{-1} \le \frac{4}{3}KL_0\rho, \quad \text{and}$$
$$\|L(v_1) - L(v_2)\|_{\mathcal{L}_0} \le 4KL_0\rho\|A^\beta(v_1 - v_2)\|,$$

for all v, v_1, $v_2 \in \mathcal{D}_\rho(u_0)$.

74.4. Show that the Main Theorem and the Shadow Theorem remain valid when M is a smooth manifold of class $C^{1,1}$. (In this case, the projection $P^o(v) = R(v)$ is assumed to be locally Lipschitz continuous on M, but not necessarily Fréchet differentiable.)

74.5. Let the Standing Hypothesis A hold and let $F \in C^1_{\text{Lip}}$. Let M be an invariant Lipschitz manifold in $V^{2\beta}$, and assume that the linear skew product semiflow π has an exponential trichotomy over M, where $\dim M = \dim \mathcal{R}(R(u))$, for all $u \in M$. Show that M is a C^1-manifold and that M is normally hyperbolic. (Hint: Modify the argument for Item (6) in Theorem 74.15.)

74.6. Let $S(t)u_0$ be a periodic solution of equation (70.1), where the Standing Hypothesis A holds and let $F \in C^1_{\text{Lip}}$. Assume that the periodic orbit Γ is normally hyperbolic in the sense defined in Section 7.3. Show that the mapping $t \to S(t)u_0$ is of class C^1 and that equation (74.6) is valid. Thus, Γ is normally hyperbolic in the sense used in Section 7.4. (Hint: See Exercise 74.5.)

74.7. Show that the methodology of Section 7.4 can be extended to study perturbations of hyperbolic stationary solutions of equation (70.1). In particular, show that Theorem 71.6 is valid when $||G||_{C^1(\Omega_r)} \leq \delta$ is replaced by $||G||_{\{A;C^1(\Omega_r)\}} \leq \hat{\delta}$. Compare the sizes of δ and $\hat{\delta}$, where r is fixed. (See Section 7.5 for another comparison.)

74.8. In addition to the assumptions given in Section 7.4, assume that $F \in C^2_{\text{Lip}}$, and that for each bounded set B in $V^{2\beta}$, there is a constant $K_2(B)$ with

$$||D^2F(u_1) - D^2F(u_2)||_{\mathcal{BL}} \leq K_2(B)||A^\beta(u_1 - u_2)||, \qquad \text{for all } u_1,\, u_2 \in B,$$

where $||D^2F||_{\mathcal{BL}}$ denote the <u>bilinear</u> norm, i.e.,

$$||D^2F||_{\mathcal{BL}} = \sup\{||D^2F\{v, w\}||_W : ||A^\beta v||,\, ||A^\beta w|| \leq 1\}.$$

74.9. Derive good estimates for the various terms from the space Σ, which arise in the proofs in Section 7.4.

Section 7.6

76.1. Prove inequality (76.6).
76.2. Prove Corollary 76.2.

7.9. Commentary.

Section 7.1. The Saddle Point Property for nonlinear problems arose at the birth of dynamical systems in the latter part of the 19th century. Variations of this result appear in the works of Lyapunov (1892) and Poincaré (1890) on ordinary differential equations. While we are unable to cite where the result first appeared in the infinite dimensional literature, it does appear in Henry (1981).

The Center Manifold Theorem for finite dimensional systems of ordinary differential equations was discovered by Pliss (1964). It was the observation of Pliss, that the main application of this theorem is the Reduction Principle and the reduced evolutionary equation given by (71.43) on the subspace $\mathcal{R}(R)$. It is the study of this reduced equation which leads to valuable information about the solutions of (70.1) near u_0. For example, in the case that $P = 0$, the question of the (nonlinear) stability of the equilibrium point

$u(t) \equiv u_0$ for equation (70.1) is completely determined by the stability (or nonstability) of the solution $\xi \equiv 0$ of equation (71.43). Furthermore, the bifurcation theory for equilibria or periodic orbits reduces to the theory of the ODE dynamics on $\mathcal{R}(R)$, see Marsden and McCracken (1976), for example. This is especially significant for infinite dimensional problems, because in many practical situations the space $\mathcal{R}(R)$ is finite dimensional, and oftentimes of very low dimension.

Some applications and other theories of center manifolds can be found, for example, in Kelly (1967); Hale (1969, 1988); Ball (1973); Sell (1977, 1978, 1979, 1981b); Chencinier and Iooss (1979); Henry (1981); Chow and Hale (1982); Kirchgässner (1982); van Gils and Vanderbauwhede (1987); Chow and Lu (1988); Mielke (1991, 1992); Vanderbauwhede and Iooss (1992). Chossat and Iooss (1994); Chow and Yi (1994); Chen, Hale, and Tan (1997); Sandstede, Scheel, and Wulff (1997); and Chow, Liu, and Yi (1999).

Section 7.2. The concept of a gradient system arises in the classical theory of mechanics, where one studies solutions of the ordinary differential equation

$$\partial_t x = -\nabla V(x),$$

where $V : \mathbb{R}^m \to \mathbb{R}$ is a smooth function. Note that for any solution $x = x(t)$ of this equation, one has $\partial_t V(x(t)) = -\|\nabla V(x)\|^2 = -\|\partial_t x\|^2$. It is hard to say where the more general concept treated in this volume originated. It was likely known at the time of Euler (1707 - 1783), and it was definitely known by Lyapunov (1892). In the context of ordinary differential equations, the results described in Theorem 72.1 and 72.3 are sometimes referred to as the LaSalle Invariance Principle, see LaSalle (1960, 1974) and LaSalle and Lefschetz (1961).

The concept of a Morse decomposition presented here is based on the theory of Morse-Smale dynamical systems, wherein the Morse sets are equilibria and/or periodic orbits. See Meyer (1968); Palis (1968); Ball and Peletier (1976); Hale and Raugel (1989, 1992bc); Reineck (1990, 1995); Pilyugin (1992, 1999); Alikakos, Bates, and Fusco (1993); Oliva, De Oliveira, and Sola-Morales (1994); Benaim and Hirsch (1995); and Brunovsky and Poláčik (1997). The formulation of the concept as used here is based on the work of Conley (1978) and Mallet-Paret (1988).

There are in the literature a growing number of theories of nonlinear dynamics which are based on the concept of a discrete Lyapunov function. Such systems do have Morse structures even though the Lyapunov function is not continuous. See Matano (1982); Hale, Magalhães, and Oliva (1984); Henry (1985); Angenent (1986); Mallet-Paret (1988); Mallet-Paret and Smith (1990), Mallet-Paret and Sell (1996a,b); Krisztin, Walther, and Wu (1999); and Krisztin and Walther (2000).

We will use here some of the notation introduced in Section 5.5 on the Cahn-Hilliard problem. The Landau-Ginzburg free energy functional

$J = J(u)$ given in equation (55.6) is used in the study of the dynamical properties of the Cahn-Hilliard equation. While this appears to convert this problem into a gradient system (H^2, J), it does not meet all the requirements of the definition given in Section 7.2. The problem occurs with the requirement (G3) that the funcional be bounded below. For the Cahn-Hilliard problem, inequality (55.16) shows that $J : H^1 \to \mathbb{R}$ is bounded below in the H^1-norm. However, it is <u>not</u> the case that $J : H^2 \to \mathbb{R}$ is bounded below in the H^2-norm. Nevertheless, the restricted flow $(\mathfrak{A}_K, J)$ is a gradient system in the sense used in Section 7.2, since $\mathfrak{A}_K$ is compact in H^2.

Section 7.4. As noted in section 74, our perturbation theory for the manifold M is formulated in the case where M is a C^2-manifold. Because of this, the tangent space T_vM, at the point $v \in M$, is a C^1-function of $v \in M$. We also require that the projections $\{P(v), Q(v), R(v)\}$ be locally Lipschitz continuous functions on M. While these two notions appear to have some overlap, they are different, even in finite dimensions. In paticular, when the tangent space T_vM varies smoothly, it does not follow that <u>every</u> normal direction to M vary smoothly, or even continuously. In the setting of an exponential trichotomy, the projectors $\{P, Q, R\}$ prescribe a <u>distinguished</u> normal space $\mathcal{R}(P(v)) + \mathcal{R}(Q(v))$ at each point $v \in M$. Of course, this normal space varies continuously, since the projectors are continuous and (45.9) holds because of the invariance of the projectors. There are many related issues. For example, does the normal space $\mathcal{R}(P(v)) + \mathcal{R}(Q(v))$ vary smoothly in v? If so, does that imply that the three projectors $P(v)$, $Q(v)$,and $R(v)$ vary smoothly? We have chosen not to address these interesting technical issues in this volume. For more insight on the issue, the reader should consult the example in Anosov (1967).

As noted above, the point of view used in Section 7.4 is based on Pliss and Sell (2000). An alternate approach to this problem can be developed by using the theory of Bates, Lu, and Zeng (1999). Related works are Jones and Titi (1996) and Jones and Shkoller (1999).

Section 7.5. To a great extent, the development of the theory of dynamical systems for the infinite dimensional problems that arise in the theory of evolutionary equations grows out of the successes researchers have had in the study of the dynamics of finite dimensional ordinary differential equations. The finite dimensional experience has led to a broad atlas of techniques, which are used to navigate the infinite dimensional world. Included here one finds, for example, in-depth descriptions of the differing solution concepts which arise in the infinite dimensional world and a growing body of wisdom on how to circumvent the problem of time-reversibility.

Nevertheless, one still encounters surprises in the infinite dimensional world, surprises which seem to have little connection with the finite dimensional experience. The theory developed in Section 7.5.2 on the role of the Bubnov-Galerkin approximations in the approximation dynamics of numerical studies was a surprise to the authors. In this case, it was the infinite dimensional experience, which in this case goes back to the pioneering work of Leray and Hopf for the Navier-Stokes equations, that suggested the importance of the new norm $||G||_{\{A;C^1(\Omega)\}}$. While the use of this new norm comes from the infinite dimensional arena, it should have applications in those finite dimensional problems, which arise in multigrid calculations.

Section 7.6. Related contributions to nonautonomous dynamics can be found in Sell (1967ab, 1971), Sacker and Sell (1978, 1980), Henry (1981), Salvi (1988, 1990), Sauer (1988), Grobbelar-van Dalsen and Sauer (1989, 1993), Meyer and Sell (1989), Raugel and Sell (1993-1994), Chow and Yi (1994), Chepyzhov and Vishik (1993, 1994, 1995), Shen and Yi (1995, 1998), and Yi (1998), for example.

Section 7.7. The poet of the Yukon, Robert Service (1874-1958) wrote the poem: *Maternity*, see Service (1940) and www.ude.net/service/service.html. If he were to publish it today, perhaps he would use the title we propose: *On the Dimension of Attractors.*

In closing, we need to mention two other areas with new major developments in dynamical systems. First there is the extensive work on the Conley index and the connection matrices in Mischaikow (1995); Mischaikow, Mrozek, and Reineck (1999); and Gedeon et al (1999). Secondly there is the arena of monotone dynamical systems and their applications, see Hirsch (1985, 1988), Poláčik and I Tereščik (1992), Fusco and Oliva (1990), and Smith (1995).

While the listing of Other Topics we have given in this Volume is not complete, we sincerely hope that the citations to the extensive literature on dynamical systems will be of value to the reader.

Additional Readings

Bates, Lu, and Zeng (1998); Hale (1988); Hale, Oliva, and Magalhães (1984); Henry (1981); Pliss (1964, 1977); Pliss and Sell (1999, 2000); Sell (1978, 1979); Temam (1988); and van Gils and Vanderbauwhede (1987).

8
INERTIAL MANIFOLDS: THE REDUCTION PRINCIPLE

In the previous chapters, we have seen several illustrations of finite dimensional structures within the infinite dimensional dynamical systems. For example, many dissipative systems have global attractors, and oftentimes, the attractor $\mathfrak{A}$ has finite Hausdorff and fractal dimensions. During the last few years it has been shown that some infinite dimensional nonlinear dissipative evolutionary equations have inertial manifolds. We will give the definition shortly. It should be noted that this discovery has had an impact on the study of the longtime behavior of the solutions of these equations for the following reasons:

- The inertial manifold $\mathfrak{M}$ is a positively invariant finite dimensional manifold in the ambient infinite dimensional phase space.
- The given evolutionary equation reduces to a finite dimensional ordinary differential equation on $\mathfrak{M}$.
- Every compact, invariant set and every attractor $\mathfrak{A}$, including the global attractor, lies in $\mathfrak{M}$.
- Every solution of the nonlinear evolutionary equation is tracked at a (fast) exponential rate by a solution on $\mathfrak{M}$. This means that there is a (large) $\eta > 0$ such that for every solution $S(t)u_0 = u(t)$ of the original evolutionary system, there is a solution $v(t)$ on $\mathfrak{M}$ such that

$$\|u(t) - v(t)\| \leq Ke^{-\eta t}, \qquad \text{for } t > 0, \tag{80.1}$$

  where K depends on $u(0)$.

While the solution $S(t)u_0$ always satisfies $S(t)u_0 \to \mathfrak{A}$, as $t \to \infty$, where $\mathfrak{A}$ is the global attractor, inequality (80.1) contains more information, since it asserts that the rate of convergence is (uniformly) exponential. In this way the longtime dynamics of the solutions of an infinite dimensional nonlinear evolutionary equation with an inertial manifold $\mathfrak{M}$ is completely described by the solutions of a finite dimensional ordinary differential equation. This finite dimensional ordinary differential equation is the result of a global reduction principle for the underlying infinite dimensional evolutionary equation. In the sequel, we assume that one of the following Standing Hypotheses is satisfied .

Standing Hypothesis A. *Let A be a positive, sectorial operator on a Banach space W with associated analytic semigroup e^{-At}. Let $V^{2\alpha}$ be the family of interpolation spaces generated by the fractional powers of A, where $V^{2\alpha} = \mathcal{D}(A^\alpha)$, for $\alpha \geq 0$. Let $\|A^\alpha u\| = \|A^\alpha u\|_W = \|u\|_{V^{2\alpha}} = \|u\|_{2\alpha}$ denote the norm on $V^{2\alpha}$. See Lemma 37.4 for more information.*

Standing Hypothesis B. *The operator A is a positive, selfadjoint, linear operator, with compact resolvent, on a Hilbert space H. Consequently A satisfies the Standing Hypothesis A. Moreover, the fractional power spaces V^α are defined for all $\alpha \in \mathbb{R}$, and equation (37.2) defines the Hilbert space structure on each V^α. Also the semigroup e^{-At} is compact, for $t > 0$. See Theorem 37.2 for more information.*

As noted in Section 4.7, there is no loss of generality in assuming that the sectorial operator A is positive.

8.1. Introduction.

We begin with a nonlinear evolutionary equation of the form

$$\partial_t u + Au = F_0(u) \tag{81.1}$$

on a Hilbert space H. We assume that the Standing Hypothesis B is satisfied. In addition we assume that the nonlinear term F_0 satisfies

$$F_0 \in C_{\mathrm{Lip}}(V^{2\beta}, H), \qquad \text{for some } 0 \leq \beta < 1, \tag{81.2}$$

see (46.12) for the definition of C_{Lip}. As noted in Section 4.7, for each $u_0 \in V^{2\beta}$, the maximally defined mild solutions $S(t)u_0$ of equation (81.1) are well-defined and uniquely determined by the initial data. We now assume that each such solution is defined for all $t \geq 0$. As a result, $S(t)u_0$ is a semiflow on $V^{2\beta}$.

A subset $\mathfrak{M} \subset H$ is said to be an **inertial manifold** for (81.1), see Figure 8.1, provided $\mathfrak{M}$ satisfies the following three properties:

1. $\mathfrak{M}$ is a finite dimensional, Lipschitz continuous manifold in $V^{2\beta} \subset H$.
2. $\mathfrak{M}$ is positively invariant; i.e., if $u_0 \in \mathfrak{M}$ then $S(t)u_0 \in \mathfrak{M}$, for all $t \geq 0$.
3. $\mathfrak{M}$ is exponentially attracting; i.e., there is a $\eta > 0$ such that for every $u_0 \in H$ there is a $K = K(u_0)$ such that

$$\mathrm{dist}_H(S(t)u_0, \mathfrak{M}) \leq Ke^{-\eta t}, \qquad t \geq 0.$$

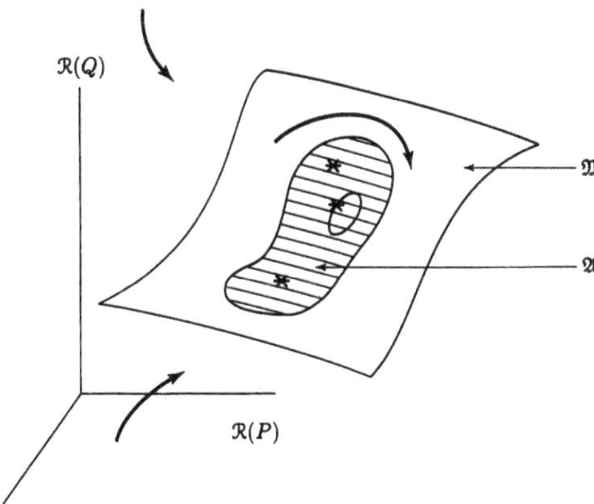

Figure 8.1. Inertial Manifold and Global Attractor

Before we turn to the theory of inertial manifolds, there are a few observations which follow directly from the definition. First note that items (2) and (3) imply that if $\mathcal{K}$ is any compact, invariant set for (81.1), then $\mathcal{K} \subset \mathfrak{M}$. In particular, if (81.1) admits a global attractor $\mathfrak{A}$, then one has $\mathfrak{A} \subset \mathfrak{M}$. We will show later that items (1) and (2) imply that nonlinear evolutionary equation (81.1) reduces to a finite system of ordinary differential equations on $\mathfrak{M}$. This proof will require, of course, more detailed knowledge about the nonlinearity F_0.

Since the nonlinearity F_0 satisfies (81.2), there exist constants $\hat{K}_0 = \hat{K}_0(r) \geq 0$ and $\hat{K}_1 = \hat{K}_1(r) \geq 0$ such that

$$\|F_0(u)\| \leq \hat{K}_0, \tag{81.3}$$

for all $u \in V^{2\beta}$ with $\|A^\beta u\| \leq r$, and

$$\|F_0(u) - F_0(v)\| \leq \hat{K}_1 \|A^\beta (u - v)\|, \tag{81.4}$$

for all $u, v \in V^{2\beta}$ with $\|A^\beta u\|$, $\|A^\beta v\| \leq r$.

As shown in Section 4.7, these assumptions on A and F_0 insure that for every $u_0 \in V^{2\beta}$ there is a maximally defined mild solution $u(t) = S(t)u_0$ of equation (81.1) in $V^{2\beta}$, defined for $0 \leq t < T$, where $T = T(u_0) = T(u_0, F)$ satisfies $0 < T \leq \infty$. For simplicity, we assume that $S(t)u_0$ generates a semiflow on $V^{2\beta}$; that is $T(u_0) = \infty$, for all $u_0 \in V^{2\beta}$. While we are especially interested in the case where there is a global attractor $\mathfrak{A}$ for this semiflow, the theory presented here will apply even in the absence of such an attractor.

We will use a preparation of the nonlinearity, as described in Lemma 47.10. That is to say, we replace F_0 with a new function $F \in C_{\mathrm{Lip}}(V^{2\beta}, H)$, where

$$F(u) = F_0(u), \qquad \text{for } u \in \Omega_\rho \stackrel{\text{def}}{=} \{u \in V^{2\beta} : \|A^\beta u\| \leq \rho\} \tag{81.5}$$

and

$$\text{Supp } F \subset \Omega_{2\rho} = \{u \in V^{2\beta} : \|A^{\beta} u\| \leq 2\rho\}. \tag{81.6}$$

As a result, there are positive constants K_0 and K_1 such that

$$\|F(u)\| \leq K_0, \qquad \text{for all } u \in V^{2\beta}, \tag{81.7}$$

and

$$\|F(u) - F(v)\| \leq K_1 \|A^{\beta}(u - v)\|, \qquad \text{for all } u, v \in V^{2\beta}. \tag{81.8}$$

One should treat ρ is a parameter in this modification of the nonlinearity F_0. It can be chosen to suit other needs, say, one is interested in studying the longtime dynamics of (81.1) only in the region Ω_ρ. For example, if there is a global attractor $\mathfrak{A}$, then one might choose $\rho > 0$ so that $\mathfrak{A} \subset \Omega_\rho$.

In the sequel we will consider the modified equation

$$\partial_t u + Au = F(u), \qquad u \in H, \tag{81.9}$$

where F satisfies (81.5) - (81.8). Notice that the prepared nonlinearity F satisfies $F \in C_{\text{Lip;Global}}(V^{2\beta}, H)$. Even if some of the maximally defined mild solutions of the original equation (81.1) fail to exist for all time $t \geq 0$, <u>all</u> the mild solutions of the modified equation (81.9) exist for all $t \geq 0$ (see Theorem 47.8).

Because of the Standing Hypothesis B, the eigenvalues of A satisfy (32.2). We let $\{e_1, e_2, e_3, \cdots\}$ denote the corresponding orthonormal basis of eigenvectors in H. Let $P = P_n$ denote the orthogonal projection of H onto Span $\{e_1, \cdots, e_n\}$. By applying P and $Q = I - P$ to (81.9) one obtains the system

$$\begin{aligned} \partial_t p + Ap &= \partial_t p + APp = PF(p+q), \\ \partial_t q + Aq &= \partial_t q + AQq = QF(p+q), \end{aligned} \tag{81.10}$$

where $p = Pu$ and $q = Qu$. The plan, which we now follow, is typical in the theory of inertial manifolds. That is, we seek an inertial manifold $\mathfrak{M}$ which can be realized as the graph of a suitable function $\Phi : PH \cap V^{2\beta} \to QH \cap V^{2\beta}$. The fact that $\mathfrak{M} = \text{Graph } \Phi$ is positively invariant means that whenever $p = p(t)$ is a solution of

$$\partial_t p + APp = PF(p + \Phi(p)), \qquad \text{for } t \geq 0, \tag{81.11}$$

then $q = q(t) = \Phi(p(t))$ is a solution of

$$\partial_t q + AQq = QF(p(t) + q), \qquad \text{for } t \geq 0, \tag{81.12}$$

and consequently, $u = u(t) = p(t) + \Phi(p(t))$ is a solution of (81.9). To put it another way, the manifold $\mathfrak{M}$ = Graph Φ is positively invariant if and only if the function $\Phi = \Phi(p(t))$ is a solution of the equation

$$\partial_t \Phi + A\Phi = QF(p(t) + \Phi), \tag{81.13}$$

for every solution $p = p(t)$ of equation (81.11). When an inertial manifold has such a representation, the finite dimensional ordinary differential equation (81.11) is said to be an **inertial form** for (81.9). The inertial form is an example of a global Reduction Principle, see Pliss (1964).

We will be looking for an inertial manifold $\mathfrak{M}$ for (81.9), where $\mathfrak{M}$ can be represented as a graph, $\mathfrak{M}$ = Graph Φ, for a suitable function Φ which satisfies

$$||A^\beta \Phi(p)|| \leq L_0, \qquad \text{for all } p \in PH \cap V^{2\beta}, \tag{81.14}$$

and for all $p_1, p_2 \in PH \cap V^{2\beta}$, one has

$$||A^\beta(\Phi(p_1) - \Phi(p_2))|| \leq L_1 ||A^\beta(p_1 - p_2)||, \tag{81.15}$$

and the Lipschitz constant L_1 is to satisfy $0 \leq L_1 \leq 1$. The objective of the next three sections is to give a proof of the following results.

Theorem 81.1. *Let the Standing Hypotheses B be satisfied and assume that F satisfies (81.5)-(81.8). Then there exist positive constants K_2 and K_3 such that, if for any integer $n \geq 1$ one has $\lambda_n \geq K_2$ and*

$$\lambda_{n+1} - \lambda_n \geq K_3(\lambda_{n+1}^\beta + \lambda_n^\beta), \tag{81.16}$$

then there is a positively invariant manifold, $\mathfrak{M}$ = Graph Φ, for (81.1) and (81.9), where $\Phi : PH \cap V^{2\beta} \to QH \cap V^{2\beta}$, and there exist L_0 and L_1 such that Φ satisfies (81.14) and (81.15). Furthermore, one has $\mathfrak{M} \subset \mathcal{D}(A)$, and for every $u_0 \in \mathfrak{M}$, the mild solution $S(t)u_0$ of equation (81.9) in $V^{2\beta}$ is a strong solution in $V^{2\beta}$, for all $t \geq 0$. Moreover, Φ satisfies the property that $(L_0, L_1) \to (0,0)$, as $(K_0, K_1) \to (0,0)$.

Inequality (81.16) is referred to as the **Spectral Gap Condition**. Next we show that $\mathfrak{M}$ is an inertial manifold, i.e., $\mathfrak{M}$ attracts all solutions at an exponential rate.

Theorem 81.2. *Let the hypotheses of Theorem 81.1 be satisfied, and let K_2, K_3, and $\mathfrak{M}$ be given by Theorem 81.1. Let n be chosen so that $\lambda_n \geq K_2$ and (81.16) is satisfied. Define μ by*

$$\mu \stackrel{\text{def}}{=} \lambda_{n+1} - K_3 \lambda_{n+1}^\beta. \tag{81.17}$$

Then for every solution $(p(t), q(t))$ satisfying equations (81.11) and (81.12) with $p(0) = p_0$, $q(0) = q_0$, and p_0, $q_0 \in V^{2\beta}$, one has

$$||A^\beta(q(t) - \Phi(p(t)))|| \leq 2||A^\beta(q_0 - \Phi(p_0))||e^{-\mu t}, \qquad \textit{for } t \geq 0. \tag{81.18}$$

In other words, $\mathfrak{M}$ is an inertial manifold for (81.9).

We also show that $\mu > 0$ is "large" in the sense that inequality (81.18) guarantees that the inertial manifold $\mathfrak{M}$ is normally hyperbolic. Specifically we show that: (1) $\mathfrak{M}$ is normally hyperbolic, and (2) $\mathfrak{M}$ has an exponential tracking property.

Theorem 81.3. *Let the hypotheses of Theorem 81.1 be satisfied, and let K_2, K_3, and $\mathfrak{M}$ be given by Theorem 81.1. Let n be chosen so that $\lambda_n \geq K_2$ and (81.16) is satisfied. Let $u(t) = p(t) + q(t)$ be a solution of (81.9) with initial condition $u(0) = u_0 = p_0 + q_0 \in \mathcal{D}(A)$. Then the following properties are satisfied, for every $u_0 \in V^{2\beta}$:*

(1) *There is a solution $\hat{u}(t)$ on the inertial manifold $\mathfrak{M}$ such that*

$$\|A^\beta(u(t) - \hat{u}(t))\| \leq 3\|A^\beta(q_0 - \Phi(p_0))\|e^{-\mu t}, \qquad t \geq 0,$$

where μ is given by (81.17).

(2) *The mapping $u_0 \to \hat{p}_0$, where $\hat{u}(0) = \hat{p}_0 + \hat{q}_0$ and $\hat{q}_0 = \Phi(p_0)$, is a continuous mapping of $V^{2\beta}$ onto $PH \cap V^{2\beta}$.*

In the next result we note that the inertial manifold $\mathfrak{M}$ is smooth whenever the nonlinearity F_0, or F, is smooth.

Theorem 81.4. *Let the hypotheses of Theorem 81.1 be satisfied, and let K_2, K_3, and $\mathfrak{M}$ be given by Theorem 81.1. Let n be chosen so that $\lambda_n \geq K_2$ and (81.16) is satisfied. Assume that the nonlinear function F_0, or F, is in $C^1_{\text{Lip}}(V^{2\beta}, H)$. Then the inertial manifold $\mathfrak{M}$ is a C^1-manifold, and Φ is in $C^1_{\text{Lip}}(PV^{2\beta}, QV^{2\beta})$. Moreover, the inertial manifold $\mathfrak{M}$ is normally hyperbolic.*

As will be seen shortly, the theory of inertial manifolds is a part of the perturbation theory of normally hyperbolic invariant manifolds. Because of this, the theory presented here has considerable overlap with the theory presented in Section 7.4. However, there is a subtle, yet very important, change in the point of view between Chapters 7 and 8. While the technical issue of compact versus noncompact manifolds is in the background, this is not what we refer to here. The change in point of view addresses a much deeper issue.

In order to describe this issue, it is convenient to turn to the linear problem

$$\partial_t u + Au = 0. \tag{81.19}$$

The inertial manifold $\mathfrak{M}$ for equation (81.9), as formulated here, is a perturbation of $\mathfrak{M}_0 = \{u = p + q \in V^{2\beta} : q \equiv 0\}$, the inertial manifold for equation (81.19), where the Standing Hypothesis B is satisfied. The linear gap condition $\lambda_{n+1} - \lambda_n > 0$ implies that the inertial manifold $\mathfrak{M}_0$ for (81.19) is normally hyperbolic (see Section 7.4). Because of the theory described in Section 7.4, it should be expected that if F satisfies (81.5) - (81.8), then for small $\epsilon > 0$, the perturbed nonlinear equation

$$\partial_t u + Au = \epsilon F(u), \tag{81.20}$$

has an invariant manifold $\mathfrak{M}_\epsilon$ and $\mathfrak{M}_\epsilon$ is "close to" $\mathfrak{M}_0$. The point to make here is that the nonlinear term $\epsilon F(u)$ is a "small" perturbation of the linear equation (81.19). The problem we face in this chapter is to show that, even for $\epsilon = 1$, equation (81.20) has an invariant manifold $\mathfrak{M}_1$. The issue here is to show that the spectral gap is "large enough" so that the nonlinear problem (81.9) has an inertial manifold. Thus there is a shift in the point of view from "the perturbation is small enough" to "the spectral gap is large enough". Concomitant with the change in the point of view, there is a special emphasis on sharp estimates. This is needed for the applications, as we will see.

8.2. The Lyapunov-Perron Method.

The Lyapunov-Perron method for the construction of inertial manifolds is a particular case of a very powerful technique used throughout the general theory of dynamical systems. In this section we first focus on the underlying motivation for this important approach. We assume throughout this section that the linear operator A and the nonlinear term $F = F(u)$ satisfy the hypotheses given in Section 8.1.

For the moment assume that $\Phi : PH \cap V^{2\beta} \to QH \cap V^{2\beta}$ is a Lipschitz continuous function, and let $p(t) = p(\Phi, p_0, t)$ denote the solution of the Initial Value Problem

$$\partial_t p + Ap = PF(p + \Phi(p)), \qquad p(0) = p_0. \tag{82.1}$$

Since (82.1) is an ordinary differential equation on $PV^{2\beta}$, and since Supp $F \subset \Omega_{2\rho}$, we see that the solution $p(t)$ is defined for all $t \in \mathbb{R}$, and the mapping

$$t \to \hat{F}(t) \stackrel{\text{def}}{=} F(p(t) + \Phi(p(t)))$$

is a (locally) Lipschitz continuous mapping of $\mathbb{R}$ into H. Furthermore, from inequality (81.7), we see that $\hat{F} \in L^\infty(\mathbb{R}; H)$.

Consider now the inhomogenous linear equation on $QH \cap V^{2\beta}$ given by

$$\partial_t q + Aq = QF(p(t) + \Phi(p(t))) = Q\hat{F}(t), \tag{82.2}$$

where $Aq = AQq$. Note that $-AQ$ generates an analytic semigroup e^{-AQt} on QH. By using equation (37.6) with $r = \beta$, we define $b = b(\tau)$, for $0 < \tau < \infty$, by

$$b(\tau) \stackrel{\text{def}}{=} \|A^\beta e^{-AQ\tau}\|_{\mathcal{L}(H)} = \begin{cases} \beta^\beta e^{-\beta}\tau^{-\beta}, & 0 < \tau \le a, \\ \lambda_{n+1}^\beta e^{-\lambda_{n+1}\tau}, & a < \tau < \infty, \end{cases} \tag{82.3}$$

where $a = \beta\lambda_{n+1}^{-1}$. Note that equation (82.2) is the same as equation (45.36), with $\lambda = 0$, $\Phi_\lambda(B, t) = e^{-AQt}$, and $g(t) = h(t) = Q\hat{F}(t)$, for $t \in \mathbb{R}$.

Lemma 82.1. *Assume that the hypotheses of Theorem 81.1 are satisfied. Then for every* $p_0 \in PV^{2\beta}$, *there is a unique mild solution* $q(t)$ *of (82.2) in* $V^{2\beta}$ *that satisfies* $\sup_{t\leq 0} \|A^\beta q(t)\| < \infty$, *and this solution is given by*

$$q(t) = \int_{-\infty}^{t} e^{-AQ(t-s)} QF(p(s) + \Phi(p(s)))\, ds, \qquad \text{for } t \in \mathbb{R}. \tag{82.4}$$

Moreover, q *is a strong solution of (82.2) in* $V^{2\beta}$ *on* $\mathbb{R}$.

Proof. The fact that there is a unique mild solution $q = q(t)$ satisfying (82.4) with $q \in L^\infty(-\infty, 0; V^{2\beta})$ follows directly from Theorem 45.7, Item (1). Since the solution $p(t)$ of (82.1) is locally Lipschitz continuous in t, it follows from inequalities (81.8) and (81.15) that $Q\hat{F}(t)$ is locally Lipschitz continuous in t, as well. It then follows from Theorem 42.9 that $q(t)$ is a strong solution of (82.2) in $V^{2\beta}$ on $\mathbb{R}$. □

We will use the value of this unique, bounded solution of (82.2) at $t = 0$ to define the **Lyapunov-Perron Transformation** $\hat{\Phi} = \mathcal{T}\Phi$; i.e.,

$$q(0) = \hat{\Phi}(p_0) = \mathcal{T}\Phi(p_0) = \int_{-\infty}^{0} e^{AQs} QF(p(s) + \Phi(p(s)))\, ds, \tag{82.5}$$

for $p_0 \in PV^{2\beta}$. An important feature underlying the Lyapunov-Perron method and the dynamical theory of solutions of (81.9) occurs in the case where the transformation $\mathcal{T}$ has a fixed-point in some space of Lipschitz continuous functions $\Phi : PV^{2\beta} \to QV^{2\beta}$, with Lipschitz coefficient $\ell < 1$. In this setting, it is easily shown that $q(t) = \Phi(p(t))$ is a solution of (82.2) and

$$u(t) \stackrel{\text{def}}{=} p(t) + \Phi(p(t))$$

is a solution to (81.9). To put it another way the graph $\mathfrak{M} = \text{Graph } \Phi$ is an invariant manifold for (81.9). Of course, one has $\dim \mathfrak{M} = \dim PV^{2\beta} = n$ and $\mathfrak{M}$ is a Lipschitz continuous manifold in $V^{2\beta}$. Our search for inertial manifolds then begins with a search for fixed point of $\mathcal{T}$.

We define $\mathcal{F}_\ell$ to be the set of $\Phi \in C_{\text{Lip}} = C_{\text{Lip}}(PV^{2\beta}, QV^{2\beta})$ that satisfy the following three properties:

(1) One has

$$\|A^\beta(\Phi(p_1) - \Phi(p_2))\| \leq \ell \|A^\beta(p_1 - p_2)\|, \qquad \text{for all } p_1, p_2 \in PV^{2\beta}. \tag{82.6}$$

(2) The sup-metric $\|A^\beta \Phi\|_\infty$ satisfies

$$\|A^\beta \Phi\|_\infty \stackrel{\text{def}}{=} \sup_{p \in PV^{2\beta}} \|A^\beta \Phi(p)\| < \infty. \tag{82.7}$$

(3) The support satisfies $\text{Supp } \Phi \subset \Omega_{2\rho}$, see (81.6).

Note that with ρ and ℓ fixed, $\mathfrak{F}_\ell$ is a closed set in the Banach space C_{Lip}.

Let $\Phi \in \mathfrak{F}_\ell$ and let $\hat{\Phi} = \mathcal{T}\Phi$ be defined by (82.5). Our first objective is to show that for an appropriate choice of ℓ, the Lyapunov-Perron Transformation $\mathcal{T}$ maps $\mathfrak{F}_\ell$ into itself and that it is a strict contraction in the sup-metric defined by (82.7). We then show that the fixed point Φ of $\mathcal{T}$ has the property that $\mathfrak{M} = \text{Graph}\ \Phi$ is an inertial manifold, and it satisfies the other properties mentioned in the theorems given above.

Let us return to the issue of the positive invariance of $\mathfrak{M} = \text{Graph}\ \Phi$, as described in Section 8.1. Let $\Phi \in \mathfrak{F}_\ell$ be the fixed point of the Lyapunov-Perron Transformation $\mathcal{T}$. We show below that $\Phi = \Phi(p(t))$ is a solution of equation (81.13) whenever $p = p(t)$ is a solution of equation (81.11). That is to say, $\mathfrak{M}$ is positively invariant. Since the mapping Φ is a Lipschitz continuous mapping on a finite dimensional space $PV^{2\beta}$, it is (strongly) differentiable with respect to $p \in PV^{2\beta}$ almost everywhere. Let $D\Phi \overset{\text{a.e.}}{=} \frac{\partial}{\partial p}\Phi$ denote this derivative, where it exists. Now the chain rule for differentiation, see Berger (1977), implies that $q(t) = \Phi(p(t))$ satisfies $\partial_t q \overset{\text{a.e.}}{=} D\Phi\, \partial_t p$, and from equations (81.11) and (81.12) with $\Phi = \Phi(p(t))$, we obtain

$$D\Phi(PF(p+\Phi) - Ap) \overset{\text{a.e.}}{=} QF(p+\Phi) - A\Phi, \qquad \text{for almost all } t \geq 0. \tag{82.8}$$

Furthermore, one has

$$\|A^\beta D\Phi(p)v\| \leq \ell\, \|A^\beta v\|, \qquad \text{for } p, v \in PV^{2\beta}. \tag{82.9}$$

Indeed, the differentiability of Φ implies that for almost all $p \in PV^{2\beta}$ and all $v \in PV^{2\beta}$, one has

$$\left[\frac{1}{h}(\Phi(p+hv) - \Phi(p)) - D\Phi(p)v\right] \overset{s}{\to} 0, \text{ in } V^{2\beta}, \qquad \text{as } h \to 0.$$

This implies that

$$\begin{aligned}\|A^\beta D\Phi(p)v\| &= \lim_{h\to 0} \left\| A^\beta \left(\frac{1}{h}(\Phi(p+hv) - \Phi(p)\right)\right\| \\ &\leq \lim_{h\to 0} \frac{1}{h}\ell\, \|A^\beta(hv)\| = \ell\, \|A^\beta v\|.\end{aligned}$$

8.3. Existence of Inertial Manifolds: Spectral Gap Condition.

The first step in this section is a calculation of a Lipschitz coefficient L for $\hat{\Phi}$, so that the Lyapunov-Perron Transformation $\mathcal{T}$ has the property that it maps $\mathfrak{F}_\ell$ into $\mathfrak{F}_L$. As we will see, this will occur when the spectral gap $\lambda_{n+1} - \lambda_n$ is large enough. For any F satisfying (81.5)-(81.8) and for any $\Phi \in \mathfrak{F}_\ell$, we define $\hat{F}$ by

$$\hat{F}(\Phi)(p) \overset{\text{def}}{=} F(p + \Phi(p)) \quad \text{and} \quad \hat{F}(t) \overset{\text{def}}{=} F(p(t) + \Phi(p(t))), \tag{83.1}$$

where $p(t)$ is some solution of equation (82.1). In the sequel we will restrict ℓ to satisfy $0 \le \ell < 1$. In this case with $p_1, p_2 \in PV^{2\beta}$, one has

$$\|\hat{F}(\Phi)(p_1) - \hat{F}(\Phi)(p_2)\| \le 2K_1\|A^\beta(p_1 - p_2)\|, \tag{83.2}$$

whenever $\Phi \in \mathcal{F}_\ell$. Also if $\Phi_1, \Phi_2 \in \mathcal{F}_\ell$ then

$$\|\hat{F}(\Phi_1)(p_1) - \hat{F}(\Phi_2)(p_2)\| \le 2K_1\|A^\beta(p_1 - p_2)\| + K_1\|A^\beta(\Phi_1 - \Phi_2)\|_\infty, \tag{83.3}$$

for $p_1, p_2 \in PV^{2\beta} = PH$. The proof of (83.3) follows from the observations that (82.6) and (83.2) imply

$$\begin{aligned}\|\hat{F}(\Phi_1)(p_1) - \hat{F}(\Phi_1)(p_2)\| &+ \|\hat{F}(\Phi_1)(p_2) - \hat{F}(\Phi_2)(p_2)\| \\ &\le 2K_1\|A^\beta(p_1 - p_2)\| + K_1\|A^\beta(\Phi_1(p_2) - \Phi_2(p_2))\|,\end{aligned}$$

and that the last term above is dominated by $K_1\|A^\beta(\Phi_1 - \Phi_2)\|_\infty$. For $\Phi \in \mathcal{F}_\ell$, let $\hat{\Phi} = \mathcal{T}\Phi$ be defined by (82.5). Then (37.7), (82.3) and (81.7) imply that

$$\begin{aligned}\|A^\beta\hat{\Phi}(p_0)\| &\le \int_{-\infty}^0 \|A^\beta e^{-AQ(-s)}\|\,\|Q\hat{F}(s)\|\,ds \\ &\le L_0 \overset{\text{def}}{=} K_0(1-\beta)^{-1}e^{-\beta}\lambda_{n+1}^{\beta-1}.\end{aligned} \tag{83.4}$$

Since Supp $F \subset \Omega_{2\rho}$ and Supp $\Phi \subset \Omega_{2\rho}$, it follows that if $\|A^\beta p_0\| > 2\rho$, then $\Phi(p_0) = 0$ and one has

$$\|A^\beta e^{-APt}p_0\| \ge e^{-\lambda_1 t}\|A^\beta p_0\| \ge \|A^\beta p_0\| > 2\rho, \qquad \text{for all } t \le 0.$$

With $p(t) = e^{-APt}p_0$, it follows that $F(p(t) + \Phi(p(t))) = 0$, for $t \le 0$. Consequently $\hat{\Phi}(p_0) = 0$ whenever $\|A^\beta p_0\| > 2\rho$, i.e., Supp $\hat{\Phi} \subset \Omega_{2\rho}$. Hence, we have the following result.

Lemma 83.1. *Let $\Phi \in \mathcal{F}_\ell$, where $0 \le \ell \le 1$. Then one has Supp $\mathcal{T}\Phi \subset \Omega_{2\rho}$ and $\|A^\beta\mathcal{T}\Phi(p)\| \le L_0$, for all $p \in PV^{2\beta}$, where L_0 is given by equation (83.4).*

Next we will study the growth of the solutions of equation (82.1), as $s \to -\infty$.

Lemma 83.2. *Let $\Phi_1, \Phi_2 \in \mathcal{F}_\ell$, where $0 \le \ell < 1$, and let $p_i(t)$ be any solutions of*

$$\partial_t p + APp = PF(p + \Phi_i(p)), \qquad \text{for } i = 1, 2. \tag{83.5}$$

Then one has

(83.6)

$$\begin{aligned}&\|A^\beta(p_1(s) - p_2(s))\| \\ &\qquad \le (\|A^\beta(p_1(t) - p_2(t))\| + K_1\lambda_n^{\beta-1}\|A^\beta(\Phi_1 - \Phi_2)\|_\infty)e^{-\gamma(s-t)},\end{aligned}$$

for $s \le t$, where

$$\gamma \stackrel{\text{def}}{=} \lambda_n + 2\, K_1\, \lambda_n^\beta. \tag{83.7}$$

Proof. Since $\delta(t) \stackrel{\text{def}}{=} p_1(t) - p_2(t)$ is a mild solution of

$$\partial_t \delta + AP\delta = P\hat{F}(\Phi_1)(p_1) - P\hat{F}(\Phi_2)(p_2),$$

one finds that

$$\delta(s) = e^{AP(t-s)}\delta(t) - \int_s^t e^{AP(\sigma-s)} \left[P\hat{F}(\Phi_1)(p_1(\sigma)) - P\hat{F}(\Phi_2)(p_2(\sigma))\right] d\sigma,$$

for $s \le t$. From inequality (83.3), one then obtains

$$\begin{aligned} \|A^\beta \delta(s)\| &\le e^{\lambda_n(t-s)}\|A^\beta \delta(t)\| \\ &\quad + \int_s^t \lambda_n^\beta e^{\lambda_n(\sigma-s)} \left[2K_1\|A^\beta\delta(\sigma)\| + K_1\|A^\beta(\Phi_1 - \Phi_2)\|_\infty\right] d\sigma \\ &\le e^{\lambda_n(t-s)}\|A^\beta\delta(t)\| + e^{\lambda_n(t-s)}K_1\lambda_n^{\beta-1}\|A^\beta(\Phi_1 - \Phi_2)\|_\infty \\ &\quad + 2K_1\lambda_n^\beta \int_s^t e^{\lambda_n(\sigma-s)}\|A^\beta\delta(\sigma)\|\, d\sigma. \end{aligned}$$

Set $w(\sigma) = e^{\lambda_n \sigma}\|A^\beta\delta(\sigma)\|$ and $g(t) = w(t) + e^{\lambda_n t}K_1\lambda_n^{\beta-1}\|A^\beta(\Phi_1 - \Phi_2)\|_\infty$. Then w satisfies $w(s) \le g(t) + 2K_1\lambda_n^\beta \int_s^t w(\sigma)\, d\sigma$, for $s \le t$. It then follows from the Gronwall inequality (Lemma D.1) that $w(s) \le e^{2K_1\lambda_n^\beta(t-s)}g(t)$, for $s \le t$. Consequently, one has

$$\|A^\beta\delta(s)\| \le e^{\gamma(t-s)}\left(\|A^\beta\delta(t)\| + K_1\lambda_n^{\beta-1}\|A^\beta(\Phi_1 - \Phi_2)\|_\infty\right),$$

for $s \le t$, which is (83.6). □

In order to show that the Lyapunov-Perron Transformation $\mathcal{T}$ maps $\mathcal{F}_\ell$ into $\mathcal{F}_L$, for an appropriate L, and to derive an estimate for L, we now assume that the eigenvalues of A satisfy the inequality

$$\lambda_{n+1} - \lambda_n > 2K_1\lambda_n^\beta, \tag{83.8}$$

and that $0 \le \ell < 1$.

Lemma 83.3. *Let $\Phi \in \mathcal{F}_\ell$, where $0 \le \ell < 1$, and assume that inequality (83.8) holds, i.e., one has $\gamma < \lambda_{n+1}$, where γ is given by (83.7). Then for all $p_1, p_2 \in PV^{2\beta}$, one has*

$$\|A^\beta(\mathcal{T}\Phi(p_1) - \mathcal{T}\Phi(p_2))\| \le L\|A^\beta(p_1 - p_2)\|, \tag{83.9}$$

where

$$L \stackrel{\text{def}}{=} 2K_1 \int_0^\infty b(\tau)e^{\gamma\tau}\,d\tau. \tag{83.10}$$

Proof. First note that due to inequality (81.16), one has $\gamma < \lambda_{n+1}$. the integral in (83.10) is finite, see (37.7). Let $p_i(s)$ denote the solutions of $\partial_t p + APp = PF(p + \Phi(p))$ with initial conditions $p_i(0) = p_{i0}$, for $i = 1, 2$. We now use (37.6), (82.3), (83.2), and (83.6) with $\Phi_1 = \Phi_2 = \Phi$ to obtain

$$\begin{aligned}
&||A^\beta(\mathcal{T}\Phi(p_{10}) - \mathcal{T}\Phi(p_{20}))|| \\
&\qquad \le \int_{-\infty}^0 ||A^\beta e^{AQs}||\,||\hat{F}(\Phi)(p_1(s)) - \hat{F}(\Phi)(p_2(s))||\,ds \\
&\qquad \le 2K_1 \int_{-\infty}^0 b(-s)||A^\beta(p_1(s) - p_2(s))||\,ds \\
&\qquad \le 2K_1 \int_{-\infty}^0 b(-s)e^{-\gamma s}\,ds||A^\beta(p_{10} - p_{20})||. \quad \square
\end{aligned}$$

Lemma 83.4. *Let $\Phi_1, \Phi_2 \in \mathcal{F}_\ell$, where $0 \le \ell < 1$ and $\gamma < \lambda_{n+1}$. Then one has*

$$||A^\beta(\mathcal{T}\Phi_1 - \mathcal{T}\Phi_2)||_\infty \le K||A^\beta(\Phi_1 - \Phi_2)||_\infty, \tag{83.11}$$

where $K = K_1\left(L\lambda_n^{\beta-1} + (1-\beta)^{-1}e^{-\beta}\lambda_{n+1}^{\beta-1}\right)$, and L is given by (83.10).

Proof. Let $p_i(s)$ be the solutions of $\partial_t p + APp = PF(p + \Phi_i(p))$ that satisfy the same initial condition $p_i(0) = p_0$, for $i = 1, 2$. Then from (83.6), with $t = 0$, one has $||A^\beta(p_1(s) - p_2(s))|| \le K_1\lambda_n^{\beta-1}e^{-\gamma s}||A^\beta(\Phi_1 - \Phi_2)||_\infty$, for $s \le 0$. Hence by using (37.7), (82.3), and (83.10), one obtains

$$\begin{aligned}
&||A^\beta(\mathcal{T}\Phi_1(p_0) - A\mathcal{T}\Phi_2(p_0))|| \\
&\quad \le \int_{-\infty}^0 ||A^\beta e^{AQs}||\,||\hat{F}(\Phi_1)(p_1) - \hat{F}(\Phi_2)(p_2)||\,ds \\
&\quad \le \int_{-\infty}^0 b(-s)\left[2K_1||A^\beta(p_1(s) - p_2(s))|| + K_1||A^\beta(\Phi_1 - \Phi_2)||_\infty\right] ds \\
&\quad \le \int_{-\infty}^0 b(-s)\left[2K_1^2\lambda_n^{\beta-1}e^{-\gamma s} + K_1\right] ds\,||A^\beta(\Phi_1 - \Phi_2)||_\infty \\
&\quad \le K_1\left(L\lambda_n^{\beta-1} + (1-\beta)^{-1}e^{-\beta}\lambda_{n+1}^{\beta-1}\right)||A^\beta(\Phi_1 - \Phi_2)||_\infty. \quad \square
\end{aligned}$$

The next step is to find conditions under which one has $L \le \ell$. As we will see, our argument will use a gap condition which is stronger than (83.8).

Lemma 83.5. *Define K_2 and K_3 by $K_2^{1-\beta} = 12K_1(1-\beta)^{-1}$ and $K_3 = 6K_1$. If $\lambda_n \geq K_2$ and the Spectral Gap Condition (81.16) holds, then (83.9) is satisfied and $L < \frac{2}{3}$. Consequently $\mathcal{T}$ maps $\mathcal{F}_\ell$ into itself, provided that ℓ satisfies $L \leq \ell < 1$.*

Proof. Since $K_3 = 6K_1$ and $\lambda_{n+1} \geq K_2 > 0$, inequality (83.9) is an immediate consequence of (81.16), (83.7) and (83.10). From (37.7) one obtains

$$
\begin{aligned}
L_1 &\stackrel{\text{def}}{=} L = 2K_1 \int_0^\infty b(\tau) e^{\gamma\tau}\, d\tau \\
&\leq 2K_1 \left[(1-\beta)^{-1} \lambda_{n+1}^{\beta-1} + \lambda_{n+1}^{\beta} (\lambda_{n+1} - \gamma)^{-1} \right].
\end{aligned}
\tag{83.12}
$$

Since $\lambda_{n+1} > \lambda_n \geq K_2$, the first term on the right-side of (83.12) is $< \frac{1}{6}$, and from (81.16) and (83.7), one finds that the last term of (83.12) is $< \frac{1}{2}$. □

In the sequel we will fix ℓ so that $\ell = L_1 < \frac{2}{3}$, where L_1 is defined by equation (83.12). The final lemma will complete the proof of the existence of fixed points of the Lyapunov-Perron operator $\mathcal{T}$.

Lemma 83.6. *Let K_2 and K_3 be given as in Lemma 83.5. If $\lambda_n \geq K_2$ and (81.16) holds, then $\mathcal{T}$ is a strict contraction on $\mathcal{F}_\ell$ with*

$$\|A(\mathcal{T}\Phi_1 - \mathcal{T}\Phi_2)\|_\infty \leq \frac{1}{6} \|A(\Phi_1 - \Phi_2)\|_\infty$$

for all $\Phi_1, \Phi_2 \in \mathcal{F}_\ell$. In particular, $\mathcal{T}$ has a unique fixed point Φ in $\mathcal{F}_\ell$.

Proof. By Lemma 83.5, $\mathcal{T}$ maps $\mathcal{F}_\ell$ into itself. From Lemma 83.4, inequality (83.11) holds. Since $L < 1$ (Lemma 83.5) and $\lambda_{n+1} > \lambda_n \geq K_2 > 0$, one has

$$
\begin{aligned}
K &= K_1 \left(L\lambda_n^{\beta-1} + (1-\beta)^{-1} e^{-\beta} \lambda_{n+1}^{\beta-1} \right) \\
&< K_1 \lambda_n^{\beta-1} + K_1 (1-\beta)^{-1} \lambda_{n+1}^{\beta-1} < \frac{1}{6}. \quad \square
\end{aligned}
$$

Proof of Theorem 81.1. Lemma 83.6 completes the proof of the existence of a Lipschitz continuous fixed point $\Phi \in \mathcal{F}_\ell$. The bounds L_0 and L_1 are given by (83.4) and (83.12). It follows from these definitions that $(L_0, L_1) \to 0$, as $(K_0, K_1) \to 0$.

Let Φ be the fixed point of $\mathcal{T}$ given by Lemma 83.6, and let $p(t) = \pi(p_0, t)$ denote the solution of equation (81.11) that satisfies $\pi(p_0, 0) = p_0$. Let $u_0 = p_0 + \Phi(p_0)$ Note that $S(t)u_0 = p(t) + \Phi(p(t))$ is strong solution of equation (81.9) in $V^{2\beta}$, by Corollary 47.3. From Lemma 82.1, $q(t) = \Phi(p(t))$ is a strong solution of equation (82.2) in $V^{2\beta}$, and by a change of variables, $q = \Phi = \mathcal{T}\Phi$ satisfies

$$\Phi(p(t)) = \int_{-\infty}^{t} e^{-AQ(t-s)} Q\hat{F}(\Phi)(p(s))\, ds. \tag{83.13}$$

By differentiating (83.13) with respect to t we see that Φ is a strong solution of equation (81.13). In other words, $\mathfrak{M} = \text{Graph } \Phi$ is an invariant manifold for (81.9). This completes the proof of Theorem 81.1. □

8.4. Exponential Attraction of Inertial Manifolds.

In this section we will prove Theorems 81.2 and 81.3. As before we assume here that the Standing Hypothesis B is satisfied and that n is chosen so that the Spectral Gap Condition (81.16) is valid and that $\lambda_n \geq K_2$, where K_2 and K_3 are given in Theorem 81.1. (Also see Lemma 83.5.) We let Φ denote the fixed point of $\mathcal{T}$ given by Theorem 81.1, and define μ by equation (81.17). Note that because of the Spectral Gap Condition (81.16), one has

$$\mu \geq \lambda_n + K_3 \lambda_n^\beta > \gamma,$$

where γ is given by (83.7) and μ is given by (81.17). It is important to note that the decay rate μ, which appears in Theorems 81.2 and 81.3 is faster than the decay rate γ occurring inside the manifold $\mathfrak{M}$ (see Lemma 83.2). As we will see, this implies that the inertial manifold $\mathfrak{M}$ is normally hyperbolic.

Proof of Theorem 81.2. It is convenient to make the change of variables

$$q = \Phi + r = \Phi(p) + r.$$

Since $\Phi : PV^{2\beta} \to QV^{2\beta}$ satisfies (82.6), with $\ell = L_1$ given by (83.12), it follows that the Fréchet derivative $D\Phi = D\Phi(p) = \frac{\partial}{\partial p}\Phi$ exists, for almost all $p \in PV^{2\beta}$, and $\|D\Phi\|_{\mathfrak{L}} \leq L_1 < 1$, where $\mathfrak{L} = \mathfrak{L}(PV^{2\beta}, QV^{2\beta})$.

We claim that in the (p, r)-variables, the system (81.10) now takes the form

$$\begin{aligned} \partial_t p + APp &\overset{\text{a.e.}}{=} PF(p + \Phi + r) \\ \partial_t r + AQr &\overset{\text{a.e.}}{=} QF(p + \Phi + r) - QF(p + \Phi) \\ &\quad - D\Phi(PF(p + \Phi + r) - PF(p + \Phi)), \end{aligned} \tag{84.1}$$

for $t \geq 0$. Indeed, the first equation in (84.1) is elementary. For the second equation, we note that $\partial_t q + AQq = QF(p + \Phi + r)$, while $\partial_t \Phi$ is given by the chain rule

$$\begin{aligned} \partial_t \Phi &\overset{\text{a.e.}}{=} D\Phi\, \partial_t p \overset{\text{a.e.}}{=} D\Phi(PF(p + \Phi + r) - Ap) \\ &\overset{\text{a.e.}}{=} D\Phi(PF(p + \Phi) - Ap) + D\Phi(PF(p + \Phi + r) - PF(p + \Phi)). \end{aligned}$$

By using equation (82.8), we then obtain

$$\partial_t \Phi + AQ\Phi \overset{\text{a.e.}}{=} QF(p + \Phi) + D\Phi(PF(p + \Phi + r) - PF(p + \Phi)),$$

which implies (84.1).

Let p_0, r_0 be fixed, where $p_0 \in PV^{2\beta} = PH$ and $r_0 \in V^{2\beta}$. Let $(p(t), r(t))$, $t \geq 0$, denote the mild solution of (84.1) in $V^{2\beta}$ with $p(0) = p_0$ and $r(0) = r_0$. From the Variation of Constants Formula for mild solutions one has

$$\begin{aligned} r(t) = e^{-AQt} r_0 + \int_0^t e^{-AQ(t-s)}[QF(p+\Phi+r) - QF(p+\Phi)]\, ds \\ - \int_0^t e^{-AQ(t-s)} D\Phi(PF(p+\Phi+r) - PF(p+\Phi))\, ds. \end{aligned}$$

By applying A^β to the last equation, setting $\lambda = \lambda_{n+1}$, using (37.6), (81.8), (82.3), and $\mu < \lambda$, one obtains

$$\|A^\beta r(t)\| \leq e^{-\mu t}\|A^\beta r_0\| + 2K_1 \int_0^t b(t-s)\|A^\beta r(s)\|\, ds. \tag{84.2}$$

This means that $\rho(t) = \|A^\beta r(t)\|$ is a subsolution of the equation

$$v(t) = e^{-\mu t}\|A^\beta r_0\| + 2K_1 \int_0^t b(t-s) v(s)\, ds, \qquad \text{for } t \geq 0. \tag{84.3}$$

By using the definitions of μ and b, see (81.17) and (82.3), inequality (81.16) and Lemma 83.5, with $\lambda = \lambda_{n+1} > K_2$ and $K_3 = 6K_1$, one can verify that

$$2K_1 \int_0^t b(t-s) e^{\mu(t-s)}\, ds \leq 2K_1 \left(\frac{\beta}{1-\beta} \lambda^{\beta-1} + \frac{\lambda^\beta}{\lambda - \mu} \right) \leq 1,$$

for all $t \geq 0$. It follows that $w(t) = 2\|A^\beta r_0\| e^{-\mu t}$ is a supersolution of (84.3), for $t \geq 0$. Since $\rho(0) < w(0)$, it follows that $\|A^\beta r(t)\| \leq w(t)$, for $t \geq 0$.

This shows that inequality (81.18) holds, for almost all $p_0 \in PV^{2\beta}$. Since this set of p_0 is dense in $PV^{2\beta}$, it follows from the continuity of the terms that, inequality (81.18) is valid, for all $p_0 \in PV^{2\beta}$. □

Proof of Theorem 81.3. Let $\mathfrak{M} = \text{Graph}\, \Phi$ be given by Theorem 81.1. Let $u(t)$ be any solution of (81.9) with $u(0) = u_0 \in V^{2\beta}$, and set $p(t) = Pu(t)$, $q(t) = Qu(t)$, $p_0 = Pu_0$ and $q_0 = Qu_0$. Let $\overline{p}(t)$ denote an unknown solution of (81.11), and set $\overline{q}(t) = \Phi(\overline{p}(t))$. Thus $\overline{u}(t) = \overline{p}(t) + q(t)$ is an unknown solution of (81.9) on the inertial manifold $\mathfrak{M} = \text{Graph}\, \Phi$. Define $r(t)$ and $\delta(t)$ by $q = \Phi(p) + r$ and $p + \delta = \overline{p}$. Then (p, r) is a solution of (84.1), with initial condition $r_0 = q_0 - \Phi(p_0)$. Also δ satisfies the finite dimensional ordinary differential equation

$$\partial_t \delta + AP\delta = -PF(p + \Phi(p) + r) + PF(p + \delta + \Phi(p + \delta)). \tag{84.4}$$

With r_0 fixed and $r_0 \neq 0$, we let $\mathcal{X}$ denote the space of continuous functions

$$\delta = \delta(u_0, t) : V^{2\beta} \times [0, \infty) \to PV^{2\beta}$$

that satisfy

$$N(\delta) \overset{\text{def}}{=} \sup_{u_0 \in V^{2\beta}} \sup_{t \geq 0} \|A^\beta \delta(u_0, t)\| \, \|A^\beta (q_0 - \Phi(p_0))\|^{-1} e^{\mu t} < \infty.$$

Note that $N(\delta)$ defines a metric $N(\delta_1 - \delta_2)$ on $\mathcal{X}$, and $\mathcal{X}$ is complete as a metric space. Furthermore, if $\delta \in \mathcal{X}$, then for $t \geq 0$ and $u_0 \in V^{2\beta}$, one has

$$\|A^\beta \delta(u_0, t)\| \leq N(\delta) \|A^\beta (q_0 - \Phi(p_0))\| e^{-\mu t}. \tag{84.5}$$

We define a Lyapunov-Perron operator $\mathcal{J}$ on $\mathcal{X}$ formally by $\hat{\delta}(u_0, t) = \mathcal{J}\delta(u_0, t)$, where

$$\mathcal{J}\delta(u_0, t) \overset{\text{def}}{=} -\int_t^\infty e^{-AP(t-s)} \big[PF(p + \Phi(p) + r) \\ - PF(p + \delta + \Phi(p + \delta))\big] \, ds,$$

for $t \geq 0$. We claim that $\mathcal{J}$ maps $\mathcal{X}$ into itself and that $\mathcal{J}$ is a strict contraction on $\mathcal{X}$. Indeed, if $\delta \in \mathcal{X}$, then from (81.8), (81.18), (84.5), and the fact that

$$\|A^\beta e^{-AP(t-s)}\| \leq \lambda_n^\beta e^{-\lambda_n (t-s)}, \qquad \text{for } t \leq s,$$

one has

$$\begin{aligned} \|A^\beta \hat{\delta}(u_0, t)\| &\leq \int_t^\infty K_1 \lambda_n^\beta e^{-\lambda_n (t-s)} (\|A^\beta r\| + 2\|A^\beta \delta\|) \, ds \\ &\leq K_1 \lambda_n^\beta \int_t^\infty e^{-\lambda_n (t-s)} e^{-\mu s} \, ds \, (2 + 2N(\delta)) \|A^\beta (q_0 - \Phi(p_0))\| \\ &\leq 2K_1 \lambda_n^\beta (\mu - \lambda_n)^{-1} (1 + N(\delta)) \|A^\beta (q_0 - \Phi(p_0))\| e^{-\mu t}, \end{aligned}$$

which implies that $\hat{\delta} \in \mathcal{X}$. By using the definition of μ in (81.17) and $K_3 = 6K_1$, one has $\lambda_n^\beta (\mu - \lambda_n)^{-1} \leq K_3^{-1}$ and consequently,

$$\|A^\beta \hat{\delta}(u_0, t)\| \leq \frac{1}{3} (1 + N(\delta)) \|A^\beta (q_0 - \Phi(p_0))\| e^{-\mu t}, \qquad \text{for } t \geq 0.$$

From the definition of $N(\delta)$, this implies that any fixed point δ of $\mathcal{J}$ must satisfy $N(\delta) \leq \frac{1}{2}$.

Next if $\delta_1, \delta_2 \in \mathcal{X}$, then from (81.8) one has

$$\begin{aligned} \|A^\beta (\hat{\delta}_1 - \hat{\delta}_2)\| &\leq \int_t^\infty 2K_1 \lambda_n^\beta e^{-\lambda_n (t-s)} \|A^\beta (\delta_1 - \delta_2)\| \, ds \\ &\leq 2K_1 \lambda_n^\beta \int_t^\infty e^{-\lambda_n (t-s)} e^{-\mu s} \, ds \, N(\delta_1 - \delta_2) \|A^\beta (q_0 - \Phi(p_0))\| \\ &\leq 2K_1 \lambda_n^\beta (\mu - \lambda_n)^{-1} N(\delta_1 - \delta_2) \|A^\beta (q_0 - \Phi(p_0))\| e^{-\mu t}. \end{aligned}$$

Once again, by using the definition of μ in (81.17), and $K_3 = 6K_1$, one obtains

$$N(\hat{\delta}_1 - \hat{\delta}_2) \le \frac{1}{3} N(\delta_1 - \delta_2).$$

Hence $\mathcal{J}$ is a strict contraction on $\mathfrak{X}$, and therefore, it has a unique fixed point $\delta \in \mathfrak{X}$. We now fix δ to be this fixed point. The equation $\delta = \mathcal{J}\delta$ then becomes

$$\delta(u_0, t) = -\int_t^{\infty} e^{-AP(t-s)} \left[PF(p + \Phi(p) + r) - PF(p + \delta + \Phi(p + \delta))\right] ds.$$

The reader should verify that one can differentiate this equation with respect to t. Furthermore, the initial value is $\delta_0 = \delta(u_0, 0)$, or

$$\delta_0 = -\int_0^{\infty} e^{AP(s)} \left[PF(p + \Phi(p) + r) - PF(p + \delta + \Phi(p + \delta))\right] ds,$$

which is a continuous function of u_0. The function $\overline{p}(t) = p(t) + \delta(u_0, t)$ is then a solution of the inertial form (81.11) with initial condition $\overline{p}_0 = p_0 + \delta_0$, and the function $\overline{u}(t) = \overline{p}(t) + \overline{q}(t)$, where $\overline{q}(t) = \Phi(\overline{p}(t))$, is a solution of (81.9) which lies on the inertial manifold $\mathfrak{M} = \text{Graph } \Phi$. Furthermore, $\overline{p}_0$ is a continuous function of u_0, and since $N(\delta) \le \frac{1}{2}$, one has

$$\|A^{\beta}(p(t) - \overline{p}(t))\| = \|A^{\beta}\delta(u_0, t)\| \le \frac{1}{2}\|A^{\beta}(q_0 - \Phi(p_0))\|e^{-\mu t}, \tag{84.6}$$

for $t \ge 0$. Next observe that

$$\begin{aligned} u(t) - \overline{u}(t) &= (p(t) - \overline{p}(t)) + (q(t) - \overline{q}(t)) \\ &= (p(t) - \overline{p}(t)) + (q(t) - \Phi(p(t))) + (\Phi(p(t)) - \Phi(\overline{p}(t))) \end{aligned} \tag{84.7}$$

By applying A^{β} to (84.7) and using (81.18), (84.6), and $0 \le \ell \le 1$, one obtains

$$\begin{aligned} \|A^{\beta}(u(t) - \overline{u}(t))\| &\le 2\|A^{\beta}(p(t) - \overline{p}(t))\| + \|A^{\beta}(q(t) - \Phi(p(t)))\| \\ &\le 3\|A^{\beta}(q_0 - \Phi(p_0))\|e^{-\mu t}, \end{aligned}$$

which completes the proof of Theorem 81.3. □

8.5. Smoothness of Inertial Manifolds.

The proof presented here of Theorem 81.4 on the smoothness of the invariant manifold $\mathfrak{M} = \text{Graph } \Phi$, where Φ is the fixed point of $\mathcal{T}$, is based on the argument in Chow, Lu and Sell (1992). Since the full argument is rather long, we will only present an outline here. This argument uses the following lemma, which is is an extension of the Contraction Mapping Theorem and is proved in Chow, Lu, and Sell (1992).

Lemma 85.1. *Let X and Y be complete metric spaces with metrics d_x and d_y, and let Λ be a locally compact Hausdorff space. Let $H : \Lambda \times X \times Y \to X \times Y$ be a continuous function satisfying the following properties:*

1. *$H(\lambda, x, y) = H_\lambda(x, y) = (F_\lambda(x), G_\lambda(x, y))$, where F_λ does not depend on y.*
2. *There is a constant k with $0 \le k < 1$ such that*
$$d_x(F_\lambda(x_1), F_\lambda(x_2)) \le k d_x(x_1, x_2), \quad x_1, x_2 \in X,$$
$$d_y(G_\lambda(x, y_1), G_\lambda(x, y_2)) \le k d_y(y_1, y_2), \quad x \in X, \quad y_1, y_2 \in Y,$$
for all $\lambda \in \Lambda$.

Then for each $\lambda \in \Lambda$, H_λ has a unique fixed point $(x(\lambda), y(\lambda)) \in X \times Y$. Furthermore, the mapping
$$\lambda \to (x(\lambda), y(\lambda))$$
is a continuous mapping of Λ into $X \times Y$. Moreover, if (x_n, y_n) is any sequence of successive approximations, i.e., $(x_{n+1}, y_{n+1}) = H_\lambda(x_n, y_n)$ for $n \ge 1$ and some (fixed) $\lambda \in \Lambda$, then
$$\lim_{n\to\infty} (x_n, y_n) = (x(\lambda), y(\lambda)).$$

Let $\mathfrak{M} = \text{Graph } \Phi$ be given as in the hypotheses of Theorem 81.4. Define $\mathcal{G}_\ell$ to be the collection of continuous mappings $\Psi = \Psi(p)$ of $PV^{2\beta}$ into the space $\mathcal{L} = \mathcal{L}(PV^{2\beta}, QV^{2\beta})$ of bounded linear operators from $PV^{2\beta}$ into $QV^{2\beta}$, such that

$$\|A^\beta \Psi(p)\hat{p}\| \le \ell \|A^\beta \hat{p}\|, \qquad \text{for all } p, \hat{p} \in PV^{2\beta}. \tag{85.1}$$

We will use the norms
$$\|\Psi(p)\|_{\mathcal{L}} \stackrel{\text{def}}{=} \sup\{\|A^\beta \Psi(p)\hat{p}\| : \|A^\beta \hat{p}\| \le 1\}$$
and
$$\|\Psi\|_\infty = \sup\{\|\Psi(p)\|_{\mathcal{L}} : p \in PV^{2\beta}\}$$
for mappings in $\mathcal{G}_\ell$. Next define $\mathcal{F}_\ell^1 \stackrel{\text{def}}{=} \mathcal{F}_\ell \cap C_F^1(PV^{2\beta}, QV^{2\beta})$. Note that if $\Phi \in \mathcal{F}_\ell^1$, then the Fréchet derivative $D\Phi$ is in $\mathcal{G}_\ell$. We now use the Lyapunov-Perron Transformation $\mathcal{T}^0 \stackrel{\text{def}}{=} \mathcal{T}$, see equation (82.5). Let $p(t) = p(t; p_0, \Phi)$ be the solution of (81.11) satisfying $p(0) = p_0$. Likewise for $\Phi \in \mathcal{F}_\ell$, $\Psi \in \mathcal{G}_\ell$, we define the operator-valued function $J(p_0, t) = J(\Phi, \Psi, p_0, t) \in \mathcal{L}(PV^{2\beta}, PV^{2\beta})$ as the solution of the linear ordinary differential equation

$$\partial_t \rho + AP\rho = PDF(u(t))(P + \Psi(p(t)))\rho \tag{85.2}$$

satisfying $J(p_0, 0) = P$, where $u(t) = p(t) + \Phi(p(t))$ and $DF(u)$ is the Fréchet derivative of F and $\|DF(u)\|_{\mathcal{L}(V^{2\beta}; H)} \le K_1$. For $\Phi \in \mathcal{F}_\ell$ and $\Psi \in \mathcal{G}_\ell$, we define $\hat{\Psi} = \hat{\Psi}(p_0) = \mathcal{T}^1(\Phi, \Psi)$ by

(85.3)
$$\hat{\Psi}(p_0) = \mathcal{T}^1(\Phi, \Psi)(p_0) = \int_{-\infty}^0 e^{AQs} QDF(u(s))(P + \Psi(p(s)))J(p_0, s)\, ds.$$

The first result is the infinitesimal version of Lemma 83.2. We leave the proof as an exercise for the reader.

Lemma 85.2. *For every $\Phi \in \mathcal{F}_\ell$ and $\Psi \in \mathcal{G}_\ell$, one has*

$$\|A^\beta J(p_0, s)\| \leq e^{-\gamma s}, \qquad \text{for } s \leq 0 \text{ and } p_0 \in PV^{2\beta}, \tag{85.4}$$

where γ is given by (83.7).

Using (82.3) and (83.12) with $\ell = L_1$, one finds that $\Psi \in \mathcal{G}_\ell$ implies $\hat{\Psi} \in \mathcal{G}_\ell$ since

$$\begin{aligned} \|A^\beta \hat{\Psi}(p_0)\hat{p}_0\| &\leq \int_{-\infty}^0 \|A^\beta e^{AQs}\| \, \|DF(u(s))(P + \Psi(p(s)))J(p_0, s)\hat{p}_0\| \, ds \\ &\leq 2K_1 \int_{-\infty}^0 b(-s)e^{-\gamma s} ds \, \|A^\beta \hat{p}_0\| \leq \ell \|A^\beta \hat{p}_0\|, \end{aligned}$$

for $p_0, \hat{p}_0 \in PV^{2\beta}$. In the next two lemmas we fix $\Phi \in \mathcal{F}_\ell$ and $\Psi_1, \Psi_2 \in \mathcal{G}_\ell$ and define $J_i = J_i(t) = J(\Phi, \Psi_i, p_0, t)$ and $\hat{\Psi}_i = \mathcal{T}^1(\Phi, \Psi_i)$ for $i = 1, 2$. The first lemma gives an estimate of $\|A^\beta(J_1 - J_2)\|$, and the second an estimate of $\|\hat{\Psi}_1 - \hat{\Psi}_2\|_\infty$.

Lemma 85.3. *With the above notation one has*

$$\|A^\beta(J_1(t) - J_2(t))\| \leq e^{-\gamma t}\|\Psi_1 - \Psi_2\|_\infty, \qquad \text{for all } t \leq 0, \tag{85.5}$$

where γ is given by (83.7).

Lemma 85.4. *With the above notation one has $\frac{3}{2}L_1 < 1$ and*

$$\|\hat{\Psi}_1 - \hat{\Psi}_2\|_\infty \leq \frac{3}{2} L_1 \|\Psi_1 - \Psi_2\|_\infty. \tag{85.6}$$

One can now prove that the mapping of $[0, \infty) \times \mathcal{F}_\ell \times \mathcal{G}_\ell$ into $\mathcal{F}_\ell \times \mathcal{G}_\ell$ given by

$$\mathcal{S} : (\lambda, \Phi, \Psi) \to (\mathcal{T}^0\Phi, \mathcal{T}^1(\Phi, \Psi))$$

satisfies the hypotheses of Lemma 85.1. First the continuity of $\mathcal{S}$ is a consequence of the Lebesgue Dominated Convergence Theorem. The contraction properties for $\mathcal{T}^0\Phi$ and $\mathcal{T}^1(\Phi, \Psi)$ follow from Lemmas 83.6 and 85.4.

For the remainder of this section, we will let (Φ, Ψ) be the unique fixed point in $\mathcal{F}_\ell \times \mathcal{G}_\ell$ given by Lemma 85.1. The component Φ is, of course, the fixed point of $\mathcal{T}^0$ proved in Lemma 83.6. What we want to show is that $\Phi \in \mathcal{F}_\ell^1$ and that $D\Phi = \Psi$. The next step is to observe that, to a limited extent, the operator $\mathcal{T}^0$ maps smooth functions onto smooth functions.

Lemma 85.5. *Let $\Phi \in \mathcal{F}_\ell$ and $\Psi \in \mathcal{G}_\ell$ and define $\hat{\Phi} = \mathcal{T}^0\Phi$ and $\hat{\Psi} = \mathcal{T}^1(\Phi, \Psi)$. If $\Phi \in \mathcal{F}_\ell^1$ and $D\Phi = \Psi$, then $\hat{\Phi} \in \mathcal{F}_\ell^1$ and $D\hat{\Phi} = \hat{\Psi}$.*

This lemma is the most significant step in the proof of the smoothness of the mapping Φ. The basic idea is to use the decomposition $\int_{-\infty}^0 =$

$\int_{-\infty}^{T} + \int_{T}^{0}$, for an appropriate T with $-\infty < T < 0$. One then uses the Leibniz argument which allows one to bring the differentiation inside the integral on the finite interval $(T, 0)$ (see Section 4.2), and one shows that the integral over the infinite interval $(-\infty, T)$ is small.

Lemma 85.5 can now be applied to a sequence of successive approximations

$$\Phi^{n+1} = \mathcal{T}^0 \Phi^n, \qquad \Psi^{n+1} = \mathcal{T}^1(\Phi^n, \Psi^n)$$

where $\Phi^0 \in \mathcal{F}_\ell^1$ and $D\Phi^0 = \Psi^0$. For definiteness we fix $\Phi^0(p) \equiv 0$ and $\Psi^0(p) \equiv 0$. One then has $\Phi^n \in \mathcal{F}_\ell^1$ and $D\Phi^n = \Psi^n$ for all $n \geq 0$. It then follows from Lemma 85.1 that

$$\begin{pmatrix} \Phi^n \\ \Psi^n \end{pmatrix} \to \begin{pmatrix} \Phi \\ \Psi \end{pmatrix}, \qquad \text{as } n \to \infty,$$

where the convergence is in the given topology on $\mathcal{F}_\ell \times \mathcal{G}_\ell$.

The final step is to show that Φ is in $\mathcal{F}_\ell^1$ and that $D\Phi = \Psi$. First note that since $\Phi^n \in \mathcal{F}_\ell^1$ and $\Psi^n = D\Phi^n$, the integral mean value theorem implies that

$$\Phi^n(p+h) - \Phi^n(p) = \int_0^1 \Psi^n(p+\theta h)\, d\theta\, h.$$

Now pass to the limit as $n \to \infty$ and note that the convergence $\Psi^n \to \Psi$ is uniform. Consequently, one has

$$\begin{aligned} \Phi(p+h) - \Phi(p) &= \int_0^1 \Psi(p+\theta h)\, d\theta\, h \\ &= \Psi(p)h + \int_0^1 [\Psi(p+\theta h) - \Psi(p)]\, d\theta\, h. \end{aligned} \tag{85.7}$$

Since $\Psi(p)$ is uniformly continuous in p, it follows that for every $\epsilon > 0$ there is a $\delta > 0$ such that

$$\|A^\beta(\Psi(p+\theta h) - \Psi(p))\| \leq \epsilon, \qquad 0 \leq \theta \leq 1,$$

whenever $\|A^\beta h\| \leq \delta$. By combining this with (85.7) we conclude that $\Phi \in \mathcal{F}_\ell^1$ and $\Psi = D\Phi$. This completes the proof that Φ is a C^1-function and that $\mathfrak{M}$ is a C^1-manifold. Furthermore, $D\Phi$ satisfies inequality (82.9), for all $p, v \in PV^{2\beta}$.

The final item, the proof that $\mathfrak{M}$ is normally hyperbolic, is left as an exercise. In this connection, see (74.6) and the proof of Item (7) in Theorem 74.15.

8.6. Applications of Inertial Manifold Theorems.

While the use of inertial[42] manifolds and inertial forms in the study of the longtime dynamics of solutions of certain evolutionary differential equations dates back to the mid 1980s, one can find precursors of this in earlier works. The earliest we are aware of is in seminal paper of Kurzweil (1966), where it is shown that an inertial manifold occurs when one introduces a small time-delay into a system of ordinary differential equations. Another precursor is the basic paper of Conway, Hoff and Smoller (1978), where they obtain a similar conclusion for systems of reaction diffusion equations with high diffusivity. Also see Mañé (1977) and Mora (1983).

Most of the known theory on the existence of inertial manifolds is based on the applications of the Spectral Gap Condition (81.16). The two primary parameters behind this inequality are the space dimension m and the parameter β. Let us look first at the case where the linear operator A is a second order uniformly elliptic operator, say $A = -\Delta$, with suitable boundary conditions. For the moment, we also assume that A is a selfadjoint operator on $\mathcal{D}(A) \subset L^2(\Omega)$, where Ω is an open bounded domain with smooth boundary in $\mathbb{R}^m$. The eigenvalues λ of A, which satisfy (32.2), then satisfy the asymptotic distribution relation

$$\lambda_n \sim c n^{\frac{2}{m}}, \qquad \text{as } n \to \infty, \tag{86.1}$$

where c is a positive constant which depends only on the domain Ω, see Courant and Hilbert (1953, 1962).

In space dimension $m = 1$ one has $(\lambda_{n+1} - \lambda_n) \sim cn$. Therefore, (81.16) is satisfied, for any $K_3 > 0$, whenever $F \in C_{\text{Lip}}(V^{2\beta}, M)$ and $0 \leq \beta < \frac{1}{2}$. The Chafee-Infante problem satisfies this condtion, but the equations of convection do not.

In space dimension $m = 2$, for a general domain Ω, one has $(\lambda_{n+1} - \lambda_n) \sim c$. Therefore, (81.16) is satisfied only when $\beta = 0$ and K_3 is small.

Nevertheless, for special geometries in $\mathbb{R}^2$, one is able to make an application of the Spectral Gap Condition. In particular, if $\Omega = Q_2 = (0, \ell_1) \times (0, \ell_2)$ is a square, and $A = -\Delta$ has either periodic, Dirichlet, or Neumann boundary conditions, then the eigenvalues assume the form

$$\lambda_k = 4\pi^2 \left(a_1^2 k_1^2 + a_2^2 k_2^2\right), \qquad \text{where } k = (k_1, k_2) \in \mathbb{Z}^2,$$

and $a_i = \ell_i^{-1}$. From a basic property of number theory, it is known that

$$\limsup_{n \to \infty} (\lambda_{n+1} - \lambda_n) = \infty, \qquad \text{whenever } a_1^2 a_2^{-2} \text{ is rational}, \tag{86.2}$$

see Richards (1982). It then follows from (86.2) that for <u>any</u> $K_3 > 0$, there exist integers $n \to \infty$ such that inequality (81.16) holds, whenever $\beta = 0$.

---

[42] As it is used in this context, the term "inertial manifold" was coined by Professor Ciprian Foias.

$m = 1$	$0 \leq \beta < \frac{3}{4}$
$m = 2$	$0 \leq \beta < \frac{1}{2}$
$m = 3$	$0 \leq \beta < \frac{1}{4}$

Table 8.1. Gap Conditions for Fourth-Order Problems

Even for the simplest geometries in $\mathbb{R}^3$, say $\Omega = Q_3 = (0,1)^3$, the Spectral Gap Condition (81.16) fails to hold unless $\beta = 0$ and K_3 is small. Nevertheless, there is a theory of inertial manifolds for scalar-valued reaction diffusion equations

$$\partial_t u - \Delta u = f(u, x), \qquad \text{for } x \in Q_3, \tag{86.3}$$

where $u = u(x,t)$ is a real-valued or complex-valued function, and $f = f(u,x)$ is sufficiently smooth, see Mallet-Paret and Sell (1988), and Mallet-Paret, Sell, and Shao (1993). Instead of using the Lyapunov-Perron method, as is done in this chapter, one uses a graph transformation method, which is similar to the method used in Section 7.4. The key feature arising in these papers is a Principle of Spatial Averaging, which is valid on appropriate domains when $1 \leq m \leq 3$. In particular, it applies to any rectangle Q_2 in $\mathbb{R}^2$, even when the ratio a_1^2/a_2^2 is irrational. While the Spectral Gap Conditon on Q_3 fails when K_3 is large, it should be noted that the eigenvalues on Q_3 do satisfy

$$\limsup_{n \to \infty} (\lambda_{n+1} - \lambda_n) > 0,$$

and this feature plays a critical role in the proof of the existence of an inertial manifold for (86.3). There are some extensions of the Principle of Spatial Averaging, see Kwean (1999). For example, an earlier theorem of Pinsky (1980) can be used to show that the Principle of Spatial Averaging holds on some triangular domains. Also, in his extension to the 3D setting of a result of Marion (1989a) on inertial manifolds for partly dissipative systems, Shao (1998) made a partial extension of the Principle of Spatial Averaging to systems of partial differential equations.

For higher-order elliptic operators, say, the biharmonic operator $A = \Delta^2$, the eigenvalues λ of A satisfy (32.2) and a stronger asymptotic distribution relation

$$\lambda_n \sim c n^{\frac{4}{m}}, \qquad \text{as } n \to \infty,$$

where c is a positive constant which depends on the domain Ω. This leads to the question: what conditions should (m, β) satisfy in order to guarantee that for any $K_3 > 0$, there are eigenvalues of A that satisfy the Spectral Gap Condition (81.16), for some n? An answer to this query is summarized in Table 8.1. In particular, we see that the Cahn-Hilliard problem and the Kuramoto-Sivashinsky problem always have inertial manifolds when $m = 1$.

The theorems developed in this chapter do have extensions to problems where the Standing Hypothesis B is replaced by the Standing Hypothesis A, and A is not selfadjoint, see Sell and You (1992). Like the theory developed here, the extension to the non-selfadjoint problem is built on properties of exponential dichotomies. The main difficulty one encounters in this extension is that the characteristic K, which arises in dichotomic inequality separating the p and the q spaces, depends on the rank n of the Bubnov-Galerkin projection $P = P_n$. The spaces $\mathcal{R}(P_n)$ and $\mathcal{R}(I - P_n)$ are generally not orthogonal, and the angle between them can become small, as $n \to \infty$.

Our description of the existence of inertial manifolds ends in space dimension $m = 3$. There is a good reason for this. A counterexample exists on the domain $Q_4 = (0,1)^4$ in $\mathbb{R}^4$, see Mallet-Paret, Sell, and Shao (1993).

8.7. Open Problems for Inertial Manifolds.

In a few locations in this chapter we left some of the details of the proofs as exercises for the reader. Rather than repeating these here, we choose to focus instead on some open problems connected with the theory of inertial manifolds. In each case, some new mathematics seems to be needed to solve the problem.

87.1. Assume that the Standing Hypothesis B is satsified, and let H denote the Hilbert space and A the corresponding sectorial operator. Consider the partially coupled system

$$\partial_t w + \mathbb{A} w = F(w), \qquad \text{where } w \in \mathbb{H} = H \times H, \tag{87.1}$$

with

$$w = \begin{pmatrix} u \\ v \end{pmatrix}, \quad \mathbb{A} = \begin{pmatrix} A & aA \\ 0 & A \end{pmatrix}, \quad F = \begin{pmatrix} f \\ g \end{pmatrix},$$

and $a \in \mathbb{R}$ is nonzero. Assume that $F \in C_{\text{Lip}}(\mathbb{H}, \mathbb{H})$ satisfies inequalities (81.7) and (81.8), where $V^{2\beta}$ is replaced by $\mathbb{H}$. Does there always exist an inertial manifold for (87.1), for any $a \in \mathbb{R}$? If so, give a proof; if not, a counterexample.

87.2. Does there exist an inertial manifold for the Navier-Stokes equations on $Q_2 = (0,1)^2$, with periodic boundary conditions?

87.3. Does there exist an inertial manifold for a reaction diffusion equation, say, the Chafee-Infante equation, on an arbitrary, smooth, bounded domain Ω in $\mathbb{R}^2$? On a polygonal domain in $\mathbb{R}^2$?

87.4. Can the Principle of Spatial Averaging be extended to systems of partial differential equations?

8.8. Commentary.

The various theories on the existence of inertial manifolds are based on one of three different approaches for the construction of invariant manifolds, namely (1) the Lyapunov-Perron method, which is based on the Variation of Constants Formula (see Section 4.2), (2) the Hadamard method, which uses a graph transformation approach (see Sections 7.4 and Chapter 8), and (3) the Sacker method, which is a method of elliptic regularization and is not treated here. For example, the Lyapunov-Perron method is used in Henry (1981), and Foias, Sell, and Temam (1986); the Hadamard method is used in Constantin, Foias, Nicolaenko,and Temam (1988), and Mallet-Paret and Sell (1987); and the Sacker method is described in Sacker (1965) for the finite dimensional case, and Fabes, Luskin, and Sell (1988) in the infinite dimensional case. Also see Sacker (1969). Other references are cited below.

Variations of the exponential tracking property, which we present in Section 8.4, are proved elsewhere. See Mañé (1977), Henry (1981), and Foias, Sell, and Titi (1988). The proof we give here differs from the arguments used in these papers.

A concept which fits somewhere between the concept of a global attractor and an inertial manifold is that of an **inertial set**. An inertial set is a positively invariant set (not necessarily a manifold), which contains the global attractor, and which has the property that it attracts all other solutions at a uniform exponential rate. For more information, see Eden, et al (1994); Eden, Foias, and Nicolaenko (1994); and Dung and Nicolaenko (1999).

There are some theories of inertial manifolds which arise in hyperbolic evolutionary equations, such as damped nonlinear wave equations and nonlinear beam equations. For more on this, see, for example, Mora and Solà-Morales (1989), Taboada and You (1992), You (1993b, 1996a,c), and Bianchi and Marzocchi (1998).

Additional Readings

Chow, Lu, and Sell (1992); Constantin, Foias, Nicolaenko, and Temam (1988); Fabes, Luskin, and Sell (1991); Foias, Nicolaenko, and Temam (1989); Foias, Sell, and Temam (1988); Foias, Sell, and Titi (1988); Jolly (1989); Mallet-Paret and Sell (1988); Mallet-Paret, Sell, and Shao (1993); Marion (1989a); Rosa and Temam (1996); Sell and Taboada (1992); Sell and You (1992); Shao (1994); Taboada and You (1992); Temam (1988); You (1993b, 1996a,c); and Sell (2001).

APPENDICES: BASICS OF FUNCTIONAL ANALYSIS

Appendix A: Banach Spaces and Fréchet Spaces.

Our purpose in this appendix is to introduce some basic notational conventions and basic theorems arising in the study of dynamics on metric spaces, with special emphasis on Banach spaces, Hilbert spaces, and Fréchet spaces.

A.1. Metric Spaces. Let W be a nonempty set. A real-valued function $\rho(x,y)$, which is defined for x, $y \in W$, is said to be a pseudometric on W if one has (1) $\rho(x,y) \geq 0$, (2) $\rho(x,x) = 0$, (3) $\rho(x,y) = \rho(y,x)$, and (4) $\rho(x,y) \leq \rho(x,z) + \rho(z,y)$, for all $x, y, z \in W$. If in addition, one has (5) $\rho(x,y) = 0$ if and only if $x = y$, then ρ is a metric on W. In this case, we say that W, or (W,ρ), is a **metric** space.

For many applications we will use Hilbert spaces. However, in some cases we will use Banach spaces, or Fréchet spaces, or subsets of such spaces. Since the completeness of the underlying space will be important for some considerations, we will be interested in the setting where W is a closed subset of a Banach space, or a Fréchet space. The metric on W will be denoted by $d_W(\cdot,\cdot) = d(\cdot,\cdot)$. For a Banach space W the norm will be denoted by $\|\cdot\|_W = \|\cdot\|$. The **dual space** of a Banach space W, which is denoted by W' or W^*, is the Banach space consisting of all bounded linear functionals on W with the operator norm.

For any set U in W, the **interior** of U is the largest open set V that satisfies $V \subset U$. We will write $V = \text{int}\, U$ for the interior of U. The **boundary** of U is denoted by ∂U and is defined as $\partial U \stackrel{\text{def}}{=} \text{Cl}_W U \setminus \text{int}\, U$. A set U in W is said to be a **neighborhood** of another set A in W provided one has $A \subset \text{int}\, U$. For any set U in W and any $\epsilon \geq 0$ we let $N_\epsilon(U)$ denote the set

$$N_\epsilon(U) \stackrel{\text{def}}{=} \{v \in W : \inf_{u \in W} d_W(u,v) \leq \epsilon\}.$$

Note that for $\epsilon > 0$, the set $N_\epsilon(U)$ is a (closed) neighborhood of U. For a Banach space W, we let $N_a(0) \stackrel{\text{def}}{=} \{u \in W : \|u\| \leq a\}$ denote the closed ball of radius a in W, where $a > 0$. A metric space (W,d) is said to be **separable** if there is a countable set in W that is dense in W.

A.2. Banach Spaces. A Banach space W may be equipped with more that one norm. Two norms $\|\cdot\|_1$ and $\|\cdot\|_2$ defined on W are said to be **equivalent** if there exist positive constants α and β such that $\alpha\|w\|_1 \leq \|w\|_2 \leq \beta\|w\|_1$, for all $w \in W$. Let V and W be complex Banach spaces. A mapping $L : V \to W$ is said to be **additive** if $L(v_1 + v_2) = Lv_1 + Lv_2$, for all $v_1, v_2 \in V$. An additive mapping L is said to be **hemilinear** if $L(\alpha v) = \bar{\alpha} Lv$, for all $v \in V$, and $\alpha \in \mathbb{C}$, and L is **linear** if $L(\alpha v) = \alpha Lv$, for all $v \in V$, and $\alpha \in \mathbb{C}$. A mapping $a = a(u, v)$ from $V \times V$ to W is said to be **sesquilinear**[43] if it is linear in u and hemilinear in v, and $a(u, v)$ is **bilinear** if it is linear in both variables. If V and W are real Banach spaces, then, of course, hemilinearity is equivalent to linearity, and sesquilinearity is equivalent to bilinearity.

Let U and V be two closed linear subspaces of a Banach space W, with the property that $W = U + V = \text{Span}(U \cup V)$. We say that U and V have **transversal intersection** in W, provided that there exist three bounded linear projections P, Q, and R, such that these projections all commute; one has $PQ = PR = QR = 0$, $I = P + Q + R$; and

$$\mathcal{R}(P + Q) = U, \quad \mathcal{R}(R + Q) = V, \quad \text{and} \quad \mathcal{R}(Q) = U \cap V,$$

where $\mathcal{R}(L)$ is used to denote the range of the operator L.

A.3. Hilbert Spaces. In the case of a Hilbert space, we will denote the inner product of two elements $u, v \in H$ by $\langle u, v\rangle = \langle u, v\rangle_H$. The Riesz Representation Theorem states that if ℓ is a bounded, linear functional on a Hilbert space H, then there is a unique $v \in H$ such that

$$\ell(u) = \langle u, v\rangle_H, \qquad \text{for all } u \in H.$$

Furthermore, the mapping of the dual space H^* into H, given by $j : \ell \to v$, is an isometric, hemilinear mapping of H^* onto H. We will refer to j, or its inverse, as the **Riesz mapping**.

A.4. Linear Operators. Most differential operators with appropriate boundary conditions can be described as unbounded linear operators in some Sobolev spaces. As a counterpart, most integral operators, including solution operators for differential equations, are bounded linear (or affine) operators.

Let V and W be two Banach spaces. The collection of all bounded, linear operators $T : V \to W$ will be denoted by $\mathcal{L} = \mathcal{L}(V, W)$. If $V = W$, we will write $\mathcal{L}(W) = \mathcal{L}(W, W)$. The **operator norm** is given by

$$\|T\| = \|T\|_{\mathcal{L}} = \|T\|_{\mathcal{L}(V,W)} \stackrel{\text{def}}{=} \sup\{\|Tv\|_W : \|v\|_W \leq 1\}.$$

An **isomorphism** T, from a Banach space V onto another Banach space W, is a mapping $T \in \mathcal{L}(V, W)$, where T is a one-to-one mapping of V onto

[43] Sesqui comes from the Latin, and it means "one and a half".

W, where the inverse satisfies $T^{-1} \in \mathcal{L}(W, V)$. A **(linear) projection** P on W is a mapping $P \in \mathcal{L}(W)$ with the property that $P^2 = P$. If P is a projection, then $Q = I - P$ is a projection, as well. In this case, Q is called the **complementary projection** for P. Note that if P is a projection, then the range and null spaces, $\mathcal{R}(P)$ and $\mathcal{N}(P)$, are closed linear subspaces of W with $W = \mathcal{R}(P) + \mathcal{N}(P)$.

Let $\mathbb{Z} = \{n \in \mathbb{R}\}$ denote the set of all integers and set $\mathbb{Z}^+ = \{n \in \mathbb{Z} : n \geq 0\}$. We will let $\ell^\infty(\mathbb{Z}, W)$, or $\ell^\infty(\mathbb{Z}^+, W)$, respectively, denote the Banach space of all sequences w_n, with $w_n \in W$ for each $n \in \mathbb{Z}$, or $n \in \mathbb{Z}^+$, respectively, such that

$$\|w_n\|_\infty \stackrel{\text{def}}{=} \sup\{\|w_n\|_W : n \in \mathbb{Z}, \text{ or } \mathbb{Z}^+\} < \infty.$$

In the case of a sequence $T_n \in \mathcal{L}(W)$, for $n \in \mathbb{Z}$, we define $T_{n,m}$ by

$$T_{n,m} \stackrel{\text{def}}{=} T_{n-1} \cdot \cdots \cdot T_{m+1} \cdot T_m, \qquad \text{for } n > m,$$

and we set $T_{m,m} = I$. The following result is proved in Henry (1981, pp 230 - 232).

Lemma A.0. *Let W be a Banach space and let T_n be a sequence in $\mathcal{L}(W)$, for $n \in \mathbb{Z}$. Assume that T_n has the property that for every $f \in B \stackrel{\text{def}}{=} \ell^\infty(\mathbb{Z}, W)$, there is a unique sequence $x = \{x_n\} \in B$ such that*

$$x_{n+1} = T_n x_n + f_n, \qquad \text{for all } n \in \mathbb{Z}.$$

Then there is a sequence of projections P_n on W, for $n \in \mathbb{Z}$, such that the following properties hold:

(1) *One has $T_n P_n = P_{n+1} T_n$, for all $n \in \mathbb{Z}$.*

(2) *The restriction $T \mid_{\mathcal{R}(P_n)}$ is an isomorphism from $\mathcal{R}(P_n)$ onto the space $\mathcal{R}(P_{n+1})$, for all $n \in \mathbb{Z}$. Since the inverses are well-defined, we set*

$$T_{n,m} w \stackrel{\text{def}}{=} T \mid^{-1}_{\mathcal{R}(P_{n-1})} \cdot \cdots \cdot T \mid^{-1}_{\mathcal{R}(P_m)} w, \qquad \text{for } n < m \text{ and } w \in \mathcal{R}(P_m).$$

Note that for $w \in \mathcal{R}(P_m)$, one has $P_m w = w$ and $T_{n,m} P_m w = T_{n,m} w$, for all $n < m$.

(3) *There exists constant $K \geq 1$ and θ with $0 < \theta < 1$, such that*

$$\|T_{n,m}(I - P_m)w\| \leq K\theta^{n-m}\|w\|, \qquad \text{for all } n \geq m \text{ and } w \in W,$$

and

$$\|T_{n,m} P_m w\| \leq K\theta^{m-n}\|w\|, \qquad \text{for all } n < m \text{ and } w \in W.$$

Because of the importance of this lemma in the study of the longtime dyamics of evolutionary equations, we add a few comments on the proof. The details can be found in Henry (1981, pp 229 - 233). The proof of this lemma is based on the observation that the linear mapping $L : B \to B$ given by $y = Lx$, where $y_n = x_{n+1} - T_n x_n$ for $n \in \mathbb{Z}$, is an isomorphism on B with a bounded inverse G which has the form $x = Gy$, where

$$x_n = \sum_{k \in \mathbb{Z}} G_{n,k+1} y_k,$$

and each coefficient $G_{n,k+1}$ is in $\mathcal{L}(W)$ with $||G_{n,k+1}||_{\mathcal{L}(W)} \leq ||G||_{\mathcal{L}(B)}$. The projection P_n is then given by $P_n = I - G_{n,n}$, for $n \in \mathbb{Z}$. Clearly, $G_{n,n}$ is the complementary projection for P_n. One then shows that $P_{m+1}T_m w = T_m P_m w$, whenever $P_m w = 0$. Next one argues that the restriction $T_m |_{\mathcal{R}(P_m)}$ is an isomorphism of $\mathcal{R}(P_m)$ onto $\mathcal{R}(P_{m+1})$. One then shows that $P_{n+1}T_n x = T_n P_n x$, for all $w \in W$. Finally one verifies the exponential growth estimates in Item (3).

Let T be a linear mapping (called operator) from $\mathcal{D}(T) \subset V$ to W, where V and W are two Banach spaces. The most important class of unbounded linear operators is closed operators. A linear operator $T : \mathcal{D}(T)(\subset V) \to W$ is said to be a **closed** operator if its graph

$$\text{Graph } (T) = \{(x, y) \in V \times W : x \in \mathcal{D}(T), y = Tx\}$$

is a closed set in the product Banach space $V \times W$. To put it another way, the operator T is closed if and only if for all sequences x_n in $\mathcal{D}(T)$ with the property that $\lim_{n\to\infty} x_n = x$ in V and $\lim_{n\to\infty} Tx_n = y$ in W, then one has $x \in \mathcal{D}(T)$ and $y = Tx$.

All the closed operators from V (which means their domains are linear subspaces of V) to W are denoted by $CLO(V, W)$. If $V = W$, we will write $CLO(W) = CLO(W, W)$. A linear operator $T : \mathcal{D}(T)(\subset V) \to W$ is called **densely defined** if $\mathcal{D}(T)$ is a dense subspace in V. Let $T^* : \mathcal{D}(T^*)(\subset W^*) \to V^*$ denote the adjoint operator of T. The concepts of adjoint operators and selfadjoint operators can be found in Naylor and Sell (1982). Note that if $T \in CLO(H, H)$, where H is a Hilbert space, satisfies $T \subset T^*$ (i.e., $\mathcal{D}(T) \subset \mathcal{D}(T^*)$ and $\langle Tx, y\rangle = \langle x, Ty\rangle$, for all $x, y \in \mathcal{D}(T)$), then T is called a **symmetric** operator. If in addition, $\mathcal{D}(T) = \mathcal{D}(T^*)$, then T is called **selfadjoint**.

Theorem A.1. *Let V and W be Banach spaces. Then the class of closed operators $CLO(V, W)$ has the following properties:*

(1) *For any closed operator $A : \mathcal{D}(A)(\subset V) \to W$, its domain $\mathcal{D}(A)$ equipped with the graph norm $||w||_{\text{Graph}(A)} \stackrel{\text{def}}{=} ||w||_V + ||Aw||_W$ becomes a Banach space, and $A \in \mathcal{L}(\mathcal{D}(A), W)$ is this sense.*
(2) *An operator A satisfies $A \in \mathcal{L}(V, W)$ if and only if $A \in CLO(V, W)$ and $\mathcal{D}(A) = V$.*

(3) *For any densely defined linear operator T (no matter closed or not), its adjoint operator T^* (if it exists) must be a closed operator. In particular, any selfadjoint operator (bounded or unbounded) on a Hilbert space must be a closed operator.*

In an infinite dimensional space, a symmetric linear operator need not be selfadjoint and need not share the spectral theory for selfadjoint operators. The following proposition is sometimes useful in that regard.

Theorem A.2. *Let T be a symmetric linear operator in a Hilbert space H. If for some $\lambda \in \mathbb{R}$, one has $\mathcal{R}(\lambda I - T) = H$, then T is a selfadjoint operator. In particular, if $\mathcal{R}(T) = H$, then T is a selfadjoint operator.*

Proof. Since $\mathcal{D}(T)$ is dense and $\mathcal{D}(T) \subset \mathcal{D}(T^*)$ by the symmetry, $\mathcal{D}(T^*)$ is dense in H. We need to prove that $\mathcal{D}(T^*) \subset \mathcal{D}(T)$, then $T^* = T$.

Owing to the fact that $\mathcal{R}(\lambda I - T^*) \subset \mathcal{R}(\lambda I - T) = H$, for any $f \in \mathcal{D}(T^*)$, there is an $g \in \mathcal{D}(T)$ such that

$$(\lambda I - T^*)f = (\lambda I - T)g = (\lambda I - T^*)g,$$

where the second equality follows from the fact $T^*|_{\mathcal{D}(T)} = T$. Hence,

$$f - g \in \mathcal{N}(\lambda I - T^*) \subset \mathcal{R}(\lambda I - T)^{\perp} = \{0\},$$

where the inclusion relation is always true for any adjoint operator, cf. Naylor and Sell (1982). Thus, $f = g$ and $f \in \mathcal{D}(T)$. Hence, $\mathcal{D}(T^*) \subset \mathcal{D}(T)$, and the proof is completed. □

Let H be a Hilbert space and $A : \mathcal{D}(A)(\subset H) \to H$ be a symmetric operator. The operator A is called **upper bounded** (resp. **lower bounded**) if there is a constant real number $\delta \in \mathbb{R}$ such that

$$\langle Au, u\rangle \leq \delta \|u\|^2 \qquad (\text{resp. } \geq \delta\|u\|^2).$$

In both cases, A is called **semibounded**.

Let W be a real Banach space and $A : \mathcal{D}(A)(\subset W) \to W'$ be a single-valued mapping, possibly nonlinear. The operator A is called **accretive** if

$$(Au_1 - Au_2, u_1 - u_2) \geq 0, \qquad \text{for all } u_1, u_2 \text{ in } \mathcal{D}(A),$$

where $(\cdot,\cdot)$ represents the functional action between elements in W' and in W. The operator A is called **maximal accretive**, or **m-accretive**, if it is accretive and there is no proper accretive extension of A. The operator A is called **demicontinuous** if it is strongly-weakly continuous in the sense that, if $\{u_n\}$ is a sequence in $\mathcal{D}(A)$ with strong limit u_0 in W, then $\{Au_n\}$ converges weakly to Au_0 in W' and $u_0 \in \mathcal{D}(A)$. The operator A is called **bounded** if it maps every bounded set in $\mathcal{D}(A)$ of W into a bounded set of W'. Finally, A is called **coercive** if

$$\lim_{\|u\|_W \to \infty} \frac{(Au, u)}{\|u\|_W^2} = +\infty.$$

Theorem A.3 (Friedrichs). *Let H be a Hilbert space. For any linear operator $A : \mathcal{D}(A)(\subset H) \to H$ that is symmetric and semibounded, there exists a selfadjoint extension A_e of A, $\mathcal{D}(A) \subset \mathcal{D}(A_e) \subset H$.*

The operator A_e, which need not be unique, is referred to as a **Friedrichs extension** of A. The proof and construction of the extension A_e can be found from Yosida (1966).

A.5. Spectrum of a Closed Operator. Let W be a Banach space and $T \in CLO(W)$ be a closed, linear operator whose domain is $\mathcal{D}(T) \subset W$. Then the complex plane $\mathbb{C}$ has a disjoint composition $\mathbb{C} = \rho(T) \cup \sigma(T)$, where $\rho(T)$ is the set of regular points and $\sigma(T)$ is the spectrum. A complex number λ satisfies $\lambda \in \rho(T)$ if and only if the linear operator $\lambda I - T : \mathcal{D}(T) \to W$ is one-to-one and onto, and the inverse operator $R(\lambda; T) \stackrel{\text{def}}{=} (\lambda I - T)^{-1}$ is a bounded linear operator on W. The operator $R(\lambda; T)$ is called the **resolvent operator**. If any of the three properties mentioned above is not satisfied by $\lambda I - T$, then this λ is in the spectrum, and we write $\lambda \in \sigma(T)$. We use $\sigma_p(T)$ to denote the subset of $\sigma(T)$, that consists of all the eigenvalues of T and is called the **point spectrum**. In other words, $\lambda \in \sigma_p(T)$ if there is an $x \in \mathcal{D}(T)$ such that $x \neq 0$ and $(\lambda I - T)x = 0$.

Here is a key difference between a bounded linear operator $K \in \mathcal{L}(W)$ and a closed operator $T \in CLO(W)$, where W is an infinite dimensional Banach space. By the Liouville Theorem and the boundedness of the spectral radius, one knows both $\sigma(K)$ and $\rho(K)$ are nonempty, for $K \in \mathcal{L}(W)$. However, for some closed operators, it is possible that $\sigma(T) = \emptyset$ or $\rho(T) = \emptyset$. Try to find such examples!

Theorem A.4. *Let W be a Banach space and $T : \mathcal{D}(T)(\subset W) \to W$ be a closed operator. Then the following properties are valid.*

(1) *$\sigma(T)$ is a closed set in $\mathbb{C}$ and $\rho(T)$ is an open set in $\mathbb{C}$.*
(2) *For any fixed $\lambda_0 \in \rho(T)$, there is a neighborhood $N_\varepsilon(\lambda_0)$ in $\mathbb{C}$ such that for any $\lambda \in N_\varepsilon(\lambda_0)$, one has $\lambda \in \rho(T)$ and the following Neuman expansion holds in $\mathcal{L}(W)$,*

$$R(\lambda; T) = \sum_{n=0}^{\infty} (\lambda_0 - \lambda)^n R(\lambda_0; T)^{n+1}, \qquad \textit{for } \lambda \in N_\epsilon(\lambda_0).$$

(3) *If T is a compact operator in $\mathcal{L}(W)$, then $\sigma(T)$ is a countable set and the only possible accumulation point of $\sigma(T)$ is $\lambda_0 = 0$. Every nonzero spectral point must be an eigenvalue of finite multiplicity.*
(4) *If T is an unbounded, selfadjoint, positive definite operator on a Hilbert space H such that for some $\lambda \in \rho(T)$, $R(\lambda; T)$ is a compact operator, then the properties described in Section 3.2 are satisfied.*

A.6. Compact and Continuous Imbeddings. Let W be a Banach space with norm $\|\cdot\|_W$, and let V be a linear subspace of W. Assume that there is a norm $\|\cdot\|_V$ on V and that V is complete in this norm, i.e., $(V, \|\cdot\|_V)$ is a Banach space. We will say that V is a **continuous imbedding** into W (and will denote this by $V \mapsto W$) provided the inclusion mapping $i(v) = v$, where

$$i : (V, \|\cdot\|_V) \to (W, \|\cdot\|_W), \tag{91.1}$$

is continuous. The continuity of the inclusion mapping means that there is a constant $C > 0$ such that

$$\|v\|_W \leq C\|v\|_V, \qquad \text{for all } v \in V. \tag{91.2}$$

We will say that V is a **compact imbedding** into W (and will denote this by $V \hookrightarrow W$) provided the inclusion mapping i given by (91.1) is a compact linear operator. In the case of a compact imbedding, inequality (91.2) is valid and every bounded set B in V is **precompact** in W, i.e., $\mathrm{Cl}_W B$ is compact in W.

A.7. Fréchet Spaces. Let F be any linear space. A metric $d(f, g)$ on F is said to be **invariant** if one has

$$d(f, g) = d(f - g, 0), \qquad \text{for all } f,\, g \in F.$$

If the last equation holds, one sometimes expresses the metric in the form $N(f - g)$, where $N(f) = d(f, 0)$, for $f \in F$. A **Fréchet space** F is a linear space with an invariant metric $d(f, g)$ and such that the metric space (F, d) is complete. Clearly, every Banach space W is a Fréchet space, with the metric $d(v, w) = \|v - w\|$.

Here is a typical example of a Fréchet space, where the metric is not a norm. Let X and Y denote two Banach spaces. Define $C(X, Y)$ as the collection of all continuous fucntions $f : X \to Y$, and let $L^\infty_{\mathrm{loc}}(X, Y)$ be the collection of all functions $f : X \to Y$ with the property that f is bounded on every bounded set B in X. Let

$$F = C_b(X, Y) \stackrel{\text{def}}{=} C(X, Y) \cap L^\infty_{\mathrm{loc}}(X, Y).$$

Then F becomes a Fréchet space, where $d(f, g) = N(f - g)$,

$$N(f) \stackrel{\text{def}}{=} \sum_{n=1}^{\infty} 2^{-n} \frac{\|f\|_n}{1 + \|f\|_n}, \qquad \text{for } f \in F, \tag{91.3}$$

and

$$\|f\|_n = \sup\{\|f(x)\|_Y : \|x\|_X \leq n\}, \qquad \text{for } n = 1, 2, \cdots. \tag{91.4}$$

With this metric, the topology on F is the **topology of uniform convergence on bounded sets** in X.

Notice that $\|f\|_n$ is a pseudonorm on F and one has $\|f\|_n \leq \|f\|_{n+1}$, for every $f \in F$ and every $n \geq 1$. The use of a countable family of pseudonorms $\|\cdot\|_n$, such as (91.4), and the definition (91.3) is viewed as a common way for the construction of a Fréchet space. The definition (91.3) itself is not sacred. Other definitions, such as

$$N_0(f) = \sum_{n=1}^{\infty} 2^{-n} \min(\|f\|_n, 1),$$

work just as well and lead to the same topology.

Let X denote any Banach space, with norm $\|\cdot\|_X$. We will distinguish between two Fréchet spaces, $L^p_{\text{loc}}(0,\infty;X)$ and $L^p_{\text{loc}}[0,\infty;X)$, for $1 \leq p \leq \infty$. The integrals of any abstract function with values in a Banach space X will refer to as the Bochner integral (see Appendix C). The space $L^p_{\text{loc}}[0,\infty;X)$ is always the smaller of the two spaces, and it consists of certain functions from $L^p_{\text{loc}}(0,\infty;X)$ that have better regularity properties at $t=0$. More precisely, for $1 \leq p < \infty$, we will let $L^p_{\text{loc}}(0,\infty;X)$ denote the collection of all functions $\phi : (0,\infty) \to X$ with the property that, for all τ and T with $0 < \tau \leq T < \infty$, one has

$$\int_{\tau}^{T} \|\phi(s)\|_X^p \, ds < \infty.$$

We let $L^p_{\text{loc}}[0,\infty;X)$ denote the collection of all functions $\phi \in L^p_{\text{loc}}(0,\infty;X)$ with the property that, for all T with $0 < T < \infty$, one has

$$\int_{0}^{T} \|\phi(s)\|_X^p \, ds < \infty.$$

Similarly for $p = \infty$, we will let $L^\infty_{\text{loc}}(0,\infty;X)$ denote the collection of all $\phi : (0,\infty) \to X$ with the property that, for all τ and T with $0 < \tau \leq T < \infty$, one has

$$\operatorname*{ess\,sup}_{\tau < s < T} \|\phi(s)\|_X < \infty.$$

We let $L^\infty_{\text{loc}}[0,\infty;X)$ denote the collection of all functions $\phi \in L^\infty_{\text{loc}}(0,\infty;X)$ with the property that, for all T with $0 < T < \infty$, one has

$$\operatorname*{ess\,sup}_{0 < s < T} \|\phi(s)\|_X < \infty.$$

For $0 \leq a < b \leq \infty$ and $1 \leq p \leq \infty$, the spaces $L^p_{\text{loc}}(a,b;X)$ and $L^p_{\text{loc}}[a,b;X)$ are defined in an analogous manner. See Dunford and Schwartz (1958) or Hille and Phillips (1957), for more details about these L^p-spaces.

For $-\infty \leq a < b \leq \infty$ we let $L^\infty(a,b;X)$ denote the usual Banach space of functions $\phi : (a,b) \to X$, where $\operatorname{ess\,sup}_{a<s<b} \|\phi(s)\|_X < \infty$, with the norm given by

$$\|\phi\|_\infty = \|\phi\|_{L^\infty(a,b;X)} \stackrel{\text{def}}{=} \operatorname*{ess\,sup}_{a<s<b} \|\phi(s)\|_X.$$

We let $C(0, \infty; H_w)$ and $C[0, \infty; H_w)$ denote the spaces of weakly continuous functions with range in H and defined, respectively, on the intervals $(0, \infty)$ and $[0, \infty)$. Similarly, $C(a, b; X)$, $C[a, b; X)$, and $C([a, b]; X)$ will denote the spaces of strongly continuous functions with values in X and defined on the respective intervals, (a, b), $[a, b)$, and $[a, b]$.

The spaces $L^p_{\text{loc}}(0, \infty; X)$ and $L^p_{\text{loc}}[0, \infty; X)$, for $1 \le p < \infty$, are examples of Fréchet spaces. This means that they are metrizable, locally convex topological linear spaces, and they are complete, see Kelley and Namioka (1976), and Dunford and Schwartz (1958). The topology on each of these spaces is generated by a countable family of pseudonorms. In particular, on the space $L^p_{\text{loc}}(0, \infty; X)$ one can use the pseudonorms $N_0 = N_1 = 0$ and

$$N_n(\phi) = \left(\int_{n^{-1}}^{n} \|\phi\|_X^p \, ds \right)^{\frac{1}{p}}, \qquad \text{for } n = 2, 3, 4, \cdots, \tag{91.5}$$

while on the space $L^p_{\text{loc}}[0, \infty; X)$ one can use

$$N_n(\phi) = \left(\int_{n}^{n+1} \|\phi\|_X^p \, ds \right)^{\frac{1}{p}}, \qquad \text{for } n = 0, 1, 2, \cdots. \tag{91.6}$$

The concepts of continuous and compact imbeddings extend directly to the Fréchet space setting. Note that $L^p_{\text{loc}}[0, \infty; X) \mapsto L^p_{\text{loc}}(0, \infty; X)$, when $1 \le p < \infty$, and

$$L^\infty(0, \infty; X) \mapsto L^\infty_{\text{loc}}[0, \infty; X) \mapsto L^\infty_{\text{loc}}(0, \infty; X).$$

There are other Fréchet spaces which arise in the study of the longtime dynamics of evolutionary equations. These are

$$\mathcal{M} \stackrel{\text{def}}{=} L^\infty(\mathbb{R}; X) \cap C(\mathbb{R}; X) \quad \text{and} \quad \mathcal{M}^p \stackrel{\text{def}}{=} L^p(\mathbb{R}; X).$$

In the case of $\mathcal{M}$, we will use the L^∞_{loc}-**topology**, which is generated by the family of pseudonorms

$$N_n(\phi) \stackrel{\text{def}}{=} \sup \{ \|\phi(s)\|_X : n \le s \le n+1 \}, \qquad n \in \mathbb{Z}. \tag{91.7}$$

The L^∞_{loc}-topology on $\mathcal{M}$ is the topology of uniform convergence on bounded sets. On $\mathcal{M}^p$, for $1 \le p < \infty$, one uses the L^p_{loc}-**topology**, which is generated by the family of pseudonorms

$$N_n(\phi) \stackrel{\text{def}}{=} \left(\int_{n}^{n+1} \|\phi(s)\|_X^p \, ds \right)^{\frac{1}{p}}, \qquad n \in \mathbb{Z}. \tag{91.8}$$

Recall that a set B in a linear topological space Z is said to be **bounded** if for every neighborhood U of the origin in Z there is an $r > 0$ such that

$B \subset rU$, where $rU = \{ru : u \in U\}$, see Kelley and Namioka (1963). In the case of the Fréchet spaces $L^p_{\text{loc}}(0, \infty; X)$ and $L^p_{\text{loc}}[0, \infty; X)$, for $1 \leq p < \infty$, it follows that a set B is bounded if and only if one has

$$\sup\{N_n(\phi) : \phi \in B\} < \infty, \qquad \text{for each } n = 0, 1, 2, \ldots, \tag{91.9}$$

where $N_n(\phi)$ is given in (91.5) or (91.6). Recall that in a metric space a set B is compact if and only if it is sequentially compact, see Naylor and Sell (1982).

It should be noted that the concept of a bounded set given above is a topological concept. It does not depend on the metric used to construct the topology. In order to illustrate this point, let X be a Banach space with norm $\|\cdot\|_X$. Define N_X by

$$N_X(u) \stackrel{\text{def}}{=} \min\{\|u\|_X, 1\}, \qquad \text{for } u \in X.$$

Note that $N_X(u - v)$ is a metric on X. Furthermore, a set B is bounded in X if and only if

$$\sup\{\|u - v\|_X : u, v \in B\} < \infty.$$

A similar characterization of boundedness in terms of N_X is not possible, since $\sup\{N_X(u-v) : u, v \in X\} \leq 1$. Thus in terms of the invariant metric N_X, every set in X has a finite N_X-diameter, while a set B in X has finite $\|\cdot\|_X$-diameter if and only if it is bounded.

A.8. Topology of Duality. Let W and W' be a Banach space and its dual space, respectively. The duality provides topologies on these spaces in a natural way. In particular, these topologies are defined as the *weakest* topologies such that <u>all</u> the bilinear mappings from $W \times W'$ into the field of scalars, $(x, f) \to f(x)$, are continuous. The associated topology on W is called the **weak** topology on W, while the topology on W' is called the **weak*** on W'. By using finite intersections, a base for the neighborhood system of the zero-element 0, in each of these spaces, is generated by the sets $U_f(\epsilon) = \{x \in W : |f(x)| < \epsilon\}$ and $V_x(\epsilon) = \{f \in W' : |f(x)| < \epsilon\}$, respectively. Each space is a locally convex linear space in the respective topologies. See Kelley and Namioka (1963), for more details.

A.9. Some Basic Theorems in Banach Spaces. Besides the very important theorems (the Hahn-Banach Theorem and the Closed Graph Theorem), there are several basic results we have used in the presentation of the text that hold in Banach spaces.

Uniform Boundedness Principle. *Let $\{T_n\}$ be a sequence of bounded, linear operators in $\mathfrak{L} = \mathfrak{L}(V, W)$, where V and W are Banach spaces. If for each $x \in V$, the sequence $\{T_n x\}$ is bounded in W, then the sequence $\{T_n\}$ is bounded in $\mathfrak{L}(V, W)$, i.e., $\sup_n \|T_n\|_{\mathfrak{L}} < \infty$.*

This Uniform Boundedness Principle is also referred to as the Banach-Steinhaus Theorem or the Resonance Theorem.

Banach-Alaoglu Theorem. *Let W be any Banach space and W' be its dual space, where W' is equipped with the operator norm. Then every bounded set in W' is weak* precompact.*

According to the latter theorem, if $\{\varphi_n\}$ is any bounded sequence in W', then there exists a subsequence $\{\varphi_{n_k}\}$ and a $\varphi \in W'$, such that $\lim_{k\to\infty} \varphi_{n_k}(v) = \varphi(v)$, for any $v \in W$.

Corollary. *Every bounded set in a reflexive Banach space, or a Hilbert space, is weakly precompact.*

Dunford-Pettis Theorem. *Let W_1 and W_2 be two Banach spaces, and assume that the intersection $W = W_1 \cap W_2$ is nonempty. Then W is a Banach space with the norm $\|u\|_W = \|u\|_{W_1} + \|u\|_{W_2}$, and its dual space is*

$$W' = W_1' + W_2' \qquad \textit{(the linear sum)}$$

whose norm is

$$\|\varphi\|_{W'} = \inf\{\|g\|_{W_1'} + \|h\|_{W_2'} \mid \varphi = g + h, g \in W_1' \text{ and } h \in W_2'\}.$$

Mazur Theorem. *Let W be a reflexive Banach space. Then every strongly closed, convex set S in W is weakly closed.*

As a consequence of the Mazur Theorem, if W is a reflexive Banach space and $f : W \to \mathbb{R}$ is a strongly continuous and convex functional, then f is weakly lower semicontinuous. That is to say, if $x_n \to x$ weakly in W, then

$$\liminf_{n\to\infty} f(x_n) \geq f(x).$$

We have seen such an example where $f(\cdot) = \|\cdot\|_W$.

The first two theorems can be found in Naylor and Sell (1982), the third one in Lions and Magenes (1972), and the fourth one in Hille and Phillips (1957).

A.10. Vitali Convergence Theorem. In certain cases, where the Lebesgue Dominated Convergence Theorem is not applicable, one can still prove the commutivity of the limit and the integral by using an alternate result. Let $\Psi = \{f_\lambda : \lambda \in \Lambda\}$ denote a family of integrable functionss on a Lebesgue measurable set E in $\mathbb{R}$, where Λ is an index set. We say that Ψ has **equi-absolutely continuous integrals** on E, if for every $\epsilon > 0$, there is a $\delta = \delta(\epsilon) > 0$ such that for every measurable subset $B \subset E$ with $\mu(B) \leq \delta$, one has

$$\left|\int_B f_\lambda \, d\mu\right| \leq \epsilon, \qquad \text{for all } \lambda \in \Lambda.$$

Note that δ does <u>not</u> depend on λ. The proof of the following result can be found in Natanson (1957).

Vitali Convergence Theorem. *Let E be a measurable set in $\mathbb{R}$ with $\mu(E) < \infty$. Let $\{f_n\}$ be a sequence of functions in $L^1(E;\mathbb{R})$. If the following two conditions are satisfied:*

(1) *$f_n \to f$, as $n \to \infty$, almost everywhere on E, and*
(2) *$\{f_n\}$ has equi-absolutely continuous integrals on E,*

then the limit function f satisfies $f \in L^1(E;\mathbb{R})$ and

$$\lim_{n\to\infty} \int_E f_n \, d\mu = \int_E f \, d\mu. \tag{91.10}$$

Appendix B: Function Spaces and Sobolev Imbedding Theorems.

In this section, we sketch the basic facts of Sobolev spaces that serve as the phase spaces for the infinite dimensional dynamical systems modeled from boundary value problems of partial differential equations. The detailed proof will be referred to the references.

Recall that a homeomorphism ϕ is said to be a **Lipschitz homeomorphism** if both ϕ and ϕ^{-1} are locally Lipschitz continuous. Let Ω be an open, bounded subset in the Euclidean space $\mathbb{R}^n$, and let $\partial\Omega$ denote the boundary of Ω and $\overline{\Omega}$ the closure of Ω. We will say that Ω is of **Lipschitz class**, or that the boundary $\partial\Omega$ is **locally Lipschitz continuous**, if for every point $x \in \partial\Omega$ there is a Lipschitz homeomorphism ϕ of a neighborhood U of x into $\mathbb{R}^n$ such that ϕ^{-1} is Lipschitz continuous with

$$\phi(\Omega \cap U) \subset \{(x_1, \dots, x_n) \in \mathbb{R}^n : x_n > 0\}$$

and

$$\phi(\partial\Omega \cap U) \subset \{(x_1, \dots, x_n) \in \mathbb{R}^n : x_n = 0\}.$$

If in addition, ϕ and ϕ^{-1} are C^k-functions, where k is a positive integer, then Ω is said to be of **class** C^k. Likewise, Ω is said to be of class $C^{k,\mu}$, where $0 < \mu \leq 1$, provided that ϕ and ϕ^{-1} are C^k functions, and the kth derivatives are Hölder continuous with exponent μ. If the exponent satisfies $\mu = 1$, then Hölder continuity is the same as Lipschitz continuity.

We say that a subset $\Omega \subset \mathbb{R}^n$ is a **domain** (in $\mathbb{R}^n$), if Ω is an open, connected set in $\mathbb{R}^n$. For a smooth function $u : \Omega \to \mathbb{R}$ we let $D^\alpha u$ denote the α-derivative

$$D^\alpha u = \frac{\partial^{|\alpha|} u}{\partial x_1^{\alpha_1} \dots \partial x_u^{\alpha_u}}$$

where $\alpha = (\alpha_1, \dots, \alpha_n)$, α_i is a nonnegative integer, for $1 \leq i \leq n$, and $|\alpha| = \alpha_1 + \cdots + \alpha_n$. Depending on the context, one needs to specify whether the derivative is a strong derivative, weak derivative, or a distributional derivative. Derivatives of vector-valued functions $u : \Omega \to \mathbb{R}^m$ are defined similarly. The vector α is sometimes referred to as a **(multi) index of differentiability.**

B.1. Spaces of Continuous Functions. We let Ω be a domain in $\mathbb{R}^n$, and let $C(\Omega) = C^0(\Omega) = C^0(\Omega, \mathbb{R}^m)$ denote the space of continuous functions $u : \Omega \to \mathbb{R}^m$. For a positive integer k, we define the spaces

$$C^k(\Omega) = \{u : D^\alpha u \in C(\Omega), 0 \leq |\alpha| \leq k\},$$

where $D^\alpha u$ are strong derivatives, and

$$C_0^k(\Omega) = \{u \in C^k(\Omega) : \text{supp } u \subset \Omega\},$$

where supp $u = \text{Cl}_\Omega\{x \in \mathbb{R}^n : u(x) \neq 0\}$ is the **support of** u. Since supp u is a closed set, it is a compact set in Ω, whenever Ω is bounded. For θ with $0 < \theta \leq 1$, we let $C^{0,\theta}(\Omega)$ denote those functions $u : \Omega \to \mathbb{R}^m$ for which there is a constant $K \geq 0$ such that

$$|u(x_1) - u(x_2)| \leq K|x_1 - x_2|^\theta, \qquad \text{for all } x_1, x_2 \in \Omega.$$

Similarly, we let $C_{\text{loc}}^{0,\theta}(\Omega)$ denote those functions $u : \Omega \to \mathbb{R}^m$ with the property that, for each compact set $\Omega_c \subset \Omega$, constant $K = K(\Omega_c) \geq 0$ such that

$$|u(x_1) - u(x_2)| \leq K|x_1 - x_2|^\theta, \qquad \text{for all } x_1, x_2 \in \Omega_c.$$

Next we define $C^k(\overline{\Omega})$ to be those functions $u \in C^k(\Omega)$ with the property that $D^\alpha u$ has a continuous extension to $\overline{\Omega}$, for each α with $0 \leq |\alpha| \leq k$. For $0 < \lambda \leq 1$, we let $C^{k,\lambda}(\overline{\Omega})$ denote those $u \in C^k(\overline{\Omega})$ with the property that $D^\alpha u$ is Hölder continuous on $\overline{\Omega}$ with exponent λ provided that $|\alpha| = k$. That is to say, $C^{k,\lambda}(\overline{\Omega})$ consists of those function $u \in C^k(\overline{\Omega})$ for which

$$\sup_{x,y\in\overline{\Omega},x\neq y} |D^\alpha u(x) - D^\alpha u(y)|\,|x - y|^{-\lambda} < \infty.$$

The exponent λ is referred to as the **Hölder exponent** of u. Note that $C^k(\overline{\Omega})$ with the norm $\|u\|_{C^k} = \max_{0\leq|\alpha|\leq k} \sup_{x\in\Omega} |D^\alpha u|$ and $C^{k,\lambda}(\overline{\Omega})$ with the norm

$$\|u\|_{C^{k,\lambda}} = \|u\|_{C^k} + \max_{|\alpha|=k} \sup_{x,y\in\overline{\Omega},x\neq y} |D^\alpha u(x) - D^\alpha u(y)|\,|x - y|^{-\lambda}$$

are separable Banach spaces, see Friedman (1969) and Adams (1975). Finally we let

$$C^\infty(\Omega) = \cap_{k\geq 0} C^k(\Omega), \qquad \text{and} \qquad C_0^\infty(\Omega) = \cap_{k\geq 0} C_0^k(\Omega).$$

B.2. Lebesgue Spaces and Sobolev Spaces. Let Ω be a domain in $\mathbb{R}^n$. The **Lebesgue spaces** $L^p(\Omega) = L^p(\Omega, \mathbb{R}^m)$, for $1 \leq p < \infty$, are defined as the collection of Lebesgue measurable functions $u : \Omega \to \mathbb{R}^m$ such that the Lebesgue integral $\int_\Omega |u|^p dx < \infty$, where $|\cdot|$ is a norm on $\mathbb{R}^m$, and the L^p-norm is

$$\|u\|_{L^p} = \left(\int_\Omega |u|^p dx \right)^{1/p}.$$

For $p = \infty$, the space $L^\infty(\Omega)$ consists of the Lebesgue measurable functions $u : \Omega \to \mathbb{R}^m$ that are bounded almost everywhere with

$$\|u\|_\infty = \operatorname{ess\,sup}\{|u(x)| : x \in \Omega\}.$$

Finally the space $L^1_{\text{loc}}(\Omega)$ is defined as the collection of all Lebesgue measurable functions $u : \Omega \to \mathbb{R}^m$ such that for every compact set $K \subset \Omega$, one has $u \in L^1(K)$.

A function $v \in L^1_{\text{loc}}(\Omega)$ is said to be the **weak derivative** of $u \in L^1_{\text{loc}}(\Omega)$, or a **deriviative in the sense of distributions** of order α, if

$$\int_\Omega v \cdot \varphi \, dx = (-1)^{|\alpha|} \int_\Omega u \cdot D^\alpha \varphi \, dx, \qquad \text{for all } \varphi \in C_0^\infty(\Omega).$$

For a weak derivative, one can write $v = D^\alpha u$. It is this concept of derivative in the sense of distributions which is used to define the Sobolev spaces $W^{k,p}(\Omega)$ and $W_0^{k,p}(\Omega)$. For $1 \leq p < \infty$ and an integer $k \geq 0$, we define the **Sobolev space** $W^{k,p}(\Omega)$ by

$$W^{k,p}(\Omega) = L^p(\Omega) \cap \{u : D^\alpha u \in L^p(\Omega), |\alpha| \leq k\}$$

equipped with a norm

$$\|u\|_{W^{k,p}} = \left(\int_\Omega \sum_{|\alpha| \leq k} |D^\alpha u|^p dx \right)^{1/p}.$$

It can be shown that $W^{k,p}(\Omega)$ is a separable Banach space. The space $W_0^{k,p}(\Omega)$ is defined as the closure of $C_0^\infty(\Omega)$ in $W^{k,p}(\Omega)$, see Adams (1975). The spaces $H^k(\Omega) \stackrel{\text{def}}{=} W^{k,2}(\Omega)$ and $H_0^k(\Omega) \stackrel{\text{def}}{=} W_0^{k,p}(\Omega)$ are Hilbert spaces with the inner-products given by

$$\langle u, v \rangle_{H^k} \stackrel{\text{def}}{=} \sum_{|\alpha| \leq k} \int_\Omega D^\alpha u \cdot \overline{D^\alpha v} \, dx.$$

The Sobolev spaces associated with the boundary $\Gamma = \partial\Omega$ can also be defined in the same way with respect to the hypersurface measure and integral, see Lions and Magenes (1972). The definitions of $W^{k.p}$ and H^k can be extended to the case where k is not an integer, see Lions and Magenes (1972) and Constantin and Foias (1988, pp 48-49).

B.3 Sobolev Inequalities. The structure of the family of Sobolev spaces on a given open, bounded domain Ω of $\mathbb{R}^n$ is fascinating and interconnected by so-called **Sobolev Imbedding Theorems** and **Sobolev inequalities**.

The Sobolev Imbedding Theorems describe properties of the Sobolev spaces $W^{k,p}(\Omega)$, where Ω is a suitable open, bounded set in $\mathbb{R}^n$. We emphasize that the description given here is valid when Ω is open, bounded, and of Lipschitz class. Generalization for unbounded domains can be found in Adams (1975) and Edmunds and Evans (1987); for nonsmooth domains, see Grisvard (1985).

For the most part, there are two major classes of Sobolev Imbedding Theorems. The difference depends on the sign of the index

$$k - \frac{n}{p}.$$

For the first class of theorems we assume that the index $k - \frac{n}{p}$ is positive.

Theorem B.1. *Let Ω be an open, bounded domain of Lipschitz class in $\mathbb{R}^n$, and assume that $k \geq 1$ is an integer that satisfies $k - \frac{n}{p} > 0$, for some p, $1 \leq p < \infty$. Then the following imbedding relations hold.*

(1) *$W^{k,p}(\Omega) \mapsto C^{N,\lambda}(\overline{\Omega})$ where $(k - N - \lambda)p \geq n$, N is a nonnegative integer, and $0 < \lambda \leq 1$. Moreover, if $(k - N - \lambda)p > n$, then this imbedding is compact, i.e., $W^{k,p}(\Omega) \hookrightarrow C^{N,\lambda}(\overline{\Omega})$.*

(2) *$W^{k,p}(\Omega) \mapsto W^{m,q}(\Omega)$, when $k - \frac{n}{p} \geq m - \frac{n}{q}$. Moreover if $k - \frac{n}{p} > m - \frac{n}{q}$ then this imbedding is compact, i.e., $W^{k,p}(\Omega) \hookrightarrow W^{m,q}(\Omega)$.*

The proof of this theorem goes through two major steps. First, one shows that these imbedding relations hold for the spaces $W_0^{k,p}(\Omega)$ and $W_0^{m,q}(\Omega)$. (Note that for $\Omega = \mathbb{R}^n$ one has $W_0^{k,p}(\mathbb{R}^n) = W^{k,p}(\mathbb{R}^n)$, see Edmunds and Evans (1987).) Secondly, one uses the extension theorem of Calderon-Stein, which states that every domain Ω of Lipschitz class has the property that there exists a bounded linear operator $L : W^{k,p}(\Omega) \to W^{k,p}(\mathbb{R}^n)$ such that $L(u)|_{x\in\Omega} = u$, for all $u \in W^{k,p}(\Omega)$, see Edmunds and Evans (1987; Theorem 4.12) or Stein (1970). Then the conclusion follows. The proof of the compact imbedding properties relies on the necessary and/or sufficient conditions for a set to be precompact in the space $L^q(\Omega)$ or $C^{k,\lambda}(\overline{\Omega})$, see Edmunds and Evans (1987).

Since $C^{N,\lambda}(\overline{\Omega}) \mapsto L^\infty(\Omega)$, when N is a nonnegative integer and $0 < \lambda \leq 1$, the following compact imbedding follows from Theorem B.1:

$$W^{k,p}(\Omega) \hookrightarrow L^\infty(\Omega), \qquad \text{whenever } k - \frac{n}{p} > 0. \tag{92.1}$$

As an application of (92.1), we see that, for $1 \leq n \leq 3$, one has $H^2(\Omega) \hookrightarrow L^\infty(\Omega)$, and for $n = 1$, $H^1(\Omega) \hookrightarrow L^\infty(\Omega)$. It is natural then to ask what happens if the index $k - \frac{n}{p}$ becomes negative, or equivalently, if $\frac{1}{p} - \frac{k}{n}$ is positive? While (92.1) is no longer valid, one has the following important result, see Adams (1975, Theorem 5.4).

Theorem B.2. *Let Ω be an open, bounded domain of Lipschitz class in $\mathbb{R}^n$, and assume that $k \geq 1$ is an integer that satisfies $k - \frac{n}{p} < 0$, for some p with $1 \leq p < \infty$. Let q satisfy*

$$p \leq q \leq \frac{np}{n - kp}.$$

Then one has the continuous imbedding $W^{k,p}(\Omega) \mapsto L^q(\Omega)$. Moreover, if $k - \frac{n}{p} = 0$, then $W^{k,p}(\Omega) \mapsto L^q(\Omega)$, for any q with $p \leq q < \infty$.

As applications of the last theorem, we find that, for $n = 3$, one has $H^1(\Omega) \mapsto L^p(\Omega)$ and $H^2(\Omega) \mapsto W^{1,p}(\Omega)$, for $2 \leq p \leq 6$. Also, for $n = 2$ one has $W^{1,2}(\Omega) = H^1(\Omega) \mapsto L^p(\Omega)$, for any p with $2 \leq p < \infty$.

The effect of Theorems B.1 and B.2 is that for an appropriate q, with $p \leq q \leq \infty$, one has $W^{k,p}(\Omega) \mapsto L^q(\Omega)$. Next we describe two **Nirenberg-Gagliardo inequalities**, see Nirenberg (1955, 1959) and Gagliardo (1958).

Theorem B.3. *Let Ω be an open, bounded domain of Lipschitz class in $\mathbb{R}^n$. Assume that $p \geq 1$, $q \geq 1$, $r \geq 1$, $0 < \theta \leq 1$ and*

$$k - \frac{n}{p} \leq \theta \left(m - \frac{n}{q} \right) - (1 - \theta)\frac{n}{r}.$$

Then there is a $C_0 > 0$ such that one has

$$\|u\|_{W^{k,p}} \leq C_0 \|u\|^{\theta}_{W^{m,q}} \|u\|^{(1-\theta)}_{L^r}, \qquad \text{for all } u \in W^{m,q}. \tag{92.2}$$

Assume instead that $0 < \theta \leq 1$, and

$$N + \lambda < \theta \left(m - \frac{n}{q} \right) - (1 - \theta)\frac{n}{r},$$

where N is a nonnegative integer and $0 < \lambda \leq 1$. Then there is a $C_1 > 0$ such that one has

$$\|u\|_{C^{N,\lambda}} \leq C_1 \|u\|^{\theta}_{W^{m,q}} \|u\|^{(1-\theta)}_{L^r}, \qquad \text{for all } u \in W^{m,q}. \tag{92.3}$$

Some other useful properties, such as the Poincaré inequality and the regularity theory for solutions of elliptic equations, can be found in Agmon (1965).

B.4. Time Dependence and Translations. For any Banach spaces V and W we will let $C(I, W)$, $C(V, W)$ and $C(V \times I, W)$ denote the spaces of continuous functions defined on, respectively, I, V, and $V \times I$, and taking values in W, where I denotes an interval in the real line $\mathbb{R}$. In this book we are primarily interested in two cases: (1) where $I = \mathbb{R}^+ = [0, \infty)$, and (2) where $I = \mathbb{R}$. In addition to these spaces, we define $C_b(V, W)$ (and $C_b(V \times I, W)$) to be the collection of all f in $C(V, W)$ (and $C(V \times I, W)$), such that for every bounded set $B \subset V$ (and every compact set $J \subset I$), there is a $K_0 \geq 0$, such that $\|f(u)\|_W \leq K_0$ (and $\|f(u,t)\|_W \leq K_0$), for all $u \in B$ (and $(u,t) \in B \times J$). The subscript b in C_b refers to the condition that f maps bounded sets in V into bounded sets in W. In this book, $C_b(V \times I, W)$ is usually written as $L^\infty_{\text{loc}}(V \times I, W) \cap C(V \times I, W)$. Of course, if V is finite dimensional, then the boundedness condition is automatically satisfied, for any continuous function, since the bounded set B has compact closure.

These spaces of continuous functions are Fréchet spaces with a metric topology which is described by **uniform convergence on bounded sets**. The metric in this case is generated by a countable family of pseudonorms $\|\cdot\|_k$ as follows: Let B_k be the closed neighborhood of the origin in V of radius k, and set $I_k = I \cap [-k,k]$. Define[44]

$$\|f\|_k \stackrel{\text{def}}{=} \sup_{u \in B_k, t \in I_k} \|f(u,t)\|_W.$$

A sequence f_n is said to converge to f - i.e., $f = \lim_{n\to\infty} f_n$ - if and only if

$$\lim_{n\to\infty} \|f - f_n\|_k = 0, \qquad \text{for all } k \geq 1.$$

It turns out that $C_b(V \times I, W)$ is a complete metric space with this metric.

We now restrict to $I = \mathbb{R}^+$, or $I = \mathbb{R}$. For each $f \in C$, where $C = C(V \times I, W)$ or $C = C(I, W)$, we define the **translate** f_τ by

$$f_\tau(u,t) \stackrel{\text{def}}{=} f(u, \tau + t), \qquad u \in V \text{ and } t, \tau \in I.$$

Note that $f_\tau \in C$ for all $\tau \in I$. Furthermore, the mapping $(f,\tau) \to f_\tau$ is a continuous mapping of $C \times I$ into C, where C has the topology defined by uniform convergence on compact sets. Next define the **(positive) orbit** and the **(full) orbit** through f by

$$\gamma^+(f) \stackrel{\text{def}}{=} \{f_\tau : \tau \in \mathbb{R}^+\}, \qquad \text{and} \qquad \gamma(f) \stackrel{\text{def}}{=} \{f_\tau : \tau \in \mathbb{R}\}.$$

Similarly the **(positive) hull** and the **(full) hull** are defined by

$$H^+(f) \stackrel{\text{def}}{=} \text{Cl}_C \gamma^+(f), \qquad \text{and} \qquad H(f) \stackrel{\text{def}}{=} \text{Cl}_C \gamma(f).$$

We will use $C_{\text{Lip}}(V \times I, W)$ to denote the space of functions $f \in C(V \times I, W)$ with the property that for any bounded set B in V, one has

$$f \in L^\infty(B \times I, W) \cap C(B \times I, W)$$

[44] We define $\|\cdot\|_k$ here for a function f in $C(V \times I, W)$. If f is independent of either u, or t, the same formula describes pseudonorms on $C(I, W)$, or $C(V, W)$, respectively.

and there is a constant $K_1 = K_1(B) \geq 0$ such that

$$\|f(u,t) - f(v,t)\|_W \leq K_1\|u - v\|_V, \qquad \text{for all } t \in I, \tag{92.4}$$

and all $u, v \in B$. In other words, each $f \in C_{\text{Lip}}(V \times I, W)$ is locally Lipschitz continuous on V. Note that $C_{\text{Lip}}(V \times I, W)$ is a linear subspace of $C(V \times I, W)$. On some occasions we will use the space $C_{\text{Lip};\alpha} = C_{\text{Lip};\alpha}(V \times I, W)$, where $0 < \alpha \leq 1$. The space $C_{\text{Lip};\alpha}$ consists of those functions $f \in C_{\text{Lip}}(V \times I, W)$ with the property that for any bounded set B in V and compact set J in I, there are constant $K_2 = K_2(B, J)$, such that

$$\|f(u,t) - f(v,s)\|_W \leq K_2(B,J)\left(\|u - v\|_V + |t - s|^\alpha\right),$$

for all $t, s \in J$ and all $u, v \in B$.

One can readily show that, with respect to the metric d on $L^p_{\text{loc}}(I, \mathbb{R}^n)$, with $1 \leq p < \infty$, one has

$$d(\varphi_{\tau_n} - \varphi_\tau) \to 0, \qquad \text{as } n \to \infty, \tag{92.5}$$

see for example, Sell (1971), Miller and Sell (1970), and Sacker and Sell (1977).

Assume that $f : \Omega \times \mathbb{R} \to \mathbb{R}$ is a real function satisfying the conditions of **Carathéodory**, i.e., one has

(C1) for $s \in \mathbb{R}$, the function $x \to f(x,s)$ is Lebesgue measurable in Ω;
(C2) for almost every $x \in \Omega$, the function $s \to f(x,s)$ is continuous in $\mathbb{R}$.

Sometimes a third condition arises, and we include it as a Carathéodory condition:

(C3) for almost every $x \in \Omega$, the function $s \to f(x,s)$ is differentiable in $\mathbb{R}$, and the partial derivative $\partial_s f(x,s)$ satisfies (C1) and (C2).

A **Carathéodory function** is a function that satisfies these three conditions. Moreover, we assume that there are positive constants c_i, for $1 \leq i \leq 4$, and an integer $p \geq 2$, such that

$$\begin{cases} |f(x,s)| \leq c_1|s|^{p-1} + c_2, \\ |\partial_s f(x,s)| \leq c_3|s|^{p-2} + c_4, \end{cases} \qquad \text{for } x \in \Omega, s \in \mathbb{R}.$$

For any Carathéodory function f, the **Nemytskii mapping** is given by $F : g(x) \to f(x, g(x))$, $x \in \Omega$. In general, the Nemytskii mapping associated with a scalar function f can be denoted N_f.

Denote the set of all measurable functions $g : \Omega \to \mathbb{R}$ by $M(\Omega)$. Let q be the number satisfying $(1/p) + (1/q) = 1$. Since $p \geq 2$, we have $1 < q \leq 2$. The proof of the next lemma can be found in DeFigueiredo (1989).

Lemma B.4. *Let $f(x,s)$ be a Carathéodory function satisfying the conditions above. Then the following statements hold.*

(1) *The Nemytskii mapping F maps $M(\Omega)$ into $M(\Omega)$.*
(2) $F \in C_{\mathrm{Lip}}(L^p(\Omega); L^q(\Omega))$.
(3) $F \in C^1_{\mathrm{F}}(L^p(\Omega); L^q(\Omega))$ *and the Fréchet derivative $DF(g)$ is*

$$DF(g)h = \partial_s f(x, g(x))h(x), \qquad \text{for all } g, h \in L^p(\Omega).$$

(4) *The potential function of f, defined by*

$$\Phi(x,s) = \int_0^s f(x,\tau)\, d\tau,$$

is also a Carathéodory function and its corresponding Nemytskii mapping N_Φ maps $L^p(\Omega)$ into $L^1(\Omega)$.

B.5. Almost Periodicity and Quasi-Periodicity. A function $f : \mathbb{R} \to \mathbb{C}$ is said to be **(Bohr) almost periodic** if $f \in C(\mathbb{R}, \mathbb{C})$ and, for every $\epsilon > 0$, the set

$$E(\epsilon, f) = \{\tau \in \mathbb{R} : |f(\tau + t) - f(t)| < \epsilon \text{ for all } t \in \mathbb{R}\}$$

is relatively dense in $\mathbb{R}$; i.e., there is an $\ell = \ell(\epsilon) > 0$ such that, every interval in $\mathbb{R}$ of length $\geq \ell$ contains at least one point of $E(\epsilon, f)$. The space of Bohr almost periodic functions generates a Hilbert space H_{ap} with the inner product

$$\langle f, g\rangle_{\mathrm{ap}} = \lim_{T\to\infty} \frac{1}{2T} \int_{-T}^{T} f(s)\overline{g}(s)\, ds.$$

The uncountable set $\{e^{i\lambda t} : \lambda \in \mathbb{R}\}$ is an orthonormal basis for H_{ap}. For each Bohr almost periodic function f, there is a countable set of frequencies $\{\lambda_n\}$ such that f has the Fourier series expansion

$$f(t) = \sum_n f_n e^{i\lambda_n t}, \qquad \text{where } f_n = \langle f, e^{i\lambda_n t}\rangle_{\mathrm{ap}}.$$

The smallest subgroup of $\mathbb{R}$ that contains this set of frequencies $\{\lambda_n\}$ is called the **module** of f, and it is denoted by $\mathcal{M}(f)$.

A vector $\omega = (\omega_1, \cdots, \omega_m) \in \mathbb{R}^m$ is said to be **independent (over the integers)**, if one has

$$N \cdot \omega = 0, \text{ for any } N \in \mathbb{Z}^m \Longleftrightarrow N = 0.$$

For any collection $\{\omega_1, \cdots, \omega_m\}$ in the module $\mathcal{M}(f)$, one has $N \cdot \omega \in \mathcal{M}(f)$, for $\omega = (\omega_1, \cdots, \omega_m)$ and all $N \in \mathbb{Z}^m$. A finite collection $\{\omega_1, \cdots, \omega_m\}$ in $\mathcal{M}(f)$ is said to be a **(modular) basis** for $\mathcal{M}(f)$, if the vector $\omega =$

$(\omega_1, \cdots, \omega_m)$ is independent and $\mathcal{M}(f) = \{N \cdot \omega : N \in \mathbb{Z}^m\}$. In this case, then integer m is called the **algebraic dimension** of $\mathcal{M}(f)$.

A Bohr almost periodic function $f : \mathbb{R} \to \mathbb{C}$ is said to be **quasi periodic** if the frequency module satisfies $\mathcal{M}(f) = \{N \cdot \omega : N \in \mathbb{Z}^m\}$, for some vector $\omega \in \mathbb{R}^m$. In this definition, we do not require that the vector ω be independent. However, if it is not independent, then there is a collection $\{\hat{\omega}_1, \cdots, \hat{\omega}_k\}$ in $\mathcal{M}(f)$, where $1 \leq k < m$ and $\hat{\omega} = (\hat{\omega}_1, \cdots, \hat{\omega}_k) \in \mathbb{R}^k$ is independent, and $\mathcal{M}(f) = \{\hat{N} \cdot \hat{\omega} : \hat{N} \in \mathbb{Z}^k\}$.

There is a second point of view for quasi periodicity. Let $f : \mathbb{R} \to \mathbb{C}$ be a quasi periodic function with $\mathcal{M}(f) = \{N \cdot \omega : N \in \mathbb{Z}^m\}$, where $\omega \in \mathbb{R}^m$ is independent. The Fourier series for f is given by

$$f(t) = \sum_{N \in \mathbb{Z}^m} f_N e^{i N \cdot \omega t}, \qquad \text{where } f_N = \langle f, e^{-i N \cdot \omega t} \rangle_{\text{ap}}.$$

In this case, there is a function $F \in C(T^m, \mathbb{C})$, where T^m is the m-dimensional torus, and a point $\theta = (\theta_1, \cdots, \theta_m) \in T^m$ such that

$$f(t) = F(\theta + \omega t), \qquad \text{for all } t \in \mathbb{R}. \tag{92.6}$$

In equation (92.6), one uses the standard group operation on T^m for the term $\theta + \omega t$, that is,

$$\theta + \omega t = (\theta_1 + \omega_1 t, \cdots, \theta_m + \omega_m t) \mod 2\pi.$$

Conversely, it $F \in C(T^m, \mathbb{C})$ and $\omega \in \mathbb{R}^m$, then any function f defined by (92.6) is quasi periodic, with $\mathcal{M}(f) = \{N \cdot \omega : N \in \mathbb{Z}^m\}$. Since $C(T^m, \mathbb{C}) \mapsto L^2(T^m, \mathbb{C}; \mu)$, where μ is the Haar measure on T^m with $\mu(T^m) = 1$, the function F has a Fourier series expansion

$$F(\theta) = \sum_{N \in \mathbb{Z}^m} F_N e^{i N \cdot \theta}, \qquad \text{where } F_N = \int_{T^m} F(\theta) e^{-i N \cdot \theta} \, d\mu(\theta).$$

Since the ergodic theorem is applicable here, one has

$$\lim_{T \to \infty} \frac{1}{2T} \int_{-T}^{T} f(t) e^{-i N \cdot \omega t} \, dt = \int_{T^m} F(\theta) e^{-i N \cdot \theta} \, d\mu(\theta).$$

Hence, the Fourier coefficients satisfy $f_N = F_N$, for all $N \in \mathbb{Z}^m$.

As noted in Sell (1971), the theory of almost periodicity admits numerous generalizations. In fact, parts of the theory hold for $f : \mathbb{R} \to U$, where U is a complete, uniform space, see Kelley (1955, p 178). In order to obtain a formulation of the Fourier series expansions, it is typical to consider the case where $f : \mathbb{R} \to H$ and H is a complex Hilbert space. In this setting, if $\{\varphi_a : a \in \mathcal{A}\}$ is an orthonormal basis for H, then an orthonormal basis for the corresponding space H_{ap} is $\{e^{i \lambda t} \varphi_a : \lambda \in \mathbb{R}, a \in \mathcal{A}\}$. As is the case for periodic functions, there are corresponding theories of real-valued almost periodic functions and almost periodic functions in real Hilbert spaces.

Appendix C: Calculus of Vector-Valued Functions.

Let W denote your favorite Banach space. In this section, we outline some basic concepts related to the calculus of vector-valued functions. We begin with the integral calculus of functions $f : I \to W$, where $I \subset \mathbb{R}$ is a Lebesgue measurable set. Later we will examine some aspects of the differentiability theory.

C.1 The Bochner Integral. In the field of infinite dimensional dynamical systems, one will deal with vector-valued functions $f : I \to W$, where I is an interval, or more genarally, a Lebesgue measurable set in $\mathbb{R}$ and W is a Banach space. These functions are also called **abstract functions**. For abstract functions, one can treat its calculus almost parallel to the calculus of functions valued in the Euclidean space $\mathbb{R}^n$ (or $\mathbb{C}^n$). However, the infinite dimensionality causes much more complexity and forces us to be scrupulous. In this section, we introduce the strong measurable functions and the Bochner integral, which is a natural generalization of Lebesgue integral.

A function $f : I \to W$ is said to be a **simple function** if there exist countably many disjoint measurable sets E_j of I, $j = 1, 2, \cdots$, with finite Lebesgue measure $\mu(E_j) < \infty$, and there exist vectors $w_j \in W$, $j = 1, 2, \cdots$, such that $f(t) = w_j$, for all $t \in E_j$ and $f(t) = 0$, for all $t \in I \backslash (\cup_j E_j)$. A function f is said to be **strongly measurable** if there is a sequence of simple functions f_n that converges strongly to f for almost all $t \in I$, i.e., one has $f \overset{\text{a.e.}}{=} \lim_{n\to\infty} f_n$. Finally a function f is said to be **weakly measurable** if for any vector w^* in the dual space W' the scalar function $t \to w^*(f(t))$ is Lebesgue measurable on I. Note that any strongly (weakly) continuous function is strongly (weakly) measurable. Also, $f : I \to W$ is said to be **almost separably valued** if there exists a set $E_0 \subset I$, with $\mu(E_0) = 0$, such that there is a countable subset of $f(I \backslash E_0)$ that is dense in $f(I \setminus E_0)$.

Except where otherwise noted, the proofs of all the results cited in this Section C.1 can be found in Hille and Phillips (1957).

Theorem C.1 (Pettis). *A vector-valued function $f : I \to W$ is strongly measurable if and only if it is weakly measurable and it is almost separably valued. In particular, if W is a separable Banach space, then an abstract function $f : I \to W$ is strongly measurable if and only if it is weakly measurable.*

For operator-valued functions $F : I \to \mathcal{L}(V, W)$ where V and W are Banach spaces, the measurability is defined as follows.

(1) An operator-valued function $F : I \to \mathcal{L}(V, W)$ is called **uniformly measurable** if there is a sequence of simple functions $\{F_n\} : I \to \mathcal{L}(V, W)$ that converges to F in the operator norm for almost all $t \in I$.

(2) $F : I \to \mathcal{L}(V, W)$ is called **strongly measurable** if for any $x \in V$, the vector-valued function $t \mapsto F(t)x$ is strongly measurable.

The following result is very important.

Proposition C.2. *If a function $F : I \to \mathcal{L}(V, W)$ is strongly continuous, then the scalar function $t \mapsto \|F(t)\|_{\mathcal{L}(V,W)}$ is Lebesgue measurable on I.*

For an abstract function $f : I = [a, b] \to W$, which is strongly continuous, one can define its **Riemann integral**, as in the case of a real, or complex, valued function, except that the convergence of the Riemann sums is in terms of the W-norm.

Let $I \subset \mathbb{R}^n$ be a measurable set with $\mu(I) < \infty$.

(1) A simple function $f : I \to W$ is called **Bochner integrable** if its norm function $\|f(t)\| : I \to \mathbb{R}$ is Lebesgue integrable, and its Bochner integral is defined by

$$\int_I f(t)dt = \sum_{j=1}^{\infty} w_j \mu(E_j).$$

(2) A general abstract function $f : I \to W$ is called **Bochner integrable**, if there exists a sequence of Bochner integrable simple functions $\{f_n\} : I \to W$ such that

(i) $\lim_{n\to\infty} f_n(t) = f(t)$ strongly in W, for almost all $t \in I$; and

(ii) $\lim_{n\to\infty} \int_I \|f_n(t) - f(t)\| dt = 0$, where the integral is in the Lebesgue sense. In this case, the Bochner integral of f on I is defined to be the limit of Bochner integrals of $f_n(\cdot)$,

$$\int_I f(t)dt = \lim_{n\to\infty} \int_I f_n(t)dt.$$

For a set I with $\mu(I) = \infty$, the Bochner integral is generalized, just as in the theory of the Lebesgue integral of real, or complex, valued functions. There is a convenient way to verify the Bochner integrability for abstract functions as follows.

Theorem C.3. *An abstract function $f : I \to W$ is Bochner integrable if and only if f is strongly measurable and $\|f(t)\|$ is Lebesgue integrable on I.*

Many properties of Lebesgue integral are also shared by Bochner integral. Some useful properties are listed here.

Theorem C.4. *Suppose that $f : I \to W$ is Bochner integrable. Then the following properties are valid:*

(1) $\| \int_I f(t)\, dt\| \le \int_I \|f(t)\|\, dt$.

(2) *$\int_I f(t)\, dt$ is strongly absolutely continuous in the sense that for any $\epsilon > 0$, there is $\delta = \delta(\epsilon) > 0$ such that if $\mu(E) < \delta$ and $E \subset I$, then $\| \int_E f(t)\, dt\| < \epsilon$.*

(3) *Let* $A \in CLO(W, Y)$. *If* f *is Bochner integrable in* W *and* Af *is Bochner integrable in* Y, *then* $\int_I f(t)dt \in \mathcal{D}(A)$ *and the following commutivity relation holds:*

$$A \int_I f(t)\, dt = \int_I Af(t)\, dt.$$

(4) *(Dominated Convergence) Let* $\{f_n\} : I \to W$ *be Bochner integrable,* $f : I \to W$, *and assume that following two conditions are satisfied: (a)* $\lim_{n\to\infty} f_n(t) = f(t)$ *strongly for almost all* $t \in I$, *and (b) there is a Lebesgue integrable function* $p(t)$ *on* I *such that*

$$\|f_n(t)\| \le p(t), \text{ for almost all } t \in I, \text{ and for all } n.$$

Then the limit function $f : I \to W$ *is Bochner integrable and*

$$\lim_{n\to\infty} \int_I f_n(t)\, dt = \int_I f(t)\, dt.$$

(5) *(Fubini Theorem) Let* $I \subset \mathbb{R}$ *and* $J \subset \mathbb{R}$ *be measurable. If* $f(t,s)$: $I \times J \to W$ *is a Bochner integrable function, then*

$$\int_{I\times J} f(t,s)\, dt\, ds = \int_I \left(\int_J f(t,s)\, ds\right) dt = \int_J \left(\int_I f(t,s)\, dt\right) ds,$$

in which the Bochner integral for function of multi-variable is defined in a similar way as in the single variable case via simple functions, but now with respect to the product measure. Moreover, the iterated integral $\int_I(\int_J f(t,s)\, ds)\, dt$ *means that for almost all* $t \in I$, $g(t) = \int_J f(t,s)ds$ *is defined and* g *is Bochner integrable on* I; *and the second iterated integral is defined similarly.*

(6) *(Tonelli Theorem) If* $f : I \times J (\subset \mathbb{R} \times \mathbb{R}) \to W$ *is jointly strongly measurable and satisfies*

$$\int_I \left(\int_J \|f(t,s)\|\, ds\right) dt < \infty, \quad \text{or} \quad \int_J \left(\int_I \|f(t,s)\|\, dt\right) ds < \infty,$$

then f *is (double) Bochner integrable on* $I \times J$ *so that, by the Fubini Theorem, one has*

$$\int_I \left(\int_J f(t,s)\, ds\right) dt = \int_J \left(\int_I f(t,s)\, dt\right) ds.$$

Let I be an interval, or more generally, a Lebesgue measurable set in $\mathbb{R}$, and let W be a Banach space. For $1 \le p < \infty$, we define $L^p(I; W)$ to be the space of strongly measurable functions $f : I \to W$ for which $\|f(\cdot)\|$ is in

$L^p(I, \mathbb{R})$. This is a Banach space with the norm $\|f\|_p = (\int_I \|f(t)\|^p \, dx)^{1/p}$. If $p = \infty$, we define $L^\infty(I; W)$ to be the space of strongly measurable functions $f : I \to W$ for which $\|f(\cdot)\|$ is in $L^\infty(I, \mathbb{R})$. This is a Banach space with the norm $\|f\|_\infty = \operatorname{ess\,sup}_{t \in I} \|f(t)\|$. Let

$$L^p_{\text{loc}}(I; W) = \{f : I \to W \,|\, f \in L^p(E; W) \text{ for every compact set } E \subset I\}.$$

For $1 \leq p < \infty$, if W is a separable Banach space, then $L^p(I; W)$ is also separable. Moreover the dual space of $L^p(I; W)$ is $L^q(I; W')$, where $\frac{1}{p} + \frac{1}{q} = 1$.

C.2. Differentiable Functions. Next we turn to the theory of the differentiability of functions $f : I \to W.$, where I is an interval in $\mathbb{R}$ and W is a given Banach space. The function f is said to be **strongly differentiable** at the point $t_0 \in I$, if there is a vector $w_0 \in W$ such that

$$f(t_0 + h) - f(t_0) = hw_0 + E(w_0, h), \qquad \text{for } t_0 + h \in I,$$

and one has

$$h^{-1}\|E(w_0, h)\| = h^{-1}\|f(t_0 + h) - f(t_0) - hw_0\| \to 0, \qquad \text{as } |h| \to 0.$$

In other words, one has that $\frac{1}{h}(f(t_0 + h) - f(t_0)) \to w_0$ (strongly) in W, as $h \to 0$. In this case one writes $w_0 = f'(t_0) = \partial_t f \,|_{t=t_0} = \partial_t f(t_0)$. We refer to $\partial_t f$ as the **(strong) derivative** of f.

The following lemma provides a condition under which the Newton-Leibniz formula holds. Let W be a Banach space. A W-valued function f defined on an interval $I = [a, b]$ is said to be **absolutely continuous**, if for each $\varepsilon > 0$ there exists $\delta(\varepsilon) > 0$ such that

$$\sum_{n=1}^{N} \|f(t_n) - f(s_n)\|_W < \varepsilon,$$

whenever $\sum_{n=1}^{N} |t_n - s_n| < \delta(\varepsilon)$, where $\{(s_n, t_n)\}$ is any finite collection of mutually disjoint intervals in $[a, b]$.

Lemma C.5. *Let W be a Banach space. Then for every strongly, absolutely continuous function $f : [a, b] \to W$ with the property that the strong derivative f' is Bochner integrable over $[a, b]$, i.e., $f' \in L^1([a, b]; W)$, one has*

$$f(t) = f(a) + \int_a^t f'(s) \, ds, \qquad \text{for } t \in [a, b]. \tag{93.1}$$

If in addition W is a reflexive Banach space, then every strongly, absolutely continuous function $f : [a, b] \to W$ is strongly differentiable a.e. on $[a, b]$ with $f' \in L^1([a, b]; W)$, and (93.1) is valid.

The equality (93.1) is called a **Newton-Leibniz formula**. The proof of Lemma C.5 is in Barbu and Precupanu (1986, Theorem 3.4). Also see Hille and Phillips (1957). An important property of strong differentiability is given in the following lemma, which is a consequence of Lemma C.5.

Lemma C.6. *If W is a Hilbert space or a reflexive Banach space and $f : [a,b] \to W$ is Lipschitz continuous, then f is strongly differentiable almost everywhere in $[a,b]$ and the strong derivative $f' \in L^1(a,b;W)$. Furthermore, the Newton-Leibniz formula (93.1) holds.*

We will also use the following result, which describes the behavior of the limit of a sequence of strongly differentiable functions.

Lemma C.7. *Let W be a Banach space and let $[0,T)$ be an interval in $\mathbb{R}$. Let $f_n : [0,T) \to W$ be a sequence of continuous fucntions where each f_n is strongly differentiable on $(0,T)$, and the derivative $\partial_t f_n$ is continuous on $(0,T)$. Assume that there exist two functions $f : [0,T) \to W$ and $g : (0,T) \to W$ with the following two properties:*

(1) *$\lim_{n\to\infty} f_n(t) = f(t)$ in W, uniformly on compact sets in $[0,T)$, and*

(2) *$\lim_{n\to\infty} \partial_t f_n(t) = g(t)$ in W, uniformly on compact sets in $(0,T)$.*

Then f is strongly differentiable on $(0,T)$, with $\partial_t f = g$ on $(0,T)$;

$$f \in C[0,T;W), \quad \partial_t f = g \in L^1_{\text{loc}}[0,T;W) \cap C(0,T;W);$$

and $f(t) = f(0) + \int_0^t g(s)\,ds$, for all $t \in [0,T)$.

C.3 Fréchet and Gâteaux Derivatives. In this section we examine two theories of differentiability for functions $f : U \to W$, where $U \subset V$ and both V and W are Banach spaces. This issue is especially interesting when V is infinite dimensional.

A function $f : U \to W$ is said to be **Fréchet differentiable** (with respect to $U \subset V$) at a point $v_0 \in U$ if there is a bounded linear operator $L \in \mathcal{L}(V,W)$ such that for all $v \in V$ with $v_0 + v \in U$, the difference function

$$E(v_0, v) \stackrel{\text{def}}{=} f(v_0 + v) - f(v_0) - Lv$$

satisfies

$$\lim_{||v||\to 0} \frac{||E(v_0,v)||}{||v||} = 0. \tag{93.2}$$

In this case one writes $Df(v_0) = L$, and $Df(v_0)$ is the **Fréchet derivative** (with respect to U) of f at v_0. In the important case where U is open in V, then one refers to $Df(v_0)$ as the **Fréchet derivative** of f at v_0.

The Gâteaux derivative is formulated in terms of the directional derivatives at a point $v_0 \in U$. In particular, a function $f : U \to W$ is said to be **Gâteaux differentiable** (with respect to $U \subset V$) at a point $v_0 \in U$ if there is a bounded linear operator $L \in \mathcal{L}(V,W)$ such that for all $v \in V$ and all $t \in \mathbb{R}$ with $v_0 + tv \in U$, the difference function

$$E(v_0, v, t) \stackrel{\text{def}}{=} f(v_0 + tv) - f(v_0) - tLv$$

satisfies

$$\lim_{|t|\to 0} \frac{\|E(v_0, v, t)\|}{|t|} = 0. \tag{93.3}$$

In this case one writes $Df(v_0) = L$, and $Df(v_0)$ is the **Gâteaux derivative** (with respect to U) of f at v_0. In the important case where U is open in V, then one refers to $Df(v_0)$ as the **Gâteaux derivative** of f at v_0.

It follows immediately from the definitions that if f is Fréchet differentiable (with respect to U) at $v_0 \in U$, then it is Gâteaux differentiable (with respect to U) at v_0, and the two derivatives at v_0 agree. Thus, if f is Fréchet differentiable, then one can use the Gâteaux derivative, which is based on directional derivatives, to calculate $Df(v_0)$.

It is well known that, by the Rademacher Theorem, a Lipschitz mapping from an open set in $\mathbb{R}^n$ into $\mathbb{R}^m$ is almost everywhere differentiable. One may ask: "Is a Lipschitz mapping from a Banach space V (or an open set in V) into a Banach space W almost everywhere Fréchet differentiable or Gâteaux differentiable?" The answer is "no" in general. Here is a relevant result.

Theorem C.8. *Let V be a separable Banach space and U be an open set in V. Let W be a Banach space with Radon-Nikodým property. Then every Lipschitz mapping $f : U \to W$ is Gâteaux differentiable, except possibly on an Aronszajn null set.*

The definition of the Radon-Nikodým property (RNP) and the definition of the Aronszajn null set can be found in Benyamini and Lindenstrauss (2000).

A mapping $f : U(\subset V) \to W$ is said to be **continuously Fréchet (or Gâteaux) differentiable** on U if f is Fréchet (or Gâteaux) differentiable (with respect to U) at each point $v_0 \in U$ and the mapping of U into $\mathcal{L}(V, W)$ given by $v_0 \to Df(v_0)$ is continuous in terms of the uniform topology[45] on $\mathcal{L}(V, W)$, i.e., one has $\|Df(v_n) - Df(v_0)\|_{\mathcal{L}(V,W)} \to 0$, whenever $v_n \overset{s}{\to} v_0 \in U$. We will let $C^1_F(U, W)$, and $C^1_G(U, W)$, denote respectively the space of all continuously Fréchet, and Gâteaux, differentiable functions on U. Similarly, we let $C^1_{F,b} = C^1_{F,b}(U, W)$ (and $C^1_{G,b} = C^1_{G,b}(U, W)$) denote those function $f \in C^1_F(U, W)$ (and $f \in C^1_G(U, W)$) such that both f and Df map bounded sets into bounded sets. The spaces $C^1_{F,b}(U, W)$ and $C^1_{G,b}(U, W)$ will have the Fréchet space topology generated by uniform C^1-convergence on bounded sets in U. That is to say, one has $F_n \to F$ in $C^1_{F,b}$, or in $C^1_{G,b}$, provided that $F_n(u) \to F(u)$ and $DF_n(u) \to DF(u)$, uniformly for u in bounded sets in U. By using a suitable countable family of pseudonorms, one can readily show that these spaces are Fréchet spaces.

---

[45] A weaker form of continuity arises when one uses the strong topology on $\mathcal{L}(V, W)$. In this case, one has instead that for each $v \in V$, the mapping of U into W given by $u_0 \to Df(u_0)v$ is continuous.

Let $f \in C_F^1(U,W) \subset C_G^1(U,W)$. Notice that if $u(s) \in U$, for $0 \le s < T \le \infty$, is a continuous function, then the derivative $Df(u(s))$ is continous for $0 \le s < T$. Also note that if $u_n \xrightarrow{s} u_0 \in U$ and $v_n \xrightarrow{s} v_0 \in V$, then one has $Df(u_n)v_n \xrightarrow{s} Df(u_0)v_0$. Indeed one has

$$\|Df(u_n)v_n - Df(u_0)v_0\| \le \|Df(u_n)(v_n - v_0)\| + \|(Df(u_n) - Df(u_0))v_0\| \to 0.$$

Since Df is continuous and $K = \{u_0\} \cup \{u_n : n \ge 1\}$ is a compact set, one has $\sup_n \|Df(u_n)\| < \infty$ and $\|(Df(u_n) - Df(u_0))v_0\| \to 0$, as $n \to \infty$.

An extensively useful formula is the **Mean Value Formula** of integral form, which has been used numerous times in this text. Let U be a convex, open set in a Banach space V, and let W be a Banach space. If $f \in C_F^1(U,W)$, then one has

$$(93.4) \quad f(u) - f(v) = \int_0^1 Df(v + s(u-v))(u-v)\, ds, \qquad \text{for any } u, v \in U,$$

where $Df(\cdot)$ stands for the Fréchet derivative of the mapping f. See Berger (1977).

Let V and W be two Banach spaces. We define $C^1_{\text{Lip}} \stackrel{\text{def}}{=} C_{\text{Lip}}(V,W) \cap C_F^1(V,W)$. Thus $F \in C^1_{\text{Lip}}$ means that F has a continuous Fréchet derivative $DF(u)$ at each point $u \in V$, and that for every bounded set B in V there are constants $k_0 = k_0(B)$ and $k_1 = k_1(B)$, such that $\|F(u)\|_W \le k_0$, for all $u \in B$, and

$$(93.5) \qquad \|F(u_1) - F(u_2)\|_W \le k_1 \|u_1 - u_2\|_V, \qquad \text{for all } u_1,\, u_1 \in B.$$

The Fréchet derivative is said to have the **Hölder property** if for every compact set $\mathcal{K}$ in V there exist constants $C = C(\mathcal{K}) > 0$ and $\theta = \theta(\mathcal{K})$, with $0 < \theta \le 1$, such that

$$\|DF(u_1) - DF(u_2)\|_{\mathcal{L}(V,W)} \le C \|u_1 - u_2\|_V^\theta, \qquad \text{for all } u_1,\, u_2 \in \mathcal{K}.$$

If one can choose $\theta(\mathcal{K}) = 1$, for every compact set $\mathcal{K}$, then we say that DF has the **Lipschitz property**.

Note that one has $C^1_{\text{Lip}} \subset C^1_{\text{F,b}}(V,W)$. Indeed if $F \in C^1_{\text{Lip}}$ and B is a bounded set in V, then F is bounded on B and inequality (93.5) implies that the derivative DF is bounded on B, as well.

C.4. Vector-Valued Sobolev Spaces. In this section we introduce the concept of vector-valued distributions and define certain Sobolev spaces of vector-valued functions. Our goal is an important result, Theorem C.9, which is often used in the theory of evolutionary equations. See the proof of Lemma 42.1, for example.

Let X be a Banach space and let $I = (a,b)$ be an interval in $\mathbb{R}$. We define $\mathcal{D}'(I;X) = \mathcal{L}(C_0^\infty(I);X)$ to be the space of all continuous, linear operators from $C_0^\infty(I) = C_0^\infty(I, \mathbb{C})$ to X. The elements T in $\mathcal{D}'(I; X)$ are referred to as X-**valued distributions** on the interval I. In particular, for any vector-valued function $x \in L^p_{\text{loc}}(a, b; X)$, where $1 \leq p < \infty$, the mapping T_x defined by

$$T_x(\varphi) = \int_a^b \varphi(t)\, x(t)\, dt, \qquad \text{for any } \varphi \in C_0^\infty(I)$$

is in $\mathcal{D}'(I; X)$, where $I = (a, b)$.

For any $T \in \mathcal{D}'(I; X)$ and any integer $k \geq 1$, the **distribution derivative** of T of **order** k is defined by

$$T^{(k)}(\varphi) = (D^k T)(\varphi) \stackrel{\text{def}}{=} (-1)^k T(\partial_t^k \varphi), \qquad \text{for } \varphi \in C_0^\infty(I).$$

The distribution derivatives $T^{(k)}$ always exist, for any $T \in \mathcal{D}'(I; X)$, and one has $T^{(k)} \in \mathcal{D}'(I; X)$, for each $k \geq 1$. For $k = 0$ we set $T^{(0)} = T$.

For any $T \in \mathcal{D}'(I; X)$, we will write $T^{(k)} \in L^p(I; X)$, for an integer $k \geq 0$, whenever there is a function $f_k \in L^p(I; X)$ such that

$$T^{(k)}(\varphi) = \int_a^b \varphi(t)\, f_k(t)\, dt, \qquad \text{for all } \varphi \in C_0^\infty(I).$$

We then define the vector-valued Sobolev space $W^{m,p}(I; X)$, for an integer $m \geq 0$ and $1 \leq p < \infty$, as

$$W^{m,p}(I; X) \stackrel{\text{def}}{=} \{T \in \mathcal{D}'(I; X) : T^{(k)} \in L^p(I; X) \text{ for } k = 0, \cdots, m\}.$$

For a bounded interval $I = (a, b)$, we define $A^{m,p}(I; X)$ to be the collection of all functions $f : [a, b] \to X$ such that the kth-strong derivatives $\partial_t^k f$ are absolutely continuous on $[a, b]$, for $0 \leq k \leq m - 1$, and $\partial_t^k f \in L^p(I; X)$, for $0 \leq k \leq m$. In the case where the interval I is unbounded, we alter the definition of $A^{m,p}(I; X)$ by requiring that the derivatives $\partial_t^k f$ be locally absolutely continuous on (a, b).

The difference between the two spaces $W^{m,p}(I; X)$ and $A^{m,p}(I; X)$ lies in the fact that, in the first case one has "distribution" derivatives, while in the second case one has "strong" derivatives. An important matter of concern is the relationship between these two spaces. The proof of the following result, which addresses this issue, can be found in Barbu and Precupanu (1986).

Theorem C.9. *Let $I = (a, b)$ be a bounded interval and let p satisfy $1 \leq p \leq \infty$. Let X be a Banach space and let $u \in L^p(I; X)$. Then for any integer $m \geq 0$, the following two statements are equivalent:*

(1) *$u \in W^{m,p}(I; X)$.*

(2) *There is a function $v \in A^{m,p}(I; X)$ such that $u(t) \stackrel{\text{a.e.}}{=} v(t)$ on I.*

Based on Theorem C.9, one may identify $W^{m,p}(I;X)$ with $A^{m,p}(I;X)$ in the sense of Item (2). For $1 \le p < \infty$, $W^{m,p}(I;X)$ is a Banach space with the norm

$$\|u\|_{m,p} \stackrel{\text{def}}{=} \left(\sum_{k=0}^{m} \int_a^b \|\partial_t^k u(t)\|_X^p \, dt \right)^{\frac{1}{p}} = \left(\sum_{k=0}^{m} \int_a^b \|D^k u(t)\|_X^p \, dt \right)^{\frac{1}{p}}.$$

Moreover, if $u \in W^{1,1}(I;X)$, where $I = (a,b)$ is a bounded interval, then from Lemma C.5, the Newton-Leibniz Formula holds for u on I, and one has

$$u(t) = u(a) + \int_a^t u'(s) \, ds, \qquad \text{for } t \in [a,b],$$

where u' is the distribution derivative of u.

Appendix D: Basic Inequalities.

In this section we state, for reference, several basic inequalities. Let a and b be nonnegative real numbers, and let p and q satisfy $p + q = pq$, where $1 \le p < \infty$ and $1 \le q < \infty$. Then for every $\lambda > 0$ one has

$$ab = (\lambda a)(\lambda^{-1} b) \le \frac{\lambda^p a^p}{p} + \frac{b^q}{q \lambda^q}, \tag{94.1}$$

see Naylor and Sell (1982; p. 549), for example. This inequality is referred to as the **Young inequality**. For $\lambda = 1$ and $p = q = 2$, it reduces to the **Cauchy inequality**, or the **Schwarz inequality**. An equivalent version of the Young inequality is

$$ab \le \epsilon a^p + C(\epsilon) b^q, \qquad \text{with } 1 < p < \infty \text{ and } \frac{1}{p} + \frac{1}{q} = 1, \tag{94.2}$$

where $\epsilon > 0$ is arbitrary and $C(\epsilon) = p^{-q}(p-1)\epsilon^{-(p-1)^{-1}}$. From the Young inequality follows the **Hölder inequality**,

$$\left| \int_\Omega u \, v \, dx \right| \le \|u\|_{L^p(\Omega)} \|v\|_{L^q(\Omega)}, \qquad \text{for any } u \in L^p(\Omega) \text{ and } v \in L^q(\Omega), \tag{94.3}$$

where $1 \le p \le \infty$ and $\frac{1}{p} + \frac{1}{q} = 1$. The Hölder inequality can be extended to multiple products with $u_j \in L^{p_j}(\Omega)$, and $\sum_{j=1}^k p_j^{-1} = 1$, $p_j \ge 1$, for $j = 1, \cdots, k$. One then has

$$\left| \int_\Omega u_1 \cdots u_k \, dx \right| \le \|u_1\|_{L^{p_1}(\Omega)} \cdots \|u_k\|_{L^{p_k}(\Omega)}. \tag{94.4}$$

A further consequence of the Hölder inequality is the **interpolation inequality**:

(94.5) $$\|u\|_{L^q(\Omega)} \leq \|u\|^{\theta}_{L^p(\Omega)}\|u\|^{1-\theta}_{L^r(\Omega)}, \qquad \text{for any } u \in L^r(\Omega),$$

where $1 \leq p \leq q \leq r$, $\frac{1}{q} = \frac{\theta}{p} + \frac{1-\theta}{r}$, and $0 \leq \theta \leq 1$.

Another very useful inequality in is the **Lions compactness inequality**. Let W, X and Y be Banach spaces with the imbeddings $W \hookrightarrow X \mapsto Y$. Then for any $\epsilon > 0$, there is a constant $C_\epsilon > 0$ such that

(94.6) $$\|v\|_X \leq \epsilon\|v\|_W + C_\epsilon\|v\|_Y, \qquad \text{for any } v \in W.$$

The proof can be found in Lions (1969). An application of (94.6) is the inequality

(94.7) $$\|u\|_{W^{k,p}} \leq \epsilon\|u\|_{W^{m,p}} + C_\epsilon\|u\|_{L^p}, \qquad \text{for any } u \in W^{m,p}(\Omega),$$

where $0 \leq k \leq m-1$, $1 \leq p < \infty$, and $\epsilon > 0$ can be arbitrarily small.

Many of the following inequalities are referred to as **Gronwall inequalities** in the literature.

Lemma D.1. *Assume that $\rho(t)$ is a nonnegative function in $L^1_{\text{loc}}[0,\infty;\mathbb{R})$ that satisfies*

$$\rho(t) \leq a + b\int_0^t e^{\alpha(t-s)}\rho(s)\,ds, \qquad \text{for all } t \geq 0,$$

where a, b, and α are constants that satisfy $a \geq 0$, $b \geq 0$, and $b+\alpha \neq 0$. Then one has

(94.8a) $$\rho(t) \leq a + ab(b+\alpha)^{-1}\left(e^{(b+\alpha)t} - 1\right), \qquad \text{for all } t \geq 0.$$

Assume that $h(t)$ is a nonnegative function in $L^1_{\text{loc}}(\mathbb{R};\mathbb{R})$ that satisfies

$$h(s) \leq a + b\int_s^t e^{\beta(\sigma-s)}h(\sigma)\,d\sigma, \qquad \text{for all } s \leq t < t_0,$$

where a, b, and β are constants that satisfy $a \geq 0$, $b \geq 0$, $b+\beta \neq 0$, and $t_0 \leq \infty$. Then one has

(94.8b) $$h(s) \leq a + ab(b+\beta)^{-1}\left(e^{(b+\beta)(t-s)} - 1\right), \qquad \text{for all } s \leq t < t_0.$$

Proof. Let $\sigma(t) = e^{-\alpha t}\rho(t)$ and $V(t) = \int_0^t \sigma(s)\,ds$. Then one has $V' \leq bV + ae^{-\alpha t}$ and $V(0) = 0$. Since $b + \alpha \neq 0$, we find that

$$V(t) \leq \int_0^t ae^{-\alpha s}e^{b(t-s)}\,ds = a(b+\alpha)^{-1}(e^{bt} - e^{-\alpha t}),$$

which implies (94.8a). A similar argument establishes (94.8b). □

Lemma D.2. *Assume that $r = r(t)$, $K = K(t)$, and $a = a(t)$ are nonnegative functions in $L^1_{\text{loc}}[0, \infty; \mathbb{R})$ that satisfy*

$$r(t) \leq a(t) + K(t) \int_0^t b(s) r(s)\, ds, \qquad \textit{for almos all } t \geq 0,$$

where $b = b(t)$ is a nonnegative continuous function in $L^\infty_{\text{loc}}[0, \infty; \mathbb{R})$. Then for almost all $t \geq 0$, one has

$$r(t) \leq a(t) + K(t) \int_0^t a(s) b(s)\, ds \exp\left(\int_0^t K(s) b(s)\, ds \right). \tag{94.9}$$

If in addition, K and a are nondecreasing, then

$$r(t) \leq a(t) \left[1 + K(t) \int_0^t b(s)\, ds \exp\left(K(t) \int_0^t b(s)\, ds \right) \right].$$

Proof. Let $p(t) = \int_0^t b(s) r(s)\, ds$. Then $p(t)$ satisfies $p(0) = 0$ and the differential inequality

$$p'(t) \leq K(t) b(t) p(t) + a(t) b(t), \qquad \text{for almost all } t \geq 0.$$

Using the multiplier $\exp \int_0^t (-1) K(s) b(s)\, ds$, one can solve this inequality to obtain

$$p(t) \leq \int_0^t a(\tau) b(\tau) \exp\left(\int_\tau^t K(s) b(s)\, ds \right) d\tau.$$

By integrating by parts, using the nonnegativity of a and b, and making the substitution $w = \int_\tau^t K(s) b(s)\, ds$, in that order, one finds that

$$\begin{aligned} &\int_0^t a(\tau) b(\tau) \exp\left(\int_\tau^t K(s) b(s)\, ds \right) d\tau \\ &\qquad \leq \int_0^t a(s) b(s)\, ds \times \exp\left(\int_0^t K(s) b(s)\, ds \right). \end{aligned}$$

Inequality (94.9) then follows from $r(t) \leq a(t) + K(t) p(t)$. $\square$

Lemma D.3 (Uniform Gronwall inequality). *Let y, g, and h be nonnegative functions in $L^1_{\text{loc}}[0, T; \mathbb{R})$, where $0 < T \leq \infty$. Assume that y is absolutely continuous on $(0, T)$ and that*

$$\partial_t y(t) \leq g(t) y(t) + h(t), \qquad \textit{almost everywhere on } (0, T). \tag{94.10}$$

Then $y \in L^\infty_{\text{loc}}(0, T; \mathbb{R})$ and one has

$$y(t) \leq y(t_0) \exp\left(\int_{t_0}^t g(s)\, ds \right) + \int_{t_0}^t \exp\left(\int_s^t g(r)\, dr \right) h(s)\, ds, \tag{94.11}$$

for $0 < t_0 < t < T$, *and*

$$y(t) \leq \left(\frac{1}{t-\tau} \int_{\tau}^{t} y(s)\, ds + \int_{\tau}^{t} h(s)\, ds \right) \exp\left(\int_{\tau}^{t} g(s)\, ds \right), \tag{94.12}$$

for $0 < t < T$, *where* $\tau = \max(0, t-1)$. *If in addition one has* $y \in C[0, T; \mathbb{R})$, *then inequality (94.11) is valid at* $t_0 = 0$.

Proof. First we observe that (94.10) can be written in the form

$$\partial_s \left(y(s) \exp\left(-\int_{t_0}^{s} g(r)\, dr \right) \right) \leq h(s) \exp\left(-\int_{t_0}^{s} g(r)\, dr \right).$$

By integrating the last inequality with respect to s from t_0 to t, one obtains (94.11). Next we fix t_0 to be in the interval (τ, t). Then (94.11) implies that

$$y(t) \leq \left(y(t_0) + \int_{\tau}^{t} h(s)\, ds \right) \exp\left(\int_{\tau}^{t} g(s)\, ds \right).$$

By integrating the last inequality with respect to t_0 from τ to t, one obtains (94.12). □

Note that inequality (94.12) has the following two forms:

$$y(t) \leq \left(\frac{1}{t} \int_{0}^{t} y(s)\, ds + \int_{0}^{t} h(s)\, ds \right) \exp\left(\int_{0}^{t} g(s)\, ds \right), \tag{94.13}$$

for $0 < t \leq 1$, and

$$y(t) \leq \left(\int_{t-1}^{t} y(s)\, ds + \int_{t-1}^{t} h(s)\, ds \right) \exp\left(\int_{t-1}^{t} g(s)\, ds \right), \tag{94.14}$$

for $1 \leq t < T$.

In the following result, we present the Gronwall-Henry inequality. This important inequality is a variation of the classical Gronwall inequality, and it arises in the study of the dynamics of solutions of nonlinear partial differential equations. For this purpose, we define

$$E_{r,c}(z) = \sum_{n=0}^{\infty} \frac{\Gamma(c)}{\Gamma(nr+c)} z^{nr},$$

$$E'_{r,c}(z) = \frac{d}{dz} E_{r,c}(z) = \sum_{n=1}^{\infty} \frac{nr\,\Gamma(c)}{\Gamma(nr+c)} z^{nr-1},$$

where r and c are positive real numbers and $\Gamma(\cdot)$ is the Gamma function.

Lemma D.4 (Gronwall-Henry inequality). *Let $v(t)$ be a nonnegative function in $L^\infty_{\text{loc}}[0,\tau;\mathbb{R})$ and $h(\cdot) \in L^1_{\text{loc}}[0,\tau;\mathbb{R})$ satisfy*

$$v(t) \leq h(t) + M\int_0^t (t-s)^{r-1}v(s)\,ds, \qquad t \in (0,\tau),$$

where $0 < \tau \leq \infty$ and $r > 0$. Then one has

$$v(t) \leq h(t) + \mu\int_0^t E'_{r,1}(\mu(t-s))h(s)\,ds, \qquad t \in (0,\tau), \tag{94.15}$$

where

$$\mu^r = M\Gamma(r). \tag{94.16}$$

If in addition, one has $h(t) \equiv at^{c-1}$, where a and c are positive constants, then there exists a positive constant $C_{r,c}$ such that

$$v(t) \leq C_{r,c}\, a\, t^{c-1}\, E_{r,c}(\mu t), \qquad t \in (0,\tau). \tag{94.17}$$

Moreover, if $h(t) = ae^{\lambda t}$, where $\lambda > \mu$, then with $\theta = \mu^r\lambda^{-r}$, one has

$$v(t) \leq a(1-\theta)^{-1}e^{\lambda t}, \qquad \text{for } t \in (0,\tau). \tag{94.18}$$

The proof of (94.15) can be found in Henry (1981, p. 188). The proofs of (94.17) and (94.18), as well as other properties used in this book, are based on the following elementary properties of the beta and gamma functions,

$$B(r,c) = \int_0^1 (1-x)^{r-1}x^{c-1}\,dx \qquad \text{and} \qquad \Gamma(r) = \int_0^\infty x^{r-1}e^{-x}\,dx,$$

where r and c are positive real numbers. One has $\Gamma(p) = (p-1)!$, for any integer $p \geq 1$, $\Gamma(r)$ is everywhere positive with a unique minimum at a point r_0, where $1 < r_0 < 2$, and $\Gamma(r)$ is strictly increasing, for $r > r_0$. Also one has

$$\int_0^t (t-s)^{r-1}e^{-\lambda(t-s)}\,ds \leq \lambda^{-r}\Gamma(r), \qquad \int_0^t (t-s)^{r-1}s^{c-1}\,ds = t^{r+c-1}B(r,c),$$

for $t > 0$, and

$$B(r,c) = \frac{\Gamma(r)\Gamma(c)}{\Gamma(r+c)} \qquad \text{and} \qquad \Gamma(r+1) = r\Gamma(r).$$

The constant $C_{r,c}$ appearing in inequality (94.17) is given by

$$C_{r,c} = \max\left(1, \max\{\Gamma(nr+1)\Gamma(nr+c)^{-1} : nr + c \leq r_0\}\right).$$

By using these properties one can show, for example, that for any r and c with $r > 0$ and $c > 0$, there exist four polynomials in z and z^r, say $P_1(z)$, $P_2(z)$, $P_3(z)$, and $P_4(z)$, such that

$$P_1(z) + P_2(z)e^z \leq E_{r,c}(z) \leq P_3(z) + P_4(z)e^z, \qquad \text{for } z \geq 1, \tag{94.19}$$

with a similar inequality valid for $E'_{r,c}(z) = \frac{d}{dz}E_{r,c}(z)$. It follows immediately from (94.19) that for any $\mu > 0$ one has

$$\lim_{t\to\infty} \frac{1}{t}\log(E_{r,c}(\mu t)) = \mu, \qquad \text{whenever } r > 0 \text{ and } c > 0. \tag{94.20}$$

The following result is a typical case of the nonlinear Gronwall inequality.

Lemma D.5. *Let a and b be positive real numbers, and let $y = y(t)$ be a nonnegative, absolutely continuous function that satisfies*

$$y' \leq a + by^3, \qquad 0 \leq t < \infty.$$

Then one has $y(t) \leq r(t)$, for $0 \leq t < T_0$, where $r(t)$ is the solution of

$$r' = a + br^3, \qquad r(0) = r_0, \tag{94.21}$$

with $r_0 \geq y(0)$. Furthermore, one has $T_0 < \infty$ and $r(t) = r(r_0, a, b, t) \to \infty$, as $t \to T_0^-$.

Appendix E: Commentary.

Detailed information about Sobolev spaces can be found in Adams (1975), Edmunds and Evans (1987), and Sobolev (1963). In order to study some PDEs with periodic conditions in space variables, one needs Sobolev spaces of periodic functions, cf. Temam (1988). For handling the boundary value problems with variable coefficients or problems on unbounded domains, one will use weighted Sobolev spaces, cf. Kufner (1985). Sobolev subspaces associated with certain additional conditions in terms of gradient, divergence, or curl operators also play important roles in dealing with many dynamics problems of fluid mechanics and mathematical physics, cf. Dautray and Lions (1990).

Appendix B: Two important facts are shown in Grisvard (1985). First, any open bounded set Ω in $\mathbb{R}^n$ has the uniform cone property if and only if its boundary $\partial\Omega$ is locally Lipschitz continuous. Second, any open, bounded, and convex set Ω in $\mathbb{R}^n$ has a locally Lipschitz continuous boundary.

Another fascinating feature of Sobolev spaces on a domain Ω is the intrinsic relations between the Sobolev spaces defined on Ω and the Sobolev spaces defined on its boundary $\partial\Omega$. These relations, called Trace Theorems, are related to Sobolev spaces with fractional orders, see Lions (1962) and Lions and Magenes (1972). These relations involve two important notions: interpolation spaces and fractional powers of operators. These two notions in turn provide a refined framework to study the regularity of weak solutions. We present these notions in Chapter 3 together with the semigroup theory of linear operators.

The Sobolev spaces with negative order can be defined in terms of certain dual spaces. Let $1 \leq p < \infty$, and $\ell \geq 0$ be an integer. The Sobolev space $W^{-\ell,q}(\Omega)$ is defined to be the dual space of $W_0^{\ell,p}(\Omega)$, where $\frac{1}{p} + \frac{1}{q} = 1$. The structure of $W^{-\ell,q}(\Omega)$ is described by the following result, where Ω is an open, bounded domain. In this case, one has $(W^{\ell,p}(\Omega))' \subsetneq W^{-\ell,q}(\Omega)$. See Section 3.8.3 where this concept is developed for the study of the Stokes equations and the Navier-Stokes equations.

Appendix D: Applications of the Uniform Gronwall inequality to nonlinear evolutionary equations was first done by Foias and Prodi (1967) in their study of the 2D Navier-Stokes equations. See Chapter 6 for this application.

Additional Readings

Adams (1975), Dautry and Lions (1990), Edmunds and Evans (1987), Fraenkel (1979), Grisvard (1985), Hille and Phillips (1957), Kufner (1985), Lions (1962), Lions and Magenes (1972), Sobolev (1963), and Stein (1970).

Bibliography

M Abounouh and O Goubet (2000), *Attractors for a damped cubic Schrödinger equation on two-dimensional thin domain*, Differential Integral Equations **13**, 311–340.

R S Adams (1975), *Sobolev Spaces*, Academic Press, New York.

S Agmon (1965), *Elliptic Boundary Value Problems*, Van Nostrand, New York.

S Agmon, A Douglis, and L Nirenberg (1959, 1964), *Estimates near the boundary for solutions of elliptic partial differential equations satisfying general boundary conditions I, II*, Comm Pure Appl Math **12, 17**, 623-727, 35–92.

N Alikakos, P W Bates, and G Fusco (1991), *Slow motion for the Cahn-Hilliard equation in one-space dimension*, J Differential Equations **90**, 81–135.

N D Alikakos, P W Bates, and G Fusco (1993), *Solutions to the nonautonomous bistable equation with specified Morse index. I Existence*, Trans Am Math Soc **340**, 641–654.

N Alikakos and G Fusco (1998), *Slow dynamics for the Cahn-Hilliard equation in higher-space dimensions: the motion of bubbles*, Arch Rational Mech Anal **141**, 1–61.

S Allen and J W Cahn (1979), *A microscopic theory for antiphase boundary motion and its application to antiphase domain coarsening*, Acta Metallurgica **27**, 1084–1095.

H Amann (1986), *Quasilinear evolution equations and parabolic systems*, Trans Am Math Soc **293**, 191–227.

H Amann (1989), *Dynamic theory of quasilinear parabolic equations III: global existence*, Math Z **202**, 219–250.

H Amann (1990), *Dynamic theory of quasilinear parabolic equations II: reaction diffusion systems*, Differential Integral Equations **3**, 13–76.

H Amann (2000), *On the strong solvability of the Navier-Stokes equations*, Preprint.

C D Andereck, S S Liu, and H L Swinney (1986), *Flow regimes in a circular Couette system with independently rotating cylinders*, J Fluid Mech **164**, 155–183.

F Andreu, J M Mazon, F Simondon, and J Toledo (1998), *Attractors for a degenerate nonlinear diffusion problem with nonlinear boundary conditions*, J Dynamics Differential Equations **10**, 347–377.

A A Andronov (1929), *Application of Poincaré's theorem on "bifurcation points" and "change in stability" to simple auto-oscillatory systems*, C R Acad Sci Paris **189**, 559–561.

S B Angenent (1986), *The Morse-Smale property for a semilinear parabolic equation*, J Differential Equations **62**, 427–442.

D V Anosov (1967), *Geodesic flows on closed Riemannian manifolds with negative curvature*, Proc Steklov Inst Math **90**, 1–209.

L Arnold (1995), *Random dynamical systems*, Lecture Notes in Math, No 1609, Springer Verlag, New York, pp. 1–43.

L Arnold (1998), *Random Dynamical Systems*, Monographs in Math, Springer Verlag, New York.

V I Arnold (1983), *Geometrical Methods in the Theory of Ordinary Differential Equations*, Springer Verlag, New York.

J Arrieta, A N Carvalho, and J K Hale (1992), *A damped hyperbolic equation with critical exponent*, Comm Partial Differential Equations **17**, 841–866.

Z Artstein (1977a), *The topological dynamics of an ordinary differential equation*, J Differential Equations **23**, 216–223.

Z Artstein (1977b), *Topological dynamics of ordinary differential equations and Kurzweil equations*, J Differential Equations **23**, 224–243.

M Aassila (1998), *On a quasilinear wave equation with strong damping*, Funk Ekvacioj **41**, 67–78.

A V Babin and S-N Chow (1998), *Uniform longtime behavior of solutions of parabolic equations depending on slow time*, J Differential Equations **150**, 264–316.

A V Babin and G R Sell (2000), *Attractors of nonautonomous parabolic equations and their symmetry properties*, J Differential Equations **160**, 1–50.

A V Babin and M I Vishik (1983a), *Attractors of partial differential evolution equations and estimates of their dimensions*, Russian Math Surveys **38**, 151–213.

A V Babin and M I Vishik (1983b), *Regular attractors of semigroups of evolutionary equations*, J Math Pures Appl **62**, 441–491.

A V Babin and M I Vishik (1990), *Attractors of partial differential equations in an unbounded domain*, Proc Roy Soc Edinburgh **116**, 221–243.

A V Babin and M I Vishik (1992), *Attractors of Evolution Equations*, English translation, Studies in Math and its Appl, No 25, North-Holland, Amsterdam.

J M Ball (1973), *Saddle point analysis for an ordinary differential equation in a Banach space and an application to dynamic buckling of a beam*, Nonlinear Elasticity, R W Dickey, ed, Academic Press, New York, pp. 93–160.

J M Ball (1974), *Continuity properties of nonlinear semigroups*, J Functional Anal **17**, 91–102.

J M Ball (1976), *Measurability and continuity conditions for nonlinear processes*, Proc Am Math Soc **55**, 353–358.

J M Ball (1978), *On the asymptotic behaviour of generalized processes with applications to nonlinear evolutionary equations*, J Differential Equations **27**, 224–265.

J M Ball (1997), *Continuity properties and global attractors of generalized semiflows and the Navier-Stokes equations*, J Nonlinear Sci **7**, 475–502.

J M Ball and L A Peletier (1976), *Stabilization of concentration profiles in catalyst particles*, J Differential Equations **20**, 356–368.

S Banach (1932), *Théorie des Opérations Linéaries*, Monografje Matematyczne, Vol 1, Warsaw.

V Barbu and Th. Precupanu (1986), *Convexity and Optimization in Banach Spaces*, D Reidel Publ, Boston.

C Bardos and L Tartar (1973), *Sur l'unicité rétrograde des équations paraboliques et quelques équations voisines*, Arch Rational Mech Anal **50**, 10–25.

P W Bates, *See Alikakos, Bates, and Fusco.*

P W Bates and P C Fife (1993), *The dynamics of nucleation for the Cahn-Hilliard equation*, SIAM J Appl Math **53**, 990–1008.

P W Bates, K Lu, and C Zeng (1998), *Existence and Persistence of Invariant Manifolds for Semiflows in Banach Space*, Memoirs Am Math Soc, 135, No 645.

R Bellman and K L Cooke (1963), *Differential Difference Equations*, Academic Press, New York.

M Benaim and M W Hirsch (1995), *Dynamics of Morse-Smale urn processes*, Ergodic Theory Dynam Systems **15**, 1005–1030.

Y Benyamini and J Lindenstrauss (2000), *Geometric Nonlinear Functional Analysis, Vol I*, Colloquium Publications, No 48, Am Math Soc, Providence.

M S Berger (1977), *Nonlinearity and Functional Analysis*, Academic Press, New York.

N P Bhatia and G P Szegö (1970), *Stability Theory of Dynamical Systems*, Springer Verlag, New York.

G Bianchi and A Marzocchi (1998), *Inertial manifold for the motion of strongly damped nonlinear elastic beams*, NoDEA Nonlinear Differential Equations **5**, 181–192.

J E Billotti and J P LaSalle (1971), *Dissipative periodic processes*, Bull Am Math Soc **77**, 1082–1088.

G D Birkhoff (1927), *Dynamical Systems*, Am Math Soc, Providence.

G D Birkhoff (1931a), *Proof of a recurrence theorem for strongly transitive systems*, Proc Natl Acad Sci USA **17**, 650–655.

G D Birkhoff (1931b), *Proof of the ergodic theorem*, Proc Natl Acad Sci USA **17**, 656–660.

R I Bogdanov (1974), *Versal deformation of a singular point of a vector field on the plane in the case of zero eigenvalues*, Functional Anal Appl **9**, 144–145.

N N Bogoliubov, *See Krylov and Bogoliubov.*

N N Bogoliubov and Y A Mitropolski (1955), *Asymptotic Methods in the Theory of Nonlinear Oscillations*, (Russian), Akad Nauk, Moscow.

H Bohr (1925–1926), *Zur Theorie der fastperiodischen Funktionen I, II, III*, Acta Math **45, 46, 47**, 29–127, 101–214, 237–281.

V G Bondarevsky (1996), *On the global regularity problem for the 3-dimensional Navier-Stokes equations on thin domains*, University of Minnesota PhD thesis.

J Bourgain (1998), *Quasi periodic solutions of Hamiltonian perturbations of 2D linear Schrödinger equations*, Ann Math **148**, 363–439.

J Bourgain (1999), *Global Solutions of Nonlinear Schrödinger Equations*, Colloquium Publ, No 46, Am Math Soc, Providence.

P Brunovský (1990), *The attractor of the scalar reaction diffusion equation is a smooth graph*, J Dynamics Differential Equations **2**, 293–323.

P Brunovský and B Fiedler (1988), *Connecting orbits in scalar reaction diffusion equations*, Dynamics Reported I (U Kirchgrader and H Ealther, ed.), Wiley Teubner, pp. 57–89.

P Brunovský and B Fiedler (1989), *Connecting orbits in scalar reaction diffusion equations II: the complete solution*, J Differential Equations **81**, 106–136.

P Brunovský and P Poláčik (1997), *The Morse-Smale structure of a generic reaction diffusion equation in higher space dimension*, J Differential Equations **135**, 129–181.

I G Bubnov (1912, 1914), *Mechanics of Shipbuilding (2 Volumes)*, St Petersburg.

I G Bubnov (1913), *A review of the works of Professor S P Timoshenko*, Proc Inst Railway Engineers, St Petersburg, **31**.

L A Caffarelli and N E Muler (1995), *An L^∞ bound for solutions of the Cahn-Hilliard equation*, Arch Rational Mech Anal **133**, 129–144.

J W Cahn, *See Allen and Cahn.*

J W Cahn (1961), *On spinodal decomposition*, Acta Metall **9**, 795–801.

J W Cahn and J E Hilliard (1958; 1959), *Free energy of a nonuniform system, I; III*, J Chem Phys **28; 31**, 258–267; 688–699.

J Carr and R L Pego (1989), *Metastable patterns in solutions of $u_t = \epsilon^2 u_{xx} - f(u)$*, Comm Pure Appl Math **42**, 523–586.

M L Cartwright (1967), *Almost periodic flows and solutions of differential equations*, Proc Lond Math Soc **17**, 355–380.

M L Cartwright (1969), *Almost periodic equations and almost periodic flows*, J Differential Equations **5**, 167–181.

M M Cavalcanti, N A Larkin, and J A Soriano (1998), *On solvability and stability of solutions of nonlinear degenerate hyperbolic equations with boundary damping*, Funk Ekvacioj **41**, 271–289.

A N Carvalho, *See Arrieta, Carvalho, and Hale.*

S S Ceron, *See Lopes and Ceron.*

N Chafee (1974), *A stability analysis for a semilinear parabolic partial differential equations*, J Differential Equations **15**, 522–540.

N Chafee and E F Infante (1974), *A bifurcation problem for a nonlinear partial differential equation of parabolic type*, Applicable Anal **4**, 17–37.

G Chen and D L Russell (1982), *A mathematical model for linear elastic systems with structural damping*, Quart J Appl Math, 433–454.

S Y Chen, C Foias, D D Holm, E Olson, E S Titi, and S Wynne (1999a), *The Camassa-Holm equations and turbulence. Predictability: quantifying uncertainty in models of complex phenomena*, Physica D **133**, 49–65.

S Y Chen, C Foias, D D Holm, E Olson, E S Titi, and S Wynne (1999b), *A connection between the Camassa-Holm equations and turbulent flows in channels and pipes*, Phys Fluids **11**, 2343–2353.

S Chen and R Triggiani (1989), *Proof of extensions of two conjectures on structural damping for elastic systems*, Pacific J Math **136**, 15–55.

X F Chen and M Kowalczyk (1996), *Existence of equilibria for the Cahn-Hilliard equation via local minimizers of perimeter*, Comm Partial Differential Equations vol 21, 1097–1123.

X Y Chen and H Matano (1989), *Convergence, asymptotic periodicity, and finite-point blowup in one-dimensional semilinear heat equations*, J Differential Equations **78**, 160–190.

X Y Chen, J K Hale, and B Tan (1997), *Invariant foliations for C^1 semigroups in Banach spaces*, J Differential Equations **139**, 283–318.

Y G Chen (1990), *Blowup solutions of a semilinear parabolic equations with the Neumann and Robin boundary conditions*, J Fac Sci Univ Tokyo, Sect IA **37**, 537–574.

A Chenciner and G Iooss (1979), *Bifurcations de tores invariant*, Arch Rational Mech Anal **69**, 109–198.

V V Chepyzhov and M I Vishik (1993), *A Hausdorff dimension estimate for the kernel sections of nonautonomous evolution equations*, Indiana Univ Math J **42**, 1057–1076.

V V Chepyzhov and M I Vishik (1994), *Attractors of nonautonomous dynamical systems and their dimensions*, J Math Pures Appl **73**, 279–333.

V V Chepyzhov and M I Vishik (1995), *Nonautonomous evolutionary equations with translation-compact symbols and their attractors*, C R Acad Sci Paris, Ser I, Math **321**, 153–158.

P R Chernoff (1975), *A note on the continuity of maps*, Proc Am Math Soc **53**, 318–320.

P R Chernoff and J E Marsden (1970), *On continuity and smoothness of group actions*, Bull Am Math Soc **76**, 1044–1049.

P R Chernoff and J E Marsden (1974), *Properties of Infinite Dimensional Hamiltonian Systems*, Lecture Notes in Math, vol 425, Springer Verlag, New York.

A I Chernykh, *See Lvov, Predtechensky, and Chernykh.*

J W Cholewa and T Dlotko (1994), *Global attractor of the Cahn-Hilliard system*, Bull Austral Math Soc **49**, 277–302.

J W Cholewa and T Dlotko (2000), *Global Attractors in Abstract Parabolic Problems*, London Math Soc Lecture Note Series, No 278, Cambridge Univ Press, Cambridge, UK.

P Chossat and G Iooss (1994), *The Couette-Taylor Problem*, Springer Verlag, New York.

S-N Chow, *See Babin and Chow.*

S-N Chow and J K Hale (1982), *Methods of Bifurcation Theory*, Springer-Verlag, New York.

S-N Chow, J K Hale, and J Mallet-Paret (1975; 1976), *Applications of generic bifurcations. I; II*, Arch Rational Mech Anal **59; 62**, 159–188; 209–235.

S-N Chow and H Leiva (1995), *Existence and roughness of the exponential dichotomy for skew product semiflows in Banach spaces*, J Differential Equations **120**, 429–477.

S-N Chow, W Liu, and Y Yi (2000), *Center manifolds for smooth invariant manifolds*, Trans Am Math Soc **352**, 5179–5211.

S-N Chow and K Lu (1988), *Invariant manifolds for flows in Banach spaces*, J Differential Equations **74**, 285–317.

S-N Chow, K Lu, and J Mallet-Paret (1994), *Floquet theory for parabolic differential equations*, J Differential Equations **109**, 147–200.

S-N Chow, K Lu, and J Mallet-Paret (1995), *Floquet bundles for scalar parabolic equations*, Arch Rational Mech Anal **129**, 245–304.

S-N Chow, K Lu, and G R Sell (1992), *Smoothness of inertial manifolds*, J Math Anal Appl **169**, 283–312.

S-N Chow and J Mallet-Paret (1977), *Integral averaging and bifurcation*, J Differential Equations **26**, 112–159.

S-N Chow and J Mallet-Paret (1978), *The Fuller index and global Hopf bifurcation*, J Differential Equations **29**, 66–85.

S-N Chow and Y Yi (1994), *Center manifold and stability for skewproduct flows*, J Dynamics Differential Equations **6**, 543–582.

K N Chueh, C C Conley, and J A Smoller (1977), *Positively invariant regions for systems of nonlinear diffusion equations*, Indiana Univ Math J **26**, 373–392.

I D Chueshov (1991), *Strong solutions and the attractor of the von Kármán equation*, Math USSR Sbornik **69**, 25–36.

P Collet, J-P Eckmann, H Epstein, and J Stubbe (1993), *A global attracting set for the Kuramoto-Sivashinsky equation*, Comm Math Phys **152**, 203–214

C C Conley, *See Chueh, Conley, and Smoller.*

C C Conley (1978), *Isolated Invariant Sets and the Morse Index*, CBMS Regional Conference, vol 89, Am Math Soc, Providence.

P Constantin and C Foias (1985), *Global Lyapunov exponents, Kaplan-Yorke formulas and the dimension of the attractor for 2D Navier-Stokes equation*, Comm Pure Appl Math **38**, 1–27.

P Constantin and C Foias (1988), *Navier-Stokes Equations*, Univ Chicago Press, Chicago.

P Constantin, C Foias, B Nicolaenko, and R Temam (1988), *Integral Manifolds and Inertial Manifolds for Dissipative Partial Differential Equations*, Appl Math Sciences, No 70, Springer Verlag, New York.

P Constantin, C Foias, and R Temam (1985), *Attractors representing turbulent flows*, No 314, Memoir Amer Math Soc **53**, Amer Math Soc, Providence.

E Conway, D Hoff, and J Smoller (1978), *Large time behavior of solutions of systems of nonlinear reaction diffusion equations*, SIAM J Appl Math **35**, 1–16.

K L Cooke, *See Bellman and Cooke.*

W A Coppel (1965), *Stability and Asymptotic Behavior of Differential Equations*, Heath, Boston.

W A Coppel (1967, 1968), *Dichotomies and reducibility I, II*, J Differential Equations **3, 4**, 500–521, 386–398.

W A Coppel (1978), *Dichotomies in Stability Theory*, Lect Notes in Math, Vol 629, Springer Verlag, New York.

W A Coppel (1984), *Dichotomies and Lyapunov functions*, J Differential Equations **52**, 58–65.

R Courant and D Hilbert (1953, 1962), *Methods of Mathematical Physics, Vol 1, 2*, Interscience Publ, New York.

C Dafermos (1971), *An invariance principle for compact processes*, J Differential Equations **9**, 239–252.

C Dafermos (1975), *Semiflows generated by compact and uniform processes*, Math Systems Theory **8**, 142–149.

P S Datti (1990), *Nonlinear wave equations in exterior domains*, Nonlinear Anal **15**, 321–331.

R Dautray and J L Lions (1990), *Mathematical Analysis and Numerical Methods for Science and Technology*, Vol 1–4, Springer Verlag, New York.

A Debussche and L Dettori (1995), *On the Cahn-Hilliard equation with a logarithmic free energy*, Nonlinear Anal **24**, 1491–1514.

D G DeFigueiredo (1989), *The Ekeland Variational Principle with Applications*, Springer Verlag, New York.

K Deimling (1985), *Nonlinear Functional Analysis*, Springer Verlag, New York.

J C F De Oliveira, *See Oliva, De Oliveira, and Sola-Morales.*

L Dettori, *See Debussche and Dettori.*

P Deuring and W von Wahl (1995), *Strong solutions of the Navier-Stokes equations in Lipschitz bounded domains*, Math Nachr **171**, 111–148.

E DiBenedetto (1983), *Continuity of weak solutions to a general porous medium equations*, Indiana Univ J Math **32**, 88–118.

T Dlotko, *See Cholewa and Dlotko.*

T Dlotko (1994), *Global attractor for the Cahn-Hilliard equation in H^2 and H^3*, J Differential Equations **113**, 381–393.

E J Doedel, *See M J Friedman and Doedel.*

C R Doering and J D Gibbon (1995), *Applied Analysis of the Navier-Stokes Equations*, Cambridge Texts in Applied Mathematics, Cambridge Univ Press, Cambridge, UK.

A Douady and J Oesterlé (1980), *Dimension de Hausdorff des attracteurs*, C R Acad Sci Paris, Ser A **290**, 1135–1138.

A Douglis, *See Agmon, Douglis, and Nirenberg.*

J Duan and V J Ervin (1998), *Dynamics of a nonlocal Kuramoto-Sivashinsky equation*, J Differential Equations **143**, 243–266.

N Dunford and J T Schwartz (1958), *Linear Operators, Parts 1, 2 and 3*, Wiley Interscience, New York.

L Dung (1997), *Dissipativity and global attractors for a class of quasilinear parabolic systems*, Comm Partial Differential Equations **22**, 413–433.

L Dung and B Nicolaenko (2001), *Exponential attractors in Banach spaces*, (to appear), J Dynamics Differential Equations **13**.

J-P Eckmann, *See Collet, Eckmann, Epstein, and Stubbe.*

A Eden, C Foias, and B Nicolaenko (1994), *Exponential attractors of optimal Lyapunov dimension for the Navier-Stokes equations*, J Dynamics Differential Equations **6**, 301–323.

A Eden, C Foias, B Nicolaenko, and R Temam (1994), *Exponential Attractors for Dissipative Evolution Equations*, Research in Applied Mathematics Series, Masson, Paris.

A Eden, C Foias, and R Temam (1991), *Local and global Lyapunov exponents*, J Dynamics Differential Equations, 133–177.

A Eden, B Michaux, and J M Rakotoson (1991), *Doubly nonlinear parabolic-type equations as dynamical systems*, J Dynamics Differential Equations **3**, 97–131.

D E Edmunds and W D Evans (1987), *Spectral Theory and Differential Operators*, Clarendon Press, Oxford.

C M Elliott and A M Stuart (1996), *Viscous Cahn-Hilliard equations II: analysis*, J Differential Equations **128**, 387–414.

C M Elliott and S Zheng (1986), *On the Cahn-Hilliard equation*, Arch Rational Mech Anal **96**, 339–357.

R Ellis (1969), *Lectures on Topological Dynamics*, Benjamin, New York.

K-J Engel and R Nagel (2000), *One-Parameter Semigroups for Linear Evolution Equations*, Springer Verlag, New York.

H Epstein, *See Collet, Eckmann, Epstein, and Stubbe.*

V J Ervin, *See Duan and Ervin.*

M Escobedo, J L Vazquez, and E Zuazua (1994), *Entropy solutions for diffusion-convection equations with partial diffusivity*, Trans Am Math Soc **343**, 829–842.

M Escobedo and E Zuazua (1991), *Large time behavior for convective diffusion equations in R^N*, J Func. Anal **100**, 119–161.

W D Evans, *See Edmunds and Evans.*

D J Eyre (1993), *Systems of Cahn-Hilliard equations*, SIAM J Appl Math **33**, 1686–1712.

E Fabes, M Luskin and G R Sell (1991), *Construction of inertial manifolds by elliptic regularization*, J Dynamics Differential Equations **89**, 355–387.

S Faedo (1949), *Un nuovo metodo per l'analisi esistenziale e quantitativa dei problemi di propagazione*, Ann Sci Norm Sup Pisa, 1–40.

H O Fattorini (1983), *The Cauchy Problem*, Encyclopedia of Mathematics and its Applications, Addison Wesley, Boston.

J Favard (1933), *Lecons sur les fonctions presque periodiques*, Gauthier Villars, Paris.

E Feireisl (1994), *Finite dimensional asymptotic behavior of some semilinear damped hyperbolic problems*, J Dynamics Differential Equations **6**, 23–35.

E Feireisl (1995), *Asymptotic behavior and attractors for a semilinear damped wave equation with supercritical exponents*, Proc Royal Soc Edinburgh **125**, 1051–1062.

E Feireisl (1997), *Longtime behavior and convergence for semilinear wave equations in R^N*, J Dynamics Differential Equations **9**, 133–155.

E Feireisl, P Laurencot, and F Simondon (1996), *Global attractors for degenerate parabolic equations on unbounded domains*, J Differential Equations **129**, 239–261.

E Feireisl and E Zuazua (1993), *Global attractors for semilinear wave equations with locally distributed nonlinear damping and critical exponent*, Comm Partial Differential Equations **18**, 1539–1555.

N Fenichel (1971), *Persistence and smoothness of invariant manifolds for flows*, Indiana Univ Math J **21**, 193–226.

N Fenichel (1979), *Geometric singular perturbation theory for ordinary differential equations*, J Differential Equations **31**, 53–98.

B Fiedler, *See Brunovský and Fiedler.*

B Fiedler (1988), *Global Bifurcation of Periodic Solutions with Symmetry*, Lecture Notes in Math, 1309, Springer Verlag, New York.

B Fiedler and J Mallet-Paret (1989), *A Poincaré Bendixson theorem for scalar reaction diffusion equations*, Arch Rational Mech Anal **107**, 325–345.

P C Fife, *See Bates and Fife.*

W E Fitzgibbon, M Parrott, and Y You (1996), *Finite dimensionality and upper semicontinuity of the global attractor of singularly perturbed Hodgkin-Huxley systems*, J Differential Equations **129**, 193–237.

I Flahaut (1991), *Attractors for the dissipative Zakharov system*, Nonlinear Anal **16**, 599–633.

C Foias, *See Chen, Foias, Holm, Olson, Titi, and Wynne; Constantin and Foias; Constantin, Foias, and Temam; Constantin, Foias, Nicolaenko, and Temam; Eden, Foias, and Nicolaenko; Eden, Foias, Nicolaenko, and Temam; Eden, Foias, and Temam; and Sz Nagy and Foias.*

C Foias, D D Holm, and E S Titi (2002), *The three dimensional viscous Camassa-Holm equations, and their relation to the Navier-Stokes equations and turbulence theory*, J Dynamics Differential Equations **14**.

C Foias and M S Jolly (1995), *On the numerical algebraic approximations of global attractors*, Nonlinearity **8**, 295–319.

C Foias, M S Jolly, I G Kevrekidis, and E S Titi (1991), *Dissipativity of numerical schemes*, Nonlinearity **4**, 591–613.

C Foias, O Manley, R Rosa, and R Temam (2001), *Navier-Stokes Equations and Turbulence*, (to appear).

C Foias, O P Manley, and R Temam (1987), *Attractors for the Bénard problem and physical bounds on their fractal dimensions*, Nonlinear Anal **11**, 939–967.

C Foias, O P Manley, R Temam, and Y M Treve (1983), *Asymptotic analysis of the Navier-Stokes equations*, Physica D **9**, 157–188.

C Foias, B Nicolaenko, G R Sell, and R Temam (1988), *Inertial manifolds for the Kuramoto-Sivashinsky equation and an estimate of their lowest dimension*, J Math Pures et Appl **67**, 197–226.

C Foias, B Nicolaenko, and R Temam (1989), *Spectral barriers and inertial manifolds for dissipative partial differential equations*, J Dynamics Differential Equations **1**, 45–73.

C Foias and G Prodi (1967), *Sur le compartement global des solutions non stationnaires des équations de Navier-Stokes en dimension 2*, Rend Sem Mat Univ Padova **39**, 1–34.

C Foias, G R Sell and R Temam (1988), *Inertial manifolds for nonlinear evolutionary equations*, J Differential Equations **73**, 309–353.

C Foias, G R Sell and E S Titi (1989), *Exponential tracking and approximation of inertial manifolds for dissipative equations*, J Dynamics Differential Equations **1**, 199–244.

C Foias and R Temam (1979), *Some analytic and geometric properties of the solutions of the Navier-Stokes equations*, J Math Pures Appl **58**, 339–368.

C Foias and R Temam (1980), *Homogeneous statistical solutions of Navier-Stokes equations*, Indiana Math J **29**, 913–957.

C Foias and R Temam (1987), *The connection between the Navier-Stokes equations, dynamical systems, and turbulence theory*, Directions in Partial Differential Equations, Academic Press, New York, pp. 55–73.

L E Fraenkel (1979), *On regularity of the boundary in the theory of Sobolev spaces*, Proc Lond Math Soc **39**, 385–427.

J Franks (1972), *Differentiably Ω-stable diffeomorphisms*, Topology **11**, 107–113.

A Friedman (1969), *Partial Differential Equations*, Holt Rinehart and Winston, New York.

M J Friedman and E J Doedel (1993), *Computational methods for global analysis of homoclinic and heteroclinic orbits: a case study*, J Dynamics Differential Equations **5**, 37–57.

P A Fuhrman (1984), *Linear Systems and Operators in Hilbert Spaces*, 2nd ed, McGraw Hill, New York.

H Fujita, *See Kato and Fujita.*

H Furstenberg (1963), *The structure of distal flows*, Am J Math **85**, 477–515.

H Furstenberg (1967), *Disjointness in ergodic theory, minimal sets, and a problem in Diophantine approximation*, Math Systems Theory **1**, 1–49.

G Fusco, *See Alikakos, Bates, and Fusco; and Alikakos and Fusco.*

G Fusco (1987), *Describing the flow on the attractor of one-dimensional reaction diffusion equations by systems of ordinary differential equations*, Dynamics of Infinite Dimensional Systems (S N Chow and J K Hale, ed.), Springer Verlag, New York, pp. 113–122.

G Fusco and J K Hale (1989), *Slow motion manifolds, dormant instability, and singular perturbations*, J Dynamics Differential Equations **1**, 75–94.

G Fusco and W M Oliva (1990), *Transversality between invariant manifolds of periodic orbits for a class of monotone dynamical systems*, J Dynamics Differential Equations **2**, 1–17.

E Gagliardo (1958), *Proprietá di alcune classi di funzioni in piu variabili*, Ricerche di Mat **7**, 102–137.

E Gagliardo (1959), *Ulteriori proprietá di alcune classi di funzioni in piu variabili*, Ricerche di Mat **8**, 24–51.

G P Galdi (1994), *An Introduction to the Mathematical Theory of the Navier-Stokes Equations, Vol I and II*, Springer Tracts in Natural Philosophy, Springer Verlag, New York.

B G Galerkin (1915), *On the electrical circuits for the approximate solution of the Laplace equation*, Vestnik Inzh **19**, 897–908.

T Gedeon, H Kokubu, K Mischaikow, H Oka, and J F Reineck (1999), *The Conley index for fast-slow systems. I One-dimensional slow variable*, J Dynamics Differential Equations **11**, 427–470.

J M Ghidaglia (1988), *Weakly damped forced Korteweg-de Vries equatinos behave as a finite dimensional dynamical system in the long time*, J Differential Equations **74**, 369–390.

J M Ghidaglia and B Heron (1987), *Dimension of the attractor associated to the Ginzburg-Landau equations*, Physica D **28**, 282–304.

J M Ghidaglia and R Temam (1987), *Attractors for damped nonlinear hyperbolic equations*, J Math Pures Appl **66**, 273–319.

J R Gibbon, *See Doering and Gibbon.*

Y Giga and R V Kohn (1987), *Characterizing blowup using similarity variables*, Indiana Univ J Math **36**, 1–44.

J A Goldstein (1985), *Semigroups of Linear Operators and Applications*, Oxford Univ Press, Oxford.

J P Gollub and H L Swinney (1975), *Onset of turbulence in a rotating fluid*, Phy Rev Letters **35**, 921–.

M Golubitsky and I Stewart (1986), *Symmetry and stability in Taylor-Couette flow*, SIAM J Math Anal **17**, 249–286.

J Goodman (1994), *Stability of the Kuramoto-Sivashinsky and related systems*, Comm Pure Appl Math **47**, 293–306.

D L Goroff (1993), *Introduction to the English translation of: Les Méthodes nouvelles de la Mécanique céleste*, New Methods of Celestial Mechanics, Am Inst Physics.

O Goubet, *See Abounouh and Goubet.*

O Goubet (1998), *Regularity of the attractors for a weakly damped nonlinear Schrödinger equation in R^2*, Adv Differential Equations **3**, 337–360.

Grigolyuk (1971), *An interview with Professor S P Timoshenko*, Stability of Bars, Plates, and Shells, Moscow.

M Grinfeld and A Novick-Cohen (1999), *The viscous Cahn-Hilliard equation: Morse decomposition and structure of the global attractor*, Trans Am Math Soc **351**, 2375–2406.

P Grisvard (1985), *Elliptic Problems in Nonsmooth Domains*, Monographs and Studies in Mathematics, 24, Pitman, Boston.

M Grobbelar-van Dalsen and N Sauer (1989), *Dynamic boundary conditions for the Navier-Stokes equations*, Proc Roy Soc Edinburgh **113**, 1–11.

M Grobbelar-van Dalsen and N Sauer (1993), *Solutions in Lebesgue spaces of the Navier-Stokes equations with dynamic boundary conditions*, Proc Roy Soc Edinburgh **123**, 745–761.

J Guckenheimer and P Holmes (1983), *Nonlinear Osillations, Dynamical Systems and Bifurcations of Vector Fields*, Springer Verlag, New York.

B Guo and F Su (1993), *The global attractors for the periodic initial value problem of generalized Kuramoto-Sivashinsky type equations in multidimensions*, J Partial Differential Equations **6**, 217–236.

J Hadamard (1901), *Sur l'iteration et les solutions asymptotiques des équations différentielles*, Bull Soc Math France **29**, 224–228.

A Hagen (1996), *Hyperbolic structures of time discretizations and the dependence on the time step*, University of Minnesota PhD Thesis.

H Haken (1981), *Chaos and order in nature*, Chaos and Order in Nature, H Haken, ed., Springer Verlag, Berlin, pp. 2–11.

H Haken (1983), *Advanced Synergetics: Instability Hierarchies of Self-Organizing Systems and Devices*, Springer Verlag, New York.

J K Hale, *See Arrieta, Carvalho, and Hale; Chen, Hale, and Tan; Chow and Hale; and Fusco and Hale.*

J K Hale (1961), *Integral manifolds of perturbed differential equations*, Ann Math **73**, 496–531.

J K Hale (1963), *A stability theorem for functional differential equations*, Proc Natl Acad Sci USA **50**, 942–946.

J K Hale (1969), *Dynamical systems and stability*, J Math Anal Appl **26**, 39–69.

J K Hale (1977), *Theory of Functional Differential Equations*, Appl Math Sci, No 3, Springer Verlag, New York.

J K Hale (1985), *Asymptotic behavior and dynamics in infinite dimensions*, Research Notes in Math, Vol 132, Pitman, Boston, pp. 1–41.

J K Hale (1988), *Asymptotic Behavior of Dissipative Systems*, Mathematical Surveys and Monographs, Vol 25, Am Math Soc, Providence.

J K Hale (1989), *Compact attractors and singular perturbations*, Advanced Topics in the Theory of Dynamical Systems (Trento, 1987), Academic Press, Boston, pp. 137–153.

J K Hale and H Kocak (1991), *Dynamics and Bifurcations*, Springer Verlag, New York.

J K Hale, J P LaSalle, and M Slemrod (1972), *Theory of a general class of dissipative processes*, J Math Anal Appl **39**, 177–191.

J K Hale, X-B Lin, and G Raugel (1988), *Upper semicontinuity of attractors and partial differential equations*, Math Comp **50**, 89–123.

J K Hale and O Lopes (1973), *Fixed point theorems and dissipative processes*, J Differential Equations **13**, 391–402.

J K Hale, L T Magalhães, and W M Oliva (1984), *An Introduction to Infinite Dimensional Dynamical Systems - Geometric Theory*, Appl Math Sci, No 47, Springer Verlag, New York.

J K Hale and G Raugel (1989), *Lower semicontinuity of attractors of gradient systems and applications*, Ann Mat Pura Appl **154**, 281–326.

J K Hale and G Raugel (1991), *Attractors for dissipative evolutionary equations*, Georgia Tech Preprint CDSNS91-72.

J K Hale and G Raugel (1992a), *Reaction diffusion equation on thin domains*, J Math Pures Appl **71**, 33–95.

J K Hale and G Raugel (1992b), *A damped hyperbolic equation on thin domains*, Trans Am Math Soc **329**, 185–219.

J K Hale and G Raugel (1992c), *Convergence in gradient-like systems and applications*, ZAMP **43**, 63–124.

J K Hale and G Raugel (1992d), *Dynamics of partial differential equations on thin domains*, Preprint.

J K Hale and G Raugel (1995), *A reaction diffusion equation on a thin L-shaped domain*, Proc Roy Soc Edinburgh, Sect A **125**, 283–327.

J K Hale and K Sakamoto (1989), *Shadow systems and attractors in reaction diffusion equations*, Appl Anal **32**, 287–303.

J K Hale and S M Verduyn Lunel (1993), *Introduction to Functional Differential Equations*, Applied Mathematical Sciences, Vol 99, Springer Verlag, New York.

G R Hall, *See Meyer and Hall.*

A Haraux (1985), *Two remarks on dissipative hyperbolic problems*, College de France Seminaire, Pitmann, pp. 161–179.

A Haraux (1988), *Attractors of asymptotically compact processes and applications to nonlinear partial differential equations*, Comm in PDE **13**, 1383–1414.

P Hartman (1964), *Ordinary Differential Equations*, Wiley, New York.

N N Hayashi and P Naumkin (1998), *Large time asymptotics of solutions to the generalized Korteweg-de Vries equations*, J Funct Anal **159**, 110–136.

D B Henry (1981), *Geometric Theory of Semilinear Parabolic Equations*, Lecture Notes in Mathematics, No 840, Springer Verlag, New York.

D B Henry (1985), *Some infinite dimensional Morse-Smale systems defined by parabolic partial differential equations*, J Differential Equations **53**, 401–458.

B Heron, *See Ghidaglia and Heron.*

M A Herrero and J Velazquez (1992), *Some results on blow up for semilinear parabolic equations*, IMA Preprint No 1000.

J G Heywood and R Rannacher (1982; 1986; 1988; 1990), *Finite-element method of the nonstationary Navier-Stokes problem I; II; III; IV*, SIAM J Numerical Anal **19; 23; 25; 27**, 275–311; 750–777; 489–512; 353–384.

D Hilbert, *See Courant and Hilbert.*

E Hille and R S Phillips (1948; 1957), *Functional Analysis and Semigroups*, Am Math Soc Colloquium Publ, vol 31, Am Math Soc, Providence.

J E Hilliard, *See Cahn and Hilliard.*

J E Hilliard (1970), *Spinodal decomposition in phase transformation*, Am Soc Metals, 497–560.

M W Hirsch, *See Benaim and Hirsch.*

M W Hirsch (1985), *Attractors for discrete-time monotone dynamical systems in strongly ordered spaces*, Lecture Notes in Math, 1167, Springer Verlag, New York.

M W Hirsch (1988), *Stability and convergence in strongly monotone dynamical systems*, J Reine Angew Math **383**, 1–53.

M W Hirsch, C C Pugh and M Shub (1977), *Invariant Manifolds*, Lecture Notes In Math, No 583, Springer Verlag, New York.

D Hoff, *See Conway, Hoff, and Smoller.*

P C Hohenberg, *See Swift and Hohenberg.*

D D Holm, *See Chen, Foias, Holm, Olson, Titi, and Wynne; and Foias, Holm, and Titi.*

P Holmes, *See Guckenheimer and Holmes.*

E Hopf (1942), *Abzweigung einer periodischen Lösung von einer stationären Lösung eines Differentialsystems*, English translation in Marsden and McCracken (1976), Ber Math Phys Suchische Akad Wiss Leipzig **94**, 1–22.

E Hopf (1951), *Über die Anfangswertaufgabe für die hydrodynamischen Grundgleichungen*, Math Nachr **4**, 213–231.

J J Hoyt (1989), *Spinodal decomposition in ternary alloys*, Acta Metall **37**, 2489–2497.

F Huang (1985), (in Chinese), Ann Differential Equations **1**.

A R Humphries, *See Stuart and Humphries.*

Ju S Il'yashenko (1992), *Global analysis of the phase portrait for the Kuramoto Sivashinsky equation*, J Dynamics Differential Equations **4**, 585–615.

E F Infante, *See Chafee and Infante.*

H Inoue (1999), *Strong solutions of two-dimensional heat convection equations with dissipative terms*, Tokyo J Math **22**, 445–472.

G Iooss, *See Chenciner and Iooss; Chossat and Iooss; and Vanderbauwhede and Iooss.*

R A Johnson (1978a), *Ergodic theory and linear differential equations*, J Differential Equations **28**, 23–34.

R A Johnson (1978b), *Concerning a theorem of Sell*, J Differential Equations **30**, 324–339.

R A Johnson (1979), *Measurable subbundles in linear skewproduct flows*, Illinois J Math **23**, 183–198.

R A Johnson (1980a), *On a Floquet theory for almost periodic two-dimensional linear systems*, J Differential Equations **37**, 184–205.

R A Johnson (1980b), *Analyticity of spectral subbundles*, J Differential Equations **35**, 366–387.

R A Johnson (1981), *A linear, almost periodic equation with an almost automorphic solution*, Proc Am Math Soc **82**, 199–295.

R A Johnson and J K Moser (1982), *The rotation number for almost periodic potentials*, Comm Math Phys **84**, 403–438.

R A Johnson, K J Palmer, and G R Sell (1987), *Ergodic properties of linear dynamical systems*, SIAM J Math Anal **18**, 1–33.

R A Johnson and G R Sell (1981), *Smoothness of spectral subbundles and reducibility of quasi periodic linear differential equations*, J Differential Equations **41**, 262–288.

M S Jolly, *See Foias and Jolly; Foias, Jolly, Kevrekidis, and Titi.*

M S Jolly (1989), *Explicit construction of an inertial manifold for a reaction diffusion equation*, J Differential Equations **78**, 220–261.

M S Jolly, R Rosa, and R Temam (2000), *Evaluation of the dimension of an inertial manifold for the Kuramoto-Sivashinsky equation*, Adv Differential Equations **5**, 31–66.

C K R T Jones (1995), *Geometric singular perturbation theory*, Dynamical Systems (Montecatini Terme, 1994), Lecture Notes in Math, 1609, Springer Verlag, Berlin, pp. 44–118.

D A Jones and S Shkoller (1999), *Persistence of invariant manifolds for nonlinear partial differential equations*, Studies Appl Math **102**, 27–67.

D A Jones and E S Titi (1996), *C^1 approximations of inertial manifolds for dissipative nonlinear equations*, J Differential Equations **127**, 54–86.

D W Kahn (1980), *Introduction to Global Analysis*, Pure Appl Math, Vol 91, Academic Press, New York.

D A Kamaev (1980), *Hyperbolic limit sets of evolutionary equations and the Galerkin method*, Russian Math Surveys **35(3)**, 239–243.

T Kato (1961), *Fractional powers of dissipative operators*, J Math Soc Japan **13**, 246–274.

T Kato (1962), *Fractional powers of dissipative operators II*, J Math Soc Japan **14**, 242–248.

T Kato (1966), *Perturbation Theory for Linear Operators*, Springer Verlag, New York.

T Kato and H Fujita (1966), *On the stationary Navier-Stokes system*, Rend Sem Mat Univ Padova **32**, 243–260.

A Katok (1980), *Lyapunov exponents, entropy and periodic orbits for diffeomorphisms*, Inst Hautes Ètudes Sci Publ Math **51**, 137–173.

J P Keener and J J Tyson (1992), *The dynamics of scroll waves in excitable media*, SIAM Review **34**, 1–39.

A Kelley (1967), *The stable, center-stable, center, center-unstable, and unstable manifolds*, J Differential Equations **3**, 546–570.

J L Kelley (1955), *General Topology*, Van Nostrand, New York.

J L Kelley and I Namioka (1976), *Linear Topological Spaces*, Graduate Texts in Math, No 36, Springer Verlag, New York.

J Kepler (1609), *Astronomia Nova.*

J Kepler (1619), *Harmonice Mundi.*

I G Kevrekidis, *See Foias, Jolly, Kevrekidis, and Titi.*

K Kirchgässner (1975), *Bifurcation in nonlinear hydrodynamic stability*, SIAM Rev **17**, 652–683.

K Kirchgässner (1982), *Wave-solutions of reversible systems and applications*, J Differential Equations **45**, 113–127.

K Kirchgässner and P Sorger (1969), *Branching analysis for the Taylor problem*, Quart J Mech Appl Math **22**, 183–209.

H Kocak, *See Hale and Kocak.*

R V Kohn, *See Giga and Kohn.*

H Kokubu, *See Gedeon, Kokubu, Mischaikow, Oka, and Reineck.*

M Kowalczyk, *See Chen and Kowalczyk.*

N Krasovskii (1963), *On the stabilization of unstable motions by additional forces when the feedback loop is incomplete*, Prikl Mat Mek **27**, 641–663.

S G Krein (1971), *Linear Differential Equations in Banach Spaces*, Translations of Math Monographs, Vol 29, Am Math Soc, Providence.

T Krisztin and H-O Walther (2001), *Unique periodic orbits for delayed positive feedback and the global attractor*, J Dynamics Differential Equations **13**, 1–57.

T Krisztin, H-O Walther, and J Wu (1999), *Shape, Smoothness and Invariant Stratification of an Attracting Set for Delayed Monotone Positive Feedback*, Fields Institute Monograph Series, Vol 11, Am Math Soc, Providence.

N M Krylov and N N Bogoliubov (1934), *The Application of Methods of Nonlinear Mechanics to the Theory of Stationary Oscillations*, (Russian), Ukraine Akad Nauk, Kiev.

N M Krylov and N N Bogoliubov (1937), *La théorie générale de la mesure dan son application à l'étude des systémes dynamiques de al mécanique non linéaire*, Ann Math **38**, 65–113.

P Krzyzanowski, *See Lukaszewicz and Krzyzanowski.*

A Kufner (1985), *Weighted Sobolev Spaces*, Wiley Interscience Publ, New York.

Y Kuramoto (1978), *Diffusion-induced chaos in reaction systems*, Supp Prog Theor Phys **64**, 346–347.

J Kurzweil (1966), *Exponentially stable integral manifolds, averaging principle and continuous dependence on a parameter*, Czechslovak Math J **16**, 463–492.

M Kwak (1992a), *Finite dimensional behavior of convective reaction diffusion equations*, J Dynamics Differential Equations **4**, 515–543.

M Kwak (1992b), *Finite dimensional inertial forms for the 2D Navier-Stokes equations*, Indiana J Math **41**, 927–981.

H Kwean (1999), *An extension of the principle of spatial averaging for inertial manifolds*, J Austral Math Soc **66**, 125–142.

O A Ladyzhenskaya (1963), *The Mathematical Theory of Viscous Incompressible Flow*, Gordon and Breach, New York.

O A Ladyzhenskaya (1972), *On the dynamical system generated by the Navier-Stokes equations*, English translation, J Soviet Math **3**, 458–479.

O A Ladyzhenskaya (1987), *On the determination of minimal global attractors for the Navier-Stokes and other partial differential equations*, English translation, Russian Math Surveys **42**, 27–73.

O A Ladyzhenskaya (1991), *Attractors for Semigroups and Evolution Equations*, Cambridge Univ Press, Cambridge.

R E La Quey, P H Mahajan, P H Rutherford, and W M Tang (1975), Phys Review Letters **34**, 391.

N A Larkin, *See Cavalcanti, Larkin, and Soriano.*

J P LaSalle, *See Billotti and LaSalle; and Hale, LaSalle, and Slemrod.*

J P LaSalle (1960), *The extent of asymptotic stability*, Proc Natl Acad Sci, USA **46**, 573–577.

J P LaSalle (1974), *Invariance principles and stability theory for nonautonomous systems*, Proc of Carathéodory International Symposium, Greek Math Soc, Athens, pp. 397–408.

J P LaSalle and S Lefschetz (1961), *Stability by Liapunov's Direct Method with Applications*, Academic Press, New York.

I Lasiecka (1999), *Finite dimensionality and compactness of attractors for von Kármán equations with nonlinear dissipation*, Nonlinear Differential Equations Appl **6**, 437–472.

Y Latushkin, T Randolph, and R Schnaubelt (1998), *Exponential dichotomy and mild solutions of nonautonomous equations in Banach spaces*, J Dynamics Differential Equations **10**, 489–510.

P Laurencot, *See Feireisl, Laurencot, and Simondon.*

P D Lax (1976), *Almost periodic solutions of the KdV equation*, SIAM Review **18**, 351–375.

S Lefschetz, *See LaSalle and Lefschetz.*

H Leiva, *See Chow and Leiva.*

H Leiva (1998), *Exponential dichotomy for a nonautonomous system of parabolic equations*, J Dynamics Differential Equations **10**, 475–488.

J Leray (1933), *Etude de diverses équations intégrales nonlinéaires et de quelques problèmes que pose l'hydrodynamique*, J Math Pures Appl **12**, 1–82.

J Leray (1934a), *Essai sur les mouvement plans d'un liquide visqueux que limitent des parois*, J Math Pures Appl **13**, 331–418.

J Leray (1934b), *Sur le mouvement d'un liquide visqueux emplissant l'espace*, Acta Math **63**, 193–248.

N Levinson (1950), *Small periodic perturbations of an autonomous system with a stable orbit*, Ann Math **52**, 727–738.

D Li and C Zhong (1998), *Global attractor for the Cahn-Hilliard system with fast growing nonlinearity*, J Differential Equations **149**, 191–210.

Y Li, D W McLaughlin, J Shatah, and S Wiggins (1996), *Persistent homoclinic orbits for a perturbed nonlinear Schrödinger equation*, Comm Pure Appl Math **49**, 1175–1255.

S T Liao (1966), *Applications to phase space structure of ergodic properties of the one-parameter transformation group induced on the tangent bundle by a differential system on a manifold I*, (In Chinese), Acta Sci Natur Univ Pekinensis **12**, 1–43.

S T Liao (1973), *An ergodic property theorem for a differential system*, Sci Sinica **16**, 1–24.

X B Lin, *See Hale, Lin, and Raugel.*

X B Lin (1990), *Using Melnikov's method to solve Šilnikov's problems*, Proc Roy Soc Edinburgh Sect A **116**, 295–325.

J Lindenstrauss, *See Benyamini and Lindenstrauss.*

J L Lions, *See Dautray and Lions.*

J L Lions (1962), *Problèmes aux Limites dans les Équations aux Dérivées Partielles*, Seminaire de Math Sup, Univ de Montréal.

J L Lions (1969), *Quelques Méthodes de Résolution des Problèmes aux Limites non Linéaires*, Gauthier Villars, Paris.

J L Lions and E Magenes (1972), *Nonhomogeneous Boundary Value Problems and Applications, Vol I, II, III*, Springer Verlag, New York.

S S Liu, *See Andereck, Liu, and Swinney.*

W Liu, *See Chow, Liu, and Yi.*

O Lopes, *See Hale and Lopes.*

O Lopes (1986), *A semilinear wave equation in one space variable with weak damping: convergence to equilibrium*, Nonlinear Anal **10**, 1491–1502.

O Lopes and S S Ceron (1984), *Existence of forced periodic solutions of dissipative semilinear hyperbolic equations and systems*, Ann Mat Pura Appl.

E N Lorenz (1963), *Deterministic nonperiodic flow*, J Atmospheric Sci **20**, 130–141.

K Lu, *See Bates, Lu, and Zeng; Chow and Lu; Chow, Lu, and Mallet-Paret; and Chow, Lu, and Sell.*

K Lu (1991), *A Grobman-Hartman theorem for scalar reaction diffusion equations*, J Differential Equations **93**, 364–394.

G Lukaszewicz and P Krzyzanowski (1997), *On the heat convection equations with dissipative terms in regions with moving boundaries*, Math Methods Appl Sci **20**, 347–368.

A Lunardi (1995), *Analytic Semigroups and Optimal Regularity in Parabolic Problems*, Birkhäuser, Boston.

M Luskin, *See Fabes, Luskin, and Sell.*

V S Lvov, A A Predtechensky, and A I Chernykh (1983), *Bifurcation and chaos in the system of Taylor vortices - laboratory and numerical experiment*, Nonlinear Dynamics and Turbulence, (G I Barenblatt, G Iooss, and D D Joseph, eds), Pitman, Lond, pp. 238–280.

A M Lyapunov (1892), *Problème géneral de la stabilité du mouvement*, Annals Math Studies, No 17 (1947), French version published first in 1907, Princeton Univ Press, Princeton, N J.

L T Magalhães, *See Hale, Magalhães, and Oliva.*

L T Magalhães (1987), *The spectrum of invariant sets for dissipative semiflows in dynamics of infinite dimensional systems*, NATO ASI Series, No. F-37, Springer Verlag, New York.

E Magenes, *See Lions and Magenes.*

P H Mahajan, *See La Quey, Mahajan, Rutherford, and Tang.*

J Mallet-Paret, *See Chow, Lu, and Mallet-Paret; Chow and Mallet-Paret and Fiedler and Mallet-Paret.*

J Mallet-Paret (1976), *Negatively invariant sets of compact maps and an extension of a theorem of Cartwright*, J Differential Equations **22**, 331–348.

J Mallet-Paret (1988), *Morse decompositions for delay differential equations*, J Differential Equations **72**, 270–315.

J Mallet-Paret and G R Sell (1988), *Inertial manifolds for reaction diffusion equations in higher space dimensions*, J Am Math Soc **1**, 805–866.

J Mallet-Paret and G R Sell (1996a), *Systems of differential delay equations: Floquet multipliers and discrete Lyapunov functions*, J Differential Equations **125**, 385–440.

J Mallet-Paret and G R Sell (1996b), *The Poincaré-Bendixson theorem for monotone cyclic feedback systems with delay*, J Differential Equations **125**, 441–489.

J Mallet-Paret, G R Sell, and Z Shao (1993), *Obstructions for the existence of normally hyperbolic inertial manifolds*, Indiana J Math **42**, 1027–1055.

J Mallet-Paret and H L Smith (1990), *The Poincaré-Bendixson theorem for monotone cyclic feedback systems*, J Dynamics Differential Equations **2**, 367–421.

R Mañé (1977), *Reduction of semilinear parabolic equations to finite dimensional C^1 flows*, Lecture Notes in Math, vol 597, Springer Verlag, New York, pp. 361–378.

R Mañé (1981), *On the dimension of the compact invariant sets of certain nonlinear maps*, Lecture Notes in Math, vol 898, Springer Verlag, New York, pp. 230–242.

R Mañé (1988), *A proof of the C^1 stability conjecture*, Inst Hautes Sci Publ Math, No 66, 161–210.

O P Manley, *See Foias, Manley, Rosa, and Temam; Foias, Manley, and Temam; and Foias, Manley, Temam, and Treve.*

O P Manley and M Marion (1992), *Attractor dimension for a simple premixed flame model*, Combustion Sci Tech **88**, 15–32.

O P Manley, M Marion, and R Temam (1993), *Equations of combustion in the presence of complex chemistry*, Indiana Univ Math J **42**, 941–967.

M Marion, *See Manley and Marion; and Manley, Marion and Temam.*

M Marion (1987), *Attractors for reaction diffusion equations: existence and estimate of their dimensions*, Applicable Analysis **25**, 101–147.

M Marion (1989a), *Inertial manifolds associated to partly dissipative reaction diffusion equations*, J Math Anal Appl **143**, 295–326.

M Marion (1989b), *Finite dimensional attractors associated to partly dissipative reaction diffusion equations*, SIAM J Math Anal **20**, 816–844.

M Marion (1991), *Attractors and turbulence for some combustion models*, IMA Volumes in Mathematics and its Applications, vol. 35, Springer Verlag, New York, pp. 213–228.

L Markus (1961), *Structurally stable differential systems*, Ann Math **73**, 1–19.

L Markus and K R Meyer (1974), *Generic Hamiltonian dynamical systems are neither integrable nor ergodic*, Memoirs Am Math Soc, No 144, Am Math Soc, Providence.

L Markus and K R Meyer (1980), *Periodic orbits and solenoids in generic Hamiltonian dynamical systems*, Am J Math **102**, 25–92.

L Markus and G R Sell (1968), *Control and capture in conservative dynamical systems*, Arch Rational Mech Anal **31**, 271–287.

L Markus and G R Sell (1974), *Control in conservative dynamical systems: Recurrence and capture in aperiodic fields*, J Differential Equations **16**, 472–505.

J E Marsden, *See Chernoff and Marsden.*

J E Marsden and M F McCracken (1976), *The Hopf Bifurcation and its Applications*, Springer Verlag, New York.

J E Marsden, T Ratiu, and G Raugel (1991), *Symplectic connections and the linearization of Hamiltonian systems*, Proc Roy Soc Edinburgh Sect A **117**, 329–380.

R H Martin (1976), *Nonlinear Operators and Differential Equations in Banach Spaces*, Wiley, New York.

A Marzocchi, *See Bianchi and Marzocchi.*

P Massatt (1983), *Limiting behavior for strongly damped nonlinear wave equations*, J Differential Equations **48**, 334–349.

J L Massera and J J Schäffer (1958, 1959a, 1959b), *Linear differential equations and functional analysis I, II, III*, Ann Math **67, 69, 69**, 517–573, 88–104, 535–574.

H Matano, *See Chen and Matano.*

H Matano (1979), *Asymptotic behavior and stability of solutions of semilinear diffusion equations*, Publ Rest Inst Math Sci **15**, 401–458.

H Matano (1082), *Nonincrease of lap number of a solution for a one dimensional semilinear parabolic equation*, J Fac Sci Univ Tokyo **23**, 401–441.

N Matsuda, *See Oeda and Matsuda.*

J M Mazon, *See Andreu, Mazon, Simondon, and Toledo.*

M F McCracken, *See Marsden and McCracken.*

R McGehee (1974), *Triple collision in the collinear three-body problem*, Invent Math **27**, 409–421.

D W McLaughlin, *See Li, McLaughlin, Shatah, and Wiggins.*

D W McLaughlin and J Shatah (1996), *Melnikov analysis for PDEs*, Dynamical Systems and Probabilistic Methods in Partial Differential Equations, Lectures in Appl Math, 31, Am Math Soc, Providence.

V K Melnikov (1963), *On the stability of the center for time periodic perturbations*, Trans Moscow Math Soc **12**, 1–57.

G P Menzala and E Zuazua (1998), *Energy decay rates for the von Kármán systems of thermoelastic plates*, Differential Integral Equations **11**, 755–770.

K R Meyer, *See Markus and Meyer.*

K R Meyer (1968), *Energy functions for Morse-Smale systems*, Am J Math **90**, 1031–1040.

K R Meyer and G R Hall (1992), *Introduction to Hamiltonian Dynamical Systems and the N-Body Problem*, Applied Mathematical Sciences, Vol 90, Springer Verlag, New York.

K R Meyer and G R Sell (1989), *Melnikov transforms, Bernoulli bundles and almost periodic perturbations*, Trans Am Math Soc **314**, 63–105.

B Michaux, *See Eden, Michaux, and Rakotoson.*

D Michelson (1996), *Stability of the Bunsen flame profile in the Kuramoto-Sivashinsky equation*, SIAM J Math Anal **27**, 765–781.

A Mielke (1991), *Hamiltonian and Lagrangian Flows on Center Manifolds With Applications to Elliptic Variational Problems*, Lecture Notes in Math, No 1489, Springer Verlag, New York.

A Mielke (1992), *On nonlinear problems of mixed type: a qualitative theory using infinite dimensional center manifolds*, J Dynamics Differential Equations **4**, 419–443.

R K Miller (1965), *Almost periodic differential equations as dynamical systems with applications to the existence of a.p. solutions*, J Differential Equations **1**, 337–345.

R K Miller and G R Sell (1970), *Volterra integral equations and topological dynamics*, Memoir Am Math Soc, No 102.

V M Millionščikov (1968), *Metric theory of linear systems of differential equations*, Math USSR Sbornik **6**, 149–158.

K Mischaikow, *See Gedeon, Kokubu, Mischaikow, Oka, and Reineck.*

K Mischaikow (1995), *Conley index theory*, Dynamical Systems (Montecatini Terme, 1994), Lecture Notes in Math, 1609, Springer Verlag, Berlin.

K Mischaikow, M Mrozek, and J F Reineck (1999), *Singular index pairs*, J Dynamics Differential Equations **11**, 399–425.

Y A Mitropolski, *See Bogoliubov and Mitropolski.*

K Mochizuki and T Motai (1995), *On energy decay-nondecay problems for wave equations with nonlinear dissipative term in* R^N, J Math Soc Japan **47**, 405–421.

K Mochizuki and T Motai (1996), *On asymptotic behaviors for wave equations with a nonlinear dissipative term in* R^N, Hokkaido Math J **25**, 119–135.

R Moeckel (1981), *Orbits of the three-body problem which pass infinitely close to triple collision*, Am J Math **103**, 1323–1341.

R Moeckel (1985), *Relative equilibria of the four body problem*, Ergodic Theory Dynamical Systems **4**, 417–435.

I Moise and R Rosa (1997), *On the regularity of the global attractor of a weakly damped, forced Korteweg-de Vries equation*, Adv Differential Equations **2**, 257–296.

X Mora (1983), *Finite dimensional attracting manifolds in reaction diffusion equations*, Nonlinear Partial Differential Equations, Contemp Math, 17, Am Math Soc, Providence, pp. 353–360.

X Mora and J Solà-Morales (1989), *Inertial manifolds of damped semilinear wave equations*, Attractors, Inertial Manifolds and their Approximation (Marseille-Luminy, 1987), RAIRO Model Math Anal Numer, vol. 23, pp. 489–505.

J K Moser, *See Johnson and Moser; and Siegel and Moser.*

J K Moser (1969), *On a theorem of Anosov*, J Differential Equations **5**, 411–440.

T Motai, *See Mochizuki and Motai.*

M Mrozek, *See Mischaikow, Mrozek, and Reineck.*

N E Muler, *See Caffarelli and Muler.*

J D Murray (1993), *Mathematical Biology*, Biomath Series No. 19, Springer Verlag, New York.

R Nagel, *See Engel and Nagel.*

M Nakao (1990), *Global existence and classical solutions to the initial-boundary value problems of the semilinear wave equations with a degenerate dissipative term*, Nonlinear Anal **15**, 115–140.

I Namioka, *See Kelley and Namioka.*

I P Natanson (1957), *Theory of Functions of Real Variables*, 2nd ed (in Russian), Nat Tech Theo Publ, Moscow.

P Naumkin, *See Hayashi and Naumkin.*

C L M H Navier (1827), *Mémoire sur les lois du mouvement des fluides*, Mem Acad Sci Inst France **6**, 389-440.

A W Naylor and G R Sell (1982), *Linear Operator Theory in Engineering and Science*, Applied Mathematical Sciences, Vol 40, Springer Verlag, New York.

J Nečas, M Ružička, and V Šverak (1996), *On Leray's self-similar solutions of the Navier-Stokes equations*, Acta Math **176**, 283–294.

V V Nemytskii and V V Stepanov (1960), *Qualitative Theory of Differential Equations*, Princeton Univ, Princeton NJ.

J von Neumann (1930), *Allgemeine Eigenwerttheorie Hermitscher Funktionaloperatoren*, Math Ann **102**, 49–131.

I Newton (1687), *Philosophiae Naturalis Principia Matematica.*

B Nicolaenko, *See Constantin, Foias, Nicolaenko, and Temam; Eden, Foias, and Nicolaenko; Eden, Foias, Nicolaenko, and Temam; Foias, Nicolaenko, Sell, and Temam; and Foias, Nicolaenko, and Temam.*

B Nicolaenko, B Sheurer, and R Temam (1985), *Some global dynamical properties of the Kuramoto-Sivashinsky equations: nonlinear stability and attractors*, Physica D **16**, 155–183.

B Nicolaenko, B Scheurer, and R Temam (1989), *Some global dynamical properties of a class of pattern formation equations*, Comm Partial Differential Equations **14**, 245–297.

L Nirenberg, *See Agmon, Douglis, and Nirenberg.*

L Nirenberg (1955), *Remarks on strongly elliptic partial differential equations*, Comm Pure Appl Math **8**, 648–674.

L Nirenberg (1959), *On elliptic partial differential equations*, Ann Scuola Norm Sup Pisa **13**, 115–162.

Y Nishiura (1982), *Global structure of bifurcating solutions of some reaction diffusion equations*, SIAM J Math Anal **13**, 555–593.

D Norman (1999), *A model for chemically reacting fluid flows: weak solutions and global attractors*, J Differential Equations **152**, 75–135.

D Norman (2000a), *Chemically reacting fluid flows: strong solutions and global attractors*, J Dynamics Differential Equations **12**, 273–307.

D Norman (2000b), *Chemically reacting flows and continuous stirred tank reactors: the large diffusivity limit*, J Dynamics Differential Equations **12**, 309–355.

A Novick-Cohen, *See Grinfeld and Novick-Cohen.*

A Novick-Cohen (1988), *Energy methods for the Cahn-Hilliard equation*, Quart Appl Math **46**, 681–690.

A Novick-Cohen (1990), *On Cahn-Hilliard type equations*, Nonlinear Analysis **15**, 797–814.

A Novick-Cohen (1998), *The Cahn-Hilliard equation: mathematical and modeling perspectives*, Adv Math Sci Appl **8**, 965–985.

A Novick-Cohen and L A Segel (1984), *Nonlinear aspects of the Cahn-Hilliard equation*, Physica D **10**, 277–298.

K Oeda and N Matsuda (1998), *Initial value problems for the heat convection equations in exterior domains*, Tokyo J Math **21**, 359–375.

J Oesterlé, *See Douady and Oesterlé.*

H Oka, *See Gedeon, Kokubu, Mischaikow, Oka, and Reineck.*

W M Oliva, *See Hale, Magalhães, and Oliva.*

W M Oliva, J C F De Oliveira, and J Sola-Morales (1994), *An infnite-dimensional Morse-Smale map*, Nonlinear Differential Equations Appl **1**, 365–387.

E Olson, *See Chen, Foias, Holm, Olson, Titi, and Wynne.*

V Oseledec (1968), *A multiplicative ergodic theorem. Lyapunov characteristic numbers for dynamical systems*, Trans Moscow Math Soc **19**, 197–231.

J Palis (1968), *On Morse-Smale dynamical systems*, Topology **8**, 385–404.

K J Palmer, *See Johnson, Palmer, and Sell.*

K J Palmer (1984), *Exponential dichotomies and transversal homoclinic points*, J Differential Equations **55**, 225–256.

M Parrott, *See Fitzgibbon, Parrott, and You.*

A Pazy (1983), *Semigroups of Linear Operators and Applications to Partial Differential Equations*, Applied Mathematical Sciences, Vol 44, Springer Verlag, New York.

R L Pego, *See Carr and Pego.*

R L Pego (1989), *Front migration in the nonlinear Cahn-Hilliard equation*, Proc Roy Soc London, Ser A **422**, 261–278.

M M Peixoto (1959), *On structural stability*, Ann Math **69**, 199–222.

L A Peletier, *See Ball and Peletier.*

L A Peletier and J Zhao (1993), *Large time behavior of solutions of the porous media equations with absorption: the fast diffusion case*, Nonlinear Anal **17**, 991–1009.

O Perron (1928), *Über Stabilität und asymptotisches Verhalten der Integrale von Differentialgleichungssystemen*, Math Zeit **29**, 129–160.

O Perron (1930a), *Über eine Matrixtransformation*, Math Zeit **32**, 465–473.

O Perron (1930b), *Die Stabilitätsfrage bei Differentialgleichungen*, Math Zeit **32**, 703–728.

Ja B Pesin (1977), *Characteristic Lyapunov exponents and smooth ergodic theory*, Uspehi Mat Nauk **32**, 55–112, 287.

R S Phillips, *See Hille and Phillips.*

S Yu Pilyugin (1992), *Introduction to Structurally Stable Systems of Differential Equations*, Birkhäuser, Boston.

S Yu Pilyugin (1999), *Shadowing in Dynamical Systems*, Lecture Notes in Mathematics, No. 1706, Springer Verlag, New York.

A M Pinsky (1980), *The eigenvalues of an equilateral triangle*, SIAM J Math Anal **11**, 819–827.

V A Pliss (1964), *A reduction principle in the theory of the stability of motion*, Izv Akad Nauk SSSR, Mat Ser **28**, 1297–1324.

V A Pliss (1966), *Nonlinear Problems of the Theory of Oscillations*, English translation, Academic Press, New York.

V A Pliss (1969), *The structure of the asymptotically stable invariant sets of structurally stable autonomous systems of differential equations*, Diff Urav **5**, 979–991.

V A Pliss (1977), *Integral Sets of Periodic Systems of Differential Equations*, Russian, Izdat Nauka, Moscow.

V A Pliss and G R Sell (1991), *Perturbations of attractors of differential equations*, J Differential Equations **92**, 100–124.

V A Pliss and G R Sell (1998), *Approximation dynamics and the stability of invariant sets*, J Differential Equations **149**, 1–51.

V A Pliss and G R Sell (1999), *Robustness of exponential dichotomies for evolutionary equations*, J Dynamics Differential Equations **11**, 471–513.

V A Pliss and G R Sell (2001a), *Perturbations of normally hyperbolic invariant manifolds with applications to the Navier-Stokes equations*, J Differential Equations **169**, 396–492.

V A Pliss and G R Sell (2001b), *Stability of normally hyperbolic sets of smooth flows*, Dokl Akad Nauk (to appear).

P Poláčik, *See Brunovský and Poláčik.*

P Poláčik and I Tereščik (1992), *Convergence to cycles as a typical asymptotic behavior in smooth strongly monotone discrete-time dynamical systems*, Arch Rational Mech Anal **116**, 339–360.

H Poincaré (1890), *Sur le problème des trois corps et les équations de la dynamique*, Acta Math **13**, 1–270.

H Poincaré (1892), *Les Méthodes Nouvelles de la Mécanique Céleste I-II-III*, Gauthier Villars, Paris.

H Poincaré (1892), *Les Méthodes Nouvelles de la Mécanique Céleste I-II-III; English Translation (1993)*, Am Inst Physics.

S D Poisson (1831), *Mémoire sur les équations génèrale de léquilibre et du mouvement des corps solides élastiques et des fluides*, J Ecole Polytechnique **13**, 1-174.

T Precupanu, *See Barbu and Precupanu.*

A A Predtechensky, *See Lvov, Predtechensky, and Chernykh.*

G Prodi, *See Foias and Prodi.*

M H Protter and H F Weinberger (1984), *Maximum Principles in Differential Equations*, Prentice Hall, Englewood Cliffs NJ.

C C Pugh, *See Hirsch, Pugh, and Shub.*

J M Rakotoson, *See Eden, Michaux, and Rakotoson.*

T Randolph, *See Latushkin, Randolph, and Schnaubelt.*

R Rannacher, *See Heywood and Rannacher.*

T Ratiu, *See Marsden, Ratiu, and Raugel.*

G Raugel, *See Hale, Lin, and Raugel; Hale and Raugel; and Marsden, Ratiu, and Raugel.*

G Raugel (1995), *Dynamics of partial differential equations on thin domains*, Dynamical Systems (Montecatini Terme, 1994), Lecture Notes in Math, 1609, Springer Verlag, Berlin.

G Raugel and G R Sell (1993a), *Navier-Stokes equations on thin 3D domains I: Global attractors and global regularity of solutions*, J Am Math Soc **6**, 503–568.

G Raugel and G R Sell (1994), *Navier-Stokes equations on thin 3D domains II: Global regularity of spatially periodic solutions*, Nonlinear Partial Differential Equations and Their Applications; Collège de France Seminar; Vol XI, eds H Brezis and J L Lions, Pitman Research Notes in Mathematics Series 299, Longman Scientific Technical, Essex UK, pp. 205–247.

G Raugel and G R Sell (1993b), *Navier-Stokes equations on thin 3D domains III: Global and local attractors*, Turbulence in Fluid Flows: A Dynamical Systems Approach, IMA Volumes in Mathematics and Its Applications, Vol 55, (G R Sell, C Foias, and R Temam, eds), Springer Verlag, New York, pp. 137–163.

G Raugel and G R Sell (2002), *Navier-Stokes equations on irregular thin domains*, Under preparation.

R Redlinger (1995), *Existence of the global attractor for a strongly coupled parabolic system arising in population dynamics*, J Differential Equations **118**, 219–252.

M Reed and B Simon, *Methods of Modern Mathematical Physics, Vol IV*, Academic Press, New York.

J F Reineck, *See Gedeon, Kokubu, Mischaikow, Oka, and Reineck; Mischaikow, Mrozek, and Reineck.*

J F Reineck (1990; 1995), *The connection matrix in Morse-Smale flows I; II*, Trans Am Math Soc **322; 347**, 523–545; 2097–2110.

I Richards (1982), *On the gaps between numbers which are the sum of two squares*, Adv Math **46**, 1–2.

C Robinson (1974), *Structural stability of vector fields*, Ann Math **99**, 154–175.

R Rosa, *See Foias, Manley, Rosa, and Temam; Jolly, Rosa, and Temam; and Moise and Rosa.*

R Rosa and R Temam (1996), *Inertial manifolds and normal hyperbolicity*, Acta Appl Math **45**, 1–50.

D Ruelle (1979), *Ergodic theory of differentiable dynamical systems*, Inst Hautes Ètudes Sci Publ Math **51**, 27–58.

D Ruelle and F Takens (1971), *On the nature of turbulence*, Comm Math Phys **20, 21**, 167–192, 343–344.

A Ruiz (1992), *Unique continuation for weak solutions of the wave equations plus a potential*, J Math Pures Appl **71**, 455–467.

D L Russell, *See Chen and Russell.*

P H Rutherford, *See La Quey, Mahajan, Rutherford, and Tang.*

M Ružička, *See Nečas, Ružička, and Šverak.*

R J Sacker (1964), *On invariant surfaces and bifurcation of periodic solutions of ordinary differential equations*, NYU Preprint No 333, October 1964.

R J Sacker (1965), *A new approach to the perturbation theory of invariant surfaces*, Comm Pure Appl Math **18**, 717–732.

R J Sacker (1969), *A perturbation theorem for invariant manifolds and Hölder continuity*, J Math Mech **18**, 705–762.

R J Sacker (1978), *Existence of dichotomies and invariant splitting for linear differential systems IV*, J Differential Equations **27**, 106–137.

R J Sacker and G R Sell (1974; 1976a; 1976b), *Existence of dichotomies and invariant splitting for linear differential systems I; II; III*, J Differential Equations **15; 22; 22**, 429–458; 478–496; 497–522.

R J Sacker and G R Sell (1977), *Lifting properties in skewproduct flows with applications to differential equations*, Memoirs Am Math Soc, No 190.

R J Sacker and G R Sell (1978), *A spectral theory for linear differential systems*, J Differential Equations **27**, 320–358.

R J Sacker and G R Sell (1980), *The spectrum of an invariant submanifold*, J Differential Equations **38**, 135–160.

R J Sacker and G R Sell (1994), *Dichotomies for linear evolutionary equations in Banach spaces*, J Differential Equations **113**, 17–67.

K Sakamoto, *See Hale and Sakamoto.*

R Salvi (1988), *Some problems for nonhomogeneous fluids with time dependent domains and convex sets*, Math Z **199**, 153–170.

R Salvi (1990), *On the Navier-Stokes equations in noncylindrical domains: on the existence and regularity*, Rend Accad Naz Sci XL Mem Mat **14**, 87–105.

B Sandstede, A Scheel, and C Wulff (1997), *Dynamics of spiral waves on unbounded domains using center manifold reductions*, J Differential Equations **141**, 122–149.

N Sauer, *See Grobbelar-van Dalsen and Sauer.*

N Sauer (1981/82), *Linear evolutionary equations in two Banach spaces*, Proc Roy Soc Edinburgh **91**, 287–303.

N Sauer (1988), *The steady state Navier-Stokes equations with rotating boundaries*, Proc Roy Soc Edinburgh **110**, 93–99.

V E Ščadilov, *See Solonnikov and Ščadilov.*

J J Schäffer, *See Massera and Schäffer.*

M Schechter (1981), *Operator Methods in Quantum Mechanics*, North Holland, Amsterdam.

A Scheel, *See Sandstede, Scheel, and Wulff.*

R Schnaubelt, *See Latushkin, Randolph, and Schnaubelt.*

J T Schwartz, *See Dunford and Schwartz.*

L A Segel, *See Novick-Cohen and Segel.*

J Selgrade (1975), *Isolated invariant sets for flows on vector bundles*, Trans Am Math Soc **203**, 359–390.

G R Sell, *See Babin and Sell; Chow, Lu, and Sell; Fabes, Luskin, and Sell; Foias, Nicolaenko, Sell, and Temam; Foias, Sell, and Temam; Foias, Sell, and Titi; Johnson, Palmer, and Sell; Johnson and Sell; Naylor and Sell; Mallet-Paret and Sell; Mallet-Paret, Sell, and Shao; Markus and Sell; Meyer and Sell; Pliss and Sell; Raugel and Sell; Sacker and Sell; and Varghese and Sell.*

G R Sell (1967), *Nonautonomous differential equations and topological dynamics I: The basic theory, and II: Limiting equations*, Trans Am Math Soc **127**, 241–283.

G R Sell (1971), *Topological Dynamics and Ordinary Differential Equations*, Van Nostrand, New York.

G R Sell (1972), *A characterization of smooth α-Lipschitz mappings on a Hilbert space*, Atti Accad Naz Lincei Rend Cl Sci Fis Mat Natur **52**, 410–416.

G R Sell (1973), *Differential equations without uniqueness and classical topological dynamics*, J Differential Equations **14**, 42–56.

G R Sell (1977), *Hyperbolic almost periodic solutions and toroidal limit sets*, Proc Natl Acad Sci USA **74**, 3124–3125.

G R Sell (1978), *The structure of a flow in the vicinity of an almost periodic motion*, J Differential Equations **27**, 359–393.

G R Sell (1979), *Bifurcation of higher dimensional tori*, Arch Rational Mech Anal **69**, 199–230.

G R Sell (1981a), *A remark on an example of R A Johnson*, Proc Am Math Soc **82**, 206–208.

G R Sell (1981b), *Hopf-Landau bifurcation near strange attractors*, Chaos and Order in Nature, Ed: H Haken, Springer Verlag, Berlin, pp. 84–91.

G R Sell (1996), *Global attractors for the three dimensional Navier-Stokes equations*, J Dynamics Differential Equations **8**, 1–33.

G R Sell (2001), *References on dynamical systems*, http://www.math.umn.edu/~sell/.

G R Sell (2002), *Approximation Dynamics: With Applications to Numerical Analysis*, NSF Regional Conference, SIAM CBMS Regional Conference Series (to appear).

G R Sell and M Taboada (1992), *Local dissipitavity and attractors for the Kuramoto-Sivashinsky equation in thin 2D domains*, Nonlinear Anal **18**, 671–687.

G R Sell and Y You (1992), *Inertial manifolds: The non-self adjoint case*, J Differential Equations **96**, 203–255.

J Serrin (1959), *Mathematical principles of classical fluid mechanics*, Handbuch der Physik, Springer Verlag, Berlin, pp. 125–263.

J Serrin (1962), *On the interior regularity of weak solutions of the Navier-Stokes equations*, Arch Rational Mech Anal **9**, 187–195.

J Serrin (1963), *The initial value problem for the Navier-Stokes equations*, Nonlinear Problems, Univ Wisconsin Press, Madison, pp. 69–98.

J Serrin (1972), *The swirling vortex*, Phil Trans Roy Soc London **271**, 325–360.

R W Service (1940), *Bar-Room Ballads*, Dodd-Mead, New York.

T Shang and G L Sivashinsky (1983), Phys J **43**, 459–466.

Z Shao, *See Mallet-Paret, Sell, and Shao.*

Z Shao (1998), *Existence of inertial manifolds for partly dissipative reaction diffusion systems in higher space dimensions*, J Differential Equations **144**, 1–43.

J Shatah, *See Li, McLaughlin, Shatah, and Wiggins; and McLaughlin and Shatah.*

W Shen and Y Yi (1995a), *Dynamics of almost periodic scalar parabolic equations*, J Differential Equations **121**, 114–136.

W Shen and Y Yi (1995b), *Asymptotic almost periodicity of scalar parabolic equations with almost periodic time dependence*, J Differential Equations **121**, 373–397.

W Shen and Y Yi (1995c), *On minimal sets of scalar parabolic equations with skew product structures*, Trans Am Math Soc **347**, 4413–4431.

W Shen and Y Yi (1996), *Ergodicity of minimal sets of scalar parabolic equations*, J Dynamics Differential Equations **8**, 299–323.

W Shen and Y Yi (1998), *Almost automorphic and almost periodic dynamics in skew product semiflows*, Memoirs Am Math Soc, No 647, vol 136.

B Sheurer, *See Nicolaenko, Sheurer, and Temam.*

S Shkoller, *See Jones and Shkoller.*

M Shub, *See Hirsch, Pugh, and Shub.*

C L Siegel and J K Moser (1971), *Lectures on Celestial Mechanics*, Springer Verlag, New York.

B Simon, *See Reed and Simon.*

F Simondon, *See Andreu, Mazon, Simondon, and Toledo; Feireisl, Laurencot, and Simondon.*

Ya G Sinai (1994), *Topics in Ergodic Theory*, Princeton Univ Press, Princeton, NJ.

G L Sivashinsky, *See Shang and Sivashinsky.*

G L Sivashinsky (1977), *Nonlinear analysis of hydrodynamic instability in laminar flames*, Acta Astronomics **4**, 1177–1206.

M Slemrod, *See Hale, LaSalle, and Slemrod.*

S Smale (1967), *Differentiable dynamical systems*, Bull Am Math Soc **73**, 747–817.

H L Smith, *See Mallet-Paret and Smith.*

H L Smith (1995), *Monotone Dynamical Systems. An Introduction to the Theory of Competitive and Cooperative Systems.*, Math Surveys and Monographs, 41, Am Math Soc, Providence.

J A Smoller, *See Chueh, Conley, and Smoller; Conway, Hoff, and Smoller.*

J A Smoller (1983), *Shock Waves and Reaction Diffusion Equations*, Grundlehren Math Wiss, Vol 258, Springer Verlag, Berlin.

S L Sobolev (1938), *On a theorem of functional analysis*, Mat Sb **46**, 471–496.

S L Sobolev (1950; 1963), *Applications of Functional Analysis in Mathematical Physics*, English translation of 1950 Russian edition, Am Math Soc, Providence.

J Solà-Morales, *See Mora and Solà-Morales; and Oliva, De Oliveira, and Solà-Morales.*

V A Solonnikov and V E Ščadilov (1973), *On a boundary value problem for a stationary system of Navier-Stokes equations*, Proc Steklov Inst Math **125**, 186–199.

P Sorger, *See Kirchgässner and Sorger.*

J A Soriano, *See Cavalcanti, Larkin, and Soriano.*

E M Stein (1970), *Singular Integrals and Differentiability Properties of Functions*, Princeton Univ Press, Princeton.

V V Stepanov, *See Nemytskii and Stepanov.*

I Stewart, *See Golubitsky and Stewart.*

G G Stokes (1845), *On the theories of the internal friction of fluids in motion*, Trans Cambridge Phil Soc **8**, 8-106.

A M Stuart, *See Elliott and Stuart.*

A M Stuart (1994), *Numerical analysis of dynamical systems*, Acta Numerica, 1994, Cambridge Univ Press, Cambridge, pp. 467–572.

A M Stuart and A R Humphries (1996), *Dynamical Systems and Numerical Analysis*, Monograph on Applied and Computational Mathematics, Cambridge University Press.

J Stubbe, *See Collet, Eckmann, Epstein, and Stubbe.*

F Su, *See Guo and Su.*

V Šverak, *See Nečas, Ružička, and Šverak.*

J Swift and P C Hohenberg (1977), *Hydrodynamic fluctuations at the convective instability*, Phys Review A **15**, 319.

H L Swinney, *See Andereck, Liu, and Swinney, and Gollub and Swinney.*

B Sz Nagy and C Foias (1970), *Harmonic Analysis of Operators on Hilbert Spaces*, North Holland, Amsterdam.

G Szegö, *See Bhatia and Szegö.*

M Taboada, *See Sell and Taboada, You and Taboada.*

M Taboada and Y You (1992), *Global attactor, inertial manifolds and stabilizaton of nonlinear damped beam equations*, IMA Preprint, No 851.

F Takens, *See Ruelle and Takens.*

F Takens (1974), *Singularites of vector fields*, Publ Math IHES **43**, 47–100.

B Tan, *See Chen, Hale, and Tan.*

H Tanabe (1979), *Equations of Evolution*, Pitman, Lond.

W M Tang, *See Lo Quey, Mahajan, Rutherford, and Tang.*

G I Taylor (1923), *Stability of a viscous fluid contained between two rotating cylinders*, Philos Trans Roy Soc London, Ser A **223**, 289–.

S Taylor (1989), *Gevrey Semigroups*, Univ Minnesota PhD Thesis.

R Temam, *See Constantin, Foias, and Temam, Constantin, Foias, Nicolaenko, and Temam; Eden, Foias, Nicolaenko, and Temam; Eden, Foias, and Temam; Foias, Manley, Rosa, and Temam; Foias, Manley, and Temam; Foias, Manley, Temam, and Treve; Foias, Nicolaenko, Sell, and Temam; Foias, Nicolaenko, and Temam; Foias and Temam; Foias, Sell, and Temam, Ghidaglia and Temam; Jolly, Rosa, and Temam; Manley, Marion and Temam; Nicolaenko, Sheurer, and Temam; and Rosa and Temam.*

R Temam (1977), *Navier-Stokes Equations*, North-Holland, Amsterdam.

R Temam (1982), *Behaviour at time $t = 0$ of the solutions of semilinear evolution equations*, J Differential Equations **43**, 73–92.

R Temam (1983), *Navier-Stokes Equations and Nonlinear Functional Analysis*, CBMS Regional Conference Series, No 41, SIAM, Philadelphia.

R Temam (1988), *Infinite Dimensional Dynamical Systems in Mechanics and Physics*, Applied Mathematical Sciences, Vol 68, Springer Verlag, New York.

R Temam (1999), *Some developments on the Navier-Stokes equations in the second half of the 20th century*, Dévelopment des Mathématiques au cours de la seconde moitié du XXème siècle, J P Pier, Managing Editor, Birkhauser, Boston.

R Temam and X Wang (1994), *Estimates on the lowest dimension of inertial manifolds for the Kuramoto-Sivashinsky equation in the general case*, Differential and Integral Equations **7**, 1095–1108.

R Temam and M Ziane (1996), *Navier-Stokes equations in three-dimensional thin domains with various boundary conditions*, Adv Differential Equations **1**, 499–546.

R Temam and M Ziane (1997), *Navier-Stokes equations in thin spherical domains*, Optimization Methods in Partial Differential Equations, Contemp Math, 209, Am Math Soc, Providence.

I Tereščik, *See Poláčik and Tereščik.*

E S Titi, *See Chen, Foias, Holm, Olson, Titi, and Wynne; Foias, Holm, and Titi; Foias, Jolly, Kevrekidis, and Titi; and Foias, Sell, and Titi*

E S Titi (1987), *On a criterion for locating stable stationary solutions to the Navier-Stokes equations*, Nonlinear Analysis, Theory, Methods and Applications **11**, 1085–1102.

E S Titi (1991), *Un critère pour l'approximation des solutions périodiques des équations de Navier-Stokes*, C R Acad Sci Paris **312**, 41–43.

J Toledo, *See Andreu, Mazon, Simondon, and Toledo.*

Y M Treve, *See Foias, Manley, Temam, and Treve.*

H Triebel (1978), *Interpolation Theory, Function Spaces and Differential Operators*, VEB Deutscher Verlag der Wissenschaften, Berlin.

R Triggiani, *See Chen and Triggiani.*

R Triggiani (1975), *On the stabilizability problem in Banach space*, J Math Anal Appl **52**, 383–403.

J J Tyson, *See Keener and Tyson.*

A Vanderbauwhede, *See van Gils and Vanderbauwhede.*

A Vanderbauwhede and G Iooss (1992), *Center manifold theory in infinite dimensions*, Dynamics Reported, vol. 1, pp. 125–163.

S A van Gils and A Vanderbauwhede (1987), *Center manifolds and contractions on a scale of Banach spaces*, J Functional Anal **72**, 209–224.

E S Van Vleck (1995), *Numerical shadowing near hyperbolic trajectories*, SIAM J Sci Computing **16**, 1177–1189.

A Varghese and G R Sell (1997), *A conservation principle and its effect on the formulation of the Na-Ca exchanger current in cardiac cells*, J Theoretical Biology **189**, 33–40.

J L Vazquez, *See Escobedo, Vazquez, and Zuazua.*

W A Veech (1965), *Almost automorphic functions on groups*, Am J Math **87**, 719–751.

J Velazquez, *See Herrero and Velazquez.*

S M Verduyn Lunel, *See Hale and Verduyn Lunel.*

M I Vishik, *See Babin and Vishik; and Chepyzhov and Vishik.*

M I Vishik (1992), *Asymptotic Behavior of Solutions of Evolution Equations*, Accademia Nazionale dei Lincei, Cambridge Univ Press.

W von Wahl, *See Deuring and von Wahl.*

W von Wahl (1985), *The Equations of Navier-Stokes and Abstract Parabolic Problems*, Vieweg and Sohn, Braunschweig.

H-O Walther, *See Krisztin and Walther; and Krisztin, Walther, and Wu.*

X Wang, *See Temam and Wang.*

H W Weinberger, *See Protter and Weinberger.*

S Wiggins, *See Li, McLaughlin, Shatah, and Wiggins.*

S Wiggins (1994), *Normally Hyperbolic Manifolds in Dynamical Systems*, Springer Verlag, New York.

W Woinowsky (1950), *The effect of axial force on the vibration of hinged bars*, J Appl Mech **17**, 35–36.

J Wu, *See Krisztin, Walther, and Wu.*

C Wulff, *See Sandstede, Scheel, and Wulff.*

S Wynne, *See Chen, Foias, Holm, Olson, Titi, and Wynne.*

Z Xia (1992), *The existence of noncollision singularities in Newtonian systems*, Ann Math **135**, 411–468.

Y Yan (1992), *Dimensions of attractors for discretizations for the Navier Stokes equations*, J Dynamics Differential Equations **4**, 275–340.

Y Yi, *See Chow, Liu, and Yi; Chow and Yi; and Shen and Yi.*

Y Yi (1998), *Almost automorphy and almost periodicity*, Almost automorphic and almost periodic dynamics in skew product semiflows; Memoirs Am Math Soc, No 647, vol 136.

K Yosida (1980), *Functional Analysis*, Springer Verlag, New York.

Y You, *See Fitzgibbon, Parrott, and You; Sell and You; and Taboada and You.*

Y You (1993a), *Spillover problem and global dynamics of nonlinear beam equations*, Differential Equations, Dynamical Systems and Control Science, K D Elworthy, W N Everitt, and E B Lee (eds), Marcel Dekker, New York, pp. 891–912.

Y You (1993b), *Inertial manifolds and stabilization of nonlinear elastic systems with structural damping*, Differential Equations with Applications to Mathematical Physics, W F Ames, E Harrell, and J V Herod (eds), Academic Press, New York, pp. 335–346.

Y You (1996a), *Inertial manifolds and stabilization of nonlinear beam equations with Balakrishnan-Taylor damping*, Abstr Appl Anal **1**, 83–102.

Y You (1996b), *Nonlinear wave equations with asymptotically monotone damping*, Theory and Applications of Nonlinear Operators of Accretive and Monotone Type (AG Kartsatos, ed.), Marcel Dekker, New York, pp. 299–311.

Y You (1996c), *Global dynamics of dissipative generalized Korteweg-de Vries equations*, Chin Ann Math **17**, 389–402.

Y You (1996d), *A global existence theorem on nonlinear evolution equations*, Proc First World Cong Nonlinear Anal (V Lakshmikantham, ed.), Gruyter, Berlin, pp. 2239–2248.

Y You (1997a), *Nonlinear wave equations with weak dissipation*, Nonlinear Evolution Equations and Infinite Dimensional Dynamical Systems (T T Li, ed.), World Scientific, Singapore, pp. 249–258.

Y You (1997b), *Global dynamics of 2D Boussinesq equations*, Nonlinear Anal **30**, 4643–4654.

Y You (1999), $L^1 - L^\infty$ *regularity of analytic semigroups and applications*, Preprint.

Y You (2000), *Global dynamics of nonlinear hyperbolic equations with nonmonotone damping*, Proceedings of Equadiff 99, Berlin, Germany, World Scientific, Singapore.

Y You and M Taboada (1994), *Global dynamics and control of comprehensive nonlinear beam equations*, Qualitative Aspects and Applications of Nonlinear Evolution Equations (H Beirao da Veiga and T T Li, ed.), World Scientific, Singapore, pp. 109–122.

C Zeng, *See Bates, Lu, and Zeng.*

J Zhao, *See Peletier and Zhao.*

S Zheng, *See Elliott and Zheng.*

C Zhong, *See Li and Zhong.*

M Ziane, *See Temam and Ziane.*

E Zuazua, *See Escobedo, Vazquez, and Zuazua; Escobedo and Zuazua; Feireisl and Zuazua; Menzala and Zuazua.*

NOTATION INDEX

A^α	Fractional power of A
A^*	Adjoint of an operator A
$AC[a,b]$	Absolutely continuous functions $f : [a,b] \to R$
A_λ	Yosida approximant $A_\lambda = \lambda A R(\lambda, A)$
$\alpha_M(B)$	Alpha limit set of a set B in M
$\mathfrak{A}$	Attractor or global attractor
$B(u,v)$	Inertial term in the Navier-Stokes equations, $B(u,v) = \mathbb{P}((u \cdot \nabla)v)$
$\mathcal{B}(\mathfrak{A})$	Basin of attraction of an attractor $\mathfrak{A}$
B	Absorbing set
$C(\alpha, \beta, \dots)$	Constant C depending on $\alpha, \beta, \dots$
$\mathbb{C}$	Complex plane
$\mathrm{Cl}_W(B)$	Closure of B in the space W
$CLO(V,W)$	Closed linear operators from a dense subspace of V to W
C^1_{F}	Space of continuously Fréchet differentible functions
$C^1_{\mathrm{F,b}}$	$F \in C^1_{\mathrm{F}}(U,W)$ with DF bounded on bounded sets
C^1_{G}	Space of continuously Gateau differentible functions
C_{Lip}	$C_{\mathrm{Lip}}(V^{2\beta} \times R^+; W)$ or $C_{\mathrm{Lip}}(V; W)$
C^1_{Lip}	$C_{\mathrm{Lip}} \cap C^1_{\mathrm{F}}$
D_i	Spatial derivative, $D_i u = \frac{\partial}{\partial x_i} u$
∂_t	Spatial derivative, $\partial_x u = \frac{\partial}{\partial x} u$
∂_x	Time derivative, $\partial_t u = \frac{\partial}{\partial t} u$
D^α	Partial differential operator, $D^\alpha u = \frac{\partial^{\lvert\alpha\rvert} u}{\partial x_1^{\alpha_1} \cdots \partial x_n^{\alpha_n}}$
$\triangle$	Laplacian operator on R^n, $\triangle = D_1^2 + \cdots + D_n^2$
∇	Gradient operator on R^n, $\nabla = (D_1, \cdots, D_n)$
$\mathrm{diam}(B)$	Diameter of a set B
$\mathcal{D}(A)$	Domain of the linear operator A
$\mathcal{D}_{V^{2\beta}}(L)$	Domain of the linear operator L in the space $V^{2\beta} = \mathcal{D}(A^\beta)$
Δ_δ	Sector in the complex plane, $\Delta_\delta = \{z \in C : \lvert\arg z\rvert < \delta\}$
(e^{At}, A)	C_0-semigroup e^{At} with infinitesimal generator A
$E_{r,c}(\mu t)$	See the Gronwall-Henry inequality
ϕ^u	Globally defined motion through u
$\gamma_M(B)$	Full orbit through B, where $B \subset M$
$\gamma_M^-(B)$	Negative orbit through B, where $B \subset M$
$\gamma^+(B)$	Positive orbit through B
$\Gamma(\cdot)$	Gamma function
H	Hilbert space
H_w	Hilbert space with the weak topology
$H_M(B)$	Hull of a set B in M
$H_M^-(B)$	Negative hull of a set B in M

$H^+(B)$	Positive hull of B
Hypothesis T	Hypothesis for a linear skew product semiflow
$\kappa(B)$	Kuratowski measure of noncompactness of B
$\mathcal{L}(V, W)$	Space of bounded linear operators from V to W
$\|L\|_{\mathcal{L}}$	Operator norm, $\|L\|_{\mathcal{L}} = \sup\{\|Lv\|_W : \|v\|_V \le 1\}$, $\mathcal{L} = \mathcal{L}(V, W)$
$\mathcal{L}(W)$	Space of bounded linear operators on W, $\mathcal{L}(W) = \mathcal{L}(W, W)$
$m \cdot t$	Semiflow $m \cdot t = \sigma(m, t) = S(t)m$
$\mathcal{M}^\infty$	$L^\infty(R, \mathcal{L}(V^{2\beta}, W)) \cap C(R, \mathcal{L}(V^{2\beta}, W))$ with the L^∞_{loc}-topology
$\mathcal{M}^p$	$L^\infty(R, \mathcal{L}(V^{2\beta}, W))$ with the L^p_{loc}-topology
$\mathfrak{M}$	Inertial manifold
$N_\epsilon(B)$	Closed ϵ-neighborhood of a set B
$\mathcal{N}(A)$	Null space, or kernel, of an operator A
Ω	Region in physical space, usually bounded
$\partial\Omega$	Boundary of Ω
$\mathbb{P}$	Leray (orthogonal) projection onto divergent-free vector fields
$\rho(A)$	Resolvent set of a linear operator A
$\mathbb{R}$	Real line, $\mathbb{R} = (-\infty, \infty)$
$\mathbb{R}^+$	Nonnegative real numbers, $\mathbb{R}^+ = [0, \infty)$
$\mathbb{R}^-$	Nonpositive real numbers, $\mathbb{R}^- = (-\infty, 0]$
$\mathbb{R}^n$	Euclidean n-space
$R(\lambda, A)$	Resolvent operator, $R(\lambda, A) = (\lambda I - A)^{-1}$
$\mathcal{R}(A)$	Range of an operator A
$S(t)u = \sigma(u, t)$	Semiflow, Nonlinear semigroup
$\sigma(A)$	Spectrum of a linear operator A
$\sigma_p(A)$	Point spectrum of A
$\sigma(u, t) = S(t)u$	Semiflow, Nonlinear semigroup
Standing Hypothesis A	Hypothesis for C_0-semigroup
Standing Hypothesis B	Hypothesis for C_0-semigroup
$T\|_V$	Restriction of a mapping T to a subspace V
$(T(t), A)$	C_0-semigroup $T(t)$ with infinitesimal generator A
$\mathcal{T}_A$	A Fréchet space topology
$\mathcal{T}^o_A$	A Fréchet space topology
$\mathcal{T}^1_A$	A Fréchet space topology
$\mathcal{T}_{\text{bo}}$	A Fréchet space topology
$\mathcal{T}^o_{\text{bo}}$	A Fréchet space topology
$\mathcal{T}^1_{\text{bo}}$	A Fréchet space topology
u_t	Derivative of u with respect to time t, $u_t = \frac{\partial u}{\partial t}$
u_x, u_{xx}	Derivatives of u with respect to space x, e.g., $u_x = \frac{\partial u}{\partial x}$
$V^{2\alpha}$	Fractional power space, $V^{2\alpha} = \mathcal{D}(A^\alpha)$
W	Banach space, or a complete metric space
W'	Dual space of a Banach space W
$\mathcal{W}$	Sequence space $\ell_\infty(Z, \mathcal{L}(W))$
Ξ	Domain in $V^{2\beta} \times C_{\text{Lip}} \times R^+$ for mild solutions
$\omega(B)$	Omega limit set of B
$\hookrightarrow$	Continuous imbedding

$\hookrightarrow$	Compact imbedding
$\|\cdot\| = \|\cdot\|_W$	Norm on a Banach space W
$\|\cdot\|_{2\alpha}$	Norm on $V^{2\alpha} = \mathcal{D}(A^\alpha)$, $\|u\|_{2\alpha} = \|A^\alpha u\|$
$\|\cdot\|_{\{A;C^0(\Omega)\}}$	Norm on C_{Lip}
$\|\cdot\|_{\{A;C^1(\Omega)\}}$	Norm on C^1_{Lip}
$\langle\cdot,\cdot\rangle = \langle\cdot,\cdot\rangle_H$	Inner product on a Hilbert space H
$\langle\cdot,\cdot\rangle_{2\alpha}$	Inner product on $V^{2\alpha} = \mathcal{D}(A^\alpha)$, $\langle u,v\rangle_{2\alpha} = \langle A^\alpha u, A^\alpha v\rangle_H$
$\langle\langle u,v\rangle\rangle = \langle u,v\rangle_{(V',V)}$	Dual product of $u \in V'$ and $v \in V$
$\overset{\text{a.e.}}{=}$	Equality valid almost everywhere (a.e)
$\overset{\text{def}}{=}$	Equal by definition
$\overset{s}{\to}$	Strong convergence
$\overset{w}{\to}$	Weak convergence

FUNCTION SPACES: BANACH SPACES

Strongly measurable functions $\varphi : \Omega \to W$,
where Ω is bounded in R^d and W is a Banach space

$L^p(\Omega; W)$	$\int_\Omega \|f\|_W^p \, dx < \infty$, $1 \le p < \infty$
$L^\infty(\Omega; W)$	$\operatorname{ess\,sup}_{x\in\Omega} \|f(x)\|_W < \infty$

FUNCTION SPACES: SOBOLEV SPACES

Strongly measurable functions $\varphi : \Omega \to W$,
where Ω is bounded in R^d

$W^{n,p} = W^{n,p}(\Omega; R^m)$	n distribution derivatives, all in L^p
$H^n = W^{n,2}(\Omega; R^m)$	n distribution derivatives, all in L^2
$H_0^n = H_0^n(\Omega; R^m)$	Closure in H^n of smooth φ with $\varphi = 0$ near $\partial\Omega$

FUNCTION SPACES: FRÉCHET SPACES

Strongly measurable functions $\varphi : [0, T) \to W$,
where $0 < T \le \infty$ and W is a Banach space

$L^p_{\text{loc}}(0,T;W)$	$\int_\sigma^\tau \|\varphi(s)\|_W^p \, ds < \infty$, for all $0 < \sigma < \tau < T$, $1 \le p < \infty$
$L^p_{\text{loc}}[0,T;W)$	$\int_0^\tau \|\varphi(s)\|_W^p \, ds < \infty$, for all $0 < \tau < T$, $1 \le p < \infty$
$L^\infty_{\text{loc}}(0,T;W)$	$\operatorname{ess\,sup}_{\sigma<s<\tau} \|\varphi(s)\|_W < \infty$, for all $0 < \sigma < \tau < T$
$L^\infty_{\text{loc}}[0,T;W)$	$\operatorname{ess\,sup}_{0<s<\tau} \|\varphi(s)\|_W < \infty$, for all $0 < \tau < T$

DIFFERENTIABILITY AND HÖLDER CONTINUITY

$C(\Omega; W)$	Continuous functions $f : \Omega \to W$
$C^{0,\theta}(\Omega; W)$	Uniformly Hölder continuous functions $f : \Omega \to W$

$C^{0,\theta}_{\rm loc}(\Omega;W)$	$f:\Omega\to W$, uniformly Hölder continuous on compact sets in Ω
$C^{0,\theta}_{\rm loc}(0,T;W)$	$C^{0,\theta}_{\rm loc}(\Omega;W)$ with $\Omega=(0,T)$
$C^{0,\theta}_{\rm loc}[0,T;W)$	$C^{0,\theta}_{\rm loc}(\Omega;W)$ with $\Omega=[0,T)$
$C_{\rm Lip}=C^{0,1}$	Space of Lipschitz continuous functions
$C_{\rm Lip;\ Global}$	$C_{\rm Lip}$ with globally bounded Lipschitz coefficient
$C_{{\rm Lip};\alpha}$	Lipschitz continuous in space, Hölder continuous in time

SUBJECT INDEX

Applied Mathematical Sciences

(continued from page ii)

60. *Ghil/Childress:* Topics in Geophysical Dynamics: Atmospheric Dynamics, Dynamo Theory and Climate Dynamics.
61. *Sattinger/Weaver:* Lie Groups and Algebras with Applications to Physics, Geometry, and Mechanics.
62. *LaSalle:* The Stability and Control of Discrete Processes.
63. *Grasman:* Asymptotic Methods of Relaxation Oscillations and Applications.
64. *Hsu:* Cell-to-Cell Mapping: A Method of Global Analysis for Nonlinear Systems.
65. *Rand/Armbruster:* Perturbation Methods, Bifurcation Theory and Computer Algebra.
66. *Hlavácek/Haslinger/Necasl/Lovísek:* Solution of Variational Inequalities in Mechanics.
67. *Cercignani:* The Boltzmann Equation and Its Applications.
68. *Temam:* Infinite-Dimensional Dynamical Systems in Mechanics and Physics, 2nd ed.
69. *Golubitsky/Stewart/Schaeffer:* Singularities and Groups in Bifurcation Theory, Vol. II.
70. *Constantin/Foias/Nicolaenko/Temam:* Integral Manifolds and Inertial Manifolds for Dissipative Partial Differential Equations.
71. *Catlin:* Estimation, Control, and the Discrete Kalman Filter.
72. *Lochak/Meunier:* Multiphase Averaging for Classical Systems.
73. *Wiggins:* Global Bifurcations and Chaos.
74. *Mawhin/Willem:* Critical Point Theory and Hamiltonian Systems.
75. *Abraham/Marsden/Ratiu:* Manifolds, Tensor Analysis, and Applications, 2nd ed.
76. *Lagerstrom:* Matched Asymptotic Expansions: Ideas and Techniques.
77. *Aldous:* Probability Approximations via the Poisson Clumping Heuristic.
78. *Dacorogna:* Direct Methods in the Calculus of Variations.
79. *Hernández-Lerma:* Adaptive Markov Processes.
80. *Lawden:* Elliptic Functions and Applications.
81. *Bluman/Kumei:* Symmetries and Differential Equations.
82. *Kress:* Linear Integral Equations, 2nd ed.
83. *Bebernes/Eberly:* Mathematical Problems from Combustion Theory.
84. *Joseph:* Fluid Dynamics of Viscoelastic Fluids.
85. *Yang:* Wave Packets and Their Bifurcations in Geophysical Fluid Dynamics.
86. *Dendrinos/Sonis:* Chaos and Socio-Spatial Dynamics.
87. *Weder:* Spectral and Scattering Theory for Wave Propagation in Perturbed Stratified Media.
88. *Bogaevski/Povzner:* Algebraic Methods in Nonlinear Perturbation Theory.
89. *O'Malley:* Singular Perturbation Methods for Ordinary Differential Equations.
90. *Meyer/Hall:* Introduction to Hamiltonian Dynamical Systems and the N-body Problem.
91. *Straughan:* The Energy Method, Stability, and Nonlinear Convection.
92. *Naber:* The Geometry of Minkowski Spacetime.
93. *Colton/Kress:* Inverse Acoustic and Electromagnetic Scattering Theory, 2nd ed.
94. *Hoppensteadt:* Analysis and Simulation of Chaotic Systems, 2nd ed.
95. *Hackbusch:* Iterative Solution of Large Sparse Systems of Equations.
96. *Marchioro/Pulvirenti:* Mathematical Theory of Incompressible Nonviscous Fluids.
97. *Lasota/Mackey:* Chaos, Fractals, and Noise: Stochastic Aspects of Dynamics, 2nd ed.
98. *de Boor/Höllig/Riemenschneider:* Box Splines.
99. *Hale/Lunel:* Introduction to Functional Differential Equations.
100. *Sirovich (ed):* Trends and Perspectives in Applied Mathematics.
101. *Nusse/Yorke:* Dynamics: Numerical Explorations, 2nd ed.
102. *Chossat/Iooss:* The Couette-Taylor Problem.
103. *Chorin:* Vorticity and Turbulence.
104. *Farkas:* Periodic Motions.
105. *Wiggins:* Normally Hyperbolic Invariant Manifolds in Dynamical Systems.
106. *Cercignani/Illner/Pulvirenti:* The Mathematical Theory of Dilute Gases.
107. *Antman:* Nonlinear Problems of Elasticity.
108. *Zeidler:* Applied Functional Analysis: Applications to Mathematical Physics.
109. *Zeidler:* Applied Functional Analysis: Main Principles and Their Applications.
110. *Diekmann/van Gils/Verduyn Lunel/Walther:* Delay Equations: Functional-, Complex-, and Nonlinear Analysis.
111. *Visintin:* Differential Models of Hysteresis.
112. *Kuznetsov:* Elements of Applied Bifurcation Theory, 2nd ed.
113. *Hislop/Sigal:* Introduction to Spectral Theory: With Applications to Schrödinger Operators.
114. *Kevorkian/Cole:* Multiple Scale and Singular Perturbation Methods.
115. *Taylor:* Partial Differential Equations I, Basic Theory.
116. *Taylor:* Partial Differential Equations II, Qualitative Studies of Linear Equations.

(continued on next page)

Applied Mathematical Sciences

(continued from previous page)

117. *Taylor:* Partial Differential Equations III, Nonlinear Equations.
118. *Godlewski/Raviart:* Numerical Approximation of Hyperbolic Systems of Conservation Laws.
119. *Wu:* Theory and Applications of Partial Functional Differential Equations.
120. *Kirsch:* An Introduction to the Mathematical Theory of Inverse Problems.
121. *Brokate/Sprekels:* Hysteresis and Phase Transitions.
122. *Gliklikh:* Global Analysis in Mathematical Physics: Geometric and Stochastic Methods.
123. *Le/Schmitt:* Global Bifurcation in Variational Inequalities: Applications to Obstacle and Unilateral Problems.
124. *Polak:* Optimization: Algorithms and Consistent Approximations.
125. *Arnold/Khesin:* Topological Methods in Hydrodynamics.
126. *Hoppensteadt/Izhikevich:* Weakly Connected Neural Networks.
127. *Isakov:* Inverse Problems for Partial Differential Equations.
128. *Li/Wiggins:* Invariant Manifolds and Fibrations for Perturbed Nonlinear Schrödinger Equations.
129. *Müller:* Analysis of Spherical Symmetries in Euclidean Spaces.
130. *Feintuch:* Robust Control Theory in Hilbert Space.
131. *Ericksen:* Introduction to the Thermodynamics of Solids, Revised ed.
132. *Ihlenburg:* Finite Element Analysis of Acoustic Scattering.
133. *Vorovich:* Nonlinear Theory of Shallow Shells.
134. *Vein/Dale:* Determinants and Their Applications in Mathematical Physics.
135. *Drew/Passman:* Theory of Multicomponent Fluids.
136. *Cioranescu/Saint Jean Paulin:* Homogenization of Reticulated Structures.
137. *Gurtin:* Configurational Forces as Basic Concepts of Continuum Physics.
138. *Haller:* Chaos Near Resonance.
139. *Sulem/Sulem:* The Nonlinear Schrödinger Equation: Self-Focusing and Wave Collapse.
140. *Cherkaev:* Variational Methods for Structural Optimization.
141. *Naber:* Topology, Geometry, and Gauge Fields: Interactions.
142. *Schmid/Henningson:* Stability and Transition in Shear Flows.
143. *Sell/You:* Dynamics of Evolutionary Equations.
144. *Nédélec:* Acoustic and Electromagnetic Equations: Integral Representations for Harmonic Problems.
145. *Newton:* The *N*-Vortex Problem: Analytical Techniques.
146. *Allaire*: Shape Optimization by the Homogenization Method.
147. *Aubert/Kornprobst:* Mathematical Problems in Image Processing: Partial Differential Equations and the Calculus of Variations.

GPSR Compliance
The European Union's (EU) General Product Safety Regulation (GPSR) is a set of rules that requires consumer products to be safe and our obligations to ensure this.

If you have any concerns about our products, you can contact us on

ProductSafety@springernature.com

In case Publisher is established outside the EU, the EU authorized representative is:

Springer Nature Customer Service Center GmbH
Europaplatz 3
69115 Heidelberg, Germany

www.ingramcontent.com/pod-product-compliance
Ingram Content Group UK Ltd.
Pitfield, Milton Keynes, MK11 3LW, UK
UKHW022317190726
13856UKWH00001B/56

* 9 7 8 1 4 7 5 7 5 0 3 8 6 *